Win the complete library of RSMeans Online data!

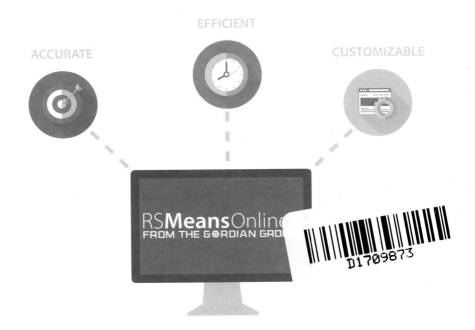

Register your book below to receive your **free** quarterly updates. **Plus** you will be entered into a quarterly drawing to win the **complete** RSMeans Online library of 2016 data!

Be sure to keep up-to-date in 2016!

Fill out the card below for RSMeans' free quarterly updates, as well as a chance to win the complete RSMeans Online library. Please provide your name, address, and email below and return this card by mail, or register online:

info.thegordiangroup.com/2016updates.html

Name _____

Email _____ Title _____

Company _____

Street _____

City/Town _____ State/Prov. ____ Zip/Postal Code _____

Win the complete library of RSMeans Online data!

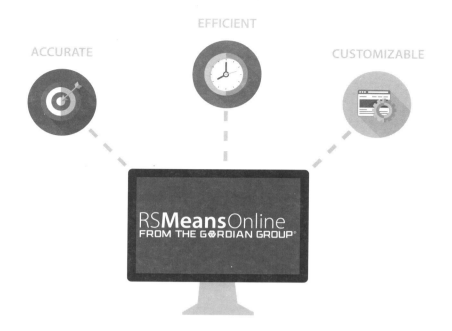

Register your book below to receive your **free** quarterly updates.
Plus you will be entered into a quarterly drawing to win the **complete** RSMeans Online library of 2016 data!

NO POSTAGE
NECESSARY
IF MAILED
IN THE
UNITED STATES

BUSINESS REPLY MAIL
FIRST-CLASS MAIL   PERMIT NO. 31   ROCKLAND, MA

POSTAGE WILL BE PAID BY ADDRESSEE

RS MEANS
1099 HINGHAM ST STE 201
ROCKLAND MA 02370-9923

Residential Cost Data

Bob Mewis, CCP, Senior Editor

RSMeans
FROM THE GORDIAN GROUP®

2016
35th annual edition

Engineering Director
Bob Mewis, CCP (*1, 2, 4, 13, 14, 31, 32, 33, 34, 35, 41, 44, 46*)

Contributing Editors
Christopher Babbitt
Adrian C. Charest, PE
Cheryl Elsmore
Wafaa Hamitou
Joseph Kelble
Charles Kibbee
Robert J. Kuchta (*8*)
Michael Landry
Thomas Lane (*6, 7*)
Genevieve Medeiros
Elisa Mello
Chris Morris (*26, 27, 28, 48*)

Melville J. Mossman, PE (*21, 22, 23*)
Marilyn Phelan, AIA (*9, 10, 11, 12*)
Stephen C. Plotner (*3, 5*)
Stephen Rosenberg
Kevin Souza
Keegan Spraker
Tim Tonello
David Yazbek

Vice President Data & Engineering
Chris Anderson

Product Manager
Andrea Sillah

Production Manager
Debbie Panarelli

Production
Sharon Larsen
Jonathan Forgit
Sheryl Rose
Mary Lou Geary

Technical Support
Gary L. Hoitt
Kathryn S. Rodriguez

Cover Design
Blaire Gaddis

Numbers in italics are the divisional responsibilities for each editor. Please contact the designated editor directly with any questions.

RSMeans
Construction Publishers & Consultants
1099 Hingham Street, Suite 201
Rockland, MA 02370
United States of America
1-877-759-4771
www.RSMeans.com

Copyright 2015 by RSMeans
All rights reserved.
Cover photo ©
Hemera Technologies/AbleStock.com/Thinkstock

Printed in the United States of America
ISSN 0896-8624
ISBN 978-1-943215-16-4

The authors, editors, and engineers of RSMeans apply diligence and judgment in locating and using reliable sources for the information published. However, RSMeans makes no express or implied warranty or guarantee in connection with the content of the information contained herein, including the accuracy, correctness, value, sufficiency, or completeness of the data, methods, and other information contained herein. RSMeans makes no express or implied warranty of merchantability or fitness for a particular purpose. RSMeans shall have no liability to any customer or third party for any loss, expense, or damage, including consequential, incidental, special, or punitive damage, including lost profits or lost revenue, caused directly or indirectly by any error or omission, or arising out of, or in connection with, the information contained herein. For the purposes of this paragraph, "RSMeans" shall include RSMeans Company LLC, and its divisions, subsidiaries, successors, parent companies, and their employees, partners, principals, agents and representatives, and any third-party providers or sources of information or data.

No part of this publication may be reproduced, stored in an information storage or retrieval system, or transmitted in any form or by any means, electronic or mechanical, including photocopying or scanning, without prior written permission of RSMeans. The cost data contained in this publication is valuable and proprietary information of RSMeans and others, and there is no transfer to you of any ownership rights in the cost data or any license granted to you to create derivative works based on or which utilize the cost data.

$219.99 per copy (in United States)
Price is subject to change without prior notice.

Related RSMeans Products and Services

This data set is aimed primarily at single-family and small multi-family housing projects costing up to $1,000,000.

The engineers at RSMeans suggest the following products and services as companion information resources to *RSMeans Residential Cost Data*:

Construction Cost Data
Square Foot Costs 2016
Residential Repair and Remodeling Costs 2016

Reference Books
Estimating Building Costs
RSMeans Estimating Handbook
Green Building: Project Planning & Estimating
How to Estimate with RSMeans Data
Plan Reading & Material Takeoff
Project Scheduling & Management for Construction
Universal Design Ideas for Style, Comfort & Safety

Seminars and In-House Training
Unit Price Estimating
RSMeans Online® Training
Practical Project Management for Construction Professionals
Scheduling with MSProject for Construction Professionals
Mechanical & Electrical Estimating

RSMeans Online Store
Visit RSMeans at www.RSMeans.com for the most reliable and current resources available on the market. Learn more about our more than 20 data sets available in Online, Book, eBook, and CD formats. Our library of reference books is also available, along with professional development seminars aimed at improving cost estimating, project management, administration, and facilities management skills.

RSMeans Electronic Data
Receive the most up-to-date cost information with RSMeans Online. This web-based service is quick, intuitive, and easy to use, giving you instant access to RSMeans' comprehensive database. Learn more at: **www.RSMeans.com/Online**.

RSMeans Custom Solutions
Building owners, facility managers, building product manufacturers, attorneys, and even insurance firms across the public and private sectors have engaged in RSMeans' custom solutions to solve their estimating needs including:

Custom Cost Engineering Solutions
Knowing the lifetime maintenance cost of your building, how much it will cost to build a specific building type in different locations across the country and globe, and if the estimate from which you are basing a major decision is accurate are all imperative for building owners and their facility managers.

Market & Custom Analytics
Taking a construction product to market requires accurate market research in order to make critical business development decisions. Once the product goes to market, one way to differentiate your product from the competition is to make any cost saving claims you have to offer.

Third-Party Legal Resources
Natural disasters can lead to reconstruction and repairs just as new developments and expansions can lead to issues of eminent domain. In these cases where one party must pay, it is plausible for legal disputes over costs and estimates to arise.

Construction Costs for Software Applications
More than 25 unit price and assemblies cost databases are available through a number of leading estimating and facilities management software partners. For more information, see "Other RSMeans Products and Services" at the back of this publication.

RSMeans data is also available to federal, state, and local government agencies as multi-year, multi-seat licenses.

For procurement construction cost data such as Job Order Contracting and Change Order Management please refer to The Gordian Group's procurement solutions.

For information on our current partners, call 1-877-759-4771.

Table of Contents

Foreword	iv
How the Cost Data Is Built: An Overview	v
Estimating with RSMeans Unit Prices	vii
How to Use the Cost Data: The Details	ix
Square Foot Cost Section	1
Assemblies Section	97
How RSMeans Assemblies Data Works	98
Unit Price Section	283
How RSMeans Unit Price Data Works	286
Reference Section	657
Construction Equipment Rental Costs	659
Crew Listings	671
Location Factors	705
Reference Tables	711
Abbreviations	732
Estimating Forms	736
Index	738
Other RSMeans Products and Services	765
Labor Trade Rates including Overhead & Profit	Inside Back Cover

Hartness Library
Vermont Technical College
One Main St.
Randolph Center, VT 05061

Foreword

Who We Are
Since 1942, RSMeans has delivered construction cost estimating information and consulting throughout North America. In 2014, RSMeans was acquired by The Gordian Group, combining two industry-leading construction cost databases. Through the RSMeans line of products and services, The Gordian Group provides innovative construction cost estimating data to organizations pursuing efficient and effective construction planning, estimating, procurement, and information solutions.

Our Offerings
With RSMeans' construction cost estimating data, contractors, architects, engineers, facility owners, and managers can utilize the most up-to-date data available in many formats to meet their specific planning and estimating needs.

When you purchase information from RSMeans, you are, in effect, hiring the services of a full-time staff of construction and engineering professionals.

Our thoroughly-experienced and highly-qualified staff works daily to collect, analyze, and disseminate comprehensive cost information to meet your needs.

These staff members have years of practical construction experience and engineering training prior to joining the firm. As a result, you can count on them not only for accurate cost figures, but also for additional background reference information that will help you create a realistic estimate.

The RSMeans organization is equipped to help you solve construction problems through its variety of data solutions.

Access our comprehensive database electronically with RSMeans Online. Quick, intuitive, easy to use, and updated continuously throughout the year, RSMeans Online is the most accurate and up-to-date cost estimating data.

This up-to-date, accurate, and localized data can be leveraged for a broad range of applications, such as Custom Cost Engineering solutions, Market and Custom Analytics, and Third-Party Legal Resources.

To ensure you are getting the most from your cost estimating data, we also offer a myriad of reference guides, training, and professional seminars (learn more at **www.RSMeans.com/Learn**).

In short, RSMeans can provide you with the tools and expertise for developing accurate and dependable construction estimates and budgets in a variety of ways.

Our Commitment
Today at RSMeans, we do more than talk about the quality of our data and the usefulness of the information. We stand behind all of our data—from historical cost indexes to construction materials and techniques—to craft current costs and predict future trends.

If you have any questions about our products or services, please call us toll-free at 1-877-759-4771. You can also visit our website at: **www.RSMeans.com**.

How the Cost Data Is Built: An Overview

A Powerful Construction Tool
You now have one of the most powerful construction tools available today. A successful project is built on the foundation of an accurate and dependable estimate. This tool will enable you to construct such an estimate.

For the casual user the information is designed to be:
- quickly and easily understood so you can get right to your estimate.
- filled with valuable information so you can understand the necessary factors that go into a cost estimate.

For the professional user, the information is designed to be:
- a handy reference that can be quickly referred to for key costs.
- a comprehensive, fully reliable source of current construction costs and productivity rates so you'll be prepared to estimate any project.
- a source for preliminary project costs, product selections, and alternate materials and methods.

To meet all of these requirements, we have organized the information into the following clearly defined sections.

Estimating with RSMeans Unit Price Cost Data*
Please refer to these steps for guidance on completing an estimate using RSMeans unit price cost data.

How to Use the Information: The Details
This section contains an in-depth explanation of how the information is arranged and how you can use it to determine a reliable construction cost estimate. It includes how we develop our cost figures and how to prepare your estimate.

Unit Prices*
All cost data has been divided into 50 divisions according to the MasterFormat system of classification and numbering.

Assemblies*
The cost data in this section has been organized in an "Assemblies" format. These assemblies are the functional elements of a building and are arranged according to the 7 elements of the UNIFORMAT II classification system. For a complete explanation of a typical "Assembly", see "How RSMeans Assemblies Data Works."

Residential Models*
Model buildings for four classes of construction—economy, average, custom, and luxury—are developed and shown with complete costs per square foot.

Commercial/Industrial/Institutional Models*
This section contains complete costs for 77 typical model buildings expressed as costs per square foot.

Green Commercial/Industrial/Institutional Models*
This section contains complete costs for 25 green model buildings expressed as costs per square foot.

References*
This section includes information on Equipment Rental Costs, Crew Listings, Historical Cost Indexes, City Cost Indexes, Location Factors, Reference Tables, and Change Orders, as well as a listing of abbreviations.

- **Equipment Rental Costs:** Included are the average costs to rent and operate hundreds of pieces of construction equipment.
- **Crew Listings:** This section lists all the crews referenced in the cost data. A crew is composed of more than one trade classification and/or the addition of power equipment to any trade classification. Power equipment is included in the cost of the crew. Costs are shown both with bare labor rates and with the installing contractor's overhead and profit added. For each, the total crew cost per eight-hour day and the composite cost per labor-hour are listed.
- **Historical Cost Indexes:** These indexes provide you with data to adjust construction costs over time.
- **City Cost Indexes:** All costs in this data set are U.S. national averages. Costs vary by region. You can adjust for this by CSI Division to over 700 locations throughout the U.S. and Canada by using this data.
- **Location Factors:** You can adjust total project costs to over 900 locations throughout the U.S. and Canada by using the weighted number, which applies across all divisions.
- **Reference Tables:** At the beginning of selected major classifications in the Unit Prices are reference numbers indicators. These numbers refer you to related information in the Reference Section. In this section, you'll find reference tables, explanations, and estimating information that support how we develop the unit price data, technical data, and estimating procedures.
- **Change Orders:** This section includes information on the factors that influence the pricing of change orders.
- **Abbreviations:** A listing of abbreviations used throughout this information, along with the terms they represent, is included.

Index (printed versions only)
A comprehensive listing of all terms and subjects will help you quickly find what you need when you are not sure where it occurs in MasterFormat.

Conclusion
This information is designed to be as comprehensive and easy to use as possible.

The Construction Specifications Institute (CSI) and Construction Specifications Canada (CSC) have produced the 2014 edition of MasterFormat®, a system of titles and numbers used extensively to organize construction information.

All unit prices in the RSMeans cost data are now arranged in the 50-division MasterFormat® 2014 system.

---

* Not all information is available in all data sets

Note: The material prices in RSMeans cost data are "contractor's prices." They are the prices that contractors can expect to pay at the lumberyards, suppliers'/distributors' warehouses, etc. Small orders of specialty items would be higher than the costs shown, while very large orders, such as truckload lots, would be less. The variation would depend on the size, timing, and negotiating power of the contractor. The labor costs are primarily for new construction or major renovation rather than repairs or minor alterations. With reasonable exercise of judgment, the figures can be used for any building work.

Estimating with RSMeans Unit Prices

Following these steps will allow you to complete an accurate estimate using RSMeans Unit Price cost data.

1. Scope Out the Project
- Think through the project and identify the CSI Divisions needed in your estimate.
- Identify the individual work tasks that will need to be covered in your estimate.
- The Unit Price data has been divided into 50 Divisions according to the CSI MasterFormat® 2014.
- In printed versions, the Unit Price Section Table of Contents on page 1 may also be helpful when scoping out your project.
- Experienced estimators find it helpful to begin with Division 2 and continue through completion. Division 1 can be estimated after the full project scope is known.

2. Quantify
- Determine the number of units required for each work task that you identified.
- Experienced estimators include an allowance for waste in their quantities. (Waste is not included in RSMeans Unit Price line items unless otherwise stated.)

3. Price the Quantities
- Use the search tools available to locate individual Unit Price line items for your estimate.
- Reference Numbers indicated within a Unit Price section refer to additional information that you may find useful.
- The crew indicates who is performing the work for that task. Crew codes are expanded in the Crew Listings in the Reference Section to include all trades and equipment that comprise the crew.
- The Daily Output is the amount of work the crew is expected to do in one day.
- The Labor-Hours value is the amount of time it will take for the crew to install one unit of work.
- The abbreviated Unit designation indicates the unit of measure upon which the crew, productivity, and prices are based.
- Bare Costs are shown for materials, labor, and equipment needed to complete the Unit Price line item. Bare costs do not include waste, project overhead, payroll insurance, payroll taxes, main office overhead, or profit.
- The Total Incl O&P cost is the billing rate or invoice amount of the installing contractor or subcontractor who performs the work for the Unit Price line item.

4. Multiply
- Multiply the total number of units needed for your project by the Total Incl O&P cost for each Unit Price line item.
- Be careful that your take off unit of measure matches the unit of measure in the Unit column.
- The price you calculate is an estimate for a completed item of work.
- Keep scoping individual tasks, determining the number of units required for those tasks, matching each task with individual Unit Price line items, and multiplying quantities by Total Incl O&P costs.
- An estimate completed in this manner is priced as if a subcontractor, or set of subcontractors, is performing the work. The estimate does not yet include Project Overhead or Estimate Summary components such as general contractor markups on subcontracted work, general contractor office overhead and profit, contingency, and location factors.

5. Project Overhead
- Include project overhead items from Division 1-General Requirements.
- These items are needed to make the job run. They are typically, but not always, provided by the general contractor. Items include, but are not limited to, field personnel, insurance, performance bond, permits, testing, temporary utilities, field office and storage facilities, temporary scaffolding and platforms, equipment mobilization and demobilization, temporary roads and sidewalks, winter protection, temporary barricades and fencing, temporary security, temporary signs, field engineering and layout, final cleaning, and commissioning.
- Each item should be quantified and matched to individual Unit Price line items in Division 1, then priced and added to your estimate.
- An alternate method of estimating project overhead costs is to apply a percentage of the total project cost, usually 5% to 15% with an average of 10% (see General Conditions).
- Include other project related expenses in your estimate such as:
  - Rented equipment not itemized in the Crew Listings
  - Rubbish handling throughout the project (see 02 41 19.19)

6. Estimate Summary
- Include sales tax as required by laws of your state or county.
- Include the general contractor's markup on self-performed work, usually 5% to 15% with an average of 10%.
- Include the general contractor's markup on subcontracted work, usually 5% to 15% with an average of 10%.
- Include the general contractor's main office overhead and profit:
  - RSMeans gives general guidelines on the general contractor's main office overhead (see section 01 31 13.60 and Reference Number R013113-50).
  - RSMeans gives no guidance on the general contractor's profit.
  - Markups will depend on the size of the general contractor's operations, projected annual revenue, the level of risk, and the level of competition in the local area and for this project in particular.
- Include a contingency, usually 3% to 5%, if appropriate.
- Adjust your estimate to the project's location by using the City Cost Indexes or the Location Factors in the Reference Section:
  - Look at the rules in "How to Use the City Cost Indexes" to see how to apply the Indexes for your location.
  - When the proper Index or Factor has been identified for the project's location, convert it to a multiplier by dividing it by 100, then multiply that multiplier by your estimated total cost. The original estimated total cost will now be adjusted up or down from the national average to a total that is appropriate for your location.

Editors' Note:
We urge you to spend time reading and understanding the supporting material.
An accurate estimate requires experience, knowledge, and careful calculation. The more you know about how we at RSMeans developed the data, the more accurate your estimate will be. In addition, it is important to take into consideration the reference material such as Equipment Listings, Crew Listings, City Cost Indexes, Location Factors, and Reference Tables.

How to Use the Cost Data: The Details

What's Behind the Numbers? The Development of Cost Data

The staff at RSMeans continually monitors developments in the construction industry in order to ensure reliable, thorough, and up-to-date cost information. While overall construction costs may vary relative to general economic conditions, price fluctuations within the industry are dependent upon many factors. Individual price variations may, in fact, be opposite to overall economic trends. Therefore, costs are constantly tracked, and complete updates are performed yearly. Also, new items are frequently added in response to changes in materials and methods.

Costs—$ (U.S.)

All costs represent U.S. national averages and are given in U.S. dollars. The RSMeans City Cost Indexes (CCI) can be used to adjust costs to a particular location. The CCI for Canada can be used to adjust U.S. national averages to local costs in Canadian dollars. No exchange rate conversion is necessary because it has already been factored in.

G The processes or products identified by the green symbol in our publications have been determined to be environmentally responsible and/or resource-efficient solely by the RSMeans engineering staff. The inclusion of the green symbol does not represent compliance with any specific industry association or standard.

Material Costs

The RSMeans staff contacts manufacturers, dealers, distributors, and contractors all across the U.S. and Canada to determine national average material costs. If you have access to current material costs for your specific location, you may wish to make adjustments to reflect differences from the national average. Included within material costs are fasteners for a normal installation. RSMeans engineers use manufacturers' recommendations, written specifications, and/or standard construction practices for size and spacing of fasteners. Adjustments to material costs may be required for your specific application or location. The manufacturer's warranty is assumed. Extended warranties are not included in the material costs. **Material costs do not include sales tax.**

Labor Costs

Labor costs are based upon a mathematical average of trade-specific wages in 30 major U.S. cities. The type of wage (union, open shop, or residential) is identified on the inside back cover of printed publications or is selected by the estimator when using the electronic products. Markups for the wages can also be found on the inside back cover of printed publications and/or under the labor references found in the electronic products.

- If wage rates in your area vary from those used, or if rate increases are expected within a given year, labor costs should be adjusted accordingly.

Labor costs reflect productivity based on actual working conditions. In addition to actual installation, these figures include time spent during a normal weekday on tasks, such as material receiving and handling, mobilization at site, site movement, breaks, and cleanup.

Productivity data is developed over an extended period so as not to be influenced by abnormal variations and reflects a typical average.

Equipment Costs

Equipment costs include not only rental, but also operating costs for equipment under normal use. The operating costs include parts and labor for routine servicing, such as repair and replacement of pumps, filters, and worn lines. Normal operating expendables, such as fuel, lubricants, tires, and electricity (where applicable), are also included. Extraordinary operating expendables with highly variable wear patterns, such as diamond bits and blades, are excluded. These costs are included under materials. Equipment rental rates are obtained from industry sources throughout North America—contractors, suppliers, dealers, manufacturers, and distributors.

Rental rates can also be treated as reimbursement costs for contractor-owned equipment. Owned equipment costs include depreciation, loan payments, interest, taxes, insurance, storage, and major repairs.

Equipment costs do not include operators' wages.

Equipment Cost/Day—The cost of equipment required for each crew is included in the Crew Listings in the Reference Section (small tools that are considered essential everyday tools are not listed out separately). The Crew Listings itemize specialized tools and heavy equipment along with labor trades. The daily cost of itemized equipment included in a crew is based on dividing the weekly bare rental rate by 5 (number of working days per week), then adding the hourly operating cost times 8 (the number of hours per day). This Equipment Cost/Day is shown in the last column of the Equipment Rental Costs in the Reference Section.

Mobilization/Demobilization—The cost to move construction equipment from an equipment yard or rental company to the job site and back again is not included in equipment costs. Mobilization (to the site) and demobilization (from the site) costs can be found in the Unit Price Section. If a piece of equipment is already at the job site, it is not appropriate to utilize mob./demob. costs again in an estimate.

Overhead and Profit

Total Cost including O&P for the installing contractor is shown in the last column of the Unit Price and/or Assemblies. This figure is the sum of the bare material cost plus 10% for profit, the bare labor cost plus total overhead and profit, and the bare equipment cost plus 10% for profit. Details for the calculation of overhead and profit on labor are shown on the inside back cover of the printed product and in the Reference Section of the electronic product.

General Conditions

Cost data in this data set is presented in two ways: Bare Costs and Total Cost including O&P (Overhead and Profit). General Conditions, or General Requirements, of the contract should also be added to the Total Cost including O&P when applicable. Costs for General Conditions are listed in Division 1 of the Unit Price Section and in the Reference Section.

General Conditions for the installing contractor may range from 0% to 10% of the Total Cost including O&P. For the general or prime contractor, costs for General Conditions may range from 5% to 15% of the Total Cost including O&P, with a figure of 10% as the most typical allowance. If applicable, the Assemblies and Models sections use costs that include the installing contractor's overhead and profit (O&P).

Factors Affecting Costs

Costs can vary depending upon a number of variables. Here's a listing of some factors that affect costs and points to consider.

Quality—The prices for materials and the workmanship upon which productivity is based represent sound construction work. They are also in line with industry standard and manufacturer specifications and are frequently used by federal, state, and local governments.

Overtime—We have made no allowance for overtime. If you anticipate premium time or work beyond normal working hours, be sure to make an appropriate adjustment to your labor costs.

Productivity—The productivity, daily output, and labor-hour figures for each line item are based on working an eight-hour day in daylight hours in moderate temperatures. For work that extends beyond normal work hours or is performed under adverse conditions, productivity may decrease.

Size of Project—The size, scope of work, and type of construction project will have a significant impact on cost. Economies of scale can reduce costs for large projects. Unit costs can often run higher for small projects.

Location—Material prices are for metropolitan areas. However, in dense urban areas, traffic and site storage limitations may increase costs. Beyond a 20-mile radius of metropolitan areas, extra trucking or transportation charges may also increase the material costs slightly. On the other hand, lower wage rates may be in effect. Be sure to consider both of these factors when preparing an estimate, particularly if the job site is located in a central city or remote rural location. In addition, highly specialized subcontract items may require travel and per-diem expenses for mechanics.

Other Factors—
- season of year
- contractor management
- weather conditions
- local union restrictions
- building code requirements
- availability of:
  - adequate energy
  - skilled labor
  - building materials
- owner's special requirements/restrictions
- safety requirements
- environmental considerations
- access

Unpredictable Factors—General business conditions influence "in-place" costs of all items. Substitute materials and construction methods may have to be employed. These may affect the installed cost and/or life cycle costs. Such factors may be difficult to evaluate and cannot necessarily be predicted on the basis of the job's location in a particular section of the country. Thus, where these factors apply, you may find significant but unavoidable cost variations for which you will have to apply a measure of judgment to your estimate.

Rounding of Costs

In printed publications only, all unit prices in excess of $5.00 have been rounded to make them easier to use and still maintain adequate precision of the results.

How Subcontracted Items Affect Costs

A considerable portion of all large construction jobs is usually subcontracted. In fact, the percentage done by subcontractors is constantly increasing and may run over 90%. Since the workers employed by these companies do nothing else but install their particular product, they soon become experts in that line. The result is, installation by these firms is accomplished so efficiently that the total in-place cost, even adding the general contractor's overhead and profit, is no more, and often less, than if the principal contractor had handled the installation. Companies that deal with construction specialties are anxious to have their product perform well and, consequently, the installation will be the best possible.

Contingencies

The allowance for contingencies generally provides for unforeseen construction difficulties. On alterations or repair jobs, 20% is not too much. If drawings are final and only field contingencies are being considered, 2% or 3% is probably sufficient, and often nothing needs to be added. Contractually, changes in plans will be covered by extras. The contractor should consider inflationary price trends and possible material shortages during the course of the job. These escalation factors are dependent upon both economic conditions and the anticipated time between the estimate and actual construction. If drawings are not complete or approved, or a budget cost is wanted, it is wise to add 5% to 10%. Contingencies, then, are a matter of judgment.

Important Estimating Considerations

The productivity, or daily output, of each craftsman or crew assumes a well-managed job where tradesmen with the proper tools and equipment, along with the appropriate construction materials, are present. Included are daily set-up and cleanup time, break time, and plan layout time. Unless otherwise indicated, time for material movement on site (for items that can be transported by hand) of up to 200' into the building and to the first or second floor is also included. If material has to be transported by other means, over greater distances, or to higher floors, an additional allowance should be considered by the estimator.

While horizontal movement is typically a sole function of distances, vertical transport introduces other variables that can significantly impact productivity. In an occupied building, the use of elevators (assuming access, size, and required protective measures are acceptable) must be understood at the time of the estimate. For new construction, hoist wait and cycle times can easily be 15 minutes and may result in scheduled access extending beyond the normal work day. Finally, all vertical transport will impose strict weight limits likely to preclude the use of any motorized material handling.

The productivity, or daily output, also assumes installation that meets manufacturer/designer/standard specifications. A time allowance for quality control checks, minor adjustments, and any task required to ensure the proper function or operation is also included. For items that require connections to services, time is included for positioning, leveling, securing the unit, and for making all the necessary connections (and start up where applicable), ensuring a complete installation. Estimating of the services themselves (electrical, plumbing, water, steam, hydraulics, dust collection, etc.) is separate.

In some cases, the estimator must consider the use of a crane and an appropriate crew for the installation of large or heavy items. For those situations where a crane is not included in the assigned crew and as part of the line item cost, equipment rental costs, mobilization and demobilization costs, and operator and support personnel costs must be considered.

Labor-Hours

The labor-hours expressed in this publication are derived by dividing the total daily labor-hours for the crew by the daily output. Based on average installation time and the assumptions listed above, the labor-hours include: direct labor, indirect labor, and nonproductive time. A typical day for a craftsman might include, but is not limited to:

- Direct Work
  - Measuring and layout
  - Preparing materials
  - Actual installation
  - Quality assurance/quality control
- Indirect Work
  - Reading plans or specifications
  - Preparing space
  - Receiving materials
  - Material movement
  - Giving or receiving instruction
  - Miscellaneous
- Non-Work
  - Chatting
  - Personal issues
  - Breaks
  - Interruptions (i.e., sickness, weather, material or equipment shortages, etc.)

If any of the items for a typical day do not apply to the particular work or project situation, the estimator should make any necessary adjustments.

Final Checklist

Estimating can be a straightforward process provided you remember the basics. Here's a checklist of some of the steps you should remember to complete before finalizing your estimate.

Did you remember to:

- factor in the City Cost Index for your locale?
- take into consideration which items have been marked up and by how much?
- mark up the entire estimate sufficiently for your purposes?
- read the background information on techniques and technical matters that could impact your project time span and cost?
- include all components of your project in the final estimate?
- double check your figures for accuracy?
- call RSMeans if you have any questions about your estimate or the data you've used? Remember, RSMeans stands behind its information. If you have any questions about your estimate, about the costs you've used from our data, or even about the technical aspects of the job that may affect your estimate, feel free to call the RSMeans editors at 1-877-759-4771.

Free Quarterly Updates

Stay up-to-date throughout the year with RSMeans' free cost data updates four times a year. Sign up online to make sure you have access to the newest data. Every quarter we provide the city cost adjustment factors for hundreds of cities and key materials.

Register at: **http://info.TheGordianGroup.com/2016Updates.html**.

Square Foot Cost Section

Table of Contents

Introduction
- General 3
- How to Use 4
- Building Classes 8
- Configurations 9
- Building Types 10
- Garage Types 12
- Building Components 13
- Exterior Wall Construction 14

Residential Cost Estimate Worksheets
- Instructions 15
- Model Residence Example 17

Economy
- Illustrations 23
- 1 Story 24
- 1-1/2 Story 26
- 2 Story 28
- Bi-Level 30
- Tri-Level 32
- Wings & Ells 34

Average
- Illustrations 37
- 1 Story 38
- 1-1/2 Story 40
- 2 Story 42
- 2-1/2 Story 44
- 3 Story 46
- Bi-Level 48
- Tri-Level 50
- Solid Wall (log home) 1 Story 52
- Solid Wall (log home) 2 Story 54
- Wings & Ells 56

Custom
- Illustrations 59
- 1 Story 60
- 1-1/2 Story 62
- 2 Story 64
- 2-1/2 Story 66
- 3 Story 68
- Bi-Level 70
- Tri-Level 72
- Wings & Ells 74

Luxury
- Illustrations 77
- 1 Story 78
- 1-1/2 Story 80
- 2 Story 82
- 2-1/2 Story 84
- 3 Story 86
- Bi-Level 88
- Tri-Level 90
- Wings & Ells 92

Residential Modifications
- Kitchen Cabinets 93
- Kitchen Countertops 93
- Vanity Bases 94
- Solid Surface Vanity Tops 94
- Fireplaces & Chimneys 94
- Windows & Skylights 94
- Dormers 94
- Appliances 95
- Breezeway 95
- Porches 95
- Finished Attic 95
- Alarm System 95
- Sauna, Prefabricated 95
- Garages 96
- Swimming Pools 96
- Wood & Coal Stoves 96
- Sidewalks 96
- Fencing 96
- Carport 96

Introduction to the Square Foot Cost Section

The Square Foot Cost Section contains costs per square foot for four classes of construction in seven building types. Costs are listed for various exterior wall materials which are typical of the class and building type. There are cost tables for wings and ells with modification tables to adjust the base cost of each class of building. Non-standard items can easily be added to the standard structures.

Cost estimating for a residence is a three-step process:
1. Identification
2. Listing dimensions
3. Calculations

Guidelines and a sample cost estimating procedure are shown on the following pages.

Identification
To properly identify a residential building, the class of construction, type, and exterior wall material must be determined. The "Building Classes" information has drawings and guidelines for determining the class of construction. There are also detailed specifications and additional drawings at the beginning of each set of tables to further aid in proper building class and type identification.

Sketches for eight types of residential buildings and their configurations are next. Definitions of living area are next to each sketch. Sketches and definitions of garage types follow.

Living Area
Base cost tables are prepared as costs per square foot of living area. The living area of a residence is that area which is suitable and normally designed for full time living. It does not include basement recreation rooms or finished attics, although these areas are often considered full time living areas by the owners.

Living area is calculated from the exterior dimensions without the need to adjust for exterior wall thickness. When calculating the living area of a 1-1/2 story, two story, three story or tri-level residence, overhangs and other differences in size and shape between floors must be considered.

Only the floor area with a ceiling height of seven feet or more in a 1-1/2 story residence is considered living area. In bi-levels and tri-levels, the areas that are below grade are considered living area, even when these areas may not be completely finished.

Base Tables and Modifications
Base cost tables show the base cost per square foot without a basement, with one full bath and one full kitchen for economy and average homes, and an additional half bath for custom and luxury models. Adjustments for finished and unfinished basements are part of the base cost tables. Adjustments for multi-family residences, additional bathrooms, townhouses, alternative roofs, and air conditioning and heating systems are listed in Modifications, Adjustments, and Alternatives tables below the base cost tables.

Costs for other modifications, adjustments, and alternatives, including garages, breezeways, and site improvements, follow the base tables.

Listing of Dimensions
To use this section, only the dimensions used to calculate the horizontal area of the building and additions, modifications, adjustments, and alternatives are needed. The dimensions, normally the length and width, can come from drawings or field measurements. For ease in calculation, consider measuring in tenths of feet, i.e., 9 ft. 6 in. = 9.5 ft., 9 ft. 4 in. = 9.3 ft.

In all cases, make a sketch of the building. Any protrusions or other variations in shape should be noted on the sketch with dimensions.

Calculations
The calculations portion of the estimate is a two-step activity:
1. The selection of appropriate costs from the tables
2. Computations

Selection of Appropriate Costs
To select the appropriate cost from the base tables, the following information is needed:
1. Class of construction
   - Economy
   - Average
   - Custom
   - Luxury
2. Type of residence
   - 1 story
   - 1-1/2 story
   - 2 story
   - 3 story
   - Bi-level
   - Tri-level
3. Occupancy
   - One family
   - Two family
   - Three family
4. Building configuration
   - Detached
   - Town/Row house
   - Semi-detached
5. Exterior wall construction
   - Wood frame
   - Brick veneer
   - Solid masonry
6. Living areas

Modifications are classified by class, type, and size.

Computations
The computation process should take the following sequence:
1. Multiply the base cost by the area.
2. Add or subtract the modifications, adjustments, and alternatives.
3. Apply the location modifier.

When selecting costs, interpolate or use the cost that most nearly matches the structure under study. This applies to size, exterior wall construction, and class.

How to Use the Residential Square Foot Cost Section

The following is a detailed explanation of a sample entry in the Residential Square Foot Cost Section. Each bold number below corresponds to the item being described on the facing page with the appropriate component of the sample entry in parentheses. Prices listed are costs that include overhead and profit of the installing contractor. Total model costs include an additional markup for general contractors' overhead and profit, and fees specific to the class of construction.

RESIDENTIAL — Average **①** — 2 Story **②**

- Simple design from standard plans
- Single family — 1 full bath, 1 kitchen
- No basement **③**
- Asphalt shingles on roof
- Hot air heat
- Gypsum wallboard interior finishes
- Materials and workmanship are average

Note: The illustration shown may contain some optional components (for example: garages and/or fireplaces) whose costs are shown in the modifications, adjustments, & alternatives below or at the end of the square foot section.

Base cost per square foot of living area

Living Area **⑤**

Exterior Wall **④**	1000	1200	1400	1600	1800	2000	2200	2600	3000	3400	3800
Wood Siding - Wood Frame	142.45	128.80	122.15	117.55	112.85	107.95 **⑥**	104.65	98.35	92.35	89.65	87.15
Brick Veneer - Wood Frame	149.30	135.15	128.05	123.25	118.20	113.05	109.55	102.75	96.45	93.50	90.80
Stucco on Wood Frame	137.75	124.40	118.00	113.65	109.20	104.40	101.25	95.30	89.55	86.90	84.55
Solid Masonry	163.05	147.85	139.95	134.60	128.90	123.40	119.30	111.65	104.60	101.25	98.15
Finished Basement, Add **⑦**	22.70	22.45	21.70	21.15	20.65	20.30	19.85 **⑧**	19.15	18.60	18.25	17.95
Unfinished Basement, Add	9.00	8.40	7.90	7.60	7.25	7.00	6.75	6.35	6.00	5.75	5.60

Modifications
Add to the total cost

Upgrade Kitchen Cabinets	**⑨** $ + 5525
Solid Surface Countertops (Included)	
Full Bath - including plumbing, wall and floor finishes	+ 7405
Half Bath - including plumbing, wall and floor finishes	+ 4331
One Car Attached Garage	+ 14,471
One Car Detached Garage	+ 19,004
Fireplace & Chimney	+ 7319

Adjustments
For multi family - add to total cost

Additional Kitchen	$ + 9600
Additional Bath	+ 7405
Additional Entry & Exit	+ 1676
Separate Heating	+ 1336
Separate Electric	+ 1945

For Townhouse/Rowhouse -
Multiply cost per square foot by

Inner Unit	**⑩** .90
End Unit	.95

Alternatives
Add to or deduct from the cost per square foot of living area

Cedar Shake Roof	+ 1.65
Clay Tile Roof	+ 3.40
Slate Roof	+ 3.75
Upgrade Walls to Skim Coat Plaster **⑪**	+ .53
Upgrade Ceilings to Textured Finish	+ .58
Air Conditioning, in Heating Ductwork	+ 2.96
In Separate Ductwork	+ 5.70
Heating Systems, Hot Water	+ 1.49
Heat Pump	+ 1.61
Electric Heat	– .63
Not Heated	– 3.28

Additional upgrades or components

Kitchen Cabinets & Countertops	Page 93
Bathroom Vanities	94
Fireplaces & Chimneys	94
Windows, Skylights & Dormers	94
Appliances	95
Breezeways & Porches	95
Finished Attic	95
Garages	96
Site Improvements **⑫**	96
Wings & Ells	56

① Class of Construction (Average)
The class of construction depends upon the design and specifications of the plan. The four classes are economy, average, custom, and luxury.

② Type of Residence (2 Story)
The building type describes the number of stories or levels in the model. The seven building types are 1 story, 1-1/2 story, 2 story, 2-1/2 story, 3 story, bi-level, and tri-level.

③ Specification Highlights (Hot Air Heat)
These specifications include information concerning the components of the model, including the number of baths, roofing types, HVAC systems, materials, and workmanship. If the components listed are not appropriate, modifications can be made by consulting the information shown below or the Assemblies section.

④ Exterior Wall System (Wood Siding–Wood Frame)
This section includes the types of exterior wall systems and the structural frame used. The exterior wall systems shown are typical of the class of construction and the building type shown.

⑤ Living Areas (2000 SF)
The living area is that area of the residence which is suitable and normally designed for full time living. It does not include basement recreation rooms or finished attics. Living area is calculated from the exterior dimensions without the need to adjust for exterior wall thickness. When calculating the living area of a 1-1/2 story, 2 story, 3 story, or tri-level residence, overhangs and other differences in size and shape between floors must be considered. Only the floor area with a ceiling height of seven feet or more in a 1-1/2 story residence is considered living area. In bi-levels and tri-levels, the areas that are below grade are considered living area, even when these areas may not be completely finished. A range of various living areas for the residential model is shown to aid in the selection of values from the matrix.

⑥ Base Costs per Square Foot of Living Area ($107.95)
Base cost tables show the cost per square foot of living area without a basement, with one full bath and one full kitchen for economy and average homes, and an additional half bath for custom and luxury models. When selecting costs, interpolate or use the cost that most nearly matches the residence under consideration for size, exterior wall system, and class of construction. Prices listed are costs that include overhead and profit of the installing contractor, a general contractor markup, and an allowance for plans that vary by class of construction. For additional information on contractor overhead and architectural fees, see the Reference Section.

⑦ Basement Types (Finished)
The two types of basements are finished and unfinished. The specifications and components for both are shown under Building Classes in the Introduction to this section.

⑧ Additional Costs for Basements ($20.30 or $7.00)
These values indicate the additional cost per square foot of living area for either a finished or an unfinished basement.

⑨ Modifications and Adjustments (Upgrade Kitchen Cabinets $5525)
Modifications and Adjustments are costs added to or subtracted from the total cost of the residence. The total cost of the residence is equal to the cost per square foot of living area times the living area. Typical modifications and adjustments include kitchens, baths, garages, and fireplaces.

⑩ Multiplier for Townhouse/Rowhouse (Inner Unit .90)
The multipliers shown adjust the base costs per square foot of living area for the common wall condition encountered in townhouses or rowhouses.

⑪ Alternatives (Skim Coat Plaster $0.53)
Alternatives are costs added to or subtracted from the base cost per square foot of living area. Typical alternatives include variations in kitchens, baths, roofing, and air conditioning and heating systems.

⑫ Additional Upgrades or Components (Wings & Ells)
Costs for additional upgrades or components, including wings or ells, breezeways, porches, finished attics, and site improvements, are shown at the end of each quality section and at the end of the Square Foot Section.

How to Use the Residential Square Foot Cost Section (Continued)

Average 2 Story

❶ Living Area - 2000 S.F.
Perimeter - 135 L.F.

			Cost Per Square Foot Of Living Area			% of Total
			Mat.	Inst.	Total	(rounded)
1	Site Work	Site preparation for slab; 4' deep trench excavation for foundation wall.		1.33	1.33	1.2%
2	Foundation	Continuous reinforced concrete footing, 10" deep x 20" wide; dampproofed and insulated 8" thick reinforced concrete block foundation wall, 4' deep; trowel finished 4" thick concrete slab on 4" crushed stone base and polyethylene vapor barrier.	3.55	4.46	8.01	7.4%
3	Framing	Exterior walls - 2" x 6" wood studs, 16" O.C.; 1/2" sheathing; gable end roof framing, 2" x 10" rafters, 16" O.C. with 1/2" plywood sheathing; 2" x 10" floor joists, 16" O.C. with bridging and 5/8" subflooring; 2" x 4" interior partitions.	7.07 ❹	9.77 ❺	16.84 ❻	15.6% ❼
4	Exterior Walls	Beveled wood siding and housewrap on insulated wood frame walls; R38 attic insulation; double hung wood windows; flush solid core doors, frame and hardware, painted finish; aluminum storm and screen doors.	13.75	5.90	19.65	18.2%
5	Roofing	25 year asphalt roof shingles; #15 felt building paper; aluminum gutters, downspouts, drip edge and flashings.	1.16	1.36	2.52	2.3%
6	Interiors	Walls & ceilings, 1/2" taped & finished gypsum wallboard, primed & painted with 2 coats of finish paint; birch faced hollow core interior doors, frames & hardware, painted finish; medium weight carpeting with pad, 40%; sheet vinyl, 15%; oak hardwood, 40%; ceramic tile, 5%; hardwood tread stairway.	14.16	14.83	28.99	26.9%
7	Specialties	Average grade kitchen cabinets and countertop; stainless steel kitchen sink; 40 gallon electric water heater.	3.55	1.19	4.74	4.4%
8	Mechanical	Three fixture bathroom: bathtub, water closet, vanity and sink; gas fired hot air heating system.	3.07	3.31	6.38	5.9%
9	Electrical	200 amp electric service; wiring, duplex & GFI receptacles, wall switches, door bell, appliance circuits, fans and communications cabling; average grade lighting fixtures	1.65	2.16	3.81	3.5%
10	Overhead	Contractor's overhead and profit and plans. ❽	8.14	7.54	15.68	14.5%
		Total	56.10	51.85	**107.95** ❾	

① Specifications
The parameters for an example dwelling from the previous pages are listed here. Included are the square foot dimensions of the proposed building. Living Area takes into account the number of floors and other factors needed to define a building's total square footage. Perimeter dimensions are defined in terms of linear feet.

② Building Type
This is a sketch of a cross section view through the dwelling. It is shown to help define the living area for the building type. For more information, see the Building Types in the Introduction.

③ Components (3 Framing)
This page contains the ten components needed to develop the complete square foot cost of the typical dwelling specified. All components are defined with a description of the materials and/or task involved. Use cost figures from each component to estimate the cost per square foot of that section of the project. The components listed here are typical of all sizes of residences from the facing page. Specific quantities of components required would vary with the size of the dwelling and the exterior wall system.

④ Materials (7.07)
This column gives the amount needed to develop the cost of materials. The figures given here are not bare costs. Ten percent has been added to bare material cost for profit.

⑤ Installation (9.77)
Installation includes labor and equipment costs. The labor rates included here incorporate the total overhead and profit costs for the installing contractor. The average mark-up used to create these figures is 72.0% over and above bare labor costs. The equipment rates include 10% for profit.

⑥ Total (16.84)
This column lists the sum of two figures. Use this total to determine the sum of material cost plus installation cost. The result is a convenient total cost for each of the ten components.

⑦ % of Total (rounded) (15.6%)
This column represents the percent of the total cost for this component group.

⑧ Overhead
The costs in components 1 through 9 include overhead and profit for the installing contractor. Item 10 is overhead and profit for the general contractor. This is typically a percentage mark-up of all other costs. The amount depends on size and type of dwelling, building class, and economic conditions. An allowance for plans or design has been included where appropriate.

⑨ Bottom Line Total (107.95)
This figure is the complete square foot cost for the construction project and equals the sum of total material and total labor costs. To determine total project cost, multiply the bottom line total by the living area.

Building Classes

Economy Class

An economy class residence is usually built from stock plans. The materials and workmanship are sufficient to satisfy building codes. Low construction cost is more important than distinctive features. The overall shape of the foundation and structure is seldom other than square or rectangular.

An unfinished basement includes a 7' high, 8" thick foundation wall composed of either concrete block or cast-in-place concrete.

Included in the finished basement cost are inexpensive paneling or drywall as the interior finish on the foundation walls, a low cost sponge backed carpeting adhered to the concrete floor, a drywall ceiling, and overhead lighting.

Average Class

An average class residence is a simple design and built from standard plans. Materials and workmanship are average but often exceed minimum building codes. There are frequently special features that give the residence some distinctive characteristics.

An unfinished basement includes a 7'-6" high, 8" thick foundation wall composed of either cast-in-place concrete or concrete block.

Included in the finished basement are plywood paneling or drywall on furring that is fastened to the foundation walls, sponge backed carpeting adhered to the concrete floor, a suspended ceiling, overhead lighting, and heating.

Custom Class

A custom class residence is usually built from plans and specifications with enough features to give the building a distinction of design. Materials and workmanship are generally above average with obvious attention given to construction details. Construction normally exceeds building code requirements.

An unfinished basement includes a 7'-6" high, 10" thick cast-in-place concrete foundation wall or a 7'-6" high, 12" thick concrete block foundation wall.

A finished basement includes painted drywall on insulated 2" x 4" wood furring as the interior finish to the concrete walls, a suspended ceiling, carpeting adhered to the concrete floor, overhead lighting, and heating.

Luxury Class

A luxury class residence is built from an architect's plan for a specific owner. It is unique both in design and workmanship. There are many special features, and construction usually exceeds all building codes. It is obvious that primary attention is placed on the owner's comfort and pleasure. Construction is supervised by an architect.

An unfinished basement includes an 8' high, 12" thick foundation wall composed of cast-in-place concrete or concrete block.

A finished basement includes painted drywall on 2" x 4" wood furring as the interior finish, a suspended ceiling, tackless carpet on wood subfloor with sleepers, overhead lighting, and heating.

Configurations

Detached House

This category of residence is a free-standing separate building with or without an attached garage. It has four complete walls.

Semi-Detached House

This category of residence has two living units side-by-side. The common wall is fireproof. Semi-detached residences can be treated as a row house with two end units. Semi-detached residences can be any of the building types.

Town/Row House

This category of residence has a number of attached units made up of inner units and end units. The units are joined by common walls. The inner units have only two exterior walls. The common walls are fireproof. The end units have three walls and a common wall. Town houses/row houses can be any of the building types.

Building Types

One Story
This is an example of a one-story dwelling. The living area of this type of residence is confined to the ground floor. The headroom in the attic is usually too low for use as a living area.

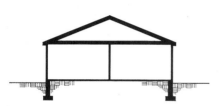

One-and-one-half Story
The living area in the upper level of this type of residence is 50% to 90% of the ground floor. This is made possible by a combination of this design's high-peaked roof and/or dormers. Only the upper level area with a ceiling height of seven feet or more is considered living area. The living area of this residence is the sum of the ground floor area plus the area on the second level with a ceiling height of seven feet or more.

One Story with Finished Attic
The main living area in this type of residence is the ground floor. The upper level or attic area has sufficient headroom for use as a living area. This is made possible by a high peaked roof. The living area in the attic is less than 50% of the ground floor. The living area of this type of residence is the ground floor area only. The finished attic is considered an adjustment.

Two Story
This type of residence has a second floor or upper level area which is equal or nearly equal to the ground floor area. The upper level of this type of residence can range from 90% to 110% of the ground floor area, depending on setbacks or overhangs. The living area is the sum of the ground floor area and the upper level floor area.

Two-and-one-half Story

This type of residence has two levels of equal or nearly equal area and a third level which has a living area that is 50% to 90% of the ground floor. This is made possible by a high peaked roof, extended wall heights, and/or dormers. Only the upper level area with a ceiling height of seven feet or more is considered living area. The living area of this residence is the sum of the ground floor area, the second floor area, and the area on the third level with a ceiling height of seven feet or more.

Bi-level

This type of residence has two living areas, one above the other. One area is about four feet below grade and the second is about four feet above grade. Both areas are equal in size. The lower level in this type of residence is designed and built to serve as a living area and not as a basement. Both levels have full ceiling heights. The living area is the sum of the lower level area and the upper level area.

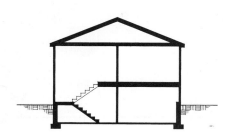

Three Story

This type of residence has three levels which are equal or nearly equal. As in the two story residence, the second and third floor areas may vary slightly depending on setbacks or overhangs. The living area is the sum of the ground floor area and the two upper level floor areas.

Tri-level

This type of residence has three levels of living area: one at grade level, one about four feet below grade, and one about four feet above grade. All levels are designed to serve as living areas. All levels have full ceiling heights. The living area is a sum of the areas of each of the three levels.

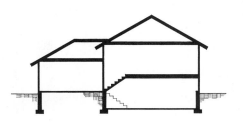

Garage Types

Attached Garage
Shares a common wall with the dwelling. Access is typically through a door between the dwelling and garage.

Basement Garage
Constructed under the roof of the dwelling but below the living area.

Built-In Garage
Constructed under the second floor living space and above the basement level of the dwelling. Reduces gross square feet of the living area.

Detached Garage
Constructed apart from the main dwelling. Shares no common area or wall with the dwelling.

Building Components

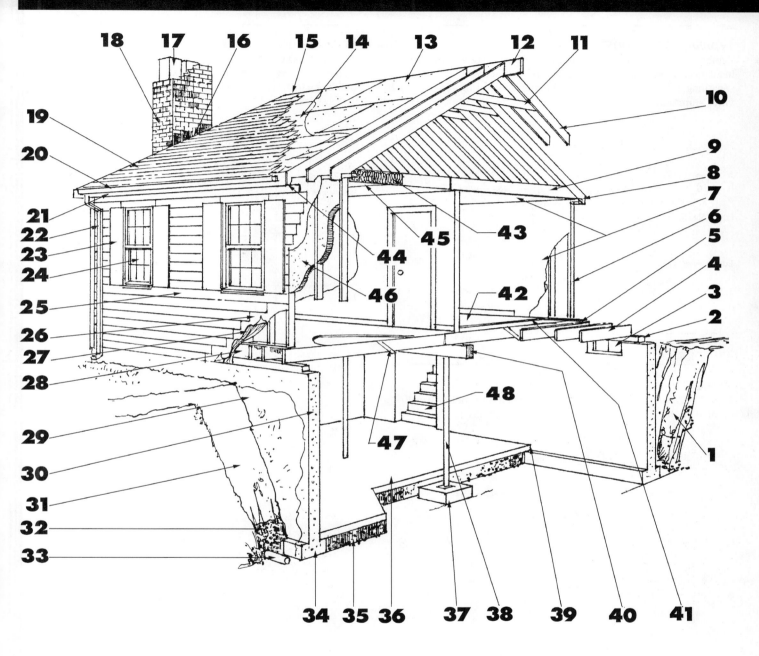

1. Excavation
2. Sill Plate
3. Basement Window
4. Floor Joist
5. Shoe Plate
6. Studs
7. Drywall
8. Plate
9. Ceiling Joists
10. Rafters
11. Collar Ties
12. Ridge Rafter
13. Roof Sheathing
14. Roof Felt
15. Roof Shingles
16. Flashing
17. Flue Lining
18. Chimney
19. Roof Shingles
20. Gutter
21. Fascia
22. Downspout
23. Shutter
24. Window
25. Wall Shingles
26. Weather Barrier
27. Wall Sheathing
28. Fire Stop
29. Dampproofing
30. Foundation Wall
31. Backfill
32. Drainage Stone
33. Drainage Tile
34. Wall Footing
35. Gravel
36. Concrete Slab
37. Column Footing
38. Pipe Column
39. Expansion Joint
40. Girder
41. Sub-floor
42. Finish Floor
43. Attic Insulation
44. Soffit
45. Ceiling Strapping
46. Wall Insulation
47. Cross Bridging
48. Bulkhead Stairs

Exterior Wall Construction

Typical Frame Construction
Typical wood frame construction consists of wood studs with insulation between them. A typical exterior surface is made up of sheathing, building paper, and exterior siding consisting of wood, vinyl, aluminum, or stucco over the wood sheathing.

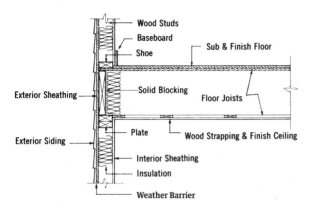

Brick Veneer
Typical brick veneer construction consists of wood studs with insulation between them. A typical exterior surface is sheathing, building paper, and an exterior of brick tied to the sheathing, with metal strips.

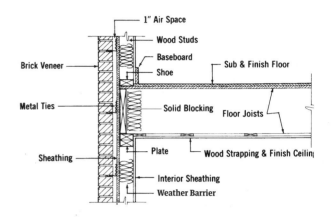

Stone
Typical solid masonry construction consists of a stone or block wall covered on the exterior with brick, stone, or other masonry.

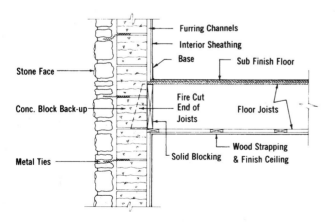

Residential Cost Estimate Worksheet

Worksheet Instructions

The residential cost estimate worksheet can be used as an outline for developing a residential construction or replacement cost. It is also useful for insurance appraisals. The design of the worksheet helps eliminate errors and omissions. To use the worksheet, follow the example below.

1. Fill out the owner's name, residence address, the estimator or appraiser's name, some type of project identifying number or code, and the date.
2. Determine from the plans, specifications, owner's description, photographs, or any other means possible the class of construction. The models in this data set use economy, average, custom, and luxury as classes. Fill in the appropriate box.
3. Fill in the appropriate box for the residence type, configuration, occupancy, and exterior wall. If you require clarification, the pages preceding this worksheet describe each of these.
4. Next, the living area of the residence must be established. The heated or air conditioned space of the residence, not including the basement, should be measured. It is easiest to break the structure up into separate components as shown in the example. The main house (A), a one-and-one-half story wing (B), and a one story wing (C). The breezeway (D), garage (E), and open covered porch (F) will be treated differently. Data entry blocks for the living area are included on the worksheet for your use. Keep each level of each component separate, and fill out the blocks as shown.
5. By using the information on the worksheet, find the model, wing, or ell in the following square foot cost pages that best matches the class, type, exterior finish, and size of the residence being estimated. Use the Modifications, Adjustments, and Alternatives to determine the adjusted cost per square foot of living area for each component.
6. For each component, multiply the cost per square foot by the living area square footage. If the residence is a townhouse/rowhouse, a multiplier should be applied based upon the configuration.
7. The second page of the residential cost estimate worksheet has space for the additional components of a house. The cost for additional bathrooms, finished attic space, breezeways, porches, fireplaces, appliance or cabinet upgrades, and garages should be added on this page. The information for each of these components is found with the model being used, or in the Modifications, Adjustments, and Alternatives pages.
8. Add the total from page one of the estimate worksheet and the items listed on page two. The sum is the adjusted total building cost.
9. Depending on the use of the final estimated cost, one of the remaining two boxes should be filled out. Any additional items or exclusions should be added or subtracted at this time. The data contained in this book is a national average. Construction costs are different throughout the country. To allow for this difference, a location factor based upon the first three digits of the residence's zip code must be applied. The location factor is a multiplier that increases or decreases the adjusted total building cost. Find the appropriate location factor and calculate the local cost. If depreciation is a concern, a dollar figure should be subtracted at this point.
10. No residence will match a model exactly. Many differences will be found. At this level of estimating, a variation of plus or minus 10% should be expected.

Adjustments Instructions

No residence matches a model exactly in shape, material, or specifications. The common differences are:
1. Two or more exterior wall systems:
   - Partial basement
   - Partly finished basement
2. Specifications or features that are between two classes
3. Crawl space instead of a basement

Examples

Below are quick examples. See pages 17-19 for complete examples of cost adjustments for these differences:

1. Residence "A" is an average one-story structure with 1,600 S.F. of living area and no basement. Three walls are wood siding on wood frame, and the fourth wall is brick veneer on wood frame. The brick veneer wall is 35% of the exterior wall area.

   Use page 38 to calculate the Base Cost per S.F. of Living Area. Wood Siding for 1,600 S.F. = $111.55 per S.F. Brick Veneer for 1,600 S.F. = $115.65 per S.F.

   .65 ($111.55) + .35 ($115.65) = $112.99 per S.F. of Living Area.

2a. Residence "B" is the same as Residence "A"; However, it has an unfinished basement under 50% of the building. To adjust the $112.99 per S.F. of living area for this partial basement, use page 38.

   $112.99 + .5($11.40) = $118.69 per S.F. of Living Area.

2b. Residence "C" is the same as Residence "A"; However, it has a full basement under the entire building. 640 S.F. or 40% of the basement area is finished.

   Using Page 38:

   $112.99 + .40 ($32.70) + .60 ($11.40) = $132.91 per S.F. of Living Area.

3. When specifications or features of a building are between classes, estimate the percent deviation, and use two tables to calculate the cost per S.F.

   A two-story residence with wood siding and 1,800 S.F. of living area has features 30% better than Average, but 70% less than Custom.

   From pages 42 and 64:

Custom 1,800 S.F. Base Cost	=	**$145.25 per S.F.**
Average 1,800 S.F. Base Cost	=	**$112.85 per S.F.**
DIFFERENCE	=	**$32.40 per S.F.**
Cost is $112.85 + .30 ($32.40)	=	**$122.57 per S.F. of Living Area**

4. To add the cost of a crawl space, use the cost of an unfinished basement as a maximum. For specific costs of components to be added or deducted, such as vapor barrier, underdrain, and floor, see the "Assemblies" section, pages 97 to 281.

Model Residence Example

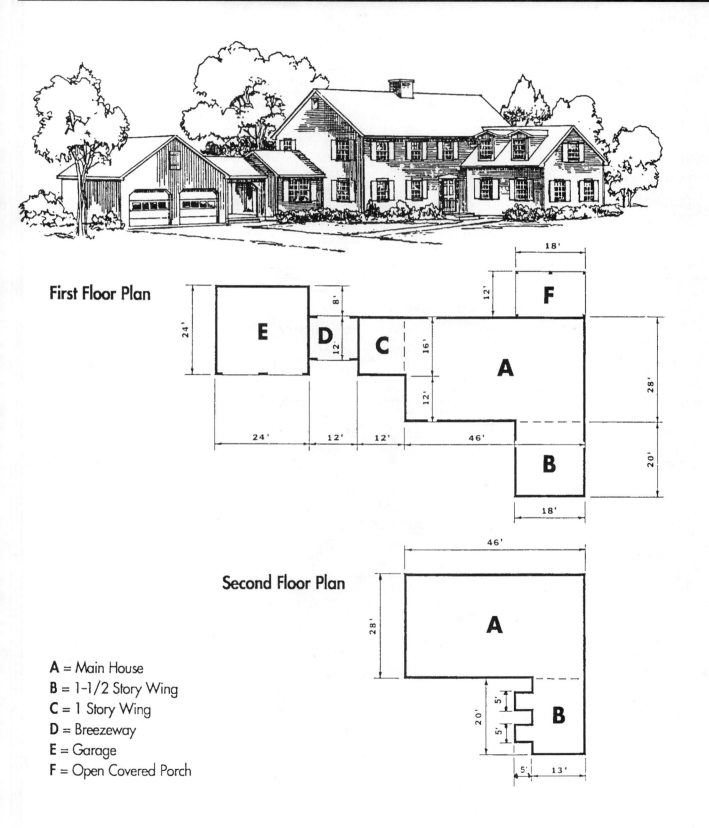

First Floor Plan

Second Floor Plan

A = Main House
B = 1-1/2 Story Wing
C = 1 Story Wing
D = Breezeway
E = Garage
F = Open Covered Porch

RESIDENTIAL COST ESTIMATE

OWNERS NAME:	**Albert Westenberg**	APPRAISER:	**Nicole Wojtowicz**
RESIDENCE ADDRESS:	**300 Sygiel Road**	PROJECT:	**# 55**
CITY, STATE, ZIP CODE:	**Three Rivers, MA 01080**	DATE:	**Jan. 1, 2016**

CLASS OF CONSTRUCTION
- ☐ ECONOMY
- ☑ AVERAGE
- ☐ CUSTOM
- ☐ LUXURY

RESIDENCE TYPE
- ☐ 1 STORY
- ☐ 1 1/2 STORY
- ☑ 2 STORY
- ☐ 2 1/2 STORY
- ☐ 3 STORY
- ☐ BI-LEVEL
- ☐ TRI-LEVEL

CONFIGURATION
- ☑ DETACHED
- ☐ TOWN/ROW HOUSE
- ☐ SEMI-DETACHED

OCCUPANCY
- ☑ ONE FAMILY
- ☐ TWO FAMILY
- ☐ THREE FAMILY
- ☐ OTHER

EXTERIOR WALL SYSTEM
- ☑ WOOD SIDING - WOOD FRAME
- ☐ BRICK VENEER - WOOD FRAME
- ☐ STUCCO ON WOOD FRAME
- ☐ PAINTED CONCRETE BLOCK
- ☐ SOLID MASONRY (AVERAGE & CUSTOM)
- ☐ STONE VENEER - WOOD FRAME
- ☐ SOLID BRICK (LUXURY)
- ☐ SOLID STONE (LUXURY)

* LIVING AREA (Main Building)		* LIVING AREA (Wing or Ell) (B)		* LIVING AREA (Wing or Ell) (C)	
First Level	1288 S.F.	First Level	360 S.F.	First Level	192 S.F.
Second level	1288 S.F.	Second level	310 S.F.	Second level	S.F.
Third Level	S.F.	Third Level	S.F.	Third Level	S.F.
Total	2576 S.F.	Total	670 S.F.	Total	192 S.F.

* Basement Area is not part of living area.

MAIN BUILDING — COSTS PER S.F. LIVING AREA

Cost per Square Foot of Living Area, from Page	**42**	$	107.95
Basement Addition: ____ % Finished, **100** % Unfinished		+	7.00
Roof Cover Adjustment: **Cedar Shake** Type, Page **42** (Add or Deduct)		( )	1.65
Central Air Conditioning: ☐ Separate Ducts ☑ Heating Ducts, Page **42**		+	2.96
Heating System Adjustment: ____ Type, Page ____ (Add or Deduct)		( )	
Main Building: Adjusted Cost per S.F. of Living Area		$ #	119.56

MAIN BUILDING TOTAL COST: $ 119.56 /S.F. x 2,576 S.F. x _____ = $ 307,987
(Cost per S.F. Living Area) (Living Area) (Town/Row House Multiplier - Use 1 for Detached) (TOTAL COST)

WING OR ELL (B) — 1-1/2 STORY — COSTS PER S.F. LIVING AREA

Cost per Square Foot of Living Area, from Page	**56**	$	97.10
Basement Addition: **100** % Finished, ____ % Unfinished		+	29.60
Roof Cover Adjustment: ____ Type, Page ____ (Add or Deduct)		( )	—
Central Air Conditioning: ☐ Separate Ducts ☑ Heating Ducts, Page **42**		+	2.96
Heating System Adjustment: ____ Type, Page ____ (Add or Deduct)		( )	—
Wing or Ell (B): Adjusted Cost per S.F. of Living Area		$ #	129.66

WING OR ELL (B) TOTAL COST: $ 129.66 /S.F. x 670 S.F. = $ 86,872

WING OR ELL (C) — 1 STORY (WOOD SIDING) — COSTS PER S.F. LIVING AREA

Cost per Square Foot of Living Area, from Page	# **56**	$ #	164.10
Basement Addition: ____ % Finished, ____ % Unfinished		+	—
Roof Cover Adjustment: ____ Type, Page ____ (Add or Deduct)		( )	—
Central Air Conditioning: ☐ Separate Ducts ☐ Heating Ducts, Page ____		+	—
Heating System Adjustment: ____ Type, Page ____ (Add or Deduct)		( )	—
Wing or Ell (C) Adjusted Cost per S.F. of Living Area		$ #	164.1

WING OR ELL (C) TOTAL COST: $ 164.10 /S.F. x 192 S.F. = $ 31,507

TOTAL THIS PAGE: 426,366

RESIDENTIAL COST ESTIMATE

Total Page 1								$	426,366
						QUANTITY	UNIT COST		
Additional Bathrooms:	2	Full,	1	Half	2 @ 7405 1 @ 4331			+	19,141
Finished Attic:	N/A	Ft. x		Ft.			S.F.		
Breezeway: ☑ Open ☐ closed		12	Ft. x	12	Ft.	144	S.F. 38.44	+	5,535
Covered Porch: ☑ Open ☐ Enclosed		18	Ft. x	12	Ft.	216	S.F. 37.09	+	8,011
Fireplace: ☑ Interior Chimney ☐ Exterior Chimney									
☑ No. of Flues (2) ☑ Additional Fireplaces				1 - 2nd Story				+	12,534
Appliances:								+	—
Kitchen Cabinets Adjustments:				(±)					—
☑ Garage ☐ Carport: 2 Car(s) Description Wood, Attached					(±)				24,571
Miscellaneous:								+	
					ADJUSTED TOTAL BUILDING COST			$	496,158

REPLACEMENT COST		
ADJUSTED TOTAL BUILDING COST	$	496,158
Site Improvements		
(A) Paving & Sidewalks	$	
(B) Landscaping	$	
(C) Fences	$	
(D) Swimming Pools	$	
(E) Miscellaneous	$	
TOTAL	$	496,158
Location Factor	x	1.060
Location Replacement Cost	$	525,927
Depreciation	-$	52,592
LOCAL DEPRECIATED COST	$	473,335

INSURANCE COST		
ADJUSTED TOTAL BUILDING COST	$	
Insurance Exclusions		
(A) Footings, sitework, Underground Piping	-$	
(B) Architects Fees	-$	
Total Building Cost Less Exclusion	$	
Location Factor	x	
LOCAL INSURABLE REPLACEMENT COST	$	

SKETCH AND ADDITIONAL CALCULATIONS

No part of this cost data may be reproduced, stored in a retrieval system, or transmitted in any form or by any means without prior written permission of RSMeans.

Did you know?

RSMeans Online gives you the same access to RSMeans' data with 24/7 access:
- Quickly locate costs in the searchable database.
- Build cost lists, estimates, and reports in minutes.
- Adjust costs to any location in the U.S. and Canada with the click of a button.

Start your free trial today at **www.RSMeansOnline.com**

RSMeans Online
FROM THE GORDIAN GROUP®

RESIDENTIAL | Economy | Illustrations

1 Story

1-1/2 Story

2 Story

Bi-Level

Tri-Level

RESIDENTIAL — Economy — 1 Story

- Mass produced from stock plans
- Single family — 1 full bath, 1 kitchen
- No basement
- Asphalt shingles on roof
- Hot air heat
- Gypsum wallboard interior finishes
- Materials and workmanship are sufficient to meet codes

Note: The illustration shown may contain some optional components (for example: garages and/or fireplaces) whose costs are shown in the modifications, adjustments, & alternatives below or at the end of the square foot section.

©Home Planners, Inc.

Base cost per square foot of living area

Exterior Wall	600	800	1000	1200	1400	1600	1800	2000	2400	2800	3200
Wood Siding - Wood Frame	140.00	126.30	116.00	107.80	100.55	96.05	93.70	90.60	84.50	80.05	77.00
Brick Veneer - Wood Frame	146.40	132.05	121.25	112.50	104.85	100.10	97.55	94.25	87.80	83.05	79.85
Stucco on Wood Frame	131.90	119.10	109.45	101.95	95.25	91.10	88.85	86.05	80.30	76.20	73.45
Painted Concrete Block	138.15	124.70	114.55	106.45	99.40	94.95	92.70	89.60	83.55	79.15	76.15
Finished Basement, Add	32.80	30.90	29.50	28.30	27.25	26.60	26.25	25.70	25.00	24.45	23.90
Unfinished Basement, Add	14.50	12.95	11.90	10.85	10.10	9.55	9.25	8.85	8.25	7.80	7.40

Modifications
Add to the total cost

Upgrade Kitchen Cabinets	$ + 1205
Solid Surface Countertops	+ 849
Full Bath - including plumbing, wall and floor finishes	+ 5924
Half Bath - including plumbing, wall and floor finishes	+ 3465
One Car Attached Garage	+ 13,419
One Car Detached Garage	+ 17,321
Fireplace & Chimney	+ 6285

Adjustments
For multi family - add to total cost

Additional Kitchen	$ + 5288
Additional Bath	+ 5924
Additional Entry & Exit	+ 1676
Separate Heating	+ 1336
Separate Electric	+ 966

For Townhouse/Rowhouse - Multiply cost per square foot by

Inner Unit	.95
End Unit	.97

Alternatives
Add to or deduct from the cost per square foot of living area

Composition Roll Roofing	– 1.15
Cedar Shake Roof	+ 3.90
Upgrade Walls and Ceilings to Skim Coat Plaster	+ .72
Upgrade Ceilings to Textured Finish	+ .58
Air Conditioning, in Heating Ductwork	+ 4.73
In Separate Ductwork	+ 6.92
Heating Systems, Hot Water	+ 1.57
Heat Pump	+ 1.33
Electric Heat	– 1.47
Not Heated	– 4.21

Additional upgrades or components

Kitchen Cabinets & Countertops	Page 93
Bathroom Vanities	94
Fireplaces & Chimneys	94
Windows, Skylights & Dormers	94
Appliances	95
Breezeways & Porches	95
Finished Attic	95
Garages	96
Site Improvements	96
Wings & Ells	34

Important: See the Reference Section for Location Factors (to adjust for your city) and Estimating Forms.

Economy 1 Story

Living Area - 1200 S.F.
Perimeter - 146 L.F.

			Cost Per Square Foot Of Living Area			% of Total
			Mat.	Inst.	Total	(rounded)
1	Site Work	Site preparation for slab; 4' deep trench excavation for foundation wall.		2.15	2.15	2.1%
2	Foundation	Continuous reinforced concrete footing, 10" deep x 20" wide; damproofed and insulated 8" thick reinforced concrete block foundation wall, 4' deep; trowel finished 4" thick concrete slab on 4" crushed stone base and polyethylene vapor barrier.	6.40	7.99	14.39	14.1%
3	Framing	Exterior walls - 2" x 4" wood studs, 16" O.C.; 1/2" sheathing; wood truss roof frame, 24" O.C. with 1/2" plywood sheating; 2" x 4" interior partitions.	5.58	7.57	13.15	12.9%
4	Exterior Walls	Metal lath reinforced stucco exterior on insulated wood frame walls; R38 attic insulation; sliding wood windows; flush solid core doors, frame and hardware, painted finish; aluminum storm and screen doors.	7.81	7.40	15.21	14.9%
5	Roofing	25 year asphalt roof shingles; #15 felt building paper; aluminum gutters, downspouts, drip edge and flashings.	2.26	2.63	4.89	4.8%
6	Interiors	Walls and ceilings, 1/2" taped and finished gypsum wallboard, primed and painted with 2 coats of finish paint; hollow core wood interior doors, frames and hardware, painted finish; lightweight carpeting with pad, 80%; sheet vinyl flooring, 20%.	10.64	12.86	23.50	23.1%
7	Specialties	Economy grade kitchen cabinets and countertops; stainless steel kitchen sink; 30 gallon electric water heater.	3.28	1.11	4.39	4.3%
8	Mechanical	Three fixture bathroom: bathtub, water closet and wall hung lavatory; gas fired hot air heating system.	4.10	3.86	7.96	7.8%
9	Electrical	100 amp electric service; wiring, duplex & GFI receptacles, wall switches, door bell, appliance circuits and communications cabling; economy grade lighting fixtures.	1.09	1.90	2.99	2.9%
10	Overhead	Contractor's overhead and profit.	6.19	7.13	13.32	13.1%
		Total	47.35	54.60	**101.95**	

For customer support on your Residential Cost Data, call 877.759.4771.

RESIDENTIAL | Economy | 1-1/2 Story

- Mass produced from stock plans
- Single family — 1 full bath, 1 kitchen
- No basement
- Asphalt shingles on roof
- Hot air heat
- Gypsum wallboard interior finishes
- Materials and workmanship are sufficient to meet codes

Note: The illustration shown may contain some optional components (for example: garages and/or fireplaces) whose costs are shown in the modifications, adjustments, & alternatives below or at the end of the square foot section.

Base cost per square foot of living area

Exterior Wall	600	800	1000	1200	1400	1600	1800	2000	2400	2800	3200
Wood Siding - Wood Frame	164.15	136.20	122.35	115.65	110.80	103.50	99.85	96.20	88.50	85.55	82.35
Brick Veneer - Wood Frame	173.40	142.90	128.50	121.50	116.35	108.55	104.65	100.75	92.50	89.30	85.85
Stucco on Wood Frame	152.60	127.75	114.60	108.35	103.90	97.20	93.90	90.55	83.50	80.80	77.95
Painted Concrete Block	161.30	134.10	120.40	113.90	109.15	101.95	98.40	94.85	87.30	84.35	81.25
Finished Basement, Add	25.20	21.40	20.45	19.75	19.25	18.45	18.05	17.65	16.90	16.55	16.10
Unfinished Basement, Add	12.80	9.80	9.00	8.45	8.00	7.45	7.10	6.80	6.15	5.90	5.55

Modifications

Add to the total cost

Upgrade Kitchen Cabinets	$ + 1205
Solid Surface Countertops	+ 849
Full Bath - including plumbing, wall and floor finishes	+ 5924
Half Bath - including plumbing, wall and floor finishes	+ 3465
One Car Attached Garage	+ 13,419
One Car Detached Garage	+ 17,321
Fireplace & Chimney	+ 6285

Adjustments

For multi family - add to total cost

Additional Kitchen	$ + 5288
Additional Bath	+ 5924
Additional Entry & Exit	+ 1676
Separate Heating	+ 1336
Separate Electric	+ 966

For Townhouse/Rowhouse - Multiply cost per square foot by

Inner Unit	.95
End Unit	.97

Alternatives

Add to or deduct from the cost per square foot of living area

Composition Roll Roofing	– .80
Cedar Shake Roof	+ 2.85
Upgrade Walls and Ceilings to Skim Coat Plaster	+ .73
Upgrade Ceilings to Textured Finish	+ .58
Air Conditioning, in Heating Ductwork	+ 3.52
In Separate Ductwork	+ 6.07
Heating Systems, Hot Water	+ 1.49
Heat Pump	+ 1.46
Electric Heat	– 1.17
Not Heated	– 3.87

Additional upgrades or components

Kitchen Cabinets & Countertops	Page 93
Bathroom Vanities	94
Fireplaces & Chimneys	94
Windows, Skylights & Dormers	94
Appliances	95
Breezeways & Porches	95
Finished Attic	95
Garages	96
Site Improvements	96
Wings & Ells	34

Economy 1-1/2 Story

Living Area - 1600 S.F.
Perimeter - 135 L.F.

			Cost Per Square Foot Of Living Area			% of Total (rounded)
			Mat.	Inst.	Total	
1	**Site Work**	Site preparation for slab; 4' deep trench excavation for foundation wall.		1.61	1.61	1.6%
2	**Foundation**	Continuous reinforced concrete footing, 10" deep x 20" wide; damproofed and insulated 8" thick reinforced concrete block foundation wall, 4' deep; trowel finished 4" thick concrete slab on 4" crushed stone base and polyethylene vapor barrier.	4.30	5.42	9.72	9.4%
3	**Framing**	Exterior walls - 2" x 4" wood studs, 16" O.C.; 1/2" sheathing; gable end roof framing, steep pitch 2" x 8" rafters, 16" O.C. with 1/2" plywood sheathing; 2" x 8" floor joists, 16" O.C. with bridging and 5/8" subflooring; 2" x 4" interior partitions.	6.27	9.92	16.19	15.6%
4	**Exterior Walls**	Beveled wood siding and housewrap on insulated wood frame walls; R38 attic insulation; sliding wood windows; flush solid core doors, frame and hardware, painted finish; aluminum storm and screen doors.	14.19	5.82	20.01	19.3%
5	**Roofing**	25 year asphalt roof shingles; #15 felt building paper; aluminum gutters, downspouts, drip edge and flashings.	1.64	1.91	3.55	3.4%
6	**Interiors**	Walls and ceilings, 1/2" taped and finished gypsum wallboard, primed and painted with 2 coats of finish paint; hollow core wood interior doors, frames and hardware, painted finish; lightweight carpeting with pad, 80%; sheet vinyl flooring, 20%; hardwood tread stairway.	12.34	13.72	26.06	25.2%
7	**Specialties**	Economy grade kitchen cabinets and countertops; stainless steel kitchen sink; 30 gallon electric water heater.	2.45	.84	3.29	3.2%
8	**Mechanical**	Three fixture bathroom: bathtub, water closet and wall hung lavatory; gas fired hot air heating system.	3.32	3.47	6.79	6.6%
9	**Electrical**	100 amp electric service; wiring, duplex & GFI receptacles, wall switches, door bell, appliance circuits and communications cabling; economy grade lighting fixtures.	1.04	1.74	2.78	2.7%
10	**Overhead**	Contractor's overhead and profit.	6.85	6.65	13.50	13.0%
		Total	52.40	51.10	**103.50**	

RESIDENTIAL | Economy | 2 Story

- Mass produced from stock plans
- Single family — 1 full bath, 1 kitchen
- No basement
- Asphalt shingles on roof
- Hot air heat
- Gypsum wallboard interior finishes
- Materials and workmanship are sufficient to meet codes

Note: The illustration shown may contain some optional components (for example: garages and/or fireplaces) whose costs are shown in the modifications, adjustments, & alternatives below or at the end of the square foot section.

Base cost per square foot of living area

Exterior Wall	Living Area										
	1000	1200	1400	1600	1800	2000	2200	2600	3000	3400	3800
Wood Siding - Wood Frame	124.95	113.20	107.50	103.65	99.70	95.35	92.45	87.00	81.65	79.30	77.10
Brick Veneer - Wood Frame	131.60	119.35	113.25	109.25	104.85	100.35	97.20	91.25	85.60	83.00	80.70
Stucco on Wood Frame	116.60	105.50	100.30	96.85	93.20	89.10	86.50	81.65	76.70	74.55	72.65
Painted Concrete Block	123.35	111.70	106.10	102.40	98.45	94.15	91.35	85.95	80.70	78.35	76.25
Finished Basement, Add	17.15	16.45	15.85	15.45	15.10	14.80	14.45	13.90	13.55	13.25	13.00
Unfinished Basement, Add	7.85	7.25	6.80	6.50	6.15	5.90	5.70	5.25	4.95	4.70	4.50

Modifications

Add to the total cost

Upgrade Kitchen Cabinets	$ + 1205
Solid Surface Countertops	+ 849
Full Bath - including plumbing, wall and floor finishes	+ 5924
Half Bath - including plumbing, wall and floor finishes	+ 3465
One Car Attached Garage	+ 13,419
One Car Detached Garage	+ 17,321
Fireplace & Chimney	+ 6945

Adjustments

For multi family - add to total cost

Additional Kitchen	$ + 5288
Additional Bath	+ 5924
Additional Entry & Exit	+ 1676
Separate Heating	+ 1336
Separate Electric	+ 966

For Townhouse/Rowhouse - Multiply cost per square foot by

Inner Unit	.93
End Unit	.96

Alternatives

Add to or deduct from the cost per square foot of living area

Composition Roll Roofing	− .55
Cedar Shake Roof	+ 1.95
Upgrade Walls and Ceilings to Skim Coat Plaster	+ .74
Upgrade Ceilings to Textured Finish	+ .58
Air Conditioning, in Heating Ductwork	+ 2.87
In Separate Ductwork	+ 5.57
Heating Systems, Hot Water	+ 1.45
Heat Pump	+ 1.55
Electric Heat	− 1.02
Not Heated	− 3.66

Additional upgrades or components

Kitchen Cabinets & Countertops	Page 93
Bathroom Vanities	94
Fireplaces & Chimneys	94
Windows, Skylights & Dormers	94
Appliances	95
Breezeways & Porches	95
Finished Attic	95
Garages	96
Site Improvements	96
Wings & Ells	34

Economy 2 Story

Living Area - 2000 S.F.
Perimeter - 135 L.F.

#	Category	Description	Mat.	Inst.	Total	% of Total (rounded)
1	Site Work	Site preparation for slab; 4' deep trench excavation for foundation wall.		1.29	1.29	1.4%
2	Foundation	Continuous reinforced concrete footing, 10" deep x 20" wide; damproofed and insulated 8" thick reinforced concrete block foundation wall, 4' deep; trowel finished 4" thick concrete slab on 4" crushed stone base and polyethylene vapor barrier.	3.45	4.34	7.79	8.2%
3	Framing	Exterior walls - 2" x 4" wood studs, 16" O.C.; 1/2" sheathing; wood truss roof frame, 24" O.C. with 1/2" plywood sheating; 2" x 8" floor joists, 16" O.C. with bridging and 5/8" subflooring; 2" x 4" interior partitions.	5.84	8.76	14.60	15.3%
4	Exterior Walls	Beveled wood siding and housewrap on insulated wood frame walls; R38 attic insulation; sliding wood windows; flush solid core doors, frame and hardware, painted finish; aluminum storm and screen doors.	14.01	5.76	19.77	20.7%
5	Roofing	25 year asphalt roof shingles; #15 felt building paper; aluminum gutters, downspouts, drip edge and flashings.	1.13	1.32	2.45	2.6%
6	Interiors	Walls and ceilings, 1/2" taped and finished gypsum wallboard, primed and painted with 2 coats of finish paint; hollow core wood interior doors, frames and hardware, painted finish; lightweight carpeting with pad, 80%; sheet vinyl flooring, 20%; hardwood tread stairway.	12.02	13.65	25.67	26.9%
7	Specialties	Economy grade kitchen cabinets and countertops; stainless steel kitchen sink; 30 gallon electric water heater.	1.97	.67	2.64	2.8%
8	Mechanical	Three fixture bathroom: bathtub, water closet and wall hung lavatory; gas fired hot air heating system.	2.83	3.23	6.06	6.4%
9	Electrical	100 amp electric service; wiring, duplex & GFI receptacles, wall switches, door bell, appliance circuits and communications cabling; economy grade lighting fixtures.	1.01	1.65	2.66	2.8%
10	Overhead	Contractor's overhead and profit.	6.34	6.08	12.42	13.0%
	Total		48.60	46.75	**95.35**	

RESIDENTIAL Economy Bi-Level

- Mass produced from stock plans
- Single family — 1 full bath, 1 kitchen
- No basement
- Asphalt shingles on roof
- Hot air heat
- Gypsum wallboard interior finishes
- Materials and workmanship are sufficient to meet codes

Note: The illustration shown may contain some optional components (for example: garages and/or fireplaces) whose costs are shown in the modifications, adjustments, & alternatives below or at the end of the square foot section.

Base cost per square foot of living area

Exterior Wall	Living Area										
	1000	1200	1400	1600	1800	2000	2200	2600	3000	3400	3800
Wood Siding - Wood Frame	115.80	104.70	99.65	96.10	92.55	**88.55**	85.95	81.10	76.15	74.05	72.15
Brick Veneer - Wood Frame	120.80	109.40	104.00	100.30	96.55	92.30	89.55	84.35	79.25	76.95	74.85
Stucco on Wood Frame	109.45	98.85	94.15	90.85	87.70	83.75	81.40	77.00	72.45	70.50	68.75
Painted Concrete Block	114.50	103.50	98.50	95.05	91.55	87.55	85.00	80.20	75.40	73.35	71.50
Finished Basement, Add	17.15	16.45	15.85	15.45	15.10	14.80	14.45	13.90	13.55	13.25	13.00
Unfinished Basement, Add	7.85	7.25	6.80	6.50	6.15	5.90	5.70	5.25	4.95	4.70	4.50

Modifications
Add to the total cost

Upgrade Kitchen Cabinets	$ + 1205
Solid Surface Countertops	+ 849
Full Bath - including plumbing, wall and floor finishes	+ 5924
Half Bath - including plumbing, wall and floor finishes	+ 3465
One Car Attached Garage	+ 13,419
One Car Detached Garage	+ 17,321
Fireplace & Chimney	+ 6285

Adjustments
For multi family - add to total cost

Additional Kitchen	$ + 5288
Additional Bath	+ 5924
Additional Entry & Exit	+ 1676
Separate Heating	+ 1336
Separate Electric	+ 966

For Townhouse/Rowhouse - Multiply cost per square foot by

Inner Unit	.94
End Unit	.97

Alternatives
Add to or deduct from the cost per square foot of living area

Composition Roll Roofing	− .55
Cedar Shake Roof	+ 1.95
Upgrade Walls and Ceilings to Skim Coat Plaster	+ .70
Upgrade Ceilings to Textured Finish	+ .58
Air Conditioning, in Heating Ductwork	+ 2.87
In Separate Ductwork	+ 5.57
Heating Systems, Hot Water	+ 1.45
Heat Pump	+ 1.55
Electric Heat	− 1.02
Not Heated	− 3.66

Additional upgrades or components

Kitchen Cabinets & Countertops	Page 93
Bathroom Vanities	94
Fireplaces & Chimneys	94
Windows, Skylights & Dormers	94
Appliances	95
Breezeways & Porches	95
Finished Attic	95
Garages	96
Site Improvements	96
Wings & Ells	34

Important: See the Reference Section for Location Factors (to adjust for your city) and Estimating Forms.

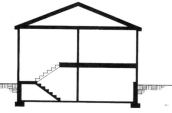

Economy Bi-Level

Living Area - 2000 S.F.
Perimeter - 135 L.F.

			Cost Per Square Foot Of Living Area			% of Total
			Mat.	Inst.	Total	(rounded)
1	Site Work	Site preparation for slab; 4' deep trench excavation for foundation wall.		1.29	1.29	1.5%
2	Foundation	Continuous reinforced concrete footing, 10" deep x 20" wide; damproofed and insulated 8" thick reinforced concrete block foundation wall, 4' deep; trowel finished 4" thick concrete slab on 4" crushed stone base and polyethylene vapor barrier.	3.45	4.34	7.79	8.8%
3	Framing	Exterior walls - 2" x 4" wood studs, 16" O.C.; 1/2" sheathing; wood truss roof frame, 24" O.C. with 1/2" plywood sheating; 2" x 8" floor joists, 16" O.C. with bridging and 5/8" subflooring; 2" x 4" interior partitions.	5.43	8.11	13.54	15.3%
4	Exterior Walls	Beveled wood siding and housewrap on insulated wood frame walls; R38 attic insulation; sliding wood windows; flush solid core doors, frame and hardware, painted finish; aluminum storm and screen doors.	11.01	4.55	15.56	17.6%
5	Roofing	25 year asphalt roof shingles; #15 felt building paper; aluminum gutters, downspouts, drip edge and flashings.	1.13	1.32	2.45	2.8%
6	Interiors	Walls and ceilings, 1/2" taped and finished gypsum wallboard, primed and painted with 2 coats of finish paint; hollow core wood interior doors, frames and hardware, painted finish; lightweight carpeting with pad, 80%; sheet vinyl flooring, 20%; hardwood tread stairway.	11.76	13.22	24.98	28.2%
7	Specialties	Economy grade kitchen cabinets and countertops; stainless steel kitchen sink; 30 gallon electric water heater.	1.97	.67	2.64	3.0%
8	Mechanical	Three fixture bathroom: bathtub, water closet and wall hung lavatory; gas fired hot air heating system.	2.83	3.23	6.06	6.8%
9	Electrical	100 amp electric service; wiring, duplex & GFI receptacles, wall switches, door bell, appliance circuits and communications cabling; economy grade lighting fixtures.	1.01	1.65	2.66	3.0%
10	Overhead	Contractor's overhead and profit.	5.81	5.77	11.58	13.1%
		Total	44.40	44.15	**88.55**	

For customer support on your Residential Cost Data, call 877.759.4771.

RESIDENTIAL | Economy | Tri-Level

- Mass produced from stock plans
- Single family — 1 full bath, 1 kitchen
- No basement
- Asphalt shingles on roof
- Hot air heat
- Gypsum wallboard interior finishes
- Materials and workmanship are sufficient to meet codes

Note: The illustration shown may contain some optional components (for example: garages and/or fireplaces) whose costs are shown in the modifications, adjustments, & alternatives below or at the end of the square foot section.

©Design Basics, Inc.

Base cost per square foot of living area

Exterior Wall	Living Area										
	1200	1500	1800	2000	2200	2400	2800	3200	3600	4000	4400
Wood Siding - Wood Frame	107.90	99.00	92.40	89.75	85.90	82.80	80.30	77.05	73.25	71.85	68.85
Brick Veneer - Wood Frame	112.60	103.30	96.25	93.40	89.40	86.10	83.50	79.95	75.95	74.45	71.35
Stucco on Wood Frame	102.00	93.65	87.55	85.10	81.50	78.70	76.35	73.40	69.85	68.55	65.80
Solid Masonry	106.65	97.90	91.35	88.75	84.95	81.90	79.50	76.25	72.45	71.15	68.25
Finished Basement, Add*	20.60	19.65	18.85	18.50	18.15	17.75	17.45	17.05	16.65	16.45	16.20
Unfinished Basement, Add*	8.60	7.90	7.25	6.95	6.70	6.40	6.15	5.80	5.50	5.35	5.15

*Basement under middle level only.

Modifications

Add to the total cost

Upgrade Kitchen Cabinets	$ + 1205
Solid Surface Countertops	+ 849
Full Bath - including plumbing, wall and floor finishes	+ 5924
Half Bath - including plumbing, wall and floor finishes	+ 3465
One Car Attached Garage	+ 13,419
One Car Detached Garage	+ 17,321
Fireplace & Chimney	+ 6285

Adjustments

For multi family - add to total cost

Additional Kitchen	$ + 5288
Additional Bath	+ 5924
Additional Entry & Exit	+ 1676
Separate Heating	+ 1336
Separate Electric	+ 966

For Townhouse/Rowhouse - Multiply cost per square foot by

Inner Unit	.93
End Unit	.96

Alternatives

Add to or deduct from the cost per square foot of living area

Composition Roll Roofing	– .80
Cedar Shake Roof	+ 2.85
Upgrade Walls and Ceilings to Skim Coat Plaster	+ .63
Upgrade Ceilings to Textured Finish	+ .58
Air Conditioning, in Heating Ductwork	+ 2.45
In Separate Ductwork	+ 5.18
Heating Systems, Hot Water	+ 1.39
Heat Pump	+ 1.61
Electric Heat	– .88
Not Heated	– 3.55

Additional upgrades or components

Kitchen Cabinets & Countertops	Page 93
Bathroom Vanities	94
Fireplaces & Chimneys	94
Windows, Skylights & Dormers	94
Appliances	95
Breezeways & Porches	95
Finished Attic	95
Garages	96
Site Improvements	96
Wings & Ells	34

Important: See the Reference Section for Location Factors (to adjust for your city) and Estimating Forms.

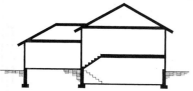

Economy Tri-Level

Living Area - 2400 S.F.
Perimeter - 163 L.F.

			Cost Per Square Foot Of Living Area			% of Total
			Mat.	Inst.	Total	(rounded)
1	Site Work	Site preparation for slab; 4' deep trench excavation for foundation wall.		1.07	1.07	1.3%
2	Foundation	Continuous reinforced concrete footing, 10" deep x 20" wide; damproofed and insulated 8" thick reinforced concrete block foundation wall, 4' deep; trowel finished 4" thick concrete slab on 4" crushed stone base and polyethylene vapor barrier.	3.84	4.71	8.55	10.3%
3	Framing	Exterior walls - 2" x 4" wood studs, 16" O.C.; 1/2" sheathing; wood truss roof frame, 24" O.C. with 1/2" plywood sheating; 2" x 8" floor joists, 16" O.C. with bridging and 5/8" subflooring; 2" x 4" interior partitions.	5.09	7.29	12.38	15.0%
4	Exterior Walls	Beveled wood siding and housewrap on insulated wood frame walls; R38 attic insulation; sliding wood windows; flush solid core doors, frame and hardware, painted finish; aluminum storm and screen doors.	9.72	3.97	13.69	16.5%
5	Roofing	25 year asphalt roof shingles; #15 felt building paper; aluminum gutters, downspouts, drip edge and flashings.	1.51	1.75	3.26	3.9%
6	Interiors	Walls and ceilings, 1/2" taped and finished gypsum wallboard, primed and painted with 2 coats of finish paint; hollow core wood interior doors, frames and hardware, painted finish; lightweight carpeting with pad, 80%; sheet vinyl flooring, 20%; hardwood tread stairway.	10.79	11.93	22.72	27.4%
7	Specialties	Economy grade kitchen cabinets and countertops; stainless steel kitchen sink; 30 gallon electric water heater.	1.63	.55	2.18	2.6%
8	Mechanical	Three fixture bathroom: bathtub, water closet and wall hung lavatory; gas fired hot air heating system.	2.51	3.07	5.58	6.7%
9	Electrical	100 amp electric service; wiring, duplex & GFI receptacles, wall switches, door bell, appliance circuits and communications cabling; economy grade lighting fixtures.	.99	1.59	2.58	3.1%
10	Overhead	Contractor's overhead and profit.	5.42	5.37	10.79	13.0%
		Total	41.50	41.30	**82.80**	

RESIDENTIAL | Economy | Wings & Ells

1 Story — Base cost per square foot of living area

Exterior Wall	Living Area							
	50	100	200	300	400	500	600	700
Wood Siding - Wood Frame	186.75	142.15	122.85	101.25	94.95	91.15	88.65	89.10
Brick Veneer - Wood Frame	201.60	152.75	131.70	107.20	100.25	96.15	93.30	93.65
Stucco on Wood Frame	168.20	128.90	111.80	93.90	88.35	84.95	82.75	83.50
Painted Concrete Block	183.85	140.05	121.15	100.10	93.90	90.20	87.70	88.20
Finished Basement, Add	50.30	41.05	37.15	30.70	29.45	28.70	28.15	27.80
Unfinished Basement, Add	27.95	20.75	17.75	12.70	11.70	11.10	10.70	10.40

1-1/2 Story — Base cost per square foot of living area

Exterior Wall	Living Area							
	100	200	300	400	500	600	700	800
Wood Siding - Wood Frame	147.45	118.70	101.00	89.60	84.35	81.70	78.40	77.45
Brick Veneer - Wood Frame	160.70	129.25	109.80	96.55	90.70	87.65	84.00	83.05
Stucco on Wood Frame	130.85	105.40	89.95	81.00	76.40	74.15	71.25	70.40
Painted Concrete Block	144.80	116.55	99.25	88.25	83.05	80.50	77.20	76.30
Finished Basement, Add	33.60	29.75	27.15	24.35	23.55	23.05	22.55	22.50
Unfinished Basement, Add	17.20	14.20	12.15	9.95	9.35	8.95	8.55	8.55

2 Story — Base cost per square foot of living area

Exterior Wall	Living Area							
	100	200	400	600	800	1000	1200	1400
Wood Siding - Wood Frame	150.80	111.95	95.05	77.40	71.95	68.60	66.40	67.15
Brick Veneer - Wood Frame	165.60	122.55	103.90	83.35	77.20	73.55	71.10	71.70
Stucco on Wood Frame	132.20	98.70	84.00	70.05	65.30	62.40	60.50	61.45
Painted Concrete Block	147.85	109.85	93.30	76.30	70.90	67.65	65.45	66.25
Finished Basement, Add	25.20	20.55	18.60	15.40	14.80	14.40	14.15	13.90
Unfinished Basement, Add	13.95	10.35	8.85	6.35	5.85	5.55	5.35	5.20

Base costs do not include bathroom or kitchen facilities. Use Modifications/Adjustments/Alternatives on pages 93–96 where appropriate.

Did you know?
RSMeans Online gives you the same access to RSMeans' data with 24/7 access:
- Quickly locate costs in the searchable database.
- Build cost lists, estimates, and reports in minutes.
- Adjust costs to any location in the U.S. and Canada with the click of a button.

Start your free trial today at **www.RSMeansOnline.com**

RSMeans Online
FROM THE GORDIAN GROUP

No part of this cost data may be reproduced, stored in a retrieval system, or transmitted in any form or by any means without prior written permission of RSMeans.

RESIDENTIAL | Average | Illustrations

1 Story

1-1/2 Story

2 Story

2-1/2 Story

Bi-Level

Tri-Level

RESIDENTIAL — Average — 1 Story

- Simple design from standard plans
- Single family — 1 full bath, 1 kitchen
- No basement
- Asphalt shingles on roof
- Hot air heat
- Gypsum wallboard interior finishes
- Materials and workmanship are average

Note: The illustration shown may contain some optional components (for example: garages and/or fireplaces) whose costs are shown in the modifications, adjustments, & alternatives below or at the end of the square foot section.

Base cost per square foot of living area

Exterior Wall	_____ Living Area _____										
	600	800	1000	1200	1400	1600	1800	2000	2400	2800	3200
Wood Siding - Wood Frame	164.45	147.45	135.05	125.20	116.95	111.55	108.60	105.00	97.95	92.85	89.45
Brick Veneer - Wood Frame	171.10	153.35	140.45	130.05	121.35	115.65	112.60	108.75	101.40	96.00	92.30
Stucco on Wood Frame	159.80	143.30	131.25	121.80	113.80	108.65	105.80	102.35	95.50	90.60	87.35
Solid Masonry	183.50	164.45	150.45	139.05	129.60	123.40	120.05	115.75	107.75	101.90	97.80
Finished Basement, Add	39.65	38.30	36.55	34.95	33.55	32.70	32.25	31.55	30.55	29.80	29.15
Unfinished Basement, Add	16.45	14.85	13.75	12.75	11.90	11.40	11.05	10.65	10.05	9.55	9.15

Modifications

Add to the total cost

Upgrade Kitchen Cabinets	$ + 5525
Solid Surface Countertops (Included)	
Full Bath - including plumbing, wall and floor finishes	+ 7405
Half Bath - including plumbing, wall and floor finishes	+ 4331
One Car Attached Garage	+ 14,471
One Car Detached Garage	+ 19,004
Fireplace & Chimney	+ 6627

Adjustments

For multi family - add to total cost

Additional Kitchen	$ + 9600
Additional Bath	+ 7405
Additional Entry & Exit	+ 1676
Separate Heating	+ 1336
Separate Electric	+ 1945

For Townhouse/Rowhouse - Multiply cost per square foot by

Inner Unit	.92
End Unit	.96

Alternatives

Add to or deduct from the cost per square foot of living area

Cedar Shake Roof	+ 3.35
Clay Tile Roof	+ 6.80
Slate Roof	+ 7.50
Upgrade Walls to Skim Coat Plaster	+ .45
Upgrade Ceilings to Textured Finish	+ .58
Air Conditioning, in Heating Ductwork	+ 4.89
In Separate Ductwork	+ 7.18
Heating Systems, Hot Water	+ 1.60
Heat Pump	+ 1.38
Electric Heat	− .81
Not Heated	− 3.52

Additional upgrades or components

Kitchen Cabinets & Countertops	Page 93
Bathroom Vanities	94
Fireplaces & Chimneys	94
Windows, Skylights & Dormers	94
Appliances	95
Breezeways & Porches	95
Finished Attic	95
Garages	96
Site Improvements	96
Wings & Ells	56

Average 1 Story

Living Area - 1600 S.F.
Perimeter - 163 L.F.

			Cost Per Square Foot Of Living Area			% of Total
			Mat.	Inst.	Total	(rounded)
1	Site Work	Site preparation for slab; 4' deep trench excavation for foundation wall.		1.66	1.66	1.5%
2	Foundation	Continuous reinforced concrete footing, 10" deep x 20" wide; damproofed and insulated 8" thick reinforced concrete block foundation wall, 4' deep; trowel finished 4" thick concrete slab on 4" crushed stone base and polyethylene vapor barrier.	5.92	7.26	13.18	11.8%
3	Framing	Exterior walls - 2" x 6" wood studs, 16" O.C.; 1/2" sheathing; gable end roof framing, 2" x 10" rafters, 16" O.C. with 1/2" plywood sheathing; 2" x 4" interior partitions.	6.73	9.48	16.21	14.5%
4	Exterior Walls	Beveled wood siding and housewrap on insulated wood frame walls; R38 attic insulation; double hung wood windows; flush solid core doors, frame and hardware, painted finish; aluminum storm and screen doors.	12.08	5.14	17.22	15.4%
5	Roofing	25 year asphalt roof shingles; #15 felt building paper; aluminum gutters, downspouts, drip edge and flashings.	2.32	2.70	5.02	4.5%
6	Interiors	Walls & ceilings, 1/2" taped & finished gypsum wallboard, primed & painted with 2 coats of finish paint; birch faced hollow core interior doors, frames & hardware, painted finish; medium weight carpeting with pad, 40%; sheet vinyl, 15%; oak hardwood, 40%; ceramic tile, 5%	11.92	12.97	24.89	22.3%
7	Specialties	Average grade kitchen cabinets and countertop; stainless steel kitchen sink; 40 gallon electric water heater.	4.44	1.49	5.93	5.3%
8	Mechanical	Three fixture bathroom: bathtub, water closet, vanity and sink; gas fired hot air heating system.	3.62	3.57	7.19	6.4%
9	Electrical	200 amp electric service; wiring, duplex & GFI receptacles, wall switches, door bell, appliance circuits, fans and communications cabling; average grade lighting fixtures.	1.74	2.31	4.05	3.6%
10	Overhead	Contractor's overhead and profit and plans.	8.28	7.92	16.20	14.5%
		Total	57.05	54.50	**111.55**	

RESIDENTIAL — Average — 1-1/2 Story

- Simple design from standard plans
- Single family — 1 full bath, 1 kitchen
- No basement
- Asphalt shingles on roof
- Hot air heat
- Gypsum wallboard interior finishes
- Materials and workmanship are average

Note: The illustration shown may contain some optional components (for example: garages and/or fireplaces) whose costs are shown in the modifications, adjustments, & alternatives below or at the end of the square foot section.

©By Designer

Base cost per square foot of living area

Exterior Wall	600	800	1000	1200	1400	1600	1800	2000	2400	2800	3200
Wood Siding - Wood Frame	187.40	155.40	139.00	130.90	125.15	116.80	112.65	108.30	99.70	96.20	92.50
Brick Veneer - Wood Frame	196.85	162.35	145.35	136.90	130.85	122.00	117.55	113.00	103.80	100.10	96.05
Stucco on Wood Frame	180.55	150.45	134.40	126.60	121.10	113.10	109.05	104.95	96.75	93.40	89.90
Solid Masonry	213.85	174.65	156.75	147.55	141.00	131.20	126.25	121.30	111.10	107.05	102.50
Finished Basement, Add	32.35	28.15	27.00	26.00	25.35	24.35	23.80	23.30	22.25	21.80	21.20
Unfinished Basement, Add	14.25	11.15	10.35	9.80	9.35	8.70	8.40	8.05	7.40	7.10	6.80

Modifications

Add to the total cost

Upgrade Kitchen Cabinets	$ + 5525
Solid Surface Countertops (Included)	
Full Bath - including plumbing, wall and floor finishes	+ 7405
Half Bath - including plumbing, wall and floor finishes	+ 4331
One Car Attached Garage	+ 14,471
One Car Detached Garage	+ 19,004
Fireplace & Chimney	+ 6627

Adjustments

For multi family - add to total cost

Additional Kitchen	$ + 9600
Additional Bath	+ 7405
Additional Entry & Exit	+ 1676
Separate Heating	+ 1336
Separate Electric	+ 1945

For Townhouse/Rowhouse - Multiply cost per square foot by

Inner Unit	.92
End Unit	.96

Alternatives

Add to or deduct from the cost per square foot of living area

Cedar Shake Roof	+ 2.40
Clay Tile Roof	+ 4.95
Slate Roof	+ 5.40
Upgrade Walls to Skim Coat Plaster	+ .52
Upgrade Ceilings to Textured Finish	+ .58
Air Conditioning, in Heating Ductwork	+ 3.71
In Separate Ductwork	+ 6.28
Heating Systems, Hot Water	+ 1.52
Heat Pump	+ 1.53
Electric Heat	– .74
Not Heated	– 3.38

Additional upgrades or components

Kitchen Cabinets & Countertops	Page 93
Bathroom Vanities	94
Fireplaces & Chimneys	94
Windows, Skylights & Dormers	94
Appliances	95
Breezeways & Porches	95
Finished Attic	95
Garages	96
Site Improvements	96
Wings & Ells	56

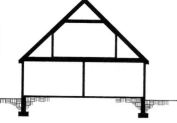

Average 1-1/2 Story

Living Area - 1800 S.F.
Perimeter - 144 L.F.

			Cost Per Square Foot Of Living Area			% of Total
			Mat.	Inst.	Total	(rounded)
1	Site Work	Site preparation for slab; 4' deep trench excavation for foundation wall.		1.47	1.47	1.3%
2	Foundation	Continuous reinforced concrete footing, 10" deep x 20" wide; damproofed and insulated 8" thick reinforced concrete block foundation wall, 4' deep; trowel finished 4" thick concrete slab on 4" crushed stone base and polyethylene vapor barrier.	4.26	5.34	9.60	8.5%
3	Framing	Exterior walls - 2" x 6" wood studs, 16" O.C.; 1/2" sheathing; gable end roof framing, steep pitch 2" x 10" rafters, 16" O.C. with 1/2" plywood sheathing; 2" x 10" floor joists, 16" O.C. with bridging and 5/8" subflooring; 2" x 4" interior partitions.	7.46	10.33	17.79	15.8%
4	Exterior Walls	Beveled wood siding and housewrap on insulated wood frame walls; R38 attic insulation; double hung wood windows; flush solid core doors, frame and hardware, painted finish; aluminum storm and screen doors.	13.19	5.65	18.84	16.7%
5	Roofing	25 year asphalt roof shingles; #15 felt building paper; aluminum gutters, downspouts, drip edge and flashings.	1.69	1.96	3.65	3.2%
6	Interiors	Walls & ceilings, 1/2" taped & finished gypsum wallboard, primed & painted with 2 coats of finish paint; birch faced hollow core interior doors, frames & hardware, painted finish; medium weight carpeting with pad, 40%; sheet vinyl, 15%; oak hardwood, 40%; ceramic tile, 5%; hardwood tread stairway.	14.24	14.74	28.98	25.7%
7	Specialties	Average grade kitchen cabinets and countertop; stainless steel kitchen sink; 40 gallon electric water heater.	3.96	1.31	5.27	4.7%
8	Mechanical	Three fixture bathroom: bathtub, water closet, vanity and sink; gas fired hot air heating system.	3.32	3.42	6.74	6.0%
9	Electrical	200 amp electric service; wiring, duplex & GFI receptacles, wall switches, door bell, appliance circuits, fans and communications cabling; average grade lighting fixtures.	1.69	2.23	3.92	3.5%
10	Overhead	Contractor's overhead and profit and plans.	8.49	7.90	16.39	14.5%
		Total	58.30	54.35	**112.65**	

RESIDENTIAL | Average | 2 Story

- Simple design from standard plans
- Single family — 1 full bath, 1 kitchen
- No basement
- Asphalt shingles on roof
- Hot air heat
- Gypsum wallboard interior finishes
- Materials and workmanship are average

Note: The illustration shown may contain some optional components (for example: garages and/or fireplaces) whose costs are shown in the modifications, adjustments, & alternatives below or at the end of the square foot section.

Base cost per square foot of living area

Exterior Wall	1000	1200	1400	1600	1800	2000	2200	2600	3000	3400	3800
Wood Siding - Wood Frame	142.45	128.80	122.15	117.55	112.85	107.95	104.65	98.35	92.35	89.65	87.15
Brick Veneer - Wood Frame	149.30	135.15	128.05	123.25	118.20	113.05	109.55	102.75	96.45	93.50	90.80
Stucco on Wood Frame	137.75	124.40	118.00	113.65	109.20	104.40	101.25	95.30	89.55	86.90	84.55
Solid Masonry	163.05	147.85	139.95	134.60	128.90	123.40	119.30	111.65	104.60	101.25	98.15
Finished Basement, Add	22.70	22.45	21.70	21.15	20.65	20.30	19.85	19.15	18.60	18.25	17.95
Unfinished Basement, Add	9.00	8.40	7.90	7.60	7.25	7.00	6.75	6.35	6.00	5.75	5.60

Modifications

Add to the total cost

Upgrade Kitchen Cabinets	$ + 5525
Solid Surface Countertops (Included)	
Full Bath - including plumbing, wall and floor finishes	+ 7405
Half Bath - including plumbing, wall and floor finishes	+ 4331
One Car Attached Garage	+ 14,471
One Car Detached Garage	+ 19,004
Fireplace & Chimney	+ 7319

Adjustments

For multi family - add to total cost

Additional Kitchen	$ + 9600
Additional Bath	+ 7405
Additional Entry & Exit	+ 1676
Separate Heating	+ 1336
Separate Electric	+ 1945

For Townhouse/Rowhouse - Multiply cost per square foot by

Inner Unit	.90
End Unit	.95

Alternatives

Add to or deduct from the cost per square foot of living area

Cedar Shake Roof	+ 1.65
Clay Tile Roof	+ 3.40
Slate Roof	+ 3.75
Upgrade Walls to Skim Coat Plaster	+ .53
Upgrade Ceilings to Textured Finish	+ .58
Air Conditioning, in Heating Ductwork	+ 2.96
In Separate Ductwork	+ 5.70
Heating Systems, Hot Water	+ 1.49
Heat Pump	+ 1.61
Electric Heat	− .63
Not Heated	− 3.28

Additional upgrades or components

Kitchen Cabinets & Countertops	Page 93
Bathroom Vanities	94
Fireplaces & Chimneys	94
Windows, Skylights & Dormers	94
Appliances	95
Breezeways & Porches	95
Finished Attic	95
Garages	96
Site Improvements	96
Wings & Ells	56

Important: See the Reference Section for Location Factors (to adjust for your city) and Estimating Forms.

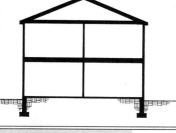

Average 2 Story

Living Area - 2000 S.F.
Perimeter - 135 L.F.

#	Category	Description	Cost Per Square Foot Of Living Area			% of Total (rounded)
			Mat.	Inst.	Total	
1	Site Work	Site preparation for slab; 4' deep trench excavation for foundation wall.		1.33	1.33	1.2%
2	Foundation	Continuous reinforced concrete footing, 10" deep x 20" wide; damproofed and insulated 8" thick reinforced concrete block foundation wall, 4' deep; trowel finished 4" thick concrete slab on 4" crushed stone base and polyethylene vapor barrier.	3.55	4.46	8.01	7.4%
3	Framing	Exterior walls - 2" x 6" wood studs, 16" O.C.; 1/2" sheathing; gable end roof framing, 2" x 10" rafters, 16" O.C. with 1/2" plywood sheathing; 2" x 10" floor joists, 16" O.C. with bridging and 5/8" subflooring; 2" x 4" interior partitions.	7.07	9.77	16.84	15.6%
4	Exterior Walls	Beveled wood siding and housewrap on insulated wood frame walls; R38 attic insulation; double hung wood windows; flush solid core doors, frame and hardware, painted finish; aluminum storm and screen doors.	13.75	5.90	19.65	18.2%
5	Roofing	25 year asphalt roof shingles; #15 felt building paper; aluminum gutters, downspouts, drip edge and flashings.	1.16	1.36	2.52	2.3%
6	Interiors	Walls & ceilings, 1/2" taped & finished gypsum wallboard, primed & painted with 2 coats of finish paint; birch faced hollow core interior doors, frames & hardware, painted finish; medium weight carpeting with pad, 40%; sheet vinyl, 15%; oak hardwood, 40%; ceramic tile, 5%; hardwood tread stairway.	14.16	14.83	28.99	26.9%
7	Specialties	Average grade kitchen cabinets and countertop; stainless steel kitchen sink; 40 gallon electric water heater.	3.55	1.19	4.74	4.4%
8	Mechanical	Three fixture bathroom: bathtub, water closet, vanity and sink; gas fired hot air heating system.	3.07	3.31	6.38	5.9%
9	Electrical	200 amp electric service; wiring, duplex & GFI receptacles, wall switches, door bell, appliance circuits, fans and communications cabling; average grade lighting fixtures	1.65	2.16	3.81	3.5%
10	Overhead	Contractor's overhead and profit and plans.	8.14	7.54	15.68	14.5%
	Total		56.10	51.85	**107.95**	

RESIDENTIAL — Average — 2-1/2 Story

- Simple design from standard plans
- Single family — 1 full bath, 1 kitchen
- No basement
- Asphalt shingles on roof
- Hot air heat
- Gypsum wallboard interior finishes
- Materials and workmanship are average

Note: The illustration shown may contain some optional components (for example: garages and/or fireplaces) whose costs are shown in the modifications, adjustments, & alternatives below or at the end of the square foot section.

Base cost per square foot of living area

Exterior Wall	____ Living Area ____										
	1200	1400	1600	1800	2000	2400	2800	3200	3600	4000	4400
Wood Siding - Wood Frame	142.15	133.10	121.45	119.05	114.50	107.40	101.85	96.20	93.35	88.20	86.55
Brick Veneer - Wood Frame	149.65	139.85	127.60	125.30	120.30	112.65	106.90	100.80	97.65	92.25	90.45
Stucco on Wood Frame	136.85	128.35	117.05	114.65	110.35	103.70	98.30	93.00	90.30	85.35	83.80
Solid Masonry	164.00	152.85	139.60	137.25	131.55	122.85	116.60	109.60	106.00	100.00	97.95
Finished Basement, Add	19.20	18.80	18.05	18.00	17.50	16.85	16.45	15.90	15.60	15.20	15.05
Unfinished Basement, Add	7.50	6.95	6.45	6.40	6.10	5.65	5.45	5.10	4.90	4.70	4.60

Modifications

Add to the total cost

Upgrade Kitchen Cabinets	$ + 5525
Solid Surface Countertops (Included)	
Full Bath - including plumbing, wall and floor finishes	+ 7405
Half Bath - including plumbing, wall and floor finishes	+ 4331
One Car Attached Garage	+ 14,471
One Car Detached Garage	+ 19,004
Fireplace & Chimney	+ 8029

Adjustments

For multi family - add to total cost

Additional Kitchen	$ + 9600
Additional Bath	+ 7405
Additional Entry & Exit	+ 1676
Separate Heating	+ 1336
Separate Electric	+ 1945

For Townhouse/Rowhouse - Multiply cost per square foot by

Inner Unit	.90
End Unit	.95

Alternatives

Add to or deduct from the cost per square foot of living area

Cedar Shake Roof	+ 1.45
Clay Tile Roof	+ 2.95
Slate Roof	+ 3.25
Upgrade Walls to Skim Coat Plaster	+ .51
Upgrade Ceilings to Textured Finish	+ .58
Air Conditioning, in Heating Ductwork	+ 2.69
In Separate Ductwork	+ 5.51
Heating Systems, Hot Water	+ 1.35
Heat Pump	+ 1.65
Electric Heat	− 1.14
Not Heated	− 3.83

Additional upgrades or components

Kitchen Cabinets & Countertops	Page 93
Bathroom Vanities	94
Fireplaces & Chimneys	94
Windows, Skylights & Dormers	94
Appliances	95
Breezeways & Porches	95
Finished Attic	95
Garages	96
Site Improvements	96
Wings & Ells	56

Average 2-1/2 Story

Living Area - 3200 S.F.
Perimeter - 150 L.F.

			Cost Per Square Foot Of Living Area			% of Total
			Mat.	Inst.	Total	(rounded)
1	Site Work	Site preparation for slab; 4' deep trench excavation for foundation wall.		.82	.82	0.9%
2	Foundation	Continuous reinforced concrete footing, 10" deep x 20" wide; damproofed and insulated 8" thick reinforced concrete block foundation wall, 4' deep; trowel finished 4" thick concrete slab on 4" crushed stone base and polyethylene vapor barrier.	2.53	3.16	5.69	5.9%
3	Framing	Exterior walls - 2" x 6" wood studs, 16" O.C.; 1/2" sheathing; gable end roof framing, steep pitch 2" x 10" rafters, 16" O.C. with 1/2" plywood sheathing; 2" x 10" floor joists, 16" O.C. with bridging and 5/8" subflooring; 2" x 4" interior partitions.	7.06	9.69	16.75	17.4%
4	Exterior Walls	Beveled wood siding and housewrap on insulated wood frame walls; R38 attic insulation; double hung wood windows; flush solid core doors, frame and hardware, painted finish; aluminum storm and screen doors.	11.79	5.01	16.80	17.5%
5	Roofing	25 year asphalt roof shingles; #15 felt building paper; aluminum gutters, downspouts, drip edge and flashings.	1.04	1.20	2.24	2.3%
6	Interiors	Walls & ceilings, 1/2" taped & finished gypsum wallboard, primed & painted with 2 coats of finish paint; birch faced hollow core interior doors, frames & hardware, painted finish; medium weight carpeting with pad, 40%; sheet vinyl, 15%; oak hardwood, 40%; ceramic tile, 5%; hardwood tread stairway.	13.98	14.33	28.31	29.4%
7	Specialties	Average grade kitchen cabinets and countertop; stainless steel kitchen sink; 40 gallon electric water heater.	2.22	.74	2.96	3.1%
8	Mechanical	Three fixture bathroom: bathtub, water closet, vanity and sink; gas fired hot air heating system.	2.28	2.95	5.23	5.4%
9	Electrical	200 amp electric service; wiring, duplex & GFI receptacles, wall switches, door bell, appliance circuits, fans and communications cabling; average grade lighting fixtures.	1.51	1.93	3.44	3.6%
10	Overhead	Contractor's overhead and profit and plans.	7.19	6.77	13.96	14.5%
		Total	49.60	46.60	**96.20**	

RESIDENTIAL — Average — 3 Story

- Simple design from standard plans
- Single family — 1 full bath, 1 kitchen
- No basement
- Asphalt shingles on roof
- Hot air heat
- Gypsum wallboard interior finishes
- Materials and workmanship are average

Note: The illustration shown may contain some optional components (for example: garages and/or fireplaces) whose costs are shown in the modifications, adjustments, & alternatives below or at the end of the square foot section.

Base cost per square foot of living area

Exterior Wall	1500	1800	2100	2500	3000	3500	4000	4500	5000	5500	6000
Wood Siding - Wood Frame	129.00	116.90	111.25	106.80	98.85	95.30	90.35	85.10	83.40	81.50	79.40
Brick Veneer - Wood Frame	135.70	123.10	117.05	112.40	103.90	100.05	94.65	89.10	87.30	85.20	82.95
Stucco on Wood Frame	124.35	112.55	107.25	102.95	95.35	92.00	87.30	82.30	80.70	78.95	77.00
Solid Masonry	149.55	135.95	129.15	123.85	114.25	109.85	103.60	97.35	95.30	92.90	90.20
Finished Basement, Add	16.45	16.25	15.75	15.40	14.85	14.50	14.10	13.75	13.60	13.40	13.20
Unfinished Basement, Add	6.10	5.70	5.35	5.15	4.80	4.55	4.35	4.10	4.00	3.90	3.80

Modifications
Add to the total cost

Upgrade Kitchen Cabinets	$ + 5525
Solid Surface Countertops (Included)	
Full Bath - including plumbing, wall and floor finishes	+ 7405
Half Bath - including plumbing, wall and floor finishes	+ 4331
One Car Attached Garage	+ 14,471
One Car Detached Garage	+ 19,004
Fireplace & Chimney	+ 8029

Adjustments
For multi family - add to total cost

Additional Kitchen	$ + 9600
Additional Bath	+ 7405
Additional Entry & Exit	+ 1676
Separate Heating	+ 1336
Separate Electric	+ 1945

For Townhouse/Rowhouse - Multiply cost per square foot by

Inner Unit	.88
End Unit	.94

Alternatives
Add to or deduct from the cost per square foot of living area

Cedar Shake Roof	+ 1.10
Clay Tile Roof	+ 2.25
Slate Roof	+ 2.50
Upgrade Walls to Skim Coat Plaster	+ .53
Upgrade Ceilings to Textured Finish	+ .58
Air Conditioning, in Heating Ductwork	+ 2.69
In Separate Ductwork	+ 5.51
Heating Systems, Hot Water	+ 1.35
Heat Pump	+ 1.65
Electric Heat	− .88
Not Heated	− 3.57

Additional upgrades or components

Kitchen Cabinets & Countertops	Page 93
Bathroom Vanities	94
Fireplaces & Chimneys	94
Windows, Skylights & Dormers	94
Appliances	95
Breezeways & Porches	95
Finished Attic	95
Garages	96
Site Improvements	96
Wings & Ells	56

Important: See the Reference Section for Location Factors (to adjust for your city) and Estimating Forms.

Average 3 Story

Living Area - 3000 S.F.
Perimeter - 135 L.F.

			Cost Per Square Foot Of Living Area			% of Total
			Mat.	Inst.	Total	(rounded)
1	Site Work	Site preparation for slab; 4' deep trench excavation for foundation wall.		.88	.88	0.9%
2	Foundation	Continuous reinforced concrete footing, 10" deep x 20" wide; damproofed and insulated 8" thick reinforced concrete block foundation wall, 4' deep; trowel finished 4" thick concrete slab on 4" crushed stone base and polyethylene vapor barrier.	2.35	2.97	5.32	5.4%
3	Framing	Exterior walls - 2" x 6" wood studs, 16" O.C.; 1/2" sheathing; gable end roof framing, 2" x 10" rafters, 16" O.C. with 1/2" plywood sheathing; 2" x 10" floor joists, 16" O.C. with bridging and 5/8" subflooring; 2" x 4" interior partitions.	6.97	9.55	16.52	16.7%
4	Exterior Walls	Beveled wood siding and housewrap on insulated wood frame walls; R38 attic insulation; double hung wood windows; flush solid core doors, frame and hardware, painted finish; aluminum storm and screen doors.	13.07	5.60	18.67	18.9%
5	Roofing	25 year asphalt roof shingles; #15 felt building paper; aluminum gutters, downspouts, drip edge and flashings.	.77	.90	1.67	1.7%
6	Interiors	Walls & ceilings, 1/2" taped & finished gypsum wallboard, primed & painted with 2 coats of finish paint; birch faced hollow core interior doors, frames & hardware, painted finish; medium weight carpeting with pad, 40%; sheet vinyl, 15%; oak hardwood, 40%; ceramic tile, 5%; hardwood tread stairway.	14.50	14.92	29.42	29.8%
7	Specialties	Average grade kitchen cabinets and countertop; stainless steel kitchen sink; 40 gallon electric water heater.	2.37	.79	3.16	3.2%
8	Mechanical	Three fixture bathroom: bathtub, water closet, vanity and sink; gas fired hot air heating system.	2.36	2.98	5.34	5.4%
9	Electrical	200 amp electric service; wiring, duplex & GFI receptacles, wall switches, door bell, appliance circuits, fans and communications cabling; average grade lighting fixtures.	1.53	1.95	3.48	3.5%
10	Overhead	Contractor's overhead and profit and plans.	7.48	6.91	14.39	14.6%
		Total	51.40	47.45	**98.85**	

For customer support on your Residential Cost Data, call 877.759.4771.

RESIDENTIAL Average Bi-Level

- Simple design from standard plans
- Single family — 1 full bath, 1 kitchen
- No basement
- Asphalt shingles on roof
- Hot air heat
- Gypsum wallboard interior finishes
- Materials and workmanship are average

Note: The illustration shown may contain some optional components (for example: garages and/or fireplaces) whose costs are shown in the modifications, adjustments, & alternatives below or at the end of the square foot section.

Base cost per square foot of living area

Exterior Wall	Living Area										
	1000	1200	1400	1600	1800	2000	2200	2600	3000	3400	3800
Wood Siding - Wood Frame	132.95	119.90	113.90	109.70	105.40	100.75	97.85	92.20	86.70	84.25	82.05
Brick Veneer - Wood Frame	138.15	124.75	118.40	114.00	109.50	104.70	101.55	95.55	89.80	87.20	84.80
Stucco on Wood Frame	129.30	116.55	110.75	106.70	102.60	98.00	95.20	89.85	84.55	82.20	80.10
Solid Masonry	148.40	134.20	127.25	122.40	117.45	112.35	108.85	102.20	95.90	92.95	90.30
Finished Basement, Add	22.70	22.45	21.70	21.15	20.65	20.30	19.85	19.15	18.60	18.25	17.95
Unfinished Basement, Add	9.00	8.40	7.90	7.60	7.25	7.00	6.75	6.35	6.00	5.75	5.60

Modifications

Add to the total cost

Upgrade Kitchen Cabinets	$ + 5525
Solid Surface Countertops (Included)	
Full Bath - including plumbing, wall and floor finishes	+ 7405
Half Bath - including plumbing, wall and floor finishes	+ 4331
One Car Attached Garage	+ 14,471
One Car Detached Garage	+ 19,004
Fireplace & Chimney	+ 6627

Adjustments

For multi family - add to total cost

Additional Kitchen	$ + 9600
Additional Bath	+ 7405
Additional Entry & Exit	+ 1676
Separate Heating	+ 1336
Separate Electric	+ 1945

For Townhouse/Rowhouse - Multiply cost per square foot by

Inner Unit	.91
End Unit	.96

Alternatives

Add to or deduct from the cost per square foot of living area

Cedar Shake Roof	+ 1.65
Clay Tile Roof	+ 3.40
Slate Roof	+ 3.75
Upgrade Walls to Skim Coat Plaster	+ .49
Upgrade Ceilings to Textured Finish	+ .58
Air Conditioning, in Heating Ductwork	+ 2.96
In Separate Ductwork	+ 5.70
Heating Systems, Hot Water	+ 1.49
Heat Pump	+ 1.61
Electric Heat	– .63
Not Heated	– 3.28

Additional upgrades or components

Kitchen Cabinets & Countertops	Page 93
Bathroom Vanities	94
Fireplaces & Chimneys	94
Windows, Skylights & Dormers	94
Appliances	95
Breezeways & Porches	95
Finished Attic	95
Garages	96
Site Improvements	96
Wings & Ells	56

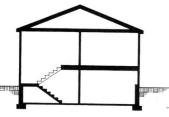

Average Bi-Level

Living Area - 2000 S.F.
Perimeter - 135 L.F.

		Cost Per Square Foot Of Living Area			% of Total
		Mat.	Inst.	Total	(rounded)
1 Site Work	Site preparation for slab; 4' deep trench excavation for foundation wall.		1.33	1.33	1.3%
2 Foundation	Continuous reinforced concrete footing, 10" deep x 20" wide; damproofed and insulated 8" thick reinforced concrete block foundation wall, 4' deep; trowel finished 4" thick concrete slab on 4" crushed stone base and polyethylene vapor barrier.	3.55	4.46	8.01	8.0%
3 Framing	Exterior walls - 2" x 6" wood studs, 16" O.C.; 1/2" sheathing; gable end roof framing, 2" x 10" rafters, 16" O.C. with 1/2" plywood sheathing; 2" x 10" floor joists, 16" O.C. with bridging and 5/8" subflooring; 2" x 4" interior partitions.	6.55	9.04	15.59	15.5%
4 Exterior Walls	Beveled wood siding and housewrap on insulated wood frame walls; R38 attic insulation; double hung wood windows; flush solid core doors, frame and hardware, painted finish; aluminum storm and screen doors.	10.82	4.66	15.48	15.4%
5 Roofing	25 year roof asphalt shingles; #15 felt building paper; aluminum gutters, downspouts, drip edge and flashings.	1.16	1.36	2.52	2.5%
6 Interiors	Walls & ceilings, 1/2" taped & finished gypsum wallboard, primed & painted with 2 coats of finish paint; birch faced hollow core interior doors, frames & hardware, painted finish; medium weight carpeting with pad, 40%; sheet vinyl, 15%; oak hardwood, 40%; ceramic tile, 5%; hardwood tread stairway.	13.89	14.39	28.28	28.1%
7 Specialties	Average grade kitchen cabinets and countertop; stainless steel kitchen sink; 40 gallon electric water heater.	3.55	1.19	4.74	4.7%
8 Mechanical	Three fixture bathroom: bathtub, water closet, vanity and sink; gas fired hot air heating system.	3.07	3.31	6.38	6.3%
9 Electrical	200 amp electric service; wiring, duplex & GFI receptacles, wall switches, door bell, appliance circuits, fans and communications cabling; average grade lighting fixtures.	1.65	2.16	3.81	3.8%
10 Overhead	Contractor's overhead and profit and plans.	7.51	7.10	14.61	14.5%
	Total	51.75	49.00	**100.75**	

RESIDENTIAL — Average — Tri-Level

- Simple design from standard plans
- Single family — 1 full bath, 1 kitchen
- No basement
- Asphalt shingles on roof
- Hot air heat
- Gypsum wallboard interior finishes
- Materials and workmanship are average

Note: The illustration shown may contain some optional components (for example: garages and/or fireplaces) whose costs are shown in the modifications, adjustments, & alternatives below or at the end of the square foot section.

©Design Basics, Inc.

Base cost per square foot of living area

Exterior Wall	1200	1500	1800	2100	2400	2700	3000	3400	3800	4200	4600
Wood Siding - Wood Frame	123.90	113.55	105.75	99.10	94.85	92.35	89.65	87.10	82.95	79.55	77.90
Brick Veneer - Wood Frame	128.75	117.90	109.75	102.70	98.30	95.60	92.70	90.05	85.75	82.20	80.45
Stucco on Wood Frame	120.50	110.45	102.95	96.55	92.45	90.05	87.50	85.00	81.05	77.80	76.20
Solid Masonry	138.05	126.40	117.35	109.65	104.75	101.85	98.55	95.75	91.05	87.10	85.20
Finished Basement, Add*	26.15	25.70	24.60	23.75	23.15	22.85	22.30	22.05	21.60	21.20	21.00
Unfinished Basement, Add*	10.05	9.25	8.55	8.05	7.70	7.50	7.20	7.05	6.70	6.45	6.35

*Basement under middle level only.

Modifications

Add to the total cost

Upgrade Kitchen Cabinets	$ + 5525
Solid Surface Countertops (Included)	
Full Bath - including plumbing, wall and floor finishes	+ 7405
Half Bath - including plumbing, wall and floor finishes	+ 4331
One Car Attached Garage	+ 14,471
One Car Detached Garage	+ 19,004
Fireplace & Chimney	+ 6627

Adjustments

For multi family - add to total cost

Additional Kitchen	$ + 9600
Additional Bath	+ 7405
Additional Entry & Exit	+ 1676
Separate Heating	+ 1336
Separate Electric	+ 1945

For Townhouse/Rowhouse - Multiply cost per square foot by

Inner Unit	.90
End Unit	.95

Alternatives

Add to or deduct from the cost per square foot of living area

Cedar Shake Roof	+ 2.40
Clay Tile Roof	+ 4.95
Slate Roof	+ 5.40
Upgrade Walls to Skim Coat Plaster	+ .43
Upgrade Ceilings to Textured Finish	+ .58
Air Conditioning, in Heating Ductwork	+ 2.49
In Separate Ductwork	+ 5.37
Heating Systems, Hot Water	+ 1.44
Heat Pump	+ 1.68
Electric Heat	− .54
Not Heated	− 3.16

Additional upgrades or components

Kitchen Cabinets & Countertops	Page 93
Bathroom Vanities	94
Fireplaces & Chimneys	94
Windows, Skylights & Dormers	94
Appliances	95
Breezeways & Porches	95
Finished Attic	95
Garages	96
Site Improvements	96
Wings & Ells	56

Important: See the Reference Section for Location Factors (to adjust for your city) and Estimating Forms.

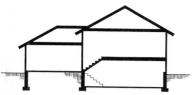

Average Tri-Level

Living Area - 2400 S.F.
Perimeter - 163 L.F.

			Cost Per Square Foot Of Living Area			% of Total
			Mat.	Inst.	Total	(rounded)
1	Site Work	Site preparation for slab; 4' deep trench excavation for foundation wall.		1.10	1.10	1.2%
2	Foundation	Continuous reinforced concrete footing, 10" deep x 20" wide; damproofed and insulated 8" thick reinforced concrete block foundation wall, 4' deep; trowel finished 4" thick concrete slab on 4" crushed stone base and polyethylene vapor barrier.	3.95	4.84	8.79	9.3%
3	Framing	Exterior walls - 2" x 6" wood studs, 16" O.C.; 1/2" sheathing; gable end roof framing, 2" x 10" rafters, 16" O.C. with 1/2" plywood sheathing; 2" x 10" floor joists, 16" O.C. with bridging and 5/8" subflooring; 2" x 4" interior partitions.	6.33	8.69	15.02	15.8%
4	Exterior Walls	Beveled wood siding and housewrap on insulated wood frame walls; R38 attic insulation; double hung wood windows; flush solid core doors, frame and hardware, painted finish; aluminum storm and screen doors.	9.57	4.08	13.65	14.4%
5	Roofing	25 year asphalt roof shingles; #15 felt building paper; aluminum gutters, downspouts, drip edge and flashings.	1.55	1.80	3.35	3.5%
6	Interiors	Walls & ceilings, 1/2" taped & finished gypsum wallboard, primed & painted with 2 coats of finish paint; birch faced hollow core interior doors, frames & hardware, painted finish; medium weight carpeting with pad, 40%; sheet vinyl, 15%; oak hardwood, 40%; ceramic tile, 5%; hardwood tread stairway.	12.74	13.04	25.78	27.2%
7	Specialties	Average grade kitchen cabinets and countertop; stainless steel kitchen sink; 40 gallon electric water heater.	2.96	1.00	3.96	4.2%
8	Mechanical	Three fixture bathroom: bathtub, water closet, vanity and sink; gas fired hot air heating system.	2.65	3.15	5.80	6.1%
9	Electrical	200 amp electric service; wiring, duplex & GFI receptacles, wall switches, door bell, appliance circuits, fans and communications cabling; average grade lighting fixtures.	1.59	2.05	3.64	3.8%
10	Overhead	Contractor's overhead and profit and plans.	7.01	6.75	13.76	14.5%
		Total	48.35	46.50	**94.85**	

RESIDENTIAL | Solid Wall | 1 Story

- Post and beam frame
- Log exterior walls
- Simple design from standard plans
- Single family — 1 full bath, 1 kitchen
- No basement
- Asphalt shingles on roof
- Hot air heat
- Gypsum wallboard interior finishes
- Materials and workmanship are average

Note: The illustration shown may contain some optional components (for example: garages and/or fireplaces) whose costs are shown in the modifications, adjustments, & alternatives below or at the end of the square foot section.

Base cost per square foot of living area

Exterior Wall	Living Area										
	600	800	1000	1200	1400	1600	1800	2000	2400	2800	3200
6" Log - Solid Wall	185.20	167.15	153.95	143.30	134.40	128.65	125.50	121.55	114.05	108.50	104.70
8" Log - Solid Wall	173.00	156.25	144.05	134.40	126.30	121.10	118.15	114.65	107.75	102.75	99.35
Finished Basement, Add	39.65	38.30	36.55	34.95	33.55	32.70	32.25	31.55	30.55	29.80	29.15
Unfinished Basement, Add	16.45	14.85	13.75	12.75	11.90	11.40	11.05	10.65	10.05	9.55	9.15

Modifications

Add to the total cost

Upgrade Kitchen Cabinets	$ + 5525
Solid Surface Countertops (Included)	
Full Bath - including plumbing, wall and floor finishes	+ 7405
Half Bath - including plumbing, wall and floor finishes	+ 4331
One Car Attached Garage	+ 14,471
One Car Detached Garage	+ 19,004
Fireplace & Chimney	+ 6627

Adjustments

For multi family - add to total cost

Additional Kitchen	$ + 9600
Additional Bath	+ 7405
Additional Entry & Exit	+ 1676
Separate Heating	+ 1336
Separate Electric	+ 1945

For Townhouse/Rowhouse - Multiply cost per square foot by

Inner Unit	.92
End Unit	.96

Alternatives

Add to or deduct from the cost per square foot of living area

Cedar Shake Roof	+ 3.35
Air Conditioning, in Heating Ductwork	+ 4.89
In Separate Ductwork	+ 7.17
Heating Systems, Hot Water	+ 1.60
Heat Pump	+ 1.37
Electric Heat	– .85
Not Heated	– 3.52

Additional upgrades or components

Kitchen Cabinets & Countertops	Page 93
Bathroom Vanities	94
Fireplaces & Chimneys	94
Windows, Skylights & Dormers	94
Appliances	95
Breezeways & Porches	95
Finished Attic	95
Garages	96
Site Improvements	96
Wings & Ells	56

Solid Wall 1 Story

Living Area - 1600 S.F.
Perimeter - 163 L.F.

		Cost Per Square Foot Of Living Area			% of Total
		Mat.	Inst.	Total	(rounded)
1 Site Work	Site preparation for slab; 4' deep trench excavation for foundation wall.		1.66	1.66	1.3%
2 Foundation	Continuous reinforced concrete footing, 10" deep x 20" wide; damproofed and insulated 8" thick reinforced concrete block foundation wall, 4' deep; trowel finished 4" thick concrete slab on 4" crushed stone base and polyethylene vapor barrier.	5.92	7.26	13.18	10.2%
3 Framing	Exterior walls - precut traditional log home, handcrafted white cedar or pine logs, delivery included; heavy timber roof framing with 2" thick tongue and groove decking, rigid insulation with 5/8" sheathing; 2" x 4" interior partitions.	26.14	15.97	42.11	32.7%
4 Exterior Walls	R38 attic insulation; double hung wood windows; flush solid core doors, frame and hardware, painted finish; aluminum storm and screen doors.	5.40	2.71	8.11	6.3%
5 Roofing	25 year asphalt roof shingles; #15 felt building paper; aluminum gutters, downspouts, drip edge and flashings.	2.32	2.70	5.02	3.9%
6 Interiors	Walls & ceilings, 1/2" taped & finished gypsum wallboard, primed & painted with 2 coats of finish paint; birch faced hollow core interior doors, frames & hardware, painted finish; medium weight carpeting with pad, 40%; sheet vinyl, 15%; oak hardwood, 40%; ceramic tile, 5%.	11.11	11.61	22.72	17.7%
7 Specialties	Average grade kitchen cabinets and countertop; stainless steel kitchen sink; 40 gallon electric water heater.	4.44	1.49	5.93	4.6%
8 Mechanical	Three fixture bathroom: bathtub, water closet, vanity and sink; gas fired hot air heating system.	3.62	3.57	7.19	5.6%
9 Electrical	200 amp electric service; wiring, duplex & GFI receptacles, wall switches, door bell, appliance circuits, fans and communications cabling; average grade lighting fixtures.	1.74	2.31	4.05	3.1%
10 Overhead	Contractor's overhead and profit and plans.	10.31	8.37	18.68	14.5%
	Total	71.00	57.65	**128.65**	

RESIDENTIAL | Solid Wall | 2 Story

- Post and beam frame
- Log exterior walls
- Simple design from standard plans
- Single family — 1 full bath, 1 kitchen
- No basement
- Asphalt shingles on roof
- Hot air heat
- Gypsum wallboard interior finishes
- Materials and workmanship are average

Note: The illustration shown may contain some optional components (for example: garages and/or fireplaces) whose costs are shown in the modifications, adjustments, & alternatives below or at the end of the square foot section.

Base cost per square foot of living area

Exterior Wall	Living Area										
	1000	1200	1400	1600	1800	2000	2200	2600	3000	3400	3800
6" Log - Solid Wall	157.80	143.40	136.15	131.25	126.05	120.85	117.15	110.25	103.80	100.80	98.00
8" Log - Solid Wall	145.20	131.70	125.25	120.80	116.25	111.40	108.25	102.10	96.25	93.65	91.20
Finished Basement, Add	22.70	22.45	21.70	21.15	20.65	20.30	19.85	19.15	18.60	18.25	17.95
Unfinished Basement, Add	9.00	8.40	7.90	7.60	7.25	7.00	6.75	6.35	6.00	5.75	5.60

Modifications

Add to the total cost

Upgrade Kitchen Cabinets	$ + 5525
Solid Surface Countertops (Included)	
Full Bath - including plumbing, wall and floor finishes	+ 7405
Half Bath - including plumbing, wall and floor finishes	+ 4331
One Car Attached Garage	+ 14,471
One Car Detached Garage	+ 19,004
Fireplace & Chimney	+ 7319

Adjustments

For multi family - add to total cost

Additional Kitchen	$ + 9600
Additional Bath	+ 7405
Additional Entry & Exit	+ 1676
Separate Heating	+ 1336
Separate Electric	+ 1945

For Townhouse/Rowhouse - Multiply cost per square foot by

Inner Unit	.92
End Unit	.96

Alternatives

Add to or deduct from the cost per square foot of living area

Cedar Shake Roof	+ 1.65
Air Conditioning, in Heating Ductwork	+ 2.96
In Separate Ductwork	+ 5.70
Heating Systems, Hot Water	+ 1.49
Heat Pump	+ 1.61
Electric Heat	– .63
Not Heated	– 3.28

Additional upgrades or components

Kitchen Cabinets & Countertops	Page 93
Bathroom Vanities	94
Fireplaces & Chimneys	94
Windows, Skylights & Dormers	94
Appliances	95
Breezeways & Porches	95
Finished Attic	95
Garages	96
Site Improvements	96
Wings & Ells	56

Solid Wall 2 Story

Living Area - 2000 S.F.
Perimeter - 135 L.F.

		Cost Per Square Foot Of Living Area			% of Total
		Mat.	Inst.	Total	(rounded)
1 Site Work	Site preparation for slab; 4' deep trench excavation for foundation wall.		1.33	1.33	1.1%
2 Foundation	Continuous reinforced concrete footing, 10" deep x 20" wide; damproofed and insulated 8" thick reinforced concrete block foundation wall, 4' deep; trowel finished 4" thick concrete slab on 4" crushed stone base and polyethylene vapor barrier.	3.55	4.46	8.01	6.6%
3 Framing	Exterior walls - precut traditional log home, handcrafted white cedar or pine logs, delivery included; heavy timber roof framing with 2" thick T. & G. decking, rigid insulation with 5/8" sheathing; heavy timber columns, beams & joists with 2" thick T. & G. decking; 2" x 4" interior partitions.	26.78	15.32	42.10	34.8%
4 Exterior Walls	R38 attic insulation; double hung wood windows; flush solid core doors, frame and hardware, painted finish; aluminum storm and screen doors.	5.40	2.86	8.26	6.8%
5 Roofing	25 year asphalt roof shingles; #15 felt building paper; aluminum gutters, downspouts, drip edge and flashings.	1.16	1.36	2.52	2.1%
6 Interiors	Walls & ceilings, 1/2" taped & finished gypsum wallboard, primed & painted with 2 coats of finish paint; birch faced hollow core interior doors, frames & hardware, painted finish; medium weight carpeting with pad, 40%; sheet vinyl, 15%; oak hardwood, 40%; ceramic tile, 5%; hardwood tread stairway.	13.10	13.06	26.16	21.6%
7 Specialties	Average grade kitchen cabinets and countertop; stainless steel kitchen sink; 40 gallon electric water heater.	3.55	1.19	4.74	3.9%
8 Mechanical	Three fixture bathroom: bathtub, water closet, vanity and sink; gas fired hot air heating system.	3.07	3.31	6.38	5.3%
9 Electrical	200 amp electric service; wiring, duplex & GFI receptacles, wall switches, door bell, appliance circuits, fans and communications cabling; average grade lighting fixtures.	1.65	2.16	3.81	3.2%
10 Overhead	Contractor's overhead and profit and plans.	9.89	7.65	17.54	14.5%
	Total	68.15	52.70	**120.85**	

RESIDENTIAL | Average | Wings & Ells

1 Story — Base cost per square foot of living area

Exterior Wall	Living Area							
	50	100	200	300	400	500	600	700
Wood Siding - Wood Frame	212.80	164.10	143.05	120.15	113.20	109.10	106.30	107.05
Brick Veneer - Wood Frame	215.40	162.30	139.45	113.50	105.95	101.50	98.50	99.05
Stucco on Wood Frame	202.30	156.50	136.65	115.85	109.25	105.35	102.80	103.65
Solid Masonry	263.55	200.30	173.20	140.30	131.30	125.95	124.10	124.00
Finished Basement, Add	63.75	52.80	47.45	38.65	36.85	35.80	35.05	34.55
Unfinished Basement, Add	30.25	22.85	19.75	14.60	13.55	12.95	12.55	12.25

1-1/2 Story — Base cost per square foot of living area

Exterior Wall	Living Area							
	100	200	300	400	500	600	700	800
Wood Siding - Wood Frame	170.75	137.90	118.20	105.95	100.00	97.10	93.45	92.35
Brick Veneer - Wood Frame	226.25	169.75	141.20	123.50	114.95	110.25	105.20	103.35
Stucco on Wood Frame	203.50	151.50	126.00	111.70	104.00	99.95	95.45	93.65
Solid Masonry	257.90	195.05	162.30	140.00	130.10	124.65	118.80	116.85
Finished Basement, Add	42.95	38.75	35.25	31.35	30.30	29.60	28.95	28.80
Unfinished Basement, Add	18.95	15.85	13.80	11.55	10.90	10.50	10.15	10.05

2 Story — Base cost per square foot of living area

Exterior Wall	Living Area							
	100	200	400	600	800	1000	1200	1400
Wood Siding - Wood Frame	170.95	128.30	109.75	91.00	84.95	81.20	78.85	79.80
Brick Veneer - Wood Frame	230.15	161.10	129.80	104.40	95.85	90.75	87.35	87.60
Stucco on Wood Frame	204.60	142.85	114.60	94.25	86.75	82.20	79.25	79.85
Solid Masonry	265.60	186.50	150.95	118.50	108.50	102.55	98.60	98.50
Finished Basement, Add	33.90	28.40	25.75	21.35	20.45	19.95	19.55	19.30
Unfinished Basement, Add	15.30	11.55	10.05	7.45	6.95	6.65	6.45	6.30

Base costs do not include bathroom or kitchen facilities. Use Modifications/Adjustments/Alternatives on pages 93–96 where appropriate.

No part of this cost data may be reproduced, stored in a retrieval system, or transmitted in any form or by any means without prior written permission of RSMeans.

Did you know?
RSMeans Online gives you the same access to RSMeans' data with 24/7 access:
- Quickly locate costs in the searchable database.
- Build cost lists, estimates, and reports in minutes.
- Adjust costs to any location in the U.S. and Canada with the click of a button.

Start your free trial today at **www.RSMeansOnline.com**

RSMeans Online
FROM THE GORDIAN GROUP®

RESIDENTIAL | Custom | Illustrations

1 Story

1-1/2 Story

2 Story

2-1/2 Story

Bi-Level

Tri-Level

RESIDENTIAL | Custom | 1 Story

- A distinct residence from designer's plans
- Single family — 1 full bath, 1 half bath, 1 kitchen
- No basement
- Asphalt shingles on roof
- Forced hot air heat/air conditioning
- Gypsum wallboard interior finishes
- Materials and workmanship are above average

Note: The illustration shown may contain some optional components (for example: garages and/or fireplaces) whose costs are shown in the modifications, adjustments, & alternatives below or at the end of the square foot section.

Base cost per square foot of living area

Exterior Wall	Living Area										
	800	1000	1200	1400	1600	1800	2000	2400	2800	3200	3600
Wood Siding - Wood Frame	206.30	185.95	170.25	157.60	148.95	143.95	138.30	128.10	120.55	115.35	109.90
Brick Veneer - Wood Frame	215.75	194.60	178.00	164.65	155.60	150.35	144.35	133.50	125.65	120.05	114.30
Stone Veneer - Wood Frame	225.60	203.50	186.05	171.95	162.40	156.95	150.55	139.20	130.85	124.95	118.75
Solid Masonry	226.30	204.10	186.65	172.55	162.95	157.45	151.00	139.60	131.25	125.25	119.15
Finished Basement, Add	57.50	57.30	54.85	52.85	51.60	50.85	49.80	48.30	47.15	46.15	45.25
Unfinished Basement, Add	25.40	23.95	22.70	21.65	20.95	20.55	20.00	19.25	18.60	18.10	17.65

Modifications

Add to the total cost

Upgrade Kitchen Cabinets	$ + 1736
Solid Surface Countertops (Included)	
Full Bath - including plumbing, wall and floor finishes	+ 8886
Half Bath - including plumbing, wall and floor finishes	+ 5197
Two Car Attached Garage	+ 28,478
Two Car Detached Garage	+ 32,548
Fireplace & Chimney	+ 6949

Adjustments

For multi family - add to total cost

Additional Kitchen	$ + 20,958
Additional Full Bath & Half Bath	+ 14,083
Additional Entry & Exit	+ 1676
Separate Heating & Air Conditioning	+ 7087
Separate Electric	+ 1945

For Townhouse/Rowhouse - Multiply cost per square foot by

Inner Unit	.90
End Unit	.95

Alternatives

Add to or deduct from the cost per square foot of living area

Cedar Shake Roof	+ 2.35
Clay Tile Roof	+ 5.85
Slate Roof	+ 6.55
Upgrade Ceilings to Textured Finish	+ .58
Air Conditioning, in Heating Ductwork	Base System
Heating Systems, Hot Water	+ 1.64
Heat Pump	+ 1.37
Electric Heat	– 2.38
Not Heated	– 4.41

Additional upgrades or components

Kitchen Cabinets & Countertops	Page 93
Bathroom Vanities	94
Fireplaces & Chimneys	94
Windows, Skylights & Dormers	94
Appliances	95
Breezeways & Porches	95
Finished Attic	95
Garages	96
Site Improvements	96
Wings & Ells	74

Custom 1 Story

Living Area - 2400 S.F.
Perimeter - 207 L.F.

#	Division	Description	Cost Per Square Foot Of Living Area			% of Total
			Mat.	Inst.	Total	(rounded)
1	Site Work	Site preparation for slab; 4' deep trench excavation for foundation wall.		1.24	1.24	1.0%
2	Foundation	Continuous reinforced concrete footing, 12" deep x 24" wide; damproofed and insulated 12" thick reinforced concrete block foundation wall, 4' deep; trowel finished 4" thick concrete slab on 4" crushed stone base and polyethylene vapor barrier.	6.33	7.64	13.97	10.9%
3	Framing	Exterior walls - 2" x 6" wood studs, 16" O.C.; 1/2" sheathing; gable end roof framing, 2" x 10" rafters, 16" O.C. with 1/2" plywood sheathing; 2" x 4" interior partitions.	6.49	9.06	15.55	12.1%
4	Exterior Walls	1" x 6" tongue and groove vertical wood siding and housewrap on insulated wood frame walls; R38 attic insulation; plastic clad double hung wood windows; raised panel exterior doors, frames and hardware, painted finish; wood storm and screen door.	11.53	4.04	15.57	12.2%
5	Roofing	Red cedar roof shingles, perfections; #15 felt building paper; aluminum gutters, downspouts and drip edge; copper flashings.	4.79	3.88	8.67	6.8%
6	Interiors	Skim coated 1/2" thick gypsum wallboard walls and ceilings, primed and painted with 2 coats; interior raised panel solid core doors, frames and hardware, painted finish; oak hardwood flooring, 70%; ceramic tile flooring, 30%.	13.43	13.09	26.52	20.7%
7	Specialties	Custom grade kitchen cabinets and countertops; double bowl kitchen sink; 75 gallon gas water heater.	8.03	1.58	9.61	7.5%
8	Mechanical	Three fixture bathroom: bathtub, water closet, vanity and sink; gas fired heating and air conditioning system.	6.84	3.65	10.49	8.2%
9	Electrical	200 amp electric service; wiring, duplex & GFI receptacles, wall switches, door bell, appliance circuits, air conditioning circuit, fans and communications cabling; custom grade lighting fixtures.	2.63	2.47	5.10	4.0%
10	Overhead	Contractor's overhead and profit and design.	12.03	9.35	21.38	16.7%
	Total		72.10	56.00	**128.10**	

For customer support on your Residential Cost Data, call 877.759.4771.

RESIDENTIAL | Custom | 1-1/2 Story

- A distinct residence from designer's plans
- Single family — 1 full bath, 1 half bath, 1 kitchen
- No basement
- Asphalt shingles on roof
- Forced hot air heat/air conditioning
- Gypsum wallboard interior finishes
- Materials and workmanship are above average

Note: The illustration shown may contain some optional components (for example: garages and/or fireplaces) whose costs are shown in the modifications, adjustments, & alternatives below or at the end of the square foot section.

Base cost per square foot of living area

Exterior Wall	Living Area										
	1000	1200	1400	1600	1800	2000	2400	2800	3200	3600	4000
Wood Siding - Wood Frame	186.50	173.10	163.45	152.15	145.60	139.35	127.45	121.90	116.80	112.95	107.65
Brick Veneer - Wood Frame	196.80	182.70	172.60	160.45	153.50	146.80	134.10	128.15	122.55	118.55	112.85
Stone Veneer - Wood Frame	207.35	192.65	182.05	169.05	161.65	154.55	140.95	134.65	128.55	124.30	118.20
Solid Masonry	208.05	193.25	182.65	169.60	162.15	155.00	141.35	135.00	128.90	124.65	118.55
Finished Basement, Add	38.15	38.45	37.40	35.85	35.00	34.25	32.70	32.00	31.15	30.70	30.05
Unfinished Basement, Add	17.05	16.30	15.75	15.00	14.55	14.10	13.30	12.90	12.45	12.30	11.90

Modifications

Add to the total cost

Upgrade Kitchen Cabinets	$ + 1736
Solid Surface Countertops (Included)	
Full Bath - including plumbing, wall and floor finishes	+ 8886
Half Bath - including plumbing, wall and floor finishes	+ 5197
Two Car Attached Garage	+ 28,478
Two Car Detached Garage	+ 32,548
Fireplace & Chimney	+ 6949

Adjustments

For multi family - add to total cost

Additional Kitchen	$ + 20,958
Additional Full Bath & Half Bath	+ 14,083
Additional Entry & Exit	+ 1676
Separate Heating & Air Conditioning	+ 7087
Separate Electric	+ 1945

For Townhouse/Rowhouse - Multiply cost per square foot by

Inner Unit	.90
End Unit	.95

Alternatives

Add to or deduct from the cost per square foot of living area

Cedar Shake Roof	+ 1.70
Clay Tile Roof	+ 4.25
Slate Roof	+ 4.70
Upgrade Ceilings to Textured Finish	+ .58
Air Conditioning, in Heating Ductwork	Base System
Heating Systems, Hot Water	+ 1.57
Heat Pump	+ 1.43
Electric Heat	− 2.10
Not Heated	− 4.07

Additional upgrades or components

Kitchen Cabinets & Countertops	Page 93
Bathroom Vanities	94
Fireplaces & Chimneys	94
Windows, Skylights & Dormers	94
Appliances	95
Breezeways & Porches	95
Finished Attic	95
Garages	96
Site Improvements	96
Wings & Ells	74

Custom 1-1/2 Story

Living Area - 2800 S.F.
Perimeter - 175 L.F.

#	Section	Description	Cost Per Square Foot Of Living Area - Mat.	Cost Per Square Foot Of Living Area - Inst.	Cost Per Square Foot Of Living Area - Total	% of Total (rounded)
1	Site Work	Site preparation for slab; 4' deep trench excavation for foundation wall.		1.07	1.07	0.9%
2	Foundation	Continuous reinforced concrete footing, 12" deep x 24" wide; damproofed and insulated 12" thick reinforced concrete block foundation wall, 4' deep; trowel finished 4" thick concrete slab on 4" crushed stone base and polyethylene vapor barrier.	4.44	5.46	9.90	8.1%
3	Framing	Exterior walls - 2" x 6" wood studs, 16" O.C.; 1/2" sheathing; gable end roof framing, steep pitch 2" x 10" rafters, 16" O.C. with 1/2" plywood sheathing; 2" x 10" floor joists, 16" O.C. with bridging and 5/8" subflooring; 2" x 4" interior partitions.	7.07	9.68	16.75	13.7%
4	Exterior Walls	1" x 6" tongue and groove vertical wood siding and housewrap on insulated wood frame walls; R38 attic insulation; plastic clad double hung wood windows; raised panel exterior doors, frames and hardware, painted finish; wood storm and screen door.	11.35	3.99	15.34	12.6%
5	Roofing	Red cedar roof shingles, perfections; #15 felt building paper; aluminum gutters, downspouts and drip edge; copper flashings.	3.47	2.81	6.28	5.2%
6	Interiors	Skim coated 1/2" thick gypsum wallboard walls and ceilings, primed and painted with 2 coats; interior raised panel solid core doors, frames and hardware, painted finish; oak hardwood flooring, 70%; ceramic tile flooring, 30%; hardwood tread stairway.	15.25	14.47	29.72	24.4%
7	Specialties	Custom grade kitchen cabinets and countertops; double bowl kitchen sink; 75 gallon gas water heater.	6.89	1.37	8.26	6.8%
8	Mechanical	Three fixture bathroom: bathtub, water closet, vanity and sink; gas fired heating and air conditioning system.	5.87	3.41	9.28	7.6%
9	Electrical	200 amp electric service; wiring, duplex & GFI receptacles, wall switches, door bell, appliance circuits, air conditioning circuit, fans and communications cabling; custom grade lighting fixtures.	2.58	2.40	4.98	4.1%
10	Overhead	Contractor's overhead and profit and design.	11.38	8.94	20.32	16.7%
		Total	68.30	53.60	**121.90**	

RESIDENTIAL | Custom | 2 Story

- A distinct residence from designer's plans
- Single family — 1 full bath, 1 half bath, 1 kitchen
- No basement
- Asphalt shingles on roof
- Forced hot air heat/air conditioning
- Gypsum wallboard interior finishes
- Materials and workmanship are above average

Note: The illustration shown may contain some optional components (for example: garages and/or fireplaces) whose costs are shown in the modifications, adjustments, & alternatives below or at the end of the square foot section.

Base cost per square foot of living area

Exterior Wall	Living Area										
	1200	1400	1600	1800	2000	2400	2800	3200	3600	4000	4400
Wood Siding - Wood Frame	170.35	159.60	152.00	145.25	138.25	128.15	119.60	114.00	110.60	107.00	104.05
Brick Veneer - Wood Frame	180.50	169.15	161.05	153.80	146.50	135.55	126.35	120.30	116.75	112.75	109.60
Stone Veneer - Wood Frame	191.00	178.95	170.50	162.60	155.05	143.30	133.35	126.85	123.05	118.60	115.35
Solid Masonry	192.10	180.00	171.40	163.50	155.90	144.05	134.05	127.55	123.65	119.25	115.95
Finished Basement, Add	30.70	30.85	30.10	29.25	28.70	27.45	26.50	25.80	25.40	24.90	24.60
Unfinished Basement, Add	13.70	13.10	12.70	12.30	12.00	11.30	10.80	10.45	10.30	10.00	9.85

Modifications

Add to the total cost

Upgrade Kitchen Cabinets	$ + 1736
Solid Surface Countertops (Included)	
Full Bath - including plumbing, wall and floor finishes	+ 8886
Half Bath - including plumbing, wall and floor finishes	+ 5197
Two Car Attached Garage	+ 28,478
Two Car Detached Garage	+ 32,548
Fireplace & Chimney	+ 7842

Adjustments

For multi family - add to total cost

Additional Kitchen	$ + 20,958
Additional Full Bath & Half Bath	+ 14,083
Additional Entry & Exit	+ 1676
Separate Heating & Air Conditioning	+ 7087
Separate Electric	+ 1945

For Townhouse/Rowhouse -
Multiply cost per square foot by

Inner Unit	.87
End Unit	.93

Alternatives

Add to or deduct from the cost per square foot of living area

Cedar Shake Roof	+ 1.20
Clay Tile Roof	+ 2.90
Slate Roof	+ 3.25
Upgrade Ceilings to Textured Finish	+ .58
Air Conditioning, in Heating Ductwork	Base System
Heating Systems, Hot Water	+ 1.52
Heat Pump	+ 1.60
Electric Heat	− 2.10
Not Heated	− 3.84

Additional upgrades or components

Kitchen Cabinets & Countertops	Page 93
Bathroom Vanities	94
Fireplaces & Chimneys	94
Windows, Skylights & Dormers	94
Appliances	95
Breezeways & Porches	95
Finished Attic	95
Garages	96
Site Improvements	96
Wings & Ells	74

Custom 2 Story

Living Area - 2800 S.F.
Perimeter - 156 L.F.

#	Category	Description	Mat.	Inst.	Total	% of Total (rounded)
1	Site Work	Site preparation for slab; 4' deep trench excavation for foundation wall.		1.07	1.07	0.9%
2	Foundation	Continuous reinforced concrete footing, 12" deep x 24" wide; damproofed and insulated 12" thick reinforced concrete block foundation wall, 4' deep; trowel finished 4" thick concrete slab on 4" crushed stone base and polyethylene vapor barrier.	3.78	4.68	8.46	7.1%
3	Framing	Exterior walls - 2" x 6" wood studs, 16" O.C.; 1/2" sheathing; gable end roof framing, 2" x 10" rafters, 16" O.C. with 1/2" plywood sheathing; 2" x 10" floor joists, 16" O.C. with bridging and 5/8" subflooring; 2" x 4" interior partitions.	6.71	9.17	15.88	13.3%
4	Exterior Walls	1" x 6" tongue and groove vertical wood siding and housewrap on insulated wood frame walls; R38 attic insulation; plastic clad double hung wood windows; raised panel exterior doors, frames and hardware, painted finish; wood storm and screen door.	12.32	4.32	16.64	13.9%
5	Roofing	Red cedar roof shingles, perfections; #15 felt building paper; aluminum gutters, downspouts and drip edge; copper flashings.	2.39	1.94	4.33	3.6%
6	Interiors	Skim coated 1/2" thick gypsum wallboard walls and ceilings, primed and painted with 2 coats; interior raised panel solid core doors, frames and hardware, painted finish; oak hardwood flooring, 70%; ceramic tile flooring, 30%; hardwood tread stairway.	15.63	14.91	30.54	25.5%
7	Specialties	Custom grade kitchen cabinets and countertops; double bowl kitchen sink; 75 gallon gas water heater.	6.89	1.37	8.26	6.9%
8	Mechanical	Three fixture bathroom: bathtub, water closet, vanity and sink; gas fired heating and air conditioning system.	6.03	3.46	9.49	7.9%
9	Electrical	200 amp electric service; wiring, duplex & GFI receptacles, wall switches, door bell, appliance circuits, air conditioning circuit, fans and communications cabling; custom grade lighting fixtures.	2.58	2.40	4.98	4.2%
10	Overhead	Contractor's overhead and profit and design.	11.27	8.68	19.95	16.7%
	Total		67.60	52.00	**119.60**	

RESIDENTIAL | Custom | 2-1/2 Story

- A distinct residence from designer's plans
- Single family — 1 full bath, 1 half bath, 1 kitchen
- No basement
- Asphalt shingles on roof
- Forced hot air heat/air conditioning
- Gypsum wallboard interior finishes
- Materials and workmanship are above average

Note: The illustration shown may contain some optional components (for example: garages and/or fireplaces) whose costs are shown in the modifications, adjustments, & alternatives below or at the end of the square foot section.

Base cost per square foot of living area

Exterior Wall	Living Area										
	1500	1800	2100	2400	2800	3200	3600	4000	4500	5000	5500
Wood Siding - Wood Frame	167.55	150.75	140.70	134.60	126.75	119.35	115.20	108.90	105.45	102.35	99.35
Brick Veneer - Wood Frame	178.45	160.75	149.65	143.05	134.85	126.70	122.15	115.35	111.60	108.20	104.85
Stone Veneer - Wood Frame	189.65	171.15	158.90	151.80	143.25	134.30	129.35	122.00	117.95	114.25	110.65
Solid Masonry	190.60	172.05	159.70	152.55	143.95	134.95	129.95	122.55	118.45	114.75	111.10
Finished Basement, Add	24.40	24.30	23.10	22.50	21.95	21.10	20.60	20.10	19.65	19.30	19.00
Unfinished Basement, Add	10.95	10.40	9.70	9.45	9.15	8.70	8.50	8.15	7.95	7.75	7.60

Modifications

Add to the total cost

Upgrade Kitchen Cabinets	$ + 1736
Solid Surface Countertops (Included)	
Full Bath - including plumbing, wall and floor finishes	+ 8886
Half Bath - including plumbing, wall and floor finishes	+ 5197
Two Car Attached Garage	+ 28,478
Two Car Detached Garage	+ 32,548
Fireplace & Chimney	+ 8856

Adjustments

For multi family - add to total cost

Additional Kitchen	$ + 20,958
Additional Full Bath & Half Bath	+ 14,083
Additional Entry & Exit	+ 1676
Separate Heating & Air Conditioning	+ 7087
Separate Electric	+ 1945

For Townhouse/Rowhouse - Multiply cost per square foot by

Inner Unit	.87
End Unit	.94

Alternatives

Add to or deduct from the cost per square foot of living area

Cedar Shake Roof	+ 1.05
Clay Tile Roof	+ 2.55
Slate Roof	+ 2.85
Upgrade Ceilings to Textured Finish	+ .58
Air Conditioning, in Heating Ductwork	Base System
Heating Systems, Hot Water	+ 1.37
Heat Pump	+ 1.65
Electric Heat	− 3.69
Not Heated	− 3.84

Additional upgrades or components

Kitchen Cabinets & Countertops	Page 93
Bathroom Vanities	94
Fireplaces & Chimneys	94
Windows, Skylights & Dormers	94
Appliances	95
Breezeways & Porches	95
Finished Attic	95
Garages	96
Site Improvements	96
Wings & Ells	74

Important: See the Reference Section for Location Factors (to adjust for your city) and Estimating Forms.

Custom 2-1/2 Story

Living Area - 3200 S.F.
Perimeter - 150 L.F.

#	Section	Description	Cost Per Square Foot Of Living Area			% of Total
			Mat.	Inst.	Total	(rounded)
1	Site Work	Site preparation for slab; 4' deep trench excavation for foundation wall.		.93	.93	0.8%
2	Foundation	Continuous reinforced concrete footing, 12" deep x 24" wide; damproofed and insulated 12" thick reinforced concrete block foundation wall, 4' deep; trowel finished 4" thick concrete slab on 4" crushed stone base and polyethylene vapor barrier.	3.12	3.89	7.01	5.9%
3	Framing	Exterior walls - 2" x 6" wood studs, 16" O.C.; 1/2" sheathing; gable end roof framing, steep pitch 2" x 10" rafters, 16" O.C. with 1/2" plywood sheathing; 2" x 10" floor joists, 16" O.C. with bridging and 5/8" subflooring; 2" x 4" interior partitions.	7.19	9.84	17.03	14.3%
4	Exterior Walls	1" x 6" tongue and groove vertical wood siding and housewrap on insulated wood frame walls; R38 attic insulation; plastic clad double hung wood windows; raised panel exterior doors, frames and hardware, painted finish; wood storm and screen door.	12.79	4.49	17.28	14.5%
5	Roofing	Red cedar roof shingles, perfections; #15 felt building paper; aluminum gutters, downspouts and drip edge; copper flashings.	2.13	1.72	3.85	3.2%
6	Interiors	Skim coated 1/2" thick gypsum wallboard walls and ceilings, primed and painted with 2 coats; interior raised panel solid core doors, frames and hardware, painted finish; oak hardwood flooring, 70%; ceramic tile flooring, 30%; hardwood tread stairway.	16.70	15.80	32.50	27.2%
7	Specialties	Custom grade kitchen cabinets and countertops; double bowl kitchen sink; 75 gallon gas water heater.	6.02	1.20	7.22	6.0%
8	Mechanical	Three fixture bathroom: bathtub, water closet, vanity and sink; gas fired heating and air conditioning system.	5.43	3.32	8.75	7.3%
9	Electrical	200 amp electric service; wiring, duplex & GFI receptacles, wall switches, door bell, appliance circuits, air conditioning circuit, fans and communications cabling; custom grade lighting fixtures.	2.55	2.35	4.90	4.1%
10	Overhead	Contractor's overhead and profit and design.	11.17	8.71	19.88	16.7%
	Total		67.10	52.25	**119.35**	

RESIDENTIAL | Custom | 3 Story

- A distinct residence from designer's plans
- Single family — 1 full bath, 1 half bath, 1 kitchen
- No basement
- Asphalt shingles on roof
- Forced hot air heat/air conditioning
- Gypsum wallboard interior finishes
- Materials and workmanship are above average

Note: The illustration shown may contain some optional components (for example: garages and/or fireplaces) whose costs are shown in the modifications, adjustments, & alternatives below or at the end of the square foot section.

Base cost per square foot of living area

Exterior Wall	Living Area										
	1500	1800	2100	2500	3000	3500	4000	4500	5000	5500	6000
Wood Siding - Wood Frame	164.40	147.95	139.65	132.60	122.30	116.95	110.65	104.15	101.75	99.15	96.50
Brick Veneer - Wood Frame	175.25	157.95	149.00	141.60	130.40	124.65	117.70	110.65	107.95	105.20	102.20
Stone Veneer - Wood Frame	186.45	168.35	158.75	150.85	138.85	132.60	124.90	117.35	114.45	111.40	108.05
Solid Masonry	187.65	169.45	159.80	151.85	139.70	133.45	125.70	118.10	115.10	112.05	108.65
Finished Basement, Add	21.35	21.35	20.55	20.00	19.15	18.60	17.90	17.40	17.15	16.90	16.60
Unfinished Basement, Add	9.65	9.15	8.70	8.45	8.00	7.70	7.35	7.10	7.00	6.80	6.65

Modifications

Add to the total cost

Upgrade Kitchen Cabinets	$ + 1736
Solid Surface Countertops (Included)	
Full Bath - including plumbing, wall and floor finishes	+ 8886
Half Bath - including plumbing, wall and floor finishes	+ 5197
Two Car Attached Garage	+ 28,478
Two Car Detached Garage	+ 32,548
Fireplace & Chimney	+ 8856

Adjustments

For multi family - add to total cost

Additional Kitchen	$ + 20,958
Additional Full Bath & Half Bath	+ 14,083
Additional Entry & Exit	+ 1676
Separate Heating & Air Conditioning	+ 7087
Separate Electric	+ 1945

For Townhouse/Rowhouse - Multiply cost per square foot by

Inner Unit	.85
End Unit	.93

Alternatives

Add to or deduct from the cost per square foot of living area

Cedar Shake Roof	+ .80
Clay Tile Roof	+ 1.95
Slate Roof	+ 2.20
Upgrade Ceilings to Textured Finish	+ .58
Air Conditioning, in Heating Ductwork	Base System
Heating Systems, Hot Water	+ 1.37
Heat Pump	+ 1.65
Electric Heat	– 3.69
Not Heated	– 3.72

Additional upgrades or components

Kitchen Cabinets & Countertops	Page 93
Bathroom Vanities	94
Fireplaces & Chimneys	94
Windows, Skylights & Dormers	94
Appliances	95
Breezeways & Porches	95
Finished Attic	95
Garages	96
Site Improvements	96
Wings & Ells	74

Important: See the Reference Section for Location Factors (to adjust for your city) and Estimating Forms.

Custom 3 Story

Living Area - 3000 S.F.
Perimeter - 135 L.F.

			Cost Per Square Foot Of Living Area			% of Total
			Mat.	Inst.	Total	(rounded)
1	Site Work	Site preparation for slab; 4' deep trench excavation for foundation wall.		.99	.99	0.8%
2	Foundation	Continuous reinforced concrete footing, 12" deep x 24" wide; damproofed and insulated 12" thick reinforced concrete block foundation wall, 4' deep; trowel finished 4" thick concrete slab on 4" crushed stone base and polyethylene vapor barrier.	2.91	3.67	6.58	5.4%
3	Framing	Exterior walls - 2" x 6" wood studs, 16" O.C.; 1/2" sheathing; gable end roof framing, 2" x 10" rafters, 16" O.C. with 1/2" plywood sheathing; 2" x 10" floor joists, 16" O.C. with bridging and 5/8" subflooring; 2" x 4" interior partitions.	7.08	9.71	16.79	13.7%
4	Exterior Walls	1" x 6" tongue and groove vertical wood siding and housewrap on insulated wood frame walls; R38 attic insulation; plastic clad double hung wood windows; raised panel exterior doors, frames and hardware, painted finish; wood storm and screen door.	14.23	5.00	19.23	15.7%
5	Roofing	Red cedar roof shingles, perfections; #15 felt building paper; aluminum gutters, downspouts and drip edge; copper flashings.	1.60	1.30	2.90	2.4%
6	Interiors	Skim coated 1/2" thick gypsum wallboard walls and ceilings, primed and painted with 2 coats; interior raised panel solid core doors, frames and hardware, painted finish; oak hardwood flooring, 70%; ceramic tile flooring, 30%; hardwood tread stairway.	17.30	16.37	33.67	27.5%
7	Specialties	Custom grade kitchen cabinets and countertops; double bowl kitchen sink; 75 gallon gas water heater.	6.42	1.28	7.70	6.3%
8	Mechanical	Three fixture bathroom: bathtub, water closet, vanity and sink; gas fired heating and air conditioning system.	5.72	3.38	9.10	7.4%
9	Electrical	200 amp electric service; wiring, duplex & GFI receptacles, wall switches, door bell, appliance circuits, air conditioning circuit, fans and communications cabling; custom grade lighting fixtures.	2.57	2.37	4.94	4.0%
10	Overhead	Contractor's overhead and profit and design.	11.57	8.83	20.40	16.7%
		Total	69.40	52.90	**122.30**	

RESIDENTIAL Custom Bi-Level

- A distinct residence from designer's plans
- Single family — 1 full bath, 1 half bath, 1 kitchen
- No basement
- Asphalt shingles on roof
- Forced hot air heat/air conditioning
- Gypsum wallboard interior finishes
- Materials and workmanship are above average

Note: The illustration shown may contain some optional components (for example: garages and/or fireplaces) whose costs are shown in the modifications, adjustments, & alternatives below or at the end of the square foot section.

Base cost per square foot of living area

Exterior Wall	Living Area										
	1200	1400	1600	1800	2000	2400	2800	3200	3600	4000	4400
Wood Siding - Wood Frame	161.20	151.00	143.80	137.45	130.85	121.45	113.45	108.20	105.10	101.80	99.05
Brick Veneer - Wood Frame	168.90	158.30	150.75	144.00	137.10	127.10	118.65	113.10	109.75	106.15	103.30
Stone Veneer - Wood Frame	176.95	165.80	157.90	150.80	143.60	132.95	123.95	118.05	114.55	110.65	107.65
Solid Masonry	177.65	166.45	158.55	151.35	144.20	133.50	124.40	118.55	115.00	111.15	108.10
Finished Basement, Add	30.70	30.85	30.10	29.25	28.70	27.45	26.50	25.80	25.40	24.90	24.60
Unfinished Basement, Add	13.70	13.10	12.70	12.30	12.00	11.30	10.80	10.45	10.30	10.00	9.85

Modifications
Add to the total cost

Upgrade Kitchen Cabinets	$ + 1736
Solid Surface Countertops (Included)	
Full Bath - including plumbing, wall and floor finishes	+ 8886
Half Bath - including plumbing, wall and floor finishes	+ 5197
Two Car Attached Garage	+ 28,478
Two Car Detached Garage	+ 32,548
Fireplace & Chimney	+ 6949

Adjustments
For multi family - add to total cost

Additional Kitchen	$ + 20,958
Additional Full Bath & Half Bath	+ 14,083
Additional Entry & Exit	+ 1676
Separate Heating & Air Conditioning	+ 7087
Separate Electric	+ 1945

For Townhouse/Rowhouse - Multiply cost per square foot by

Inner Unit	.89
End Unit	.95

Alternatives
Add to or deduct from the cost per square foot of living area

Cedar Shake Roof	+ 1.20
Clay Tile Roof	+ 2.90
Slate Roof	+ 3.25
Upgrade Ceilings to Textured Finish	+ .58
Air Conditioning, in Heating Ductwork	Base System
Heating Systems, Hot Water	+ 1.52
Heat Pump	+ 1.60
Electric Heat	– 2.10
Not Heated	– 3.72

Additional upgrades or components

Kitchen Cabinets & Countertops	Page 93
Bathroom Vanities	94
Fireplaces & Chimneys	94
Windows, Skylights & Dormers	94
Appliances	95
Breezeways & Porches	95
Finished Attic	95
Garages	96
Site Improvements	96
Wings & Ells	74

Important: See the Reference Section for Location Factors (to adjust for your city) and Estimating Forms.

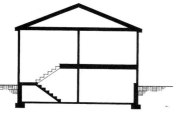

Custom Bi-Level

Living Area - 2800 S.F.
Perimeter - 156 L.F.

		Cost Per Square Foot Of Living Area			% of Total
		Mat.	Inst.	Total	(rounded)
1 Site Work	Site preparation for slab; 4' deep trench excavation for foundation wall.		1.07	1.07	0.9%
2 Foundation	Continuous reinforced concrete footing, 12" deep x 24" wide; damproofed and insulated 12" thick reinforced concrete block foundation wall, 4' deep; trowel finished 4" thick concrete slab on 4" crushed stone base and polyethylene vapor barrier.	3.78	4.68	8.46	7.5%
3 Framing	Exterior walls - 2" x 6" wood studs, 16" O.C.; 1/2" sheathing; gable end roof framing, 2" x 10" rafters, 16" O.C. with 1/2" plywood sheathing; 2" x 10" floor joists, 16" O.C. with bridging and 5/8" subflooring; 2" x 4" interior partitions.	6.30	8.56	14.86	13.1%
4 Exterior Walls	1" x 6" tongue and groove vertical wood siding and housewrap on insulated wood frame walls; R38 attic insulation; plastic clad double hung wood windows; raised panel exterior doors, frames and hardware, painted finish; wood storm and screen door.	9.78	3.43	13.21	11.6%
5 Roofing	Red cedar roof shingles, perfections; #15 felt building paper; aluminum gutters, downspouts and drip edge; copper flashings.	2.39	1.94	4.33	3.8%
6 Interiors	Skim coated 1/2" thick gypsum wallboard walls and ceilings, primed and painted with 2 coats; interior raised panel solid core doors, frames and hardware, painted finish; oak hardwood flooring, 70%; ceramic tile flooring, 30%; hardwood tread stairway.	15.39	14.51	29.90	26.4%
7 Specialties	Custom grade kitchen cabinets and countertops; double bowl kitchen sink; 75 gallon gas water heater.	6.89	1.37	8.26	7.3%
8 Mechanical	Three fixture bathroom: bathtub, water closet, vanity and sink; gas fired heating and air conditioning system.	6.03	3.46	9.49	8.4%
9 Electrical	200 amp electric service; wiring, duplex & GFI receptacles, wall switches, door bell, appliance circuits, air conditioning circuit, fans and communications cabling; custom grade lighting fixtures.	2.58	2.40	4.98	4.4%
10 Overhead	Contractor's overhead and profit and design.	10.61	8.28	18.89	16.7%
	Total	63.75	49.70	**113.45**	

RESIDENTIAL   Custom   Tri-Level

- A distinct residence from designer's plans
- Single family — 1 full bath, 1 half bath, 1 kitchen
- No basement
- Asphalt shingles on roof
- Forced hot air heat/air conditioning
- Gypsum wallboard interior finishes
- Materials and workmanship are above average

Note: The illustration shown may contain some optional components (for example: garages and/or fireplaces) whose costs are shown in the modifications, adjustments, & alternatives below or at the end of the square foot section.

Base cost per square foot of living area

Exterior Wall	Living Area										
	1200	1500	1800	2100	2400	2800	3200	3600	4000	4500	5000
Wood Siding - Wood Frame	166.60	150.65	138.90	129.20	122.70	117.90	112.35	106.60	104.05	98.60	95.60
Brick Veneer - Wood Frame	174.35	157.65	145.20	134.95	128.15	123.15	117.20	111.10	108.40	102.60	99.35
Stone Veneer - Wood Frame	182.40	164.95	151.80	140.95	133.70	128.50	122.15	115.75	112.95	106.75	103.25
Solid Masonry	183.05	165.60	152.35	141.50	134.25	128.95	122.60	116.10	113.30	107.10	103.60
Finished Basement, Add*	38.40	38.30	36.60	35.30	34.40	33.80	32.95	32.15	31.85	31.10	30.60
Unfinished Basement, Add*	16.90	16.00	15.15	14.45	13.95	13.65	13.25	12.80	12.60	12.25	12.00

*Basement under middle level only.

Modifications

Add to the total cost

Upgrade Kitchen Cabinets	$ + 1736
Solid Surface Countertops (Included)	
Full Bath - including plumbing, wall and floor finishes	+ 8886
Half Bath - including plumbing, wall and floor finishes	+ 5197
Two Car Attached Garage	+ 28,478
Two Car Detached Garage	+ 32,548
Fireplace & Chimney	+ 6949

Adjustments

For multi family - add to total cost

Additional Kitchen	$ + 20,958
Additional Full Bath & Half Bath	+ 14,083
Additional Entry & Exit	+ 1676
Separate Heating & Air Conditioning	+ 7087
Separate Electric	+ 1945

For Townhouse/Rowhouse - Multiply cost per square foot by

Inner Unit	.87
End Unit	.94

Alternatives

Add to or deduct from the cost per square foot of living area

Cedar Shake Roof	+ 1.70
Clay Tile Roof	+ 4.25
Slate Roof	+ 4.70
Upgrade Ceilings to Textured Finish	+ .58
Air Conditioning, in Heating Ductwork	Base System
Heating Systems, Hot Water	+ 1.47
Heat Pump	+ 1.67
Electric Heat	– 1.87
Not Heated	– 3.72

Additional upgrades or components

Kitchen Cabinets & Countertops	Page 93
Bathroom Vanities	94
Fireplaces & Chimneys	94
Windows, Skylights & Dormers	94
Appliances	95
Breezeways & Porches	95
Finished Attic	95
Garages	96
Site Improvements	96
Wings & Ells	74

Important: See the Reference Section for Location Factors (to adjust for your city) and Estimating Forms.

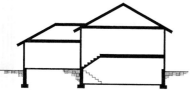

Custom Tri-Level

Living Area - 3200 S.F.
Perimeter - 198 L.F.

		Cost Per Square Foot Of Living Area			% of Total
		Mat.	Inst.	Total	(rounded)
1 Site Work	Site preparation for slab; 4' deep trench excavation for foundation wall.		.93	.93	0.8%
2 Foundation	Continuous reinforced concrete footing, 12" deep x 24" wide; damproofed and insulated 12" thick reinforced concrete block foundation wall, 4' deep; trowel finished 4" thick concrete slab on 4" crushed stone base and polyethylene vapor barrier.	4.43	5.39	9.82	8.7%
3 Framing	Exterior walls - 2" x 6" wood studs, 16" O.C.; 1/2" sheathing; gable end roof framing, 2" x 10" rafters, 16" O.C. with 1/2" plywood sheathing; 2" x 10" floor joists, 16" O.C. with bridging and 5/8" subflooring; 2" x 4" interior partitions.	6.27	8.61	14.88	13.2%
4 Exterior Walls	1" x 6" tongue and groove vertical wood siding and housewrap on insulated wood frame walls; R38 attic insulation; plastic clad double hung wood windows; raised panel exterior doors, frames and hardware, painted finish; wood storm and screen door.	9.69	3.41	13.10	11.7%
5 Roofing	Red cedar roof shingles, perfections; #15 felt building paper; aluminum gutters, downspouts and drip edge; copper flashings.	3.19	2.58	5.77	5.1%
6 Interiors	Skim coated 1/2" thick gypsum wallboard walls and ceilings, primed and painted with 2 coats; interior raised panel solid core doors, frames and hardware, painted finish; oak hardwood flooring, 70%; ceramic tile flooring, 30%; hardwood tread stairway.	14.51	13.77	28.28	25.2%
7 Specialties	Custom grade kitchen cabinets and countertops; double bowl kitchen sink; 75 gallon gas water heater.	6.02	1.20	7.22	6.4%
8 Mechanical	Three fixture bathroom: bathtub, water closet, vanity and sink; gas fired heating and air conditioning system.	5.43	3.32	8.75	7.8%
9 Electrical	200 amp electric service; wiring, duplex & GFI receptacles, wall switches, door bell, appliance circuits, air conditioning circuit, fans and communications cabling; custom grade lighting fixtures.	2.55	2.35	4.90	4.4%
10 Overhead	Contractor's overhead and profit and design.	10.41	8.29	18.70	16.6%
	Total	62.50	49.85	**112.35**	

RESIDENTIAL | Custom | Wings & Ells

1 Story — Base cost per square foot of living area

Exterior Wall	Living Area							
	50	100	200	300	400	500	600	700
Wood Siding - Wood Frame	244.80	190.45	166.90	141.20	133.45	128.80	125.75	126.60
Brick Veneer - Wood Frame	269.30	207.95	181.45	150.85	142.20	137.00	133.50	134.05
Stone Veneer - Wood Frame	294.60	226.00	196.50	160.95	151.25	145.40	141.55	141.80
Solid Masonry	300.65	230.35	200.15	163.35	153.40	147.45	143.45	143.65
Finished Basement, Add	93.20	79.40	71.85	59.30	56.80	55.25	54.25	53.55
Unfinished Basement, Add	69.05	47.80	38.00	29.25	26.90	25.50	24.55	23.90

1-1/2 Story — Base cost per square foot of living area

Exterior Wall	Living Area							
	100	200	300	400	500	600	700	800
Wood Siding - Wood Frame	194.60	159.85	138.60	125.65	119.30	116.15	112.20	111.10
Brick Veneer - Wood Frame	216.50	177.35	153.20	136.95	129.75	126.05	121.60	120.35
Stone Veneer - Wood Frame	239.10	195.40	168.25	148.70	140.65	136.30	131.25	129.95
Solid Masonry	244.45	199.75	171.90	151.55	143.20	138.75	133.55	132.20
Finished Basement, Add	62.25	57.50	52.50	47.00	45.50	44.45	43.50	43.40
Unfinished Basement, Add	41.35	31.55	26.90	22.90	21.55	20.60	19.80	19.50

2 Story — Base cost per square foot of living area

Exterior Wall	Living Area							
	100	200	400	600	800	1000	1200	1400
Wood Siding - Wood Frame	193.85	148.00	127.95	108.20	101.60	97.65	95.05	96.20
Brick Veneer - Wood Frame	218.40	165.50	142.55	117.90	110.35	105.80	102.80	103.70
Stone Veneer - Wood Frame	243.70	183.60	157.60	127.95	119.40	114.25	110.85	111.45
Solid Masonry	249.75	187.90	161.25	130.30	121.55	116.25	112.75	113.30
Finished Basement, Add	46.70	39.70	35.95	29.65	28.40	27.65	27.20	26.75
Unfinished Basement, Add	34.55	23.90	19.05	14.60	13.45	12.80	12.35	11.95

Base costs do not include bathroom or kitchen facilities. Use Modifications/Adjustments/Alternatives on pages 93–96 where appropriate.

Did you know?
RSMeans Online gives you the same access to RSMeans' data with 24/7 access:
- Quickly locate costs in the searchable database.
- Build cost lists, estimates, and reports in minutes.
- Adjust costs to any location in the U.S. and Canada with the click of a button.

Start your free trial today at **www.RSMeansOnline.com**

RSMeans Online
FROM THE GORDIAN GROUP

No part of this cost data may be reproduced, stored in a retrieval system, or transmitted in any form or by any means without prior written permission of RSMeans.

RESIDENTIAL | Luxury | Illustrations

1 Story

1-1/2 Story

2 Story

2-1/2 Story

Bi-Level

Tri-Level

RESIDENTIAL | Luxury | 1 Story

- Unique residence built from an architect's plan
- Single family — 1 full bath, 1 half bath, 1 kitchen
- No basement
- Cedar shakes on roof
- Forced hot air heat/air conditioning
- Gypsum wallboard interior finishes
- Many special features
- Extraordinary materials and workmanship

©Home Planners, Inc.

Note: The illustration shown may contain some optional components (for example: garages and/or fireplaces) whose costs are shown in the modifications, adjustments, & alternatives below or at the end of the square foot section.

Base cost per square foot of living area

Exterior Wall	\multicolumn{11}{c}{Living Area}
	1000	1200	1400	1600	1800	2000	2400	2800	3200	3600	4000
Wood Siding - Wood Frame	220.40	201.30	186.05	175.60	169.55	162.65	150.40	141.40	135.15	128.75	123.55
Brick Veneer - Wood Frame	230.05	209.90	193.95	183.00	176.65	169.35	156.45	147.05	140.35	133.60	128.10
Solid Brick	242.05	220.75	203.80	192.20	185.60	177.70	164.10	154.05	146.85	139.65	133.75
Solid Stone	250.70	228.60	210.95	198.90	192.00	183.75	169.60	159.15	151.60	144.05	137.85
Finished Basement, Add	57.80	62.25	59.80	58.15	57.25	55.90	54.05	52.65	51.35	50.25	49.35
Unfinished Basement, Add	26.00	24.40	23.15	22.35	21.85	21.15	20.20	19.45	18.85	18.25	17.80

Modifications
Add to the total cost

Upgrade Kitchen Cabinets	$ + 2445
Solid Surface Countertops (Included)	
Full Bath - including plumbing, wall and floor finishes	+ 10,663
Half Bath - including plumbing, wall and floor finishes	+ 6237
Two Car Attached Garage	+ 32,582
Two Car Detached Garage	+ 36,973
Fireplace & Chimney	+ 10,012

Adjustments
For multi family - add to total cost

Additional Kitchen	$ + 28,496
Additional Full Bath & Half Bath	+ 16,900
Additional Entry & Exit	+ 2306
Separate Heating & Air Conditioning	+ 7087
Separate Electric	+ 1945

For Townhouse/Rowhouse - Multiply cost per square foot by

Inner Unit	.90
End Unit	.95

Alternatives
Add to or deduct from the cost per square foot of living area

Heavyweight Asphalt Shingles	– 2.35
Clay Tile Roof	+ 3.50
Slate Roof	+ 4.15
Upgrade Ceilings to Textured Finish	+ .58
Air Conditioning, in Heating Ductwork	Base System
Heating Systems, Hot Water	+ 1.77
Heat Pump	+ 1.47
Electric Heat	– 2.10
Not Heated	– 4.80

Additional upgrades or components

Kitchen Cabinets & Countertops	Page 93
Bathroom Vanities	94
Fireplaces & Chimneys	94
Windows, Skylights & Dormers	94
Appliances	95
Breezeways & Porches	95
Finished Attic	95
Garages	96
Site Improvements	96
Wings & Ells	92

Important: See the Reference Section for Location Factors (to adjust for your city) and Estimating Forms.

Luxury 1 Story

Living Area - 2800 S.F.
Perimeter - 219 L.F.

#	Category	Description	Cost Per Square Foot Of Living Area			% of Total
			Mat.	Inst.	Total	(rounded)
1	Site Work	Site preparation for slab; 4' deep trench excavation for foundation wall.		1.15	1.15	0.8%
2	Foundation	Continuous reinforced concrete footing, 12" deep x 24" wide; damproofed and insulated 12" thick reinforced concrete block foundation wall, 4' deep; trowel finished 6" thick concrete slab on 4" crushed stone base and polyethylene vapor barrier.	7.41	7.93	15.34	10.8%
3	Framing	Exterior walls - 2" x 6" wood studs, 16" O.C.; 1/2" sheathing; gable end roof framing, 2" x 10" rafters, 16" O.C. with 1/2" plywood sheathing; 2" x 4" interior partitions.	6.81	9.47	16.28	11.5%
4	Exterior Walls	1" x 6" tongue and groove vertical wood siding and housewrap on insulated wood frame walls; R38 attic insulation; metal clad double hung wood windows; raised panel exterior doors, frame and hardware, painted finish; wood storm and screen door.	12.22	4.05	16.27	11.5%
5	Roofing	Red cedar roof shingles, perfections; #15 felt building paper; aluminum gutters, downspouts and drip edge; copper flashings.	5.18	4.19	9.37	6.6%
6	Interiors	Skim coated 1/2" thick gypsum wallboard walls and ceilings, primed and painted with 2 coats; interior raised panel solid core doors, frames and hardware, painted finish; oak hardwood flooring, 70%; ceramic tile flooring, 30%.	13.54	14.38	27.92	19.7%
7	Specialties	Luxury grade kitchen cabinets and countertops; double bowl kitchen sink; 75 gallon gas water heater.	9.12	1.78	10.90	7.7%
8	Mechanical	Three fixture bathroom: bathtub, water closet, vanity and sink; gas fired heating and air conditioning system.	6.77	3.80	10.57	7.5%
9	Electrical	200 amp electric service; wiring, duplex & GFI receptacles, wall switches, dimmer switches, door bell, appliance circuits, air conditioning circuit, fans and communications cabling; luxury grade lighting fixtures.	3.62	2.61	6.23	4.4%
10	Overhead	Contractor's overhead and profit and architect's fees.	15.53	11.84	27.37	19.4%
		Total	80.20	61.20	**141.40**	

RESIDENTIAL — Luxury — 1-1/2 Story

- Unique residence built from an architect's plan
- Single family — 1 full bath, 1 half bath, 1 kitchen
- No basement
- Cedar shakes on roof
- Forced hot air heat/air conditioning
- Gypsum wallboard interior finishes
- Many special features
- Extraordinary materials and workmanship

Note: The illustration shown may contain some optional components (for example: garages and/or fireplaces) whose costs are shown in the modifications, adjustments, & alternatives below or at the end of the square foot section.

Base cost per square foot of living area

Exterior Wall	Living Area										
	1000	1200	1400	1600	1800	2000	2400	2800	3200	3600	4000
Wood Siding - Wood Frame	221.85	205.10	193.30	179.85	171.80	164.15	150.00	143.20	137.05	132.45	126.15
Brick Veneer - Wood Frame	233.25	215.85	203.50	189.10	180.60	172.55	157.40	150.20	143.55	138.65	131.95
Solid Brick	246.85	228.65	215.65	200.15	191.00	182.45	166.20	158.55	151.20	146.10	138.80
Solid Stone	257.65	238.75	225.30	208.90	199.30	190.30	173.20	165.10	157.30	151.95	144.30
Finished Basement, Add	40.55	44.05	42.80	40.90	39.80	38.90	36.95	36.00	35.00	34.45	33.65
Unfinished Basement, Add	18.70	17.80	17.20	16.25	15.70	15.25	14.25	13.75	13.25	12.95	12.55

Modifications

Add to the total cost

Upgrade Kitchen Cabinets	$ + 2445
Solid Surface Countertops (Included)	
Full Bath - including plumbing, wall and floor finishes	+ 10,663
Half Bath - including plumbing, wall and floor finishes	+ 6237
Two Car Attached Garage	+ 32,582
Two Car Detached Garage	+ 36,973
Fireplace & Chimney	+ 10,012

Adjustments

For multi family - add to total cost

Additional Kitchen	$ + 28,496
Additional Full Bath & Half Bath	+ 16,900
Additional Entry & Exit	+ 2306
Separate Heating & Air Conditioning	+ 7087
Separate Electric	+ 1945

For Townhouse/Rowhouse - Multiply cost per square foot by

Inner Unit	.90
End Unit	.95

Alternatives

Add to or deduct from the cost per square foot of living area

Heavyweight Asphalt Shingles	– 1.70
Clay Tile Roof	+ 2.50
Slate Roof	+ 3
Upgrade Ceilings to Textured Finish	+ .58
Air Conditioning, in Heating Ductwork	Base System
Heating Systems, Hot Water	+ 1.69
Heat Pump	+ 1.64
Electric Heat	– 2.10
Not Heated	– 4.43

Additional upgrades or components

Kitchen Cabinets & Countertops	Page 93
Bathroom Vanities	94
Fireplaces & Chimneys	94
Windows, Skylights & Dormers	94
Appliances	95
Breezeways & Porches	95
Finished Attic	95
Garages	96
Site Improvements	96
Wings & Ells	92

Luxury 1-1/2 Story

Living Area - 2800 S.F.
Perimeter - 175 L.F.

#	Section	Description	Cost Per Square Foot Of Living Area			% of Total
			Mat.	Inst.	Total	(rounded)
1	Site Work	Site preparation for slab; 4' deep trench excavation for foundation wall.		1.15	1.15	0.8%
2	Foundation	Continuous reinforced concrete footing, 12" deep x 24" wide; damproofed and insulated 12" thick reinforced concrete block foundation wall, 4' deep; trowel finished 6" thick concrete slab on 4" crushed stone base and polyethylene vapor barrier.	5.38	5.99	11.37	7.9%
3	Framing	Exterior walls - 2" x 6" wood studs, 16" O.C.; 1/2" sheathing; gable end roof framing, steep pitch 2" x 10" rafters, 16" O.C. with 1/2" plywood sheathing; 2" x 12" floor joists, 16" O.C. with bridging and 5/8" subflooring; 2" x 4" interior partitions.	7.89	10.51	18.40	12.8%
4	Exterior Walls	1" x 6" tongue and groove vertical wood siding and housewrap on insulated wood frame walls; R38 attic insulation; metal clad double hung wood windows; raised panel exterior doors, frame and hardware, painted finish; wood storm and screen door.	13.11	4.40	17.51	12.2%
5	Roofing	Red cedar roof shingles, perfections; #15 felt building paper; aluminum gutters, downspouts and drip edge; copper flashings.	3.75	3.04	6.79	4.7%
6	Interiors	Skim coated 1/2" thick gypsum wallboard walls and ceilings, primed and painted with 2 coats; interior raised panel solid core doors, frames and hardware, painted finish; oak hardwood flooring, 70%; ceramic tile flooring, 30%; hardwood tread stairway.	16.00	16.56	32.56	22.7%
7	Specialties	Luxury grade kitchen cabinets and countertops; double bowl kitchen sink; 75 gallon gas water heater.	9.12	1.78	10.90	7.6%
8	Mechanical	Three fixture bathroom: bathtub, water closet, vanity and sink; gas fired heating and air conditioning system.	6.77	3.80	10.57	7.4%
9	Electrical	200 amp electric service; wiring, duplex & GFI receptacles, wall switches, dimmer switches, door bell, appliance circuits, air conditioning circuit, fans and communications cabling; luxury grade lighting fixtures.	3.62	2.61	6.23	4.4%
10	Overhead	Contractor's overhead and profit and architect's fees.	15.76	11.96	27.72	19.4%
	Total		81.40	61.80	**143.20**	

RESIDENTIAL | Luxury | 2 Story

- Unique residence built from an architect's plan
- Single family — 1 full bath, 1 half bath, 1 kitchen
- No basement
- Cedar shakes on roof
- Forced hot air heat/air conditioning
- Gypsum wallboard interior finishes
- Many special features
- Extraordinary materials and workmanship

Note: The illustration shown may contain some optional components (for example: garages and/or fireplaces) whose costs are shown in the modifications, adjustments, & alternatives below or at the end of the square foot section.

Base cost per square foot of living area

Exterior Wall	Living Area										
	1200	1400	1600	1800	2000	2400	2800	3200	3600	4000	4400
Wood Siding - Wood Frame	201.55	188.55	179.20	171.15	162.75	150.60	140.45	133.70	129.70	125.35	121.85
Brick Veneer - Wood Frame	212.95	199.15	189.30	180.65	171.95	158.95	147.95	140.80	136.50	131.75	128.05
Solid Brick	228.10	213.35	202.85	193.45	184.25	170.05	158.05	150.25	145.70	140.35	136.40
Solid Stone	237.60	222.25	211.40	201.45	191.95	177.00	164.40	156.15	151.35	145.70	141.55
Finished Basement, Add	32.55	35.40	34.45	33.40	32.70	31.20	29.95	29.15	28.60	28.00	27.60
Unfinished Basement, Add	15.00	14.35	13.85	13.35	13.00	12.20	11.55	11.20	10.90	10.60	10.40

Modifications

Add to the total cost

Upgrade Kitchen Cabinets	$ + 2445
Solid Surface Countertops (Included)	
Full Bath - including plumbing, wall and floor finishes	+ 10,663
Half Bath - including plumbing, wall and floor finishes	+ 6237
Two Car Attached Garage	+ 32,582
Two Car Detached Garage	+ 36,973
Fireplace & Chimney	+ 10,974

Adjustments

For multi family - add to total cost

Additional Kitchen	$ + 28,496
Additional Full Bath & Half Bath	+ 16,900
Additional Entry & Exit	+ 2306
Separate Heating & Air Conditioning	+ 7087
Separate Electric	+ 1945

For Townhouse/Rowhouse - Multiply cost per square foot by

Inner Unit	.86
End Unit	.93

Alternatives

Add to or deduct from the cost per square foot of living area

Heavyweight Asphalt Shingles	– 1.20
Clay Tile Roof	+ 1.75
Slate Roof	+ 2.10
Upgrade Ceilings to Textured Finish	+ .58
Air Conditioning, in Heating Ductwork	Base System
Heating Systems, Hot Water	+ 1.64
Heat Pump	+ 1.73
Electric Heat	– 1.90
Not Heated	– 4.19

Additional upgrades or components

Kitchen Cabinets & Countertops	Page 93
Bathroom Vanities	94
Fireplaces & Chimneys	94
Windows, Skylights & Dormers	94
Appliances	95
Breezeways & Porches	95
Finished Attic	95
Garages	96
Site Improvements	96
Wings & Ells	92

Important: See the Reference Section for Location Factors (to adjust for your city) and Estimating Forms.

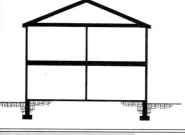

Luxury 2 Story

Living Area - 3200 S.F.
Perimeter - 163 L.F.

#	Category	Description	Cost Per Square Foot Of Living Area			% of Total
			Mat.	Inst.	Total	(rounded)
1	Site Work	Site preparation for slab; 4' deep trench excavation for foundation wall.		1.01	1.01	0.8%
2	Foundation	Continuous reinforced concrete footing, 12" deep x 24" wide; damproofed and insulated 12" thick reinforced concrete block foundation wall, 4' deep; trowel finished 6" thick concrete slab on 4" crushed stone base and polyethylene vapor barrier.	4.37	4.87	9.24	6.9%
3	Framing	Exterior walls - 2" x 6" wood studs, 16" O.C.; 1/2" sheathing; gable end roof framing, 2" x 10" rafters, 16" O.C. with 1/2" plywood sheathing; 2" x 12" floor joists, 16" O.C. with bridging and 5/8" subflooring; 2" x 4" interior partitions.	7.48	9.80	17.28	12.9%
4	Exterior Walls	1" x 6" tongue and groove vertical wood siding and housewrap on insulated wood frame walls; R38 attic insulation; metal clad double hung wood windows; raised panel exterior doors, frame and hardware, painted finish; wood storm and screen door.	13.29	4.47	17.76	13.3%
5	Roofing	Red cedar roof shingles, perfections; #15 felt building paper; aluminum gutters, downspouts and drip edge; copper flashings.	2.59	2.10	4.69	3.5%
6	Interiors	Skim coated 1/2" thick gypsum wallboard walls and ceilings, primed and painted with 2 coats; interior raised panel solid core doors, frames and hardware, painted finish; oak hardwood flooring, 70%; ceramic tile flooring, 30%; hardwood tread stairway.	15.88	16.58	32.46	24.3%
7	Specialties	Luxury grade kitchen cabinets and countertops; double bowl kitchen sink; 75 gallon gas water heater.	7.97	1.56	9.53	7.1%
8	Mechanical	Three fixture bathroom: bathtub, water closet, vanity and sink; gas fired heating and air conditioning system.	6.09	3.65	9.74	7.3%
9	Electrical	200 amp electric service; wiring, duplex & GFI receptacles, wall switches, dimmer switches, door bell, appliance circuits, air conditioning circuit, fans and communications cabling; luxury grade lighting fixtures.	3.59	2.55	6.14	4.6%
10	Overhead	Contractor's overhead and profit and architect's fees.	14.69	11.16	25.85	19.3%
	Total		75.95	57.75	**133.70**	

RESIDENTIAL — Luxury — 2-1/2 Story

- Unique residence built from an architect's plan
- Single family — 1 full bath, 1 half bath, 1 kitchen
- No basement
- Cedar shakes on roof
- Forced hot air heat/air conditioning
- Gypsum wallboard interior finishes
- Many special features
- Extraordinary materials and workmanship

Note: The illustration shown may contain some optional components (for example: garages and/or fireplaces) whose costs are shown in the modifications, adjustments, & alternatives below or at the end of the square foot section.

©Larry W. Garnett & Associates, Inc

Base cost per square foot of living area

Exterior Wall	1500	1800	2100	2500	3000	3500	4000	4500	5000	5500	6000
Wood Siding - Wood Frame	197.10	177.35	165.15	156.25	144.25	135.60	127.45	123.35	119.75	116.15	112.30
Brick Veneer - Wood Frame	209.20	188.55	175.20	165.70	152.80	143.45	134.70	130.20	126.30	122.35	118.30
Solid Brick	224.75	202.85	188.00	177.80	163.80	153.40	143.95	138.95	134.60	130.30	125.95
Solid Stone	235.40	212.75	196.80	186.05	171.30	160.20	150.25	145.00	140.35	135.70	131.15
Finished Basement, Add	26.00	28.00	26.45	25.60	24.40	23.40	22.65	22.15	21.70	21.35	21.00
Unfinished Basement, Add	12.10	11.45	10.65	10.20	9.60	9.10	8.70	8.45	8.25	8.05	7.90

Modifications

Add to the total cost

Upgrade Kitchen Cabinets	$ + 2445
Solid Surface Countertops (Included)	
Full Bath - including plumbing, wall and floor finishes	+ 10,663
Half Bath - including plumbing, wall and floor finishes	+ 6237
Two Car Attached Garage	+ 32,582
Two Car Detached Garage	+ 36,973
Fireplace & Chimney	+ 12,011

Adjustments

For multi family - add to total cost

Additional Kitchen	$ + 28,496
Additional Full Bath & Half Bath	+ 16,900
Additional Entry & Exit	+ 2306
Separate Heating & Air Conditioning	+ 7087
Separate Electric	+ 1945

For Townhouse/Rowhouse - Multiply cost per square foot by

Inner Unit	.86
End Unit	.93

Alternatives

Add to or deduct from the cost per square foot of living area

Heavyweight Asphalt Shingles	– 1.05
Clay Tile Roof	+ 1.50
Slate Roof	+ 1.80
Upgrade Ceilings to Textured Finish	+ .58
Air Conditioning, in Heating Ductwork	Base System
Heating Systems, Hot Water	+ 1.48
Heat Pump	+ 1.78
Electric Heat	– 3.73
Not Heated	– 4.19

Additional upgrades or components

Kitchen Cabinets & Countertops	Page 93
Bathroom Vanities	94
Fireplaces & Chimneys	94
Windows, Skylights & Dormers	94
Appliances	95
Breezeways & Porches	95
Finished Attic	95
Garages	96
Site Improvements	96
Wings & Ells	92

Luxury 2-1/2 Story

Living Area - 3000 S.F.
Perimeter - 148 L.F.

#	Category	Description	Cost Per Square Foot Of Living Area			% of Total
			Mat.	Inst.	Total	(rounded)
1	Site Work	Site preparation for slab; 4' deep trench excavation for foundation wall.		1.07	1.07	0.7%
2	Foundation	Continuous reinforced concrete footing, 12" deep x 24" wide; dampproofed and insulated 12" thick reinforced concrete block foundation wall, 4' deep; trowel finished 6" thick concrete slab on 4" crushed stone base and polyethylene vapor barrier.	3.83	4.42	8.25	5.7%
3	Framing	Exterior walls - 2" x 6" wood studs, 16" O.C.; 1/2" sheathing; gable end roof framing, steep pitch 2" x 10" rafters, 16" O.C. with 1/2" plywood sheathing; 2" x 12" floor joists, 16" O.C. with bridging and 5/8" subflooring; 2" x 4" interior partitions.	8.30	10.86	19.16	13.3%
4	Exterior Walls	1" x 6" tongue and groove vertical wood siding and housewrap on insulated wood frame walls; R38 attic insulation; metal clad double hung wood windows; raised panel exterior doors, frame and hardware, painted finish; wood storm and screen door.	15.42	5.19	20.61	14.3%
5	Roofing	Red cedar roof shingles, perfections; #15 felt building paper; aluminum gutters, downspouts and drip edge; copper flashings.	2.31	1.87	4.18	2.9%
6	Interiors	Skim coated 1/2" thick gypsum wallboard walls and ceilings, primed and painted with 2 coats; interior raised panel solid core doors, frames and hardware, painted finish; oak hardwood flooring, 70%; ceramic tile flooring, 30%; hardwood tread stairway.	18.12	18.44	36.56	25.3%
7	Specialties	Luxury grade kitchen cabinets and countertops; double bowl kitchen sink; 75 gallon gas water heater.	8.50	1.67	10.17	7.1%
8	Mechanical	Three fixture bathroom: bathtub, water closet, vanity and sink; gas fired heating and air conditioning system.	6.40	3.73	10.13	7.0%
9	Electrical	200 amp electric service; wiring, duplex & GFI receptacles, wall switches, dimmer switches, door bell, appliance circuits, air conditioning circuit, fans and communications cabling; luxury grade lighting fixtures.	3.61	2.57	6.18	4.3%
10	Overhead	Contractor's overhead and profit and architect's fees.	15.96	11.98	27.94	19.4%
	Total		82.45	61.80	**144.25**	

RESIDENTIAL | Luxury | 3 Story

- Unique residence built from an architect's plan
- Single family — 1 full bath, 1 half bath, 1 kitchen
- No basement
- Cedar shakes on roof
- Forced hot air heat/air conditioning
- Gypsum wallboard interior finishes
- Many special features
- Extraordinary materials and workmanship

Note: The illustration shown may contain some optional components (for example: garages and/or fireplaces) whose costs are shown in the modifications, adjustments, & alternatives below or at the end of the square foot section.

Base cost per square foot of living area

Exterior Wall	\multicolumn{11}{c}{Living Area}
	1500	1800	2100	2500	3000	3500	4000	4500	5000	5500	6000
Wood Siding - Wood Frame	193.60	174.25	164.20	155.70	143.40	137.05	129.55	122.05	119.10	116.05	112.90
Brick Veneer - Wood Frame	205.65	185.40	174.60	165.70	152.50	145.65	137.35	129.30	126.00	122.80	119.20
Solid Brick	222.20	200.75	188.90	179.35	164.90	157.40	148.05	139.15	135.60	131.95	127.85
Solid Stone	232.05	209.90	197.45	187.45	172.25	164.35	154.45	145.05	141.30	137.40	133.00
Finished Basement, Add	22.70	24.55	23.55	22.85	21.80	21.15	20.30	19.65	19.35	19.10	18.65
Unfinished Basement, Add	10.60	10.00	9.55	9.25	8.70	8.35	7.95	7.60	7.45	7.25	7.05

Modifications

Add to the total cost

Upgrade Kitchen Cabinets	$ + 2445
Solid Surface Countertops (Included)	
Full Bath - including plumbing, wall and floor finishes	+ 10,663
Half Bath - including plumbing, wall and floor finishes	+ 6237
Two Car Attached Garage	+ 32,582
Two Car Detached Garage	+ 36,973
Fireplace & Chimney	+ 12,011

Adjustments

For multi family - add to total cost

Additional Kitchen	$ + 28,496
Additional Full Bath & Half Bath	+ 16,900
Additional Entry & Exit	+ 2306
Separate Heating & Air Conditioning	+ 7087
Separate Electric	+ 1945

For Townhouse/Rowhouse - Multiply cost per square foot by

Inner Unit	.84
End Unit	.92

Alternatives

Add to or deduct from the cost per square foot of living area

Heavyweight Asphalt Shingles	– .80
Clay Tile Roof	+ 1.15
Slate Roof	+ 1.40
Upgrade Ceilings to Textured Finish	+ .58
Air Conditioning, in Heating Ductwork	Base System
Heating Systems, Hot Water	+ 1.48
Heat Pump	+ 1.78
Electric Heat	– 3.73
Not Heated	– 4.07

Additional upgrades or components

Kitchen Cabinets & Countertops	Page 93
Bathroom Vanities	94
Fireplaces & Chimneys	94
Windows, Skylights & Dormers	94
Appliances	95
Breezeways & Porches	95
Finished Attic	95
Garages	96
Site Improvements	96
Wings & Ells	92

Luxury 3 Story

Living Area - 3000 S.F.
Perimeter - 135 L.F.

			Cost Per Square Foot Of Living Area			% of Total
			Mat.	Inst.	Total	(rounded)
1	Site Work	Site preparation for slab; 4' deep trench excavation for foundation wall.		1.07	1.07	0.7%
2	Foundation	Continuous reinforced concrete footing, 12" deep x 24" wide; damproofed and insulated 12" thick reinforced concrete block foundation wall, 4' deep; trowel finished 6" thick concrete slab on 4" crushed stone base and polyethylene vapor barrier.	3.45	4.02	7.47	5.2%
3	Framing	Exterior walls - 2" x 6" wood studs, 16" O.C.; 1/2" sheathing; gable end roof framing, 2" x 10" rafters, 16" O.C. with 1/2" plywood sheathing; 2" x 12" floor joists, 16" O.C. with bridging and 5/8" subflooring; 2" x 4" interior partitions.	8.14	10.59	18.73	13.1%
4	Exterior Walls	1" x 6" tongue and groove vertical wood siding and housewrap on insulated wood frame walls; R38 attic insulation; metal clad double hung wood windows; raised panel exterior doors, frame and hardware, painted finish; wood storm and screen door.	16.44	5.54	21.98	15.3%
5	Roofing	Red cedar roof shingles, perfections; #15 felt building paper; aluminum gutters, downspouts and drip edge; copper flashings.	1.73	1.40	3.13	2.2%
6	Interiors	Skim coated 1/2" thick gypsum wallboard walls and ceilings, primed and painted with 2 coats; interior raised panel solid core doors, frames and hardware, painted finish; oak hardwood flooring, 70%; ceramic tile flooring, 30%; hardwood tread stairway.	18.20	18.61	36.81	25.7%
7	Specialties	Luxury grade kitchen cabinets and countertops; double bowl kitchen sink; 75 gallon gas water heater.	8.50	1.67	10.17	7.1%
8	Mechanical	Three fixture bathroom: bathtub, water closet, vanity and sink; gas fired heating and air conditioning system.	6.40	3.73	10.13	7.1%
9	Electrical	200 amp electric service; wiring, duplex & GFI receptacles, wall switches, dimmer switches, door bell, appliance circuits, air conditioning circuit, fans and communications cabling; luxury grade lighting fixtures.	3.61	2.57	6.18	4.3%
10	Overhead	Contractor's overhead and profit and architect's fees.	15.93	11.80	27.73	19.3%
		Total	82.40	61.00	**143.40**	

RESIDENTIAL | Luxury | Bi-Level

- Unique residence built from an architect's plan
- Single family — 1 full bath, 1 half bath, 1 kitchen
- No basement
- Cedar shakes on roof
- Forced hot air heat/air conditioning
- Gypsum wallboard interior finishes
- Many special features
- Extraordinary materials and workmanship

Note: The illustration shown may contain some optional components (for example: garages and/or fireplaces) whose costs are shown in the modifications, adjustments, & alternatives below or at the end of the square foot section.

Base cost per square foot of living area

Exterior Wall	Living Area										
	1200	1400	1600	1800	2000	2400	2800	3200	3600	4000	4400
Wood Siding - Wood Frame	190.85	178.55	169.65	162.15	154.10	142.85	133.35	127.05	123.30	119.30	116.05
Brick Veneer - Wood Frame	199.50	186.70	177.40	169.45	161.10	149.15	139.15	132.40	128.45	124.25	120.80
Solid Brick	210.80	197.20	187.50	178.90	170.25	157.40	146.60	139.45	135.25	130.55	126.95
Solid Stone	218.20	204.20	194.15	185.15	176.30	162.85	151.55	144.10	139.70	134.80	131.05
Finished Basement, Add	32.55	35.40	34.45	33.40	32.70	31.20	29.95	29.15	28.60	28.00	27.60
Unfinished Basement, Add	15.00	14.35	13.85	13.35	13.00	12.20	11.55	11.20	10.90	10.60	10.40

Modifications

Add to the total cost

Upgrade Kitchen Cabinets	$ + 2445
Solid Surface Countertops (Included)	
Full Bath - including plumbing, wall and floor finishes	+ 10,663
Half Bath - including plumbing, wall and floor finishes	+ 6237
Two Car Attached Garage	+ 32,582
Two Car Detached Garage	+ 36,973
Fireplace & Chimney	+ 10,012

Adjustments

For multi family - add to total cost

Additional Kitchen	$ + 28,496
Additional Full Bath & Half Bath	+ 16,900
Additional Entry & Exit	+ 2306
Separate Heating & Air Conditioning	+ 7087
Separate Electric	+ 1945

For Townhouse/Rowhouse - Multiply cost per square foot by

Inner Unit	.89
End Unit	.94

Alternatives

Add to or deduct from the cost per square foot of living area

Heavyweight Asphalt Shingles	– 1.20
Clay Tile Roof	+ 1.75
Slate Roof	+ 2.10
Upgrade Ceilings to Textured Finish	+ .58
Air Conditioning, in Heating Ductwork	Base System
Heating Systems, Hot Water	+ 1.64
Heat Pump	+ 1.73
Electric Heat	– 1.90
Not Heated	– 4.19

Additional upgrades or components

Kitchen Cabinets & Countertops	Page 93
Bathroom Vanities	94
Fireplaces & Chimneys	94
Windows, Skylights & Dormers	94
Appliances	95
Breezeways & Porches	95
Finished Attic	95
Garages	96
Site Improvements	96
Wings & Ells	92

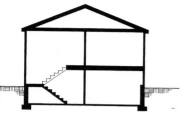

Luxury Bi-Level

Living Area - 3200 S.F.
Perimeter - 163 L.F.

			Cost Per Square Foot Of Living Area			% of Total
			Mat.	Inst.	Total	(rounded)
1	Site Work	Site preparation for slab; 4' deep trench excavation for foundation wall.		1.01	1.01	0.8%
2	Foundation	Continuous reinforced concrete footing, 12" deep x 24" wide; damproofed and insulated 12" thick reinforced concrete block foundation wall, 4' deep; trowel finished 6" thick concrete slab on 4" crushed stone base and polyethylene vapor barrier.	4.37	4.87	9.24	7.3%
3	Framing	Exterior walls - 2" x 6" wood studs, 16" O.C.; 1/2" sheathing; gable end roof framing, 2" x 10" rafters, 16" O.C. with 1/2" plywood sheathing; 2" x 12" floor joists, 16" O.C. with bridging and 5/8" subflooring; 2" x 4" interior partitions.	7.05	9.19	16.24	12.8%
4	Exterior Walls	1" x 6" tongue and groove vertical wood siding and housewrap on insulated wood frame walls; R38 attic insulation; metal clad double hung wood windows; raised panel exterior doors, frame and hardware, painted finish; wood storm and screen door.	10.52	3.54	14.06	11.1%
5	Roofing	Red cedar roof shingles, perfections: #15 felt building paper; aluminum gutters, downspouts and drip edge; copper flashings.	2.59	2.10	4.69	3.7%
6	Interiors	Skim coated 1/2" thick gypsum wallboard walls and ceilings, primed and painted with 2 coats; interior raised panel solid core doors, frames and hardware, painted finish; oak hardwood flooring, 70%; ceramic tile flooring, 30%; hardwood tread stairway.	15.64	16.17	31.81	25.0%
7	Specialties	Luxury grade kitchen cabinets and countertops; double bowl kitchen sink; 75 gallon gas water heater.	7.97	1.56	9.53	7.5%
8	Mechanical	Three fixture bathroom: bathtub, water closet, vanity and sink; gas fired heating and air conditioning system.	6.09	3.65	9.74	7.7%
9	Electrical	200 amp electric service; wiring, duplex & GFI receptacles, wall switches, dimmer switches, door bell, appliance circuits, air conditioning circuit, fans and communications cabling; luxury grade lighting fixtures	3.59	2.55	6.14	4.8%
10	Overhead	Contractor's overhead and profit and architect's fees.	13.88	10.71	24.59	19.4%
		Total	71.70	55.35	**127.05**	

RESIDENTIAL | Luxury | Tri-Level

- Unique residence built from an architect's plan
- Single family — 1 full bath, 1 half bath, 1 kitchen
- No basement
- Cedar shakes on roof
- Forced hot air heat/air conditioning
- Gypsum wallboard interior finishes
- Many special features
- Extraordinary materials and workmanship

Note: The illustration shown may contain some optional components (for example: garages and/or fireplaces) whose costs are shown in the modifications, adjustments, & alternatives below or at the end of the square foot section.

©Home Planners, Inc.

Base cost per square foot of living area

Exterior Wall	Living Area										
	1500	1800	2100	2400	2800	3200	3600	4000	4500	5000	5500
Wood Siding - Wood Frame	177.85	163.65	152.00	144.30	138.45	131.85	125.00	121.80	115.55	111.95	108.00
Brick Veneer - Wood Frame	185.75	170.75	158.50	150.35	144.20	137.20	130.05	126.65	120.00	116.25	112.05
Solid Brick	195.85	179.90	166.75	158.10	151.70	144.10	136.40	132.90	125.75	121.60	117.15
Solid Stone	202.70	186.15	172.50	163.45	156.80	148.85	140.80	137.15	129.65	125.35	120.60
Finished Basement, Add*	38.55	41.50	39.85	38.80	38.10	37.00	36.05	35.65	34.70	34.05	33.50
Unfinished Basement, Add*	17.30	16.30	15.40	14.85	14.50	14.00	13.50	13.25	12.80	12.45	12.20

*Basement under middle level only.

Modifications

Add to the total cost

Upgrade Kitchen Cabinets	$ + 2445
Solid Surface Countertops (Included)	
Full Bath - including plumbing, wall and floor finishes	+ 10,663
Half Bath - including plumbing, wall and floor finishes	+ 6237
Two Car Attached Garage	+ 32,582
Two Car Detached Garage	+ 36,973
Fireplace & Chimney	+ 10,012

Adjustments

For multi family - add to total cost

Additional Kitchen	$ + 28,496
Additional Full Bath & Half Bath	+ 16,900
Additional Entry & Exit	+ 2306
Separate Heating & Air Conditioning	+ 7087
Separate Electric	+ 1945

For Townhouse/Rowhouse - Multiply cost per square foot by

Inner Unit	.86
End Unit	.93

Alternatives

Add to or deduct from the cost per square foot of living area

Heavyweight Asphalt Shingles	– 1.70
Clay Tile Roof	+ 2.50
Slate Roof	+ 3
Upgrade Ceilings to Textured Finish	+ .58
Air Conditioning, in Heating Ductwork	Base System
Heating Systems, Hot Water	+ 1.59
Heat Pump	+ 1.80
Electric Heat	– 1.67
Not Heated	– 4.07

Additional upgrades or components

Kitchen Cabinets & Countertops	Page 93
Bathroom Vanities	94
Fireplaces & Chimneys	94
Windows, Skylights & Dormers	94
Appliances	95
Breezeways & Porches	95
Finished Attic	95
Garages	96
Site Improvements	96
Wings & Ells	92

Luxury Tri-Level

Living Area - 3600 S.F.
Perimeter - 207 L.F.

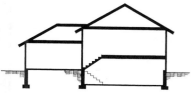

			Cost Per Square Foot Of Living Area			% of Total
			Mat.	Inst.	Total	(rounded)
1	Site Work	Site preparation for slab; 4' deep trench excavation for foundation wall.		.89	.89	0.7%
2	Foundation	Continuous reinforced concrete footing, 12" deep x 24" wide; damproofed and insulated 12" thick reinforced concrete block foundation wall, 4' deep; trowel finished 6" thick concrete slab on 4" crushed stone base and polyethylene vapor barrier.	5.18	5.61	10.79	8.6%
3	Framing	Exterior walls - 2" x 6" wood studs, 16" O.C.; 1/2" sheathing; gable end roof framing, 2" x 10" rafters, 16" O.C. with 1/2" plywood sheathing; 2" x 12" floor joists, 16" O.C. with bridging and 5/8" subflooring; 2" x 4" interior partitions.	6.84	9.04	15.88	12.7%
4	Exterior Walls	1" x 6" tongue and groove vertical wood siding and housewrap on insulated wood frame walls; R38 attic insulation; metal clad double hung wood windows; raised panel exterior doors, frame and hardware, painted finish; wood storm and screen door.	10.37	3.47	13.84	11.1%
5	Roofing	Red cedar roof shingles, perfections; #15 felt building paper; aluminum gutters, downspouts and drip edge; copper flashings.	3.45	2.79	6.24	5.0%
6	Interiors	Skim coated 1/2" thick gypsum wallboard walls and ceilings, primed and painted with 2 coats; interior raised panel solid core doors, frames and hardware, painted finish; oak hardwood flooring, 70%; ceramic tile flooring, 30%; hardwood tread stairway.	14.54	15.00	29.54	23.6%
7	Specialties	Luxury grade kitchen cabinets and countertops; double bowl kitchen sink; 75 gallon gas water heater.	7.07	1.38	8.45	6.8%
8	Mechanical	Three fixture bathroom: bathtub, water closet, vanity and sink; gas fired heating and air conditioning system.	5.57	3.55	9.12	7.3%
9	Electrical	200 amp electric service; wiring, duplex & GFI receptacles, wall switches, dimmer switches, door bell, appliance circuits, air conditioning circuit, fans and communications cabling; luxury grade lighting fixtures.	3.56	2.49	6.05	4.8%
10	Overhead	Contractor's overhead and profit and architect's fees.	13.57	10.63	24.20	19.4%
		Total	70.15	54.85	**125.00**	

RESIDENTIAL | Luxury | Wings & Ells

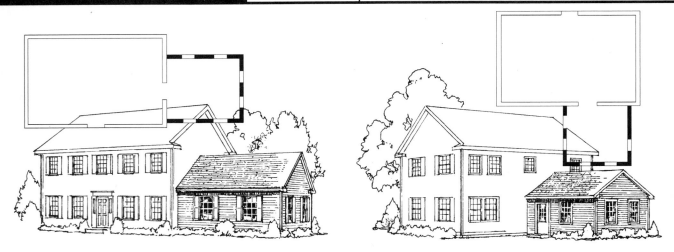

1 Story — Base cost per square foot of living area

Exterior Wall	Living Area							
	50	100	200	300	400	500	600	700
Wood Siding - Wood Frame	270.40	209.55	183.20	154.30	145.70	140.45	137.00	137.95
Brick Veneer - Wood Frame	298.50	229.60	199.90	165.50	155.70	149.80	145.95	146.50
Solid Brick	341.85	260.60	225.70	182.70	171.15	164.25	159.70	159.80
Solid Stone	362.85	275.55	238.25	191.00	178.75	171.30	166.30	166.20
Finished Basement, Add	103.40	93.80	84.25	68.25	65.05	63.15	61.85	60.95
Unfinished Basement, Add	54.05	41.70	36.55	28.05	26.30	25.30	24.60	24.15

1-1/2 Story — Base cost per square foot of living area

Exterior Wall	Living Area							
	100	200	300	400	500	600	700	800
Wood Siding - Wood Frame	214.50	175.55	151.75	137.20	130.05	126.50	122.15	120.85
Brick Veneer - Wood Frame	239.60	195.65	168.55	150.25	142.10	137.90	132.90	131.55
Solid Brick	278.30	226.60	194.35	170.40	160.60	155.45	149.50	148.00
Solid Stone	297.05	241.60	206.85	180.15	169.60	163.90	157.55	155.95
Finished Basement, Add	68.15	67.45	60.95	54.00	52.05	50.75	49.60	49.45
Unfinished Basement, Add	34.40	29.25	25.80	22.05	21.05	20.35	19.75	19.65

2 Story — Base cost per square foot of living area

Exterior Wall	Living Area							
	100	200	400	600	800	1000	1200	1400
Wood Siding - Wood Frame	210.85	159.40	136.95	114.75	107.35	103.00	100.00	101.30
Brick Veneer - Wood Frame	238.95	179.45	153.70	125.90	117.45	112.35	108.95	109.90
Solid Brick	282.30	210.45	179.55	143.10	132.90	126.80	122.70	123.15
Solid Stone	303.30	225.50	192.05	151.40	140.45	133.75	129.30	129.55
Finished Basement, Add	51.70	47.00	42.20	34.15	32.60	31.60	30.95	30.55
Unfinished Basement, Add	27.00	20.85	18.30	14.00	13.15	12.65	12.30	12.05

Base costs do not include bathroom or kitchen facilities. Use Modifications/Adjustments/Alternatives on pages 93–96 where appropriate.

RESIDENTIAL — Modifications/Adjustments/Alternatives

Kitchen cabinets - Base units, hardwood (Cost per Unit)

	Economy	Average	Custom	Luxury
24" deep, 35" high,				
One top drawer,				
One door below				
12" wide	$ 251	$ 335	$ 446	$ 586
15" wide	263	350	466	613
18" wide	285	380	505	665
21" wide	293	390	519	683
24" wide	338	450	599	788
Four drawers				
12" wide	263	350	466	613
15" wide	270	360	479	630
18" wide	293	390	519	683
24" wide	319	425	565	744
Two top drawers,				
Two doors below				
27" wide	360	480	638	840
30" wide	394	525	698	919
33" wide	409	545	725	954
36" wide	424	565	751	989
42" wide	443	590	785	1033
48" wide	476	635	845	1111
Range or sink base				
(Cost per unit)				
Two doors below				
30" wide	334	445	592	779
33" wide	356	475	632	831
36" wide	371	495	658	866
42" wide	386	515	685	901
48" wide	409	545	725	954
Corner Base Cabinet				
(Cost per unit)				
36" wide	593	790	1051	1383
Lazy Susan *(Cost per unit)*				
With revolving door	750	1000	1330	1750

Kitchen cabinets - Wall cabinets, hardwood (Cost per Unit)

	Economy	Average	Custom	Luxury
12" deep, 2 doors				
12" high				
30" wide	$ 229	$ 305	$ 406	$ 534
36" wide	270	360	479	630
15" high				
30" wide	233	310	412	543
33" wide	285	380	505	665
36" wide	278	370	492	648
24" high				
30" wide	304	405	539	709
36" wide	334	445	592	779
42" wide	375	500	665	875
30" high, 1 door				
12" wide	225	300	399	525
15" wide	236	315	419	551
18" wide	255	340	452	595
24" wide	296	395	525	691
30" high, 2 doors				
27" wide	323	430	572	753
30" wide	338	450	599	788
36" wide	383	510	678	893
42" wide	413	550	732	963
48" wide	461	615	818	1076
Corner wall, 30" high				
24" wide	334	445	592	779
30" wide	356	475	632	831
36" wide	401	535	712	936
Broom closet				
84" high, 24" deep				
18" wide	623	830	1104	1453
Oven Cabinet				
84" high, 24" deep				
27" wide	938	1250	1663	2188

Kitchen countertops (Cost per L.F.)

	Economy	Average	Custom	Luxury
Solid Surface				
24" wide, no backsplash	$ 112	$ 149	$ 198	$ 261
with backsplash	121	161	214	282
Stock plastic laminate, 24" wide				
with backsplash	26	34	45	60
Custom plastic laminate, no splash				
7/8" thick, alum. molding	36	48	65	85
1-1/4" thick, no splash	42	55	74	97
Marble				
1/2" - 3/4" thick w/splash	54	72	96	126
Maple, laminated				
1-1/2" thick w/splash	87	116	154	203
Stainless steel				
(per S.F.)	143	190	253	333
Cutting blocks, recessed				
16" x 20" x 1" (each)	120	160	213	280

RESIDENTIAL Modifications/Adjustments/Alternatives

Vanity bases (Cost per Unit)

	Economy	Average	Custom	Luxury
2 door, 30" high, 21" deep				
24" wide	$ 293	$ 390	$ 519	$ 683
30" wide	353	470	625	823
36" wide	353	470	625	823
48" wide	454	605	805	1059

Solid surface vanity tops (Cost Each)

	Economy	Average	Custom	Luxury
Center bowl				
22" x 25"	$ 425	$ 459	$ 496	$ 536
22" x 31"	490	529	571	617
22" x 37"	565	610	659	712
22" x 49"	700	756	816	881

Fireplaces & Chimneys (Cost per Unit)

	1-1/2 Story	2 Story	3 Story
Economy (prefab metal)			
Exterior chimney & 1 fireplace	$ 6284	$ 6944	$ 7617
Interior chimney & 1 fireplace	6022	6695	7004
Average (masonry)			
Exterior chimney & 1 fireplace	6625	7320	8029
Interior chimney & 1 fireplace	6348	7057	7384
For more than 1 flue, add	451	767	1286
For more than 1 fireplace, add	4445	4445	4445
Custom (masonry)			
Exterior chimney & 1 fireplace	6947	7842	8855
Interior chimney & 1 fireplace	6514	7374	7947
For more than 1 flue, add	545	943	1579
For more than 1 fireplace, add	4990	4990	4990
Luxury (masonry)			
Exterior chimney & 1 fireplace	10012	10975	12010
Interior chimney & 1 fireplace	9552	10443	11054
For more than 1 flue, add	826	1380	1926
For more than 1 fireplace, add	7891	7891	7891

Windows and Skylights (Cost Each)

	Economy	Average	Custom	Luxury
Fixed Picture Windows				
3'-6" x 4'-0"	$ 364	$ 484	$ 645	$ 849
4'-0" x 6'-0"	634	845	1125	1480
5'-0" x 6'-0"	719	958	1275	1678
6'-0" x 6'-0"	761	1015	1350	1776
Bay/Bow Windows				
8'-0" x 5'-0"	874	1165	1550	2039
10'-0" x 5'-0"	930	1240	1650	2171
10'-0" x 6'-0"	1466	1954	2600	3421
12'-0" x 6'-0"	1903	2537	3375	4441
Palladian Windows				
3'-2" x 6'-4"		1484	1975	2599
4'-0" x 6'-0"		1729	2300	3026
5'-5" x 6'-10"		2255	3000	3947
8'-0" x 6'-0"		2669	3550	4671
Skylights				
46" x 21-1/2"	611	940	1140	1254
46" x 28"	627	965	1180	1298
57" x 44"	692	1065	1195	1315

Dormers (Cost/S.F. of plan area)

	Economy	Average	Custom	Luxury
Framing and Roofing Only				
Gable dormer, 2" x 6" roof frame	$ 32	$ 36	$ 39	$ 64
2" x 8" roof frame	33	37	40	67
Shed dormer, 2" x 6" roof frame	20	24	27	41
2" x 8" roof frame	22	25	28	42
2" x 10" roof frame	24	27	29	43

RESIDENTIAL Modifications/Adjustments/Alternatives

Appliances *(Cost per Unit)*

	Economy	Average	Custom	Luxury
Range				
30" free standing, 1 oven	$ 580	$ 1415	$ 1833	$ 2250
2 oven	1450	2300	2725	3150
30" built-in, 1 oven	1075	1625	1900	2175
2 oven	1700	1955	2083	2210
21" free standing				
1 oven	620	713	759	805
Counter Top Ranges				
4 burner standard	465	1245	1635	2025
As above with griddle	1750	3100	3775	4450
Microwave Oven	252	526	663	800
Compactor				
4 to 1 compaction	850	1138	1282	1425
Deep Freeze				
15 to 23 C.F.	690	870	960	1050
30 C.F.	825	1013	1107	1200
Dehumidifier, portable, auto.				
15 pint	340	391	417	442
30 pint	365	420	448	475
Washing Machine, automatic	770	1560	1955	2350
Water Heater				
Electric, glass lined				
30 gal.	1025	1238	1344	1450
80 gal.	1950	2413	2644	2875
Water Heater, Gas, glass lined				
30 gal.	1800	2188	2382	2575
50 gal.	2075	2550	2788	3025
Dishwasher, built-in				
2 cycles	490	643	719	795
4 or more cycles	630	785	1218	1650
Dryer, automatic	790	1470	1810	2150
Garage Door Opener	525	628	679	730
Garbage Disposal	166	248	289	330
Heater, Electric, built-in				
1250 watt ceiling type	249	317	351	385
1250 watt wall type	330	353	364	375
Wall type w/blower				
1500 watt	310	357	380	403
3000 watt	590	679	723	767
Hood For Range, 2 speed				
30" wide	207	666	896	1125
42" wide	297	1336	1856	2375
Humidifier, portable				
7 gal. per day	167	192	205	217
15 gal. per day	244	281	299	317
Ice Maker, automatic				
13 lb. per day	1275	1467	1563	1658
51 lb. per day	2000	2300	2450	2600
Refrigerator, no frost				
10-12 C.F.	585	655	690	725
14-16 C.F.	710	778	812	845
18-20 C.F.	895	1473	1762	2050
21-29 C.F.	1350	2900	3675	4450
Sump Pump, 1/3 H.P.	325	455	520	585

Breezeway *(Cost per S.F.)*

Class	Type	Area (S.F.)			
		50	100	150	200
Economy	Open	$ 45.56	$ 37.94	$ 33.37	$ 31.09
	Enclosed	169.12	120.38	99.08	85.31
Average	Open	53.56	43.74	38.44	35.79
	Enclosed	173.08	125.37	103.98	93.05
Custom	Open	64.49	53.03	46.74	43.60
	Enclosed	229.22	165.75	137.08	122.50
Luxury	Open	72.27	59.64	52.96	49.62
	Enclosed	271.61	193.74	158.78	141.05

Porches *(Cost per S.F.)*

Class	Type	Area (S.F.)				
		25	50	100	200	300
Economy	Open	$ 82.76	$ 62.05	$ 46.70	$ 35.93	$ 32.36
	Enclosed	179.86	142.87	94.32	71.52	62.33
Average	Open	87.04	66.68	49.01	37.09	33.13
	Enclosed	194.30	154.71	99.66	73.60	63.33
Custom	Open	147.77	106.05	81.46	60.76	59.16
	Enclosed	237.28	191.29	126.52	95.90	91.70
Luxury	Open	165.33	118.86	90.28	65.50	59.44
	Enclosed	275.50	248.06	160.22	116.92	97.01

Finished attic *(Cost per S.F.)*

Class	Area (S.F.)				
	400	500	600	800	1000
Economy	$ 22.01	$ 21.27	$ 20.39	$ 20.05	$ 19.30
Average	34.12	33.38	32.56	32.15	31.25
Custom	43.75	42.77	41.79	41.15	40.31
Luxury	56.15	54.79	53.51	52.23	51.37

Alarm system *(Cost per System)*

	Burglar Alarm	Smoke Detector
Economy	$ 425	$ 70
Average	485	81
Custom	851	199
Luxury	1500	258

Sauna, prefabricated
(Cost per unit, including heater and controls—7' high)

Size	Cost
6' x 4'	$ 5900
6' x 5'	6575
6' x 6'	7050
6' x 9'	9175
8' x 10'	11,700
8' x 12'	13,800
10' x 12'	14,000

RESIDENTIAL — Modifications/Adjustments/Alternatives

Garages *

(Costs include exterior wall systems comparable with the quality of the residence. Included in the cost is an allowance for one personnel door, manual overhead door(s) and electrical fixture.)

Class	Detached			Attached			Built-in		Basement	
	One Car	Two Car	Three Car	One Car	Two Car	Three Car	One Car	Two Car	One Car	Two Car
Economy										
Wood	$17,321	$26,477	$35,632	$13,419	$23,095	$32,251	$-1909	$-3819	$1811	$2520
Masonry	23,431	34,123	44,814	17,242	28,455	39,147	-2660	-5319		
Average										
Wood	19,004	28,582	38,161	14,471	24,571	34,150	-2116	-4232	2060	3018
Masonry	23,406	34,091	44,776	17,226	28,432	39,118	-2657	-4222		
Custom										
Wood	21,204	32,548	43,892	16,454	28,478	39,822	-3390	-3269	2955	4808
Masonry	25,570	38,012	50,454	19,186	32,307	44,749	-3926	-4341		
Luxury										
Wood	23,737	36,973	50,208	18,667	32,582	45,817	-3495	-3479	3997	6393
Masonry	30,739	45,736	60,733	23,048	38,724	53,721	-4355	-5199		

*See the Introduction to this section for definitions of garage types.

Swimming pools (Cost per S.F.)

Residential		
In-ground		$ 39.50 - 94.00
Deck equipment		1.30
Paint pool, preparation & 3 coats (epoxy)		5.25
Rubber base paint		4.59
Pool Cover		1.86
Swimming Pool Heaters		
(not including wiring, external piping, base or pad)		
Gas		
155 MBH		$ 3250.00
190 MBH		3825.00
500 MBH		13,700.
Electric		
15 KW	7200 gallon pool	2850.00
24 KW	9600 gallon pool	3325.00
54 KW	24,000 gallon pool	5200.00

Wood and coal stoves

Wood Only	
Free Standing (minimum)	$ 2125
Fireplace Insert (minimum)	1849
Coal Only	
Free Standing	$ 2090
Fireplace Insert	2288
Wood and Coal	
Free Standing	$ 4299
Fireplace Insert	4405

Sidewalks (Cost per S.F.)

Concrete, 3000 psi with wire mesh	4" thick	$ 3.96
	5" thick	4.82
	6" thick	5.41
Precast concrete patio blocks (natural)	2" thick	6.60
Precast concrete patio blocks (colors)	2" thick	6.95
Flagstone, bluestone	1" thick	19.60
Flagstone, bluestone	1-1/2" thick	27.25
Slate (natural, irregular)	3/4" thick	18.45
Slate (random rectangular)	1/2" thick	29.00
Seeding		
Fine grading & seeding includes lime, fertilizer & seed	per S.Y.	3.03
Lawn Sprinkler System	per S.F.	1.10

Fencing (Cost per L.F.)

Chain Link, 4' high, galvanized	$ 15.60
Gate, 4' high (each)	212.00
Cedar Picket, 3' high, 2 rail	15.15
Gate (each)	234.00
3 Rail, 4' high	18.80
Gate (each)	252.00
Cedar Stockade, 3 Rail, 6' high	18.25
Gate (each)	255.00
Board & Battens, 2 sides 6' high, pine	28.00
6' high, cedar	36.00
No. 1 Cedar, basketweave, 6' high	37.00
Gate, 6' high (each)	288.00

Carport (Cost per S.F.)

Economy	$ 10.07
Average	15.07
Custom	22.15
Luxury	25.89

Assemblies Section

Table of Contents

Table No.	Page
1 Site Work	
1-04 Footing Excavation	102
1-08 Foundation Excavation	104
1-12 Utility Trenching	106
1-16 Sidewalk	108
1-20 Driveway	110
1-24 Septic	112
1-60 Chain Link Fence	114
1-64 Wood Fence	115
2 Foundations	
2-04 Footing	118
2-08 Block Wall	120
2-12 Concrete Wall	122
2-16 Wood Wall Foundation	124
2-20 Floor Slab	126
3 Framing	
3-02 Floor (Wood)	130
3-04 Floor (Wood)	132
3-06 Floor (Wood)	134
3-08 Exterior Wall	136
3-12 Gable End Roof	138
3-16 Truss Roof	140
3-20 Hip Roof	142
3-24 Gambrel Roof	144
3-28 Mansard Roof	146
3-32 Shed/Flat Roof	148
3-40 Gable Dormer	150
3-44 Shed Dormer	152
3-48 Partition	154
4 Exterior Walls	
4-02 Masonry Block Wall	158
4-04 Brick/Stone Veneer	160
4-08 Wood Siding	162
4-12 Shingle Siding	164

Table No.	Page
4-16 Metal & Plastic Siding	166
4-20 Insulation	168
4-28 Double Hung Window	170
4-32 Casement Window	172
4-36 Awning Window	174
4-40 Sliding Window	176
4-44 Bow/Bay Window	178
4-48 Fixed Window	180
4-52 Entrance Door	182
4-53 Sliding Door	184
4-56 Residential Overhead Door	186
4-58 Aluminum Window	188
4-60 Storm Door & Window	190
4-64 Shutters/Blinds	191
5 Roofing	
5-04 Gable End Roofing	194
5-08 Hip Roof Roofing	196
5-12 Gambrel Roofing	198
5-16 Mansard Roofing	200
5-20 Shed Roofing	202
5-24 Gable Dormer Roofing	204
5-28 Shed Dormer Roofing	206
5-32 Skylight/Skywindow	208
5-34 Built-up Roofing	210
6 Interiors	
6-04 Drywall & Thincoat Wall	214
6-08 Drywall & Thincoat Ceiling	216
6-12 Plaster & Stucco Wall	218
6-16 Plaster & Stucco Ceiling	220
6-18 Suspended Ceiling	222
6-20 Interior Door	224
6-24 Closet Door	226
6-60 Carpet	228

Table No.	Page
6-64 Flooring	229
6-90 Stairways	230
7 Specialties	
7-08 Kitchen	234
7-12 Appliances	236
7-16 Bath Accessories	237
7-24 Masonry Fireplace	238
7-30 Prefabricated Fireplace	240
7-32 Greenhouse	242
7-36 Swimming Pool	243
7-40 Wood Deck	244
8 Mechanical	
8-04 Two Fixture Lavatory	248
8-12 Three Fixture Bathroom	250
8-16 Three Fixture Bathroom	252
8-20 Three Fixture Bathroom	254
8-24 Three Fixture Bathroom	256
8-28 Three Fixture Bathroom	258
8-32 Three Fixture Bathroom	260
8-36 Four Fixture Bathroom	262
8-40 Four Fixture Bathroom	264
8-44 Five Fixture Bathroom	266
8-60 Gas Fired Heating/Cooling	268
8-64 Oil Fired Heating/Cooling	270
8-68 Hot Water Heating	272
8-80 Rooftop Heating/Cooling	274
9 Electrical	
9-10 Electric Service	278
9-20 Electric Heating	279
9-30 Wiring Devices	280
9-40 Light Fixtures	281

How RSMeans Assemblies Data Works

The following is a detailed explanation of a sample Assemblies Cost Table. Included are an illustration and accompanying system descriptions. Additionally, related systems and price sheets may be included. Next to each bold number below is the item being described with the appropriate component of the sample entry in parentheses. General contractors should add an additional markup to the figures shown in the Assemblies section. Note: Throughout this section, the words assembly and system will be used interchangeably.

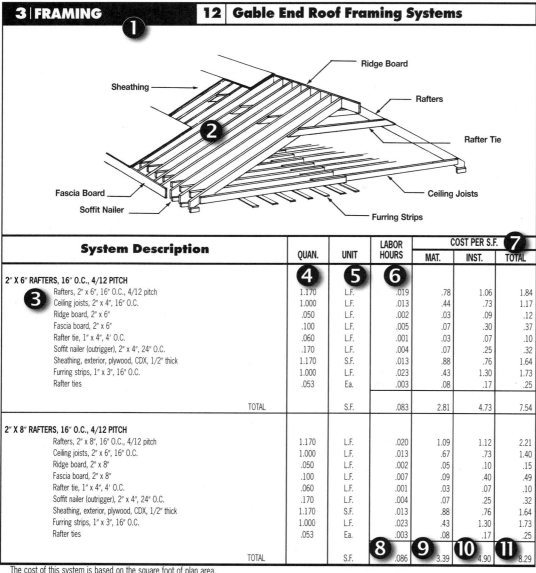

3	FRAMING ❶		12	Gable End Roof Framing Systems				
System Description ❸			QUAN. ❹	UNIT ❺	LABOR HOURS ❻	COST PER S.F. ❼		
						MAT.	INST.	TOTAL
2" X 6" RAFTERS, 16" O.C., 4/12 PITCH								
Rafters, 2" x 6", 16" O.C., 4/12 pitch			1.170	L.F.	.019	.78	1.06	1.84
Ceiling joists, 2" x 4", 16" O.C.			1.000	L.F.	.013	.44	.73	1.17
Ridge board, 2" x 6"			.050	L.F.	.002	.03	.09	.12
Fascia board, 2" x 6"			.100	L.F.	.005	.07	.30	.37
Rafter tie, 1" x 4", 4' O.C.			.060	L.F.	.001	.03	.07	.10
Soffit nailer (outrigger), 2" x 4", 24" O.C.			.170	L.F.	.004	.07	.25	.32
Sheathing, exterior, plywood, CDX, 1/2" thick			1.170	S.F.	.013	.88	.76	1.64
Furring strips, 1" x 3", 16" O.C.			1.000	L.F.	.023	.43	1.30	1.73
Rafter ties			.053	Ea.	.003	.08	.17	.25
TOTAL				S.F.	.083	2.81	4.73	7.54
2" X 8" RAFTERS, 16" O.C., 4/12 PITCH								
Rafters, 2" x 8", 16" O.C., 4/12 pitch			1.170	L.F.	.020	1.09	1.12	2.21
Ceiling joists, 2" x 6", 16" O.C.			1.000	L.F.	.013	.67	.73	1.40
Ridge board, 2" x 8"			.050	L.F.	.002	.05	.10	.15
Fascia board, 2" x 8"			.100	L.F.	.007	.09	.40	.49
Rafter tie, 1" x 4", 4' O.C.			.060	L.F.	.001	.03	.07	.10
Soffit nailer (outrigger), 2" x 4", 24" O.C.			.170	L.F.	.004	.07	.25	.32
Sheathing, exterior, plywood, CDX, 1/2" thick			1.170	S.F.	.013	.88	.76	1.64
Furring strips, 1" x 3", 16" O.C.			1.000	L.F.	.023	.43	1.30	1.73
Rafter ties			.053	Ea.	.003	.08	.17	.25
TOTAL				S.F.	.086 ❽	3.39 ❾	4.90 ❿	8.29 ⓫

The cost of this system is based on the square foot of plan area. All quantities have been adjusted accordingly.

Description	QUAN. ⓬	UNIT	LABOR HOURS	COST PER S.F.		
				MAT.	INST.	TOTAL

**① System Identification
(3 Framing 12)**

Each Assemblies section has been assigned a unique identification number, component category, system number, and system description.

② Illustration

Included with most assemblies is an illustration with individual components labeled. Elements involved in the total system function are shown.

**③ System Description
(2" x 6" Rafters, 16" O.C., 4/12 Pitch)**

The components of a typical system are listed separately to show what has been included in the development of the total system price. Each assembly includes a brief outline of any special conditions to be used when pricing a system. Alternative components can also be found. Simply insert any chosen new element into the chart to develop a custom system.

④ Quantities for Each Component

Each material in a system is shown with the quantity required for the system unit. For example, there are 1.170 L.F. of rafter per S.F. of plan area.

⑤ Unit of Measure for Each Component

The abbreviated designation indicates the unit of measure as defined by industry standards, upon which the individual component has been priced. In this example, items are priced by the linear foot (L.F.) or the square foot (S.F.).

⑥ Labor-Hours

This is the amount of time it takes to install the quantity of the individual component.

⑦ Unit of Measure (Cost per S.F.)

In the three right-hand columns, each cost figure is adjusted to agree with the unit of measure for the entire system. In this case, cost per S.F. is the common unit of measure.

⑧ Labor-Hours (.086)

The labor-hours column shows the amount of time necessary to install the system per the unit of measure. For example, it takes .086 labor-hours to install one square foot (plan area) of this roof framing system.

⑨ Materials (3.39)

This column contains the material cost of each element. These cost figures include 10% for profit.

⑩ Installation (4.90)

This column contains labor and equipment costs. Labor rates include bare cost and the installing contractor's overhead and profit. On the average, the labor cost will be 69.5% over the bare labor cost. Equipment costs include 10% for profit.

⑪ Totals (8.29)

The figure in this column is the sum of the material and installation costs.

⑫ Work Sheet

Using the selective price sheet, it is possible to create estimates with alternative items for any number of systems.

Did you know?
RSMeans Online gives you the same access to RSMeans' data with 24/7 access:
- Quickly locate costs in the searchable database.
- Build cost lists, estimates, and reports in minutes.
- Adjust costs to any location in the U.S. and Canada with the click of a button.

Start your free trial today at **www.RSMeansOnline.com**

RSMeans Online
FROM THE GORDIAN GROUP®

No part of this cost data may be reproduced, stored in a retrieval system, or transmitted in any form or by any means without prior written permission of RSMeans.

1 | SITEWORK
04 | Footing Excavation Systems

System Description	QUAN.	UNIT	LABOR HOURS	COST EACH MAT.	COST EACH INST.	COST EACH TOTAL
BUILDING, 24' X 38', 4' DEEP						
Cut & chip light trees to 6" diam.	.190	Acre	9.120		760	760
Excavator, hydraulic, crawler mtd., 1 C.Y. cap. = 100 C.Y./hr.	174.000	C.Y.	3.480		370.62	370.62
Backfill, dozer, 4" lifts, no compaction	87.000	C.Y.	.580		145.29	145.29
Rough grade, dozer, 30' from building	87.000	C.Y.	.580		145.29	145.29
Mobilize and demobilize equipment	4.000	Ea.	12.000		970	970
TOTAL		Ea.	25.760		2,391.20	2,391.20
BUILDING, 26' X 46', 4' DEEP						
Cut & chip light trees to 6" diam.	.210	Acre	10.080		840	840
Excavator, hydraulic, crawler mtd., 1 C.Y. cap. = 100 C.Y./hr.	201.000	C.Y.	4.020		428.13	428.13
Backfill, dozer, 4" lifts, no compaction	100.000	C.Y.	.667		167	167
Rough grade, dozer, 30' from building	100.000	C.Y.	.667		167	167
Mobilize and demobilize equipment	4.000	Ea.	12.000		970	970
TOTAL		Ea.	27.434		2,572.13	2,572.13
BUILDING, 26' X 60', 4' DEEP						
Cut & chip light trees to 6" diam.	.240	Acre	11.520		960	960
Excavator, hydraulic, crawler mtd., 1 C.Y. cap. = 100 C.Y./hr.	240.000	C.Y.	4.800		511.20	511.20
Backfill, dozer, 4" lifts, no compaction	120.000	C.Y.	.800		200.40	200.40
Rough grade, dozer, 30' from building	120.000	C.Y.	.800		200.40	200.40
Mobilize and demobilize equipment	4.000	Ea.	12.000		970	970
TOTAL		Ea.	29.920		2,842	2,842
BUILDING, 30' X 66', 4' DEEP						
Cut & chip light trees to 6" diam.	.260	Acre	12.480		1,040	1,040
Excavator, hydraulic, crawler mtd., 1 C.Y. cap. = 100 C.Y./hr.	268.000	C.Y.	5.360		570.84	570.84
Backfill, dozer, 4" lifts, no compaction	134.000	C.Y.	.894		223.78	223.78
Rough grade, dozer, 30' from building	134.000	C.Y.	.894		223.78	223.78
Mobilize and demobilize equipment	4.000	Ea.	12.000		970	970
TOTAL		Ea.	31.628		3,028.40	3,028.40

The costs in this system are on a cost each basis.
Quantities are based on 1'-0" clearance on each side of footing.

Description	QUAN.	UNIT	LABOR HOURS	COST EACH MAT.	COST EACH INST.	COST EACH TOTAL

Footing Excavation Price Sheet	QUAN.	UNIT	LABOR HOURS	COST EACH		
				MAT.	INST.	TOTAL
Clear and grub, medium brush, 30' from building, 24' x 38'	.190	Acre	9.120		760	760
26' x 46'	.210	Acre	10.080		840	840
26' x 60'	.240	Acre	11.520		960	960
30' x 66'	.260	Acre	12.480		1,050	1,050
Light trees, to 6" dia. cut & chip, 24' x 38'	.190	Acre	9.120		760	760
26' x 46'	.210	Acre	10.080		840	840
26' x 60'	.240	Acre	11.520		960	960
30' x 66'	.260	Acre	12.480		1,050	1,050
Medium trees, to 10" dia. cut & chip, 24' x 38'	.190	Acre	13.029		1,075	1,075
26' x 46'	.210	Acre	14.400		1,200	1,200
26' x 60'	.240	Acre	16.457		1,350	1,350
30' x 66'	.260	Acre	17.829		1,475	1,475
Excavation, footing, 24' x 38', 2' deep	68.000	C.Y.	.906		145	145
4' deep	174.000	C.Y.	2.319		370	370
8' deep	384.000	C.Y.	5.119		820	820
26' x 46', 2' deep	79.000	C.Y.	1.053		169	169
4' deep	201.000	C.Y.	2.679		430	430
8' deep	404.000	C.Y.	5.385		860	860
26' x 60', 2' deep	94.000	C.Y.	1.253		200	200
4' deep	240.000	C.Y.	3.199		510	510
8' deep	483.000	C.Y.	6.438		1,025	1,025
30' x 66', 2' deep	105.000	C.Y.	1.400		224	224
4' deep	268.000	C.Y.	3.572		570	570
8' deep	539.000	C.Y.	7.185		1,150	1,150
Backfill, 24' x 38', 2" lifts, no compaction	34.000	C.Y.	.227		57	57
Compaction, air tamped, add	34.000	C.Y.	2.267		465	465
4" lifts, no compaction	87.000	C.Y.	.580		145	145
Compaction, air tamped, add	87.000	C.Y.	5.800		1,200	1,200
8" lifts, no compaction	192.000	C.Y.	1.281		320	320
Compaction, air tamped, add	192.000	C.Y.	12.801		2,650	2,650
26' x 46', 2" lifts, no compaction	40.000	C.Y.	.267		66.50	66.50
Compaction, air tamped, add	40.000	C.Y.	2.667		550	550
4" lifts, no compaction	100.000	C.Y.	.667		167	167
Compaction, air tamped, add	100.000	C.Y.	6.667		1,375	1,375
8" lifts, no compaction	202.000	C.Y.	1.347		340	340
Compaction, air tamped, add	202.000	C.Y.	13.467		2,800	2,800
26' x 60', 2" lifts, no compaction	47.000	C.Y.	.313		78.50	78.50
Compaction, air tamped, add	47.000	C.Y.	3.133		650	650
4" lifts, no compaction	120.000	C.Y.	.800		201	201
Compaction, air tamped, add	120.000	C.Y.	8.000		1,650	1,650
8" lifts, no compaction	242.000	C.Y.	1.614		405	405
Compaction, air tamped, add	242.000	C.Y.	16.134		3,325	3,325
30' x 66', 2" lifts, no compaction	53.000	C.Y.	.354		88.50	88.50
Compaction, air tamped, add	53.000	C.Y.	3.534		730	730
4" lifts, no compaction	134.000	C.Y.	.894		225	225
Compaction, air tamped, add	134.000	C.Y.	8.934		1,850	1,850
8" lifts, no compaction	269.000	C.Y.	1.794		450	450
Compaction, air tamped, add	269.000	C.Y.	17.934		3,700	3,700
Rough grade, 30' from building, 24' x 38'	87.000	C.Y.	.580		145	145
26' x 46'	100.000	C.Y.	.667		167	167
26' x 60'	120.000	C.Y.	.800		201	201
30' x 66'	134.000	C.Y.	.894		225	225
Mobilize and demobilize equipment	4.000	Ea.	12.000		970	970

1 | SITEWORK   08 | Foundation Excavation Systems

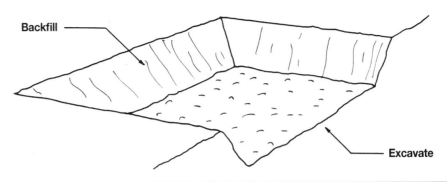

System Description	QUAN.	UNIT	LABOR HOURS	COST EACH		
				MAT.	INST.	TOTAL
BUILDING, 24' X 38', 8' DEEP						
Medium clearing	.190	Acre	2.027		295.45	295.45
Excavate, track loader, 1-1/2 C.Y. bucket	550.000	C.Y.	7.860		1,023	1,023
Backfill, dozer, 8" lifts, no compaction	180.000	C.Y.	1.201		300.60	300.60
Rough grade, dozer, 30' from building	280.000	C.Y.	1.868		467.60	467.60
Mobilize and demobilize equipment	4.000	Ea.	12.000		970	970
TOTAL		Ea.	24.956		3,056.65	3,056.65
BUILDING, 26' X 46', 8' DEEP						
Medium clearing	.210	Acre	2.240		326.55	326.55
Excavate, track loader, 1-1/2 C.Y. bucket	672.000	C.Y.	9.603		1,249.92	1,249.92
Backfill, dozer, 8" lifts, no compaction	220.000	C.Y.	1.467		367.40	367.40
Rough grade, dozer, 30' from building	340.000	C.Y.	2.268		567.80	567.80
Mobilize and demobilize equipment	4.000	Ea.	12.000		970	970
TOTAL		Ea.	27.578		3,481.67	3,481.67
BUILDING, 26' X 60', 8' DEEP						
Medium clearing	.240	Acre	2.560		373.20	373.20
Excavate, track loader, 1-1/2 C.Y. bucket	829.000	C.Y.	11.846		1,541.94	1,541.94
Backfill, dozer, 8" lifts, no compaction	270.000	C.Y.	1.801		450.90	450.90
Rough grade, dozer, 30' from building	420.000	C.Y.	2.801		701.40	701.40
Mobilize and demobilzie equipment	4.000	Ea.	12.000		970	970
TOTAL		Ea.	31.008		4,037.44	4,037.44
BUILDING, 30' X 66', 8' DEEP						
Medium clearing	.260	Acre	2.773		404.30	404.30
Excavate, track loader, 1-1/2 C.Y. bucket	990.000	C.Y.	14.147		1,841.40	1,841.40
Backfill dozer, 8" lifts, no compaction	320.000	C.Y.	2.134		534.40	534.40
Rough grade, dozer, 30' from building	500.000	C.Y.	3.335		835	835
Mobilize and demobilize equipment	4.000	Ea.	12.000		970	970
TOTAL		Ea.	34.389		4,585.10	4,585.10

The costs in this system are on a cost each basis.
Quantities are based on 1'-0" clearance beyond footing projection.

Description	QUAN.	UNIT	LABOR HOURS	COST EACH		
				MAT.	INST.	TOTAL

For customer support on your Residential Cost Data, call 877.759.4771.

Foundation Excavation Price Sheet	QUAN.	UNIT	LABOR HOURS	COST EACH		
				MAT.	INST.	TOTAL
Clear & grub, medium brush, 30' from building, 24' x 38'	.190	Acre	2.027		296	296
26' x 46'	.210	Acre	2.240		325	325
26' x 60'	.240	Acre	2.560		375	375
30' x 66'	.260	Acre	2.773		405	405
Light trees, to 6" dia. cut & chip, 24' x 38'	.190	Acre	9.120		760	760
26' x 46'	.210	Acre	10.080		840	840
26' x 60'	.240	Acre	11.520		960	960
30' x 66'	.260	Acre	12.480		1,050	1,050
Medium trees, to 10" dia. cut & chip, 24' x 38'	.190	Acre	13.029		1,075	1,075
26' x 46'	.210	Acre	14.400		1,200	1,200
26' x 60'	.240	Acre	16.457		1,350	1,350
30' x 66'	.260	Acre	17.829		1,475	1,475
Excavation, basement, 24' x 38', 2' deep	98.000	C.Y.	1.400		183	183
4' deep	220.000	C.Y.	3.144		410	410
8' deep	550.000	C.Y.	7.860		1,025	1,025
26' x 46', 2' deep	123.000	C.Y.	1.758		229	229
4' deep	274.000	C.Y.	3.915		510	510
8' deep	672.000	C.Y.	9.603		1,250	1,250
26' x 60', 2' deep	157.000	C.Y.	2.244		292	292
4' deep	345.000	C.Y.	4.930		640	640
8' deep	829.000	C.Y.	11.846		1,550	1,550
30' x 66', 2' deep	192.000	C.Y.	2.744		355	355
4' deep	419.000	C.Y.	5.988		780	780
8' deep	990.000	C.Y.	14.147		1,850	1,850
Backfill, 24' x 38', 2" lifts, no compaction	32.000	C.Y.	.213		53.50	53.50
Compaction, air tamped, add	32.000	C.Y.	2.133		440	440
4" lifts, no compaction	72.000	C.Y.	.480		120	120
Compaction, air tamped, add	72.000	C.Y.	4.800		990	990
8" lifts, no compaction	180.000	C.Y.	1.201		300	300
Compaction, air tamped, add	180.000	C.Y.	12.001		2,475	2,475
26' x 46', 2" lifts, no compaction	40.000	C.Y.	.267		66.50	66.50
Compaction, air tamped, add	40.000	C.Y.	2.667		550	550
4" lifts, no compaction	90.000	C.Y.	.600		150	150
Compaction, air tamped, add	90.000	C.Y.	6.000		1,250	1,250
8" lifts, no compaction	220.000	C.Y.	1.467		370	370
Compacton, air tamped, add	220.000	C.Y.	14.667		3,025	3,025
26' x 60', 2" lifts, no compaction	50.000	C.Y.	.334		83.50	83.50
Compaction, air tamped, add	50.000	C.Y.	3.334		690	690
4" lifts, no compaction	110.000	C.Y.	.734		184	184
Compaction, air tamped, add	110.000	C.Y.	7.334		1,525	1,525
8" lifts, no compaction	270.000	C.Y.	1.801		450	450
Compaction, air tamped, add	270.000	C.Y.	18.001		3,725	3,725
30' x 66', 2" lifts, no compaction	60.000	C.Y.	.400		101	101
Compaction, air tamped, add	60.000	C.Y.	4.000		825	825
4" lifts, no compaction	130.000	C.Y.	.867		217	217
Compaction, air tamped, add	130.000	C.Y.	8.667		1,800	1,800
8" lifts, no compaction	320.000	C.Y.	2.134		535	535
Compaction, air tamped, add	320.000	C.Y.	21.334		4,425	4,425
Rough grade, 30' from building, 24' x 38'	280.000	C.Y.	1.868		470	470
26' x 46'	340.000	C.Y.	2.268		570	570
26' x 60'	420.000	C.Y.	2.801		705	705
30' x 66'	500.000	C.Y.	3.335		835	835
Mobilize and demobilize equipment	4.000	Ea.	12.000		970	970

1 | SITEWORK
12 | Utility Trenching Systems

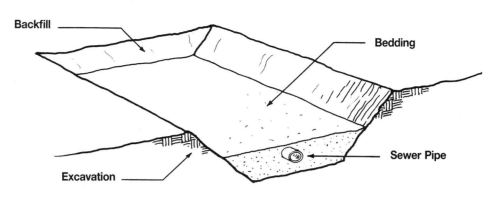

System Description	QUAN.	UNIT	LABOR HOURS	COST PER L.F. MAT.	COST PER L.F. INST.	COST PER L.F. TOTAL
2' DEEP						
Excavation, backhoe	.296	C.Y.	.032		2.36	2.36
Alternate pricing method, 4" deep	.111	C.Y.	.044	4.05	2.13	6.18
Utility, sewer, 6" cast iron	1.000	L.F.	.283	30.53	15.78	46.31
Compaction in 12" layers, hand tamp, add to above	.185	C.Y.	.044		1.79	1.79
TOTAL		L.F.	.403	34.58	22.06	56.64
4' DEEP						
Excavation, backhoe	.889	C.Y.	.095		7.09	7.09
Alternate pricing method, 4" deep	.111	C.Y.	.044	4.05	2.13	6.18
Utility, sewer, 6" cast iron	1.000	L.F.	.283	30.53	15.78	46.31
Compaction in 12" layers, hand tamp, add to above	.778	C.Y.	.183		7.55	7.55
TOTAL		L.F.	.605	34.58	32.55	67.13
6' DEEP						
Excavation, backhoe	1.770	C.Y.	.189		14.12	14.12
Alternate pricing method, 4" deep	.111	C.Y.	.044	4.05	2.13	6.18
Utility, sewer, 6" cast iron	1.000	L.F.	.283	30.53	15.78	46.31
Compaction in 12" layers, hand tamp, add to above	1.660	C.Y.	.391		16.10	16.10
TOTAL		L.F.	.907	34.58	48.13	82.71
8' DEEP						
Excavation, backhoe	2.960	C.Y.	.316		23.62	23.62
Alternate pricing method, 4" deep	.111	C.Y.	.044	4.05	2.13	6.18
Utility, sewer, 6" cast iron	1.000	L.F.	.283	30.53	15.78	46.31
Compaction in 12" layers, hand tamp, add to above	2.850	C.Y.	.671		27.65	27.65
TOTAL		L.F.	1.314	34.58	69.18	103.76

The costs in this system are based on a cost per linear foot of trench, and based on 2' wide at bottom of trench up to 6' deep.

Description	QUAN.	UNIT	LABOR HOURS	COST PER L.F. MAT.	COST PER L.F. INST.	COST PER L.F. TOTAL

Utility Trenching Price Sheet

Utility Trenching Price Sheet	QUAN.	UNIT	LABOR HOURS	COST PER UNIT MAT.	COST PER UNIT INST.	COST PER UNIT TOTAL
Excavation, bottom of trench 2' wide, 2' deep	.296	C.Y.	.032		2.36	2.36
4' deep	.889	C.Y.	.095		7.10	7.10
6' deep	1.770	C.Y.	.142		10.90	10.90
8' deep	2.960	C.Y.	.105		20	20
Bedding, sand, bottom of trench 2' wide, no compaction, pipe, 2" diameter	.070	C.Y.	.028	2.56	1.34	3.90
4" diameter	.084	C.Y.	.034	3.07	1.62	4.69
6" diameter	.105	C.Y.	.042	3.83	2.02	5.85
8" diameter	.122	C.Y.	.049	4.45	2.34	6.79
Compacted, pipe, 2" diameter	.074	C.Y.	.030	2.70	1.42	4.12
4" diameter	.092	C.Y.	.037	3.36	1.77	5.13
6" diameter	.111	C.Y.	.044	4.05	2.13	6.18
8" diameter	.129	C.Y.	.052	4.71	2.47	7.18
3/4" stone, bottom of trench 2' wide, pipe, 4" diameter	.082	C.Y.	.033	2.99	1.57	4.56
6" diameter	.099	C.Y.	.040	3.61	1.90	5.51
3/8" stone, bottom of trench 2' wide, pipe, 4" diameter	.084	C.Y.	.034	3.07	1.62	4.69
6" diameter	.102	C.Y.	.041	3.72	1.96	5.68
Utilities, drainage & sewerage, corrugated plastic, 6" diameter	1.000	L.F.	.069	3.61	2.91	6.52
8" diameter	1.000	L.F.	.072	6.75	3.04	9.79
Concrete, non-reinforced, 6" diameter	1.000	L.F.	.181	8	9.50	17.50
8" diameter	1.000	L.F.	.214	8.75	11.30	20.05
PVC, SDR 35, 4" diameter	1.000	L.F.	.064	1.64	2.71	4.35
6" diameter	1.000	L.F.	.069	3.61	2.91	6.52
8" diameter	1.000	L.F.	.072	6.75	3.04	9.79
Gas & service, polyethylene, 1-1/4" diameter	1.000	L.F.	.059	1.78	2.97	4.75
Steel sched.40, 1" diameter	1.000	L.F.	.107	5.50	6.75	12.25
2" diameter	1.000	L.F.	.114	8.65	7.25	15.90
Sub-drainage, PVC, perforated, 3" diameter	1.000	L.F.	.064	1.64	2.71	4.35
4" diameter	1.000	L.F.	.064	1.64	2.71	4.35
5" diameter	1.000	L.F.	.069	3.61	2.91	6.52
6" diameter	1.000	L.F.	.069	3.61	2.91	6.52
Water service, copper, type K, 3/4"	1.000	L.F.	.083	6.90	5.35	12.25
1" diameter	1.000	L.F.	.093	8.95	6	14.95
PVC, 3/4"	1.000	L.F.	.121	4.50	7.80	12.30
1" diameter	1.000	L.F.	.134	5.05	8.60	13.65
Backfill, bottom of trench 2' wide no compact, 2' deep, pipe, 2" diameter	.226	L.F.	.053		2.19	2.19
4" diameter	.212	L.F.	.050		2.06	2.06
6" diameter	.185	L.F.	.044		1.79	1.79
4' deep, pipe, 2" diameter	.819	C.Y.	.193		7.95	7.95
4" diameter	.805	C.Y.	.189		7.80	7.80
6" diameter	.778	C.Y.	.183		7.55	7.55
6' deep, pipe, 2" diameter	1.700	C.Y.	.400		16.50	16.50
4" diameter	1.690	C.Y.	.398		16.40	16.40
6" diameter	1.660	C.Y.	.391		16.10	16.10
8' deep, pipe, 2" diameter	2.890	C.Y.	.680		28	28
4" diameter	2.870	C.Y.	.675		28	28
6" diameter	2.850	C.Y.	.671		27.50	27.50

1 | SITEWORK — 16 | Sidewalk Systems

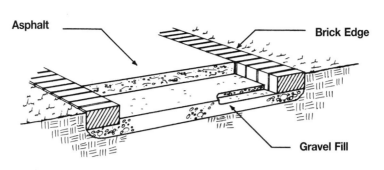

System Description	QUAN.	UNIT	LABOR HOURS	COST PER S.F. MAT.	COST PER S.F. INST.	COST PER S.F. TOTAL
ASPHALT SIDEWALK SYSTEM, 3' WIDE WALK						
Gravel fill, 4" deep	1.000	S.F.	.001	.32	.05	.37
Compact fill	.012	C.Y.			.02	.02
Handgrade	1.000	S.F.	.004		.17	.17
Walking surface, bituminous paving, 2" thick	1.000	S.F.	.007	.89	.36	1.25
Edging, brick, laid on edge	.670	L.F.	.079	1.90	3.92	5.82
TOTAL		S.F.	.091	3.11	4.52	7.63
CONCRETE SIDEWALK SYSTEM, 3' WIDE WALK						
Gravel fill, 4" deep	1.000	S.F.	.001	.32	.05	.37
Compact fill	.012	C.Y.			.02	.02
Handgrade	1.000	S.F.	.004		.17	.17
Walking surface, concrete, 4" thick	1.000	S.F.	.040	1.96	2	3.96
Edging, brick, laid on edge	.670	L.F.	.079	1.90	3.92	5.82
TOTAL		S.F.	.124	4.18	6.16	10.34
PAVERS, BRICK SIDEWALK SYSTEM, 3' WIDE WALK						
Sand base fill, 4" deep	1.000	S.F.	.001	.50	.09	.59
Compact fill	.012	C.Y.			.02	.02
Handgrade	1.000	S.F.	.004		.17	.17
Walking surface, brick pavers	1.000	S.F.	.160	3.14	7.85	10.99
Edging, redwood, untreated, 1" x 4"	.670	L.F.	.032	1.66	1.83	3.49
TOTAL		S.F.	.197	5.30	9.96	15.26

The costs in this system are based on a cost per square foot of sidewalk area. Concrete used is 3000 p.s.i.

Description	QUAN.	UNIT	LABOR HOURS	COST PER S.F. MAT.	COST PER S.F. INST.	COST PER S.F. TOTAL

Sidewalk Price Sheet

Sidewalk Price Sheet	QUAN.	UNIT	LABOR HOURS	COST PER S.F. MAT.	COST PER S.F. INST.	COST PER S.F. TOTAL
Base, crushed stone, 3" deep	1.000	S.F.	.001	.29	.10	.39
6" deep	1.000	S.F.	.001	.58	.11	.69
9" deep	1.000	S.F.	.002	.84	.15	.99
12" deep	1.000	S.F.	.002	1.48	.18	1.66
Bank run gravel, 6" deep	1.000	S.F.	.001	.48	.08	.56
9" deep	1.000	S.F.	.001	.71	.11	.82
12" deep	1.000	S.F.	.001	.97	.13	1.10
Compact base, 3" deep	.009	C.Y.	.001		.01	.01
6" deep	.019	C.Y.	.001		.03	.03
9" deep	.028	C.Y.	.001		.05	.05
Handgrade	1.000	S.F.	.004		.17	.17
Surface, brick, pavers dry joints, laid flat, running bond	1.000	S.F.	.160	3.14	7.85	10.99
Basket weave	1.000	S.F.	.168	3.97	8.30	12.27
Herringbone	1.000	S.F.	.174	3.97	8.55	12.52
Laid on edge, running bond	1.000	S.F.	.229	4.33	11.25	15.58
Mortar jts. laid flat, running bond	1.000	S.F.	.192	3.77	9.40	13.17
Basket weave	1.000	S.F.	.202	4.76	9.95	14.71
Herringbone	1.000	S.F.	.209	4.76	10.25	15.01
Laid on edge, running bond	1.000	S.F.	.274	5.20	13.50	18.70
Bituminous paving, 1-1/2" thick	1.000	S.F.	.006	.66	.26	.92
2" thick	1.000	S.F.	.007	.89	.36	1.25
2-1/2" thick	1.000	S.F.	.008	1.12	.39	1.51
Sand finish, 3/4" thick	1.000	S.F.	.001	.36	.13	.49
1" thick	1.000	S.F.	.001	.44	.15	.59
Concrete, reinforced, broom finish, 4" thick	1.000	S.F.	.040	1.96	2	3.96
5" thick	1.000	S.F.	.044	2.62	2.20	4.82
6" thick	1.000	S.F.	.047	3.06	2.35	5.41
Crushed stone, white marble, 3" thick	1.000	S.F.	.009	.46	.39	.85
Bluestone, 3" thick	1.000	S.F.	.009	.13	.39	.52
Flagging, bluestone, 1"	1.000	S.F.	.198	9.90	9.70	19.60
1-1/2"	1.000	S.F.	.188	18	9.25	27.25
Slate, natural cleft, 3/4"	1.000	S.F.	.174	9.90	8.55	18.45
Random rect., 1/2"	1.000	S.F.	.152	21.50	7.50	29
Granite blocks	1.000	S.F.	.174	16.50	8.55	25.05
Edging, corrugated aluminum, 4", 3' wide walk	.666	L.F.	.008	1.61	.47	2.08
4' wide walk	.500	L.F.	.006	1.21	.35	1.56
6", 3' wide walk	.666	L.F.	.010	2.02	.55	2.57
4' wide walk	.500	L.F.	.007	1.52	.42	1.94
Redwood-cedar-cypress, 1" x 4", 3' wide walk	.666	L.F.	.021	.85	1.21	2.06
4' wide walk	.500	L.F.	.016	.64	.91	1.55
2" x 4", 3' wide walk	.666	L.F.	.032	1.66	1.83	3.49
4' wide walk	.500	L.F.	.024	1.25	1.38	2.63
Brick, dry joints, 3' wide walk	.666	L.F.	.079	1.90	3.92	5.82
4' wide walk	.500	L.F.	.059	1.42	2.93	4.35
Mortar joints, 3' wide walk	.666	L.F.	.095	2.28	4.70	6.98
4' wide walk	.500	L.F.	.071	1.70	3.51	5.21

1 | SITEWORK 20 | Driveway Systems

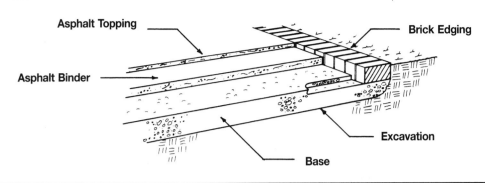

System Description	QUAN.	UNIT	LABOR HOURS	COST PER S.F.		
				MAT.	INST.	TOTAL
ASPHALT DRIVEWAY TO 10' WIDE						
Excavation, driveway to 10' wide, 6" deep	.019	C.Y.			.04	.04
Base, 6" crushed stone	1.000	S.F.	.001	.58	.11	.69
Handgrade base	1.000	S.F.	.004		.17	.17
2" thick base	1.000	S.F.	.002	.89	.20	1.09
1" topping	1.000	S.F.	.001	.44	.15	.59
Edging, brick pavers	.200	L.F.	.024	.57	1.17	1.74
TOTAL		S.F.	.032	2.48	1.84	4.32
CONCRETE DRIVEWAY TO 10' WIDE						
Excavation, driveway to 10' wide, 6" deep	.019	C.Y.			.04	.04
Base, 6" crushed stone	1.000	S.F.	.001	.58	.11	.69
Handgrade base	1.000	S.F.	.004		.17	.17
Surface, concrete, 4" thick	1.000	S.F.	.040	1.96	2	3.96
Edging, brick pavers	.200	L.F.	.024	.57	1.17	1.74
TOTAL		S.F.	.069	3.11	3.49	6.60
PAVERS, BRICK DRIVEWAY TO 10' WIDE						
Excavation, driveway to 10' wide, 6" deep	.019	C.Y.			.04	.04
Base, 6" sand	1.000	S.F.	.001	.79	.15	.94
Handgrade base	1.000	S.F.	.004		.17	.17
Surface, pavers, brick laid flat, running bond	1.000	S.F.	.160	3.14	7.85	10.99
Edging, redwood, untreated, 2" x 4"	.200	L.F.	.010	.50	.55	1.05
TOTAL		S.F.	.175	4.43	8.76	13.19

Description	QUAN.	UNIT	LABOR HOURS	COST PER S.F.		
				MAT.	INST.	TOTAL

Driveway Price Sheet	QUAN.	UNIT	LABOR HOURS	COST PER S.F. MAT.	COST PER S.F. INST.	COST PER S.F. TOTAL
Excavation, by machine, 10' wide, 6" deep	.019	C.Y.	.001		.04	.04
12" deep	.037	C.Y.	.001		.08	.08
18" deep	.055	C.Y.	.001		.12	.12
20' wide, 6" deep	.019	C.Y.	.001		.04	.04
12" deep	.037	C.Y.	.001		.08	.08
18" deep	.055	C.Y.	.001		.12	.12
Base, crushed stone, 10' wide, 3" deep	1.000	S.F.	.001	.29	.06	.35
6" deep	1.000	S.F.	.001	.58	.11	.69
9" deep	1.000	S.F.	.002	.84	.15	.99
20' wide, 3" deep	1.000	S.F.	.001	.29	.06	.35
6" deep	1.000	S.F.	.001	.58	.11	.69
9" deep	1.000	S.F.	.002	.84	.15	.99
Bank run gravel, 10' wide, 3" deep	1.000	S.F.	.001	.24	.05	.29
6" deep	1.000	S.F.	.001	.48	.08	.56
9" deep	1.000	S.F.	.001	.71	.11	.82
20' wide, 3" deep	1.000	S.F.	.001	.24	.05	.29
6" deep	1.000	S.F.	.001	.48	.08	.56
9" deep	1.000	S.F.	.001	.71	.11	.82
Handgrade, 10' wide	1.000	S.F.	.004		.17	.17
20' wide	1.000	S.F.	.004		.17	.17
Surface, asphalt, 10' wide, 3/4" topping, 1" base	1.000	S.F.	.002	1.05	.28	1.33
2" base	1.000	S.F.	.003	1.25	.33	1.58
1" topping, 1" base	1.000	S.F.	.002	1.13	.30	1.43
2" base	1.000	S.F.	.003	1.33	.35	1.68
20' wide, 3/4" topping, 1" base	1.000	S.F.	.002	1.05	.28	1.33
2" base	1.000	S.F.	.003	1.25	.33	1.58
1" topping, 1" base	1.000	S.F.	.002	1.13	.30	1.43
2" base	1.000	S.F.	.003	1.33	.35	1.68
Concrete, 10' wide, 4" thick	1.000	S.F.	.040	1.96	2	3.96
6" thick	1.000	S.F.	.047	3.06	2.35	5.41
20' wide, 4" thick	1.000	S.F.	.040	1.96	2	3.96
6" thick	1.000	S.F.	.047	3.06	2.35	5.41
Paver, brick 10' wide dry joints, running bond, laid flat	1.000	S.F.	.160	3.14	7.85	10.99
Laid on edge	1.000	S.F.	.229	4.33	11.25	15.58
Mortar joints, laid flat	1.000	S.F.	.192	3.77	9.40	13.17
Laid on edge	1.000	S.F.	.274	5.20	13.50	18.70
20' wide, running bond, dry jts., laid flat	1.000	S.F.	.160	3.14	7.85	10.99
Laid on edge	1.000	S.F.	.229	4.33	11.25	15.58
Mortar joints, laid flat	1.000	S.F.	.192	3.77	9.40	13.17
Laid on edge	1.000	S.F.	.274	5.20	13.50	18.70
Crushed stone, 10' wide, white marble, 3"	1.000	S.F.	.009	.46	.39	.85
Bluestone, 3"	1.000	S.F.	.009	.13	.39	.52
20' wide, white marble, 3"	1.000	S.F.	.009	.46	.39	.85
Bluestone, 3"	1.000	S.F.	.009	.13	.39	.52
Soil cement, 10' wide	1.000	S.F.	.007	.36	1.06	1.42
20' wide	1.000	S.F.	.007	.36	1.06	1.42
Granite blocks, 10' wide	1.000	S.F.	.174	16.50	8.55	25.05
20' wide	1.000	S.F.	.174	16.50	8.55	25.05
Asphalt block, solid 1-1/4" thick	1.000	S.F.	.119	6.15	5.85	12
Solid 3" thick	1.000	S.F.	.123	8.60	6.05	14.65
Edging, brick, 10' wide	.200	L.F.	.024	.57	1.17	1.74
20' wide	.100	L.F.	.012	.28	.59	.87
Redwood, untreated 2" x 4", 10' wide	.200	L.F.	.010	.50	.55	1.05
20' wide	.100	L.F.	.005	.25	.28	.53
Granite, 4 1/2" x 12" straight, 10' wide	.200	L.F.	.032	1.54	2.02	3.56
20' wide	.100	L.F.	.016	.77	1.02	1.79
Finishes, asphalt sealer, 10' wide	1.000	S.F.	.023	1.02	.97	1.99
20' wide	1.000	S.F.	.023	1.02	.97	1.99
Concrete, exposed aggregate 10' wide	1.000	S.F.	.013	.22	.66	.88
20' wide	1.000	S.F.	.013	.22	.66	.88

1 | SITEWORK — 24 | Septic Systems

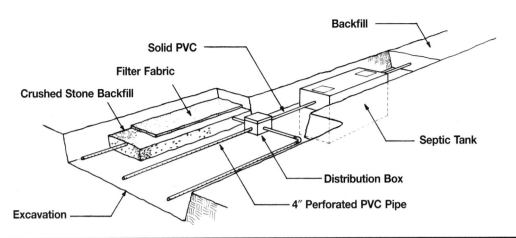

System Description	QUAN.	UNIT	LABOR HOURS	COST EACH MAT.	COST EACH INST.	COST EACH TOTAL
SEPTIC SYSTEM WITH 600 S.F. LEACHING FIELD, 1000 GALLON TANK						
Mobilization	2.000	Ea.	16.000		1,370	1,370
Tank, precast, 1000 gallon	1.000	Ea.	3.500	1,125	176.30	1,301.30
Effluent filter	1.000	Ea.	1.000	47.50	57.50	105
Distribution box, precast	1.000	Ea.	1.000	90	41.50	131.50
Flow leveler	3.000	Ea.	.480	7.77	19.80	27.57
PVC pipe, 4" diameter	25.000	L.F.	1.600	41	67.75	108.75
Tee	1.000	Ea.	1.000	14.80	57.50	72.30
Elbow	2.000	Ea.	1.333	21.60	77	98.60
Viewport cap	1.000	Ea.	.333	8.45	19.15	27.60
Filter fabric	67.000	S.Y.	.447	121.94	18.76	140.70
Detectable marking tape	1.600	C.L.F.	.085	10.56	3.52	14.08
PVC perforated pipe, 4" diameter	135.000	L.F.	8.640	221.40	365.85	587.25
Excavation	160.000	C.Y.	17.654		1,681.60	1,681.60
Backfill	133.000	L.C.Y.	2.660		264.67	264.67
Spoil	55.000	L.C.Y.	3.056		330	330
Compaction	113.000	E.C.Y.	15.066		696.08	696.08
Stone fill	39.000	C.Y.	6.240	1,638	393.12	2,031.12
TOTAL		Ea.	80.094	3,348.02	5,640.10	8,988.12
SEPTIC SYSTEM WITH 750 S.F. LEACHING FIELD, 1500 GALLON TANK						
Mobilization	2.000	Ea.	16.000		1,370	1,370
Tank, precast, 1500 gallon	1.000	Ea.	4.000	1,750	201	1,951
Effluent filter	1.000	Ea.	1.000	47.50	57.50	105
Distribution box, precast	1.000	Ea.	1.000	90	41.50	131.50
Flow leveler	3.000	Ea.	.480	7.77	19.80	27.57
PVC pipe, 4" diameter	25.000	L.F.	1.600	41	67.75	108.75
Tee	125.000	Ea.	1.000	14.80	57.50	72.30
Elbow	2.000	Ea.	1.333	21.60	77	98.60
Viewport cap	1.000	Ea.	.333	8.45	19.15	27.60
Filter fabric	84.000	S.Y.	.560	152.88	23.52	176.40
Detectable marking tape	1.900	C.L.F.	.101	12.54	4.18	16.72
PVC perforated pipe, 4" diameter	165.000	L.F.	10.560	270.60	447.15	717.75
Excavation	199.000	C.Y.	21.958		2,091.49	2,091.49
Backfill	162.000	L.C.Y.	3.240		322.38	322.38
Spoil	73.500	L.C.Y.	4.084		441.01	441.01
Compaction	137.000	E.C.Y.	18.266		843.92	843.92
Stone fill	48.500	C.Y.	7.760	2,037	488.88	2,525.88
TOTAL		Ea.	93.275	4,454.14	6,573.73	11,027.87

The costs in this system include all necessary piping and excavation.

1 | SITEWORK — 24 | Septic Systems

System Description	QUAN.	UNIT	LABOR HOURS	COST EACH MAT.	COST EACH INST.	COST EACH TOTAL
SEPTIC SYSTEM WITH 1000 S.F. LEACHING FIELD, 1500 GALLON TANK						
Mobilization	2.000	Ea.	16.000		1,370	1,370
Tank, precast, 1500 gallon	1.000	Ea.	4.000	1,750	201	1,951
Effluent filter	1.000	Ea.	1.000	47.50	57.50	105
Distribution box, precast	1.000	Ea.	1.000	90	41.50	131.50
Flow leveler	4.000	Ea.	.640	10.36	26.40	36.76
PVC pipe, 4" diameter	25.000	L.F.	1.600	41	67.75	108.75
Tee	1.000	Ea.	1.000	14.80	57.50	72.30
Elbow	4.000	Ea.	2.667	43.20	154	197.20
Viewport cap	1.000	Ea.	.333	8.45	19.15	27.60
Filter fabric	111.000	S.Y.	.740	202.02	31.08	233.10
Detectable marking tape	2.400	C.L.F.	.128	15.84	5.28	21.12
PVC perforated pipe, 4" diameter	215.000	L.F.	13.760	352.60	582.65	935.25
Excavation	229.000	C.Y.	25.268		2,406.79	2,406.79
Backfill	178.000	L.C.Y.	3.560		354.22	354.22
Spoil	91.500	L.C.Y.	5.084		549.01	549.01
Compaction	151.000	E.C.Y.	20.133		930.16	930.16
Stone fill	64.000	C.Y.	10.240	2,688	645.12	3,333.12
TOTAL		Ea.	107.153	5,263.77	7,499.11	12,762.88

The costs in this system include all necessary piping and excavation.

Septic Systems Price Sheet	QUAN.	UNIT	LABOR HOURS	COST EACH MAT.	COST EACH INST.	COST EACH TOTAL
Tank, precast concrete, 1000 gallon	1.000	Ea.	3.500	1,125	176	1,301
1500 gallon	1.000	Ea.	4.000	1,750	201	1,951
Distribution box, concrete, 5 outlets	1.000	Ea.	1.000	90	41.50	131.50
12 outlets	1.000	Ea.	2.000	550	82.50	632.50
4" pipe, PVC, solid	25.000	L.F.	1.600	41	68	109
Tank and field excavation, 600 S.F. field	160.000	C.Y.	17.654		1,675	1,675
750 S.F. field	199.000	C.Y.	21.958		2,075	2,075
1000 S.F. field	229.000	C.Y.	25.268		2,400	2,400
Tank excavation only, 1000 gallon tank	20.000	C.Y.	2.206		210	210
1500 gallon tank	26.000	C.Y.	2.869		273	273
Backfill, crushed stone, 600 S.F. field	39.000	C.Y.	12.160	1,650	395	2,045
750 S.F. field	48.500	C.Y.	22.400	2,025	490	2,515
1000 S.F. field	64.000	C.Y.	.240	2,700	645	3,345
Backfill with excavated material, 600 S.F. field	133.000	L.C.Y.	.400		265	265
750 S.F. field	162.000	L.C.Y.	.367		325	325
1000 S.F. field	178.000	L.C.Y.	.280		355	355
Filter fabric, 600 S.F. field	67.000	S.Y.	2.376	122	18.75	140.75
750 S.F. field	84.000	S.Y.	4.860	153	23.50	176.50
1000 S.F. field	111.000	S.Y.	.740	202	31	233
4" pipe, PVC, perforated, 600 S.F. field	135.000	L.F.	9.280	221	365	586
750 S.F. field	165.000	L.F.	16.960	271	445	716
1000 S.F. field	215.000	L.F.	1.939	355	585	940
Pipe fittings, PVC, 600 S.F. field	2.000	Ea.	3.879	21.50	77	98.50
750 S.F. field	2.000	Ea.		21.50	77	98.50
1000 S.F. field	4.000	Ea.		43	154	197
Mobilization	2.000	Ea.	16.000		1,375	1,375
Effluent filter	1.000	Ea.	1.000	47.50	57.50	105
Flow leveler, 600 S.F. field	3.000	Ea.	.480	7.75	19.80	27.55
750 S.F. field	3.000	Ea.	.480	7.75	19.80	27.55
1000 S.F. field	4.000	Ea.	.640	10.35	26.50	36.85
Viewport cap	1.000	Ea.	.333	8.45	19.15	27.60
Detectable marking tape, 600 S.F. field	1.600	C.L.F.	.085	10.55	3.52	14.07
750 S.F. field	1.900	C.L.F.	.101	12.55	4.18	16.73
1000 S.F. field	2.400	C.L.F.	.128	15.85	5.30	21.15

1 | SITEWORK — 60 | Chain Link Fence Systems

System Description	QUAN.	UNIT	LABOR HOURS	COST PER UNIT MAT.	COST PER UNIT INST.	COST PER UNIT TOTAL
Chain link fence						
Galv. 9ga. wire, 1-5/8"post 10' O.C., 1-3/8"top rail, 2"corner post, 3' hi	1.000	L.F.	.130	10.05	5.50	15.55
4' high	1.000	L.F.	.141	9.60	6	15.60
6' high	1.000	L.F.	.209	13.35	8.85	22.20
Add for gate 3' wide 1-3/8" frame 3' high	1.000	Ea.	2.000	103	84.50	187.50
4' high	1.000	Ea.	2.400	110	102	212
6' high	1.000	Ea.	2.400	132	102	234
Add for gate 4' wide 1-3/8" frame 3' high	1.000	Ea.	2.667	105	113	218
4' high	1.000	Ea.	2.667	114	113	227
6' high	1.000	Ea.	3.000	139	127	266
Alum. 9ga. wire, 1-5/8"post, 10' O.C., 1-3/8"top rail, 2"corner post, 3' high	1.000	L.F.	.130	8.40	5.50	13.90
4' high	1.000	L.F.	.141	8.80	6	14.80
6' high	1.000	L.F.	.209	11.45	8.85	20.30
Add for gate 3' wide 1-3/8" frame 3' high	1.000	Ea.	2.000	140	84.50	224.50
4' high	1.000	Ea.	2.400	150	102	252
6' high	1.000	Ea.	2.400	190	102	292
Add for gate 4' wide 1-3/8" frame 3' high	1.000	Ea.	2.400	145	102	247
4' high	1.000	Ea.	2.667	150	113	263
6' high	1.000	Ea.	3.000	202	127	329
Vinyl 9ga. wire, 1-5/8"post 10' O.C., 1-3/8"top rail, 2"corner post, 3' hi	1.000	L.F.	.130	8.15	5.50	13.65
4' high	1.000	L.F.	.141	8	6	14
6' high	1.000	L.F.	.209	10.65	8.85	19.50
Add for gate 3' wide 1-3/8" frame 3' high	1.000	Ea.	2.000	112	84.50	196.50
4' high	1.000	Ea.	2.400	118	102	220
6' high	1.000	Ea.	2.400	146	102	248
Add for gate 4' wide 1-3/8" frame 3' high	1.000	Ea.	2.400	115	102	217
4' high	1.000	Ea.	2.667	121	113	234
6' high	1.000	Ea.	3.000	156	127	283
Tennis court, chain link fence, 10' high						
Galv. 11ga. wire, 2"post 10' O.C., 1-3/8"top rail, 2-1/2"corner post	1.000	L.F.	.253	8.50	10.70	19.20
Add for gate 3' wide 1-3/8" frame	1.000	Ea.	2.400	242	102	344
Alum. 11ga. wire, 2"post 10' O.C., 1-3/8"top rail, 2-1/2"corner post	1.000	L.F.	.253	10.20	10.70	20.90
Add for gate 3' wide 1-3/8" frame	1.000	Ea.	2.400	156	102	258
Vinyl 11ga. wire, 2"post 10' O.C., 1-3/8"top rail, 2-1/2"corner post	1.000	L.F.	.253	8.85	10.70	19.55
Add for gate 3' wide 1-3/8" frame	1.000	Ea.	2.400	310	102	412
Railings, commercial						
Aluminum balcony rail, 1-1/2" posts with pickets	1.000	L.F.	.164	81	11.75	92.75
With expanded metal panels	1.000	L.F.	.164	106	11.75	117.75
With porcelain enamel panel inserts	1.000	L.F.	.164	99	11.75	110.75
Mild steel, ornamental rounded top rail	1.000	L.F.	.164	92	11.75	103.75
As above, but pitch down stairs	1.000	L.F.	.183	120	13.10	133.10
Steel pipe, welded, 1-1/2" round, painted	1.000	L.F.	.160	30	11.40	41.40
Galvanized	1.000	L.F.	.160	42	11.40	53.40
Residential, stock units, mild steel, deluxe	1.000	L.F.	.102	17.25	7.25	24.50
Economy	1.000	L.F.	.102	12.85	7.25	20.10

1 | SITEWORK   64 | Wood Fence Systems

System Description	QUAN.	UNIT	LABOR HOURS	COST PER UNIT		
				MAT.	INST.	TOTAL
Basketweave, 3/8"x4" boards, 2"x4" stringers on spreaders, 4"x4" posts						
No. 1 cedar, 6' high	1.000	L.F.	.150	28.50	8.35	36.85
Treated pine, 6' high	1.000	L.F.	.160	38	8.90	46.90
Board fence, 1"x4" boards, 2"x4" rails, 4"x4" posts						
Preservative treated, 2 rail, 3' high	1.000	L.F.	.166	10.40	9.20	19.60
4' high	1.000	L.F.	.178	12.20	9.90	22.10
3 rail, 5' high	1.000	L.F.	.185	13.30	10.25	23.55
6' high	1.000	L.F.	.192	15	10.70	25.70
Western cedar, No. 1, 2 rail, 3' high	1.000	L.F.	.166	11.45	9.20	20.65
3 rail, 4' high	1.000	L.F.	.178	12.60	9.90	22.50
5' high	1.000	L.F.	.185	14.15	10.25	24.40
6' high	1.000	L.F.	.192	15.10	10.70	25.80
No. 1 cedar, 2 rail, 3' high	1.000	L.F.	.166	14.50	9.20	23.70
4' high	1.000	L.F.	.178	15.70	9.90	25.60
3 rail, 5' high	1.000	L.F.	.185	19.05	10.25	29.30
6' high	1.000	L.F.	.192	21	10.70	31.70
Shadow box, 1"x6" boards, 2"x4" rails, 4"x4" posts						
Fir, pine or spruce, treated, 3 rail, 6' high	1.000	L.F.	.160	19.30	8.90	28.20
No. 1 cedar, 3 rail, 4' high	1.000	L.F.	.185	20	10.25	30.25
6' high	1.000	L.F.	.192	25.50	10.70	36.20
Open rail, split rails, No. 1 cedar, 2 rail, 3' high	1.000	L.F.	.150	14.05	8.35	22.40
3 rail, 4' high	1.000	L.F.	.160	14.50	8.90	23.40
No. 2 cedar, 2 rail, 3' high	1.000	L.F.	.150	12.90	8.35	21.25
3 rail, 4' high	1.000	L.F.	.160	10.25	8.90	19.15
Open rail, rustic rails, No. 1 cedar, 2 rail, 3' high	1.000	L.F.	.150	11.20	8.35	19.55
3 rail, 4' high	1.000	L.F.	.160	10.75	8.90	19.65
No. 2 cedar, 2 rail, 3' high	1.000	L.F.	.150	10.50	8.35	18.85
3 rail, 4' high	1.000	L.F.	.160	8.75	8.90	17.65
Rustic picket, molded pine pickets, 2 rail, 3' high	1.000	L.F.	.171	9.25	9.50	18.75
3 rail, 4' high	1.000	L.F.	.197	10.65	10.95	21.60
No. 1 cedar, 2 rail, 3' high	1.000	L.F.	.171	10.70	9.50	20.20
3 rail, 4' high	1.000	L.F.	.197	12.30	10.95	23.25
Picket fence, fir, pine or spruce, preserved, treated						
2 rail, 3' high	1.000	L.F.	.171	8.60	9.50	18.10
3 rail, 4' high	1.000	L.F.	.185	9.80	10.25	20.05
Western cedar, 2 rail, 3' high	1.000	L.F.	.171	9.90	9.50	19.40
3 rail, 4' high	1.000	L.F.	.185	10	10.25	20.25
No. 1 cedar, 2 rail, 3' high	1.000	L.F.	.171	15.05	9.50	24.55
3 rail, 4' high	1.000	L.F.	.185	19.60	10.25	29.85
Stockade, No. 1 cedar, 3-1/4" rails, 6' high	1.000	L.F.	.150	14.45	8.35	22.80
8' high	1.000	L.F.	.155	19.90	8.60	28.50
No. 2 cedar, treated rails, 6' high	1.000	L.F.	.150	14.80	8.35	23.15
Treated pine, treated rails, 6' high	1.000	L.F.	.150	15.35	8.35	23.70
Gates, No. 2 cedar, picket, 3'-6" wide 4' high	1.000	Ea.	2.667	87	149	236
No. 2 cedar, rustic round, 3' wide, 3' high	1.000	Ea.	2.667	112	149	261
No. 2 cedar, stockade screen, 3'-6" wide, 6' high	1.000	Ea.	3.000	98.50	168	266.50
General, wood, 3'-6" wide, 4' high	1.000	Ea.	2.400	96	135	231
6' high	1.000	Ea.	3.000	120	168	288

Division 2 - Foundations

Did you know?
RSMeans Online gives you the same access to RSMeans' data with 24/7 access:
- Quickly locate costs in the searchable database.
- Build cost lists, estimates, and reports in minutes.
- Adjust costs to any location in the U.S. and Canada with the click of a button.

Start your free trial today at **www.RSMeansOnline.com**

RSMeans Online
FROM THE GORDIAN GROUP

No part of this cost data may be reproduced, stored in a retrieval system, or transmitted in any form or by any means without prior written permission of RSMeans.

2 FOUNDATIONS — 04 Footing Systems

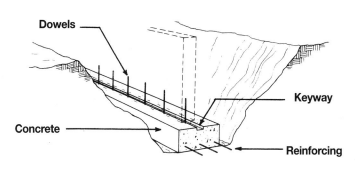

System Description	QUAN.	UNIT	LABOR HOURS	COST PER L.F. MAT.	COST PER L.F. INST.	COST PER L.F. TOTAL
8" THICK BY 18" WIDE FOOTING						
Concrete, 3000 psi	.040	C.Y.		4.72		4.72
Place concrete, direct chute	.040	C.Y.	.016		.72	.72
Forms, footing, 4 uses	1.330	SFCA	.103	1.02	5.05	6.07
Reinforcing, 1/2" diameter bars, 2 each	1.380	Lb.	.011	.73	.63	1.36
Keyway, 2" x 4", beveled, 4 uses	1.000	L.F.	.015	.23	.86	1.09
Dowels, 1/2" diameter bars, 2' long, 6' O.C.	.166	Ea.	.006	.13	.33	.46
TOTAL		L.F.	.151	6.83	7.59	14.42
12" THICK BY 24" WIDE FOOTING						
Concrete, 3000 psi	.070	C.Y.		8.26		8.26
Place concrete, direct chute	.070	C.Y.	.028		1.26	1.26
Forms, footing, 4 uses	2.000	SFCA	.155	1.54	7.60	9.14
Reinforcing, 1/2" diameter bars, 2 each	1.380	Lb.	.011	.73	.63	1.36
Keyway, 2" x 4", beveled, 4 uses	1.000	L.F.	.015	.23	.86	1.09
Dowels, 1/2" diameter bars, 2' long, 6' O.C.	.166	Ea.	.006	.13	.33	.46
TOTAL		L.F.	.215	10.89	10.68	21.57
12" THICK BY 36" WIDE FOOTING						
Concrete, 3000 psi	.110	C.Y.		12.98		12.98
Place concrete, direct chute	.110	C.Y.	.044		1.98	1.98
Forms, footing, 4 uses	2.000	SFCA	.155	1.54	7.60	9.14
Reinforcing, 1/2" diameter bars, 2 each	1.380	Lb.	.011	.73	.63	1.36
Keyway, 2" x 4", beveled, 4 uses	1.000	L.F.	.015	.23	.86	1.09
Dowels, 1/2" diameter bars, 2' long, 6' O.C.	.166	Ea.	.006	.13	.33	.46
TOTAL		L.F.	.231	15.61	11.40	27.01

The footing costs in this system are on a cost per linear foot basis.

Description	QUAN.	UNIT	LABOR HOURS	COST PER S.F. MAT.	COST PER S.F. INST.	COST PER S.F. TOTAL

Footing Price Sheet	QUAN.	UNIT	LABOR HOURS	COST PER L.F. MAT.	COST PER L.F. INST.	COST PER L.F. TOTAL
Concrete, 8" thick by 18" wide footing						
2000 psi concrete	.040	C.Y.		4.52		4.52
2500 psi concrete	.040	C.Y.		4.60		4.60
3000 psi concrete	.040	C.Y.		4.72		4.72
3500 psi concrete	.040	C.Y.		4.84		4.84
4000 psi concrete	.040	C.Y.		4.96		4.96
12" thick by 24" wide footing						
2000 psi concrete	.070	C.Y.		7.90		7.90
2500 psi concrete	.070	C.Y.		8.05		8.05
3000 psi concrete	.070	C.Y.		8.25		8.25
3500 psi concrete	.070	C.Y.		8.45		8.45
4000 psi concrete	.070	C.Y.		8.70		8.70
12" thick by 36" wide footing						
2000 psi concrete	.110	C.Y.		12.45		12.45
2500 psi concrete	.110	C.Y.		12.65		12.65
3000 psi concrete	.110	C.Y.		13		13
3500 psi concrete	.110	C.Y.		13.30		13.30
4000 psi concrete	.110	C.Y.		13.65		13.65
Place concrete, 8" thick by 18" wide footing, direct chute	.040	C.Y.	.016		.72	.72
Pumped concrete	.040	C.Y.	.017		1	1
Crane & bucket	.040	C.Y.	.032		2.07	2.07
12" thick by 24" wide footing, direct chute	.070	C.Y.	.028		1.26	1.26
Pumped concrete	.070	C.Y.	.030		1.76	1.76
Crane & bucket	.070	C.Y.	.056		3.63	3.63
12" thick by 36" wide footing, direct chute	.110	C.Y.	.044		1.98	1.98
Pumped concrete	.110	C.Y.	.047		2.76	2.76
Crane & bucket	.110	C.Y.	.088		5.70	5.70
Forms, 8" thick footing, 1 use	1.330	SFCA	.140	3.15	6.85	10
4 uses	1.330	SFCA	.103	1.02	5.05	6.07
12" thick footing, 1 use	2.000	SFCA	.211	4.74	10.30	15.04
4 uses	2.000	SFCA	.155	1.54	7.60	9.14
Reinforcing, 3/8" diameter bar, 1 each	.400	Lb.	.003	.21	.18	.39
2 each	.800	Lb.	.006	.42	.37	.79
3 each	1.200	Lb.	.009	.64	.55	1.19
1/2" diameter bar, 1 each	.700	Lb.	.005	.37	.32	.69
2 each	1.380	Lb.	.011	.73	.63	1.36
3 each	2.100	Lb.	.016	1.11	.97	2.08
5/8" diameter bar, 1 each	1.040	Lb.	.008	.55	.48	1.03
2 each	2.080	Lb.	.016	1.10	.96	2.06
Keyway, beveled, 2" x 4", 1 use	1.000	L.F.	.030	.46	1.72	2.18
2 uses	1.000	L.F.	.023	.35	1.29	1.64
2" x 6", 1 use	1.000	L.F.	.032	.66	1.82	2.48
2 uses	1.000	L.F.	.024	.50	1.37	1.87
Dowels, 2 feet long, 6' O.C., 3/8" bar	.166	Ea.	.005	.07	.31	.38
1/2" bar	.166	Ea.	.006	.13	.33	.46
5/8" bar	.166	Ea.	.006	.20	.37	.57
3/4" bar	.166	Ea.	.006	.20	.37	.57

2 | FOUNDATIONS — 08 | Block Wall Systems

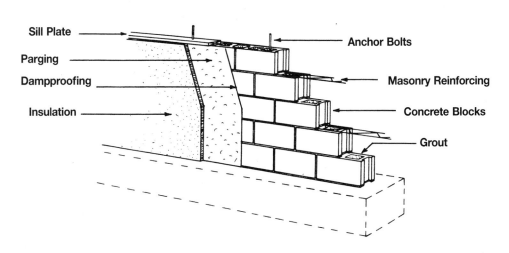

System Description	QUAN.	UNIT	LABOR HOURS	COST PER S.F. MAT.	COST PER S.F. INST.	COST PER S.F. TOTAL
8" WALL, GROUTED, FULL HEIGHT						
Concrete block, 8" x 16" x 8"	1.000	S.F.	.094	3.56	4.72	8.28
Masonry reinforcing, every second course	.750	L.F.	.002	.21	.11	.32
Parging, plastering with portland cement plaster, 1 coat	1.000	S.F.	.014	.26	.76	1.02
Dampproofing, bituminous coating, 1 coat	1.000	S.F.	.012	.25	.64	.89
Insulation, 1" rigid polystyrene	1.000	S.F.	.010	.62	.57	1.19
Grout, solid, pumped	1.000	S.F.	.059	1.31	2.91	4.22
Anchor bolts, 1/2" diameter, 8" long, 4' O.C.	.060	Ea.	.004	.09	.20	.29
Sill plate, 2" x 4", treated	.250	L.F.	.007	.14	.41	.55
TOTAL		S.F.	.202	6.44	10.32	16.76
12" WALL, GROUTED, FULL HEIGHT						
Concrete block, 8" x 16" x 12"	1.000	S.F.	.160	5.35	7.85	13.20
Masonry reinforcing, every second course	.750	L.F.	.003	.14	.17	.31
Parging, plastering with portland cement plaster, 1 coat	1.000	S.F.	.014	.26	.76	1.02
Dampproofing, bituminous coating, 1 coat	1.000	S.F.	.012	.25	.64	.89
Insulation, 1" rigid polystyrene	1.000	S.F.	.010	.62	.57	1.19
Grout, solid, pumped	1.000	S.F.	.063	2.15	3.08	5.23
Anchor bolts, 1/2" diameter, 8" long, 4' O.C.	.060	Ea.	.004	.09	.20	.29
Sill plate, 2" x 4", treated	.250	L.F.	.007	.14	.41	.55
TOTAL		S.F.	.273	9	13.68	22.68

The costs in this system are based on a square foot of wall. Do not subtract for window or door openings.

Description	QUAN.	UNIT	LABOR HOURS	COST PER S.F. MAT.	COST PER S.F. INST.	COST PER S.F. TOTAL

Block Wall Price Sheet

Block Wall Price Sheet	QUAN.	UNIT	LABOR HOURS	COST PER S.F. MAT.	COST PER S.F. INST.	COST PER S.F. TOTAL
Concrete, block, 8" x 16" x, 6" thick	1.000	S.F.	.089	3.32	4.41	7.73
8" thick	1.000	S.F.	.093	3.56	4.72	8.28
10" thick	1.000	S.F.	.095	4.07	5.75	9.82
12" thick	1.000	S.F.	.122	5.35	7.85	13.20
Solid block, 8" x 16" x, 6" thick	1.000	S.F.	.091	3.38	4.56	7.94
8" thick	1.000	S.F.	.096	4.72	4.83	9.55
10" thick	1.000	S.F.	.096	4.72	4.83	9.55
12" thick	1.000	S.F.	.126	6.75	6.75	13.50
Masonry reinforcing, wire strips, to 8" wide, every course	1.500	L.F.	.004	.42	.21	.63
Every 2nd course	.750	L.F.	.002	.21	.11	.32
Every 3rd course	.500	L.F.	.001	.14	.07	.21
Every 4th course	.400	L.F.	.001	.11	.06	.17
Wire strips to 12" wide, every course	1.500	L.F.	.006	.27	.33	.60
Every 2nd course	.750	L.F.	.003	.14	.17	.31
Every 3rd course	.500	L.F.	.002	.09	.11	.20
Every 4th course	.400	L.F.	.002	.07	.09	.16
Parging, plastering with portland cement plaster, 1 coat	1.000	S.F.	.014	.26	.76	1.02
2 coats	1.000	S.F.	.022	.40	1.16	1.56
Dampproofing, bituminous, brushed on, 1 coat	1.000	S.F.	.012	.25	.64	.89
2 coats	1.000	S.F.	.016	.49	.85	1.34
Sprayed on, 1 coat	1.000	S.F.	.010	.25	.51	.76
2 coats	1.000	S.F.	.016	.48	.85	1.33
Troweled on, 1/16" thick	1.000	S.F.	.016	.39	.85	1.24
1/8" thick	1.000	S.F.	.020	.70	1.06	1.76
1/2" thick	1.000	S.F.	.023	2.27	1.21	3.48
Insulation, rigid, fiberglass, 1.5#/C.F., unfaced						
1-1/2" thick R 6.2	1.000	S.F.	.008	.42	.45	.87
2" thick R 8.5	1.000	S.F.	.008	.54	.45	.99
3" thick R 13	1.000	S.F.	.010	.62	.57	1.19
Perlite, 1" thick R 2.77	1.000	S.F.	.010	.48	.57	1.05
2" thick R 5.55	1.000	S.F.	.011	.85	.62	1.47
Polystyrene, extruded, 1" thick R 5.4	1.000	S.F.	.010	.62	.57	1.19
2" thick R 10.8	1.000	S.F.	.011	1.74	.62	2.36
Molded 1" thick R 3.85	1.000	S.F.	.010	.29	.57	.86
2" thick R 7.7	1.000	S.F.	.011	.86	.62	1.48
Grout, concrete block cores, 6" thick	1.000	S.F.	.044	.98	2.19	3.17
8" thick	1.000	S.F.	.059	1.31	2.91	4.22
10" thick	1.000	S.F.	.061	1.73	2.99	4.72
12" thick	1.000	S.F.	.063	2.15	3.08	5.23
Anchor bolts, 2' on center, 1/2" diameter, 8" long	.120	Ea.	.005	.18	.39	.57
12" long	.120	Ea.	.005	.19	.40	.59
3/4" diameter, 8" long	.120	Ea.	.006	.50	.41	.91
12" long	.120	Ea.	.006	.63	.42	1.05
4' on center, 1/2" diameter, 8" long	.060	Ea.	.002	.09	.20	.29
12" long	.060	Ea.	.003	.10	.20	.30
3/4" diameter, 8" long	.060	Ea.	.003	.25	.20	.45
12" long	.060	Ea.	.003	.32	.21	.53
Sill plates, treated, 2" x 4"	.250	L.F.	.007	.14	.41	.55
4" x 4"	.250	L.F.	.007	.37	.39	.76

2 | FOUNDATIONS — 12 | Concrete Wall Systems

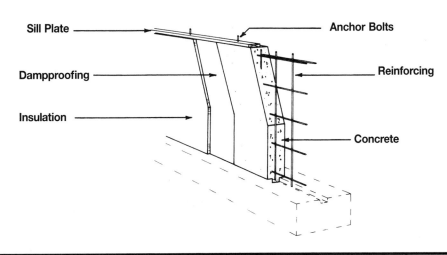

System Description	QUAN.	UNIT	LABOR HOURS	COST PER S.F. MAT.	COST PER S.F. INST.	COST PER S.F. TOTAL
8" THICK, POURED CONCRETE WALL						
Concrete, 8" thick, 3000 psi	.025	C.Y.		2.95		2.95
Forms, prefabricated plywood, up to 8' high	2.000	SFCA	.120	1.98	5.96	7.94
Reinforcing, light	.670	Lb.	.004	.36	.21	.57
Placing concrete, direct chute	.025	C.Y.	.013		.60	.60
Dampproofing, brushed on, 2 coats	1.000	S.F.	.016	.49	.85	1.34
Rigid insulation, 1" polystyrene	1.000	S.F.	.010	.62	.57	1.19
Anchor bolts, 1/2" diameter, 12" long, 4' O.C.	.060	Ea.	.004	.10	.20	.30
Sill plates, 2" x 4", treated	.250	L.F.	.007	.14	.41	.55
TOTAL		S.F.	.174	6.64	8.80	15.44
12" THICK, POURED CONCRETE WALL						
Concrete, 12" thick, 3000 psi	.040	C.Y.		4.72		4.72
Forms, prefabricated plywood, up to 8' high	2.000	SFCA	.120	1.98	5.96	7.94
Reinforcing, light	1.000	Lb.	.005	.53	.32	.85
Placing concrete, direct chute	.040	C.Y.	.019		.87	.87
Dampproofing, brushed on, 2 coats	1.000	S.F.	.016	.49	.85	1.34
Rigid insulation, 1" polystyrene	1.000	S.F.	.010	.62	.57	1.19
Anchor bolts, 1/2" diameter, 12" long, 4' O.C.	.060	Ea.	.004	.10	.20	.30
Sill plates, 2" x 4" treated	.250	L.F.	.007	.14	.41	.55
TOTAL		S.F.	.181	8.58	9.18	17.76

The costs in this system are based on sq. ft. of wall. Do not subtract for window and door openings. The costs assume a 4' high wall.

Description	QUAN.	UNIT	LABOR HOURS	COST PER S.F. MAT.	COST PER S.F. INST.	COST PER S.F. TOTAL

Concrete Wall Price Sheet	QUAN.	UNIT	LABOR HOURS	COST PER S.F.		
				MAT.	INST.	TOTAL
Formwork, prefabricated plywood, up to 8' high	2.000	SFCA	.081	1.98	5.95	7.93
Over 8' to 16' high	2.000	SFCA	.076	2.08	7.95	10.03
Job built forms, 1 use per month	2.000	SFCA	.320	5.90	12.90	18.80
4 uses per month	2.000	SFCA	.221	2.20	9.45	11.65
Reinforcing, 8" wall, light reinforcing	.670	Lb.	.004	.36	.21	.57
Heavy reinforcing	1.500	Lb.	.008	.80	.48	1.28
10" wall, light reinforcing	.850	Lb.	.005	.45	.27	.72
Heavy reinforcing	2.000	Lb.	.011	1.06	.64	1.70
12" wall light reinforcing	1.000	Lb.	.005	.53	.32	.85
Heavy reinforcing	2.250	Lb.	.012	1.19	.72	1.91
Placing concrete, 8" wall, direct chute	.025	C.Y.	.013		.60	.60
Pumped concrete	.025	C.Y.	.016		.95	.95
Crane & bucket	.025	C.Y.	.023		1.46	1.46
10" wall, direct chute	.030	C.Y.	.016		.71	.71
Pumped concrete	.030	C.Y.	.019		1.13	1.13
Crane & bucket	.030	C.Y.	.027		1.75	1.75
12" wall, direct chute	.040	C.Y.	.019		.87	.87
Pumped concrete	.040	C.Y.	.023		1.38	1.38
Crane & bucket	.040	C.Y.	.032		2.07	2.07
Dampproofing, bituminous, brushed on, 1 coat	1.000	S.F.	.012	.25	.64	.89
2 coats	1.000	S.F.	.016	.49	.85	1.34
Sprayed on, 1 coat	1.000	S.F.	.010	.25	.51	.76
2 coats	1.000	S.F.	.016	.48	.85	1.33
Troweled on, 1/16" thick	1.000	S.F.	.016	.39	.85	1.24
1/8" thick	1.000	S.F.	.020	.70	1.06	1.76
1/2" thick	1.000	S.F.	.023	2.27	1.21	3.48
Insulation rigid, fiberglass, 1.5#/C.F., unfaced						
1-1/2" thick, R 6.2	1.000	S.F.	.008	.42	.45	.87
2" thick, R 8.3	1.000	S.F.	.008	.54	.45	.99
3" thick, R 12.4	1.000	S.F.	.010	.62	.57	1.19
Perlite, 1" thick R 2.77	1.000	S.F.	.010	.48	.57	1.05
2" thick R 5.55	1.000	S.F.	.011	.85	.62	1.47
Polystyrene, extruded, 1" thick R 5.40	1.000	S.F.	.010	.62	.57	1.19
2" thick R 10.8	1.000	S.F.	.011	1.74	.62	2.36
Molded, 1" thick R 3.85	1.000	S.F.	.010	.29	.57	.86
2" thick R 7.70	1.000	S.F.	.011	.86	.62	1.48
Anchor bolts, 2' on center, 1/2" diameter, 8" long	.120	Ea.	.005	.18	.39	.57
12" long	.120	Ea.	.005	.19	.40	.59
3/4" diameter, 8" long	.120	Ea.	.006	.50	.41	.91
12" long	.120	Ea.	.006	.63	.42	1.05
Sill plates, treated lumber, 2" x 4"	.250	L.F.	.007	.14	.41	.55
4" x 4"	.250	L.F.	.007	.37	.39	.76

2 | FOUNDATIONS — 16 | Wood Wall Foundation Systems

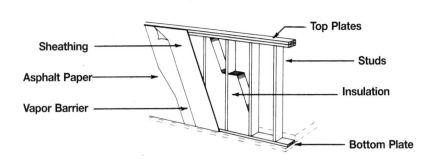

System Description	QUAN.	UNIT	LABOR HOURS	COST PER S.F. MAT.	COST PER S.F. INST.	COST PER S.F. TOTAL
2" X 4" STUDS, 16" O.C., WALL						
Studs, 2" x 4", 16" O.C., treated	1.000	L.F.	.015	.58	.83	1.41
Plates, double top plate, single bottom plate, treated, 2" x 4"	.750	L.F.	.011	.44	.62	1.06
Sheathing, 1/2", exterior grade, CDX, treated	1.000	S.F.	.014	.90	.81	1.71
Asphalt paper, 15# roll	1.100	S.F.	.002	.07	.13	.20
Vapor barrier, 4 mil polyethylene	1.000	S.F.	.002	.04	.12	.16
Fiberglass insulation, 3-1/2" thick	1.000	S.F.	.007	.36	.40	.76
TOTAL		S.F.	.051	2.39	2.91	5.30
2" X 6" STUDS, 16" O.C., WALL						
Studs, 2" x 6", 16" O.C., treated	1.000	L.F.	.016	.85	.91	1.76
Plates, double top plate, single bottom plate, treated, 2" x 6"	.750	L.F.	.012	.64	.68	1.32
Sheathing, 5/8" exterior grade, CDX, treated	1.000	S.F.	.015	1.38	.86	2.24
Asphalt paper, 15# roll	1.100	S.F.	.002	.07	.13	.20
Vapor barrier, 4 mil polyethylene	1.000	S.F.	.002	.04	.12	.16
Fiberglass insulation, 6" thick	1.000	S.F.	.007	.48	.40	.88
TOTAL		S.F.	.054	3.46	3.10	6.56
2" X 8" STUDS, 16" O.C., WALL						
Studs, 2" x 8", 16" O.C. treated	1.000	L.F.	.018	1.16	1.01	2.17
Plates, double top plate, single bottom plate, treated, 2" x 8"	.750	L.F.	.013	.87	.76	1.63
Sheathing, 3/4" exterior grade, CDX, treated	1.000	S.F.	.016	1.46	.93	2.39
Asphalt paper, 15# roll	1.100	S.F.	.002	.07	.13	.20
Vapor barrier, 4 mil polyethylene	1.000	S.F.	.002	.04	.12	.16
Fiberglass insulation, 9" thick	1.000	S.F.	.006	.80	.34	1.14
TOTAL		S.F.	.057	4.40	3.29	7.69

The costs in this system are based on a sq. ft. of wall area. Do not subtract for window or door openings. The costs assume a 4' high wall.

Description	QUAN.	UNIT	LABOR HOURS	COST PER S.F. MAT.	COST PER S.F. INST.	COST PER S.F. TOTAL

Wood Wall Foundation Price Sheet	QUAN.	UNIT	LABOR HOURS	COST PER S.F.		
				MAT.	INST.	TOTAL
Studs, treated, 2" x 4", 12" O.C.	1.250	L.F.	.018	.73	1.04	1.77
16" O.C.	1.000	L.F.	.015	.58	.83	1.41
2" x 6", 12" O.C.	1.250	L.F.	.020	1.06	1.14	2.20
16" O.C.	1.000	L.F.	.016	.85	.91	1.76
2" x 8", 12" O.C.	1.250	L.F.	.022	1.45	1.26	2.71
16" O.C.	1.000	L.F.	.018	1.16	1.01	2.17
Plates, treated double top single bottom, 2" x 4"	.750	L.F.	.011	.44	.62	1.06
2" x 6"	.750	L.F.	.012	.64	.68	1.32
2" x 8"	.750	L.F.	.013	.87	.76	1.63
Sheathing, treated exterior grade CDX, 1/2" thick	1.000	S.F.	.014	.90	.81	1.71
5/8" thick	1.000	S.F.	.015	1.38	.86	2.24
3/4" thick	1.000	S.F.	.016	1.46	.93	2.39
Asphalt paper, 15# roll	1.100	S.F.	.002	.07	.13	.20
Vapor barrier, polyethylene, 4 mil	1.000	S.F.	.002	.03	.12	.15
10 mil	1.000	S.F.	.002	.09	.12	.21
Insulation, rigid, fiberglass, 1.5#/C.F., unfaced	1.000	S.F.	.008	.32	.45	.77
1-1/2" thick, R 6.2	1.000	S.F.	.008	.42	.45	.87
2" thick, R 8.3	1.000	S.F.	.008	.54	.45	.99
3" thick, R 12.4	1.000	S.F.	.010	.63	.58	1.21
Perlite 1" thick, R 2.77	1.000	S.F.	.010	.48	.57	1.05
2" thick, R 5.55	1.000	S.F.	.011	.85	.62	1.47
Polystyrene, extruded, 1" thick, R 5.40	1.000	S.F.	.010	.62	.57	1.19
2" thick, R 10.8	1.000	S.F.	.011	1.74	.62	2.36
Molded 1" thick, R 3.85	1.000	S.F.	.010	.29	.57	.86
2" thick, R 7.7	1.000	S.F.	.011	.86	.62	1.48
Non rigid, batts, fiberglass, paper backed, 3-1/2" thick roll, R 11	1.000	S.F.	.005	.36	.40	.76
6", R 19	1.000	S.F.	.006	.48	.40	.88
9", R 30	1.000	S.F.	.006	.80	.34	1.14
12", R 38	1.000	S.F.	.006	1.16	.34	1.50
Mineral fiber, paper backed, 3-1/2", R 13	1.000	S.F.	.005	.74	.28	1.02
6", R 19	1.000	S.F.	.005	1.16	.28	1.44
10", R 30	1.000	S.F.	.006	1.53	.34	1.87

2 | FOUNDATIONS — 20 | Floor Slab Systems

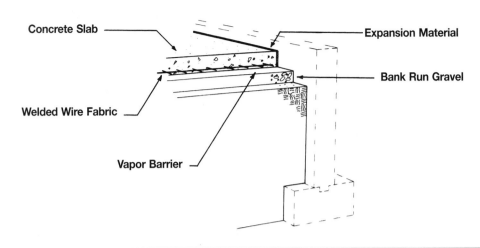

System Description	QUAN.	UNIT	LABOR HOURS	COST PER S.F. MAT.	COST PER S.F. INST.	COST PER S.F. TOTAL
4" THICK SLAB						
Concrete, 4" thick, 3000 psi concrete	.012	C.Y.		1.42		1.42
Place concrete, direct chute	.012	C.Y.	.005		.24	.24
Bank run gravel, 4" deep	1.000	S.F.	.001	.36	.06	.42
Polyethylene vapor barrier, .006" thick	1.000	S.F.	.002	.04	.12	.16
Edge forms, expansion material	.100	L.F.	.005	.04	.26	.30
Welded wire fabric, 6 x 6, 10/10 (W1.4/W1.4)	1.100	S.F.	.005	.17	.30	.47
Steel trowel finish	1.000	S.F.	.014		.71	.71
TOTAL		S.F.	.032	2.03	1.69	3.72
6" THICK SLAB						
Concrete, 6" thick, 3000 psi concrete	.019	C.Y.		2.24		2.24
Place concrete, direct chute	.019	C.Y.	.008		.37	.37
Bank run gravel, 4" deep	1.000	S.F.	.001	.36	.06	.42
Polyethylene vapor barrier, .006" thick	1.000	S.F.	.002	.04	.12	.16
Edge forms, expansion material	.100	L.F.	.005	.04	.26	.30
Welded wire fabric, 6 x 6, 10/10 (W1.4/W1.4)	1.100	S.F.	.005	.17	.30	.47
Steel trowel finish	1.000	S.F.	.014		.71	.71
TOTAL		S.F.	.035	2.85	1.82	4.67

The slab costs in this section are based on a cost per square foot of floor area.

Description	QUAN.	UNIT	LABOR HOURS	COST PER S.F. MAT.	COST PER S.F. INST.	COST PER S.F. TOTAL

Floor Slab Price Sheet

	QUAN.	UNIT	LABOR HOURS	COST PER S.F. MAT.	COST PER S.F. INST.	COST PER S.F. TOTAL
Concrete, 4" thick slab, 2000 psi concrete	.012	C.Y.		1.36		1.36
2500 psi concrete	.012	C.Y.		1.38		1.38
3000 psi concrete	.012	C.Y.		1.42		1.42
3500 psi concrete	.012	C.Y.		1.45		1.45
4000 psi concrete	.012	C.Y.		1.49		1.49
4500 psi concrete	.012	C.Y.		1.52		1.52
5" thick slab, 2000 psi concrete	.015	C.Y.		1.70		1.70
2500 psi concrete	.015	C.Y.		1.73		1.73
3000 psi concrete	.015	C.Y.		1.77		1.77
3500 psi concrete	.015	C.Y.		1.82		1.82
4000 psi concrete	.015	C.Y.		1.86		1.86
4500 psi concrete	.015	C.Y.		1.91		1.91
6" thick slab, 2000 psi concrete	.019	C.Y.		2.15		2.15
2500 psi concrete	.019	C.Y.		2.19		2.19
3000 psi concrete	.019	C.Y.		2.24		2.24
3500 psi concrete	.019	C.Y.		2.30		2.30
4000 psi concrete	.019	C.Y.		2.36		2.36
4500 psi concrete	.019	C.Y.		2.41		2.41
Place concrete, 4" slab, direct chute	.012	C.Y.	.005		.24	.24
Pumped concrete	.012	C.Y.	.006		.34	.34
Crane & bucket	.012	C.Y.	.008		.51	.51
5" slab, direct chute	.015	C.Y.	.007		.30	.30
Pumped concrete	.015	C.Y.	.007		.43	.43
Crane & bucket	.015	C.Y.	.010		.63	.63
6" slab, direct chute	.019	C.Y.	.008		.37	.37
Pumped concrete	.019	C.Y.	.009		.55	.55
Crane & bucket	.019	C.Y.	.012		.80	.80
Gravel, bank run, 4" deep	1.000	S.F.	.001	.36	.06	.42
6" deep	1.000	S.F.	.001	.48	.08	.56
9" deep	1.000	S.F.	.001	.71	.11	.82
12" deep	1.000	S.F.	.001	.97	.13	1.10
3/4" crushed stone, 3" deep	1.000	S.F.	.001	.29	.06	.35
6" deep	1.000	S.F.	.001	.58	.11	.69
9" deep	1.000	S.F.	.002	.84	.15	.99
12" deep	1.000	S.F.	.002	1.48	.18	1.66
Vapor barrier polyethylene, .004" thick	1.000	S.F.	.002	.03	.12	.15
.006" thick	1.000	S.F.	.002	.04	.12	.16
Edge forms, expansion material, 4" thick slab	.100	L.F.	.004	.02	.17	.19
6" thick slab	.100	L.F.	.005	.04	.26	.30
Welded wire fabric 6 x 6, 10/10 (W1.4/W1.4)	1.100	S.F.	.005	.17	.30	.47
6 x 6, 6/6 (W2.9/W2.9)	1.100	S.F.	.006	.28	.36	.64
4 x 4, 10/10 (W1.4/W1.4)	1.100	S.F.	.006	.24	.34	.58
Finish concrete, screed finish	1.000	S.F.	.009		.29	.29
Float finish	1.000	S.F.	.011		.29	.29
Steel trowel, for resilient floor	1.000	S.F.	.013		.92	.92
For finished floor	1.000	S.F.	.015		.71	.71

Division 3 - Framing

Did you know?
RSMeans Online gives you the same access to RSMeans' data with 24/7 access:
- Quickly locate costs in the searchable database.
- Build cost lists, estimates, and reports in minutes.
- Adjust costs to any location in the U.S. and Canada with the click of a button.

Start your free trial today at **www.RSMeansOnline.com**

RSMeans Online
FROM THE GORDIAN GROUP®

No part of this cost data may be reproduced, stored in a retrieval system, or transmitted in any form or by any means without prior written permission of RSMeans.

3 | FRAMING — 02 | Floor Framing Systems

Diagram labels: Box Sill, Sheathing, Bridging, Furring, Girder, Box Sill, Wood Joists

System Description	QUAN.	UNIT	LABOR HOURS	MAT.	INST.	TOTAL
2" X 8", 16" O.C.						
Wood joists, 2" x 8", 16" O.C.	1.000	L.F.	.015	.93	.83	1.76
Bridging, 1" x 3", 6' O.C.	.080	Pr.	.005	.06	.28	.34
Box sills, 2" x 8"	.150	L.F.	.002	.14	.12	.26
Concrete filled steel column, 4" diameter	.125	L.F.	.002	.27	.14	.41
Girder, built up from three 2" x 8"	.125	L.F.	.013	.35	.75	1.10
Sheathing, plywood, subfloor, 5/8" CDX	1.000	S.F.	.012	.84	.67	1.51
Furring, 1" x 3", 16" O.C.	1.000	L.F.	.023	.43	1.30	1.73
Joist hangers	.036	Ea.	.002	.05	.10	.15
TOTAL		S.F.	.074	3.07	4.19	7.26
2" X 10", 16" O.C.						
Wood joists, 2" x 10", 16" OC	1.000	L.F.	.018	1.47	1.01	2.48
Bridging, 1" x 3", 6' OC	.080	Pr.	.005	.06	.28	.34
Box sills, 2" x 10"	.150	L.F.	.003	.22	.15	.37
Concrete filled steel column, 4" diameter	.125	L.F.	.002	.27	.14	.41
5/8" thick	.125	L.F.	.014	.55	.81	1.36
Sheathing, plywood, subfloor, 5/8" CDX	1.000	S.F.	.012	.84	.67	1.51
Furring, 1" x 3", 16" OC	1.000	L.F.	.023	.43	1.30	1.73
Joist hangers	.036	Ea.	.002	.05	.10	.15
TOTAL		S.F.	.079	3.89	4.46	8.35
2" X 12", 16" O.C.						
Wood joists, 2" x 12", 16" O.C.	1.000	L.F.	.018	1.87	1.04	2.91
Bridging, 1" x 3", 6' O.C.	.080	Pr.	.005	.06	.28	.34
Box sills, 2" x 12"	.150	L.F.	.003	.28	.16	.44
Concrete filled steel column, 4" diameter	.125	L.F.	.002	.27	.14	.41
Girder, built up from three 2" x 12"	.125	L.F.	.015	.71	.86	1.57
Sheathing, plywood, subfloor, 5/8" CDX	1.000	S.F.	.012	.84	.67	1.51
Furring, 1" x 3", 16" O.C.	1.000	L.F.	.023	.43	1.30	1.73
Joist hangers	.036	Ea.	.002	.05	.10	.15
TOTAL		S.F.	.080	4.51	4.55	9.06

Floor costs on this page are given on a cost per square foot basis.

Description	QUAN.	UNIT	LABOR HOURS	MAT.	INST.	TOTAL

Floor Framing Price Sheet (Wood)

	QUAN.	UNIT	LABOR HOURS	COST PER S.F. MAT.	COST PER S.F. INST.	COST PER S.F. TOTAL
Joists, #2 or better, pine, 2" x 4", 12" O.C.	1.250	L.F.	.016	.55	.91	1.46
16" O.C.	1.000	L.F.	.013	.44	.73	1.17
2" x 6", 12" O.C.	1.250	L.F.	.016	.84	.91	1.75
16" O.C.	1.000	L.F.	.013	.67	.73	1.40
2" x 8", 12" O.C.	1.250	L.F.	.018	1.16	1.04	2.20
16" O.C.	1.000	L.F.	.015	.93	.83	1.76
2" x 10", 12" O.C.	1.250	L.F.	.022	1.84	1.26	3.10
16" O.C.	1.000	L.F.	.018	1.47	1.01	2.48
2" x 12", 12" O.C.	1.250	L.F.	.023	2.34	1.30	3.64
16" O.C.	1.000	L.F.	.018	1.87	1.04	2.91
Bridging, wood 1" x 3", joists 12" O.C.	.100	Pr.	.006	.07	.35	.42
16" O.C.	.080	Pr.	.005	.06	.28	.34
Metal, galvanized, joists 12" O.C.	.100	Pr.	.006	.10	.35	.45
16" O.C.	.080	Pr.	.005	.08	.28	.36
Compression type, joists 12" O.C.	.100	Pr.	.004	.10	.23	.33
16" O.C.	.080	Pr.	.003	.08	.18	.26
Box sills, #2 or better, 2" x 4"	.150	L.F.	.002	.07	.11	.18
2" x 6"	.150	L.F.	.002	.10	.11	.21
2" x 8"	.150	L.F.	.002	.14	.12	.26
2" x 10"	.150	L.F.	.003	.22	.15	.37
2" x 12"	.150	L.F.	.003	.28	.16	.44
Girders, including lally columns, 3 pieces spiked together, 2" x 8"	.125	L.F.	.015	.62	.89	1.51
2" x 10"	.125	L.F.	.016	.82	.95	1.77
2" x 12"	.125	L.F.	.017	.98	1	1.98
Solid girders, 3" x 8"	.040	L.F.	.004	.39	.23	.62
3" x 10"	.040	L.F.	.004	.43	.25	.68
3" x 12"	.040	L.F.	.004	.46	.25	.71
4" x 8"	.040	L.F.	.004	.40	.25	.65
4" x 10"	.040	L.F.	.004	.48	.27	.75
4" x 12"	.040	L.F.	.004	.51	.28	.79
Steel girders, bolted & including fabrication, wide flange shapes						
12" deep, 14#/l.f.	.040	L.F.	.003	1.04	.26	1.30
10" deep, 15#/l.f.	.040	L.F.	.003	1.04	.26	1.30
8" deep, 10#/l.f.	.040	L.F.	.003	.70	.26	.96
6" deep, 9#/l.f.	.040	L.F.	.003	.63	.26	.89
5" deep, 16#/l.f.	.040	L.F.	.003	1.04	.26	1.30
Sheathing, plywood exterior grade CDX, 1/2" thick	1.000	S.F.	.011	.75	.65	1.40
5/8" thick	1.000	S.F.	.012	.84	.67	1.51
3/4" thick	1.000	S.F.	.013	1	.73	1.73
Boards, 1" x 8" laid regular	1.000	S.F.	.016	2.07	.91	2.98
Laid diagonal	1.000	S.F.	.019	2.07	1.07	3.14
1" x 10" laid regular	1.000	S.F.	.015	2.11	.83	2.94
Laid diagonal	1.000	S.F.	.018	2.11	1.01	3.12
Furring, 1" x 3", 12" O.C.	1.250	L.F.	.029	.54	1.63	2.17
16" O.C.	1.000	L.F.	.023	.43	1.30	1.73
24" O.C.	.750	L.F.	.017	.32	.98	1.30

3 | FRAMING — 04 Floor Framing Systems

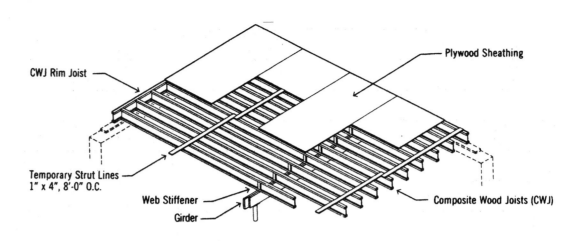

System Description	QUAN.	UNIT	LABOR HOURS	COST PER S.F. MAT.	COST PER S.F. INST.	COST PER S.F. TOTAL
9-1/2" COMPOSITE WOOD JOISTS, 16" O.C.						
CWJ, 9-1/2", 16" O.C., 15' span	1.000	L.F.	.018	2.05	1	3.05
Temp. strut line, 1" x 4", 8' O.C.	.160	L.F.	.003	.08	.18	.26
CWJ rim joist, 9-1/2"	.150	L.F.	.003	.31	.15	.46
Concrete filled steel column, 4" diameter	.125	L.F.	.002	.27	.14	.41
Girder, built up from three 2" x 8"	.125	L.F.	.013	.35	.75	1.10
Sheathing, plywood, subfloor, 5/8" CDX	1.000	S.F.	.012	.84	.67	1.51
TOTAL		S.F.	.051	3.90	2.89	6.79
11-1/2" COMPOSITE WOOD JOISTS, 16" O.C.						
CWJ, 11-1/2", 16" O.C., 18' span	1.000	L.F.	.018	2.25	1.03	3.28
Temp. strut line, 1" x 4", 8' O.C.	.160	L.F.	.003	.08	.18	.26
CWJ rim joist, 11-1/2"	.150	L.F.	.003	.34	.15	.49
Concrete filled steel column, 4" diameter	.125	L.F.	.002	.27	.14	.41
Girder, built up from three 2" x 10"	.125	L.F.	.014	.55	.81	1.36
Sheathing, plywood, subfloor, 5/8" CDX	1.000	S.F.	.012	.84	.67	1.51
TOTAL		S.F.	.052	4.33	2.98	7.31
14" COMPOSITE WOOD JOISTS, 16" O.C.						
CWJ, 14", 16" O.C., 22' span	1.000	L.F.	.020	2.35	1.10	3.45
Temp. strut line, 1" x 4", 8' O.C.	.160	L.F.	.003	.08	.18	.26
CWJ rim joist, 14"	.150	L.F.	.003	.35	.17	.52
Concrete filled steel column, 4" diameter	.600	L.F.	.002	.27	.14	.41
Girder, built up from three 2" x 12"	.600	L.F.	.015	.71	.86	1.57
Sheathing, plywood, subfloor, 5/8" CDX	1.000	S.F.	.012	.84	.67	1.51
TOTAL		S.F.	.055	4.60	3.12	7.72

Floor costs on this page are given on a cost per square foot basis.

Description	QUAN.	UNIT	LABOR HOURS	COST PER S.F. MAT.	COST PER S.F. INST.	COST PER S.F. TOTAL

Floor Framing Price Sheet (Wood)	QUAN.	UNIT	LABOR HOURS	COST PER S.F. MAT.	COST PER S.F. INST.	COST PER S.F. TOTAL
Composite wood joist 9-1/2" deep, 12" O.C.	1.250	L.F.	.022	2.56	1.25	3.81
16" O.C.	1.000	L.F.	.018	2.05	1	3.05
11-1/2" deep, 12" O.C.	1.250	L.F.	.023	2.81	1.28	4.09
16" O.C.	1.000	L.F.	.018	2.25	1.03	3.28
14" deep, 12" O.C.	1.250	L.F.	.024	2.94	1.38	4.32
16" O.C.	1.000	L.F.	.020	2.35	1.10	3.45
16" deep, 12" O.C.	1.250	L.F.	.026	4.94	1.47	6.41
16" O.C.	1.000	L.F.	.021	3.95	1.18	5.13
CWJ rim joist, 9-1/2"	.150	L.F.	.003	.31	.15	.46
11-1/2"	.150	L.F.	.003	.34	.15	.49
14"	.150	L.F.	.003	.35	.17	.52
16"	.150	L.F.	.003	.59	.18	.77
Girders, including lally columns, 3 pieces spiked together, 2" x 8"	.125	L.F.	.015	.62	.89	1.51
2" x 10"	.125	L.F.	.016	.82	.95	1.77
2" x 12"	.125	L.F.	.017	.98	1	1.98
Solid girders, 3" x 8"	.040	L.F.	.004	.39	.23	.62
3" x 10"	.040	L.F.	.004	.43	.25	.68
3" x 12"	.040	L.F.	.004	.46	.25	.71
4" x 8"	.040	L.F.	.004	.40	.25	.65
4" x 10"	.040	L.F.	.004	.48	.27	.75
4" x 12"	.040	L.F.	.004	.51	.28	.79
Steel girders, bolted & including fabrication, wide flange shapes						
12" deep, 14#/l.f.	.040	L.F.	.061	24	6.10	30.10
10" deep, 15#/l.f.	.040	L.F.	.067	26	6.65	32.65
8" deep, 10#/l.f.	.040	L.F.	.067	17.50	6.65	24.15
6" deep, 9#/l.f.	.040	L.F.	.067	15.75	6.65	22.40
5" deep, 16#/l.f.	.040	L.F.	.064	25	6.45	31.45
Sheathing, plywood exterior grade CDX, 1/2" thick	1.000	S.F.	.011	.75	.65	1.40
5/8" thick	1.000	S.F.	.012	.84	.67	1.51
3/4" thick	1.000	S.F.	.013	1	.73	1.73
Boards, 1" x 8" laid regular	1.000	S.F.	.016	2.07	.91	2.98
Laid diagonal	1.000	S.F.	.019	2.07	1.07	3.14
1" x 10" laid regular	1.000	S.F.	.015	2.11	.83	2.94
Laid diagonal	1.000	S.F.	.018	2.11	1.01	3.12
Furring, 1" x 3", 12" O.C.	1.250	L.F.	.029	.54	1.63	2.17
16" O.C.	1.000	L.F.	.023	.43	1.30	1.73
24" O.C.	.750	L.F.	.017	.32	.98	1.30

3 | FRAMING   06 | Floor Framing Systems

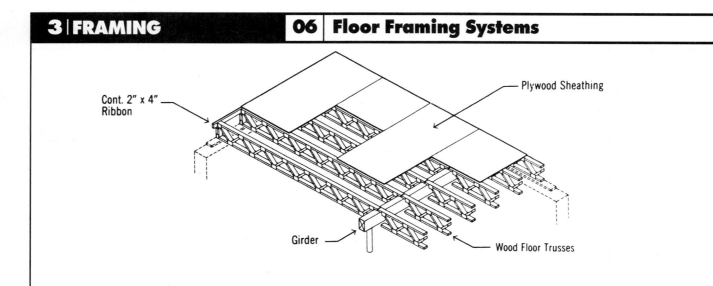

System Description	QUAN.	UNIT	LABOR HOURS	COST PER S.F. MAT.	COST PER S.F. INST.	COST PER S.F. TOTAL
12" OPEN WEB JOISTS, 16" O.C.						
OWJ 12", 16" O.C., 21' span	1.000	L.F.	.018	3.65	1.03	4.68
Continuous ribbing, 2" x 4"	.150	L.F.	.002	.07	.11	.18
Concrete filled steel column, 4" diameter	.125	L.F.	.002	.27	.14	.41
Girder, built up from three 2" x 8"	.125	L.F.	.013	.35	.75	1.10
Sheathing, plywood, subfloor, 5/8" CDX	1.000	S.F.	.012	.84	.67	1.51
Furring, 1" x 3", 16" O.C.	1.000	L.F.	.023	.43	1.30	1.73
TOTAL		S.F.	.070	5.61	4	9.61
14" OPEN WEB WOOD JOISTS, 16" O.C.						
OWJ 14", 16" O.C., 22' span	1.000	L.F.	.020	4	1.10	5.10
Continuous ribbing, 2" x 4"	.150	L.F.	.002	.07	.11	.18
Concrete filled steel column, 4" diameter	.125	L.F.	.002	.27	.14	.41
Girder, built up from three 2" x 10"	.125	L.F.	.014	.55	.81	1.36
Sheathing, plywood, subfloor, 5/8" CDX	1.000	S.F.	.012	.84	.67	1.51
Furring, 1" x 3", 16" O.C.	1.000	L.F.	.023	.43	1.30	1.73
TOTAL		S.F.	.073	6.16	4.13	10.29
16" OPEN WEB WOOD JOISTS, 16" O.C.						
OWJ 16", 16" O.C., 24' span	1.000	L.F.	.021	4.18	1.18	5.36
Continuous ribbing, 2" x 4"	.150	L.F.	.002	.07	.11	.18
Concrete filled steel column, 4" diameter	.125	L.F.	.002	.27	.14	.41
Girder, built up from three 2" x 12"	.125	L.F.	.015	.71	.86	1.57
Sheathing, plywood, subfloor, 5/8" CDX	1.000	S.F.	.012	.84	.67	1.51
Furring, 1" x 3", 16" O.C.	1.000	L.F.	.023	.43	1.30	1.73
TOTAL		S.F.	.075	6.50	4.26	10.76

Floor costs on this page are given on a cost per square foot basis.

Description	QUAN.	UNIT	LABOR HOURS	COST PER S.F. MAT.	COST PER S.F. INST.	COST PER S.F. TOTAL

Floor Framing Price Sheet (Wood)	QUAN.	UNIT	LABOR HOURS	COST PER S.F.		
				MAT.	INST.	TOTAL
Open web joists, 12" deep, 12" O.C.	1.250	L.F.	.023	4.56	1.28	5.84
16" O.C.	1.000	L.F.	.018	3.65	1.03	4.68
14" deep, 12" O.C.	1.250	L.F.	.024	5	1.38	6.38
16" O.C.	1.000	L.F.	.020	4	1.10	5.10
16" deep, 12" O.C.	1.250	L.F.	.026	5.20	1.47	6.67
16" O.C.	1.000	L.F.	.021	4.18	1.18	5.36
18" deep, 12" O.C.	1.250	L.F.	.027	5.30	1.53	6.83
16" O.C.	1.000	L.F.	.022	4.25	1.23	5.48
Continuous ribbing, 2" x 4"	.150	L.F.	.002	.07	.11	.18
2" x 6"	.150	L.F.	.002	.10	.11	.21
2" x 8"	.150	L.F.	.002	.14	.12	.26
2" x 10"	.150	L.F.	.003	.22	.15	.37
2" x 12"	.150	L.F.	.003	.28	.16	.44
Girders, including lally columns, 3 pieces spiked together, 2" x 8"	.125	L.F.	.015	.62	.89	1.51
2" x 10"	.125	L.F.	.016	.82	.95	1.77
2" x 12"	.125	L.F.	.017	.98	1	1.98
Solid girders, 3" x 8"	.040	L.F.	.004	.39	.23	.62
3" x 10"	.040	L.F.	.004	.43	.25	.68
3" x 12"	.040	L.F.	.004	.46	.25	.71
4" x 8"	.040	L.F.	.004	.40	.25	.65
4" x 10"	.040	L.F.	.004	.48	.27	.75
4" x 12"	.040	L.F.	.004	.51	.28	.79
Steel girders, bolted & including fabrication, wide flange shapes						
12" deep, 14#/l.f.	.040	L.F.	.061	24	6.10	30.10
10" deep, 15#/l.f.	.040	L.F.	.067	26	6.65	32.65
8" deep, 10#/l.f.	.040	L.F.	.067	17.50	6.65	24.15
6" deep, 9#/l.f.	.040	L.F.	.067	15.75	6.65	22.40
5" deep, 16#/l.f.	.040	L.F.	.064	25	6.45	31.45
Sheathing, plywood exterior grade CDX, 1/2" thick	1.000	S.F.	.011	.75	.65	1.40
5/8" thick	1.000	S.F.	.012	.84	.67	1.51
3/4" thick	1.000	S.F.	.013	1	.73	1.73
Boards, 1" x 8" laid regular	1.000	S.F.	.016	2.07	.91	2.98
Laid diagonal	1.000	S.F.	.019	2.07	1.07	3.14
1" x 10" laid regular	1.000	S.F.	.015	2.11	.83	2.94
Laid diagonal	1.000	S.F.	.018	2.11	1.01	3.12
Furring, 1" x 3", 12" O.C.	1.250	L.F.	.029	.54	1.63	2.17
16" O.C.	1.000	L.F.	.023	.43	1.30	1.73
24" O.C.	.750	L.F.	.017	.32	.98	1.30

3 | FRAMING 08 | Exterior Wall Framing Systems

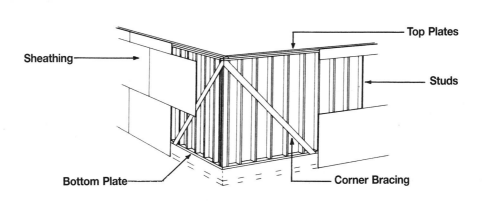

System Description	QUAN.	UNIT	LABOR HOURS	COST PER S.F. MAT.	COST PER S.F. INST.	COST PER S.F. TOTAL
2" X 4", 16" O.C.						
2" x 4" studs, 16" O.C.	1.000	L.F.	.015	.43	.83	1.26
Plates, 2" x 4", double top, single bottom	.375	L.F.	.005	.16	.31	.47
Corner bracing, let-in, 1" x 6"	.063	L.F.	.003	.05	.19	.24
Sheathing, 1/2" plywood, CDX	1.000	S.F.	.011	.75	.65	1.40
Framing connectors, holddowns	.013	Ea.	.013	.36	.74	1.10
TOTAL		S.F.	.047	1.75	2.72	4.47
2" X 4", 24" O.C.						
2" x 4" studs, 24" O.C.	.750	L.F.	.011	.32	.62	.94
Plates, 2" x 4", double top, single bottom	.375	L.F.	.005	.16	.31	.47
Corner bracing, let-in, 1" x 6"	.063	L.F.	.002	.05	.12	.17
Sheathing, 1/2" plywood, CDX	1.000	S.F.	.011	.75	.65	1.40
Framing connectors, holddowns	.013	Ea.	.013	.36	.74	1.10
TOTAL		S.F.	.042	1.64	2.44	4.08
2" X 6", 16" O.C.						
2" x 6" studs, 16" O.C.	1.000	L.F.	.016	.67	.91	1.58
Plates, 2" x 6", double top, single bottom	.375	L.F.	.006	.25	.34	.59
Corner bracing, let-in, 1" x 6"	.063	L.F.	.003	.05	.19	.24
Sheathing, 1/2" plywood, CDX	1.000	S.F.	.014	.75	.81	1.56
Framing connectors, holddowns	.013	Ea.	.013	.36	.74	1.10
TOTAL		S.F.	.052	2.08	2.99	5.07
2" X 6", 24" O.C.						
2" x 6" studs, 24" O.C.	.750	L.F.	.012	.50	.68	1.18
Plates, 2" x 6", double top, single bottom	.375	L.F.	.006	.25	.34	.59
Corner bracing, let-in, 1" x 6"	.063	L.F.	.002	.05	.12	.17
Sheathing, 1/2" plywood, CDX	1.000	S.F.	.011	.75	.65	1.40
Framing connectors, holddowns	.013	Ea.	.013	.36	.74	1.10
TOTAL		S.F.	.044	1.91	2.53	4.44

The wall costs on this page are given in cost per square foot of wall.
For window and door openings see below.

Description	QUAN.	UNIT	LABOR HOURS	COST PER S.F. MAT.	COST PER S.F. INST.	COST PER S.F. TOTAL

Exterior Wall Framing Price Sheet	QUAN.	UNIT	LABOR HOURS	COST PER S.F.		
				MAT.	INST.	TOTAL
Studs, #2 or better, 2" x 4", 12" O.C.	1.250	L.F.	.018	.54	1.04	1.58
16" O.C.	1.000	L.F.	.015	.43	.83	1.26
24" O.C.	.750	L.F.	.011	.32	.62	.94
32" O.C.	.600	L.F.	.009	.26	.50	.76
2" x 6", 12" O.C.	1.250	L.F.	.020	.84	1.14	1.98
16" O.C.	1.000	L.F.	.016	.67	.91	1.58
24" O.C.	.750	L.F.	.012	.50	.68	1.18
32" O.C.	.600	L.F.	.010	.40	.55	.95
2" x 8", 12" O.C.	1.250	L.F.	.025	1.58	1.43	3.01
16" O.C.	1.000	L.F.	.020	1.26	1.14	2.40
24" O.C.	.750	L.F.	.015	.95	.86	1.81
32" O.C.	.600	L.F.	.012	.76	.68	1.44
Plates, #2 or better, double top, single bottom, 2" x 4"	.375	L.F.	.005	.16	.31	.47
2" x 6"	.375	L.F.	.006	.25	.34	.59
2" x 8"	.375	L.F.	.008	.47	.43	.90
Corner bracing, let-in 1" x 6" boards, studs, 12" O.C.	.070	L.F.	.004	.05	.21	.26
16" O.C.	.063	L.F.	.003	.05	.19	.24
24" O.C.	.063	L.F.	.002	.05	.12	.17
32" O.C.	.057	L.F.	.002	.04	.11	.15
Let-in steel ("T" shape), studs, 12" O.C.	.070	L.F.	.001	.06	.05	.11
16" O.C.	.063	L.F.	.001	.06	.05	.11
24" O.C.	.063	L.F.	.001	.06	.05	.11
32" O.C.	.057	L.F.	.001	.05	.04	.09
Sheathing, plywood CDX, 3/8" thick	1.000	S.F.	.010	.68	.60	1.28
1/2" thick	1.000	S.F.	.011	.75	.65	1.40
5/8" thick	1.000	S.F.	.012	.84	.70	1.54
3/4" thick	1.000	S.F.	.013	1	.76	1.76
Boards, 1" x 6", laid regular	1.000	S.F.	.025	1.69	1.40	3.09
Laid diagonal	1.000	S.F.	.027	1.69	1.55	3.24
1" x 8", laid regular	1.000	S.F.	.021	2.07	1.19	3.26
Laid diagonal	1.000	S.F.	.025	2.07	1.40	3.47
Wood fiber, regular, no vapor barrier, 1/2" thick	1.000	S.F.	.013	.70	.76	1.46
5/8" thick	1.000	S.F.	.013	.84	.76	1.60
Asphalt impregnated 25/32" thick	1.000	S.F.	.013	.36	.76	1.12
1/2" thick	1.000	S.F.	.013	.28	.76	1.04
Polystyrene, regular, 3/4" thick	1.000	S.F.	.010	.62	.57	1.19
2" thick	1.000	S.F.	.011	1.74	.62	2.36
Fiberglass, foil faced, 1" thick	1.000	S.F.	.008	.99	.45	1.44
2" thick	1.000	S.F.	.009	1.86	.51	2.37

Window & Door Openings	QUAN.	UNIT	LABOR HOURS	COST EACH		
				MAT.	INST.	TOTAL
The following costs are to be added to the total costs of the wall for each opening. Do not subtract the area of the openings.						
Headers, 2" x 6" double, 2' long	4.000	L.F.	.178	2.68	10.10	12.78
3' long	6.000	L.F.	.267	4.02	15.10	19.12
4' long	8.000	L.F.	.356	5.35	20	25.35
5' long	10.000	L.F.	.444	6.70	25	31.70
2" x 8" double, 4' long	8.000	L.F.	.376	7.45	21.50	28.95
5' long	10.000	L.F.	.471	9.30	26.50	35.80
6' long	12.000	L.F.	.565	11.15	32	43.15
8' long	16.000	L.F.	.753	14.90	42.50	57.40
2" x 10" double, 4' long	8.000	L.F.	.400	11.75	22.50	34.25
6' long	12.000	L.F.	.600	17.65	34	51.65
8' long	16.000	L.F.	.800	23.50	45.50	69
10' long	20.000	L.F.	1.000	29.50	57	86.50
2" x 12" double, 8' long	16.000	L.F.	.853	30	48.50	78.50
12' long	24.000	L.F.	1.280	45	72.50	117.50

3 | FRAMING — 12 | Gable End Roof Framing Systems

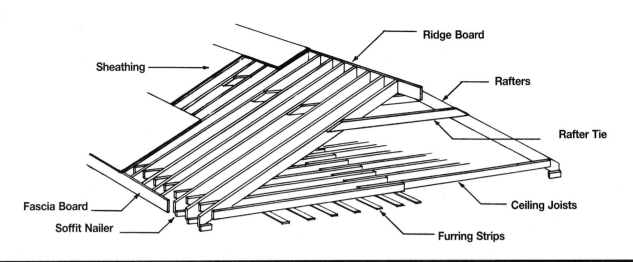

System Description	QUAN.	UNIT	LABOR HOURS	COST PER S.F. MAT.	COST PER S.F. INST.	COST PER S.F. TOTAL
2″ X 6″ RAFTERS, 16″ O.C., 4/12 PITCH						
Rafters, 2″ x 6″, 16″ O.C., 4/12 pitch	1.170	L.F.	.019	.78	1.06	1.84
Ceiling joists, 2″ x 4″, 16″ O.C.	1.000	L.F.	.013	.44	.73	1.17
Ridge board, 2″ x 6″	.050	L.F.	.002	.03	.09	.12
Fascia board, 2″ x 6″	.100	L.F.	.005	.07	.30	.37
Rafter tie, 1″ x 4″, 4′ O.C.	.060	L.F.	.001	.03	.07	.10
Soffit nailer (outrigger), 2″ x 4″, 24″ O.C.	.170	L.F.	.004	.07	.25	.32
Sheathing, exterior, plywood, CDX, 1/2″ thick	1.170	S.F.	.013	.88	.76	1.64
Furring strips, 1″ x 3″, 16″ O.C.	1.000	L.F.	.023	.43	1.30	1.73
Rafter ties	.053	Ea.	.003	.08	.17	.25
TOTAL		S.F.	.083	2.81	4.73	7.54
2″ X 8″ RAFTERS, 16″ O.C., 4/12 PITCH						
Rafters, 2″ x 8″, 16″ O.C., 4/12 pitch	1.170	L.F.	.020	1.09	1.12	2.21
Ceiling joists, 2″ x 6″, 16″ O.C.	1.000	L.F.	.013	.67	.73	1.40
Ridge board, 2″ x 8″	.050	L.F.	.002	.05	.10	.15
Fascia board, 2″ x 8″	.100	L.F.	.007	.09	.40	.49
Rafter tie, 1″ x 4″, 4′ O.C.	.060	L.F.	.001	.03	.07	.10
Soffit nailer (outrigger), 2″ x 4″, 24″ O.C.	.170	L.F.	.004	.07	.25	.32
Sheathing, exterior, plywood, CDX, 1/2″ thick	1.170	S.F.	.013	.88	.76	1.64
Furring strips, 1″ x 3″, 16″ O.C.	1.000	L.F.	.023	.43	1.30	1.73
Rafter ties	.053	Ea.	.003	.08	.17	.25
TOTAL		S.F.	.086	3.39	4.90	8.29

The cost of this system is based on the square foot of plan area.
All quantities have been adjusted accordingly.

Description	QUAN.	UNIT	LABOR HOURS	COST PER S.F. MAT.	COST PER S.F. INST.	COST PER S.F. TOTAL

Gable End Roof Framing Price Sheet

	QUAN.	UNIT	LABOR HOURS	COST PER S.F. MAT.	COST PER S.F. INST.	COST PER S.F. TOTAL
Rafters, #2 or better, 16" O.C., 2" x 6", 4/12 pitch	1.170	L.F.	.019	.78	1.06	1.84
8/12 pitch	1.330	L.F.	.027	.89	1.52	2.41
2" x 8", 4/12 pitch	1.170	L.F.	.020	1.09	1.12	2.21
8/12 pitch	1.330	L.F.	.028	1.24	1.61	2.85
2" x 10", 4/12 pitch	1.170	L.F.	.030	1.72	1.68	3.40
8/12 pitch	1.330	L.F.	.043	1.96	2.43	4.39
24" O.C., 2" x 6", 4/12 pitch	.940	L.F.	.015	.63	.86	1.49
8/12 pitch	1.060	L.F.	.021	.71	1.21	1.92
2" x 8", 4/12 pitch	.940	L.F.	.016	.87	.90	1.77
8/12 pitch	1.060	L.F.	.023	.99	1.28	2.27
2" x 10", 4/12 pitch	.940	L.F.	.024	1.38	1.35	2.73
8/12 pitch	1.060	L.F.	.034	1.56	1.94	3.50
Ceiling joist, #2 or better, 2" x 4", 16" O.C.	1.000	L.F.	.013	.44	.73	1.17
24" O.C.	.750	L.F.	.010	.33	.55	.88
2" x 6", 16" O.C.	1.000	L.F.	.013	.67	.73	1.40
24" O.C.	.750	L.F.	.010	.50	.55	1.05
2" x 8", 16" O.C.	1.000	L.F.	.015	.93	.83	1.76
24" O.C.	.750	L.F.	.011	.70	.62	1.32
2" x 10", 16" O.C.	1.000	L.F.	.018	1.47	1.01	2.48
24" O.C.	.750	L.F.	.013	1.10	.76	1.86
Ridge board, #2 or better, 1" x 6"	.050	L.F.	.001	.04	.08	.12
1" x 8"	.050	L.F.	.001	.07	.08	.15
1" x 10"	.050	L.F.	.002	.08	.09	.17
2" x 6"	.050	L.F.	.002	.03	.09	.12
2" x 8"	.050	L.F.	.002	.05	.10	.15
2" x 10"	.050	L.F.	.002	.07	.11	.18
Fascia board, #2 or better, 1" x 6"	.100	L.F.	.004	.05	.22	.27
1" x 8"	.100	L.F.	.005	.06	.26	.32
1" x 10"	.100	L.F.	.005	.07	.29	.36
2" x 6"	.100	L.F.	.006	.07	.32	.39
2" x 8"	.100	L.F.	.007	.09	.40	.49
2" x 10"	.100	L.F.	.004	.29	.20	.49
Rafter tie, #2 or better, 4' O.C., 1" x 4"	.060	L.F.	.001	.03	.07	.10
1" x 6"	.060	L.F.	.001	.03	.08	.11
2" x 4"	.060	L.F.	.002	.04	.09	.13
2" x 6"	.060	L.F.	.002	.05	.11	.16
Soffit nailer (outrigger), 2" x 4", 16" O.C.	.220	L.F.	.006	.10	.32	.42
24" O.C.	.170	L.F.	.004	.07	.25	.32
2" x 6", 16" O.C.	.220	L.F.	.006	.11	.37	.48
24" O.C.	.170	L.F.	.005	.09	.29	.38
Sheathing, plywood CDX, 4/12 pitch, 3/8" thick.	1.170	S.F.	.012	.80	.70	1.50
1/2" thick	1.170	S.F.	.013	.88	.76	1.64
5/8" thick	1.170	S.F.	.014	.98	.82	1.80
8/12 pitch, 3/8"	1.330	S.F.	.014	.90	.80	1.70
1/2" thick	1.330	S.F.	.015	1	.86	1.86
5/8" thick	1.330	S.F.	.016	1.12	.93	2.05
Boards, 4/12 pitch roof, 1" x 6"	1.170	S.F.	.026	1.98	1.46	3.44
1" x 8"	1.170	S.F.	.021	2.42	1.22	3.64
8/12 pitch roof, 1" x 6"	1.330	S.F.	.029	2.25	1.66	3.91
1" x 8"	1.330	S.F.	.024	2.75	1.38	4.13
Furring, 1" x 3", 12" O.C.	1.200	L.F.	.027	.52	1.56	2.08
16" O.C.	1.000	L.F.	.023	.43	1.30	1.73
24" O.C.	.800	L.F.	.018	.34	1.04	1.38

3 | FRAMING 16 | Truss Roof Framing Systems

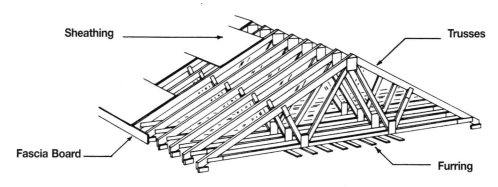

System Description	QUAN.	UNIT	LABOR HOURS	COST PER S.F. MAT.	COST PER S.F. INST.	COST PER S.F. TOTAL
TRUSS, 16" O.C., 4/12 PITCH, 1' OVERHANG, 26' SPAN						
Truss, 40# loading, 16" O.C., 4/12 pitch, 26' span	.030	Ea.	.021	2.64	1.46	4.10
Fascia board, 2" x 6"	.100	L.F.	.005	.07	.30	.37
Sheathing, exterior, plywood, CDX, 1/2" thick	1.170	S.F.	.013	.88	.76	1.64
Furring, 1" x 3", 16" O.C.	1.000	L.F.	.023	.43	1.30	1.73
Rafter ties	.053	Ea.	.003	.08	.17	.25
TOTAL		S.F.	.065	4.10	3.99	8.09
TRUSS, 16" O.C., 8/12 PITCH, 1' OVERHANG, 26' SPAN						
Truss, 40# loading, 16" O.C., 8/12 pitch, 26' span	.030	Ea.	.023	3.39	1.61	5
Fascia board, 2" x 6"	.100	L.F.	.005	.07	.30	.37
Sheathing, exterior, plywood, CDX, 1/2" thick	1.330	S.F.	.015	1	.86	1.86
Furring, 1" x 3", 16" O.C.	1.000	L.F.	.023	.43	1.30	1.73
Rafter ties	.053	Ea.	.003	.08	.17	.25
TOTAL		S.F.	.069	4.97	4.24	9.21
TRUSS, 24" O.C., 4/12 PITCH, 1' OVERHANG, 26' SPAN						
Truss, 40# loading, 24" O.C., 4/12 pitch, 26' span	.020	Ea.	.014	1.76	.97	2.73
Fascia board, 2" x 6"	.100	L.F.	.005	.07	.30	.37
Sheathing, exterior, plywood, CDX, 1/2" thick	1.170	S.F.	.013	.88	.76	1.64
Furring, 1" x 3", 16" O.C.	1.000	L.F.	.023	.43	1.30	1.73
Rafter ties	.035	Ea.	.002	.06	.11	.17
TOTAL		S.F.	.057	3.20	3.44	6.64
TRUSS, 24" O.C., 8/12 PITCH, 1' OVERHANG, 26' SPAN						
Truss, 40# loading, 24" O.C., 8/12 pitch, 26' span	.020	Ea.	.015	2.26	1.07	3.33
Fascia board, 2" x 6"	.100	L.F.	.005	.07	.30	.37
Sheathing, exterior, plywood, CDX, 1/2" thick	1.330	S.F.	.015	1	.86	1.86
Furring, 1" x 3", 16" O.C.	1.000	L.F.	.023	.43	1.30	1.73
Rafter ties	.035	Ea.	.002	.06	.11	.17
TOTAL		S.F.	.060	3.82	3.64	7.46

The cost of this system is based on the square foot of plan area.
A one foot overhang is included.

Description	QUAN.	UNIT	LABOR HOURS	COST PER S.F. MAT.	COST PER S.F. INST.	COST PER S.F. TOTAL

Truss Roof Framing Price Sheet	QUAN.	UNIT	LABOR HOURS	COST PER S.F.		
				MAT.	INST.	TOTAL
Truss, 40# loading, including 1' overhang, 4/12 pitch, 24' span, 16" O.C.	.033	Ea.	.022	2.79	1.52	4.31
24" O.C.	.022	Ea.	.015	1.86	1.02	2.88
26' span, 16" O.C.	.030	Ea.	.021	2.64	1.46	4.10
24" O.C.	.020	Ea.	.014	1.76	.97	2.73
28' span, 16" O.C.	.027	Ea.	.020	2.75	1.41	4.16
24" O.C.	.019	Ea.	.014	1.94	.99	2.93
32' span, 16" O.C.	.024	Ea.	.019	2.90	1.33	4.23
24" O.C.	.016	Ea.	.013	1.94	.89	2.83
36' span, 16" O.C.	.022	Ea.	.019	3.08	1.33	4.41
24" O.C.	.015	Ea.	.013	2.10	.91	3.01
8/12 pitch, 24' span, 16" O.C.	.033	Ea.	.024	3.53	1.65	5.18
24" O.C.	.022	Ea.	.016	2.35	1.10	3.45
26' span, 16" O.C.	.030	Ea.	.023	3.39	1.61	5
24" O.C.	.020	Ea.	.015	2.26	1.07	3.33
28' span, 16" O.C.	.027	Ea.	.022	3.48	1.52	5
24" O.C.	.019	Ea.	.016	2.45	1.07	3.52
32' span, 16" O.C.	.024	Ea.	.021	3.50	1.48	4.98
24" O.C.	.016	Ea.	.014	2.34	.99	3.33
36' span, 16" O.C.	.022	Ea.	.021	3.96	1.49	5.45
24" O.C.	.015	Ea.	.015	2.70	1.01	3.71
Fascia board, #2 or better, 1" x 6"	.100	L.F.	.004	.05	.22	.27
1" x 8"	.100	L.F.	.005	.06	.26	.32
1" x 10"	.100	L.F.	.005	.07	.29	.36
2" x 6"	.100	L.F.	.006	.07	.32	.39
2" x 8"	.100	L.F.	.007	.09	.40	.49
2" x 10"	.100	L.F.	.009	.15	.51	.66
Sheathing, plywood CDX, 4/12 pitch, 3/8" thick	1.170	S.F.	.012	.80	.70	1.50
1/2" thick	1.170	S.F.	.013	.88	.76	1.64
5/8" thick	1.170	S.F.	.014	.98	.82	1.80
8/12 pitch, 3/8" thick	1.330	S.F.	.014	.90	.80	1.70
1/2" thick	1.330	S.F.	.015	1	.86	1.86
5/8" thick	1.330	S.F.	.016	1.12	.93	2.05
Boards, 4/12 pitch, 1" x 6"	1.170	S.F.	.026	1.98	1.46	3.44
1" x 8"	1.170	S.F.	.021	2.42	1.22	3.64
8/12 pitch, 1" x 6"	1.330	S.F.	.029	2.25	1.66	3.91
1" x 8"	1.330	S.F.	.024	2.75	1.38	4.13
Furring, 1" x 3", 12" O.C.	1.200	L.F.	.027	.52	1.56	2.08
16" O.C.	1.000	L.F.	.023	.43	1.30	1.73
24" O.C.	.800	L.F.	.018	.34	1.04	1.38

3 | FRAMING — 20 | Hip Roof Framing Systems

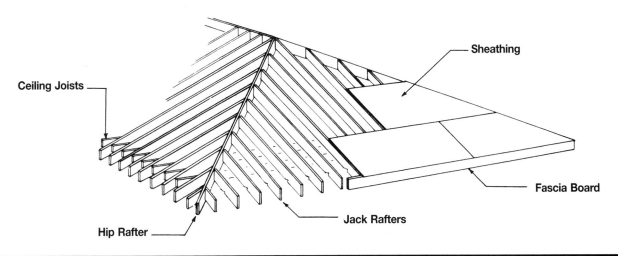

System Description	QUAN.	UNIT	LABOR HOURS	COST PER S.F. MAT.	COST PER S.F. INST.	COST PER S.F. TOTAL
2" X 6", 16" O.C., 4/12 PITCH						
Hip rafters, 2" x 8", 4/12 pitch	.160	L.F.	.004	.15	.20	.35
Jack rafters, 2" x 6", 16" O.C., 4/12 pitch	1.430	L.F.	.038	.96	2.16	3.12
Ceiling joists, 2" x 6", 16" O.C.	1.000	L.F.	.013	.67	.73	1.40
Fascia board, 2" x 8"	.220	L.F.	.016	.20	.89	1.09
Soffit nailer (outrigger), 2" x 4", 24" O.C.	.220	L.F.	.006	.10	.32	.42
Sheathing, 1/2" exterior plywood, CDX	1.570	S.F.	.018	1.18	1.02	2.20
Furring strips, 1" x 3", 16" O.C.	1.000	L.F.	.023	.43	1.30	1.73
Rafter ties	.070	Ea.	.004	.11	.22	.33
TOTAL		S.F.	.122	3.80	6.84	10.64
2" X 8", 16" O.C., 4/12 PITCH						
Hip rafters, 2" x 10", 4/12 pitch	.160	L.F.	.004	.24	.25	.49
Jack rafters, 2" x 8", 16" O.C., 4/12 pitch	1.430	L.F.	.047	1.33	2.65	3.98
Ceiling joists, 2" x 6", 16" O.C.	1.000	L.F.	.013	.67	.73	1.40
Fascia board, 2" x 8"	.220	L.F.	.012	.16	.69	.85
Soffit nailer (outrigger), 2" x 4", 24" O.C.	.220	L.F.	.006	.10	.32	.42
Sheathing, 1/2" exterior plywood, CDX	1.570	S.F.	.018	1.18	1.02	2.20
Furring strips, 1" x 3", 16" O.C.	1.000	L.F.	.023	.43	1.30	1.73
Rafter ties	.070	Ea.	.004	.11	.22	.33
TOTAL		S.F.	.127	4.22	7.18	11.40

The cost of this system is based on S.F. of plan area. Measurement is area under the hip roof only. See gable roof system for added costs.

Description	QUAN.	UNIT	LABOR HOURS	COST PER S.F. MAT.	COST PER S.F. INST.	COST PER S.F. TOTAL

Hip Roof Framing Price Sheet	QUAN.	UNIT	LABOR HOURS	COST PER S.F.		
				MAT.	INST.	TOTAL
Hip rafters, #2 or better, 2" x 6", 4/12 pitch	.160	L.F.	.003	.11	.19	.30
8/12 pitch	.210	L.F.	.006	.14	.33	.47
2" x 8", 4/12 pitch	.160	L.F.	.004	.15	.20	.35
8/12 pitch	.210	L.F.	.006	.20	.35	.55
2" x 10", 4/12 pitch	.160	L.F.	.004	.24	.25	.49
8/12 pitch roof	.210	L.F.	.008	.31	.43	.74
Jack rafters, #2 or better, 16" O.C., 2" x 6", 4/12 pitch	1.430	L.F.	.038	.96	2.16	3.12
8/12 pitch	1.800	L.F.	.061	1.21	3.44	4.65
2" x 8", 4/12 pitch	1.430	L.F.	.047	1.33	2.65	3.98
8/12 pitch	1.800	L.F.	.075	1.67	4.25	5.92
2" x 10", 4/12 pitch	1.430	L.F.	.051	2.10	2.89	4.99
8/12 pitch	1.800	L.F.	.082	2.65	4.66	7.31
24" O.C., 2" x 6", 4/12 pitch	1.150	L.F.	.031	.77	1.74	2.51
8/12 pitch	1.440	L.F.	.048	.96	2.75	3.71
2" x 8", 4/12 pitch	1.150	L.F.	.038	1.07	2.13	3.20
8/12 pitch	1.440	L.F.	.060	1.34	3.40	4.74
2" x 10", 4/12 pitch	1.150	L.F.	.041	1.69	2.32	4.01
8/12 pitch	1.440	L.F.	.066	2.12	3.73	5.85
Ceiling joists, #2 or better, 2" x 4", 16" O.C.	1.000	L.F.	.013	.44	.73	1.17
24" O.C.	.750	L.F.	.010	.33	.55	.88
2" x 6", 16" O.C.	1.000	L.F.	.013	.67	.73	1.40
24" O.C.	.750	L.F.	.010	.50	.55	1.05
2" x 8", 16" O.C.	1.000	L.F.	.015	.93	.83	1.76
24" O.C.	.750	L.F.	.011	.70	.62	1.32
2" x 10", 16" O.C.	1.000	L.F.	.018	1.47	1.01	2.48
24" O.C.	.750	L.F.	.013	1.10	.76	1.86
Fascia board, #2 or better, 1" x 6"	.220	L.F.	.009	.11	.49	.60
1" x 8"	.220	L.F.	.010	.13	.57	.70
1" x 10"	.220	L.F.	.011	.15	.64	.79
2" x 6"	.220	L.F.	.013	.16	.71	.87
2" x 8"	.220	L.F.	.016	.20	.89	1.09
2" x 10"	.220	L.F.	.020	.32	1.11	1.43
Soffit nailer (outrigger), 2" x 4", 16" O.C.	.280	L.F.	.007	.12	.41	.53
24" O.C.	.220	L.F.	.006	.10	.32	.42
2" x 8", 16" O.C.	.280	L.F.	.007	.20	.38	.58
24" O.C.	.220	L.F.	.005	.16	.31	.47
Sheathing, plywood CDX, 4/12 pitch, 3/8" thick	1.570	S.F.	.016	1.07	.94	2.01
1/2" thick	1.570	S.F.	.018	1.18	1.02	2.20
5/8" thick	1.570	S.F.	.019	1.32	1.10	2.42
8/12 pitch, 3/8" thick	1.900	S.F.	.020	1.29	1.14	2.43
1/2" thick	1.900	S.F.	.022	1.43	1.24	2.67
5/8" thick	1.900	S.F.	.023	1.60	1.33	2.93
Boards, 4/12 pitch, 1" x 6" boards	1.450	S.F.	.032	2.45	1.81	4.26
1" x 8" boards	1.450	S.F.	.027	3	1.51	4.51
8/12 pitch, 1" x 6" boards	1.750	S.F.	.039	2.96	2.19	5.15
1" x 8" boards	1.750	S.F.	.032	3.62	1.82	5.44
Furring, 1" x 3", 12" O.C.	1.200	L.F.	.027	.52	1.56	2.08
16" O.C.	1.000	L.F.	.023	.43	1.30	1.73
24" O.C.	.800	L.F.	.018	.34	1.04	1.38

3 | FRAMING — 24 | Gambrel Roof Framing Systems

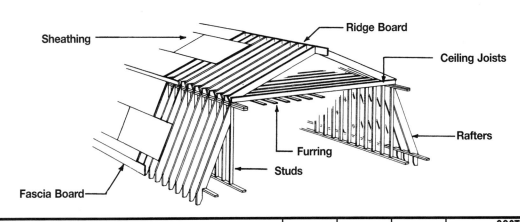

System Description	QUAN.	UNIT	LABOR HOURS	COST PER S.F. MAT.	COST PER S.F. INST.	COST PER S.F. TOTAL
2" X 6" RAFTERS, 16" O.C.						
Roof rafters, 2" x 6", 16" O.C.	1.430	L.F.	.029	.96	1.63	2.59
Ceiling joists, 2" x 6", 16" O.C.	.710	L.F.	.009	.48	.52	1
Stud wall, 2" x 4", 16" O.C., including plates	.790	L.F.	.012	.35	.70	1.05
Furring strips, 1" x 3", 16" O.C.	.710	L.F.	.016	.31	.92	1.23
Ridge board, 2" x 8"	.050	L.F.	.002	.05	.10	.15
Fascia board, 2" x 6"	.100	L.F.	.006	.07	.32	.39
Sheathing, exterior grade plywood, 1/2" thick	1.450	S.F.	.017	1.09	.94	2.03
Rafter ties	.106	Ea.	.006	.17	.33	.50
TOTAL		S.F.	.097	3.48	5.46	8.94
2" X 8" RAFTERS, 16" O.C.						
Roof rafters, 2" x 8", 16" O.C.	1.430	L.F.	.031	1.33	1.73	3.06
Ceiling joists, 2" x 6", 16" O.C.	.710	L.F.	.009	.48	.52	1
Stud wall, 2" x 4", 16" O.C., including plates	.790	L.F.	.012	.35	.70	1.05
Furring strips, 1" x 3", 16" O.C.	.710	L.F.	.016	.31	.92	1.23
Ridge board, 2" x 8"	.050	L.F.	.002	.05	.10	.15
Fascia board, 2" x 8"	.100	L.F.	.007	.09	.40	.49
Sheathing, exterior grade plywood, 1/2" thick	1.450	S.F.	.017	1.09	.94	2.03
Rafter ties	.106	Ea.	.006	.17	.33	.50
TOTAL		S.F.	.100	3.87	5.64	9.51

The cost of this system is based on the square foot of plan area on the first floor.

Description	QUAN.	UNIT	LABOR HOURS	COST PER S.F. MAT.	COST PER S.F. INST.	COST PER S.F. TOTAL

Gambrel Roof Framing Price Sheet

	QUAN.	UNIT	LABOR HOURS	COST PER S.F. MAT.	COST PER S.F. INST.	COST PER S.F. TOTAL
Roof rafters, #2 or better, 2" x 6", 16" O.C.	1.430	L.F.	.029	.96	1.63	2.59
24" O.C.	1.140	L.F.	.023	.76	1.30	2.06
2" x 8", 16" O.C.	1.430	L.F.	.031	1.33	1.73	3.06
24" O.C.	1.140	L.F.	.024	1.06	1.38	2.44
2" x 10", 16" O.C.	1.430	L.F.	.046	2.10	2.62	4.72
24" O.C.	1.140	L.F.	.037	1.68	2.09	3.77
Ceiling joist, #2 or better, 2" x 4", 16" O.C.	.710	L.F.	.009	.31	.52	.83
24" O.C.	.570	L.F.	.007	.25	.42	.67
2" x 6", 16" O.C.	.710	L.F.	.009	.48	.52	1
24" O.C.	.570	L.F.	.007	.38	.42	.80
2" x 8", 16" O.C.	.710	L.F.	.010	.66	.59	1.25
24" O.C.	.570	L.F.	.008	.53	.47	1
Stud wall, #2 or better, 2" x 4", 16" O.C.	.790	L.F.	.012	.35	.70	1.05
24" O.C.	.630	L.F.	.010	.28	.56	.84
2" x 6", 16" O.C.	.790	L.F.	.014	.53	.81	1.34
24" O.C.	.630	L.F.	.011	.42	.64	1.06
Furring, 1" x 3", 16" O.C.	.710	L.F.	.016	.31	.92	1.23
24" O.C.	.590	L.F.	.013	.25	.77	1.02
Ridge board, #2 or better, 1" x 6"	.050	L.F.	.001	.04	.08	.12
1" x 8"	.050	L.F.	.001	.07	.08	.15
1" x 10"	.050	L.F.	.002	.08	.09	.17
2" x 6"	.050	L.F.	.002	.03	.09	.12
2" x 8"	.050	L.F.	.002	.05	.10	.15
2" x 10"	.050	L.F.	.002	.07	.11	.18
Fascia board, #2 or better, 1" x 6"	.100	L.F.	.004	.05	.22	.27
1" x 8"	.100	L.F.	.005	.06	.26	.32
1" x 10"	.100	L.F.	.005	.07	.29	.36
2" x 6"	.100	L.F.	.006	.07	.32	.39
2" x 8"	.100	L.F.	.007	.09	.40	.49
2" x 10"	.100	L.F.	.009	.15	.51	.66
Sheathing, plywood, exterior grade CDX, 3/8" thick	1.450	S.F.	.015	.99	.87	1.86
1/2" thick	1.450	S.F.	.017	1.09	.94	2.03
5/8" thick	1.450	S.F.	.018	1.22	1.02	2.24
3/4" thick	1.450	S.F.	.019	1.45	1.10	2.55
Boards, 1" x 6", laid regular	1.450	S.F.	.032	2.45	1.81	4.26
Laid diagonal	1.450	S.F.	.036	2.45	2.03	4.48
1" x 8", laid regular	1.450	S.F.	.027	3	1.51	4.51
Laid diagonal	1.450	S.F.	.032	3	1.81	4.81

3 | FRAMING — 28 | Mansard Roof Framing Systems

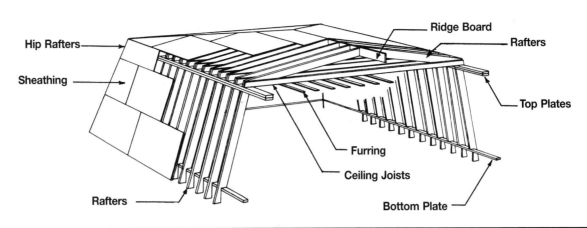

System Description	QUAN.	UNIT	LABOR HOURS	COST PER S.F. MAT.	COST PER S.F. INST.	COST PER S.F. TOTAL
2" X 6" RAFTERS, 16" O.C.						
Roof rafters, 2" x 6", 16" O.C.	1.210	L.F.	.033	.81	1.86	2.67
Rafter plates, 2" x 6", double top, single bottom	.364	L.F.	.010	.24	.56	.80
Ceiling joists, 2" x 4", 16" O.C.	.920	L.F.	.012	.40	.67	1.07
Hip rafter, 2" x 6"	.070	L.F.	.002	.05	.12	.17
Jack rafter, 2" x 6", 16" O.C.	1.000	L.F.	.039	.67	2.21	2.88
Ridge board, 2" x 6"	.018	L.F.	.001	.01	.03	.04
Sheathing, exterior grade plywood, 1/2" thick	2.210	S.F.	.025	1.66	1.44	3.10
Furring strips, 1" x 3", 16" O.C.	.920	L.F.	.021	.40	1.20	1.60
Rafter ties	.140	Ea.	.008	.22	.44	.66
TOTAL		S.F.	.151	4.46	8.53	12.99
2" X 8" RAFTERS, 16" O.C.						
Roof rafters, 2" x 8", 16" O.C.	1.210	L.F.	.036	1.13	2.03	3.16
Rafter plates, 2" x 8", double top, single bottom	.364	L.F.	.011	.34	.61	.95
Ceiling joists, 2" x 6", 16" O.C.	.920	L.F.	.012	.62	.67	1.29
Hip rafter, 2" x 8"	.070	L.F.	.002	.07	.14	.21
Jack rafter, 2" x 8", 16" O.C.	1.000	L.F.	.048	.93	2.71	3.64
Ridge board, 2" x 8"	.018	L.F.	.001	.02	.04	.06
Sheathing, exterior grade plywood, 1/2" thick	2.210	S.F.	.025	1.66	1.44	3.10
Furring strips, 1" x 3", 16" O.C.	.920	L.F.	.021	.40	1.20	1.60
Rafter ties	.140	Ea.	.008	.22	.44	.66
TOTAL		S.F.	.164	5.39	9.28	14.67

The cost of this system is based on the square foot of plan area.

Description	QUAN.	UNIT	LABOR HOURS	COST PER S.F. MAT.	COST PER S.F. INST.	COST PER S.F. TOTAL

Mansard Roof Framing Price Sheet

			LABOR	COST PER S.F.		
	QUAN.	UNIT	HOURS	MAT.	INST.	TOTAL
Roof rafters, #2 or better, 2" x 6", 16" O.C.	1.210	L.F.	.033	.81	1.86	2.67
24" O.C.	.970	L.F.	.026	.65	1.49	2.14
2" x 8", 16" O.C.	1.210	L.F.	.036	1.13	2.03	3.16
24" O.C.	.970	L.F.	.029	.90	1.63	2.53
2" x 10", 16" O.C.	1.210	L.F.	.046	1.78	2.59	4.37
24" O.C.	.970	L.F.	.037	1.43	2.08	3.51
Rafter plates, #2 or better double top single bottom, 2" x 6"	.364	L.F.	.010	.24	.56	.80
2" x 8"	.364	L.F.	.011	.34	.61	.95
2" x 10"	.364	L.F.	.014	.54	.78	1.32
Ceiling joist, #2 or better, 2" x 4", 16" O.C.	.920	L.F.	.012	.40	.67	1.07
24" O.C.	.740	L.F.	.009	.33	.54	.87
2" x 6", 16" O.C.	.920	L.F.	.012	.62	.67	1.29
24" O.C.	.740	L.F.	.009	.50	.54	1.04
2" x 8", 16" O.C.	.920	L.F.	.013	.86	.76	1.62
24" O.C.	.740	L.F.	.011	.69	.61	1.30
Hip rafter, #2 or better, 2" x 6"	.070	L.F.	.002	.05	.12	.17
2" x 8"	.070	L.F.	.002	.07	.14	.21
2" x 10"	.070	L.F.	.003	.10	.17	.27
Jack rafter, #2 or better, 2" x 6", 16" O.C.	1.000	L.F.	.039	.67	2.21	2.88
24" O.C.	.800	L.F.	.031	.54	1.77	2.31
2" x 8", 16" O.C.	1.000	L.F.	.048	.93	2.71	3.64
24" O.C.	.800	L.F.	.038	.74	2.17	2.91
Ridge board, #2 or better, 1" x 6"	.018	L.F.	.001	.01	.03	.04
1" x 8"	.018	L.F.	.001	.02	.03	.05
1" x 10"	.018	L.F.	.001	.03	.03	.06
2" x 6"	.018	L.F.	.001	.01	.03	.04
2" x 8"	.018	L.F.	.001	.02	.04	.06
2" x 10"	.018	L.F.	.001	.03	.04	.07
Sheathing, plywood exterior grade CDX, 3/8" thick	2.210	S.F.	.023	1.50	1.33	2.83
1/2" thick	2.210	S.F.	.025	1.66	1.44	3.10
5/8" thick	2.210	S.F.	.027	1.86	1.55	3.41
3/4" thick	2.210	S.F.	.029	2.21	1.68	3.89
Boards, 1" x 6", laid regular	2.210	S.F.	.049	3.73	2.76	6.49
Laid diagonal	2.210	S.F.	.054	3.73	3.09	6.82
1" x 8", laid regular	2.210	S.F.	.040	4.57	2.30	6.87
Laid diagonal	2.210	S.F.	.049	4.57	2.76	7.33
Furring, 1" x 3", 12" O.C.	1.150	L.F.	.026	.49	1.50	1.99
24" O.C.	.740	L.F.	.017	.32	.96	1.28

3 | FRAMING — 32 | Shed/Flat Roof Framing Systems

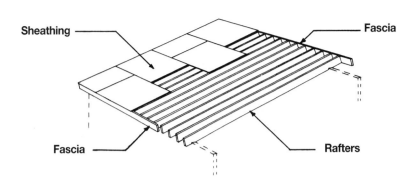

System Description	QUAN.	UNIT	LABOR HOURS	COST PER S.F. MAT.	COST PER S.F. INST.	COST PER S.F. TOTAL
2" X 6", 16" O.C., 4/12 PITCH						
Rafters, 2" x 6", 16" O.C., 4/12 pitch	1.170	L.F.	.019	.78	1.06	1.84
Fascia, 2" x 6"	.100	L.F.	.006	.07	.32	.39
Bridging, 1" x 3", 6' O.C.	.080	Pr.	.005	.06	.28	.34
Sheathing, exterior grade plywood, 1/2" thick	1.230	S.F.	.014	.92	.80	1.72
Rafter ties	.053	Ea.	.003	.08	.17	.25
TOTAL		S.F.	.047	1.91	2.63	4.54
2" X 6", 24" O.C., 4/12 PITCH						
Rafters, 2" x 6", 24" O.C., 4/12 pitch	.940	L.F.	.015	.63	.86	1.49
Fascia, 2" x 6"	.100	L.F.	.006	.07	.32	.39
Bridging, 1" x 3", 6' O.C.	.060	Pr.	.004	.04	.21	.25
Sheathing, exterior grade plywood, 1/2" thick	1.230	S.F.	.014	.92	.80	1.72
Rafter ties	.035	Ea.	.002	.06	.11	.17
TOTAL		S.F.	.041	1.72	2.30	4.02
2" X 8", 16" O.C., 4/12 PITCH						
Rafters, 2" x 8", 16" O.C., 4/12 pitch	1.170	L.F.	.020	1.09	1.12	2.21
Fascia, 2" x 8"	.100	L.F.	.007	.09	.40	.49
Bridging, 1" x 3", 6' O.C.	.080	Pr.	.005	.06	.28	.34
Sheathing, exterior grade plywood, 1/2" thick	1.230	S.F.	.014	.92	.80	1.72
Rafter ties	.053	Ea.	.003	.08	.17	.25
TOTAL		S.F.	.049	2.24	2.77	5.01
2" X 8", 24" O.C., 4/12 PITCH						
Rafters, 2" x 8", 24" O.C., 4/12 pitch	.940	L.F.	.016	.87	.90	1.77
Fascia, 2" x 8"	.100	L.F.	.007	.09	.40	.49
Bridging, 1" x 3", 6' O.C.	.060	Pr.	.004	.04	.21	.25
Sheathing, exterior grade plywood, 1/2" thick	1.230	S.F.	.014	.92	.80	1.72
Rafter ties	.035	Ea.	.002	.06	.11	.17
TOTAL		S.F.	.043	1.98	2.42	4.40

The cost of this system is based on the square foot of plan area.
A 1' overhang is assumed. No ceiling joists or furring are included.

Description	QUAN.	UNIT	LABOR HOURS	COST PER S.F. MAT.	COST PER S.F. INST.	COST PER S.F. TOTAL

Shed/Flat Roof Framing Price Sheet

Shed/Flat Roof Framing Price Sheet	QUAN.	UNIT	LABOR HOURS	COST PER S.F. MAT.	COST PER S.F. INST.	COST PER S.F. TOTAL
Rafters, #2 or better, 16" O.C., 2" x 4", 0 - 4/12 pitch	1.170	L.F.	.014	.58	.79	1.37
5/12 - 8/12 pitch	1.330	L.F.	.020	.67	1.14	1.81
2" x 6", 0 - 4/12 pitch	1.170	L.F.	.019	.78	1.06	1.84
5/12 - 8/12 pitch	1.330	L.F.	.027	.89	1.52	2.41
2" x 8", 0 - 4/12 pitch	1.170	L.F.	.020	1.09	1.12	2.21
5/12 - 8/12 pitch	1.330	L.F.	.028	1.24	1.61	2.85
2" x 10", 0 - 4/12 pitch	1.170	L.F.	.030	1.72	1.68	3.40
5/12 - 8/12 pitch	1.330	L.F.	.043	1.96	2.43	4.39
24" O.C., 2" x 4", 0 - 4/12 pitch	.940	L.F.	.011	.48	.65	1.13
5/12 - 8/12 pitch	1.060	L.F.	.021	.71	1.21	1.92
2" x 6", 0 - 4/12 pitch	.940	L.F.	.015	.63	.86	1.49
5/12 - 8/12 pitch	1.060	L.F.	.021	.71	1.21	1.92
2" x 8", 0 - 4/12 pitch	.940	L.F.	.016	.87	.90	1.77
5/12 - 8/12 pitch	1.060	L.F.	.023	.99	1.28	2.27
2" x 10", 0 - 4/12 pitch	.940	L.F.	.024	1.38	1.35	2.73
5/12 - 8/12 pitch	1.060	L.F.	.034	1.56	1.94	3.50
Fascia, #2 or better,, 1" x 4"	.100	L.F.	.003	.04	.16	.20
1" x 6"	.100	L.F.	.004	.05	.22	.27
1" x 8"	.100	L.F.	.005	.06	.26	.32
1" x 10"	.100	L.F.	.005	.07	.29	.36
2" x 4"	.100	L.F.	.005	.06	.27	.33
2" x 6"	.100	L.F.	.006	.07	.32	.39
2" x 8"	.100	L.F.	.007	.09	.40	.49
2" x 10"	.100	L.F.	.009	.15	.51	.66
Bridging, wood 6' O.C., 1" x 3", rafters, 16" O.C.	.080	Pr.	.005	.06	.28	.34
24" O.C.	.060	Pr.	.004	.04	.21	.25
Metal, galvanized, rafters, 16" O.C.	.080	Pr.	.005	.08	.28	.36
24" O.C.	.060	Pr.	.003	.13	.19	.32
Compression type, rafters, 16" O.C.	.080	Pr.	.003	.08	.18	.26
24" O.C.	.060	Pr.	.002	.06	.14	.20
Sheathing, plywood, exterior grade, 3/8" thick, flat 0 - 4/12 pitch	1.230	S.F.	.013	.84	.74	1.58
5/12 - 8/12 pitch	1.330	S.F.	.014	.90	.80	1.70
1/2" thick, flat 0 - 4/12 pitch	1.230	S.F.	.014	.92	.80	1.72
5/12 - 8/12 pitch	1.330	S.F.	.015	1	.86	1.86
5/8" thick, flat 0 - 4/12 pitch	1.230	S.F.	.015	1.03	.86	1.89
5/12 - 8/12 pitch	1.330	S.F.	.016	1.12	.93	2.05
3/4" thick, flat 0 - 4/12 pitch	1.230	S.F.	.016	1.23	.93	2.16
5/12 - 8/12 pitch	1.330	S.F.	.018	1.33	1.01	2.34
Boards, 1" x 6", laid regular, flat 0 - 4/12 pitch	1.230	S.F.	.027	2.08	1.54	3.62
5/12 - 8/12 pitch	1.330	S.F.	.041	2.25	2.33	4.58
Laid diagonal, flat 0 - 4/12 pitch	1.230	S.F.	.030	2.08	1.72	3.80
5/12 - 8/12 pitch	1.330	S.F.	.044	2.25	2.51	4.76
1" x 8", laid regular, flat 0 - 4/12 pitch	1.230	S.F.	.022	2.55	1.28	3.83
5/12 - 8/12 pitch	1.330	S.F.	.034	2.75	1.90	4.65
Laid diagonal, flat 0 - 4/12 pitch	1.230	S.F.	.027	2.55	1.54	4.09
5/12 - 8/12 pitch	1.330	S.F.	.044	2.25	2.51	4.76

3 | FRAMING

40 | Gable Dormer Framing Systems

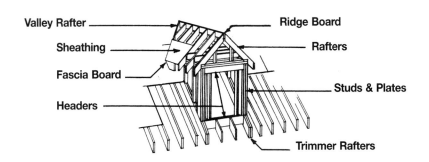

System Description	QUAN.	UNIT	LABOR HOURS	COST PER S.F. MAT.	COST PER S.F. INST.	COST PER S.F. TOTAL
2" X 6", 16" O.C.						
Dormer rafter, 2" x 6", 16" O.C.	1.330	L.F.	.036	.89	2.05	2.94
Ridge board, 2" x 6"	.280	L.F.	.009	.19	.51	.70
Trimmer rafters, 2" x 6"	.880	L.F.	.014	.59	.80	1.39
Wall studs & plates, 2" x 4", 16" O.C.	3.160	L.F.	.056	1.39	3.16	4.55
Fascia, 2" x 6"	.220	L.F.	.012	.16	.69	.85
Valley rafter, 2" x 6", 16" O.C.	.280	L.F.	.009	.19	.50	.69
Cripple rafter, 2" x 6", 16" O.C.	.560	L.F.	.022	.38	1.24	1.62
Headers, 2" x 6", doubled	.670	L.F.	.030	.45	1.69	2.14
Ceiling joist, 2" x 4", 16" O.C.	1.000	L.F.	.013	.44	.73	1.17
Sheathing, exterior grade plywood, 1/2" thick	3.610	S.F.	.041	2.71	2.35	5.06
TOTAL		S.F.	.242	7.39	13.72	21.11
2" X 8", 16" O.C.						
Dormer rafter, 2" x 8", 16" O.C.	1.330	L.F.	.039	1.24	2.23	3.47
Ridge board, 2" x 8"	.280	L.F.	.010	.26	.57	.83
Trimmer rafter, 2" x 8"	.880	L.F.	.015	.82	.84	1.66
Wall studs & plates, 2" x 4", 16" O.C.	3.160	L.F.	.056	1.39	3.16	4.55
Fascia, 2" x 8"	.220	L.F.	.016	.20	.89	1.09
Valley rafter, 2" x 8", 16" O.C.	.280	L.F.	.010	.26	.54	.80
Cripple rafter, 2" x 8", 16" O.C.	.560	L.F.	.027	.52	1.52	2.04
Headers, 2" x 8", doubled	.670	L.F.	.032	.62	1.79	2.41
Ceiling joist, 2" x 4", 16" O.C.	1.000	L.F.	.013	.44	.73	1.17
Sheathing,, exterior grade plywood, 1/2" thick	3.610	S.F.	.041	2.71	2.35	5.06
TOTAL		S.F.	.259	8.46	14.62	23.08

The cost in this system is based on the square foot of plan area.
The measurement being the plan area of the dormer only.

Description	QUAN.	UNIT	LABOR HOURS	COST PER S.F. MAT.	COST PER S.F. INST.	COST PER S.F. TOTAL

Gable Dormer Framing Price Sheet

	QUAN.	UNIT	LABOR HOURS	COST PER S.F. MAT.	COST PER S.F. INST.	COST PER S.F. TOTAL
Dormer rafters, #2 or better, 2" x 4", 16" O.C.	1.330	L.F.	.029	.71	1.64	2.35
24" O.C.	1.060	L.F.	.023	.57	1.31	1.88
2" x 6", 16" O.C.	1.330	L.F.	.036	.89	2.05	2.94
24" O.C.	1.060	L.F.	.029	.71	1.63	2.34
2" x 8", 16" O.C.	1.330	L.F.	.039	1.24	2.23	3.47
24" O.C.	1.060	L.F.	.031	.99	1.78	2.77
Ridge board, #2 or better, 1" x 4"	.280	L.F.	.006	.17	.34	.51
1" x 6"	.280	L.F.	.007	.22	.42	.64
1" x 8"	.280	L.F.	.008	.36	.46	.82
2" x 4"	.280	L.F.	.007	.15	.41	.56
2" x 6"	.280	L.F.	.009	.19	.51	.70
2" x 8"	.280	L.F.	.010	.26	.57	.83
Trimmer rafters, #2 or better, 2" x 4"	.880	L.F.	.011	.47	.64	1.11
2" x 6"	.880	L.F.	.014	.59	.80	1.39
2" x 8"	.880	L.F.	.015	.82	.84	1.66
2" x 10"	.880	L.F.	.022	1.29	1.27	2.56
Wall studs & plates, #2 or better, 2" x 4" studs, 16" O.C.	3.160	L.F.	.056	1.39	3.16	4.55
24" O.C.	2.800	L.F.	.050	1.23	2.80	4.03
2" x 6" studs, 16" O.C.	3.160	L.F.	.063	2.12	3.60	5.72
24" O.C.	2.800	L.F.	.056	1.88	3.19	5.07
Fascia, #2 or better, 1" x 4"	.220	L.F.	.006	.08	.36	.44
1" x 6"	.220	L.F.	.008	.10	.44	.54
1" x 8"	.220	L.F.	.009	.12	.52	.64
2" x 4"	.220	L.F.	.011	.14	.61	.75
2" x 6"	.220	L.F.	.014	.18	.77	.95
2" x 8"	.220	L.F.	.016	.20	.89	1.09
Valley rafter, #2 or better, 2" x 4"	.280	L.F.	.007	.15	.40	.55
2" x 6"	.280	L.F.	.009	.19	.50	.69
2" x 8"	.280	L.F.	.010	.26	.54	.80
2" x 10"	.280	L.F.	.012	.41	.67	1.08
Cripple rafter, #2 or better, 2" x 4", 16" O.C.	.560	L.F.	.018	.30	.99	1.29
24" O.C.	.450	L.F.	.014	.24	.80	1.04
2" x 6", 16" O.C.	.560	L.F.	.022	.38	1.24	1.62
24" O.C.	.450	L.F.	.018	.30	.99	1.29
2" x 8", 16" O.C.	.560	L.F.	.027	.52	1.52	2.04
24" O.C.	.450	L.F.	.021	.42	1.22	1.64
Headers, #2 or better double header, 2" x 4"	.670	L.F.	.024	.36	1.36	1.72
2" x 6"	.670	L.F.	.030	.45	1.69	2.14
2" x 8"	.670	L.F.	.032	.62	1.79	2.41
2" x 10"	.670	L.F.	.034	.98	1.90	2.88
Ceiling joist, #2 or better, 2" x 4", 16" O.C.	1.000	L.F.	.013	.44	.73	1.17
24" O.C.	.800	L.F.	.010	.35	.58	.93
2" x 6", 16" O.C.	1.000	L.F.	.013	.67	.73	1.40
24" O.C.	.800	L.F.	.010	.54	.58	1.12
Sheathing, plywood exterior grade, 3/8" thick	3.610	S.F.	.038	2.45	2.17	4.62
1/2" thick	3.610	S.F.	.041	2.71	2.35	5.06
5/8" thick	3.610	S.F.	.044	3.03	2.53	5.56
3/4" thick	3.610	S.F.	.048	3.61	2.74	6.35
Boards, 1" x 6", laid regular	3.610	S.F.	.089	6.10	5.05	11.15
Laid diagonal	3.610	S.F.	.099	6.10	5.60	11.70
1" x 8", laid regular	3.610	S.F.	.076	7.45	4.30	11.75
Laid diagonal	3.610	S.F.	.089	7.45	5.05	12.50

3 | FRAMING — 44 Shed Dormer Framing Systems

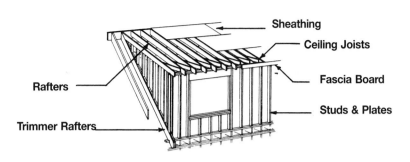

System Description	QUAN.	UNIT	LABOR HOURS	COST PER S.F. MAT.	COST PER S.F. INST.	COST PER S.F. TOTAL
2" X 6" RAFTERS, 16" O.C.						
Dormer rafter, 2" x 6", 16" O.C.	1.080	L.F.	.029	.72	1.66	2.38
Trimmer rafter, 2" x 6"	.400	L.F.	.006	.27	.36	.63
Studs & plates, 2" x 4", 16" O.C.	2.750	L.F.	.049	1.21	2.75	3.96
Fascia, 2" x 6"	.250	L.F.	.014	.18	.77	.95
Ceiling joist, 2" x 4", 16" O.C.	1.000	L.F.	.013	.44	.73	1.17
Sheathing, exterior grade plywood, CDX, 1/2" thick	2.940	S.F.	.034	2.21	1.91	4.12
TOTAL		S.F.	.145	5.03	8.18	13.21
2" X 8" RAFTERS, 16" O.C.						
Dormer rafter, 2" x 8", 16" O.C.	1.080	L.F.	.032	1	1.81	2.81
Trimmer rafter, 2" x 8"	.400	L.F.	.007	.37	.38	.75
Studs & plates, 2" x 4", 16" O.C.	2.750	L.F.	.049	1.21	2.75	3.96
Fascia, 2" x 8"	.250	L.F.	.018	.23	1.01	1.24
Ceiling joist, 2" x 6", 16" O.C.	1.000	L.F.	.013	.67	.73	1.40
Sheathing, exterior grade plywood, CDX, 1/2" thick	2.940	S.F.	.034	2.21	1.91	4.12
TOTAL		S.F.	.153	5.69	8.59	14.28
2" X 10" RAFTERS, 16" O.C.						
Dormer rafter, 2" x 10", 16" O.C.	1.080	L.F.	.041	1.59	2.31	3.90
Trimmer rafter, 2" x 10"	.400	L.F.	.010	.59	.58	1.17
Studs & plates, 2" x 4", 16" O.C.	2.750	L.F.	.049	1.21	2.75	3.96
Fascia, 2" x 10"	.250	L.F.	.022	.37	1.26	1.63
Ceiling joist, 2" x 6", 16" O.C.	1.000	L.F.	.013	.67	.73	1.40
Sheathing, exterior grade plywood, CDX, 1/2" thick	2.940	S.F.	.034	2.21	1.91	4.12
TOTAL		S.F.	.169	6.64	9.54	16.18

The cost in this system is based on the square foot of plan area.
The measurement is the plan area of the dormer only.

Description	QUAN.	UNIT	LABOR HOURS	COST PER S.F. MAT.	COST PER S.F. INST.	COST PER S.F. TOTAL

Shed Dormer Framing Price Sheet

	QUAN.	UNIT	LABOR HOURS	COST PER S.F. MAT.	COST PER S.F. INST.	COST PER S.F. TOTAL
Dormer rafters, #2 or better, 2" x 4", 16" O.C.	1.080	L.F.	.023	.58	1.33	1.91
24" O.C.	.860	L.F.	.019	.46	1.06	1.52
2" x 6", 16" O.C.	1.080	L.F.	.029	.72	1.66	2.38
24" O.C.	.860	L.F.	.023	.58	1.32	1.90
2" x 8", 16" O.C.	1.080	L.F.	.032	1	1.81	2.81
24" O.C.	.860	L.F.	.025	.80	1.44	2.24
2" x 10", 16" O.C.	1.080	L.F.	.041	1.59	2.31	3.90
24" O.C.	.860	L.F.	.032	1.26	1.84	3.10
Trimmer rafter, #2 or better, 2" x 4"	.400	L.F.	.005	.21	.29	.50
2" x 6"	.400	L.F.	.006	.27	.36	.63
2" x 8"	.400	L.F.	.007	.37	.38	.75
2" x 10"	.400	L.F.	.010	.59	.58	1.17
Studs & plates, #2 or better, 2" x 4", 16" O.C.	2.750	L.F.	.049	1.21	2.75	3.96
24" O.C.	2.200	L.F.	.039	.97	2.20	3.17
2" x 6", 16" O.C.	2.750	L.F.	.055	1.84	3.14	4.98
24" O.C.	2.200	L.F.	.044	1.47	2.51	3.98
Fascia, #2 or better, 1" x 4"	.250	L.F.	.006	.08	.36	.44
1" x 6"	.250	L.F.	.008	.10	.44	.54
1" x 8"	.250	L.F.	.009	.12	.52	.64
2" x 4"	.250	L.F.	.011	.14	.61	.75
2" x 6"	.250	L.F.	.014	.18	.77	.95
2" x 8"	.250	L.F.	.018	.23	1.01	1.24
Ceiling joist, #2 or better, 2" x 4", 16" O.C.	1.000	L.F.	.013	.44	.73	1.17
24" O.C.	.800	L.F.	.010	.35	.58	.93
2" x 6", 16" O.C.	1.000	L.F.	.013	.67	.73	1.40
24" O.C.	.800	L.F.	.010	.54	.58	1.12
2" x 8", 16" O.C.	1.000	L.F.	.015	.93	.83	1.76
24" O.C.	.800	L.F.	.012	.74	.66	1.40
Sheathing, plywood exterior grade, 3/8" thick	2.940	S.F.	.031	2	1.76	3.76
1/2" thick	2.940	S.F.	.034	2.21	1.91	4.12
5/8" thick	2.940	S.F.	.036	2.47	2.06	4.53
3/4" thick	2.940	S.F.	.039	2.94	2.23	5.17
Boards, 1" x 6", laid regular	2.940	S.F.	.072	4.97	4.12	9.09
Laid diagonal	2.940	S.F.	.080	4.97	4.56	9.53
1" x 8", laid regular	2.940	S.F.	.062	6.10	3.50	9.60
Laid diagonal	2.940	S.F.	.072	6.10	4.12	10.22

Window Openings

	QUAN.	UNIT	LABOR HOURS	COST EACH MAT.	COST EACH INST.	COST EACH TOTAL

The following are to be added to the total cost of the dormers for window openings. Do not subtract window area from the stud wall quantities.

	QUAN.	UNIT	LABOR HOURS	MAT.	INST.	TOTAL
Headers, 2" x 6" doubled, 2' long	4.000	L.F.	.178	2.68	10.10	12.78
3' long	6.000	L.F.	.267	4.02	15.10	19.12
4' long	8.000	L.F.	.356	5.35	20	25.35
5' long	10.000	L.F.	.444	6.70	25	31.70
2" x 8" doubled, 4' long	8.000	L.F.	.376	7.45	21.50	28.95
5' long	10.000	L.F.	.471	9.30	26.50	35.80
6' long	12.000	L.F.	.565	11.15	32	43.15
8' long	16.000	L.F.	.753	14.90	42.50	57.40
2" x 10" doubled, 4' long	8.000	L.F.	.400	11.75	22.50	34.25
6' long	12.000	L.F.	.600	17.65	34	51.65
8' long	16.000	L.F.	.800	23.50	45.50	69
10' long	20.000	L.F.	1.000	29.50	57	86.50

3 | FRAMING   48 | Partition Framing Systems

System Description	QUAN.	UNIT	LABOR HOURS	COST PER S.F.		
				MAT.	INST.	TOTAL
2" X 4", 16" O.C.						
2" x 4" studs, #2 or better, 16" O.C.	1.000	L.F.	.015	.43	.83	1.26
Plates, double top, single bottom	.375	L.F.	.005	.16	.31	.47
Cross bracing, let-in, 1" x 6"	.080	L.F.	.004	.06	.24	.30
TOTAL		S.F.	.024	.65	1.38	2.03
2" X 4", 24" O.C.						
2" x 4" studs, #2 or better, 24" O.C.	.800	L.F.	.012	.34	.66	1
Plates, double top, single bottom	.375	L.F.	.005	.16	.31	.47
Cross bracing, let-in, 1" x 6"	.080	L.F.	.003	.06	.16	.22
TOTAL		S.F.	.020	.56	1.13	1.69
2" X 6", 16" O.C.						
2" x 6" studs, #2 or better, 16" O.C.	1.000	L.F.	.016	.67	.91	1.58
Plates, double top, single bottom	.375	L.F.	.006	.25	.34	.59
Cross bracing, let-in, 1" x 6"	.080	L.F.	.004	.06	.24	.30
TOTAL		S.F.	.026	.98	1.49	2.47
2" X 6", 24" O.C.						
2" x 6" studs, #2 or better, 24" O.C.	.800	L.F.	.013	.54	.73	1.27
Plates, double top, single bottom	.375	L.F.	.006	.25	.34	.59
Cross bracing, let-in, 1" x 6"	.080	L.F.	.003	.06	.16	.22
TOTAL		S.F.	.022	.85	1.23	2.08

The costs in this system are based on a square foot of wall area. Do not subtract for door or window openings.

Description	QUAN.	UNIT	LABOR HOURS	COST PER S.F.		
				MAT.	INST.	TOTAL

Partition Framing Price Sheet	QUAN.	UNIT	LABOR HOURS	COST PER S.F.		
				MAT.	INST.	TOTAL
Wood studs, #2 or better, 2" x 4", 12" O.C.	1.250	L.F.	.018	.54	1.04	1.58
16" O.C.	1.000	L.F.	.015	.43	.83	1.26
24" O.C.	.800	L.F.	.012	.34	.66	1
32" O.C.	.650	L.F.	.009	.28	.54	.82
2" x 6", 12" O.C.	1.250	L.F.	.020	.84	1.14	1.98
16" O.C.	1.000	L.F.	.016	.67	.91	1.58
24" O.C.	.800	L.F.	.013	.54	.73	1.27
32" O.C.	.650	L.F.	.010	.44	.59	1.03
Plates, #2 or better double top single bottom, 2" x 4"	.375	L.F.	.005	.16	.31	.47
2" x 6"	.375	L.F.	.006	.25	.34	.59
2" x 8"	.375	L.F.	.005	.35	.31	.66
Cross bracing, let-in, 1" x 6" boards studs, 12" O.C.	.080	L.F.	.005	.08	.30	.38
16" O.C.	.080	L.F.	.004	.06	.24	.30
24" O.C.	.080	L.F.	.003	.06	.16	.22
32" O.C.	.080	L.F.	.002	.05	.13	.18
Let-in steel (T shaped) studs, 12" O.C.	.080	L.F.	.001	.09	.08	.17
16" O.C.	.080	L.F.	.001	.07	.06	.13
24" O.C.	.080	L.F.	.001	.07	.06	.13
32" O.C.	.080	L.F.	.001	.06	.05	.11
Steel straps studs, 12" O.C.	.080	L.F.	.001	.10	.06	.16
16" O.C.	.080	L.F.	.001	.09	.06	.15
24" O.C.	.080	L.F.	.001	.09	.06	.15
32" O.C.	.080	L.F.	.001	.09	.05	.14
Metal studs, load bearing 24" O.C., 20 ga. galv., 2-1/2" wide	1.000	S.F.	.015	.70	.85	1.55
3-5/8" wide	1.000	S.F.	.015	.84	.87	1.71
4" wide	1.000	S.F.	.016	.87	.88	1.75
6" wide	1.000	S.F.	.016	1.12	.90	2.02
16 ga., 2-1/2" wide	1.000	S.F.	.017	.82	.97	1.79
3-5/8" wide	1.000	S.F.	.017	.98	.99	1.97
4" wide	1.000	S.F.	.018	1.03	1.01	2.04
6" wide	1.000	S.F.	.018	1.30	1.03	2.33
Non-load bearing 24" O.C., 25 ga. galv., 1-5/8" wide	1.000	S.F.	.011	.20	.60	.80
2-1/2" wide	1.000	S.F.	.011	.25	.61	.86
3-5/8" wide	1.000	S.F.	.011	.30	.61	.91
4" wide	1.000	S.F.	.011	.33	.61	.94
6" wide	1.000	S.F.	.011	.40	.63	1.03
20 ga., 2-1/2" wide	1.000	S.F.	.013	.33	.76	1.09
3-5/8" wide	1.000	S.F.	.014	.37	.77	1.14
4" wide	1.000	S.F.	.014	.44	.77	1.21
6" wide	1.000	S.F.	.014	.52	.78	1.30

Window & Door Openings	QUAN.	UNIT	LABOR HOURS	COST EACH		
				MAT.	INST.	TOTAL
The following costs are to be added to the total costs of the walls. Do not subtract openings from total wall area.						
Headers, 2" x 6" double, 2' long	4.000	L.F.	.178	2.68	10.10	12.78
3' long	6.000	L.F.	.267	4.02	15.10	19.12
4' long	8.000	L.F.	.356	5.35	20	25.35
5' long	10.000	L.F.	.444	6.70	25	31.70
2" x 8" double, 4' long	8.000	L.F.	.376	7.45	21.50	28.95
5' long	10.000	L.F.	.471	9.30	26.50	35.80
6' long	12.000	L.F.	.565	11.15	32	43.15
8' long	16.000	L.F.	.753	14.90	42.50	57.40
2" x 10" double, 4' long	8.000	L.F.	.400	11.75	22.50	34.25
6' long	12.000	L.F.	.600	17.65	34	51.65
8' long	16.000	L.F.	.800	23.50	45.50	69
10' long	20.000	L.F.	1.000	29.50	57	86.50
2" x 12" double, 8' long	16.000	L.F.	.853	30	48.50	78.50
12' long	24.000	L.F.	1.280	45	72.50	117.50

Did you know?
RSMeans Online gives you the same access to RSMeans' data with 24/7 access:
- Quickly locate costs in the searchable database.
- Build cost lists, estimates, and reports in minutes.
- Adjust costs to any location in the U.S. and Canada with the click of a button.

Start your free trial today at **www.RSMeansOnline.com**

RSMeans Online
FROM THE GORDIAN GROUP

No part of this cost data may be reproduced, stored in a retrieval system, or transmitted in any form or by any means without prior written permission of RSMeans.

4 | EXTERIOR WALLS   02 | Block Masonry Systems

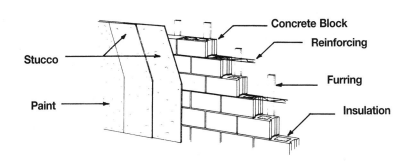

System Description	QUAN.	UNIT	LABOR HOURS	COST PER S.F. MAT.	COST PER S.F. INST.	COST PER S.F. TOTAL
6" THICK CONCRETE BLOCK WALL						
6" thick concrete block, 6" x 8" x 16"	1.000	S.F.	.100	2.53	5	7.53
Masonry reinforcing, truss strips every other course	.625	L.F.	.002	.18	.09	.27
Furring, 1" x 3", 16" O.C.	1.000	L.F.	.016	.46	.92	1.38
Masonry insulation, poured perlite	1.000	S.F.	.013	1.93	.76	2.69
Stucco, 2 coats	1.000	S.F.	.069	.24	3.58	3.82
Masonry paint, 2 coats	1.000	S.F.	.016	.24	.76	1
TOTAL		S.F.	.216	5.58	11.11	16.69
8" THICK CONCRETE BLOCK WALL						
8" thick concrete block, 8" x 8" x 16"	1.000	S.F.	.107	2.71	5.35	8.06
Masonry reinforcing, truss strips every other course	.625	L.F.	.002	.18	.09	.27
Furring, 1" x 3", 16" O.C.	1.000	L.F.	.016	.46	.92	1.38
Masonry insulation, poured perlite	1.000	S.F.	.018	2.55	1	3.55
Stucco, 2 coats	1.000	S.F.	.069	.24	3.58	3.82
Masonry paint, 2 coats	1.000	S.F.	.016	.24	.76	1
TOTAL		S.F.	.228	6.38	11.70	18.08
12" THICK CONCRETE BLOCK WALL						
12" thick concrete block, 12" x 8" x 16"	1.000	S.F.	.141	4.48	6.95	11.43
Masonry reinforcing, truss strips every other course	.625	L.F.	.003	.11	.14	.25
Furring, 1" x 3", 16" O.C.	1.000	L.F.	.016	.46	.92	1.38
Masonry insulation, poured perlite	1.000	S.F.	.026	3.77	1.48	5.25
Stucco, 2 coats	1.000	S.F.	.069	.24	3.58	3.82
Masonry paint, 2 coats	1.000	S.F.	.016	.24	.76	1
TOTAL		S.F.	.271	9.30	13.83	23.13

Costs for this system are based on a square foot of wall area. Do not subtract for window openings.

Description	QUAN.	UNIT	LABOR HOURS	COST PER S.F. MAT.	COST PER S.F. INST.	COST PER S.F. TOTAL

Masonry Block Price Sheet

	QUAN.	UNIT	LABOR HOURS	COST PER S.F. MAT.	COST PER S.F. INST.	COST PER S.F. TOTAL
Block concrete, 8" x 16" regular, 4" thick	1.000	S.F.	.093	1.92	4.66	6.58
6" thick	1.000	S.F.	.100	2.53	5	7.53
8" thick	1.000	S.F.	.107	2.71	5.35	8.06
10" thick	1.000	S.F.	.111	3.23	5.55	8.78
12" thick	1.000	S.F.	.141	4.48	6.95	11.43
Solid block, 4" thick	1.000	S.F.	.096	2.19	4.83	7.02
6" thick	1.000	S.F.	.104	2.59	5.20	7.79
8" thick	1.000	S.F.	.111	3.87	5.55	9.42
10" thick	1.000	S.F.	.133	5.25	6.55	11.80
12" thick	1.000	S.F.	.148	5.85	7.25	13.10
Lightweight, 4" thick	1.000	S.F.	.093	1.92	4.66	6.58
6" thick	1.000	S.F.	.100	2.53	5	7.53
8" thick	1.000	S.F.	.107	2.71	5.35	8.06
10" thick	1.000	S.F.	.111	3.23	5.55	8.78
12" thick	1.000	S.F.	.141	4.48	6.95	11.43
Split rib profile, 4" thick	1.000	S.F.	.116	4.38	5.80	10.18
6" thick	1.000	S.F.	.123	4.96	6.15	11.11
8" thick	1.000	S.F.	.131	5.60	6.70	12.30
10" thick	1.000	S.F.	.157	5.90	7.75	13.65
12" thick	1.000	S.F.	.175	6.55	8.60	15.15
Masonry reinforcing, wire truss strips, every course, 8" block	1.375	L.F.	.004	.39	.19	.58
12" block	1.375	L.F.	.006	.25	.30	.55
Every other course, 8" block	.625	L.F.	.002	.18	.09	.27
12" block	.625	L.F.	.003	.11	.14	.25
Furring, wood, 1" x 3", 12" O.C.	1.250	L.F.	.020	.58	1.15	1.73
16" O.C.	1.000	L.F.	.016	.46	.92	1.38
24" O.C.	.800	L.F.	.013	.37	.74	1.11
32" O.C.	.640	L.F.	.010	.29	.59	.88
Steel, 3/4" channels, 12" O.C.	1.250	L.F.	.034	.42	1.82	2.24
16" O.C.	1.000	L.F.	.030	.38	1.62	2
24" O.C.	.800	L.F.	.023	.25	1.22	1.47
32" O.C.	.640	L.F.	.018	.20	.98	1.18
Masonry insulation, vermiculite or perlite poured 4" thick	1.000	S.F.	.009	1.25	.49	1.74
6" thick	1.000	S.F.	.013	1.91	.75	2.66
8" thick	1.000	S.F.	.018	2.55	1	3.55
10" thick	1.000	S.F.	.021	3.09	1.21	4.30
12" thick	1.000	S.F.	.026	3.77	1.48	5.25
Block inserts polystyrene, 6" thick	1.000	S.F.		1.32		1.32
8" thick	1.000	S.F.		1.49		1.49
10" thick	1.000	S.F.		1.54		1.54
12" thick	1.000	S.F.		1.71		1.71
Stucco, 1 coat	1.000	S.F.	.057	.20	2.95	3.15
2 coats	1.000	S.F.	.069	.24	3.58	3.82
3 coats	1.000	S.F.	.081	.28	4.21	4.49
Painting, 1 coat	1.000	S.F.	.011	.15	.52	.67
2 coats	1.000	S.F.	.016	.24	.76	1
Primer & 1 coat	1.000	S.F.	.013	.22	.62	.84
2 coats	1.000	S.F.	.018	.31	.86	1.17
Lath, metal lath expanded 2.5 lb/S.Y., painted	1.000	S.F.	.010	.45	.56	1.01
Galvanized	1.000	S.F.	.012	.50	.62	1.12

4 | EXTERIOR WALLS 04 | Brick/Stone Veneer Systems

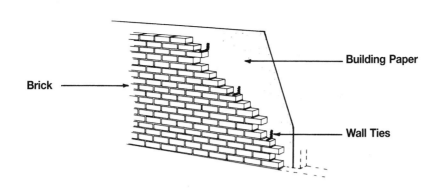

System Description	QUAN.	UNIT	LABOR HOURS	COST PER S.F. MAT.	COST PER S.F. INST.	COST PER S.F. TOTAL
SELECT COMMON BRICK						
Brick, select common, running bond	1.000	S.F.	.174	4.77	8.70	13.47
Wall ties, 7/8" x 7", 22 gauge	1.000	Ea.	.008	.17	.41	.58
Building paper, spunbonded polypropylene	1.100	S.F.	.002	.17	.12	.29
Trim, pine, painted	.125	L.F.	.004	.07	.23	.30
TOTAL		S.F.	.188	5.18	9.46	14.64
RED FACED COMMON BRICK						
Brick, common, red faced, running bond	1.000	S.F.	.182	4.32	9.10	13.42
Wall ties, 7/8" x 7", 22 gauge	1.000	Ea.	.008	.17	.41	.58
Building paper, spunbonded polypropylene	1.100	S.F.	.002	.17	.12	.29
Trim, pine, painted	.125	L.F.	.004	.07	.23	.30
TOTAL		S.F.	.196	4.73	9.86	14.59
BUFF OR GREY FACE BRICK						
Brick, buff or grey	1.000	S.F.	.182	4.56	9.10	13.66
Wall ties, 7/8" x 7", 22 gauge	1.000	Ea.	.008	.17	.41	.58
Building paper, spunbonded polypropylene	1.100	S.F.	.002	.17	.12	.29
Trim, pine, painted	.125	L.F.	.004	.07	.23	.30
TOTAL		S.F.	.196	4.97	9.86	14.83
STONE WORK, ROUGH STONE, AVERAGE						
Field stone veneer	1.000	S.F.	.223	9.28	11.19	20.47
Wall ties, 7/8" x 7", 22 gauge	1.000	Ea.	.008	.17	.41	.58
Building paper, spunbonded polypropylene	1.000	S.F.	.002	.17	.12	.29
Trim, pine, painted	.125	L.F.	.004	.07	.23	.30
TOTAL		S.F.	.237	9.69	11.95	21.64

The costs in this system are based on a square foot of wall area. Do not subtract area for window & door openings.

Description	QUAN.	UNIT	LABOR HOURS	COST PER S.F. MAT.	COST PER S.F. INST.	COST PER S.F. TOTAL

Brick/Stone Veneer Price Sheet	QUAN.	UNIT	LABOR HOURS	COST PER S.F.		
				MAT.	INST.	TOTAL
Brick						
Select common, running bond	1.000	S.F.	.174	4.77	8.70	13.47
Red faced, running bond	1.000	S.F.	.182	4.32	9.10	13.42
Buff or grey faced, running bond	1.000	S.F.	.182	4.56	9.10	13.66
Header every 6th course	1.000	S.F.	.216	5.05	10.85	15.90
English bond	1.000	S.F.	.286	6.45	14.35	20.80
Flemish bond	1.000	S.F.	.195	4.56	9.80	14.36
Common bond	1.000	S.F.	.267	5.75	13.35	19.10
Stack bond	1.000	S.F.	.182	4.56	9.10	13.66
Jumbo, running bond	1.000	S.F.	.092	5.35	4.61	9.96
Norman, running bond	1.000	S.F.	.125	6.65	6.25	12.90
Norwegian, running bond	1.000	S.F.	.107	5.90	5.35	11.25
Economy, running bond	1.000	S.F.	.129	4.66	6.45	11.11
Engineer, running bond	1.000	S.F.	.154	4.14	7.70	11.84
Roman, running bond	1.000	S.F.	.160	7.90	8	15.90
Utility, running bond	1.000	S.F.	.089	5.25	5.55	10.80
Glazed, running bond	1.000	S.F.	.190	13.40	9.55	22.95
Stone work, rough stone, average	1.000	S.F.	.179	9.30	11.20	20.50
Maximum	1.000	S.F.	.267	13.85	16.70	30.55
Wall ties, galvanized, corrugated 7/8" x 7", 22 gauge	1.000	Ea.	.008	.17	.41	.58
16 gauge	1.000	Ea.	.008	.32	.41	.73
Cavity wall, every 3rd course 6" long Z type, 1/4" diameter	1.330	L.F.	.010	.56	.53	1.09
3/16" diameter	1.330	L.F.	.010	.44	.53	.97
8" long, Z type, 1/4" diameter	1.330	L.F.	.010	.69	.53	1.22
3/16" diameter	1.330	L.F.	.010	.38	.53	.91
Building paper, aluminum and kraft laminated foil, 1 side	1.000	S.F.	.002	.13	.12	.25
2 sides	1.000	S.F.	.002	.15	.12	.27
#15 asphalt paper	1.100	S.F.	.002	.07	.13	.20
Polyethylene, .002" thick	1.000	S.F.	.002	.02	.12	.14
.004" thick	1.000	S.F.	.002	.03	.12	.15
.006" thick	1.000	S.F.	.002	.04	.12	.16
.010" thick	1.000	S.F.	.002	.09	.12	.21
Trim, 1" x 4", cedar	.125	L.F.	.005	.16	.28	.44
Fir	.125	L.F.	.005	.12	.28	.40
Redwood	.125	L.F.	.005	.16	.28	.44
White pine	.125	L.F.	.005	.12	.28	.40

4 | EXTERIOR WALLS   08 | Wood Siding Systems

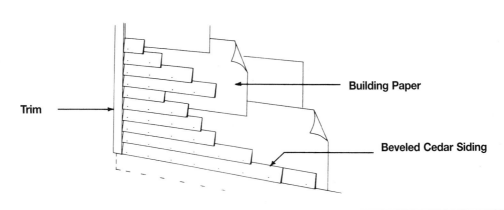

System Description	QUAN.	UNIT	LABOR HOURS	COST PER S.F. MAT.	COST PER S.F. INST.	COST PER S.F. TOTAL
1/2" X 6" BEVELED CEDAR SIDING, "A" GRADE						
1/2" x 6" beveled cedar siding	1.000	S.F.	.027	4.68	1.54	6.22
Building wrap, spundbonded polypropylene	1.100	S.F.	.002	.17	.12	.29
Trim, cedar	.125	L.F.	.005	.16	.28	.44
Paint, primer & 2 coats	1.000	S.F.	.017	.23	.81	1.04
TOTAL		S.F.	.051	5.24	2.75	7.99
1/2" X 8" BEVELED CEDAR SIDING, "A" GRADE						
1/2" x 8" beveled cedar siding	1.000	S.F.	.024	7.15	1.38	8.53
Building wrap, spundbonded polypropylene	1.100	S.F.	.002	.17	.12	.29
Trim, cedar	.125	L.F.	.005	.16	.28	.44
Paint, primer & 2 coats	1.000	S.F.	.017	.23	.81	1.04
TOTAL		S.F.	.048	7.71	2.59	10.30
1" X 4" TONGUE & GROOVE, REDWOOD, VERTICAL GRAIN						
Redwood, clear, vertical grain, 1" x 10"	1.000	S.F.	.020	5.41	1.11	6.52
Building wrap, spunbonded polypropylene	1.100	S.F.	.002	.17	.12	.29
Trim, redwood	.125	L.F.	.005	.16	.28	.44
Sealer, 1 coat, stain, 1 coat	1.000	S.F.	.013	.15	.63	.78
TOTAL		S.F.	.040	5.89	2.14	8.03
1" X 6" TONGUE & GROOVE, REDWOOD, VERTICAL GRAIN						
Redwood, clear, vertical grain, 1" x 10"	1.000	S.F.	.020	5.57	1.14	6.71
Building wrap, spunbonded polypropylene	1.100	S.F.	.002	.17	.12	.29
Trim, redwood	.125	L.F.	.005	.16	.28	.44
Sealer, 1 coat, stain, 1 coat	1.000	S.F.	.013	.15	.63	.78
TOTAL		S.F.	.040	6.05	2.17	8.22

The costs in this system are based on a square foot of wall area.
Do not subtract area for door or window openings.

Description	QUAN.	UNIT	LABOR HOURS	COST PER S.F. MAT.	COST PER S.F. INST.	COST PER S.F. TOTAL

Wood Siding Price Sheet

	QUAN.	UNIT	LABOR HOURS	COST PER S.F. MAT.	COST PER S.F. INST.	COST PER S.F. TOTAL
Siding, beveled cedar, "A" grade, 1/2" x 6"	1.000	S.F.	.028	4.68	1.54	6.22
1/2" x 8"	1.000	S.F.	.023	7.15	1.38	8.53
"B" grade, 1/2" x 6"	1.000	S.F.	.032	5.20	1.71	6.91
1/2" x 8"	1.000	S.F.	.029	7.95	1.53	9.48
Clear grade, 1/2" x 6"	1.000	S.F.	.028	5.85	1.93	7.78
1/2" x 8"	1.000	S.F.	.023	8.95	1.73	10.68
Redwood, clear vertical grain, 1/2" x 6"	1.000	S.F.	.036	5.25	1.54	6.79
1/2" x 8"	1.000	S.F.	.032	5.70	1.38	7.08
Clear all heart vertical grain, 1/2" x 6"	1.000	S.F.	.028	5.85	1.71	7.56
1/2" x 8"	1.000	S.F.	.023	6.35	1.53	7.88
Siding board & batten, cedar, "B" grade, 1" x 10"	1.000	S.F.	.031	4.93	1.08	6.01
1" x 12"	1.000	S.F.	.031	4.93	1.08	6.01
Redwood, clear vertical grain, 1" x 6"	1.000	S.F.	.043	5.30	1.54	6.84
1" x 8"	1.000	S.F.	.018	5.70	1.38	7.08
White pine, #2 & better, 1" x 10"	1.000	S.F.	.029	2.41	1.38	3.79
1" x 12"	1.000	S.F.	.029	2.41	1.38	3.79
Siding vertical, tongue & groove, cedar "B" grade, 1" x 4"	1.000	S.F.	.033	5.25	1.11	6.36
1" x 6"	1.000	S.F.	.024	5.40	1.14	6.54
1" x 8"	1.000	S.F.	.024	5.55	1.17	6.72
1" x 10"	1.000	S.F.	.021	5.70	1.21	6.91
"A" grade, 1" x 4"	1.000	S.F.	.033	4.79	1.02	5.81
1" x 6"	1.000	S.F.	.024	4.91	1.04	5.95
1" x 8"	1.000	S.F.	.024	5.05	1.07	6.12
1" x 10"	1.000	S.F.	.021	5.20	1.10	6.30
Clear vertical grain, 1" x 4"	1.000	S.F.	.033	4.42	.94	5.36
1" x 6"	1.000	S.F.	.024	4.52	.96	5.48
1" x 8"	1.000	S.F.	.024	4.63	.98	5.61
1" x 10"	1.000	S.F.	.021	4.75	1.01	5.76
Redwood, clear vertical grain, 1" x 4"	1.000	S.F.	.033	5.40	1.11	6.51
1" x 6"	1.000	S.F.	.024	5.55	1.14	6.69
1" x 8"	1.000	S.F.	.024	5.75	1.17	6.92
1" x 10"	1.000	S.F.	.021	5.90	1.21	7.11
Clear all heart vertical grain, 1" x 4"	1.000	S.F.	.033	4.96	1.02	5.98
1" x 6"	1.000	S.F.	.024	5.10	1.04	6.14
1" x 8"	1.000	S.F.	.024	5.20	1.07	6.27
1" x 10"	1.000	S.F.	.021	5.35	1.10	6.45
White pine, 1" x 10"	1.000	S.F.	.024	3.50	1.21	4.71
Siding plywood, texture 1-11 cedar, 3/8" thick	1.000	S.F.	.024	1.38	1.35	2.73
5/8" thick	1.000	S.F.	.024	2.97	1.35	4.32
Redwood, 3/8" thick	1.000	S.F.	.024	1.38	1.35	2.73
5/8" thick	1.000	S.F.	.024	2.11	1.35	3.46
Fir, 3/8" thick	1.000	S.F.	.024	.99	1.35	2.34
5/8" thick	1.000	S.F.	.024	1.61	1.35	2.96
Southern yellow pine, 3/8" thick	1.000	S.F.	.024	.99	1.35	2.34
5/8" thick	1.000	S.F.	.024	1.51	1.35	2.86
Paper, #15 asphalt felt	1.100	S.F.	.002	.07	.13	.20
Trim, cedar	.125	L.F.	.005	.16	.28	.44
Fir	.125	L.F.	.005	.12	.28	.40
Redwood	.125	L.F.	.005	.16	.28	.44
White pine	.125	L.F.	.005	.12	.28	.40
Painting, primer, & 1 coat	1.000	S.F.	.013	.15	.63	.78
2 coats	1.000	S.F.	.017	.23	.81	1.04
Stain, sealer, & 1 coat	1.000	S.F.	.017	.16	.82	.98
2 coats	1.000	S.F.	.019	.25	.89	1.14

4 | EXTERIOR WALLS — 12 | Shingle Siding Systems

System Description	QUAN.	UNIT	LABOR HOURS	COST PER S.F. MAT.	COST PER S.F. INST.	COST PER S.F. TOTAL
WHITE CEDAR SHINGLES, 5" EXPOSURE						
White cedar shingles, 16" long, grade "A", 5" exposure	1.000	S.F.	.033	2.12	1.89	4.01
Building wrap, spunbonded polypropylene	1.100	S.F.	.002	.17	.12	.29
Trim, cedar	.125	S.F.	.005	.16	.28	.44
Paint, primer & 1 coat	1.000	S.F.	.017	.16	.82	.98
TOTAL		S.F.	.057	2.61	3.11	5.72
RESQUARED & REBUTTED PERFECTIONS, 5-1/2" EXPOSURE						
Resquared & rebutted perfections, 5-1/2" exposure	1.000	S.F.	.027	3.10	1.51	4.61
Building wrap, spunbonded polypropylene	1.100	S.F.	.002	.17	.12	.29
Trim, cedar	.125	S.F.	.005	.16	.28	.44
Stain, sealer & 1 coat	1.000	S.F.	.017	.16	.82	.98
TOTAL		S.F.	.051	3.59	2.73	6.32
HAND-SPLIT SHAKES, 8-1/2" EXPOSURE						
Hand-split red cedar shakes, 18" long, 8-1/2" exposure	1.000	S.F.	.040	1.97	2.27	4.24
Building wrap, spunbonded polypropylene	1.100	S.F.	.002	.17	.12	.29
Trim, cedar	.125	S.F.	.005	.16	.28	.44
Stain, sealer & 1 coat	1.000	S.F.	.017	.16	.82	.98
TOTAL		S.F.	.064	2.46	3.49	5.95

The costs in this system are based on a square foot of wall area.
Do not subtract area for door or window openings.

Description	QUAN.	UNIT	LABOR HOURS	COST PER S.F. MAT.	COST PER S.F. INST.	COST PER S.F. TOTAL

Shingle Siding Price Sheet

Shingle Siding Price Sheet	QUAN.	UNIT	LABOR HOURS	COST PER S.F. MAT.	COST PER S.F. INST.	COST PER S.F. TOTAL
Shingles wood, white cedar 16" long, "A" grade, 5" exposure	1.000	S.F.	.033	2.12	1.89	4.01
7" exposure	1.000	S.F.	.030	1.91	1.70	3.61
8-1/2" exposure	1.000	S.F.	.032	1.21	1.82	3.03
10" exposure	1.000	S.F.	.028	1.06	1.59	2.65
"B" grade, 5" exposure	1.000	S.F.	.040	1.82	2.27	4.09
7" exposure	1.000	S.F.	.028	1.27	1.59	2.86
8-1/2" exposure	1.000	S.F.	.024	1.09	1.36	2.45
10" exposure	1.000	S.F.	.020	.91	1.14	2.05
Fire retardant, "A" grade, 5" exposure	1.000	S.F.	.033	2.74	1.89	4.63
7" exposure	1.000	S.F.	.028	1.70	1.59	3.29
8-1/2" exposure	1.000	S.F.	.032	1.81	1.82	3.63
10" exposure	1.000	S.F.	.025	1.41	1.41	2.82
Fire retardant, 5" exposure	1.000	S.F.	.029	3.35	1.65	5
7" exposure	1.000	S.F.	.036	2.63	2.02	4.65
8-1/2" exposure	1.000	S.F.	.032	2.36	1.82	4.18
10" exposure	1.000	S.F.	.025	1.84	1.41	3.25
Resquared & rebutted, 5-1/2" exposure	1.000	S.F.	.027	3.10	1.51	4.61
7" exposure	1.000	S.F.	.024	2.79	1.36	4.15
8-1/2" exposure	1.000	S.F.	.021	2.48	1.21	3.69
10" exposure	1.000	S.F.	.019	2.17	1.06	3.23
Fire retardant, 5" exposure	1.000	S.F.	.027	3.72	1.51	5.23
7" exposure	1.000	S.F.	.024	3.34	1.36	4.70
8-1/2" exposure	1.000	S.F.	.021	2.97	1.21	4.18
10" exposure	1.000	S.F.	.023	2.03	1.30	3.33
Hand-split, red cedar, 24" long, 7" exposure	1.000	S.F.	.045	4.41	2.55	6.96
8-1/2" exposure	1.000	S.F.	.038	3.78	2.18	5.96
10" exposure	1.000	S.F.	.032	3.15	1.82	4.97
12" exposure	1.000	S.F.	.026	2.52	1.46	3.98
Fire retardant, 7" exposure	1.000	S.F.	.045	5.25	2.55	7.80
8-1/2" exposure	1.000	S.F.	.038	4.52	2.18	6.70
10" exposure	1.000	S.F.	.032	3.77	1.82	5.59
12" exposure	1.000	S.F.	.026	3.01	1.46	4.47
18" long, 5" exposure	1.000	S.F.	.068	3.35	3.86	7.21
7" exposure	1.000	S.F.	.048	2.36	2.72	5.08
8-1/2" exposure	1.000	S.F.	.040	1.97	2.27	4.24
10" exposure	1.000	S.F.	.036	1.77	2.04	3.81
Fire retardant, 5" exposure	1.000	S.F.	.068	4.40	3.86	8.26
7" exposure	1.000	S.F.	.048	3.10	2.72	5.82
8-1/2" exposure	1.000	S.F.	.040	2.59	2.27	4.86
10" exposure	1.000	S.F.	.036	2.32	2.04	4.36
Paper, #15 asphalt felt	1.100	S.F.	.002	.06	.12	.18
Trim, cedar	.125	S.F.	.005	.16	.28	.44
Fir	.125	S.F.	.005	.12	.28	.40
Redwood	.125	S.F.	.005	.16	.28	.44
White pine	.125	S.F.	.005	.12	.28	.40
Painting, primer, & 1 coat	1.000	S.F.	.013	.15	.63	.78
2 coats	1.000	S.F.	.017	.23	.81	1.04
Staining, sealer, & 1 coat	1.000	S.F.	.017	.16	.82	.98
2 coats	1.000	S.F.	.019	.25	.89	1.14

4 | EXTERIOR WALLS | 16 | Metal & Plastic Siding Systems

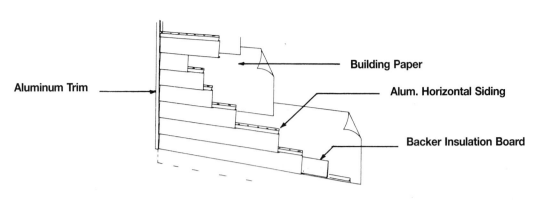

System Description	QUAN.	UNIT	LABOR HOURS	COST PER S.F. MAT.	COST PER S.F. INST.	COST PER S.F. TOTAL
ALUMINUM CLAPBOARD SIDING, 8" WIDE, WHITE						
Aluminum horizontal siding, 8" clapboard	1.000	S.F.	.031	3	1.76	4.76
Backer, insulation board	1.000	S.F.	.008	.42	.45	.87
Trim, aluminum	.600	L.F.	.016	1.12	.89	2.01
Building wrap, spunbonded polypropylene	1.100	S.F.	.002	.17	.12	.29
TOTAL		S.F.	.057	4.71	3.22	7.93
ALUMINUM VERTICAL BOARD & BATTEN, WHITE						
Aluminum vertical board & batten	1.000	S.F.	.027	2.56	1.54	4.10
Backer insulation board	1.000	S.F.	.008	.42	.45	.87
Trim, aluminum	.600	L.F.	.016	1.12	.89	2.01
Building wrap, spunbonded polypropylene	1.100	S.F.	.002	.17	.12	.29
TOTAL		S.F.	.053	4.27	3	7.27
VINYL CLAPBOARD SIDING, 8" WIDE, WHITE						
Vinyl siding, clabboard profile, smooth texture, .042 thick, single 8	1.000	S.F.	.032	.87	1.83	2.70
Backer, insulation board	1.000	S.F.	.008	.42	.45	.87
Vinyl siding, access., outside corner, woodgrain, 4" face, 3/4" pocket	.600	L.F.	.014	1.37	.78	2.15
Building wrap, spunbonded polypropylene	1.100	S.F.	.002	.17	.12	.29
TOTAL		S.F.	.056	2.83	3.18	6.01
VINYL VERTICAL BOARD & BATTEN, WHITE						
Vinyl siding, vertical pattern, .046 thick, double 5	1.000	S.F.	.029	1.72	1.65	3.37
Backer, insulation board	1.000	S.F.	.008	.42	.45	.87
Vinyl siding, access., outside corner, woodgrain, 4" face, 3/4" pocket	.600	L.F.	.014	1.37	.78	2.15
Building wrap, spunbonded polypropylene	1.100	S.F.	.002	.17	.12	.29
TOTAL		S.F.	.053	3.68	3	6.68

The costs in this system are on a square foot of wall basis.
Subtract openings from wall area.

Description	QUAN.	UNIT	LABOR HOURS	COST PER S.F. MAT.	COST PER S.F. INST.	COST PER S.F. TOTAL

Metal & Plastic Siding Price Sheet

Description	QUAN.	UNIT	LABOR HOURS	COST PER S.F. MAT.	COST PER S.F. INST.	COST PER S.F. TOTAL
Siding, aluminum, .024" thick, smooth, 8" wide, white	1.000	S.F.	.031	3	1.76	4.76
Color	1.000	S.F.	.031	3.17	1.76	4.93
Double 4" pattern, 8" wide, white	1.000	S.F.	.031	3.08	1.76	4.84
Color	1.000	S.F.	.031	3.25	1.76	5.01
Double 5" pattern, 10" wide, white	1.000	S.F.	.029	2.99	1.65	4.64
Color	1.000	S.F.	.029	3.16	1.65	4.81
Embossed, single, 8" wide, white	1.000	S.F.	.031	2.99	1.76	4.75
Color	1.000	S.F.	.031	3.16	1.76	4.92
Double 4" pattern, 8" wide, white	1.000	S.F.	.031	3.07	1.76	4.83
Color	1.000	S.F.	.031	3.24	1.76	5
Double 5" pattern, 10" wide, white	1.000	S.F.	.029	3.05	1.65	4.70
Color	1.000	S.F.	.029	3.22	1.65	4.87
Alum siding with insulation board, smooth, 8" wide, white	1.000	S.F.	.031	2.70	1.76	4.46
Color	1.000	S.F.	.031	2.87	1.76	4.63
Double 4" pattern, 8" wide, white	1.000	S.F.	.031	2.66	1.76	4.42
Color	1.000	S.F.	.031	2.83	1.76	4.59
Double 5" pattern, 10" wide, white	1.000	S.F.	.029	2.66	1.65	4.31
Color	1.000	S.F.	.029	2.83	1.65	4.48
Embossed, single, 8" wide, white	1.000	S.F.	.031	3.10	1.76	4.86
Color	1.000	S.F.	.031	3.27	1.76	5.03
Double 4" pattern, 8" wide, white	1.000	S.F.	.031	3.14	1.76	4.90
Color	1.000	S.F.	.031	3.31	1.76	5.07
Double 5" pattern, 10" wide, white	1.000	S.F.	.029	3.14	1.65	4.79
Color	1.000	S.F.	.029	3.31	1.65	4.96
Aluminum, shake finish, 10" wide, white	1.000	S.F.	.029	3.37	1.65	5.02
Color	1.000	S.F.	.029	3.54	1.65	5.19
Aluminum, vertical, 12" wide, white	1.000	S.F.	.027	2.56	1.54	4.10
Color	1.000	S.F.	.027	2.73	1.54	4.27
Vinyl siding, 8" wide, smooth, white	1.000	S.F.	.032	.87	1.83	2.70
Color	1.000	S.F.	.032	1.02	1.83	2.85
10" wide, Dutch lap, smooth, white	1.000	S.F.	.029	1.17	1.65	2.82
Color	1.000	S.F.	.029	1.32	1.65	2.97
Double 4" pattern, 8" wide, white	1.000	S.F.	.032	.87	1.83	2.70
Color	1.000	S.F.	.032	1.02	1.83	2.85
Double 5" pattern, 10" wide, white	1.000	S.F.	.029	.87	1.65	2.52
Color	1.000	S.F.	.029	1.02	1.65	2.67
Embossed, single, 8" wide, white	1.000	S.F.	.032	1.49	1.83	3.32
Color	1.000	S.F.	.032	1.64	1.83	3.47
10" wide, white	1.000	S.F.	.029	1.79	1.65	3.44
Color	1.000	S.F.	.029	1.94	1.65	3.59
Double 4" pattern, 8" wide, white	1.000	S.F.	.032	1.17	1.83	3
Color	1.000	S.F.	.032	1.32	1.83	3.15
Double 5" pattern, 10" wide, white	1.000	S.F.	.029	1.17	1.65	2.82
Color	1.000	S.F.	.029	1.32	1.65	2.97
Vinyl, shake finish, 10" wide, white	1.000	S.F.	.029	3.98	2.27	6.25
Color	1.000	S.F.	.029	4.13	2.27	6.40
Vinyl, vertical, double 5" pattern, 10" wide, white	1.000	S.F.	.029	1.72	1.65	3.37
Color	1.000	S.F.	.029	1.87	1.65	3.52
Backer board, installed in siding panels 8" or 10" wide	1.000	S.F.	.008	.42	.45	.87
4' x 8' sheets, polystyrene, 3/4" thick	1.000	S.F.	.010	.62	.57	1.19
4' x 8' fiberboard, plain	1.000	S.F.	.008	.42	.45	.87
Trim, aluminum, white	.600	L.F.	.016	1.12	.89	2.01
Color	.600	L.F.	.016	1.21	.89	2.10
Vinyl, white	.600	L.F.	.014	1.37	.78	2.15
Color	.600	L.F.	.014	1.57	.78	2.35
Paper, #15 asphalt felt	1.100	S.F.	.002	.07	.13	.20
Kraft paper, plain	1.100	S.F.	.002	.14	.13	.27
Foil backed	1.100	S.F.	.002	.17	.13	.30

4 | EXTERIOR WALLS — 20 | Insulation Systems

Description	QUAN.	UNIT	LABOR HOURS	COST PER S.F. MAT.	COST PER S.F. INST.	COST PER S.F. TOTAL
Poured insulation, cellulose fiber, R3.8 per inch (1" thick)	1.000	S.F.	.003	.06	.19	.25
Fiberglass, R4.0 per inch (1" thick)	1.000	S.F.	.003	.05	.19	.24
Mineral wool, R3.0 per inch (1" thick)	1.000	S.F.	.003	.04	.19	.23
Polystyrene, R4.0 per inch (1" thick)	1.000	S.F.	.003	.15	.19	.34
Vermiculite, R2.7 per inch (1" thick)	1.000	S.F.	.003	.48	.19	.67
Perlite, R2.7 per inch (1" thick)	1.000	S.F.	.003	.48	.19	.67
Reflective insulation, aluminum foil reinforced with scrim	1.000	S.F.	.004	.16	.24	.40
Reinforced with woven polyolefin	1.000	S.F.	.004	.24	.24	.48
With single bubble air space, R8.8	1.000	S.F.	.005	.29	.31	.60
With double bubble air space, R9.8	1.000	S.F.	.005	.34	.31	.65
Rigid insulation, fiberglass, unfaced,						
1-1/2" thick, R6.2	1.000	S.F.	.008	.42	.45	.87
2" thick, R8.3	1.000	S.F.	.008	.54	.45	.99
2-1/2" thick, R10.3	1.000	S.F.	.010	.62	.57	1.19
3" thick, R12.4	1.000	S.F.	.010	.62	.57	1.19
Foil faced, 1" thick, R4.3	1.000	S.F.	.008	.99	.45	1.44
1-1/2" thick, R6.2	1.000	S.F.	.008	1.49	.45	1.94
2" thick, R8.7	1.000	S.F.	.009	1.86	.51	2.37
2-1/2" thick, R10.9	1.000	S.F.	.010	2.18	.57	2.75
3" thick, R13.0	1.000	S.F.	.010	2.40	.57	2.97
Perlite, 1" thick R2.77	1.000	S.F.	.010	.48	.57	1.05
2" thick R5.55	1.000	S.F.	.011	.85	.62	1.47
Polystyrene, extruded, blue, 2.2#/C.F., 3/4" thick R4	1.000	S.F.	.010	.62	.57	1.19
1-1/2" thick R8.1	1.000	S.F.	.011	1.24	.62	1.86
2" thick R10.8	1.000	S.F.	.011	1.74	.62	2.36
Molded bead board, white, 1" thick R3.85	1.000	S.F.	.010	.29	.57	.86
1-1/2" thick, R5.6	1.000	S.F.	.011	.57	.62	1.19
2" thick, R7.7	1.000	S.F.	.011	.86	.62	1.48
Non-rigid insulation, batts						
Fiberglass, kraft faced, 3-1/2" thick, R13, 11" wide	1.000	S.F.	.005	.36	.40	.76
15" wide	1.000	S.F.	.005	.36	.40	.76
23" wide	1.000	S.F.	.005	.36	.40	.76
6" thick, R19, 11" wide	1.000	S.F.	.006	.48	.40	.88
15" wide	1.000	S.F.	.006	.48	.40	.88
23" wide	1.000	S.F.	.006	.48	.40	.88
9" thick, R30, 15" wide	1.000	S.F.	.006	.80	.34	1.14
23" wide	1.000	S.F.	.006	.80	.34	1.14
12" thick, R38, 15" wide	1.000	S.F.	.006	1.16	.34	1.50
23" wide	1.000	S.F.	.006	1.16	.34	1.50
Fiberglass, foil faced, 3-1/2" thick, R13, 15" wide	1.000	S.F.	.005	.52	.34	.86
23" wide	1.000	S.F.	.005	.52	.34	.86
6" thick, R19, 15" thick	1.000	S.F.	.005	.68	.28	.96
23" wide	1.000	S.F.	.005	.68	.28	.96
9" thick, R30, 15" wide	1.000	S.F.	.006	1.03	.34	1.37
23" wide	1.000	S.F.	.006	1.03	.34	1.37

Insulation Systems	QUAN.	UNIT	LABOR HOURS	COST PER S.F.		
				MAT.	INST.	TOTAL
Non-rigid insulation batts						
Fiberglass unfaced, 3-1/2" thick, R13, 15" wide	1.000	S.F.	.005	.36	.28	.64
23" wide	1.000	S.F.	.005	.36	.28	.64
6" thick, R19, 15" wide	1.000	S.F.	.006	.43	.34	.77
23" wide	1.000	S.F.	.006	.43	.34	.77
9" thick, R19, 15" wide	1.000	S.F.	.007	.66	.40	1.06
23" wide	1.000	S.F.	.007	.66	.40	1.06
12" thick, R38, 15" wide	1.000	S.F.	.007	.83	.40	1.23
23" wide	1.000	S.F.	.007	.83	.40	1.23
Mineral fiber batts, 3" thick, R11	1.000	S.F.	.005	.74	.28	1.02
3-1/2" thick, R13	1.000	S.F.	.005	.74	.28	1.02
6" thick, R19	1.000	S.F.	.005	1.16	.28	1.44
6-1/2" thick, R22	1.000	S.F.	.005	1.16	.28	1.44
10" thick, R30	1.000	S.F.	.006	1.53	.34	1.87

4 | EXTERIOR WALLS — 28 | Double Hung Window Systems

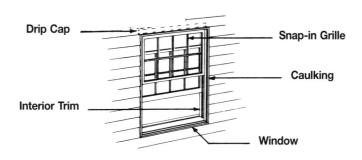

System Description	QUAN.	UNIT	LABOR HOURS	COST EACH MAT.	COST EACH INST.	COST EACH TOTAL
BUILDER'S QUALITY WOOD WINDOW 2' X 3', DOUBLE HUNG						
Window, primed, builder's quality, 2' x 3', insulating glass	1.000	Ea.	.800	256	45.50	301.50
Trim, interior casing	11.000	L.F.	.352	17.05	20.02	37.07
Paint, interior & exterior, primer & 2 coats	2.000	Face	1.778	2.66	84	86.66
Caulking	10.000	L.F.	.278	2.50	15	17.50
Snap-in grille	1.000	Set	.333	56.50	18.90	75.40
Drip cap, metal	2.000	L.F.	.040	1.26	2.28	3.54
TOTAL		Ea.	3.581	335.97	185.70	521.67
PLASTIC CLAD WOOD WINDOW 3' X 4', DOUBLE HUNG						
Window, plastic clad, premium, 3' x 4', insulating glass	1.000	Ea.	.889	435	50.50	485.50
Trim, interior casing	15.000	L.F.	.480	23.25	27.30	50.55
Paint, interior, primer & 2 coats	1.000	Face	.889	1.33	42	43.33
Caulking	14.000	L.F.	.389	3.50	21	24.50
Snap-in grille	1.000	Set	.333	56.50	18.90	75.40
TOTAL		Ea.	2.980	519.58	159.70	679.28
METAL CLAD WOOD WINDOW, 3' X 5', DOUBLE HUNG						
Window, metal clad, deluxe, 3' x 5', insulating glass	1.000	Ea.	1.000	415	57	472
Trim, interior casing	17.000	L.F.	.544	26.35	30.94	57.29
Paint, interior, primer & 2 coats	1.000	Face	.889	1.33	42	43.33
Caulking	16.000	L.F.	.444	4	24	28
Snap-in grille	1.000	Set	.235	144	13.35	157.35
Drip cap, metal	3.000	L.F.	.060	1.89	3.42	5.31
TOTAL		Ea.	3.172	592.57	170.71	763.28

The cost of this system is on a cost per each window basis.

Description	QUAN.	UNIT	LABOR HOURS	COST EACH MAT.	COST EACH INST.	COST EACH TOTAL

Double Hung Window Price Sheet

Description	QUAN.	UNIT	LABOR HOURS	COST EACH MAT.	COST EACH INST.	COST EACH TOTAL
Windows, double-hung, builder's quality, 2' x 3', single glass	1.000	Ea.	.800	231	45.50	276.50
Insulating glass	1.000	Ea.	.800	256	45.50	301.50
3' x 4', single glass	1.000	Ea.	.889	305	50.50	355.50
Insulating glass	1.000	Ea.	.889	320	50.50	370.50
4' x 4'-6", single glass	1.000	Ea.	1.000	355	57	412
Insulating glass	1.000	Ea.	1.000	380	57	437
Plastic clad premium insulating glass, 2'-6" x 3'	1.000	Ea.	.800	340	45.50	385.50
3' x 3'-6"	1.000	Ea.	.800	370	45.50	415.50
3' x 4'	1.000	Ea.	.889	435	50.50	485.50
3' x 4'-6"	1.000	Ea.	.889	450	50.50	500.50
3' x 5'	1.000	Ea.	1.000	485	57	542
3'-6" x 6'	1.000	Ea.	1.000	540	57	597
Metal clad deluxe insulating glass, 2'-6" x 3'	1.000	Ea.	.800	310	45.50	355.50
3' x 3'-6"	1.000	Ea.	.800	350	45.50	395.50
3' x 4'	1.000	Ea.	.889	370	50.50	420.50
3' x 4'-6"	1.000	Ea.	.889	385	50.50	435.50
3' x 5'	1.000	Ea.	1.000	415	57	472
3'-6" x 6'	1.000	Ea.	1.000	505	57	562
Trim, interior casing, window 2' x 3'	11.000	L.F.	.367	17.05	20	37.05
2'-6" x 3'	12.000	L.F.	.400	18.60	22	40.60
3' x 3'-6"	14.000	L.F.	.467	21.50	25.50	47
3' x 4'	15.000	L.F.	.500	23.50	27.50	51
3' x 4'-6"	16.000	L.F.	.533	25	29	54
3' x 5'	17.000	L.F.	.567	26.50	31	57.50
3'-6" x 6'	20.000	L.F.	.667	31	36.50	67.50
4' x 4'-6"	18.000	L.F.	.600	28	33	61
Paint or stain, interior or exterior, 2' x 3' window, 1 coat	1.000	Face	.444	.51	21	21.51
2 coats	1.000	Face	.727	1.02	34.50	35.52
Primer & 1 coat	1.000	Face	.727	.88	34.50	35.38
Primer & 2 coats	1.000	Face	.889	1.33	42	43.33
3' x 4' window, 1 coat	1.000	Face	.667	1.05	31.50	32.55
2 coats	1.000	Face	.667	1.81	31.50	33.31
Primer & 1 coat	1.000	Face	.727	1.36	34.50	35.86
Primer & 2 coats	1.000	Face	.889	1.33	42	43.33
4' x 4'-6" window, 1 coat	1.000	Face	.667	1.05	31.50	32.55
2 coats	1.000	Face	.667	1.81	31.50	33.31
Primer & 1 coat	1.000	Face	.727	1.36	34.50	35.86
Primer & 2 coats	1.000	Face	.889	1.33	42	43.33
Caulking, window, 2' x 3'	10.000	L.F.	.323	2.50	15	17.50
2'-6" x 3'	11.000	L.F.	.355	2.75	16.50	19.25
3' x 3'-6"	13.000	L.F.	.419	3.25	19.50	22.75
3' x 4'	14.000	L.F.	.452	3.50	21	24.50
3' x 4'-6"	15.000	L.F.	.484	3.75	22.50	26.25
3' x 5'	16.000	L.F.	.516	4	24	28
3'-6" x 6'	19.000	L.F.	.613	4.75	28.50	33.25
4' x 4'-6"	17.000	L.F.	.548	4.25	25.50	29.75
Grilles, glass size to, 16" x 24" per sash	1.000	Set	.333	56.50	18.90	75.40
32" x 32" per sash	1.000	Set	.235	144	13.35	157.35
Drip cap, aluminum, 2' long	2.000	L.F.	.040	1.26	2.28	3.54
3' long	3.000	L.F.	.060	1.89	3.42	5.31
4' long	4.000	L.F.	.080	2.52	4.56	7.08
Wood, 2' long	2.000	L.F.	.067	3.10	3.64	6.74
3' long	3.000	L.F.	.100	4.65	5.45	10.10
4' long	4.000	L.F.	.133	6.20	7.30	13.50

4 | EXTERIOR WALLS 32 | Casement Window Systems

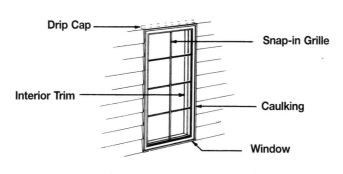

System Description	QUAN.	UNIT	LABOR HOURS	COST EACH		
				MAT.	INST.	TOTAL
BUILDER'S QUALITY WINDOW, WOOD, 2' BY 3', CASEMENT						
Window, primed, builder's quality, 2' x 3', insulating glass	1.000	Ea.	.800	285	45.50	330.50
Trim, interior casing	11.000	L.F.	.352	17.05	20.02	37.07
Paint, interior & exterior, primer & 2 coats	2.000	Face	1.778	2.66	84	86.66
Caulking	10.000	L.F.	.278	2.50	15	17.50
Snap-in grille	1.000	Ea.	.267	35	15.15	50.15
Drip cap, metal	2.000	L.F.	.040	1.26	2.28	3.54
TOTAL		Ea.	3.515	343.47	181.95	525.42
PLASTIC CLAD WOOD WINDOW, 2' X 4', CASEMENT						
Window, plastic clad, premium, 2' x 4', insulating glass	1.000	Ea.	.889	350	50.50	400.50
Trim, interior casing	13.000	L.F.	.416	20.15	23.66	43.81
Paint, interior, primer & 2 coats	1.000	Ea.	.889	1.33	42	43.33
Caulking	12.000	L.F.	.333	3	18	21
Snap-in grille	1.000	Ea.	.267	35	15.15	50.15
TOTAL		Ea.	2.794	409.48	149.31	558.79
METAL CLAD WOOD WINDOW, 2' X 5', CASEMENT						
Window, metal clad, deluxe, 2' x 5', insulating glass	1.000	Ea.	1.000	365	57	422
Trim, interior casing	15.000	L.F.	.480	23.25	27.30	50.55
Paint, interior, primer & 2 coats	1.000	Ea.	.889	1.33	42	43.33
Caulking	14.000	L.F.	.389	3.50	21	24.50
Snap-in grille	1.000	Ea.	.250	47.50	14.20	61.70
Drip cap, metal	12.000	L.F.	.040	1.26	2.28	3.54
TOTAL		Ea.	3.048	441.84	163.78	605.62

The cost of this system is on a cost per each window basis.

Description	QUAN.	UNIT	LABOR HOURS	COST EACH		
				MAT.	INST.	TOTAL

Casement Window Price Sheet

	QUAN.	UNIT	LABOR HOURS	COST EACH MAT.	COST EACH INST.	COST EACH TOTAL
Window, casement, builders quality, 2' x 3', double insulated glass	1.000	Ea.	.800	300	45.50	345.50
Insulating glass	1.000	Ea.	.800	285	45.50	330.50
2' x 4'-6", double insulated glass	1.000	Ea.	.727	980	41.50	1,021.50
Insulating glass	1.000	Ea.	.727	590	41.50	631.50
2' x 6', double insulated glass	1.000	Ea.	.889	485	57	542
Insulating glass	1.000	Ea.	.889	485	57	542
Plastic clad premium insulating glass, 2' x 3'	1.000	Ea.	.800	297	45.50	342.50
2' x 4'	1.000	Ea.	.889	450	50.50	500.50
2' x 5'	1.000	Ea.	1.000	610	57	667
2' x 6'	1.000	Ea.	1.000	420	57	477
Metal clad deluxe insulating glass, 2' x 3'	1.000	Ea.	.800	315	45.50	360.50
2' x 4'	1.000	Ea.	.889	310	50.50	360.50
2' x 5'	1.000	Ea.	1.000	355	57	412
2' x 6'	1.000	Ea.	1.000	405	57	462
Trim, interior casing, window 2' x 3'	11.000	L.F.	.367	17.05	20	37.05
2' x 4'	13.000	L.F.	.433	20	23.50	43.50
2' x 4'-6"	14.000	L.F.	.467	21.50	25.50	47
2' x 5'	15.000	L.F.	.500	23.50	27.50	51
2' x 6'	17.000	L.F.	.567	26.50	31	57.50
Paint or stain, interior or exterior, 2' x 3' window, 1 coat	1.000	Face	.444	.51	21	21.51
2 coats	1.000	Face	.727	1.02	34.50	35.52
Primer & 1 coat	1.000	Face	.727	.88	34.50	35.38
Primer & 2 coats	1.000	Face	.889	1.33	42	43.33
2' x 4' window, 1 coat	1.000	Face	.444	.51	21	21.51
2 coats	1.000	Face	.727	1.02	34.50	35.52
Primer & 1 coat	1.000	Face	.727	.88	34.50	35.38
Primer & 2 coats	1.000	Face	.889	1.33	42	43.33
2' x 6' window, 1 coat	1.000	Face	.667	1.05	31.50	32.55
2 coats	1.000	Face	.667	1.81	31.50	33.31
Primer & 1 coat	1.000	Face	.727	1.36	34.50	35.86
Primer & 2 coats	1.000	Face	.889	1.33	42	43.33
Caulking, window, 2' x 3'	10.000	L.F.	.323	2.50	15	17.50
2' x 4'	12.000	L.F.	.387	3	18	21
2' x 4'-6"	13.000	L.F.	.419	3.25	19.50	22.75
2' x 5'	14.000	L.F.	.452	3.50	21	24.50
2' x 6'	16.000	L.F.	.516	4	24	28
Grilles, glass size, to 20" x 36"	1.000	Ea.	.267	35	15.15	50.15
To 20" x 56"	1.000	Ea.	.250	47.50	14.20	61.70
Drip cap, metal, 2' long	2.000	L.F.	.040	1.26	2.28	3.54
Wood, 2' long	2.000	L.F.	.067	3.10	3.64	6.74

4 | EXTERIOR WALLS 36 | Awning Window Systems

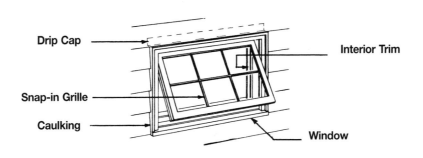

System Description	QUAN.	UNIT	LABOR HOURS	COST EACH		
				MAT.	INST.	TOTAL
BUILDER'S QUALITY WINDOW, WOOD, 34" X 22", AWNING						
Window, 34" x 22", insulating glass	1.000	Ea.	.800	299	45.50	344.50
Trim, interior casing	10.500	L.F.	.336	16.28	19.11	35.39
Paint, interior & exterior, primer & 2 coats	2.000	Face	1.778	2.66	84	86.66
Caulking	9.500	L.F.	.264	2.38	14.25	16.63
Snap-in grille	1.000	Ea.	.267	31	15.15	46.15
Drip cap, metal	3.000	L.F.	.060	1.89	3.42	5.31
TOTAL		Ea.	3.505	353.21	181.43	534.64
PLASTIC CLAD WOOD WINDOW, 40" X 28", AWNING						
Window, plastic clad, premium, 40" x 28", insulating glass	1.000	Ea.	.889	375	50.50	425.50
Trim interior casing	13.500	L.F.	.432	20.93	24.57	45.50
Paint, interior, primer & 2 coats	1.000	Face	.889	1.33	42	43.33
Caulking	12.500	L.F.	.347	3.13	18.75	21.88
Snap-in grille	1.000	Ea.	.267	31	15.15	46.15
TOTAL		Ea.	2.824	431.39	150.97	582.36
METAL CLAD WOOD WINDOW, 48" X 36", AWNING						
Window, metal clad, deluxe, 48" x 36", insulating glass	1.000	Ea.	1.000	410	57	467
Trim, interior casing	15.000	L.F.	.480	23.25	27.30	50.55
Paint, interior, primer & 2 coats	1.000	Face	.889	1.33	42	43.33
Caulking	14.000	L.F.	.389	3.50	21	24.50
Snap-in grille	1.000	Ea.	.250	44	14.20	58.20
Drip cap, metal	4.000	L.F.	.080	2.52	4.56	7.08
TOTAL		Ea.	3.088	484.60	166.06	650.66

The cost of this system is on a cost per each window basis.

Description	QUAN.	UNIT	LABOR HOURS	COST EACH		
				MAT.	INST.	TOTAL

Awning Window Price Sheet

	QUAN.	UNIT	LABOR HOURS	COST EACH MAT.	COST EACH INST.	COST EACH TOTAL
Windows, awning, builder's quality, 34" x 22", insulated glass	1.000	Ea.	.800	298	45.50	343.50
Low E glass	1.000	Ea.	.800	299	45.50	344.50
40" x 28", insulated glass	1.000	Ea.	.889	345	50.50	395.50
Low E glass	1.000	Ea.	.889	375	50.50	425.50
48" x 36", insulated glass	1.000	Ea.	1.000	515	57	572
Low E glass	1.000	Ea.	1.000	540	57	597
Plastic clad premium insulating glass, 34" x 22"	1.000	Ea.	.800	293	45.50	338.50
40" x 22"	1.000	Ea.	.800	320	45.50	365.50
36" x 28"	1.000	Ea.	.889	335	50.50	385.50
36" x 36"	1.000	Ea.	.889	375	50.50	425.50
48" x 28"	1.000	Ea.	1.000	405	57	462
60" x 36"	1.000	Ea.	1.000	550	57	607
Metal clad deluxe insulating glass, 34" x 22"	1.000	Ea.	.800	273	50.50	323.50
40" x 22"	1.000	Ea.	.800	320	50.50	370.50
36" x 25"	1.000	Ea.	.889	305	50.50	355.50
40" x 30"	1.000	Ea.	.889	375	50.50	425.50
48" x 28"	1.000	Ea.	1.000	385	57	442
60" x 36"	1.000	Ea.	1.000	410	57	467
Trim, interior casing window, 34" x 22"	10.500	L.F.	.350	16.30	19.10	35.40
40" x 22"	11.500	L.F.	.383	17.85	21	38.85
36" x 28"	12.500	L.F.	.417	19.40	23	42.40
40" x 28"	13.500	L.F.	.450	21	24.50	45.50
48" x 28"	14.500	L.F.	.483	22.50	26.50	49
48" x 36"	15.000	L.F.	.500	23.50	27.50	51
Paint or stain, interior or exterior, 34" x 22", 1 coat	1.000	Face	.444	.51	21	21.51
2 coats	1.000	Face	.727	1.02	34.50	35.52
Primer & 1 coat	1.000	Face	.727	.88	34.50	35.38
Primer & 2 coats	1.000	Face	.889	1.33	42	43.33
36" x 28", 1 coat	1.000	Face	.444	.51	21	21.51
2 coats	1.000	Face	.727	1.02	34.50	35.52
Primer & 1 coat	1.000	Face	.727	.88	34.50	35.38
Primer & 2 coats	1.000	Face	.889	1.33	42	43.33
48" x 36", 1 coat	1.000	Face	.667	1.05	31.50	32.55
2 coats	1.000	Face	.667	1.81	31.50	33.31
Primer & 1 coat	1.000	Face	.727	1.36	34.50	35.86
Primer & 2 coats	1.000	Face	.889	1.33	42	43.33
Caulking, window, 34" x 22"	9.500	L.F.	.306	2.38	14.25	16.63
40" x 22"	10.500	L.F.	.339	2.63	15.75	18.38
36" x 28"	11.500	L.F.	.371	2.88	17.25	20.13
40" x 28"	12.500	L.F.	.403	3.13	18.75	21.88
48" x 28"	13.500	L.F.	.436	3.38	20.50	23.88
48" x 36"	14.000	L.F.	.452	3.50	21	24.50
Grilles, glass size, to 28" by 16"	1.000	Ea.	.267	31	15.15	46.15
To 44" by 24"	1.000	Ea.	.250	44	14.20	58.20
Drip cap, aluminum, 3' long	3.000	L.F.	.060	1.89	3.42	5.31
3'-6" long	3.500	L.F.	.070	2.21	3.99	6.20
4' long	4.000	L.F.	.080	2.52	4.56	7.08
Wood, 3' long	3.000	L.F.	.100	4.65	5.45	10.10
3'-6" long	3.500	L.F.	.117	5.45	6.35	11.80
4' long	4.000	L.F.	.133	6.20	7.30	13.50

4 | EXTERIOR WALLS — 40 | Sliding Window Systems

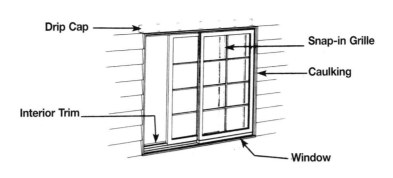

System Description	QUAN.	UNIT	LABOR HOURS	COST EACH MAT.	COST EACH INST.	COST EACH TOTAL
BUILDER'S QUALITY WOOD WINDOW, 3' X 2', SLIDING						
Window, primed, builder's quality, 3' x 3', insul. glass	1.000	Ea.	.800	340	45.50	385.50
Trim, interior casing	11.000	L.F.	.352	17.05	20.02	37.07
Paint, interior & exterior, primer & 2 coats	2.000	Face	1.778	2.66	84	86.66
Caulking	10.000	L.F.	.278	2.50	15	17.50
Snap-in grille	1.000	Set	.333	39	18.90	57.90
Drip cap, metal	3.000	L.F.	.060	1.89	3.42	5.31
TOTAL		Ea.	3.601	403.10	186.84	589.94
PLASTIC CLAD WOOD WINDOW, 4' X 3'-6", SLIDING						
Window, plastic clad, premium, 4' x 3'-6", insulating glass	1.000	Ea.	.889	755	50.50	805.50
Trim, interior casing	16.000	L.F.	.512	24.80	29.12	53.92
Paint, interior, primer & 2 coats	1.000	Face	.889	1.33	42	43.33
Caulking	17.000	L.F.	.472	4.25	25.50	29.75
Snap-in grille	1.000	Set	.333	39	18.90	57.90
TOTAL		Ea.	3.095	824.38	166.02	990.40
METAL CLAD WOOD WINDOW, 6' X 5', SLIDING						
Window, metal clad, deluxe, 6' x 5', insulating glass	1.000	Ea.	1.000	800	57	857
Trim, interior casing	23.000	L.F.	.736	35.65	41.86	77.51
Paint, interior, primer & 2 coats	1.000	Face	.889	1.33	42	43.33
Caulking	22.000	L.F.	.611	5.50	33	38.50
Snap-in grille	1.000	Set	.364	47.50	20.50	68
Drip cap, metal	6.000	L.F.	.120	3.78	6.84	10.62
TOTAL		Ea.	3.720	893.76	201.20	1,094.96

The cost of this system is on a cost per each window basis.

Description	QUAN.	UNIT	LABOR HOURS	COST EACH MAT.	COST EACH INST.	COST EACH TOTAL

Sliding Window Price Sheet

	QUAN.	UNIT	LABOR HOURS	COST EACH MAT.	COST EACH INST.	COST EACH TOTAL
Windows, sliding, builder's quality, 3' x 3', single glass	1.000	Ea.	.800	310	45.50	355.50
Insulating glass	1.000	Ea.	.800	340	45.50	385.50
4' x 3'-6", single glass	1.000	Ea.	.889	390	50.50	440.50
Insulating glass	1.000	Ea.	.889	400	50.50	450.50
6' x 5', single glass	1.000	Ea.	1.000	535	57	592
Insulating glass	1.000	Ea.	1.000	580	57	637
Plastic clad premium insulating glass, 3' x 3'	1.000	Ea.	.800	665	45.50	710.50
4' x 3'-6"	1.000	Ea.	.889	755	50.50	805.50
5' x 4'	1.000	Ea.	.889	990	50.50	1,040.50
6' x 5'	1.000	Ea.	1.000	1,250	57	1,307
Metal clad deluxe insulating glass, 3' x 3'	1.000	Ea.	.800	360	45.50	405.50
4' x 3'-6"	1.000	Ea.	.889	440	50.50	490.50
5' x 4'	1.000	Ea.	.889	525	50.50	575.50
6' x 5'	1.000	Ea.	1.000	800	57	857
Trim, interior casing, window 3' x 2'	11.000	L.F.	.367	17.05	20	37.05
3' x 3'	13.000	L.F.	.433	20	23.50	43.50
4' x 3'-6"	16.000	L.F.	.533	25	29	54
5' x 4'	19.000	L.F.	.633	29.50	34.50	64
6' x 5'	23.000	L.F.	.767	35.50	42	77.50
Paint or stain, interior or exterior, 3' x 2' window, 1 coat	1.000	Face	.444	.51	21	21.51
2 coats	1.000	Face	.727	1.02	34.50	35.52
Primer & 1 coat	1.000	Face	.727	.88	34.50	35.38
Primer & 2 coats	1.000	Face	.889	1.33	42	43.33
4' x 3'-6" window, 1 coat	1.000	Face	.667	1.05	31.50	32.55
2 coats	1.000	Face	.667	1.81	31.50	33.31
Primer & 1 coat	1.000	Face	.727	1.36	34.50	35.86
Primer & 2 coats	1.000	Face	.889	1.33	42	43.33
6' x 5' window, 1 coat	1.000	Face	.889	2.88	42	44.88
2 coats	1.000	Face	1.333	5.25	63	68.25
Primer & 1 coat	1.000	Face	1.333	4.71	63	67.71
Primer & 2 coats	1.000	Face	1.600	7.10	75.50	82.60
Caulking, window, 3' x 2'	10.000	L.F.	.323	2.50	15	17.50
3' x 3'	12.000	L.F.	.387	3	18	21
4' x 3'-6"	15.000	L.F.	.484	3.75	22.50	26.25
5' x 4'	18.000	L.F.	.581	4.50	27	31.50
6' x 5'	22.000	L.F.	.710	5.50	33	38.50
Grilles, glass size, to 14" x 36"	1.000	Set	.333	39	18.90	57.90
To 36" x 36"	1.000	Set	.364	47.50	20.50	68
Drip cap, aluminum, 3' long	3.000	L.F.	.060	1.89	3.42	5.31
4' long	4.000	L.F.	.080	2.52	4.56	7.08
5' long	5.000	L.F.	.100	3.15	5.70	8.85
6' long	6.000	L.F.	.120	3.78	6.85	10.63
Wood, 3' long	3.000	L.F.	.100	4.65	5.45	10.10
4' long	4.000	L.F.	.133	6.20	7.30	13.50
5' long	5.000	L.F.	.167	7.75	9.10	16.85
6' long	6.000	L.F.	.200	9.30	10.90	20.20

4 | EXTERIOR WALLS — 44 | Bow/Bay Window Systems

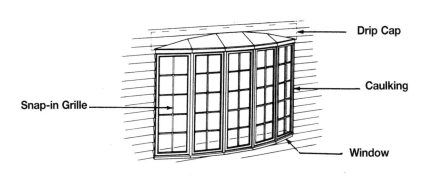

System Description	QUAN.	UNIT	LABOR HOURS	COST EACH MAT.	COST EACH INST.	COST EACH TOTAL
AWNING TYPE BOW WINDOW, BUILDER'S QUALITY, 8' X 5'						
Window, primed, builder's quality, 8' x 5', insulating glass	1.000	Ea.	1.600	1,450	91	1,541
Trim, interior casing	27.000	L.F.	.864	41.85	49.14	90.99
Paint, interior & exterior, primer & 1 coat	2.000	Face	3.200	14.20	151	165.20
Caulking	26.000	L.F.	.722	6.50	39	45.50
Snap-in grilles	1.000	Set	1.067	140	60.60	200.60
TOTAL		Ea.	7.453	1,652.55	390.74	2,043.29
CASEMENT TYPE BOW WINDOW, PLASTIC CLAD, 10' X 6'						
Window, plastic clad, premium, 10' x 6', insulating glass	1.000	Ea.	2.286	2,875	130	3,005
Trim, interior casing	33.000	L.F.	1.056	51.15	60.06	111.21
Paint, interior, primer & 1 coat	1.000	Face	1.778	2.66	84	86.66
Caulking	32.000	L.F.	.889	8	48	56
Snap-in grilles	1.000	Set	1.333	175	75.75	250.75
TOTAL		Ea.	7.342	3,111.81	397.81	3,509.62
DOUBLE HUNG TYPE, METAL CLAD, 9' X 5'						
Window, metal clad, deluxe, 9' x 5', insulating glass	1.000	Ea.	2.667	1,625	151	1,776
Trim, interior casing	29.000	L.F.	.928	44.95	52.78	97.73
Paint, interior, primer & 1 coat	1.000	Face	1.778	2.66	84	86.66
Caulking	28.000	L.F.	.778	7	42	49
Snap-in grilles	1.000	Set	1.067	140	60.60	200.60
TOTAL		Ea.	7.218	1,819.61	390.38	2,209.99

The cost of this system is on a cost per each window basis.

Description	QUAN.	UNIT	LABOR HOURS	COST EACH MAT.	COST EACH INST.	COST EACH TOTAL

Bow/Bay Window Price Sheet

Description	QUAN.	UNIT	LABOR HOURS	COST EACH MAT.	COST EACH INST.	TOTAL
Windows, bow awning type, builder's quality, 8' x 5', insulating glass	1.000	Ea.	1.600	1,650	91	1,741
Low E glass	1.000	Ea.	1.600	1,450	91	1,541
12' x 6', insulating glass	1.000	Ea.	2.667	1,500	151	1,651
Low E glass	1.000	Ea.	2.667	1,600	151	1,751
Plastic clad premium insulating glass, 6' x 4'	1.000	Ea.	1.600	1,125	91	1,216
9' x 4'	1.000	Ea.	2.000	1,500	114	1,614
10' x 5'	1.000	Ea.	2.286	2,475	130	2,605
12' x 6'	1.000	Ea.	2.667	3,225	151	3,376
Metal clad deluxe insulating glass, 6' x 4'	1.000	Ea.	1.600	1,325	91	1,416
9' x 4'	1.000	Ea.	2.000	1,700	114	1,814
10' x 5'	1.000	Ea.	2.286	2,325	130	2,455
12' x 6'	1.000	Ea.	2.667	2,950	151	3,101
Bow casement type, builder's quality, 8' x 5', single glass	1.000	Ea.	1.600	2,100	91	2,191
Insulating glass	1.000	Ea.	1.600	2,550	91	2,641
12' x 6', single glass	1.000	Ea.	2.667	2,625	151	2,776
Insulating glass	1.000	Ea.	2.667	3,500	151	3,651
Plastic clad premium insulating glass, 8' x 5'	1.000	Ea.	1.600	1,925	91	2,016
10' x 5'	1.000	Ea.	2.000	2,525	114	2,639
10' x 6'	1.000	Ea.	2.286	2,875	130	3,005
12' x 6'	1.000	Ea.	2.667	3,525	151	3,676
Metal clad deluxe insulating glass, 8' x 5'	1.000	Ea.	1.600	1,875	91	1,966
10' x 5'	1.000	Ea.	2.000	2,000	114	2,114
10' x 6'	1.000	Ea.	2.286	2,375	130	2,505
12' x 6'	1.000	Ea.	2.667	3,225	151	3,376
Bow, double hung type, builder's quality, 8' x 4', single glass	1.000	Ea.	1.600	1,475	91	1,566
Insulating glass	1.000	Ea.	1.600	1,600	91	1,691
9' x 5', single glass	1.000	Ea.	2.667	1,600	151	1,751
Insulating glass	1.000	Ea.	2.667	1,675	151	1,826
Plastic clad premium insulating glass, 7' x 4'	1.000	Ea.	1.600	1,525	91	1,616
8' x 4'	1.000	Ea.	2.000	1,550	114	1,664
8' x 5'	1.000	Ea.	2.286	1,650	130	1,780
9' x 5'	1.000	Ea.	2.667	1,675	151	1,826
Metal clad deluxe insulating glass, 7' x 4'	1.000	Ea.	1.600	1,425	91	1,516
8' x 4'	1.000	Ea.	2.000	1,475	114	1,589
8' x 5'	1.000	Ea.	2.286	1,525	130	1,655
9' x 5'	1.000	Ea.	2.667	1,625	151	1,776
Trim, interior casing, window 7' x 4'	1.000	Ea.	.767	35.50	42	77.50
8' x 5'	1.000	Ea.	.900	42	49	91
10' x 6'	1.000	Ea.	1.100	51	60	111
12' x 6'	1.000	Ea.	1.233	57.50	67.50	125
Paint or stain, interior, or exterior, 7' x 4' window, 1 coat	1.000	Face	.889	2.88	42	44.88
Primer & 1 coat	1.000	Face	1.333	4.71	63	67.71
8' x 5' window, 1 coat	1.000	Face	.889	2.88	42	44.88
Primer & 1 coat	1.000	Face	1.333	4.71	63	67.71
10' x 6' window, 1 coat	1.000	Face	1.333	2.10	63	65.10
Primer & 1 coat	1.000	Face	1.778	2.66	84	86.66
12' x 6' window, 1 coat	1.000	Face	1.778	5.75	84	89.75
Primer & 1 coat	1.000	Face	2.667	9.40	126	135.40
Drip cap, vinyl moulded window, 7' long	1.000	Ea.	.533	125	24.50	149.50
8' long	1.000	Ea.	.533	143	28	171
Caulking, window, 7' x 4'	1.000	Ea.	.710	5.50	33	38.50
8' x 5'	1.000	Ea.	.839	6.50	39	45.50
10' x 6'	1.000	Ea.	1.032	8	48	56
12' x 6'	1.000	Ea.	1.161	9	54	63
Grilles, window, 7' x 4'	1.000	Set	.800	105	45.50	150.50
8' x 5'	1.000	Set	1.067	140	60.50	200.50
10' x 6'	1.000	Set	1.333	175	76	251
12' x 6'	1.000	Set	1.600	210	91	301

4 | EXTERIOR WALLS — 48 Fixed Window Systems

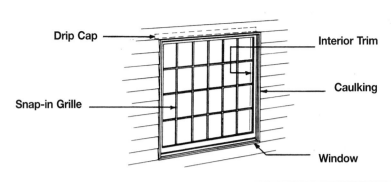

System Description	QUAN.	UNIT	LABOR HOURS	COST EACH MAT.	COST EACH INST.	COST EACH TOTAL
BUILDER'S QUALITY PICTURE WINDOW, 4' X 4'						
Window, primed, builder's quality, 3'-0" x 4', insulating glass	1.000	Ea.	1.333	475	75.50	550.50
Trim, interior casing	17.000	L.F.	.544	26.35	30.94	57.29
Paint, interior & exterior, primer & 2 coats	2.000	Face	1.778	2.66	84	86.66
Caulking	16.000	L.F.	.444	4	24	28
Snap-in grille	1.000	Ea.	.267	132	15.15	147.15
Drip cap, metal	4.000	L.F.	.080	2.52	4.56	7.08
TOTAL		Ea.	4.446	642.53	234.15	876.68
PLASTIC CLAD WOOD WINDOW, 4'-6" X 6'-6"						
Window, plastic clad, prem., 4'-6" x 6'-6", insul. glass	1.000	Ea.	1.455	1,050	82.50	1,132.50
Trim, interior casing	23.000	L.F.	.736	35.65	41.86	77.51
Paint, interior, primer & 2 coats	1.000	Face	.889	1.33	42	43.33
Caulking	22.000	L.F.	.611	5.50	33	38.50
Snap-in grille	1.000	Ea.	.267	132	15.15	147.15
TOTAL		Ea.	3.958	1,224.48	214.51	1,438.99
METAL CLAD WOOD WINDOW, 6'-6" X 6'-6"						
Window, metal clad, deluxe, 6'-0" x 6'-0", insulating glass	1.000	Ea.	1.600	780	91	871
Trim interior casing	27.000	L.F.	.864	41.85	49.14	90.99
Paint, interior, primer & 2 coats	1.000	Face	1.600	7.10	75.50	82.60
Caulking	26.000	L.F.	.722	6.50	39	45.50
Snap-in grille	1.000	Ea.	.267	132	15.15	147.15
Drip cap, metal	6.500	L.F.	.130	4.10	7.41	11.51
TOTAL		Ea.	5.183	971.55	277.20	1,248.75

The cost of this system is on a cost per each window basis.

Description	QUAN.	UNIT	LABOR HOURS	COST EACH MAT.	COST EACH INST.	COST EACH TOTAL

Fixed Window Price Sheet

Fixed Window Price Sheet	QUAN.	UNIT	LABOR HOURS	COST EACH MAT.	COST EACH INST.	COST EACH TOTAL
Window-picture, builder's quality, 4' x 4', single glass	1.000	Ea.	1.333	465	75.50	540.50
Insulating glass	1.000	Ea.	1.333	475	75.50	550.50
4' x 4'-6", single glass	1.000	Ea.	1.455	605	82.50	687.50
Insulating glass	1.000	Ea.	1.455	585	82.50	667.50
5' x 4', single glass	1.000	Ea.	1.455	640	82.50	722.50
Insulating glass	1.000	Ea.	1.455	665	82.50	747.50
6' x 4'-6", single glass	1.000	Ea.	1.600	700	91	791
Insulating glass	1.000	Ea.	1.600	705	91	796
Plastic clad premium insulating glass, 4' x 4'	1.000	Ea.	1.333	570	75.50	645.50
4'-6" x 6'-6"	1.000	Ea.	1.455	1,050	82.50	1,132.50
5'-6" x 6'-6"	1.000	Ea.	1.600	1,175	91	1,266
6'-6" x 6'-6"	1.000	Ea.	1.600	1,250	91	1,341
Metal clad deluxe insulating glass, 4' x 4'	1.000	Ea.	1.333	420	75.50	495.50
4'-6" x 6'-6"	1.000	Ea.	1.455	615	82.50	697.50
5'-6" x 6'-6"	1.000	Ea.	1.600	680	91	771
6'-6" x 6'-6"	1.000	Ea.	1.600	780	91	871
Trim, interior casing, window 4' x 4'	17.000	L.F.	.567	26.50	31	57.50
4'-6" x 4'-6"	19.000	L.F.	.633	29.50	34.50	64
5'-0" x 4'-0"	19.000	L.F.	.633	29.50	34.50	64
4'-6" x 6'-6"	23.000	L.F.	.767	35.50	42	77.50
5'-6" x 6'-6"	25.000	L.F.	.833	39	45.50	84.50
6'-6" x 6'-6"	27.000	L.F.	.900	42	49	91
Paint or stain, interior or exterior, 4' x 4' window, 1 coat	1.000	Face	.667	1.05	31.50	32.55
2 coats	1.000	Face	.667	1.81	31.50	33.31
Primer & 1 coat	1.000	Face	.727	1.36	34.50	35.86
Primer & 2 coats	1.000	Face	.889	1.33	42	43.33
4'-6" x 6'-6" window, 1 coat	1.000	Face	.667	1.05	31.50	32.55
2 coats	1.000	Face	.667	1.81	31.50	33.31
Primer & 1 coat	1.000	Face	.727	1.36	34.50	35.86
Primer & 2 coats	1.000	Face	.889	1.33	42	43.33
6'-6" x 6'-6" window, 1 coat	1.000	Face	.889	2.88	42	44.88
2 coats	1.000	Face	1.333	5.25	63	68.25
Primer & 1 coat	1.000	Face	1.333	4.71	63	67.71
Primer & 2 coats	1.000	Face	1.600	7.10	75.50	82.60
Caulking, window, 4' x 4'	1.000	Ea.	.516	4	24	28
4'-6" x 4'-6"	1.000	Ea.	.581	4.50	27	31.50
5'-0" x 4'-0"	1.000	Ea.	.581	4.50	27	31.50
4'-6" x 6'-6"	1.000	Ea.	.710	5.50	33	38.50
5'-6" x 6'-6"	1.000	Ea.	.774	6	36	42
6'-6" x 6'-6"	1.000	Ea.	.839	6.50	39	45.50
Grilles, glass size, to 48" x 48"	1.000	Ea.	.267	132	15.15	147.15
To 60" x 68"	1.000	Ea.	.286	202	16.20	218.20
Drip cap, aluminum, 4' long	4.000	L.F.	.080	2.52	4.56	7.08
4'-6" long	4.500	L.F.	.090	2.84	5.15	7.99
5' long	5.000	L.F.	.100	3.15	5.70	8.85
6' long	6.000	L.F.	.120	3.78	6.85	10.63
Wood, 4' long	4.000	L.F.	.133	6.20	7.30	13.50
4'-6" long	4.500	L.F.	.150	7	8.20	15.20
5' long	5.000	L.F.	.167	7.75	9.10	16.85
6' long	6.000	L.F.	.200	9.30	10.90	20.20

4 | EXTERIOR WALLS   52 | Entrance Door Systems

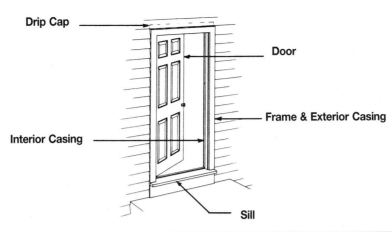

System Description	QUAN.	UNIT	LABOR HOURS	COST EACH MAT.	COST EACH INST.	COST EACH TOTAL
COLONIAL, 6 PANEL, 3' X 6'-8", WOOD						
Door, 3' x 6'-8" x 1-3/4" thick, pine, 6 panel colonial	1.000	Ea.	1.067	490	60.50	550.50
Frame, 5-13/16" deep, incl. exterior casing & drip cap	17.000	L.F.	.725	147.05	41.14	188.19
Interior casing, 2-1/2" wide	18.000	L.F.	.576	27.90	32.76	60.66
Sill, 8/4 x 8" deep	3.000	L.F.	.480	67.50	27.30	94.80
Butt hinges, brass, 4-1/2" x 4-1/2"	1.500	Pr.		33.75		33.75
Average quality	1.000	Ea.	.571	55	32.50	87.50
Weatherstripping, metal, spring type, bronze	1.000	Set	1.053	25.50	59.50	85
Paint, interior & exterior, primer & 2 coats	2.000	Face	1.778	14.10	84	98.10
TOTAL		Ea.	6.250	860.80	337.70	1,198.50
SOLID CORE BIRCH, FLUSH, 3' X 6'-8"						
Door, 3' x 6'-8", 1-3/4" thick, birch, flush solid core	1.000	Ea.	1.067	130	60.50	190.50
Frame, 5-13/16" deep, incl. exterior casing & drip cap	17.000	L.F.	.725	147.05	41.14	188.19
Interior casing, 2-1/2" wide	18.000	L.F.	.576	27.90	32.76	60.66
Sill, 8/4 x 8" deep	3.000	L.F.	.480	67.50	27.30	94.80
Butt hinges, brass, 4-1/2" x 4-1/2"	1.500	Pr.		33.75		33.75
Average quality	1.000	Ea.	.571	55	32.50	87.50
Weatherstripping, metal, spring type, bronze	1.000	Set	1.053	25.50	59.50	85
Paint, Interior & exterior, primer & 2 coats	2.000	Face	1.778	13.20	84	97.20
TOTAL		Ea.	6.250	499.90	337.70	837.60

These systems are on a cost per each door basis.

Description	QUAN.	UNIT	LABOR HOURS	COST EACH MAT.	COST EACH INST.	COST EACH TOTAL

Entrance Door Price Sheet

Entrance Door Price Sheet	QUAN.	UNIT	LABOR HOURS	COST EACH MAT.	COST EACH INST.	TOTAL
Door exterior wood 1-3/4" thick, pine, dutch door, 2'-8" x 6'-8" minimum	1.000	Ea.	1.333	910	75.50	985.50
Maximum	1.000	Ea.	1.600	990	91	1,081
3'-0" x 6'-8", minimum	1.000	Ea.	1.333	865	75.50	940.50
Maximum	1.000	Ea.	1.600	1,075	91	1,166
Colonial, 6 panel, 2'-8" x 6'-8"	1.000	Ea.	1.000	605	57	662
3'-0" x 6'-8"	1.000	Ea.	1.067	490	60.50	550.50
8 panel, 2'-6" x 6'-8"	1.000	Ea.	1.000	660	57	717
3'-0" x 6'-8"	1.000	Ea.	1.067	625	60.50	685.50
Flush, birch, solid core, 2'-8" x 6'-8"	1.000	Ea.	1.000	125	57	182
3'-0" x 6'-8"	1.000	Ea.	1.067	130	60.50	190.50
Porch door, 2'-8" x 6'-8"	1.000	Ea.	1.000	690	57	747
3'-0" x 6'-8"	1.000	Ea.	1.067	655	60.50	715.50
Hand carved mahogany, 2'-8" x 6'-8"	1.000	Ea.	1.067	1,550	60.50	1,610.50
3'-0" x 6'-8"	1.000	Ea.	1.067	1,525	60.50	1,585.50
Rosewood, 2'-8" x 6'-8"	1.000	Ea.	1.067	825	57.50	882.50
3'-0" x 6-8"	1.000	Ea.	1.067	770	60.50	830.50
Door, metal clad wood 1-3/8" thick raised panel, 2'-8" x 6'-8"	1.000	Ea.	1.067	345	53.50	398.50
3'-0" x 6'-8"	1.000	Ea.	1.067	300	60.50	360.50
Deluxe metal door, 3'-0" x 6'-8"	1.000	Ea.	1.231	300	60.50	360.50
3'-0" x 6'-8"	1.000	Ea.	1.231	300	60.50	360.50
Frame, pine, including exterior trim & drip cap, 5/4, x 4-9/16" deep	17.000	L.F.	.725	124	41	165
5-13/16" deep	17.000	L.F.	.725	147	41	188
6-9/16" deep	17.000	L.F.	.725	164	41	205
Safety glass lites, add	1.000	Ea.		87		87
Interior casing, 2'-8" x 6'-8" door	18.000	L.F.	.600	28	33	61
3'-0" x 6'-8" door	19.000	L.F.	.633	29.50	34.50	64
Sill, oak, 8/4 x 8" deep	3.000	L.F.	.480	67.50	27.50	95
8/4 x 10" deep	3.000	L.F.	.533	87	30.50	117.50
Butt hinges, steel plated, 4-1/2" x 4-1/2", plain	1.500	Pr.		34		34
Ball bearing	1.500	Pr.		61		61
Bronze, 4-1/2" x 4-1/2", plain	1.500	Pr.		46.50		46.50
Ball bearing	1.500	Pr.		64		64
Lockset, minimum	1.000	Ea.	.571	55	32.50	87.50
Maximum	1.000	Ea.	1.000	242	57	299
Weatherstripping, metal, interlocking, zinc	1.000	Set	2.667	48.50	151	199.50
Bronze	1.000	Set	2.667	61.50	151	212.50
Spring type, bronze	1.000	Set	1.053	25.50	59.50	85
Rubber, minimum	1.000	Set	1.053	10.05	59.50	69.55
Maximum	1.000	Set	1.143	12.25	65	77.25
Felt minimum	1.000	Set	.571	4.29	32.50	36.79
Maximum	1.000	Set	.615	4.90	35	39.90
Paint or stain, flush door, interior or exterior, 1 coat	2.000	Face	.941	4.92	44	48.92
2 coats	2.000	Face	1.455	9.85	69	78.85
Primer & 1 coat	2.000	Face	1.455	8.55	69	77.55
Primer & 2 coats	2.000	Face	1.778	13.20	84	97.20
Paneled door, interior & exterior, 1 coat	2.000	Face	1.143	5.25	54	59.25
2 coats	2.000	Face	2.000	10.50	94	104.50
Primer & 1 coat	2.000	Face	1.455	9.10	69	78.10
Primer & 2 coats	2.000	Face	1.778	14.10	84	98.10

For customer support on your Residential Cost Data, call 877.759.4771.

4 | EXTERIOR WALLS — 53 | Sliding Door Systems

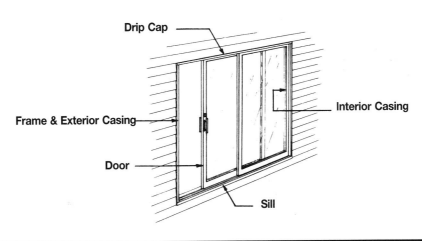

System Description	QUAN.	UNIT	LABOR HOURS	COST EACH		
				MAT.	INST.	TOTAL
WOOD SLIDING DOOR, 8' WIDE, PREMIUM						
Wood, 5/8" thick tempered insul. glass, 8' wide, premium	1.000	Ea.	5.333	2,025	305	2,330
Interior casing	22.000	L.F.	.704	34.10	40.04	74.14
Exterior casing	22.000	L.F.	.704	34.10	40.04	74.14
Sill, oak, 8/4 x 8" deep	8.000	L.F.	1.280	180	72.80	252.80
Drip cap	8.000	L.F.	.160	5.04	9.12	14.16
Paint, interior & exterior, primer & 2 coats	2.000	Face	2.816	20.24	132.88	153.12
TOTAL		Ea.	10.997	2,298.48	599.88	2,898.36
ALUMINUM SLIDING DOOR, 8' WIDE, PREMIUM						
Aluminum, 5/8" tempered insul. glass, 8' wide, premium	1.000	Ea.	5.333	1,825	305	2,130
Interior casing	22.000	L.F.	.704	34.10	40.04	74.14
Exterior casing	22.000	L.F.	.704	34.10	40.04	74.14
Sill, oak, 8/4 x 8" deep	8.000	L.F.	1.280	180	72.80	252.80
Drip cap	8.000	L.F.	.160	5.04	9.12	14.16
Paint, interior & exterior, primer & 2 coats	2.000	Face	2.816	20.24	132.88	153.12
TOTAL		Ea.	10.997	2,098.48	599.88	2,698.36

The cost of this system is on a cost per each door basis.

Description	QUAN.	UNIT	LABOR HOURS	COST EACH		
				MAT.	INST.	TOTAL

Sliding Door Price Sheet	QUAN.	UNIT	LABOR HOURS	COST PER UNIT		
				MAT.	INST.	TOTAL
Sliding door, wood, 5/8" thick, tempered insul. glass, 6' wide, premium	1.000	Ea.	4.000	1,600	227	1,827
Economy	1.000	Ea.	4.000	1,275	227	1,502
8' wide, wood premium	1.000	Ea.	5.333	2,025	305	2,330
Economy	1.000	Ea.	5.333	1,650	305	1,955
12' wide, wood premium	1.000	Ea.	6.400	3,300	365	3,665
Economy	1.000	Ea.	6.400	2,700	365	3,065
Aluminum, 5/8" thick, tempered insul. glass, 6' wide, premium	1.000	Ea.	4.000	1,775	227	2,002
Economy	1.000	Ea.	4.000	900	227	1,127
8' wide, premium	1.000	Ea.	5.333	1,825	305	2,130
Economy	1.000	Ea.	5.333	1,575	305	1,880
12' wide, premium	1.000	Ea.	6.400	3,300	365	3,665
Economy	1.000	Ea.	6.400	1,700	365	2,065
Interior casing, 6' wide door	20.000	L.F.	.667	31	36.50	67.50
8' wide door	22.000	L.F.	.733	34	40	74
12' wide door	26.000	L.F.	.867	40.50	47.50	88
Exterior casing, 6' wide door	20.000	L.F.	.667	31	36.50	67.50
8' wide door	22.000	L.F.	.733	34	40	74
12' wide door	26.000	L.F.	.867	40.50	47.50	88
Sill, oak, 8/4 x 8" deep, 6' wide door	6.000	L.F.	.960	135	54.50	189.50
8' wide door	8.000	L.F.	1.280	180	73	253
12' wide door	12.000	L.F.	1.920	270	109	379
8/4 x 10" deep, 6' wide door	6.000	L.F.	1.067	174	60.50	234.50
8' wide door	8.000	L.F.	1.422	232	81	313
12' wide door	12.000	L.F.	2.133	350	121	471
Drip cap, 6' wide door	6.000	L.F.	.120	3.78	6.85	10.63
8' wide door	8.000	L.F.	.160	5.05	9.10	14.15
12' wide door	12.000	L.F.	.240	7.55	13.70	21.25
Paint or stain, interior & exterior, 6' wide door, 1 coat	2.000	Face	1.600	6.40	75	81.40
2 coats	2.000	Face	1.600	6.40	75	81.40
Primer & 1 coat	2.000	Face	1.778	12	84	96
Primer & 2 coats	2.000	Face	2.560	18.40	121	139.40
8' wide door, 1 coat	2.000	Face	1.760	7.05	82.50	89.55
2 coats	2.000	Face	1.760	7.05	82.50	89.55
Primer & 1 coat	2.000	Face	1.955	13.20	92.50	105.70
Primer & 2 coats	2.000	Face	2.816	20	133	153
12' wide door, 1 coat	2.000	Face	2.080	8.30	98	106.30
2 coats	2.000	Face	2.080	8.30	98	106.30
Primer & 1 coat	2.000	Face	2.311	15.60	109	124.60
Primer & 2 coats	2.000	Face	3.328	24	157	181
Aluminum door, trim only, interior & exterior, 6' door, 1 coat	2.000	Face	.800	3.20	37.50	40.70
2 coats	2.000	Face	.800	3.20	37.50	40.70
Primer & 1 coat	2.000	Face	.889	6	42	48
Primer & 2 coats	2.000	Face	1.280	9.20	60.50	69.70
8' wide door, 1 coat	2.000	Face	.880	3.52	41.50	45.02
2 coats	2.000	Face	.880	3.52	41.50	45.02
Primer & 1 coat	2.000	Face	.978	6.60	46	52.60
Primer & 2 coats	2.000	Face	1.408	10.10	66.50	76.60
12' wide door, 1 coat	2.000	Face	1.040	4.16	49	53.16
2 coats	2.000	Face	1.040	4.16	49	53.16
Primer & 1 coat	2.000	Face	1.155	7.80	54.50	62.30
Primer & 2 coats	2.000	Face	1.664	11.95	78.50	90.45

4 | EXTERIOR WALLS — 56 | Residential Garage Door Systems

System Description	QUAN.	UNIT	LABOR HOURS	COST EACH MAT.	COST EACH INST.	COST EACH TOTAL
OVERHEAD, SECTIONAL GARAGE DOOR, 9' X 7'						
Wood, overhead sectional door, std., incl. hardware, 9' x 7'	1.000	Ea.	2.000	1,075	114	1,189
Jamb & header blocking, 2" x 6"	25.000	L.F.	.901	16.75	51.25	68
Exterior trim	25.000	L.F.	.800	38.75	45.50	84.25
Paint, interior & exterior, primer & 2 coats	2.000	Face	3.556	28.20	168	196.20
Weatherstripping, molding type	1.000	Set	.736	35.65	41.86	77.51
Drip cap	9.000	L.F.	.180	5.67	10.26	15.93
TOTAL		Ea.	8.173	1,200.02	430.87	1,630.89
OVERHEAD, SECTIONAL GARAGE DOOR, 16' X 7'						
Wood, overhead sectional, std., incl. hardware, 16' x 7'	1.000	Ea.	2.667	1,800	151	1,951
Jamb & header blocking, 2" x 6"	30.000	L.F.	1.081	20.10	61.50	81.60
Exterior trim	30.000	L.F.	.960	46.50	54.60	101.10
Paint, interior & exterior, primer & 2 coats	2.000	Face	5.333	42.30	252	294.30
Weatherstripping, molding type	1.000	Set	.960	46.50	54.60	101.10
Drip cap	16.000	L.F.	.320	10.08	18.24	28.32
TOTAL		Ea.	11.321	1,965.48	591.94	2,557.42
OVERHEAD, SWING-UP TYPE, GARAGE DOOR, 16' X 7'						
Wood, overhead, swing-up, std., incl. hardware, 16' x 7'	1.000	Ea.	2.667	990	151	1,141
Jamb & header blocking, 2" x 6"	30.000	L.F.	1.081	20.10	61.50	81.60
Exterior trim	30.000	L.F.	.960	46.50	54.60	101.10
Paint, interior & exterior, primer & 2 coats	2.000	Face	5.333	42.30	252	294.30
Weatherstripping, molding type	1.000	Set	.960	46.50	54.60	101.10
Drip cap	16.000	L.F.	.320	10.08	18.24	28.32
TOTAL		Ea.	11.321	1,155.48	591.94	1,747.42

This system is on a cost per each door basis.

Description	QUAN.	UNIT	LABOR HOURS	COST EACH MAT.	COST EACH INST.	COST EACH TOTAL

Resi Garage Door Price Sheet

Description	QUAN.	UNIT	LABOR HOURS	COST EACH MAT.	COST EACH INST.	COST EACH TOTAL
Overhead, sectional, including hardware, fiberglass, 9' x 7', standard	1.000	Ea.	3.030	1,050	182	1,232
Deluxe	1.000	Ea.	3.030	1,250	182	1,432
16' x 7', standard	1.000	Ea.	2.667	1,725	151	1,876
Deluxe	1.000	Ea.	2.667	2,350	151	2,501
Hardboard, 9' x 7', standard	1.000	Ea.	2.000	755	114	869
Deluxe	1.000	Ea.	2.000	900	114	1,014
16' x 7', standard	1.000	Ea.	2.667	1,350	151	1,501
Deluxe	1.000	Ea.	2.667	1,575	151	1,726
Metal, 9' x 7', standard	1.000	Ea.	3.030	915	114	1,029
Deluxe	1.000	Ea.	2.000	1,050	151	1,201
16' x 7', standard	1.000	Ea.	5.333	1,125	151	1,276
Deluxe	1.000	Ea.	2.667	1,550	182	1,732
Wood, 9' x 7', standard	1.000	Ea.	2.000	1,075	114	1,189
Deluxe	1.000	Ea.	2.000	2,350	114	2,464
16' x 7', standard	1.000	Ea.	2.667	1,800	151	1,951
Deluxe	1.000	Ea.	2.667	3,325	151	3,476
Overhead swing-up type including hardware, fiberglass, 9' x 7', standard	1.000	Ea.	2.000	1,100	114	1,214
Deluxe	1.000	Ea.	2.000	1,200	114	1,314
16' x 7', standard	1.000	Ea.	2.667	1,375	151	1,526
Deluxe	1.000	Ea.	2.667	1,750	151	1,901
Hardboard, 9' x 7', standard	1.000	Ea.	2.000	605	114	719
Deluxe	1.000	Ea.	2.000	715	114	829
16' x 7', standard	1.000	Ea.	2.667	735	151	886
Deluxe	1.000	Ea.	2.667	935	151	1,086
Metal, 9' x 7', standard	1.000	Ea.	2.000	660	114	774
Deluxe	1.000	Ea.	2.000	1,075	114	1,189
16' x 7', standard	1.000	Ea.	2.667	880	151	1,031
Deluxe	1.000	Ea.	2.667	1,200	151	1,351
Wood, 9' x 7', standard	1.000	Ea.	2.000	770	114	884
Deluxe	1.000	Ea.	2.000	1,250	114	1,364
16' x 7', standard	1.000	Ea.	2.667	990	151	1,141
Deluxe	1.000	Ea.	2.667	2,300	151	2,451
Jamb & header blocking, 2" x 6", 9' x 7' door	25.000	L.F.	.901	16.75	51.50	68.25
16' x 7' door	30.000	L.F.	1.081	20	61.50	81.50
2" x 8", 9' x 7' door	25.000	L.F.	1.000	23.50	57	80.50
16' x 7' door	30.000	L.F.	1.200	28	68	96
Exterior trim, 9' x 7' door	25.000	L.F.	.833	39	45.50	84.50
16' x 7' door	30.000	L.F.	1.000	46.50	54.50	101
Paint or stain, interior & exterior, 9' x 7' door, 1 coat	1.000	Face	2.286	10.50	108	118.50
2 coats	1.000	Face	4.000	21	188	209
Primer & 1 coat	1.000	Face	2.909	18.25	138	156.25
Primer & 2 coats	1.000	Face	3.556	28	168	196
16' x 7' door, 1 coat	1.000	Face	3.429	15.80	162	177.80
2 coats	1.000	Face	6.000	31.50	282	313.50
Primer & 1 coat	1.000	Face	4.364	27.50	207	234.50
Primer & 2 coats	1.000	Face	5.333	42.50	252	294.50
Weatherstripping, molding type, 9' x 7' door	1.000	Set	.767	35.50	42	77.50
16' x 7' door	1.000	Set	1.000	46.50	54.50	101
Drip cap, 9' door	9.000	L.F.	.180	5.65	10.25	15.90
16' door	16.000	L.F.	.320	10.10	18.25	28.35
Garage door opener, economy	1.000	Ea.	1.000	470	57	527
Deluxe, including remote control	1.000	Ea.	1.000	675	57	732

4 | EXTERIOR WALLS 58 | Aluminum Window Systems

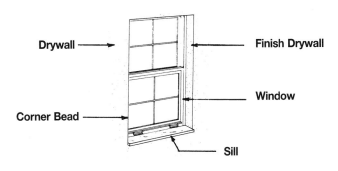

System Description	QUAN.	UNIT	LABOR HOURS	COST EACH MAT.	COST EACH INST.	COST EACH TOTAL
SINGLE HUNG, 2' X 3' OPENING						
Window, 2' x 3' opening, enameled, insulating glass	1.000	Ea.	1.600	277	105	382
Blocking, 1" x 3" furring strip nailers	10.000	L.F.	.146	4.30	8.30	12.60
Drywall, 1/2" thick, standard	5.000	S.F.	.040	1.80	2.25	4.05
Corner bead, 1" x 1", galvanized steel	8.000	L.F.	.160	1.36	9.12	10.48
Finish drywall, tape and finish corners inside and outside	16.000	L.F.	.269	1.76	15.36	17.12
Sill, slate	2.000	L.F.	.400	26.80	19.70	46.50
TOTAL		Ea.	2.615	313.02	159.73	472.75
SLIDING, 3' X 2' OPENING						
Window, 3' x 2' opening, enameled, insulating glass	1.000	Ea.	1.600	255	105	360
Blocking, 1" x 3" furring strip nailers	10.000	L.F.	.146	4.30	8.30	12.60
Drywall, 1/2" thick, standard	5.000	S.F.	.040	1.80	2.25	4.05
Corner bead, 1" x 1", galvanized steel	7.000	L.F.	.140	1.19	7.98	9.17
Finish drywall, tape and finish corners inside and outside	14.000	L.F.	.236	1.54	13.44	14.98
Sill, slate	3.000	L.F.	.600	40.20	29.55	69.75
TOTAL		Ea.	2.762	304.03	166.52	470.55
AWNING, 3'-1" X 3'-2"						
Window, 3'-1" x 3'-2" opening, enameled, insul. glass	1.000	Ea.	1.600	425	105	530
Blocking, 1" x 3" furring strip, nailers	12.500	L.F.	.182	5.38	10.38	15.76
Drywall, 1/2" thick, standard	4.500	S.F.	.036	1.62	2.03	3.65
Corner bead, 1" x 1", galvanized steel	9.250	L.F.	.185	1.57	10.55	12.12
Finish drywall, tape and finish corners, inside and outside	18.500	L.F.	.312	2.04	17.76	19.80
Sill, slate	3.250	L.F.	.650	43.55	32.01	75.56
TOTAL		Ea.	2.965	479.16	177.73	656.89

Description	QUAN.	UNIT	LABOR HOURS	COST PER S.F. MAT.	COST PER S.F. INST.	COST PER S.F. TOTAL

Aluminum Window Price Sheet	QUAN.	UNIT	LABOR HOURS	COST EACH MAT.	COST EACH INST.	TOTAL
Window, aluminum, awning, 3'-1" x 3'-2", standard glass	1.000	Ea.	1.600	390	105	495
Insulating glass	1.000	Ea.	1.600	425	105	530
4'-5" x 5'-3", standard glass	1.000	Ea.	2.000	440	131	571
Insulating glass	1.000	Ea.	2.000	525	131	656
Casement, 3'-1" x 3'-2", standard glass	1.000	Ea.	1.600	410	105	515
Insulating glass	1.000	Ea.	1.600	565	105	670
Single hung, 2' x 3', standard glass	1.000	Ea.	1.600	229	105	334
Insulating glass	1.000	Ea.	1.600	277	105	382
2'-8" x 6'-8", standard glass	1.000	Ea.	2.000	400	131	531
Insulating glass	1.000	Ea.	2.000	520	131	651
3'-4" x 5'-0", standard glass	1.000	Ea.	1.778	330	116	446
Insulating glass	1.000	Ea.	1.778	365	116	481
Sliding, 3' x 2', standard glass	1.000	Ea.	1.600	239	105	344
Insulating glass	1.000	Ea.	1.600	255	105	360
5' x 3', standard glass	1.000	Ea.	1.778	365	116	481
Insulating glass	1.000	Ea.	1.778	425	116	541
8' x 4', standard glass	1.000	Ea.	2.667	385	174	559
Insulating glass	1.000	Ea.	2.667	620	174	794
Blocking, 1" x 3" furring, opening 3' x 2'	10.000	L.F.	.146	4.30	8.30	12.60
3' x 3'	12.500	L.F.	.182	5.40	10.40	15.80
3' x 5'	16.000	L.F.	.233	6.90	13.30	20.20
4' x 4'	16.000	L.F.	.233	6.90	13.30	20.20
4' x 5'	18.000	L.F.	.262	7.75	14.95	22.70
4' x 6'	20.000	L.F.	.291	8.60	16.60	25.20
4' x 8'	24.000	L.F.	.349	10.30	19.90	30.20
6'-8" x 2'-8"	19.000	L.F.	.276	8.15	15.75	23.90
Drywall, 1/2" thick, standard, opening 3' x 2'	5.000	S.F.	.040	1.80	2.25	4.05
3' x 3'	6.000	S.F.	.048	2.16	2.70	4.86
3' x 5'	8.000	S.F.	.064	2.88	3.60	6.48
4' x 4'	8.000	S.F.	.064	2.88	3.60	6.48
4' x 5'	9.000	S.F.	.072	3.24	4.05	7.29
4' x 6'	10.000	S.F.	.080	3.60	4.50	8.10
4' x 8'	12.000	S.F.	.096	4.32	5.40	9.72
6'-8" x 2'	9.500	S.F.	.076	3.42	4.28	7.70
Corner bead, 1" x 1", galvanized steel, opening 3' x 2'	7.000	L.F.	.140	1.19	8	9.19
3' x 3'	9.000	L.F.	.180	1.53	10.25	11.78
3' x 5'	11.000	L.F.	.220	1.87	12.55	14.42
4' x 4'	12.000	L.F.	.240	2.04	13.70	15.74
4' x 5'	13.000	L.F.	.260	2.21	14.80	17.01
4' x 6'	14.000	L.F.	.280	2.38	15.95	18.33
4' x 8'	16.000	L.F.	.320	2.72	18.25	20.97
6'-8" x 2'	15.000	L.F.	.300	2.55	17.10	19.65
Tape and finish corners, inside and outside, opening 3' x 2'	14.000	L.F.	.204	1.54	13.45	14.99
3' x 3'	18.000	L.F.	.262	1.98	17.30	19.28
3' x 5'	22.000	L.F.	.320	2.42	21	23.42
4' x 4'	24.000	L.F.	.349	2.64	23	25.64
4' x 5'	26.000	L.F.	.378	2.86	25	27.86
4' x 6'	28.000	L.F.	.407	3.08	27	30.08
4' x 8'	32.000	L.F.	.466	3.52	30.50	34.02
6'-8" x 2'	30.000	L.F.	.437	3.30	29	32.30
Sill, slate, 2' long	2.000	L.F.	.400	27	19.70	46.70
3' long	3.000	L.F.	.600	40	29.50	69.50
4' long	4.000	L.F.	.800	53.50	39.50	93
Wood, 1-5/8" x 6-1/4", 2' long	2.000	L.F.	.128	11.70	7.25	18.95
3' long	3.000	L.F.	.192	17.55	10.90	28.45
4' long	4.000	L.F.	.256	23.50	14.50	38

4 | EXTERIOR WALLS — 60 | Storm Door & Window Systems

System Description	QUAN.	UNIT	LABOR HOURS	COST EACH MAT.	COST EACH INST.	COST EACH TOTAL
Storm door, aluminum, combination, storm & screen, anodized, 2'-6" x 6'-8"	1.000	Ea.	1.067	226	60.50	286.50
2'-8" x 6'-8"	1.000	Ea.	1.143	228	65	293
3'-0" x 6'-8"	1.000	Ea.	1.143	197	65	262
Mill finish, 2'-6" x 6'-8"	1.000	Ea.	1.067	264	60.50	324.50
2'-8" x 6'-8"	1.000	Ea.	1.143	264	65	329
3'-0" x 6'-8"	1.000	Ea.	1.143	289	65	354
Painted, 2'-6" x 6'-8"	1.000	Ea.	1.067	325	60.50	385.50
2'-8" x 6'-8"	1.000	Ea.	1.143	315	65	380
3'-0" x 6'-8"	1.000	Ea.	1.143	340	65	405
Wood, combination, storm & screen, crossbuck, 2'-6" x 6'-9"	1.000	Ea.	1.455	340	82.50	422.50
2'-8" x 6'-9"	1.000	Ea.	1.600	330	91	421
3'-0" x 6'-9"	1.000	Ea.	1.778	340	101	441
Full lite, 2'-6" x 6'-9"	1.000	Ea.	1.455	350	82.50	432.50
2'-8" x 6'-9"	1.000	Ea.	1.600	350	91	441
3'-0" x 6'-9"	1.000	Ea.	1.778	355	101	456
Windows, aluminum, combination storm & screen, basement, 1'-10" x 1'-0"	1.000	Ea.	.533	39.50	30.50	70
2'-9" x 1'-6"	1.000	Ea.	.533	43	30.50	73.50
3'-4" x 2'-0"	1.000	Ea.	.533	50.50	30.50	81
Double hung, anodized, 2'-0" x 3'-5"	1.000	Ea.	.533	107	30.50	137.50
2'-6" x 5'-0"	1.000	Ea.	.571	129	32.50	161.50
4'-0" x 6'-0"	1.000	Ea.	.640	259	36.50	295.50
Painted, 2'-0" x 3'-5"	1.000	Ea.	.533	123	30.50	153.50
2'-6" x 5'-0"	1.000	Ea.	.571	190	32.50	222.50
4'-0" x 6'-0"	1.000	Ea.	.640	315	36.50	351.50
Fixed window, anodized, 4'-6" x 4'-6"	1.000	Ea.	.640	146	36.50	182.50
5'-8" x 4'-6"	1.000	Ea.	.800	165	45.50	210.50
Painted, 4'-6" x 4'-6"	1.000	Ea.	.640	146	36.50	182.50
5'-8" x 4'-6"	1.000	Ea.	.800	169	45.50	214.50

4 | EXTERIOR WALLS 64 | Shutters/Blinds Systems

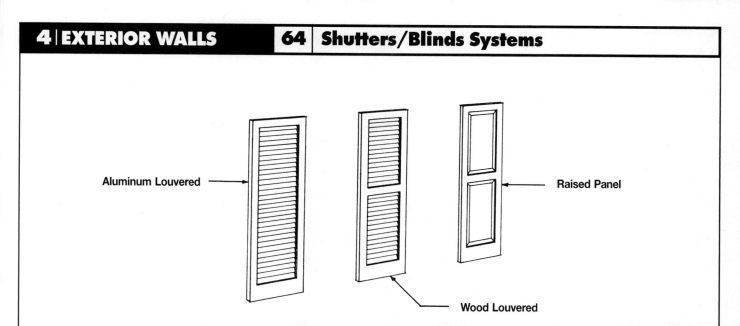

System Description	QUAN.	UNIT	LABOR HOURS	COST PER PAIR		
				MAT.	INST.	TOTAL
Shutters, exterior blinds, aluminum, louvered, 1'-4" wide, 3"-0" long	1.000	Set	.800	220	45.50	265.50
4'-0" long	1.000	Set	.800	264	45.50	309.50
5'-4" long	1.000	Set	.800	310	45.50	355.50
6'-8" long	1.000	Set	.889	390	50.50	440.50
Wood, louvered, 1'-2" wide, 3'-3" long	1.000	Set	.800	242	45.50	287.50
4'-7" long	1.000	Set	.800	297	45.50	342.50
5'-3" long	1.000	Set	.800	360	45.50	405.50
1'-6" wide, 3'-3" long	1.000	Set	.800	268	45.50	313.50
4'-7" long	1.000	Set	.800	355	45.50	400.50
Polystyrene, louvered, 1'-2" wide, 3'-3" long	1.000	Set	.800	39.50	45.50	85
4'-7" long	1.000	Set	.800	50.50	45.50	96
5'-3" long	1.000	Set	.800	65	45.50	110.50
6'-8" long	1.000	Set	.889	75	50.50	125.50
Vinyl, louvered, 1'-2" wide, 4'-7" long	1.000	Set	.720	63.50	41	104.50
1'-4" x 6'-8" long	1.000	Set	.889	83.50	50.50	134

Division 5 - Roofing

Did you know?
RSMeans Online gives you the same access to RSMeans' data with 24/7 access:
- Quickly locate costs in the searchable database.
- Build cost lists, estimates, and reports in minutes.
- Adjust costs to any location in the U.S. and Canada with the click of a button.

Start your free trial today at **www.RSMeansOnline.com**

RSMeans Online
FROM THE GORDIAN GROUP

No part of this cost data may be reproduced, stored in a retrieval system, or transmitted in any form or by any means without prior written permission of RSMeans.

5 | ROOFING   04 | Gable End Roofing Systems

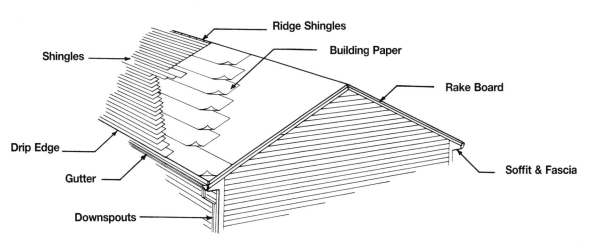

System Description	QUAN.	UNIT	LABOR HOURS	COST PER S.F.		
				MAT.	INST.	TOTAL
ASPHALT, ROOF SHINGLES, CLASS A						
Shingles, inorganic class A, 210-235 lb./sq., 4/12 pitch	1.160	S.F.	.017	1.07	.92	1.99
Drip edge, metal, 5" wide	.150	L.F.	.003	.10	.17	.27
Building paper, #15 felt	1.300	S.F.	.002	.08	.09	.17
Ridge shingles, asphalt	.042	L.F.	.001	.10	.05	.15
Soffit & fascia, white painted aluminum, 1' overhang	.083	L.F.	.012	.41	.68	1.09
Rake trim, 1" x 6"	.040	L.F.	.002	.05	.09	.14
Rake trim, prime and paint	.040	L.F.	.002	.01	.08	.09
Gutter, seamless, aluminum painted	.083	L.F.	.005	.25	.34	.59
Downspouts, aluminum painted	.035	L.F.	.002	.08	.10	.18
Ridge vent	.042	L.F.	.002	.11	.11	.22
TOTAL		S.F.	.048	2.26	2.63	4.89
WOOD, CEDAR SHINGLES NO. 1 PERFECTIONS, 18" LONG						
Shingles, wood, cedar, No. 1 perfections, 4/12 pitch	1.160	S.F.	.035	3.28	1.98	5.26
Drip edge, metal, 5" wide	.150	L.F.	.003	.10	.17	.27
Building paper, #15 felt	1.300	S.F.	.002	.08	.09	.17
Ridge shingles, cedar	.042	L.F.	.001	.21	.07	.28
Soffit & fascia, white painted aluminum, 1' overhang	.083	L.F.	.012	.41	.68	1.09
Rake trim, 1" x 6"	.040	L.F.	.002	.05	.09	.14
Rake trim, prime and paint	.040	L.F.	.002	.01	.08	.09
Gutter, seamless, aluminum, painted	.083	L.F.	.005	.25	.34	.59
Downspouts, aluminum, painted	.035	L.F.	.002	.08	.10	.18
Ridge vent	.042	L.F.	.002	.11	.11	.22
TOTAL		S.F.	.066	4.58	3.71	8.29

The prices in these systems are based on a square foot of plan area.
All quantities have been adjusted accordingly.

Description	QUAN.	UNIT	LABOR HOURS	COST PER S.F.		
				MAT.	INST.	TOTAL

Gable End Roofing Price Sheet

	QUAN.	UNIT	LABOR HOURS	COST PER S.F. MAT.	COST PER S.F. INST.	COST PER S.F. TOTAL
Shingles, asphalt, inorganic, class A, 210-235 lb./sq., 4/12 pitch	1.160	S.F.	.017	1.07	.92	1.99
8/12 pitch	1.330	S.F.	.019	1.16	1	2.16
Laminated, multi-layered, 240-260 lb./sq., 4/12 pitch	1.160	S.F.	.021	1.44	1.13	2.57
8/12 pitch	1.330	S.F.	.023	1.56	1.22	2.78
Premium laminated, multi-layered, 260-300 lb./sq., 4/12 pitch	1.160	S.F.	.027	2.09	1.45	3.54
8/12 pitch	1.330	S.F.	.030	2.26	1.57	3.83
Clay tile, Spanish tile, red, 4/12 pitch	1.160	S.F.	.053	6.60	2.78	9.38
8/12 pitch	1.330	S.F.	.058	7.15	3.02	10.17
Mission tile, red, 4/12 pitch	1.160	S.F.	.083	5.40	2.78	8.18
8/12 pitch	1.330	S.F.	.090	5.85	3.02	8.87
French tile, red, 4/12 pitch	1.160	S.F.	.071	15	2.54	17.54
8/12 pitch	1.330	S.F.	.077	16.25	2.76	19.01
Slate, Buckingham, Virginia, black, 4/12 pitch	1.160	S.F.	.055	7.15	2.92	10.07
8/12 pitch	1.330	S.F.	.059	7.75	3.16	10.91
Vermont, black or grey, 4/12 pitch	1.160	S.F.	.055	6.30	2.92	9.22
8/12 pitch	1.330	S.F.	.059	6.85	3.16	10.01
Wood, No. 1 red cedar, 5X, 16" long, 5" exposure, 4/12 pitch	1.160	S.F.	.038	3.72	2.18	5.90
8/12 pitch	1.330	S.F.	.042	4.03	2.37	6.40
Fire retardant, 4/12 pitch	1.160	S.F.	.038	4.46	2.18	6.64
8/12 pitch	1.330	S.F.	.042	4.83	2.37	7.20
18" long, No.1 perfections, 5" exposure, 4/12 pitch	1.160	S.F.	.035	3.28	1.98	5.26
8/12 pitch	1.330	S.F.	.038	3.55	2.15	5.70
Fire retardant, 4/12 pitch	1.160	S.F.	.035	4.02	1.98	6
8/12 pitch	1.330	S.F.	.038	4.35	2.15	6.50
Resquared & rebutted, 18" long, 6" exposure, 4/12 pitch	1.160	S.F.	.032	3.72	1.81	5.53
8/12 pitch	1.330	S.F.	.035	4.03	1.96	5.99
Fire retardant, 4/12 pitch	1.160	S.F.	.032	4.46	1.81	6.27
8/12 pitch	1.330	S.F.	.035	4.83	1.96	6.79
Wood shakes hand split, 24" long, 10" exposure, 4/12 pitch	1.160	S.F.	.038	3.78	2.18	5.96
8/12 pitch	1.330	S.F.	.042	4.10	2.37	6.47
Fire retardant, 4/12 pitch	1.160	S.F.	.038	4.52	2.18	6.70
8/12 pitch	1.330	S.F.	.042	4.90	2.37	7.27
18" long, 8" exposure, 4/12 pitch	1.160	S.F.	.048	2.36	2.72	5.08
8/12 pitch	1.330	S.F.	.052	2.56	2.95	5.51
Fire retardant, 4/12 pitch	1.160	S.F.	.048	3.10	2.72	5.82
8/12 pitch	1.330	S.F.	.052	3.36	2.95	6.31
Drip edge, metal, 5" wide	.150	L.F.	.003	.10	.17	.27
8" wide	.150	L.F.	.003	.14	.17	.31
Building paper, #15 asphalt felt	1.300	S.F.	.002	.08	.09	.17
Ridge shingles, asphalt	.042	L.F.	.001	.10	.05	.15
Clay	.042	L.F.	.002	.23	.27	.50
Slate	.042	L.F.	.002	.46	.09	.55
Wood, shingles	.042	L.F.	.001	.21	.07	.28
Shakes	.042	L.F.	.001	.21	.07	.28
		L.F.				
Soffit & fascia, aluminum, vented, 1' overhang	.083	L.F.	.012	.41	.68	1.09
2' overhang	.083	L.F.	.013	.59	.76	1.35
Vinyl, vented, 1' overhang	.083	L.F.	.011	.42	.63	1.05
2' overhang	.083	L.F.	.012	.57	.76	1.33
Wood, board fascia, plywood soffit, 1' overhang	.083	L.F.	.004	.02	.17	.19
2' overhang	.083	L.F.	.006	.04	.26	.30
Rake trim, painted, 1" x 6"	.040	L.F.	.004	.06	.17	.23
1" x 8"	.040	L.F.	.004	.24	.16	.40
Gutter, 5" box, aluminum, seamless, painted	.083	L.F.	.006	.25	.34	.59
Vinyl	.083	L.F.	.006	.15	.33	.48
Downspout, 2" x 3", aluminum, one story house	.035	L.F.	.001	.05	.09	.14
Two story house	.060	L.F.	.003	.09	.16	.25
Vinyl, one story house	.035	L.F.	.002	.08	.10	.18
Two story house	.060	L.F.	.003	.09	.16	.25

For customer support on your Residential Cost Data, call 877.759.4771.

5 | ROOFING — 08 | Hip Roof Roofing Systems

System Description	QUAN.	UNIT	LABOR HOURS	COST PER S.F. MAT.	COST PER S.F. INST.	COST PER S.F. TOTAL
ASPHALT, ROOF SHINGLES, CLASS A						
Shingles, inorganic, class A, 210-235 lb./sq. 4/12 pitch	1.570	S.F.	.023	1.42	1.23	2.65
Drip edge, metal, 5" wide	.122	L.F.	.002	.08	.14	.22
Building paper, #15 asphalt felt	1.800	S.F.	.002	.11	.13	.24
Ridge shingles, asphalt	.075	L.F.	.002	.18	.10	.28
Soffit & fascia, white painted aluminum, 1' overhang	.120	L.F.	.017	.59	.99	1.58
Gutter, seamless, aluminum, painted	.120	L.F.	.008	.37	.48	.85
Downspouts, aluminum, painted	.035	L.F.	.002	.08	.10	.18
Ridge vent	.028	L.F.	.001	.07	.07	.14
TOTAL		S.F.	.057	2.90	3.24	6.14
WOOD, CEDAR SHINGLES, NO. 1 PERFECTIONS, 18" LONG						
Shingles, red cedar, No. 1 perfections, 5" exp., 4/12 pitch	1.570	S.F.	.047	4.37	2.64	7.01
Drip edge, metal, 5" wide	.122	L.F.	.002	.08	.14	.22
Building paper, #15 asphalt felt	1.800	S.F.	.002	.11	.13	.24
Ridge shingles, wood, cedar	.075	L.F.	.002	.37	.12	.49
Soffit & fascia, white painted aluminum, 1' overhang	.120	L.F.	.017	.59	.99	1.58
Gutter, seamless, aluminum, painted	.120	L.F.	.008	.37	.48	.85
Downspouts, aluminum, painted	.035	L.F.	.002	.08	.10	.18
Ridge vent	.028	L.F.	.001	.07	.07	.14
TOTAL		S.F.	.081	6.04	4.67	10.71

The prices in these systems are based on a square foot of plan area.
All quantities have been adjusted accordingly.

Description	QUAN.	UNIT	LABOR HOURS	COST PER S.F. MAT.	COST PER S.F. INST.	COST PER S.F. TOTAL

Hip Roof - Roofing Price Sheet

Hip Roof - Roofing Price Sheet	QUAN.	UNIT	LABOR HOURS	COST PER S.F. MAT.	COST PER S.F. INST.	COST PER S.F. TOTAL
Shingles, asphalt, inorganic, class A, 210-235 lb./sq., 4/12 pitch	1.570	S.F.	.023	1.42	1.23	2.65
8/12 pitch	1.850	S.F.	.028	1.69	1.46	3.15
Laminated, multi-layered, 240-260 lb./sq., 4/12 pitch	1.570	S.F.	.028	1.92	1.50	3.42
8/12 pitch	1.850	S.F.	.034	2.28	1.79	4.07
Prem. laminated, multi-layered, 260-300 lb./sq., 4/12 pitch	1.570	S.F.	.037	2.78	1.94	4.72
8/12 pitch	1.850	S.F.	.043	3.31	2.30	5.61
Clay tile, Spanish tile, red, 4/12 pitch	1.570	S.F.	.071	8.80	3.71	12.51
8/12 pitch	1.850	S.F.	.084	10.45	4.41	14.86
Mission tile, red, 4/12 pitch	1.570	S.F.	.111	7.20	3.71	10.91
8/12 pitch	1.850	S.F.	.132	8.55	4.41	12.96
French tile, red, 4/12 pitch	1.570	S.F.	.095	20	3.39	23.39
8/12 pitch	1.850	S.F.	.113	24	4.03	28.03
Slate, Buckingham, Virginia, black, 4/12 pitch	1.570	S.F.	.073	9.50	3.89	13.39
8/12 pitch	1.850	S.F.	.087	11.30	4.62	15.92
Vermont, black or grey, 4/12 pitch	1.570	S.F.	.073	8.40	3.89	12.29
8/12 pitch	1.850	S.F.	.087	10	4.62	14.62
Wood, red cedar, No.1 5X, 16" long, 5" exposure, 4/12 pitch	1.570	S.F.	.051	4.96	2.91	7.87
8/12 pitch	1.850	S.F.	.061	5.90	3.46	9.36
Fire retardant, 4/12 pitch	1.570	S.F.	.051	5.95	2.91	8.86
8/12 pitch	1.850	S.F.	.061	7.05	3.46	10.51
18" long, No.1 perfections, 5" exposure, 4/12 pitch	1.570	S.F.	.047	4.37	2.64	7.01
8/12 pitch	1.850	S.F.	.055	5.20	3.14	8.34
Fire retardant, 4/12 pitch	1.570	S.F.	.047	5.35	2.64	7.99
8/12 pitch	1.850	S.F.	.055	6.35	3.14	9.49
Resquared & rebutted, 18" long, 6" exposure, 4/12 pitch	1.570	S.F.	.043	4.96	2.42	7.38
8/12 pitch	1.850	S.F.	.051	5.90	2.87	8.77
Fire retardant, 4/12 pitch	1.570	S.F.	.043	5.95	2.42	8.37
8/12 pitch	1.850	S.F.	.051	7.05	2.87	9.92
Wood shakes hand split, 24" long, 10" exposure, 4/12 pitch	1.570	S.F.	.051	5.05	2.91	7.96
8/12 pitch	1.850	S.F.	.061	6	3.46	9.46
Fire retardant, 4/12 pitch	1.570	S.F.	.051	6	2.91	8.91
8/12 pitch	1.850	S.F.	.061	7.15	3.46	10.61
18" long, 8" exposure, 4/12 pitch	1.570	S.F.	.064	3.15	3.63	6.78
8/12 pitch	1.850	S.F.	.076	3.74	4.31	8.05
Fire retardant, 4/12 pitch	1.570	S.F.	.064	4.13	3.63	7.76
8/12 pitch	1.850	S.F.	.076	4.91	4.31	9.22
Drip edge, metal, 5" wide	.122	L.F.	.002	.08	.14	.22
8" wide	.122	L.F.	.002	.11	.14	.25
Building paper, #15 asphalt felt	1.800	S.F.	.002	.11	.13	.24
Ridge shingles, asphalt	.075	L.F.	.002	.18	.10	.28
Clay	.075	L.F.	.003	.41	.48	.89
Slate	.075	L.F.	.003	.83	.16	.99
Wood, shingles	.075	L.F.	.002	.37	.12	.49
Shakes	.075	L.F.	.002	.37	.12	.49
Soffit & fascia, aluminum, vented, 1' overhang	.120	L.F.	.017	.59	.99	1.58
2' overhang	.120	L.F.	.019	.86	1.09	1.95
Vinyl, vented, 1' overhang	.120	L.F.	.016	.61	.91	1.52
2' overhang	.120	L.F.	.017	.82	1.09	1.91
Wood, board fascia, plywood soffit, 1' overhang	.120	L.F.	.004	.02	.17	.19
2' overhang	.120	L.F.	.006	.04	.26	.30
Gutter, 5" box, aluminum, seamless, painted	.120	L.F.	.008	.37	.48	.85
Vinyl	.120	L.F.	.009	.21	.47	.68
Downspout, 2" x 3", aluminum, one story house	.035	L.F.	.002	.08	.10	.18
Two story house	.060	L.F.	.003	.09	.16	.25
Vinyl, one story house	.035	L.F.	.001	.05	.09	.14
Two story house	.060	L.F.	.003	.09	.16	.25

5 | ROOFING   12 | Gambrel Roofing Systems

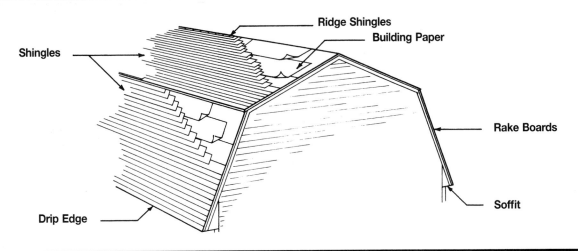

System Description	QUAN.	UNIT	LABOR HOURS	COST PER S.F. MAT.	COST PER S.F. INST.	COST PER S.F. TOTAL
ASPHALT, ROOF SHINGLES, CLASS A						
Shingles, asphalt, inorganic, class A, 210-235 lb./sq.	1.450	S.F.	.022	1.34	1.16	2.50
Drip edge, metal, 5" wide	.146	L.F.	.003	.09	.17	.26
Building paper, #15 asphalt felt	1.500	S.F.	.002	.09	.11	.20
Ridge shingles, asphalt	.042	L.F.	.001	.10	.05	.15
Soffit & fascia, painted aluminum, 1' overhang	.083	L.F.	.012	.41	.68	1.09
Rake trim, 1" x 6"	.063	L.F.	.003	.08	.14	.22
Rake trim, prime and paint	.063	L.F.	.003	.02	.13	.15
Gutter, seamless, aluminum, painted	.083	L.F.	.005	.25	.34	.59
Downspouts, aluminum, painted	.042	L.F.	.002	.09	.12	.21
Ridge vent	.042	L.F.	.002	.11	.11	.22
TOTAL		S.F.	.055	2.58	3.01	5.59
WOOD, CEDAR SHINGLES, NO. 1 PERFECTIONS, 18" LONG						
Shingles, wood, red cedar, No. 1 perfections, 5" exposure	1.450	S.F.	.044	4.10	2.48	6.58
Drip edge, metal, 5" wide	.146	L.F.	.003	.09	.17	.26
Building paper, #15 asphalt felt	1.500	S.F.	.002	.09	.11	.20
Ridge shingles, wood	.042	L.F.	.001	.21	.07	.28
Soffit & fascia, white painted aluminum, 1' overhang	.083	L.F.	.012	.41	.68	1.09
Rake trim, 1" x 6"	.063	L.F.	.003	.08	.14	.22
Rake trim, prime and paint	.063	L.F.	.001	.02	.06	.08
Gutter, seamless, aluminum, painted	.083	L.F.	.005	.25	.34	.59
Downspouts, aluminum, painted	.042	L.F.	.002	.09	.12	.21
Ridge vent	.042	L.F.	.002	.11	.11	.22
TOTAL		S.F.	.075	5.45	4.28	9.73

The prices in this system are based on a square foot of plan area.
All quantities have been adjusted accordingly.

Description	QUAN.	UNIT	LABOR HOURS	COST PER S.F. MAT.	COST PER S.F. INST.	COST PER S.F. TOTAL

Gambrel Roofing Price Sheet	QUAN.	UNIT	LABOR HOURS	COST PER S.F.		
				MAT.	INST.	TOTAL
Shingles, asphalt, standard, inorganic, class A, 210-235 lb./sq.	1.450	S.F.	.022	1.34	1.16	2.50
Laminated, multi-layered, 240-260 lb./sq.	1.450	S.F.	.027	1.80	1.41	3.21
Premium laminated, multi-layered, 260-300 lb./sq.	1.450	S.F.	.034	2.61	1.82	4.43
Slate, Buckingham, Virginia, black	1.450	S.F.	.069	8.95	3.65	12.60
Vermont, black or grey	1.450	S.F.	.069	7.90	3.65	11.55
Wood, red cedar, No.1 5X, 16" long, 5" exposure, plain	1.450	S.F.	.048	4.65	2.73	7.38
Fire retardant	1.450	S.F.	.048	5.55	2.73	8.28
18" long, No.1 perfections, 6" exposure, plain	1.450	S.F.	.044	4.10	2.48	6.58
Fire retardant	1.450	S.F.	.044	5	2.48	7.48
Resquared & rebutted, 18" long, 6" exposure, plain	1.450	S.F.	.040	4.65	2.27	6.92
Fire retardant	1.450	S.F.	.040	5.55	2.27	7.82
Shakes, hand split, 24" long, 10" exposure, plain	1.450	S.F.	.048	4.73	2.73	7.46
Fire retardant	1.450	S.F.	.048	5.65	2.73	8.38
18" long, 8" exposure, plain	1.450	S.F.	.060	2.96	3.41	6.37
Fire retardant	1.450	S.F.	.060	3.88	3.41	7.29
Drip edge, metal, 5" wide	.146	L.F.	.003	.09	.17	.26
8" wide	.146	L.F.	.003	.13	.17	.30
Building paper, #15 asphalt felt	1.500	S.F.	.002	.09	.11	.20
Ridge shingles, asphalt	.042	L.F.	.001	.10	.05	.15
Slate	.042	L.F.	.002	.46	.09	.55
Wood, shingles	.042	L.F.	.001	.21	.07	.28
Soffit & fascia, aluminum, vented, 1' overhang	.083	L.F.	.012	.41	.68	1.09
2' overhang	.083	L.F.	.013	.59	.76	1.35
Vinyl vented, 1' overhang	.083	L.F.	.011	.42	.63	1.05
2' overhang	.083	L.F.	.012	.57	.76	1.33
Wood board fascia, plywood soffit, 1' overhang	.083	L.F.	.004	.02	.17	.19
2' overhang	.083	L.F.	.006	.04	.26	.30
Rake trim, painted, 1" x 6"	.063	L.F.	.006	.10	.27	.37
1" x 8"	.063	L.F.	.007	.12	.36	.48
Gutter, 5" box, aluminum, seamless, painted	.083	L.F.	.006	.25	.34	.59
Vinyl	.083	L.F.	.006	.15	.33	.48
Downspout 2" x 3", aluminum, one story house	.042	L.F.	.002	.06	.11	.17
Two story house	.070	L.F.	.003	.11	.19	.30
Vinyl, one story house	.042	L.F.	.002	.06	.11	.17
Two story house	.070	L.F.	.003	.11	.19	.30

5 | ROOFING   16 | Mansard Roofing Systems

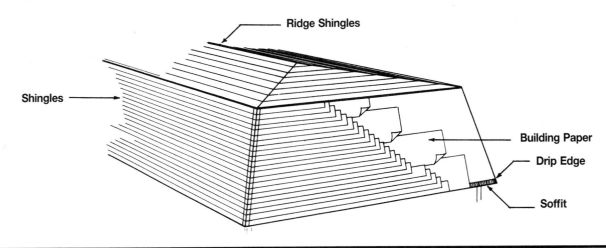

System Description	QUAN.	UNIT	LABOR HOURS	COST PER S.F. MAT.	COST PER S.F. INST.	COST PER S.F. TOTAL
ASPHALT, ROOF SHINGLES, CLASS A						
Shingles, standard inorganic class A 210-235 lb./sq.	2.210	S.F.	.032	1.96	1.69	3.65
Drip edge, metal, 5" wide	.122	L.F.	.002	.08	.14	.22
Building paper, #15 asphalt felt	2.300	S.F.	.003	.13	.17	.30
Ridge shingles, asphalt	.090	L.F.	.002	.22	.12	.34
Soffit & fascia, white painted aluminum, 1' overhang	.122	L.F.	.018	.60	1.01	1.61
Gutter, seamless, aluminum, painted	.122	L.F.	.008	.37	.49	.86
Downspouts, aluminum, painted	.042	L.F.	.002	.09	.12	.21
Ridge vent	.028	L.F.	.001	.07	.07	.14
TOTAL		S.F.	.068	3.52	3.81	7.33
WOOD, CEDAR SHINGLES, NO. 1 PERFECTIONS, 18" LONG						
Shingles, wood, red cedar, No. 1 perfections, 5" exposure	2.210	S.F.	.064	6.01	3.63	9.64
Drip edge, metal, 5" wide	.122	L.F.	.002	.08	.14	.22
Building paper, #15 asphalt felt	2.300	S.F.	.003	.13	.17	.30
Ridge shingles, wood	.090	L.F.	.003	.45	.15	.60
Soffit & fascia, white painted aluminum, 1' overhang	.122	L.F.	.018	.60	1.01	1.61
Gutter, seamless, aluminum, painted	.122	L.F.	.008	.37	.49	.86
Downspouts, aluminum, painted	.042	L.F.	.002	.09	.12	.21
Ridge vent	.028	L.F.	.001	.07	.07	.14
TOTAL		S.F.	.101	7.80	5.78	13.58

The prices in these systems are based on a square foot of plan area.
All quantities have been adjusted accordingly.

Description	QUAN.	UNIT	LABOR HOURS	COST PER S.F. MAT.	COST PER S.F. INST.	COST PER S.F. TOTAL

Mansard Roofing Price Sheet

	QUAN.	UNIT	LABOR HOURS	COST PER S.F. MAT.	COST PER S.F. INST.	COST PER S.F. TOTAL
Shingles, asphalt, standard, inorganic, class A, 210-235 lb./sq.	2.210	S.F.	.032	1.96	1.69	3.65
Laminated, multi-layered, 240-260 lb./sq.	2.210	S.F.	.039	2.64	2.07	4.71
Premium laminated, multi-layered, 260-300 lb./sq.	2.210	S.F.	.050	3.83	2.66	6.49
Slate Buckingham, Virginia, black	2.210	S.F.	.101	13.10	5.35	18.45
Vermont, black or grey	2.210	S.F.	.101	11.55	5.35	16.90
Wood, red cedar, No.1 5X, 16" long, 5" exposure, plain	2.210	S.F.	.070	6.80	4	10.80
Fire retardant	2.210	S.F.	.070	8.15	4	12.15
18" long, No.1 perfections 6" exposure, plain	2.210	S.F.	.064	6	3.63	9.63
Fire retardant	2.210	S.F.	.064	7.35	3.63	10.98
Resquared & rebutted, 18" long, 6" exposure, plain	2.210	S.F.	.059	6.80	3.32	10.12
Fire retardant	2.210	S.F.	.059	8.15	3.32	11.47
Shakes, hand split, 24" long 10" exposure, plain	2.210	S.F.	.070	6.95	4	10.95
Fire retardant	2.210	S.F.	.070	8.30	4	12.30
18" long, 8" exposure, plain	2.210	S.F.	.088	4.33	4.99	9.32
Fire retardant	2.210	S.F.	.088	5.70	4.99	10.69
Drip edge, metal, 5" wide	.122	S.F.	.002	.08	.14	.22
8" wide	.122	S.F.	.002	.11	.14	.25
Building paper, #15 asphalt felt	2.300	S.F.	.003	.13	.17	.30
Ridge shingles, asphalt	.090	L.F.	.002	.22	.12	.34
Slate	.090	L.F.	.004	.99	.19	1.18
Wood, shingles	.090	L.F.	.003	.45	.15	.60
Soffit & fascia, aluminum vented, 1' overhang	.122	L.F.	.018	.60	1.01	1.61
2' overhang	.122	L.F.	.020	.87	1.11	1.98
Vinyl vented, 1' overhang	.122	L.F.	.016	.62	.92	1.54
2' overhang	.122	L.F.	.018	.84	1.11	1.95
Wood board fascia, plywood soffit, 1' overhang	.122	L.F.	.013	.37	.69	1.06
2' overhang	.122	L.F.	.019	.52	1.05	1.57
Gutter, 5" box, aluminum, seamless, painted	.122	L.F.	.008	.37	.49	.86
Vinyl	.122	L.F.	.009	.21	.48	.69
Downspout 2" x 3", aluminum, one story house	.042	L.F.	.002	.06	.11	.17
Two story house	.070	L.F.	.003	.11	.18	.29
Vinyl, one story house	.042	L.F.	.002	.06	.11	.17
Two story house	.070	L.F.	.003	.11	.18	.29

5 | ROOFING — 20 | Shed Roofing Systems

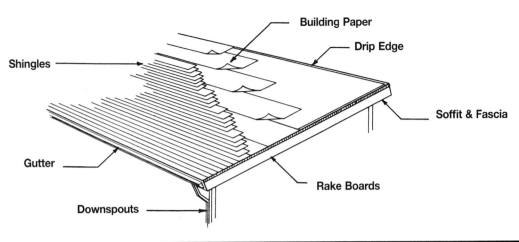

System Description	QUAN.	UNIT	LABOR HOURS	COST PER S.F. MAT.	COST PER S.F. INST.	COST PER S.F. TOTAL
ASPHALT, ROOF SHINGLES, CLASS A						
Shingles, inorganic class A 210-235 lb./sq. 4/12 pitch	1.230	S.F.	.019	1.16	1	2.16
Drip edge, metal, 5" wide	.100	L.F.	.002	.07	.11	.18
Building paper, #15 asphalt felt	1.300	S.F.	.002	.08	.09	.17
Soffit & fascia, white painted aluminum, 1' overhang	.080	L.F.	.012	.39	.66	1.05
Rake trim, 1" x 6"	.043	L.F.	.002	.05	.10	.15
Rake trim, prime and paint	.043	L.F.	.002	.01	.09	.10
Gutter, seamless, aluminum, painted	.040	L.F.	.003	.12	.16	.28
Downspouts, painted aluminum	.020	L.F.	.001	.05	.06	.11
TOTAL		S.F.	.043	1.93	2.27	4.20
WOOD, CEDAR SHINGLES, NO. 1 PERFECTIONS, 18" LONG						
Shingles, red cedar, No. 1 perfections, 5" exp., 4/12 pitch	1.230	S.F.	.035	3.28	1.98	5.26
Drip edge, metal, 5" wide	.100	L.F.	.002	.07	.11	.18
Building paper, #15 asphalt felt	1.300	S.F.	.002	.08	.09	.17
Soffit & fascia, white painted aluminum, 1' overhang	.080	L.F.	.012	.39	.66	1.05
Rake trim, 1" x 6"	.043	L.F.	.002	.05	.10	.15
Rake trim, prime and paint	.043	L.F.	.001	.01	.04	.05
Gutter, seamless, aluminum, painted	.040	L.F.	.003	.12	.16	.28
Downspouts, painted aluminum	.020	L.F.	.001	.05	.06	.11
TOTAL		S.F.	.058	4.05	3.20	7.25

The prices in these systems are based on a square foot of plan area. All quantities have been adjusted accordingly.

Description	QUAN.	UNIT	LABOR HOURS	COST PER S.F. MAT.	COST PER S.F. INST.	COST PER S.F. TOTAL

Shed Roofing Price Sheet	QUAN.	UNIT	LABOR HOURS	COST PER S.F.		
				MAT.	INST.	TOTAL
Shingles, asphalt, inorganic, class A, 210-235 lb./sq., 4/12 pitch	1.230	S.F.	.017	1.07	.92	1.99
8/12 pitch	1.330	S.F.	.019	1.16	1	2.16
Laminated, multi-layered, 240-260 lb./sq. 4/12 pitch	1.230	S.F.	.021	1.44	1.13	2.57
8/12 pitch	1.330	S.F.	.023	1.56	1.22	2.78
Premium laminated, multi-layered, 260-300 lb./sq. 4/12 pitch	1.230	S.F.	.027	2.09	1.45	3.54
8/12 pitch	1.330	S.F.	.030	2.26	1.57	3.83
Clay tile, Spanish tile, red, 4/12 pitch	1.230	S.F.	.053	6.60	2.78	9.38
8/12 pitch	1.330	S.F.	.058	7.15	3.02	10.17
Mission tile, red, 4/12 pitch	1.230	S.F.	.083	5.40	2.78	8.18
8/12 pitch	1.330	S.F.	.090	5.85	3.02	8.87
French tile, red, 4/12 pitch	1.230	S.F.	.071	15	2.54	17.54
8/12 pitch	1.330	S.F.	.077	16.25	2.76	19.01
Slate, Buckingham, Virginia, black, 4/12 pitch	1.230	S.F.	.055	7.15	2.92	10.07
8/12 pitch	1.330	S.F.	.059	7.75	3.16	10.91
Vermont, black or grey, 4/12 pitch	1.230	S.F.	.055	6.30	2.92	9.22
8/12 pitch	1.330	S.F.	.059	6.85	3.16	10.01
Wood, red cedar, No.1 5X, 16" long, 5" exposure, 4/12 pitch	1.230	S.F.	.038	3.72	2.18	5.90
8/12 pitch	1.330	S.F.	.042	4.03	2.37	6.40
Fire retardant, 4/12 pitch	1.230	S.F.	.038	4.46	2.18	6.64
8/12 pitch	1.330	S.F.	.042	4.83	2.37	7.20
18" long, 6" exposure, 4/12 pitch	1.230	S.F.	.035	3.28	1.98	5.26
8/12 pitch	1.330	S.F.	.038	3.55	2.15	5.70
Fire retardant, 4/12 pitch	1.230	S.F.	.035	4.02	1.98	6
8/12 pitch	1.330	S.F.	.038	4.35	2.15	6.50
Resquared & rebutted, 18" long, 6" exposure, 4/12 pitch	1.230	S.F.	.032	3.72	1.81	5.53
8/12 pitch	1.330	S.F.	.035	4.03	1.96	5.99
Fire retardant, 4/12 pitch	1.230	S.F.	.032	4.46	1.81	6.27
8/12 pitch	1.330	S.F.	.035	4.83	1.96	6.79
Wood shakes, hand split, 24" long, 10" exposure, 4/12 pitch	1.230	S.F.	.038	3.78	2.18	5.96
8/12 pitch	1.330	S.F.	.042	4.10	2.37	6.47
Fire retardant, 4/12 pitch	1.230	S.F.	.038	4.52	2.18	6.70
8/12 pitch	1.330	S.F.	.042	4.90	2.37	7.27
18" long, 8" exposure, 4/12 pitch	1.230	S.F.	.048	2.36	2.72	5.08
8/12 pitch	1.330	S.F.	.052	2.56	2.95	5.51
Fire retardant, 4/12 pitch	1.230	S.F.	.048	3.10	2.72	5.82
8/12 pitch	1.330	S.F.	.052	3.36	2.95	6.31
Drip edge, metal, 5" wide	.100	L.F.	.002	.07	.11	.18
8" wide	.100	L.F.	.002	.09	.11	.20
Building paper, #15 asphalt felt	1.300	S.F.	.002	.08	.09	.17
Soffit & fascia, aluminum vented, 1' overhang	.080	L.F.	.012	.39	.66	1.05
2' overhang	.080	L.F.	.013	.57	.73	1.30
Vinyl vented, 1' overhang	.080	L.F.	.011	.41	.60	1.01
2' overhang	.080	L.F.	.012	.55	.73	1.28
Wood board fascia, plywood soffit, 1' overhang	.080	L.F.	.010	.24	.51	.75
2' overhang	.080	L.F.	.014	.35	.77	1.12
Rake, trim, painted, 1" x 6"	.043	L.F.	.004	.06	.19	.25
1" x 8"	.043	L.F.	.004	.06	.19	.25
Gutter, 5" box, aluminum, seamless, painted	.040	L.F.	.003	.12	.16	.28
Vinyl	.040	L.F.	.003	.07	.16	.23
Downspout 2" x 3", aluminum, one story house	.020	L.F.	.001	.03	.05	.08
Two story house	.020	L.F.	.001	.05	.09	.14
Vinyl, one story house	.020	L.F.	.001	.03	.05	.08
Two story house	.020	L.F.	.001	.05	.09	.14

5 | ROOFING
24 | Gable Dormer Roofing Systems

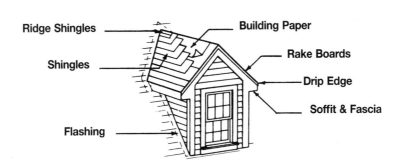

System Description	QUAN.	UNIT	LABOR HOURS	COST PER S.F. MAT.	COST PER S.F. INST.	COST PER S.F. TOTAL
ASPHALT, ROOF SHINGLES, CLASS A						
Shingles, standard inorganic class A 210-235 lb./sq	1.400	S.F.	.020	1.25	1.08	2.33
Drip edge, metal, 5" wide	.220	L.F.	.004	.14	.25	.39
Building paper, #15 asphalt felt	1.500	S.F.	.002	.09	.11	.20
Ridge shingles, asphalt	.280	L.F.	.007	.68	.36	1.04
Soffit & fascia, aluminum, vented	.220	L.F.	.032	1.08	1.82	2.90
Flashing, aluminum, mill finish, .013" thick	1.500	S.F.	.083	1.37	4.38	5.75
TOTAL		S.F.	.148	4.61	8	12.61
WOOD, CEDAR, NO. 1 PERFECTIONS						
Shingles, red cedar, No.1 perfections, 18" long, 5" exp.	1.400	S.F.	.041	3.82	2.31	6.13
Drip edge, metal, 5" wide	.220	L.F.	.004	.14	.25	.39
Building paper, #15 asphalt felt	1.500	S.F.	.002	.09	.11	.20
Ridge shingles, wood	.280	L.F.	.008	1.39	.45	1.84
Soffit & fascia, aluminum, vented	.220	L.F.	.032	1.08	1.82	2.90
Flashing, aluminum, mill finish, .013" thick	1.500	S.F.	.083	1.37	4.38	5.75
TOTAL		S.F.	.170	7.89	9.32	17.21
SLATE, BUCKINGHAM, BLACK						
Shingles, Buckingham, Virginia, black	1.400	S.F.	.064	8.33	3.40	11.73
Drip edge, metal, 5" wide	.220	L.F.	.004	.14	.25	.39
Building paper, #15 asphalt felt	1.500	S.F.	.002	.09	.11	.20
Ridge shingles, slate	.280	L.F.	.011	3.08	.59	3.67
Soffit & fascia, aluminum, vented	.220	L.F.	.032	1.08	1.82	2.90
Flashing, copper, 16 oz.	1.500	S.F.	.104	12.98	5.54	18.52
TOTAL		S.F.	.217	25.70	11.71	37.41

The prices in these systems are based on a square foot of plan area under the dormer roof.

Description	QUAN.	UNIT	LABOR HOURS	COST PER S.F. MAT.	COST PER S.F. INST.	COST PER S.F. TOTAL

Gable Dormer Roofing Price Sheet

	QUAN.	UNIT	LABOR HOURS	COST PER S.F. MAT.	COST PER S.F. INST.	COST PER S.F. TOTAL
Shingles, asphalt, standard, inorganic, class A, 210-235 lb./sq.	1.400	S.F.	.020	1.25	1.08	2.33
Laminated, multi-layered, 240-260 lb./sq.	1.400	S.F.	.025	1.68	1.32	3
Premium laminated, multi-layered, 260-300 lb./sq.	1.400	S.F.	.032	2.44	1.69	4.13
Clay tile, Spanish tile, red	1.400	S.F.	.062	7.70	3.25	10.95
Mission tile, red	1.400	S.F.	.097	6.30	3.25	9.55
French tile, red	1.400	S.F.	.083	17.50	2.97	20.47
Slate Buckingham, Virginia, black	1.400	S.F.	.064	8.35	3.40	11.75
Vermont, black or grey	1.400	S.F.	.064	7.35	3.40	10.75
Wood, red cedar, No.1 5X, 16" long, 5" exposure	1.400	S.F.	.045	4.34	2.55	6.89
Fire retardant	1.400	S.F.	.045	5.20	2.55	7.75
18" long, No.1 perfections, 5" exposure	1.400	S.F.	.041	3.82	2.31	6.13
Fire retardant	1.400	S.F.	.041	4.68	2.31	6.99
Resquared & rebutted, 18" long, 5" exposure	1.400	S.F.	.037	4.34	2.11	6.45
Fire retardant	1.400	S.F.	.037	5.20	2.11	7.31
Shakes hand split, 24" long, 10" exposure	1.400	S.F.	.045	4.41	2.55	6.96
Fire retardant	1.400	S.F.	.045	5.25	2.55	7.80
18" long, 8" exposure	1.400	S.F.	.056	2.76	3.18	5.94
Fire retardant	1.400	S.F.	.056	3.62	3.18	6.80
Drip edge, metal, 5" wide	.220	L.F.	.004	.14	.25	.39
8" wide	.220	L.F.	.004	.20	.25	.45
Building paper, #15 asphalt felt	1.500	S.F.	.002	.09	.11	.20
Ridge shingles, asphalt	.280	L.F.	.007	.68	.36	1.04
Clay	.280	L.F.	.011	1.51	1.78	3.29
Slate	.280	L.F.	.011	3.08	.59	3.67
Wood	.280	L.F.	.008	1.39	.45	1.84
Soffit & fascia, aluminum, vented	.220	L.F.	.032	1.08	1.82	2.90
Vinyl, vented	.220	L.F.	.029	1.12	1.66	2.78
Wood, board fascia, plywood soffit	.220	L.F.	.026	.69	1.39	2.08
Flashing, aluminum, .013" thick	1.500	S.F.	.083	1.37	4.38	5.75
.032" thick	1.500	S.F.	.083	2.24	4.38	6.62
.040" thick	1.500	S.F.	.083	3.81	4.38	8.19
.050" thick	1.500	S.F.	.083	4.35	4.38	8.73
Copper, 16 oz.	1.500	S.F.	.104	13	5.55	18.55
20 oz.	1.500	S.F.	.109	17.05	5.80	22.85
24 oz.	1.500	S.F.	.114	24	6.05	30.05
32 oz.	1.500	S.F.	.120	31	6.35	37.35

5 | ROOFING — 28 | Shed Dormer Roofing Systems

System Description	QUAN.	UNIT	LABOR HOURS	COST PER S.F. MAT.	COST PER S.F. INST.	COST PER S.F. TOTAL
ASPHALT, ROOF SHINGLES, CLASS A						
Shingles, standard inorganic class A 210-235 lb./sq.	1.100	S.F.	.016	.98	.85	1.83
Drip edge, aluminum, 5" wide	.250	L.F.	.005	.16	.29	.45
Building paper, #15 asphalt felt	1.200	S.F.	.002	.07	.09	.16
Soffit & fascia, aluminum, vented, 1' overhang	.250	L.F.	.036	1.23	2.06	3.29
Flashing, aluminum, mill finish, 0.013" thick	.800	L.F.	.044	.73	2.34	3.07
TOTAL		S.F.	.103	3.17	5.63	8.80
WOOD, CEDAR, NO. 1 PERFECTIONS, 18" LONG						
Shingles, wood, red cedar, #1 perfections, 5" exposure	1.100	S.F.	.032	3	1.82	4.82
Drip edge, aluminum, 5" wide	.250	L.F.	.005	.16	.29	.45
Building paper, #15 asphalt felt	1.200	S.F.	.002	.07	.09	.16
Soffit & fascia, aluminum, vented, 1' overhang	.250	L.F.	.036	1.23	2.06	3.29
Flashing, aluminum, mill finish, 0.013" thick	.800	L.F.	.044	.73	2.34	3.07
TOTAL		S.F.	.119	5.19	6.60	11.79
SLATE, BUCKINGHAM, BLACK						
Shingles, slate, Buckingham, black	1.100	S.F.	.050	6.55	2.67	9.22
Drip edge, aluminum, 5" wide	.250	L.F.	.005	.16	.29	.45
Building paper, #15 asphalt felt	1.200	S.F.	.002	.07	.09	.16
Soffit & fascia, aluminum, vented, 1' overhang	.250	L.F.	.036	1.23	2.06	3.29
Flashing, copper, 16 oz.	.800	L.F.	.056	6.92	2.95	9.87
TOTAL		S.F.	.149	14.93	8.06	22.99

The prices in this system are based on a square foot of plan area under the dormer roof.

Description	QUAN.	UNIT	LABOR HOURS	COST PER S.F. MAT.	COST PER S.F. INST.	COST PER S.F. TOTAL

Shed Dormer Roofing Price Sheet

	QUAN.	UNIT	LABOR HOURS	COST PER S.F. MAT.	COST PER S.F. INST.	COST PER S.F. TOTAL
Shingles, asphalt, standard, inorganic, class A, 210-235 lb./sq.	1.100	S.F.	.016	.98	.85	1.83
Laminated, multi-layered, 240-260 lb./sq.	1.100	S.F.	.020	1.32	1.03	2.35
Premium laminated, multi-layered, 260-300 lb./sq.	1.100	S.F.	.025	1.91	1.33	3.24
Clay tile, Spanish tile, red	1.100	S.F.	.049	6.05	2.55	8.60
Mission tile, red	1.100	S.F.	.077	4.95	2.55	7.50
French tile, red	1.100	S.F.	.065	13.75	2.33	16.08
Slate Buckingham, Virginia, black	1.100	S.F.	.050	6.55	2.67	9.22
Vermont, black or grey	1.100	S.F.	.050	5.80	2.67	8.47
Wood, red cedar, No. 1 5X, 16" long, 5" exposure	1.100	S.F.	.035	3.41	2	5.41
Fire retardant	1.100	S.F.	.035	4.09	2	6.09
18" long, No.1 perfections, 5" exposure	1.100	S.F.	.032	3	1.82	4.82
Fire retardant	1.100	S.F.	.032	3.68	1.82	5.50
Resquared & rebutted, 18" long, 5" exposure	1.100	S.F.	.029	3.41	1.66	5.07
Fire retardant	1.100	S.F.	.029	4.09	1.66	5.75
Shakes hand split, 24" long, 10" exposure	1.100	S.F.	.035	3.47	2	5.47
Fire retardant	1.100	S.F.	.035	4.15	2	6.15
18" long, 8" exposure	1.100	S.F.	.044	2.17	2.50	4.67
Fire retardant	1.100	S.F.	.044	2.85	2.50	5.35
Drip edge, metal, 5" wide	.250	L.F.	.005	.16	.29	.45
8" wide	.250	L.F.	.005	.23	.29	.52
Building paper, #15 asphalt felt	1.200	S.F.	.002	.07	.09	.16
Soffit & fascia, aluminum, vented	.250	L.F.	.036	1.23	2.06	3.29
Vinyl, vented	.250	L.F.	.033	1.28	1.89	3.17
Wood, board fascia, plywood soffit	.250	L.F.	.030	.79	1.59	2.38
Flashing, aluminum, .013" thick	.800	L.F.	.044	.73	2.34	3.07
.032" thick	.800	L.F.	.044	1.19	2.34	3.53
.040" thick	.800	L.F.	.044	2.03	2.34	4.37
.050" thick	.800	L.F.	.044	2.32	2.34	4.66
Copper, 16 oz.	.800	L.F.	.056	6.90	2.95	9.85
20 oz.	.800	L.F.	.058	9.10	3.08	12.18
24 oz.	.800	L.F.	.061	12.85	3.23	16.08
32 oz.	.800	L.F.	.064	16.40	3.39	19.79

5 | ROOFING 32 | Skylight/Skywindow Systems

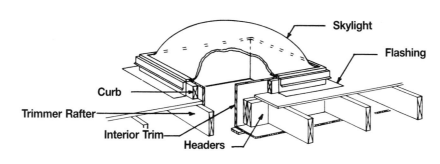

System Description	QUAN.	UNIT	LABOR HOURS	COST EACH MAT.	COST EACH INST.	COST EACH TOTAL
SKYLIGHT, FIXED, 32" X 32"						
Skylight, fixed bubble, insulating, 32" x 32"	1.000	Ea.	1.422	227.55	74.31	301.86
Trimmer rafters, 2" x 6"	28.000	L.F.	.448	18.76	25.48	44.24
Headers, 2" x 6"	6.000	L.F.	.267	4.02	15.12	19.14
Curb, 2" x 4"	12.000	L.F.	.154	5.28	8.76	14.04
Flashing, aluminum, .013" thick	13.500	S.F.	.745	12.29	39.42	51.71
Moldings, casing, ogee, 11/16" x 2-1/2", pine	12.000	L.F.	.384	18.60	21.84	40.44
Trim primer coat, oil base, brushwork	12.000	L.F.	.148	.36	6.96	7.32
Trim paint, 1 coat, brushwork	12.000	L.F.	.148	.72	6.96	7.68
TOTAL		Ea.	3.716	287.58	198.85	486.43
SKYLIGHT, FIXED, 48" X 48"						
Skylight, fixed bubble, insulating, 48" x 48"	1.000	Ea.	1.296	520	67.52	587.52
Trimmer rafters, 2" x 6"	28.000	L.F.	.448	18.76	25.48	44.24
Headers, 2" x 6"	8.000	L.F.	.356	5.36	20.16	25.52
Curb, 2" x 4"	16.000	L.F.	.205	7.04	11.68	18.72
Flashing, aluminum, .013" thick	16.000	S.F.	.883	14.56	46.72	61.28
Moldings, casing, ogee, 11/16" x 2-1/2", pine	16.000	L.F.	.512	24.80	29.12	53.92
Trim primer coat, oil base, brushwork	16.000	L.F.	.197	.48	9.28	9.76
Trim paint, 1 coat, brushwork	16.000	L.F.	.197	.96	9.28	10.24
TOTAL		Ea.	4.094	591.96	219.24	811.20
SKYWINDOW, OPERATING, 24" X 48"						
Skywindow, operating, thermopane glass, 24" x 48"	1.000	Ea.	3.200	645	167	812
Trimmer rafters, 2" x 6"	28.000	L.F.	.448	18.76	25.48	44.24
Headers, 2" x 6"	8.000	L.F.	.267	4.02	15.12	19.14
Curb, 2" x 4"	14.000	L.F.	.179	6.16	10.22	16.38
Flashing, aluminum, .013" thick	14.000	S.F.	.772	12.74	40.88	53.62
Moldings, casing, ogee, 11/16" x 2-1/2", pine	14.000	L.F.	.448	21.70	25.48	47.18
Trim primer coat, oil base, brushwork	14.000	L.F.	.172	.42	8.12	8.54
Trim paint, 1 coat, brushwork	14.000	L.F.	.172	.84	8.12	8.96
TOTAL		Ea.	5.658	709.64	300.42	1,010.06

The prices in these systems are on a cost each basis.

Description	QUAN.	UNIT	LABOR HOURS	COST EACH MAT.	COST EACH INST.	COST EACH TOTAL

For customer support on your Residential Cost Data, call 877.759.4771.

Skylight/Skywindow Price Sheet	QUAN.	UNIT	LABOR HOURS	COST EACH		
				MAT.	INST.	TOTAL
Skylight, fixed bubble insulating, 24" x 24"	1.000	Ea.	.800	128	42	170
32" x 32"	1.000	Ea.	1.422	228	74.50	302.50
32" x 48"	1.000	Ea.	.864	345	45	390
48" x 48"	1.000	Ea.	1.296	520	67.50	587.50
Ventilating bubble insulating, 36" x 36"	1.000	Ea.	2.667	530	139	669
52" x 52"	1.000	Ea.	2.667	735	139	874
28" x 52"	1.000	Ea.	3.200	540	167	707
36" x 52"	1.000	Ea.	3.200	595	167	762
Skywindow, operating, thermopane glass, 24" x 48"	1.000	Ea.	3.200	645	167	812
32" x 48"	1.000	Ea.	3.556	675	185	860
Trimmer rafters, 2" x 6"	28.000	L.F.	.448	18.75	25.50	44.25
2" x 8"	28.000	L.F.	.472	26	27	53
2" x 10"	28.000	L.F.	.711	41	40.50	81.50
Headers, 24" window, 2" x 6"	4.000	L.F.	.178	2.68	10.10	12.78
2" x 8"	4.000	L.F.	.188	3.72	10.70	14.42
2" x 10"	4.000	L.F.	.200	5.90	11.35	17.25
32" window, 2" x 6"	6.000	L.F.	.267	4.02	15.10	19.12
2" x 8"	6.000	L.F.	.282	5.60	16	21.60
2" x 10"	6.000	L.F.	.300	8.80	17.05	25.85
48" window, 2" x 6"	8.000	L.F.	.356	5.35	20	25.35
2" x 8"	8.000	L.F.	.376	7.45	21.50	28.95
2" x 10"	8.000	L.F.	.400	11.75	22.50	34.25
Curb, 2" x 4", skylight, 24" x 24"	8.000	L.F.	.102	3.52	5.85	9.37
32" x 32"	12.000	L.F.	.154	5.30	8.75	14.05
32" x 48"	14.000	L.F.	.179	6.15	10.20	16.35
48" x 48"	16.000	L.F.	.205	7.05	11.70	18.75
Flashing, aluminum .013" thick, skylight, 24" x 24"	9.000	S.F.	.497	8.20	26.50	34.70
32" x 32"	13.500	S.F.	.745	12.30	39.50	51.80
32" x 48"	14.000	S.F.	.772	12.75	41	53.75
48" x 48"	16.000	S.F.	.883	14.55	46.50	61.05
Copper 16 oz., skylight, 24" x 24"	9.000	S.F.	.626	78	33	111
32" x 32"	13.500	S.F.	.939	117	50	167
32" x 48"	14.000	S.F.	.974	121	51.50	172.50
48" x 48"	16.000	S.F.	1.113	138	59	197
Trim, interior casing painted, 24" x 24"	8.000	L.F.	.347	13.50	18.30	31.80
32" x 32"	12.000	L.F.	.520	20.50	27.50	48
32" x 48"	14.000	L.F.	.607	23.50	32	55.50
48" x 48"	16.000	L.F.	.693	27	36.50	63.50

5 | ROOFING
34 | Built-up Roofing Systems

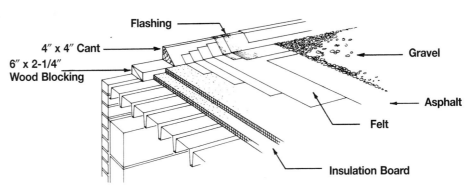

System Description	QUAN.	UNIT	LABOR HOURS	COST PER S.F. MAT.	COST PER S.F. INST.	COST PER S.F. TOTAL
ASPHALT, ORGANIC, 4-PLY, INSULATED DECK						
Membrane, asphalt, 4-plies #15 felt, gravel surfacing	1.000	S.F.	.025	1.43	1.56	2.99
Insulation board, 2-layers of 1-1/16" glass fiber	2.000	S.F.	.012	2.22	.66	2.88
Roof deck insulation, fastening alternatives, coated screws, 4" long	1.000	S.F	.003	.12	.14	.26
Wood blocking, 2" x 6"	.040	L.F.	.004	.08	.25	.33
Treated 4" x 4" cant strip	.040	L.F.	.001	.08	.06	.14
Flashing, aluminum, 0.040" thick	.050	S.F.	.003	.13	.15	.28
TOTAL		S.F.	.048	4.06	2.82	6.88
ASPHALT, INORGANIC, 3-PLY, INSULATED DECK						
Membrane, asphalt, 3-plies type IV glass felt, gravel surfacing	1.000	S.F.	.028	1.47	1.71	3.18
Insulation board, 2-layers of 1-1/16" glass fiber	2.000	S.F.	.012	2.22	.66	2.88
Roof deck insulation, fastening alternatives, coated screws, 4" long	1.000	S.F	.003	.12	.14	.26
Wood blocking, 2" x 6"	.040	L.F.	.004	.08	.25	.33
Treated 4" x 4" cant strip	.040	L.F.	.001	.08	.06	.14
Flashing, aluminum, 0.040" thick	.050	S.F.	.003	.13	.15	.28
TOTAL		S.F.	.051	4.10	2.97	7.07
COAL TAR, ORGANIC, 4-PLY, INSULATED DECK						
Membrane, coal tar, 4-plies #15 felt, gravel surfacing	1.000	S.F.	.027	2.20	1.64	3.84
Insulation board, 2-layers of 1-1/16" glass fiber	2.000	S.F.	.012	2.22	.66	2.88
Roof deck insulation, fastening alternatives, coated screws, 4" long	1.000	S.F	.003	.12	.14	.26
Wood blocking, 2" x 6"	.040	L.F.	.004	.08	.25	.33
Treated 4" x 4" cant strip	.040	L.F.	.001	.08	.06	.14
Flashing, aluminum, 0.040" thick	.050	S.F.	.003	.13	.15	.28
TOTAL		S.F.	.050	4.83	2.90	7.73
COAL TAR, INORGANIC, 3-PLY, INSULATED DECK						
Membrane, coal tar, 3-plies type IV glass felt, gravel surfacing	1.000	S.F.	.029	1.81	1.81	3.62
Insulation board, 2-layers of 1-1/16" glass fiber	2.000	S.F.	.012	2.22	.66	2.88
Roof deck insulation, fastening alternatives, coated screws, 4" long	1.000	S.F	.003	.12	.14	.26
Wood blocking, 2" x 6"	.040	L.F.	.004	.08	.25	.33
Treated 4" x 4" cant strip	.040	L.F.	.001	.08	.06	.14
Flashing, aluminum, 0.040" thick	.050	S.F.	.003	.13	.15	.28
TOTAL		S.F.	.052	4.44	3.07	7.51

Built-Up Roofing Price Sheet	QUAN.	UNIT	LABOR HOURS	COST PER S.F.		
				MAT.	INST.	TOTAL
Membrane, asphalt, 4-plies #15 organic felt, gravel surfacing	1.000	S.F.	.025	1.43	1.56	2.99
Asphalt base sheet & 3-plies #15 asphalt felt	1.000	S.F.	.025	1.09	1.56	2.65
3-plies type IV glass fiber felt	1.000	S.F.	.028	1.47	1.71	3.18
4-plies type IV glass fiber felt	1.000	S.F.	.028	1.82	1.71	3.53
Coal tar, 4-plies #15 organic felt, gravel surfacing	1.000	S.F.	.027			
4-plies tarred felt	1.000	S.F.	.027	2.20	1.64	3.84
3-plies type IV glass fiber felt	1.000	S.F.	.029	1.81	1.81	3.62
4-plies type IV glass fiber felt	1.000	S.F.	.027	2.52	1.64	4.16
Roll, asphalt, 1-ply #15 organic felt, 2-plies mineral surfaced	1.000	S.F.	.021	.85	1.27	2.12
3-plies type IV glass fiber, 1-ply mineral surfaced	1.000	S.F.	.022	1.45	1.38	2.83
Insulation boards, glass fiber, 1-1/16" thick	1.000	S.F.	.008	1.23	.47	1.70
2-1/16" thick	1.000	S.F.	.010	1.72	.55	2.27
2-7/16" thick	1.000	S.F.	.010	1.97	.55	2.52
Expanded perlite, 1" thick	1.000	S.F.	.010	.70	.55	1.25
1-1/2" thick	1.000	S.F.	.010	.98	.55	1.53
2" thick	1.000	S.F.	.011	1.29	.61	1.90
Fiberboard, 1" thick	1.000	S.F.	.010	.73	.55	1.28
1-1/2" thick	1.000	S.F.	.010	1.04	.55	1.59
2" thick	1.000	S.F.	.010	1.33	.55	1.88
Extruded polystyrene, 15 PSI compressive strength, 2" thick R10	1.000	S.F.	.006	.98	.40	1.38
3" thick R15	1.000	S.F.	.008	1.84	.47	2.31
4" thick R20	1.000	S.F.	.008	2.43	.47	2.90
Tapered for drainage	1.000	S.F.	.005	.80	.36	1.16
40 PSI compressive strength, 1" thick R5	1.000	S.F.	.005	.70	.36	1.06
2" thick R10	1.000	S.F.	.006	1.23	.40	1.63
3" thick R15	1.000	S.F.	.008	1.72	.47	2.19
4" thick R20	1.000	S.F.	.008	2.22	.47	2.69
Fiberboard high density, 1/2" thick R1.3	1.000	S.F.	.008	.45	.47	.92
1" thick R2.5	1.000	S.F.	.010	.74	.55	1.29
1 1/2" thick R3.8	1.000	S.F.	.010	1.07	.55	1.62
Polyisocyanurate, 1 1/2" thick	1.000	S.F.	.006	.82	.40	1.22
2" thick	1.000	S.F.	.007	1.02	.44	1.46
3 1/2" thick	1.000	S.F.	.008	2.21	.47	2.68
Tapered for drainage	1.000	S.F.	.006	.77	.37	1.14
Expanded polystyrene, 1" thick	1.000	S.F.	.005	.41	.36	.77
2" thick R10	1.000	S.F.	.006	.69	.40	1.09
3" thick	1.000	S.F.	.006	.98	.40	1.38
Wood blocking, treated, 6" x 2" & 4" x 4" cant	.040	L.F.	.002	.14	.14	.28
6" x 4-1/2" & 4" x 4" cant	.040	L.F.	.005	.20	.31	.51
6" x 5" & 4" x 4" cant	.040	L.F.	.007	.24	.39	.63
Flashing, aluminum, 0.019" thick	.050	S.F.	.003	.07	.15	.22
0.032" thick	.050	S.F.	.003	.07	.15	.22
0.040" thick	.050	S.F.	.003	.13	.15	.28
Copper sheets, 16 oz., under 500 lbs.	.050	S.F.	.003	.43	.18	.61
Over 500 lbs.	.050	S.F.	.003	.43	.14	.57
20 oz., under 500 lbs.	.050	S.F.	.004	.57	.19	.76
Over 500 lbs.	.050	S.F.	.003	.54	.15	.69
Stainless steel, 32 gauge	.050	S.F.	.003	.18	.14	.32
28 gauge	.050	S.F.	.003	.23	.14	.37
26 gauge	.050	S.F.	.003	.25	.14	.39
24 gauge	.050	S.F.	.003	.27	.14	.41

Did you know?
RSMeans Online gives you the same access to RSMeans' data with 24/7 access:
- Quickly locate costs in the searchable database.
- Build cost lists, estimates, and reports in minutes.
- Adjust costs to any location in the U.S. and Canada with the click of a button.

Start your free trial today at **www.RSMeansOnline.com**

RSMeans Online
FROM THE GORDIAN GROUP®

No part of this cost data may be reproduced, stored in a retrieval system, or transmitted in any form or by any means without prior written permission of RSMeans.

6 | INTERIORS — 04 Drywall & Thincoat Wall Systems

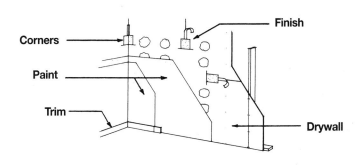

System Description	QUAN.	UNIT	LABOR HOURS	MAT.	INST.	TOTAL
1/2" DRYWALL, TAPED & FINISHED						
Gypsum wallboard, 1/2" thick, standard	1.000	S.F.	.008	.36	.45	.81
Finish, taped & finished joints	1.000	S.F.	.008	.05	.45	.50
Corners, taped & finished, 32 L.F. per 12' x 12' room	.083	L.F.	.002	.01	.09	.10
Painting, primer & 2 coats	1.000	S.F.	.011	.21	.50	.71
Paint trim, to 6" wide, primer + 1 coat enamel	.125	L.F.	.001	.02	.06	.08
Moldings, base, ogee profile, 9/16" x 4-1/2, red oak	.125	L.F.	.005	.49	.26	.75
TOTAL		S.F.	.035	1.14	1.81	2.95
THINCOAT, SKIM-COAT, ON 1/2" BACKER DRYWALL						
Gypsum wallboard, 1/2" thick, thincoat backer	1.000	S.F.	.008	.36	.45	.81
Thincoat plaster	1.000	S.F.	.011	.12	.58	.70
Corners, taped & finished, 32 L.F. per 12' x 12' room	.083	L.F.	.002	.01	.09	.10
Painting, primer & 2 coats	1.000	S.F.	.011	.21	.50	.71
Paint trim, to 6" wide, primer + 1 coat enamel	.125	L.F.	.001	.02	.06	.08
Moldings, base, ogee profile, 9/16" x 4-1/2, red oak	.125	L.F.	.005	.49	.26	.75
TOTAL		S.F.	.038	1.21	1.94	3.15
5/8" DRYWALL, TAPED & FINISHED						
Gypsum wallboard, 5/8" thick, standard	1.000	S.F.	.008	.37	.45	.82
Finish, taped & finished joints	1.000	S.F.	.008	.05	.45	.50
Corners, taped & finished, 32 L.F. per 12' x 12' room	.083	L.F.	.002	.01	.09	.10
Painting, primer & 2 coats	1.000	S.F.	.011	.21	.50	.71
Moldings, base, ogee profile, 9/16" x 4-1/2, red oak	.125	L.F.	.005	.49	.26	.75
Paint trim, to 6" wide, primer + 1 coat enamel	.125	L.F.	.001	.02	.06	.08
TOTAL		S.F.	.035	1.15	1.81	2.96

The costs in this system are based on a square foot of wall.
Do not deduct for openings.

Description	QUAN.	UNIT	LABOR HOURS	MAT.	INST.	TOTAL

Drywall & Thincoat Wall Price Sheet

	QUAN.	UNIT	LABOR HOURS	COST PER S.F. MAT.	INST.	TOTAL
Gypsum wallboard, 1/2" thick, standard	1.000	S.F.	.008	.36	.45	.81
Fire resistant	1.000	S.F.	.008	.40	.45	.85
Water resistant	1.000	S.F.	.008	.45	.45	.90
5/8" thick, standard	1.000	S.F.	.008	.37	.45	.82
Fire resistant	1.000	S.F.	.008	.39	.45	.84
Water resistant	1.000	S.F.	.008	.47	.45	.92
Gypsum wallboard backer for thincoat system, 1/2" thick	1.000	S.F.	.008	.36	.45	.81
5/8" thick	1.000	S.F.	.008	.37	.45	.82
Gypsum wallboard, taped & finished	1.000	S.F.	.008	.05	.45	.50
Texture spray	1.000	S.F.	.010	.04	.54	.58
Thincoat plaster, including tape	1.000	S.F.	.011	.12	.58	.70
Gypsum wallboard corners, taped & finished, 32 L.F. per 4' x 4' room	.250	L.F.	.004	.03	.24	.27
6' x 6' room	.110	L.F.	.002	.01	.11	.12
10' x 10' room	.100	L.F.	.001	.01	.10	.11
12' x 12' room	.083	L.F.	.001	.01	.08	.09
16' x 16' room	.063	L.F.	.001	.01	.06	.07
Thincoat system, 32 L.F. per 4' x 4' room	.250	L.F.	.003	.03	.15	.18
6' x 6' room	.110	L.F.	.001	.01	.06	.07
10' x 10' room	.100	L.F.	.001	.01	.05	.06
12' x 12' room	.083	L.F.	.001	.01	.04	.05
16' x 16' room	.063	L.F.	.001	.01	.03	.04
Painting, primer, & 1 coat	1.000	S.F.	.008	.14	.39	.53
& 2 coats	1.000	S.F.	.011	.21	.50	.71
Wallpaper, $7/double roll	1.000	S.F.	.013	.65	.59	1.24
$17/double roll	1.000	S.F.	.015	1.24	.71	1.95
$40/double roll	1.000	S.F.	.018	1.89	.87	2.76
Wallcovering, medium weight vinyl		S.F.	.017	.99	.79	1.78
Tile, ceramic adhesive thin set, 4 1/4" x 4 1/4" tiles	1.000	S.F.	.084	2.74	3.78	6.52
6" x 6" tiles	1.000	S.F.	.080	3.80	4.11	7.91
Pregrouted sheets	1.000	S.F.	.067	5.95	2.99	8.94
Trim, painted or stained, baseboard	.125	L.F.	.006	.51	.32	.83
Base shoe	.125	L.F.	.005	.06	.30	.36
Chair rail	.125	L.F.	.005	.24	.27	.51
Cornice molding	.125	L.F.	.004	.15	.27	.42
Cove base, vinyl	.125	L.F.	.003	.16	.16	.32
Paneling, not including furring or trim						
Plywood, prefinished, 1/4" thick, 4' x 8' sheets, vert. grooves						
Birch faced, minimum	1.000	S.F.	.032	1.60	1.82	3.42
Average	1.000	S.F.	.038	1.32	2.16	3.48
Maximum	1.000	S.F.	.046	1.38	2.59	3.97
Mahogany, African	1.000	S.F.	.040	3.08	2.27	5.35
Philippine (lauan)	1.000	S.F.	.032	.72	1.82	2.54
Oak or cherry, minimum	1.000	S.F.	.032	1.54	1.82	3.36
Maximum	1.000	S.F.	.040	2.31	2.27	4.58
Rosewood	1.000	S.F.	.050	3.25	2.84	6.09
Teak	1.000	S.F.	.040	3.25	2.27	5.52
Chestnut	1.000	S.F.	.043	5.35	2.42	7.77
Pecan	1.000	S.F.	.040	2.81	2.27	5.08
Walnut, minimum	1.000	S.F.	.032	2.64	1.82	4.46
Maximum	1.000	S.F.	.040	5.80	2.27	8.07

6 | INTERIORS — 08 Drywall & Thincoat Ceiling Systems

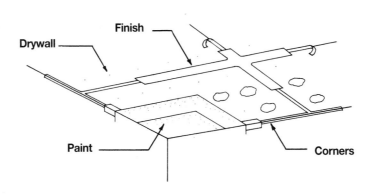

System Description	QUAN.	UNIT	LABOR HOURS	COST PER S.F. MAT.	COST PER S.F. INST.	COST PER S.F. TOTAL
1/2" GYPSUM WALLBOARD, TAPED & FINISHED						
Gypsum wallboard, 1/2" thick, standard	1.000	S.F.	.008	.36	.45	.81
Finish, taped & finished	1.000	S.F.	.008	.05	.45	.50
Corners, taped & finished, 12' x 12' room	.333	L.F.	.006	.04	.32	.36
Paint, primer & 2 coats	1.000	S.F.	.011	.21	.50	.71
TOTAL		S.F.	.033	.66	1.72	2.38
THINCOAT, SKIM COAT ON 1/2" GYPSUM WALLBOARD						
Gypsum wallboard, 1/2" thick, thincoat backer	1.000	S.F.	.008	.36	.45	.81
Thincoat plaster	1.000	S.F.	.011	.12	.58	.70
Corners, taped & finished, 12' x 12' room	.333	L.F.	.006	.04	.32	.36
Paint, primer & 2 coats	1.000	S.F.	.011	.21	.50	.71
TOTAL		S.F.	.036	.73	1.85	2.58
WATER-RESISTANT GYPSUM WALLBOARD, 1/2" THICK, TAPED & FINISHED						
Gypsum wallboard, 1/2" thick, water-resistant	1.000	S.F.	.008	.45	.45	.90
Finish, taped & finished	1.000	S.F.	.008	.05	.45	.50
Corners, taped & finished, 12' x 12' room	.333	L.F.	.006	.04	.32	.36
Paint, primer & 2 coats	1.000	S.F.	.011	.21	.50	.71
TOTAL		S.F.	.033	.75	1.72	2.47
5/8" GYPSUM WALLBOARD, TAPED & FINISHED						
Gypsum wallboard, 5/8" thick, standard	1.000	S.F.	.008	.37	.45	.82
Finish, taped & finished	1.000	S.F.	.008	.05	.45	.50
Corners, taped & finished, 12' x 12' room	.333	L.F.	.006	.04	.32	.36
Paint, primer & 2 coats	1.000	S.F.	.011	.21	.50	.71
TOTAL		S.F.	.033	.67	1.72	2.39

The costs in this system are based on a square foot of ceiling.

Description	QUAN.	UNIT	LABOR HOURS	COST PER S.F. MAT.	COST PER S.F. INST.	COST PER S.F. TOTAL

For customer support on your Residential Cost Data, call 877.759.4771.

Drywall & Thincoat Ceilings Price Sheet

	QUAN.	UNIT	LABOR HOURS	COST PER S.F. MAT.	COST PER S.F. INST.	COST PER S.F. TOTAL
Gypsum wallboard ceilings, 1/2" thick, standard	1.000	S.F.	.008	.36	.45	.81
Fire resistant	1.000	S.F.	.008	.40	.45	.85
Water resistant	1.000	S.F.	.008	.45	.45	.90
5/8" thick, standard	1.000	S.F.	.008	.37	.45	.82
Fire resistant	1.000	S.F.	.008	.39	.45	.84
Water resistant	1.000	S.F.	.008	.47	.45	.92
Gypsum wallboard backer for thincoat ceiling system, 1/2" thick	1.000	S.F.	.016	.76	.90	1.66
5/8" thick	1.000	S.F.	.016	.77	.90	1.67
Gypsum wallboard ceilings, taped & finished	1.000	S.F.	.008	.05	.45	.50
Texture spray	1.000	S.F.	.010	.04	.54	.58
Thincoat plaster	1.000	S.F.	.011	.12	.58	.70
Corners taped & finished, 4' x 4' room	1.000	L.F.	.015	.11	.96	1.07
6' x 6' room	.667	L.F.	.010	.07	.64	.71
10' x 10' room	.400	L.F.	.006	.04	.38	.42
12' x 12' room	.333	L.F.	.005	.04	.32	.36
16' x 16' room	.250	L.F.	.003	.02	.18	.20
Thincoat system, 4' x 4' room	1.000	L.F.	.011	.12	.58	.70
6' x 6' room	.667	L.F.	.007	.08	.39	.47
10' x 10' room	.400	L.F.	.004	.05	.24	.29
12' x 12' room	.333	L.F.	.004	.04	.19	.23
16' x 16' room	.250	L.F.	.002	.02	.11	.13
Painting, primer & 1 coat	1.000	S.F.	.008	.14	.39	.53
& 2 coats	1.000	S.F.	.011	.21	.50	.71
Wallpaper, double roll, solid pattern, avg. workmanship	1.000	S.F.	.013	.65	.59	1.24
Basic pattern, avg. workmanship	1.000	S.F.	.015	1.24	.71	1.95
Basic pattern, quality workmanship	1.000	S.F.	.018	1.89	.87	2.76
Tile, ceramic adhesive thin set, 4 1/4" x 4 1/4" tiles	1.000	S.F.	.084	2.74	3.78	6.52
6" x 6" tiles	1.000	S.F.	.080	3.80	4.11	7.91
Pregrouted sheets	1.000	S.F.	.067	5.95	2.99	8.94

6 | INTERIORS — 12 | Plaster & Stucco Wall Systems

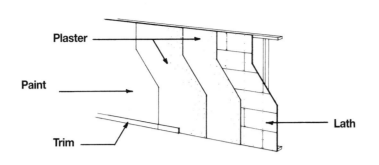

System Description	QUAN.	UNIT	LABOR HOURS	COST PER S.F. MAT.	COST PER S.F. INST.	COST PER S.F. TOTAL
PLASTER ON GYPSUM LATH						
Plaster, gypsum or perlite, 2 coats	1.000	S.F.	.053	.45	2.78	3.23
Lath, 3/8" gypsum	1.000	S.F.	.010	.37	.56	.93
Corners, expanded metal, 32 L.F. per 12' x 12' room	.083	L.F.	.002	.01	.09	.10
Painting, primer & 2 coats	1.000	S.F.	.011	.21	.50	.71
Paint trim, to 6" wide, primer + 1 coat enamel	.125	L.F.	.001	.02	.06	.08
Moldings, base, ogee profile, 9/16" x 4-1/2, red oak	.125	L.F.	.005	.49	.26	.75
TOTAL		S.F.	.082	1.55	4.25	5.80
PLASTER ON METAL LATH						
Plaster, gypsum or perlite, 2 coats	1.000	S.F.	.053	.45	2.78	3.23
Lath, 2.5 Lb. diamond, metal	1.000	S.F.	.010	.45	.56	1.01
Corners, expanded metal, 32 L.F. per 12' x 12' room	.083	L.F.	.002	.01	.09	.10
Painting, primer & 2 coats	1.000	S.F.	.011	.21	.50	.71
Paint trim, to 6" wide, primer + 1 coat enamel	.125	L.F.	.001	.02	.06	.08
Moldings, base, ogee profile, 9/16" x 4-1/2, red oak	.125	L.F.	.005	.49	.26	.75
TOTAL		S.F.	.082	1.63	4.25	5.88
STUCCO ON METAL LATH						
Stucco, 2 coats	1.000	S.F.	.041	.28	2.13	2.41
Lath, 2.5 Lb. diamond, metal	1.000	S.F.	.010	.45	.56	1.01
Corners, expanded metal, 32 L.F. per 12' x 12' room	.083	L.F.	.002	.01	.09	.10
Painting, primer & 2 coats	1.000	S.F.	.011	.21	.50	.71
Paint trim, to 6" wide, primer + 1 coat enamel	.125	L.F.	.001	.02	.06	.08
Moldings, base, ogee profile, 9/16" x 4-1/2, red oak	.125	L.F.	.005	.49	.26	.75
TOTAL		S.F.	.070	1.46	3.60	5.06

The costs in these systems are based on a per square foot of wall area.
Do not deduct for openings.

Description	QUAN.	UNIT	LABOR HOURS	COST PER S.F. MAT.	COST PER S.F. INST.	COST PER S.F. TOTAL

Plaster & Stucco Wall Price Sheet

	QUAN.	UNIT	LABOR HOURS	COST PER S.F. MAT.	COST PER S.F. INST.	COST PER S.F. TOTAL
Plaster, gypsum or perlite, 2 coats	1.000	S.F.	.053	.45	2.78	3.23
3 coats	1.000	S.F.	.065	.65	3.36	4.01
Lath, gypsum, standard, 3/8" thick	1.000	S.F.	.010	.37	.56	.93
Fire resistant, 3/8" thick	1.000	S.F.	.013	.30	.68	.98
1/2" thick	1.000	S.F.	.014	.36	.73	1.09
Metal, diamond, 2.5 Lb.	1.000	S.F.	.010	.45	.56	1.01
3.4 Lb.	1.000	S.F.	.012	.48	.63	1.11
Rib, 2.75 Lb.	1.000	S.F.	.012	.38	.63	1.01
3.4 Lb.	1.000	S.F.	.013	.50	.68	1.18
Corners, expanded metal, 32 L.F. per 4' x 4' room	.250	L.F.	.005	.04	.29	.33
6' x 6' room	.110	L.F.	.002	.02	.13	.15
10' x 10' room	.100	L.F.	.002	.02	.11	.13
12' x 12' room	.083	L.F.	.002	.01	.09	.10
16' x 16' room	.063	L.F.	.001	.01	.07	.08
Painting, primer & 1 coats	1.000	S.F.	.008	.14	.39	.53
Primer & 2 coats	1.000	S.F.	.011	.21	.50	.71
Wallpaper, low price double roll	1.000	S.F.	.013	.65	.59	1.24
Medium price double roll	1.000	S.F.	.015	1.24	.71	1.95
High price double roll	1.000	S.F.	.018	1.89	.87	2.76
Tile, ceramic thin set, 4-1/4" x 4-1/4" tiles	1.000	S.F.	.084	2.74	3.78	6.52
6" x 6" tiles	1.000	S.F.	.080	3.80	4.11	7.91
Pregrouted sheets	1.000	S.F.	.067	5.95	2.99	8.94
Trim, painted or stained, baseboard	.125	L.F.	.006	.51	.32	.83
Base shoe	.125	L.F.	.005	.06	.30	.36
Chair rail	.125	L.F.	.005	.24	.27	.51
Cornice molding	.125	L.F.	.004	.15	.27	.42
Cove base, vinyl	.125	L.F.	.003	.16	.16	.32
Paneling not including furring or trim						
Plywood, prefinished, 1/4" thick, 4' x 8' sheets, vert. grooves						
Birch faced, minimum	1.000	S.F.	.032	1.60	1.82	3.42
Average	1.000	S.F.	.038	1.32	2.16	3.48
Maximum	1.000	S.F.	.046	1.38	2.59	3.97
Mahogany, African	1.000	S.F.	.040	3.08	2.27	5.35
Philippine (lauan)	1.000	S.F.	.032	.72	1.82	2.54
Oak or cherry, minimum	1.000	S.F.	.032	1.54	1.82	3.36
Maximum	1.000	S.F.	.040	2.31	2.27	4.58
Rosewood	1.000	S.F.	.050	3.25	2.84	6.09
Teak	1.000	S.F.	.040	3.25	2.27	5.52
Chestnut	1.000	S.F.	.043	5.35	2.42	7.77
Pecan	1.000	S.F.	.040	2.81	2.27	5.08
Walnut, minimum	1.000	S.F.	.032	2.64	1.82	4.46
Maximum	1.000	S.F.	.040	5.80	2.27	8.07

6 | INTERIORS — 16 | Plaster & Stucco Ceiling Systems

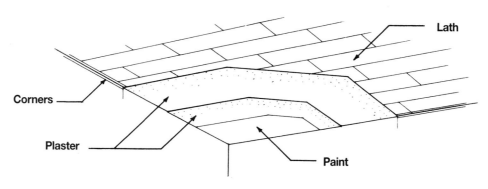

System Description	QUAN.	UNIT	LABOR HOURS	COST PER S.F. MAT.	COST PER S.F. INST.	COST PER S.F. TOTAL
PLASTER ON GYPSUM LATH						
Plaster, gypsum or perlite, 2 coats	1.000	S.F.	.061	.45	3.16	3.61
Gypsum lath, plain or perforated, nailed, 3/8" thick	1.000	S.F.	.010	.37	.56	.93
Gypsum lath, ceiling installation adder	1.000	S.F.	.004		.22	.22
Corners, expanded metal, 12' x 12' room	.330	L.F.	.007	.06	.38	.44
Painting, primer & 2 coats	1.000	S.F.	.011	.21	.50	.71
TOTAL		S.F.	.093	1.09	4.82	5.91
PLASTER ON METAL LATH						
Plaster, gypsum or perlite, 2 coats	1.000	S.F.	.061	.45	3.16	3.61
Lath, 2.5 Lb. diamond, metal	1.000	S.F.	.012	.45	.63	1.08
Corners, expanded metal, 12' x 12' room	.330	L.F.	.007	.06	.38	.44
Painting, primer & 2 coats	1.000	S.F.	.011	.21	.50	.71
TOTAL		S.F.	.091	1.17	4.67	5.84
STUCCO ON GYPSUM LATH						
Stucco, 2 coats	1.000	S.F.	.041	.28	2.13	2.41
Gypsum lath, plain or perforated, nailed, 3/8" thick	1.000	S.F.	.010	.37	.56	.93
Gypsum lath, ceiling installation adder	1.000	S.F.	.004		.22	.22
Corners, expanded metal, 12' x 12' room	.330	L.F.	.007	.06	.38	.44
Painting, primer & 2 coats	1.000	S.F.	.011	.21	.50	.71
TOTAL		S.F.	.073	.92	3.79	4.71
STUCCO ON METAL LATH						
Stucco, 2 coats	1.000	S.F.	.041	.28	2.13	2.41
Lath, 2.5 Lb. diamond, metal	1.000	S.F.	.012	.45	.63	1.08
Corners, expanded metal, 12' x 12' room	.330	L.F.	.007	.06	.38	.44
Painting, primer & 2 coats	1.000	S.F.	.011	.21	.50	.71
TOTAL		S.F.	.071	1	3.64	4.64

The costs in these systems are based on a square foot of ceiling area.

Description	QUAN.	UNIT	LABOR HOURS	COST PER S.F. MAT.	COST PER S.F. INST.	COST PER S.F. TOTAL

Plaster & Stucco Ceiling Price Sheet

	QUAN.	UNIT	LABOR HOURS	COST PER S.F. MAT.	COST PER S.F. INST.	COST PER S.F. TOTAL
Plaster, gypsum or perlite, 2 coats	1.000	S.F.	.061	.45	3.16	3.61
3 coats	1.000	S.F.	.065	.65	3.36	4.01
Lath, gypsum, standard, 3/8" thick	1.000	S.F.	.014	.37	.78	1.15
Fire resistant, 3/8" thick	1.000	S.F.	.017	.30	.90	1.20
1/2" thick	1.000	S.F.	.018	.36	.95	1.31
Metal, diamond, 2.5 Lb.	1.000	S.F.	.012	.45	.63	1.08
3.4 Lb.	1.000	S.F.	.015	.48	.79	1.27
Rib, 2.75 Lb.	1.000	S.F.	.012	.38	.63	1.01
3.4 Lb.	1.000	S.F.	.013	.50	.68	1.18
Corners expanded metal, 4' x 4' room	1.000	L.F.	.020	.17	1.14	1.31
6' x 6' room	.667	L.F.	.013	.11	.76	.87
10' x 10' room	.400	L.F.	.008	.07	.46	.53
12' x 12' room	.333	L.F.	.007	.06	.38	.44
16' x 16' room	.250	L.F.	.004	.03	.21	.24
Painting, primer & 1 coat	1.000	S.F.	.008	.14	.39	.53
Primer & 2 coats	1.000	S.F.	.011	.21	.50	.71

6 | INTERIORS 18 | Suspended Ceiling Systems

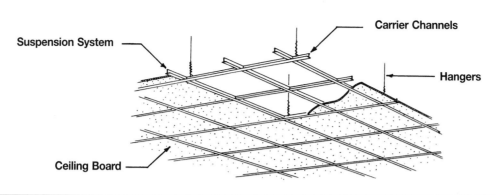

System Description	QUAN.	UNIT	LABOR HOURS	COST PER S.F. MAT.	COST PER S.F. INST.	COST PER S.F. TOTAL
2' X 2' GRID, FILM FACED FIBERGLASS, 5/8" THICK						
Suspension system, 2' x 2' grid, T bar	1.000	S.F.	.012	1.09	.70	1.79
Ceiling board, film faced fiberglass, 5/8" thick	1.000	S.F.	.013	1.38	.73	2.11
Carrier channels, 1-1/2" x 3/4"	1.000	S.F.	.017	.13	.97	1.10
Hangers, #12 wire	1.000	S.F.	.002		.10	.10
TOTAL		S.F.	.044	2.60	2.50	5.10
2' X 4' GRID, FILM FACED FIBERGLASS, 5/8" THICK						
Suspension system, 2' x 4' grid, T bar	1.000	S.F.	.010	.84	.57	1.41
Ceiling board, film faced fiberglass, 5/8" thick	1.000	S.F.	.013	1.38	.73	2.11
Carrier channels, 1-1/2" x 3/4"	1.000	S.F.	.017	.13	.97	1.10
Hangers, #12 wire	1.000	S.F.	.002		.10	.10
TOTAL		S.F.	.042	2.35	2.37	4.72
2' X 2' GRID, MINERAL FIBER, REVEAL EDGE, 1" THICK						
Suspension system, 2' x 2' grid, T bar	1.000	S.F.	.012	1.09	.70	1.79
Ceiling board, mineral fiber, reveal edge, 1" thick	1.000	S.F.	.013	2.20	.76	2.96
Carrier channels, 1-1/2" x 3/4"	1.000	S.F.	.017	.13	.97	1.10
Hangers, #12 wire	1.000	S.F.	.002		.10	.10
TOTAL		S.F.	.044	3.42	2.53	5.95
2' X 4' GRID, MINERAL FIBER, REVEAL EDGE, 1" THICK						
Suspension system, 2' x 4' grid, T bar	1.000	S.F.	.010	.84	.57	1.41
Ceiling board, mineral fiber, reveal edge, 1" thick	1.000	S.F.	.013	2.20	.76	2.96
Carrier channels, 1-1/2" x 3/4"	1.000	S.F.	.017	.13	.97	1.10
Hangers, #12 wire	1.000	S.F.	.002		.10	.10
TOTAL		S.F.	.042	3.17	2.40	5.57

Description	QUAN.	UNIT	LABOR HOURS	COST PER S.F. MAT.	COST PER S.F. INST.	COST PER S.F. TOTAL

Suspended Ceiling Price Sheet

	QUAN.	UNIT	LABOR HOURS	COST PER S.F. MAT.	COST PER S.F. INST.	COST PER S.F. TOTAL
Suspension systems, T bar, 2' x 2' grid	1.000	S.F.	.012	1.09	.70	1.79
2' x 4' grid	1.000	S.F.	.010	.84	.57	1.41
Concealed Z bar, 12" module	1.000	S.F.	.015	.95	.87	1.82
Ceiling boards, fiberglass, film faced, 2' x 2' or 2' x 4', 5/8" thick	1.000	S.F.	.013	1.38	.73	2.11
3/4" thick	1.000	S.F.	.013	2.89	.76	3.65
3" thick thermal R11	1.000	S.F.	.018	2.88	1.01	3.89
Glass cloth faced, 3/4" thick	1.000	S.F.	.016	2.99	.91	3.90
1" thick	1.000	S.F.	.016	3.64	.94	4.58
1-1/2" thick, nubby face	1.000	S.F.	.017	2.86	.96	3.82
Mineral fiber boards, 5/8" thick, aluminum face 2' x 2'	1.000	S.F.	.013	4.05	.76	4.81
2' x 4'	1.000	S.F.	.012	4.05	.70	4.75
Standard faced, 2' x 2' or 2' x 4'	1.000	S.F.	.012	.75	.67	1.42
Plastic coated face, 2' x 2' or 2' x 4'	1.000	S.F.	.020	2.76	1.14	3.90
Fire rated, 2 hour rating, 5/8" thick	1.000	S.F.	.012	1.25	.67	1.92
Tegular edge, 2' x 2' or 2' x 4', 5/8" thick, fine textured	1.000	S.F.	.013	1.14	.97	2.11
Rough textured	1.000	S.F.	.015	1.35	.97	2.32
3/4" thick, fine textured	1.000	S.F.	.016	2.29	1.01	3.30
Rough textured	1.000	S.F.	.018	1.55	1.01	2.56
Luminous panels, prismatic, acrylic	1.000	S.F.	.020	2.90	1.14	4.04
Polystyrene	1.000	S.F.	.020	1.78	1.14	2.92
Flat or ribbed, acrylic	1.000	S.F.	.020	4.37	1.14	5.51
Polystyrene	1.000	S.F.	.020	2.52	1.14	3.66
Drop pan, white, acrylic	1.000	S.F.	.020	6.25	1.14	7.39
Polystyrene	1.000	S.F.	.020	4.90	1.14	6.04
Carrier channels, 4'-0" on center, 3/4" x 1-1/2"	1.000	S.F.	.017	.13	.97	1.10
1-1/2" x 3-1/2"	1.000	S.F.	.017	.35	.97	1.32
Hangers, #12 wire	1.000	S.F.	.002		.10	.10

6 | INTERIORS 20 | Interior Door Systems

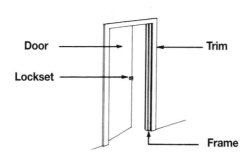

System Description	QUAN.	UNIT	LABOR HOURS	COST EACH		
				MAT.	INST.	TOTAL
LAUAN, FLUSH DOOR, HOLLOW CORE						
Door, flush, lauan, hollow core, 2'-8" wide x 6'-8" high	1.000	Ea.	.889	44	50.50	94.50
Frame, pine, 4-5/8" jamb	17.000	L.F.	.725	91.80	41.14	132.94
Moldings, casing, ogee, 11/16" x 2-1/2", pine	34.000	L.F.	1.088	52.70	61.88	114.58
Paint trim, to 6" wide, primer + 1 coat enamel	34.000	L.F.	.340	4.76	15.98	20.74
Butt hinges, chrome, 3-1/2" x 3-1/2"	1.500	Pr.		48.75		48.75
Lockset, passage	1.000	Ea.	.500	33	28.50	61.50
Prime door & frame, oil, brushwork	2.000	Face	1.600	8.14	76	84.14
Paint door and frame, oil, 2 coats	2.000	Face	2.667	9.98	126	135.98
TOTAL		Ea.	7.809	293.13	400	693.13
BIRCH, FLUSH DOOR, HOLLOW CORE						
Door, flush, birch, hollow core, 2'-8" wide x 6'-8" high	1.000	Ea.	.889	55.50	50.50	106
Frame, pine, 4-5/8" jamb	17.000	L.F.	.725	91.80	41.14	132.94
Moldings, casing, ogee, 11/16" x 2-1/2", pine	34.000	L.F.	1.088	52.70	61.88	114.58
Butt hinges, chrome, 3-1/2" x 3-1/2"	1.500	Pr.		48.75		48.75
Lockset, passage	1.000	Ea.	.500	33	28.50	61.50
Prime door & frame, oil, brushwork	2.000	Face	1.600	8.14	76	84.14
Paint door and frame, oil, 2 coats	2.000	Face	2.667	9.98	126	135.98
TOTAL		Ea.	7.469	299.87	384.02	683.89
RAISED PANEL, SOLID, PINE DOOR						
Door, pine, raised panel, 2'-8" wide x 6'-8" high	1.000	Ea.	.889	204	50.50	254.50
Frame, pine, 4-5/8" jamb	17.000	L.F.	.725	91.80	41.14	132.94
Moldings, casing, ogee, 11/16" x 2-1/2", pine	34.000	L.F.	1.088	52.70	61.88	114.58
Butt hinges, bronze, 3-1/2" x 3-1/2"	1.500	Pr.		57		57
Lockset, passage	1.000	Ea.	.500	33	28.50	61.50
Prime door & frame, oil, brushwork	2.000	Face	1.600	8.14	76	84.14
Paint door and frame, oil, 2 coats	2.000	Face	2.667	9.98	126	135.98
TOTAL		Ea.	7.469	456.62	384.02	840.64

The costs in these systems are based on a cost per each door.

Description	QUAN.	UNIT	LABOR HOURS	COST EACH		
				MAT.	INST.	TOTAL

For customer support on your Residential Cost Data, call 877.759.4771.

Interior Door Price Sheet

			LABOR	COST EACH		
	QUAN.	UNIT	HOURS	MAT.	INST.	TOTAL
Door, hollow core, lauan 1-3/8" thick, 6'-8" high x 1'-6" wide	1.000	Ea.	.889	39	50.50	89.50
2'-0" wide	1.000	Ea.	.889	38.50	50.50	89
2'-6" wide	1.000	Ea.	.889	42.50	50.50	93
2'-8" wide	1.000	Ea.	.889	44	50.50	94.50
3'-0" wide	1.000	Ea.	.941	46.50	53.50	100
Birch 1-3/8" thick, 6'-8" high x 1'-6" wide	1.000	Ea.	.889	45.50	50.50	96
2'-0" wide	1.000	Ea.	.889	46.50	50.50	97
2'-6" wide	1.000	Ea.	.889	55	50.50	105.50
2'-8" wide	1.000	Ea.	.889	55.50	50.50	106
3'-0" wide	1.000	Ea.	.941	57.50	53.50	111
Louvered pine 1-3/8" thick, 6'-8" high x 1'-6" wide	1.000	Ea.	.842	132	48	180
2'-0" wide	1.000	Ea.	.889	144	50.50	194.50
2'-6" wide	1.000	Ea.	.889	165	50.50	215.50
2'-8" wide	1.000	Ea.	.889	177	50.50	227.50
3'-0" wide	1.000	Ea.	.941	192	53.50	245.50
Paneled pine 1-3/8" thick, 6'-8" high x 1'-6" wide	1.000	Ea.	.842	141	48	189
2'-0" wide	1.000	Ea.	.889	168	50.50	218.50
2'-6" wide	1.000	Ea.	.889	200	50.50	250.50
2'-8" wide	1.000	Ea.	.889	204	50.50	254.50
3'-0" wide	1.000	Ea.	.941	224	53.50	277.50
Frame, pine, 1'-6" thru 2'-0" wide door, 3-5/8" deep	16.000	L.F.	.683	78	38.50	116.50
4-5/8" deep	16.000	L.F.	.683	86.50	38.50	125
5-5/8" deep	16.000	L.F.	.683	82.50	38.50	121
2'-6" thru 3'0" wide door, 3-5/8" deep	17.000	L.F.	.725	83	41	124
4-5/8" deep	17.000	L.F.	.725	92	41	133
5-5/8" deep	17.000	L.F.	.725	87.50	41	128.50
Trim, casing, painted, both sides, 1'-6" thru 2'-6" wide door	32.000	L.F.	1.855	52.50	95.50	148
2'-6" thru 3'-0" wide door	34.000	L.F.	1.971	56	101	157
Butt hinges 3-1/2" x 3-1/2", steel plated, chrome	1.500	Pr.		49		49
Bronze	1.500	Pr.		57		57
Locksets, passage, minimum	1.000	Ea.	.500	33	28.50	61.50
Maximum	1.000	Ea.	.575	38	33	71
Privacy, miniumum	1.000	Ea.	.625	41.50	35.50	77
Maximum	1.000	Ea.	.675	44.50	38.50	83
Paint 2 sides, primer & 2 cts., flush door, 1'-6" to 2'-0" wide	2.000	Face	5.547	25.50	262	287.50
2'-6" thru 3'-0" wide	2.000	Face	6.933	31.50	330	361.50
Louvered door, 1'-6" thru 2'-0" wide	2.000	Face	6.400	23	300	323
2'-6" thru 3'-0" wide	2.000	Face	8.000	29	380	409
Paneled door, 1'-6" thru 2'-0" wide	2.000	Face	6.400	23	300	323
2'-6" thru 3'-0" wide	2.000	Face	8.000	29	380	409

6 | INTERIORS — 24 | Closet Door Systems

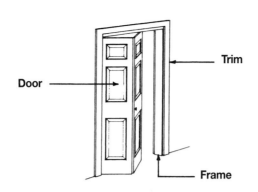

System Description	QUAN.	UNIT	LABOR HOURS	COST EACH MAT.	COST EACH INST.	COST EACH TOTAL
BI-PASSING, FLUSH, LAUAN, HOLLOW CORE, 4'-0" X 6'-8"						
Door, flush, lauan, hollow core, 4'-0" x 6'-8" opening	1.000	Ea.	1.333	194	75.50	269.50
Frame, pine, 4-5/8" jamb	18.000	L.F.	.768	97.20	43.56	140.76
Moldings, casing, ogee, 11/16" x 2-1/2", pine	36.000	L.F.	1.152	55.80	65.52	121.32
Prime door & frame, oil, brushwork	2.000	Face	1.600	8.14	76	84.14
Paint door and frame, oil, 2 coats	2.000	Face	2.667	9.98	126	135.98
TOTAL		Ea.	7.520	365.12	386.58	751.70
BI-PASSING, FLUSH, BIRCH, HOLLOW CORE, 6'-0" X 6'-8"						
Door, flush, birch, hollow core, 6'-0" x 6'-8" opening	1.000	Ea.	1.600	310	91	401
Frame, pine, 4-5/8" jamb	19.000	L.F.	.811	102.60	45.98	148.58
Moldings, casing, ogee, 11/16" x 2-1/2", pine	38.000	L.F.	1.216	58.90	69.16	128.06
Prime door & frame, oil, brushwork	2.000	Face	2.000	10.18	95	105.18
Paint door and frame, oil, 2 coats	2.000	Face	3.333	12.48	157.50	169.98
TOTAL		Ea.	8.960	494.16	458.64	952.80
BI-FOLD, PINE, PANELED, 3'-0" X 6'-8"						
Door, pine, paneled, 3'-0" x 6'-8" opening	1.000	Ea.	1.231	231	70	301
Frame, pine, 4-5/8" jamb	17.000	L.F.	.725	91.80	41.14	132.94
Moldings, casing, ogee, 11/16" x 2-1/2", pine	34.000	L.F.	1.088	52.70	61.88	114.58
Prime door & frame, oil, brushwork	2.000	Face	1.600	8.14	76	84.14
Paint door and frame, oil, 2 coats	2.000	Face	2.667	9.98	126	135.98
TOTAL		Ea.	7.311	393.62	375.02	768.64
BI-FOLD, PINE, LOUVERED, 6'-0" X 6'-8"						
Door, pine, louvered, 6'-0" x 6'-8" opening	1.000	Ea.	1.600	320	91	411
Frame, pine, 4-5/8" jamb	19.000	L.F.	.811	102.60	45.98	148.58
Moldings, casing, ogee, 11/16" x 2-1/2", pine	38.000	L.F.	1.216	58.90	69.16	128.06
Prime door & frame, oil, brushwork	2.500	Face	2.000	10.18	95	105.18
Paint door and frame, oil, 2 coats	2.500	Face	3.333	12.48	157.50	169.98
TOTAL		Ea.	8.960	504.16	458.64	962.80

The costs in this system are based on a cost per each door.

Description	QUAN.	UNIT	LABOR HOURS	COST EACH MAT.	COST EACH INST.	COST EACH TOTAL

Closet Door Price Sheet	QUAN.	UNIT	LABOR HOURS	COST EACH		
				MAT.	INST.	TOTAL
Doors, bi-passing, pine, louvered, 4'-0" x 6'-8" opening	1.000	Ea.	1.333	530	75.50	605.50
6'-0" x 6'-8" opening	1.000	Ea.	1.600	690	91	781
Paneled, 4'-0" x 6'-8" opening	1.000	Ea.	1.333	520	75.50	595.50
6'-0" x 6'-8" opening	1.000	Ea.	1.600	805	91	896
Flush, birch, hollow core, 4'-0" x 6'-8" opening	1.000	Ea.	1.333	266	75.50	341.50
6'-0" x 6'-8" opening	1.000	Ea.	1.600	310	91	401
Flush, lauan, hollow core, 4'-0" x 6'-8" opening	1.000	Ea.	1.333	194	75.50	269.50
6'-0" x 6'-8" opening	1.000	Ea.	1.600	230	91	321
Bi-fold, pine, louvered, 3'-0" x 6'-8" opening	1.000	Ea.	1.231	231	70	301
6'-0" x 6'-8" opening	1.000	Ea.	1.600	320	91	411
Paneled, 3'-0" x 6'-8" opening	1.000	Ea.	1.231	231	70	301
6'-0" x 6'-8" opening	1.000	Ea.	1.600	320	91	411
Flush, birch, hollow core, 3'-0" x 6'-8" opening	1.000	Ea.	1.231	78.50	70	148.50
6'-0" x 6'-8" opening	1.000	Ea.	1.600	141	91	232
Flush, lauan, hollow core, 3'-0" x 6'8" opening	1.000	Ea.	1.231	310	70	380
6'-0" x 6'-8" opening	1.000	Ea.	1.600	440	91	531
Frame pine, 3'-0" door, 3-5/8" deep	17.000	L.F.	.725	83	41	124
4-5/8" deep	17.000	L.F.	.725	92	41	133
5-5/8" deep	17.000	L.F.	.725	87.50	41	128.50
4'-0" door, 3-5/8" deep	18.000	L.F.	.768	88	43.50	131.50
4-5/8" deep	18.000	L.F.	.768	97	43.50	140.50
5-5/8" deep	18.000	L.F.	.768	92.50	43.50	136
6'-0" door, 3-5/8" deep	19.000	L.F.	.811	92.50	46	138.50
4-5/8" deep	19.000	L.F.	.811	103	46	149
5-5/8" deep	19.000	L.F.	.811	98	46	144
Trim both sides, painted 3'-0" x 6'-8" door	34.000	L.F.	1.971	56	101	157
4'-0" x 6'-8" door	36.000	L.F.	2.086	59	107	166
6'-0" x 6'-8" door	38.000	L.F.	2.203	62.50	113	175.50
Paint 2 sides, primer & 2 cts., flush door & frame, 3' x 6'-8" opng	2.000	Face	2.914	13.60	152	165.60
4'-0" x 6'-8" opening	2.000	Face	3.886	18.10	202	220.10
6'-0" x 6'-8" opening	2.000	Face	4.857	22.50	253	275.50
Paneled door & frame, 3'-0" x 6'-8" opening	2.000	Face	6.000	21.50	284	305.50
4'-0" x 6'-8" opening	2.000	Face	8.000	29	380	409
6'-0" x 6'-8" opening	2.000	Face	10.000	36	475	511
Louvered door & frame, 3'-0" x 6'-8" opening	2.000	Face	6.000	21.50	284	305.50
4'-0" x 6'-8" opening	2.000	Face	8.000	29	380	409
6'-0" x 6'-8" opening	2.000	Face	10.000	36	475	511

6 | INTERIORS — 60 | Carpet Systems

System Description	QUAN.	UNIT	LABOR HOURS	COST PER S.F. MAT.	COST PER S.F. INST.	COST PER S.F. TOTAL
Carpet, direct glue-down, nylon, level loop, 26 oz.	1.000	S.F.	.018	2.61	.60	3.21
32 oz.	1.000	S.F.	.018	4.83	.60	5.43
40 oz.	1.000	S.F.	.018	5.50	.60	6.10
Nylon, plush, 20 oz.	1.000	S.F.	.018	2.24	.60	2.84
24 oz.	1.000	S.F.	.018	2.33	.60	2.93
30 oz.	1.000	S.F.	.018	3.78	.60	4.38
42 oz.	1.000	S.F.	.022	5.50	.64	6.14
48 oz.	1.000	S.F.	.022	6.20	.64	6.84
54 oz.	1.000	S.F.	.022	6.75	.64	7.39
Olefin, 15 oz.	1.000	S.F.	.018	1.84	.60	2.44
22 oz.	1.000	S.F.	.018	1.74	.60	2.34
Tile, foam backed, needle punch	1.000	S.F.	.014	4.75	.71	5.46
Tufted loop or shag	1.000	S.F.	.014	3.62	.71	4.33
Wool, 36 oz., level loop	1.000	S.F.	.018	12.30	.64	12.94
32 oz., patterned	1.000	S.F.	.020	11.15	.64	11.79
48 oz., patterned	1.000	S.F.	.020	12.30	.64	12.94
Padding, sponge rubber cushion, minimum	1.000	S.F.	.006	.56	.30	.86
Maximum	1.000	S.F.	.006	1.07	.30	1.37
Felt, 32 oz. to 56 oz., minimum	1.000	S.F.	.006	.69	.30	.99
Maximum	1.000	S.F.	.006	1.14	.30	1.44
Bonded urethane, 3/8" thick, minimum	1.000	S.F.	.006	.73	.30	1.03
Maximum	1.000	S.F.	.006	.98	.30	1.28
Prime urethane, 1/4" thick, minimum	1.000	S.F.	.006	.40	.30	.70
Maximum	1.000	S.F.	.006	.61	.30	.91
Stairs, for stairs, add to above carpet prices	1.000	Riser	.267		13.45	13.45
Underlayment plywood, 3/8" thick	1.000	S.F.	.011	1.09	.61	1.70
1/2" thick	1.000	S.F.	.011	1.29	.63	1.92
5/8" thick	1.000	S.F.	.011	1.42	.65	2.07
3/4" thick	1.000	S.F.	.012	1.63	.70	2.33
Particle board, 3/8" thick	1.000	S.F.	.011	.45	.61	1.06
1/2" thick	1.000	S.F.	.011	.50	.63	1.13
5/8" thick	1.000	S.F.	.011	.59	.65	1.24
3/4" thick	1.000	S.F.	.012	.75	.70	1.45
Hardboard, 4' x 4', 0.215" thick	1.000	S.F.	.011	.69	.61	1.30

6 | INTERIORS — 64 | Flooring Systems

System Description	QUAN.	UNIT	LABOR HOURS	COST PER S.F. MAT.	COST PER S.F. INST.	COST PER S.F. TOTAL
Resilient flooring, asphalt tile on concrete, 1/8" thick						
Color group B	1.000	S.F.	.020	1.31	1.01	2.32
Color group C & D	1.000	S.F.	.020	1.43	1.01	2.44
Asphalt tile on wood subfloor, 1/8" thick						
Color group B	1.000	S.F.	.020	1.60	1.01	2.61
Color group C & D	1.000	S.F.	.020	1.72	1.01	2.73
Vinyl composition tile, 12" x 12", 1/16" thick	1.000	S.F.	.016	1.31	.81	2.12
Embossed	1.000	S.F.	.016	2.43	.81	3.24
Marbleized	1.000	S.F.	.016	2.43	.81	3.24
Plain	1.000	S.F.	.016	3.14	.81	3.95
.080" thick, embossed	1.000	S.F.	.016	1.57	.81	2.38
Marbleized	1.000	S.F.	.016	2.79	.81	3.60
Plain	1.000	S.F.	.016	2.60	.81	3.41
1/8" thick, marbleized	1.000	S.F.	.016	2.68	.81	3.49
Plain	1.000	S.F.	.016	1.71	.81	2.52
Vinyl tile, 12" x 12", .050" thick, minimum	1.000	S.F.	.016	4.08	.81	4.89
Maximum	1.000	S.F.	.016	5.70	.81	6.51
1/8" thick, minimum	1.000	S.F.	.016	7.70	.81	8.51
Maximum	1.000	S.F.	.016	3.58	.81	4.39
1/8" thick, solid colors	1.000	S.F.	.016	6.50	.81	7.31
Florentine pattern	1.000	S.F.	.016	6.95	.81	7.76
Marbleized or travertine pattern	1.000	S.F.	.016	6.85	.81	7.66
Vinyl sheet goods, backed, .070" thick, minimum	1.000	S.F.	.032	4.06	1.61	5.67
Maximum	1.000	S.F.	.040	4.77	2.02	6.79
.093" thick, minimum	1.000	S.F.	.035	4.53	1.75	6.28
Maximum	1.000	S.F.	.040	7.10	2.02	9.12
.125" thick, minimum	1.000	S.F.	.035	4.26	1.75	6.01
Maximum	1.000	S.F.	.040	8.25	2.02	10.27
Wood, oak, finished in place, 25/32" x 2-1/2" clear	1.000	S.F.	.074	4.65	3.79	8.44
Select	1.000	S.F.	.074	5.60	3.79	9.39
No. 1 common	1.000	S.F.	.074	5.50	3.79	9.29
Prefinished, oak, 2-1/2" wide	1.000	S.F.	.047	5.30	2.67	7.97
3-1/4" wide	1.000	S.F.	.043	5.75	2.45	8.20
Ranch plank, oak, random width	1.000	S.F.	.055	7.85	3.13	10.98
Parquet, 5/16" thick, finished in place, oak, minimum	1.000	S.F.	.077	6.95	3.96	10.91
Maximum	1.000	S.F.	.107	11.65	5.65	17.30
Teak, minimum	1.000	S.F.	.077	7.95	3.96	11.91
Maximum	1.000	S.F.	.107	12.75	5.65	18.40
Sleepers, treated, 16" O.C., 1" x 2"	1.000	S.F.	.007	.30	.39	.69
1" x 3"	1.000	S.F.	.008	.49	.45	.94
2" x 4"	1.000	S.F.	.011	.62	.61	1.23
2" x 6"	1.000	S.F.	.012	.94	.70	1.64
Subfloor, plywood, 1/2" thick	1.000	S.F.	.011	.75	.61	1.36
5/8" thick	1.000	S.F.	.012	.84	.67	1.51
3/4" thick	1.000	S.F.	.013	1	.73	1.73
Ceramic tile, color group 2, 1" x 1"	1.000	S.F.	.087	5.70	3.93	9.63
2" x 2" or 2" x 1"	1.000	S.F.	.084	7.05	3.78	10.83
Color group 1, 8" x 8"	1.000	S.F.	.064	5.10	2.39	7.49
12" x 12"	1.000	S.F.	.049	6.70	2.48	9.18
16" x 16"	1.000	S.F.	.029	8.15	2.57	10.72

6 | INTERIORS   90 | Stairways

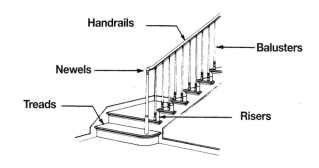

System Description	QUAN.	UNIT	LABOR HOURS	COST EACH MAT.	COST EACH INST.	COST EACH TOTAL
7 RISERS, OAK TREADS, BOX STAIRS						
Treads, oak, 1-1/4" x 10" wide, 3' long	6.000	Ea.	2.667	636	150	786
Risers, 3/4" thick, beech	7.000	Ea.	.672	196.35	38.22	234.57
30" primed pine balusters	12.000	Ea.	1.000	39.60	56.76	96.36
Newels, 3" wide, plain, paint grade, square	2.000	Ea.	2.286	111	130	241
Handrails, oak laminated	7.000	L.F.	.933	280	52.85	332.85
Stringers, 2" x 10", 3 each	21.000	L.F.	.306	9.03	17.43	26.46
TOTAL		Ea.	7.864	1,271.98	445.26	1,717.24
14 RISERS, OAK TREADS, BOX STAIRS						
Treads, oak, 1-1/4" x 10" wide, 3' long	13.000	Ea.	5.778	1,378	325	1,703
Risers, 3/4" thick, beech	14.000	Ea.	1.344	392.70	76.44	469.14
30" primed pine balusters	26.000	Ea.	2.167	85.80	122.98	208.78
Newels, 3" wide, plain, paint grade, square	2.000	Ea.	2.286	111	130	241
Handrails, oak, laminated	14.000	L.F.	1.867	560	105.70	665.70
Stair stringers, 2" x 10"	42.000	L.F.	5.169	61.74	294	355.74
TOTAL		Ea.	18.611	2,589.24	1,054.12	3,643.36
14 RISERS, PINE TREADS, BOX STAIRS						
Treads, pine, 9-1/2" x 3/4" thick	13.000	Ea.	5.778	220.35	325	545.35
Risers, 3/4" thick, pine	14.000	Ea.	1.344	150.36	76.44	226.80
30" primed pine balusters	26.000	Ea.	2.167	85.80	122.98	208.78
Newels, 3" wide, plain, paint grade, square	2.000	Ea.	2.286	111	130	241
Handrails, oak, laminated	14.000	L.F.	1.867	560	105.70	665.70
Stair stringers, 2" x 10"	42.000	L.F.	5.169	61.74	294	355.74
TOTAL		Ea.	18.611	1,189.25	1,054.12	2,243.37

Description	QUAN.	UNIT	LABOR HOURS	COST EACH MAT.	COST EACH INST.	COST EACH TOTAL

6 | INTERIORS — 90 | Stairways

Stairway Price Sheet	QUAN.	UNIT	LABOR HOURS	COST EACH		
				MAT.	INST.	TOTAL
Treads, oak, 1-1/16" x 9-1/2", 3' long, 7 riser stair	6.000	Ea.	2.667	635	150	785
14 riser stair	13.000	Ea.	5.778	1,375	325	1,700
1-1/16" x 11-1/2", 3' long, 7 riser stair	6.000	Ea.	2.667	695	150	845
14 riser stair	13.000	Ea.	5.778	1,500	325	1,825
Pine, 3/4" x 9-1/2", 3' long, 7 riser stair	6.000	Ea.	2.667	102	150	252
14 riser stair	13.000	Ea.	5.778	220	325	545
3/4" x 11-1/4", 3' long, 7 riser stair	6.000	Ea.	2.667	113	150	263
14 riser stair	13.000	Ea.	5.778	245	325	570
Risers, oak, 3/4" x 7-1/2" high, 7 riser stair	7.000	Ea.	2.625	150	38	188
14 riser stair	14.000	Ea.	5.250	300	76.50	376.50
Beech, 3/4" x 7-1/2" high, 7 riser stair	7.000	Ea.	2.625	196	38	234
14 riser stair	14.000	Ea.	5.250	395	76.50	471.50
Baluster, turned, 30" high, primed pine, 7 riser stair	12.000	Ea.	3.429	39.50	57	96.50
14 riser stair	26.000	Ea.	7.428	86	123	209
30" birch, 7 riser stair	12.000	Ea.	3.429	36	51.50	87.50
14 riser stair	26.000	Ea.	7.428	78	112	190
42" pine, 7 riser stair	12.000	Ea.	3.556	58.50	57	115.50
14 riser stair	26.000	Ea.	7.704	126	123	249
42" birch, 7 riser stair	12.000	Ea.	3.556	58.50	57	115.50
14 riser stair	26.000	Ea.	7.704	126	123	249
Newels, 3-1/4" wide, starting, 7 riser stair	2.000	Ea.	2.286	111	130	241
14 riser stair	2.000	Ea.	2.286	111	130	241
Landing, 7 riser stair	2.000	Ea.	3.200	139	182	321
14 riser stair	2.000	Ea.	3.200	139	182	321
Handrails, oak, laminated, 7 riser stair	7.000	L.F.	.933	280	53	333
14 riser stair	14.000	L.F.	1.867	560	106	666
Stringers, fir, 2" x 10" 7 riser stair	21.000	L.F.	2.585	31	147	178
14 riser stair	42.000	L.F.	5.169	61.50	294	355.50
2" x 12", 7 riser stair	21.000	L.F.	2.585	39.50	147	186.50
14 riser stair	42.000	L.F.	5.169	78.50	294	372.50

Special Stairways	QUAN.	UNIT	LABOR HOURS	COST EACH		
				MAT.	INST.	TOTAL
Basement stairs, open risers	1.000	Flight	4.000	850	227	1,077
Spiral stairs, oak, 4'-6" diameter, prefabricated, 9' high	1.000	Flight	10.667	3,775	605	4,380
Aluminum, 5'-0" diameter stock unit	1.000	Flight	9.956	8,200	710	8,910
Custom unit	1.000	Flight	9.956	14,400	710	15,110
Cast iron, 4'-0" diameter, minimum	1.000	Flight	9.956	10,200	710	10,910
Maximum	1.000	Flight	17.920	17,900	1,300	19,200
Steel, industrial, pre-erected, 3'-6" wide, bar rail	1.000	Flight	7.724	8,475	775	9,250
Picket rail	1.000	Flight	7.724	9,525	775	10,300

No part of this cost data may be reproduced, stored in a retrieval system, or transmitted in any form or by any means without prior written permission of RSMeans.

Did you know?
RSMeans Online gives you the same access to RSMeans' data with 24/7 access:
- Quickly locate costs in the searchable database.
- Build cost lists, estimates, and reports in minutes.
- Adjust costs to any location in the U.S. and Canada with the click of a button.

Start your free trial today at **www.RSMeansOnline.com**

RSMeans Online
FROM THE GORDIAN GROUP®

7 | SPECIALTIES 08 | Kitchen Systems

System Description	QUAN.	UNIT	LABOR HOURS	COST PER L.F. MAT.	COST PER L.F. INST.	COST PER L.F. TOTAL
KITCHEN, ECONOMY GRADE						
Top cabinets, economy grade	1.000	L.F.	.171	64	9.76	73.76
Bottom cabinets, economy grade	1.000	L.F.	.256	96	14.64	110.64
Square edge, plastic face countertop	1.000	L.F.	.267	37	15.15	52.15
Blocking, wood, 2" x 4"	1.000	L.F.	.032	.44	1.82	2.26
Soffit, framing, wood, 2" x 4"	4.000	L.F.	.071	1.76	4.04	5.80
Soffit drywall	2.000	S.F.	.047	.86	2.70	3.56
Drywall painting	2.000	S.F.	.013	.12	.76	.88
TOTAL		L.F.	.857	200.18	48.87	249.05
AVERAGE GRADE						
Top cabinets, average grade	1.000	L.F.	.213	80	12.20	92.20
Bottom cabinets, average grade	1.000	L.F.	.320	120	18.30	138.30
Solid surface countertop, solid color	1.000	L.F.	.800	81	45.50	126.50
Blocking, wood, 2" x 4"	1.000	L.F.	.032	.44	1.82	2.26
Soffit framing, wood, 2" x 4"	4.000	L.F.	.071	1.76	4.04	5.80
Soffit drywall	2.000	S.F.	.047	.86	2.70	3.56
Drywall painting	2.000	S.F.	.013	.12	.76	.88
TOTAL		L.F.	1.496	284.18	85.32	369.50
CUSTOM GRADE						
Top cabinets, custom grade	1.000	L.F.	.256	178	14.60	192.60
Bottom cabinets, custom grade	1.000	L.F.	.384	267	21.90	288.90
Solid surface countertop, premium patterned color	1.000	L.F.	1.067	140	60.50	200.50
Blocking, wood, 2" x 4"	1.000	L.F.	.032	.44	1.82	2.26
Soffit framing, wood, 2" x 4"	4.000	L.F.	.071	1.76	4.04	5.80
Soffit drywall	2.000	S.F.	.047	.86	2.70	3.56
Drywall painting	2.000	S.F.	.013	.12	.76	.88
TOTAL		L.F.	1.870	588.18	106.32	694.50

Description	QUAN.	UNIT	LABOR HOURS	COST PER L.F. MAT.	COST PER L.F. INST.	COST PER L.F. TOTAL

Kitchen Price Sheet	QUAN.	UNIT	LABOR HOURS	COST PER L.F. MAT.	COST PER L.F. INST.	COST PER L.F. TOTAL
Top cabinets, economy grade	1.000	L.F.	.171	64	9.75	73.75
Average grade	1.000	L.F.	.213	80	12.20	92.20
Custom grade	1.000	L.F.	.256	178	14.60	192.60
Bottom cabinets, economy grade	1.000	L.F.	.256	96	14.65	110.65
Average grade	1.000	L.F.	.320	120	18.30	138.30
Custom grade	1.000	L.F.	.384	267	22	289
Counter top, laminated plastic, 7/8" thick, no splash	1.000	L.F.	.267	33.50	15.15	48.65
With backsplash	1.000	L.F.	.267	32.50	15.15	47.65
1-1/4" thick, no splash	1.000	L.F.	.286	39.50	16.20	55.70
With backsplash	1.000	L.F.	.286	47.50	16.20	63.70
Post formed, laminated plastic	1.000	L.F.	.267	11.40	15.15	26.55
Ceramic tile, with backsplash		L.F.	.427	18.80	9.10	27.90
Marble, with backsplash, minimum	1.000	L.F.	.471	46.50	25.50	72
Maximum	1.000	L.F.	.615	117	33	150
Maple, solid laminated, no backsplash	1.000	L.F.	.286	84.50	16.20	100.70
With backsplash	1.000	L.F.	.286	100	16.20	116.20
Solid Surface, with backsplash, minimum	1.000	L.F.	.842	89	48	137
Maximum	1.000	L.F.	1.067	140	60.50	200.50
Blocking, wood, 2" x 4"	1.000	L.F.	.032	.44	1.82	2.26
2" x 6"	1.000	L.F.	.036	.67	2.05	2.72
2" x 8"	1.000	L.F.	.040	.93	2.27	3.20
Soffit framing, wood, 2" x 3"	4.000	L.F.	.064	1.64	3.64	5.28
2" x 4"	4.000	L.F.	.071	1.76	4.04	5.80
Soffit, drywall, painted	2.000	S.F.	.060	.98	3.46	4.44
Paneling, standard	2.000	S.F.	.064	3.20	3.64	6.84
Deluxe	2.000	S.F.	.091	2.76	5.20	7.96
Sinks, porcelain on cast iron, single bowl, 21" x 24"	1.000	Ea.	10.334	595	600	1,195
21" x 30"	1.000	Ea.	10.334	870	600	1,470
Double bowl, 20" x 32"	1.000	Ea.	10.810	675	630	1,305
Stainless steel, single bowl, 16" x 20"	1.000	Ea.	10.334	935	600	1,535
22" x 25"	1.000	Ea.	10.334	1,000	600	1,600
Double bowl, 20" x 32"	1.000	Ea.	10.810	840	630	1,470

7 | SPECIALTIES — 12 | Appliance Systems

Appliance Price Sheet

	QUAN.	UNIT	LABOR HOURS	MAT.	INST.	TOTAL
Range, free standing, minimum	1.000	Ea.	3.600	595	187	782
Maximum	1.000	Ea.	6.000	2,150	286	2,436
Built-in, minimum	1.000	Ea.	3.333	1,075	202	1,277
Maximum	1.000	Ea.	10.000	1,800	575	2,375
Counter top range, 4-burner, minimum	1.000	Ea.	3.333	465	202	667
Maximum	1.000	Ea.	4.667	1,950	282	2,232
Compactor, built-in, minimum	1.000	Ea.	2.215	780	128	908
Maximum	1.000	Ea.	3.282	1,300	188	1,488
Dishwasher, built-in, minimum	1.000	Ea.	6.735	550	430	980
Maximum	1.000	Ea.	9.235	715	590	1,305
Garbage disposer, minimum	1.000	Ea.	2.810	201	180	381
Maximum	1.000	Ea.	2.810	365	180	545
Microwave oven, minimum	1.000	Ea.	2.615	153	158	311
Maximum	1.000	Ea.	4.615	580	278	858
Range hood, ducted, minimum	1.000	Ea.	4.658	126	272	398
Maximum	1.000	Ea.	5.991	975	350	1,325
Ductless, minimum	1.000	Ea.	2.615	135	152	287
Maximum	1.000	Ea.	3.948	985	229	1,214
Refrigerator, 16 cu.ft., minimum	1.000	Ea.	2.000	650	82.50	732.50
Maximum	1.000	Ea.	3.200	1,050	132	1,182
16 cu.ft. with icemaker, minimum	1.000	Ea.	4.210	880	216	1,096
Maximum	1.000	Ea.	5.410	1,275	265	1,540
19 cu.ft., minimum	1.000	Ea.	2.667	615	110	725
Maximum	1.000	Ea.	4.667	1,075	193	1,268
19 cu.ft. with icemaker, minimum	1.000	Ea.	5.143	870	254	1,124
Maximum	1.000	Ea.	7.143	1,325	335	1,660
Sinks, porcelain on cast iron single bowl, 21" x 24"	1.000	Ea.	10.334	595	600	1,195
21" x 30"	1.000	Ea.	10.334	870	600	1,470
Double bowl, 20" x 32"	1.000	Ea.	10.810	675	630	1,305
Stainless steel, single bowl 16" x 20"	1.000	Ea.	10.334	935	600	1,535
22" x 25"	1.000	Ea.	10.334	1,000	600	1,600
Double bowl, 20" x 32"	1.000	Ea.	10.810	840	630	1,470
Water heater, electric, 30 gallon	1.000	Ea.	3.636	995	234	1,229
40 gallon	1.000	Ea.	4.000	1,075	258	1,333
Gas, 30 gallon	1.000	Ea.	4.000	1,775	258	2,033
75 gallon	1.000	Ea.	5.333	2,675	345	3,020
Wall, packaged terminal heater/air conditioner cabinet, wall sleeve, louver, electric heat, thermostat, manual changeover, 208V		Ea.				
6000 BTUH cooling, 8800 BTU heating	1.000	Ea.	2.667	910	159	1,069
9000 BTUH cooling, 13,900 BTU heating	1.000	Ea.	3.200	1,350	190	1,540
12,000 BTUH cooling, 13,900 BTU heating	1.000	Ea.	4.000	1,500	238	1,738
15,000 BTUH cooling, 13,900 BTU heating	1.000	Ea.	5.333	1,625	315	1,940

7 | SPECIALTIES   16 | Bath Accessories Systems

System Description	QUAN.	UNIT	LABOR HOURS	COST EACH MAT.	COST EACH INST.	COST EACH TOTAL
Curtain rods, stainless, 1" diameter, 3' long	1.000	Ea.	.615	31	35	66
5' long	1.000	Ea.	.615	31	35	66
Grab bar, 1" diameter, 12" long	1.000	Ea.	.283	28.50	16.05	44.55
36" long	1.000	Ea.	.340	34.50	19.15	53.65
1-1/4" diameter, 12" long	1.000	Ea.	.333	33.50	18.90	52.40
36" long	1.000	Ea.	.400	40.50	22.50	63
1-1/2" diameter, 12" long	1.000	Ea.	.383	38.50	21.50	60
36" long	1.000	Ea.	.460	46.50	26	72.50
Mirror, 18" x 24"	1.000	Ea.	.400	54.50	22.50	77
72" x 24"	1.000	Ea.	1.333	335	75.50	410.50
Medicine chest with mirror, 18" x 24"	1.000	Ea.	.400	218	22.50	240.50
36" x 24"	1.000	Ea.	.600	325	34	359
Toilet tissue dispenser, surface mounted, minimum	1.000	Ea.	.267	21	15.15	36.15
Maximum	1.000	Ea.	.400	31.50	22.50	54
Flush mounted, minimum	1.000	Ea.	.293	23	16.65	39.65
Maximum	1.000	Ea.	.427	33.50	24	57.50
Towel bar, 18" long, minimum	1.000	Ea.	.278	36.50	15.80	52.30
Maximum	1.000	Ea.	.348	45.50	19.75	65.25
24" long, minimum	1.000	Ea.	.313	41	17.80	58.80
Maximum	1.000	Ea.	.383	50	21.50	71.50
36" long, minimum	1.000	Ea.	.381	73	21.50	94.50
Maximum	1.000	Ea.	.419	80.50	23.50	104

7 | SPECIALTIES

24 | Masonry Fireplace Systems

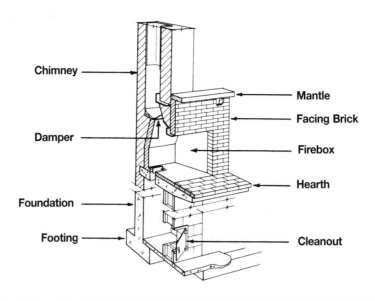

System Description	QUAN.	UNIT	LABOR HOURS	COST EACH MAT.	COST EACH INST.	COST EACH TOTAL
MASONRY FIREPLACE						
Footing, 8" thick, concrete, 4' x 7'	.700	C.Y.	2.800	131.60	148.57	280.17
Foundation, concrete block, 32" x 60" x 4' deep	1.000	Ea.	5.275	199.20	264.60	463.80
Fireplace, brick firebox, 30" x 29" opening	1.000	Ea.	40.000	630	1,975	2,605
Damper, cast iron, 30" opening	1.000	Ea.	1.333	137	72	209
Facing brick, standard size brick, 6' x 5'	30.000	S.F.	5.217	143.10	261	404.10
Hearth, standard size brick, 3' x 6'	1.000	Ea.	8.000	233	395	628
Chimney, standard size brick, 8" x 12" flue, one story house	12.000	V.L.F.	12.000	516	588	1,104
Mantle, 4" x 8", wood	6.000	L.F.	1.333	54.90	75.60	130.50
Cleanout, cast iron, 8" x 8"	1.000	Ea.	.667	46.50	36	82.50
TOTAL		Ea.	76.625	2,091.30	3,815.77	5,907.07

The costs in this system are on a cost each basis.

Description	QUAN.	UNIT	LABOR HOURS	COST EACH MAT.	COST EACH INST.	COST EACH TOTAL

7 | SPECIALTIES   24 | Masonry Fireplace Systems

Masonry Fireplace Price Sheet

	QUAN.	UNIT	LABOR HOURS	COST EACH MAT.	COST EACH INST.	TOTAL
Footing 8" thick, 3' x 6'	.440	C.Y.	1.326	82.50	93.50	176
4' x 7'	.700	C.Y.	2.110	132	149	281
5' x 8'	1.000	C.Y.	3.014	188	212	400
1' thick, 3' x 6'	.670	C.Y.	2.020	126	142	268
4' x 7'	1.030	C.Y.	3.105	194	218	412
5' x 8'	1.480	C.Y.	4.461	278	310	588
Foundation-concrete block, 24" x 48", 4' deep	1.000	Ea.	4.267	159	212	371
8' deep	1.000	Ea.	8.533	320	425	745
24" x 60", 4' deep	1.000	Ea.	4.978	186	247	433
8' deep	1.000	Ea.	9.956	370	495	865
32" x 48", 4' deep	1.000	Ea.	4.711	176	234	410
8' deep	1.000	Ea.	9.422	350	465	815
32" x 60", 4' deep	1.000	Ea.	5.333	199	265	464
8' deep	1.000	Ea.	10.845	405	540	945
32" x 72", 4' deep	1.000	Ea.	6.133	229	305	534
8' deep	1.000	Ea.	12.267	460	610	1,070
Fireplace, brick firebox 30" x 29" opening	1.000	Ea.	40.000	630	1,975	2,605
48" x 30" opening	1.000	Ea.	60.000	945	2,975	3,920
Steel fire box with registers, 25" opening	1.000	Ea.	26.667	1,150	1,325	2,475
48" opening	1.000	Ea.	44.000	1,925	2,200	4,125
Damper, cast iron, 30" opening	1.000	Ea.	1.333	137	72	209
36" opening	1.000	Ea.	1.556	160	84	244
Steel, 30" opening	1.000	Ea.	1.333	131	72	203
36" opening	1.000	Ea.	1.556	153	84	237
Facing for fireplace, standard size brick, 6' x 5'	30.000	S.F.	5.217	143	261	404
7' x 5'	35.000	S.F.	6.087	167	305	472
8' x 6'	48.000	S.F.	8.348	229	420	649
Fieldstone, 6' x 5'	30.000	S.F.	5.217	425	261	686
7' x 5'	35.000	S.F.	6.087	495	305	800
8' x 6'	48.000	S.F.	8.348	680	420	1,100
Sheetrock on metal, studs, 6' x 5'	30.000	S.F.	.980	26	55.50	81.50
7' x 5'	35.000	S.F.	1.143	30	65	95
8' x 6'	48.000	S.F.	1.568	41.50	89	130.50
Hearth, standard size brick, 3' x 6'	1.000	Ea.	8.000	233	395	628
3' x 7'	1.000	Ea.	9.280	270	460	730
3' x 8'	1.000	Ea.	10.640	310	525	835
Stone, 3' x 6'	1.000	Ea.	8.000	242	395	637
3' x 7'	1.000	Ea.	9.280	280	460	740
3' x 8'	1.000	Ea.	10.640	320	525	845
Chimney, standard size brick, 8" x 12" flue, one story house	12.000	V.L.F.	12.000	515	590	1,105
Two story house	20.000	V.L.F.	20.000	860	980	1,840
Mantle wood, beams, 4" x 8"	6.000	L.F.	1.333	55	75.50	130.50
4" x 10"	6.000	L.F.	1.371	71.50	77.50	149
Ornate, prefabricated, 6' x 3'-6" opening, minimum	1.000	Ea.	1.600	480	91	571
Maximum	1.000	Ea.	1.600	650	91	741
Cleanout, door and frame, cast iron, 8" x 8"	1.000	Ea.	.667	46.50	36	82.50
12" x 12"	1.000	Ea.	.800	107	43	150

7 | SPECIALTIES 30 | Prefabricated Fireplace Systems

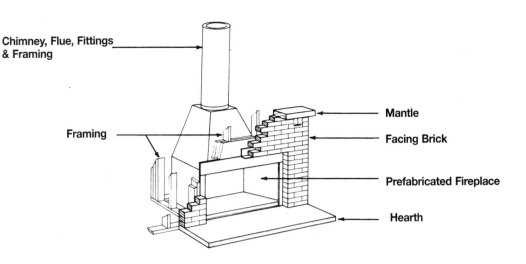

System Description	QUAN.	UNIT	LABOR HOURS	COST EACH MAT.	COST EACH INST.	COST EACH TOTAL
PREFABRICATED FIREPLACE						
Prefabricated fireplace, metal, painted	1.000	Ea.	6.154	1,625	350	1,975
Framing, 2" x 4" studs, 6' x 5'	35.000	L.F.	.509	15.05	29.05	44.10
Fire resistant gypsum drywall, unfinished	40.000	S.F.	.320	16	18	34
Drywall finishing adder	40.000	S.F.	.320	2	18	20
Facing, brick, standard size brick, 6' x 5'	30.000	S.F.	5.217	143.10	261	404.10
Hearth, standard size brick, 3' x 6'	1.000	Ea.	8.000	233	395	628
Chimney, one story house, framing, 2" x 4" studs	80.000	L.F.	1.164	34.40	66.40	100.80
Sheathing, plywood, 5/8" thick	32.000	S.F.	.758	95.04	43.20	138.24
Flue, 10" metal, insulated pipe	12.000	V.L.F.	4.000	570	226.80	796.80
Fittings, ceiling support	1.000	Ea.	.667	156	38	194
Fittings, joist shield	1.000	Ea.	.727	770	41.50	811.50
Fittings, roof flashing	1.000	Ea.	.667	440	38	478
Mantle beam, wood, 4" x 8"	6.000	L.F.	1.333	54.90	75.60	130.50
TOTAL		Ea.	29.836	4,154.49	1,600.55	5,755.04

The costs in this system are on a cost each basis.

Description	QUAN.	UNIT	LABOR HOURS	COST EACH MAT.	COST EACH INST.	COST EACH TOTAL

7 | SPECIALTIES | 30 | Prefabricated Fireplace Systems

Prefabricated Fireplace Price Sheet

	QUAN.	UNIT	LABOR HOURS	MAT.	INST.	TOTAL
Prefabricated fireplace, minimum	1.000	Ea.	6.154	1,625	350	1,975
Average	1.000	Ea.	8.000	1,875	455	2,330
Maximum	1.000	Ea.	8.889	3,350	505	3,855
Framing, 2" x 4" studs, fireplace, 6' x 5'	35.000	L.F.	.509	15.05	29	44.05
7' x 5'	40.000	L.F.	.582	17.20	33	50.20
8' x 6'	45.000	L.F.	.655	19.35	37.50	56.85
Sheetrock, 1/2" thick, fireplace, 6' x 5'	40.000	S.F.	.640	18	36	54
7' x 5'	45.000	S.F.	.720	20.50	40.50	61
8' x 6'	50.000	S.F.	.800	22.50	45	67.50
Facing for fireplace, brick, 6' x 5'	30.000	S.F.	5.217	143	261	404
7' x 5'	35.000	S.F.	6.087	167	305	472
8' x 6'	48.000	S.F.	8.348	229	420	649
Fieldstone, 6' x 5'	30.000	S.F.	5.217	405	261	666
7' x 5'	35.000	S.F.	6.087	475	305	780
8' x 6'	48.000	S.F.	8.348	650	420	1,070
Hearth, standard size brick, 3' x 6'	1.000	Ea.	8.000	233	395	628
3' x 7'	1.000	Ea.	9.280	270	460	730
3' x 8'	1.000	Ea.	10.640	310	525	835
Stone, 3' x 6'	1.000	Ea.	8.000	242	395	637
3' x 7'	1.000	Ea.	9.280	280	460	740
3' x 8'	1.000	Ea.	10.640	320	525	845
Chimney, framing, 2" x 4", one story house	80.000	L.F.	1.164	34.50	66.50	101
Two story house	120.000	L.F.	1.746	51.50	99.50	151
Sheathing, plywood, 5/8" thick	32.000	S.F.	.758	95	43	138
Stucco on plywood	32.000	S.F.	1.125	57	62.50	119.50
Flue, 10" metal pipe, insulated, one story house	12.000	V.L.F.	4.000	570	227	797
Two story house	20.000	V.L.F.	6.667	950	380	1,330
Fittings, ceiling support	1.000	Ea.	.667	156	38	194
Fittings joist sheild, one story house	1.000	Ea.	.667	770	41.50	811.50
Two story house	2.000	Ea.	1.333	1,550	83	1,633
Fittings roof flashing	1.000	Ea.	.667	440	38	478
Mantle, wood beam, 4" x 8"	6.000	L.F.	1.333	55	75.50	130.50
4" x 10"	6.000	L.F.	1.371	71.50	77.50	149
Ornate prefabricated, 6' x 3'-6" opening, minimum	1.000	Ea.	1.600	480	91	571
Maximum	1.000	Ea.	1.600	650	91	741

7 | SPECIALTIES — 32 | Greenhouse Systems

System Description	QUAN.	UNIT	LABOR HOURS	COST EACH MAT.	COST EACH INST.	COST EACH TOTAL
Economy, lean to, shell only, not including 2' stub wall, fndtn, flrs, heat						
4' x 16'	1.000	Ea.	26.212	2,650	1,475	4,125
4' x 24'	1.000	Ea.	30.259	3,050	1,700	4,750
6' x 10'	1.000	Ea.	16.552	1,925	940	2,865
6' x 16'	1.000	Ea.	23.034	2,675	1,300	3,975
6' x 24'	1.000	Ea.	29.793	3,450	1,700	5,150
8' x 10'	1.000	Ea.	22.069	2,550	1,250	3,800
8' x 16'	1.000	Ea.	38.400	4,450	2,175	6,625
8' x 24'	1.000	Ea.	49.655	5,750	2,825	8,575
Free standing, 8' x 8'	1.000	Ea.	17.356	1,750	985	2,735
8' x 16'	1.000	Ea.	30.211	3,075	1,725	4,800
8' x 24'	1.000	Ea.	39.051	3,950	2,225	6,175
10' x 10'	1.000	Ea.	18.824	4,500	1,075	5,575
10' x 16'	1.000	Ea.	24.095	5,750	1,375	7,125
10' x 24'	1.000	Ea.	31.624	7,550	1,800	9,350
14' x 10'	1.000	Ea.	20.741	6,650	1,175	7,825
14' x 16'	1.000	Ea.	24.889	7,975	1,400	9,375
14' x 24'	1.000	Ea.	33.349	10,700	1,900	12,600
Standard, lean to, shell only, not incl. 2' stub wall, fndtn, flrs, heat 4'x10'	1.000	Ea.	28.235	2,850	1,600	4,450
4' x 16'	1.000	Ea.	39.341	3,975	2,225	6,200
4' x 24'	1.000	Ea.	45.412	4,575	2,550	7,125
6' x 10'	1.000	Ea.	24.827	2,875	1,400	4,275
6' x 16'	1.000	Ea.	34.538	4,000	1,950	5,950
6' x 24'	1.000	Ea.	44.689	5,175	2,525	7,700
8' x 10'	1.000	Ea.	33.103	3,850	1,875	5,725
8' x 16'	1.000	Ea.	57.600	6,675	3,275	9,950
8' x 24'	1.000	Ea.	74.482	8,650	4,225	12,875
Free standing, 8' x 8'	1.000	Ea.	26.034	2,650	1,475	4,125
8' x 16'	1.000	Ea.	45.316	4,600	2,575	7,175
8' x 24'	1.000	Ea.	58.577	5,950	3,325	9,275
10' x 10'	1.000	Ea.	28.236	6,750	1,600	8,350
10' x 16'	1.000	Ea.	36.142	8,650	2,050	10,700
10' x 24'	1.000	Ea.	47.436	11,300	2,700	14,000
14' x 10'	1.000	Ea.	31.112	9,975	1,775	11,750
14' x 16'	1.000	Ea.	37.334	12,000	2,125	14,125
14' x 24'	1.000	Ea.	50.030	16,000	2,825	18,825
Deluxe, lean to, shell only, not incl. 2' stub wall, fndtn, flrs or heat, 4'x10'	1.000	Ea.	20.645	4,650	1,175	5,825
4' x 16'	1.000	Ea.	33.032	7,425	1,900	9,325
4' x 24'	1.000	Ea.	49.548	11,100	2,825	13,925
6' x 10'	1.000	Ea.	30.968	6,950	1,775	8,725
6' x 16'	1.000	Ea.	49.548	11,100	2,825	13,925
6' x 24'	1.000	Ea.	74.323	16,700	4,250	20,950
8' x 10'	1.000	Ea.	41.290	9,275	2,350	11,625
8' x 16'	1.000	Ea.	66.065	14,800	3,775	18,575
8' x 24'	1.000	Ea.	99.097	22,300	5,675	27,975
Freestanding, 8' x 8'	1.000	Ea.	18.618	5,975	1,050	7,025
8' x 16'	1.000	Ea.	37.236	12,000	2,100	14,100
8' x 24'	1.000	Ea.	55.855	18,000	3,175	21,175
10' x 10'	1.000	Ea.	29.091	9,350	1,650	11,000
10' x 16'	1.000	Ea.	46.546	15,000	2,650	17,650
10' x 24'	1.000	Ea.	69.818	22,400	3,950	26,350
14' x 10'	1.000	Ea.	40.727	13,100	2,300	15,400
14' x 16'	1.000	Ea.	65.164	20,900	3,700	24,600
14' x 24'	1.000	Ea.	97.746	31,400	5,550	36,950

7 | SPECIALTIES — 36 | Swimming Pool Systems

System Description	QUAN.	UNIT	LABOR HOURS	COST EACH MAT.	COST EACH INST.	COST EACH TOTAL
Swimming pools, vinyl lined, metal sides, sand bottom, 12' x 28'	1.000	Ea.	50.177	7,650	2,950	10,600
12' x 32'	1.000	Ea.	55.366	8,450	3,225	11,675
12' x 36'	1.000	Ea.	60.061	9,175	3,500	12,675
16' x 32'	1.000	Ea.	66.798	10,200	3,900	14,100
16' x 36'	1.000	Ea.	71.190	10,900	4,175	15,075
16' x 40'	1.000	Ea.	74.703	11,400	4,350	15,750
20' x 36'	1.000	Ea.	77.860	11,900	4,550	16,450
20' x 40'	1.000	Ea.	82.135	12,500	4,800	17,300
20' x 44'	1.000	Ea.	90.348	13,800	5,275	19,075
24' x 40'	1.000	Ea.	98.562	15,000	5,750	20,750
24' x 44'	1.000	Ea.	108.418	16,600	6,325	22,925
24' x 48'	1.000	Ea.	118.274	18,100	6,900	25,000
Vinyl lined, concrete sides, 12' x 28'	1.000	Ea.	79.447	12,100	4,625	16,725
12' x 32'	1.000	Ea.	88.818	13,600	5,200	18,800
12' x 36'	1.000	Ea.	97.656	14,900	5,700	20,600
16' x 32'	1.000	Ea.	111.393	17,000	6,525	23,525
16' x 36'	1.000	Ea.	121.354	18,500	7,075	25,575
16' x 40'	1.000	Ea.	130.445	19,900	7,600	27,500
28' x 36'	1.000	Ea.	140.585	21,500	8,200	29,700
20' x 40'	1.000	Ea.	149.336	22,800	8,700	31,500
20' x 44'	1.000	Ea.	164.270	25,100	9,575	34,675
24' x 40'	1.000	Ea.	179.203	27,400	10,500	37,900
24' x 44'	1.000	Ea.	197.124	30,100	11,500	41,600
24' x 48'	1.000	Ea.	215.044	32,800	12,500	45,300
Gunite, bottom and sides, 12' x 28'	1.000	Ea.	129.767	17,600	7,550	25,150
12' x 32'	1.000	Ea.	142.164	19,300	8,275	27,575
12' x 36'	1.000	Ea.	153.028	20,800	8,900	29,700
16' x 32'	1.000	Ea.	167.743	22,800	9,775	32,575
16' x 36'	1.000	Ea.	176.421	24,000	10,300	34,300
16' x 40'	1.000	Ea.	182.368	24,800	10,600	35,400
20' x 36'	1.000	Ea.	187.949	25,500	10,900	36,400
20' x 40'	1.000	Ea.	179.200	34,000	10,400	44,400
20' x 44'	1.000	Ea.	197.120	37,400	11,500	48,900
24' x 40'	1.000	Ea.	215.040	40,800	12,500	53,300
24' x 44'	1.000	Ea.	273.244	37,100	15,900	53,000
24' x 48'	1.000	Ea.	298.077	40,500	17,400	57,900

7 | SPECIALTIES 40 | Wood Deck Systems

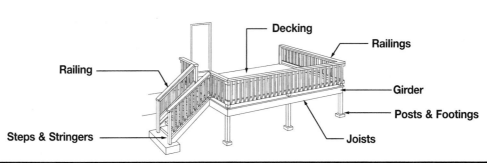

System Description	QUAN.	UNIT	LABOR HOURS	COST PER S.F. MAT.	COST PER S.F. INST.	COST PER S.F. TOTAL
8' X 12' DECK, PRESSURE TREATED LUMBER, JOISTS 16" O.C.						
4" x 4" Posts	.500	L.F.	.021	.74	1.17	1.91
Ledger, bolted 4' O.C.	.125	L.F.	.005	.21	.30	.51
Joists, 2" x 10", 16" O.C.	1.080	L.F.	.019	1.63	1.09	2.72
5/4" x 6" decking	1.000	S.F.	.025	2.48	1.42	3.90
2" x 12" stair stringers	.375	L.F.	.046	.68	2.63	3.31
5/4" x 6" stair treads	.500	L.F.	.050	.58	2.85	3.43
4' x 4" handrail posts	.210	L.F.	.009	.31	.49	.80
2" x 6" handrail cap	.520	L.F.	.014	.45	.79	1.24
2" x 4" baluster support	1.040	L.F.	.028	.60	1.57	2.17
2" x 2" balusters	3.850	L.F.	.093	1.42	5.31	6.73
Post footings	.042	Ea.	.134	1.11	6.59	7.70
Concrete step	.010	Ea.	.016	.78	.88	1.66
Joist hangers	.210	Ea.	.010	.32	.58	.90
TOTAL		S.F.	.470	11.31	25.67	36.98
12' X 16' DECK, PRESSURE TREATED LUMBER, JOISTS 24" O.C.						
4" x 4" Posts	.312	L.F.	.013	.46	.73	1.19
Ledger, bolted 4' O.C.	.083	L.F.	.003	.11	.19	.30
Joists, 2" x 8", 16" O.C.	.810	L.F.	.014	1.22	.82	2.04
5/4" x 6" decking	1.000	S.F.	.025	2.48	1.42	3.90
2" x 12" stair stringers	.188	L.F.	.023	.34	1.32	1.66
5/4" x 6" stair treads	.250	L.F.	.025	1.22	1.43	2.65
4" x 4" handrail posts	.125	L.F.	.005	.19	.29	.48
2" x 6" handrail cap	.323	L.F.	.009	.28	.49	.77
2" x 4" baluster support	.650	L.F.	.017	.38	.98	1.36
2" x 2" balusters	2.440	L.F.	.059	.90	3.37	4.27
Post footings	.026	Ea.	.083	.69	4.08	4.77
Concrete step	.005	Ea.	.008	.39	.45	.84
Joist hangers	.135	Ea.	.007	.20	.37	.57
Deck support girder	.083	L.F.	.002	.25	.14	.39
TOTAL		S.F.	.293	9.11	16.08	25.19

The costs in this system are on a square foot basis.

7 | SPECIALTIES — 40 | Wood Deck Systems

System Description	QUAN.	UNIT	LABOR HOURS	COST PER S.F. MAT.	COST PER S.F. INST.	COST PER S.F. TOTAL
12' X 24' DECK, REDWOOD OR CEDAR, JOISTS 16" O.C.						
4" x 4" redwood posts	.291	L.F.	.012	2.07	.68	2.75
Redwood joists, 2" x 8", 16" O.C.	.001	M.B.F.	.010	5.50	.57	6.07
5/4" x 6" redwood decking	1.000	S.F.	.025	5.30	1.42	6.72
Stair stringers	.125	L.F.	.015	.23	.88	1.11
Stringer trim	.083	L.F.	.003	.44	.19	.63
Stair treads	.135	S.F.	.003	.72	.19	.91
Handrail supports	.083	L.F.	.003	.59	.19	.78
Handrail cap	.152	L.F.	.004	1.25	.23	1.48
Baluster supports	.416	L.F.	.011	3.43	.63	4.06
Balusters	1.875	L.F.	.050	2.42	2.83	5.25
Post footings	.024	Ea.	.077	.64	3.77	4.41
Concrete step	.003	Ea.	.005	.23	.26	.49
Joist hangers	.132	Ea.	.006	.20	.36	.56
Deck support girder		M.B.F.	.003	1.50	.16	1.66
TOTAL		S.F.	.227	24.52	12.36	36.88

The costs in this system are on a square foot basis.

Wood Deck Price Sheet	QUAN.	UNIT	LABOR HOURS	COST PER S.F. MAT.	COST PER S.F. INST.	COST PER S.F. TOTAL
Decking, treated lumber, 1" x 4"	3.430	L.F.	.031	8.15	5.65	13.80
2" x 4"	3.430	L.F.	.033	6.70	5.20	11.90
2" x 6"	2.200	L.F.	.041	4.07	3.12	7.19
5/4" x 6"	2.200	L.F.	.027	5.45	3.12	8.57
Redwood or cedar, 1" x 4"	3.430	L.F.	.035	7.20	5.65	12.85
2" x 4"	3.430	L.F.	.036	14.60	5.20	19.80
2" x 6"	2.200	L.F.	.028	16.85	3.12	19.97
5/4" x 6"	2.200	L.F.	.027	10.75	3.12	13.87
Joists for deck, treated lumber, 2" x 8", 16" O.C.	1.000	L.F.	.015	1.18	.83	2.01
24" O.C.	.750	L.F.	.012	.89	.62	1.51
2" x 10", 16" O.C.	1.000	L.F.	.018	1.51	1.01	2.52
24" O.C.	.750	L.F.	.014	1.13	.76	1.89
Redwood or cedar, 2" x 8", 16" O.C.	1.000	L.F.	.015	5	.83	5.83
24" O.C.	.750	L.F.	.012	3.75	.62	4.37
2" x 10", 16" O.C.	1.000	L.F.	.018	8.30	1.01	9.31
24" O.C.	.750	L.F.	.014	6.25	.76	7.01
Girder for joists, treated lumber, 2" x 10", 8' x 12' deck	.250	L.F.	.002	.38	.21	.59
12' x 16' deck	.167	L.F.	.001	.25	.14	.39
12' x 24' deck	.167	L.F.	.001	.25	.14	.39
Redwood or cedar, 2" x 10", 8' x 12' deck	.250	L.F.	.002	2.08	.25	2.33
12' x 16' deck	.167	L.F.	.001	1.39	.17	1.56
12' x 24' deck	.167	L.F.	.001	1.39	.17	1.56
Posts, 4" x 4", including concrete footing, 8' x 12' deck	.500	L.F.	.022	1.83	7.60	9.43
12' x 16' deck	.312	L.F.	.017	1.15	4.81	5.96
12' x 24' deck	.291	L.F.	.017	1.07	4.45	5.52
Stairs 2" x 10" stringers, treated lumber, 8' x 12' deck	1.000	Set	.020	1.51	4.55	6.06
12' x 16' deck	1.000	Set	.012	.76	2.28	3.04
12' x 24' deck	1.000	Set	.008	.53	1.59	2.12
Redwood or cedar, 8' x 12' deck	1.000	Set	.040	3.95	4.55	8.50
12' x 16' deck	1.000	Set	.020	1.98	2.28	4.26
12' x 24' deck	1.000	Set	.012	1.38	1.59	2.97
Railings 2" x 4", treated lumber, 8' x 12' deck	.520	L.F.	.026	2.47	7.65	10.12
12' x 16' deck	.323	L.F.	.017	1.56	4.84	6.40
12' x 24' deck	.210	L.F.	.014	1.12	3.54	4.66
Redwood or cedar, 8' x 12' deck	1.000	L.F.	.009	10.60	8.15	18.75

7 | SPECIALTIES — 40 | Wood Deck Systems

Wood Deck Price Sheet

	QUAN.	UNIT	LABOR HOURS	COST PER S.F. MAT.	COST PER S.F. INST.	COST PER S.F. TOTAL
12' x 16' deck	.670	L.F.	.006	6.60	5.15	11.75
12' x 24' deck	.540	L.F.	.005	4.70	3.79	8.49
Alternative decking, wood/plastic composite, 5/4" x 6"	2.200	L.F.	.055	7.60	3.12	10.72
1" x 4" square edge fir	3.430	L.F.	.100	8.25	5.65	13.90
1" x 4" tongue and groove fir	3.430	L.F.	.122	5.50	6.95	12.45
1" x 4" mahogany	3.430	L.F.	.100	7.60	5.65	13.25
5/4" x 6" PVC	2.200	L.F.	.064	9	3.63	12.63

Did you know?

RSMeans Online gives you the same access to RSMeans' data with 24/7 access:
- Quickly locate costs in the searchable database.
- Build cost lists, estimates, and reports in minutes.
- Adjust costs to any location in the U.S. and Canada with the click of a button.

Start your free trial today at **www.RSMeansOnline.com**

RSMeans Online
FROM THE GORDIAN GROUP®

No part of this cost data may be reproduced, stored in a retrieval system, or transmitted in any form or by any means without prior written permission of RSMeans.

8 | MECHANICAL — 04 | Two Fixture Lavatory Systems

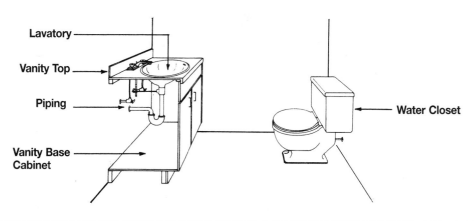

System Description	QUAN.	UNIT	LABOR HOURS	COST EACH MAT.	COST EACH INST.	COST EACH TOTAL
LAVATORY INSTALLED WITH VANITY, PLUMBING IN 2 WALLS						
Water closet, floor mounted, 2 piece, close coupled, white	1.000	Ea.	3.019	253	175	428
Rough-in, vent, 2" diameter DWV piping	1.000	Ea.	.955	53	55.40	108.40
Waste, 4" diameter DWV piping	1.000	Ea.	.828	64.50	48	112.50
Supply, 1/2" diameter type "L" copper supply piping	1.000	Ea.	.593	24.30	38.10	62.40
Lavatory, 20" x 18", P.E. cast iron white	1.000	Ea.	2.500	345	145	490
Rough-in, vent, 1-1/2" diameter DWV piping	1.000	Ea.	.901	50.60	52.20	102.80
Waste, 2" diameter DWV piping	1.000	Ea.	.955	53	55.40	108.40
Supply, 1/2" diameter type "L" copper supply piping	1.000	Ea.	.988	40.50	63.50	104
Piping, supply, 1/2" diameter type "L" copper supply piping	10.000	L.F.	.988	40.50	63.50	104
Waste, 4" diameter DWV piping	7.000	L.F.	1.931	150.50	112	262.50
Vent, 2" diameter DWV piping	12.000	L.F.	2.866	159	166.20	325.20
Vanity base cabinet, 2 door, 30" wide	1.000	Ea.	1.000	415	57	472
Vanity top, plastic & laminated, square edge	2.670	L.F.	.712	117.48	40.45	157.93
TOTAL		Ea.	18.236	1,766.38	1,071.75	2,838.13
LAVATORY WITH WALL-HUNG LAVATORY, PLUMBING IN 2 WALLS						
Water closet, floor mounted, 2 piece close coupled, white	1.000	Ea.	3.019	253	175	428
Rough-in, vent, 2" diameter DWV piping	1.000	Ea.	.955	53	55.40	108.40
Waste, 4" diameter DWV piping	1.000	Ea.	.828	64.50	48	112.50
Supply, 1/2" diameter type "L" copper supply piping	1.000	Ea.	.593	24.30	38.10	62.40
Lavatory, 20" x 18", P.E. cast iron, wall hung, white	1.000	Ea.	2.000	291	116	407
Rough-in, vent, 1-1/2" diameter DWV piping	1.000	Ea.	.901	50.60	52.20	102.80
Waste, 2" diameter DWV piping	1.000	Ea.	.955	53	55.40	108.40
Supply, 1/2" diameter type "L" copper supply piping	1.000	Ea.	.988	40.50	63.50	104
Piping, supply, 1/2" diameter type "L" copper supply piping	10.000	L.F.	.988	40.50	63.50	104
Waste, 4" diameter DWV piping	7.000	L.F.	1.931	150.50	112	262.50
Vent, 2" diameter DWV piping	12.000	L.F.	2.866	159	166.20	325.20
Carrier, steel for studs, no arms	1.000	Ea.	1.143	105	73.50	178.50
TOTAL		Ea.	17.167	1,284.90	1,018.80	2,303.70

Description	QUAN.	UNIT	LABOR HOURS	COST EACH MAT.	COST EACH INST.	COST EACH TOTAL

Two Fixture Lavatory Price Sheet	QUAN.	UNIT	LABOR HOURS	COST EACH		
				MAT.	INST.	TOTAL
Water closet, close coupled standard 2 piece, white	1.000	Ea.	3.019	253	175	428
Color	1.000	Ea.	3.019	460	175	635
One piece elongated bowl, white	1.000	Ea.	3.019	805	175	980
Color	1.000	Ea.	3.019	1,075	175	1,250
Low profile, one piece elongated bowl, white	1.000	Ea.	3.019	820	175	995
Color	1.000	Ea.	3.019	1,075	175	1,250
Rough-in for water closet						
1/2" copper supply, 4" cast iron waste, 2" cast iron vent	1.000	Ea.	2.376	142	142	284
4" PVC waste, 2" PVC vent	1.000	Ea.	2.678	83.50	159	242.50
4" copper waste, 2" copper vent	1.000	Ea.	2.520	229	154	383
3" cast iron waste, 1-1/2" cast iron vent	1.000	Ea.	2.244	126	134	260
3" PVC waste, 1-1/2" PVC vent	1.000	Ea.	2.388	74.50	148	222.50
3" copper waste, 1-1/2" copper vent	1.000	Ea.	2.524	227	150	377
1/2" PVC supply, 4" PVC waste, 2" PVC vent	1.000	Ea.	2.974	105	178	283
3" PVC waste, 1-1/2" PVC vent	1.000	Ea.	2.684	96	167	263
1/2" steel supply, 4" cast iron waste, 2" cast iron vent	1.000	Ea.	2.545	138	153	291
4" cast iron waste, 2" steel vent	1.000	Ea.	2.590	123	155	278
4" PVC waste, 2" PVC vent	1.000	Ea.	2.847	80	170	250
Lavatory, vanity top mounted, P.E. on cast iron 20" x 18" white	1.000	Ea.	2.500	345	145	490
Color	1.000	Ea.	2.500	505	145	650
Steel, enameled 10" x 17" white	1.000	Ea.	2.759	165	160	325
Color	1.000	Ea.	2.500	169	145	314
Vitreous china 20" x 16", white	1.000	Ea.	2.963	260	172	432
Color	1.000	Ea.	2.963	260	172	432
Wall hung, P.E. on cast iron, 20" x 18", white	1.000	Ea.	2.000	291	116	407
Color	1.000	Ea.	2.000	305	116	421
Vitreous china 19" x 17", white	1.000	Ea.	2.286	162	133	295
Color	1.000	Ea.	2.286	184	133	317
Rough-in supply waste and vent for lavatory						
1/2" copper supply, 2" cast iron waste, 1-1/2" cast iron vent	1.000	Ea.	2.844	144	171	315
2" PVC waste, 1-1/2" PVC vent	1.000	Ea.	2.962	89	184	273
2" copper waste, 1-1/2" copper vent	1.000	Ea.	2.308	155	149	304
1-1/2" PVC waste, 1-1/4" PVC vent	1.000	Ea.	2.639	88	170	258
1-1/2" copper waste, 1-1/4" copper vent	1.000	Ea.	2.114	131	136	267
1/2" PVC supply, 2" PVC waste, 1-1/2" PVC vent	1.000	Ea.	3.456	125	216	341
1-1/2" PVC waste, 1-1/4" PVC vent	1.000	Ea.	3.133	124	202	326
1/2" steel supply, 2" cast iron waste, 1-1/2" cast iron vent	1.000	Ea.	3.126	139	190	329
2" cast iron waste, 2" steel vent	1.000	Ea.	3.225	125	195	320
2" PVC waste, 1-1/2" PVC vent	1.000	Ea.	3.244	83.50	202	285.50
1-1/2" PVC waste, 1-1/4" PVC vent	1.000	Ea.	2.921	82.50	188	270.50
Piping, supply, 1/2" copper, type "L"	10.000	L.F.	.988	40.50	63.50	104
1/2" steel	10.000	L.F.	1.270	35	82	117
1/2" PVC	10.000	L.F.	1.482	76	95.50	171.50
Waste, 4" cast iron	7.000	L.F.	1.931	151	112	263
4" copper	7.000	L.F.	2.800	325	161	486
4" PVC	7.000	L.F.	2.333	79	135	214
Vent, 2" cast iron	12.000	L.F.	2.866	159	166	325
2" copper	12.000	L.F.	2.182	195	140	335
2" PVC	12.000	L.F.	3.254	75.50	188	263.50
2" steel	12.000	Ea.	3.000	112	174	286
Vanity base cabinet, 2 door, 24" x 30"	1.000	Ea.	1.000	415	57	472
24" x 36"	1.000	Ea.	1.200	400	68	468
Vanity top, laminated plastic, square edge 25" x 32"	2.670	L.F.	.712	117	40.50	157.50
25" x 38"	3.170	L.F.	.845	139	48	187
Post formed, laminated plastic, 25" x 32"	2.670	L.F.	.712	30.50	40.50	71
25" x 38"	3.170	L.F.	.845	36	48	84
Cultured marble, 25" x 32" with bowl	1.000	Ea.	2.500	209	145	354
25" x 38" with bowl	1.000	Ea.	2.500	243	145	388
Carrier for lavatory, steel for studs	1.000	Ea.	1.143	105	73.50	178.50
Wood 2" x 8" blocking	1.330	L.F.	.053	1.24	3.02	4.26

8 | MECHANICAL — 12 | Three Fixture Bathroom Systems

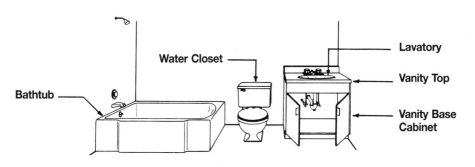

System Description	QUAN.	UNIT	LABOR HOURS	MAT.	INST.	TOTAL
BATHROOM INSTALLED WITH VANITY						
Water closet, floor mounted, 2 piece, close coupled, white	1.000	Ea.	3.019	253	175	428
Rough-in, waste, 4" diameter DWV piping	1.000	Ea.	.828	64.50	48	112.50
Vent, 2" diameter DWV piping	1.000	Ea.	.955	53	55.40	108.40
Supply, 1/2" diameter type "L" copper supply piping	1.000	Ea.	.593	24.30	38.10	62.40
Lavatory, 20" x 18", P.E. cast iron with accessories, white	1.000	Ea.	2.500	345	145	490
Rough-in, supply, 1/2" diameter type "L" copper supply piping	1.000	Ea.	.988	40.50	63.50	104
Waste, 1-1/2" diameter DWV piping	1.000	Ea.	1.803	101.20	104.40	205.60
Bathtub, P.E. cast iron, 5' long with accessories, white	1.000	Ea.	3.636	1,250	211	1,461
Rough-in, waste, 4" diameter DWV piping	1.000	Ea.	.828	64.50	48	112.50
Vent, 1-1/2" diameter DWV piping	1.000	Ea.	.593	49.60	38.20	87.80
Supply, 1/2" diameter type "L" copper supply piping	1.000	Ea.	.988	40.50	63.50	104
Piping, supply, 1/2" diameter type "L" copper supply piping	20.000	L.F.	1.975	81	127	208
Waste, 4" diameter DWV piping	9.000	L.F.	2.483	193.50	144	337.50
Vent, 2" diameter DWV piping	6.000	L.F.	1.500	55.80	87	142.80
Vanity base cabinet, 2 door, 30" wide	1.000	Ea.	1.000	415	57	472
Vanity top, plastic laminated square edge	2.670	L.F.	.712	98.79	40.45	139.24
TOTAL		Ea.	24.401	3,130.19	1,445.55	4,575.74
BATHROOM WITH WALL HUNG LAVATORY						
Water closet, floor mounted, 2 piece, close coupled, white	1.000	Ea.	3.019	253	175	428
Rough-in, vent, 2" diameter DWV piping	1.000	Ea.	.955	53	55.40	108.40
Waste, 4" diameter DWV piping	1.000	Ea.	.828	64.50	48	112.50
Supply, 1/2" diameter type "L" copper supply piping	1.000	Ea.	.593	24.30	38.10	62.40
Lavatory, 20" x 18" P.E. cast iron, wall hung, white	1.000	Ea.	2.000	291	116	407
Rough-in, waste, 1-1/2" diameter DWV piping	1.000	Ea.	1.803	101.20	104.40	205.60
Supply, 1/2" diameter type "L" copper supply piping	1.000	Ea.	.988	40.50	63.50	104
Bathtub, P.E. cast iron, 5' long with accessories, white	1.000	Ea.	3.636	1,250	211	1,461
Rough-in, waste, 4" diameter DWV piping	1.000	Ea.	.828	64.50	48	112.50
Supply, 1/2" diameter type "L" copper supply piping	1.000	Ea.	.988	40.50	63.50	104
Vent, 1-1/2" diameter DWV piping	1.000	Ea.	1.482	124	95.50	219.50
Piping, supply, 1/2" diameter type "L" copper supply piping	20.000	L.F.	1.975	81	127	208
Waste, 4" diameter DWV piping	9.000	L.F.	2.483	193.50	144	337.50
Vent, 2" diameter DWV piping	6.000	L.F.	1.500	55.80	87	142.80
Carrier, steel, for studs, no arms	1.000	Ea.	1.143	105	73.50	178.50
TOTAL		Ea.	24.221	2,741.80	1,449.90	4,191.70

The costs in this system are a cost each basis, all necessary piping is included.

Three Fixture Bathroom Price Sheet	QUAN.	UNIT	LABOR HOURS	COST EACH MAT.	COST EACH INST.	COST EACH TOTAL
Water closet, close coupled standard 2 piece, white	1.000	Ea.	3.019	253	175	428
Color	1.000	Ea.	3.019	460	175	635
One piece, elongated bowl, white	1.000	Ea.	3.019	805	175	980
Color	1.000	Ea.	3.019	1,075	175	1,250
Low profile, one piece elongated bowl, white	1.000	Ea.	3.019	820	175	995
Color	1.000	Ea.	3.019	1,075	175	1,250
Rough-in, for water closet						
1/2" copper supply, 4" cast iron waste, 2" cast iron vent	1.000	Ea.	2.376	142	142	284
4" PVC/DWV waste, 2" PVC vent	1.000	Ea.	2.678	83.50	159	242.50
4" copper waste, 2" copper vent	1.000	Ea.	2.520	229	154	383
3" cast iron waste, 1-1/2" cast iron vent	1.000	Ea.	2.244	126	134	260
3" PVC waste, 1-1/2" PVC vent	1.000	Ea.	2.388	74.50	148	222.50
3" copper waste, 1-1/2" copper vent	1.000	Ea.	2.014	161	124	285
1/2" PVC supply, 4" PVC waste, 2" PVC vent	1.000	Ea.	2.974	105	178	283
3" PVC waste, 1-1/2" PVC supply	1.000	Ea.	2.684	96	167	263
1/2" steel supply, 4" cast iron waste, 2" cast iron vent	1.000	Ea.	2.545	138	153	291
4" cast iron waste, 2" steel vent	1.000	Ea.	2.590	123	155	278
4" PVC waste, 2" PVC vent	1.000	Ea.	2.847	80	170	250
Lavatory, wall hung, P.E. cast iron 20" x 18", white	1.000	Ea.	2.000	291	116	407
Color	1.000	Ea.	2.000	305	116	421
Vitreous china 19" x 17", white	1.000	Ea.	2.286	162	133	295
Color	1.000	Ea.	2.286	184	133	317
Lavatory, for vanity top, P.E. cast iron 20" x 18", white	1.000	Ea.	2.500	345	145	490
Color	1.000	Ea.	2.500	505	145	650
Steel, enameled 20" x 17", white	1.000	Ea.	2.759	165	160	325
Color	1.000	Ea.	2.500	169	145	314
Vitreous china 20" x 16", white	1.000	Ea.	2.963	260	172	432
Color	1.000	Ea.	2.963	260	172	432
Rough-in, for lavatory						
1/2" copper supply, 1-1/2" C.I. waste, 1-1/2" C.I. vent	1.000	Ea.	2.791	142	168	310
1-1/2" PVC waste, 1-1/4" PVC vent	1.000	Ea.	2.639	88	170	258
1/2" steel supply, 1-1/4" cast iron waste, 1-1/4" steel vent	1.000	Ea.	2.890	111	176	287
1-1/4" PVC waste, 1-1/4" PVC vent	1.000	Ea.	2.794	83.50	180	263.50
1/2" PVC supply, 1-1/2" PVC waste, 1-1/2" PVC vent	1.000	Ea.	3.260	123	210	333
Bathtub, P.E. cast iron, 5' long corner with fittings, white	1.000	Ea.	3.636	1,250	211	1,461
Color	1.000	Ea.	3.636	1,550	211	1,761
Rough-in, for bathtub						
1/2" copper supply, 4" cast iron waste, 1-1/2" copper vent	1.000	Ea.	2.409	155	150	305
4" PVC waste, 1-1/2" PVC vent	1.000	Ea.	2.877	97.50	179	276.50
1/2" steel supply, 4" cast iron waste, 1-1/2" steel vent	1.000	Ea.	2.898	128	176	304
4" PVC waste, 1-1/2" PVC vent	1.000	Ea.	3.159	92	197	289
1/2" PVC supply, 4" PVC waste, 1-1/2" PVC vent	1.000	Ea.	3.371	133	211	344
Piping, supply 1/2" copper	20.000	L.F.	1.975	81	127	208
1/2" steel	20.000	L.F.	2.540	70	164	234
1/2" PVC	20.000	L.F.	2.963	152	191	343
Piping, waste, 4" cast iron no hub	9.000	L.F.	2.483	194	144	338
4" PVC/DWV	9.000	L.F.	3.000	101	174	275
4" copper/DWV	9.000	L.F.	3.600	420	207	627
Piping, vent 2" cast iron no hub	6.000	L.F.	1.433	79.50	83	162.50
2" copper/DWV	6.000	L.F.	1.091	97.50	70	167.50
2" PVC/DWV	6.000	L.F.	1.627	38	94	132
2" steel, galvanized	6.000	L.F.	1.500	56	87	143
Vanity base cabinet, 2 door, 24" x 30"	1.000	Ea.	1.000	415	57	472
24" x 36"	1.000	Ea.	1.200	400	68	468
Vanity top, laminated plastic square edge 25" x 32"	2.670	L.F.	.712	99	40.50	139.50
25" x 38"	3.160	L.F.	.843	117	48	165
Cultured marble, 25" x 32", with bowl	1.000	Ea.	2.500	209	145	354
25" x 38", with bowl	1.000	Ea.	2.500	243	145	388
Carrier, for lavatory, steel for studs, no arms	1.000	Ea.	1.143	105	73.50	178.50
Wood, 2" x 8" blocking	1.300	L.F.	.052	1.21	2.95	4.16

8 | MECHANICAL   16 | Three Fixture Bathroom Systems

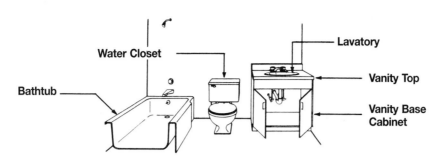

System Description	QUAN.	UNIT	LABOR HOURS	COST EACH MAT.	COST EACH INST.	COST EACH TOTAL
BATHROOM WITH LAVATORY INSTALLED IN VANITY						
Water closet, floor mounted, 2 piece, close coupled, white	1.000	Ea.	3.019	253	175	428
Rough-in, waste, 4" diameter DWV piping	1.000	Ea.	.828	64.50	48	112.50
Vent, 2" diameter DWV piping	1.000	Ea.	.955	53	55.40	108.40
Supply, 1/2" diameter type "L" copper supply piping	1.000	Ea.	.593	24.30	38.10	62.40
Lavatory, 20" x 18", P.E. cast iron with accessories, white	1.000	Ea.	2.500	345	145	490
Rough-in, waste, 1-1/2" diameter DWV piping	1.000	Ea.	1.803	101.20	104.40	205.60
Supply, 1/2" diameter type "L" copper supply piping	1.000	Ea.	.988	40.50	63.50	104
Bathtub, P.E. cast iron 5' long with accessories, white	1.000	Ea.	3.636	1,250	211	1,461
Rough-in, waste, 4" diameter DWV piping	1.000	Ea.	.828	64.50	48	112.50
Vent, 1-1/2" diameter DWV piping	1.000	Ea.	.593	49.60	38.20	87.80
Supply, 1/2" diameter type "L" copper supply piping	1.000	Ea.	.988	40.50	63.50	104
Piping, supply, 1/2" diameter type "L" copper supply piping	10.000	L.F.	.988	40.50	63.50	104
Waste, 4" diameter DWV piping	6.000	L.F.	1.655	129	96	225
Vent, 2" diameter DWV piping	6.000	L.F.	1.500	55.80	87	142.80
Vanity base cabinet, 2 door, 30" wide	1.000	Ea.	1.000	415	57	472
Vanity top, plastic laminated square edge	2.670	L.F.	.712	98.79	40.45	139.24
TOTAL		Ea.	22.586	3,025.19	1,334.05	4,359.24
BATHROOM WITH WALL HUNG LAVATORY						
Water closet, floor mounted, 2 piece, close coupled, white	1.000	Ea.	3.019	253	175	428
Rough-in, vent, 2" diameter DWV piping	1.000	Ea.	.955	53	55.40	108.40
Waste, 4" diameter DWV piping	1.000	Ea.	.828	64.50	48	112.50
Supply, 1/2" diameter type "L" copper supply piping	1.000	Ea.	.593	24.30	38.10	62.40
Lavatory, 20" x 18" P.E. cast iron, wall hung, white	1.000	Ea.	2.000	291	116	407
Rough-in, waste, 1-1/2" diameter DWV piping	1.000	Ea.	1.803	101.20	104.40	205.60
Supply, 1/2" diameter type "L" copper supply piping	1.000	Ea.	.988	40.50	63.50	104
Bathtub, P.E. cast iron, 5' long with accessories, white	1.000	Ea.	3.636	1,250	211	1,461
Rough-in, waste, 4" diameter DWV piping	1.000	Ea.	.828	64.50	48	112.50
Supply, 1/2" diameter type "L" copper supply piping	1.000	Ea.	.988	40.50	63.50	104
Vent, 1-1/2" diameter DWV piping	1.000	Ea.	.593	49.60	38.20	87.80
Piping, supply, 1/2" diameter type "L" copper supply piping	10.000	L.F.	.988	40.50	63.50	104
Waste, 4" diameter DWV piping	6.000	L.F.	1.655	129	96	225
Vent, 2" diameter DWV piping	6.000	L.F.	1.500	55.80	87	142.80
Carrier, steel, for studs, no arms	1.000	Ea.	1.143	105	73.50	178.50
TOTAL		Ea.	21.517	2,562.40	1,281.10	3,843.50

The costs in this system are on a cost each basis. All necessary piping is included.

Three Fixture Bathroom Price Sheet

	QUAN.	UNIT	LABOR HOURS	COST EACH MAT.	COST EACH INST.	COST EACH TOTAL
Water closet, close coupled standard 2 piece, white	1.000	Ea.	3.019	253	175	428
Color	1.000	Ea.	3.019	460	175	635
One piece elongated bowl, white	1.000	Ea.	3.019	805	175	980
Color	1.000	Ea.	3.019	1,075	175	1,250
Low profile, one piece elongated bowl, white	1.000	Ea.	3.019	820	175	995
Color	1.000	Ea.	3.019	1,075	175	1,250
Rough-in for water closet						
1/2" copper supply, 4" cast iron waste, 2" cast iron vent	1.000	Ea.	2.376	142	142	284
4" PVC/DWV waste, 2" PVC vent	1.000	Ea.	2.678	83.50	159	242.50
4" carrier waste, 2" copper vent	1.000	Ea.	2.520	229	154	383
3" cast iron waste, 1-1/2" cast iron vent	1.000	Ea.	2.244	126	134	260
3" PVC waste, 1-1/2" PVC vent	1.000	Ea.	2.388	74.50	148	222.50
3" copper waste, 1-1/2" copper vent	1.000	Ea.	2.014	161	124	285
1/2" PVC supply, 4" PVC waste, 2" PVC vent	1.000	Ea.	2.974	105	178	283
3" PVC waste, 1-1/2" PVC supply	1.000	Ea.	2.684	96	167	263
1/2" steel supply, 4" cast iron waste, 2" cast iron vent	1.000	Ea.	2.545	138	153	291
4" cast iron waste, 2" steel vent	1.000	Ea.	2.590	123	155	278
4" PVC waste, 2" PVC vent	1.000	Ea.	2.847	80	170	250
Lavatory, wall hung, PE cast iron 20" x 18", white	1.000	Ea.	2.000	291	116	407
Color	1.000	Ea.	2.000	305	116	421
Vitreous china 19" x 17", white	1.000	Ea.	2.286	162	133	295
Color	1.000	Ea.	2.286	184	133	317
Lavatory, for vanity top, PE cast iron 20" x 18", white	1.000	Ea.	2.500	345	145	490
Color	1.000	Ea.	2.500	505	145	650
Steel enameled 20" x 17", white	1.000	Ea.	2.759	165	160	325
Color	1.000	Ea.	2.500	169	145	314
Vitreous china 20" x 16", white	1.000	Ea.	2.963	260	172	432
Color	1.000	Ea.	2.963	260	172	432
Rough-in for lavatory						
1/2" copper supply, 1-1/2" cast iron waste, 1-1/2" cast iron vent	1.000	Ea.	2.791	142	168	310
1-1/2" PVC waste, 1-1/4" PVC vent	1.000	Ea.	2.639	88	170	258
1/2" steel supply, 1-1/4" cast iron waste, 1-1/4" steel vent	1.000	Ea.	2.890	111	176	287
1-1/4" PVC waste, 1-1/4" PVC vent	1.000	Ea.	2.794	83.50	180	263.50
1/2" PVC supply, 1-1/2" PVC waste, 1-1/2" PVC vent	1.000	Ea.	3.260	123	210	333
Bathtub, PE cast iron, 5' long corner with fittings, white	1.000	Ea.	3.636	1,250	211	1,461
Color	1.000	Ea.	3.636	1,550	211	1,761
Rough-in for bathtub						
1/2" copper supply, 4" cast iron waste, 1-1/2" copper vent	1.000	Ea.	2.409	155	150	305
4" PVC waste, 1/2" PVC vent	1.000	Ea.	2.877	97.50	179	276.50
1/2" steel supply, 4" cast iron waste, 1-1/2" steel vent	1.000	Ea.	2.898	128	176	304
4" PVC waste, 1-1/2" PVC vent	1.000	Ea.	3.159	92	197	289
1/2" PVC supply, 4" PVC waste, 1-1/2" PVC vent	1.000	Ea.	3.371	133	211	344
Piping supply, 1/2" copper	10.000	L.F.	.988	40.50	63.50	104
1/2" steel	10.000	L.F.	1.270	35	82	117
1/2" PVC	10.000	L.F.	1.482	76	95.50	171.50
Piping waste, 4" cast iron no hub	6.000	L.F.	1.655	129	96	225
4" PVC/DWV	6.000	L.F.	2.000	67.50	116	183.50
4" copper/DWV	6.000	L.F.	2.400	279	138	417
Piping vent 2" cast iron no hub	6.000	L.F.	1.433	79.50	83	162.50
2" copper/DWV	6.000	L.F.	1.091	97.50	70	167.50
2" PVC/DWV	6.000	L.F.	1.627	38	94	132
2" steel, galvanized	6.000	L.F.	1.500	56	87	143
Vanity base cabinet, 2 door, 24" x 30"	1.000	Ea.	1.000	415	57	472
24" x 36"	1.000	Ea.	1.200	400	68	468
Vanity top, laminated plastic square edge 25" x 32"	2.670	L.F.	.712	99	40.50	139.50
25" x 38"	3.160	L.F.	.843	117	48	165
Cultured marble, 25" x 32", with bowl	1.000	Ea.	2.500	209	145	354
25" x 38", with bowl	1.000	Ea.	2.500	243	145	388
Carrier, for lavatory, steel for studs, no arms	1.000	Ea.	1.143	105	73.50	178.50
Wood, 2" x 8" blocking	1.300	L.F.	.052	1.21	2.95	4.16

For customer support on your Residential Cost Data, call 877.759.4771.

8 | MECHANICAL — 20 | Three Fixture Bathroom Systems

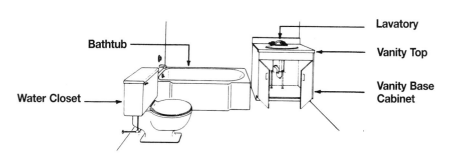

System Description	QUAN.	UNIT	LABOR HOURS	COST EACH MAT.	COST EACH INST.	COST EACH TOTAL
BATHROOM WITH LAVATORY INSTALLED IN VANITY						
Water closet, floor mounted, 2 piece, close coupled, white	1.000	Ea.	3.019	253	175	428
Rough-in, vent, 2" diameter DWV piping	1.000	Ea.	.955	53	55.40	108.40
Waste, 4" diameter DWV piping	1.000	Ea.	.828	64.50	48	112.50
Supply, 1/2" diameter type "L" copper supply piping	1.000	Ea.	.593	24.30	38.10	62.40
Lavatory, 20" x 18", PE cast iron with accessories, white	1.000	Ea.	2.500	345	145	490
Rough-in, vent, 1-1/2" diameter DWV piping	1.000	Ea.	1.803	101.20	104.40	205.60
Supply, 1/2" diameter type "L" copper supply piping	1.000	Ea.	.988	40.50	63.50	104
Bathtub, P.E. cast iron, 5' long with accessories, white	1.000	Ea.	3.636	1,250	211	1,461
Rough-in, waste, 4" diameter DWV piping	1.000	Ea.	.828	64.50	48	112.50
Supply, 1/2" diameter type "L" copper supply piping	1.000	Ea.	.988	40.50	63.50	104
Vent, 1-1/2" diameter DWV piping	1.000	Ea.	.593	49.60	38.20	87.80
Piping, supply, 1/2" diameter type "L" copper supply piping	32.000	L.F.	3.161	129.60	203.20	332.80
Waste, 4" diameter DWV piping	12.000	L.F.	3.310	258	192	450
Vent, 2" diameter DWV piping	6.000	L.F.	1.500	55.80	87	142.80
Vanity base cabinet, 2 door, 30" wide	1.000	Ea.	1.000	415	57	472
Vanity top, plastic laminated square edge	2.670	L.F.	.712	98.79	40.45	139.24
TOTAL		Ea.	26.414	3,243.29	1,569.75	4,813.04
BATHROOM WITH WALL HUNG LAVATORY						
Water closet, floor mounted, 2 piece, close coupled, white	1.000	Ea.	3.019	253	175	428
Rough-in, vent, 2" diameter DWV piping	1.000	Ea.	.955	53	55.40	108.40
Waste, 4" diameter DWV piping	1.000	Ea.	.828	64.50	48	112.50
Supply, 1/2" diameter type "L" copper supply piping	1.000	Ea.	.593	24.30	38.10	62.40
Lavatory, 20" x 18" P.E. cast iron, wall hung, white	1.000	Ea.	2.000	291	116	407
Rough-in, waste, 1-1/2" diameter DWV piping	1.000	Ea.	1.803	101.20	104.40	205.60
Supply, 1/2" diameter type "L" copper supply piping	1.000	Ea.	.988	40.50	63.50	104
Bathtub, P.E. cast iron, 5' long with accessories, white	1.000	Ea.	3.636	1,250	211	1,461
Rough-in, waste, 4" diameter DWV piping	1.000	Ea.	.828	64.50	48	112.50
Supply, 1/2" diameter type "L" copper supply piping	1.000	Ea.	.988	40.50	63.50	104
Vent, 1-1/2" diameter DWV piping	1.000	Ea.	.593	49.60	38.20	87.80
Piping, supply, 1/2" diameter type "L" copper supply piping	32.000	L.F.	3.161	129.60	203.20	332.80
Waste, 4" diameter DWV piping	12.000	L.F.	3.310	258	192	450
Vent, 2" diameter DWV piping	6.000	L.F.	1.500	55.80	87	142.80
Carrier steel, for studs, no arms	1.000	Ea.	1.143	105	73.50	178.50
TOTAL		Ea.	25.345	2,780.50	1,516.80	4,297.30

The costs in this system are on a cost each basis. All necessary piping is included.

Three Fixture Bathroom Price Sheet

	QUAN.	UNIT	LABOR HOURS	COST EACH MAT.	COST EACH INST.	COST EACH TOTAL
Water closet, close coupled, standard 2 piece, white	1.000	Ea.	3.019	253	175	428
Color	1.000	Ea.	3.019	460	175	635
One piece, elongated bowl, white	1.000	Ea.	3.019	805	175	980
Color	1.000	Ea.	3.019	1,075	175	1,250
Low profile, one piece, elongated bowl, white	1.000	Ea.	3.019	820	175	995
Color	1.000	Ea.	3.019	1,075	175	1,250
Rough-in, for water closet						
1/2" copper supply, 4" cast iron waste, 2" cast iron vent	1.000	Ea.	2.376	142	142	284
4" PVC/DWV waste, 2" PVC vent	1.000	Ea.	2.678	83.50	159	242.50
4" copper waste, 2" copper vent	1.000	Ea.	2.520	229	154	383
3" cast iron waste, 1-1/2" cast iron vent	1.000	Ea.	2.244	126	134	260
3" PVC waste, 1-1/2" PVC vent	1.000	Ea.	2.388	74.50	148	222.50
3" copper waste, 1-1/2" copper vent	1.000	Ea.	2.014	161	124	285
1/2" PVC supply, 4" PVC waste, 2" PVC vent	1.000	Ea.	2.974	105	178	283
3" PVC waste, 1-1/2" PVC supply	1.000	Ea.	2.684	96	167	263
1/2" steel supply, 4" cast iron waste, 2" cast iron vent	1.000	Ea.	2.545	138	153	291
4" cast iron waste, 2" steel vent	1.000	Ea.	2.590	123	155	278
4" PVC waste, 2" PVC vent	1.000	Ea.	2.847	80	170	250
Lavatory wall hung, P.E. cast iron, 20" x 18", white	1.000	Ea.	2.000	291	116	407
Color	1.000	Ea.	2.000	305	116	421
Vitreous china, 19" x 17", white	1.000	Ea.	2.286	162	133	295
Color	1.000	Ea.	2.286	184	133	317
Lavatory, for vanity top, P.E., cast iron, 20" x 18", white	1.000	Ea.	2.500	345	145	490
Color	1.000	Ea.	2.500	505	145	650
Steel, enameled, 20" x 17", white	1.000	Ea.	2.759	165	160	325
Color	1.000	Ea.	2.500	169	145	314
Vitreous china, 20" x 16", white	1.000	Ea.	2.963	260	172	432
Color	1.000	Ea.	2.963	260	172	432
Rough-in, for lavatory						
1/2" copper supply, 1-1/2" C.I. waste, 1-1/2" C.I. vent	1.000	Ea.	2.791	142	168	310
1-1/2" PVC waste, 1-1/4" PVC vent	1.000	Ea.	2.639	88	170	258
1/2" steel supply, 1-1/4" cast iron waste, 1-1/4" steel vent	1.000	Ea.	2.890	111	176	287
1-1/4" PVC waste, 1-1/4" PVC vent	1.000	Ea.	2.794	83.50	180	263.50
1/2" PVC supply, 1-1/2" PVC waste, 1-1/2" PVC vent	1.000	Ea.	3.260	123	210	333
Bathtub, P.E. cast iron, 5' long corner with fittings, white	1.000	Ea.	3.636	1,250	211	1,461
Color	1.000	Ea.	3.636	1,550	211	1,761
Rough-in, for bathtub						
1/2" copper supply, 4" cast iron waste, 1-1/2" copper vent	1.000	Ea.	2.409	155	150	305
4" PVC waste, 1/2" PVC vent	1.000	Ea.	2.877	97.50	179	276.50
1/2" steel supply, 4" cast iron waste, 1-1/2" steel vent	1.000	Ea.	2.898	128	176	304
4" PVC waste, 1-1/2" PVC vent	1.000	Ea.	3.159	92	197	289
1/2" PVC supply, 4" PVC waste, 1-1/2" PVC vent	1.000	Ea.	3.371	133	211	344
Piping, supply, 1/2" copper	32.000	L.F.	3.161	130	203	333
1/2" steel	32.000	L.F.	4.063	112	262	374
1/2" PVC	32.000	L.F.	4.741	243	305	548
Piping, waste, 4" cast iron no hub	12.000	L.F.	3.310	258	192	450
4" PVC/DWV	12.000	L.F.	4.000	135	232	367
4" copper/DWV	12.000	L.F.	4.800	560	276	836
Piping, vent, 2" cast iron no hub	6.000	L.F.	1.433	79.50	83	162.50
2" copper/DWV	6.000	L.F.	1.091	97.50	70	167.50
2" PVC/DWV	6.000	L.F.	1.627	38	94	132
2" steel, galvanized	6.000	L.F.	1.500	56	87	143
Vanity base cabinet, 2 door, 24" x 30"	1.000	Ea.	1.000	415	57	472
24" x 36"	1.000	Ea.	1.200	400	68	468
Vanity top, laminated plastic square edge, 25" x 32"	2.670	L.F.	.712	99	40.50	139.50
25" x 38"	3.160	L.F.	.843	117	48	165
Cultured marble, 25" x 32", with bowl	1.000	Ea.	2.500	209	145	354
25" x 38", with bowl	1.000	Ea.	2.500	243	145	388
Carrier, for lavatory, steel for studs, no arms	1.000	Ea.	1.143	105	73.50	178.50
Wood, 2" x 8" blocking	1.300	L.F.	.052	1.21	2.95	4.16

For customer support on your Residential Cost Data, call 877.759.4771.

8 | MECHANICAL — 24 | Three Fixture Bathroom Systems

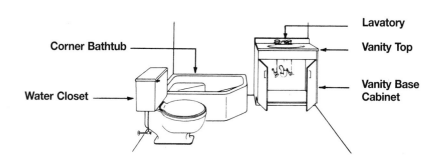

System Description	QUAN.	UNIT	LABOR HOURS	MAT.	INST.	TOTAL
BATHROOM WITH LAVATORY INSTALLED IN VANITY						
Water closet, floor mounted, 2 piece, close coupled, white	1.000	Ea.	3.019	253	175	428
Rough-in, vent, 2" diameter DWV piping	1.000	Ea.	.955	53	55.40	108.40
Waste, 4" diameter DWV piping	1.000	Ea.	.828	64.50	48	112.50
Supply, 1/2" diameter type "L" copper supply piping	1.000	Ea.	.593	24.30	38.10	62.40
Lavatory, 20" x 18", P.E. cast iron with fittings, white	1.000	Ea.	2.500	345	145	490
Rough-in, waste, 1-1/2" diameter DWV piping	1.000	Ea.	1.803	101.20	104.40	205.60
Supply, 1/2" diameter type "L" copper supply piping	1.000	Ea.	.988	40.50	63.50	104
Bathtub, P.E. cast iron, corner with fittings, white	1.000	Ea.	3.636	2,925	211	3,136
Rough-in, waste, 4" diameter DWV piping	1.000	Ea.	.828	64.50	48	112.50
Supply, 1/2" diameter type "L" copper supply piping	1.000	Ea.	.988	40.50	63.50	104
Vent, 1-1/2" diameter DWV piping	1.000	Ea.	.593	49.60	38.20	87.80
Piping, supply, 1/2" diameter type "L" copper supply piping	32.000	L.F.	3.161	129.60	203.20	332.80
Waste, 4" diameter DWV piping	12.000	L.F.	3.310	258	192	450
Vent, 2" diameter DWV piping	6.000	L.F.	1.500	55.80	87	142.80
Vanity base cabinet, 2 door, 30" wide	1.000	Ea.	1.000	415	57	472
Vanity top, plastic laminated, square edge	2.670	L.F.	.712	117.48	40.45	157.93
TOTAL		Ea.	26.414	4,936.98	1,569.75	6,506.73
BATHROOM WITH WALL HUNG LAVATORY						
Water closet, floor mounted, 2 piece, close coupled, white	1.000	Ea.	3.019	253	175	428
Rough-in, vent, 2" diameter DWV piping	1.000	Ea.	.955	53	55.40	108.40
Waste, 4" diameter DWV piping	1.000	Ea.	.828	64.50	48	112.50
Supply, 1/2" diameter type "L" copper supply piping	1.000	Ea.	.593	24.30	38.10	62.40
Lavatory, 20" x 18", P.E. cast iron, with fittings, white	1.000	Ea.	2.000	291	116	407
Rough-in, waste, 1-1/2" diameter DWV piping	1.000	Ea.	1.803	101.20	104.40	205.60
Supply, 1/2" diameter type "L" copper supply piping	1.000	Ea.	.988	40.50	63.50	104
Bathtub, P.E. cast iron, corner, with fittings, white	1.000	Ea.	3.636	2,925	211	3,136
Rough-in, waste, 4" diameter DWV piping	1.000	Ea.	.828	64.50	48	112.50
Supply, 1/2" diameter type "L" copper supply piping	1.000	Ea.	.988	40.50	63.50	104
Vent, 1-1/2" diameter DWV piping	1.000	Ea.	.593	49.60	38.20	87.80
Piping, supply, 1/2" diameter type "L" copper supply piping	32.000	L.F.	3.161	129.60	203.20	332.80
Waste, 4" diameter DWV piping	12.000	L.F.	3.310	258	192	450
Vent, 2" diameter DWV piping	6.000	L.F.	1.500	55.80	87	142.80
Carrier, steel, for studs, no arms	1.000	Ea.	1.143	105	73.50	178.50
TOTAL		Ea.	25.345	4,455.50	1,516.80	5,972.30

The costs in this system are on a cost each basis. All necessary piping is included.

Three Fixture Bathroom Price Sheet	QUAN.	UNIT	LABOR HOURS	COST EACH		
				MAT.	INST.	TOTAL
Water closet, close coupled, standard 2 piece, white	1.000	Ea.	3.019	253	175	428
Color	1.000	Ea.	3.019	460	175	635
One piece elongated bowl, white	1.000	Ea.	3.019	805	175	980
Color	1.000	Ea.	3.019	1,075	175	1,250
Low profile, one piece elongated bowl, white	1.000	Ea.	3.019	820	175	995
Color	1.000	Ea.	3.019	1,075	175	1,250
Rough-in, for water closet						
1/2" copper supply, 4" cast iron waste, 2" cast iron vent	1.000	Ea.	2.376	142	142	284
4" PVC/DWV waste, 2" PVC vent	1.000	Ea.	2.678	83.50	159	242.50
4" copper waste, 2" copper vent	1.000	Ea.	2.520	229	154	383
3" cast iron waste, 1-1/2" cast iron vent	1.000	Ea.	2.244	126	134	260
3" PVC waste, 1-1/2" PVC vent	1.000	Ea.	2.388	74.50	148	222.50
3" copper waste, 1-1/2" copper vent	1.000	Ea.	2.014	161	124	285
1/2" PVC supply, 4" PVC waste, 2" PVC vent	1.000	Ea.	2.974	105	178	283
3" PVC waste, 1-1/2" PVC supply	1.000	Ea.	2.684	96	167	263
1/2" steel supply, 4" cast iron waste, 2" cast iron vent	1.000	Ea.	2.545	138	153	291
4" cast iron waste, 2" steel vent	1.000	Ea.	2.590	123	155	278
4" PVC waste, 2" PVC vent	1.000	Ea.	2.847	80	170	250
Lavatory, wall hung P.E. cast iron 20" x 18", white	1.000	Ea.	2.000	291	116	407
Color	1.000	Ea.	2.000	305	116	421
Vitreous china 19" x 17", white	1.000	Ea.	2.286	162	133	295
Color	1.000	Ea.	2.286	184	133	317
Lavatory, for vanity top, P.E., cast iron, 20" x 18", white	1.000	Ea.	2.500	345.	145.	490
Color	1.000	Ea.	2.500	505	145	650
Steel enameled 20" x 17", white	1.000	Ea.	2.759	165	160	325
Color	1.000	Ea.	2.500	169	145	314
Vitreous china 20" x 16", white	1.000	Ea.	2.963	260	172	432
Color	1.000	Ea.	2.963	260	172	432
Rough-in, for lavatory						
1/2" copper supply, 1-1/2" cast iron waste, 1-1/2" cast iron vent	1.000	Ea.	2.791	142	168	310
1-1/2" PVC waste, 1-1/4" PVC vent	1.000	Ea.	2.639	88	170	258
1/2" steel supply, 1-1/4" cast iron waste, 1-1/4" steel vent	1.000	Ea.	2.890	111	176	287
1-1/4" PVC waste, 1-1/4" PVC vent	1.000	Ea.	2.794	83.50	180	263.50
1/2" PVC supply, 1-1/2" PVC waste, 1-1/2" PVC vent	1.000	Ea.	3.260	123	210	333
Bathtub, P.E. cast iron, corner with fittings, white	1.000	Ea.	3.636	2,925	211	3,136
Color	1.000	Ea.	4.000	3,350	232	3,582
Rough-in, for bathtub						
1/2" copper supply, 4" cast iron waste, 1-1/2" copper vent	1.000	Ea.	2.409	155	150	305
4" PVC waste, 1-1/2" PVC vent	1.000	Ea.	2.877	97.50	179	276.50
1/2" steel supply, 4" cast iron waste, 1-1/2" steel vent	1.000	Ea.	2.898	128	176	304
4" PVC waste, 1-1/2" PVC vent	1.000	Ea.	3.159	92	197	289
1/2" PVC supply, 4" PVC waste, 1-1/2" PVC vent	1.000	Ea.	3.371	133	211	344
Piping, supply, 1/2" copper	32.000	L.F.	3.161	130	203	333
1/2" steel	32.000	L.F.	4.063	112	262	374
1/2" PVC	32.000	L.F.	4.741	243	305	548
Piping, waste, 4" cast iron, no hub	12.000	L.F.	3.310	258	192	450
4" PVC/DWV	12.000	L.F.	4.000	135	232	367
4" copper/DWV	12.000	L.F.	4.800	560	276	836
Piping, vent 2" cast iron, no hub	6.000	L.F.	1.433	79.50	83	162.50
2" copper/DWV	6.000	L.F.	1.091	97.50	70	167.50
2" PVC/DWV	6.000	L.F.	1.627	38	94	132
2" steel, galvanized	6.000	L.F.	1.500	56	87	143
Vanity base cabinet, 2 door, 24" x 30"	1.000	Ea.	1.000	415	57	472
24" x 36"	1.000	Ea.	1.200	400	68	468
Vanity top, laminated plastic square edge 25" x 32"	2.670	L.F.	.712	117	40.50	157.50
25" x 38"	3.160	L.F.	.843	139	48	187
Cultured marble, 25" x 32", with bowl	1.000	Ea.	2.500	209	145	354
25" x 38", with bowl	1.000	Ea.	2.500	243	145	388
Carrier, for lavatory, steel for studs, no arms	1.000	Ea.	1.143	105	73.50	178.50
Wood, 2" x 8" blocking	1.300	L.F.	.053	1.24	3.02	4.26

For customer support on your Residential Cost Data, call 877.759.4771.

8 | MECHANICAL — 28 | Three Fixture Bathroom Systems

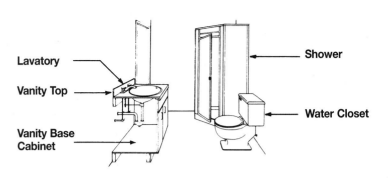

System Description	QUAN.	UNIT	LABOR HOURS	COST EACH MAT.	COST EACH INST.	COST EACH TOTAL
BATHROOM WITH SHOWER, LAVATORY INSTALLED IN VANITY						
Water closet, floor mounted, 2 piece, close coupled, white	1.000	Ea.	3.019	253	175	428
Rough-in, vent, 2" diameter DWV piping	1.000	Ea.	.955	53	55.40	108.40
Waste, 4" diameter DWV piping	1.000	Ea.	.828	64.50	48	112.50
Supply, 1/2" diameter type "L" copper supply piping	1.000	Ea.	.593	24.30	38.10	62.40
Lavatory, 20" x 18" P.E. cast iron with fittings, white	1.000	Ea.	2.500	345	145	490
Rough-in, waste, 1-1/2" diameter DWV piping	1.000	Ea.	1.803	101.20	104.40	205.60
Supply, 1/2" diameter type "L" copper supply piping	1.000	Ea.	.988	40.50	63.50	104
Shower, steel enameled, stone base, corner, white	1.000	Ea.	3.200	1,275	186	1,461
Shower mixing valve	1.000	Ea.	1.333	145	86	231
Shower door	1.000	Ea.	1.000	455	63	518
Rough-in, vent, 1-1/2" diameter DWV piping	1.000	Ea.	.225	12.65	13.05	25.70
Waste, 2" diameter DWV piping	1.000	Ea.	1.433	79.50	83.10	162.60
Supply, 1/2" diameter type "L" copper supply piping	1.000	Ea.	1.580	64.80	101.60	166.40
Piping, supply, 1/2" diameter type "L" copper supply piping	36.000	L.F.	4.148	170.10	266.70	436.80
Waste, 4" diameter DWV piping	7.000	L.F.	2.759	215	160	375
Vent, 2" diameter DWV piping	6.000	L.F.	2.250	83.70	130.50	214.20
Vanity base 2 door, 30" wide	1.000	Ea.	1.000	415	57	472
Vanity top, plastic laminated, square edge	2.170	L.F.	.712	86.78	40.45	127.23
TOTAL		Ea.	30.326	3,884.03	1,816.80	5,700.83
BATHROOM WITH SHOWER, WALL HUNG LAVATORY						
Water closet, floor mounted, close coupled	1.000	Ea.	3.019	253	175	428
Rough-in, vent, 2" diameter DWV piping	1.000	Ea.	.955	53	55.40	108.40
Waste, 4" diameter DWV piping	1.000	Ea.	.828	64.50	48	112.50
Supply, 1/2" diameter type "L" copper supply piping	1.000	Ea.	.593	24.30	38.10	62.40
Lavatory, 20" x 18" P.E. cast iron with fittings, white	1.000	Ea.	2.000	291	116	407
Rough-in, waste, 1-1/2" diameter DWV piping	1.000	Ea.	1.803	101.20	104.40	205.60
Supply, 1/2" diameter type "L" copper supply piping	1.000	Ea.	.988	40.50	63.50	104
Shower, steel enameled, stone base, white	1.000	Ea.	3.200	1,275	186	1,461
Mixing valve	1.000	Ea.	1.333	145	86	231
Shower door	1.000	Ea.	1.000	455	63	518
Rough-in, vent, 1-1/2" diameter DWV piping	1.000	Ea.	.225	12.65	13.05	25.70
Waste, 2" diameter DWV piping	1.000	Ea.	1.433	79.50	83.10	162.60
Supply, 1/2" diameter type "L" copper supply piping	1.000	Ea.	1.580	64.80	101.60	166.40
Piping, supply, 1/2" diameter type "L" copper supply piping	36.000	L.F.	4.148	170.10	266.70	436.80
Waste, 4" diameter DWV piping	7.000	L.F.	2.759	215	160	375
Vent, 2" diameter DWV piping	6.000	L.F.	2.250	83.70	130.50	214.20
Carrier, steel, for studs, no arms	1.000	Ea.	1.143	105	73.50	178.50
TOTAL		Ea.	29.257	3,433.25	1,763.85	5,197.10

The costs in this system are on a cost each basis. All necessary piping is included.

Three Fixture Bathroom Price Sheet

Description	QUAN.	UNIT	LABOR HOURS	COST EACH MAT.	COST EACH INST.	COST EACH TOTAL
Water closet, close coupled, standard 2 piece, white	1.000	Ea.	3.019	253	175	428
Color	1.000	Ea.	3.019	460	175	635
One piece elongated bowl, white	1.000	Ea.	3.019	805	175	980
Color	1.000	Ea.	3.019	1,075	175	1,250
Low profile, one piece elongated bowl, white	1.000	Ea.	3.019	820	175	995
Color	1.000	Ea.	3.019	1,075	175	1,250
Rough-in, for water closet						
1/2" copper supply, 4" cast iron waste, 2" cast iron vent	1.000	Ea.	2.376	142	142	284
4" PVC/DWV waste, 2" PVC vent	1.000	Ea.	2.678	83.50	159	242.50
4" copper waste, 2" copper vent	1.000	Ea.	2.520	229	154	383
3" cast iron waste, 1-1/2" cast iron vent	1.000	Ea.	2.244	126	134	260
3" PVC waste, 1-1/2" PVC vent	1.000	Ea.	2.388	74.50	148	222.50
3" copper waste, 1-1/2" copper vent	1.000	Ea.	2.014	161	124	285
1/2" PVC supply, 4" PVC waste, 2" PVC vent	1.000	Ea.	2.974	105	178	283
3" PVC waste, 1-1/2" PVC supply	1.000	Ea.	2.684	96	167	263
1/2" steel supply, 4" cast iron waste, 2" cast iron vent	1.000	Ea.	2.545	138	153	291
4" cast iron waste, 2" steel vent	1.000	Ea.	2.590	123	155	278
4" PVC waste, 2" PVC vent	1.000	Ea.	2.847	80	170	250
Lavatory, wall hung, P.E. cast iron 20" x 18", white	1.000	Ea.	2.000	291	116	407
Color	1.000	Ea.	2.000	305	116	421
Vitreous china 19" x 17", white	1.000	Ea.	2.286	162	133	295
Color	1.000	Ea.	2.286	184	133	317
Lavatory, for vanity top, P.E. cast iron 20" x 18", white	1.000	Ea.	2.500	345	145	490
Color	1.000	Ea.	2.500	505	145	650
Steel enameled 20" x 17", white	1.000	Ea.	2.759	165	160	325
Color	1.000	Ea.	2.500	169	145	314
Vitreous china 20" x 16", white	1.000	Ea.	2.963	260	172	432
Color	1.000	Ea.	2.963	260	172	432
Rough-in, for lavatory						
1/2" copper supply, 1-1/2" cast iron waste, 1-1/2" cast iron vent	1.000	Ea.	2.791	142	168	310
1-1/2" PVC waste, 1-1/2" PVC vent	1.000	Ea.	2.639	88	170	258
1/2" steel supply, 1-1/4" cast iron waste, 1-1/4" steel vent	1.000	Ea.	2.890	111	176	287
1-1/4" PVC waste, 1-1/4" PVC vent	1.000	Ea.	2.921	82.50	188	270.50
1/2" PVC supply, 1-1/2" PVC waste, 1-1/2" PVC vent	1.000	Ea.	3.260	123	210	333
Shower, steel enameled stone base, 32" x 32", white	1.000	Ea.	8.000	1,275	186	1,461
Color	1.000	Ea.	7.822	1,950	170	2,120
36" x 36" white	1.000	Ea.	8.889	1,625	193	1,818
Color	1.000	Ea.	8.889	2,225	193	2,418
Rough-in, for shower						
1/2" copper supply, 4" cast iron waste, 1-1/2" copper vent	1.000	Ea.	3.238	157	198	355
4" PVC waste, 1-1/2" PVC vent	1.000	Ea.	3.429	108	210	318
1/2" steel supply, 4" cast iron waste, 1-1/2" steel vent	1.000	Ea.	3.665	143	226	369
4" PVC waste, 1-1/2" PVC vent	1.000	Ea.	3.881	99.50	240	339.50
1/2" PVC supply, 4" PVC waste, 1-1/2" PVC vent	1.000	Ea.	4.219	165	261	426
Piping, supply, 1/2" copper	36.000	L.F.	4.148	170	267	437
1/2" steel	36.000	L.F.	5.333	147	345	492
1/2" PVC	36.000	L.F.	6.222	320	400	720
Piping, waste, 4" cast iron no hub	7.000	L.F.	2.759	215	160	375
4" PVC/DWV	7.000	L.F.	3.333	113	194	307
4" copper/DWV	7.000	L.F.	4.000	465	230	695
Piping, vent, 2" cast iron no hub	6.000	L.F.	2.149	119	125	244
2" copper/DWV	6.000	L.F.	1.636	146	105	251
2" PVC/DWV	6.000	L.F.	2.441	56.50	141	197.50
2" steel, galvanized	6.000	L.F.	2.250	83.50	131	214.50
Vanity base cabinet, 2 door, 24" x 30"	1.000	Ea.	1.000	415	57	472
24" x 36"	1.000	Ea.	1.200	400	68	468
Vanity top, laminated plastic square edge, 25" x 32"	2.170	L.F.	.712	87	40.50	127.50
25" x 38"	2.670	L.F.	.845	103	48	151
Carrier, for lavatory, steel for studs, no arms	1.000	Ea.	1.143	105	73.50	178.50
Wood, 2" x 8" blocking	1.300	L.F.	.052	1.21	2.95	4.16

For customer support on your Residential Cost Data, call 877.759.4771.

8 | MECHANICAL 32 | Three Fixture Bathroom Systems

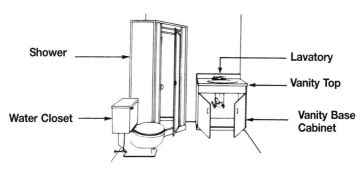

System Description	QUAN.	UNIT	LABOR HOURS	COST EACH MAT.	COST EACH INST.	COST EACH TOTAL
BATHROOM WITH LAVATORY INSTALLED IN VANITY						
Water closet, floor mounted, 2 piece, close coupled, white	1.000	Ea.	3.019	253	175	428
Rough-in, vent, 2" diameter DWV piping	1.000	Ea.	.955	53	55.40	108.40
Waste, 4" diameter DWV piping	1.000	Ea.	.828	64.50	48	112.50
Supply, 1/2" diameter type "L" copper supply piping	1.000	Ea.	.593	24.30	38.10	62.40
Lavatory, 20" x 18", P.E. cast iron with fittings, white	1.000	Ea.	2.500	345	145	490
Rough-in, waste, 1-1/2" diameter DWV piping	1.000	Ea.	1.803	101.20	104.40	205.60
Supply, 1/2" diameter type "L" copper supply piping	1.000	Ea.	.988	40.50	63.50	104
Shower, steel enameled, stone base, corner, white	1.000	Ea.	3.200	1,275	186	1,461
Mixing valve	1.000	Ea.	1.333	145	86	231
Shower door	1.000	Ea.	1.000	455	63	518
Rough-in, vent, 1-1/2" diameter DWV piping	1.000	Ea.	.225	12.65	13.05	25.70
Waste, 2" diameter DWV piping	1.000	Ea.	1.433	79.50	83.10	162.60
Supply, 1/2" diameter type "L" copper supply piping	1.000	Ea.	1.580	64.80	101.60	166.40
Piping, supply, 1/2" diameter type "L" copper supply piping	36.000	L.F.	3.556	145.80	228.60	374.40
Waste, 4" diameter DWV piping	7.000	L.F.	1.931	150.50	112	262.50
Vent, 2" diameter DWV piping	6.000	L.F.	1.500	55.80	87	142.80
Vanity base, 2 door, 30" wide	1.000	Ea.	1.000	415	57	472
Vanity top, plastic laminated, square edge	2.670	L.F.	.712	98.79	40.45	139.24
TOTAL		Ea.	28.156	3,779.34	1,687.20	5,466.54
BATHROOM, WITH WALL HUNG LAVATORY						
Water closet, floor mounted, 2 piece, close coupled, white	1.000	Ea.	3.019	253	175	428
Rough-in, vent, 2" diameter DWV piping	1.000	Ea.	.955	53	55.40	108.40
Waste, 4" diameter DWV piping	1.000	Ea.	.828	64.50	48	112.50
Supply, 1/2" diameter type "L" copper supply piping	1.000	Ea.	.593	24.30	38.10	62.40
Lavatory, wall hung, 20" x 18" P.E. cast iron with fittings, white	1.000	Ea.	2.000	291	116	407
Rough-in, waste, 1-1/2" diameter DWV piping	1.000	Ea.	1.803	101.20	104.40	205.60
Supply, 1/2" diameter type "L" copper supply piping	1.000	Ea.	.988	40.50	63.50	104
Shower, steel enameled, stone base, corner, white	1.000	Ea.	3.200	1,275	186	1,461
Mixing valve	1.000	Ea.	1.333	145	86	231
Shower door	1.000	Ea.	1.000	455	63	518
Rough-in, waste, 1-1/2" diameter DWV piping	1.000	Ea.	.225	12.65	13.05	25.70
Waste, 2" diameter DWV piping	1.000	Ea.	1.433	79.50	83.10	162.60
Supply, 1/2" diameter type "L" copper supply piping	1.000	Ea.	1.580	64.80	101.60	166.40
Piping, supply, 1/2" diameter type "L" copper supply piping	36.000	L.F.	3.556	145.80	228.60	374.40
Waste, 4" diameter DWV piping	7.000	L.F.	1.931	150.50	112	262.50
Vent, 2" diameter DWV piping	6.000	L.F.	1.500	55.80	87	142.80
Carrier, steel, for studs, no arms	1.000	Ea.	1.143	105	73.50	178.50
TOTAL		Ea.	27.087	3,316.55	1,634.25	4,950.80

The costs in this system are on a cost each basis. All necessary piping is included.

Three Fixture Bathroom Price Sheet	QUAN.	UNIT	LABOR HOURS	COST EACH MAT.	COST EACH INST.	COST EACH TOTAL
Water closet, close coupled, standard 2 piece, white	1.000	Ea.	3.019	253	175	428
Color	1.000	Ea.	3.019	460	175	635
One piece elongated bowl, white	1.000	Ea.	3.019	805	175	980
Color	1.000	Ea.	3.019	1,075	175	1,250
Low profile one piece elongated bowl, white	1.000	Ea.	3.019	820	175	995
Color	1.000	Ea.	3.623	1,075	175	1,250
Rough-in, for water closet						
1/2" copper supply, 4" cast iron waste, 2" cast iron vent	1.000	Ea.	2.376	142	142	284
4" P.V.C./DWV waste, 2" PVC vent	1.000	Ea.	2.678	83.50	159	242.50
4" copper waste, 2" copper vent	1.000	Ea.	2.520	229	154	383
3" cast iron waste, 1-1/2" cast iron vent	1.000	Ea.	2.244	126	134	260
3" PVC waste, 1-1/2" PVC vent	1.000	Ea.	2.388	74.50	148	222.50
3" copper waste, 1-1/2" copper vent	1.000	Ea.	2.014	161	124	285
1/2" P.V.C. supply, 4" P.V.C. waste, 2" P.V.C. vent	1.000	Ea.	2.974	105	178	283
3" P.V.C. waste, 1-1/2" P.V.C. vent	1.000	Ea.	2.684	96	167	263
1/2" steel supply, 4" cast iron waste, 2" cast iron vent	1.000	Ea.	2.545	138	153	291
4" cast iron waste, 2" steel vent	1.000	Ea.	2.590	123	155	278
4" P.V.C. waste, 2" P.V.C. vent	1.000	Ea.	2.847	80	170	250
Lavatory, wall hung P.E. cast iron 20" x 18", white	1.000	Ea.	2.000	291	116	407
Color	1.000	Ea.	2.000	305	116	421
Vitreous china 19" x 17", white	1.000	Ea.	2.286	162	133	295
Color	1.000	Ea.	2.286	184	133	317
Lavatory, for vanity top P.E. cast iron 20" x 18", white	1.000	Ea.	2.500	345	145	490
Color	1.000	Ea.	2.500	505	145	650
Steel enameled 20" x 17", white	1.000	Ea.	2.759	165	160	325
Color	1.000	Ea.	2.500	169	145	314
Vitreous china 20" x 16", white	1.000	Ea.	2.963	260	172	432
Color	1.000	Ea.	2.963	260	172	432
Rough-in, for lavatory						
1/2" copper supply, 1-1/2" cast iron waste, 1-1/2" cast iron vent	1.000	Ea.	2.791	142	168	310
1-1/2" P.V.C. waste, 1-1/2" P.V.C. vent	1.000	Ea.	2.639	88	170	258
1/2" steel supply, 1-1/2" cast iron waste, 1-1/4" steel vent	1.000	Ea.	2.890	111	176	287
1-1/2" P.V.C. waste, 1-1/4" P.V.C. vent	1.000	Ea.	2.921	82.50	188	270.50
1/2" P.V.C. supply, 1-1/2" P.V.C. waste, 1-1/2" P.V.C. vent	1.000	Ea.	3.260	123	210	333
Shower, steel enameled stone base, 32" x 32", white	1.000	Ea.	8.000	1,275	186	1,461
Color	1.000	Ea.	7.822	1,950	170	2,120
36" x 36", white	1.000	Ea.	8.889	1,625	193	1,818
Color	1.000	Ea.	8.889	2,225	193	2,418
Rough-in, for shower						
1/2" copper supply, 2" cast iron waste, 1-1/2" copper vent	1.000	Ea.	3.161	157	194	351
2" P.V.C. waste, 1-1/2" P.V.C. vent	1.000	Ea.	3.429	108	210	318
1/2" steel supply, 2" cast iron waste, 1-1/2" steel vent	1.000	Ea.	3.887	192	239	431
2" P.V.C. waste, 1-1/2" P.V.C. vent	1.000	Ea.	3.881	99.50	240	339.50
1/2" P.V.C. supply, 2" P.V.C. waste, 1-1/2" P.V.C. vent	1.000	Ea.	4.219	165	261	426
Piping, supply, 1/2" copper	36.000	L.F.	3.556	146	229	375
1/2" steel	36.000	L.F.	4.571	126	295	421
1/2" P.V.C.	36.000	L.F.	5.333	274	345	619
Waste, 4" cast iron, no hub	7.000	L.F.	1.931	151	112	263
4" P.V.C./DWV	7.000	L.F.	2.333	79	135	214
4" copper/DWV	7.000	L.F.	2.800	325	161	486
Vent, 2" cast iron, no hub	6.000	L.F.	1.091	97.50	70	167.50
2" copper/DWV	6.000	L.F.	1.091	97.50	70	167.50
2" P.V.C./DWV	6.000	L.F.	1.627	38	94	132
2" steel, galvanized	6.000	L.F.	1.500	56	87	143
Vanity base cabinet, 2 door, 24" x 30"	1.000	Ea.	1.000	415	57	472
24" x 36"	1.000	Ea.	1.200	400	68	468
Vanity top, laminated plastic square edge, 25" x 32"	2.670	L.F.	.712	99	40.50	139.50
25" x 38"	3.170	L.F.	.845	117	48	165
Carrier, for lavatory, steel, for studs, no arms	1.000	Ea.	1.143	105	73.50	178.50
Wood, 2" x 8" blocking	1.300	L.F.	.052	1.21	2.95	4.16

For customer support on your Residential Cost Data, call 877.759.4771.

8 | MECHANICAL — 36 | Four Fixture Bathroom Systems

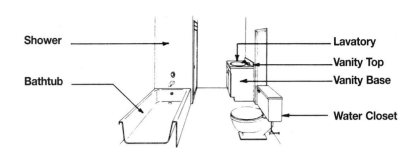

System Description	QUAN.	UNIT	LABOR HOURS	COST EACH MAT.	COST EACH INST.	COST EACH TOTAL
BATHROOM WITH LAVATORY INSTALLED IN VANITY						
Water closet, floor mounted, 2 piece, close coupled, white	1.000	Ea.	3.019	253	175	428
Rough-in, vent, 2" diameter DWV piping	1.000	Ea.	.955	53	55.40	108.40
Waste, 4" diameter DWV piping	1.000	Ea.	.828	64.50	48	112.50
Supply, 1/2" diameter type "L" copper supply piping	1.000	Ea.	.593	24.30	38.10	62.40
Lavatory, 20" x 18" P.E. cast iron with fittings, white	1.000	Ea.	2.500	345	145	490
Shower, steel, enameled, stone base, corner, white	1.000	Ea.	3.333	1,900	193	2,093
Mixing valve	1.000	Ea.	1.333	145	86	231
Shower door	1.000	Ea.	1.000	455	63	518
Rough-in, waste, 1-1/2" diameter DWV piping	2.000	Ea.	4.507	253	261	514
Supply, 1/2" diameter type "L" copper supply piping	2.000	Ea.	3.161	129.60	203.20	332.80
Bathtub, P.E. cast iron, 5' long with fittings, white	1.000	Ea.	3.636	1,250	211	1,461
Rough-in, waste, 4" diameter DWV piping	1.000	Ea.	.828	64.50	48	112.50
Supply, 1/2" diameter type "L" copper supply piping	1.000	Ea.	.988	40.50	63.50	104
Vent, 1-1/2" diameter DWV piping	1.000	Ea.	.593	49.60	38.20	87.80
Piping, supply, 1/2" diameter type "L" copper supply piping	42.000	L.F.	4.148	170.10	266.70	436.80
Waste, 4" diameter DWV piping	10.000	L.F.	2.759	215	160	375
Vent, 2" diameter DWV piping	13.000	L.F.	3.250	120.90	188.50	309.40
Vanity base, 2 doors, 30" wide	1.000	Ea.	1.000	415	57	472
Vanity top, plastic laminated, square edge	2.670	L.F.	.712	98.79	40.45	139.24
TOTAL		Ea.	39.143	6,046.79	2,341.05	8,387.84
BATHROOM WITH WALL HUNG LAVATORY						
Water closet, floor mounted, 2 piece, close coupled, white	1.000	Ea.	3.019	253	175	428
Rough-in, vent, 2" diameter DWV piping	1.000	Ea.	.955	53	55.40	108.40
Waste, 4" diameter DWV piping	1.000	Ea.	.828	64.50	48	112.50
Supply, 1/2" diameter type "L" copper supply piping	1.000	Ea.	.593	24.30	38.10	62.40
Lavatory, 20" x 18" P.E. cast iron with fittings, white	1.000	Ea.	2.000	291	116	407
Shower, steel enameled, stone base, corner, white	1.000	Ea.	3.333	1,900	193	2,093
Mixing valve	1.000	Ea.	1.333	145	86	231
Shower door	1.000	Ea.	1.000	455	63	518
Rough-in, waste, 1-1/2" diameter DWV piping	2.000	Ea.	4.507	253	261	514
Supply, 1/2" diameter type "L" copper supply piping	2.000	Ea.	3.161	129.60	203.20	332.80
Bathtub, P.E. cast iron, 5' long with fittings, white	1.000	Ea.	3.636	1,250	211	1,461
Rough-in, waste, 4" diameter DWV piping	1.000	Ea.	.828	64.50	48	112.50
Supply, 1/2" diameter type "L" copper supply piping	1.000	Ea.	.988	40.50	63.50	104
Vent, 1-1/2" diameter copper DWV piping	1.000	Ea.	.593	49.60	38.20	87.80
Piping, supply, 1/2" diameter type "L" copper supply piping	42.000	L.F.	4.148	170.10	266.70	436.80
Waste, 4" diameter DWV piping	10.000	L.F.	2.759	215	160	375
Vent, 2" diameter DWV piping	13.000	L.F.	3.250	120.90	188.50	309.40
Carrier, steel, for studs, no arms	1.000	Ea.	1.143	105	73.50	178.50
TOTAL		Ea.	38.074	5,584	2,288.10	7,872.10

The costs in this system are on a cost each basis. All necessary piping is included.

Four Fixture Bathroom Price Sheet	QUAN.	UNIT	LABOR HOURS	COST EACH MAT.	COST EACH INST.	COST EACH TOTAL
Water closet, close coupled, standard 2 piece, white	1.000	Ea.	3.019	253	175	428
Color	1.000	Ea.	3.019	460	175	635
One piece elongated bowl, white	1.000	Ea.	3.019	805	175	980
Color	1.000	Ea.	3.019	1,075	175	1,250
Low profile, one piece elongated bowl, white	1.000	Ea.	3.019	820	175	995
Color	1.000	Ea.	3.019	1,075	175	1,250
1/2" copper supply, 4" cast iron waste, 2" cast iron vent	1.000	Ea.	2.376	142	142	284
4" PVC/DWV waste, 2" PVC vent	1.000	Ea.	2.678	83.50	159	242.50
4" copper waste, 2" copper vent	1.000	Ea.	2.520	229	154	383
3" cast iron waste, 1-1/2" cast iron vent	1.000	Ea.	2.244	126	134	260
3" P.V.C. waste, 1-1/2" P.V.C. vent	1.000	Ea.	2.388	74.50	148	222.50
3" copper waste, 1-1/2" copper vent	1.000	Ea.	2.014	161	124	285
1/2" P.V.C. supply, 4" P.V.C. waste, 2" P.V.C. vent	1.000	Ea.	2.974	105	178	283
3" P.V.C. waste, 1-1/2" P.V.C. vent	1.000	Ea.	2.684	96	167	263
1/2" steel supply, 4" cast iron waste, 2" cast iron vent	1.000	Ea.	2.545	138	153	291
4" cast iron waste, 2" steel vent	1.000	Ea.	2.590	123	155	278
4" P.V.C. waste, 2" P.V.C. vent	1.000	Ea.	2.847	80	170	250
Lavatory, wall hung P.E. cast iron 20" x 18", white	1.000	Ea.	2.000	291	116	407
Color	1.000	Ea.	2.000	305	116	421
Vitreous china 19" x 17", white	1.000	Ea.	2.286	162	133	295
Color	1.000	Ea.	2.286	184	133	317
Lavatory for vanity top, P.E. cast iron 20" x 18", white	1.000	Ea.	2.500	345	145	490
Color	1.000	Ea.	2.500	505	145	650
Steel enameled, 20" x 17", white	1.000	Ea.	2.759	165	160	325
Color	1.000	Ea.	2.500	169	145	314
Vitreous china 20" x 16", white	1.000	Ea.	2.963	260	172	432
Color	1.000	Ea.	2.963	260	172	432
Shower, steel enameled stone base, 36" square, white	1.000	Ea.	8.889	1,900	193	2,093
Color	1.000	Ea.	8.889	1,950	193	2,143
Rough-in, for lavatory or shower						
1/2" copper supply, 1-1/2" cast iron waste, 1-1/2" cast iron vent	1.000	Ea.	3.834	191	232	423
1-1/2" P.V.C. waste, 1-1/4" P.V.C. vent	1.000	Ea.	3.675	124	236	360
1/2" steel supply, 1-1/4" cast iron waste, 1-1/4" steel vent	1.000	Ea.	4.103	157	251	408
1-1/4" P.V.C. waste, 1-1/4" P.V.C. vent	1.000	Ea.	3.937	116	254	370
1/2" P.V.C. supply, 1-1/2" P.V.C. waste, 1-1/2" P.V.C. vent	1.000	Ea.	4.592	180	296	476
Bathtub, P.E. cast iron, 5' long with fittings, white	1.000	Ea.	3.636	1,250	211	1,461
Color	1.000	Ea.	3.636	1,550	211	1,761
Steel, enameled 5' long with fittings, white	1.000	Ea.	2.909	550	169	719
Color	1.000	Ea.	2.909	550	169	719
Rough-in, for bathtub						
1/2" copper supply, 4" cast iron waste, 1-1/2" copper vent	1.000	Ea.	2.409	155	150	305
4" P.V.C. waste, 1-1/2" P.V.C. vent	1.000	Ea.	2.877	97.50	179	276.50
1/2" steel supply, 4" cast iron waste, 1-1/2" steel vent	1.000	Ea.	2.898	128	176	304
4" P.V.C. waste, 1-1/2" P.V.C. vent	1.000	Ea.	3.159	92	197	289
1/2" P.V.C. supply, 4" P.V.C. waste, 1-1/2" P.V.C. vent	1.000	Ea.	3.371	133	211	344
Piping, supply, 1/2" copper	42.000	L.F.	4.148	170	267	437
1/2" steel	42.000	L.F.	5.333	147	345	492
1/2" P.V.C.	42.000	L.F.	6.222	320	400	720
Waste, 4" cast iron, no hub	10.000	L.F.	2.759	215	160	375
4" P.V.C./DWV	10.000	L.F.	3.333	113	194	307
4" copper/DWV	10.000	Ea.	4.000	465	230	695
Vent 2" cast iron, no hub	13.000	L.F.	3.105	172	180	352
2" copper/DWV	13.000	L.F.	2.364	211	152	363
2" P.V.C./DWV	13.000	L.F.	3.525	82	204	286
2" steel, galvanized	13.000	L.F.	3.250	121	189	310
Vanity base cabinet, 2 doors, 30" wide	1.000	Ea.	1.000	415	57	472
Vanity top, plastic laminated, square edge	2.670	L.F.	.712	99	40.50	139.50
Carrier, steel for studs, no arms	1.000	Ea.	1.143	105	73.50	178.50
Wood, 2" x 8" blocking	1.300	L.F.	.052	1.21	2.95	4.16

8 | MECHANICAL — 40 | Four Fixture Bathroom Systems

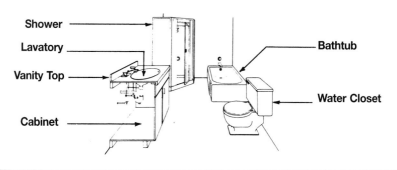

System Description	QUAN.	UNIT	LABOR HOURS	MAT.	INST.	TOTAL
BATHROOM WITH LAVATORY INSTALLED IN VANITY						
Water closet, floor mounted, 2 piece, close coupled, white	1.000	Ea.	3.019	253	175	428
Rough-in, vent, 2" diameter DWV piping	1.000	Ea.	.955	53	55.40	108.40
Waste, 4" diameter DWV piping	1.000	Ea.	.828	64.50	48	112.50
Supply, 1/2" diameter type "L" copper supply piping	1.000	Ea.	.593	24.30	38.10	62.40
Lavatory, 20" x 18" P.E. cast iron with fittings, white	1.000	Ea.	2.500	345	145	490
Shower, steel, enameled, stone base, corner, white	1.000	Ea.	3.333	1,900	193	2,093
Mixing valve	1.000	Ea.	1.333	145	86	231
Shower door	1.000	Ea.	1.000	455	63	518
Rough-in, waste, 1-1/2" diameter DWV piping	2.000	Ea.	4.507	253	261	514
Supply, 1/2" diameter type "L" copper supply piping	2.000	Ea.	3.161	129.60	203.20	332.80
Bathtub, P.E. cast iron, 5' long with fittings, white	1.000	Ea.	3.636	1,250	211	1,461
Rough-in, waste, 4" diameter DWV piping	1.000	Ea.	.828	64.50	48	112.50
Supply, 1/2" diameter type "L" copper supply piping	1.000	Ea.	.988	40.50	63.50	104
Vent, 1-1/2" diameter DWV piping	1.000	Ea.	.593	49.60	38.20	87.80
Piping, supply, 1/2" diameter type "L" copper supply piping	42.000	L.F.	4.939	202.50	317.50	520
Waste, 4" diameter DWV piping	10.000	L.F.	4.138	322.50	240	562.50
Vent, 2" diameter DWV piping	13.000	L.F.	4.500	167.40	261	428.40
Vanity base, 2 doors, 30" wide	1.000	Ea.	1.000	415	57	472
Vanity top, plastic laminated, square edge	2.670	L.F.	.712	86.78	40.45	127.23
TOTAL		Ea.	42.563	6,221.18	2,544.35	8,765.53
BATHROOM WITH WALL HUNG LAVATORY						
Water closet, floor mounted, 2 piece, close coupled, white	1.000	Ea.	3.019	253	175	428
Rough-in, vent, 2" diameter DWV piping	1.000	Ea.	.955	53	55.40	108.40
Waste, 4" diameter DWV piping	1.000	Ea.	.828	64.50	48	112.50
Supply, 1/2" diameter type "L" copper supply piping	1.000	Ea.	.593	24.30	38.10	62.40
Lavatory, 20" x 18" P.E. cast iron with fittings, white	1.000	Ea.	2.000	291	116	407
Shower, steel enameled, stone base, corner, white	1.000	Ea.	3.333	1,900	193	2,093
Mixing valve	1.000	Ea.	1.333	145	86	231
Shower door	1.000	Ea.	1.000	455	63	518
Rough-in, waste, 1-1/2" diameter DWV piping	2.000	Ea.	4.507	253	261	514
Supply, 1/2" diameter type "L" copper supply piping	2.000	Ea.	3.161	129.60	203.20	332.80
Bathtub, P.E. cast iron, 5" long with fittings, white	1.000	Ea.	3.636	1,250	211	1,461
Rough-in, waste, 4" diameter DWV piping	1.000	Ea.	.828	64.50	48	112.50
Supply, 1/2" diameter type "L" copper supply piping	1.000	Ea.	.988	40.50	63.50	104
Vent, 1-1/2" diameter DWV piping	1.000	Ea.	.593	49.60	38.20	87.80
Piping, supply, 1/2" diameter type "L" copper supply piping	42.000	L.F.	4.939	202.50	317.50	520
Waste, 4" diameter DWV piping	10.000	L.F.	4.138	322.50	240	562.50
Vent, 2" diameter DWV piping	13.000	L.F.	4.500	167.40	261	428.40
Carrier, steel for studs, no arms	1.000	Ea.	1.143	105	73.50	178.50
TOTAL		Ea.	41.494	5,770.40	2,491.40	8,261.80

The costs in this system are on a cost each basis. All necessary piping is included.

Four Fixture Bathroom Price Sheet

	QUAN.	UNIT	LABOR HOURS	COST EACH MAT.	COST EACH INST.	COST EACH TOTAL
Water closet, close coupled, standard 2 piece, white	1.000	Ea.	3.019	253	175	428
Color	1.000	Ea.	3.019	460	175	635
One piece, elongated bowl, white	1.000	Ea.	3.019	805	175	980
Color	1.000	Ea.	3.019	1,075	175	1,250
Low profile, one piece elongated bowl, white	1.000	Ea.	3.019	820	175	995
Color	1.000	Ea.	3.019	1,075	175	1,250
Rough-in, for water closet						
1/2" copper supply, 4" cast iron waste, 2" cast iron vent	1.000	Ea.	2.376	142	142	284
4" PVC/DWV waste, 2" PVC vent	1.000	Ea.	2.678	83.50	159	242.50
4" copper waste, 2" copper vent	1.000	Ea.	2.520	229	154	383
3" cast iron waste, 1-1/2" cast iron vent	1.000	Ea.	2.244	126	134	260
3" PVC waste, 1-1/2" PVC vent	1.000	Ea.	2.388	74.50	148	222.50
3" PVC waste, 1-1/2" PVC vent	1.000	Ea.	2.014	161	124	285
1/2" PVC supply, 4" PVC waste, 2" PVC vent	1.000	Ea.	2.974	105	178	283
3" PVC waste, 1-1/2" PVC vent	1.000	Ea.	2.684	96	167	263
1/2" steel supply, 4" cast iron waste, 2" cast iron vent	1.000	Ea.	2.545	138	153	291
4" cast iron waste, 2" steel vent	1.000	Ea.	2.590	123	155	278
4" PVC waste, 2" PVC vent	1.000	Ea.	2.847	80	170	250
Lavatory wall hung, P.E. cast iron 20" x 18", white	1.000	Ea.	2.000	291	116	407
Color	1.000	Ea.	2.000	305	116	421
Vitreous china 19" x 17", white	1.000	Ea.	2.286	162	133	295
Color	1.000	Ea.	2.286	184	133	317
Lavatory for vanity top, P.E. cast iron, 20" x 18", white	1.000	Ea.	2.500	345	145	490
Color	1.000	Ea.	2.500	505	145	650
Steel, enameled 20" x 17", white	1.000	Ea.	2.759	165	160	325
Color	1.000	Ea.	2.500	169	145	314
Vitreous china 20" x 16", white	1.000	Ea.	2.963	260	172	432
Color	1.000	Ea.	2.963	260	172	432
Shower, steel enameled, stone base 36" square, white	1.000	Ea.	8.889	1,900	193	2,093
Color	1.000	Ea.	8.889	1,950	193	2,143
Rough-in, for lavatory and shower						
1/2" copper supply, 1-1/2" cast iron waste, 1-1/2" cast iron vent	1.000	Ea.	7.668	385	465	850
1-1/2" PVC waste, 1-1/4" PVC vent	1.000	Ea.	7.352	248	475	723
1/2" steel supply, 1-1/4" cast iron waste, 1-1/4" steel vent	1.000	Ea.	8.205	315	500	815
1-1/4" PVC waste, 1-1/4" PVC vent	1.000	Ea.	7.873	233	505	738
1/2" PVC supply, 1-1/2" PVC waste, 1-1/2" PVC vent	1.000	Ea.	9.185	360	590	950
Bathtub, P.E. cast iron, 5' long with fittings, white	1.000	Ea.	3.636	1,250	211	1,461
Color	1.000	Ea.	3.636	1,550	211	1,761
Steel enameled, 5' long with fittings, white	1.000	Ea.	2.909	550	169	719
Color	1.000	Ea.	2.909	550	169	719
Rough-in, for bathtub						
1/2" copper supply, 4" cast iron waste, 1-1/2" copper vent	1.000	Ea.	2.409	155	150	305
4" PVC waste, 1-1/2" PVC vent	1.000	Ea.	2.877	97.50	179	276.50
1/2" steel supply, 4" cast iron waste, 1-1/2" steel vent	1.000	Ea.	2.898	128	176	304
4" PVC waste, 1-1/2" PVC vent	1.000	Ea.	3.159	92	197	289
1/2" PVC supply, 4" PVC waste, 1-1/2" PVC vent	1.000	Ea.	3.371	133	211	344
Piping supply, 1/2" copper	42.000	L.F.	4.148	170	267	437
1/2" steel	42.000	L.F.	5.333	147	345	492
1/2" PVC	42.000	L.F.	6.222	320	400	720
Piping, waste, 4" cast iron, no hub	10.000	L.F.	3.586	280	208	488
4" PVC/DWV	10.000	L.F.	4.333	146	252	398
4" copper/DWV	10.000	L.F.	5.200	605	299	904
Piping, vent, 2" cast iron, no hub	13.000	L.F.	3.105	172	180	352
2" copper/DWV	13.000	L.F.	2.364	211	152	363
2" PVC/DWV	13.000	L.F.	3.525	82	204	286
2" steel, galvanized	13.000	L.F.	3.250	121	189	310
Vanity base cabinet, 2 doors, 30" wide	1.000	Ea.	1.000	415	57	472
Vanity top, plastic laminated, square edge	3.160	L.F.	.843	117	48	165
Carrier, steel, for studs, no arms	1.000	Ea.	1.143	105	73.50	178.50
Wood, 2" x 8" blocking	1.300	L.F.	.052	1.21	2.95	4.16

8 | MECHANICAL   44 | Five Fixture Bathroom Systems

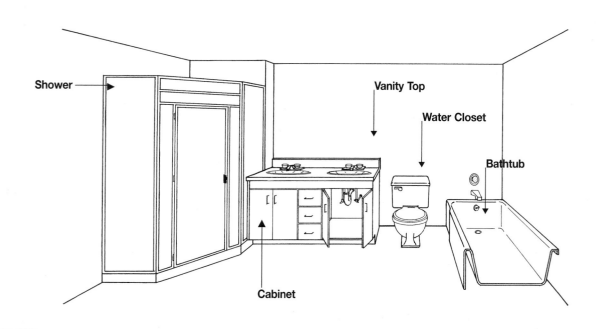

System Description	QUAN.	UNIT	LABOR HOURS	COST EACH MAT.	COST EACH INST.	COST EACH TOTAL
BATHROOM WITH SHOWER, BATHTUB, LAVATORIES IN VANITY						
Water closet, floor mounted, 1 piece, white	1.000	Ea.	3.019	820	175	995
Rough-in, vent, 2" diameter DWV piping	1.000	Ea.	.955	53	55.40	108.40
Waste, 4" diameter DWV piping	1.000	Ea.	.828	64.50	48	112.50
Supply, 1/2" diameter type "L" copper supply piping	1.000	Ea.	.593	24.30	38.10	62.40
Lavatory, 20" x 16", vitreous china oval, with fittings, white	2.000	Ea.	5.926	520	344	864
Shower, steel enameled, stone base, corner, white	1.000	Ea.	3.333	1,900	193	2,093
Mixing valve	1.000	Ea.	1.333	145	86	231
Shower door	1.000	Ea.	1.000	455	63	518
Rough-in, waste, 1-1/2" diameter DWV piping	3.000	Ea.	5.408	303.60	313.20	616.80
Supply, 1/2" diameter type "L" copper supply piping	3.000	Ea.	2.963	121.50	190.50	312
Bathtub, P.E. cast iron, 5' long with fittings, white	1.000	Ea.	3.636	1,250	211	1,461
Rough-in, waste, 4" diameter DWV piping	1.000	Ea.	1.103	86	64	150
Supply, 1/2" diameter type "L" copper supply piping	1.000	Ea.	.988	40.50	63.50	104
Vent, 1-1/2" diameter copper DWV piping	1.000	Ea.	.593	49.60	38.20	87.80
Piping, supply, 1/2" diameter type "L" copper supply piping	42.000	L.F.	4.148	170.10	266.70	436.80
Waste, 4" diameter DWV piping	10.000	L.F.	2.759	215	160	375
Vent, 2" diameter DWV piping	13.000	L.F.	3.250	120.90	188.50	309.40
Vanity base, 2 door, 24" x 48"	1.000	Ea.	1.400	525	79.50	604.50
Vanity top, plastic laminated, square edge	4.170	L.F.	1.112	154.29	63.18	217.47
TOTAL		Ea.	44.347	7,018.29	2,640.78	9,659.07

The costs in this system are on a cost each basis. All necessary piping is included.

Description	QUAN.	UNIT	LABOR HOURS	COST EACH MAT.	COST EACH INST.	COST EACH TOTAL

Five Fixture Bathroom Price Sheet	QUAN.	UNIT	LABOR HOURS	COST EACH MAT.	COST EACH INST.	COST EACH TOTAL
Water closet, close coupled, standard 2 piece, white	1.000	Ea.	3.019	253	175	428
Color	1.000	Ea.	3.019	460	175	635
One piece elongated bowl, white	1.000	Ea.	3.019	805	175	980
Color	1.000	Ea.	3.019	1,075	175	1,250
Low profile, one piece elongated bowl, white	1.000	Ea.	3.019	820	175	995
Color	1.000	Ea.	3.019	1,075	175	1,250
Rough-in, supply, waste and vent for water closet						
1/2" copper supply, 4" cast iron waste, 2" cast iron vent	1.000	Ea.	2.376	142	142	284
4" P.V.C./DWV waste, 2" P.V.C. vent	1.000	Ea.	2.678	83.50	159	242.50
4" copper waste, 2" copper vent	1.000	Ea.	2.520	229	154	383
3" cast iron waste, 1-1/2" cast iron vent	1.000	Ea.	2.244	126	134	260
3" P.V.C. waste, 1-1/2" P.V.C. vent	1.000	Ea.	2.388	74.50	148	222.50
3" copper waste, 1-1/2" copper vent	1.000	Ea.	2.014	161	124	285
1/2" P.V.C. supply, 4" P.V.C. waste, 2" P.V.C. vent	1.000	Ea.	2.974	105	178	283
3" P.V.C. waste, 1-1/2" P.V.C. supply	1.000	Ea.	2.684	96	167	263
1/2" steel supply, 4" cast iron waste, 2" cast iron vent	1.000	Ea.	2.545	138	153	291
4" cast iron waste, 2" steel vent	1.000	Ea.	2.590	123	155	278
4" P.V.C. waste, 2" P.V.C. vent	1.000	Ea.	2.847	80	170	250
Lavatory, wall hung, P.E. cast iron 20" x 18", white	2.000	Ea.	4.000	580	232	812
Color	2.000	Ea.	4.000	610	232	842
Vitreous china, 19" x 17", white	2.000	Ea.	4.571	325	266	591
Color	2.000	Ea.	4.571	370	266	636
Lavatory, for vanity top, P.E. cast iron, 20" x 18", white	2.000	Ea.	5.000	690	290	980
Color	2.000	Ea.	5.000	1,000	290	1,290
Steel enameled 20" x 17", white	2.000	Ea.	5.517	330	320	650
Color	2.000	Ea.	5.000	340	290	630
Vitreous china 20" x 16", white	2.000	Ea.	5.926	520	345	865
Color	2.000	Ea.	5.926	520	345	865
Shower, steel enameled, stone base 36" square, white	1.000	Ea.	8.889	1,900	193	2,093
Color	1.000	Ea.	8.889	1,950	193	2,143
Rough-in, for lavatory or shower						
1/2" copper supply, 1-1/2" cast iron waste, 1-1/2" cast iron vent	3.000	Ea.	8.371	425	505	930
1-1/2" P.V.C. waste, 1-1/4" P.V.C. vent	3.000	Ea.	7.916	264	510	774
1/2" steel supply, 1-1/4" cast iron waste, 1-1/4" steel vent	3.000	Ea.	8.670	335	525	860
1-1/4" P.V.C. waste, 1-1/4" P.V.C. vent	3.000	Ea.	8.381	250	540	790
1/2" P.V.C. supply, 1-1/2" P.V.C. waste, 1-1/2" P.V.C. vent	3.000	Ea.	9.778	370	630	1,000
Bathtub, P.E. cast iron 5' long with fittings, white	1.000	Ea.	3.636	1,250	211	1,461
Color	1.000	Ea.	3.636	1,550	211	1,761
Steel, enameled 5' long with fittings, white	1.000	Ea.	2.909	550	169	719
Color	1.000	Ea.	2.909	550	169	719
Rough-in, for bathtub						
1/2" copper supply, 4" cast iron waste, 1-1/2" copper vent	1.000	Ea.	2.684	176	166	342
4" P.V.C. waste, 1-1/2" P.V.C. vent	1.000	Ea.	3.210	109	198	307
1/2" steel supply, 4" cast iron waste, 1-1/2" steel vent	1.000	Ea.	3.173	150	192	342
4" P.V.C. waste, 1-1/2" P.V.C. vent	1.000	Ea.	3.492	103	217	320
1/2" P.V.C. supply, 4" P.V.C. waste, 1-1/2" P.V.C. vent	1.000	Ea.	3.704	144	230	374
Piping, supply, 1/2" copper	42.000	L.F.	4.148	170	267	437
1/2" steel	42.000	L.F.	5.333	147	345	492
1/2" P.V.C.	42.000	L.F.	6.222	320	400	720
Piping, waste, 4" cast iron, no hub	10.000	L.F.	2.759	215	160	375
4" P.V.C./DWV	10.000	L.F.	3.333	113	194	307
4" copper/DWV	10.000	L.F.	4.000	465	230	695
Piping, vent, 2" cast iron, no hub	13.000	L.F.	3.105	172	180	352
2" copper/DWV	13.000	L.F.	2.364	211	152	363
2" P.V.C./DWV	13.000	L.F.	3.525	82	204	286
2" steel, galvanized	13.000	L.F.	3.250	121	189	310
Vanity base cabinet, 2 doors, 24" x 48"	1.000	Ea.	1.400	525	79.50	604.50
Vanity top, plastic laminated, square edge	4.170	L.F.	1.112	154	63	217
Carrier, steel, for studs, no arms	1.000	Ea.	1.143	105	73.50	178.50
Wood, 2" x 8" blocking	1.300	L.F.	.052	1.21	2.95	4.16

For customer support on your Residential Cost Data, call 877.759.4771.

8 | MECHANICAL — 60 | Gas Fired Heating/Cooling Systems

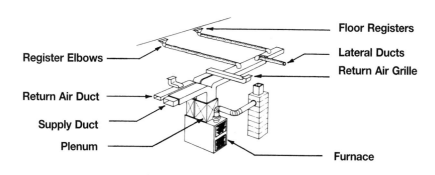

System Description	QUAN.	UNIT	LABOR HOURS	COST PER SYSTEM MAT.	COST PER SYSTEM INST.	COST PER SYSTEM TOTAL
HEATING ONLY, GAS FIRED HOT AIR, ONE ZONE, 1200 S.F. BUILDING						
Furnace, gas, up flow	1.000	Ea.	5.000	760	284	1,044
Intermittent pilot	1.000	Ea.		168		168
Supply duct, rigid fiberglass	176.000	S.F.	12.068	153.12	711.04	864.16
Return duct, sheet metal, galvanized	158.000	Lb.	16.137	99.54	948	1,047.54
Lateral ducts, 6" flexible fiberglass	144.000	L.F.	8.862	465.12	502.56	967.68
Register, elbows	12.000	Ea.	6.400	195	366	561
Floor registers, enameled steel	12.000	Ea.	3.000	123.60	189	312.60
Floor grille, return air	2.000	Ea.	.727	58	46	104
Thermostat	1.000	Ea.	1.000	58.50	66	124.50
Plenum	1.000	Ea.	1.000	100	57	157
TOTAL		System	54.194	2,180.88	3,169.60	5,350.48
HEATING/COOLING, GAS FIRED FORCED AIR, ONE ZONE, 1200 S.F. BUILDING						
Furnace, including plenum, compressor, coil	1.000	Ea.	14.720	5,957	837.20	6,794.20
Intermittent pilot	1.000	Ea.		168		168
Supply duct, rigid fiberglass	176.000	S.F.	12.068	153.12	711.04	864.16
Return duct, sheet metal, galvanized	158.000	Lb.	16.137	99.54	948	1,047.54
Lateral duct, 6" flexible fiberglass	144.000	L.F.	8.862	465.12	502.56	967.68
Register elbows	12.000	Ea.	6.400	195	366	561
Floor registers, enameled steel	12.000	Ea.	3.000	123.60	189	312.60
Floor grille return air	2.000	Ea.	.727	58	46	104
Thermostat	1.000	Ea.	1.000	58.50	66	124.50
Refrigeration piping, 25 ft. (pre-charged)	1.000	Ea.		305		305
TOTAL		System	62.914	7,582.88	3,665.80	11,248.68

The costs in these systems are based on complete system basis. For larger buildings use the price sheet on the opposite page.

Description	QUAN.	UNIT	LABOR HOURS	COST PER SYSTEM MAT.	COST PER SYSTEM INST.	COST PER SYSTEM TOTAL

Gas Fired Heating/Cooling Price Sheet

Description	QUAN.	UNIT	LABOR HOURS	COST EACH MAT.	COST EACH INST.	TOTAL
Furnace, heating only, 100 MBH, area to 1200 S.F.	1.000	Ea.	5.000	760	284	1,044
120 MBH, area to 1500 S.F.	1.000	Ea.	5.000	760	284	1,044
160 MBH, area to 2000 S.F.	1.000	Ea.	5.714	1,500	325	1,825
200 MBH, area to 2400 S.F.	1.000	Ea.	6.154	3,250	350	3,600
Heating/cooling, 100 MBH heat, 36 MBH cool, to 1200 S.F.	1.000	Ea.	16.000	6,475	910	7,385
120 MBH heat, 42 MBH cool, to 1500 S.F.	1.000	Ea.	18.462	6,925	1,075	8,000
144 MBH heat, 47 MBH cool, to 2000 S.F.	1.000	Ea.	20.000	8,000	1,175	9,175
200 MBH heat, 60 MBH cool, to 2400 S.F.	1.000	Ea.	34.286	8,400	2,025	10,425
Intermittent pilot, 100 MBH furnace	1.000	Ea.		168		168
200 MBH furnace	1.000	Ea.		168		168
Supply duct, rectangular, area to 1200 S.F., rigid fiberglass	176.000	S.F.	12.068	153	710	863
Sheet metal insulated	228.000	Lb.	31.331	1,025	3,175	4,200
Area to 1500 S.F., rigid fiberglass	176.000	S.F.	12.068	153	710	863
Sheet metal insulated	228.000	Lb.	31.331	1,025	3,175	4,200
Area to 2400 S.F., rigid fiberglass	205.000	S.F.	14.057	178	830	1,008
Sheet metal insulated	271.000	Lb.	37.048	1,200	3,725	4,925
Round flexible, insulated 6" diameter, to 1200 S.F.	156.000	L.F.	9.600	505	545	1,050
To 1500 S.F.	184.000	L.F.	11.323	595	640	1,235
8" diameter, to 2000 S.F.	269.000	L.F.	23.911	1,025	1,350	2,375
To 2400 S.F.	248.000	L.F.	22.045	950	1,250	2,200
Return duct, sheet metal galvanized, to 1500 S.F.	158.000	Lb.	16.137	99.50	950	1,049.50
To 2400 S.F.	191.000	Lb.	19.507	120	1,150	1,270
Lateral ducts, flexible round 6" insulated, to 1200 S.F.	144.000	L.F.	8.862	465	505	970
To 1500 S.F.	172.000	L.F.	10.585	555	600	1,155
To 2000 S.F.	261.000	L.F.	16.062	845	910	1,755
To 2400 S.F.	300.000	L.F.	18.462	970	1,050	2,020
Spiral steel insulated, to 1200 S.F.	144.000	L.F.	20.067	1,600	3,125	4,725
To 1500 S.F.	172.000	L.F.	23.952	1,900	3,725	5,625
To 2000 S.F.	261.000	L.F.	36.352	2,875	5,650	8,525
To 2400 S.F.	300.000	L.F.	41.825	3,300	6,500	9,800
Rectangular sheet metal galvanized insulated, to 1200 S.F.	228.000	Lb.	39.056	1,875	4,900	6,775
To 1500 S.F.	344.000	Lb.	53.966	2,275	6,275	8,550
To 2000 S.F.	522.000	Lb.	81.926	3,450	9,550	13,000
To 2400 S.F.	600.000	Lb.	94.189	3,975	11,000	14,975
Register elbows, to 1500 S.F.	12.000	Ea.	6.400	195	365	560
To 2400 S.F.	14.000	Ea.	7.460	228	425	653
Floor registers, enameled steel w/damper, to 1500 S.F.	12.000	Ea.	3.000	124	189	313
To 2400 S.F.	14.000	Ea.	4.308	169	272	441
Return air grille, area to 1500 S.F. 12" x 12"	2.000	Ea.	.727	58	46	104
Area to 2400 S.F. 8" x 16"	2.000	Ea.	.444	39.50	28	67.50
Area to 2400 S.F. 8" x 16"	2.000	Ea.	.727	58	46	104
16" x 16"	1.000	Ea.	.364	37.50	23	60.50
Thermostat, manual, 1 set back	1.000	Ea.	1.000	58.50	66	124.50
Electric, timed, 1 set back	1.000	Ea.	1.000	41.50	66	107.50
2 set back	1.000	Ea.	1.000	209	66	275
Plenum, heating only, 100 M.B.H.	1.000	Ea.	1.000	100	57	157
120 MBH	1.000	Ea.	1.000	100	57	157
160 MBH	1.000	Ea.	1.000	100	57	157
200 MBH	1.000	Ea.	1.000	100	57	157
Refrigeration piping, 3/8"	25.000	L.F.		33		33
3/4"	25.000	L.F.		73.50		73.50
7/8"	25.000	L.F.		85.50		85.50
Refrigerant piping, 25 ft. (precharged)	1.000	Ea.		305		305
Diffusers, ceiling, 6" diameter, to 1500 S.F.	10.000	Ea.	4.444	104	280	384
To 2400 S.F.	12.000	Ea.	6.000	132	380	512
Floor, aluminum, adjustable, 2-1/4" x 12" to 1500 S.F.	12.000	Ea.	3.000	90	189	279
To 2400 S.F.	14.000	Ea.	3.500	105	221	326
Side wall, aluminum, adjustable, 8" x 4", to 1500 S.F.	12.000	Ea.	3.000	221	189	410
5" x 10" to 2400 S.F.	12.000	Ea.	3.692	294	233	527

For customer support on your Residential Cost Data, call 877.759.4771.

8 | MECHANICAL — 64 | Oil Fired Heating/Cooling Systems

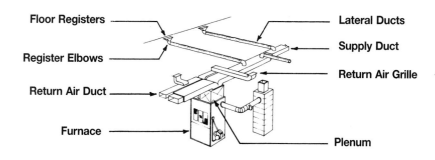

System Description	QUAN.	UNIT	LABOR HOURS	COST PER SYSTEM MAT.	COST PER SYSTEM INST.	COST PER SYSTEM TOTAL
HEATING ONLY, OIL FIRED HOT AIR, ONE ZONE, 1200 S.F. BUILDING						
Furnace, oil fired, atomizing gun type burner	1.000	Ea.	4.571	2,750	259	3,009
3/8" diameter copper supply pipe	1.000	Ea.	2.759	104.10	177	281.10
Shut off valve	1.000	Ea.	.333	15.60	21.50	37.10
Oil tank, 275 gallon, on legs	1.000	Ea.	3.200	540	190	730
Supply duct, rigid fiberglass	176.000	S.F.	12.068	153.12	711.04	864.16
Return duct, sheet metal, galvanized	158.000	Lb.	16.137	99.54	948	1,047.54
Lateral ducts, 6" flexible fiberglass	144.000	L.F.	8.862	465.12	502.56	967.68
Register elbows	12.000	Ea.	6.400	195	366	561
Floor register, enameled steel	12.000	Ea.	3.000	123.60	189	312.60
Floor grille, return air	2.000	Ea.	.727	58	46	104
Thermostat	1.000	Ea.	1.000	58.50	66	124.50
TOTAL		System	59.057	4,562.58	3,476.10	8,038.68
HEATING/COOLING, OIL FIRED, FORCED AIR, ONE ZONE, 1200 S.F. BUILDING						
Furnace, including plenum, compressor, coil	1.000	Ea.	16.000	6,925	910	7,835
3/8" diameter copper supply pipe	1.000	Ea.	2.759	104.10	177	281.10
Shut off valve	1.000	Ea.	.333	15.60	21.50	37.10
Oil tank, 275 gallon on legs	1.000	Ea.	3.200	540	190	730
Supply duct, rigid fiberglass	176.000	S.F.	12.068	153.12	711.04	864.16
Return duct, sheet metal, galvanized	158.000	Lb.	16.137	99.54	948	1,047.54
Lateral ducts, 6" flexible fiberglass	144.000	L.F.	8.862	465.12	502.56	967.68
Register elbows	12.000	Ea.	6.400	195	366	561
Floor registers, enameled steel	12.000	Ea.	3.000	123.60	189	312.60
Floor grille, return air	2.000	Ea.	.727	58	46	104
Refrigeration piping (precharged)	25.000	L.F.		305		305
TOTAL		System	69.486	8,984.08	4,061.10	13,045.18

Description	QUAN.	UNIT	LABOR HOURS	COST EACH MAT.	COST EACH INST.	COST EACH TOTAL

Oil Fired Heating/Cooling Price Sheet

Description	QUAN.	UNIT	LABOR HOURS	COST EACH MAT.	COST EACH INST.	TOTAL
Furnace, heating, 95.2 MBH, area to 1200 S.F.	1.000	Ea.	4.706	2,825	267	3,092
123.2 MBH, area to 1500 S.F.	1.000	Ea.	5.000	2,875	284	3,159
151.2 MBH, area to 2000 S.F.	1.000	Ea.	5.333	3,000	305	3,305
200 MBH, area to 2400 S.F.	1.000	Ea.	6.154	3,325	350	3,675
Heating/cooling, 95.2 MBH heat, 36 MBH cool, to 1200 S.F.	1.000	Ea.	16.000	6,925	910	7,835
112 MBH heat, 42 MBH cool, to 1500 S.F.	1.000	Ea.	24.000	10,400	1,375	11,775
151 MBH heat, 47 MBH cool, to 2000 S.F.	1.000	Ea.	20.800	9,000	1,175	10,175
184.8 MBH heat, 60 MBH cool, to 2400 S.F.	1.000	Ea.	24.000	4,450	1,400	5,850
Oil piping to furnace, 3/8" dia., copper	1.000	Ea.	3.412	258	217	475
Oil tank, on legs above ground, 275 gallons	1.000	Ea.	3.200	540	190	730
550 gallons	1.000	Ea.	5.926	4,325	350	4,675
Below ground, 275 gallons	1.000	Ea.	3.200	540	190	730
550 gallons	1.000	Ea.	5.926	4,325	350	4,675
1000 gallons	1.000	Ea.	6.400	7,100	380	7,480
Supply duct, rectangular, area to 1200 S.F., rigid fiberglass	176.000	S.F.	12.068	153	710	863
Sheet metal, insulated	228.000	Lb.	31.331	1,025	3,175	4,200
Area to 1500 S.F., rigid fiberglass	176.000	S.F.	12.068	153	710	863
Sheet metal, insulated	228.000	Lb.	31.331	1,025	3,175	4,200
Area to 2400 S.F., rigid fiberglass	205.000	S.F.	14.057	178	830	1,008
Sheet metal, insulated	271.000	Lb.	37.048	1,200	3,725	4,925
Round flexible, insulated, 6" diameter to 1200 S.F.	156.000	L.F.	9.600	505	545	1,050
To 1500 S.F.	184.000	L.F.	11.323	595	640	1,235
8" diameter to 2000 S.F.	269.000	L.F.	23.911	1,025	1,350	2,375
To 2400 S.F.	269.000	L.F.	22.045	950	1,250	2,200
Return duct, sheet metal galvanized, to 1500 S.F.	158.000	Lb.	16.137	99.50	950	1,049.50
To 2400 S.F.	191.000	Lb.	19.507	120	1,150	1,270
Lateral ducts, flexible round, 6", insulated to 1200 S.F.	144.000	L.F.	8.862	465	505	970
To 1500 S.F.	172.000	L.F.	10.585	555	600	1,155
To 2000 S.F.	261.000	L.F.	16.062	845	910	1,755
To 2400 S.F.	300.000	L.F.	18.462	970	1,050	2,020
Spiral steel, insulated to 1200 S.F.	144.000	L.F.	20.067	1,600	3,125	4,725
To 1500 S.F.	172.000	L.F.	23.952	1,900	3,725	5,625
To 2000 S.F.	261.000	L.F.	36.352	2,875	5,650	8,525
To 2400 S.F.	300.000	L.F.	41.825	3,300	6,500	9,800
Rectangular sheet metal galvanized insulated, to 1200 S.F.	288.000	Lb.	45.183	1,900	5,275	7,175
To 1500 S.F.	344.000	Lb.	53.966	2,275	6,275	8,550
To 2000 S.F.	522.000	Lb.	81.926	3,450	9,550	13,000
To 2400 S.F.	600.000	Lb.	94.189	3,975	11,000	14,975
Register elbows, to 1500 S.F.	12.000	Ea.	6.400	195	365	560
To 2400 S.F.	14.000	Ea.	7.470	228	425	653
Floor registers, enameled steel w/damper, to 1500 S.F.	12.000	Ea.	3.000	124	189	313
To 2400 S.F.	14.000	Ea.	4.308	169	272	441
Return air grille, area to 1500 S.F., 12" x 12"	2.000	Ea.	.727	58	46	104
12" x 24"	1.000	Ea.	.444	39.50	28	67.50
Area to 2400 S.F., 8" x 16"	2.000	Ea.	.727	58	46	104
16" x 16"	1.000	Ea.	.364	37.50	23	60.50
Thermostat, manual, 1 set back	1.000	Ea.	1.000	58.50	66	124.50
Electric, timed, 1 set back	1.000	Ea.	1.000	41.50	66	107.50
2 set back	1.000	Ea.	1.000	209	66	275
Refrigeration piping, 3/8"	25.000	L.F.		33		33
3/4"	25.000	L.F.		73.50		73.50
Diffusers, ceiling, 6" diameter, to 1500 S.F.	10.000	Ea.	4.444	104	280	384
To 2400 S.F.	12.000	Ea.	6.000	132	380	512
Floor, aluminum, adjustable, 2-1/4" x 12" to 1500 S.F.	12.000	Ea.	3.000	90	189	279
To 2400 S.F.	14.000	Ea.	3.500	105	221	326
Side wall, aluminum, adjustable, 8" x 4", to 1500 S.F.	12.000	Ea.	3.000	221	189	410
5" x 10" to 2400 S.F.	12.000	Ea.	3.692	294	233	527

8 | MECHANICAL 68 | Hot Water Heating Systems

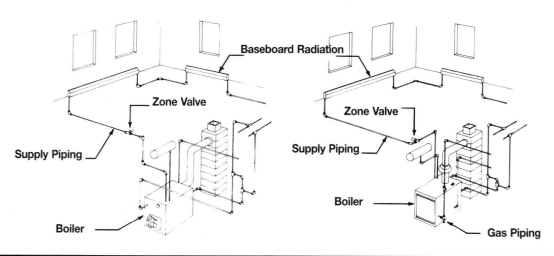

System Description	QUAN.	UNIT	LABOR HOURS	COST EACH		
				MAT.	INST.	TOTAL
OIL FIRED HOT WATER HEATING SYSTEM, AREA TO 1200 S.F.						
Boiler package, oil fired, 97 MBH, area to 1200 S.F. building	1.000	Ea.	15.000	2,000	860	2,860
3/8" diameter copper supply pipe	1.000	Ea.	2.759	104.10	177	281.10
Shut off valve	1.000	Ea.	.333	15.60	21.50	37.10
Oil tank, 275 gallon, with black iron filler pipe	1.000	Ea.	3.200	540	190	730
Supply piping, 3/4" copper tubing	176.000	L.F.	18.526	1,012	1,196.80	2,208.80
Supply fittings, copper 3/4"	36.000	Ea.	15.158	100.08	972	1,072.08
Supply valves, 3/4"	2.000	Ea.	.800	228	52	280
Baseboard radiation, 3/4"	106.000	L.F.	35.333	731.40	2,098.80	2,830.20
Zone valve	1.000	Ea.	.400	188	26.50	214.50
TOTAL		Ea.	91.509	4,919.18	5,594.60	10,513.78
OIL FIRED HOT WATER HEATING SYSTEM, AREA TO 2400 S.F.						
Boiler package, oil fired, 225 MBH, area to 2400 S.F. building	1.000	Ea.	25.105	7,700	1,425	9,125
3/8" diameter copper supply pipe	1.000	Ea.	2.759	104.10	177	281.10
Shut off valve	1.000	Ea.	.333	15.60	21.50	37.10
Oil tank, 550 gallon, with black iron pipe filler pipe	1.000	Ea.	5.926	4,325	350	4,675
Supply piping, 3/4" copper tubing	228.000	L.F.	23.999	1,311	1,550.40	2,861.40
Supply fittings, copper	46.000	Ea.	19.368	127.88	1,242	1,369.88
Supply valves	2.000	Ea.	.800	228	52	280
Baseboard radiation	212.000	L.F.	70.666	1,462.80	4,197.60	5,660.40
Zone valve	1.000	Ea.	.400	188	26.50	214.50
TOTAL		Ea.	149.356	15,462.38	9,042	24,504.38

The costs in this system are on a cost each basis. The costs represent total cost for the system based on a gross square foot of plan area.

Description	QUAN.	UNIT	LABOR HOURS	COST EACH		
				MAT.	INST.	TOTAL

Hot Water Heating Price Sheet	QUAN.	UNIT	LABOR HOURS	COST EACH		
				MAT.	INST.	TOTAL
Boiler, oil fired, 97 MBH, area to 1200 S.F.	1.000	Ea.	15.000	2,000	860	2,860
118 MBH, area to 1500 S.F.	1.000	Ea.	16.506	2,200	945	3,145
161 MBH, area to 2000 S.F.	1.000	Ea.	18.405	2,475	1,050	3,525
215 MBH, area to 2400 S.F.	1.000	Ea.	19.704	4,100	2,250	6,350
Oil piping, (valve & filter), 3/8" copper	1.000	Ea.	3.289	120	199	319
1/4" copper	1.000	Ea.	3.242	140	207	347
Oil tank, filler pipe and cap on legs, 275 gallon	1.000	Ea.	3.200	540	190	730
550 gallon	1.000	Ea.	5.926	4,325	350	4,675
Buried underground, 275 gallon	1.000	Ea.	3.200	540	190	730
550 gallon	1.000	Ea.	5.926	4,325	350	4,675
1000 gallon	1.000	Ea.	6.400	7,100	380	7,480
Supply piping copper, area to 1200 S.F., 1/2" tubing	176.000	L.F.	17.384	715	1,125	1,840
3/4" tubing	176.000	L.F.	18.526	1,000	1,200	2,200
Area to 1500 S.F., 1/2" tubing	186.000	L.F.	18.371	755	1,175	1,930
3/4" tubing	186.000	L.F.	19.578	1,075	1,275	2,350
Area to 2000 S.F., 1/2" tubing	204.000	L.F.	20.149	825	1,300	2,125
3/4" tubing	204.000	L.F.	21.473	1,175	1,375	2,550
Area to 2400 S.F., 1/2" tubing	228.000	L.F.	22.520	925	1,450	2,375
3/4" tubing	228.000	L.F.	23.999	1,300	1,550	2,850
Supply pipe fittings copper, area to 1200 S.F., 1/2"	36.000	Ea.	14.400	44.50	935	979.50
3/4"	36.000	Ea.	15.158	100	970	1,070
Area to 1500 S.F., 1/2"	40.000	Ea.	16.000	49.50	1,050	1,099.50
3/4"	40.000	Ea.	16.842	111	1,075	1,186
Area to 2000 S.F., 1/2"	44.000	Ea.	17.600	54.50	1,150	1,204.50
3/4"	44.000	Ea.	18.526	122	1,200	1,322
Area to 2400, S.F., 1/2"	46.000	Ea.	18.400	57	1,200	1,257
3/4"	46.000	Ea.	19.368	128	1,250	1,378
Supply valves, 1/2" pipe size	2.000	Ea.	.667	222	43	265
3/4"	2.000	Ea.	.800	228	52	280
Baseboard radiation, area to 1200 S.F., 1/2" tubing	106.000	L.F.	28.267	1,300	1,675	2,975
3/4" tubing	106.000	L.F.	35.333	730	2,100	2,830
Area to 1500 S.F., 1/2" tubing	134.000	L.F.	35.734	1,650	2,125	3,775
3/4" tubing	134.000	L.F.	44.666	925	2,650	3,575
Area to 2000 S.F., 1/2" tubing	178.000	L.F.	47.467	2,200	2,825	5,025
3/4" tubing	178.000	L.F.	59.333	1,225	3,525	4,750
Area to 2400 S.F., 1/2" tubing	212.000	L.F.	56.534	2,600	3,350	5,950
3/4" tubing	212.000	L.F.	70.666	1,475	4,200	5,675
Zone valves, 1/2" tubing	1.000	Ea.	.400	188	26.50	214.50
3/4" tubing	1.000	Ea.	.400	189	26.50	215.50

8 | MECHANICAL — 80 | Rooftop Systems

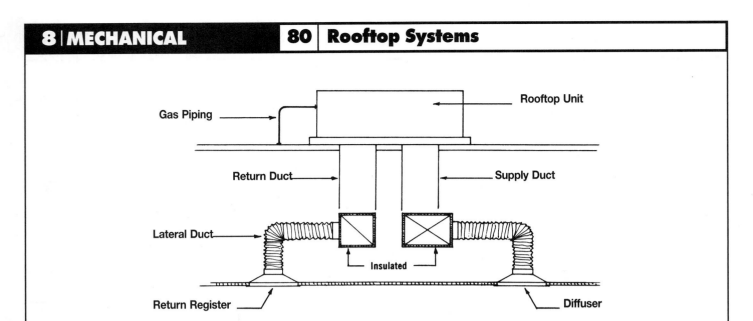

System Description	QUAN.	UNIT	LABOR HOURS	MAT.	INST.	TOTAL
ROOFTOP HEATING/COOLING UNIT, AREA TO 2000 S.F.						
Rooftop unit, single zone, electric cool, gas heat, to 2000 S.F.	1.000	Ea.	28.521	4,850	1,700	6,550
Gas piping	34.500	L.F.	5.207	163.19	334.65	497.84
Duct, supply and return, galvanized steel	38.000	Lb.	3.881	23.94	228	251.94
Insulation, ductwork	33.000	S.F.	6.286	164.67	338.25	502.92
Lateral duct, flexible duct 12" diameter, insulated	72.000	L.F.	11.520	388.80	655.20	1,044
Diffusers	4.000	Ea.	4.571	1,184	288	1,472
Return registers	1.000	Ea.	.727	74	46	120
TOTAL		Ea.	60.713	6,848.60	3,590.10	10,438.70
ROOFTOP HEATING/COOLING UNIT, AREA TO 5000 S.F.						
Rooftop unit, single zone, electric cool, gas heat, to 5000 S.F.	1.000	Ea.	42.032	14,400	2,400	16,800
Gas piping	86.250	L.F.	13.019	407.96	836.63	1,244.59
Duct supply and return, galvanized steel	95.000	Lb.	9.702	59.85	570	629.85
Insulation, ductwork	82.000	S.F.	15.619	409.18	840.50	1,249.68
Lateral duct, flexible duct, 12" diameter, insulated	180.000	L.F.	28.800	972	1,638	2,610
Diffusers	10.000	Ea.	11.429	2,960	720	3,680
Return registers	3.000	Ea.	2.182	222	138	360
TOTAL		Ea.	122.783	19,430.99	7,143.13	26,574.12

Description	QUAN.	UNIT	LABOR HOURS	MAT.	INST.	TOTAL

Rooftop Price Sheet	QUAN.	UNIT	LABOR HOURS	COST EACH		
				MAT.	INST.	TOTAL
Rooftop unit, single zone, electric cool, gas heat to 2000 S.F.	1.000	Ea.	28.521	4,850	1,700	6,550
Area to 3000 S.F.	1.000	Ea.	35.982	10,000	2,050	12,050
Area to 5000 S.F.	1.000	Ea.	42.032	14,400	2,400	16,800
Area to 10000 S.F.	1.000	Ea.	68.376	39,200	4,075	43,275
Gas piping, area 2000 through 4000 S.F.	34.500	L.F.	5.207	163	335	498
Area 5000 to 10000 S.F.	86.250	L.F.	13.019	410	835	1,245
Duct, supply and return, galvanized steel, to 2000 S.F.	38.000	Lb.	3.881	24	228	252
Area to 3000 S.F.	57.000	Lb.	5.821	36	340	376
Area to 5000 S.F.	95.000	Lb.	9.702	60	570	630
Area to 10000 S.F.	190.000	Lb.	19.405	120	1,150	1,270
Rigid fiberglass, area to 2000 S.F.	33.000	S.F.	2.263	28.50	133	161.50
Area to 3000 S.F.	49.000	S.F.	3.360	42.50	198	240.50
Area to 5000 S.F.	82.000	S.F.	5.623	71.50	330	401.50
Area to 10000 S.F.	164.000	S.F.	11.245	143	665	808
Insulation, supply and return, blanket type, area to 2000 S.F.	33.000	S.F.	1.508	165	340	505
Area to 3000 S.F.	49.000	S.F.	2.240	245	500	745
Area to 5000 S.F.	82.000	S.F.	3.748	410	840	1,250
Area to 10000 S.F.	164.000	S.F.	7.496	820	1,675	2,495
Lateral ducts, flexible round, 12" insulated, to 2000 S.F.	72.000	L.F.	11.520	390	655	1,045
Area to 3000 S.F.	108.000	L.F.	17.280	585	985	1,570
Area to 5000 S.F.	180.000	L.F.	28.800	970	1,650	2,620
Area to 10000 S.F.	360.000	L.F.	57.600	1,950	3,275	5,225
Rectangular, galvanized steel, to 2000 S.F.	239.000	Lb.	24.409	151	1,425	1,576
Area to 3000 S.F.	360.000	Lb.	36.767	227	2,150	2,377
Area to 5000 S.F.	599.000	Lb.	61.176	375	3,600	3,975
Area to 10000 S.F.	998.000	Lb.	101.926	630	6,000	6,630
Diffusers, ceiling, 1 to 4 way blow, 24" x 24", to 2000 S.F.	4.000	Ea.	4.571	1,175	288	1,463
Area to 3000 S.F.	6.000	Ea.	6.857	1,775	430	2,205
Area to 5000 S.F.	10.000	Ea.	11.429	2,950	720	3,670
Area to 10000 S.F.	20.000	Ea.	22.857	5,925	1,450	7,375
Return grilles, 24" x 24", to 2000 S.F.	1.000	Ea.	.727	74	46	120
Area to 3000 S.F.	2.000	Ea.	1.455	148	92	240
Area to 5000 S.F.	3.000	Ea.	2.182	222	138	360
Area to 10000 S.F.	5.000	Ea.	3.636	370	230	600

Division 9 - Electrical

9 | ELECTRICAL 10 | Electric Service Systems

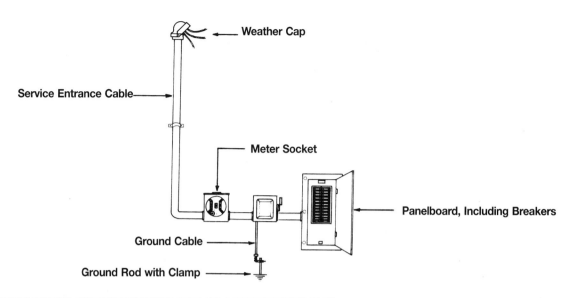

System Description	QUAN.	UNIT	LABOR HOURS	COST EACH MAT.	COST EACH INST.	COST EACH TOTAL
100 AMP SERVICE						
Weather cap	1.000	Ea.	.667	8.90	40	48.90
Service entrance cable	10.000	L.F.	.762	38.10	46	84.10
Meter socket	1.000	Ea.	2.500	47.50	151	198.50
Ground rod with clamp	1.000	Ea.	1.455	22	88	110
Ground cable	5.000	L.F.	.250	7.35	15.10	22.45
Panel board, 12 circuit	1.000	Ea.	6.667	102	400	502
TOTAL		Ea.	12.301	225.85	740.10	965.95
200 AMP SERVICE						
Weather cap	1.000	Ea.	1.000	23	60.50	83.50
Service entrance cable	10.000	L.F.	1.143	36	69	105
Meter socket	1.000	Ea.	4.211	101	254	355
Ground rod with clamp	1.000	Ea.	1.818	42.50	110	152.50
Ground cable	10.000	L.F.	.500	14.70	30.20	44.90
3/4" EMT	5.000	L.F.	.308	5.50	18.55	24.05
Panel board, 24 circuit	1.000	Ea.	12.308	510	670	1,180
TOTAL		Ea.	21.288	732.70	1,212.25	1,944.95
400 AMP SERVICE						
Weather cap	1.000	Ea.	2.963	239	179	418
Service entrance cable	180.000	L.F.	5.760	680.40	347.40	1,027.80
Meter socket	1.000	Ea.	4.211	101	254	355
Ground rod with clamp	1.000	Ea.	2.000	56	121	177
Ground cable	20.000	L.F.	.485	48	29.20	77.20
3/4" Greenfield	20.000	L.F.	1.000	12.80	60.40	73.20
Current transformer cabinet	1.000	Ea.	6.154	154	370	524
Panel board, 42 circuit	1.000	Ea.	33.333	3,875	2,000	5,875
TOTAL		Ea.	55.906	5,166.20	3,361	8,527.20

9 | ELECTRICAL — 20 | Electric Perimeter Heating Systems

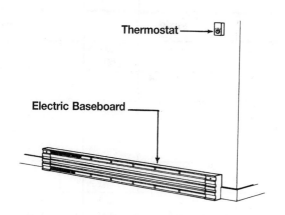

System Description	QUAN.	UNIT	LABOR HOURS	COST EACH MAT.	COST EACH INST.	COST EACH TOTAL
4' BASEBOARD HEATER						
Electric baseboard heater, 4' long	1.000	Ea.	1.194	39	72	111
Thermostat, integral	1.000	Ea.	.500	29	30	59
Romex, 12-3 with ground	40.000	L.F.	1.600	20.40	96.40	116.80
Panel board breaker, 20 Amp	1.000	Ea.	.300	27.15	18.15	45.30
TOTAL		Ea.	3.594	115.55	216.55	332.10
6' BASEBOARD HEATER						
Electric baseboard heater, 6' long	1.000	Ea.	1.600	56	96.50	152.50
Thermostat, integral	1.000	Ea.	.500	29	30	59
Romex, 12-3 with ground	40.000	L.F.	1.600	20.40	96.40	116.80
Panel board breaker, 20 Amp	1.000	Ea.	.400	36.20	24.20	60.40
TOTAL		Ea.	4.100	141.60	247.10	388.70
8' BASEBOARD HEATER						
Electric baseboard heater, 8' long	1.000	Ea.	2.000	61.50	121	182.50
Thermostat, integral	1.000	Ea.	.500	29	30	59
Romex, 12-3 with ground	40.000	L.F.	1.600	20.40	96.40	116.80
Panel board breaker, 20 Amp	1.000	Ea.	.500	45.25	30.25	75.50
TOTAL		Ea.	4.600	156.15	277.65	433.80
10' BASEBOARD HEATER						
Electric baseboard heater, 10' long	1.000	Ea.	2.424	198	146	344
Thermostat, integral	1.000	Ea.	.500	29	30	59
Romex, 12-3 with ground	40.000	L.F.	1.600	20.40	96.40	116.80
Panel board breaker, 20 Amp	1.000	Ea.	.750	67.88	45.38	113.26
TOTAL		Ea.	5.274	315.28	317.78	633.06

The costs in this system are on a cost each basis and include all necessary conduit fittings.

Description	QUAN.	UNIT	LABOR HOURS	COST EACH MAT.	COST EACH INST.	COST EACH TOTAL

9 | ELECTRICAL — 30 | Wiring Device Systems

System Description	QUAN.	UNIT	LABOR HOURS	COST EACH MAT.	COST EACH INST.	COST EACH TOTAL
Air conditioning receptacles						
Using non-metallic sheathed cable	1.000	Ea.	.800	21.50	48.50	70
Using BX cable	1.000	Ea.	.964	28	58	86
Using EMT conduit	1.000	Ea.	1.194	42	72	114
Disposal wiring						
Using non-metallic sheathed cable	1.000	Ea.	.889	26.50	53.50	80
Using BX cable	1.000	Ea.	1.067	32.50	64.50	97
Using EMT conduit	1.000	Ea.	1.333	48	80.50	128.50
Dryer circuit						
Using non-metallic sheathed cable	1.000	Ea.	1.455	35.50	88	123.50
Using BX cable	1.000	Ea.	1.739	42	105	147
Using EMT conduit	1.000	Ea.	2.162	54	130	184
Duplex receptacles						
Using non-metallic sheathed cable	1.000	Ea.	.615	21.50	37	58.50
Using BX cable	1.000	Ea.	.741	28	44.50	72.50
Using EMT conduit	1.000	Ea.	.920	42	55.50	97.50
Exhaust fan wiring						
Using non-metallic sheathed cable	1.000	Ea.	.800	16.20	48.50	64.70
Using BX cable	1.000	Ea.	.964	23	58	81
Using EMT conduit	1.000	Ea.	1.194	37	72	109
Furnace circuit & switch						
Using non-metallic sheathed cable	1.000	Ea.	1.333	29.50	80.50	110
Using BX cable	1.000	Ea.	1.600	39	96.50	135.50
Using EMT conduit	1.000	Ea.	2.000	49	121	170
Ground fault						
Using non-metallic sheathed cable	1.000	Ea.	1.000	59.50	60.50	120
Using BX cable	1.000	Ea.	1.212	68.50	73	141.50
Using EMT conduit	1.000	Ea.	1.481	83.50	89.50	173
Heater circuits						
Using non-metallic sheathed cable	1.000	Ea.	1.000	21	60.50	81.50
Using BX cable	1.000	Ea.	1.212	26.50	73	99.50
Using EMT conduit	1.000	Ea.	1.481	39.50	89.50	129
Lighting wiring						
Using non-metallic sheathed cable	1.000	Ea.	.500	21.50	30	51.50
Using BX cable	1.000	Ea.	.602	26	36.50	62.50
Using EMT conduit	1.000	Ea.	.748	38	45	83
Range circuits						
Using non-metallic sheathed cable	1.000	Ea.	2.000	80.50	121	201.50
Using BX cable	1.000	Ea.	2.424	113	146	259
Using EMT conduit	1.000	Ea.	2.963	95.50	179	274.50
Switches, single pole						
Using non-metallic sheathed cable	1.000	Ea.	.500	16.20	30	46.20
Using BX cable	1.000	Ea.	.602	23	36.50	59.50
Using EMT conduit	1.000	Ea.	.748	37	45	82
Switches, 3-way						
Using non-metallic sheathed cable	1.000	Ea.	.667	19.75	40	59.75
Using BX cable	1.000	Ea.	.800	26	48.50	74.50
Using EMT conduit	1.000	Ea.	1.333	46.50	80.50	127
Water heater						
Using non-metallic sheathed cable	1.000	Ea.	1.600	28.50	96.50	125
Using BX cable	1.000	Ea.	1.905	40.50	115	155.50
Using EMT conduit	1.000	Ea.	2.353	45.50	142	187.50
Weatherproof receptacle						
Using non-metallic sheathed cable	1.000	Ea.	1.333	171	80.50	251.50
Using BX cable	1.000	Ea.	1.600	176	96.50	272.50
Using EMT conduit	1.000	Ea.	2.000	190	121	311

9 | ELECTRICAL — 40 | Light Fixture Systems

System Description	QUAN.	UNIT	LABOR HOURS	COST EACH MAT.	COST EACH INST.	COST EACH TOTAL
Fluorescent strip, 4' long, 1 light, average	1.000	Ea.	.941	25.50	57	82.50
Deluxe	1.000	Ea.	1.129	30.50	68.50	99
2 lights, average	1.000	Ea.	1.000	45.50	60.50	106
Deluxe	1.000	Ea.	1.200	54.50	72.50	127
8' long, 1 light, average	1.000	Ea.	1.194	55.50	72	127.50
Deluxe	1.000	Ea.	1.433	66.50	86.50	153
2 lights, average	1.000	Ea.	1.290	67	78	145
Deluxe	1.000	Ea.	1.548	80.50	93.50	174
Surface mounted, 4' x 1', economy	1.000	Ea.	.914	56.50	55	111.50
Average	1.000	Ea.	1.143	70.50	69	139.50
Deluxe	1.000	Ea.	1.371	84.50	83	167.50
4' x 2', economy	1.000	Ea.	1.208	71	73	144
Average	1.000	Ea.	1.509	89	91	180
Deluxe	1.000	Ea.	1.811	107	109	216
Recessed, 4'x 1', 2 lamps, economy	1.000	Ea.	1.123	44	67.50	111.50
Average	1.000	Ea.	1.404	55	84.50	139.50
Deluxe	1.000	Ea.	1.684	66	101	167
4' x 2', 4' lamps, economy	1.000	Ea.	1.362	53	82.50	135.50
Average	1.000	Ea.	1.702	66.50	103	169.50
Deluxe	1.000	Ea.	2.043	80	124	204
Incandescent, exterior, 150W, single spot	1.000	Ea.	.500	35	30	65
Double spot	1.000	Ea.	1.167	97	70	167
Recessed, 100W, economy	1.000	Ea.	.800	63	48.50	111.50
Average	1.000	Ea.	1.000	78.50	60.50	139
Deluxe	1.000	Ea.	1.200	94	72.50	166.50
150W, economy	1.000	Ea.	.800	91	48.50	139.50
Average	1.000	Ea.	1.000	114	60.50	174.50
Deluxe	1.000	Ea.	1.200	137	72.50	209.50
Surface mounted, 60W, economy	1.000	Ea.	.800	49	48.50	97.50
Average	1.000	Ea.	1.000	55	60.50	115.50
Deluxe	1.000	Ea.	1.194	77.50	72	149.50
Metal halide, recessed 2' x 2' 250W	1.000	Ea.	2.500	350	151	501
2' x 2', 400W	1.000	Ea.	2.759	400	166	566
Surface mounted, 2' x 2', 250W	1.000	Ea.	2.963	380	179	559
2' x 2', 400W	1.000	Ea.	3.333	445	201	646
High bay, single, unit, 400W	1.000	Ea.	3.478	450	210	660
Twin unit, 400W	1.000	Ea.	5.000	905	300	1,205
Low bay, 250W	1.000	Ea.	2.500	400	151	551

Unit Price Section

Table of Contents

Sect. No.		Page
	General Requirements	
01 11	Summary of Work	292
01 21	Allowances	292
01 31	Project Management and Coordination	293
01 41	Regulatory Requirements	293
01 54	Construction Aids	293
01 56	Temporary Barriers and Enclosures	296
01 71	Examination and Preparation	297
01 74	Cleaning and Waste Management	297
01 76	Protecting Installed Construction	297
	Existing Conditions	
02 21	Surveys	300
02 32	Geotechnical Investigations	300
02 41	Demolition	300
02 83	Lead Remediation	304
02 85	Mold Remediation	306
	Concrete	
03 01	Maintenance of Concrete	308
03 05	Common Work Results for Concrete	308
03 11	Concrete Forming	308
03 15	Concrete Accessories	310
03 21	Reinforcement Bars	312
03 22	Fabric and Grid Reinforcing	312
03 23	Stressed Tendon Reinforcing	313
03 24	Fibrous Reinforcing	313
03 30	Cast-In-Place Concrete	314
03 31	Structural Concrete	315
03 35	Concrete Finishing	317
03 39	Concrete Curing	318
03 41	Precast Structural Concrete	318
03 48	Precast Concrete Specialties	318
03 54	Cast Underlayment	319
03 63	Epoxy Grouting	319
03 82	Concrete Boring	319
	Masonry	
04 01	Maintenance of Masonry	322
04 05	Common Work Results for Masonry	322
04 21	Clay Unit Masonry	325
04 22	Concrete Unit Masonry	327
04 23	Glass Unit Masonry	329
04 24	Adobe Unit Masonry	329
04 27	Multiple-Wythe Unit Masonry	330
04 41	Dry-Placed Stone	330
04 43	Stone Masonry	330
04 51	Flue Liner Masonry	332
04 57	Masonry Fireplaces	332
04 72	Cast Stone Masonry	333
	Metals	
05 05	Common Work Results for Metals	336
05 12	Structural Steel Framing	339
05 31	Steel Decking	340
05 41	Structural Metal Stud Framing	341
05 42	Cold-Formed Metal Joist Framing	344
05 44	Cold-Formed Metal Trusses	350
05 51	Metal Stairs	351
05 52	Metal Railings	351
05 58	Formed Metal Fabrications	352
05 71	Decorative Metal Stairs	353
05 75	Decorative Formed Metal	353
	Wood, Plastics & Composites	
06 05	Common Work Results for Wood, Plastics, and Composites	356
06 11	Wood Framing	370
06 12	Structural Panels	381
06 13	Heavy Timber Construction	382
06 15	Wood Decking	384
06 16	Sheathing	384
06 17	Shop-Fabricated Structural Wood	387
06 18	Glued-Laminated Construction	388
06 22	Millwork	391
06 25	Prefinished Paneling	402
06 26	Board Paneling	404
06 43	Wood Stairs and Railings	404
06 44	Ornamental Woodwork	406
06 48	Wood Frames	408
06 49	Wood Screens and Exterior Wood Shutters	409
06 51	Structural Plastic Shapes and Plates	410
06 63	Plastic Railings	411
06 65	Plastic Trim	411
06 80	Composite Fabrications	412
06 81	Composite Railings	412
	Thermal & Moisture Protection	
07 01	Operation and Maint. of Thermal and Moisture Protection	416
07 05	Common Work Results for Thermal and Moisture Protection	416
07 11	Dampproofing	417
07 19	Water Repellents	418
07 21	Thermal Insulation	418
07 22	Roof and Deck Insulation	421
07 24	Exterior Insulation and Finish Systems	423
07 25	Weather Barriers	423
07 26	Vapor Retarders	424
07 27	Air Barriers	424
07 31	Shingles and Shakes	424
07 32	Roof Tiles	426
07 33	Natural Roof Coverings	427
07 41	Roof Panels	427
07 42	Wall Panels	428
07 46	Siding	429
07 51	Built-Up Bituminous Roofing	431
07 52	Modified Bituminous Membrane Roofing	432
07 57	Coated Foamed Roofing	433
07 58	Roll Roofing	433
07 61	Sheet Metal Roofing	434
07 62	Sheet Metal Flashing and Trim	434
07 65	Flexible Flashing	435
07 71	Roof Specialties	436
07 72	Roof Accessories	440
07 76	Roof Pavers	440
07 91	Preformed Joint Seals	441
07 92	Joint Sealants	442
	Openings	
08 01	Operation and Maintenance of Openings	444
08 05	Common Work Results for Openings	445
08 11	Metal Doors and Frames	446
08 12	Metal Frames	446
08 13	Metal Doors	447
08 14	Wood Doors	448
08 16	Composite Doors	454
08 17	Integrated Door Opening Assemblies	455
08 31	Access Doors and Panels	456
08 32	Sliding Glass Doors	457
08 36	Panel Doors	457
08 51	Metal Windows	459
08 52	Wood Windows	460
08 53	Plastic Windows	468
08 54	Composite Windows	471
08 61	Roof Windows	471
08 62	Unit Skylights	472
08 71	Door Hardware	473
08 75	Window Hardware	477
08 79	Hardware Accessories	477
08 81	Glass Glazing	478
08 83	Mirrors	479
08 87	Glazing Surface Films	479
08 91	Louvers	479
08 95	Vents	480
	Finishes	
09 01	Maintenance of Finishes	482
09 05	Common Work Results for Finishes	482
09 22	Supports for Plaster and Gypsum Board	483
09 23	Gypsum Plastering	486
09 24	Cement Plastering	487
09 25	Other Plastering	487
09 26	Veneer Plastering	487
09 28	Backing Boards and Underlayments	488
09 29	Gypsum Board	488
09 30	Tiling	492
09 34	Waterproofing-Membrane Tiling	494
09 51	Acoustical Ceilings	495
09 53	Acoustical Ceiling Suspension Assemblies	496
09 61	Flooring Treatment	496
09 62	Specialty Flooring	496
09 63	Masonry Flooring	497
09 64	Wood Flooring	498
09 65	Resilient Flooring	499
09 66	Terrazzo Flooring	500
09 68	Carpeting	501
09 72	Wall Coverings	502
09 91	Painting	503
09 93	Staining and Transparent Finishing	517
09 96	High-Performance Coatings	517
	Specialties	
10 28	Toilet, Bath, and Laundry Accessories	520
10 31	Manufactured Fireplaces	521
10 32	Fireplace Specialties	522
10 35	Stoves	523
10 44	Fire Protection Specialties	523
10 55	Postal Specialties	523
10 56	Storage Assemblies	523
10 57	Wardrobe and Closet Specialties	524
10 73	Protective Covers	524
10 74	Manufactured Exterior Specialties	525
10 75	Flagpoles	526
	Equipment	
11 30	Residential Equipment	528
11 32	Unit Kitchens	530
11 41	Foodservice Storage Equipment	530
11 81	Facility Maintenance Equipment	531
	Furnishings	
12 21	Window Blinds	534
12 22	Curtains and Drapes	534
12 23	Interior Shutters	535
12 24	Window Shades	535
12 32	Manufactured Wood Casework	536
12 34	Manufactured Plastic Casework	538
12 36	Countertops	538
	Special Construction	
13 11	Swimming Pools	542
13 12	Fountains	543
13 17	Tubs and Pools	543
13 24	Special Activity Rooms	543
13 34	Fabricated Engineered Structures	544
	Conveying Equipment	
14 21	Electric Traction Elevators	546
14 42	Wheelchair Lifts	546
	Fire Suppression	
21 05	Common Work Results for Fire Suppression	548

Table of Contents (cont.)

Sect. No.		Page
21 11	Facility Fire-Suppression Water-Service Piping	548
21 12	Fire-Suppression Standpipes	551
21 13	Fire-Suppression Sprinkler Systems	552

Plumbing

22 05	Common Work Results for Plumbing	556
22 07	Plumbing Insulation	559
22 11	Facility Water Distribution	559
22 13	Facility Sanitary Sewerage	567
22 14	Facility Storm Drainage	570
22 31	Domestic Water Softeners	570
22 33	Electric Domestic Water Heaters	571
22 34	Fuel-Fired Domestic Water Heaters	571
22 41	Residential Plumbing Fixtures	571
22 42	Commercial Plumbing Fixtures	575
22 51	Swimming Pool Plumbing Systems	576

Heating Ventilation Air Conditioning

23 05	Common Work Results for HVAC	578
23 07	HVAC Insulation	578
23 09	Instrumentation and Control for HVAC	579
23 13	Facility Fuel-Storage Tanks	579
23 21	Hydronic Piping and Pumps	580
23 23	Refrigerant Piping	580
23 31	HVAC Ducts and Casings	581
23 33	Air Duct Accessories	582
23 34	HVAC Fans	582
23 37	Air Outlets and Inlets	583
23 41	Particulate Air Filtration	584
23 42	Gas-Phase Air Filtration	585
23 43	Electronic Air Cleaners	585
23 51	Breechings, Chimneys, and Stacks	585
23 52	Heating Boilers	586
23 54	Furnaces	587
23 62	Packaged Compressor and Condenser Units	589
23 74	Packaged Outdoor HVAC Equipment	590
23 81	Decentralized Unitary HVAC Equipment	590
23 82	Convection Heating and Cooling Units	591
23 83	Radiant Heating Units	592

Electrical

26 05	Common Work Results for Electrical	596
26 24	Switchboards and Panelboards	605
26 27	Low-Voltage Distribution Equipment	606
26 28	Low-Voltage Circuit Protective Devices	608
26 32	Packaged Generator Assemblies	608
26 33	Battery Equipment	609
26 36	Transfer Switches	609
26 41	Facility Lightning Protection	609
26 51	Interior Lighting	609
26 55	Special Purpose Lighting	611
26 56	Exterior Lighting	611
26 61	Lighting Systems and Accessories	612

Communications

27 41	Audio-Video Systems	614

Electronic Safety & Security

28 16	Intrusion Detection	616
28 31	Fire Detection and Alarm	616

Earthwork

31 05	Common Work Results for Earthwork	618
31 11	Clearing and Grubbing	618
31 13	Selective Tree and Shrub Removal and Trimming	618
31 14	Earth Stripping and Stockpiling	618
31 22	Grading	619
31 23	Excavation and Fill	619
31 25	Erosion and Sedimentation Controls	631
31 31	Soil Treatment	631

Exterior Improvements

32 01	Operation and Maintenance of Exterior Improvements	634
32 06	Schedules for Exterior Improvements	634
32 11	Base Courses	634
32 12	Flexible Paving	635
32 13	Rigid Paving	636
32 14	Unit Paving	636
32 16	Curbs, Gutters, Sidewalks, and Driveways	637
32 31	Fences and Gates	638
32 32	Retaining Walls	640
32 33	Site Furnishings	642
32 84	Planting Irrigation	642
32 91	Planting Preparation	642
32 92	Turf and Grasses	643
32 93	Plants	644
32 94	Planting Accessories	646
32 96	Transplanting	646

Utilities

33 05	Common Work Results for Utilities	648
33 11	Water Utility Distribution Piping	648
33 12	Water Utility Distribution Equipment	649
33 21	Water Supply Wells	649
33 31	Sanitary Utility Sewerage Piping	650
33 36	Utility Septic Tanks	650
33 41	Storm Utility Drainage Piping	651
33 44	Storm Utility Water Drains	652
33 46	Subdrainage	652
33 49	Storm Drainage Structures	652
33 51	Natural-Gas Distribution	652

Electrical Power Generation

48 15	Wind Energy Electrical Power Generation Equipment	656

How RSMeans Unit Price Data Works

All RSMeans unit price data is organized in the same way.

It is important to understand the structure, so that you can find information easily and use it correctly.

03 30 Cast-In-Place Concrete
03 30 53 – Miscellaneous Cast-In-Place Concrete

03 30 53.40 Concrete In Place		Crew	Daily Output	Labor-Hours	Unit	Material	2016 Bare Costs Labor	Equipment	Total	Total Incl O&P
0010	**CONCRETE IN PLACE**									
0020	Including forms (4 uses), Grade 60 rebar, concrete (Portland cement									
0050	Type I), placement and finishing unless otherwise indicated									
0500	Chimney foundations (5000 psi), over 5 C.Y.	C-14C	32.22	3.476	C.Y.	155	109	.97	264.97	355
0510	(3500 psi), under 5 C.Y.	"	23.71	4.724	"	182	149	1.32	332.32	450
3540	Equipment pad (3000 psi), 3' x 3' x 6" thick	C-14H	45	1.067	Ea.	47.50	35	.70	83.20	111
3550	4' x 4' x 6" thick		30	1.600		70.50	52	1.06	123.56	166
3560	5' x 5' x 8" thick		18	2.667		124	87	1.76	212.76	284
3570	6' x 6' x 8" thick		14	3.429		167	112	2.26	281.26	375
3580	8' x 8' x 10" thick		8	6		360	196	3.96	559.96	725
3590	10' x 10' x 12" thick		5	9.600		615	315	6.35	936.35	1,200
3800	Footings (3000 psi), spread under 1 C.Y.	C-14C	28	4	C.Y.	171	126	1.12	298.12	400
3825	1 C.Y. to 5 C.Y.		43	2.605		207	82	.73	289.73	365
3850	Over 5 C.Y.		75	1.493		190	47	.42	237.42	288

❶ Line Numbers
RSMeans **Line Numbers** consist of 12 characters, which identify a unique location in the database for each task. The first 6 or 8 digits conform to the Construction Specifications Institute MasterFormat® 2014. The remainder of the digits are a further breakdown by RSMeans in order to arrange items in understandable groups of similar tasks. Line numbers are consistent across all RSMeans publications, so a line number in any RSMeans product will always refer to the same unit of work.

❷ Descriptions
RSMeans **Descriptions** are shown in a hierarchical structure to make them readable. In order to read a complete description, read up through the indents to the top of the section. Include everything that is above and to the left that is not contradicted by information below. For instance, the complete description for line 03 30 53.40 3550 is "Concrete in place, including forms (4 uses), Grade 60 rebar, concrete (Portland cement Type 1), placement and finishing unless otherwise indicated; Equipment pad (3000 psi), 4' x 4' x 6" thick."

❸ RSMeans Data
When using **RSMeans data**, it is important to read through an entire section to ensure that you use the data that most closely matches your work. Note that sometimes there is additional information shown in the section that may improve your price. There are frequently lines that further describe, add to, or adjust data for specific situations.

❹ Reference Information
RSMeans engineers have created **reference** information to assist you in your estimate. **If** there is information that applies to a section, it will be indicated at the start of the section.

❺ Crews
Crews include labor and/or equipment necessary to accomplish each task. In this case, Crew C-14H is used.

RSMeans selects a crew to represent the workers and equipment that are typically used for that task. In this case, Crew C-14H consists of one carpenter foreman (outside), two carpenters, one rodman, one laborer, one cement finisher, and one gas engine vibrator. Details of all crews can be found in the reference section.

Crews - Residential

Crew No.	Bare Costs		Incl. Subs O&P		Cost Per Labor-Hour	
Crew C-14H	Hr.	Daily	Hr.	Daily	Bare Costs	Incl. O&P
1 Carpenter Foreman (outside)	$35.90	$287.20	$60.10	$480.80	$32.63	$54.44
2 Carpenters	33.90	542.40	56.75	908.00		
1 Rodman (reinf.)	35.50	284.00	59.95	479.60		
1 Laborer	24.65	197.20	41.25	330.00		
1 Cement Finisher	31.95	255.60	51.85	414.80		
1 Gas Engine Vibrator		31.60		34.76	0.66	0.72
48 L.H., Daily Totals		$1598.00		$2647.96	$33.29	$55.17

6. Daily Output

The **Daily Output** is the amount of work that the crew can do in a normal 8-hour workday, including mobilization, layout, movement of materials, and cleanup. In this case, crew C-14H can install thirty 4' x 4' x 6" thick concrete pads in a day. Daily output is variable and based on many factors, including the size of the job, location, and environmental conditions. RSMeans data represents work done in daylight (or adequate lighting) and temperate conditions.

7. Labor-Hours

The figure in the **Labor-Hours** column is the amount of labor required to perform one unit of work–in this case the amount of labor required to construct one 4' x 4' equipment pad. This figure is calculated by dividing the number of hours of labor in the crew by the daily output (48 labor hours divided by 30 pads = 1.6 hours of labor per pad). Multiply 1.600 times 60 to see the value in minutes: 60 x 1.6 = 96 minutes. Note: the labor-hour figure is not dependent on the crew size. A change in crew size will result in a corresponding change in daily output, but the labor-hours per unit of work will not change.

8. Unit of Measure

All RSMeans unit cost data includes the typical **Unit of Measure** used for estimating that item. For concrete-in-place the typical unit is cubic yards (C.Y.) or each (Ea.). For installing broadloom carpet it is square yard, and for gypsum board it is square foot. The estimator needs to take special care that the unit in the data matches the unit in the take-off. Unit conversions may be found in the Reference Section.

9. Bare Costs

Bare Costs are the costs of materials, labor, and equipment that the installing contractor pays. They represent the cost, in U.S. dollars, for one unit of work. They do not include any markups for profit or labor burden.

10. Bare Total

The **Total column** represents the total bare cost for the installing contractor, in U.S. dollars. In this case, the sum of $70.50 for material + $52.00 for labor + $1.06 for equipment is $123.56.

11. Total Incl O&P

The **Total Incl O&P column** is the total cost, including overhead and profit, that the installing contractor will charge the customer. This represents the cost of materials plus 10% profit, the cost of labor plus labor burden and 10% profit, and the cost of equipment plus 10% profit. It does not include the general contractor's overhead and profit. Note: See the inside back cover of the printed product or the reference section of the electronic product for details of how RSMeans calculates labor burden.

National Average

*The data in RSMeans print publications represents a "national average" cost. This data should be modified to the project location using the **City Cost Indexes** or **Location Factors tables** found in the Reference Section. Use the location factors to adjust estimate totals if the project covers multiple trades. Use the city cost indexes (CCI) for single trade projects or projects where a more detailed analysis is required. All figures in the two tables are derived from the same research. The last row of data in the CCI, the weighted average, is the same as the numbers reported for each location in the location factor table.*

How RSMeans Unit Price Data Works (Continued)

							01/01/16	RESI
Project Name: Pre-Engineered Steel Building			Architect: As Shown					
Location:	Anywhere, USA							
Line Number	Description	Qty	Unit	Material	Labor	Equipment	SubContract	Estimate Total
03 30 53.40 3940	Strip footing, 12" x 24", reinforced	34	C.Y.	$4,896.00	$2,499.00	$22.10	$0.00	
03 30 53.40 3950	Strip footing, 12" x 36", reinforced	15	C.Y.	$2,070.00	$885.00	$7.80	$0.00	
03 11 13.65 3000	Concrete slab edge forms	500	L.F.	$165.00	$780.00	$0.00	$0.00	
03 22 11.10 0200	Welded wire fabric reinforcing	150	C.S.F.	$2,632.50	$2,745.00	$0.00	$0.00	
03 31 13.35 0300	Ready mix concrete, 4000 psi for slab on grade	278	C.Y.	$31,414.00	$0.00	$0.00	$0.00	
03 31 13.70 4300	Place, strike off & consolidate concrete slab	278	C.Y.	$0.00	$3,183.10	$161.24	$0.00	
03 35 13.30 0250	Machine float & trowel concrete slab	15,000	S.F.	$0.00	$6,150.00	$450.00	$0.00	
03 15 16.20 0140	Cut control joints in concrete slab	950	L.F.	$47.50	$266.00	$85.50	$0.00	
03 39 23.13 0300	Sprayed concrete curing membrane	150	C.S.F.	$1,905.00	$622.50	$0.00	$0.00	
Division 03	**Subtotal**			**$43,130.00**	**$17,130.60**	**$726.64**	**$0.00**	**$60,987.24**
08 36 13.10 2650	Manual 10' x 10' steel sectional overhead door	8	Ea.	$9,800.00	$2,400.00	$0.00	$0.00	
08 36 13.10 2860	Insulation and steel back panel for OH door	800	S.F.	$3,800.00	$0.00	$0.00	$0.00	
Division 08	**Subtotal**			**$13,600.00**	**$2,400.00**	**$0.00**	**$0.00**	**$16,000.00**
13 34 19.50 1100	Pre-Engineered Steel Building, 100' x 150' x 24'	15,000	SF Flr.	$0.00	$0.00	$0.00	$293,250.00	
13 34 19.50 6050	Framing for PESB door opening, 3' x 7'	4	Opng.	$0.00	$0.00	$0.00	$1,860.00	
13 34 19.50 6100	Framing for PESB door opening, 10' x 10'	8	Opng.	$0.00	$0.00	$0.00	$8,200.00	
13 34 19.50 6200	Framing for PESB window opening, 4' x 3'	6	Opng.	$0.00	$0.00	$0.00	$2,820.00	
13 34 19.50 5750	PESB door, 3' x 7', single leaf	4	Opng.	$2,380.00	$456.00	$0.00	$0.00	
13 34 19.50 7750	PESB sliding window, 4' x 3' with screen	6	Opng.	$2,430.00	$264.00	$67.80	$0.00	
13 34 19.50 6550	PESB gutter, eave type, 26 ga., painted	300	L.F.	$2,040.00	$534.00	$0.00	$0.00	
13 34 19.50 8650	PESB roof vent, 12" wide x 10' long	15	Ea.	$540.00	$2,145.00	$0.00	$0.00	
13 34 19.50 6900	PESB insulation, vinyl faced, 4" thick	27,400	S.F.	$12,330.00	$6,576.00	$0.00	$0.00	
Division 13	**Subtotal**			**$19,720.00**	**$9,975.00**	**$67.80**	**$306,130.00**	**$335,892.80**
			Subtotal	$76,450.00	$29,505.60	$794.44	$306,130.00	$412,880.04
Division 01	General Requirements @ 7%			5,351.50	2,065.39	55.61	21,429.10	
			Estimate Subtotal	$81,801.50	$31,570.99	$850.05	$327,559.10	$412,880.04
			Sales Tax @ 5%	4,090.08		42.50	8,188.98	
			Subtotal A	85,891.58	31,570.99	892.55	335,748.08	
			GC O & P	8,589.16	21,468.27	89.26	33,574.81	
			Subtotal B	94,480.73	53,039.27	981.81	369,322.89	$517,824.69
			Contingency @ 5%					25,891.23
			Subtotal C					$543,715.93
			Bond @ $12/1000 +10% O&P					7,177.05
			Subtotal D					$550,892.98
			Location Adjustment Factor		102.30			12,670.54
			Grand Total					**$563,563.52**

This estimate is based on an interactive spreadsheet. A copy of this spreadsheet is located on the RSMeans website at www.RSMeans.com/2016extras. You are free to download it and adjust it to your methodology.

Sample Estimate

This sample demonstrates the elements of an estimate, including a tally of the RSMeans data lines and a summary of the markups on a contractor's work to arrive at a total cost to the owner. The RSMeans Location Factor is added at the bottom of the estimate to adjust the cost of the work to a specific location.

1. Work Performed

The body of the estimate shows the RSMeans data selected, including the line number, a brief description of each item, its take-off unit and quantity, and the bare costs of materials, labor, and equipment. This estimate also includes a column titled "SubContract." This data is taken from the RSMeans column "Total Incl O&P" and represents the total that a subcontractor would charge a general contractor for the work, including the sub's markup for overhead and profit.

2. Division 1, General Requirements

This is the first division numerically but the last division estimated. Division 1 includes project-wide needs provided by the general contractor. These requirements vary by project but may include temporary facilities and utilities, security, testing, project cleanup, etc. For small projects a percentage can be used, typically between 5% and 15% of project cost. For large projects the costs may be itemized and priced individually.

3. Sales Tax

If the work is subject to state or local sales taxes, the amount must be added to the estimate. Sales tax may be added to material costs, equipment costs, and subcontracted work. In this case, sales tax was added in all three categories. It was assumed that approximately half the subcontracted work would be material cost, so the tax was applied to 50% of the subcontract total.

4. GC O&P

This entry represents the general contractor's markup on material, labor, equipment, and subcontractor costs. RSMeans' standard markup on materials, equipment, and subcontracted work is 10%. In this estimate, the markup on the labor performed by the GC's workers uses "Skilled Workers Average" shown in Column F on the table "Installing Contractor's Overhead & Profit," which can be found on the inside-back cover of the printed product or in the Reference Section of the electronic product.

5. Contingency

A factor for contingency may be added to any estimate to represent the cost of unknowns that may occur between the time that the estimate is performed and the time the project is constructed. The amount of the allowance will depend on the stage of design at which the estimate is done and the contractor's assessment of the risk involved. Refer to section 01 21 16.50 for contingency allowances.

6. Bonds

Bond costs should be added to the estimate. The figures here represent a typical performance bond, ensuring the owner that if the general contractor does not complete the obligations in the construction contract the bonding company will pay the cost for completion of the work.

7. Location Adjustment

RSMeans published data is based on national average costs. If necessary, adjust the total cost of the project using a location factor from the "Location Factor" table or the "City Cost Index" table. Use location factors if the work is general, covering multiple trades. If the work is by a single trade (e.g., masonry) use the more specific data found in the "City Cost Indexes."

Estimating Tips

01 20 00 Price and Payment Procedures

- Allowances that should be added to estimates to cover contingencies and job conditions that are not included in the national average material and labor costs are shown in section 01 21.
- When estimating historic preservation projects (depending on the condition of the existing structure and the owner's requirements), a 15%–20% contingency or allowance is recommended, regardless of the stage of the drawings.

01 30 00 Administrative Requirements

- Before determining a final cost estimate, it is a good practice to review all the items listed in Subdivisions 01 31 and 01 32 to make final adjustments for items that may need customizing to specific job conditions.
- Requirements for initial and periodic submittals can represent a significant cost to the General Requirements of a job. Thoroughly check the submittal specifications when estimating a project to determine any costs that should be included.

01 40 00 Quality Requirements

- All projects will require some degree of Quality Control. This cost is not included in the unit cost of construction listed in each division. Depending upon the terms of the contract, the various costs of inspection and testing can be the responsibility of either the owner or the contractor. Be sure to include the required costs in your estimate.

01 50 00 Temporary Facilities and Controls

- Barricades, access roads, safety nets, scaffolding, security, and many more requirements for the execution of a safe project are elements of direct cost. These costs can easily be overlooked when preparing an estimate. When looking through the major classifications of this subdivision, determine which items apply to each division in your estimate.
- Construction Equipment Rental Costs can be found in the Reference Section in section 01 54 33. Operators' wages are not included in equipment rental costs.
- Equipment mobilization and demobilization costs are not included in equipment rental costs and must be considered separately.

- The cost of small tools provided by the installing contractor for his workers is covered in the "Overhead" column on the "Installing Contractor's Overhead and Profit" table that lists labor trades, base rates and markups and, therefore, is included in the "Total Incl. O&P" cost of any Unit Price line item.

01 70 00 Execution and Closeout Requirements

- When preparing an estimate, thoroughly read the specifications to determine the requirements for Contract Closeout. Final cleaning, record documentation, operation and maintenance data, warranties and bonds, and spare parts and maintenance materials can all be elements of cost for the completion of a contract. Do not overlook these in your estimate.

Reference Numbers

Reference numbers are shown at the beginning of some major classifications. These numbers refer to related items in the Reference Section. The reference information may be an estimating procedure, an alternate pricing method, or technical information.

Note: Not all subdivisions listed here necessarily appear. ■

Did you know?

RSMeans Online gives you the same access to RSMeans' data with 24/7 access:

- Quickly locate costs in the searchable database.
- Build cost lists, estimates, and reports in minutes.
- Adjust costs to any location in the U.S. and Canada with the click of a button.

Start your free trial today at **www.RSMeansOnline.com**

RSMeans Online
FROM THE GORDIAN GROUP

No part of this cost data may be reproduced, stored in a retrieval system, or transmitted in any form or by any means without prior written permission of RSMeans.

01 11 Summary of Work

01 11 31 – Professional Consultants

01 11 31.10 Architectural Fees

		Crew	Daily Output	Labor-Hours	Unit	Material	2016 Bare Costs Labor	Equipment	Total	Total Incl O&P
0010	**ARCHITECTURAL FEES** R011110-10									
0020	For new construction									
0060	Minimum				Project				4.90%	4.90%
0090	Maximum								16%	16%
0100	For alteration work, to $500,000, add to new construction fee								50%	50%
0150	Over $500,000, add to new construction fee								25%	25%

01 11 31.20 Construction Management Fees

		Crew	Daily Output	Labor-Hours	Unit	Material	Labor	Equipment	Total	Total Incl O&P
0010	**CONSTRUCTION MANAGEMENT FEES**									
0060	For work to $100,000				Project				10%	10%
0070	To $250,000								9%	9%
0090	To $1,000,000								6%	6%

01 11 31.75 Renderings

		Crew	Daily Output	Labor-Hours	Unit	Material	Labor	Equipment	Total	Total Incl O&P
0010	**RENDERINGS** Color, matted, 20" x 30", eye level,									
0050	Average				Ea.	2,875			2,875	3,175

01 21 Allowances

01 21 16 – Contingency Allowances

01 21 16.50 Contingencies

		Crew	Daily Output	Labor-Hours	Unit	Material	Labor	Equipment	Total	Total Incl O&P
0010	**CONTINGENCIES**, Add to estimate									
0020	Conceptual stage				Project				20%	20%
0150	Final working drawing stage				"				3%	3%

01 21 55 – Job Conditions Allowance

01 21 55.50 Job Conditions

		Crew	Daily Output	Labor-Hours	Unit	Material	Labor	Equipment	Total	Total Incl O&P
0010	**JOB CONDITIONS** Modifications to applicable									
8000	Remove and reset contents of small room	1 Clab	6.50	1.231	Room		30.50		30.50	51
8010	Average room		4.70	1.702			42		42	70
8020	Large room		3.50	2.286			56.50		56.50	94.50
8030	Extra large room		2.40	3.333			82		82	138

01 21 63 – Taxes

01 21 63.10 Taxes

		Crew	Daily Output	Labor-Hours	Unit	Material	Labor	Equipment	Total	Total Incl O&P
0010	**TAXES** R012909-80									
0020	Sales tax, State, average				%	5.08%				
0050	Maximum R012909-85					7.50%				
0200	Social Security, on first $118,500 of wages						7.65%			
0300	Unemployment, combined Federal and State, minimum						.60%			
0350	Average						9.60%			
0400	Maximum						12%			

01 31 Project Management and Coordination

01 31 13 – Project Coordination

01 31 13.30 Insurance

		Crew	Daily Output	Labor-Hours	Unit	Material	2016 Bare Costs Labor	Equipment	Total	Total Incl O&P
0010	**INSURANCE** R013113-40									
0020	Builders risk, standard, minimum				Job				.24%	.24%
0050	Maximum R013113-60								.64%	.64%
0200	All-risk type, minimum								.25%	.25%
0250	Maximum								.62%	.62%
0400	Contractor's equipment floater, minimum				Value				.50%	.50%
0450	Maximum				"				1.50%	1.50%
0600	Public liability, average				Job				2.02%	2.02%
0800	Workers' compensation & employer's liability, average									
0850	by trade, carpentry, general				Payroll		14.37%			
0900	Clerical						.48%			
0950	Concrete						12.56%			
1000	Electrical						5.64%			
1050	Excavation						9.34%			
1100	Glazing						13.08%			
1150	Insulation						11.56%			
1200	Lathing						8.63%			
1250	Masonry						13.73%			
1300	Painting & decorating						11.66%			
1350	Pile driving						14.63%			
1400	Plastering						10.49%			
1450	Plumbing						6.94%			
1500	Roofing						31.44%			
1550	Sheet metal work (HVAC)						8.84%			
1600	Steel erection, structural						27.45%			
1650	Tile work, interior ceramic						8.82%			
1700	Waterproofing, brush or hand caulking						6.76%			
1800	Wrecking						22.92%			
2000	Range of 35 trades in 50 states, excl. wrecking & clerical, min.						1.42%			
2100	Average						12.61%			
2200	Maximum						107.12%			

01 41 Regulatory Requirements

01 41 26 – Permit Requirements

01 41 26.50 Permits

		Crew	Daily Output	Labor-Hours	Unit	Material	Labor	Equipment	Total	Total Incl O&P
0010	**PERMITS**									
0020	Rule of thumb, most cities, minimum				Job				.50%	.50%
0100	Maximum				"				2%	2%

01 54 Construction Aids

01 54 16 – Temporary Hoists

01 54 16.50 Weekly Forklift Crew

		Crew	Daily Output	Labor-Hours	Unit	Material	Labor	Equipment	Total	Total Incl O&P
0010	**WEEKLY FORKLIFT CREW**									
0100	All-terrain forklift, 45' lift, 35' reach, 9000 lb. capacity	A-3P	.20	40	Week	1,350	2,675		4,025	5,200

01 54 Construction Aids

01 54 19 – Temporary Cranes

01 54 19.50 Daily Crane Crews

		Crew	Daily Output	Labor-Hours	Unit	Material	2016 Bare Costs Labor	2016 Bare Costs Equipment	Total	Total Incl O&P
0010	**DAILY CRANE CREWS** for small jobs, portal to portal									
0100	12-ton truck-mounted hydraulic crane	A-3H	1	8	Day		288	875	1,163	1,450
0900	If crane is needed on a Saturday, Sunday or Holiday									
0910	At time-and-a-half, add				Day		50%			
0920	At double time, add				"		100%			

01 54 23 – Temporary Scaffolding and Platforms

01 54 23.60 Pump Staging

		Crew	Daily Output	Labor-Hours	Unit	Material	Labor	Equipment	Total	Total Incl O&P
0010	**PUMP STAGING**, Aluminum									
1300	System in place, 50' working height, per use based on 50 uses	2 Carp	84.80	.189	C.S.F.	6.95	6.40		13.35	18.35
1400	100 uses	R015423-20	84.80	.189		3.49	6.40		9.89	14.55
1500	150 uses		84.80	.189		2.34	6.40		8.74	13.25

01 54 23.70 Scaffolding

		Crew	Daily Output	Labor-Hours	Unit	Material	Labor	Equipment	Total	Total Incl O&P
0010	**SCAFFOLDING** R015423-10									
0015	Steel tube, regular, no plank, labor only to erect & dismantle									
0091	Building exterior, wall face, 1 to 5 stories, 6'-4" x 5' frames	3 Clab	8	3	C.S.F.		74		74	124
0201	6 to 12 stories	4 Clab	8	4			98.50		98.50	165
0310	13 to 20 stories	5 Carp	8	5			170		170	284
0461	Building interior, walls face area, up to 16' high	3 Clab	12	2			49.50		49.50	82.50
0561	16' to 40' high		10	2.400			59		59	99
0801	Building interior floor area, up to 30' high		150	.160	C.C.F.		3.94		3.94	6.60
0901	Over 30' high	4 Clab	160	.200	"		4.93		4.93	8.25
0906	Complete system for face of walls, no plank, material only rent/mo				C.S.F.	35.50			35.50	39
0908	Interior spaces, no plank, material only rent/mo				C.C.F.	4.30			4.30	4.73
0910	Steel tubular, heavy duty shoring, buy									
0920	Frames 5' high 2' wide				Ea.	81.50			81.50	90
0925	5' high 4' wide					93			93	102
0930	6' high 2' wide					93.50			93.50	103
0935	6' high 4' wide					109			109	120
0940	Accessories									
0945	Cross braces				Ea.	15.50			15.50	17.05
0950	U-head, 8" x 8"					19.10			19.10	21
0955	J-head, 4" x 8"					13.90			13.90	15.30
0960	Base plate, 8" x 8"					15.50			15.50	17.05
0965	Leveling jack					33.50			33.50	36.50
1000	Steel tubular, regular, buy									
1100	Frames 3' high 5' wide				Ea.	89			89	97.50
1150	5' high 5' wide					104			104	114
1200	6'-4" high 5' wide					147			147	162
1350	7'-6" high 6' wide					158			158	173
1500	Accessories cross braces					18.10			18.10	19.90
1550	Guardrail post					16.40			16.40	18.05
1600	Guardrail 7' section					8.70			8.70	9.60
1650	Screw jacks & plates					21.50			21.50	24
1700	Sidearm brackets					30.50			30.50	33.50
1750	8" casters					31			31	34
1800	Plank 2" x 10" x 16'-0"					59.50			59.50	65.50
1900	Stairway section					286			286	315
1910	Stairway starter bar					32			32	35
1920	Stairway inside handrail					58.50			58.50	64
1930	Stairway outside handrail					86.50			86.50	95
1940	Walk-thru frame guardrail					41.50			41.50	46

01 54 Construction Aids

01 54 23 – Temporary Scaffolding and Platforms

01 54 23.70 Scaffolding		Crew	Daily Output	Labor-Hours	Unit	Material	2016 Bare Costs Labor	Equipment	Total	Total Incl O&P
2000	Steel tubular, regular, rent/mo.									
2100	Frames 3' high 5' wide				Ea.	5			5	5.50
2150	5' high 5' wide					5			5	5.50
2200	6'-4" high 5' wide					5.35			5.35	5.90
2250	7'-6" high 6' wide					10			10	11
2500	Accessories, cross braces					1			1	1.10
2550	Guardrail post					1			1	1.10
2600	Guardrail 7' section					1			1	1.10
2650	Screw jacks & plates					2			2	2.20
2700	Sidearm brackets					2			2	2.20
2750	8" casters					8			8	8.80
2800	Outrigger for rolling tower					3			3	3.30
2850	Plank 2" x 10" x 16'-0"					10			10	11
2900	Stairway section					35			35	38.50
2940	Walk-thru frame guardrail				↓	2.50			2.50	2.75
3000	Steel tubular, heavy duty shoring, rent/mo.									
3250	5' high 2' & 4' wide				Ea.	8.50			8.50	9.35
3300	6' high 2' & 4' wide					8.50			8.50	9.35
3500	Accessories, cross braces					1			1	1.10
3600	U - head, 8" x 8"					2.50			2.50	2.75
3650	J - head, 4" x 8"					2.50			2.50	2.75
3700	Base plate, 8" x 8"					1			1	1.10
3750	Leveling jack					2.50			2.50	2.75
5700	Planks, 2" x 10" x 16'-0", labor only to erect & remove to 50' H	3 Carp	72	.333			11.30		11.30	18.90
5800	Over 50' high	4 Carp	80	.400	↓		13.55		13.55	22.50

01 54 23.80 Staging Aids

		Crew	Daily Output	Labor-Hours	Unit	Material	Labor	Equipment	Total	Total Incl O&P
0010	**STAGING AIDS** and fall protection equipment									
0100	Sidewall staging bracket, tubular, buy				Ea.	51			51	56
0110	Cost each per day, based on 250 days use				Day	.20			.20	.22
0200	Guard post, buy				Ea.	52			52	57
0210	Cost each per day, based on 250 days use				Day	.21			.21	.23
0300	End guard chains, buy per pair				Pair	38			38	41.50
0310	Cost per set per day, based on 250 days use				Day	.20			.20	.22
1010	Cost each per day, based on 250 days use				"	.04			.04	.04
1100	Wood bracket, buy				Ea.	21			21	23.50
1110	Cost each per day, based on 250 days use				Day	.08			.08	.09
2010	Cost per pair per day, based on 250 days use				"	.49			.49	.53
2100	Steel siderail jack, buy per pair				Pair	97.50			97.50	107
2110	Cost per pair per day, based on 250 days use				Day	.39			.39	.43
3010	Cost each per day, based on 250 days use				"	.24			.24	.26
3100	Aluminum scaffolding plank, 20" wide x 24' long, buy				Ea.	790			790	870
3110	Cost each per day, based on 250 days use				Day	3.16			3.16	3.48
4010	Cost each per day, based on 250 days use				"	.70			.70	.77
4100	Rope for safety line, 5/8" x 100' nylon, buy				Ea.	56			56	61.50
4110	Cost each per day, based on 250 days use				Day	.22			.22	.25
4200	Permanent U-Bolt roof anchor, buy				Ea.	35.50			35.50	39.50
4300	Temporary (one use) roof ridge anchor, buy				"	32			32	35.50
5000	Installation (setup and removal) of staging aids									
5010	Sidewall staging bracket	2 Carp	64	.250	Ea.		8.50		8.50	14.20
5020	Guard post with 2 wood rails	"	64	.250			8.50		8.50	14.20
5030	End guard chains, set	1 Carp	64	.125			4.24		4.24	7.10
5100	Roof shingling bracket		96	.083	↓		2.82		2.82	4.73

01 54 Construction Aids

01 54 23 – Temporary Scaffolding and Platforms

01 54 23.80 Staging Aids

		Crew	Daily Output	Labor-Hours	Unit	Material	2016 Bare Costs Labor	2016 Bare Costs Equipment	Total	Total Incl O&P
5200	Ladder jack	1 Carp	64	.125	Ea.		4.24		4.24	7.10
5300	Wood plank, 2" x 10" x 16'	2 Carp	80	.200			6.80		6.80	11.35
5310	Aluminum scaffold plank, 20" x 24'	"	40	.400			13.55		13.55	22.50
5410	Safety rope	1 Carp	40	.200			6.80		6.80	11.35
5420	Permanent U-Bolt roof anchor (install only)	2 Carp	40	.400			13.55		13.55	22.50
5430	Temporary roof ridge anchor (install only)	1 Carp	64	.125			4.24		4.24	7.10

01 54 36 – Equipment Mobilization

01 54 36.50 Mobilization

		Crew	Daily Output	Labor-Hours	Unit	Material	Labor	Equipment	Total	Total Incl O&P
0010	**MOBILIZATION** (Use line item again for demobilization) R015436-50									
0015	Up to 25 mi. haul dist. (50 mi. RT for mob/demob crew)									
1200	Small equipment, placed in rear of, or towed by pickup truck	A-3A	4	2	Ea.		68	39	107	155
1300	Equipment hauled on 3-ton capacity towed trailer	A-3Q	2.67	3			102	67.50	169.50	243
1400	20-ton capacity	B-34U	2	8			258	236	494	685
1500	40-ton capacity	B-34N	2	8			264	375	639	850
1700	Crane, truck-mounted, up to 75 ton (driver only)	1 Eqhv	4	2			72		72	119
2500	For each additional 5 miles haul distance, add						10%	10%		
3000	For large pieces of equipment, allow for assembly/knockdown									
3100	For mob/demob of micro-tunneling equip, see Section 33 05 23.19									

01 56 Temporary Barriers and Enclosures

01 56 13 – Temporary Air Barriers

01 56 13.60 Tarpaulins

		Crew	Daily Output	Labor-Hours	Unit	Material	Labor	Equipment	Total	Total Incl O&P
0010	**TARPAULINS**									
0020	Cotton duck, 10 oz. to 13.13 oz. per S.Y., 6' x 8'				S.F.	.88			.88	.97
0050	30' x 30'					.64			.64	.70
0200	Reinforced polyethylene 3 mils thick, white					.04			.04	.04
0300	4 mils thick, white, clear or black					.11			.11	.12
0730	Polyester reinforced w/integral fastening system 11 mils thick					.22			.22	.24

01 56 16 – Temporary Dust Barriers

01 56 16.10 Dust Barriers, Temporary

		Crew	Daily Output	Labor-Hours	Unit	Material	Labor	Equipment	Total	Total Incl O&P
0010	**DUST BARRIERS, TEMPORARY**									
0020	Spring loaded telescoping pole & head, to 12', erect and dismantle	1 Clab	240	.033	Ea.		.82		.82	1.37
0025	Cost per day (based upon 250 days)				Day	.26			.26	.29
0030	To 21', erect and dismantle	1 Clab	240	.033	Ea.		.82		.82	1.37
0035	Cost per day (based upon 250 days)				Day	.54			.54	.60
0040	Accessories, caution tape reel, erect and dismantle	1 Clab	480	.017	Ea.		.41		.41	.69
0045	Cost per day (based upon 250 days)				Day	.36			.36	.40
0060	Foam rail and connector, erect and dismantle	1 Clab	240	.033	Ea.		.82		.82	1.37
0065	Cost per day (based upon 250 days)				Day	.12			.12	.13
0070	Caution tape	1 Clab	384	.021	C.L.F.	3.13	.51		3.64	4.30
0080	Zipper, standard duty		60	.133	Ea.	8	3.29		11.29	14.30
0090	Heavy duty		48	.167	"	9.75	4.11		13.86	17.65
0100	Polyethylene sheet, 4 mil		37	.216	Sq.	2.55	5.35		7.90	11.70
0110	6 mil		37	.216	"	3.70	5.35		9.05	12.95
1000	Dust partition, 6 mil polyethylene, 1" x 3" frame	2 Carp	2000	.008	S.F.	.30	.27		.57	.77
1080	2" x 4" frame	"	2000	.008	"	.33	.27		.60	.81

01 71 Examination and Preparation

01 71 23 – Field Engineering

01 71 23.13 Construction Layout

		Crew	Daily Output	Labor-Hours	Unit	Material	2016 Bare Costs Labor	2016 Bare Costs Equipment	Total	Total Incl O&P
0010	**CONSTRUCTION LAYOUT**									
1100	Crew for layout of building, trenching or pipe laying, 2 person crew	A-6	1	16	Day		540	43.50	583.50	950
1200	3 person crew	A-7	1	24	"		865	43.50	908.50	1,500

01 74 Cleaning and Waste Management

01 74 13 – Progress Cleaning

01 74 13.20 Cleaning Up

		Crew	Daily Output	Labor-Hours	Unit	Material	Labor	Equipment	Total	Total Incl O&P
0010	**CLEANING UP**									
0020	After job completion, allow, minimum				Job				.30%	.30%
0040	Maximum				"				1%	1%

01 76 Protecting Installed Construction

01 76 13 – Temporary Protection of Installed Construction

01 76 13.20 Temporary Protection

		Crew	Daily Output	Labor-Hours	Unit	Material	Labor	Equipment	Total	Total Incl O&P
0010	**TEMPORARY PROTECTION**									
0020	Flooring, 1/8" tempered hardboard, taped seams	2 Carp	1500	.011	S.F.	.42	.36		.78	1.07
0030	Peel away carpet protection	1 Clab	3200	.003	"	.13	.06		.19	.24

Division Notes

Estimating Tips
02 30 00 Subsurface Investigation
In preparing estimates on structures involving earthwork or foundations, all information concerning soil characteristics should be obtained. Look particularly for hazardous waste, evidence of prior dumping of debris, and previous stream beds.

02 40 00 Demolition and Structure Moving
The costs shown for selective demolition do not include rubbish handling or disposal. These items should be estimated separately using RSMeans data or other sources.
- Historic preservation often requires that the contractor remove materials from the existing structure, rehab them, and replace them. The estimator must be aware of any related measures and precautions that must be taken when doing selective demolition and cutting and patching. Requirements may include special handling and storage, as well as security.
- In addition to Subdivision 02 41 00, you can find selective demolition items in each division. Example: Roofing demolition is in Division 7.
- Absent of any other specific reference, an approximate demolish-in-place cost can be obtained by halving the new-install labor cost. To remove for reuse, allow the entire new-install labor figure.

02 40 00 Building Deconstruction
This section provides costs for the careful dismantling and recycling of most of low-rise building materials.

02 50 00 Containment of Hazardous Waste
This section addresses on-site hazardous waste disposal costs.

02 80 00 Hazardous Material Disposal/ Remediation
This subdivision includes information on hazardous waste handling, asbestos remediation, lead remediation, and mold remediation. See reference R028213-20 and R028319-60 for further guidance in using these unit price lines.

02 90 00 Monitoring Chemical Sampling, Testing Analysis
This section provides costs for on-site sampling and testing hazardous waste.

Reference Numbers
Reference numbers are shown at the beginning of some major classifications. These numbers refer to related items in the Reference Section. The reference information may be an estimating procedure, an alternate pricing method, or technical information.

Note: Not all subdivisions listed here necessarily appear. ∎

No part of this cost data may be reproduced, stored in a retrieval system, or transmitted in any form or by any means without prior written permission of RSMeans.

Did you know?
RSMeans Online gives you the same access to RSMeans' data with 24/7 access:
- Quickly locate costs in the searchable database.
- Build cost lists, estimates, and reports in minutes.
- Adjust costs to any location in the U.S. and Canada with the click of a button.

Start your free trial today at **www.RSMeansOnline.com**

RSMeans Online
FROM THE GORDIAN GROUP

02 21 Surveys

02 21 13 – Site Surveys

02 21 13.09 Topographical Surveys

		Crew	Daily Output	Labor-Hours	Unit	Material	2016 Bare Costs Labor	Equipment	Total	Total Incl O&P
0010	**TOPOGRAPHICAL SURVEYS**									
0020	Topographical surveying, conventional, minimum	A-7	3.30	7.273	Acre	20	262	13.25	295.25	470
0100	Maximum	A-8	.60	53.333	"	60	1,875	73	2,008	3,275

02 21 13.13 Boundary and Survey Markers

0010	**BOUNDARY AND SURVEY MARKERS**									
0300	Lot location and lines, large quantities, minimum	A-7	2	12	Acre	35	430	22	487	785
0320	Average	"	1.25	19.200		55	690	35	780	1,250
0400	Small quantities, maximum	A-8	1	32		75	1,125	44	1,244	2,000
0600	Monuments, 3' long	A-7	10	2.400	Ea.	40	86.50	4.37	130.87	193
0800	Property lines, perimeter, cleared land	"	1000	.024	L.F.	.05	.86	.04	.95	1.55
0900	Wooded land	A-8	875	.037	"	.07	1.29	.05	1.41	2.28

02 32 Geotechnical Investigations

02 32 13 – Subsurface Drilling and Sampling

02 32 13.10 Boring and Exploratory Drilling

		Crew	Daily Output	Labor-Hours	Unit	Material	Labor	Equipment	Total	Total Incl O&P
0010	**BORING AND EXPLORATORY DRILLING**									
0020	Borings, initial field stake out & determination of elevations	A-6	1	16	Day		540	43.50	583.50	950
0100	Drawings showing boring details				Total		335		335	425
0200	Report and recommendations from P.E.						775		775	970
0300	Mobilization and demobilization	B-55	4	4			109	273	382	480
0350	For over 100 miles, per added mile		450	.036	Mile		.97	2.43	3.40	4.28
0600	Auger holes in earth, no samples, 2-1/2" diameter		78.60	.204	L.F.		5.55	13.90	19.45	24.50
0800	Cased borings in earth, with samples, 2-1/2" diameter		55.50	.288	"	15	7.85	19.70	42.55	51
1400	Borings, earth, drill rig and crew with truck mounted auger		1	16	Day		435	1,100	1,535	1,925
1500	For inner city borings add, minimum								10%	10%
1510	Maximum								20%	20%

02 41 Demolition

02 41 13 – Selective Site Demolition

02 41 13.17 Demolish, Remove Pavement and Curb

			Crew	Daily Output	Labor-Hours	Unit	Material	Labor	Equipment	Total	Total Incl O&P
0010	**DEMOLISH, REMOVE PAVEMENT AND CURB**	R024119-10									
5010	Pavement removal, bituminous roads, up to 3" thick		B-38	690	.035	S.Y.		.96	.79	1.75	2.48
5050	4" to 6" thick			420	.057			1.58	1.30	2.88	4.07
5100	Bituminous driveways			640	.038			1.04	.85	1.89	2.67
5200	Concrete to 6" thick, hydraulic hammer, mesh reinforced			255	.094			2.61	2.14	4.75	6.70
5300	Rod reinforced			200	.120			3.33	2.73	6.06	8.55
5600	With hand held air equipment, bituminous, to 6" thick		B-39	1900	.025	S.F.		.63	.12	.75	1.19
5700	Concrete to 6" thick, no reinforcing			1600	.030			.75	.14	.89	1.41
5800	Mesh reinforced			1400	.034			.86	.16	1.02	1.61
5900	Rod reinforced			765	.063			1.57	.30	1.87	2.95
6000	Curbs, concrete, plain		B-6	360	.067	L.F.		1.85	1.02	2.87	4.20
6100	Reinforced			275	.087			2.42	1.33	3.75	5.50
6200	Granite			360	.067			1.85	1.02	2.87	4.20
6300	Bituminous			528	.045			1.26	.69	1.95	2.86

02 41 13.23 Utility Line Removal

		Crew	Daily Output	Labor-Hours	Unit	Material	Labor	Equipment	Total	Total Incl O&P
0010	**UTILITY LINE REMOVAL**									
0015	No hauling, abandon catch basin or manhole	B-6	7	3.429	Ea.		95	52	147	216
0020	Remove existing catch basin or manhole, masonry		4	6			166	91.50	257.50	380
0030	Catch basin or manhole frames and covers, stored		13	1.846			51	28	79	117

02 41 Demolition

02 41 13 – Selective Site Demolition

02 41 13.23 Utility Line Removal

		Crew	Daily Output	Labor-Hours	Unit	Material	2016 Bare Costs Labor	2016 Bare Costs Equipment	Total	Total Incl O&P
0040	Remove and reset	B-6	7	3.429	Ea.		95	52	147	216
2900	Pipe removal, sewer/water, no excavation, 12" diameter	↓	175	.137	L.F.		3.80	2.09	5.89	8.65
2930	15"-18" diameter	B-12Z	150	.160			4.55	10.55	15.10	19.20
2960	21"-24" diameter	"	120	.200	↓		5.70	13.20	18.90	24

02 41 13.30 Minor Site Demolition

		Crew	Daily Output	Labor-Hours	Unit	Material	Labor	Equipment	Total	Total Incl O&P
0010	**MINOR SITE DEMOLITION** R024119-10									
1000	Masonry walls, block, solid	B-5	1800	.022	C.F.		.60	.79	1.39	1.88
1200	Brick, solid		900	.044			1.21	1.58	2.79	3.75
1400	Stone, with mortar		900	.044			1.21	1.58	2.79	3.75
1500	Dry set	↓	1500	.027	↓		.72	.95	1.67	2.26
4000	Sidewalk removal, bituminous, 2" thick	B-6	350	.069	S.Y.		1.90	1.04	2.94	4.32
4010	2-1/2" thick		325	.074			2.05	1.12	3.17	4.65
4050	Brick, set in mortar		185	.130			3.60	1.98	5.58	8.15
4100	Concrete, plain, 4"		160	.150			4.16	2.28	6.44	9.45
4110	Plain, 5"		140	.171			4.75	2.61	7.36	10.75
4120	Plain, 6"		120	.200			5.55	3.05	8.60	12.60
4200	Mesh reinforced, concrete, 4"		150	.160			4.44	2.44	6.88	10.10
4210	5" thick		131	.183			5.10	2.79	7.89	11.50
4220	6" thick	↓	112	.214	↓		5.95	3.26	9.21	13.50
4300	Slab on grade removal, plain	B-5	45	.889	C.Y.		24	31.50	55.50	75.50
4310	Mesh reinforced		33	1.212			33	43	76	103
4320	Rod reinforced	↓	25	1.600			43.50	57	100.50	135
4400	For congested sites or small quantities, add up to								200%	200%
4450	For disposal on site, add	B-11A	232	.069			2.07	6	8.07	10.05
4500	To 5 miles, add	B-34D	76	.105	↓		3.23	9.75	12.98	16

02 41 13.60 Selective Demolition Fencing

		Crew	Daily Output	Labor-Hours	Unit	Material	Labor	Equipment	Total	Total Incl O&P
0010	**SELECTIVE DEMOLITION FENCING** R024119-10									
1600	Fencing, barbed wire, 3 strand	2 Clab	430	.037	L.F.		.92		.92	1.53
1650	5 strand	"	280	.057			1.41		1.41	2.36
1700	Chain link, posts & fabric, 8' to 10' high, remove only	B-6	445	.054	↓		1.50	.82	2.32	3.39

02 41 16 – Structure Demolition

02 41 16.13 Building Demolition

		Crew	Daily Output	Labor-Hours	Unit	Material	Labor	Equipment	Total	Total Incl O&P
0010	**BUILDING DEMOLITION** Large urban projects, incl. 20 mi. haul R024119-10									
0500	Small bldgs, or single bldgs, no salvage included, steel	B-3	14800	.003	C.F.		.09	.17	.26	.34
0600	Concrete	"	11300	.004	"		.12	.23	.35	.45
0605	Concrete, plain	B-5	33	1.212	C.Y.		33	43	76	103
0610	Reinforced		25	1.600			43.50	57	100.50	135
0615	Concrete walls		34	1.176			32	42	74	99.50
0620	Elevated slabs	↓	26	1.538	↓		42	55	97	130
0650	Masonry	B-3	14800	.003	C.F.		.09	.17	.26	.34
0700	Wood		14800	.003	"		.09	.17	.26	.34
1000	Demoliton single family house, one story, wood 1600 S.F.		1	48	Ea.		1,375	2,575	3,950	5,125
1020	3200 S.F.		.50	96			2,750	5,150	7,900	10,300
1200	Demoliton two family house, two story, wood 2400 S.F.		.67	71.964			2,075	3,875	5,950	7,675
1220	4200 S.F.		.38	128			3,675	6,875	10,550	13,700
1300	Demoliton three family house, three story, wood 3200 S.F.		.50	96			2,750	5,150	7,900	10,300
1320	5400 S.F.	↓	.30	160	↓		4,600	8,600	13,200	17,100

02 41 Demolition

02 41 16 – Structure Demolition

02 41 16.17 Building Demolition Footings and Foundations	Crew	Daily Output	Labor-Hours	Unit	Material	2016 Bare Costs Labor	2016 Bare Costs Equipment	Total	Total Incl O&P
0010 **BUILDING DEMOLITION FOOTINGS AND FOUNDATIONS** R024119-10									
0200 Floors, concrete slab on grade,									
0240 4" thick, plain concrete	B-13L	5000	.003	S.F.		.12	.41	.53	.64
0280 Reinforced, wire mesh		4000	.004			.14	.51	.65	.80
0300 Rods		4500	.004			.13	.46	.59	.71
0400 6" thick, plain concrete		4000	.004			.14	.51	.65	.80
0420 Reinforced, wire mesh		3200	.005			.18	.64	.82	1.01
0440 Rods		3600	.004			.16	.57	.73	.89
1000 Footings, concrete, 1' thick, 2' wide		300	.053	L.F.		1.92	6.85	8.77	10.70
1080 1'-6" thick, 2' wide		250	.064			2.31	8.20	10.51	12.80
1120 3' wide		200	.080			2.88	10.25	13.13	16.05
1200 Average reinforcing, add								10%	10%
2000 Walls, block, 4" thick	B-13L	8000	.002	S.F.		.07	.26	.33	.40
2040 6" thick		6000	.003			.10	.34	.44	.54
2080 8" thick		4000	.004			.14	.51	.65	.80
2100 12" thick		3000	.005			.19	.68	.87	1.07
2400 Concrete, plain concrete, 6" thick		4000	.004			.14	.51	.65	.80
2420 8" thick		3500	.005			.16	.59	.75	.91
2440 10" thick		3000	.005			.19	.68	.87	1.07
2500 12" thick		2500	.006			.23	.82	1.05	1.28
2600 For average reinforcing, add								10%	10%
4000 For congested sites or small quantities, add up to								200%	200%
4200 Add for disposal, on site	B-11A	232	.069	C.Y.		2.07	6	8.07	10.05
4250 To five miles	B-30	220	.109	"		3.51	10.95	14.46	17.85

02 41 19 – Selective Demolition

02 41 19.13 Selective Building Demolition

0010 **SELECTIVE BUILDING DEMOLITION**									
0020 Costs related to selective demolition of specific building components									
0025 are included under Common Work Results (XX 05)									
0030 in the component's appropriate division.									

02 41 19.16 Selective Demolition, Cutout

	Crew	Daily Output	Labor-Hours	Unit	Material	Labor	Equipment	Total	Total Incl O&P
0010 **SELECTIVE DEMOLITION, CUTOUT** R024119-10									
0020 Concrete, elev. slab, light reinforcement, under 6 C.F.	B-9	65	.615	C.F.		15.40	3.52	18.92	30
0050 Light reinforcing, over 6 C.F.		75	.533	"		13.35	3.05	16.40	26
0200 Slab on grade to 6" thick, not reinforced, under 8 S.F.		85	.471	S.F.		11.80	2.69	14.49	22.50
0250 8 – 16 S.F.		175	.229	"		5.75	1.31	7.06	11.05
0255 For over 16 S.F. see Line 02 41 16.17 0400									
0600 Walls, not reinforced, under 6 C.F.	B-9	60	.667	C.F.		16.70	3.81	20.51	32
0650 6 – 12 C.F.	"	80	.500	"		12.55	2.86	15.41	24
0655 For over 12 C.F. see Line 02 41 16.17 2500									
1000 Concrete, elevated slab, bar reinforced, under 6 C.F.	B-9	45	.889	C.F.		22.50	5.10	27.60	43
1050 Bar reinforced, over 6 C.F.		50	.800	"		20	4.58	24.58	38.50
1200 Slab on grade to 6" thick, bar reinforced, under 8 S.F.		75	.533	S.F.		13.35	3.05	16.40	26
1250 8 – 16 S.F.		150	.267	"		6.70	1.53	8.23	12.90
1255 For over 16 S.F. see Line 02 41 16.17 0440									
1400 Walls, bar reinforced, under 6 C.F.	B-9	50	.800	C.F.		20	4.58	24.58	38.50
1450 6 – 12 C.F.	"	70	.571	"		14.30	3.27	17.57	27.50
1455 For over 12 C.F. see Lines 02 41 16.17 2500 and 2600									
2000 Brick, to 4 S.F. opening, not including toothing									
2040 4" thick	B-9	30	1.333	Ea.		33.50	7.65	41.15	64.50
2060 8" thick		18	2.222			55.50	12.70	68.20	107

02 41 Demolition

02 41 19 – Selective Demolition

02 41 19.16 Selective Demolition, Cutout

		Crew	Daily Output	Labor-Hours	Unit	Material	2016 Bare Costs Labor	2016 Bare Costs Equipment	Total	Total Incl O&P
2080	12" thick	B-9	10	4	Ea.		100	23	123	193
2400	Concrete block, to 4 S.F. opening, 2" thick		35	1.143			28.50	6.55	35.05	55
2420	4" thick		30	1.333			33.50	7.65	41.15	64.50
2440	8" thick		27	1.481			37	8.45	45.45	71.50
2460	12" thick		24	1.667			42	9.55	51.55	80.50
2600	Gypsum block, to 4 S.F. opening, 2" thick		80	.500			12.55	2.86	15.41	24
2620	4" thick		70	.571			14.30	3.27	17.57	27.50
2640	8" thick		55	.727			18.20	4.16	22.36	35
2800	Terra cotta, to 4 S.F. opening, 4" thick		70	.571			14.30	3.27	17.57	27.50
2840	8" thick		65	.615			15.40	3.52	18.92	30
2880	12" thick		50	.800			20	4.58	24.58	38.50
3000	Toothing masonry cutouts, brick, soft old mortar	1 Brhe	40	.200	V.L.F.		5.30		5.30	8.85
3100	Hard mortar		30	.267			7.10		7.10	11.85
3200	Block, soft old mortar		70	.114			3.04		3.04	5.05
3400	Hard mortar		50	.160			4.26		4.26	7.10
6000	Walls, interior, not including re-framing,									
6010	openings to 5 S.F.									
6100	Drywall to 5/8" thick	1 Clab	24	.333	Ea.		8.20		8.20	13.75
6200	Paneling to 3/4" thick		20	.400			9.85		9.85	16.50
6300	Plaster, on gypsum lath		20	.400			9.85		9.85	16.50
6340	On wire lath		14	.571			14.10		14.10	23.50
7000	Wood frame, not including re-framing, openings to 5 S.F.									
7200	Floors, sheathing and flooring to 2" thick	1 Clab	5	1.600	Ea.		39.50		39.50	66
7310	Roofs, sheathing to 1" thick, not including roofing		6	1.333			33		33	55
7410	Walls, sheathing to 1" thick, not including siding		7	1.143			28		28	47

02 41 19.19 Selective Demolition

		Crew	Daily Output	Labor-Hours	Unit	Material	Labor	Equipment	Total	Total Incl O&P
0010	**SELECTIVE DEMOLITION**, Rubbish Handling R024119-10									
0020	The following are to be added to the demolition prices									
0600	Dumpster, weekly rental, 1 dump/week, 6 C.Y. capacity (2 Tons)				Week	415			415	455
0700	10 C.Y. capacity (3 Tons)					480			480	530
0725	20 C.Y. capacity (5 Tons) R024119-20					565			565	625
0800	30 C.Y. capacity (7 Tons)					730			730	800
0840	40 C.Y. capacity (10 Tons)					775			775	850
2000	Load, haul, dump and return, 0 – 50' haul, hand carried	2 Clab	24	.667	C.Y.		16.45		16.45	27.50
2005	Wheeled		37	.432			10.65		10.65	17.85
2040	0 – 100' haul, hand carried		16.50	.970			24		24	40
2045	Wheeled		25	.640			15.80		15.80	26.50
2080	Haul and return, add per each extra 100' haul, hand carried		35.50	.451			11.10		11.10	18.60
2085	Wheeled		54	.296			7.30		7.30	12.20
2120	For travel in elevators, up to 10 floors, add		140	.114			2.82		2.82	4.71
2130	0 – 50' haul, incl. up to 5 riser stair, hand carried		23	.696			17.15		17.15	28.50
2135	Wheeled		35	.457			11.25		11.25	18.85
2140	6 – 10 riser stairs, hand carried		22	.727			17.95		17.95	30
2145	Wheeled		34	.471			11.60		11.60	19.40
2150	11 – 20 riser stairs, hand carried		20	.800			19.70		19.70	33
2155	Wheeled		31	.516			12.70		12.70	21.50
2160	21 – 40 riser stairs, hand carried		16	1			24.50		24.50	41.50
2165	Wheeled		24	.667			16.45		16.45	27.50
2170	0 – 100' haul, incl. 5 riser stair, hand carried		15	1.067			26.50		26.50	44
2175	Wheeled		23	.696			17.15		17.15	28.50
2180	6 – 10 riser stair, hand carried		14	1.143			28		28	47
2185	Wheeled		21	.762			18.80		18.80	31.50

02 41 Demolition

02 41 19 – Selective Demolition

02 41 19.19 Selective Demolition

		Crew	Daily Output	Labor-Hours	Unit	Material	2016 Bare Costs Labor	Equipment	Total	Total Incl O&P
2190	11 – 20 riser stair, hand carried	2 Clab	12	1.333	C.Y.		33		33	55
2195	Wheeled		18	.889			22		22	36.50
2200	21 – 40 riser stair, hand carried		8	2			49.50		49.50	82.50
2205	Wheeled		12	1.333			33		33	55
2210	Haul and return, add per each extra 100' haul, hand carried		35.50	.451			11.10		11.10	18.60
2215	Wheeled		54	.296			7.30		7.30	12.20
2220	For each additional flight of stairs, up to 5 risers, add		550	.029	Flight		.72		.72	1.20
2225	6 – 10 risers, add		275	.058			1.43		1.43	2.40
2230	11 – 20 risers, add		138	.116			2.86		2.86	4.78
2235	21 – 40 risers, add		69	.232			5.70		5.70	9.55
3000	Loading & trucking, including 2 mile haul, chute loaded	B-16	45	.711	C.Y.		18.95	15.35	34.30	48.50
3040	Hand loading truck, 50' haul	"	48	.667			17.75	14.40	32.15	45.50
3080	Machine loading truck	B-17	120	.267			7.60	6.50	14.10	19.80
5000	Haul, per mile, up to 8 C.Y. truck	B-34B	1165	.007			.21	.59	.80	1
5100	Over 8 C.Y. truck	"	1550	.005			.16	.45	.61	.75

02 41 19.20 Selective Demolition, Dump Charges

		Crew	Daily Output	Labor-Hours	Unit	Material	Labor	Equipment	Total	Total Incl O&P
0010	**SELECTIVE DEMOLITION, DUMP CHARGES** R024119-10									
0020	Dump charges, typical urban city, tipping fees only									
0100	Building construction materials				Ton	74			74	81
0200	Trees, brush, lumber					63			63	69.50
0300	Rubbish only					63			63	69.50
0500	Reclamation station, usual charge					74			74	81

02 41 19.21 Selective Demolition, Gutting

		Crew	Daily Output	Labor-Hours	Unit	Material	Labor	Equipment	Total	Total Incl O&P
0010	**SELECTIVE DEMOLITION, GUTTING** R024119-10									
0020	Building interior, including disposal, dumpster fees not included									
0500	Residential building									
0560	Minimum	B-16	400	.080	SF Flr.		2.13	1.73	3.86	5.45
0580	Maximum	"	360	.089	"		2.37	1.92	4.29	6.05
0900	Commercial building									
1000	Minimum	B-16	350	.091	SF Flr.		2.44	1.97	4.41	6.25
1020	Maximum	"	250	.128	"		3.41	2.76	6.17	8.75

02 83 Lead Remediation

02 83 19 – Lead-Based Paint Remediation

02 83 19.22 Preparation of Lead Containment Area

		Crew	Daily Output	Labor-Hours	Unit	Material	Labor	Equipment	Total	Total Incl O&P
0010	**PREPARATION OF LEAD CONTAINMENT AREA**									
0020	Lead abatement work area, test kit, per swab	1 Skwk	16	.500	Ea.	3.12	17.10		20.22	32
0025	For dust barriers see Section 01 56 16.10									
0050	Caution sign	1 Skwk	48	.167	Ea.	8.35	5.70		14.05	18.75
0100	Pre-cleaning, HEPA vacuum and wet wipe, floor and wall surfaces	3 Skwk	5000	.005	S.F.	.01	.16		.17	.30
0105	Ceiling, 6' - 11' high		4100	.006		.08	.20		.28	.43
0108	12' - 15' high		3550	.007		.07	.23		.30	.47
0115	Over 15' high		3000	.008		.09	.27		.36	.56
0500	Cover surfaces with polyethylene sheeting									
0550	Floors, each layer, 6 mil	3 Skwk	8000	.003	S.F.	.04	.10		.14	.21
0560	Walls, each layer, 6 mil	"	6000	.004	"	.04	.14		.18	.27
0570	For heights above 12', add						20%			
2400	Post abatement cleaning of protective sheeting, HEPA vacuum & wet wipe	3 Skwk	5000	.005	S.F.	.01	.16		.17	.30
2450	Doff, bag and seal protective sheeting		12000	.002		.01	.07		.08	.12
2500	Post abatement cleaning, HEPA vacuum & wet wipe		5000	.005		.01	.16		.17	.30

02 83 Lead Remediation

02 83 19 – Lead-Based Paint Remediation

02 83 19.23 Encapsulation of Lead-Based Paint

		Crew	Daily Output	Labor-Hours	Unit	Material	2016 Bare Costs Labor	Equipment	Total	Total Incl O&P
0010	**ENCAPSULATION OF LEAD-BASED PAINT**									
0020	Interior, brushwork, trim, under 6"	1 Pord	240	.033	L.F.	2.30	.95		3.25	4.10
0030	6" to 12" wide		180	.044		3	1.27		4.27	5.40
0040	Balustrades		300	.027		2	.76		2.76	3.46
0050	Pipe to 4" diameter		500	.016		1.20	.46		1.66	2.08
0060	To 8" diameter		375	.021		1.50	.61		2.11	2.66
0070	To 12" diameter		250	.032		2.20	.92		3.12	3.93
0080	To 16" diameter		170	.047		3.30	1.35		4.65	5.85
0090	Cabinets, ornate design		200	.040	S.F.	3	1.15		4.15	5.20
0100	Simple design		250	.032	"	2.30	.92		3.22	4.04
0110	Doors, 3' x 7', both sides, incl. frame & trim									
0120	Flush	1 Pord	6	1.333	Ea.	28	38		66	94
0130	French, 10 – 15 lite		3	2.667		6	76.50		82.50	133
0140	Panel		4	2		35	57.50		92.50	133
0150	Louvered		2.75	2.909		32	83.50		115.50	172
0160	Windows, per interior side, per 15 S.F.									
0170	1 to 6 lite	1 Pord	14	.571	Ea.	20	16.35		36.35	49
0180	7 to 10 lite		7.50	1.067		22	30.50		52.50	74.50
0190	12 lite		5.75	1.391		30	40		70	98.50
0200	Radiators		8	1		70	28.50		98.50	124
0210	Grilles, vents		275	.029	S.F.	2	.83		2.83	3.57
0220	Walls, roller, drywall or plaster		1000	.008		.60	.23		.83	1.04
0230	With spunbonded reinforcing fabric		720	.011		.70	.32		1.02	1.29
0240	Wood		800	.010		.70	.29		.99	1.24
0250	Ceilings, roller, drywall or plaster		900	.009		.70	.25		.95	1.19
0260	Wood		700	.011		.80	.33		1.13	1.42
0270	Exterior, brushwork, gutters and downspouts		300	.027	L.F.	1.90	.76		2.66	3.35
0280	Columns		400	.020	S.F.	1.40	.57		1.97	2.48
0290	Spray, siding		600	.013	"	1	.38		1.38	1.73
0300	Miscellaneous									
0310	Electrical conduit, brushwork, to 2" diameter	1 Pord	500	.016	L.F.	1.20	.46		1.66	2.08
0320	Brick, block or concrete, spray		500	.016	S.F.	1.20	.46		1.66	2.08
0330	Steel, flat surfaces and tanks to 12"		500	.016		1.20	.46		1.66	2.08
0340	Beams, brushwork		400	.020		1.40	.57		1.97	2.48
0350	Trusses		400	.020		1.40	.57		1.97	2.48

02 83 19.26 Removal of Lead-Based Paint

		Crew	Daily Output	Labor-Hours	Unit	Material	2016 Bare Costs Labor	Equipment	Total	Total Incl O&P
0010	**REMOVAL OF LEAD-BASED PAINT**									
0011	By chemicals, per application									
0050	Baseboard, to 6" wide	1 Pord	64	.125	L.F.	.80	3.58		4.38	6.80
0070	To 12" wide		32	.250	"	1.47	7.15		8.62	13.40
0200	Balustrades, one side		28	.286	S.F.	1.50	8.20		9.70	15.15
1400	Cabinets, simple design		32	.250		1.37	7.15		8.52	13.30
1420	Ornate design		25	.320		1.60	9.15		10.75	16.85
1600	Cornice, simple design		60	.133		1.60	3.82		5.42	8.05
1620	Ornate design		20	.400		5.50	11.45		16.95	25
2800	Doors, one side, flush		84	.095		1.73	2.73		4.46	6.40
2820	Two panel		80	.100		1.37	2.87		4.24	6.25
2840	Four panel		45	.178		1.43	5.10		6.53	9.95
2880	For trim, one side, add		64	.125	L.F.	.77	3.58		4.35	6.75
3000	Fence, picket, one side		30	.267	S.F.	1.37	7.65		9.02	14.10
3200	Grilles, one side, simple design		30	.267		1.37	7.65		9.02	14.10
3220	Ornate design		25	.320		1.47	9.15		10.62	16.70

02 83 Lead Remediation

02 83 19 – Lead-Based Paint Remediation

02 83 19.26 Removal of Lead-Based Paint		Crew	Daily Output	Labor-Hours	Unit	Material	2016 Bare Costs Labor	Equipment	Total	Total Incl O&P
3240	Handrails	1 Pord	90	.089	L.F.	1.37	2.55		3.92	5.70
4400	Pipes, to 4" diameter		90	.089		1.73	2.55		4.28	6.10
4420	To 8" diameter		50	.160		3.47	4.58		8.05	11.35
4440	To 12" diameter		36	.222		5.20	6.35		11.55	16.20
4460	To 16" diameter		20	.400	↓	6.95	11.45		18.40	26.50
4500	For hangers, add		40	.200	Ea.	2.50	5.75		8.25	12.20
4800	Siding		90	.089	S.F.	1.30	2.55		3.85	5.65
5000	Trusses, open		55	.145	SF Face	2	4.17		6.17	9.05
6200	Windows, one side only, double hung, 1/1 light, 24" x 48" high		4	2	Ea.	24	57.50		81.50	121
6220	30" x 60" high		3	2.667		32	76.50		108.50	161
6240	36" x 72" high		2.50	3.200		38	91.50		129.50	193
6280	40" x 80" high		2	4		47	115		162	241
6400	Colonial window, 6/6 light, 24" x 48" high		2	4		47	115		162	241
6420	30" x 60" high		1.50	5.333		62	153		215	320
6440	36" x 72" high		1	8		93	229		322	480
6480	40" x 80" high		1	8		93	229		322	480
6600	8/8 light, 24" x 48" high		2	4		47	115		162	241
6620	40" x 80" high		1	8		93	229		322	480
6800	12/12 light, 24" x 48" high		1	8		93	229		322	480
6820	40" x 80" high		.75	10.667	↓	125	305		430	645
6840	Window frame & trim items, included in pricing above									

02 85 Mold Remediation

02 85 33 – Removal and Disposal of Materials with Mold

02 85 33.50 Demolition in Mold Contaminated Area		Crew	Daily Output	Labor-Hours	Unit	Material	2016 Bare Costs Labor	Equipment	Total	Total Incl O&P
0010	**DEMOLITION IN MOLD CONTAMINATED AREA**									
0200	Ceiling, including suspension system, plaster and lath	A-9	2100	.030	S.F.	.09	1.08		1.17	1.93
0210	Finished plaster, leaving wire lath		585	.109		.33	3.86		4.19	6.90
0220	Suspended acoustical tile		3500	.018		.05	.65		.70	1.16
0230	Concealed tile grid system		3000	.021		.06	.75		.81	1.35
0240	Metal pan grid system		1500	.043		.13	1.51		1.64	2.70
0250	Gypsum board		2500	.026		.08	.90		.98	1.61
0255	Plywood		2500	.026	↓	.08	.90		.98	1.61
0260	Lighting fixtures up to 2' x 4'	↓	72	.889	Ea.	2.65	31.50		34.15	56.50
0400	Partitions, non load bearing									
0410	Plaster, lath, and studs	A-9	690	.093	S.F.	.91	3.28		4.19	6.55
0450	Gypsum board and studs		1390	.046		.14	1.63		1.77	2.91
0465	Carpet & pad		1390	.046		.14	1.63		1.77	2.91
0600	Pipe insulation, air cell type, up to 4" diameter pipe		900	.071	L.F.	.21	2.51		2.72	4.49
0610	4" to 8" diameter pipe		800	.080		.24	2.82		3.06	5.05
0620	10" to 12" diameter pipe		700	.091		.27	3.23		3.50	5.80
0630	14" to 16" diameter pipe		550	.116	↓	.35	4.11		4.46	7.35
0650	Over 16" diameter pipe		650	.098	S.F.	.29	3.48		3.77	6.20
9000	For type B (supplied air) respirator equipment, add				%				10%	10%

Estimating Tips

General

- Carefully check all the plans and specifications. Concrete often appears on drawings other than structural drawings, including mechanical and electrical drawings for equipment pads. The cost of cutting and patching is often difficult to estimate. See Subdivision 03 81 for Concrete Cutting, Subdivision 02 41 19.16 for Cutout Demolition, Subdivision 03 05 05.10 for Concrete Demolition, and Subdivision 02 41 19.19 for Rubbish Handling (handling, loading and hauling of debris).
- Always obtain concrete prices from suppliers near the job site. A volume discount can often be negotiated, depending upon competition in the area. Remember to add for waste, particularly for slabs and footings on grade.

03 10 00 Concrete Forming and Accessories

- A primary cost for concrete construction is forming. Most jobs today are constructed with prefabricated forms. The selection of the forms best suited for the job and the total square feet of forms required for efficient concrete forming and placing are key elements in estimating concrete construction. Enough forms must be available for erection to make efficient use of the concrete placing equipment and crew.
- Concrete accessories for forming and placing depend upon the systems used. Study the plans and specifications to ensure that all special accessory requirements have been included in the cost estimate, such as anchor bolts, inserts, and hangers.
- Included within costs for forms-in-place are all necessary bracing and shoring.

03 20 00 Concrete Reinforcing

- Ascertain that the reinforcing steel supplier has included all accessories, cutting, bending, and an allowance for lapping, splicing, and waste. A good rule of thumb is 10% for lapping, splicing, and waste. Also, 10% waste should be allowed for welded wire fabric.
- The unit price items in the subdivisions for Reinforcing In Place, Glass Fiber Reinforcing, and Welded Wire Fabric include the labor to install accessories such as beam and slab bolsters, high chairs, and bar ties and tie wire. The material cost for these accessories is not included; they may be obtained from the Accessories Subdivisions.

03 30 00 Cast-In-Place Concrete

- When estimating structural concrete, pay particular attention to requirements for concrete additives, curing methods, and surface treatments. Special consideration for climate, hot or cold, must be included in your estimate. Be sure to include requirements for concrete placing equipment, and concrete finishing.
- For accurate concrete estimating, the estimator must consider each of the following major components individually: forms, reinforcing steel, ready-mix concrete, placement of the concrete, and finishing of the top surface. For faster estimating, Subdivision 03 30 53.40 for Concrete-In-Place can be used; here, various items of concrete work are presented that include the costs of all five major components (unless specifically stated otherwise).

03 40 00 Precast Concrete
03 50 00 Cast Decks and Underlayment

- The cost of hauling precast concrete structural members is often an important factor. For this reason, it is important to get a quote from the nearest supplier. It may become economically feasible to set up precasting beds on the site if the hauling costs are prohibitive.

Reference Numbers

Reference numbers are shown at the beginning of some major classifications. These numbers refer to related items in the Reference Section. The reference information may be an estimating procedure, an alternate pricing method, or technical information.

Note: Not all subdivisions listed here necessarily appear. ■

No part of this cost data may be reproduced, stored in a retrieval system, or transmitted in any form or by any means without prior written permission of RSMeans.

Did you know?

RSMeans Online gives you the same access to RSMeans' data with 24/7 access:

- Quickly locate costs in the searchable database.
- Build cost lists, estimates, and reports in minutes.
- Adjust costs to any location in the U.S. and Canada with the click of a button.

Start your free trial today at **www.RSMeansOnline.com**

03 01 Maintenance of Concrete

03 01 30 – Maintenance of Cast-In-Place Concrete

03 01 30.64 Floor Patching		Crew	Daily Output	Labor-Hours	Unit	Material	2016 Bare Costs Labor	Equipment	Total	Total Incl O&P
0010	**FLOOR PATCHING**									
0012	Floor patching, 1/4" thick, small areas, regular	1 Cefi	170	.047	S.F.	3.07	1.50		4.57	5.80
0100	Epoxy	"	100	.080	"	7.50	2.56		10.06	12.40

03 05 Common Work Results for Concrete

03 05 13 – Basic Concrete Materials

03 05 13.85 Winter Protection		Crew	Daily Output	Labor-Hours	Unit	Material	2016 Bare Costs Labor	Equipment	Total	Total Incl O&P
0010	**WINTER PROTECTION**									
0012	For heated ready mix, add				C.Y.	4.45			4.45	4.90
0100	Temporary heat to protect concrete, 24 hours	2 Clab	50	.320	M.S.F.	200	7.90		207.90	232
0200	Temporary shelter for slab on grade, wood frame/polyethylene sheeting									
0201	Build or remove, light framing for short spans	2 Carp	10	1.600	M.S.F.	292	54		346	410
0210	Large framing for long spans	"	3	5.333	"	380	181		561	725
0710	Electrically, heated pads, 15 watts/S.F., 20 uses				S.F.	.55			.55	.61

03 11 Concrete Forming

03 11 13 – Structural Cast-In-Place Concrete Forming

03 11 13.25 Forms In Place, Columns

			Crew	Daily Output	Labor-Hours	Unit	Material	Labor	Equipment	Total	Total Incl O&P
0010	**FORMS IN PLACE, COLUMNS**										
1500	Round fiber tube, recycled paper, 1 use, 8" diameter	G	C-1	155	.206	L.F.	1.82	6.05		7.87	12.15
1550	10" diameter	G		155	.206		2.46	6.05		8.51	12.85
1600	12" diameter	G		150	.213		2.84	6.25		9.09	13.60
1650	14" diameter	G		145	.221		4.02	6.45		10.47	15.25
1700	16" diameter	G		140	.229		4.97	6.70		11.67	16.70
1720	18" diameter	G		140	.229		5.80	6.70		12.50	17.65
5000	Job-built plywood, 8" x 8" columns, 1 use			165	.194	SFCA	2.71	5.70		8.41	12.55
5500	12" x 12" columns, 1 use			180	.178	"	2.59	5.20		7.79	11.60
7400	Steel framed plywood, based on 50 uses of purchased										
7420	forms, and 4 uses of bracing lumber										
7500	8" x 8" column		C-1	340	.094	SFCA	2.17	2.76		4.93	7
7550	10" x 10"			350	.091		1.89	2.68		4.57	6.55
7600	12" x 12"			370	.086		1.60	2.53		4.13	6

03 11 13.45 Forms In Place, Footings

		Crew	Daily Output	Labor-Hours	Unit	Material	Labor	Equipment	Total	Total Incl O&P
0010	**FORMS IN PLACE, FOOTINGS**									
0020	Continuous wall, plywood, 1 use	C-1	375	.085	SFCA	6.80	2.50		9.30	11.70
0150	4 use	"	485	.066	"	2.21	1.93		4.14	5.70
1500	Keyway, 4 use, tapered wood, 2" x 4"	1 Carp	530	.015	L.F.	.21	.51		.72	1.09
1550	2" x 6"	"	500	.016	"	.30	.54		.84	1.24
5000	Spread footings, job-built lumber, 1 use	C-1	305	.105	SFCA	2.15	3.07		5.22	7.50
5150	4 use	"	414	.077	"	.70	2.26		2.96	4.57

03 11 13.50 Forms In Place, Grade Beam

		Crew	Daily Output	Labor-Hours	Unit	Material	Labor	Equipment	Total	Total Incl O&P
0010	**FORMS IN PLACE, GRADE BEAM**									
0020	Job-built plywood, 1 use	C-2	530	.091	SFCA	3.06	2.68		5.74	7.90
0150	4 use	"	605	.079	"	1	2.35		3.35	5.05

03 11 13.65 Forms In Place, Slab On Grade

			Crew	Daily Output	Labor-Hours	Unit	Material	Labor	Equipment	Total	Total Incl O&P
0010	**FORMS IN PLACE, SLAB ON GRADE**										
1000	Bulkhead forms w/keyway, wood, 6" high, 1 use		C-1	510	.063	L.F.	.97	1.84		2.81	4.15
1400	Bulkhead form for slab, 4-1/2" high, exp metal, incl keyway & stakes	G		1200	.027		.91	.78		1.69	2.31

03 11 Concrete Forming

03 11 13 – Structural Cast-In-Place Concrete Forming

03 11 13.65 Forms In Place, Slab On Grade

			Crew	Daily Output	Labor-Hours	Unit	Material	2016 Bare Costs Labor	Equipment	Total	Total Incl O&P
1410	5-1/2" high	G	C-1	1100	.029	L.F.	1.15	.85		2	2.70
1420	7-1/2" high	G		960	.033		1.26	.98		2.24	3.03
1430	9-1/2" high	G		840	.038		1.38	1.12		2.50	3.39
2000	Curb forms, wood, 6" to 12" high, on grade, 1 use			215	.149	SFCA	2.64	4.36		7	10.20
2150	4 use			275	.116	"	.86	3.41		4.27	6.65
3000	Edge forms, wood, 4 use, on grade, to 6" high			600	.053	L.F.	.33	1.56		1.89	2.99
3050	7" to 12" high			435	.074	SFCA	.83	2.16		2.99	4.52
4000	For slab blockouts, to 12" high, 1 use			200	.160	L.F.	.78	4.69		5.47	8.70
4100	Plastic (extruded), to 6" high, multiple use, on grade			800	.040	"	8.20	1.17		9.37	10.95
8760	Void form, corrugated fiberboard, 4" x 12", 4' long	G		3000	.011	S.F.	3.33	.31		3.64	4.19
8770	6" x 12", 4' long			3000	.011		4.10	.31		4.41	5.05
8780	1/4" thick hardboard protective cover for void form		2 Carp	1500	.011		.63	.36		.99	1.30

03 11 13.85 Forms In Place, Walls

			Crew	Daily Output	Labor-Hours	Unit	Material	Labor	Equipment	Total	Total Incl O&P
0010	**FORMS IN PLACE, WALLS**										
0100	Box out for wall openings, to 16" thick, to 10 S.F.		C-2	24	2	Ea.	25.50	59.50		85	128
0150	Over 10 S.F. (use perimeter)		"	280	.171	L.F.	2.20	5.10		7.30	10.95
0250	Brick shelf, 4" w, add to wall forms, use wall area above shelf										
0260	1 use		C-2	240	.200	SFCA	2.37	5.95		8.32	12.55
0350	4 use			300	.160	"	.95	4.74		5.69	9
0500	Bulkhead, wood with keyway, 1 use, 2 piece			265	.181	L.F.	2.07	5.35		7.42	11.30
0600	Bulkhead forms with keyway, 1 piece expanded metal, 8" wall	G	C-1	1000	.032		1.26	.94		2.20	2.96
0610	10" wall	G	"	800	.040		1.38	1.17		2.55	3.49
2000	Wall, job-built plywood, to 8' high, 1 use		C-2	370	.130	SFCA	2.68	3.85		6.53	9.40
2050	2 use			435	.110		1.70	3.27		4.97	7.35
2100	3 use			495	.097		1.24	2.87		4.11	6.20
2150	4 use			505	.095		1	2.82		3.82	5.85
2400	Over 8' to 16' high, 1 use			280	.171		2.96	5.10		8.06	11.80
2450	2 use			345	.139		1.29	4.12		5.41	8.30
2500	3 use			375	.128		.92	3.79		4.71	7.35
2550	4 use			395	.122		.76	3.60		4.36	6.90
7800	Modular prefabricated plywood, based on 20 uses of purchased										
7820	forms, and 4 uses of bracing lumber										
7860	To 8' high		C-2	800	.060	SFCA	.90	1.78		2.68	3.97
8060	Over 8' to 16' high		"	600	.080	"	.95	2.37		3.32	5

03 11 19 – Insulating Concrete Forming

03 11 19.10 Insulating Forms, Left In Place

			Crew	Daily Output	Labor-Hours	Unit	Material	Labor	Equipment	Total	Total Incl O&P
0010	**INSULATING FORMS, LEFT IN PLACE**										
0020	S.F. is for exterior face, but includes forms for both faces (total R22)										
2000	4" wall, straight block, 16" x 48" (5.33 S.F.)	G	2 Carp	90	.178	Ea.	20	6.05		26.05	32
2010	90 corner block, exterior 16" x 38" x 22" (6.67 S.F.)	G		75	.213		23.50	7.25		30.75	38
2020	45 corner block, exterior 16" x 34" x 18" (5.78 S.F.)	G		75	.213		23.50	7.25		30.75	38
2100	6" wall, straight block, 16" x 48" (5.33 S.F.)	G		90	.178		19.70	6.05		25.75	31.50
2110	90 corner block, exterior 16" x 32" x 24" (6.22 S.F.)	G		75	.213		25	7.25		32.25	39.50
2120	45 corner block, exterior 16" x 26" x 18" (4.89 S.F.)	G		75	.213		22.50	7.25		29.75	36.50
2130	Brick ledge block, 16" x 48" (5.33 S.F.)	G		80	.200		25.50	6.80		32.30	39.50
2140	Taper top block, 16" x 48" (5.33 S.F.)	G		80	.200		24.50	6.80		31.30	38.50
2200	8" wall, straight block, 16" x 48" (5.33 S.F.)	G		90	.178		20.50	6.05		26.55	32.50
2210	90 corner block, exterior 16" x 34" x 26" (6.67 S.F.)	G		75	.213		27	7.25		34.25	42
2220	45 corner block, exterior 16" x 28" x 20" (5.33 S.F.)	G		75	.213		24	7.25		31.25	38
2230	Brick ledge block, 16" x 48" (5.33 S.F.)	G		80	.200		26	6.80		32.80	40.50
2240	Taper top block, 16" x 48" (5.33 S.F.)	G		80	.200		25	6.80		31.80	39

For customer support on your Residential Cost Data, call 877.759.4771.

03 11 Concrete Forming

03 11 23 – Permanent Stair Forming

03 11 23.75 Forms In Place, Stairs

		Crew	Daily Output	Labor-Hours	Unit	Material	2016 Bare Costs Labor	2016 Bare Costs Equipment	Total	Total Incl O&P
0010	**FORMS IN PLACE, STAIRS**									
0015	(Slant length x width), 1 use	C-2	165	.291	S.F.	5.80	8.60		14.40	21
0150	4 use		190	.253		2.04	7.50		9.54	14.80
2000	Stairs, cast on sloping ground (length x width), 1 use		220	.218		2.22	6.45		8.67	13.30
2025	2 use		232	.207		1.22	6.15		7.37	11.65
2050	3 use		244	.197		.89	5.85		6.74	10.75
2100	4 use		256	.188		.72	5.55		6.27	10.15

03 15 Concrete Accessories

03 15 05 – Concrete Forming Accessories

03 15 05.12 Chamfer Strips

		Crew	Daily Output	Labor-Hours	Unit	Material	Labor	Equipment	Total	Total Incl O&P
0010	**CHAMFER STRIPS**									
5000	Wood, 1/2" wide	1 Carp	535	.015	L.F.	.14	.51		.65	1
5200	3/4" wide		525	.015		.16	.52		.68	1.04
5400	1" wide		515	.016		.29	.53		.82	1.20

03 15 05.15 Column Form Accessories

					Unit	Material	Labor	Equipment	Total	Total Incl O&P
0010	**COLUMN FORM ACCESSORIES**									
1000	Column clamps, adjustable to 24" x 24", buy	G			Set	145			145	160
1100	Rent per month	G			"	14.55			14.55	16

03 15 05.75 Sleeves and Chases

		Crew	Daily Output	Labor-Hours	Unit	Material	Labor	Equipment	Total	Total Incl O&P
0010	**SLEEVES AND CHASES**									
0100	Plastic, 1 use, 12" long, 2" diameter	1 Carp	100	.080	Ea.	2.23	2.71		4.94	7
0150	4" diameter		90	.089		6.25	3.01		9.26	11.95
0200	6" diameter		75	.107		11.05	3.62		14.67	18.20

03 15 05.80 Snap Ties

					Unit	Material	Labor	Equipment	Total	Total Incl O&P
0010	**SNAP TIES**, 8-1/4" L&W (Lumber and wedge)									
0100	2250 lb., w/flat washer, 8" wall	G			C	93			93	102
0150	10" wall	G				135			135	149
0200	12" wall	G				140			140	154
0500	With plastic cone, 8" wall	G				82			82	90
0550	10" wall	G				85			85	93.50
0600	12" wall	G				92			92	101

03 15 05.95 Wall and Foundation Form Accessories

					Unit	Material	Labor	Equipment	Total	Total Incl O&P
0010	**WALL AND FOUNDATION FORM ACCESSORIES**									
4000	Nail stakes, 3/4" diameter, 18" long	G			Ea.	3.40			3.40	3.74
4050	24" long	G				4.39			4.39	4.83
4200	30" long	G				5.65			5.65	6.20
4250	36" long	G				6.80			6.80	7.45

03 15 13 – Waterstops

03 15 13.50 Waterstops

		Crew	Daily Output	Labor-Hours	Unit	Material	Labor	Equipment	Total	Total Incl O&P
0010	**WATERSTOPS**, PVC and Rubber									
0020	PVC, ribbed 3/16" thick, 4" wide	1 Carp	155	.052	L.F.	1.32	1.75		3.07	4.38
0050	6" wide		145	.055		2.29	1.87		4.16	5.65
0500	With center bulb, 6" wide, 3/16" thick		135	.059		2.02	2.01		4.03	5.60
0550	3/8" thick		130	.062		3.79	2.09		5.88	7.65
0600	9" wide x 3/8" thick		125	.064		6.15	2.17		8.32	10.40

03 15 Concrete Accessories

03 15 16 – Concrete Construction Joints

03 15 16.20 Control Joints, Saw Cut

		Crew	Daily Output	Labor-Hours	Unit	Material	2016 Bare Costs Labor	Equipment	Total	Total Incl O&P
0010	**CONTROL JOINTS, SAW CUT**									
0100	Sawcut control joints in green concrete									
0120	1" depth	C-27	2000	.008	L.F.	.04	.26	.08	.38	.54
0140	1-1/2" depth		1800	.009		.05	.28	.09	.42	.62
0160	2" depth	↓	1600	.010	↓	.07	.32	.11	.50	.72
0180	Sawcut joint reservoir in cured concrete									
0182	3/8" wide x 3/4" deep, with single saw blade	C-27	1000	.016	L.F.	.05	.51	.17	.73	1.08
0184	1/2" wide x 1" deep, with double saw blades		900	.018		.10	.57	.19	.86	1.24
0186	3/4" wide x 1-1/2" deep, with double saw blades	↓	800	.020		.21	.64	.21	1.06	1.50
0190	Water blast joint to wash away laitance, 2 passes	C-29	2500	.003			.08	.03	.11	.16
0200	Air blast joint to blow out debris and air dry, 2 passes	C-28	2000	.004	↓		.13	.01	.14	.22
0300	For backer rod, see Section 07 91 23.10									
0340	For joint sealant, see Section 03 15 16.30 or 07 92 13.20									

03 15 16.30 Expansion Joints

			Crew	Daily Output	Labor-Hours	Unit	Material	2016 Bare Costs Labor	Equipment	Total	Total Incl O&P
0010	**EXPANSION JOINTS**										
0020	Keyed, cold, 24 ga., incl. stakes, 3-1/2" high	G	1 Carp	200	.040	L.F.	.86	1.36		2.22	3.22
0050	4-1/2" high	G		200	.040		.91	1.36		2.27	3.27
0100	5-1/2" high	G		195	.041		1.15	1.39		2.54	3.60
2000	Premolded, bituminous fiber, 1/2" x 6"			375	.021		.40	.72		1.12	1.65
2050	1" x 12"			300	.027		1.97	.90		2.87	3.68
2140	Concrete expansion joint, recycled paper and fiber, 1/2" x 6"	G		390	.021		.43	.70		1.13	1.63
2150	1/2" x 12"	G		360	.022		.85	.75		1.60	2.20
2500	Neoprene sponge, closed cell, 1/2" x 6"			375	.021		2.41	.72		3.13	3.86
2550	1" x 12"		↓	300	.027	↓	8.40	.90		9.30	10.75
5000	For installation in walls, add							75%			
5250	For installation in boxouts, add							25%			

03 15 19 – Cast-In Concrete Anchors

03 15 19.05 Anchor Bolt Accessories

		Crew	Daily Output	Labor-Hours	Unit	Material	2016 Bare Costs Labor	Equipment	Total	Total Incl O&P
0010	**ANCHOR BOLT ACCESSORIES**									
0015	For anchor bolts set in fresh concrete, see Section 03 15 19.10									
8150	Anchor bolt sleeve, plastic, 1" diam. bolts	1 Carp	60	.133	Ea.	12.95	4.52		17.47	22
8500	1-1/2" diameter		28	.286		19.20	9.70		28.90	37
8600	2" diameter		24	.333		19.45	11.30		30.75	40.50
8650	3" diameter		20	.400		35.50	13.55		49.05	61.50

03 15 19.10 Anchor Bolts

			Crew	Daily Output	Labor-Hours	Unit	Material	2016 Bare Costs Labor	Equipment	Total	Total Incl O&P
0010	**ANCHOR BOLTS**										
0015	Made from recycled materials										
0025	Single bolts installed in fresh concrete, no templates										
0030	Hooked w/nut and washer, 1/2" diameter, 8" long	G	1 Carp	132	.061	Ea.	1.32	2.05		3.37	4.90
0040	12" long	G		131	.061		1.47	2.07		3.54	5.10
0070	3/4" diameter, 8" long	G		127	.063		3.81	2.14		5.95	7.75
0080	12" long	G	↓	125	.064	↓	4.76	2.17		6.93	8.90

03 15 19.30 Inserts

			Crew	Daily Output	Labor-Hours	Unit	Material	2016 Bare Costs Labor	Equipment	Total	Total Incl O&P
0010	**INSERTS**										
1000	Inserts, slotted nut type for 3/4" bolts, 4" long	G	1 Carp	84	.095	Ea.	20	3.23		23.23	28
2100	6" long	G		84	.095		23	3.23		26.23	30.50
2150	8" long	G		84	.095		30.50	3.23		33.73	39
2200	Slotted, strap type, 4" long	G		84	.095		22	3.23		25.23	29.50
2300	8" long	G	↓	84	.095	↓	32.50	3.23		35.73	41.50
9950	For galvanized inserts, add						30%				

03 21 Reinforcement Bars

03 21 11 – Plain Steel Reinforcement Bars

03 21 11.60 Reinforcing In Place

			Crew	Daily Output	Labor-Hours	Unit	Material	2016 Bare Costs Labor	Equipment	Total	Total Incl O&P
0010	**REINFORCING IN PLACE**, 50-60 ton lots, A615 Grade 60										
0020	Includes labor, but not material cost, to install accessories										
0030	Made from recycled materials										
0502	Footings, #4 to #7	G	4 Rodm	4200	.008	Lb.	.48	.27		.75	.99
0550	#8 to #18	G		3.60	8.889	Ton	960	315		1,275	1,575
0602	Slab on grade, #3 to #7	G		4200	.008	Lb.	.48	.27		.75	.99
0702	Walls, #3 to #7	G		6000	.005	"	.48	.19		.67	.85
0750	#8 to #18	G		4	8	Ton	960	284		1,244	1,525
0900	For other than 50 – 60 ton lots										
1000	Under 10 ton job, #3 to #7, add						25%	10%			
1010	#8 to #18, add						20%	10%			
1050	10 – 50 ton job, #3 to #7, add						10%				
1060	#8 to #18, add						5%				
1100	60 – 100 ton job, #3 to #7, deduct						5%				
1110	#8 to #18, deduct						10%				
1150	Over 100 ton job, #3 to #7, deduct						10%				
1160	#8 to #18, deduct						15%				
2400	Dowels, 2 feet long, deformed, #3	G	2 Rodm	520	.031	Ea.	.40	1.09		1.49	2.28
2410	#4	G		480	.033		.71	1.18		1.89	2.78
2420	#5	G		435	.037		1.10	1.31		2.41	3.42
2430	#6	G		360	.044		1.59	1.58		3.17	4.40
2600	Dowel sleeves for CIP concrete, 2-part system										
2610	Sleeve base, plastic, for 5/8" smooth dowel sleeve, fasten to edge form		1 Rodm	200	.040	Ea.	.53	1.42		1.95	2.98
2615	Sleeve, plastic, 12" long, for 5/8" smooth dowel, snap onto base			400	.020		1.33	.71		2.04	2.66
2620	Sleeve base, for 3/4" smooth dowel sleeve			175	.046		.53	1.62		2.15	3.32
2625	Sleeve, 12" long, for 3/4" smooth dowel			350	.023		1.61	.81		2.42	3.14
2630	Sleeve base, for 1" smooth dowel sleeve			150	.053		.68	1.89		2.57	3.95
2635	Sleeve, 12" long, for 1" smooth dowel			300	.027		1.42	.95		2.37	3.16
2700	Dowel caps, visual warning only, plastic, #3 to #8		2 Rodm	800	.020		.30	.71		1.01	1.53
2720	#8 to #18			750	.021		.74	.76		1.50	2.10
2750	Impalement protective, plastic, #4 to #9			800	.020		1.66	.71		2.37	3.03

03 22 Fabric and Grid Reinforcing

03 22 11 – Plain Welded Wire Fabric Reinforcing

03 22 11.10 Welded Wire Fabric

			Crew	Daily Output	Labor-Hours	Unit	Material	Labor	Equipment	Total	Total Incl O&P
0011	**WELDED WIRE FABRIC**, 6 x 6 - W1.4 x W1.4 (10 x 10)	G	2 Rodm	3500	.005	S.F.	.14	.16		.30	.42
0301	6 x 6 - W2.9 x W2.9 (6 x 6) 42 lb. per C.S.F.	G		2900	.006		.23	.20		.43	.58
0501	4 x 4 - W1.4 x W1.4 (10 x 10) 31 lb. per C.S.F.	G		3100	.005		.20	.18		.38	.53
0750	Rolls										
0901	2 x 2 - #12 galv. for gunite reinforcing	G	2 Rodm	650	.025	S.F.	.63	.87		1.50	2.17

03 22 13 – Galvanized Welded Wire Fabric Reinforcing

03 22 13.10 Galvanized Welded Wire Fabric

		Unit	Material			Total	Total Incl O&P
0010	**GALVANIZED WELDED WIRE FABRIC**						
0100	Add to plain welded wire pricing for galvanized welded wire	Lb.	.23			.23	.25

03 22 16 – Epoxy-Coated Welded Wire Fabric Reinforcing

03 22 16.10 Epoxy-Coated Welded Wire Fabric

		Unit	Material			Total	Total Incl O&P
0010	**EPOXY-COATED WELDED WIRE FABRIC**						
0100	Add to plain welded wire pricing for epoxy-coated welded wire	Lb.	.22			.22	.24

03 23 Stressed Tendon Reinforcing

03 23 05 – Prestressing Tendons

03 23 05.50 Prestressing Steel

		Crew	Daily Output	Labor-Hours	Unit	Material	2016 Bare Costs Labor	Equipment	Total	Total Incl O&P
0010	**PRESTRESSING STEEL**									
3000	Slabs on grade, 0.5-inch diam. non-bonded strands, HDPE sheathed,									
3050	attached dead-end anchors, loose stressing-end anchors									
3100	25' x 30' slab, strands @ 36" O.C., placing	2 Rodm	2940	.005	S.F.	.60	.19		.79	.99
3105	Stressing	C-4A	3750	.004			.15	.01	.16	.27
3110	42" O.C., placing	2 Rodm	3200	.005		.53	.18		.71	.89
3115	Stressing	C-4A	4040	.004			.14	.01	.15	.25
3120	48" O.C., placing	2 Rodm	3510	.005		.47	.16		.63	.78
3125	Stressing	C-4A	4390	.004			.13	.01	.14	.23
3150	25' x 40' slab, strands @ 36" O.C., placing	2 Rodm	3370	.005		.58	.17		.75	.92
3155	Stressing	C-4A	4360	.004			.13	.01	.14	.23
3160	42" O.C., placing	2 Rodm	3760	.004		.50	.15		.65	.81
3165	Stressing	C-4A	4820	.003			.12	.01	.13	.21
3170	48" O.C., placing	2 Rodm	4090	.004		.45	.14		.59	.72
3175	Stressing	C-4A	5190	.003			.11	.01	.12	.19
3200	30' x 30' slab, strands @ 36" O.C., placing	2 Rodm	3260	.005		.58	.17		.75	.93
3205	Stressing	C-4A	4190	.004			.14	.01	.15	.24
3210	42" O.C., placing	2 Rodm	3530	.005		.52	.16		.68	.85
3215	Stressing	C-4A	4500	.004			.13	.01	.14	.22
3220	48" O.C., placing	2 Rodm	3840	.004		.47	.15		.62	.76
3225	Stressing	C-4A	4850	.003			.12	.01	.13	.21
3230	30' x 40' slab, strands @ 36" O.C., placing	2 Rodm	3780	.004		.56	.15		.71	.87
3235	Stressing	C-4A	4920	.003			.12	.01	.13	.20
3240	42" O.C., placing	2 Rodm	4190	.004		.49	.14		.63	.77
3245	Stressing	C-4A	5410	.003			.11	.01	.12	.19
3250	48" O.C., placing	2 Rodm	4520	.004		.45	.13		.58	.70
3255	Stressing	C-4A	5790	.003			.10	.01	.11	.18
3260	30' x 50' slab, strands @ 36" O.C., placing	2 Rodm	4300	.004		.53	.13		.66	.81
3265	Stressing	C-4A	5650	.003			.10	.01	.11	.18
3270	42" O.C., placing	2 Rodm	4720	.003		.47	.12		.59	.72
3275	Stressing	C-4A	6150	.003			.09	.01	.10	.17
3280	48" O.C., placing	2 Rodm	5240	.003		.42	.11		.53	.64
3285	Stressing	C-4A	6760	.002			.08	.01	.09	.15

03 24 Fibrous Reinforcing

03 24 05 – Reinforcing Fibers

03 24 05.30 Synthetic Fibers

					Unit	Material			Total	Total Incl O&P
0010	**SYNTHETIC FIBERS**									
0100	Synthetic fibers, add to concrete				Lb.	4.40			4.40	4.84
0110	1-1/2 lb. per C.Y.				C.Y.	6.80			6.80	7.50

03 24 05.70 Steel Fibers

						Unit	Material			Total	Total Incl O&P
0010	**STEEL FIBERS**										
0140	ASTM A850, Type V, continuously deformed, 1-1/2" long x 0.045" diam.										
0150	Add to price of ready mix concrete	G				Lb.	1.13			1.13	1.24
0205	Alternate pricing, dosing at 5 lb. per C.Y., add to price of RMC	G				C.Y.	5.65			5.65	6.20
0210	10 lb. per C.Y.	G					11.30			11.30	12.45
0215	15 lb. per C.Y.	G					16.95			16.95	18.65
0220	20 lb. per C.Y.	G					22.50			22.50	25
0225	25 lb. per C.Y.	G					28.50			28.50	31
0230	30 lb. per C.Y.	G					34			34	37.50
0235	35 lb. per C.Y.	G					39.50			39.50	43.50

03 24 Fibrous Reinforcing

03 24 05 – Reinforcing Fibers

03 24 05.70 Steel Fibers		Crew	Daily Output	Labor-Hours	Unit	Material	2016 Bare Costs Labor	Equipment	Total	Total Incl O&P
0240	40 lb. per C.Y.	G			C.Y.	45			45	49.50
0250	50 lb. per C.Y.	G				56.50			56.50	62
0275	75 lb. per C.Y.	G				85			85	93
0300	100 lb. per C.Y.	G				113			113	124

03 30 Cast-In-Place Concrete

03 30 53 – Miscellaneous Cast-In-Place Concrete

03 30 53.40 Concrete In Place

		Crew	Daily Output	Labor-Hours	Unit	Material	Labor	Equipment	Total	Total Incl O&P
0010	**CONCRETE IN PLACE**									
0020	Including forms (4 uses), Grade 60 rebar, concrete (Portland cement									
0050	Type I), placement and finishing unless otherwise indicated									
0500	Chimney foundations (5000 psi), over 5 C.Y.	C-14C	32.22	3.476	C.Y.	155	109	.97	264.97	355
0510	(3500 psi), under 5 C.Y.	"	23.71	4.724	"	182	149	1.32	332.32	450
3540	Equipment pad (3000 psi), 3' x 3' x 6" thick	C-14H	45	1.067	Ea.	47.50	35	.70	83.20	111
3550	4' x 4' x 6" thick		30	1.600		70.50	52	1.06	123.56	166
3560	5' x 5' x 8" thick		18	2.667		124	87	1.76	212.76	284
3570	6' x 6' x 8" thick		14	3.429		167	112	2.26	281.26	375
3580	8' x 8' x 10" thick		8	6		360	196	3.96	559.96	725
3590	10' x 10' x 12" thick		5	9.600		615	315	6.35	936.35	1,200
3800	Footings (3000 psi), spread under 1 C.Y.	C-14C	28	4	C.Y.	171	126	1.12	298.12	400
3825	1 C.Y. to 5 C.Y.		43	2.605		207	82	.73	289.73	365
3850	Over 5 C.Y.		75	1.493		190	47	.42	237.42	288
3900	Footings, strip (3000 psi), 18" x 9", unreinforced	C-14L	40	2.400		131	74	.79	205.79	268
3920	18" x 9", reinforced	C-14C	35	3.200		154	101	.90	255.90	340
3925	20" x 10", unreinforced	C-14L	45	2.133		127	66	.70	193.70	251
3930	20" x 10", reinforced	C-14C	40	2.800		146	88	.78	234.78	310
3935	24" x 12", unreinforced	C-14L	55	1.745		125	54	.58	179.58	229
3940	24" x 12", reinforced	C-14C	48	2.333		144	73.50	.65	218.15	283
3945	36" x 12", unreinforced	C-14L	70	1.371		121	42.50	.45	163.95	204
3950	36" x 12", reinforced	C-14C	60	1.867		138	59	.52	197.52	251
4000	Foundation mat (3000 psi), under 10 C.Y.		38.67	2.896		209	91	.81	300.81	385
4050	Over 20 C.Y.		56.40	1.986		184	62.50	.56	247.06	310
4520	Handicap access ramp (4000 psi), railing both sides, 3' wide	C-14H	14.58	3.292	L.F.	315	107	2.17	424.17	530
4525	5' wide		12.22	3.928		325	128	2.59	455.59	575
4530	With 6" curb and rails both sides, 3' wide		8.55	5.614		325	183	3.71	511.71	670
4535	5' wide		7.31	6.566		330	214	4.33	548.33	725
4650	Slab on grade (3500 psi), not including finish, 4" thick	C-14E	60.75	1.449	C.Y.	129	46.50	.52	176.02	220
4700	6" thick	"	92	.957	"	124	30.50	.34	154.84	188
4701	Thickened slab edge (3500 psi), for slab on grade poured									
4702	monolithically with slab; depth is in addition to slab thickness;									
4703	formed vertical outside edge, earthen bottom and inside slope									
4705	8" deep x 8" wide bottom, unreinforced	C-14L	2190	.044	L.F.	3.57	1.35	.01	4.93	6.20
4710	8" x 8", reinforced	C-14C	1670	.067		5.85	2.11	.02	7.98	9.95
4715	12" deep x 12" wide bottom, unreinforced	C-14L	1800	.053		7.30	1.64	.02	8.96	10.75
4720	12" x 12", reinforced	C-14C	1310	.086		11.40	2.69	.02	14.11	17.10
4725	16" deep x 16" wide bottom, unreinforced	C-14L	1440	.067		12.30	2.05	.02	14.37	17
4730	16" x 16", reinforced	C-14C	1120	.100		17.25	3.15	.03	20.43	24
4735	20" deep x 20" wide bottom, unreinforced	C-14L	1150	.083		18.70	2.57	.03	21.30	25
4740	20" x 20", reinforced	C-14C	920	.122		25	3.83	.03	28.86	34
4745	24" deep x 24" wide bottom, unreinforced	C-14L	930	.103		26.50	3.18	.03	29.71	34.50
4750	24" x 24", reinforced	C-14C	740	.151		34.50	4.77	.04	39.31	46

03 30 Cast-In-Place Concrete

03 30 53 – Miscellaneous Cast-In-Place Concrete

03 30 53.40 Concrete In Place

		Crew	Daily Output	Labor-Hours	Unit	Material	2016 Bare Costs Labor	2016 Bare Costs Equipment	Total	Total Incl O&P
4751	Slab on grade (3500 psi), incl. troweled finish, not incl. forms									
4760	or reinforcing, over 10,000 S.F., 4" thick	C-14F	3425	.021	S.F.	1.43	.63	.01	2.07	2.60
4820	6" thick	"	3350	.021	"	2.08	.64	.01	2.73	3.35
5000	Slab on grade (3000 psi), incl. broom finish, not incl. forms									
5001	or reinforcing, 4" thick	C-14G	2873	.019	S.F.	1.39	.57	.01	1.97	2.47
5010	6" thick		2590	.022		2.18	.63	.01	2.82	3.44
5020	8" thick		2320	.024		2.84	.70	.01	3.55	4.29
6800	Stairs (3500 psi), not including safety treads, free standing, 3'-6" wide	C-14H	83	.578	LF Nose	5.70	18.85	.38	24.93	38
6850	Cast on ground		125	.384	"	4.73	12.55	.25	17.53	26.50
7000	Stair landings, free standing		200	.240	S.F.	4.61	7.85	.16	12.62	18.25
7050	Cast on ground		475	.101	"	3.61	3.30	.07	6.98	9.55

03 31 Structural Concrete

03 31 13 – Heavyweight Structural Concrete

03 31 13.25 Concrete, Hand Mix

		Crew	Daily Output	Labor-Hours	Unit	Material	Labor	Equipment	Total	Total Incl O&P
0010	**CONCRETE, HAND MIX** for small quantities or remote areas									
0050	Includes bulk local aggregate, bulk sand, bagged Portland									
0060	cement (Type I) and water, using gas powered cement mixer									
0125	2500 psi	C-30	135	.059	C.F.	3.50	1.46	1.28	6.24	7.70
0130	3000 psi		135	.059		3.75	1.46	1.28	6.49	7.95
0135	3500 psi		135	.059		3.89	1.46	1.28	6.63	8.15
0140	4000 psi		135	.059		4.06	1.46	1.28	6.80	8.30
0145	4500 psi		135	.059		4.26	1.46	1.28	7	8.55
0150	5000 psi		135	.059		4.53	1.46	1.28	7.27	8.85
0300	Using pre-bagged dry mix and wheelbarrow (80-lb. bag = 0.6 C.F.)									
0340	4000 psi	1 Clab	48	.167	C.F.	6.50	4.11		10.61	14.05

03 31 13.30 Concrete, Volumetric Site-Mixed

		Crew	Daily Output	Labor-Hours	Unit	Material	Labor	Equipment	Total	Total Incl O&P
0010	**CONCRETE, VOLUMETRIC SITE-MIXED**									
0015	Mixed on-site in volumetric truck									
0020	Includes local aggregate, sand, Portland cement (Type I) and water									
0025	Excludes all additives and treatments									
0100	3000 psi, 1 C.Y. mixed and discharged				C.Y.	193			193	213
0110	2 C.Y.					147			147	162
0120	3 C.Y.					131			131	144
0130	4 C.Y.					118			118	130
0140	5 C.Y.					110			110	121
0200	For truck holding/waiting time past first 2 on-site hours, add				Hr.	88			88	97
0210	For trip charge beyond first 20 miles, each way, add				Mile	3.50			3.50	3.85
0220	For each additional increase of 500 psi, add				Ea.	4.17			4.17	4.59

03 31 13.35 Heavyweight Concrete, Ready Mix

		Crew	Daily Output	Labor-Hours	Unit	Material	Labor	Equipment	Total	Total Incl O&P
0010	**HEAVYWEIGHT CONCRETE, READY MIX**, delivered									
0012	Includes local aggregate, sand, Portland cement (Type I) and water									
0015	Excludes all additives and treatments									
0020	2000 psi				C.Y.	102			102	113
0100	2500 psi					105			105	115
0150	3000 psi					107			107	118
0200	3500 psi					110			110	121
0300	4000 psi					113			113	124
0350	4500 psi					116			116	127
0400	5000 psi					119			119	131

03 31 Structural Concrete

03 31 13 – Heavyweight Structural Concrete

03 31 13.35 Heavyweight Concrete, Ready Mix

		Crew	Daily Output	Labor-Hours	Unit	Material	2016 Bare Costs Labor	2016 Bare Costs Equipment	Total	Total Incl O&P
0411	6000 psi				C.Y.	122			122	134
0412	8000 psi					129			129	141
0413	10,000 psi					135			135	149
0414	12,000 psi					142			142	156
1000	For high early strength (Portland cement Type III), add					10%				
1300	For winter concrete (hot water), add					4.45			4.45	4.90
1410	For mid-range water reducer, add					3.33			3.33	3.66
1420	For high-range water reducer/superplasticizer, add					6			6	6.60
1430	For retarder, add					3.23			3.23	3.55
1440	For non-Chloride accelerator, add					5.70			5.70	6.25
1450	For Chloride accelerator, per 1%, add					3			3	3.30
1460	For fiber reinforcing, synthetic (1 lb./C.Y.), add					6.70			6.70	7.35
1500	For Saturday delivery, add					7.50			7.50	8.25
1510	For truck holding/waiting time past 1st hour per load, add				Hr.	96.50			96.50	106
1520	For short load (less than 4 C.Y.), add per load				Ea.	74.50			74.50	82
2000	For all lightweight aggregate, add				C.Y.	45%				

03 31 13.70 Placing Concrete

		Crew	Daily Output	Labor-Hours	Unit	Material	2016 Bare Costs Labor	2016 Bare Costs Equipment	Total	Total Incl O&P
0010	**PLACING CONCRETE**									
0020	Includes labor and equipment to place, level (strike off) and consolidate									
1900	Footings, continuous, shallow, direct chute	C-6	120	.400	C.Y.		10.50	.53	11.03	18.05
1950	Pumped	C-20	150	.427			11.60	5.25	16.85	25
2000	With crane and bucket	C-7	90	.800			22	13.45	35.45	52
2400	Footings, spread, under 1 C.Y., direct chute	C-6	55	.873			23	1.15	24.15	39.50
2600	Over 5 C.Y., direct chute		120	.400			10.50	.53	11.03	18.05
2900	Foundation mats, over 20 C.Y., direct chute		350	.137			3.59	.18	3.77	6.20
4300	Slab on grade, up to 6" thick, direct chute		110	.436			11.45	.58	12.03	19.65
4350	Pumped	C-20	130	.492			13.35	6.10	19.45	28.50
4400	With crane and bucket	C-7	110	.655			18.10	11	29.10	42
4900	Walls, 8" thick, direct chute	C-6	90	.533			13.95	.70	14.65	24
4950	Pumped	C-20	100	.640			17.35	7.90	25.25	37.50
5000	With crane and bucket	C-7	80	.900			25	15.10	40.10	58
5050	12" thick, direct chute	C-6	100	.480			12.60	.63	13.23	21.50
5100	Pumped	C-20	110	.582			15.80	7.20	23	34.50
5200	With crane and bucket	C-7	90	.800			22	13.45	35.45	52
5600	Wheeled concrete dumping, add to placing costs above									
5610	Walking cart, 50' haul, add	C-18	32	.281	C.Y.		7	1.88	8.88	13.75
5620	150' haul, add		24	.375			9.35	2.50	11.85	18.35
5700	250' haul, add		18	.500			12.45	3.34	15.79	24.50
5800	Riding cart, 50' haul, add	C-19	80	.113			2.80	1.25	4.05	6.05
5810	150' haul, add		60	.150			3.73	1.66	5.39	8.10
5900	250' haul, add		45	.200			4.97	2.22	7.19	10.75

03 35 Concrete Finishing

03 35 13 – High-Tolerance Concrete Floor Finishing

03 35 13.30 Finishing Floors, High Tolerance

		Daily Output	Labor-Hours	Unit	Material	2016 Bare Costs Labor	2016 Bare Costs Equipment	Total	Total Incl O&P
		Crew							
0010	**FINISHING FLOORS, HIGH TOLERANCE**								
0012	Finishing of fresh concrete flatwork requires that concrete								
0013	first be placed, struck off & consolidated								
0015	Basic finishing for various unspecified flatwork								
0100	Bull float only	C-10	4000	.006	S.F.		.18		.18	.29
0125	Bull float & manual float		2000	.012			.35		.35	.58
0150	Bull float, manual float, & broom finish, w/edging & joints		1850	.013			.38		.38	.63
0200	Bull float, manual float & manual steel trowel		1265	.019			.56		.56	.92
0210	For specified Random Access Floors in ACI Classes 1, 2, 3 and 4 to achieve								
0215	Composite Overall Floor Flatness and Levelness values up to FF35/FL25								
0250	Bull float, machine float & machine trowel (walk-behind)	C-10C	1715	.014	S.F.		.41	.03	.44	.71
0300	Power screed, bull float, machine float & trowel (walk-behind)	C-10D	2400	.010			.30	.05	.35	.53
0350	Power screed, bull float, machine float & trowel (ride-on)	C-10E	4000	.006			.18	.06	.24	.36

03 35 23 – Exposed Aggregate Concrete Finishing

03 35 23.30 Finishing Floors, Exposed Aggregate

		Crew	Daily Output	Labor-Hours	Unit	Material	Labor	Equipment	Total	Total Incl O&P
0010	**FINISHING FLOORS, EXPOSED AGGREGATE**									
1600	Exposed local aggregate finish, seeded on fresh concrete, 3 lb. per S.F.	1 Cefi	625	.013	S.F.	.20	.41		.61	.88
1650	4 lb. per S.F.	"	465	.017	"	.33	.55		.88	1.26

03 35 29 – Tooled Concrete Finishing

03 35 29.30 Finishing Floors, Tooled

		Crew	Daily Output	Labor-Hours	Unit	Material	Labor	Equipment	Total	Total Incl O&P
0010	**FINISHING FLOORS, TOOLED**									
4400	Stair finish, fresh concrete, float finish	1 Cefi	275	.029	S.F.		.93		.93	1.51
4500	Steel trowel finish	"	200	.040	"		1.28		1.28	2.07

03 35 29.60 Finishing Walls

		Crew	Daily Output	Labor-Hours	Unit	Material	Labor	Equipment	Total	Total Incl O&P
0010	**FINISHING WALLS**									
0020	Break ties and patch voids	1 Cefi	540	.015	S.F.	.04	.47		.51	.82
0050	Burlap rub with grout	"	450	.018		.04	.57		.61	.97
0300	Bush hammer, green concrete	B-39	1000	.048			1.20	.23	1.43	2.26
0350	Cured concrete	"	650	.074			1.84	.35	2.19	3.48
0500	Acid etch	1 Cefi	575	.014		.14	.44		.58	.88
0850	Grind form fins flush	1 Clab	700	.011	L.F.		.28		.28	.47

03 35 33 – Stamped Concrete Finishing

03 35 33.50 Slab Texture Stamping

		Crew	Daily Output	Labor-Hours	Unit	Material	Labor	Equipment	Total	Total Incl O&P
0010	**SLAB TEXTURE STAMPING**									
0050	Stamping requires that concrete first be placed, struck off, consolidated,									
0060	bull floated and free of bleed water. Decorative stamping tasks include:									
0100	Step 1 - first application of dry shake colored hardener	1 Cefi	6400	.001	S.F.	.44	.04		.48	.55
0110	Step 2 - bull float		6400	.001			.04		.04	.06
0130	Step 3 - second application of dry shake colored hardener		6400	.001		.22	.04		.26	.30
0140	Step 4 - bull float, manual float & steel trowel	3 Cefi	1280	.019			.60		.60	.97
0150	Step 5 - application of dry shake colored release agent	1 Cefi	6400	.001		.09	.04		.13	.16
0160	Step 6 - place, tamp & remove mats	3 Cefi	2400	.010		.88	.32		1.20	1.49
0170	Step 7 - touch up edges, mat joints & simulated grout lines	1 Cefi	1280	.006			.20		.20	.32
0300	Alternate stamping estimating method includes all tasks above	4 Cefi	800	.040		1.63	1.28		2.91	3.87
0400	Step 8 - pressure wash @ 3000 psi after 24 hours	1 Cefi	1600	.005			.16		.16	.26
0500	Step 9 - roll 2 coats cure/seal compound when dry	"	800	.010		.63	.32		.95	1.21

03 39 Concrete Curing

03 39 13 – Water Concrete Curing

03 39 13.50 Water Curing

		Crew	Daily Output	Labor-Hours	Unit	Material	2016 Bare Costs Labor	Equipment	Total	Total Incl O&P
0011	**WATER CURING**									
0020	With burlap, 4 uses assumed, 7.5 oz.	2 Clab	5500	.003	S.F.	.14	.07		.21	.27
0101	10 oz.	"	5500	.003	"	.25	.07		.32	.40

03 39 23 – Membrane Concrete Curing

03 39 23.13 Chemical Compound Membrane Concrete Curing

0010	**CHEMICAL COMPOUND MEMBRANE CONCRETE CURING**									
0301	Sprayed membrane curing compound	2 Clab	9500	.002	S.F.	.13	.04		.17	.21

03 39 23.23 Sheet Membrane Concrete Curing

0010	**SHEET MEMBRANE CONCRETE CURING**									
0201	Curing blanket, burlap/poly, 2-ply	2 Clab	7000	.002	S.F.	.29	.06		.35	.41

03 41 Precast Structural Concrete

03 41 23 – Precast Concrete Stairs

03 41 23.50 Precast Stairs

		Crew	Daily Output	Labor-Hours	Unit	Material	Labor	Equipment	Total	Total Incl O&P
0010	**PRECAST STAIRS**									
0020	Precast concrete treads on steel stringers, 3' wide	C-12	75	.640	Riser	135	21	8.80	164.80	193
0300	Front entrance, 5' wide with 48" platform, 2 risers		16	3	Flight	485	99	41	625	740
0350	5 risers		12	4		765	132	55	952	1,125
0500	6' wide, 2 risers		15	3.200		535	106	44	685	815
0550	5 risers		11	4.364		845	144	60	1,049	1,225
0700	7' wide, 2 risers	↓	14	3.429		685	113	47	845	990
1200	Basement entrance stairwell, 6 steps, incl. steel bulkhead door	B-51	22	2.182		1,450	56.50	11.30	1,517.80	1,675
1250	14 steps	"	11	4.364	↓	2,400	113	22.50	2,535.50	2,850

03 48 Precast Concrete Specialties

03 48 43 – Precast Concrete Trim

03 48 43.40 Precast Lintels

		Crew	Daily Output	Labor-Hours	Unit	Material	Labor	Equipment	Total	Total Incl O&P
0010	**PRECAST LINTELS**, smooth gray, prestressed, stock units only									
0800	4" wide, 8" high, x 4' long	D-10	28	1.143	Ea.	25	37	17.10	79.10	108
0850	8' long		24	1.333		63	43	19.95	125.95	164
1000	6" wide, 8" high, x 4' long		26	1.231		37	40	18.45	95.45	128
1050	10' long	↓	22	1.455	↓	92.50	47	22	161.50	205

03 48 43.90 Precast Window Sills

0010	**PRECAST WINDOW SILLS**									
0600	Precast concrete, 4" tapers to 3", 9" wide	D-1	70	.229	L.F.	14.65	6.75		21.40	27.50
0650	11" wide	"	60	.267	"	24	7.85		31.85	39

03 54 Cast Underlayment

03 54 13 – Gypsum Cement Underlayment

03 54 13.50 Poured Gypsum Underlayment

		Crew	Daily Output	Labor-Hours	Unit	Material	2016 Bare Costs Labor	Equipment	Total	Total Incl O&P
0010	POURED GYPSUM UNDERLAYMENT									
0400	Underlayment, gypsum based, self-leveling 2500 psi, pumped, 1/2" thick	C-8	24000	.002	S.F.	.36	.07	.03	.46	.54
0500	3/4" thick		20000	.003		.54	.08	.04	.66	.77
0600	1" thick		16000	.004		.72	.10	.05	.87	1.02
1400	Hand placed, 1/2" thick	C-18	450	.020		.36	.50	.13	.99	1.38
1500	3/4" thick	"	300	.030		.54	.75	.20	1.49	2.07

03 63 Epoxy Grouting

03 63 05 – Grouting of Dowels and Fasteners

03 63 05.10 Epoxy Only

		Crew	Daily Output	Labor-Hours	Unit	Material	2016 Bare Costs Labor	Equipment	Total	Total Incl O&P
0010	EPOXY ONLY									
1500	Chemical anchoring, epoxy cartridge, excludes layout, drilling, fastener									
1530	For fastener 3/4" diam. x 6" embedment	2 Skwk	72	.222	Ea.	4.84	7.60		12.44	18.05
1535	1" diam. x 8" embedment		66	.242		7.25	8.30		15.55	22
1540	1-1/4" diam. x 10" embedment		60	.267		14.50	9.10		23.60	31.50
1545	1-3/4" diam. x 12" embedment		54	.296		24	10.15		34.15	43.50
1550	14" embedment		48	.333		29	11.40		40.40	51
1555	2" diam. x 12" embedment		42	.381		38.50	13.05		51.55	64.50
1560	18" embedment		32	.500		48.50	17.10		65.60	81.50

03 82 Concrete Boring

03 82 16 – Concrete Drilling

03 82 16.10 Concrete Impact Drilling

		Crew	Daily Output	Labor-Hours	Unit	Material	2016 Bare Costs Labor	Equipment	Total	Total Incl O&P
0010	CONCRETE IMPACT DRILLING									
0020	Includes bit cost, layout and set-up time, no anchors									
0050	Up to 4" deep in concrete/brick floors/walls									
0100	Holes, 1/4" diameter	1 Carp	75	.107	Ea.	.07	3.62		3.69	6.15
0150	For each additional inch of depth in same hole, add		430	.019		.02	.63		.65	1.08
0200	3/8" diameter		63	.127		.06	4.30		4.36	7.25
0250	For each additional inch of depth in same hole, add		340	.024		.01	.80		.81	1.36
0300	1/2" diameter		50	.160		.06	5.40		5.46	9.15
0350	For each additional inch of depth in same hole, add		250	.032		.02	1.08		1.10	1.84
0400	5/8" diameter		48	.167		.10	5.65		5.75	9.55
0450	For each additional inch of depth in same hole, add		240	.033		.03	1.13		1.16	1.92
0500	3/4" diameter		45	.178		.13	6.05		6.18	10.25
0550	For each additional inch of depth in same hole, add		220	.036		.03	1.23		1.26	2.10
0600	7/8" diameter		43	.186		.18	6.30		6.48	10.75
0650	For each additional inch of depth in same hole, add		210	.038		.04	1.29		1.33	2.21
0700	1" diameter		40	.200		.16	6.80		6.96	11.50
0750	For each additional inch of depth in same hole, add		190	.042		.04	1.43		1.47	2.43
0800	1-1/4" diameter		38	.211		.27	7.15		7.42	12.25
0850	For each additional inch of depth in same hole, add		180	.044		.07	1.51		1.58	2.59
0900	1-1/2" diameter		35	.229		.41	7.75		8.16	13.40
0950	For each additional inch of depth in same hole, add		165	.048		.10	1.64		1.74	2.86
1000	For ceiling installations, add						40%			

Division Notes

Estimating Tips
04 05 00 Common Work Results for Masonry

- The terms mortar and grout are often used interchangeably, and incorrectly. Mortar is used to bed masonry units, seal the entry of air and moisture, provide architectural appearance, and allow for size variations in the units. Grout is used primarily in reinforced masonry construction and is used to bond the masonry to the reinforcing steel. Common mortar types are M(2500 psi), S(1800 psi), N(750 psi), and O(350 psi), and conform to ASTM C270. Grout is either fine or coarse and conforms to ASTM C476, and in-place strengths generally exceed 2500 psi. Mortar and grout are different components of masonry construction and are placed by entirely different methods. An estimator should be aware of their unique uses and costs.
- Mortar is included in all assembled masonry line items. The mortar cost, part of the assembled masonry material cost, includes all ingredients, all labor, and all equipment required. Please see reference number R040513-10.
- Waste, specifically the loss/droppings of mortar and the breakage of brick and block, is included in all unit cost lines that include mortar and masonry units in this division. A factor of 25% is added for mortar and 3% for brick and concrete masonry units.
- Scaffolding or staging is not included in any of the Division 4 costs. Refer to Subdivision 01 54 23 for scaffolding and staging costs.

04 20 00 Unit Masonry

- The most common types of unit masonry are brick and concrete masonry. The major classifications of brick are building brick (ASTM C62), facing brick (ASTM C216), glazed brick, fire brick, and pavers. Many varieties of texture and appearance can exist within these classifications, and the estimator would be wise to check local custom and availability within the project area. For repair and remodeling jobs, matching the existing brick may be the most important criteria.
- Brick and concrete block are priced by the piece and then converted into a price per square foot of wall. Openings less than two square feet are generally ignored by the estimator because any savings in units used is offset by the cutting and trimming required.
- It is often difficult and expensive to find and purchase small lots of historic brick. Costs can vary widely. Many design issues affect costs, selection of mortar mix, and repairs or replacement of masonry materials. Cleaning techniques must be reflected in the estimate.
- All masonry walls, whether interior or exterior, require bracing. The cost of bracing walls during construction should be included by the estimator, and this bracing must remain in place until permanent bracing is complete. Permanent bracing of masonry walls is accomplished by masonry itself, in the form of pilasters or abutting wall corners, or by anchoring the walls to the structural frame. Accessories in the form of anchors, anchor slots, and ties are used, but their supply and installation can be by different trades. For instance, anchor slots on spandrel beams and columns are supplied and welded in place by the steel fabricator, but the ties from the slots into the masonry are installed by the bricklayer. Regardless of the installation method, the estimator must be certain that these accessories are accounted for in pricing.

Reference Numbers

Reference numbers are shown at the beginning of some major classifications. These numbers refer to related items in the Reference Section. The reference information may be an estimating procedure, an alternate pricing method, or technical information.

Note: Not all subdivisions listed here necessarily appear. ■

Did you know?

RSMeans Online gives you the same access to RSMeans' data with 24/7 access:
- Quickly locate costs in the searchable database.
- Build cost lists, estimates, and reports in minutes.
- Adjust costs to any location in the U.S. and Canada with the click of a button.

Start your free trial today at **www.RSMeansOnline.com**

No part of this cost data may be reproduced, stored in a retrieval system, or transmitted in any form or by any means without prior written permission of RSMeans.

04 01 Maintenance of Masonry

04 01 20 – Maintenance of Unit Masonry

04 01 20.20 Pointing Masonry

		Crew	Daily Output	Labor-Hours	Unit	Material	2016 Bare Costs Labor	2016 Bare Costs Equipment	Total	Total Incl O&P
0010	**POINTING MASONRY**									
0300	Cut and repoint brick, hard mortar, running bond	1 Bric	80	.100	S.F.	.58	3.24		3.82	6.05
0320	Common bond		77	.104		.58	3.37		3.95	6.25
0360	Flemish bond		70	.114		.61	3.70		4.31	6.80
0400	English bond		65	.123		.61	3.99		4.60	7.30
0600	Soft old mortar, running bond		100	.080		.58	2.59		3.17	4.96
0620	Common bond		96	.083		.58	2.70		3.28	5.15
0640	Flemish bond		90	.089		.61	2.88		3.49	5.45
0680	English bond		82	.098		.61	3.16		3.77	5.90
0700	Stonework, hard mortar		140	.057	L.F.	.77	1.85		2.62	3.94
0720	Soft old mortar		160	.050	"	.77	1.62		2.39	3.55
1000	Repoint, mask and grout method, running bond		95	.084	S.F.	.77	2.73		3.50	5.40
1020	Common bond		90	.089		.77	2.88		3.65	5.65
1040	Flemish bond		86	.093		.81	3.01		3.82	5.90
1060	English bond		77	.104		.81	3.37		4.18	6.50
2000	Scrub coat, sand grout on walls, thin mix, brushed		120	.067		3.22	2.16		5.38	7.15
2020	Troweled		98	.082		4.49	2.64		7.13	9.35

04 01 20.30 Pointing CMU

		Crew	Daily Output	Labor-Hours	Unit	Material	Labor	Equipment	Total	Total Incl O&P
0010	**POINTING CMU**									
0300	Cut and repoint block, hard mortar, running bond	1 Bric	190	.042	S.F.	.24	1.36		1.60	2.53
0310	Stacked bond		200	.040		.24	1.30		1.54	2.42
0600	Soft old mortar, running bond		230	.035		.24	1.13		1.37	2.14
0610	Stacked bond		245	.033		.24	1.06		1.30	2.02

04 01 30 – Unit Masonry Cleaning

04 01 30.60 Brick Washing

		Crew	Daily Output	Labor-Hours	Unit	Material	Labor	Equipment	Total	Total Incl O&P
0010	**BRICK WASHING** R040130-10									
0012	Acid cleanser, smooth brick surface	1 Bric	560	.014	S.F.	.05	.46		.51	.82
0050	Rough brick		400	.020		.06	.65		.71	1.15
0060	Stone, acid wash		600	.013		.08	.43		.51	.80
1000	Muriatic acid, price per gallon in 5 gallon lots				Gal.	9.45			9.45	10.35

04 05 Common Work Results for Masonry

04 05 05 – Selective Demolition for Masonry

04 05 05.10 Selective Demolition

		Crew	Daily Output	Labor-Hours	Unit	Material	Labor	Equipment	Total	Total Incl O&P
0010	**SELECTIVE DEMOLITION** R024119-10									
0200	Bond beams, 8" block with #4 bar	2 Clab	32	.500	L.F.		12.35		12.35	20.50
0300	Concrete block walls, unreinforced, 2" thick		1200	.013	S.F.		.33		.33	.55
0310	4" thick		1150	.014			.34		.34	.57
0320	6" thick		1100	.015			.36		.36	.60
0330	8" thick		1050	.015			.38		.38	.63
0340	10" thick		1000	.016			.39		.39	.66
0360	12" thick		950	.017			.42		.42	.69
0380	Reinforced alternate courses, 2" thick		1130	.014			.35		.35	.58
0390	4" thick		1080	.015			.37		.37	.61
0400	6" thick		1035	.015			.38		.38	.64
0410	8" thick		990	.016			.40		.40	.67
0420	10" thick		940	.017			.42		.42	.70
0430	12" thick		890	.018			.44		.44	.74
0440	Reinforced alternate courses & vertically 48" OC, 4" thick		900	.018			.44		.44	.73
0450	6" thick		850	.019			.46		.46	.78

04 05 Common Work Results for Masonry

04 05 05 – Selective Demolition for Masonry

04 05 05.10 Selective Demolition		Crew	Daily Output	Labor-Hours	Unit	Material	2016 Bare Costs Labor	2016 Bare Costs Equipment	Total	Total Incl O&P
0460	8" thick	2 Clab	800	.020	S.F.		.49		.49	.83
0480	10" thick		750	.021			.53		.53	.88
0490	12" thick		700	.023			.56		.56	.94
1000	Chimney, 16" x 16", soft old mortar	1 Clab	55	.145	C.F.		3.59		3.59	6
1020	Hard mortar		40	.200			4.93		4.93	8.25
1030	16" x 20", soft old mortar		55	.145			3.59		3.59	6
1040	Hard mortar		40	.200			4.93		4.93	8.25
1050	16" x 24", soft old mortar		55	.145			3.59		3.59	6
1060	Hard mortar		40	.200			4.93		4.93	8.25
1080	20" x 20", soft old mortar		55	.145			3.59		3.59	6
1100	Hard mortar		40	.200			4.93		4.93	8.25
1110	20" x 24", soft old mortar		55	.145			3.59		3.59	6
1120	Hard mortar		40	.200			4.93		4.93	8.25
1140	20" x 32", soft old mortar		55	.145			3.59		3.59	6
1160	Hard mortar		40	.200			4.93		4.93	8.25
1200	48" x 48", soft old mortar		55	.145			3.59		3.59	6
1220	Hard mortar		40	.200			4.93		4.93	8.25
1250	Metal, high temp steel jacket, 24" diameter	E-2	130	.369	V.L.F.		13.30	11.70	25	37
1260	60" diameter	"	60	.800			29	25.50	54.50	80
1280	Flue lining, up to 12" x 12"	1 Clab	200	.040			.99		.99	1.65
1282	Up to 24" x 24"		150	.053			1.31		1.31	2.20
2000	Columns, 8" x 8", soft old mortar		48	.167			4.11		4.11	6.90
2020	Hard mortar		40	.200			4.93		4.93	8.25
2060	16" x 16", soft old mortar		16	.500			12.35		12.35	20.50
2100	Hard mortar		14	.571			14.10		14.10	23.50
2140	24" x 24", soft old mortar		8	1			24.50		24.50	41.50
2160	Hard mortar		6	1.333			33		33	55
2200	36" x 36", soft old mortar		4	2			49.50		49.50	82.50
2220	Hard mortar		3	2.667			65.50		65.50	110
2230	Alternate pricing method, soft old mortar		30	.267	C.F.		6.55		6.55	11
2240	Hard mortar		23	.348	"		8.55		8.55	14.35
3000	Copings, precast or masonry, to 8" wide									
3020	Soft old mortar	1 Clab	180	.044	L.F.		1.10		1.10	1.83
3040	Hard mortar	"	160	.050	"		1.23		1.23	2.06
3100	To 12" wide									
3120	Soft old mortar	1 Clab	160	.050	L.F.		1.23		1.23	2.06
3140	Hard mortar	"	140	.057	"		1.41		1.41	2.36
4000	Fireplace, brick, 30" x 24" opening									
4020	Soft old mortar	1 Clab	2	4	Ea.		98.50		98.50	165
4040	Hard mortar		1.25	6.400			158		158	264
4100	Stone, soft old mortar		1.50	5.333			131		131	220
4120	Hard mortar		1	8			197		197	330
5000	Veneers, brick, soft old mortar		140	.057	S.F.		1.41		1.41	2.36
5020	Hard mortar		125	.064			1.58		1.58	2.64
5050	Glass block, up to 4" thick		500	.016			.39		.39	.66
5100	Granite and marble, 2" thick		180	.044			1.10		1.10	1.83
5120	4" thick		170	.047			1.16		1.16	1.94
5140	Stone, 4" thick		180	.044			1.10		1.10	1.83
5160	8" thick		175	.046			1.13		1.13	1.89
5400	Alternate pricing method, stone, 4" thick		60	.133	C.F.		3.29		3.29	5.50
5420	8" thick		85	.094	"		2.32		2.32	3.88

04 05 Common Work Results for Masonry

04 05 13 – Masonry Mortaring

04 05 13.10 Cement

		Crew	Daily Output	Labor-Hours	Unit	Material	2016 Bare Costs Labor	Equipment	Total	Total Incl O&P
0010	**CEMENT** R040513-10									
0100	Masonry, 70 lb. bag, T.L. lots				Bag	12			12	13.20
0150	L.T.L. lots					12.75			12.75	14
0200	White, 70 lb. bag, T.L. lots					15.70			15.70	17.30
0250	L.T.L. lots					16.55			16.55	18.20

04 05 16 – Masonry Grouting

04 05 16.30 Grouting

		Crew	Daily Output	Labor-Hours	Unit	Material	Labor	Equipment	Total	Total Incl O&P
0010	**GROUTING**									
0011	Bond beams & lintels, 8" deep, 6" thick, 0.15 C.F. per L.F.	D-4	1480	.027	L.F.	.69	.74	.09	1.52	2.09
0020	8" thick, 0.2 C.F. per L.F.		1400	.029		1.11	.78	.10	1.99	2.63
0050	10" thick, 0.25 C.F. per L.F.		1200	.033		1.16	.91	.11	2.18	2.91
0060	12" thick, 0.3 C.F. per L.F.		1040	.038		1.39	1.05	.13	2.57	3.43
0200	Concrete block cores, solid, 4" thk., by hand, 0.067 C.F./S.F. of wall	D-8	1100	.036	S.F.	.31	1.09		1.40	2.16
0210	6" thick, pumped, 0.175 C.F. per S.F.	D-4	720	.056		.81	1.52	.19	2.52	3.64
0250	8" thick, pumped, 0.258 C.F. per S.F.		680	.059		1.19	1.61	.20	3	4.22
0300	10" thick, pumped, 0.340 C.F. per S.F.		660	.061		1.57	1.66	.20	3.43	4.72
0350	12" thick, pumped, 0.422 C.F. per S.F.		640	.063		1.95	1.71	.21	3.87	5.25

04 05 19 – Masonry Anchorage and Reinforcing

04 05 19.05 Anchor Bolts

		Crew	Daily Output	Labor-Hours	Unit	Material	Labor	Equipment	Total	Total Incl O&P
0010	**ANCHOR BOLTS**									
0015	Installed in fresh grout in CMU bond beams or filled cores, no templates									
0020	Hooked, with nut and washer, 1/2" diam., 8" long	1 Bric	132	.061	Ea.	1.32	1.96		3.28	4.73
0030	12" long		131	.061		1.47	1.98		3.45	4.92
0040	5/8" diameter, 8" long		129	.062		3.09	2.01		5.10	6.75
0050	12" long		127	.063		3.81	2.04		5.85	7.60
0060	3/4" diameter, 8" long		127	.063		3.81	2.04		5.85	7.60
0070	12" long		125	.064		4.76	2.07		6.83	8.70

04 05 19.16 Masonry Anchors

		Crew	Daily Output	Labor-Hours	Unit	Material	Labor	Equipment	Total	Total Incl O&P
0010	**MASONRY ANCHORS**									
0020	For brick veneer, galv., corrugated, 7/8" x 7", 22 Ga.	1 Bric	10.50	.762	C	15.15	24.50		39.65	57.50
0100	24 Ga.		10.50	.762		9.90	24.50		34.40	52
0150	16 Ga.		10.50	.762		29	24.50		53.50	72.50
0200	Buck anchors, galv., corrugated, 16 ga., 2" bend, 8" x 2"		10.50	.762		56.50	24.50		81	103
0250	8" x 3"		10.50	.762		58.50	24.50		83	106
0660	Cavity wall, Z-type, galvanized, 6" long, 1/8" diam.		10.50	.762		25	24.50		49.50	68.50
0670	3/16" diameter		10.50	.762		30.50	24.50		55	74.50
0680	1/4" diameter		10.50	.762		39	24.50		63.50	84
0850	8" long, 3/16" diameter		10.50	.762		26.50	24.50		51	70
0855	1/4" diameter		10.50	.762		48	24.50		72.50	94
1000	Rectangular type, galvanized, 1/4" diameter, 2" x 6"		10.50	.762		73	24.50		97.50	122
1050	4" x 6"		10.50	.762		88	24.50		112.50	138
1100	3/16" diameter, 2" x 6"		10.50	.762		45.50	24.50		70	91
1150	4" x 6"		10.50	.762		52	24.50		76.50	98.50
1500	Rigid partition anchors, plain, 8" long, 1" x 1/8"		10.50	.762		243	24.50		267.50	310
1550	1" x 1/4"		10.50	.762		285	24.50		309.50	355
1580	1-1/2" x 1/8"		10.50	.762		267	24.50		291.50	335
1600	1-1/2" x 1/4"		10.50	.762		335	24.50		359.50	405
1650	2" x 1/8"		10.50	.762		315	24.50		339.50	385
1700	2" x 1/4"		10.50	.762		415	24.50		439.50	495

04 05 Common Work Results for Masonry

04 05 19 – Masonry Anchorage and Reinforcing

04 05 19.26 Masonry Reinforcing Bars

			Crew	Daily Output	Labor-Hours	Unit	Material	2016 Bare Costs Labor	Equipment	Total	Total Incl O&P
0010	**MASONRY REINFORCING BARS**	R040519-50									
0015	Steel bars A615, placed horiz., #3 & #4 bars		1 Bric	450	.018	Lb.	.48	.58		1.06	1.49
0050	Placed vertical, #3 & #4 bars			350	.023		.48	.74		1.22	1.76
0060	#5 & #6 bars			650	.012		.48	.40		.88	1.19
0200	Joint reinforcing, regular truss, to 6" wide, mill std galvanized			30	.267	C.L.F.	22.50	8.65		31.15	39.50
0250	12" wide			20	.400		26.50	12.95		39.45	50.50
0400	Cavity truss with drip section, to 6" wide			30	.267		23	8.65		31.65	40
0450	12" wide			20	.400		26.50	12.95		39.45	51

04 21 Clay Unit Masonry

04 21 13 – Brick Masonry

04 21 13.13 Brick Veneer Masonry

			Crew	Daily Output	Labor-Hours	Unit	Material	Labor	Equipment	Total	Total Incl O&P
0010	**BRICK VENEER MASONRY**, T.L. lots, excl. scaff., grout & reinforcing										
0015	Material costs incl. 3% brick and 25% mortar waste	R042110-10									
2000	Standard, sel. common, 4" x 2-2/3" x 8", (6.75/S.F.)	R042110-20	D-8	230	.174	S.F.	4.34	5.25		9.59	13.45
2020	Red, 4" x 2-2/3" x 8", running bond			220	.182		3.93	5.45		9.38	13.40
2050	Full header every 6th course (7.88/S.F.)	R042110-50		185	.216		4.58	6.50		11.08	15.90
2100	English, full header every 2nd course (10.13/S.F.)			140	.286		5.85	8.60		14.45	21
2150	Flemish, alternate header every course (9.00/S.F.)			150	.267		5.20	8		13.20	19.10
2200	Flemish, alt. header every 6th course (7.13/S.F.)			205	.195		4.15	5.85		10	14.35
2250	Full headers throughout (13.50/S.F.)			105	.381		7.80	11.45		19.25	27.50
2300	Rowlock course (13.50/S.F.)			100	.400		7.80	12.05		19.85	28.50
2350	Rowlock stretcher (4.50/S.F.)			310	.129		2.64	3.88		6.52	9.35
2400	Soldier course (6.75/S.F.)			200	.200		3.93	6		9.93	14.35
2450	Sailor course (4.50/S.F.)			290	.138		2.64	4.15		6.79	9.80
2600	Buff or gray face, running bond, (6.75/S.F.)			220	.182		4.15	5.45		9.60	13.65
2700	Glazed face brick, running bond			210	.190		12.20	5.75		17.95	23
2750	Full header every 6th course (7.88/S.F.)			170	.235		14.20	7.10		21.30	27.50
3000	Jumbo, 6" x 4" x 12" running bond (3.00/S.F.)			435	.092		4.85	2.77		7.62	9.95
3050	Norman, 4" x 2-2/3" x 12" running bond, (4.5/S.F.)			320	.125		6.05	3.76		9.81	12.90
3100	Norwegian, 4" x 3-1/5" x 12" (3.75/S.F.)			375	.107		5.35	3.21		8.56	11.25
3150	Economy, 4" x 4" x 8" (4.50/S.F.)			310	.129		4.24	3.88		8.12	11.10
3200	Engineer, 4" x 3-1/5" x 8" (5.63/S.F.)			260	.154		3.76	4.63		8.39	11.85
3250	Roman, 4" x 2" x 12" (6.00/S.F.)			250	.160		7.20	4.81		12.01	15.90
3300	SCR, 6" x 2-2/3" x 12" (4.50/S.F.)			310	.129		6.20	3.88		10.08	13.30
3350	Utility, 4" x 4" x 12" (3.00/S.F.)			360	.111		4.78	3.34		8.12	10.80
3360	For less than truck load lots, add					M	15%				
3400	For cavity wall construction, add							15%			
3450	For stacked bond, add							10%			
3500	For interior veneer construction, add							15%			
3550	For curved walls, add							30%			

04 21 13.14 Thin Brick Veneer

		Crew	Daily Output	Labor-Hours	Unit	Material	Labor	Equipment	Total	Total Incl O&P
0010	**THIN BRICK VENEER**									
0015	Material costs incl. 3% brick and 25% mortar waste									
0020	On & incl. metal panel support sys, modular, 2-2/3" x 5/8" x 8", red	D-7	92	.174	S.F.	9.35	4.83		14.18	18.05
0100	Closure, 4" x 5/8" x 8"		110	.145		9.10	4.04		13.14	16.55
0110	Norman, 2-2/3" x 5/8" x 12"		110	.145		9.15	4.04		13.19	16.60
0120	Utility, 4" x 5/8" x 12"		125	.128		8.85	3.55		12.40	15.50
0130	Emperor, 4" x 3/4" x 16"		175	.091		10	2.54		12.54	15.10
0140	Super emperor, 8" x 3/4" x 16"		195	.082		10.30	2.28		12.58	15.05

04 21 Clay Unit Masonry

04 21 13 – Brick Masonry

04 21 13.14 Thin Brick Veneer

		Crew	Daily Output	Labor-Hours	Unit	Material	2016 Bare Costs Labor	Equipment	Total	Total Incl O&P
0150	For L shaped corners with 4" return, add				L.F.	9.25			9.25	10.20
0200	On masonry/plaster back-up, modular, 2-2/3" x 5/8" x 8", red	D-7	137	.117	S.F.	4.36	3.24		7.60	10.05
0210	Closure, 4" x 5/8" x 8"		165	.097		4.12	2.69		6.81	8.90
0220	Norman, 2-2/3" x 5/8" x 12"		165	.097		4.16	2.69		6.85	8.95
0230	Utility, 4" x 5/8" x 12"		185	.086		3.88	2.40		6.28	8.15
0240	Emperor, 4" x 3/4" x 16"		260	.062		5	1.71		6.71	8.25
0250	Super emperor, 8" x 3/4" x 16"		285	.056		5.35	1.56		6.91	8.35
0260	For L shaped corners with 4" return, add				L.F.	9.25			9.25	10.20
0270	For embedment into pre-cast concrete panels, add				S.F.	14.40			14.40	15.85

04 21 13.15 Chimney

		Crew	Daily Output	Labor-Hours	Unit	Material	Labor	Equipment	Total	Total Incl O&P
0010	**CHIMNEY**, excludes foundation, scaffolding, grout and reinforcing									
0100	Brick, 16" x 16", 8" flue	D-1	18.20	.879	V.L.F.	25	26		51	70.50
0150	16" x 20" with one 8" x 12" flue		16	1		39	29.50		68.50	92
0200	16" x 24" with two 8" x 8" flues		14	1.143		57	33.50		90.50	119
0250	20" x 20" with one 12" x 12" flue		13.70	1.168		46.50	34.50		81	109
0300	20" x 24" with two 8" x 12" flues		12	1.333		65	39.50		104.50	137
0350	20" x 32" with two 12" x 12" flues		10	1.600		81.50	47		128.50	169

04 21 13.18 Columns

		Crew	Daily Output	Labor-Hours	Unit	Material	Labor	Equipment	Total	Total Incl O&P
0010	**COLUMNS**, solid, excludes scaffolding, grout and reinforcing									
0050	Brick, 8" x 8", 9 brick per V.L.F.	D-1	56	.286	V.L.F.	5	8.45		13.45	19.55
0100	12" x 8", 13.5 brick per V.L.F.		37	.432		7.50	12.75		20.25	30
0200	12" x 12", 20 brick per V.L.F.		25	.640		11.10	18.90		30	44
0300	16" x 12", 27 brick per V.L.F.		19	.842		15	25		40	58
0400	16" x 16", 36 brick per V.L.F.		14	1.143		20	33.50		53.50	78
0500	20" x 16", 45 brick per V.L.F.		11	1.455		25	43		68	99
0600	20" x 20", 56 brick per V.L.F.		9	1.778		31	52.50		83.50	122
0700	24" x 20", 68 brick per V.L.F.		7	2.286		38	67.50		105.50	154
0800	24" x 24", 81 brick per V.L.F.		6	2.667		45	78.50		123.50	181
1000	36" x 36", 182 brick per V.L.F.		3	5.333		101	157		258	375

04 21 13.30 Oversized Brick

		Crew	Daily Output	Labor-Hours	Unit	Material	Labor	Equipment	Total	Total Incl O&P
0010	**OVERSIZED BRICK**, excludes scaffolding, grout and reinforcing									
0100	Veneer, 4" x 2.25" x 16"	D-8	387	.103	S.F.	5.30	3.11		8.41	11.05
0102	8" x 2.25" x 16", multicell		265	.151		16.65	4.54		21.19	26
0105	4" x 2.75" x 16"		412	.097		5.50	2.92		8.42	10.95
0107	8" x 2.75" x 16", multicell		295	.136		16.70	4.08		20.78	25
0110	4" x 4" x 16"		460	.087		3.61	2.62		6.23	8.35
0120	4" x 8" x 16"		533	.075		4.25	2.26		6.51	8.45
0122	4" x 8" x 16" multicell		327	.122		15.65	3.68		19.33	23.50
0125	Loadbearing, 6" x 4" x 16", grouted and reinforced		387	.103		10.70	3.11		13.81	16.95
0130	8" x 4" x 16", grouted and reinforced		327	.122		11.75	3.68		15.43	19.10
0132	10" x 4" x 16", grouted and reinforced		327	.122		23.50	3.68		27.18	32
0135	6" x 8" x 16", grouted and reinforced		440	.091		13.90	2.73		16.63	19.85
0140	8" x 8" x 16", grouted and reinforced		400	.100		14.85	3.01		17.86	21.50
0145	Curtainwall/reinforced veneer, 6" x 4" x 16"		387	.103		15.25	3.11		18.36	22
0150	8" x 4" x 16"		327	.122		18.55	3.68		22.23	26.50
0152	10" x 4" x 16"		327	.122		26	3.68		29.68	35
0155	6" x 8" x 16"		440	.091		18.95	2.73		21.68	25.50
0160	8" x 8" x 16"		400	.100		26.50	3.01		29.51	34
0200	For 1 to 3 slots in face, add					15%				
0210	For 4 to 7 slots in face, add					25%				
0220	For bond beams, add					20%				
0230	For bullnose shapes, add					20%				

04 21 Clay Unit Masonry

04 21 13 – Brick Masonry

04 21 13.30 Oversized Brick

		Crew	Daily Output	Labor-Hours	Unit	Material	2016 Bare Costs Labor	Equipment	Total	Total Incl O&P
0240	For open end knockout, add				S.F.	10%				
0250	For white or gray color group, add					10%				
0260	For 135 degree corner, add					250%				

04 21 13.35 Common Building Brick

			Crew	Daily Output	Labor-Hours	Unit	Material	Labor	Equipment	Total	Total Incl O&P
0010	**COMMON BUILDING BRICK**, C62, TL lots, material only	R042110-20									
0020	Standard					M	495			495	545
0050	Select					"	520			520	570

04 21 13.45 Face Brick

0010	**FACE BRICK** Material Only, C216, TL lots	R042110-20									
0300	Standard modular, 4" x 2-2/3" x 8"					M	460			460	505
2170	For less than truck load lots, add						15			15	16.50
2180	For buff or gray brick, add						16			16	17.60

04 22 Concrete Unit Masonry

04 22 10 – Concrete Masonry Units

04 22 10.11 Autoclave Aerated Concrete Block

			Crew	Daily Output	Labor-Hours	Unit	Material	Labor	Equipment	Total	Total Incl O&P
0010	**AUTOCLAVE AERATED CONCRETE BLOCK**, excl. scaffolding, grout & reinforcing										
0050	Solid, 4" x 8" x 24", incl. mortar	G	D-8	600	.067	S.F.	1.55	2.01		3.56	5.05
0060	6" x 8" x 24"	G		600	.067		2.34	2.01		4.35	5.90
0070	8" x 8" x 24"	G		575	.070		3.11	2.09		5.20	6.90
0080	10" x 8" x 24"	G		575	.070		3.80	2.09		5.89	7.65
0090	12" x 8" x 24"	G		550	.073		4.66	2.19		6.85	8.80

04 22 10.14 Concrete Block, Back-Up

			Crew	Daily Output	Labor-Hours	Unit	Material	Labor	Equipment	Total	Total Incl O&P
0010	**CONCRETE BLOCK, BACK-UP**, C90, 2000 psi	R042210-20									
0020	Normal weight, 8" x 16" units, tooled joint 1 side										
0050	Not-reinforced, 2000 psi, 2" thick		D-8	475	.084	S.F.	1.53	2.53		4.06	5.90
0200	4" thick			460	.087		1.84	2.62		4.46	6.40
0300	6" thick			440	.091		2.40	2.73		5.13	7.20
0350	8" thick			400	.100		2.56	3.01		5.57	7.80
0400	10" thick			330	.121		3.03	3.65		6.68	9.45
0450	12" thick		D-9	310	.155		4.17	4.57		8.74	12.20
1000	Reinforced, alternate courses, 4" thick		D-8	450	.089		2	2.67		4.67	6.65
1100	6" thick			430	.093		2.56	2.80		5.36	7.50
1150	8" thick			395	.101		2.76	3.05		5.81	8.15
1200	10" thick			320	.125		3.19	3.76		6.95	9.75
1250	12" thick		D-9	300	.160		4.34	4.72		9.06	12.65

04 22 10.16 Concrete Block, Bond Beam

		Crew	Daily Output	Labor-Hours	Unit	Material	Labor	Equipment	Total	Total Incl O&P
0010	**CONCRETE BLOCK, BOND BEAM**, C90, 2000 psi									
0020	Not including grout or reinforcing									
0125	Regular block, 6" thick	D-8	584	.068	L.F.	2.73	2.06		4.79	6.45
0130	8" high, 8" thick	"	565	.071		2.82	2.13		4.95	6.65
0150	12" thick	D-9	510	.094		3.83	2.78		6.61	8.85
0525	Lightweight, 6" thick	D-8	592	.068		2.85	2.03		4.88	6.55

04 22 10.19 Concrete Block, Insulation Inserts

		Crew	Daily Output	Labor-Hours	Unit	Material	Labor	Equipment	Total	Total Incl O&P
0010	**CONCRETE BLOCK, INSULATION INSERTS**									
0100	Styrofoam, plant installed, add to block prices									
0200	8" x 16" units, 6" thick				S.F.	1.20			1.20	1.32
0250	8" thick					1.35			1.35	1.49
0300	10" thick					1.40			1.40	1.54
0350	12" thick					1.55			1.55	1.71

04 22 Concrete Unit Masonry

04 22 10 – Concrete Masonry Units

04 22 10.19 Concrete Block, Insulation Inserts	Crew	Daily Output	Labor-Hours	Unit	Material	2016 Bare Costs Labor	Equipment	Total	Total Incl O&P
0500 8" x 8" units, 8" thick				S.F.	1.20			1.20	1.32
0550 12" thick					1.40			1.40	1.54
04 22 10.23 Concrete Block, Decorative									
0010 **CONCRETE BLOCK, DECORATIVE**, C90, 2000 psi									
5000 Split rib profile units, 1" deep ribs, 8 ribs									
5100 8" x 16" x 4" thick	D-8	345	.116	S.F.	3.98	3.49		7.47	10.20
5150 6" thick		325	.123		4.50	3.70		8.20	11.10
5200 8" thick		300	.133		5.10	4.01		9.11	12.30
5250 12" thick	D-9	275	.175		5.95	5.15		11.10	15.15
5400 For special deeper colors, 4" thick, add					1.29			1.29	1.42
5450 12" thick, add					1.33			1.33	1.47
5600 For white, 4" thick, add					1.29			1.29	1.42
5650 6" thick, add					1.31			1.31	1.44
5700 8" thick, add					1.34			1.34	1.48
5750 12" thick, add					1.39			1.39	1.53
04 22 10.24 Concrete Block, Exterior									
0010 **CONCRETE BLOCK, EXTERIOR**, C90, 2000 psi									
0020 Reinforced alt courses, tooled joints 2 sides									
0100 Normal weight, 8" x 16" x 6" thick	D-8	395	.101	S.F.	2.35	3.05		5.40	7.70
0200 8" thick		360	.111		3.62	3.34		6.96	9.55
0250 10" thick		290	.138		4.19	4.15		8.34	11.50
0300 12" thick	D-9	250	.192		4.89	5.65		10.54	14.85
04 22 10.26 Concrete Block Foundation Wall									
0010 **CONCRETE BLOCK FOUNDATION WALL**, C90/C145									
0050 Normal-weight, cut joints, horiz joint reinf, no vert reinf.									
0200 Hollow, 8" x 16" x 6" thick	D-8	455	.088	S.F.	3.02	2.64		5.66	7.75
0250 8" thick		425	.094		3.24	2.83		6.07	8.30
0300 10" thick		350	.114		3.70	3.44		7.14	9.80
0350 12" thick	D-9	300	.160		4.88	4.72		9.60	13.20
0500 Solid, 8" x 16" block, 6" thick	D-8	440	.091		3.08	2.73		5.81	7.95
0550 8" thick	"	415	.096		4.29	2.90		7.19	9.55
0600 12" thick	D-9	350	.137		6.15	4.05		10.20	13.50
04 22 10.32 Concrete Block, Lintels									
0010 **CONCRETE BLOCK, LINTELS**, C90, normal weight									
0100 Including grout and horizontal reinforcing									
0200 8" x 8" x 8", 1 #4 bar	D-4	300	.133	L.F.	3.87	3.65	.45	7.97	10.85
0250 2 #4 bars		295	.136		4.09	3.71	.46	8.26	11.20
0400 8" x 16" x 8", 1 #4 bar		275	.145		3.90	3.98	.49	8.37	11.50
0450 2 #4 bars		270	.148		4.12	4.05	.50	8.67	11.85
1000 12" x 8" x 8", 1 #4 bar		275	.145		5.30	3.98	.49	9.77	13.05
1100 2 #4 bars		270	.148		5.50	4.05	.50	10.05	13.35
1150 2 #5 bars		270	.148		5.75	4.05	.50	10.30	13.65
1200 2 #6 bars		265	.151		6.05	4.13	.51	10.69	14.10
1500 12" x 16" x 8", 1 #4 bar		250	.160		5.95	4.38	.54	10.87	14.45
1600 2 #3 bars		245	.163		6	4.47	.55	11.02	14.65
1650 2 #4 bars		245	.163		6.15	4.47	.55	11.17	14.85
1700 2 #5 bars		240	.167		6.40	4.56	.56	11.52	15.25
04 22 10.33 Lintel Block									
0010 **LINTEL BLOCK**									
3481 Lintel block 6" x 8" x 8"	D-1	300	.053	Ea.	1.32	1.57		2.89	4.07
3501 6" x 16" x 8"		275	.058		2.07	1.72		3.79	5.15

04 22 Concrete Unit Masonry

04 22 10 – Concrete Masonry Units

04 22 10.33 Lintel Block

		Crew	Daily Output	Labor-Hours	Unit	Material	2016 Bare Costs Labor	Equipment	Total	Total Incl O&P
3521	8" x 8" x 8"	D-1	275	.058	Ea.	1.19	1.72		2.91	4.17
3561	8" x 16" x 8"		250	.064		2.09	1.89		3.98	5.45

04 22 10.34 Concrete Block, Partitions

		Crew	Daily Output	Labor-Hours	Unit	Material	Labor	Equipment	Total	Total Incl O&P
0010	**CONCRETE BLOCK, PARTITIONS**, excludes scaffolding									
1000	Lightweight block, tooled joints, 2 sides, hollow									
1100	Not reinforced, 8" x 16" x 4" thick	D-8	440	.091	S.F.	1.85	2.73		4.58	6.60
1150	6" thick		410	.098		2.63	2.93		5.56	7.80
1200	8" thick		385	.104		3.20	3.13		6.33	8.70
1250	10" thick		370	.108		3.88	3.25		7.13	9.65
1300	12" thick	D-9	350	.137		4.09	4.05		8.14	11.25
4000	Regular block, tooled joints, 2 sides, hollow									
4100	Not reinforced, 8" x 16" x 4" thick	D-8	430	.093	S.F.	1.74	2.80		4.54	6.60
4150	6" thick		400	.100		2.30	3.01		5.31	7.55
4200	8" thick		375	.107		2.47	3.21		5.68	8.05
4250	10" thick		360	.111		2.93	3.34		6.27	8.80
4300	12" thick	D-9	340	.141		4.08	4.16		8.24	11.45

04 23 Glass Unit Masonry

04 23 13 – Vertical Glass Unit Masonry

04 23 13.10 Glass Block

		Crew	Daily Output	Labor-Hours	Unit	Material	Labor	Equipment	Total	Total Incl O&P
0010	**GLASS BLOCK**									
0100	Plain, 4" thick, under 1,000 S.F., 6" x 6"	D-8	115	.348	S.F.	27.50	10.45		37.95	48
0150	8" x 8"		160	.250		16	7.50		23.50	30
0160	end block		160	.250		60.50	7.50		68	79
0170	90 deg corner		160	.250		56	7.50		63.50	74
0180	45 deg corner		160	.250		46	7.50		53.50	63
0200	12" x 12"		175	.229		24	6.90		30.90	38
0210	4" x 8"		160	.250		30	7.50		37.50	45.50
0220	6" x 8"		160	.250		21.50	7.50		29	36
0700	For solar reflective blocks, add					100%				
1000	Thinline, plain, 3-1/8" thick, under 1,000 S.F., 6" x 6"	D-8	115	.348	S.F.	22.50	10.45		32.95	42
1050	8" x 8"		160	.250		12.55	7.50		20.05	26.50
1400	For cleaning block after installation (both sides), add		1000	.040		.16	1.20		1.36	2.19

04 24 Adobe Unit Masonry

04 24 16 – Manufactured Adobe Unit Masonry

04 24 16.06 Adobe Brick

			Crew	Daily Output	Labor-Hours	Unit	Material	Labor	Equipment	Total	Total Incl O&P
0010	**ADOBE BRICK**, Semi-stabilized, with cement mortar										
0060	Brick, 10" x 4" x 14", 2.6/S.F.	G	D-8	560	.071	S.F.	4.60	2.15		6.75	8.65
0080	12" x 4" x 16", 2.3/S.F.	G		580	.069		6.90	2.07		8.97	11.05
0100	10" x 4" x 16", 2.3/S.F.	G		590	.068		6.45	2.04		8.49	10.50
0120	8" x 4" x 16", 2.3/S.F.	G		560	.071		4.94	2.15		7.09	9.05
0140	4" x 4" x 16", 2.3/S.F.	G		540	.074		4.82	2.23		7.05	9
0160	6" x 4" x 16", 2.3/S.F.	G		540	.074		4.48	2.23		6.71	8.65
0180	4" x 4" x 12", 3.0/S.F.	G		520	.077		5.20	2.31		7.51	9.55
0200	8" x 4" x 12", 3.0/S.F.	G		520	.077		4.24	2.31		6.55	8.55

04 27 Multiple-Wythe Unit Masonry

04 27 10 – Multiple-Wythe Masonry

04 27 10.10 Cornices		Crew	Daily Output	Labor-Hours	Unit	Material	2016 Bare Costs Labor	Equipment	Total	Total Incl O&P
0010	**CORNICES**									
0110	Face bricks, 12 brick/S.F.	D-1	30	.533	SF Face	5.85	15.75		21.60	32.50
0150	15 brick/S.F.	"	23	.696	"	7	20.50		27.50	41.50
04 27 10.30 Brick Walls										
0010	**BRICK WALLS**, including mortar, excludes scaffolding									
0800	Face brick, 4" thick wall, 6.75 brick/S.F.	D-8	215	.186	S.F.	3.86	5.60		9.46	13.60
0850	Common brick, 4" thick wall, 6.75 brick/S.F.		240	.167		4.12	5		9.12	12.90
0900	8" thick, 13.50 bricks per S.F.		135	.296		8.55	8.90		17.45	24.50
1000	12" thick, 20.25 bricks per S.F.		95	.421		12.85	12.65		25.50	35
1050	16" thick, 27.00 bricks per S.F.		75	.533		17.40	16.05		33.45	45.50
1200	Reinforced, face brick, 4" thick wall, 6.75 brick/S.F.		210	.190		4.02	5.75		9.77	13.95
1220	Common brick, 4" thick wall, 6.75 brick/S.F.		235	.170		4.28	5.10		9.38	13.25
1250	8" thick, 13.50 bricks per S.F.		130	.308		8.85	9.25		18.10	25
1300	12" thick, 20.25 bricks per S.F.		90	.444		13.35	13.35		26.70	37
1350	16" thick, 27.00 bricks per S.F.		70	.571		18.05	17.20		35.25	48.50
04 27 10.40 Steps										
0010	**STEPS**									
0012	Entry steps, select common brick	D-1	.30	53.333	M	520	1,575		2,095	3,200

04 41 Dry-Placed Stone

04 41 10 – Dry Placed Stone

04 41 10.10 Rough Stone Wall			Crew	Daily Output	Labor-Hours	Unit	Material	2016 Bare Costs Labor	Equipment	Total	Total Incl O&P
0011	**ROUGH STONE WALL**, Dry										
0012	Dry laid (no mortar), under 18" thick	G	D-1	60	.267	C.F.	12.65	7.85		20.50	27
0100	Random fieldstone, under 18" thick	G	D-12	60	.533		12.65	15.75		28.40	40
0150	Over 18" thick	G	"	63	.508		15.15	15		30.15	41.50
0500	Field stone veneer	G	D-8	120	.333	S.F.	12.60	10.05		22.65	30.50
0510	Valley stone veneer	G		120	.333		12.60	10.05		22.65	30.50
0520	River stone veneer	G		120	.333		12.60	10.05		22.65	30.50
0600	Rubble stone walls, in mortar bed, up to 18" thick	G	D-11	75	.320	C.F.	15.25	9.75		25	33

04 43 Stone Masonry

04 43 10 – Masonry with Natural and Processed Stone

04 43 10.45 Granite		Crew	Daily Output	Labor-Hours	Unit	Material	2016 Bare Costs Labor	Equipment	Total	Total Incl O&P
0010	**GRANITE**, cut to size									
2500	Steps, copings, etc., finished on more than one surface									
2550	Low price, gray, light gray, etc.	D-10	50	.640	C.F.	91	20.50	9.60	121.10	145
2575	Medium price, pink, brown, etc.		50	.640		118	20.50	9.60	148.10	175
2600	High price, red, black, etc.		50	.640		146	20.50	9.60	176.10	205
2800	Pavers, 4" x 4" x 4" blocks, split face and joints									
2850	Low price, gray, light gray, etc.	D-11	80	.300	S.F.	13.10	9.15		22.25	29.50
2875	Medium price, pinks, browns, etc.		80	.300		21	9.15		30.15	38.50
2900	High price, red, black, etc.		80	.300		29	9.15		38.15	47.50
5000	Reclaimed or Antique									
5010	Treads, up to 12" wide	D-10	100	.320	L.F.	42.50	10.35	4.79	57.64	69.50
5020	Up to 18" wide		100	.320		38.50	10.35	4.79	53.64	65
5030	Capstone, size varies		50	.640		30.50	20.50	9.60	60.60	78.50
5040	Posts		30	1.067	V.L.F.	30.50	34.50	16	81	109

04 43 Stone Masonry

04 43 10 – Masonry with Natural and Processed Stone

04 43 10.55 Limestone

		Crew	Daily Output	Labor-Hours	Unit	Material	2016 Bare Costs Labor	Equipment	Total	Total Incl O&P
0010	**LIMESTONE**, cut to size									
0020	Veneer facing panels									
0500	Texture finish, light stick, 4-1/2" thick, 5' x 12'	D-4	300	.133	S.F.	20.50	3.65	.45	24.60	29
0750	5" thick, 5' x 14' panels	D-10	275	.116		21.50	3.77	1.74	27.01	32
1000	Sugarcube finish, 2" Thick, 3' x 5' panels		275	.116		29	3.77	1.74	34.51	40
1050	3" Thick, 4' x 9' panels		275	.116		25.50	3.77	1.74	31.01	36.50
1200	4" Thick, 5' x 11' panels		275	.116		31.50	3.77	1.74	37.01	42.50
1400	Sugarcube, textured finish, 4-1/2" thick, 5' x 12'		275	.116		32.50	3.77	1.74	38.01	44
1450	5" thick, 5' x 14' panels		275	.116		34	3.77	1.74	39.51	45
2000	Coping, sugarcube finish, top & 2 sides		30	1.067	C.F.	68.50	34.50	16	119	151
2100	Sills, lintels, jambs, trim, stops, sugarcube finish, simple		20	1.600		68.50	52	24	144.50	188
2150	Detailed		20	1.600		68.50	52	24	144.50	188
2300	Steps, extra hard, 14" wide, 6" rise		50	.640	L.F.	25	20.50	9.60	55.10	72.50
3000	Quoins, plain finish, 6" x 12" x 12"	D-12	25	1.280	Ea.	41	38		79	109
3050	6" x 16" x 24"	"	25	1.280	"	55	38		93	124

04 43 10.60 Marble

		Crew	Daily Output	Labor-Hours	Unit	Material	Labor	Equipment	Total	Total Incl O&P
0011	**MARBLE**, ashlar, split face, 4" + or - thick, random									
0040	Lengths 1' to 4' & heights 2" to 7-1/2", average	D-8	175	.229	S.F.	18.30	6.90		25.20	31.50
0100	Base, polished, 3/4" or 7/8" thick, polished, 6" high	D-10	65	.492	L.F.	12	15.95	7.35	35.30	48
1000	Facing, polished finish, cut to size, 3/4" to 7/8" thick									
1050	Carrara or equal	D-10	130	.246	S.F.	22.50	7.95	3.69	34.14	42.50
1100	Arabescato or equal	"	130	.246	"	39.50	7.95	3.69	51.14	61
2200	Window sills, 6" x 3/4" thick	D-1	85	.188	L.F.	11.10	5.55		16.65	21.50
2500	Flooring, polished tiles, 12" x 12" x 3/8" thick									
2510	Thin set, Giallo Solare or equal	D-11	90	.267	S.F.	17.50	8.15		25.65	33
2600	Sky Blue or equal		90	.267		16	8.15		24.15	31
2700	Mortar bed, Giallo Solare or equal		65	.369		17.50	11.25		28.75	38
2740	Sky Blue or equal		65	.369		16	11.25		27.25	36.50
2780	Travertine, 3/8" thick, Sierra or equal	D-10	130	.246		9.25	7.95	3.69	20.89	27.50
2790	Silver or equal	"	130	.246		25.50	7.95	3.69	37.14	45.50
3500	Thresholds, 3' long, 7/8" thick, 4" to 5" wide, plain	D-12	24	1.333	Ea.	35.50	39.50		75	105
3550	Beveled		24	1.333	"	71.50	39.50		111	144
3700	Window stools, polished, 7/8" thick, 5" wide		85	.376	L.F.	21.50	11.10		32.60	42.50

04 43 10.75 Sandstone or Brownstone

		Crew	Daily Output	Labor-Hours	Unit	Material	Labor	Equipment	Total	Total Incl O&P
0011	**SANDSTONE OR BROWNSTONE**									
0100	Sawed face veneer, 2-1/2" thick, to 2' x 4' panels	D-10	130	.246	S.F.	19.35	7.95	3.69	30.99	39
0150	4" thick, to 3'-6" x 8'panels		100	.320		19.35	10.35	4.79	34.49	44
0300	Split face, random sizes		100	.320		14.15	10.35	4.79	29.29	38
0350	Cut stone trim (limestone)									
0360	Ribbon stone, 4" thick, 5' pieces	D-8	120	.333	Ea.	160	10.05		170.05	193
0370	Cove stone, 4" thick, 5' pieces		105	.381		161	11.45		172.45	196
0380	Cornice stone, 10" to 12" wide		90	.444		199	13.35		212.35	242
0390	Band stone, 4" thick, 5' pieces		145	.276		109	8.30		117.30	134
0410	Window and door trim, 3" to 4" wide		160	.250		93.50	7.50		101	116
0420	Key stone, 18" long		60	.667		93.50	20		113.50	137

04 43 10.80 Slate

		Crew	Daily Output	Labor-Hours	Unit	Material	Labor	Equipment	Total	Total Incl O&P
0010	**SLATE**									
3100	Stair landings, 1" thick, black, clear	D-1	65	.246	S.F.	21	7.25		28.25	35.50
3200	Ribbon	"	65	.246	"	23.50	7.25		30.75	37.50
3500	Stair treads, sand finish, 1" thick x 12" wide									
3600	3 L.F. to 6 L.F.	D-10	120	.267	L.F.	25	8.65	3.99	37.64	46

04 43 Stone Masonry

04 43 10 – Masonry with Natural and Processed Stone

04 43 10.80 Slate

		Crew	Daily Output	Labor-Hours	Unit	Material	2016 Bare Costs Labor	Equipment	Total	Total Incl O&P
3700	Ribbon, sand finish, 1" thick x 12" wide									
3750	To 6 L.F.	D-10	120	.267	L.F.	21	8.65	3.99	33.64	42

04 43 10.85 Window Sill

		Crew	Daily Output	Labor-Hours	Unit	Material	2016 Bare Costs Labor	Equipment	Total	Total Incl O&P
0010	**WINDOW SILL**									
0020	Bluestone, thermal top, 10" wide, 1-1/2" thick	D-1	85	.188	S.F.	10	5.55		15.55	20.50
0050	2" thick		75	.213	"	10	6.30		16.30	21.50
0100	Cut stone, 5" x 8" plain		48	.333	L.F.	12.55	9.85		22.40	30
0200	Face brick on edge, brick, 8" wide		80	.200		5.25	5.90		11.15	15.65
0400	Marble, 9" wide, 1" thick		85	.188		8.95	5.55		14.50	19.10
0900	Slate, colored, unfading, honed, 12" wide, 1" thick		85	.188		8.50	5.55		14.05	18.60
0950	2" thick		70	.229		8.50	6.75		15.25	20.50

04 51 Flue Liner Masonry

04 51 10 – Clay Flue Lining

04 51 10.10 Flue Lining

		Crew	Daily Output	Labor-Hours	Unit	Material	2016 Bare Costs Labor	Equipment	Total	Total Incl O&P
0010	**FLUE LINING**, including mortar									
0020	Clay, 8" x 8"	D-1	125	.128	V.L.F.	5.70	3.78		9.48	12.55
0100	8" x 12"		103	.155		8.45	4.58		13.03	16.95
0200	12" x 12"		93	.172		10.80	5.10		15.90	20.50
0300	12" x 18"		84	.190		22	5.60		27.60	33.50
0400	18" x 18"		75	.213		28	6.30		34.30	41.50
0500	20" x 20"		66	.242		42.50	7.15		49.65	58.50
0600	24" x 24"		56	.286		54.50	8.45		62.95	74
1000	Round, 18" diameter		66	.242		38.50	7.15		45.65	54.50
1100	24" diameter		47	.340		74.50	10.05		84.55	99

04 57 Masonry Fireplaces

04 57 10 – Brick or Stone Fireplaces

04 57 10.10 Fireplace

		Crew	Daily Output	Labor-Hours	Unit	Material	2016 Bare Costs Labor	Equipment	Total	Total Incl O&P
0010	**FIREPLACE**									
0100	Brick fireplace, not incl. foundations or chimneys									
0110	30" x 29" opening, incl. chamber, plain brickwork	D-1	.40	40	Ea.	570	1,175		1,745	2,600
0200	Fireplace box only (110 brick)	"	2	8	"	162	236		398	575
0300	For elaborate brickwork and details, add					35%	35%			
0400	For hearth, brick & stone, add	D-1	2	8	Ea.	212	236		448	630
0410	For steel, damper, cleanouts, add		4	4		18	118		136	217
0600	Plain brickwork, incl. metal circulator		.50	32		900	945		1,845	2,575
0800	Face brick only, standard size, 8" x 2-2/3" x 4"		.30	53.333	M	600	1,575		2,175	3,275
0900	Stone fireplace, fieldstone, add				SF Face	8.50			8.50	9.35
1000	Cut stone, add				"	8			8	8.80

04 72 Cast Stone Masonry

04 72 10 – Cast Stone Masonry Features

04 72 10.10 Coping

		Crew	Daily Output	Labor-Hours	Unit	Material	2016 Bare Costs Labor	2016 Bare Costs Equipment	Total	Total Incl O&P
0010	**COPING**, stock units									
0050	Precast concrete, 10" wide, 4" tapers to 3-1/2", 8" wall	D-1	75	.213	L.F.	17.45	6.30		23.75	29.50
0100	12" wide, 3-1/2" tapers to 3", 10" wall		70	.229		18.80	6.75		25.55	32
0110	14" wide, 4" tapers to 3-1/2", 12" wall		65	.246		21.50	7.25		28.75	35.50
0150	16" wide, 4" tapers to 3-1/2", 14" wall		60	.267		23	7.85		30.85	38.50
0300	Limestone for 12" wall, 4" thick		90	.178		14.70	5.25		19.95	25
0350	6" thick		80	.200		22	5.90		27.90	34.50
0500	Marble, to 4" thick, no wash, 9" wide		90	.178		12.85	5.25		18.10	23
0550	12" wide		80	.200		18	5.90		23.90	29.50
0700	Terra cotta, 9" wide		90	.178		6.45	5.25		11.70	15.85
0750	12" wide		80	.200		8.80	5.90		14.70	19.55
0800	Aluminum, for 12" wall		80	.200		8.90	5.90		14.80	19.65

04 72 20 – Cultured Stone Veneer

04 72 20.10 Cultured Stone Veneer Components

		Crew	Daily Output	Labor-Hours	Unit	Material	2016 Bare Costs Labor	2016 Bare Costs Equipment	Total	Total Incl O&P
0010	**CULTURED STONE VENEER COMPONENTS**									
0110	On wood frame and sheathing substrate, random sized cobbles, corner stones	D-8	70	.571	V.L.F.	10.25	17.20		27.45	40
0120	Field stones		140	.286	S.F.	7.45	8.60		16.05	22.50
0130	Random sized flats, corner stones		70	.571	V.L.F.	10.55	17.20		27.75	40
0140	Field stones		140	.286	S.F.	8.75	8.60		17.35	24
0150	Horizontal lined ledgestones, corner stones		75	.533	V.L.F.	10.25	16.05		26.30	38
0160	Field stones		150	.267	S.F.	7.45	8		15.45	21.50
0170	Random shaped flats, corner stones		65	.615	V.L.F.	10.25	18.50		28.75	42.50
0180	Field stones		150	.267	S.F.	7.45	8		15.45	21.50
0190	Random shaped/textured face, corner stones		65	.615	V.L.F.	10.25	18.50		28.75	42.50
0200	Field stones		130	.308	S.F.	7.45	9.25		16.70	23.50
0210	Random shaped river rock, corner stones		65	.615	V.L.F.	10.25	18.50		28.75	42.50
0220	Field stones		130	.308	S.F.	7.45	9.25		16.70	23.50
0240	On concrete or CMU substrate, random sized cobbles, corner stones		70	.571	V.L.F.	9.60	17.20		26.80	39
0250	Field stones		140	.286	S.F.	7.10	8.60		15.70	22
0260	Random sized flats, corner stones		70	.571	V.L.F.	9.90	17.20		27.10	39.50
0270	Field stones		140	.286	S.F.	8.45	8.60		17.05	23.50
0280	Horizontal lined ledgestones, corner stones		75	.533	V.L.F.	9.60	16.05		25.65	37
0290	Field stones		150	.267	S.F.	7.10	8		15.10	21
0300	Random shaped flats, corner stones		70	.571	V.L.F.	9.60	17.20		26.80	39
0310	Field stones		140	.286	S.F.	7.10	8.60		15.70	22
0320	Random shaped/textured face, corner stones		65	.615	V.L.F.	9.60	18.50		28.10	41.50
0330	Field stones		130	.308	S.F.	7.10	9.25		16.35	23.50
0340	Random shaped river rock, corner stones		65	.615	V.L.F.	9.60	18.50		28.10	41.50
0350	Field stones		130	.308	S.F.	7.10	9.25		16.35	23.50
0360	Cultured stone veneer, #15 felt weather resistant barrier	1 Clab	3700	.002	Sq.	5.30	.05		5.35	5.95
0390	Water table or window sill, 18" long	1 Bric	80	.100	Ea.	9.80	3.24		13.04	16.20

Division Notes

		CREW	DAILY OUTPUT	LABOR-HOURS	UNIT	BARE COSTS MAT.	LABOR	EQUIP.	TOTAL	TOTAL INCL O&P

Estimating Tips

05 05 00 Common Work Results for Metals

- Nuts, bolts, washers, connection angles, and plates can add a significant amount to both the tonnage of a structural steel job and the estimated cost. As a rule of thumb, add 10% to the total weight to account for these accessories.
- Type 2 steel construction, commonly referred to as "simple construction," consists generally of field-bolted connections with lateral bracing supplied by other elements of the building, such as masonry walls or x-bracing. The estimator should be aware, however, that shop connections may be accomplished by welding or bolting. The method may be particular to the fabrication shop and may have an impact on the estimated cost.

05 10 00 Structural Steel

- Steel items can be obtained from two sources: a fabrication shop or a metals service center. Fabrication shops can fabricate items under more controlled conditions than can crews in the field. They are also more efficient and can produce items more economically. Metal service centers serve as a source of long mill shapes to both fabrication shops and contractors.
- Most line items in this structural steel subdivision, and most items in 05 50 00 Metal Fabrications, are indicated as being shop fabricated. The bare material cost for these shop fabricated items is the "Invoice Cost" from the shop and includes the mill base price of steel plus mill extras, transportation to the shop, shop drawings and detailing where warranted, shop fabrication and handling, sandblasting and a shop coat of primer paint, all necessary structural bolts, and delivery to the job site. The bare labor cost and bare equipment cost for these shop fabricated items is for field installation or erection.
- Line items in Subdivision 05 12 23.40 Lightweight Framing, and other items scattered in Division 5, are indicated as being field fabricated. The bare material cost for these field fabricated items is the "Invoice Cost" from the metals service center and includes the mill base price of steel plus mill extras, transportation to the metals service center, material handling, and delivery of long lengths of mill shapes to the job site. Material costs for structural bolts and welding rods should be added to the estimate. The bare labor cost and bare equipment cost for these items is for both field fabrication and field installation or erection, and include time for cutting, welding and drilling in the fabricated metal items. Drilling into concrete and fasteners to fasten field fabricated items to other work are not included and should be added to the estimate.

05 20 00 Steel Joist Framing

- In any given project the total weight of open web steel joists is determined by the loads to be supported and the design. However, economies can be realized in minimizing the amount of labor used to place the joists. This is done by maximizing the joist spacing, and therefore minimizing the number of joists required to be installed on the job. Certain spacings and locations may be required by the design, but in other cases maximizing the spacing and keeping it as uniform as possible will keep the costs down.

05 30 00 Steel Decking

- The takeoff and estimating of metal deck involves more than simply the area of the floor or roof and the type of deck specified or shown on the drawings. Many different sizes and types of openings may exist. Small openings for individual pipes or conduits may be drilled after the floor/roof is installed, but larger openings may require special deck lengths as well as reinforcing or structural support. The estimator should determine who will be supplying this reinforcing. Additionally, some deck terminations are part of the deck package, such as screed angles and pour stops, and others will be part of the steel contract, such as angles attached to structural members and cast-in-place angles and plates. The estimator must ensure that all pieces are accounted for in the complete estimate.

05 50 00 Metal Fabrications

- The most economical steel stairs are those that use common materials, standard details, and most importantly, a uniform and relatively simple method of field assembly. Commonly available A36/A992 channels and plates are very good choices for the main stringers of the stairs, as are angles and tees for the carrier members. Risers and treads are usually made by specialty shops, and it is most economical to use a typical detail in as many places as possible. The stairs should be pre-assembled and shipped directly to the site. The field connections should be simple and straightforward to be accomplished efficiently, and with minimum equipment and labor.

Reference Numbers

Reference numbers are shown at the beginning of some major classifications. These numbers refer to related items in the Reference Section. The reference information may be an estimating procedure, an alternate pricing method, or technical information.

Note: Not all subdivisions listed here necessarily appear. ■

No part of this cost data may be reproduced, stored in a retrieval system, or transmitted in any form or by any means without prior written permission of RSMeans.

05 05 Common Work Results for Metals

05 05 19 – Post-Installed Concrete Anchors

05 05 19.10 Chemical Anchors		Crew	Daily Output	Labor-Hours	Unit	Material	2016 Bare Costs Labor	Equipment	Total	Total Incl O&P
0010	**CHEMICAL ANCHORS**									
0020	Includes layout & drilling									
1430	Chemical anchor, w/rod & epoxy cartridge, 3/4" diam. x 9-1/2" long	B-89A	27	.593	Ea.	9.25	17.45	4.33	31.03	44
1435	1" diameter x 11-3/4" long		24	.667		16.95	19.60	4.87	41.42	57
1440	1-1/4" diameter x 14" long		21	.762		34	22.50	5.55	62.05	81
1445	1-3/4" diameter x 15" long		20	.800		63.50	23.50	5.85	92.85	116
1450	18" long		17	.941		76.50	27.50	6.90	110.90	138
1455	2" diameter x 18" long		16	1		101	29.50	7.30	137.80	169
1460	24" long		15	1.067		132	31.50	7.80	171.30	206

05 05 19.20 Expansion Anchors			Crew	Daily Output	Labor-Hours	Unit	Material	Labor	Equipment	Total	Total Incl O&P
0010	**EXPANSION ANCHORS**										
0100	Anchors for concrete, brick or stone, no layout and drilling										
0200	Expansion shields, zinc, 1/4" diameter, 1-5/16" long, single	G	1 Carp	90	.089	Ea.	.44	3.01		3.45	5.55
0300	1-3/8" long, double	G		85	.094		.55	3.19		3.74	5.95
0400	3/8" diameter, 1-1/2" long, single	G		85	.094		.66	3.19		3.85	6.10
0500	2" long, double	G		80	.100		1.20	3.39		4.59	7
0600	1/2" diameter, 2-1/16" long, single	G		80	.100		1.21	3.39		4.60	7.05
0700	2-1/2" long, double	G		75	.107		1.94	3.62		5.56	8.20
0800	5/8" diameter, 2-5/8" long, single	G		75	.107		2.01	3.62		5.63	8.25
0900	2-3/4" long, double	G		70	.114		2.74	3.87		6.61	9.50
1000	3/4" diameter, 2-3/4" long, single	G		70	.114		3.18	3.87		7.05	10
1100	3-15/16" long, double	G		65	.123		5.10	4.17		9.27	12.60
2100	Hollow wall anchors for gypsum wall board, plaster or tile										
2500	3/16" diameter, short	G	1 Carp	150	.053	Ea.	.46	1.81		2.27	3.54
3000	Toggle bolts, bright steel, 1/8" diameter, 2" long	G		85	.094		.21	3.19		3.40	5.60
3100	4" long	G		80	.100		.27	3.39		3.66	6
3200	3/16" diameter, 3" long	G		80	.100		.28	3.39		3.67	6
3300	6" long	G		75	.107		.39	3.62		4.01	6.50
3400	1/4" diameter, 3" long	G		75	.107		.38	3.62		4	6.45
3500	6" long	G		70	.114		.55	3.87		4.42	7.10
3600	3/8" diameter, 3" long	G		70	.114		.93	3.87		4.80	7.50
3700	6" long	G		60	.133		1.43	4.52		5.95	9.10
3800	1/2" diameter, 4" long	G		60	.133		2	4.52		6.52	9.75
3900	6" long	G		50	.160		2.59	5.40		7.99	11.95
4000	Nailing anchors										
4100	Nylon nailing anchor, 1/4" diameter, 1" long		1 Carp	3.20	2.500	C	22	85		107	166
4200	1-1/2" long			2.80	2.857		25.50	97		122.50	190
4300	2" long			2.40	3.333		30	113		143	222
4400	Metal nailing anchor, 1/4" diameter, 1" long	G		3.20	2.500		18.60	85		103.60	163
4500	1-1/2" long	G		2.80	2.857		24.50	97		121.50	189
4600	2" long	G		2.40	3.333		29	113		142	221
5000	Screw anchors for concrete, masonry,										
5100	stone & tile, no layout or drilling included										
5700	Lag screw shields, 1/4" diameter, short	G	1 Carp	90	.089	Ea.	.37	3.01		3.38	5.45
5800	Long	G		85	.094		.41	3.19		3.60	5.80
5900	3/8" diameter, short	G		85	.094		.71	3.19		3.90	6.15
6000	Long	G		80	.100		.94	3.39		4.33	6.75
6100	1/2" diameter, short	G		80	.100		1.01	3.39		4.40	6.80
6200	Long	G		75	.107		1.31	3.62		4.93	7.50
6300	5/8" diameter, short	G		70	.114		1.57	3.87		5.44	8.25
6400	Long	G		65	.123		2	4.17		6.17	9.20
6600	Lead, #6 & #8, 3/4" long	G		260	.031		.19	1.04		1.23	1.96

05 05 Common Work Results for Metals

05 05 19 – Post-Installed Concrete Anchors

05 05 19.20 Expansion Anchors

		Crew	Daily Output	Labor-Hours	Unit	Material	2016 Bare Costs Labor	Equipment	Total	Total Incl O&P
6700	#10 - #14, 1-1/2" long [G]	1 Carp	200	.040	Ea.	.39	1.36		1.75	2.70
6800	#16 & #18, 1-1/2" long [G]		160	.050		.42	1.70		2.12	3.30
6900	Plastic, #6 & #8, 3/4" long		260	.031		.04	1.04		1.08	1.79
7000	#8 & #10, 7/8" long		240	.033		.04	1.13		1.17	1.93
7100	#10 & #12, 1" long		220	.036		.05	1.23		1.28	2.12
7200	#14 & #16, 1-1/2" long		160	.050		.08	1.70		1.78	2.93
8950	Self-drilling concrete screw, hex washer head, 3/16" diam. x 1-3/4" long [G]		300	.027		.21	.90		1.11	1.74
8960	2-1/4" long [G]		250	.032		.24	1.08		1.32	2.08
8970	Phillips flat head, 3/16" diam. x 1-3/4" long [G]		300	.027		.21	.90		1.11	1.74
8980	2-1/4" long [G]		250	.032		.24	1.08		1.32	2.08

05 05 21 – Fastening Methods for Metal

05 05 21.15 Drilling Steel

		Crew	Daily Output	Labor-Hours	Unit	Material	Labor	Equipment	Total	Total Incl O&P
0010	**DRILLING STEEL**									
1910	Drilling & layout for steel, up to 1/4" deep, no anchor									
1920	Holes, 1/4" diameter	1 Sswk	112	.071	Ea.	.10	2.55		2.65	4.78
1925	For each additional 1/4" depth, add		336	.024		.10	.85		.95	1.67
1930	3/8" diameter		104	.077		.09	2.74		2.83	5.15
1935	For each additional 1/4" depth, add		312	.026		.09	.91		1	1.78
1940	1/2" diameter		96	.083		.11	2.97		3.08	5.60
1945	For each additional 1/4" depth, add		288	.028		.11	.99		1.10	1.95
1950	5/8" diameter		88	.091		.16	3.24		3.40	6.15
1955	For each additional 1/4" depth, add		264	.030		.16	1.08		1.24	2.16
1960	3/4" diameter		80	.100		.20	3.57		3.77	6.75
1965	For each additional 1/4" depth, add		240	.033		.20	1.19		1.39	2.40
1970	7/8" diameter		72	.111		.26	3.96		4.22	7.55
1975	For each additional 1/4" depth, add		216	.037		.26	1.32		1.58	2.71
1980	1" diameter		64	.125		.23	4.46		4.69	8.45
1985	For each additional 1/4" depth, add		192	.042		.23	1.49		1.72	2.98
1990	For drilling up, add					40%				

05 05 23 – Metal Fastenings

05 05 23.10 Bolts and Hex Nuts

		Crew	Daily Output	Labor-Hours	Unit	Material	Labor	Equipment	Total	Total Incl O&P
0010	**BOLTS & HEX NUTS**, Steel, A307									
0100	1/4" diameter, 1/2" long [G]	1 Sswk	140	.057	Ea.	.06	2.04		2.10	3.81
0200	1" long [G]		140	.057		.07	2.04		2.11	3.82
0300	2" long [G]		130	.062		.10	2.19		2.29	4.13
0400	3" long [G]		130	.062		.15	2.19		2.34	4.19
0500	4" long [G]		120	.067		.17	2.38		2.55	4.54
0600	3/8" diameter, 1" long [G]		130	.062		.14	2.19		2.33	4.18
0700	2" long [G]		130	.062		.18	2.19		2.37	4.22
0800	3" long [G]		120	.067		.24	2.38		2.62	4.62
0900	4" long [G]		120	.067		.30	2.38		2.68	4.69
1000	5" long [G]		115	.070		.38	2.48		2.86	4.96
1100	1/2" diameter, 1-1/2" long [G]		120	.067		.40	2.38		2.78	4.80
1200	2" long [G]		120	.067		.46	2.38		2.84	4.87
1300	4" long [G]		115	.070		.75	2.48		3.23	5.40
1400	6" long [G]		110	.073		1.05	2.59		3.64	5.90
1500	8" long [G]		105	.076		1.38	2.72		4.10	6.50
1600	5/8" diameter, 1-1/2" long [G]		120	.067		.85	2.38		3.23	5.30
1700	2" long [G]		120	.067		.94	2.38		3.32	5.40
1800	4" long [G]		115	.070		1.34	2.48		3.82	6.05
1900	6" long [G]		110	.073		1.72	2.59		4.31	6.65
2000	8" long [G]		105	.076		2.55	2.72		5.27	7.80

05 05 Common Work Results for Metals

05 05 23 – Metal Fastenings

05 05 23.10 Bolts and Hex Nuts		Crew	Daily Output	Labor-Hours	Unit	Material	2016 Bare Costs Labor	Equipment	Total	Total Incl O&P
2100	10" long [G]	1 Sswk	100	.080	Ea.	3.20	2.85		6.05	8.75
2200	3/4" diameter, 2" long [G]		120	.067		1.15	2.38		3.53	5.65
2300	4" long [G]		110	.073		1.65	2.59		4.24	6.55
2400	6" long [G]		105	.076		2.12	2.72		4.84	7.30
2500	8" long [G]		95	.084		3.20	3		6.20	9
2600	10" long [G]		85	.094		4.20	3.36		7.56	10.75
2700	12" long [G]		80	.100		4.92	3.57		8.49	11.95
2800	1" diameter, 3" long [G]		105	.076		2.69	2.72		5.41	7.95
2900	6" long [G]		90	.089		3.94	3.17		7.11	10.15
3000	12" long [G]		75	.107		7.10	3.80		10.90	14.80
3100	For galvanized, add					75%				
3200	For stainless, add					350%				

05 05 23.30 Lag Screws

		Crew	Daily Output	Labor-Hours	Unit	Material	Labor	Equipment	Total	Total Incl O&P
0010	**LAG SCREWS**									
0020	Steel, 1/4" diameter, 2" long [G]	1 Carp	200	.040	Ea.	.09	1.36		1.45	2.37
0100	3/8" diameter, 3" long [G]		150	.053		.32	1.81		2.13	3.38
0200	1/2" diameter, 3" long [G]		130	.062		.68	2.09		2.77	4.24
0300	5/8" diameter, 3" long [G]		120	.067		1.22	2.26		3.48	5.10

05 05 23.50 Powder Actuated Tools and Fasteners

		Crew	Daily Output	Labor-Hours	Unit	Material	Labor	Equipment	Total	Total Incl O&P
0010	**POWDER ACTUATED TOOLS & FASTENERS**									
0020	Stud driver, .22 caliber, single shot				Ea.	149			149	164
0100	.27 caliber, semi automatic, strip				"	450			450	495
0300	Powder load, single shot, .22 cal, power level 2, brown				C	5.50			5.50	6.05
0400	Strip, .27 cal, power level 4, red					7.90			7.90	8.70
0600	Drive pin, .300 x 3/4" long [G]	1 Carp	4.80	1.667		4.23	56.50		60.73	99
0700	.300 x 3" long with washer [G]	"	4	2		13	68		81	128

05 05 23.55 Rivets

		Crew	Daily Output	Labor-Hours	Unit	Material	Labor	Equipment	Total	Total Incl O&P
0010	**RIVETS**									
0100	Aluminum rivet & mandrel, 1/2" grip length x 1/8" diameter [G]	1 Carp	4.80	1.667	C	8.05	56.50		64.55	103
0200	3/16" diameter [G]		4	2		11.90	68		79.90	127
0300	Aluminum rivet, steel mandrel, 1/8" diameter [G]		4.80	1.667		10.30	56.50		66.80	106
0400	3/16" diameter [G]		4	2		16.50	68		84.50	132
0500	Copper rivet, steel mandrel, 1/8" diameter [G]		4.80	1.667		9.50	56.50		66	105
0800	Stainless rivet & mandrel, 1/8" diameter [G]		4.80	1.667		25.50	56.50		82	123
0900	3/16" diameter [G]		4	2		40.50	68		108.50	159
1000	Stainless rivet, steel mandrel, 1/8" diameter [G]		4.80	1.667		15.70	56.50		72.20	112
1100	3/16" diameter [G]		4	2		25.50	68		93.50	142
1200	Steel rivet and mandrel, 1/8" diameter [G]		4.80	1.667		7.75	56.50		64.25	103
1300	3/16" diameter [G]		4	2		12.55	68		80.55	128
1400	Hand riveting tool, standard				Ea.	76.50			76.50	84
1500	Deluxe					385			385	425
1600	Power riveting tool, standard					500			500	555
1700	Deluxe					1,650			1,650	1,825

05 12 Structural Steel Framing

05 12 23 – Structural Steel for Buildings

05 12 23.10 Ceiling Supports

		Crew	Daily Output	Labor-Hours	Unit	Material	2016 Bare Costs Labor	Equipment	Total	Total Incl O&P
0010	**CEILING SUPPORTS**									
1000	Entrance door/folding partition supports, shop fabricated G	E-4	60	.533	L.F.	26.50	19.30	2.45	48.25	67
1100	Linear accelerator door supports G		14	2.286		121	82.50	10.50	214	297
1200	Lintels or shelf angles, hung, exterior hot dipped galv. G		267	.120		18.10	4.33	.55	22.98	28.50
1250	Two coats primer paint instead of galv. G		267	.120		15.65	4.33	.55	20.53	26
1400	Monitor support, ceiling hung, expansion bolted G		4	8	Ea.	420	289	37	746	1,025
1450	Hung from pre-set inserts G		6	5.333		450	193	24.50	667.50	875
1600	Motor supports for overhead doors G		4	8		214	289	37	540	805
1700	Partition support for heavy folding partitions, without pocket G		24	1.333	L.F.	60.50	48	6.15	114.65	162
1750	Supports at pocket only G		12	2.667		121	96.50	12.25	229.75	325
2000	Rolling grilles & fire door supports G		34	.941		51.50	34	4.33	89.83	124
2100	Spider-leg light supports, expansion bolted to ceiling slab G		8	4	Ea.	172	145	18.40	335.40	475
2150	Hung from pre-set inserts G		12	2.667	"	186	96.50	12.25	294.75	395
2400	Toilet partition support G		36	.889	L.F.	60.50	32	4.09	96.59	130
2500	X-ray travel gantry support G		12	2.667	"	207	96.50	12.25	315.75	420

05 12 23.15 Columns, Lightweight

		Crew	Daily Output	Labor-Hours	Unit	Material	Labor	Equipment	Total	Total Incl O&P
0010	**COLUMNS, LIGHTWEIGHT**									
8000	Lally columns, to 8', 3-1/2" diameter	2 Carp	24	.667	Ea.	66	22.50		88.50	111
8080	4" diameter	"	20	.800	"	81	27		108	135

05 12 23.17 Columns, Structural

		Crew	Daily Output	Labor-Hours	Unit	Material	Labor	Equipment	Total	Total Incl O&P
0010	**COLUMNS, STRUCTURAL**									
0015	Made from recycled materials									
0020	Shop fab'd for 100-ton, 1-2 story project, bolted connections									
0800	Steel, concrete filled, extra strong pipe, 3-1/2" diameter	E-2	660	.073	L.F.	43.50	2.62	2.30	48.42	55.50
0830	4" diameter		780	.062		48.50	2.22	1.95	52.67	59.50
0890	5" diameter		1020	.047		58	1.70	1.49	61.19	68.50
0930	6" diameter		1200	.040		77	1.44	1.27	79.71	88.50
0940	8" diameter		1100	.044		77	1.57	1.38	79.95	89
1100	For galvanizing, add				Lb.	.25			.25	.28
1300	For web ties, angles, etc., add per added lb.	1 Sswk	945	.008		1.33	.30		1.63	2.01
1500	Steel pipe, extra strong, no concrete, 3" to 5" diameter G	E-2	16000	.003		1.33	.11	.10	1.54	1.76
1600	6" to 12" diameter G		14000	.003		1.33	.12	.11	1.56	1.80
5100	Structural tubing, rect., 5" to 6" wide, light section G		8000	.006		1.33	.22	.19	1.74	2.06
5200	Heavy section G		12000	.004		1.33	.14	.13	1.60	1.86
8090	For projects 75 to 99 tons, add				%	10%				
8092	50 to 74 tons, add					20%				
8094	25 to 49 tons, add					30%	10%			
8096	10 to 24 tons, add					50%	25%			
8098	2 to 9 tons, add					75%	50%			
8099	Less than 2 tons, add					100%	100%			

05 12 23.45 Lintels

		Crew	Daily Output	Labor-Hours	Unit	Material	Labor	Equipment	Total	Total Incl O&P
0010	**LINTELS**									
0015	Made from recycled materials									
0020	Plain steel angles, shop fabricated, under 500 lb. G	1 Bric	550	.015	Lb.	1.02	.47		1.49	1.91
0100	500 to 1000 lb. G		640	.013	"	.99	.41		1.40	1.77
2000	Steel angles, 3-1/2" x 3", 1/4" thick, 2'-6" long G		47	.170	Ea.	14.30	5.50		19.80	25
2100	4'-6" long G		26	.308		26	9.95		35.95	45
2600	4" x 3-1/2", 1/4" thick, 5'-0" long G		21	.381		33	12.35		45.35	56.50
2700	9'-0" long G		12	.667		59	21.50		80.50	101

05 12 Structural Steel Framing

05 12 23 – Structural Steel for Buildings

05 12 23.65 Plates

		Crew	Daily Output	Labor-Hours	Unit	Material	2016 Bare Costs Labor	Equipment	Total	Total Incl O&P
0010	**PLATES**									
0015	Made from recycled materials									
0020	For connections & stiffener plates, shop fabricated									
0050	1/8" thick (5.1 lb./S.F.)	G			S.F.	6.75			6.75	7.45
0100	1/4" thick (10.2 lb./S.F.)	G				13.50			13.50	14.85
0300	3/8" thick (15.3 lb./S.F.)	G				20.50			20.50	22.50
0400	1/2" thick (20.4 lb./S.F.)	G				27			27	29.50
0450	3/4" thick (30.6 lb./S.F.)	G				40.50			40.50	44.50
0500	1" thick (40.8 lb./S.F.)	G				54			54	59.50
2000	Steel plate, warehouse prices, no shop fabrication									
2100	1/4" thick (10.2 lb./S.F.)	G			S.F.	6.85			6.85	7.55

05 12 23.79 Structural Steel

		Crew	Daily Output	Labor-Hours	Unit	Material	2016 Bare Costs Labor	Equipment	Total	Total Incl O&P
0010	**STRUCTURAL STEEL**									
0020	Shop fab'd for 100-ton, 1-2 story project, bolted conn's.									
0050	Beams, W 6 x 9	G	E-2	720	.067	L.F.	14.30	2.40	2.11	18.81	22.50
0100	W 8 x 10	G		720	.067		15.90	2.40	2.11	20.41	24
0200	Columns, W 6 x 15	G		540	.089		26	3.20	2.82	32.02	37.50
0250	W 8 x 31	G		540	.089		53.50	3.20	2.82	59.52	67.50
7990	For projects 75 to 99 tons, add					All	10%				
7992	50 to 75 tons, add						20%				
7994	25 to 49 tons, add						30%	10%			
7996	10 to 24 tons, add						50%	25%			
7998	2 to 9 tons, add						75%	50%			
7999	Less than 2 tons, add						100%	100%			

05 31 Steel Decking

05 31 23 – Steel Roof Decking

05 31 23.50 Roof Decking

		Crew	Daily Output	Labor-Hours	Unit	Material	2016 Bare Costs Labor	Equipment	Total	Total Incl O&P
0010	**ROOF DECKING**									
0015	Made from recycled materials									
2100	Open type, 1-1/2" deep, Type B, wide rib, galv., 22 ga., under 50 sq.	G	E-4	4500	.007	S.F.	2.07	.26	.03	2.36	2.79
2600	20 ga., under 50 squares	G		3865	.008		2.42	.30	.04	2.76	3.25
2900	18 ga., under 50 squares	G		3800	.008		3.12	.30	.04	3.46	4.03
3050	16 ga., under 50 squares	G		3700	.009		4.21	.31	.04	4.56	5.25

05 31 33 – Steel Form Decking

05 31 33.50 Form Decking

		Crew	Daily Output	Labor-Hours	Unit	Material	2016 Bare Costs Labor	Equipment	Total	Total Incl O&P
0010	**FORM DECKING**									
0015	Made from recycled materials									
6100	Slab form, steel, 28 ga., 9/16" deep, Type UFS, uncoated	G	E-4	4000	.008	S.F.	1.56	.29	.04	1.89	2.29
6200	Galvanized	G		4000	.008		1.38	.29	.04	1.71	2.09
6220	24 ga., 1" deep, Type UF1X, uncoated	G		3900	.008		1.50	.30	.04	1.84	2.23
6240	Galvanized	G		3900	.008		1.76	.30	.04	2.10	2.52
6300	24 ga., 1-5/16" deep, Type UFX, uncoated	G		3800	.008		1.59	.30	.04	1.93	2.35
6400	Galvanized	G		3800	.008		1.87	.30	.04	2.21	2.66
6500	22 ga., 1-5/16" deep, uncoated	G		3700	.009		2.02	.31	.04	2.37	2.83
6600	Galvanized	G		3700	.009		2.06	.31	.04	2.41	2.88
6700	22 ga., 2" deep, uncoated	G		3600	.009		2.62	.32	.04	2.98	3.52
6800	Galvanized	G		3600	.009		2.57	.32	.04	2.93	3.47

05 41 Structural Metal Stud Framing

05 41 13 – Load-Bearing Metal Stud Framing

05 41 13.05 Bracing

		Crew	Daily Output	Labor-Hours	Unit	Material	2016 Bare Costs Labor	Equipment	Total	Total Incl O&P
0010	**BRACING**, shear wall X-bracing, per 10' x 10' bay, one face									
0015	Made of recycled materials									
0120	Metal strap, 20 ga. x 4" wide G	2 Carp	18	.889	Ea.	18.90	30		48.90	71.50
0130	6" wide G		18	.889		31.50	30		61.50	85
0160	18 ga. x 4" wide G		16	1		33	34		67	93
0170	6" wide G		16	1		48.50	34		82.50	111
0410	Continuous strap bracing, per horizontal row on both faces									
0420	Metal strap, 20 ga. x 2" wide, studs 12" O.C. G	1 Carp	7	1.143	C.L.F.	57	38.50		95.50	128
0430	16" O.C. G		8	1		57	34		91	120
0440	24" O.C. G		10	.800		57	27		84	109
0450	18 ga. x 2" wide, studs 12" O.C. G		6	1.333		79	45		124	163
0460	16" O.C. G		7	1.143		79	38.50		117.50	152
0470	24" O.C. G		8	1		79	34		113	144

05 41 13.10 Bridging

		Crew	Daily Output	Labor-Hours	Unit	Material	2016 Bare Costs Labor	Equipment	Total	Total Incl O&P
0010	**BRIDGING**, solid between studs w/1-1/4" leg track, per stud bay									
0015	Made from recycled materials									
0200	Studs 12" O.C., 18 ga. x 2-1/2" wide G	1 Carp	125	.064	Ea.	.95	2.17		3.12	4.68
0210	3-5/8" wide G		120	.067		1.16	2.26		3.42	5.05
0220	4" wide G		120	.067		1.22	2.26		3.48	5.10
0230	6" wide G		115	.070		1.60	2.36		3.96	5.70
0240	8" wide G		110	.073		1.97	2.47		4.44	6.30
0300	16 ga. x 2-1/2" wide G		115	.070		1.21	2.36		3.57	5.30
0310	3-5/8" wide G		110	.073		1.47	2.47		3.94	5.75
0320	4" wide G		110	.073		1.57	2.47		4.04	5.85
0330	6" wide G		105	.076		2.01	2.58		4.59	6.55
0340	8" wide G		100	.080		2.51	2.71		5.22	7.30
1200	Studs 16" O.C., 18 ga. x 2-1/2" wide G		125	.064		1.22	2.17		3.39	4.97
1210	3-5/8" wide G		120	.067		1.49	2.26		3.75	5.40
1220	4" wide G		120	.067		1.57	2.26		3.83	5.50
1230	6" wide G		115	.070		2.05	2.36		4.41	6.20
1240	8" wide G		110	.073		2.52	2.47		4.99	6.90
1300	16 ga. x 2-1/2" wide G		115	.070		1.55	2.36		3.91	5.65
1310	3-5/8" wide G		110	.073		1.88	2.47		4.35	6.20
1320	4" wide G		110	.073		2.01	2.47		4.48	6.35
1330	6" wide G		105	.076		2.57	2.58		5.15	7.15
1340	8" wide G		100	.080		3.22	2.71		5.93	8.10
2200	Studs 24" O.C., 18 ga. x 2-1/2" wide G		125	.064		1.77	2.17		3.94	5.55
2210	3-5/8" wide G		120	.067		2.15	2.26		4.41	6.15
2220	4" wide G		120	.067		2.27	2.26		4.53	6.25
2230	6" wide G		115	.070		2.96	2.36		5.32	7.20
2240	8" wide G		110	.073		3.65	2.47		6.12	8.15
2300	16 ga. x 2-1/2" wide G		115	.070		2.24	2.36		4.60	6.40
2310	3-5/8" wide G		110	.073		2.72	2.47		5.19	7.10
2320	4" wide G		110	.073		2.91	2.47		5.38	7.35
2330	6" wide G		105	.076		3.72	2.58		6.30	8.40
2340	8" wide G		100	.080		4.65	2.71		7.36	9.65
3000	Continuous bridging, per row									
3100	16 ga. x 1-1/2" channel thru studs 12" O.C. G	1 Carp	6	1.333	C.L.F.	51.50	45		96.50	133
3110	16" O.C. G		7	1.143		51.50	38.50		90	122
3120	24" O.C. G		8.80	.909		51.50	31		82.50	109
4100	2" x 2" angle x 18 ga., studs 12" O.C. G		7	1.143		80.50	38.50		119	154
4110	16" O.C. G		9	.889		80.50	30		110.50	139

05 41 Structural Metal Stud Framing

05 41 13 – Load-Bearing Metal Stud Framing

05 41 13.10 Bridging

			Crew	Daily Output	Labor-Hours	Unit	Material	2016 Bare Costs Labor	Equipment	Total	Total Incl O&P
4120		24" O.C.	1 Carp	12	.667	C.L.F.	80.50	22.50		103	127
4200		16 ga., studs 12" O.C.		5	1.600		101	54		155	202
4210		16" O.C.		7	1.143		101	38.50		139.50	176
4220		24" O.C.		10	.800		101	27		128	157

05 41 13.25 Framing, Boxed Headers/Beams

			Crew	Daily Output	Labor-Hours	Unit	Material	Labor	Equipment	Total	Total Incl O&P
0010	**FRAMING, BOXED HEADERS/BEAMS**										
0015	Made from recycled materials										
0200	Double, 18 ga. x 6" deep		2 Carp	220	.073	L.F.	5.50	2.47		7.97	10.20
0210	8" deep			210	.076		6.05	2.58		8.63	10.95
0220	10" deep			200	.080		7.40	2.71		10.11	12.70
0230	12" deep			190	.084		8.05	2.85		10.90	13.65
0300	16 ga. x 8" deep			180	.089		7	3.01		10.01	12.75
0310	10" deep			170	.094		8.45	3.19		11.64	14.60
0320	12" deep			160	.100		9.15	3.39		12.54	15.80
0400	14 ga. x 10" deep			140	.114		9.75	3.87		13.62	17.20
0410	12" deep			130	.123		10.65	4.17		14.82	18.75
1210	Triple, 18 ga. x 8" deep			170	.094		8.75	3.19		11.94	15
1220	10" deep			165	.097		10.60	3.29		13.89	17.15
1230	12" deep			160	.100		11.60	3.39		14.99	18.45
1300	16 ga. x 8" deep			145	.110		10.15	3.74		13.89	17.40
1310	10" deep			140	.114		12.15	3.87		16.02	19.85
1320	12" deep			135	.119		13.25	4.02		17.27	21.50
1400	14 ga. x 10" deep			115	.139		13.25	4.72		17.97	22.50
1410	12" deep			110	.145		14.65	4.93		19.58	24.50

05 41 13.30 Framing, Stud Walls

			Crew	Daily Output	Labor-Hours	Unit	Material	Labor	Equipment	Total	Total Incl O&P
0010	**FRAMING, STUD WALLS** w/top & bottom track, no openings,										
0020	Headers, beams, bridging or bracing										
0025	Made from recycled materials										
4100	8' high walls, 18 ga. x 2-1/2" wide, studs 12" O.C.		2 Carp	54	.296	L.F.	9.10	10.05		19.15	27
4110	16" O.C.			77	.208		7.25	7.05		14.30	19.80
4120	24" O.C.			107	.150		5.45	5.05		10.50	14.50
4130	3-5/8" wide, studs 12" O.C.			53	.302		10.85	10.25		21.10	29
4140	16" O.C.			76	.211		8.70	7.15		15.85	21.50
4150	24" O.C.			105	.152		6.55	5.15		11.70	15.85
4160	4" wide, studs 12" O.C.			52	.308		11.30	10.45		21.75	30
4170	16" O.C.			74	.216		9.05	7.35		16.40	22
4180	24" O.C.			103	.155		6.80	5.25		12.05	16.30
4190	6" wide, studs 12" O.C.			51	.314		14.40	10.65		25.05	33.50
4200	16" O.C.			73	.219		11.55	7.45		19	25
4210	24" O.C.			101	.158		8.70	5.35		14.05	18.60
4220	8" wide, studs 12" O.C.			50	.320		17.45	10.85		28.30	37.50
4230	16" O.C.			72	.222		14	7.55		21.55	28
4240	24" O.C.			100	.160		10.60	5.40		16	21
4300	16 ga. x 2-1/2" wide, studs 12" O.C.			47	.340		10.85	11.55		22.40	31.50
4310	16" O.C.			68	.235		8.60	8		16.60	23
4320	24" O.C.			94	.170		6.35	5.75		12.10	16.60
4330	3-5/8" wide, studs 12" O.C.			46	.348		12.95	11.80		24.75	34
4340	16" O.C.			66	.242		10.25	8.20		18.45	25
4350	24" O.C.			92	.174		7.55	5.90		13.45	18.15
4360	4" wide, studs 12" O.C.			45	.356		13.55	12.05		25.60	35
4370	16" O.C.			65	.246		10.75	8.35		19.10	26
4380	24" O.C.			90	.178		7.95	6.05		14	18.80

05 41 Structural Metal Stud Framing

05 41 13 – Load-Bearing Metal Stud Framing

05 41 13.30 Framing, Stud Walls

			Crew	Daily Output	Labor-Hours	Unit	Material	2016 Bare Costs Labor	Equipment	Total	Total Incl O&P
4390	6" wide, studs 12" O.C.	G	2 Carp	44	.364	L.F.	17.05	12.35		29.40	39.50
4400	16" O.C.	G		64	.250		13.55	8.50		22.05	29
4410	24" O.C.	G		88	.182		10.05	6.15		16.20	21.50
4420	8" wide, studs 12" O.C.	G		43	.372		21	12.60		33.60	44
4430	16" O.C.	G		63	.254		16.60	8.60		25.20	32.50
4440	24" O.C.	G		86	.186		12.30	6.30		18.60	24
5100	10' high walls, 18 ga. x 2-1/2" wide, studs 12" O.C.	G		54	.296		10.90	10.05		20.95	29
5110	16" O.C.	G		77	.208		8.65	7.05		15.70	21.50
5120	24" O.C.	G		107	.150		6.35	5.05		11.40	15.50
5130	3-5/8" wide, studs 12" O.C.	G		53	.302		13	10.25		23.25	31.50
5140	16" O.C.	G		76	.211		10.30	7.15		17.45	23.50
5150	24" O.C.	G		105	.152		7.60	5.15		12.75	17
5160	4" wide, studs 12" O.C.	G		52	.308		13.55	10.45		24	32.50
5170	16" O.C.	G		74	.216		10.75	7.35		18.10	24
5180	24" O.C.	G		103	.155		7.95	5.25		13.20	17.50
5190	6" wide, studs 12" O.C.	G		51	.314		17.25	10.65		27.90	37
5200	16" O.C.	G		73	.219		13.70	7.45		21.15	27.50
5210	24" O.C.	G		101	.158		10.15	5.35		15.50	20
5220	8" wide, studs 12" O.C.	G		50	.320		21	10.85		31.85	41
5230	16" O.C.	G		72	.222		16.55	7.55		24.10	31
5240	24" O.C.	G		100	.160		12.30	5.40		17.70	22.50
5300	16 ga. x 2-1/2" wide, studs 12" O.C.	G		47	.340		13.10	11.55		24.65	33.50
5310	16" O.C.	G		68	.235		10.30	8		18.30	24.50
5320	24" O.C.	G		94	.170		7.45	5.75		13.20	17.85
5330	3-5/8" wide, studs 12" O.C.	G		46	.348		15.60	11.80		27.40	37
5340	16" O.C.	G		66	.242		12.25	8.20		20.45	27.50
5350	24" O.C.	G		92	.174		8.90	5.90		14.80	19.65
5360	4" wide, studs 12" O.C.	G		45	.356		16.35	12.05		28.40	38
5370	16" O.C.	G		65	.246		12.85	8.35		21.20	28
5380	24" O.C.	G		90	.178		9.35	6.05		15.40	20.50
5390	6" wide, studs 12" O.C.	G		44	.364		20.50	12.35		32.85	43
5400	16" O.C.	G		64	.250		16.15	8.50		24.65	32
5410	24" O.C.	G		88	.182		11.80	6.15		17.95	23.50
5420	8" wide, studs 12" O.C.	G		43	.372		25	12.60		37.60	48.50
5430	16" O.C.	G		63	.254		19.80	8.60		28.40	36.50
5440	24" O.C.	G		86	.186		14.45	6.30		20.75	26.50
6190	12' high walls, 18 ga. x 6" wide, studs 12" O.C.	G		41	.390		20	13.25		33.25	44
6200	16" O.C.	G		58	.276		15.80	9.35		25.15	33
6210	24" O.C.	G		81	.198		11.55	6.70		18.25	24
6220	8" wide, studs 12" O.C.	G		40	.400		24.50	13.55		38.05	49
6230	16" O.C.	G		57	.281		19.15	9.50		28.65	37
6240	24" O.C.	G		80	.200		14	6.80		20.80	27
6390	16 ga. x 6" wide, studs 12" O.C.	G		35	.457		24	15.50		39.50	52.50
6400	16" O.C.	G		51	.314		18.80	10.65		29.45	38.50
6410	24" O.C.	G		70	.229		13.55	7.75		21.30	28
6420	8" wide, studs 12" O.C.	G		34	.471		29.50	15.95		45.45	59
6430	16" O.C.	G		50	.320		23	10.85		33.85	43.50
6440	24" O.C.	G		69	.232		16.60	7.85		24.45	31.50
6530	14 ga. x 3-5/8" wide, studs 12" O.C.	G		34	.471		22.50	15.95		38.45	51.50
6540	16" O.C.	G		48	.333		17.75	11.30		29.05	38.50
6550	24" O.C.	G		65	.246		12.75	8.35		21.10	28
6560	4" wide, studs 12" O.C.	G		33	.485		24	16.45		40.45	54
6570	16" O.C.	G		47	.340		18.75	11.55		30.30	40

05 41 Structural Metal Stud Framing

05 41 13 – Load-Bearing Metal Stud Framing

05 41 13.30 Framing, Stud Walls

			Crew	Daily Output	Labor-Hours	Unit	Material	2016 Bare Costs Labor	Equipment	Total	Total Incl O&P
6580	24" O.C.	G	2 Carp	64	.250	L.F.	13.50	8.50		22	29
6730	12 ga. x 3-5/8" wide, studs 12" O.C.	G		31	.516		31.50	17.50		49	64
6740	16" O.C.	G		43	.372		24.50	12.60		37.10	47.50
6750	24" O.C.	G		59	.271		17.15	9.20		26.35	34.50
6760	4" wide, studs 12" O.C.	G		30	.533		33.50	18.10		51.60	67.50
6770	16" O.C.	G		42	.381		26	12.90		38.90	50
6780	24" O.C.	G		58	.276		18.30	9.35		27.65	35.50
7390	16' high walls, 16 ga. x 6" wide, studs 12" O.C.	G		33	.485		31	16.45		47.45	61.50
7400	16" O.C.	G		48	.333		24	11.30		35.30	45.50
7410	24" O.C.	G		67	.239		17.05	8.10		25.15	32.50
7420	8" wide, studs 12" O.C.	G		32	.500		38	16.95		54.95	70.50
7430	16" O.C.	G		47	.340		29.50	11.55		41.05	52
7440	24" O.C.	G		66	.242		21	8.20		29.20	37
7560	14 ga. x 4" wide, studs 12" O.C.	G		31	.516		31	17.50		48.50	63.50
7570	16" O.C.	G		45	.356		24	12.05		36.05	46.50
7580	24" O.C.	G		61	.262		17	8.90		25.90	33.50
7590	6" wide, studs 12" O.C.	G		30	.533		39	18.10		57.10	73.50
7600	16" O.C.	G		44	.364		30	12.35		42.35	53.50
7610	24" O.C.	G		60	.267		21.50	9.05		30.55	38.50
7760	12 ga. x 4" wide, studs 12" O.C.	G		29	.552		44	18.70		62.70	79.50
7770	16" O.C.	G		40	.400		33.50	13.55		47.05	59.50
7780	24" O.C.	G		55	.291		23.50	9.85		33.35	42
7790	6" wide, studs 12" O.C.	G		28	.571		55.50	19.35		74.85	93.50
7800	16" O.C.	G		39	.410		42.50	13.90		56.40	70
7810	24" O.C.	G		54	.296		29.50	10.05		39.55	49.50
8590	20' high walls, 14 ga. x 6" wide, studs 12" O.C.	G		29	.552		48	18.70		66.70	84
8600	16" O.C.	G		42	.381		37	12.90		49.90	62
8610	24" O.C.	G		57	.281		26	9.50		35.50	44.50
8620	8" wide, studs 12" O.C.	G		28	.571		52	19.35		71.35	89.50
8630	16" O.C.	G		41	.390		40	13.25		53.25	66
8640	24" O.C.	G		56	.286		28.50	9.70		38.20	47
8790	12 ga. x 6" wide, studs 12" O.C.	G		27	.593		68	20		88	109
8800	16" O.C.	G		37	.432		52	14.65		66.65	82
8810	24" O.C.	G		51	.314		36	10.65		46.65	57.50
8820	8" wide, studs 12" O.C.	G		26	.615		82.50	21		103.50	126
8830	16" O.C.	G		36	.444		63	15.05		78.05	94.50
8840	24" O.C.	G		50	.320		43.50	10.85		54.35	66

05 42 Cold-Formed Metal Joist Framing

05 42 13 – Cold-Formed Metal Floor Joist Framing

05 42 13.05 Bracing

			Crew	Daily Output	Labor-Hours	Unit	Material	Labor	Equipment	Total	Total Incl O&P
0010	**BRACING**, continuous, per row, top & bottom										
0015	Made from recycled materials										
0120	Flat strap, 20 ga. x 2" wide, joists at 12" O.C.	G	1 Carp	4.67	1.713	C.L.F.	60	58		118	163
0130	16" O.C.	G		5.33	1.501		57.50	51		108.50	149
0140	24" O.C.	G		6.66	1.201		55.50	40.50		96	129
0150	18 ga. x 2" wide, joists at 12" O.C.	G		4	2		78.50	68		146.50	201
0160	16" O.C.	G		4.67	1.713		77	58		135	182
0170	24" O.C.	G		5.33	1.501		75.50	51		126.50	168

05 42 Cold-Formed Metal Joist Framing

05 42 13 – Cold-Formed Metal Floor Joist Framing

05 42 13.10 Bridging

			Crew	Daily Output	Labor-Hours	Unit	Material	2016 Bare Costs Labor	Equipment	Total	Total Incl O&P
0010	**BRIDGING**, solid between joists w/1-1/4" leg track, per joist bay										
0015	Made from recycled materials										
0230	Joists 12" O.C., 18 ga. track x 6" wide	G	1 Carp	80	.100	Ea.	1.60	3.39		4.99	7.45
0240	8" wide	G		75	.107		1.97	3.62		5.59	8.20
0250	10" wide	G		70	.114		2.46	3.87		6.33	9.20
0260	12" wide	G		65	.123		2.78	4.17		6.95	10.05
0330	16 ga. track x 6" wide	G		70	.114		2.01	3.87		5.88	8.70
0340	8" wide	G		65	.123		2.51	4.17		6.68	9.75
0350	10" wide	G		60	.133		3.10	4.52		7.62	10.95
0360	12" wide	G		55	.145		3.58	4.93		8.51	12.20
0440	14 ga. track x 8" wide	G		60	.133		3.13	4.52		7.65	11
0450	10" wide	G		55	.145		3.89	4.93		8.82	12.55
0460	12" wide	G		50	.160		4.49	5.40		9.89	14.05
0550	12 ga. track x 10" wide	G		45	.178		5.70	6.05		11.75	16.35
0560	12" wide	G		40	.200		5.80	6.80		12.60	17.75
1230	16" O.C., 18 ga. track x 6" wide	G		80	.100		2.05	3.39		5.44	7.95
1240	8" wide	G		75	.107		2.52	3.62		6.14	8.85
1250	10" wide	G		70	.114		3.15	3.87		7.02	9.95
1260	12" wide	G		65	.123		3.56	4.17		7.73	10.90
1330	16 ga. track x 6" wide	G		70	.114		2.57	3.87		6.44	9.35
1340	8" wide	G		65	.123		3.22	4.17		7.39	10.55
1350	10" wide	G		60	.133		3.98	4.52		8.50	11.90
1360	12" wide	G		55	.145		4.59	4.93		9.52	13.30
1440	14 ga. track x 8" wide	G		60	.133		4.01	4.52		8.53	11.95
1450	10" wide	G		55	.145		4.98	4.93		9.91	13.75
1460	12" wide	G		50	.160		5.75	5.40		11.15	15.45
1550	12 ga. track x 10" wide	G		45	.178		7.30	6.05		13.35	18.10
1560	12" wide	G		40	.200		7.45	6.80		14.25	19.55
2230	24" O.C., 18 ga. track x 6" wide	G		80	.100		2.96	3.39		6.35	8.95
2240	8" wide	G		75	.107		3.65	3.62		7.27	10.05
2250	10" wide	G		70	.114		4.56	3.87		8.43	11.50
2260	12" wide	G		65	.123		5.15	4.17		9.32	12.65
2330	16 ga. track x 6" wide	G		70	.114		3.72	3.87		7.59	10.60
2340	8" wide	G		65	.123		4.65	4.17		8.82	12.10
2350	10" wide	G		60	.133		5.75	4.52		10.27	13.90
2360	12" wide	G		55	.145		6.65	4.93		11.58	15.55
2440	14 ga. track x 8" wide	G		60	.133		5.80	4.52		10.32	13.95
2450	10" wide	G		55	.145		7.20	4.93		12.13	16.20
2460	12" wide	G		50	.160		8.35	5.40		13.75	18.25
2550	12 ga. track x 10" wide	G		45	.178		10.55	6.05		16.60	21.50
2560	12" wide	G		40	.200		10.80	6.80		17.60	23

05 42 13.25 Framing, Band Joist

			Crew	Daily Output	Labor-Hours	Unit	Material	2016 Bare Costs Labor	Equipment	Total	Total Incl O&P
0010	**FRAMING, BAND JOIST** (track) fastened to bearing wall										
0015	Made from recycled materials										
0220	18 ga. track x 6" deep	G	2 Carp	1000	.016	L.F.	1.30	.54		1.84	2.34
0230	8" deep	G		920	.017		1.61	.59		2.20	2.76
0240	10" deep	G		860	.019		2.01	.63		2.64	3.27
0320	16 ga. track x 6" deep	G		900	.018		1.64	.60		2.24	2.81
0330	8" deep	G		840	.019		2.05	.65		2.70	3.33
0340	10" deep	G		780	.021		2.53	.70		3.23	3.94
0350	12" deep	G		740	.022		2.92	.73		3.65	4.44
0430	14 ga. track x 8" deep	G		750	.021		2.55	.72		3.27	4.02

05 42 Cold-Formed Metal Joist Framing

05 42 13 – Cold-Formed Metal Floor Joist Framing

05 42 13.25 Framing, Band Joist

			Crew	Daily Output	Labor-Hours	Unit	Material	2016 Bare Costs Labor	Equipment	Total	Total Incl O&P
0440	10" deep	G	2 Carp	720	.022	L.F.	3.17	.75		3.92	4.75
0450	12" deep	G		700	.023		3.66	.78		4.44	5.35
0540	12 ga. track x 10" deep	G		670	.024		4.64	.81		5.45	6.45
0550	12" deep	G		650	.025		4.75	.83		5.58	6.60

05 42 13.30 Framing, Boxed Headers/Beams

			Crew	Daily Output	Labor-Hours	Unit	Material	Labor	Equipment	Total	Total Incl O&P
0010	**FRAMING, BOXED HEADERS/BEAMS**										
0015	Made from recycled materials										
0200	Double, 18 ga. x 6" deep	G	2 Carp	220	.073	L.F.	5.50	2.47		7.97	10.20
0210	8" deep	G		210	.076		6.05	2.58		8.63	10.95
0220	10" deep	G		200	.080		7.40	2.71		10.11	12.70
0230	12" deep	G		190	.084		8.05	2.85		10.90	13.65
0300	16 ga. x 8" deep	G		180	.089		7	3.01		10.01	12.75
0310	10" deep	G		170	.094		8.45	3.19		11.64	14.60
0320	12" deep	G		160	.100		9.15	3.39		12.54	15.80
0400	14 ga. x 10" deep	G		140	.114		9.75	3.87		13.62	17.20
0410	12" deep	G		130	.123		10.65	4.17		14.82	18.75
0500	12 ga. x 10" deep	G		110	.145		12.85	4.93		17.78	22.50
0510	12" deep	G		100	.160		14.15	5.40		19.55	24.50
1210	Triple, 18 ga. x 8" deep	G		170	.094		8.75	3.19		11.94	15
1220	10" deep	G		165	.097		10.60	3.29		13.89	17.15
1230	12" deep	G		160	.100		11.60	3.39		14.99	18.45
1300	16 ga. x 8" deep	G		145	.110		10.15	3.74		13.89	17.40
1310	10" deep	G		140	.114		12.15	3.87		16.02	19.85
1320	12" deep	G		135	.119		13.25	4.02		17.27	21.50
1400	14 ga. x 10" deep	G		115	.139		14.10	4.72		18.82	23.50
1410	12" deep	G		110	.145		15.50	4.93		20.43	25.50
1500	12 ga. x 10" deep	G		90	.178		18.75	6.05		24.80	30.50
1510	12" deep	G		85	.188		21	6.40		27.40	33.50

05 42 13.40 Framing, Joists

			Crew	Daily Output	Labor-Hours	Unit	Material	Labor	Equipment	Total	Total Incl O&P
0010	**FRAMING, JOISTS**, no band joists (track), web stiffeners, headers,										
0020	Beams, bridging or bracing										
0025	Made from recycled materials										
0030	Joists (2" flange) and fasteners, materials only										
0220	18 ga. x 6" deep	G				L.F.	1.69			1.69	1.86
0230	8" deep	G					1.98			1.98	2.18
0240	10" deep	G					2.34			2.34	2.58
0320	16 ga. x 6" deep	G					2.07			2.07	2.28
0330	8" deep	G					2.47			2.47	2.71
0340	10" deep	G					2.89			2.89	3.18
0350	12" deep	G					3.28			3.28	3.60
0430	14 ga. x 8" deep	G					3.10			3.10	3.41
0440	10" deep	G					3.57			3.57	3.93
0450	12" deep	G					4.06			4.06	4.47
0540	12 ga. x 10" deep	G					5.20			5.20	5.70
0550	12" deep	G					5.90			5.90	6.50
1010	Installation of joists to band joists, beams & headers, labor only										
1220	18 ga. x 6" deep	G	2 Carp	110	.145	Ea.		4.93		4.93	8.25
1230	8" deep	G		90	.178			6.05		6.05	10.10
1240	10" deep	G		80	.200			6.80		6.80	11.35
1320	16 ga. x 6" deep	G		95	.168			5.70		5.70	9.55
1330	8" deep	G		70	.229			7.75		7.75	12.95
1340	10" deep	G		60	.267			9.05		9.05	15.15

05 42 Cold-Formed Metal Joist Framing

05 42 13 – Cold-Formed Metal Floor Joist Framing

05 42 13.40 Framing, Joists		Crew	Daily Output	Labor-Hours	Unit	Material	2016 Bare Costs Labor	Equipment	Total	Total Incl O&P
1350	12" deep	2 Carp	55	.291	Ea.		9.85		9.85	16.50
1430	14 ga. x 8" deep		65	.246			8.35		8.35	13.95
1440	10" deep		45	.356			12.05		12.05	20
1450	12" deep		35	.457			15.50		15.50	26
1540	12 ga. x 10" deep		40	.400			13.55		13.55	22.50
1550	12" deep		30	.533			18.10		18.10	30.50

05 42 13.45 Framing, Web Stiffeners

0010	**FRAMING, WEB STIFFENERS** at joist bearing, fabricated from										
0020	Stud piece (1-5/8" flange) to stiffen joist (2" flange)										
0025	Made from recycled materials										
2120	For 6" deep joist, with 18 ga. x 2-1/2" stud	G	1 Carp	120	.067	Ea.	.91	2.26		3.17	4.78
2130	3-5/8" stud	G		110	.073		1.08	2.47		3.55	5.30
2140	4" stud	G		105	.076		1.12	2.58		3.70	5.55
2150	6" stud	G		100	.080		1.42	2.71		4.13	6.10
2160	8" stud	G		95	.084		1.71	2.85		4.56	6.65
2220	8" deep joist, with 2-1/2" stud	G		120	.067		1.22	2.26		3.48	5.10
2230	3-5/8" stud	G		110	.073		1.45	2.47		3.92	5.70
2240	4" stud	G		105	.076		1.50	2.58		4.08	5.95
2250	6" stud	G		100	.080		1.90	2.71		4.61	6.65
2260	8" stud	G		95	.084		2.29	2.85		5.14	7.30
2320	10" deep joist, with 2-1/2" stud	G		110	.073		1.51	2.47		3.98	5.80
2330	3-5/8" stud	G		100	.080		1.79	2.71		4.50	6.50
2340	4" stud	G		95	.084		1.86	2.85		4.71	6.85
2350	6" stud	G		90	.089		2.36	3.01		5.37	7.65
2360	8" stud	G		85	.094		2.84	3.19		6.03	8.45
2420	12" deep joist, with 2-1/2" stud	G		110	.073		1.82	2.47		4.29	6.15
2430	3-5/8" stud	G		100	.080		2.16	2.71		4.87	6.90
2440	4" stud	G		95	.084		2.24	2.85		5.09	7.25
2450	6" stud	G		90	.089		2.84	3.01		5.85	8.15
2460	8" stud	G		85	.094		3.42	3.19		6.61	9.10
3130	For 6" deep joist, with 16 ga. x 3-5/8" stud	G		100	.080		1.34	2.71		4.05	6
3140	4" stud	G		95	.084		1.40	2.85		4.25	6.30
3150	6" stud	G		90	.089		1.75	3.01		4.76	7
3160	8" stud	G		85	.094		2.14	3.19		5.33	7.70
3230	8" deep joist, with 3-5/8" stud	G		100	.080		1.80	2.71		4.51	6.50
3240	4" stud	G		95	.084		1.88	2.85		4.73	6.85
3250	6" stud	G		90	.089		2.35	3.01		5.36	7.65
3260	8" stud	G		85	.094		2.87	3.19		6.06	8.50
3330	10" deep joist, with 3-5/8" stud	G		85	.094		2.22	3.19		5.41	7.80
3340	4" stud	G		80	.100		2.32	3.39		5.71	8.25
3350	6" stud	G		75	.107		2.91	3.62		6.53	9.25
3360	8" stud	G		70	.114		3.55	3.87		7.42	10.40
3430	12" deep joist, with 3-5/8" stud	G		85	.094		2.68	3.19		5.87	8.30
3440	4" stud	G		80	.100		2.80	3.39		6.19	8.80
3450	6" stud	G		75	.107		3.50	3.62		7.12	9.90
3460	8" stud	G		70	.114		4.28	3.87		8.15	11.20
4230	For 8" deep joist, with 14 ga. x 3-5/8" stud	G		90	.089		2.22	3.01		5.23	7.50
4240	4" stud	G		85	.094		2.35	3.19		5.54	7.95
4250	6" stud	G		80	.100		2.95	3.39		6.34	8.95
4260	8" stud	G		75	.107		3.15	3.62		6.77	9.50
4330	10" deep joist, with 3-5/8" stud	G		75	.107		2.76	3.62		6.38	9.10
4340	4" stud	G		70	.114		2.91	3.87		6.78	9.70

05 42 Cold-Formed Metal Joist Framing

05 42 13 – Cold-Formed Metal Floor Joist Framing

05 42 13.45 Framing, Web Stiffeners

		Crew	Daily Output	Labor-Hours	Unit	Material	2016 Bare Costs Labor	Equipment	Total	Total Incl O&P
4350	6" stud	1 Carp G	65	.123	Ea.	3.65	4.17		7.82	11
4360	8" stud	G	60	.133		3.90	4.52		8.42	11.85
4430	12" deep joist, with 3-5/8" stud	G	75	.107		3.32	3.62		6.94	9.70
4440	4" stud	G	70	.114		3.50	3.87		7.37	10.35
4450	6" stud	G	65	.123		4.40	4.17		8.57	11.85
4460	8" stud	G	60	.133		4.70	4.52		9.22	12.70
5330	For 10" deep joist, with 12 ga. x 3-5/8" stud	G	65	.123		3.97	4.17		8.14	11.35
5340	4" stud	G	60	.133		4.23	4.52		8.75	12.20
5350	6" stud	G	55	.145		5.35	4.93		10.28	14.15
5360	8" stud	G	50	.160		6.45	5.40		11.85	16.20
5430	12" deep joist, with 3-5/8" stud	G	65	.123		4.78	4.17		8.95	12.25
5440	4" stud	G	60	.133		5.10	4.52		9.62	13.15
5450	6" stud	G	55	.145		6.45	4.93		11.38	15.35
5460	8" stud	G	50	.160		7.80	5.40		13.20	17.65

05 42 23 – Cold-Formed Metal Roof Joist Framing

05 42 23.05 Framing, Bracing

		Crew	Daily Output	Labor-Hours	Unit	Material	Labor	Equipment	Total	Incl O&P
0010	**FRAMING, BRACING**									
0015	Made from recycled materials									
0020	Continuous bracing, per row									
0100	16 ga. x 1-1/2" channel thru rafters/trusses @ 16" O.C.	1 Carp G	4.50	1.778	C.L.F.	51.50	60.50		112	158
0120	24" O.C.	G	6	1.333		51.50	45		96.50	133
0300	2" x 2" angle x 18 ga., rafters/trusses @ 16" O.C.	G	6	1.333		80.50	45		125.50	164
0320	24" O.C.	G	8	1		80.50	34		114.50	146
0400	16 ga., rafters/trusses @ 16" O.C.	G	4.50	1.778		101	60.50		161.50	212
0420	24" O.C.	G	6.50	1.231		101	41.50		142.50	181

05 42 23.10 Framing, Bridging

		Crew	Daily Output	Labor-Hours	Unit	Material	Labor	Equipment	Total	Incl O&P
0010	**FRAMING, BRIDGING**									
0015	Made from recycled materials									
0020	Solid, between rafters w/1-1/4" leg track, per rafter bay									
1200	Rafters 16" O.C., 18 ga. x 4" deep	1 Carp G	60	.133	Ea.	1.57	4.52		6.09	9.25
1210	6" deep	G	57	.140		2.05	4.76		6.81	10.20
1220	8" deep	G	55	.145		2.52	4.93		7.45	11.05
1230	10" deep	G	52	.154		3.15	5.20		8.35	12.20
1240	12" deep	G	50	.160		3.56	5.40		8.96	13
2200	24" O.C., 18 ga. x 4" deep	G	60	.133		2.27	4.52		6.79	10.05
2210	6" deep	G	57	.140		2.96	4.76		7.72	11.20
2220	8" deep	G	55	.145		3.65	4.93		8.58	12.25
2230	10" deep	G	52	.154		4.56	5.20		9.76	13.75
2240	12" deep	G	50	.160		5.15	5.40		10.55	14.75

05 42 23.50 Framing, Parapets

		Crew	Daily Output	Labor-Hours	Unit	Material	Labor	Equipment	Total	Incl O&P
0010	**FRAMING, PARAPETS**									
0015	Made from recycled materials									
0100	3' high installed on 1st story, 18 ga. x 4" wide studs, 12" O.C.	2 Carp G	100	.160	L.F.	5.70	5.40		11.10	15.35
0110	16" O.C.	G	150	.107		4.85	3.62		8.47	11.40
0120	24" O.C.	G	200	.080		4.01	2.71		6.72	8.95
0200	6" wide studs, 12" O.C.	G	100	.160		7.30	5.40		12.70	17.15
0210	16" O.C.	G	150	.107		6.25	3.62		9.87	12.90
0220	24" O.C.	G	200	.080		5.15	2.71		7.86	10.25
1100	Installed on 2nd story, 18 ga. x 4" wide studs, 12" O.C.	G	95	.168		5.70	5.70		11.40	15.80
1110	16" O.C.	G	145	.110		4.85	3.74		8.59	11.60
1120	24" O.C.	G	190	.084		4.01	2.85		6.86	9.20
1200	6" wide studs, 12" O.C.	G	95	.168		7.30	5.70		13	17.60

05 42 Cold-Formed Metal Joist Framing

05 42 23 – Cold-Formed Metal Roof Joist Framing

	05 42 23.50 Framing, Parapets		Crew	Daily Output	Labor-Hours	Unit	Material	2016 Bare Costs Labor	Equipment	Total	Total Incl O&P
1210	16" O.C.	G	2 Carp	145	.110	L.F.	6.25	3.74		9.99	13.10
1220	24" O.C.	G		190	.084		5.15	2.85		8	10.50
2100	Installed on gable, 18 ga. x 4" wide studs, 12" O.C.	G		85	.188		5.70	6.40		12.10	16.95
2110	16" O.C.	G		130	.123		4.85	4.17		9.02	12.35
2120	24" O.C.	G		170	.094		4.01	3.19		7.20	9.75
2200	6" wide studs, 12" O.C.	G		85	.188		7.30	6.40		13.70	18.75
2210	16" O.C.	G		130	.123		6.25	4.17		10.42	13.85
2220	24" O.C.	G		170	.094		5.15	3.19		8.34	11.05

	05 42 23.60 Framing, Roof Rafters		Crew	Daily Output	Labor-Hours	Unit	Material	Labor	Equipment	Total	Total Incl O&P
0010	**FRAMING, ROOF RAFTERS**										
0015	Made from recycled materials										
0100	Boxed ridge beam, double, 18 ga. x 6" deep	G	2 Carp	160	.100	L.F.	5.50	3.39		8.89	11.75
0110	8" deep	G		150	.107		6.05	3.62		9.67	12.70
0120	10" deep	G		140	.114		7.40	3.87		11.27	14.65
0130	12" deep	G		130	.123		8.05	4.17		12.22	15.85
0200	16 ga. x 6" deep	G		150	.107		6.20	3.62		9.82	12.90
0210	8" deep	G		140	.114		7	3.87		10.87	14.20
0220	10" deep	G		130	.123		8.45	4.17		12.62	16.25
0230	12" deep	G		120	.133		9.15	4.52		13.67	17.65
1100	Rafters, 2" flange, material only, 18 ga. x 6" deep	G					1.69			1.69	1.86
1110	8" deep	G					1.98			1.98	2.18
1120	10" deep	G					2.34			2.34	2.58
1130	12" deep	G					2.70			2.70	2.97
1200	16 ga. x 6" deep	G					2.07			2.07	2.28
1210	8" deep	G					2.47			2.47	2.71
1220	10" deep	G					2.89			2.89	3.18
1230	12" deep	G					3.28			3.28	3.60
2100	Installation only, ordinary rafter to 4:12 pitch, 18 ga. x 6" deep		2 Carp	35	.457	Ea.		15.50		15.50	26
2110	8" deep			30	.533			18.10		18.10	30.50
2120	10" deep			25	.640			21.50		21.50	36.50
2130	12" deep			20	.800			27		27	45.50
2200	16 ga. x 6" deep			30	.533			18.10		18.10	30.50
2210	8" deep			25	.640			21.50		21.50	36.50
2220	10" deep			20	.800			27		27	45.50
2230	12" deep			15	1.067			36		36	60.50
8100	Add to labor, ordinary rafters on steep roofs							25%			
8110	Dormers & complex roofs							50%			
8200	Hip & valley rafters to 4:12 pitch							25%			
8210	Steep roofs							50%			
8220	Dormers & complex roofs							75%			
8300	Hip & valley jack rafters to 4:12 pitch							50%			
8310	Steep roofs							75%			
8320	Dormers & complex roofs							100%			

	05 42 23.70 Framing, Soffits and Canopies		Crew	Daily Output	Labor-Hours	Unit	Material	Labor	Equipment	Total	Total Incl O&P
0010	**FRAMING, SOFFITS & CANOPIES**										
0015	Made from recycled materials										
0130	Continuous ledger track @ wall, studs @ 16" O.C., 18 ga. x 4" wide	G	2 Carp	535	.030	L.F.	1.05	1.01		2.06	2.85
0140	6" wide	G		500	.032		1.36	1.08		2.44	3.32
0150	8" wide	G		465	.034		1.68	1.17		2.85	3.80
0160	10" wide	G		430	.037		2.10	1.26		3.36	4.42
0230	Studs @ 24" O.C., 18 ga. x 4" wide	G		800	.020		1	.68		1.68	2.24
0240	6" wide	G		750	.021		1.30	.72		2.02	2.64

05 42 Cold-Formed Metal Joist Framing

05 42 23 – Cold-Formed Metal Roof Joist Framing

05 42 23.70 Framing, Soffits and Canopies

			Crew	Daily Output	Labor-Hours	Unit	Material	2016 Bare Costs Labor	2016 Bare Costs Equipment	Total	Total Incl O&P
0250	8" wide	G	2 Carp	700	.023	L.F.	1.61	.78		2.39	3.07
0260	10" wide	G		650	.025		2.01	.83		2.84	3.61
1000	Horizontal soffit and canopy members, material only										
1030	1-5/8" flange studs, 18 ga. x 4" deep	G				L.F.	1.34			1.34	1.48
1040	6" deep	G					1.70			1.70	1.87
1050	8" deep	G					2.05			2.05	2.26
1140	2" flange joists, 18 ga. x 6" deep	G					1.93			1.93	2.13
1150	8" deep	G					2.27			2.27	2.49
1160	10" deep	G					2.68			2.68	2.94
4030	Installation only, 18 ga., 1-5/8" flange x 4" deep		2 Carp	130	.123	Ea.		4.17		4.17	7
4040	6" deep			110	.145			4.93		4.93	8.25
4050	8" deep			90	.178			6.05		6.05	10.10
4140	2" flange, 18 ga. x 6" deep			110	.145			4.93		4.93	8.25
4150	8" deep			90	.178			6.05		6.05	10.10
4160	10" deep			80	.200			6.80		6.80	11.35
6010	Clips to attach fascia to rafter tails, 2" x 2" x 18 ga. angle	G	1 Carp	120	.067		.95	2.26		3.21	4.82
6020	16 ga. angle	G	"	100	.080		1.20	2.71		3.91	5.85

05 44 Cold-Formed Metal Trusses

05 44 13 – Cold-Formed Metal Roof Trusses

05 44 13.60 Framing, Roof Trusses

			Crew	Daily Output	Labor-Hours	Unit	Material	Labor	Equipment	Total	Total Incl O&P
0010	**FRAMING, ROOF TRUSSES**										
0015	Made from recycled materials										
0020	Fabrication of trusses on ground, Fink (W) or King Post, to 4:12 pitch										
0120	18 ga. x 4" chords, 16' span	G	2 Carp	12	1.333	Ea.	62.50	45		107.50	145
0130	20' span	G		11	1.455		78.50	49.50		128	169
0140	24' span	G		11	1.455		94	49.50		143.50	186
0150	28' span	G		10	1.600		110	54		164	212
0160	32' span	G		10	1.600		125	54		179	229
0250	6" chords, 28' span	G		9	1.778		139	60.50		199.50	254
0260	32' span	G		9	1.778		159	60.50		219.50	276
0270	36' span	G		8	2		179	68		247	310
0280	40' span	G		8	2		199	68		267	335
1120	5:12 to 8:12 pitch, 18 ga. x 4" chords, 16' span	G		10	1.600		71.50	54		125.50	170
1130	20' span	G		9	1.778		89.50	60.50		150	200
1140	24' span	G		9	1.778		108	60.50		168.50	219
1150	28' span	G		8	2		125	68		193	252
1160	32' span	G		8	2		143	68		211	272
1250	6" chords, 28' span	G		7	2.286		159	77.50		236.50	305
1260	32' span	G		7	2.286		182	77.50		259.50	330
1270	36' span	G		6	2.667		204	90.50		294.50	375
1280	40' span	G		6	2.667		227	90.50		317.50	400
2120	9:12 to 12:12 pitch, 18 ga. x 4" chords, 16' span	G		8	2		89.50	68		157.50	213
2130	20' span	G		7	2.286		112	77.50		189.50	253
2140	24' span	G		7	2.286		134	77.50		211.50	278
2150	28' span	G		6	2.667		157	90.50		247.50	325
2160	32' span	G		6	2.667		179	90.50		269.50	350
2250	6" chords, 28' span	G		5	3.200		199	108		307	400
2260	32' span	G		5	3.200		227	108		335	430
2270	36' span	G		4	4		256	136		392	510
2280	40' span	G		4	4		284	136		420	535

05 44 Cold-Formed Metal Trusses

05 44 13 – Cold-Formed Metal Roof Trusses

05 44 13.60 Framing, Roof Trusses

		Crew	Daily Output	Labor-Hours	Unit	Material	2016 Bare Costs Labor	2016 Bare Costs Equipment	Total	Total Incl O&P
5120	Erection only of roof trusses, to 4:12 pitch, 16' span	F-6	48	.833	Ea.		25.50	13.70	39.20	57.50
5130	20' span		46	.870			26.50	14.30	40.80	60.50
5140	24' span		44	.909			28	14.95	42.95	63
5150	28' span		42	.952			29	15.70	44.70	66
5160	32' span		40	1			30.50	16.45	46.95	69
5170	36' span		38	1.053			32	17.35	49.35	73
5180	40' span		36	1.111			34	18.30	52.30	77
5220	5:12 to 8:12 pitch, 16' span		42	.952			29	15.70	44.70	66
5230	20' span		40	1			30.50	16.45	46.95	69
5240	24' span		38	1.053			32	17.35	49.35	73
5250	28' span		36	1.111			34	18.30	52.30	77
5260	32' span		34	1.176			36	19.35	55.35	81.50
5270	36' span		32	1.250			38.50	20.50	59	86.50
5280	40' span		30	1.333			41	22	63	92
5320	9:12 to 12:12 pitch, 16' span		36	1.111			34	18.30	52.30	77
5330	20' span		34	1.176			36	19.35	55.35	81.50
5340	24' span		32	1.250			38.50	20.50	59	86.50
5350	28' span		30	1.333			41	22	63	92
5360	32' span		28	1.429			44	23.50	67.50	99
5370	36' span		26	1.538			47	25.50	72.50	107
5380	40' span		24	1.667			51	27.50	78.50	115

05 51 Metal Stairs

05 51 13 – Metal Pan Stairs

05 51 13.50 Pan Stairs

			Crew	Daily Output	Labor-Hours	Unit	Material	Labor	Equipment	Total	Total Incl O&P
0010	**PAN STAIRS**, shop fabricated, steel stringers										
0015	Made from recycled materials										
1700	Pre-erected, steel pan tread, 3'-6" wide, 2 line pipe rail	G	E-2	87	.552	Riser	550	19.90	17.45	587.35	660
1800	With flat bar picket rail	G	"	87	.552	"	615	19.90	17.45	652.35	735

05 51 23 – Metal Fire Escapes

05 51 23.50 Fire Escape Stairs

		Crew	Daily Output	Labor-Hours	Unit	Material	Labor	Equipment	Total	Total Incl O&P
0010	**FIRE ESCAPE STAIRS**, portable									
0100	Portable ladder					Ea.	122			122	135

05 52 Metal Railings

05 52 13 – Pipe and Tube Railings

05 52 13.50 Railings, Pipe

			Crew	Daily Output	Labor-Hours	Unit	Material	Labor	Equipment	Total	Total Incl O&P
0010	**RAILINGS, PIPE**, shop fab'd, 3'-6" high, posts @ 5' O.C.										
0015	Made from recycled materials										
0020	Aluminum, 2 rail, satin finish, 1-1/4" diameter	G	E-4	160	.200	L.F.	50.50	7.25	.92	58.67	70
0030	Clear anodized	G		160	.200		62	7.25	.92	70.17	82.50
0040	Dark anodized	G		160	.200		69	7.25	.92	77.17	90.50
0080	1-1/2" diameter, satin finish	G		160	.200		59	7.25	.92	67.17	79.50
0090	Clear anodized	G		160	.200		66.50	7.25	.92	74.67	87.50
0100	Dark anodized	G		160	.200		73	7.25	.92	81.17	95
0140	Aluminum, 3 rail, 1-1/4" diam., satin finish	G		137	.234		65	8.45	1.07	74.52	88
0150	Clear anodized	G		137	.234		81.50	8.45	1.07	91.02	106
0160	Dark anodized	G		137	.234		90	8.45	1.07	99.52	115
0200	1-1/2" diameter, satin finish	G		137	.234		77.50	8.45	1.07	87.02	102

05 52 Metal Railings

05 52 13 – Pipe and Tube Railings

05 52 13.50 Railings, Pipe

			Crew	Daily Output	Labor-Hours	Unit	Material	2016 Bare Costs Labor	Equipment	Total	Total Incl O&P
0210	Clear anodized	G	E-4	137	.234	L.F.	88.50	8.45	1.07	98.02	114
0220	Dark anodized	G		137	.234		96.50	8.45	1.07	106.02	123
0500	Steel, 2 rail, on stairs, primed, 1-1/4" diameter	G		160	.200		25	7.25	.92	33.17	42
0520	1-1/2" diameter	G		160	.200		27	7.25	.92	35.17	44.50
0540	Galvanized, 1-1/4" diameter	G		160	.200		34	7.25	.92	42.17	52
0560	1-1/2" diameter	G		160	.200		38.50	7.25	.92	46.67	56.50
0580	Steel, 3 rail, primed, 1-1/4" diameter	G		137	.234		37.50	8.45	1.07	47.02	57.50
0600	1-1/2" diameter	G		137	.234		39	8.45	1.07	48.52	59.50
0620	Galvanized, 1-1/4" diameter	G		137	.234		52.50	8.45	1.07	62.02	74
0640	1-1/2" diameter	G		137	.234		60.50	8.45	1.07	70.02	83
0700	Stainless steel, 2 rail, 1-1/4" diam. #4 finish	G		137	.234		110	8.45	1.07	119.52	138
0720	High polish	G		137	.234		178	8.45	1.07	187.52	213
0740	Mirror polish	G		137	.234		223	8.45	1.07	232.52	262
0760	Stainless steel, 3 rail, 1-1/2" diam., #4 finish	G		120	.267		166	9.65	1.23	176.88	202
0770	High polish	G		120	.267		275	9.65	1.23	285.88	325
0780	Mirror finish	G		120	.267		335	9.65	1.23	345.88	390
0900	Wall rail, alum. pipe, 1-1/4" diam., satin finish	G		213	.150		24.50	5.45	.69	30.64	37
0905	Clear anodized	G		213	.150		30	5.45	.69	36.14	43.50
0910	Dark anodized	G		213	.150		35.50	5.45	.69	41.64	49.50
0915	1-1/2" diameter, satin finish	G		213	.150		27	5.45	.69	33.14	40.50
0920	Clear anodized	G		213	.150		34	5.45	.69	40.14	47.50
0925	Dark anodized	G		213	.150		42	5.45	.69	48.14	56.50
0930	Steel pipe, 1-1/4" diameter, primed	G		213	.150		15	5.45	.69	21.14	27
0935	Galvanized	G		213	.150		22	5.45	.69	28.14	34.50
0940	1-1/2" diameter	G		176	.182		15.50	6.55	.84	22.89	30
0945	Galvanized	G		213	.150		22	5.45	.69	28.14	34.50
0955	Stainless steel pipe, 1-1/2" diam., #4 finish	G		107	.299		88	10.80	1.38	100.18	118
0960	High polish	G		107	.299		179	10.80	1.38	191.18	218
0965	Mirror polish	G		107	.299		212	10.80	1.38	224.18	254
2000	2-line pipe rail (1-1/2" T&B) with 1/2" pickets @ 4-1/2" O.C.,										
2005	attached handrail on brackets										
2010	42" high aluminum, satin finish, straight & level	G	E-4	120	.267	L.F.	257	9.65	1.23	267.88	300
2050	42" high steel, primed, straight & level	G	"	120	.267		127	9.65	1.23	137.88	159
4000	For curved and level rails, add						10%	10%			
4100	For sloped rails for stairs, add						30%	30%			

05 58 Formed Metal Fabrications

05 58 25 – Formed Lamp Posts

05 58 25.40 Lamp Posts

			Crew	Daily Output	Labor-Hours	Unit	Material	Labor	Equipment	Total	Total Incl O&P
0010	**LAMP POSTS**										
0020	Aluminum, 7' high, stock units, post only	G	1 Carp	16	.500	Ea.	85	16.95		101.95	122
0100	Mild steel, plain	G	"	16	.500	"	73.50	16.95		90.45	110

05 71 Decorative Metal Stairs

05 71 13 – Fabricated Metal Spiral Stairs

05 71 13.50 Spiral Stairs

		Crew	Daily Output	Labor-Hours	Unit	Material	2016 Bare Costs Labor	Equipment	Total	Total Incl O&P
0010	**SPIRAL STAIRS**									
1805	Shop fabricated, custom ordered									
1810	Aluminum, 5'-0" diameter, plain units	G E-4	45	.711	Riser	530	25.50	3.27	558.77	635
1820	Fancy units	G	45	.711		925	25.50	3.27	953.77	1,075
1900	Cast iron, 4'-0" diameter, plain units	G	45	.711		665	25.50	3.27	693.77	780
1920	Fancy units	G	25	1.280		1,150	46.50	5.90	1,202.40	1,375
3100	Spiral stair kits, 12 stacking risers to fit exact floor height									
3110	Steel, flat metal treads, primed, 3'-6" diameter	G 2 Carp	1.60	10	Flight	1,300	340		1,640	2,000
3120	4'-0" diameter	G	1.45	11.034		1,475	375		1,850	2,225
3130	4'-6" diameter	G	1.35	11.852		1,625	400		2,025	2,475
3140	5'-0" diameter	G	1.25	12.800		1,775	435		2,210	2,700
3310	Checkered plate tread, primed, 3'-6" diameter	G	1.45	11.034		1,550	375		1,925	2,325
3320	4'-0" diameter	G	1.35	11.852		1,725	400		2,125	2,575
3330	4'-6" diameter	G	1.25	12.800		1,900	435		2,335	2,800
3340	5'-0" diameter	G	1.15	13.913		2,075	470		2,545	3,075
3510	Red oak covers on flat metal treads, 3'-6" diameter		1.35	11.852		2,350	400		2,750	3,275
3520	4'-0" diameter		1.25	12.800		2,800	435		3,235	3,800
3530	4'-6" diameter		1.15	13.913		3,025	470		3,495	4,125
3540	5'-0" diameter		1.05	15.238		3,250	515		3,765	4,450

05 75 Decorative Formed Metal

05 75 13 – Columns

05 75 13.10 Aluminum Columns

		Crew	Daily Output	Labor-Hours	Unit	Material	Labor	Equipment	Total	Total Incl O&P
0010	**ALUMINUM COLUMNS**									
0015	Made from recycled materials									
0020	Aluminum, extruded, stock units, no cap or base, 6" diameter	G E-4	240	.133	L.F.	11.85	4.82	.61	17.28	22.50
0100	8" diameter	G "	170	.188	"	15.05	6.80	.87	22.72	30
0410	Caps and bases, plain, 6" diameter	G			Set	21			21	23
0420	8" diameter	G			"	26			26	28.50
0500	For square columns, add to column prices above				L.F.	50%				
0700	Residential, flat, 8' high, plain	G E-4	20	1.600	Ea.	97.50	58	7.35	162.85	221
0720	Fancy	G	20	1.600		190	58	7.35	255.35	320
0740	Corner type, plain	G	20	1.600		168	58	7.35	233.35	299
0760	Fancy	G	20	1.600		330	58	7.35	395.35	480

05 75 13.20 Columns, Ornamental

		Crew	Daily Output	Labor-Hours	Unit	Material	Labor	Equipment	Total	Total Incl O&P
0010	**COLUMNS, ORNAMENTAL**, shop fabricated									
6400	Mild steel, flat, 9" wide, stock units, painted, plain	G E-4	160	.200	V.L.F.	9.15	7.25	.92	17.32	24.50
6450	Fancy	G	160	.200		17.75	7.25	.92	25.92	34
6500	Corner columns, painted, plain	G	160	.200		15.75	7.25	.92	23.92	31.50
6550	Fancy	G	160	.200		31	7.25	.92	39.17	49

Division Notes

	CREW	DAILY OUTPUT	LABOR-HOURS	UNIT	BARE COSTS				TOTAL INCL O&P
					MAT.	LABOR	EQUIP.	TOTAL	

Estimating Tips

06 05 00 Common Work Results for Wood, Plastics, and Composites

- Common to any wood-framed structure are the accessory connector items such as screws, nails, adhesives, hangers, connector plates, straps, angles, and hold-downs. For typical wood-framed buildings, such as residential projects, the aggregate total for these items can be significant, especially in areas where seismic loading is a concern. For floor and wall framing, the material cost is based on 10 to 25 lbs. of accessory connectors per MBF. Hold-downs, hangers, and other connectors should be taken off by the piece.

  Included with material costs are fasteners for a normal installation. RSMeans engineers use manufacturer's recommendations, written specifications, and/or standard construction practice for size and spacing of fasteners. Prices for various fasteners are shown for informational purposes only. Adjustments should be made if unusual fastening conditions exist.

06 10 00 Carpentry

- Lumber is a traded commodity and therefore sensitive to supply and demand in the marketplace. Even in "budgetary" estimating of wood-framed projects, it is advisable to call local suppliers for the latest market pricing.

- Common quantity units for wood-framed projects are "thousand board feet" (MBF). A board foot is a volume of wood, 1" x 1' x 1', or 144 cubic inches. Board-foot quantities are generally calculated using nominal material dimensions—dressed sizes are ignored. Board foot per lineal foot of any stick of lumber can be calculated by dividing the nominal cross-sectional area by 12. As an example, 2,000 lineal feet of 2 x 12 equates to 4 MBF by dividing the nominal area, 2 x 12, by 12, which equals 2, and multiplying by 2,000 to give 4,000 board feet. This simple rule applies to all nominal dimensioned lumber.

- Waste is an issue of concern at the quantity takeoff for any area of construction. Framing lumber is sold in even foot lengths, i.e., 8', 10', 12', 14', 16' and, depending on spans, wall heights, and the grade of lumber, waste is inevitable. A rule of thumb for lumber waste is 5%–10% depending on material quality and the complexity of the framing.

- Wood in various forms and shapes is used in many projects, even where the main structural framing is steel, concrete, or masonry. Plywood as a back-up partition material and 2x boards used as blocking and cant strips around roof edges are two common examples. The estimator should ensure that the costs of all wood materials are included in the final estimate.

06 20 00 Finish Carpentry

- It is necessary to consider the grade of workmanship when estimating labor costs for erecting millwork and interior finish. In practice, there are three grades: premium, custom, and economy. The RSMeans daily output for base and case moldings is in the range of 200 to 250 L.F. per carpenter per day. This is appropriate for most average custom-grade projects. For premium projects, an adjustment to productivity of 25%–50% should be made, depending on the complexity of the job.

Reference Numbers

Reference numbers are shown at the beginning of some major classifications. These numbers refer to related items in the Reference Section. The reference information may be an estimating procedure, an alternate pricing method, or technical information.

Note: Not all subdivisions listed here necessarily appear. ■

No part of this cost data may be reproduced, stored in a retrieval system, or transmitted in any form or by any means without prior written permission of RSMeans.

Did you know?

RSMeans Online gives you the same access to RSMeans' data with 24/7 access:

- Quickly locate costs in the searchable database.
- Build cost lists, estimates, and reports in minutes.
- Adjust costs to any location in the U.S. and Canada with the click of a button.

Start your free trial today at **www.RSMeansOnline.com**

RSMeans Online
FROM THE GORDIAN GROUP

06 05 Common Work Results for Wood, Plastics, and Composites

06 05 05 – Selective Demolition for Wood, Plastics, and Composites

06 05 05.10 Selective Demolition Wood Framing		Crew	Daily Output	Labor-Hours	Unit	Material	2016 Bare Costs Labor	Equipment	Total	Total Incl O&P
0010	**SELECTIVE DEMOLITION WOOD FRAMING** R024119-10									
0100	Timber connector, nailed, small	1 Clab	96	.083	Ea.		2.05		2.05	3.44
0110	Medium		60	.133			3.29		3.29	5.50
0120	Large		48	.167			4.11		4.11	6.90
0130	Bolted, small		48	.167			4.11		4.11	6.90
0140	Medium		32	.250			6.15		6.15	10.30
0150	Large		24	.333			8.20		8.20	13.75
2958	Beams, 2" x 6"	2 Clab	1100	.015	L.F.		.36		.36	.60
2960	2" x 8"		825	.019			.48		.48	.80
2965	2" x 10"		665	.024			.59		.59	.99
2970	2" x 12"		550	.029			.72		.72	1.20
2972	2" x 14"		470	.034			.84		.84	1.40
2975	4" x 8"	B-1	413	.058			1.47		1.47	2.46
2980	4" x 10"		330	.073			1.84		1.84	3.08
2985	4" x 12"		275	.087			2.21		2.21	3.70
3000	6" x 8"		275	.087			2.21		2.21	3.70
3040	6" x 10"		220	.109			2.76		2.76	4.62
3080	6" x 12"		185	.130			3.28		3.28	5.50
3120	8" x 12"		140	.171			4.34		4.34	7.25
3160	10" x 12"		110	.218			5.50		5.50	9.25
3162	Alternate pricing method		1.10	21.818	M.B.F.		550		550	925
3170	Blocking, in 16" OC wall framing, 2" x 4"	1 Clab	600	.013	L.F.		.33		.33	.55
3172	2" x 6"		400	.020			.49		.49	.83
3174	In 24" OC wall framing, 2" x 4"		600	.013			.33		.33	.55
3176	2" x 6"		400	.020			.49		.49	.83
3178	Alt method, wood blocking removal from wood framing		.40	20	M.B.F.		495		495	825
3179	Wood blocking removal from steel framing		.36	22.222	"		550		550	915
3180	Bracing, let in, 1" x 3", studs 16" OC		1050	.008	L.F.		.19		.19	.31
3181	Studs 24" OC		1080	.007			.18		.18	.31
3182	1" x 4", studs 16" OC		1050	.008			.19		.19	.31
3183	Studs 24" OC		1080	.007			.18		.18	.31
3184	1" x 6", studs 16" OC		1050	.008			.19		.19	.31
3185	Studs 24" OC		1080	.007			.18		.18	.31
3186	2" x 3", studs 16" OC		800	.010			.25		.25	.41
3187	Studs 24" OC		830	.010			.24		.24	.40
3188	2" x 4", studs 16" OC		800	.010			.25		.25	.41
3189	Studs 24" OC		830	.010			.24		.24	.40
3190	2" x 6", studs 16" OC		800	.010			.25		.25	.41
3191	Studs 24" OC		830	.010			.24		.24	.40
3192	2" x 8", studs 16" OC		800	.010			.25		.25	.41
3193	Studs 24" OC		830	.010			.24		.24	.40
3194	"T" shaped metal bracing, studs at 16" OC		1060	.008			.19		.19	.31
3195	Studs at 24" OC		1200	.007			.16		.16	.28
3196	Metal straps, studs at 16" OC		1200	.007			.16		.16	.28
3197	Studs at 24" OC		1240	.006			.16		.16	.27
3200	Columns, round, 8' to 14' tall		40	.200	Ea.		4.93		4.93	8.25
3202	Dimensional lumber sizes	2 Clab	1.10	14.545	M.B.F.		360		360	600
3250	Blocking, between joists	1 Clab	320	.025	Ea.		.62		.62	1.03
3252	Bridging, metal strap, between joists		320	.025	Pr.		.62		.62	1.03
3254	Wood, between joists		320	.025	"		.62		.62	1.03
3260	Door buck, studs, header & access., 8' high 2" x 4" wall, 3' wide		32	.250	Ea.		6.15		6.15	10.30
3261	4' wide		32	.250			6.15		6.15	10.30
3262	5' wide		32	.250			6.15		6.15	10.30

06 05 Common Work Results for Wood, Plastics, and Composites

06 05 05 – Selective Demolition for Wood, Plastics, and Composites

06 05 05.10 Selective Demolition Wood Framing

		Crew	Daily Output	Labor-Hours	Unit	Material	2016 Bare Costs Labor	2016 Bare Costs Equipment	Total	Total Incl O&P
3263	6' wide	1 Clab	32	.250	Ea.		6.15		6.15	10.30
3264	8' wide		30	.267			6.55		6.55	11
3265	10' wide		30	.267			6.55		6.55	11
3266	12' wide		30	.267			6.55		6.55	11
3267	2" x 6" wall, 3' wide		32	.250			6.15		6.15	10.30
3268	4' wide		32	.250			6.15		6.15	10.30
3269	5' wide		32	.250			6.15		6.15	10.30
3270	6' wide		32	.250			6.15		6.15	10.30
3271	8' wide		30	.267			6.55		6.55	11
3272	10' wide		30	.267			6.55		6.55	11
3273	12' wide		30	.267			6.55		6.55	11
3274	Window buck, studs, header & access, 8' high 2" x 4" wall, 2' wide		24	.333			8.20		8.20	13.75
3275	3' wide		24	.333			8.20		8.20	13.75
3276	4' wide		24	.333			8.20		8.20	13.75
3277	5' wide		24	.333			8.20		8.20	13.75
3278	6' wide		24	.333			8.20		8.20	13.75
3279	7' wide		24	.333			8.20		8.20	13.75
3280	8' wide		22	.364			8.95		8.95	15
3281	10' wide		22	.364			8.95		8.95	15
3282	12' wide		22	.364			8.95		8.95	15
3283	2" x 6" wall, 2' wide		24	.333			8.20		8.20	13.75
3284	3' wide		24	.333			8.20		8.20	13.75
3285	4' wide		24	.333			8.20		8.20	13.75
3286	5' wide		24	.333			8.20		8.20	13.75
3287	6' wide		24	.333			8.20		8.20	13.75
3288	7' wide		24	.333			8.20		8.20	13.75
3289	8' wide		22	.364			8.95		8.95	15
3290	10' wide		22	.364			8.95		8.95	15
3291	12' wide		22	.364			8.95		8.95	15
3360	Deck or porch decking		825	.010	L.F.		.24		.24	.40
3400	Fascia boards, 1" x 6"		500	.016			.39		.39	.66
3440	1" x 8"		450	.018			.44		.44	.73
3480	1" x 10"		400	.020			.49		.49	.83
3490	2" x 6"		450	.018			.44		.44	.73
3500	2" x 8"		400	.020			.49		.49	.83
3510	2" x 10"		350	.023			.56		.56	.94
3610	Furring, on wood walls or ceiling		4000	.002	S.F.		.05		.05	.08
3620	On masonry or concrete walls or ceiling		1200	.007	"		.16		.16	.28
3800	Headers over openings, 2 @ 2" x 6"		110	.073	L.F.		1.79		1.79	3
3840	2 @ 2" x 8"		100	.080			1.97		1.97	3.30
3880	2 @ 2" x 10"		90	.089			2.19		2.19	3.67
3885	Alternate pricing method		.26	30.651	M.B.F.		755		755	1,275
3920	Joists, 1" x 4"		1250	.006	L.F.		.16		.16	.26
3930	1" x 6"		1135	.007			.17		.17	.29
3940	1" x 8"		1000	.008			.20		.20	.33
3950	1" x 10"		895	.009			.22		.22	.37
3960	1" x 12"		765	.010			.26		.26	.43
4200	2" x 4"	2 Clab	1000	.016			.39		.39	.66
4230	2" x 6"		970	.016			.41		.41	.68
4240	2" x 8"		940	.017			.42		.42	.70
4250	2" x 10"		910	.018			.43		.43	.73
4280	2" x 12"		880	.018			.45		.45	.75
4281	2" x 14"		850	.019			.46		.46	.78

For customer support on your Residential Cost Data, call 877.759.4771.

06 05 Common Work Results for Wood, Plastics, and Composites

06 05 05 – Selective Demolition for Wood, Plastics, and Composites

06 05 05.10 Selective Demolition Wood Framing		Crew	Daily Output	Labor-Hours	Unit	Material	2016 Bare Costs Labor	2016 Bare Costs Equipment	Total	Total Incl O&P
4282	Composite joists, 9-1/2"	2 Clab	960	.017	L.F.		.41		.41	.69
4283	11-7/8"		930	.017			.42		.42	.71
4284	14"		897	.018			.44		.44	.74
4285	16"		865	.019			.46		.46	.76
4290	Wood joists, alternate pricing method		1.50	10.667	M.B.F.		263		263	440
4500	Open web joist, 12" deep		500	.032	L.F.		.79		.79	1.32
4505	14" deep		475	.034			.83		.83	1.39
4510	16" deep		450	.036			.88		.88	1.47
4520	18" deep		425	.038			.93		.93	1.55
4530	24" deep		400	.040			.99		.99	1.65
4550	Ledger strips, 1" x 2"	1 Clab	1200	.007			.16		.16	.28
4560	1" x 3"		1200	.007			.16		.16	.28
4570	1" x 4"		1200	.007			.16		.16	.28
4580	2" x 2"		1100	.007			.18		.18	.30
4590	2" x 4"		1000	.008			.20		.20	.33
4600	2" x 6"		1000	.008			.20		.20	.33
4601	2" x 8" or 2" x 10"		800	.010			.25		.25	.41
4602	4" x 6"		600	.013			.33		.33	.55
4604	4" x 8"		450	.018			.44		.44	.73
5400	Posts, 4" x 4"	2 Clab	800	.020			.49		.49	.83
5405	4" x 6"		550	.029			.72		.72	1.20
5410	4" x 8"		440	.036			.90		.90	1.50
5425	4" x 10"		390	.041			1.01		1.01	1.69
5430	4" x 12"		350	.046			1.13		1.13	1.89
5440	6" x 6"		400	.040			.99		.99	1.65
5445	6" x 8"		350	.046			1.13		1.13	1.89
5450	6" x 10"		320	.050			1.23		1.23	2.06
5455	6" x 12"		290	.055			1.36		1.36	2.28
5480	8" x 8"		300	.053			1.31		1.31	2.20
5500	10" x 10"		240	.067			1.64		1.64	2.75
5660	Tongue and groove floor planks		2	8	M.B.F.		197		197	330
5682	Rafters, ordinary, 16" OC, 2" x 4"		880	.018	S.F.		.45		.45	.75
5683	2" x 6"		840	.019			.47		.47	.79
5684	2" x 8"		820	.020			.48		.48	.80
5685	2" x 10"		820	.020			.48		.48	.80
5686	2" x 12"		810	.020			.49		.49	.81
5687	24" OC, 2" x 4"		1170	.014			.34		.34	.56
5688	2" x 6"		1117	.014			.35		.35	.59
5689	2" x 8"		1091	.015			.36		.36	.61
5690	2" x 10"		1091	.015			.36		.36	.61
5691	2" x 12"		1077	.015			.37		.37	.61
5795	Rafters, ordinary, 2" x 4" (alternate method)		862	.019	L.F.		.46		.46	.77
5800	2" x 6" (alternate method)		850	.019			.46		.46	.78
5840	2" x 8" (alternate method)		837	.019			.47		.47	.79
5855	2" x 10" (alternate method)		825	.019			.48		.48	.80
5865	2" x 12" (alternate method)		812	.020			.49		.49	.81
5870	Sill plate, 2" x 4"	1 Clab	1170	.007			.17		.17	.28
5871	2" x 6"		780	.010			.25		.25	.42
5872	2" x 8"		586	.014			.34		.34	.56
5873	Alternate pricing method		.78	10.256	M.B.F.		253		253	425
5885	Ridge board, 1" x 4"	2 Clab	900	.018	L.F.		.44		.44	.73
5886	1" x 6"		875	.018			.45		.45	.75
5887	1" x 8"		850	.019			.46		.46	.78

06 05 Common Work Results for Wood, Plastics, and Composites

06 05 05 – Selective Demolition for Wood, Plastics, and Composites

06 05 05.10 Selective Demolition Wood Framing		Crew	Daily Output	Labor-Hours	Unit	Material	2016 Bare Costs Labor	Equipment	Total	Total Incl O&P
5888	1" x 10"	2 Clab	825	.019	L.F.		.48		.48	.80
5889	1" x 12"		800	.020			.49		.49	.83
5890	2" x 4"		900	.018			.44		.44	.73
5892	2" x 6"		875	.018			.45		.45	.75
5894	2" x 8"		850	.019			.46		.46	.78
5896	2" x 10"		825	.019			.48		.48	.80
5898	2" x 12"		800	.020			.49		.49	.83
6050	Rafter tie, 1" x 4"		1250	.013			.32		.32	.53
6052	1" x 6"		1135	.014			.35		.35	.58
6054	2" x 4"		1000	.016			.39		.39	.66
6056	2" x 6"		970	.016			.41		.41	.68
6070	Sleepers, on concrete, 1" x 2"	1 Clab	4700	.002			.04		.04	.07
6075	1" x 3"		4000	.002			.05		.05	.08
6080	2" x 4"		3000	.003			.07		.07	.11
6085	2" x 6"		2600	.003			.08		.08	.13
6086	Sheathing from roof, 5/16"	2 Clab	1600	.010	S.F.		.25		.25	.41
6088	3/8"		1525	.010			.26		.26	.43
6090	1/2"		1400	.011			.28		.28	.47
6092	5/8"		1300	.012			.30		.30	.51
6094	3/4"		1200	.013			.33		.33	.55
6096	Board sheathing from roof		1400	.011			.28		.28	.47
6100	Sheathing, from walls, 1/4"		1200	.013			.33		.33	.55
6110	5/16"		1175	.014			.34		.34	.56
6120	3/8"		1150	.014			.34		.34	.57
6130	1/2"		1125	.014			.35		.35	.59
6140	5/8"		1100	.015			.36		.36	.60
6150	3/4"		1075	.015			.37		.37	.61
6152	Board sheathing from walls		1500	.011			.26		.26	.44
6158	Subfloor/roof deck, with boards		2200	.007			.18		.18	.30
6159	Subfloor/roof deck, with tongue & groove boards		2000	.008			.20		.20	.33
6160	Plywood, 1/2" thick		768	.021			.51		.51	.86
6162	5/8" thick		760	.021			.52		.52	.87
6164	3/4" thick		750	.021			.53		.53	.88
6165	1-1/8" thick		720	.022			.55		.55	.92
6166	Underlayment, particle board, 3/8" thick	1 Clab	780	.010			.25		.25	.42
6168	1/2" thick		768	.010			.26		.26	.43
6170	5/8" thick		760	.011			.26		.26	.43
6172	3/4" thick		750	.011			.26		.26	.44
6200	Stairs and stringers, straight run	2 Clab	40	.400	Riser		9.85		9.85	16.50
6240	With platforms, winders or curves	"	26	.615	"		15.15		15.15	25.50
6300	Components, tread	1 Clab	110	.073	Ea.		1.79		1.79	3
6320	Riser		80	.100	"		2.47		2.47	4.13
6390	Stringer, 2" x 10"		260	.031	L.F.		.76		.76	1.27
6400	2" x 12"		260	.031			.76		.76	1.27
6410	3" x 10"		250	.032			.79		.79	1.32
6420	3" x 12"		250	.032			.79		.79	1.32
6590	Wood studs, 2" x 3"	2 Clab	3076	.005			.13		.13	.21
6600	2" x 4"		2000	.008			.20		.20	.33
6640	2" x 6"		1600	.010			.25		.25	.41
6720	Wall framing, including studs, plates and blocking, 2" x 4"	1 Clab	600	.013	S.F.		.33		.33	.55
6740	2" x 6"		480	.017	"		.41		.41	.69
6750	Headers, 2" x 4"		1125	.007	L.F.		.18		.18	.29
6755	2" x 6"		1125	.007			.18		.18	.29

06 05 Common Work Results for Wood, Plastics, and Composites

06 05 05 – Selective Demolition for Wood, Plastics, and Composites

06 05 05.10 Selective Demolition Wood Framing

		Crew	Daily Output	Labor-Hours	Unit	Material	2016 Bare Costs Labor	Equipment	Total	Total Incl O&P
6760	2" x 8"	1 Clab	1050	.008	L.F.		.19		.19	.31
6765	2" x 10"		1050	.008			.19		.19	.31
6770	2" x 12"		1000	.008			.20		.20	.33
6780	4" x 10"		525	.015			.38		.38	.63
6785	4" x 12"		500	.016			.39		.39	.66
6790	6" x 8"		560	.014			.35		.35	.59
6795	6" x 10"		525	.015			.38		.38	.63
6797	6" x 12"		500	.016			.39		.39	.66
7000	Trusses									
7050	12' span	2 Clab	74	.216	Ea.		5.35		5.35	8.90
7150	24' span	F-3	66	.606			18.60	10	28.60	42
7200	26' span		64	.625			19.15	10.30	29.45	43.50
7250	28' span		62	.645			19.80	10.60	30.40	44.50
7300	30' span		58	.690			21	11.35	32.35	48
7350	32' span		56	.714			22	11.75	33.75	49.50
7400	34' span		54	.741			22.50	12.20	34.70	51.50
7450	36' span		52	.769			23.50	12.65	36.15	53.50
8000	Soffit, T & G wood	1 Clab	520	.015	S.F.		.38		.38	.63
8010	Hardboard, vinyl or aluminum	"	640	.013			.31		.31	.52
8030	Plywood	2 Carp	315	.051			1.72		1.72	2.88
9500	See Section 02 41 19.19 for rubbish handling									

06 05 05.20 Selective Demolition Millwork and Trim

		Crew	Daily Output	Labor-Hours	Unit	Material	Labor	Equipment	Total	Total Incl O&P
0010	**SELECTIVE DEMOLITION MILLWORK AND TRIM** R024119-10									
1000	Cabinets, wood, base cabinets, per L.F.	2 Clab	80	.200	L.F.		4.93		4.93	8.25
1020	Wall cabinets, per L.F.	"	80	.200	"		4.93		4.93	8.25
1060	Remove and reset, base cabinets	2 Carp	18	.889	Ea.		30		30	50.50
1070	Wall cabinets		20	.800			27		27	45.50
1072	Oven cabinet, 7' high		11	1.455			49.50		49.50	82.50
1074	Cabinet door, up to 2' high	1 Clab	66	.121			2.99		2.99	5
1076	2' - 4' high	"	46	.174			4.29		4.29	7.15
1100	Steel, painted, base cabinets	2 Clab	60	.267	L.F.		6.55		6.55	11
1120	Wall cabinets		60	.267	"		6.55		6.55	11
1200	Casework, large area		320	.050	S.F.		1.23		1.23	2.06
1220	Selective		200	.080	"		1.97		1.97	3.30
1500	Counter top, straight runs		200	.080	L.F.		1.97		1.97	3.30
1510	L, U or C shapes		120	.133	"		3.29		3.29	5.50
2000	Paneling, 4' x 8' sheets		2000	.008	S.F.		.20		.20	.33
2100	Boards, 1" x 4"		700	.023			.56		.56	.94
2120	1" x 6"		750	.021			.53		.53	.88
2140	1" x 8"		800	.020			.49		.49	.83
3000	Trim, baseboard, to 6" wide		1200	.013	L.F.		.33		.33	.55
3040	Greater than 6" and up to 12" wide		1000	.016			.39		.39	.66
3080	Remove and reset, minimum	2 Carp	400	.040			1.36		1.36	2.27
3090	Maximum	"	300	.053			1.81		1.81	3.03
3100	Ceiling trim	2 Clab	1000	.016			.39		.39	.66
3120	Chair rail		1200	.013			.33		.33	.55
3140	Railings with balusters		240	.067			1.64		1.64	2.75
3160	Wainscoting		700	.023	S.F.		.56		.56	.94
4000	Curtain rod	1 Clab	80	.100	L.F.		2.47		2.47	4.13

06 05 Common Work Results for Wood, Plastics, and Composites

06 05 23 – Wood, Plastic, and Composite Fastenings

06 05 23.10 Nails		Crew	Daily Output	Labor-Hours	Unit	Material	2016 Bare Costs Labor	Equipment	Total	Total Incl O&P
0010	**NAILS**, material only, based upon 50# box purchase									
0020	Copper nails, plain				Lb.	10.10			10.10	11.15
0400	Stainless steel, plain					9.05			9.05	10
0500	Box, 3d to 20d, bright					1.40			1.40	1.54
0520	Galvanized					1.70			1.70	1.87
0600	Common, 3d to 60d, plain					1.37			1.37	1.51
0700	Galvanized					2			2	2.20
0800	Aluminum					10.90			10.90	11.95
1000	Annular or spiral thread, 4d to 60d, plain					2.58			2.58	2.84
1200	Galvanized					3.10			3.10	3.41
1400	Drywall nails, plain					1.40			1.40	1.54
1600	Galvanized					2.10			2.10	2.31
1800	Finish nails, 4d to 10d, plain					1.20			1.20	1.32
2000	Galvanized					1.83			1.83	2.01
2100	Aluminum					5.80			5.80	6.40
2300	Flooring nails, hardened steel, 2d to 10d, plain					3.33			3.33	3.66
2400	Galvanized					4			4	4.40
2500	Gypsum lath nails, 1-1/8", 13 ga. flathead, blued					2.05			2.05	2.26
2600	Masonry nails, hardened steel, 3/4" to 3" long, plain					1.70			1.70	1.87
2700	Galvanized					3.25			3.25	3.58
2900	Roofing nails, threaded, galvanized					1.75			1.75	1.93
3100	Aluminum					4.80			4.80	5.30
3300	Compressed lead head, threaded, galvanized					2.45			2.45	2.70
3600	Siding nails, plain shank, galvanized					2			2	2.20
3800	Aluminum					5.95			5.95	6.55
5000	Add to prices above for cement coating					.12			.12	.13
5200	Zinc or tin plating					.15			.15	.17
5500	Vinyl coated sinkers, 8d to 16d					2.40			2.40	2.64

06 05 23.50 Wood Screws		Crew	Daily Output	Labor-Hours	Unit	Material	Labor	Equipment	Total	Total Incl O&P
0010	**WOOD SCREWS**									
0020	#8 x 1" long, steel				C	2.75			2.75	3.03
0100	Brass					12.25			12.25	13.50
0200	#8, 2" long, steel					4.25			4.25	4.68
0300	Brass					22.50			22.50	24.50
0400	#10, 1" long, steel					3.45			3.45	3.80
0500	Brass					15.60			15.60	17.15
0600	#10, 2" long, steel					5.25			5.25	5.80
0700	Brass					27			27	29.50
0800	#10, 3" long, steel					8.90			8.90	9.80
1000	#12, 2" long, steel					7.35			7.35	8.10
1100	Brass					35			35	38
1500	#12, 3" long, steel					11.20			11.20	12.30
2000	#12, 4" long, steel					19.20			19.20	21

06 05 23.60 Timber Connectors		Crew	Daily Output	Labor-Hours	Unit	Material	Labor	Equipment	Total	Total Incl O&P
0010	**TIMBER CONNECTORS**									
0020	Add up cost of each part for total cost of connection									
0100	Connector plates, steel, with bolts, straight	2 Carp	75	.213	Ea.	30.50	7.25		37.75	45.50
0110	Tee, 7 ga.		50	.320		35	10.85		45.85	56.50
0120	T- Strap, 14 ga., 12" x 8" x 2"		50	.320		35	10.85		45.85	56.50
0150	Anchor plates, 7 ga., 9" x 7"		75	.213		30.50	7.25		37.75	45.50
0200	Bolts, machine, sq. hd. with nut & washer, 1/2" diameter, 4" long	1 Carp	140	.057		.75	1.94		2.69	4.07
0300	7-1/2" long		130	.062		1.37	2.09		3.46	5

06 05 Common Work Results for Wood, Plastics, and Composites

06 05 23 – Wood, Plastic, and Composite Fastenings

06 05 23.60 Timber Connectors		Crew	Daily Output	Labor-Hours	Unit	Material	2016 Bare Costs Labor	Equipment	Total	Total Incl O&P
0500	3/4" diameter, 7-1/2" long	1 Carp	130	.062	Ea.	3.20	2.09		5.29	7
0610	Machine bolts, w/nut, washer, 3/4" diam., 15" L, HD's & beam hangers		95	.084		5.95	2.85		8.80	11.35
0720	Machine bolts, sq. hd. w/nut & wash		150	.053	Lb.	3.44	1.81		5.25	6.80
0800	Drilling bolt holes in timber, 1/2" diameter		450	.018	Inch		.60		.60	1.01
0900	1" diameter		350	.023	"		.78		.78	1.30
1100	Framing anchor, angle, 3" x 3" x 1-1/2", 12 ga		175	.046	Ea.	2.56	1.55		4.11	5.40
1150	Framing anchors, 18 ga., 4-1/2" x 2-3/4"		175	.046		2.56	1.55		4.11	5.40
1160	Framing anchors, 18 ga., 4-1/2" x 3"		175	.046		2.56	1.55		4.11	5.40
1170	Clip anchors plates, 18 ga., 12" x 1-1/8"		175	.046		2.56	1.55		4.11	5.40
1250	Holdowns, 3 ga. base, 10 ga. body		8	1		25	34		59	84.50
1260	Holdowns, 7 ga. 11-1/16" x 3-1/4"		8	1		25	34		59	84.50
1270	Holdowns, 7 ga. 14-3/8" x 3-1/8"		8	1		25	34		59	84.50
1275	Holdowns, 12 ga. 8" x 2-1/2"		8	1		25	34		59	84.50
1300	Joist and beam hangers, 18 ga. galv., for 2" x 4" joist		175	.046		.72	1.55		2.27	3.38
1400	2" x 6" to 2" x 10" joist		165	.048		1.36	1.64		3	4.25
1600	16 ga. galv., 3" x 6" to 3" x 10" joist		160	.050		2.92	1.70		4.62	6.05
1700	3" x 10" to 3" x 14" joist		160	.050		4.88	1.70		6.58	8.20
1800	4" x 6" to 4" x 10" joist		155	.052		3.04	1.75		4.79	6.25
1900	4" x 10" to 4" x 14" joist		155	.052		4.96	1.75		6.71	8.40
2000	Two-2" x 6" to two-2" x 10" joists		150	.053		4.16	1.81		5.97	7.60
2100	Two-2" x 10" to two-2" x 14" joists		150	.053		4.64	1.81		6.45	8.15
2300	3/16" thick, 6" x 8" joist		145	.055		65	1.87		66.87	74.50
2400	6" x 10" joist		140	.057		67.50	1.94		69.44	77.50
2500	6" x 12" joist		135	.059		70.50	2.01		72.51	81
2700	1/4" thick, 6" x 14" joist		130	.062		73	2.09		75.09	84
2900	Plywood clips, extruded aluminum H clip, for 3/4" panels					.22			.22	.24
3000	Galvanized 18 ga. back-up clip					.20			.20	.22
3200	Post framing, 16 ga. galv. for 4" x 4" base, 2 piece	1 Carp	130	.062		16.35	2.09		18.44	21.50
3300	Cap		130	.062		22.50	2.09		24.59	28.50
3500	Rafter anchors, 18 ga. galv., 1-1/2" wide, 5-1/4" long		145	.055		.48	1.87		2.35	3.66
3600	10-3/4" long		145	.055		1.44	1.87		3.31	4.71
3800	Shear plates, 2-5/8" diameter		120	.067		2.44	2.26		4.70	6.45
3900	4" diameter		115	.070		5.70	2.36		8.06	10.25
4000	Sill anchors, embedded in concrete or block, 25-1/2" long		115	.070		12.70	2.36		15.06	17.95
4100	Spike grids, 3" x 6"		120	.067		.94	2.26		3.20	4.81
4400	Split rings, 2-1/2" diameter		120	.067		2.02	2.26		4.28	6
4500	4" diameter		110	.073		2.98	2.47		5.45	7.40
4550	Tie plate, 20 ga., 7" x 3 1/8"		110	.073		2.98	2.47		5.45	7.40
4560	Tie plate, 20 ga., 5" x 4 1/8"		110	.073		2.98	2.47		5.45	7.40
4575	Twist straps, 18 ga., 12" x 1 1/4"		110	.073		2.98	2.47		5.45	7.40
4580	Twist straps, 18 ga., 16" x 1 1/4"		110	.073		2.98	2.47		5.45	7.40
4600	Strap ties, 20 ga., 2-1/16" wide, 12 13/16" long		180	.044		.93	1.51		2.44	3.54
4700	Strap ties, 16 ga., 1-3/8" wide, 12" long		180	.044		.93	1.51		2.44	3.54
4800	21-5/8" x 1-1/4"		160	.050		2.94	1.70		4.64	6.05
5000	Toothed rings, 2-5/8" or 4" diameter		90	.089		1.79	3.01		4.80	7
5200	Truss plates, nailed, 20 ga., up to 32' span		17	.471	Truss	12.85	15.95		28.80	40.50
5400	Washers, 2" x 2" x 1/8"				Ea.	.38			.38	.42
5500	3" x 3" x 3/16"				"	1.04			1.04	1.14
6000	Angles and gussets, painted									
6012	7 ga., 3-1/4" x 3-1/4" x 2-1/2" long	1 Carp	1.90	4.211	C	1,075	143		1,218	1,450
6014	3-1/4" x 3-1/4" x 5" long		1.90	4.211		2,125	143		2,268	2,575
6016	3-1/4" x 3-1/4" x 7-1/2" long		1.85	4.324		4,000	147		4,147	4,650
6018	5-3/4" x 5-3/4" x 2-1/2" long		1.85	4.324		2,550	147		2,697	3,075

06 05 Common Work Results for Wood, Plastics, and Composites

06 05 23 – Wood, Plastic, and Composite Fastenings

06 05 23.60 Timber Connectors		Crew	Daily Output	Labor-Hours	Unit	Material	2016 Bare Costs Labor	Equipment	Total	Total Incl O&P
6020	5-3/4" x 5-3/4" x 5" long	1 Carp	1.85	4.324	C	4,150	147		4,297	4,825
6022	5-3/4" x 5-3/4" x 7-1/2" long		1.80	4.444		6,150	151		6,301	7,025
6024	3 ga., 4-1/4" x 4-1/4" x 3" long		1.85	4.324		2,750	147		2,897	3,275
6026	4-1/4" x 4-1/4" x 6" long		1.85	4.324		6,000	147		6,147	6,850
6028	4-1/4" x 4-1/4" x 9" long		1.80	4.444		6,725	151		6,876	7,650
6030	7-1/4" x 7-1/4" x 3" long		1.80	4.444		4,800	151		4,951	5,525
6032	7-1/4" x 7-1/4" x 6" long		1.80	4.444		6,475	151		6,626	7,375
6034	7-1/4" x 7-1/4" x 9" long		1.75	4.571		14,600	155		14,755	16,300
6036	Gussets									
6038	7 ga., 8-1/8" x 8-1/8" x 2-3/4" long	1 Carp	1.80	4.444	C	4,625	151		4,776	5,325
6040	3 ga., 9-3/4" x 9-3/4" x 3-1/4" long	"	1.80	4.444	"	6,400	151		6,551	7,300
6101	Beam hangers, polymer painted									
6102	Bolted, 3 ga., (W x H x L)									
6104	3-1/4" x 9" x 12" top flange	1 Carp	1	8	C	19,600	271		19,871	22,100
6106	5-1/4" x 9" x 12" top flange		1	8		20,400	271		20,671	23,000
6108	5-1/4" x 11" x 11-3/4" top flange		1	8		23,200	271		23,471	26,000
6110	6-7/8" x 9" x 12" top flange		1	8		21,200	271		21,471	23,800
6112	6-7/8" x 11" x 13-1/2" top flange		1	8		24,400	271		24,671	27,300
6114	8-7/8" x 11" x 15-1/2" top flange		1	8		26,100	271		26,371	29,200
6116	Nailed, 3 ga., (W x H x L)									
6118	3-1/4" x 10-1/2" x 10" top flange	1 Carp	1.80	4.444	C	20,400	151		20,551	22,700
6120	3-1/4" x 10-1/2" x 12" top flange		1.80	4.444		20,400	151		20,551	22,700
6122	5-1/4" x 9-1/2" x 10" top flange		1.80	4.444		20,900	151		21,051	23,300
6124	5-1/4" x 9-1/2" x 12" top flange		1.80	4.444		20,900	151		21,051	23,300
6128	6-7/8" x 8-1/2" x 12" top flange		1.80	4.444		21,400	151		21,551	23,800
6134	Saddle hangers, glu-lam (W x H x L)									
6136	3-1/4" x 10-1/2" x 5-1/4" x 6" saddle	1 Carp	.50	16	C	15,300	540		15,840	17,700
6138	3-1/4" x 10-1/2" x 6-7/8" x 6" saddle		.50	16		16,100	540		16,640	18,600
6140	3-1/4" x 10-1/2" x 8-7/8" x 6" saddle		.50	16		16,900	540		17,440	19,500
6142	3-1/4" x 19-1/2" x 5-1/4" x 10-1/8" saddle		.40	20		15,300	680		15,980	17,900
6144	3-1/4" x 19-1/2" x 6-7/8" x 10-1/8" saddle		.40	20		16,100	680		16,780	18,800
6146	3-1/4" x 19-1/2" x 8-7/8" x 10-1/8" saddle		.40	20		16,900	680		17,580	19,700
6148	5-1/4" x 9-1/2" x 5-1/4" x 12" saddle		.50	16		18,200	540		18,740	21,000
6150	5-1/4" x 9-1/2" x 6-7/8" x 9" saddle		.50	16		20,000	540		20,540	22,900
6152	5-1/4" x 10-1/2" x spec x 12" saddle		.50	16		21,800	540		22,340	24,800
6154	5-1/4" x 18" x 5-1/4" x 12-1/8" saddle		.40	20		18,200	680		18,880	21,200
6156	5-1/4" x 18" x 6-7/8" x 12-1/8" saddle		.40	20		20,000	680		20,680	23,100
6158	5-1/4" x 18" x spec x 12-1/8" saddle		.40	20		21,800	680		22,480	25,000
6160	6-7/8" x 8-1/2" x 6-7/8" x 12" saddle		.50	16		21,800	540		22,340	24,800
6162	6-7/8" x 8-1/2" x 8-7/8" x 12" saddle		.50	16		22,600	540		23,140	25,700
6164	6-7/8" x 10-1/2" x spec x 12" saddle		.50	16		21,800	540		22,340	24,800
6166	6-7/8" x 18" x 6-7/8" x 13-3/4" saddle		.40	20		21,800	680		22,480	25,000
6168	6-7/8" x 18" x 8-7/8" x 13-3/4" saddle		.40	20		22,600	680		23,280	25,900
6170	6-7/8" x 18" x spec x 13-3/4" saddle		.40	20		24,300	680		24,980	27,900
6172	8-7/8" x 18" x spec x 15-3/4" saddle		.40	20		38,000	680		38,680	42,900
6201	Beam and purlin hangers, galvanized, 12 ga.									
6202	Purlin or joist size, 3" x 8"	1 Carp	1.70	4.706	C	2,025	160		2,185	2,500
6204	3" x 10"		1.70	4.706		2,175	160		2,335	2,675
6206	3" x 12"		1.65	4.848		2,500	164		2,664	3,025
6208	3" x 14"		1.65	4.848		2,675	164		2,839	3,200
6210	3" x 16"		1.65	4.848		2,825	164		2,989	3,375
6212	4" x 8"		1.65	4.848		2,025	164		2,189	2,500
6214	4" x 10"		1.65	4.848		2,200	164		2,364	2,675

For customer support on your Residential Cost Data, call 877.759.4771.

06 05 Common Work Results for Wood, Plastics, and Composites

06 05 23 – Wood, Plastic, and Composite Fastenings

06 05 23.60 Timber Connectors		Crew	Daily Output	Labor-Hours	Unit	Material	2016 Bare Costs Labor	Equipment	Total	Total Incl O&P
6216	4" x 12"	1 Carp	1.60	5	C	2,600	170		2,770	3,125
6218	4" x 14"		1.60	5		2,750	170		2,920	3,300
6220	4" x 16"		1.60	5		2,925	170		3,095	3,475
6222	6" x 8"		1.60	5		2,625	170		2,795	3,150
6224	6" x 10"		1.55	5.161		2,675	175		2,850	3,250
6226	6" x 12"		1.55	5.161		4,575	175		4,750	5,325
6228	6" x 14"		1.50	5.333		4,850	181		5,031	5,625
6230	6" x 16"		1.50	5.333		5,125	181		5,306	5,950
6250	Beam seats									
6252	Beam size, 5-1/4" wide									
6254	5" x 7" x 1/4"	1 Carp	1.80	4.444	C	7,700	151		7,851	8,725
6256	6" x 7" x 3/8"		1.80	4.444		8,650	151		8,801	9,750
6258	7" x 7" x 3/8"		1.80	4.444		9,250	151		9,401	10,500
6260	8" x 7" x 3/8"		1.80	4.444		10,900	151		11,051	12,300
6262	Beam size, 6-7/8" wide									
6264	5" x 9" x 1/4"	1 Carp	1.80	4.444	C	9,200	151		9,351	10,400
6266	6" x 9" x 3/8"		1.80	4.444		12,000	151		12,151	13,500
6268	7" x 9" x 3/8"		1.80	4.444		12,100	151		12,251	13,600
6270	8" x 9" x 3/8"		1.80	4.444		14,400	151		14,551	16,100
6272	Special beams, over 6-7/8" wide									
6274	5" x 10" x 3/8"	1 Carp	1.80	4.444	C	12,400	151		12,551	13,900
6276	6" x 10" x 3/8"		1.80	4.444		14,500	151		14,651	16,300
6278	7" x 10" x 3/8"		1.80	4.444		15,200	151		15,351	17,000
6280	8" x 10" x 3/8"		1.75	4.571		16,300	155		16,455	18,200
6282	5-1/4" x 12" x 5/16"		1.75	4.571		12,600	155		12,755	14,200
6284	6-1/2" x 12" x 3/8"		1.75	4.571		20,900	155		21,055	23,300
6286	5-1/4" x 16" x 5/16"		1.70	4.706		18,600	160		18,760	20,700
6288	6-1/2" x 16" x 3/8"		1.70	4.706		24,200	160		24,360	27,000
6290	5-1/4" x 20" x 5/16"		1.70	4.706		21,800	160		21,960	24,300
6292	6-1/2" x 20" x 3/8"		1.65	4.848		28,500	164		28,664	31,600
6300	Column bases									
6302	4 x 4, 16 ga.	1 Carp	1.80	4.444	C	775	151		926	1,100
6306	7 ga.		1.80	4.444		2,850	151		3,001	3,375
6308	4 x 6, 16 ga.		1.80	4.444		1,800	151		1,951	2,250
6312	7 ga.		1.80	4.444		2,975	151		3,126	3,525
6314	6 x 6, 16 ga.		1.75	4.571		2,050	155		2,205	2,500
6318	7 ga.		1.75	4.571		4,050	155		4,205	4,700
6320	6 x 8, 7 ga.		1.70	4.706		3,150	160		3,310	3,750
6322	6 x 10, 7 ga.		1.70	4.706		3,400	160		3,560	4,000
6324	6 x 12, 7 ga.		1.70	4.706		3,675	160		3,835	4,325
6326	8 x 8, 7 ga.		1.65	4.848		6,200	164		6,364	7,100
6330	8 x 10, 7 ga.		1.65	4.848		7,425	164		7,589	8,450
6332	8 x 12, 7 ga.		1.60	5		8,075	170		8,245	9,150
6334	10 x 10, 3 ga.		1.60	5		8,225	170		8,395	9,325
6336	10 x 12, 3 ga.		1.60	5		9,475	170		9,645	10,700
6338	12 x 12, 3 ga.		1.55	5.161		10,300	175		10,475	11,600
6350	Column caps, painted, 3 ga.									
6352	3-1/4" x 3-5/8"	1 Carp	1.80	4.444	C	10,400	151		10,551	11,800
6354	3-1/4" x 5-1/2"		1.80	4.444		10,400	151		10,551	11,800
6356	3-5/8" x 3-5/8"		1.80	4.444		8,550	151		8,701	9,650
6358	3-5/8" x 5-1/2"		1.80	4.444		8,550	151		8,701	9,650
6360	5-1/4" x 5-1/2"		1.75	4.571		11,200	155		11,355	12,600
6362	5-1/4" x 7-1/2"		1.75	4.571		11,200	155		11,355	12,600

06 05 Common Work Results for Wood, Plastics, and Composites

06 05 23 – Wood, Plastic, and Composite Fastenings

06 05 23.60 Timber Connectors

		Crew	Daily Output	Labor-Hours	Unit	Material	2016 Bare Costs Labor	Equipment	Total	Total Incl O&P
6364	5-1/2" x 3-5/8"	1 Carp	1.75	4.571	C	12,000	155		12,155	13,600
6366	5-1/2" x 5-1/2"		1.75	4.571		12,000	155		12,155	13,600
6368	5-1/2" x 7-1/2"		1.70	4.706		12,000	160		12,160	13,600
6370	6-7/8" x 5-1/2"		1.70	4.706		12,600	160		12,760	14,100
6372	6-7/8" x 6-7/8"		1.70	4.706		12,600	160		12,760	14,100
6374	6-7/8" x 7-1/2"		1.70	4.706		12,600	160		12,760	14,100
6376	7-1/2" x 5-1/2"		1.65	4.848		13,100	164		13,264	14,700
6378	7-1/2" x 7-1/2"		1.65	4.848		13,100	164		13,264	14,700
6380	8-7/8" x 5-1/2"		1.60	5		13,900	170		14,070	15,600
6382	8-7/8" x 7-1/2"		1.60	5		13,900	170		14,070	15,600
6384	9-1/2" x 5-1/2"		1.60	5		18,800	170		18,970	20,900
6400	Floor tie anchors, polymer paint									
6402	10 ga., 3" x 37-1/2"	1 Carp	1.80	4.444	C	5,000	151		5,151	5,750
6404	3-1/2" x 45-1/2"		1.75	4.571		5,250	155		5,405	6,025
6406	3 ga., 3-1/2" x 56"		1.70	4.706		9,225	160		9,385	10,500
6410	Girder hangers									
6412	6" wall thickness, 4" x 6"	1 Carp	1.80	4.444	C	2,800	151		2,951	3,350
6414	4" x 8"		1.80	4.444		3,125	151		3,276	3,700
6416	8" wall thickness, 4" x 6"		1.80	4.444		3,250	151		3,401	3,825
6418	4" x 8"		1.80	4.444		3,250	151		3,401	3,825
6420	Hinge connections, polymer painted									
6422	3/4" thick top plate									
6424	5-1/4" x 12" w/5" x 5" top	1 Carp	1	8	C	35,800	271		36,071	39,900
6426	5-1/4" x 15" w/6" x 6" top		.80	10		38,100	340		38,440	42,500
6428	5-1/4" x 18" w/7" x 7" top		.70	11.429		40,100	385		40,485	44,800
6430	5-1/4" x 26" w/9" x 9" top		.60	13.333		42,600	450		43,050	47,700
6432	1" thick top plate									
6434	6-7/8" x 14" w/5" x 5" top	1 Carp	.80	10	C	43,700	340		44,040	48,600
6436	6-7/8" x 17" w/6" x 6" top		.80	10		48,600	340		48,940	54,000
6438	6-7/8" x 21" w/7" x 7" top		.70	11.429		53,000	385		53,385	59,000
6440	6-7/8" x 31" w/9" x 9" top		.60	13.333		58,000	450		58,450	64,500
6442	1-1/4" thick top plate									
6444	8-7/8" x 16" w/5" x 5" top	1 Carp	.60	13.333	C	54,500	450		54,950	61,000
6446	8-7/8" x 21" w/6" x 6" top		.50	16		60,000	540		60,540	67,000
6448	8-7/8" x 26" w/7" x 7" top		.40	20		68,000	680		68,680	75,500
6450	8-7/8" x 39" w/9" x 9" top		.30	26.667		84,500	905		85,405	94,500
6460	Holddowns									
6462	Embedded along edge									
6464	26" long, 12 ga.	1 Carp	.90	8.889	C	1,375	300		1,675	2,000
6466	35" long, 12 ga.		.85	9.412		1,800	320		2,120	2,525
6468	35" long, 10 ga.		.85	9.412		1,425	320		1,745	2,100
6470	Embedded away from edge									
6472	Medium duty, 12 ga.									
6474	18-1/2" long	1 Carp	.95	8.421	C	825	285		1,110	1,375
6476	23-3/4" long		.90	8.889		960	300		1,260	1,550
6478	28" long		.85	9.412		980	320		1,300	1,600
6480	35" long		.85	9.412		1,350	320		1,670	2,000
6482	Heavy duty, 10 ga.									
6484	28" long	1 Carp	.85	9.412	C	1,750	320		2,070	2,450
6486	35" long	"	.85	9.412	"	1,900	320		2,220	2,625
6490	Surface mounted (W x H)									
6492	2-1/2" x 5-3/4", 7 ga.	1 Carp	1	8	C	2,000	271		2,271	2,650
6494	2-1/2" x 8", 12 ga.		1	8		1,225	271		1,496	1,800

For customer support on your Residential Cost Data, call 877.759.4771.

06 05 Common Work Results for Wood, Plastics, and Composites

06 05 23 – Wood, Plastic, and Composite Fastenings

06 05 23.60 Timber Connectors		Crew	Daily Output	Labor-Hours	Unit	Material	2016 Bare Costs Labor	Equipment	Total	Total Incl O&P
6496	2-7/8" x 6-3/8", 7 ga.	1 Carp	1	8	C	4,425	271		4,696	5,325
6498	2-7/8" x 12-1/2", 3 ga.		1	8		4,550	271		4,821	5,450
6500	3-3/16" x 9-3/8", 10 ga.		1	8		3,025	271		3,296	3,775
6502	3-1/2" x 11-5/8", 3 ga.		1	8		5,525	271		5,796	6,525
6504	3-1/2" x 14-3/4", 3 ga.		1	8		7,000	271		7,271	8,150
6506	3-1/2" x 16-1/2", 3 ga.		1	8		8,450	271		8,721	9,725
6508	3-1/2" x 20-1/2", 3 ga.		.90	8.889		8,650	300		8,950	10,000
6510	3-1/2" x 24-1/2", 3 ga.		.90	8.889		10,900	300		11,200	12,500
6512	4-1/4" x 20-3/4", 3 ga.		.90	8.889		7,450	300		7,750	8,675
6520	Joist hangers									
6522	Sloped, field adjustable, 18 ga.									
6524	2" x 6"	1 Carp	1.65	4.848	C	545	164		709	875
6526	2" x 8"		1.65	4.848		975	164		1,139	1,350
6528	2" x 10" and up		1.65	4.848		1,625	164		1,789	2,075
6530	3" x 10" and up		1.60	5		1,225	170		1,395	1,625
6532	4" x 10" and up		1.55	5.161		1,475	175		1,650	1,925
6536	Skewed 45°, 16 ga.									
6538	2" x 4"	1 Carp	1.75	4.571	C	870	155		1,025	1,225
6540	2" x 6" or 2" x 8"		1.65	4.848		880	164		1,044	1,250
6542	2" x 10" or 2" x 12"		1.65	4.848		1,025	164		1,189	1,400
6544	2" x 14" or 2" x 16"		1.60	5		1,825	170		1,995	2,275
6546	(2) 2" x 6" or (2) 2" x 8"		1.60	5		1,650	170		1,820	2,075
6548	(2) 2" x 10" or (2) 2" x 12"		1.55	5.161		1,750	175		1,925	2,225
6550	(2) 2" x 14" or (2) 2" x 16"		1.50	5.333		2,775	181		2,956	3,350
6552	4" x 6" or 4" x 8"		1.60	5		1,375	170		1,545	1,775
6554	4" x 10" or 4" x 12"		1.55	5.161		1,600	175		1,775	2,075
6556	4" x 14" or 4" x 16"		1.55	5.161		2,475	175		2,650	3,025
6560	Skewed 45°, 14 ga.									
6562	(2) 2" x 6" or (2) 2" x 8"	1 Carp	1.60	5	C	1,900	170		2,070	2,350
6564	(2) 2" x 10" or (2) 2" x 12"		1.55	5.161		2,625	175		2,800	3,175
6566	(2) 2" x 14" or (2) 2" x 16"		1.50	5.333		3,850	181		4,031	4,525
6568	4" x 6" or 4" x 8"		1.60	5		2,250	170		2,420	2,750
6570	4" x 10" or 4" x 12"		1.55	5.161		2,375	175		2,550	2,900
6572	4" x 14" or 4" x 16"		1.55	5.161		3,175	175		3,350	3,800
6590	Joist hangers, heavy duty 12 ga., galvanized									
6592	2" x 4"	1 Carp	1.75	4.571	C	1,275	155		1,430	1,650
6594	2" x 6"		1.65	4.848		1,400	164		1,564	1,800
6595	2" x 6", 16 ga.		1.65	4.848		1,325	164		1,489	1,725
6596	2" x 8"		1.65	4.848		2,100	164		2,264	2,600
6597	2" x 8", 16 ga.		1.65	4.848		2,000	164		2,164	2,475
6598	2" x 10"		1.65	4.848		2,175	164		2,339	2,650
6600	2" x 12"		1.65	4.848		2,650	164		2,814	3,200
6602	2" x 14"		1.65	4.848		2,775	164		2,939	3,325
6604	2" x 16"		1.65	4.848		2,925	164		3,089	3,500
6606	3" x 4"		1.65	4.848		1,775	164		1,939	2,225
6608	3" x 6"		1.65	4.848		2,375	164		2,539	2,875
6610	3" x 8"		1.65	4.848		2,400	164		2,564	2,925
6612	3" x 10"		1.60	5		2,775	170		2,945	3,325
6614	3" x 12"		1.60	5		3,325	170		3,495	3,950
6616	3" x 14"		1.60	5		3,900	170		4,070	4,575
6618	3" x 16"		1.60	5		4,300	170		4,470	5,025
6620	(2) 2" x 4"		1.75	4.571		2,075	155		2,230	2,525
6622	(2) 2" x 6"		1.60	5		2,500	170		2,670	3,000

06 05 Common Work Results for Wood, Plastics, and Composites
06 05 23 – Wood, Plastic, and Composite Fastenings

06 05 23.60 Timber Connectors		Crew	Daily Output	Labor-Hours	Unit	Material	2016 Bare Costs Labor	Equipment	Total	Total Incl O&P
6624	(2) 2" x 8"	1 Carp	1.60	5	C	2,550	170		2,720	3,075
6626	(2) 2" x 10"		1.55	5.161		2,750	175		2,925	3,325
6628	(2) 2" x 12"		1.55	5.161		3,525	175		3,700	4,175
6630	(2) 2" x 14"		1.50	5.333		3,550	181		3,731	4,225
6632	(2) 2" x 16"		1.50	5.333		3,600	181		3,781	4,250
6634	4" x 4"		1.65	4.848		1,500	164		1,664	1,925
6636	4" x 6"		1.60	5		1,650	170		1,820	2,100
6638	4" x 8"		1.60	5		1,900	170		2,070	2,375
6640	4" x 10"		1.55	5.161		2,350	175		2,525	2,875
6642	4" x 12"		1.55	5.161		2,500	175		2,675	3,025
6644	4" x 14"		1.55	5.161		2,975	175		3,150	3,575
6646	4" x 16"		1.55	5.161		3,275	175		3,450	3,900
6648	(3) 2" x 10"		1.50	5.333		3,600	181		3,781	4,275
6650	(3) 2" x 12"		1.50	5.333		4,025	181		4,206	4,725
6652	(3) 2" x 14"		1.45	5.517		4,325	187		4,512	5,075
6654	(3) 2" x 16"		1.45	5.517		4,400	187		4,587	5,150
6656	6" x 6"		1.60	5		1,950	170		2,120	2,400
6658	6" x 8"		1.60	5		2,000	170		2,170	2,475
6660	6" x 10"		1.55	5.161		2,400	175		2,575	2,925
6662	6" x 12"		1.55	5.161		2,725	175		2,900	3,300
6664	6" x 14"		1.50	5.333		3,425	181		3,606	4,050
6666	6" x 16"		1.50	5.333		4,025	181		4,206	4,725
6690	Knee braces, galvanized, 12 ga.									
6692	Beam depth, 10" x 15" x 5' long	1 Carp	1.80	4.444	C	5,225	151		5,376	6,000
6694	15" x 22-1/2" x 7' long		1.70	4.706		6,000	160		6,160	6,875
6696	22-1/2" x 28-1/2" x 8' long		1.60	5		6,450	170		6,620	7,350
6698	28-1/2" x 36" x 10' long		1.55	5.161		6,725	175		6,900	7,700
6700	36" x 42" x 12' long		1.50	5.333		7,425	181		7,606	8,450
6710	Mudsill anchors									
6714	2" x 4" or 3" x 4"	1 Carp	115	.070	C	1,350	2.36		1,352.36	1,475
6716	2" x 6" or 3" x 6"		115	.070		1,350	2.36		1,352.36	1,475
6718	Block wall, 13-1/4" long		115	.070		85	2.36		87.36	97.50
6720	21-1/4" long		115	.070		126	2.36		128.36	143
6730	Post bases, 12 ga. galvanized									
6732	Adjustable, 3-9/16" x 3-9/16"	1 Carp	1.30	6.154	C	1,075	209		1,284	1,525
6734	3-9/16" x 5-1/2"		1.30	6.154		2,025	209		2,234	2,575
6736	4" x 4"		1.30	6.154		930	209		1,139	1,375
6738	4" x 6"		1.30	6.154		2,700	209		2,909	3,325
6740	5-1/2" x 5-1/2"		1.30	6.154		3,325	209		3,534	4,000
6742	6" x 6"		1.30	6.154		3,325	209		3,534	4,000
6744	Elevated, 3-9/16" x 3-1/4"		1.30	6.154		1,150	209		1,359	1,625
6746	5-1/2" x 3-5/16"		1.30	6.154		1,650	209		1,859	2,150
6748	5-1/2" x 5"		1.30	6.154		2,450	209		2,659	3,050
6750	Regular, 3-9/16" x 3-3/8"		1.30	6.154		880	209		1,089	1,325
6752	4" x 3-3/8"		1.30	6.154		1,250	209		1,459	1,725
6754	18 ga., 5-1/4" x 3-1/8"		1.30	6.154		1,300	209		1,509	1,775
6755	5-1/2" x 3-3/8"		1.30	6.154		1,300	209		1,509	1,775
6756	5-1/2" x 5-3/8"		1.30	6.154		1,875	209		2,084	2,400
6758	6" x 3-3/8"		1.30	6.154		2,250	209		2,459	2,825
6760	6" x 5-3/8"		1.30	6.154		2,500	209		2,709	3,100
6762	Post combination cap/bases									
6764	3-9/16" x 3-9/16"	1 Carp	1.20	6.667	C	460	226		686	885
6766	3-9/16" x 5-1/2"		1.20	6.667		1,050	226		1,276	1,525

06 05 Common Work Results for Wood, Plastics, and Composites

06 05 23 – Wood, Plastic, and Composite Fastenings

06 05 23.60 Timber Connectors		Crew	Daily Output	Labor-Hours	Unit	Material	2016 Bare Costs Labor	Equipment	Total	Total Incl O&P
6768	4" x 4"	1 Carp	1.20	6.667	C	2,075	226		2,301	2,675
6770	5-1/2" x 5-1/2"		1.20	6.667		1,150	226		1,376	1,650
6772	6" x 6"		1.20	6.667		4,150	226		4,376	4,950
6774	7-1/2" x 7-1/2"		1.20	6.667		4,550	226		4,776	5,400
6776	8" x 8"		1.20	6.667		4,725	226		4,951	5,575
6790	Post-beam connection caps									
6792	Beam size 3-9/16"									
6794	12 ga. post, 4" x 4"	1 Carp	1	8	C	2,900	271		3,171	3,625
6796	4" x 6"		1	8		3,850	271		4,121	4,700
6798	4" x 8"		1	8		5,700	271		5,971	6,700
6800	16 ga. post, 4" x 4"		1	8		1,200	271		1,471	1,750
6802	4" x 6"		1	8		2,000	271		2,271	2,625
6804	4" x 8"		1	8		3,350	271		3,621	4,125
6805	18 ga. post, 2-7/8" x 3"		1	8		3,350	271		3,621	4,125
6806	Beam size 5-1/2"									
6808	12 ga. post, 6" x 4"	1 Carp	1	8	C	3,450	271		3,721	4,250
6810	6" x 6"		1	8		5,450	271		5,721	6,450
6812	6" x 8"		1	8		3,750	271		4,021	4,575
6816	16 ga. post, 6" x 4"		1	8		1,875	271		2,146	2,525
6818	6" x 6"		1	8		2,000	271		2,271	2,625
6820	Beam size 7-1/2"									
6822	12 ga. post, 8" x 4"	1 Carp	1	8	C	4,775	271		5,046	5,700
6824	8" x 6"		1	8		5,025	271		5,296	5,975
6826	8" x 8"		1	8		7,550	271		7,821	8,750
6840	Purlin anchors, embedded									
6842	Heavy duty, 10 ga.									
6844	Straight, 28" long	1 Carp	1.60	5	C	1,425	170		1,595	1,850
6846	35" long		1.50	5.333		1,775	181		1,956	2,250
6848	Twisted, 28" long		1.60	5		1,425	170		1,595	1,850
6850	35" long		1.50	5.333		1,775	181		1,956	2,250
6852	Regular duty, 12 ga.									
6854	Straight, 18-1/2" long	1 Carp	1.80	4.444	C	820	151		971	1,150
6856	23-3/4" long		1.70	4.706		1,025	160		1,185	1,400
6858	29" long		1.60	5		1,050	170		1,220	1,425
6860	35" long		1.50	5.333		1,450	181		1,631	1,900
6862	Twisted, 18" long		1.80	4.444		820	151		971	1,150
6866	28" long		1.60	5		980	170		1,150	1,350
6868	35" long		1.50	5.333		1,450	181		1,631	1,900
6870	Straight, plastic coated									
6872	23-1/2" long	1 Carp	1.60	5	C	2,025	170		2,195	2,500
6874	26-7/8" long		1.60	5		2,375	170		2,545	2,875
6876	32-1/2" long		1.50	5.333		2,525	181		2,706	3,075
6878	35-7/8" long		1.50	5.333		2,625	181		2,806	3,200
6890	Purlin hangers, painted									
6892	12 ga., 2" x 6"	1 Carp	1.80	4.444	C	1,850	151		2,001	2,300
6894	2" x 8"		1.80	4.444		2,025	151		2,176	2,475
6896	2" x 10"		1.80	4.444		2,175	151		2,326	2,650
6898	2" x 12"		1.75	4.571		2,325	155		2,480	2,825
6900	2" x 14"		1.75	4.571		2,500	155		2,655	3,000
6902	2" x 16"		1.75	4.571		2,650	155		2,805	3,175
6904	3" x 6"		1.70	4.706		1,875	160		2,035	2,325
6906	3" x 8"		1.70	4.706		2,025	160		2,185	2,500
6908	3" x 10"		1.70	4.706		2,175	160		2,335	2,675

06 05 Common Work Results for Wood, Plastics, and Composites

06 05 23 – Wood, Plastic, and Composite Fastenings

06 05 23.60 Timber Connectors		Crew	Daily Output	Labor-Hours	Unit	Material	2016 Bare Costs Labor	Equipment	Total	Total Incl O&P
6910	3" x 12"	1 Carp	1.65	4.848	C	2,500	164		2,664	3,025
6912	3" x 14"		1.65	4.848		2,675	164		2,839	3,200
6914	3" x 16"		1.65	4.848		2,825	164		2,989	3,375
6916	4" x 6"		1.65	4.848		1,875	164		2,039	2,325
6918	4" x 8"		1.65	4.848		2,025	164		2,189	2,500
6920	4" x 10"		1.65	4.848		2,200	164		2,364	2,675
6922	4" x 12"		1.60	5		2,600	170		2,770	3,125
6924	4" x 14"		1.60	5		2,750	170		2,920	3,300
6926	4" x 16"		1.60	5		2,925	170		3,095	3,475
6928	6" x 6"		1.60	5		2,475	170		2,645	3,000
6930	6" x 8"		1.60	5		2,625	170		2,795	3,150
6932	6" x 10"		1.55	5.161		2,675	175		2,850	3,250
6934	double 2" x 6"		1.70	4.706		2,025	160		2,185	2,500
6936	double 2" x 8"		1.70	4.706		2,175	160		2,335	2,675
6938	double 2" x 10"		1.70	4.706		2,350	160		2,510	2,850
6940	double 2" x 12"		1.65	4.848		2,500	164		2,664	3,025
6942	double 2" x 14"		1.65	4.848		2,675	164		2,839	3,200
6944	double 2" x 16"		1.65	4.848		2,825	164		2,989	3,375
6960	11 ga., 4" x 6"		1.65	4.848		3,700	164		3,864	4,350
6962	4" x 8"		1.65	4.848		3,975	164		4,139	4,650
6964	4" x 10"		1.65	4.848		4,250	164		4,414	4,950
6966	6" x 6"		1.60	5		3,750	170		3,920	4,400
6968	6" x 8"		1.60	5		4,025	170		4,195	4,700
6970	6" x 10"		1.55	5.161		4,300	175		4,475	5,025
6972	6" x 12"		1.55	5.161		4,575	175		4,750	5,325
6974	6" x 14"		1.55	5.161		4,850	175		5,025	5,625
6976	6" x 16"		1.50	5.333		5,125	181		5,306	5,950
6978	7 ga., 8" x 6"		1.60	5		4,075	170		4,245	4,750
6980	8" x 8"		1.60	5		4,350	170		4,520	5,050
6982	8" x 10"		1.55	5.161		4,625	175		4,800	5,375
6984	8" x 12"		1.55	5.161		4,900	175		5,075	5,700
6986	8" x 14"		1.50	5.333		5,175	181		5,356	6,000
6988	8" x 16"		1.50	5.333		5,450	181		5,631	6,300
7000	Strap connectors, galvanized									
7002	12 ga., 2-1/16" x 36"	1 Carp	1.55	5.161	C	1,175	175		1,350	1,600
7004	2-1/16" x 47"		1.50	5.333		1,625	181		1,806	2,100
7005	10 ga., 2-1/16" x 72"		1.50	5.333		1,725	181		1,906	2,200
7006	7 ga., 2-1/16" x 34"		1.55	5.161		2,925	175		3,100	3,525
7008	2-1/16" x 45"		1.50	5.333		3,825	181		4,006	4,500
7010	3 ga., 3" x 32"		1.55	5.161		4,925	175		5,100	5,725
7012	3" x 41"		1.55	5.161		5,125	175		5,300	5,925
7014	3" x 50"		1.50	5.333		7,825	181		8,006	8,900
7016	3" x 59"		1.50	5.333		9,475	181		9,656	10,700
7018	3-1/2" x 68"		1.45	5.517		9,650	187		9,837	10,900
7030	Tension ties									
7032	19-1/8" long, 16 ga., 3/4" anchor bolt	1 Carp	1.80	4.444	C	1,350	151		1,501	1,725
7034	20" long, 12 ga., 1/2" anchor bolt		1.80	4.444		1,750	151		1,901	2,175
7036	20" long, 12 ga., 3/4" anchor bolt		1.80	4.444		1,750	151		1,901	2,175
7038	27-3/4" long, 12 ga., 3/4" anchor bolt		1.75	4.571		3,100	155		3,255	3,675
7050	Truss connectors, galvanized									
7052	Adjustable hanger									
7054	18 ga., 2" x 6"	1 Carp	1.65	4.848	C	530	164		694	855
7056	4" x 6"		1.65	4.848		695	164		859	1,050

06 05 Common Work Results for Wood, Plastics, and Composites

06 05 23 – Wood, Plastic, and Composite Fastenings

06 05 23.60 Timber Connectors		Crew	Daily Output	Labor-Hours	Unit	Material	2016 Bare Costs Labor	Equipment	Total	Total Incl O&P
7058	16 ga., 4" x 10"	1 Carp	1.60	5	C	1,025	170		1,195	1,400
7060	(2) 2" x 10"	"	1.60	5	"	1,025	170		1,195	1,400
7062	Connectors to plate									
7064	16 ga., 2" x 4" plate	1 Carp	1.80	4.444	C	520	151		671	820
7066	2" x 6" plate	"	1.80	4.444	"	670	151		821	990
7068	Hip jack connector									
7070	14 ga.	1 Carp	1.50	5.333	C	2,750	181		2,931	3,325

06 05 23.80 Metal Bracing

0010	**METAL BRACING**									
0302	Let-in, "T" shaped, 22 ga. galv. steel, studs at 16" O.C.	1 Carp	580	.014	L.F.	.83	.47		1.30	1.69
0402	Studs at 24" O.C.		600	.013		.83	.45		1.28	1.67
0502	Steel straps, 16 ga. galv. steel, studs at 16" O.C.		600	.013		1.07	.45		1.52	1.94
0602	Studs at 24" O.C.		620	.013		1.07	.44		1.51	1.91

06 11 Wood Framing

06 11 10 – Framing with Dimensional, Engineered or Composite Lumber

06 11 10.01 Forest Stewardship Council Certification

0010	**FOREST STEWARDSHIP COUNCIL CERTIFICATION**									
0020	For Forest Stewardship Council (FSC) cert dimension lumber, add G					65%				

06 11 10.02 Blocking

0010	**BLOCKING**									
1790	Bolted to concrete									
1798	Ledger board, 2" x 4"	1 Carp	180	.044	L.F.	4.69	1.51		6.20	7.65
1800	2" x 6"		160	.050		4.70	1.70		6.40	8
1810	4" x 6"		140	.057		8.25	1.94		10.19	12.35
1820	4" x 8"		120	.067		8.30	2.26		10.56	12.95
1950	Miscellaneous, to wood construction									
2000	2" x 4"	1 Carp	250	.032	L.F.	.40	1.08		1.48	2.26
2005	Pneumatic nailed		305	.026		.40	.89		1.29	1.93
2050	2" x 6"		222	.036		.61	1.22		1.83	2.72
2055	Pneumatic nailed		271	.030		.62	1		1.62	2.36
2100	2" x 8"		200	.040		.85	1.36		2.21	3.20
2105	Pneumatic nailed		244	.033		.85	1.11		1.96	2.80
2150	2" x 10"		178	.045		1.34	1.52		2.86	4.02
2155	Pneumatic nailed		217	.037		1.34	1.25		2.59	3.57
2200	2" x 12"		151	.053		1.70	1.80		3.50	4.88
2205	Pneumatic nailed		185	.043		1.71	1.47		3.18	4.33
2300	To steel construction									
2320	2" x 4"	1 Carp	208	.038	L.F.	.40	1.30		1.70	2.62
2340	2" x 6"		180	.044		.61	1.51		2.12	3.19
2360	2" x 8"		158	.051		.85	1.72		2.57	3.80
2380	2" x 10"		136	.059		1.34	1.99		3.33	4.81
2400	2" x 12"		109	.073		1.70	2.49		4.19	6.05

06 11 10.04 Wood Bracing

0010	**WOOD BRACING**									
0012	Let-in, with 1" x 6" boards, studs @ 16" O.C.	1 Carp	150	.053	L.F.	.71	1.81		2.52	3.81
0202	Studs @ 24" O.C.	"	230	.035	"	.71	1.18		1.89	2.75

06 11 Wood Framing

06 11 10 – Framing with Dimensional, Engineered or Composite Lumber

06 11 10.06 Bridging		Crew	Daily Output	Labor-Hours	Unit	Material	2016 Bare Costs Labor	Equipment	Total	Total Incl O&P
0010	**BRIDGING**									
0012	Wood, for joists 16" O.C., 1" x 3"	1 Carp	130	.062	Pr.	.64	2.09		2.73	4.20
0017	Pneumatic nailed		170	.047		.69	1.60		2.29	3.43
0102	2" x 3" bridging		130	.062		.66	2.09		2.75	4.22
0107	Pneumatic nailed		170	.047		.66	1.60		2.26	3.40
0302	Steel, galvanized, 18 ga., for 2" x 10" joists at 12" O.C.		130	.062		.94	2.09		3.03	4.52
0352	16" O.C.		135	.059		.94	2.01		2.95	4.40
0402	24" O.C.		140	.057		1.94	1.94		3.88	5.35
0602	For 2" x 14" joists at 16" O.C.		130	.062		1.43	2.09		3.52	5.05
0902	Compression type, 16" O.C., 2" x 8" joists		200	.040		.94	1.36		2.30	3.31
1002	2" x 12" joists		200	.040		.94	1.36		2.30	3.31

06 11 10.10 Beam and Girder Framing		Crew	Daily Output	Labor-Hours	Unit	Material	2016 Bare Costs Labor	Equipment	Total	Total Incl O&P
0010	**BEAM AND GIRDER FRAMING** R061110-30									
1000	Single, 2" x 6"	2 Carp	700	.023	L.F.	.61	.78		1.39	1.97
1005	Pneumatic nailed		812	.020		.62	.67		1.29	1.80
1020	2" x 8"		650	.025		.85	.83		1.68	2.33
1025	Pneumatic nailed		754	.021		.85	.72		1.57	2.14
1040	2" x 10"		600	.027		1.34	.90		2.24	2.98
1045	Pneumatic nailed		696	.023		1.34	.78		2.12	2.78
1060	2" x 12"		550	.029		1.70	.99		2.69	3.52
1065	Pneumatic nailed		638	.025		1.71	.85		2.56	3.30
1080	2" x 14"		500	.032		2.15	1.08		3.23	4.18
1085	Pneumatic nailed		580	.028		2.16	.94		3.10	3.94
1100	3" x 8"		550	.029		2.72	.99		3.71	4.64
1120	3" x 10"		500	.032		3.43	1.08		4.51	5.60
1140	3" x 12"		450	.036		4.11	1.21		5.32	6.55
1160	3" x 14"		400	.040		4.80	1.36		6.16	7.55
1170	4" x 6"	F-3	1100	.036		2.90	1.12	.60	4.62	5.70
1180	4" x 8"		1000	.040		2.96	1.23	.66	4.85	6.05
1200	4" x 10"		950	.042		4.70	1.29	.69	6.68	8.05
1220	4" x 12"		900	.044		5.20	1.36	.73	7.29	8.80
1240	4" x 14"		850	.047		6.05	1.44	.77	8.26	9.95
1250	6" x 8"		525	.076		7.50	2.34	1.25	11.09	13.55
1260	6" x 10"		500	.080		6.55	2.45	1.32	10.32	12.75
1290	8" x 12"		300	.133		15.70	4.09	2.19	21.98	26.50
2000	Double, 2" x 6"	2 Carp	625	.026		1.22	.87		2.09	2.79
2005	Pneumatic nailed		725	.022		1.23	.75		1.98	2.60
2020	2" x 8"		575	.028		1.70	.94		2.64	3.45
2025	Pneumatic nailed		667	.024		1.71	.81		2.52	3.24
2040	2" x 10"		550	.029		2.67	.99		3.66	4.59
2045	Pneumatic nailed		638	.025		2.69	.85		3.54	4.38
2060	2" x 12"		525	.030		3.40	1.03		4.43	5.45
2065	Pneumatic nailed		610	.026		3.42	.89		4.31	5.25
2080	2" x 14"		475	.034		4.30	1.14		5.44	6.65
2085	Pneumatic nailed		551	.029		4.32	.98		5.30	6.40
3000	Triple, 2" x 6"		550	.029		1.83	.99		2.82	3.67
3005	Pneumatic nailed		638	.025		1.85	.85		2.70	3.45
3020	2" x 8"		525	.030		2.54	1.03		3.57	4.53
3025	Pneumatic nailed		609	.026		2.56	.89		3.45	4.31
3040	2" x 10"		500	.032		4.01	1.08		5.09	6.25
3045	Pneumatic nailed		580	.028		4.03	.94		4.97	6
3060	2" x 12"		475	.034		5.10	1.14		6.24	7.50

06 11 Wood Framing

06 11 10 – Framing with Dimensional, Engineered or Composite Lumber

		Crew	Daily Output	Labor-Hours	Unit	Material	2016 Bare Costs Labor	Equipment	Total	Total Incl O&P
06 11 10.10 Beam and Girder Framing										
3065	Pneumatic nailed	2 Carp	551	.029	L.F.	5.15	.98		6.13	7.30
3080	2" x 14"		450	.036		7.15	1.21		8.36	9.85
3085	Pneumatic nailed		522	.031		6.50	1.04		7.54	8.85
06 11 10.12 Ceiling Framing										
0010	**CEILING FRAMING**									
6000	Suspended, 2" x 3"	2 Carp	1000	.016	L.F.	.38	.54		.92	1.32
6050	2" x 4"		900	.018		.40	.60		1	1.45
6100	2" x 6"		800	.020		.61	.68		1.29	1.81
6150	2" x 8"		650	.025		.85	.83		1.68	2.33
06 11 10.14 Posts and Columns										
0010	**POSTS AND COLUMNS**									
0100	4" x 4"	2 Carp	390	.041	L.F.	1.73	1.39		3.12	4.24
0150	4" x 6"		275	.058		2.90	1.97		4.87	6.50
0200	4" x 8"		220	.073		2.96	2.47		5.43	7.40
0250	6" x 6"		215	.074		5.05	2.52		7.57	9.80
0300	6" x 8"		175	.091		7.50	3.10		10.60	13.45
0350	6" x 10"		150	.107		6.55	3.62		10.17	13.25
06 11 10.18 Joist Framing										
0010	**JOIST FRAMING**									
2000	Joists, 2" x 4"	2 Carp	1250	.013	L.F.	.40	.43		.83	1.17
2005	Pneumatic nailed		1438	.011		.40	.38		.78	1.07
2100	2" x 6"		1250	.013		.61	.43		1.04	1.40
2105	Pneumatic nailed		1438	.011		.62	.38		1	1.31
2150	2" x 8"		1100	.015		.85	.49		1.34	1.76
2155	Pneumatic nailed		1265	.013		.85	.43		1.28	1.66
2200	2" x 10"		900	.018		1.34	.60		1.94	2.48
2205	Pneumatic nailed		1035	.015		1.34	.52		1.86	2.36
2250	2" x 12"		875	.018		1.70	.62		2.32	2.91
2255	Pneumatic nailed		1006	.016		1.71	.54		2.25	2.78
2300	2" x 14"		770	.021		2.15	.70		2.85	3.54
2305	Pneumatic nailed		886	.018		2.16	.61		2.77	3.39
2350	3" x 6"		925	.017		1.89	.59		2.48	3.06
2400	3" x 10"		780	.021		3.43	.70		4.13	4.93
2450	3" x 12"		600	.027		4.11	.90		5.01	6.05
2500	4" x 6"		800	.020		2.90	.68		3.58	4.33
2550	4" x 10"		600	.027		4.70	.90		5.60	6.65
2600	4" x 12"		450	.036		5.20	1.21		6.41	7.70
2605	Sister joist, 2" x 6"		800	.020		.61	.68		1.29	1.81
2606	Pneumatic nailed		960	.017		.62	.57		1.19	1.63
3000	Composite wood joist 9-1/2" deep		.90	17.778	M.L.F.	1,850	605		2,455	3,050
3010	11-1/2" deep		.88	18.182		2,050	615		2,665	3,275
3020	14" deep		.82	19.512		2,125	660		2,785	3,450
3030	16" deep		.78	20.513		3,575	695		4,270	5,125
4000	Open web joist 12" deep		.88	18.182		3,325	615		3,940	4,675
4002	Per linear foot		880	.018	L.F.	3.32	.62		3.94	4.68
4004	Treated, per linear foot		880	.018	"	4.21	.62		4.83	5.65
4010	14" deep		.82	19.512	M.L.F.	3,625	660		4,285	5,100
4012	Per linear foot		820	.020	L.F.	3.63	.66		4.29	5.10
4014	Treated, per linear foot		820	.020	"	4.66	.66		5.32	6.25
4020	16" deep		.78	20.513	M.L.F.	3,800	695		4,495	5,350
4022	Per linear foot		780	.021	L.F.	3.79	.70		4.49	5.35
4024	Treated, per linear foot		780	.021	"	4.97	.70		5.67	6.60

06 11 Wood Framing

06 11 10 – Framing with Dimensional, Engineered or Composite Lumber

06 11 10.18 Joist Framing

		Crew	Daily Output	Labor-Hours	Unit	Material	2016 Bare Costs Labor	Equipment	Total	Total Incl O&P
4030	18" deep	2 Carp	.74	21.622	M.L.F.	3,850	735		4,585	5,475
4032	Per linear foot		740	.022	L.F.	3.85	.73		4.58	5.45
4034	Treated, per linear foot		740	.022	"	5.20	.73		5.93	6.95
6000	Composite rim joist, 1-1/4" x 9-1/2"		.90	17.778	M.L.F.	1,975	605		2,580	3,175
6010	1-1/4" x 11-1/2"		.88	18.182		2,275	615		2,890	3,525
6020	1-1/4" x 14-1/2"		.82	19.512		2,750	660		3,410	4,125
6030	1-1/4" x 16-1/2"		.78	20.513		3,175	695		3,870	4,650

06 11 10.24 Miscellaneous Framing

		Crew	Daily Output	Labor-Hours	Unit	Material	Labor	Equipment	Total	Total Incl O&P
0010	**MISCELLANEOUS FRAMING**									
2000	Firestops, 2" x 4"	2 Carp	780	.021	L.F.	.40	.70		1.10	1.60
2005	Pneumatic nailed		952	.017		.40	.57		.97	1.39
2100	2" x 6"		600	.027		.61	.90		1.51	2.18
2105	Pneumatic nailed		732	.022		.62	.74		1.36	1.92
5000	Nailers, treated, wood construction, 2" x 4"		800	.020		.52	.68		1.20	1.72
5005	Pneumatic nailed		960	.017		.53	.57		1.10	1.53
5100	2" x 6"		750	.021		.77	.72		1.49	2.06
5105	Pneumatic nailed		900	.018		.78	.60		1.38	1.87
5120	2" x 8"		700	.023		1.06	.78		1.84	2.46
5125	Pneumatic nailed		840	.019		1.06	.65		1.71	2.25
5200	Steel construction, 2" x 4"		750	.021		.52	.72		1.24	1.79
5220	2" x 6"		700	.023		.77	.78		1.55	2.15
5240	2" x 8"		650	.025		1.06	.83		1.89	2.56
7000	Rough bucks, treated, for doors or windows, 2" x 6"		400	.040		.77	1.36		2.13	3.12
7005	Pneumatic nailed		480	.033		.78	1.13		1.91	2.75
7100	2" x 8"		380	.042		1.06	1.43		2.49	3.55
7105	Pneumatic nailed		456	.035		1.06	1.19		2.25	3.16
8000	Stair stringers, 2" x 10"		130	.123		1.34	4.17		5.51	8.45
8100	2" x 12"		130	.123		1.70	4.17		5.87	8.85
8150	3" x 10"		125	.128		3.43	4.34		7.77	11
8200	3" x 12"		125	.128		4.11	4.34		8.45	11.75
8870	Laminated structural lumber, 1-1/4" x 11-1/2"		130	.123		2.26	4.17		6.43	9.50
8880	1-1/4" x 14-1/2"		130	.123		2.72	4.17		6.89	10

06 11 10.26 Partitions

		Crew	Daily Output	Labor-Hours	Unit	Material	Labor	Equipment	Total	Total Incl O&P
0010	**PARTITIONS**									
0020	Single bottom and double top plate, no waste, std. & better lumber									
0180	2" x 4" studs, 8' high, studs 12" O.C.	2 Carp	80	.200	L.F.	4.37	6.80		11.17	16.15
0185	12" O.C., pneumatic nailed		96	.167		4.40	5.65		10.05	14.30
0200	16" O.C.		100	.160		3.58	5.40		8.98	13.05
0205	16" O.C., pneumatic nailed		120	.133		3.60	4.52		8.12	11.50
0300	24" O.C.		125	.128		2.78	4.34		7.12	10.30
0305	24" O.C., pneumatic nailed		150	.107		2.80	3.62		6.42	9.15
0380	10' high, studs 12" O.C.		80	.200		5.15	6.80		11.95	17.05
0385	12" O.C., pneumatic nailed		96	.167		5.20	5.65		10.85	15.15
0400	16" O.C.		100	.160		4.17	5.40		9.57	13.70
0405	16" O.C., pneumatic nailed		120	.133		4.20	4.52		8.72	12.15
0500	24" O.C.		125	.128		3.18	4.34		7.52	10.75
0505	24" O.C., pneumatic nailed		150	.107		3.20	3.62		6.82	9.55
0580	12' high, studs 12" O.C.		65	.246		5.95	8.35		14.30	20.50
0585	12" O.C., pneumatic nailed		78	.205		6	6.95		12.95	18.25
0600	16" O.C.		80	.200		4.77	6.80		11.57	16.60
0605	16" O.C., pneumatic nailed		96	.167		4.80	5.65		10.45	14.75
0700	24" O.C.		100	.160		3.58	5.40		8.98	13.05

06 11 Wood Framing

06 11 10 – Framing with Dimensional, Engineered or Composite Lumber

06 11 10.26 Partitions

		Crew	Daily Output	Labor-Hours	Unit	Material	2016 Bare Costs Labor	Equipment	Total	Total Incl O&P
0705	24" O.C., pneumatic nailed	2 Carp	120	.133	L.F.	3.60	4.52		8.12	11.50
0780	2" x 6" studs, 8' high, studs 12" O.C.		70	.229		6.70	7.75		14.45	20.50
0785	12" O.C., pneumatic nailed		84	.190		6.75	6.45		13.20	18.25
0800	16" O.C.		90	.178		5.50	6.05		11.55	16.15
0805	16" O.C., pneumatic nailed		108	.148		5.55	5		10.55	14.50
0900	24" O.C.		115	.139		4.28	4.72		9	12.60
0905	24" O.C., pneumatic nailed		138	.116		4.31	3.93		8.24	11.35
0980	10' high, studs 12" O.C.		70	.229		7.95	7.75		15.70	21.50
0985	12" O.C., pneumatic nailed		84	.190		8	6.45		14.45	19.60
1000	16" O.C.		90	.178		6.40	6.05		12.45	17.15
1005	16" O.C., pneumatic nailed		108	.148		6.45	5		11.45	15.50
1100	24" O.C.		115	.139		4.89	4.72		9.61	13.30
1105	24" O.C., pneumatic nailed		138	.116		4.92	3.93		8.85	12
1180	12' high, studs 12" O.C.		55	.291		9.15	9.85		19	26.50
1185	12" O.C., pneumatic nailed		66	.242		9.25	8.20		17.45	24
1200	16" O.C.		70	.229		7.35	7.75		15.10	21
1205	16" O.C., pneumatic nailed		84	.190		7.40	6.45		13.85	18.90
1300	24" O.C.		90	.178		5.50	6.05		11.55	16.15
1305	24" O.C., pneumatic nailed		108	.148		5.55	5		10.55	14.50
1400	For horizontal blocking, 2" x 4", add		600	.027		.40	.90		1.30	1.95
1500	2" x 6", add		600	.027		.61	.90		1.51	2.18
1600	For openings, add		250	.064			2.17		2.17	3.63
1702	Headers for above openings, material only, add				B.F.	.70			.70	.77

06 11 10.28 Porch or Deck Framing

		Crew	Daily Output	Labor-Hours	Unit	Material	2016 Bare Costs Labor	Equipment	Total	Total Incl O&P
0010	**PORCH OR DECK FRAMING**									
0100	Treated lumber, posts or columns, 4" x 4"	2 Carp	390	.041	L.F.	1.35	1.39		2.74	3.81
0110	4" x 6"		275	.058		2.02	1.97		3.99	5.50
0120	4" x 8"		220	.073		3.87	2.47		6.34	8.40
0130	Girder, single, 4" x 4"		675	.024		1.35	.80		2.15	2.83
0140	4" x 6"		600	.027		2.02	.90		2.92	3.73
0150	4" x 8"		525	.030		3.87	1.03		4.90	6
0160	Double, 2" x 4"		625	.026		1.07	.87		1.94	2.63
0170	2" x 6"		600	.027		1.59	.90		2.49	3.25
0180	2" x 8"		575	.028		2.17	.94		3.11	3.96
0190	2" x 10"		550	.029		2.76	.99		3.75	4.68
0200	2" x 12"		525	.030		3.87	1.03		4.90	6
0210	Triple, 2" x 4"		575	.028		1.61	.94		2.55	3.35
0220	2" x 6"		550	.029		2.38	.99		3.37	4.27
0230	2" x 8"		525	.030		3.25	1.03		4.28	5.30
0240	2" x 10"		500	.032		4.13	1.08		5.21	6.35
0250	2" x 12"		475	.034		5.80	1.14		6.94	8.30
0260	Ledger, bolted 4' O.C., 2" x 4"		400	.040		.69	1.36		2.05	3.02
0270	2" x 6"		395	.041		.93	1.37		2.30	3.33
0280	2" x 8"		390	.041		1.21	1.39		2.60	3.66
0290	2" x 10"		385	.042		1.50	1.41		2.91	4.01
0300	2" x 12"		380	.042		2.05	1.43		3.48	4.64
0310	Joists, 2" x 4"		1250	.013		.53	.43		.96	1.32
0320	2" x 6"		1250	.013		.79	.43		1.22	1.60
0330	2" x 8"		1100	.015		1.08	.49		1.57	2.01
0340	2" x 10"		900	.018		1.37	.60		1.97	2.52
0350	2" x 12"		875	.018		1.64	.62		2.26	2.85
0360	Railings and trim, 1" x 4"	1 Carp	300	.027		.46	.90		1.36	2.02

06 11 Wood Framing

06 11 10 – Framing with Dimensional, Engineered or Composite Lumber

	06 11 10.28 Porch or Deck Framing	Crew	Daily Output	Labor-Hours	Unit	Material	2016 Bare Costs Labor	Equipment	Total	Total Incl O&P
0370	2" x 2"	1 Carp	300	.027	L.F.	.33	.90		1.23	1.88
0380	2" x 4"		300	.027		.53	.90		1.43	2.09
0390	2" x 6"		300	.027		.78	.90		1.68	2.37
0400	Decking, 1" x 4"		275	.029	S.F.	2.16	.99		3.15	4.03
0410	2" x 4"		300	.027		1.78	.90		2.68	3.46
0420	2" x 6"		320	.025		1.68	.85		2.53	3.27
0430	5/4" x 6"		320	.025		2.25	.85		3.10	3.90
0440	Balusters, square, 2" x 2"	2 Carp	660	.024	L.F.	.34	.82		1.16	1.75
0450	Turned, 2" x 2"		420	.038		.45	1.29		1.74	2.65
0460	Stair stringer, 2" x 10"		130	.123		1.37	4.17		5.54	8.50
0470	2" x 12"		130	.123		1.64	4.17		5.81	8.80
0480	Stair treads, 1" x 4"		140	.114		2.17	3.87		6.04	8.90
0490	2" x 4"		140	.114		.53	3.87		4.40	7.10
0500	2" x 6"		160	.100		.80	3.39		4.19	6.60
0510	5/4" x 6"		160	.100		1.05	3.39		4.44	6.85
0520	Turned handrail post, 4" x 4"		64	.250	Ea.	31	8.50		39.50	48
0530	Lattice panel, 4' x 8', 1/2"		1600	.010	S.F.	.81	.34		1.15	1.46
0535	3/4"		1600	.010	"	1.21	.34		1.55	1.90
0540	Cedar, posts or columns, 4" x 4"		390	.041	L.F.	3.96	1.39		5.35	6.70
0550	4" x 6"		275	.058		7.70	1.97		9.67	11.75
0560	4" x 8"		220	.073		10.45	2.47		12.92	15.60
0800	Decking, 1" x 4"		550	.029		1.91	.99		2.90	3.75
0810	2" x 4"		600	.027		3.86	.90		4.76	5.75
0820	2" x 6"		640	.025		6.95	.85		7.80	9.05
0830	5/4" x 6"		640	.025		4.45	.85		5.30	6.30
0840	Railings and trim, 1" x 4"		600	.027		1.91	.90		2.81	3.61
0860	2" x 4"		600	.027		3.86	.90		4.76	5.75
0870	2" x 6"		600	.027		6.95	.90		7.85	9.15
0920	Stair treads, 1" x 4"		140	.114		1.91	3.87		5.78	8.60
0930	2" x 4"		140	.114		3.86	3.87		7.73	10.75
0940	2" x 6"		160	.100		6.95	3.39		10.34	13.35
0950	5/4" x 6"		160	.100		4.45	3.39		7.84	10.60
0980	Redwood, posts or columns, 4" x 4"		390	.041		6.45	1.39		7.84	9.45
0990	4" x 6"		275	.058		12.60	1.97		14.57	17.15
1000	4" x 8"		220	.073		23.50	2.47		25.97	30
1240	Decking, 1" x 4"	1 Carp	275	.029	S.F.	3.97	.99		4.96	6
1260	2" x 6"		340	.024		7.50	.80		8.30	9.60
1270	5/4" x 6"		320	.025		4.83	.85		5.68	6.70
1280	Railings and trim, 1" x 4"	2 Carp	600	.027	L.F.	1.17	.90		2.07	2.80
1310	2" x 6"		600	.027		7.50	.90		8.40	9.75
1420	Alternative decking, wood/plastic composite, 5/4" x 6" G		640	.025		3.14	.85		3.99	4.87
1440	1" x 4" square edge fir		550	.029		2.18	.99		3.17	4.05
1450	1" x 4" tongue and groove fir		450	.036		1.46	1.21		2.67	3.62
1460	1" x 4" mahogany		550	.029		2.01	.99		3	3.86
1462	5/4" x 6" PVC		550	.029		3.71	.99		4.70	5.75
1465	Framing, porch or deck, alt deck fastening, screws, add	1 Carp	240	.033	S.F.		1.13		1.13	1.89
1470	Accessories, joist hangers, 2" x 4"		160	.050	Ea.	.72	1.70		2.42	3.63
1480	2" x 6" through 2" x 12"		150	.053		1.36	1.81		3.17	4.53
1530	Post footing, incl excav, backfill, tube form & concrete, 4' deep, 8" dia	F-7	12	2.667		12.65	78		90.65	145
1540	10" diameter		11	2.909		18.40	85		103.40	164
1550	12" diameter		10	3.200		24	93.50		117.50	184

06 11 Wood Framing

06 11 10 – Framing with Dimensional, Engineered or Composite Lumber

06 11 10.30 Roof Framing		Crew	Daily Output	Labor-Hours	Unit	Material	2016 Bare Costs Labor	Equipment	Total	Total Incl O&P
0010	**ROOF FRAMING**									
1900	Rough fascia, 2" x 6"	2 Carp	250	.064	L.F.	.61	2.17		2.78	4.30
2000	2" x 8"		225	.071		.85	2.41		3.26	4.97
2100	2" x 10"		180	.089		1.34	3.01		4.35	6.50
5002	Rafters, to 4 in 12 pitch, 2" x 6", ordinary		1000	.016		.61	.54		1.15	1.58
5021	On steep roofs		800	.020		.61	.68		1.29	1.81
5041	On dormers or complex roofs		590	.027		.61	.92		1.53	2.21
5062	2" x 8", ordinary		950	.017		.85	.57		1.42	1.89
5081	On steep roofs		750	.021		.85	.72		1.57	2.14
5101	On dormers or complex roofs		540	.030		.85	1		1.85	2.61
5122	2" x 10", ordinary		630	.025		1.34	.86		2.20	2.91
5141	On steep roofs		495	.032		1.34	1.10		2.44	3.30
5161	On dormers or complex roofs		425	.038		1.34	1.28		2.62	3.61
5182	2" x 12", ordinary		575	.028		1.70	.94		2.64	3.45
5201	On steep roofs		455	.035		1.70	1.19		2.89	3.87
5221	On dormers or complex roofs		395	.041		1.70	1.37		3.07	4.17
5250	Composite rafter, 9-1/2" deep		575	.028		1.85	.94		2.79	3.62
5260	11-1/2" deep		575	.028		2.04	.94		2.98	3.83
5301	Hip and valley rafters, 2" x 6", ordinary		760	.021		.61	.71		1.32	1.86
5321	On steep roofs		585	.027		.61	.93		1.54	2.22
5341	On dormers or complex roofs		510	.031		.61	1.06		1.67	2.45
5361	2" x 8", ordinary		720	.022		.85	.75		1.60	2.19
5381	On steep roofs		545	.029		.85	1		1.85	2.60
5401	On dormers or complex roofs		470	.034		.85	1.15		2	2.86
5421	2" x 10", ordinary		570	.028		1.34	.95		2.29	3.06
5441	On steep roofs		440	.036		1.34	1.23		2.57	3.53
5461	On dormers or complex roofs		380	.042		1.34	1.43		2.77	3.86
5481	2" x 12", ordinary		525	.030		1.70	1.03		2.73	3.60
5501	On steep roofs		410	.039		1.70	1.32		3.02	4.08
5521	On dormers or complex roofs		355	.045		1.70	1.53		3.23	4.43
5541	Hip and valley jacks, 2" x 6", ordinary		600	.027		.61	.90		1.51	2.18
5561	On steep roofs		475	.034		.61	1.14		1.75	2.58
5581	On dormers or complex roofs		410	.039		.61	1.32		1.93	2.88
5601	2" x 8", ordinary		490	.033		.85	1.11		1.96	2.78
5621	On steep roofs		385	.042		.85	1.41		2.26	3.29
5641	On dormers or complex roofs		335	.048		.85	1.62		2.47	3.64
5661	2" x 10", ordinary		450	.036		1.34	1.21		2.55	3.49
5681	On steep roofs		350	.046		1.34	1.55		2.89	4.06
5701	On dormers or complex roofs		305	.052		1.34	1.78		3.12	4.45
5721	2" x 12", ordinary		375	.043		1.70	1.45		3.15	4.29
5741	On steep roofs		295	.054		1.70	1.84		3.54	4.95
5762	On dormers or complex roofs		255	.063		1.70	2.13		3.83	5.45
5781	Rafter tie, 1" x 4", #3		800	.020		.46	.68		1.14	1.65
5791	2" x 4", #3		800	.020		.40	.68		1.08	1.58
5801	Ridge board, #2 or better, 1" x 6"		600	.027		.71	.90		1.61	2.29
5821	1" x 8"		550	.029		1.18	.99		2.17	2.95
5841	1" x 10"		500	.032		1.53	1.08		2.61	3.51
5861	2" x 6"		500	.032		.61	1.08		1.69	2.49
5881	2" x 8"		450	.036		.85	1.21		2.06	2.95
5901	2" x 10"		400	.040		1.34	1.36		2.70	3.74
5921	Roof cants, split, 4" x 4"		650	.025		1.73	.83		2.56	3.31
5941	6" x 6"		600	.027		5.05	.90		5.95	7.10

06 11 Wood Framing

06 11 10 – Framing with Dimensional, Engineered or Composite Lumber

06 11 10.30 Roof Framing		Crew	Daily Output	Labor-Hours	Unit	Material	2016 Bare Costs Labor	Equipment	Total	Total Incl O&P
5961	Roof curbs, untreated, 2" x 6"	2 Carp	520	.031	L.F.	.61	1.04		1.65	2.42
5981	2" x 12"		400	.040		1.70	1.36		3.06	4.14
6001	Sister rafters, 2" x 6"		800	.020		.61	.68		1.29	1.81
6021	2" x 8"		640	.025		.83	.85		1.68	2.34
6041	2" x 10"		535	.030		1.34	1.01		2.35	3.17
6061	2" x 12"		455	.035		1.70	1.19		2.89	3.87

06 11 10.32 Sill and Ledger Framing										
0010	**SILL AND LEDGER FRAMING**									
0020	Extruded polystyrene sill sealer, 5-1/2" wide	1 Carp	1600	.005	L.F.	.14	.17		.31	.43
2002	Ledgers, nailed, 2" x 4"	2 Carp	755	.021		.40	.72		1.12	1.64
2052	2" x 6"		600	.027		.61	.90		1.51	2.18
2102	Bolted, not including bolts, 3" x 6"		325	.049		1.88	1.67		3.55	4.86
2152	3" x 12"		233	.069		4.08	2.33		6.41	8.40
2602	Mud sills, redwood, construction grade, 2" x 4"		895	.018		2.26	.61		2.87	3.50
2622	2" x 6"		780	.021		3.40	.70		4.10	4.90
4002	Sills, 2" x 4"		600	.027		.39	.90		1.29	1.94
4052	2" x 6"		550	.029		.60	.99		1.59	2.31
4082	2" x 8"		500	.032		.83	1.08		1.91	2.74
4101	2" x 10"		450	.036		1.32	1.21		2.53	3.47
4121	2" x 12"		400	.040		1.68	1.36		3.04	4.12
4202	Treated, 2" x 4"		550	.029		.52	.99		1.51	2.22
4222	2" x 6"		500	.032		.76	1.08		1.84	2.66
4242	2" x 8"		450	.036		1.04	1.21		2.25	3.17
4261	2" x 10"		400	.040		1.33	1.36		2.69	3.73
4281	2" x 12"		350	.046		1.88	1.55		3.43	4.65
4402	4" x 4"		450	.036		1.31	1.21		2.52	3.46
4422	4" x 6"		350	.046		1.96	1.55		3.51	4.75
4462	4" x 8"		300	.053		3.79	1.81		5.60	7.20
4480	4" x 10"		260	.062		4.73	2.09		6.82	8.70

06 11 10.34 Sleepers										
0010	**SLEEPERS**									
0100	On concrete, treated, 1" x 2"	2 Carp	2350	.007	L.F.	.27	.23		.50	.69
0150	1" x 3"		2000	.008		.44	.27		.71	.94
0200	2" x 4"		1500	.011		.56	.36		.92	1.23
0250	2" x 6"		1300	.012		.86	.42		1.28	1.64

06 11 10.36 Soffit and Canopy Framing										
0010	**SOFFIT AND CANOPY FRAMING**									
1002	Canopy or soffit framing, 1" x 4"	2 Carp	900	.018	L.F.	.46	.60		1.06	1.52
1021	1" x 6"		850	.019		.71	.64		1.35	1.85
1042	1" x 8"		750	.021		1.18	.72		1.90	2.51
1102	2" x 4"		620	.026		.40	.88		1.28	1.90
1121	2" x 6"		560	.029		.61	.97		1.58	2.29
1142	2" x 8"		500	.032		.85	1.08		1.93	2.75
1202	3" x 4"		500	.032		1.14	1.08		2.22	3.08
1221	3" x 6"		400	.040		1.89	1.36		3.25	4.35
1242	3" x 10"		300	.053		3.43	1.81		5.24	6.80

06 11 10.38 Treated Lumber Framing Material										
0010	**TREATED LUMBER FRAMING MATERIAL**									
0100	2" x 4"				M.B.F.	775			775	850
0110	2" x 6"					765			765	840
0120	2" x 8"					780			780	860
0130	2" x 10"					795			795	875

06 11 Wood Framing

06 11 10 – Framing with Dimensional, Engineered or Composite Lumber

06 11 10.38 Treated Lumber Framing Material

		Crew	Daily Output	Labor-Hours	Unit	Material	2016 Bare Costs Labor	Equipment	Total	Total Incl O&P
0140	2" x 12"				M.B.F.	940			940	1,025
0200	4" x 4"					980			980	1,075
0210	4" x 6"					980			980	1,075
0220	4" x 8"					1,425			1,425	1,550

06 11 10.40 Wall Framing

		Crew	Daily Output	Labor-Hours	Unit	Material	2016 Bare Costs Labor	Equipment	Total	Total Incl O&P
0010	**WALL FRAMING** R061110-30									
0100	Door buck, studs, header, access, 8' H, 2" x 4" wall, 3' W	1 Carp	32	.250	Ea.	16.65	8.50		25.15	32.50
0110	4' wide		32	.250		17.90	8.50		26.40	34
0120	5' wide		32	.250		21.50	8.50		30	37.50
0130	6' wide		32	.250		23	8.50		31.50	39.50
0140	8' wide		30	.267		34.50	9.05		43.55	53
0150	10' wide		30	.267		47.50	9.05		56.55	67
0160	12' wide		30	.267		65	9.05		74.05	86.50
0170	2" x 6" wall, 3' wide		32	.250		23.50	8.50		32	40
0180	4' wide		32	.250		24.50	8.50		33	41
0190	5' wide		32	.250		28.50	8.50		37	45.50
0200	6' wide		32	.250		30	8.50		38.50	47
0210	8' wide		30	.267		41.50	9.05		50.55	60.50
0220	10' wide		30	.267		54	9.05		63.05	74.50
0230	12' wide		30	.267		72	9.05		81.05	94
0240	Window buck, studs, header & access, 8' high 2" x 4" wall, 2' wide		24	.333		17.70	11.30		29	38.50
0250	3' wide		24	.333		20.50	11.30		31.80	41.50
0260	4' wide		24	.333		22.50	11.30		33.80	44
0270	5' wide		24	.333		26.50	11.30		37.80	48
0280	6' wide		24	.333		29	11.30		40.30	51
0290	7' wide		24	.333		38.50	11.30		49.80	61.50
0300	8' wide		22	.364		43	12.35		55.35	67.50
0310	10' wide		22	.364		57	12.35		69.35	83
0320	12' wide		22	.364		77	12.35		89.35	105
0330	2" x 6" wall, 2' wide		24	.333		26	11.30		37.30	48
0340	3' wide		24	.333		29	11.30		40.30	51
0350	4' wide		24	.333		32	11.30		43.30	54
0360	5' wide		24	.333		36	11.30		47.30	58.50
0370	6' wide		24	.333		39.50	11.30		50.80	62
0380	7' wide		24	.333		50	11.30		61.30	74
0390	8' wide		22	.364		54.50	12.35		66.85	80
0400	10' wide		22	.364		69	12.35		81.35	96.50
0410	12' wide		22	.364		90	12.35		102.35	120
2002	Headers over openings, 2" x 6"	2 Carp	360	.044	L.F.	.61	1.51		2.12	3.19
2007	2" x 6", pneumatic nailed		432	.037		.62	1.26		1.88	2.78
2052	2" x 8"		340	.047		.85	1.60		2.45	3.60
2057	2" x 8", pneumatic nailed		408	.039		.85	1.33		2.18	3.17
2101	2" x 10"		320	.050		1.34	1.70		3.04	4.31
2106	2" x 10", pneumatic nailed		384	.042		1.34	1.41		2.75	3.84
2152	2" x 12"		300	.053		1.70	1.81		3.51	4.90
2157	2" x 12", pneumatic nailed		360	.044		1.71	1.51		3.22	4.40
2180	4" x 8"		260	.062		2.96	2.09		5.05	6.75
2185	4" x 8", pneumatic nailed		312	.051		2.97	1.74		4.71	6.20
2191	4" x 10"		240	.067		4.70	2.26		6.96	8.95
2196	4" x 10", pneumatic nailed		288	.056		4.72	1.88		6.60	8.35
2202	4" x 12"		190	.084		5.20	2.85		8.05	10.50
2207	4" x 12", pneumatic nailed		228	.070		5.20	2.38		7.58	9.75

06 11 Wood Framing

06 11 10 – Framing with Dimensional, Engineered or Composite Lumber

06 11 10.40 Wall Framing		Crew	Daily Output	Labor-Hours	Unit	Material	2016 Bare Costs Labor	Equipment	Total	Total Incl O&P
2241	6" x 10"	2 Carp	165	.097	L.F.	6.55	3.29		9.84	12.70
2246	6" x 10", pneumatic nailed		198	.081		6.60	2.74		9.34	11.85
2251	6" x 12"		140	.114		8.25	3.87		12.12	15.55
2256	6" x 12", pneumatic nailed		168	.095		8.25	3.23		11.48	14.50
5002	Plates, untreated, 2" x 3"		850	.019		.38	.64		1.02	1.48
5007	2" x 3", pneumatic nailed		1020	.016		.38	.53		.91	1.31
5022	2" x 4"		800	.020		.40	.68		1.08	1.58
5027	2" x 4", pneumatic nailed		960	.017		.40	.57		.97	1.39
5041	2" x 6"		750	.021		.61	.72		1.33	1.88
5045	2" x 6", pneumatic nailed		900	.018		.62	.60		1.22	1.69
5061	Treated, 2" x 3"		850	.019		.45	.64		1.09	1.57
5066	2" x 3", treated, pneumatic nailed		1020	.016		.45	.53		.98	1.39
5081	2" x 4"		800	.020		.52	.68		1.20	1.72
5086	2" x 4", treated, pneumatic nailed		960	.017		.53	.57		1.10	1.53
5101	2" x 6"		750	.021		.77	.72		1.49	2.06
5106	2" x 6", treated, pneumatic nailed		900	.018		.78	.60		1.38	1.87
5122	Studs, 8' high wall, 2" x 3"		1200	.013		.38	.45		.83	1.17
5127	2" x 3", pneumatic nailed		1440	.011		.38	.38		.76	1.05
5142	2" x 4"		1100	.015		.39	.49		.88	1.26
5147	2" x 4", pneumatic nailed		1320	.012		.40	.41		.81	1.13
5162	2" x 6"		1000	.016		.61	.54		1.15	1.58
5167	2" x 6", pneumatic nailed		1200	.013		.62	.45		1.07	1.44
5182	3" x 4"		800	.020		1.14	.68		1.82	2.40
5187	3" x 4", pneumatic nailed		960	.017		1.15	.57		1.72	2.21
5201	Installed on second story, 2" x 3"		1170	.014		.38	.46		.84	1.19
5206	2" x 3", pneumatic nailed		1200	.013		.38	.45		.83	1.18
5221	2" x 4"		1015	.016		.40	.53		.93	1.33
5226	2" x 4", pneumatic nailed		1080	.015		.40	.50		.90	1.28
5241	2" x 6"		890	.018		.61	.61		1.22	1.69
5246	2" x 6", pneumatic nailed		1020	.016		.62	.53		1.15	1.57
5261	3" x 4"		800	.020		1.14	.68		1.82	2.40
5266	3" x 4", pneumatic nailed		960	.017		1.15	.57		1.72	2.21
5281	Installed on dormer or gable, 2" x 3"		1045	.015		.38	.52		.90	1.28
5286	2" x 3", pneumatic nailed		1254	.013		.38	.43		.81	1.14
5301	2" x 4"		905	.018		.40	.60		1	1.44
5306	2" x 4", pneumatic nailed		1086	.015		.40	.50		.90	1.28
5321	2" x 6"		800	.020		.61	.68		1.29	1.81
5326	2" x 6", pneumatic nailed		960	.017		.62	.57		1.19	1.63
5341	3" x 4"		700	.023		1.14	.78		1.92	2.56
5346	3" x 4", pneumatic nailed		840	.019		1.15	.65		1.80	2.34
5361	6' high wall, 2" x 3"		970	.016		.38	.56		.94	1.35
5366	2" x 3", pneumatic nailed		1164	.014		.38	.47		.85	1.20
5381	2" x 4"		850	.019		.40	.64		1.04	1.51
5386	2" x 4", pneumatic nailed		1020	.016		.40	.53		.93	1.33
5401	2" x 6"		740	.022		.61	.73		1.34	1.90
5406	2" x 6", pneumatic nailed		888	.018		.62	.61		1.23	1.70
5421	3" x 4"		600	.027		1.14	.90		2.04	2.77
5426	3" x 4", pneumatic nailed		720	.022		1.15	.75		1.90	2.52
5441	Installed on second story, 2" x 3"		950	.017		.38	.57		.95	1.38
5446	2" x 3", pneumatic nailed		1140	.014		.38	.48		.86	1.22
5461	2" x 4"		810	.020		.40	.67		1.07	1.56
5466	2" x 4", pneumatic nailed		972	.016		.40	.56		.96	1.37
5481	2" x 6"		700	.023		.61	.78		1.39	1.97

For customer support on your Residential Cost Data, call 877.759.4771.

06 11 Wood Framing

06 11 10 – Framing with Dimensional, Engineered or Composite Lumber

06 11 10.40 Wall Framing

		Crew	Daily Output	Labor-Hours	Unit	Material	2016 Bare Costs Labor	Equipment	Total	Total Incl O&P
5486	2" x 6", pneumatic nailed	2 Carp	840	.019	L.F.	.62	.65		1.27	1.76
5501	3" x 4"		550	.029		1.14	.99		2.13	2.91
5506	3" x 4", pneumatic nailed		660	.024		1.15	.82		1.97	2.64
5521	Installed on dormer or gable, 2" x 3"		850	.019		.38	.64		1.02	1.48
5526	2" x 3", pneumatic nailed		1020	.016		.38	.53		.91	1.31
5541	2" x 4"		720	.022		.40	.75		1.15	1.70
5546	2" x 4", pneumatic nailed		864	.019		.40	.63		1.03	1.49
5561	2" x 6"		620	.026		.61	.88		1.49	2.13
5566	2" x 6", pneumatic nailed		744	.022		.62	.73		1.35	1.90
5581	3" x 4"		480	.033		1.14	1.13		2.27	3.15
5586	3" x 4", pneumatic nailed		576	.028		1.15	.94		2.09	2.84
5601	3' high wall, 2" x 3"		740	.022		.38	.73		1.11	1.64
5606	2" x 3", pneumatic nailed		888	.018		.38	.61		.99	1.44
5621	2" x 4"		640	.025		.40	.85		1.25	1.86
5626	2" x 4", pneumatic nailed		768	.021		.40	.71		1.11	1.62
5641	2" x 6"		550	.029		.61	.99		1.60	2.32
5646	2" x 6", pneumatic nailed		660	.024		.62	.82		1.44	2.06
5661	3" x 4"		440	.036		1.14	1.23		2.37	3.32
5666	3" x 4", pneumatic nailed		528	.030		1.15	1.03		2.18	2.98
5681	Installed on second story, 2" x 3"		700	.023		.38	.78		1.16	1.71
5686	2" x 3", pneumatic nailed		840	.019		.38	.65		1.03	1.50
5701	2" x 4"		610	.026		.40	.89		1.29	1.93
5706	2" x 4", pneumatic nailed		732	.022		.40	.74		1.14	1.68
5721	2" x 6"		520	.031		.61	1.04		1.65	2.42
5726	2" x 6", pneumatic nailed		624	.026		.62	.87		1.49	2.14
5741	3" x 4"		430	.037		1.14	1.26		2.40	3.37
5746	3" x 4", pneumatic nailed		516	.031		1.15	1.05		2.20	3.02
5761	Installed on dormer or gable, 2" x 3"		625	.026		.38	.87		1.25	1.86
5766	2" x 3", pneumatic nailed		750	.021		.38	.72		1.10	1.63
5781	2" x 4"		545	.029		.40	1		1.40	2.11
5786	2" x 4", pneumatic nailed		654	.024		.40	.83		1.23	1.83
5801	2" x 6"		465	.034		.61	1.17		1.78	2.62
5806	2" x 6", pneumatic nailed		558	.029		.62	.97		1.59	2.31
5821	3" x 4"		380	.042		1.14	1.43		2.57	3.65
5826	3" x 4", pneumatic nailed		456	.035		1.15	1.19		2.34	3.25
8250	For second story & above, add						5%			
8300	For dormer & gable, add						15%			

06 11 10.42 Furring

		Crew	Daily Output	Labor-Hours	Unit	Material	2016 Bare Costs Labor	Equipment	Total	Total Incl O&P
0010	**FURRING**									
0012	Wood strips, 1" x 2", on walls, on wood	1 Carp	550	.015	L.F.	.24	.49		.73	1.10
0015	On wood, pneumatic nailed		710	.011		.24	.38		.62	.91
0300	On masonry		495	.016		.26	.55		.81	1.21
0400	On concrete		260	.031		.26	1.04		1.30	2.04
0600	1" x 3", on walls, on wood		550	.015		.39	.49		.88	1.26
0605	On wood, pneumatic nailed		710	.011		.39	.38		.77	1.07
0700	On masonry		495	.016		.42	.55		.97	1.38
0800	On concrete		260	.031		.42	1.04		1.46	2.21
0850	On ceilings, on wood		350	.023		.39	.78		1.17	1.73
0855	On wood, pneumatic nailed		450	.018		.39	.60		.99	1.44
0900	On masonry		320	.025		.42	.85		1.27	1.88
0950	On concrete		210	.038		.42	1.29		1.71	2.62

06 11 Wood Framing

06 11 10 – Framing with Dimensional, Engineered or Composite Lumber

06 11 10.44 Grounds

		Crew	Daily Output	Labor-Hours	Unit	Material	2016 Bare Costs Labor	Equipment	Total	Total Incl O&P
0010	**GROUNDS**									
0020	For casework, 1" x 2" wood strips, on wood	1 Carp	330	.024	L.F.	.24	.82		1.06	1.65
0100	On masonry		285	.028		.26	.95		1.21	1.88
0200	On concrete		250	.032		.26	1.08		1.34	2.11
0400	For plaster, 3/4" deep, on wood		450	.018		.24	.60		.84	1.28
0500	On masonry		225	.036		.26	1.21		1.47	2.31
0600	On concrete		175	.046		.26	1.55		1.81	2.88
0700	On metal lath		200	.040		.26	1.36		1.62	2.56

06 12 Structural Panels

06 12 10 – Structural Insulated Panels

06 12 10.10 OSB Faced Panels

			Crew	Daily Output	Labor-Hours	Unit	Material	Labor	Equipment	Total	Total Incl O&P
0010	**OSB FACED PANELS**										
0100	Structural insul. panels, 7/16" OSB both faces, EPS insul, 3-5/8" T	G	F-3	2075	.019	S.F.	3.65	.59	.32	4.56	5.35
0110	5-5/8" thick	G		1725	.023		4.10	.71	.38	5.19	6.10
0120	7-3/8" thick	G		1425	.028		4.45	.86	.46	5.77	6.85
0130	9-3/8" thick	G		1125	.036		4.75	1.09	.59	6.43	7.70
0140	7/16" OSB one face, EPS insul, 3-5/8" thick	G		2175	.018		3.75	.56	.30	4.61	5.40
0150	5-5/8" thick	G		1825	.022		4.35	.67	.36	5.38	6.30
0160	7-3/8" thick	G		1525	.026		4.85	.80	.43	6.08	7.20
0170	9-3/8" thick	G		1225	.033		5.35	1	.54	6.89	8.15
0190	7/16" OSB - 1/2" GWB faces, EPS insul, 3-5/8" T	G		2075	.019		3.35	.59	.32	4.26	5.05
0200	5-5/8" thick	G		1725	.023		4	.71	.38	5.09	6
0210	7-3/8" thick	G		1425	.028		4.50	.86	.46	5.82	6.90
0220	9-3/8" thick	G		1125	.036		5.10	1.09	.59	6.78	8.05
0240	7/16" OSB - 1/2" MRGWB faces, EPS insul, 3-5/8" T	G		2075	.019		3.45	.59	.32	4.36	5.15
0250	5-5/8" thick	G		1725	.023		4.10	.71	.38	5.19	6.10
0260	7-3/8" thick	G		1425	.028		4.55	.86	.46	5.87	6.95
0270	9-3/8" thick	G		1125	.036		5.20	1.09	.59	6.88	8.15
0300	For 1/2" GWB added to OSB skin, add	G					1.35			1.35	1.49
0310	For 1/2" MRGWB added to OSB skin, add	G					1.35			1.35	1.49
0320	For one T1-11 skin, add to OSB-OSB	G					1.95			1.95	2.15
0330	For one 19/32" CDX skin, add to OSB-OSB	G					1.50			1.50	1.65
0500	Structural insulated panel, 7/16" OSB both sides, straw core										
0510	4-3/8" T, walls (w/sill, splines, plates)	G	F-6	2400	.017	S.F.	7.55	.51	.27	8.33	9.45
0520	Floors (w/splines)	G		2400	.017		7.55	.51	.27	8.33	9.45
0530	Roof (w/splines)	G		2400	.017		7.55	.51	.27	8.33	9.45
0550	7-7/8" T, walls (w/sill, splines, plates)	G		2400	.017		11.40	.51	.27	12.18	13.70
0560	Floors (w/splines)	G		2400	.017		11.40	.51	.27	12.18	13.70
0570	Roof (w/splines)	G		2400	.017		11.40	.51	.27	12.18	13.70

06 12 19 – Composite Shearwall Panels

06 12 19.10 Steel and Wood Composite Shearwall Panels

		Crew	Daily Output	Labor-Hours	Unit	Material	Labor	Equipment	Total	Total Incl O&P
0010	**STEEL & WOOD COMPOSITE SHEARWALL PANELS**									
0020	Anchor bolts, 36" long (must be placed in wet concrete)	1 Carp	150	.053	Ea.	34	1.81		35.81	40.50
0030	On concrete, 2 x 4 & 2 x 6 walls, 7'-10' high, 360 lb. shear, 12" wide	2 Carp	8	2		460	68		528	620
0040	715 lb. shear, 15" wide		8	2		510	68		578	675
0050	1860 lb. shear, 18" wide		8	2		525	68		593	695
0060	2780 lb. shear, 21" wide		8	2		560	68		628	730
0070	3790 lb. shear, 24" wide		8	2		640	68		708	815
0080	2 x 6 walls, 11' to 13' high, 1180 lb. shear, 18" wide		6	2.667		645	90.50		735.50	860

06 12 Structural Panels

06 12 19 – Composite Shearwall Panels

06 12 19.10 Steel and Wood Composite Shearwall Panels		Crew	Daily Output	Labor-Hours	Unit	Material	2016 Bare Costs Labor	Equipment	Total	Total Incl O&P
0090	1555 lb. shear, 21" wide	2 Carp	6	2.667	Ea.	720	90.50		810.50	940
0100	2280 lb. shear, 24" wide	↓	6	2.667	↓	805	90.50		895.50	1,025
0110	For installing above on wood floor frame, add									
0120	Coupler nuts, threaded rods, bolts, shear transfer plate kit	1 Carp	16	.500	Ea.	65	16.95		81.95	100
0130	Framing anchors, angle (2 required)	"	96	.083	"	2.56	2.82		5.38	7.55
0140	For blocking see Section 06 11 10.02									
0150	For installing above, first floor to second floor, wood floor frame, add									
0160	Add stack option to first floor wall panel				Ea.	71.50			71.50	78.50
0170	Threaded rods, bolts, shear transfer plate kit	1 Carp	16	.500		75	16.95		91.95	111
0180	Framing anchors, angle (2 required)	"	96	.083	↓	2.56	2.82		5.38	7.55
0190	For blocking see section 06 11 10.02									
0200	For installing stacked panels, balloon framing									
0210	Add stack option to first floor wall panel				Ea.	71.50			71.50	78.50
0220	Threaded rods, bolts kit	1 Carp	16	.500	"	45	16.95		61.95	78

06 13 Heavy Timber Construction

06 13 13 – Log Construction

06 13 13.10 Log Structures

		Crew	Daily Output	Labor-Hours	Unit	Material	Labor	Equipment	Total	Total Incl O&P
0010	**LOG STRUCTURES**									
0020	Exterior walls, pine, D logs, with double T & G, 6" x 6"	2 Carp	500	.032	L.F.	4.74	1.08		5.82	7
0030	6" x 8"		375	.043		4.94	1.45		6.39	7.85
0040	8" x 6"		375	.043		4.74	1.45		6.19	7.60
0050	8" x 7"		322	.050		4.74	1.68		6.42	8
0060	Square/rectangular logs, with double T & G, 6" x 6"		500	.032		4.74	1.08		5.82	7
0070	6" x 8"		375	.043		4.74	1.45		6.19	7.60
0080	8" x 6"		375	.043		4.74	1.45		6.19	7.60
0090	8" x 7"		322	.050		4.74	1.68		6.42	8
0100	Round logs, with double T & G, 6" x 8"		375	.043		4.74	1.45		6.19	7.60
0110	8" x 7"		322	.050	↓	4.74	1.68		6.42	8
0120	Log siding, ship lapped, 2" x 6"		225	.071	S.F.	3.05	2.41		5.46	7.40
0130	2" x 8"		200	.080		2.56	2.71		5.27	7.35
0140	2" x 12"	↓	180	.089	↓	2.92	3.01		5.93	8.25
0150	Foam sealant strip, 3/8" x 3/8"	1 Carp	1920	.004	L.F.	.24	.14		.38	.50
0152	Chinking, 2" - 3" wide joint, 1/4" to 3/8" deep		600	.013		1.30	.45		1.75	2.19
0154	Caulking, 1/4" to 1/2" joint		900	.009		.50	.30		.80	1.04
0156	Backer rod, 1/4"	↓	900	.009	↓	.20	.30		.50	.72
0157	Penetrating wood preservative	1 Pord	2000	.004	S.F.	.15	.11		.26	.35
0158	Insect treatment	"	4000	.002	"	.34	.06		.40	.47
0160	Upper floor framing, pine, posts/columns, 4" x 6"	2 Carp	750	.021	L.F.	2.88	.72		3.60	4.38
0180	4" x 8"		562	.028		2.93	.97		3.90	4.85
0190	6" x 6"		500	.032		5.05	1.08		6.13	7.35
0200	6" x 8"		375	.043		7.45	1.45		8.90	10.60
0210	8" x 8"		281	.057		8.95	1.93		10.88	13.10
0220	8" x 10"		225	.071		11.20	2.41		13.61	16.35
0230	Beams, 4" x 8"		562	.028		2.93	.97		3.90	4.85
0240	4" x 10"		449	.036		4.67	1.21		5.88	7.15
0250	4" x 12"		375	.043		5.15	1.45		6.60	8.10
0260	6" x 8"		375	.043		7.45	1.45		8.90	10.60
0270	6" x 10"		300	.053		6.50	1.81		8.31	10.20
0280	6" x 12"		250	.064		8.15	2.17		10.32	12.65
0290	8" x 10"	↓	225	.071	↓	11.20	2.41		13.61	16.35

06 13 Heavy Timber Construction

06 13 13 – Log Construction

06 13 13.10 Log Structures

		Crew	Daily Output	Labor-Hours	Unit	Material	2016 Bare Costs Labor	Equipment	Total	Total Incl O&P
0300	8" x 12"	2 Carp	188	.085	L.F.	15.60	2.89		18.49	22
0310	Joists, 4" x 8"		562	.028		2.93	.97		3.90	4.85
0320	4" x 10"		449	.036		4.67	1.21		5.88	7.15
0330	4" x 12"		375	.043		5.15	1.45		6.60	8.10
0340	6" x 8"		375	.043		7.45	1.45		8.90	10.60
0350	6" x 10"		300	.053		6.50	1.81		8.31	10.20
0360	6" x 12"		250	.064		8.15	2.17		10.32	12.65
0370	8" x 10"		225	.071		11.20	2.41		13.61	16.35
0380	8" x 12"		188	.085		13.45	2.89		16.34	19.65
0390	Decking, 1" x 6" T & G		964	.017	S.F.	1.66	.56		2.22	2.76
0400	1" x 8" T & G		700	.023		1.49	.78		2.27	2.94
0410	2" x 6" T & G		482	.033		3.47	1.13		4.60	5.70
0420	Gable end roof framing, rafters, 4" x 8"		562	.028	L.F.	2.93	.97		3.90	4.85
0430	4" x 10"		449	.036		4.67	1.21		5.88	7.15
0450	4" x 12"		375	.043		5.15	1.45		6.60	8.10
0460	6" x 8"		375	.043		7.45	1.45		8.90	10.60
0470	6" x 10"		300	.053		6.50	1.81		8.31	10.20
0480	6" x 12"		250	.064		8.15	2.17		10.32	12.65
0490	8" x 10"		225	.071		11.20	2.41		13.61	16.35
0500	8" x 12"		188	.085		13.45	2.89		16.34	19.65
0510	Purlins, 4" x 8"		562	.028		2.93	.97		3.90	4.85
0520	6" x 8"		375	.043		7.45	1.45		8.90	10.60
0530	Roof decking, 1" x 6" T & G		640	.025	S.F.	1.66	.85		2.51	3.24
0540	1" x 8" T & G		430	.037		1.49	1.26		2.75	3.75
0550	2" x 6" T & G		320	.050		3.47	1.70		5.17	6.65

06 13 23 – Heavy Timber Framing

06 13 23.10 Heavy Framing

		Crew	Daily Output	Labor-Hours	Unit	Material	Labor	Equipment	Total	Total Incl O&P
0010	**HEAVY FRAMING**									
0020	Beams, single 6" x 10"	2 Carp	1.10	14.545	M.B.F.	1,525	495		2,020	2,500
0100	Single 8" x 16"		1.20	13.333	"	2,075	450		2,525	3,050
0202	Built from 2" lumber, multiple 2" x 14"		900	.018	B.F.	.91	.60		1.51	2.01
0212	Built from 3" lumber, multiple 3" x 6"		700	.023		1.25	.78		2.03	2.68
0222	Multiple 3" x 8"		800	.020		1.35	.68		2.03	2.63
0232	Multiple 3" x 10"		900	.018		1.36	.60		1.96	2.51
0242	Multiple 3" x 12"		1000	.016		1.36	.54		1.90	2.41
0252	Built from 4" lumber, multiple 4" x 6"		800	.020		1.44	.68		2.12	2.72
0262	Multiple 4" x 8"		900	.018		1.10	.60		1.70	2.22
0272	Multiple 4" x 10"		1000	.016		1.40	.54		1.94	2.45
0282	Multiple 4" x 12"		1100	.015		1.29	.49		1.78	2.25
0292	Columns, structural grade, 1500f, 4" x 4"		450	.036	L.F.	1.66	1.21		2.87	3.85
0302	6" x 6"		225	.071		4.27	2.41		6.68	8.75
0402	8" x 8"		240	.067		6.90	2.26		9.16	11.35
0502	10" x 10"		90	.178		12.45	6.05		18.50	24
0602	12" x 12"		70	.229		17.50	7.75		25.25	32
0802	Floor planks, 2" thick, T & G, 2" x 6"		1050	.015	B.F.	1.59	.52		2.11	2.61
0902	2" x 10"		1100	.015		1.60	.49		2.09	2.59
1102	3" thick, 3" x 6"		1050	.015		1.59	.52		2.11	2.61
1202	3" x 10"		1100	.015		1.59	.49		2.08	2.58
1402	Girders, structural grade, 12" x 12"		800	.020		1.46	.68		2.14	2.74
1502	10" x 16"		1000	.016		2.50	.54		3.04	3.66
2302	Roof purlins, 4" thick, structural grade		1050	.015		1.10	.52		1.62	2.07

06 15 Wood Decking

06 15 16 – Wood Roof Decking

06 15 16.10 Solid Wood Roof Decking

		Crew	Daily Output	Labor-Hours	Unit	Material	2016 Bare Costs Labor	Equipment	Total	Total Incl O&P
0010	**SOLID WOOD ROOF DECKING**									
0350	Cedar planks, 2" thick	2 Carp	350	.046	S.F.	6.45	1.55		8	9.70
0400	3" thick		320	.050		9.70	1.70		11.40	13.50
0500	4" thick		250	.064		12.90	2.17		15.07	17.85
0550	6" thick		200	.080		19.35	2.71		22.06	26
0650	Douglas fir, 2" thick		350	.046		2.73	1.55		4.28	5.60
0700	3" thick		320	.050		4.10	1.70		5.80	7.35
0800	4" thick		250	.064		5.45	2.17		7.62	9.65
0850	6" thick		200	.080		8.20	2.71		10.91	13.55
0950	Hemlock, 2" thick		350	.046		2.78	1.55		4.33	5.65
1000	3" thick		320	.050		4.17	1.70		5.87	7.45
1100	4" thick		250	.064		5.55	2.17		7.72	9.75
1150	6" thick		200	.080		8.35	2.71		11.06	13.70
1250	Western white spruce, 2" thick		350	.046		1.77	1.55		3.32	4.54
1300	3" thick		320	.050		2.66	1.70		4.36	5.75
1400	4" thick		250	.064		3.55	2.17		5.72	7.55
1450	6" thick		200	.080		5.30	2.71		8.01	10.40

06 15 23 – Laminated Wood Decking

06 15 23.10 Laminated Roof Deck

		Crew	Daily Output	Labor-Hours	Unit	Material	Labor	Equipment	Total	Total Incl O&P
0010	**LAMINATED ROOF DECK**									
0020	Pine or hemlock, 3" thick	2 Carp	425	.038	S.F.	6.25	1.28		7.53	9.05
0100	4" thick		325	.049		8.15	1.67		9.82	11.75
0300	Cedar, 3" thick		425	.038		7.10	1.28		8.38	9.95
0400	4" thick		325	.049		9.50	1.67		11.17	13.25
0600	Fir, 3" thick		425	.038		5.50	1.28		6.78	8.20
0700	4" thick		325	.049		7.50	1.67		9.17	11.05

06 16 Sheathing

06 16 13 – Insulating Sheathing

06 16 13.10 Insulating Sheathing

			Crew	Daily Output	Labor-Hours	Unit	Material	Labor	Equipment	Total	Total Incl O&P
0010	**INSULATING SHEATHING**										
0020	Expanded polystyrene, 1#/C.F. density, 3/4" thick R2.89	G	2 Carp	1400	.011	S.F.	.35	.39		.74	1.04
0030	1" thick R3.85	G		1300	.012		.42	.42		.84	1.16
0040	2" thick R7.69	G		1200	.013		.68	.45		1.13	1.51
0050	Extruded polystyrene, 15 PSI compressive strength, 1" thick, R5	G		1300	.012		.76	.42		1.18	1.53
0060	2" thick, R10	G		1200	.013		.94	.45		1.39	1.79
0070	Polyisocyanurate, 2#/C.F. density, 3/4" thick	G		1400	.011		.62	.39		1.01	1.33
0080	1" thick	G		1300	.012		.63	.42		1.05	1.39
0090	1-1/2" thick	G		1250	.013		.80	.43		1.23	1.61
0100	2" thick	G		1200	.013		.98	.45		1.43	1.84

06 16 23 – Subflooring

06 16 23.10 Subfloor

		Crew	Daily Output	Labor-Hours	Unit	Material	Labor	Equipment	Total	Total Incl O&P
0010	**SUBFLOOR**									
0011	Plywood, CDX, 1/2" thick	2 Carp	1500	.011	SF Flr.	.68	.36		1.04	1.36
0015	Pneumatic nailed		1860	.009		.68	.29		.97	1.24
0017	Pneumatic nailed		1860	.009		.68	.29		.97	1.24
0102	5/8" thick		1350	.012		.76	.40		1.16	1.51
0107	Pneumatic nailed		1674	.010		.76	.32		1.08	1.38
0202	3/4" thick		1250	.013		.91	.43		1.34	1.73
0207	Pneumatic nailed		1550	.010		.91	.35		1.26	1.59

06 16 Sheathing

06 16 23 – Subflooring

06 16 23.10 Subfloor

		Crew	Daily Output	Labor-Hours	Unit	Material	2016 Bare Costs Labor	Equipment	Total	Total Incl O&P
0302	1-1/8" thick, 2-4-1 including underlayment	2 Carp	1050	.015	SF Flr.	2.03	.52		2.55	3.09
0440	With boards, 1" x 6", S4S, laid regular		900	.018		1.54	.60		2.14	2.70
0452	1" x 8", laid regular		1000	.016		1.88	.54		2.42	2.98
0462	Laid diagonal		850	.019		1.88	.64		2.52	3.14
0502	1" x 10", laid regular		1100	.015		1.92	.49		2.41	2.94
0602	Laid diagonal		900	.018		1.92	.60		2.52	3.12
1500	OSB, 5/8" thick		1330	.012	S.F.	.54	.41		.95	1.27
1600	3/4" thick		1230	.013	"	.68	.44		1.12	1.49
8990	Subfloor adhesive, 3/8" bead	1 Carp	2300	.003	L.F.	.12	.12		.24	.33

06 16 26 – Underlayment

06 16 26.10 Wood Product Underlayment

		Crew	Daily Output	Labor-Hours	Unit	Material	Labor	Equipment	Total	Total Incl O&P
0010	**WOOD PRODUCT UNDERLAYMENT**									
0015	Plywood, underlayment grade, 1/4" thick	2 Carp	1500	.011	S.F.	.89	.36		1.25	1.59
0018	Pneumatic nailed		1860	.009		.89	.29		1.18	1.47
0030	3/8" thick		1500	.011		.99	.36		1.35	1.70
0070	Pneumatic nailed		1860	.009		.99	.29		1.28	1.58
0102	1/2" thick		1450	.011		1.17	.37		1.54	1.92
0107	Pneumatic nailed		1798	.009		1.17	.30		1.47	1.80
0202	5/8" thick		1400	.011		1.29	.39		1.68	2.07
0207	Pneumatic nailed		1736	.009		1.29	.31		1.60	1.94
0302	3/4" thick		1300	.012		1.48	.42		1.90	2.33
0306	Pneumatic nailed		1612	.010		1.48	.34		1.82	2.19
0502	Particle board, 3/8" thick		1500	.011		.41	.36		.77	1.06
0507	Pneumatic nailed		1860	.009		.41	.29		.70	.94
0602	1/2" thick		1450	.011		.45	.37		.82	1.13
0607	Pneumatic nailed		1798	.009		.45	.30		.75	1.01
0802	5/8" thick		1400	.011		.54	.39		.93	1.24
0807	Pneumatic nailed		1736	.009		.54	.31		.85	1.11
0902	3/4" thick		1300	.012		.68	.42		1.10	1.45
0907	Pneumatic nailed		1612	.010		.68	.34		1.02	1.31
1100	Hardboard, underlayment grade, 4' x 4', .215" thick [G]		1500	.011		.63	.36		.99	1.30

06 16 33 – Wood Board Sheathing

06 16 33.10 Board Sheathing

		Crew	Daily Output	Labor-Hours	Unit	Material	Labor	Equipment	Total	Total Incl O&P
0009	**BOARD SHEATHING**									
0010	Roof, 1" x 6" boards, laid horizontal	2 Carp	725	.022	S.F.	1.54	.75		2.29	2.94
0020	On steep roof		520	.031		1.54	1.04		2.58	3.44
0040	On dormers, hips, & valleys		480	.033		1.54	1.13		2.67	3.58
0050	Laid diagonal		650	.025		1.54	.83		2.37	3.09
0070	1" x 8" boards, laid horizontal		875	.018		1.88	.62		2.50	3.11
0080	On steep roof		635	.025		1.88	.85		2.73	3.50
0090	On dormers, hips, & valleys		580	.028		1.92	.94		2.86	3.68
0100	Laid diagonal		725	.022		1.88	.75		2.63	3.32
0110	Skip sheathing, 1" x 4", 7" OC	1 Carp	1200	.007		.58	.23		.81	1.01
0120	1" x 6", 9" OC		1450	.006		.71	.19		.90	1.09
0180	Tongue and groove sheathing/decking, 1" x 6"		1000	.008		1.82	.27		2.09	2.45
0190	2" x 6"		1000	.008		3.81	.27		4.08	4.64
0200	Walls, 1" x 6" boards, laid regular	2 Carp	650	.025		1.54	.83		2.37	3.09
0210	Laid diagonal		585	.027		1.54	.93		2.47	3.24
0220	1" x 8" boards, laid regular		765	.021		1.88	.71		2.59	3.26
0230	Laid diagonal		650	.025		1.88	.83		2.71	3.47

06 16 Sheathing

06 16 36 – Wood Panel Product Sheathing

06 16 36.10 Sheathing		Crew	Daily Output	Labor-Hours	Unit	Material	2016 Bare Costs Labor	Equipment	Total	Total Incl O&P
0010	**SHEATHING** R061636-20									
0012	Plywood on roofs, CDX									
0032	5/16" thick	2 Carp	1600	.010	S.F.	.62	.34		.96	1.26
0037	Pneumatic nailed		1952	.008		.62	.28		.90	1.16
0052	3/8" thick		1525	.010		.62	.36		.98	1.28
0057	Pneumatic nailed		1860	.009		.62	.29		.91	1.17
0102	1/2" thick		1400	.011		.68	.39		1.07	1.40
0103	Pneumatic nailed		1708	.009		.68	.32		1	1.28
0202	5/8" thick		1300	.012		.76	.42		1.18	1.54
0207	Pneumatic nailed		1586	.010		.76	.34		1.10	1.41
0302	3/4" thick		1200	.013		.91	.45		1.36	1.76
0307	Pneumatic nailed		1464	.011		.91	.37		1.28	1.62
0502	Plywood on walls, with exterior CDX, 3/8" thick		1200	.013		.62	.45		1.07	1.44
0507	Pneumatic nailed		1488	.011		.62	.36		.98	1.29
0603	1/2" thick		1125	.014		.68	.48		1.16	1.56
0608	Pneumatic nailed		1395	.011		.68	.39		1.07	1.40
0702	5/8" thick		1050	.015		.76	.52		1.28	1.70
0707	Pneumatic nailed		1302	.012		.76	.42		1.18	1.54
0803	3/4" thick		975	.016		.91	.56		1.47	1.93
0808	Pneumatic nailed		1209	.013		.91	.45		1.36	1.75
1000	For shear wall construction, add						20%			
1200	For structural 1 exterior plywood, add				S.F.	10%				
3000	Wood fiber, regular, no vapor barrier, 1/2" thick	2 Carp	1200	.013		.64	.45		1.09	1.46
3100	5/8" thick		1200	.013		.76	.45		1.21	1.60
3300	No vapor barrier, in colors, 1/2" thick		1200	.013		.77	.45		1.22	1.61
3400	5/8" thick		1200	.013		.81	.45		1.26	1.65
3600	With vapor barrier one side, white, 1/2" thick		1200	.013		.63	.45		1.08	1.45
3700	Vapor barrier 2 sides, 1/2" thick		1200	.013		.86	.45		1.31	1.71
3800	Asphalt impregnated, 25/32" thick		1200	.013		.33	.45		.78	1.12
3850	Intermediate, 1/2" thick		1200	.013		.25	.45		.70	1.04
4500	Oriented strand board, on roof, 7/16" thick G		1460	.011		.50	.37		.87	1.17
4505	Pneumatic nailed G		1780	.009		.50	.30		.80	1.06
4550	1/2" thick G		1400	.011		.50	.39		.89	1.20
4555	Pneumatic nailed G		1736	.009		.50	.31		.81	1.07
4600	5/8" thick G		1300	.012		.69	.42		1.11	1.46
4605	Pneumatic nailed G		1586	.010		.69	.34		1.03	1.33
4610	On walls, 7/16" thick G		1200	.013		.50	.45		.95	1.31
4615	Pneumatic nailed G		1488	.011		.50	.36		.86	1.16
4620	1/2" thick G		1195	.013		.50	.45		.95	1.31
4625	Pneumatic nailed G		1325	.012		.50	.41		.91	1.24
4630	5/8" thick G		1050	.015		.69	.52		1.21	1.62
4635	Pneumatic nailed G		1302	.012		.69	.42		1.11	1.46
4700	Oriented strand board, factory laminated W.R. barrier, on roof, 1/2" thick G		1400	.011		.78	.39		1.17	1.51
4705	Pneumatic nailed G		1736	.009		.78	.31		1.09	1.38
4720	5/8" thick G		1300	.012		.93	.42		1.35	1.72
4725	Pneumatic nailed G		1586	.010		.93	.34		1.27	1.59
4730	5/8" thick, T&G G		1150	.014		.93	.47		1.40	1.81
4735	Pneumatic nailed, T&G G		1400	.011		.93	.39		1.32	1.67
4740	On walls, 7/16" thick G		1200	.013		.66	.45		1.11	1.49
4745	Pneumatic nailed G		1488	.011		.66	.36		1.02	1.34
4750	1/2" thick G		1195	.013		.78	.45		1.23	1.62
4755	Pneumatic nailed G		1325	.012		.78	.41		1.19	1.55

06 16 Sheathing

06 16 36 – Wood Panel Product Sheathing

06 16 36.10 Sheathing		Crew	Daily Output	Labor-Hours	Unit	Material	2016 Bare Costs Labor	Equipment	Total	Total Incl O&P
4800	Joint sealant tape, 3-1/2"	2 Carp	7600	.002	L.F.	.31	.07		.38	.46
4810	Joint sealant tape, 6"	↓	7600	.002	"	.44	.07		.51	.60

06 16 43 – Gypsum Sheathing

06 16 43.10 Gypsum Sheathing

		Crew	Daily Output	Labor-Hours	Unit	Material	Labor	Equipment	Total	Incl O&P
0010	**GYPSUM SHEATHING**									
0020	Gypsum, weatherproof, 1/2" thick	2 Carp	1125	.014	S.F.	.46	.48		.94	1.32
0040	With embedded glass mats	"	1100	.015	"	.70	.49		1.19	1.60

06 17 Shop-Fabricated Structural Wood

06 17 33 – Wood I-Joists

06 17 33.10 Wood and Composite I-Joists

		Crew	Daily Output	Labor-Hours	Unit	Material	Labor	Equipment	Total	Incl O&P
0010	**WOOD AND COMPOSITE I-JOISTS**									
0100	Plywood webs, incl. bridging & blocking, panels 24" O.C.									
1200	15' to 24' span, 50 psf live load	F-5	2400	.013	SF Flr.	2.09	.39		2.48	2.96
1300	55 psf live load		2250	.014		2.30	.42		2.72	3.24
1400	24' to 30' span, 45 psf live load		2600	.012		2.40	.36		2.76	3.25
1500	55 psf live load	↓	2400	.013	↓	4.05	.39		4.44	5.10

06 17 53 – Shop-Fabricated Wood Trusses

06 17 53.10 Roof Trusses

		Crew	Daily Output	Labor-Hours	Unit	Material	Labor	Equipment	Total	Incl O&P
0010	**ROOF TRUSSES**									
5000	Common wood, 2" x 4" metal plate connected, 24" O.C., 4/12 slope									
5010	1' overhang, 12' span	F-5	55	.582	Ea.	41	17.05		58.05	74
5050	20' span	F-6	62	.645		70.50	19.75	10.60	100.85	122
5100	24' span		60	.667		76.50	20.50	10.95	107.95	131
5150	26' span		57	.702		80	21.50	11.55	113.05	137
5200	28' span		53	.755		93	23	12.40	128.40	154
5240	30' span		51	.784		100	24	12.90	136.90	164
5250	32' span		50	.800		110	24.50	13.15	147.65	177
5280	34' span		48	.833		114	25.50	13.70	153.20	183
5350	8/12 pitch, 1' overhang, 20' span		57	.702		77.50	21.50	11.55	110.55	134
5400	24' span		55	.727		97.50	22.50	11.95	131.95	157
5450	26' span		52	.769		103	23.50	12.65	139.15	166
5500	28' span		49	.816		117	25	13.45	155.45	185
5550	32' span		45	.889		133	27	14.65	174.65	208
5600	36' span		41	.976		164	30	16.05	210.05	248
5650	38' span		40	1		184	30.50	16.45	230.95	271
5700	40' span	↓	40	1	↓	187	30.50	16.45	233.95	275

06 18 Glued-Laminated Construction

06 18 13 – Glued-Laminated Beams

06 18 13.10 Laminated Beams		Crew	Daily Output	Labor-Hours	Unit	Material	2016 Bare Costs Labor	Equipment	Total	Total Incl O&P
0010	**LAMINATED BEAMS**									
0050	3-1/2" x 18"	F-3	480	.083	L.F.	31	2.56	1.37	34.93	40
0100	5-1/4" x 11-7/8"		450	.089		30.50	2.73	1.46	34.69	39.50
0150	5-1/4" x 16"		360	.111		41	3.41	1.83	46.24	52.50
0200	5-1/4" x 18"		290	.138		46	4.23	2.27	52.50	60.50
0250	5-1/4" x 24"		220	.182		61.50	5.60	2.99	70.09	80
0300	7" x 11-7/8"		320	.125		43.50	3.83	2.06	49.39	56
0350	7" x 16"		260	.154		59	4.72	2.53	66.25	75.50
0400	7" x 18"	↓	210	.190		68	5.85	3.14	76.99	88.50
0500	For premium appearance, add to S.F. prices					5%				
0550	For industrial type, deduct					15%				
0600	For stain and varnish, add					5%				
0650	For 3/4" laminations, add					25%				

06 18 13.20 Laminated Framing

		Crew	Daily Output	Labor-Hours	Unit	Material	Labor	Equipment	Total	Total Incl O&P
0010	**LAMINATED FRAMING**									
0020	30 lb., short term live load, 15 lb. dead load									
0200	Straight roof beams, 20' clear span, beams 8' O.C.	F-3	2560	.016	SF Flr.	2.08	.48	.26	2.82	3.37
0300	Beams 16' O.C.		3200	.013		1.52	.38	.21	2.11	2.54
0500	40' clear span, beams 8' O.C.		3200	.013		3.95	.38	.21	4.54	5.20
0600	Beams 16' O.C.	↓	3840	.010		3.25	.32	.17	3.74	4.31
0800	60' clear span, beams 8' O.C.	F-4	2880	.014		6.75	.43	.39	7.57	8.60
0900	Beams 16' O.C.	"	3840	.010		5.05	.32	.29	5.66	6.40
1100	Tudor arches, 30' to 40' clear span, frames 8' O.C.	F-3	1680	.024		8.80	.73	.39	9.92	11.35
1200	Frames 16' O.C.	"	2240	.018		6.90	.55	.29	7.74	8.85
1400	50' to 60' clear span, frames 8' O.C.	F-4	2200	.018		9.50	.56	.51	10.57	11.95
1500	Frames 16' O.C.		2640	.015		8.10	.46	.43	8.99	10.15
1700	Radial arches, 60' clear span, frames 8' O.C.		1920	.021		8.85	.64	.58	10.07	11.45
1800	Frames 16' O.C.		2880	.014		7	.43	.39	7.82	8.85
2000	100' clear span, frames 8' O.C.		1600	.025		9.15	.77	.70	10.62	12.10
2100	Frames 16' O.C.		2400	.017		8.05	.51	.47	9.03	10.20
2300	120' clear span, frames 8' O.C.		1440	.028		12.15	.85	.78	13.78	15.65
2400	Frames 16' O.C.	↓	1920	.021		11.10	.64	.58	12.32	13.90
2600	Bowstring trusses, 20' O.C., 40' clear span	F-3	2400	.017		5.50	.51	.27	6.28	7.20
2700	60' clear span	F-4	3600	.011		4.95	.34	.31	5.60	6.35
2800	100' clear span		4000	.010		7	.31	.28	7.59	8.50
2900	120' clear span	↓	3600	.011		7.45	.34	.31	8.10	9.10
3100	For premium appearance, add to S.F. prices					5%				
3300	For industrial type, deduct					15%				
3500	For stain and varnish, add					5%				
3900	For 3/4" laminations, add to straight					25%				
4100	Add to curved					15%				
4300	Alternate pricing method: (use nominal footage of									
4310	components). Straight beams, camber less than 6"	F-3	3.50	11.429	M.B.F.	2,925	350	188	3,463	4,000
4400	Columns, including hardware		2	20		3,125	615	330	4,070	4,825
4600	Curved members, radius over 32'		2.50	16		3,200	490	263	3,953	4,625
4700	Radius 10' to 32'	↓	3	13.333		3,175	410	219	3,804	4,425
4900	For complicated shapes, add maximum					100%				
5100	For pressure treating, add to straight					35%				
5200	Add to curved					45%				
6000	Laminated veneer members, southern pine or western species									
6050	1-3/4" wide x 5-1/2" deep	2 Carp	480	.033	L.F.	3.44	1.13		4.57	5.65
6100	9-1/2" deep	↓	480	.033	↓	4.74	1.13		5.87	7.10

06 18 Glued-Laminated Construction

06 18 13 – Glued-Laminated Beams

06 18 13.20 Laminated Framing		Crew	Daily Output	Labor-Hours	Unit	Material	2016 Bare Costs Labor	Equipment	Total	Total Incl O&P
6150	14" deep	2 Carp	450	.036	L.F.	7.85	1.21		9.06	10.65
6200	18" deep		450	.036		10.65	1.21		11.86	13.75
6300	Parallel strand members, southern pine or western species									
6350	1-3/4" wide x 9-1/4" deep	2 Carp	480	.033	L.F.	5.10	1.13		6.23	7.50
6400	11-1/4" deep		450	.036		5.75	1.21		6.96	8.35
6450	14" deep		400	.040		7.85	1.36		9.21	10.90
6500	3-1/2" wide x 9-1/4" deep		480	.033		16.30	1.13		17.43	19.85
6550	11-1/4" deep		450	.036		20.50	1.21		21.71	24.50
6600	14" deep		400	.040		24	1.36		25.36	29
6650	7" wide x 9-1/4" deep		450	.036		34.50	1.21		35.71	40
6700	11-1/4" deep		420	.038		43.50	1.29		44.79	49.50
6750	14" deep		400	.040		52	1.36		53.36	59.50
8000	Straight beams									
8102	20' span									
8104	3-1/8" x 9"	F-3	30	1.333	Ea.	137	41	22	200	244
8106	X 10-1/2"		30	1.333		160	41	22	223	269
8108	X 12"		30	1.333		183	41	22	246	294
8110	X 13-1/2"		30	1.333		205	41	22	268	320
8112	X 15"		29	1.379		228	42.50	22.50	293	345
8114	5-1/8" x 10-1/2"		30	1.333		262	41	22	325	380
8116	X 12"		30	1.333		299	41	22	362	425
8118	X 13-1/2"		30	1.333		335	41	22	398	465
8120	X 15"		29	1.379		375	42.50	22.50	440	505
8122	X 16-1/2"		29	1.379		410	42.50	22.50	475	550
8124	X 18"		29	1.379		450	42.50	22.50	515	590
8126	X 19-1/2"		29	1.379		485	42.50	22.50	550	630
8128	X 21"		28	1.429		525	44	23.50	592.50	675
8130	X 22-1/2"		28	1.429		560	44	23.50	627.50	715
8132	X 24"		28	1.429		600	44	23.50	667.50	760
8134	6-3/4" x 12"		29	1.379		395	42.50	22.50	460	530
8136	X 13-1/2"		29	1.379		445	42.50	22.50	510	585
8138	X 15"		29	1.379		495	42.50	22.50	560	635
8140	X 16-1/2"		28	1.429		540	44	23.50	607.50	695
8142	X 18"		28	1.429		590	44	23.50	657.50	750
8144	X 19-1/2"		28	1.429		640	44	23.50	707.50	805
8146	X 21"		27	1.481		690	45.50	24.50	760	865
8148	X 22-1/2"		27	1.481		740	45.50	24.50	810	920
8150	X 24"		27	1.481		790	45.50	24.50	860	970
8152	X 25-1/2"		27	1.481		840	45.50	24.50	910	1,025
8154	X 27"		26	1.538		885	47	25.50	957.50	1,075
8156	X 28-1/2"		26	1.538		935	47	25.50	1,007.50	1,125
8158	X 30"		26	1.538		985	47	25.50	1,057.50	1,175
8200	30' span									
8250	3-1/8" x 9"	F-3	30	1.333	Ea.	205	41	22	268	320
8252	X 10-1/2"		30	1.333		240	41	22	303	355
8254	X 12"		30	1.333		274	41	22	337	395
8256	X 13-1/2"		30	1.333		310	41	22	373	435
8258	X 15"		29	1.379		340	42.50	22.50	405	470
8260	5-1/8" x 10-1/2"		30	1.333		395	41	22	458	525
8262	X 12"		30	1.333		450	41	22	513	590
8264	X 13-1/2"		30	1.333		505	41	22	568	650
8266	X 15"		29	1.379		560	42.50	22.50	625	710
8268	X 16-1/2"		29	1.379		615	42.50	22.50	680	775

06 18 Glued-Laminated Construction

06 18 13 – Glued-Laminated Beams

06 18 13.20 Laminated Framing		Crew	Daily Output	Labor-Hours	Unit	Material	2016 Bare Costs Labor	Equipment	Total	Total Incl O&P
8270	X 18"	F-3	29	1.379	Ea.	675	42.50	22.50	740	835
8272	X 19-1/2"		29	1.379		730	42.50	22.50	795	900
8274	X 21"		28	1.429		785	44	23.50	852.50	965
8276	X 22-1/2"		28	1.429		840	44	23.50	907.50	1,025
8278	X 24"		28	1.429		900	44	23.50	967.50	1,100
8280	6-3/4" x 12"		29	1.379		590	42.50	22.50	655	745
8282	X 13-1/2"		29	1.379		665	42.50	22.50	730	825
8284	X 15"		29	1.379		740	42.50	22.50	805	910
8286	X 16-1/2"		28	1.429		815	44	23.50	882.50	995
8288	X 18"		28	1.429		885	44	23.50	952.50	1,075
8290	X 19-1/2"		28	1.429		960	44	23.50	1,027.50	1,150
8292	X 21"		27	1.481		1,025	45.50	24.50	1,095	1,250
8294	X 22-1/2"		27	1.481		1,100	45.50	24.50	1,170	1,325
8296	X 24"		27	1.481		1,175	45.50	24.50	1,245	1,400
8298	X 25-1/2"		27	1.481		1,250	45.50	24.50	1,320	1,475
8300	X 27"		26	1.538		1,325	47	25.50	1,397.50	1,575
8302	X 28-1/2"		26	1.538		1,400	47	25.50	1,472.50	1,650
8304	X 30"		26	1.538		1,475	47	25.50	1,547.50	1,725
8400	40' span									
8402	3-1/8" x 9"	F-3	30	1.333	Ea.	274	41	22	337	395
8404	X 10-1/2"		30	1.333		320	41	22	383	445
8406	X 12"		30	1.333		365	41	22	428	495
8408	X 13-1/2"		30	1.333		410	41	22	473	545
8410	X 15"		29	1.379		455	42.50	22.50	520	595
8412	5-1/8" x 10-1/2"		30	1.333		525	41	22	588	670
8414	X 12"		30	1.333		600	41	22	663	755
8416	X 13-1/2"		30	1.333		675	41	22	738	835
8418	X 15"		29	1.379		750	42.50	22.50	815	920
8420	X 16-1/2"		29	1.379		825	42.50	22.50	890	1,000
8422	X 18"		29	1.379		900	42.50	22.50	965	1,075
8424	X 19-1/2"		29	1.379		975	42.50	22.50	1,040	1,175
8426	X 21"		28	1.429		1,050	44	23.50	1,117.50	1,250
8428	X 22-1/2"		28	1.429		1,125	44	23.50	1,192.50	1,325
8430	X 24"		28	1.429		1,200	44	23.50	1,267.50	1,425
8432	6-3/4" x 12"		29	1.379		790	42.50	22.50	855	960
8434	X 13-1/2"		29	1.379		885	42.50	22.50	950	1,075
8436	X 15"		29	1.379		985	42.50	22.50	1,050	1,175
8438	X 16-1/2"		28	1.429		1,075	44	23.50	1,142.50	1,300
8440	X 18"		28	1.429		1,175	44	23.50	1,242.50	1,400
8442	X 19-1/2"		28	1.429		1,275	44	23.50	1,342.50	1,500
8444	X 21"		27	1.481		1,375	45.50	24.50	1,445	1,625
8446	X 22-1/2"		27	1.481		1,475	45.50	24.50	1,545	1,725
8448	X 24"		27	1.481		1,575	45.50	24.50	1,645	1,825
8450	X 25-1/2"		27	1.481		1,675	45.50	24.50	1,745	1,950
8452	X 27"		26	1.538		1,775	47	25.50	1,847.50	2,050
8454	X 28-1/2"		26	1.538		1,875	47	25.50	1,947.50	2,150
8456	X 30"		26	1.538		1,975	47	25.50	2,047.50	2,275

For customer support on your Residential Cost Data, call 877.759.4771.

06 22 Millwork

06 22 13 – Standard Pattern Wood Trim

06 22 13.10 Millwork

		Crew	Daily Output	Labor-Hours	Unit	Material	2016 Bare Costs Labor	Equipment	Total	Total Incl O&P
0010	**MILLWORK**									
0020	Rule of thumb, milled material equals rough lumber cost x 3									
1020	1" x 12", custom birch				L.F.	4.71			4.71	5.20
1040	Cedar					7.40			7.40	8.15
1060	Oak					4.47			4.47	4.92
1080	Redwood					4.71			4.71	5.20
1100	Southern yellow pine					4.02			4.02	4.42
1120	Sugar pine					4.10			4.10	4.51
1140	Teak					31.50			31.50	35
1160	Walnut					7.40			7.40	8.15
1180	White pine					5.20			5.20	5.75

06 22 13.15 Moldings, Base

		Crew	Daily Output	Labor-Hours	Unit	Material	Labor	Equipment	Total	Total Incl O&P
0010	**MOLDINGS, BASE**									
5100	Classic profile, 5/8" x 5-1/2", finger jointed and primed	1 Carp	250	.032	L.F.	1.49	1.08		2.57	3.46
5105	Poplar		240	.033		1.57	1.13		2.70	3.61
5110	Red oak		220	.036		2.41	1.23		3.64	4.71
5115	Maple		220	.036		3.64	1.23		4.87	6.05
5120	Cherry		220	.036		4.32	1.23		5.55	6.80
5125	3/4 x 7-1/2", finger jointed and primed		250	.032		1.89	1.08		2.97	3.90
5130	Poplar		240	.033		2.51	1.13		3.64	4.65
5135	Red oak		220	.036		3.50	1.23		4.73	5.90
5140	Maple		220	.036		4.84	1.23		6.07	7.35
5145	Cherry		220	.036		5.75	1.23		6.98	8.40
5150	Modern profile, 5/8" x 3-1/2", finger jointed and primed		250	.032		.92	1.08		2	2.83
5155	Poplar		240	.033		1	1.13		2.13	2.99
5160	Red oak		220	.036		1.65	1.23		2.88	3.87
5165	Maple		220	.036		2.55	1.23		3.78	4.86
5170	Cherry		220	.036		2.56	1.23		3.79	4.87
5175	Ogee profile, 7/16" x 3", finger jointed and primed		250	.032		.63	1.08		1.71	2.51
5180	Poplar		240	.033		.66	1.13		1.79	2.61
5185	Red oak		220	.036		.93	1.23		2.16	3.08
5200	9/16" x 3-1/2", finger jointed and primed		250	.032		.66	1.08		1.74	2.54
5205	Pine		240	.033		1.13	1.13		2.26	3.13
5210	Red oak		220	.036		2.38	1.23		3.61	4.68
5215	9/16" x 4-1/2", red oak		220	.036		3.56	1.23		4.79	6
5220	5/8" x 3-1/2", finger jointed and primed		250	.032		.91	1.08		1.99	2.82
5225	Poplar		240	.033		1	1.13		2.13	2.99
5230	Red oak		220	.036		1.65	1.23		2.88	3.87
5235	Maple		220	.036		2.55	1.23		3.78	4.86
5240	Cherry		220	.036		2.75	1.23		3.98	5.10
5245	5/8" x 4", finger jointed and primed		250	.032		1.17	1.08		2.25	3.10
5250	Poplar		240	.033		1.29	1.13		2.42	3.31
5255	Red oak		220	.036		1.85	1.23		3.08	4.09
5260	Maple		220	.036		2.85	1.23		4.08	5.20
5265	Cherry		220	.036		2.99	1.23		4.22	5.35
5270	Rectangular profile, oak, 3/8" x 1-1/4"		260	.031		1.20	1.04		2.24	3.07
5275	1/2" x 2-1/2"		255	.031		1.94	1.06		3	3.91
5280	1/2" x 3-1/2"		250	.032		2.36	1.08		3.44	4.41
5285	1" x 6"		240	.033		4.21	1.13		5.34	6.50
5290	1" x 8"		240	.033		5.80	1.13		6.93	8.30
5295	Pine, 3/8" x 1-3/4"		260	.031		.46	1.04		1.50	2.25
5300	7/16" x 2-1/2"		255	.031		.74	1.06		1.80	2.59

06 22 Millwork

06 22 13 – Standard Pattern Wood Trim

06 22 13.15 Moldings, Base

		Crew	Daily Output	Labor-Hours	Unit	Material	2016 Bare Costs Labor	Equipment	Total	Total Incl O&P
5305	1" x 6"	1 Carp	240	.033	L.F.	.83	1.13		1.96	2.80
5310	1" x 8"		240	.033		1.06	1.13		2.19	3.05
5315	Shoe, 1/2" x 3/4", primed		260	.031		.50	1.04		1.54	2.30
5320	Pine		240	.033		.30	1.13		1.43	2.22
5325	Poplar		240	.033		.39	1.13		1.52	2.32
5330	Red oak		220	.036		.53	1.23		1.76	2.64
5335	Maple		220	.036		.70	1.23		1.93	2.83
5340	Cherry		220	.036		.80	1.23		2.03	2.94
5345	11/16" x 1-1/2", pine		240	.033		.70	1.13		1.83	2.66
5350	Caps, 11/16" x 1-3/8", pine		240	.033		.55	1.13		1.68	2.49
5355	3/4" x 1-3/4", finger jointed and primed		260	.031		.71	1.04		1.75	2.53
5360	Poplar		240	.033		1.01	1.13		2.14	3
5365	Red oak		220	.036		1.21	1.23		2.44	3.39
5370	Maple		220	.036		1.58	1.23		2.81	3.80
5375	Cherry		220	.036		3.20	1.23		4.43	5.60
5380	Combination base & shoe, 9/16" x 3-1/2" & 1/2" x 3/4", pine		125	.064		1.43	2.17		3.60	5.20
5385	Three piece oak, 6" high		80	.100		5.95	3.39		9.34	12.25
5390	Including 3/4" x 1" base shoe		70	.114		6.10	3.87		9.97	13.25
5395	Flooring cant strip, 3/4" x 3/4", pre-finished pine		260	.031		.51	1.04		1.55	2.31
5400	For pre-finished, stain and clear coat, add					.53			.53	.58
5405	Clear coat only, add					.45			.45	.50

06 22 13.30 Moldings, Casings

		Crew	Daily Output	Labor-Hours	Unit	Material	2016 Bare Costs Labor	Equipment	Total	Total Incl O&P
0010	**MOLDINGS, CASINGS**									
0085	Apron, 9/16" x 2-1/2", pine	1 Carp	250	.032	L.F.	1.63	1.08		2.71	3.61
0090	5/8" x 2-1/2", pine		250	.032		1.63	1.08		2.71	3.61
0110	5/8" x 3-1/2", pine		220	.036		1.86	1.23		3.09	4.10
0300	Band, 11/16" x 1-1/8", pine		270	.030		.75	1		1.75	2.50
0310	11/16" x 1-1/2", finger jointed and primed		270	.030		.76	1		1.76	2.51
0320	Pine		270	.030		.96	1		1.96	2.73
0330	11/16" x 1-3/4", finger jointed and primed		270	.030		.95	1		1.95	2.72
0350	Pine		270	.030		1.16	1		2.16	2.95
0355	Beaded, 3/4" x 3-1/2", finger jointed and primed		220	.036		1.11	1.23		2.34	3.28
0360	Poplar		220	.036		1.11	1.23		2.34	3.28
0365	Red oak		220	.036		1.65	1.23		2.88	3.87
0370	Maple		220	.036		2.55	1.23		3.78	4.86
0375	Cherry		220	.036		3.09	1.23		4.32	5.45
0380	3/4" x 4", finger jointed and primed		220	.036		1.17	1.23		2.40	3.34
0385	Poplar		220	.036		1.45	1.23		2.68	3.65
0390	Red oak		220	.036		2.05	1.23		3.28	4.31
0395	Maple		220	.036		2.65	1.23		3.88	4.97
0400	Cherry		220	.036		3.33	1.23		4.56	5.70
0405	3/4" x 5-1/2", finger jointed and primed		200	.040		1.17	1.36		2.53	3.55
0410	Poplar		200	.040		1.93	1.36		3.29	4.39
0415	Red oak		200	.040		2.84	1.36		4.20	5.40
0420	Maple		200	.040		3.73	1.36		5.09	6.35
0425	Cherry		200	.040		4.31	1.36		5.67	7
0430	Classic profile, 3/4" x 2-3/4", finger jointed and primed		250	.032		.80	1.08		1.88	2.70
0435	Poplar		250	.032		1.02	1.08		2.10	2.94
0440	Red oak		250	.032		1.48	1.08		2.56	3.45
0445	Maple		250	.032		2.09	1.08		3.17	4.12
0450	Cherry		250	.032		2.39	1.08		3.47	4.45
0455	Fluted, 3/4" x 3-1/2", poplar		220	.036		1.12	1.23		2.35	3.29

06 22 Millwork

06 22 13 – Standard Pattern Wood Trim

06 22 13.30 Moldings, Casings

		Crew	Daily Output	Labor-Hours	Unit	Material	2016 Bare Costs Labor	Equipment	Total	Total Incl O&P
0460	Red oak	1 Carp	220	.036	L.F.	1.66	1.23		2.89	3.88
0465	Maple		220	.036		2.53	1.23		3.76	4.84
0470	Cherry		220	.036		3.09	1.23		4.32	5.45
0475	3/4" x 4", poplar		220	.036		1.44	1.23		2.67	3.64
0480	Red oak		220	.036		2.03	1.23		3.26	4.29
0485	Maple		220	.036		2.65	1.23		3.88	4.97
0490	Cherry		220	.036		3.33	1.23		4.56	5.70
0495	3/4" x 5-1/2", poplar		200	.040		1.93	1.36		3.29	4.39
0500	Red oak		200	.040		2.86	1.36		4.22	5.40
0505	Maple		200	.040		3.73	1.36		5.09	6.35
0510	Cherry		200	.040		4.32	1.36		5.68	7
0515	3/4" x 7-1/2", poplar		190	.042		1.22	1.43		2.65	3.73
0520	Red oak		190	.042		3.92	1.43		5.35	6.70
0525	Maple		190	.042		5.60	1.43		7.03	8.55
0530	Cherry		190	.042		6.70	1.43		8.13	9.75
0535	3/4" x 9-1/2", poplar		180	.044		4.13	1.51		5.64	7.05
0540	Red oak		180	.044		6.55	1.51		8.06	9.70
0545	Maple		180	.044		8.95	1.51		10.46	12.30
0550	Cherry		180	.044		9.75	1.51		11.26	13.20
0555	Modern profile, 9/16" x 2-1/4", poplar		250	.032		.59	1.08		1.67	2.47
0560	Red oak		250	.032		.79	1.08		1.87	2.69
0565	11/16" x 2-1/2", finger jointed & primed		250	.032		.84	1.08		1.92	2.74
0570	Pine		250	.032		1.28	1.08		2.36	3.23
0575	3/4" x 2-1/2", poplar		250	.032		.90	1.08		1.98	2.81
0580	Red oak		250	.032		1.21	1.08		2.29	3.15
0585	Maple		250	.032		1.81	1.08		2.89	3.81
0590	Cherry		250	.032		2.30	1.08		3.38	4.35
0595	Mullion, 5/16" x 2", pine		270	.030		.86	1		1.86	2.62
0600	9/16" x 2-1/2", fingerjointed and primed		250	.032		.98	1.08		2.06	2.90
0605	Pine		250	.032		1.28	1.08		2.36	3.23
0610	Red oak		250	.032		2.86	1.08		3.94	4.96
0615	1-1/16" x 3-3/4", red oak		220	.036		6.95	1.23		8.18	9.65
0620	Ogee, 7/16" x 2-1/2", poplar		250	.032		.58	1.08		1.66	2.46
0625	Red oak		250	.032		.79	1.08		1.87	2.69
0630	9/16" x 2-1/4", finger jointed and primed		250	.032		.51	1.08		1.59	2.38
0635	Poplar		250	.032		.59	1.08		1.67	2.47
0640	Red oak		250	.032		.78	1.08		1.86	2.68
0645	11/16" x 2-1/2", finger jointed and primed		250	.032		.72	1.08		1.80	2.61
0700	Pine		250	.032		1.41	1.08		2.49	3.37
0701	Red oak		250	.032		2.70	1.08		3.78	4.79
0730	11/16" x 3-1/2", finger jointed and primed		220	.036		1.11	1.23		2.34	3.28
0750	Pine		220	.036		1.72	1.23		2.95	3.95
0755	3/4" x 2-1/2", finger jointed and primed		250	.032		.93	1.08		2.01	2.84
0760	Poplar		250	.032		.89	1.08		1.97	2.80
0765	Red oak		250	.032		1.20	1.08		2.28	3.14
0770	Maple		250	.032		1.80	1.08		2.88	3.80
0775	Cherry		250	.032		2.27	1.08		3.35	4.31
0780	3/4" x 3-1/2", finger jointed and primed		220	.036		.92	1.23		2.15	3.07
0785	Poplar		220	.036		1.12	1.23		2.35	3.29
0790	Red oak		220	.036		1.66	1.23		2.89	3.88
0795	Maple		220	.036		2.56	1.23		3.79	4.87
0800	Cherry		220	.036		3.11	1.23		4.34	5.50
4700	Square profile, 1" x 1", teak		215	.037		2.26	1.26		3.52	4.59

06 22 Millwork

06 22 13 – Standard Pattern Wood Trim

06 22 13.30 Moldings, Casings

		Crew	Daily Output	Labor-Hours	Unit	Material	2016 Bare Costs Labor	Equipment	Total	Total Incl O&P
4800	Rectangular profile, 1" x 3", teak	1 Carp	200	.040	L.F.	6.20	1.36		7.56	9.10

06 22 13.35 Moldings, Ceilings

		Crew	Daily Output	Labor-Hours	Unit	Material	Labor	Equipment	Total	Total Incl O&P
0010	**MOLDINGS, CEILINGS**									
0600	Bed, 9/16" x 1-3/4", pine	1 Carp	270	.030	L.F.	1.06	1		2.06	2.84
0650	9/16" x 2", pine		270	.030		1.14	1		2.14	2.93
0710	9/16" x 1-3/4", oak		270	.030		1.86	1		2.86	3.72
1200	Cornice, 9/16" x 1-3/4", pine		270	.030		.96	1		1.96	2.73
1300	9/16" x 2-1/4", pine		265	.030		1.27	1.02		2.29	3.10
1350	Cove, 1/2" x 2-1/4", poplar		265	.030		1.04	1.02		2.06	2.85
1360	Red oak		265	.030		1.37	1.02		2.39	3.21
1370	Hard maple		265	.030		1.86	1.02		2.88	3.75
1380	Cherry		265	.030		2.28	1.02		3.30	4.21
2400	9/16" x 1-3/4", pine		270	.030		.97	1		1.97	2.74
2401	Oak		270	.030		1.26	1		2.26	3.06
2500	11/16" x 2-3/4", pine		265	.030		1.80	1.02		2.82	3.69
2510	Crown, 5/8" x 5/8", poplar		300	.027		.44	.90		1.34	1.99
2520	Red oak		300	.027		.52	.90		1.42	2.08
2530	Hard maple		300	.027		.70	.90		1.60	2.28
2540	Cherry		300	.027		.74	.90		1.64	2.32
2600	9/16" x 3-5/8", pine		250	.032		2.02	1.08		3.10	4.04
2700	11/16" x 4-1/4", pine		250	.032		2.90	1.08		3.98	5
2705	Oak		250	.032		6.40	1.08		7.48	8.85
2710	3/4" x 1-3/4", poplar		270	.030		.71	1		1.71	2.46
2720	Red oak		270	.030		1.05	1		2.05	2.83
2730	Hard maple		270	.030		1.39	1		2.39	3.21
2740	Cherry		270	.030		1.71	1		2.71	3.56
2750	3/4" x 2", poplar		270	.030		.94	1		1.94	2.71
2760	Red oak		270	.030		1.26	1		2.26	3.06
2770	Hard maple		270	.030		1.68	1		2.68	3.53
2780	Cherry		270	.030		1.88	1		2.88	3.75
2790	3/4" x 2-3/4", poplar		265	.030		1.02	1.02		2.04	2.83
2800	Red oak		265	.030		1.57	1.02		2.59	3.43
2810	Hard maple		265	.030		2.37	1.02		3.39	4.31
2820	Cherry		265	.030		2.40	1.02		3.42	4.35
2830	3/4" x 3-1/2", poplar		250	.032		1.30	1.08		2.38	3.25
2840	Red oak		250	.032		1.96	1.08		3.04	3.97
2850	Hard maple		250	.032		2.56	1.08		3.64	4.63
2860	Cherry		250	.032		3	1.08		4.08	5.10
2870	FJP poplar		250	.032		.92	1.08		2	2.83
2880	3/4" x 5", poplar		245	.033		1.93	1.11		3.04	3.97
2890	Red oak		245	.033		2.85	1.11		3.96	4.98
2900	Hard maple		245	.033		3.74	1.11		4.85	5.95
2910	Cherry		245	.033		4.32	1.11		5.43	6.60
2920	FJP poplar		245	.033		1.32	1.11		2.43	3.30
2930	3/4" x 6-1/4", poplar		240	.033		2.29	1.13		3.42	4.41
2940	Red oak		240	.033		3.46	1.13		4.59	5.70
2950	Hard maple		240	.033		4.52	1.13		5.65	6.85
2960	Cherry		240	.033		5.35	1.13		6.48	7.80
2970	7/8" x 8-3/4", poplar		220	.036		3.56	1.23		4.79	6
2980	Red oak		220	.036		4.97	1.23		6.20	7.50
2990	Hard maple		220	.036		6.75	1.23		7.98	9.50
3000	Cherry		220	.036		7.60	1.23		8.83	10.45

06 22 Millwork

06 22 13 – Standard Pattern Wood Trim

06 22 13.35 Moldings, Ceilings

		Crew	Daily Output	Labor-Hours	Unit	Material	2016 Bare Costs Labor	Equipment	Total	Total Incl O&P
3010	1" x 7-1/4", poplar	1 Carp	220	.036	L.F.	3.77	1.23		5	6.20
3020	Red oak		220	.036		5.85	1.23		7.08	8.50
3030	Hard maple		220	.036		8.15	1.23		9.38	11
3040	Cherry		220	.036		9.50	1.23		10.73	12.50
3050	1-1/16" x 4-1/4", poplar		250	.032		2.29	1.08		3.37	4.33
3060	Red oak		250	.032		2.69	1.08		3.77	4.77
3070	Hard maple		250	.032		3.56	1.08		4.64	5.75
3080	Cherry		250	.032		5.10	1.08		6.18	7.40
3090	Dentil crown, 3/4" x 5", poplar		250	.032		1.93	1.08		3.01	3.94
3100	Red oak		250	.032		2.85	1.08		3.93	4.95
3110	Hard maple		250	.032		3.74	1.08		4.82	5.95
3120	Cherry		250	.032		4.32	1.08		5.40	6.55
3130	Dentil piece for above, 1/2" x 1/2", poplar		300	.027		2.73	.90		3.63	4.51
3140	Red oak		300	.027		3.14	.90		4.04	4.96
3150	Hard maple		300	.027		3.57	.90		4.47	5.45
3160	Cherry		300	.027		3.89	.90		4.79	5.80

06 22 13.40 Moldings, Exterior

		Crew	Daily Output	Labor-Hours	Unit	Material	Labor	Equipment	Total	Total Incl O&P
0010	**MOLDINGS, EXTERIOR**									
0100	Band board, cedar, rough sawn, 1" x 2"	1 Carp	300	.027	L.F.	.55	.90		1.45	2.11
0110	1" x 3"		300	.027		.81	.90		1.71	2.40
0120	1" x 4"		250	.032		1.08	1.08		2.16	3.01
0130	1" x 6"		250	.032		1.63	1.08		2.71	3.61
0140	1" x 8"		225	.036		2.17	1.21		3.38	4.41
0150	1" x 10"		225	.036		2.70	1.21		3.91	4.99
0160	1" x 12"		200	.040		3.24	1.36		4.60	5.85
0240	STK, 1" x 2"		300	.027		.53	.90		1.43	2.10
0250	1" x 3"		300	.027		.59	.90		1.49	2.15
0260	1" x 4"		250	.032		.71	1.08		1.79	2.60
0270	1" x 6"		250	.032		1.13	1.08		2.21	3.06
0280	1" x 8"		225	.036		1.63	1.21		2.84	3.81
0290	1" x 10"		225	.036		2.12	1.21		3.33	4.35
0300	1" x 12"		200	.040		3.62	1.36		4.98	6.25
0310	Pine, #2, 1" x 2"		300	.027		.25	.90		1.15	1.79
0320	1" x 3"		300	.027		.40	.90		1.30	1.95
0330	1" x 4"		250	.032		.49	1.08		1.57	2.36
0340	1" x 6"		250	.032		.73	1.08		1.81	2.63
0350	1" x 8"		225	.036		1.20	1.21		2.41	3.34
0360	1" x 10"		225	.036		1.57	1.21		2.78	3.74
0370	1" x 12"		200	.040		1.92	1.36		3.28	4.38
0380	D & better, 1" x 2"		300	.027		.43	.90		1.33	1.99
0390	1" x 3"		300	.027		.66	.90		1.56	2.24
0400	1" x 4"		250	.032		.89	1.08		1.97	2.80
0410	1" x 6"		250	.032		1.12	1.08		2.20	3.05
0420	1" x 8"		225	.036		1.62	1.21		2.83	3.80
0430	1" x 10"		225	.036		2.03	1.21		3.24	4.25
0440	1" x 12"		200	.040		2.63	1.36		3.99	5.15
0450	Redwood, clear all heart, 1" x 2"		300	.027		.63	.90		1.53	2.21
0460	1" x 3"		300	.027		.94	.90		1.84	2.55
0470	1" x 4"		250	.032		1.19	1.08		2.27	3.13
0480	1" x 6"		252	.032		1.80	1.08		2.88	3.78
0490	1" x 8"		225	.036		2.36	1.21		3.57	4.62
0500	1" x 10"		225	.036		3.84	1.21		5.05	6.25

06 22 Millwork

06 22 13 – Standard Pattern Wood Trim

06 22 13.40 Moldings, Exterior

		Crew	Daily Output	Labor-Hours	Unit	Material	2016 Bare Costs Labor	Equipment	Total	Total Incl O&P
0510	1" x 12"	1 Carp	200	.040	L.F.	4.75	1.36		6.11	7.50
0530	Corner board, cedar, rough sawn, 1" x 2"		225	.036		.55	1.21		1.76	2.62
0540	1" x 3"		225	.036		.81	1.21		2.02	2.91
0550	1" x 4"		200	.040		1.08	1.36		2.44	3.46
0560	1" x 6"		200	.040		1.63	1.36		2.99	4.06
0570	1" x 8"		200	.040		2.17	1.36		3.53	4.66
0580	1" x 10"		175	.046		2.70	1.55		4.25	5.55
0590	1" x 12"		175	.046		3.24	1.55		4.79	6.15
0670	STK, 1" x 2"		225	.036		.53	1.21		1.74	2.61
0680	1" x 3"		225	.036		.59	1.21		1.80	2.66
0690	1" x 4"		200	.040		.69	1.36		2.05	3.03
0700	1" x 6"		200	.040		1.13	1.36		2.49	3.51
0710	1" x 8"		200	.040		1.63	1.36		2.99	4.06
0720	1" x 10"		175	.046		2.12	1.55		3.67	4.92
0730	1" x 12"		175	.046		3.62	1.55		5.17	6.55
0740	Pine, #2, 1" x 2"		225	.036		.25	1.21		1.46	2.30
0750	1" x 3"		225	.036		.40	1.21		1.61	2.46
0760	1" x 4"		200	.040		.49	1.36		1.85	2.81
0770	1" x 6"		200	.040		.73	1.36		2.09	3.08
0780	1" x 8"		200	.040		1.20	1.36		2.56	3.59
0790	1" x 10"		175	.046		1.57	1.55		3.12	4.31
0800	1" x 12"		175	.046		1.92	1.55		3.47	4.70
0810	D & better, 1" x 2"		225	.036		.43	1.21		1.64	2.50
0820	1" x 3"		225	.036		.66	1.21		1.87	2.75
0830	1" x 4"		200	.040		.89	1.36		2.25	3.25
0840	1" x 6"		200	.040		1.12	1.36		2.48	3.50
0850	1" x 8"		200	.040		1.62	1.36		2.98	4.05
0860	1" x 10"		175	.046		2.03	1.55		3.58	4.82
0870	1" x 12"		175	.046		2.63	1.55		4.18	5.50
0880	Redwood, clear all heart, 1" x 2"		225	.036		.63	1.21		1.84	2.72
0890	1" x 3"		225	.036		.94	1.21		2.15	3.06
0900	1" x 4"		200	.040		1.19	1.36		2.55	3.58
0910	1" x 6"		200	.040		1.80	1.36		3.16	4.25
0920	1" x 8"		200	.040		2.36	1.36		3.72	4.87
0930	1" x 10"		175	.046		3.84	1.55		5.39	6.80
0940	1" x 12"		175	.046		4.75	1.55		6.30	7.85
0950	Cornice board, cedar, rough sawn, 1" x 2"		330	.024		.55	.82		1.37	1.98
0960	1" x 3"		290	.028		.81	.94		1.75	2.46
0970	1" x 4"		250	.032		1.08	1.08		2.16	3.01
0980	1" x 6"		250	.032		1.63	1.08		2.71	3.61
0990	1" x 8"		200	.040		2.17	1.36		3.53	4.66
1000	1" x 10"		180	.044		2.70	1.51		4.21	5.50
1010	1" x 12"		180	.044		3.24	1.51		4.75	6.10
1020	STK, 1" x 2"		330	.024		.53	.82		1.35	1.97
1030	1" x 3"		290	.028		.59	.94		1.53	2.21
1040	1" x 4"		250	.032		.71	1.08		1.79	2.60
1050	1" x 6"		250	.032		1.13	1.08		2.21	3.06
1060	1" x 8"		200	.040		1.63	1.36		2.99	4.06
1070	1" x 10"		180	.044		2.12	1.51		3.63	4.85
1080	1" x 12"		180	.044		3.62	1.51		5.13	6.50
1500	Pine, #2, 1" x 2"		330	.024		.25	.82		1.07	1.66
1510	1" x 3"		290	.028		.27	.94		1.21	1.87
1600	1" x 4"		250	.032		.49	1.08		1.57	2.36

06 22 Millwork

06 22 13 – Standard Pattern Wood Trim

06 22 13.40 Moldings, Exterior		Crew	Daily Output	Labor-Hours	Unit	Material	2016 Bare Costs Labor	Equipment	Total	Total Incl O&P
1700	1" x 6"	1 Carp	250	.032	L.F.	.73	1.08		1.81	2.63
1800	1" x 8"		200	.040		1.20	1.36		2.56	3.59
1900	1" x 10"		180	.044		1.57	1.51		3.08	4.24
2000	1" x 12"		180	.044		1.92	1.51		3.43	4.63
2020	D & better, 1" x 2"		330	.024		.43	.82		1.25	1.86
2030	1" x 3"		290	.028		.66	.94		1.60	2.30
2040	1" x 4"		250	.032		.89	1.08		1.97	2.80
2050	1" x 6"		250	.032		1.12	1.08		2.20	3.05
2060	1" x 8"		200	.040		1.62	1.36		2.98	4.05
2070	1" x 10"		180	.044		2.03	1.51		3.54	4.75
2080	1" x 12"		180	.044		2.63	1.51		4.14	5.40
2090	Redwood, clear all heart, 1" x 2"		330	.024		.63	.82		1.45	2.08
2100	1" x 3"		290	.028		.94	.94		1.88	2.61
2110	1" x 4"		250	.032		1.19	1.08		2.27	3.13
2120	1" x 6"		250	.032		1.80	1.08		2.88	3.80
2130	1" x 8"		200	.040		2.36	1.36		3.72	4.87
2140	1" x 10"		180	.044		3.84	1.51		5.35	6.75
2150	1" x 12"		180	.044		4.75	1.51		6.26	7.75
2160	3 piece, 1" x 2", 1" x 4", 1" x 6", rough sawn cedar		80	.100		3.27	3.39		6.66	9.30
2180	STK cedar		80	.100		2.37	3.39		5.76	8.30
2200	#2 pine		80	.100		1.47	3.39		4.86	7.30
2210	D & better pine		80	.100		2.44	3.39		5.83	8.40
2220	Clear all heart redwood		80	.100		3.62	3.39		7.01	9.70
2230	1" x 8", 1" x 10", 1" x 12", rough sawn cedar		65	.123		8.10	4.17		12.27	15.90
2240	STK cedar		65	.123		7.35	4.17		11.52	15.10
2300	#2 pine		65	.123		4.66	4.17		8.83	12.15
2320	D & better pine		65	.123		6.25	4.17		10.42	13.90
2330	Clear all heart redwood		65	.123		10.90	4.17		15.07	19
2340	Door/window casing, cedar, rough sawn, 1" x 2"		275	.029		.55	.99		1.54	2.25
2350	1" x 3"		275	.029		.81	.99		1.80	2.54
2360	1" x 4"		250	.032		1.08	1.08		2.16	3.01
2370	1" x 6"		250	.032		1.63	1.08		2.71	3.61
2380	1" x 8"		230	.035		2.17	1.18		3.35	4.36
2390	1" x 10"		230	.035		2.70	1.18		3.88	4.94
2395	1" x 12"		210	.038		3.24	1.29		4.53	5.70
2410	STK, 1" x 2"		275	.029		.53	.99		1.52	2.24
2420	1" x 3"		275	.029		.59	.99		1.58	2.29
2430	1" x 4"		250	.032		.71	1.08		1.79	2.60
2440	1" x 6"		250	.032		1.13	1.08		2.21	3.06
2450	1" x 8"		230	.035		1.63	1.18		2.81	3.76
2460	1" x 10"		230	.035		2.12	1.18		3.30	4.30
2470	1" x 12"		210	.038		3.62	1.29		4.91	6.15
2550	Pine, #2, 1" x 2"		275	.029		.25	.99		1.24	1.93
2560	1" x 3"		275	.029		.40	.99		1.39	2.09
2570	1" x 4"		250	.032		.49	1.08		1.57	2.36
2580	1" x 6"		250	.032		.73	1.08		1.81	2.63
2590	1" x 8"		230	.035		1.20	1.18		2.38	3.29
2600	1" x 10"		230	.035		1.57	1.18		2.75	3.69
2610	1" x 12"		210	.038		1.92	1.29		3.21	4.27
2620	Pine, D & better, 1" x 2"		275	.029		.43	.99		1.42	2.13
2630	1" x 3"		275	.029		.66	.99		1.65	2.38
2640	1" x 4"		250	.032		.89	1.08		1.97	2.80
2650	1" x 6"		250	.032		1.12	1.08		2.20	3.05

For customer support on your Residential Cost Data, call 877.759.4771.

06 22 Millwork

06 22 13 – Standard Pattern Wood Trim

06 22 13.40 Moldings, Exterior		Crew	Daily Output	Labor-Hours	Unit	Material	2016 Bare Costs Labor	Equipment	Total	Total Incl O&P
2660	1" x 8"	1 Carp	230	.035	L.F.	1.62	1.18		2.80	3.75
2670	1" x 10"		230	.035		2.03	1.18		3.21	4.20
2680	1" x 12"		210	.038		2.63	1.29		3.92	5.05
2690	Redwood, clear all heart, 1" x 2"		275	.029		.63	.99		1.62	2.35
2695	1" x 3"		275	.029		.94	.99		1.93	2.69
2710	1" x 4"		250	.032		1.19	1.08		2.27	3.13
2715	1" x 6"		250	.032		1.80	1.08		2.88	3.80
2730	1" x 8"		230	.035		2.36	1.18		3.54	4.57
2740	1" x 10"		230	.035		3.84	1.18		5.02	6.20
2750	1" x 12"		210	.038		4.75	1.29		6.04	7.40
3500	Bellyband, pine, 11/16" x 4-1/4"		250	.032		3.23	1.08		4.31	5.35
3610	Brickmold, pine, 1-1/4" x 2"		200	.040		2.43	1.36		3.79	4.95
3620	FJP, 1-1/4" x 2"		200	.040		1.09	1.36		2.45	3.47
5100	Fascia, cedar, rough sawn, 1" x 2"		275	.029		.55	.99		1.54	2.25
5110	1" x 3"		275	.029		.81	.99		1.80	2.54
5120	1" x 4"		250	.032		1.08	1.08		2.16	3.01
5200	1" x 6"		250	.032		1.63	1.08		2.71	3.61
5300	1" x 8"		230	.035		2.17	1.18		3.35	4.36
5310	1" x 10"		230	.035		2.70	1.18		3.88	4.94
5320	1" x 12"		210	.038		3.24	1.29		4.53	5.70
5400	2" x 4"		220	.036		1.07	1.23		2.30	3.23
5500	2" x 6"		220	.036		1.60	1.23		2.83	3.82
5600	2" x 8"		200	.040		2.14	1.36		3.50	4.62
5700	2" x 10"		180	.044		2.66	1.51		4.17	5.45
5800	2" x 12"		170	.047		6.35	1.60		7.95	9.60
6120	STK, 1" x 2"		275	.029		.53	.99		1.52	2.24
6130	1" x 3"		275	.029		.59	.99		1.58	2.29
6140	1" x 4"		250	.032		.71	1.08		1.79	2.60
6150	1" x 6"		250	.032		1.13	1.08		2.21	3.06
6160	1" x 8"		230	.035		1.63	1.18		2.81	3.76
6170	1" x 10"		230	.035		2.12	1.18		3.30	4.30
6180	1" x 12"		210	.038		3.62	1.29		4.91	6.15
6185	2" x 2"		260	.031		.65	1.04		1.69	2.47
6190	Pine, #2, 1" x 2"		275	.029		.25	.99		1.24	1.93
6200	1" x 3"		275	.029		.40	.99		1.39	2.09
6210	1" x 4"		250	.032		.49	1.08		1.57	2.36
6220	1" x 6"		250	.032		.73	1.08		1.81	2.63
6230	1" x 8"		230	.035		1.20	1.18		2.38	3.29
6240	1" x 10"		230	.035		1.57	1.18		2.75	3.69
6250	1" x 12"		210	.038		1.92	1.29		3.21	4.27
6260	D & better, 1" x 2"		275	.029		.43	.99		1.42	2.13
6270	1" x 3"		275	.029		.66	.99		1.65	2.38
6280	1" x 4"		250	.032		.89	1.08		1.97	2.80
6290	1" x 6"		250	.032		1.12	1.08		2.20	3.05
6300	1" x 8"		230	.035		1.62	1.18		2.80	3.75
6310	1" x 10"		230	.035		2.03	1.18		3.21	4.20
6312	1" x 12"		210	.038		2.63	1.29		3.92	5.05
6330	Southern yellow, 1-1/4" x 5"		240	.033		2.43	1.13		3.56	4.56
6340	1-1/4" x 6"		240	.033		2.52	1.13		3.65	4.66
6350	1-1/4" x 8"		215	.037		3.43	1.26		4.69	5.90
6360	1-1/4" x 12"		190	.042		5.05	1.43		6.48	7.95
6370	Redwood, clear all heart, 1" x 2"		275	.029		.63	.99		1.62	2.35
6380	1" x 3"		275	.029		1.19	.99		2.18	2.96

06 22 Millwork

06 22 13 – Standard Pattern Wood Trim

	06 22 13.40 Moldings, Exterior	Crew	Daily Output	Labor-Hours	Unit	Material	Labor	Equipment	Total	Total Incl O&P
6390	1" x 4"	1 Carp	250	.032	L.F.	1.19	1.08		2.27	3.13
6400	1" x 6"		250	.032		1.80	1.08		2.88	3.80
6410	1" x 8"		230	.035		2.36	1.18		3.54	4.57
6420	1" x 10"		230	.035		3.84	1.18		5.02	6.20
6430	1" x 12"		210	.038		4.75	1.29		6.04	7.40
6440	1-1/4" x 5"		240	.033		1.88	1.13		3.01	3.96
6450	1-1/4" x 6"		240	.033		2.25	1.13		3.38	4.36
6460	1-1/4" x 8"		215	.037		3.53	1.26		4.79	6
6470	1-1/4" x 12"		190	.042		7.10	1.43		8.53	10.25
6580	Frieze, cedar, rough sawn, 1" x 2"		275	.029		.55	.99		1.54	2.25
6590	1" x 3"		275	.029		.81	.99		1.80	2.54
6600	1" x 4"		250	.032		1.08	1.08		2.16	3.01
6610	1" x 6"		250	.032		1.63	1.08		2.71	3.61
6620	1" x 8"		250	.032		2.17	1.08		3.25	4.21
6630	1" x 10"		225	.036		2.70	1.21		3.91	4.99
6640	1" x 12"		200	.040		3.21	1.36		4.57	5.80
6650	STK, 1" x 2"		275	.029		.53	.99		1.52	2.24
6660	1" x 3"		275	.029		.59	.99		1.58	2.29
6670	1" x 4"		250	.032		.71	1.08		1.79	2.60
6680	1" x 6"		250	.032		1.13	1.08		2.21	3.06
6690	1" x 8"		250	.032		1.63	1.08		2.71	3.61
6700	1" x 10"		225	.036		2.12	1.21		3.33	4.35
6710	1" x 12"		200	.040		3.62	1.36		4.98	6.25
6790	Pine, #2, 1" x 2"		275	.029		.25	.99		1.24	1.93
6800	1" x 3"		275	.029		.40	.99		1.39	2.09
6810	1" x 4"		250	.032		.49	1.08		1.57	2.36
6820	1" x 6"		250	.032		.73	1.08		1.81	2.63
6830	1" x 8"		250	.032		1.20	1.08		2.28	3.14
6840	1" x 10"		225	.036		1.57	1.21		2.78	3.74
6850	1" x 12"		200	.040		1.92	1.36		3.28	4.38
6860	D & better, 1" x 2"		275	.029		.43	.99		1.42	2.13
6870	1" x 3"		275	.029		.66	.99		1.65	2.38
6880	1" x 4"		250	.032		.89	1.08		1.97	2.80
6890	1" x 6"		250	.032		1.12	1.08		2.20	3.05
6900	1" x 8"		250	.032		1.62	1.08		2.70	3.60
6910	1" x 10"		225	.036		2.03	1.21		3.24	4.25
6920	1" x 12"		200	.040		2.63	1.36		3.99	5.15
6930	Redwood, clear all heart, 1" x 2"		275	.029		.63	.99		1.62	2.35
6940	1" x 3"		275	.029		.94	.99		1.93	2.69
6950	1" x 4"		250	.032		1.19	1.08		2.27	3.13
6960	1" x 6"		250	.032		1.80	1.08		2.88	3.80
6970	1" x 8"		250	.032		2.36	1.08		3.44	4.42
6980	1" x 10"		225	.036		3.84	1.21		5.05	6.25
6990	1" x 12"		200	.040		4.75	1.36		6.11	7.50
7000	Grounds, 1" x 1", cedar, rough sawn		300	.027		.28	.90		1.18	1.82
7010	STK		300	.027		.33	.90		1.23	1.87
7020	Pine, #2		300	.027		.16	.90		1.06	1.68
7030	D & better		300	.027		.27	.90		1.17	1.80
7050	Redwood		300	.027		.39	.90		1.29	1.93
7060	Rake/verge board, cedar, rough sawn, 1" x 2"		225	.036		.55	1.21		1.76	2.62
7070	1" x 3"		225	.036		.81	1.21		2.02	2.91
7080	1" x 4"		200	.040		1.08	1.36		2.44	3.46
7090	1" x 6"		200	.040		1.63	1.36		2.99	4.06

For customer support on your Residential Cost Data, call 877.759.4771.

06 22 Millwork
06 22 13 – Standard Pattern Wood Trim

	06 22 13.40 Moldings, Exterior	Crew	Daily Output	Labor-Hours	Unit	Material	2016 Bare Costs Labor	Equipment	Total	Total Incl O&P
7100	1" x 8"	1 Carp	190	.042	L.F.	2.17	1.43		3.60	4.78
7110	1" x 10"		190	.042		2.70	1.43		4.13	5.35
7120	1" x 12"		180	.044		3.24	1.51		4.75	6.10
7130	STK, 1" x 2"		225	.036		.53	1.21		1.74	2.61
7140	1" x 3"		225	.036		.59	1.21		1.80	2.66
7150	1" x 4"		200	.040		.71	1.36		2.07	3.05
7160	1" x 6"		200	.040		1.13	1.36		2.49	3.51
7170	1" x 8"		190	.042		1.63	1.43		3.06	4.18
7180	1" x 10"		190	.042		2.12	1.43		3.55	4.72
7190	1" x 12"		180	.044		3.62	1.51		5.13	6.50
7200	Pine, #2, 1" x 2"		225	.036		.25	1.21		1.46	2.30
7210	1" x 3"		225	.036		.40	1.21		1.61	2.46
7220	1" x 4"		200	.040		.49	1.36		1.85	2.81
7230	1" x 6"		200	.040		.73	1.36		2.09	3.08
7240	1" x 8"		190	.042		1.20	1.43		2.63	3.71
7250	1" x 10"		190	.042		1.57	1.43		3	4.11
7260	1" x 12"		180	.044		1.92	1.51		3.43	4.63
7340	D & better, 1" x 2"		225	.036		.43	1.21		1.64	2.50
7350	1" x 3"		225	.036		.66	1.21		1.87	2.75
7360	1" x 4"		200	.040		.89	1.36		2.25	3.25
7370	1" x 6"		200	.040		1.12	1.36		2.48	3.50
7380	1" x 8"		190	.042		1.62	1.43		3.05	4.17
7390	1" x 10"		190	.042		2.03	1.43		3.46	4.62
7400	1" x 12"		180	.044		2.63	1.51		4.14	5.40
7410	Redwood, clear all heart, 1" x 2"		225	.036		.63	1.21		1.84	2.72
7420	1" x 3"		225	.036		.94	1.21		2.15	3.06
7430	1" x 4"		200	.040		1.19	1.36		2.55	3.58
7440	1" x 6"		200	.040		1.80	1.36		3.16	4.25
7450	1" x 8"		190	.042		2.36	1.43		3.79	4.99
7460	1" x 10"		190	.042		3.84	1.43		5.27	6.60
7470	1" x 12"		180	.044		4.75	1.51		6.26	7.75
7480	2" x 4"		200	.040		2.29	1.36		3.65	4.79
7490	2" x 6"		182	.044		3.43	1.49		4.92	6.25
7500	2" x 8"		165	.048		4.56	1.64		6.20	7.75
7630	Soffit, cedar, rough sawn, 1" x 2"	2 Carp	440	.036		.55	1.23		1.78	2.66
7640	1" x 3"		440	.036		.81	1.23		2.04	2.95
7650	1" x 4"		420	.038		1.08	1.29		2.37	3.35
7660	1" x 6"		420	.038		1.63	1.29		2.92	3.95
7670	1" x 8"		420	.038		2.17	1.29		3.46	4.55
7680	1" x 10"		400	.040		2.70	1.36		4.06	5.25
7690	1" x 12"		400	.040		3.24	1.36		4.60	5.85
7700	STK, 1" x 2"		440	.036		.53	1.23		1.76	2.65
7710	1" x 3"		440	.036		.59	1.23		1.82	2.70
7720	1" x 4"		420	.038		.71	1.29		2	2.94
7730	1" x 6"		420	.038		1.13	1.29		2.42	3.40
7740	1" x 8"		420	.038		1.63	1.29		2.92	3.95
7750	1" x 10"		400	.040		2.12	1.36		3.48	4.60
7760	1" x 12"		400	.040		3.62	1.36		4.98	6.25
7770	Pine, #2, 1" x 2"		440	.036		.25	1.23		1.48	2.34
7780	1" x 3"		440	.036		.40	1.23		1.63	2.50
7790	1" x 4"		420	.038		.49	1.29		1.78	2.70
7800	1" x 6"		420	.038		.73	1.29		2.02	2.97
7810	1" x 8"		420	.038		1.20	1.29		2.49	3.48

06 22 Millwork

06 22 13 – Standard Pattern Wood Trim

06 22 13.40 Moldings, Exterior		Crew	Daily Output	Labor-Hours	Unit	Material	2016 Bare Costs Labor	Equipment	Total	Total Incl O&P
7820	1" x 10"	2 Carp	400	.040	L.F.	1.57	1.36		2.93	3.99
7830	1" x 12"		400	.040		1.92	1.36		3.28	4.38
7840	D & better, 1" x 2"		440	.036		.43	1.23		1.66	2.54
7850	1" x 3"		440	.036		.66	1.23		1.89	2.79
7860	1" x 4"		420	.038		.89	1.29		2.18	3.14
7870	1" x 6"		420	.038		1.12	1.29		2.41	3.39
7880	1" x 8"		420	.038		1.62	1.29		2.91	3.94
7890	1" x 10"		400	.040		2.03	1.36		3.39	4.50
7900	1" x 12"		400	.040		2.63	1.36		3.99	5.15
7910	Redwood, clear all heart, 1" x 2"		440	.036		.63	1.23		1.86	2.76
7920	1" x 3"		440	.036		.94	1.23		2.17	3.10
7930	1" x 4"		420	.038		1.19	1.29		2.48	3.47
7940	1" x 6"		420	.038		1.80	1.29		3.09	4.14
7950	1" x 8"		420	.038		2.36	1.29		3.65	4.76
7960	1" x 10"		400	.040		3.84	1.36		5.20	6.50
7970	1" x 12"		400	.040		4.75	1.36		6.11	7.50
8050	Trim, crown molding, pine, 11/16" x 4-1/4"	1 Carp	250	.032		4.54	1.08		5.62	6.80
8060	Back band, 11/16" x 1-1/16"		250	.032		.99	1.08		2.07	2.91
8070	Insect screen frame stock, 1-1/16" x 1-3/4"		395	.020		2.39	.69		3.08	3.78
8080	Dentils, 2-1/2" x 2-1/2" x 4", 6" O.C.		30	.267		1.23	9.05		10.28	16.50
8100	Fluted, 5-1/2"		165	.048		5.05	1.64		6.69	8.30
8110	Stucco bead, 1-3/8" x 1-5/8"		250	.032		2.50	1.08		3.58	4.57

06 22 13.45 Moldings, Trim		Crew	Daily Output	Labor-Hours	Unit	Material	Labor	Equipment	Total	Total Incl O&P
0010	**MOLDINGS, TRIM**									
0200	Astragal, stock pine, 11/16" x 1-3/4"	1 Carp	255	.031	L.F.	1.42	1.06		2.48	3.34
0250	1-5/16" x 2-3/16"		240	.033		2.10	1.13		3.23	4.20
0800	Chair rail, stock pine, 5/8" x 2-1/2"		270	.030		1.58	1		2.58	3.42
0900	5/8" x 3-1/2"		240	.033		2.40	1.13		3.53	4.53
1000	Closet pole, stock pine, 1-1/8" diameter		200	.040		1.16	1.36		2.52	3.55
1100	Fir, 1-5/8" diameter		200	.040		2.20	1.36		3.56	4.69
1150	Corner, inside, 5/16" x 1"		225	.036		.33	1.21		1.54	2.38
1160	Outside, 1-1/16" x 1-1/16"		240	.033		1.31	1.13		2.44	3.33
1161	1-5/16" x 1-5/16"		240	.033		1.55	1.13		2.68	3.60
3300	Half round, stock pine, 1/4" x 1/2"		270	.030		.24	1		1.24	1.94
3350	1/2" x 1"		255	.031		.73	1.06		1.79	2.58
3400	Handrail, fir, single piece, stock, hardware not included									
3450	1-1/2" x 1-3/4"	1 Carp	80	.100	L.F.	2.56	3.39		5.95	8.50
3470	Pine, 1-1/2" x 1-3/4"		80	.100		2.40	3.39		5.79	8.35
3500	1-1/2" x 2-1/2"		76	.105		2.46	3.57		6.03	8.65
3600	Lattice, stock pine, 1/4" x 1-1/8"		270	.030		.35	1		1.35	2.07
3700	1/4" x 1-3/4"		250	.032		.92	1.08		2	2.83
3800	Miscellaneous, custom, pine, 1" x 1"		270	.030		.43	1		1.43	2.16
3850	1" x 2"		265	.030		.87	1.02		1.89	2.67
3900	1" x 3"		240	.033		1.30	1.13		2.43	3.32
4100	Birch or oak, nominal 1" x 1"		240	.033		.39	1.13		1.52	2.32
4200	Nominal 1" x 3"		215	.037		1.18	1.26		2.44	3.41
4400	Walnut, nominal 1" x 1"		215	.037		.62	1.26		1.88	2.79
4500	Nominal 1" x 3"		200	.040		1.86	1.36		3.22	4.31
4700	Teak, nominal 1" x 1"		215	.037		2.64	1.26		3.90	5
4800	Nominal 1" x 3"		200	.040		7.90	1.36		9.26	10.95
4900	Quarter round, stock pine, 1/4" x 1/4"		275	.029		.24	.99		1.23	1.91
4950	3/4" x 3/4"		255	.031		.50	1.06		1.56	2.33

06 22 Millwork

06 22 13 – Standard Pattern Wood Trim

06 22 13.45 Moldings, Trim

	06 22 13.45 Moldings, Trim	Crew	Daily Output	Labor-Hours	Unit	Material	2016 Bare Costs Labor	Equipment	Total	Total Incl O&P
5600	Wainscot moldings, 1-1/8" x 9/16", 2' high, minimum	1 Carp	76	.105	S.F.	11.65	3.57		15.22	18.75
5700	Maximum		65	.123	"	15.80	4.17		19.97	24.50

06 22 13.50 Moldings, Window and Door

		Crew	Daily Output	Labor-Hours	Unit	Material	Labor	Equipment	Total	Total Incl O&P
0010	**MOLDINGS, WINDOW AND DOOR**									
2800	Door moldings, stock, decorative, 1-1/8" wide, plain	1 Carp	17	.471	Set	47.50	15.95		63.45	79
2900	Detailed		17	.471	"	91.50	15.95		107.45	128
2960	Clear pine door jamb, no stops, 11/16" x 4-9/16"		240	.033	L.F.	5.20	1.13		6.33	7.60
3150	Door trim set, 1 head and 2 sides, pine, 2-1/2 wide		12	.667	Opng.	24	22.50		46.50	64
3170	3-1/2" wide		11	.727	"	29	24.50		53.50	73.50
3250	Glass beads, stock pine, 3/8" x 1/2"		275	.029	L.F.	.33	.99		1.32	2.01
3270	3/8" x 7/8"		270	.030		.42	1		1.42	2.14
4850	Parting bead, stock pine, 3/8" x 3/4"		275	.029		.44	.99		1.43	2.13
4870	1/2" x 3/4"		255	.031		.42	1.06		1.48	2.24
5000	Stool caps, stock pine, 11/16" x 3-1/2"		200	.040		2.07	1.36		3.43	4.55
5100	1-1/16" x 3-1/4"		150	.053		3.24	1.81		5.05	6.60
5300	Threshold, oak, 3' long, inside, 5/8" x 3-5/8"		32	.250	Ea.	6.50	8.50		15	21.50
5400	Outside, 1-1/2" x 7-5/8"		16	.500	"	45	16.95		61.95	78
5900	Window trim sets, including casings, header, stops,									
5910	stool and apron, 2-1/2" wide, FJP	1 Carp	13	.615	Opng.	31.50	21		52.50	69.50
5950	Pine		10	.800		37	27		64	86.50
6000	Oak		6	1.333		64.50	45		109.50	147

06 22 13.60 Moldings, Soffits

		Crew	Daily Output	Labor-Hours	Unit	Material	Labor	Equipment	Total	Total Incl O&P
0010	**MOLDINGS, SOFFITS**									
0200	Soffits, pine, 1" x 4"	2 Carp	420	.038	L.F.	.46	1.29		1.75	2.67
0210	1" x 6"		420	.038		.71	1.29		2	2.94
0220	1" x 8"		420	.038		1.17	1.29		2.46	3.45
0230	1" x 10"		400	.040		1.53	1.36		2.89	3.95
0240	1" x 12"		400	.040		1.88	1.36		3.24	4.34
0250	STK cedar, 1" x 4"		420	.038		.68	1.29		1.97	2.91
0260	1" x 6"		420	.038		1.10	1.29		2.39	3.37
0270	1" x 8"		420	.038		1.60	1.29		2.89	3.92
0280	1" x 10"		400	.040		2.08	1.36		3.44	4.56
0290	1" x 12"		400	.040		3.58	1.36		4.94	6.20
1000	Exterior AC plywood, 1/4" thick		400	.040	S.F.	.95	1.36		2.31	3.32
1050	3/8" thick		400	.040		.99	1.36		2.35	3.36
1100	1/2" thick		400	.040		1.17	1.36		2.53	3.56
1150	Polyvinyl chloride, white, solid	1 Carp	230	.035		2.10	1.18		3.28	4.28
1160	Perforated	"	230	.035		2.10	1.18		3.28	4.28
1170	Accessories, "J" channel 5/8"	2 Carp	700	.023	L.F.	.49	.78		1.27	1.84

06 25 Prefinished Paneling

06 25 13 – Prefinished Hardboard Paneling

06 25 13.10 Paneling, Hardboard

			Crew	Daily Output	Labor-Hours	Unit	Material	Labor	Equipment	Total	Total Incl O&P
0010	**PANELING, HARDBOARD**										
0050	Not incl. furring or trim, hardboard, tempered, 1/8" thick	G	2 Carp	500	.032	S.F.	.42	1.08		1.50	2.28
0100	1/4" thick	G		500	.032		.65	1.08		1.73	2.54
0300	Tempered pegboard, 1/8" thick	G		500	.032		.42	1.08		1.50	2.28
0400	1/4" thick	G		500	.032		.68	1.08		1.76	2.57
0600	Untempered hardboard, natural finish, 1/8" thick	G		500	.032		.43	1.08		1.51	2.29
0700	1/4" thick	G		500	.032		.55	1.08		1.63	2.43

06 25 Prefinished Paneling

06 25 13 – Prefinished Hardboard Paneling

06 25 13.10 Paneling, Hardboard		Crew	Daily Output	Labor-Hours	Unit	Material	2016 Bare Costs Labor	Equipment	Total	Total Incl O&P
0900	Untempered pegboard, 1/8" thick	G	2 Carp	500	.032	S.F.	.46	1.08		1.54	2.33
1000	1/4" thick	G		500	.032		.48	1.08		1.56	2.35
1200	Plastic faced hardboard, 1/8" thick	G		500	.032		.66	1.08		1.74	2.55
1300	1/4" thick	G		500	.032		.89	1.08		1.97	2.80
1500	Plastic faced pegboard, 1/8" thick	G		500	.032		.67	1.08		1.75	2.56
1600	1/4" thick	G		500	.032		.85	1.08		1.93	2.76
1800	Wood grained, plain or grooved, 1/8" thick	G		500	.032		.69	1.08		1.77	2.58
1900	1/4" thick	G		425	.038		1.40	1.28		2.68	3.68
2100	Moldings, wood grained MDF			500	.032	L.F.	.41	1.08		1.49	2.27
2200	Pine			425	.038	"	1.40	1.28		2.68	3.68

06 25 16 – Prefinished Plywood Paneling

06 25 16.10 Paneling, Plywood

		Crew	Daily Output	Labor-Hours	Unit	Material	Labor	Equipment	Total	Total Incl O&P
0010	**PANELING, PLYWOOD**									
2400	Plywood, prefinished, 1/4" thick, 4' x 8' sheets									
2410	with vertical grooves. Birch faced, economy	2 Carp	500	.032	S.F.	1.45	1.08		2.53	3.42
2420	Average		420	.038		1.20	1.29		2.49	3.48
2430	Custom		350	.046		1.25	1.55		2.80	3.97
2600	Mahogany, African		400	.040		2.80	1.36		4.16	5.35
2700	Philippine (Lauan)		500	.032		.65	1.08		1.73	2.54
2900	Oak		500	.032		1.40	1.08		2.48	3.36
3000	Cherry		400	.040		2.10	1.36		3.46	4.58
3200	Rosewood		320	.050		2.95	1.70		4.65	6.10
3400	Teak		400	.040		2.95	1.36		4.31	5.50
3600	Chestnut		375	.043		4.85	1.45		6.30	7.75
3800	Pecan		400	.040		2.55	1.36		3.91	5.10
3900	Walnut, average		500	.032		2.40	1.08		3.48	4.46
3950	Custom		400	.040		5.25	1.36		6.61	8.05
4000	Plywood, prefinished, 3/4" thick, stock grades, economy		320	.050		1.60	1.70		3.30	4.60
4100	Average		224	.071		4.70	2.42		7.12	9.20
4300	Architectural grade, custom		224	.071		5.20	2.42		7.62	9.75
4400	Luxury		160	.100		5.20	3.39		8.59	11.40
4600	Plywood, "A" face, birch, VC, 1/2" thick, natural		450	.036		2.05	1.21		3.26	4.28
4700	Select		450	.036		2.15	1.21		3.36	4.39
4900	Veneer core, 3/4" thick, natural		320	.050		2.24	1.70		3.94	5.30
5000	Select		320	.050		2.44	1.70		4.14	5.50
5200	Lumber core, 3/4" thick, natural		320	.050		3.05	1.70		4.75	6.20
5500	Plywood, knotty pine, 1/4" thick, A2 grade		450	.036		1.70	1.21		2.91	3.89
5600	A3 grade		450	.036		2.15	1.21		3.36	4.39
5800	3/4" thick, veneer core, A2 grade		320	.050		2.20	1.70		3.90	5.25
5900	A3 grade		320	.050		2.45	1.70		4.15	5.55
6100	Aromatic cedar, 1/4" thick, plywood		400	.040		2.25	1.36		3.61	4.75
6200	1/4" thick, particle board		400	.040		1.15	1.36		2.51	3.54

06 25 26 – Panel System

06 25 26.10 Panel Systems

		Crew	Daily Output	Labor-Hours	Unit	Material	Labor	Equipment	Total	Total Incl O&P
0010	**PANEL SYSTEMS**									
0100	Raised panel, eng. wood core w/wood veneer, std., paint grade	2 Carp	300	.053	S.F.	11.20	1.81		13.01	15.35
0110	Oak veneer		300	.053		25.50	1.81		27.31	31.50
0120	Maple veneer		300	.053		32.50	1.81		34.31	39
0130	Cherry veneer		300	.053		37	1.81		38.81	43.50
0300	Class I fire rated, paint grade		300	.053		13	1.81		14.81	17.35
0310	Oak veneer		300	.053		30	1.81		31.81	36
0320	Maple veneer		300	.053		40.50	1.81		42.31	47.50

06 25 Prefinished Paneling

06 25 26 – Panel System

06 25 26.10 Panel Systems		Crew	Daily Output	Labor-Hours	Unit	Material	2016 Bare Costs Labor	2016 Bare Costs Equipment	Total	Total Incl O&P
0330	Cherry veneer	2 Carp	300	.053	S.F.	49	1.81		50.81	56.50
0510	Beadboard, 5/8" MDF, standard, primed		300	.053		8.45	1.81		10.26	12.35
0520	Oak veneer, unfinished		300	.053		13.40	1.81		15.21	17.80
0530	Maple veneer, unfinished		300	.053		14.60	1.81		16.41	19.10
0610	Rustic paneling, 5/8" MDF, standard, maple veneer, unfinished	▼	300	.053	▼	18.20	1.81		20.01	23

06 26 Board Paneling

06 26 13 – Profile Board Paneling

06 26 13.10 Paneling, Boards

		Crew	Daily Output	Labor-Hours	Unit	Material	Labor	Equipment	Total	Total Incl O&P
0010	**PANELING, BOARDS**									
6400	Wood board paneling, 3/4" thick, knotty pine	2 Carp	300	.053	S.F.	1.95	1.81		3.76	5.20
6500	Rough sawn cedar		300	.053		3.20	1.81		5.01	6.55
6700	Redwood, clear, 1" x 4" boards		300	.053		4.96	1.81		6.77	8.50
6900	Aromatic cedar, closet lining, boards	▼	275	.058	▼	2.40	1.97		4.37	5.95

06 43 Wood Stairs and Railings

06 43 13 – Wood Stairs

06 43 13.20 Prefabricated Wood Stairs

		Crew	Daily Output	Labor-Hours	Unit	Material	Labor	Equipment	Total	Total Incl O&P
0010	**PREFABRICATED WOOD STAIRS**									
0100	Box stairs, prefabricated, 3'-0" wide									
0110	Oak treads, up to 14 risers	2 Carp	39	.410	Riser	92.50	13.90		106.40	126
0600	With pine treads for carpet, up to 14 risers	"	39	.410	"	59.50	13.90		73.40	89
1100	For 4' wide stairs, add				Flight	25%				
1550	Stairs, prefabricated stair handrail with balusters	1 Carp	30	.267	L.F.	78.50	9.05		87.55	101
1700	Basement stairs, prefabricated, pine treads									
1710	Pine risers, 3' wide, up to 14 risers	2 Carp	52	.308	Riser	59.50	10.45		69.95	83
4000	Residential, wood, oak treads, prefabricated		1.50	10.667	Flight	1,200	360		1,560	1,925
4200	Built in place	▼	.44	36.364	"	2,175	1,225		3,400	4,450
4400	Spiral, oak, 4'-6" diameter, unfinished, prefabricated,									
4500	incl. railing, 9' high	2 Carp	1.50	10.667	Flight	3,425	360		3,785	4,375

06 43 13.40 Wood Stair Parts

		Crew	Daily Output	Labor-Hours	Unit	Material	Labor	Equipment	Total	Total Incl O&P
0010	**WOOD STAIR PARTS**									
0020	Pin top balusters, 1-1/4", oak, 34"	1 Carp	96	.083	Ea.	5.10	2.82		7.92	10.35
0030	38"		96	.083		5.85	2.82		8.67	11.20
0040	42"		96	.083		6.35	2.82		9.17	11.75
0050	Poplar, 34"		96	.083		3.05	2.82		5.87	8.10
0060	38"		96	.083		3.88	2.82		6.70	9
0070	42"		96	.083		8.85	2.82		11.67	14.45
0080	Maple, 34"		96	.083		4.90	2.82		7.72	10.15
0090	38"		96	.083		5.60	2.82		8.42	10.95
0100	42"		96	.083		6.50	2.82		9.32	11.90
0130	Primed, 34"		96	.083		3	2.82		5.82	8.05
0140	38"		96	.083		3.62	2.82		6.44	8.70
0150	42"		96	.083		4.42	2.82		7.24	9.60
0180	Box top balusters, 1-1/4", oak, 34"		60	.133		8.95	4.52		13.47	17.40
0190	38"		60	.133		9.95	4.52		14.47	18.50
0200	42"		60	.133		10.95	4.52		15.47	19.60
0210	Poplar, 34"		60	.133		6.25	4.52		10.77	14.45
0220	38"	▼	60	.133	▼	6.95	4.52		11.47	15.20

06 43 Wood Stairs and Railings

06 43 13 - Wood Stairs

06 43 13.40 Wood Stair Parts		Crew	Daily Output	Labor-Hours	Unit	Material	2016 Bare Costs Labor	Equipment	Total	Total Incl O&P
0230	42"	1 Carp	60	.133	Ea.	7.50	4.52		12.02	15.80
0240	Maple, 34"		60	.133		8.25	4.52		12.77	16.65
0250	38"		60	.133		9	4.52		13.52	17.45
0260	42"		60	.133		10	4.52		14.52	18.55
0290	Primed, 34"		60	.133		7	4.52		11.52	15.25
0300	38"		60	.133		8	4.52		12.52	16.35
0310	42"		60	.133		8.35	4.52		12.87	16.75
0340	Square balusters, cut from lineal stock, pine, 1-1/16" x 1-1/16"		180	.044	L.F.	1.50	1.51		3.01	4.17
0350	1-5/16" x 1-5/16"		180	.044		2.10	1.51		3.61	4.83
0360	1-5/8" x 1-5/8"		180	.044		3.30	1.51		4.81	6.15
0370	Turned newel, oak, 3-1/2" square, 48" high		8	1	Ea.	92	34		126	158
0380	62" high		8	1		92	34		126	158
0390	Poplar, 3-1/2" square, 48" high		8	1		54	34		88	117
0400	62" high		8	1		68	34		102	132
0410	Maple, 3-1/2" square, 48" high		8	1		72	34		106	136
0420	62" high		8	1		92	34		126	158
0430	Square newel, oak, 3-1/2" square, 48" high		8	1		54	34		88	117
0440	58" high		8	1		68	34		102	132
0450	Poplar, 3-1/2" square, 48" high		8	1		35	34		69	95.50
0460	58" high		8	1		42	34		76	103
0470	Maple, 3" square, 48" high		8	1		52	34		86	114
0480	58" high		8	1		64	34		98	128
0490	Railings, oak, economy		96	.083	L.F.	8.65	2.82		11.47	14.25
0500	Average		96	.083		13	2.82		15.82	19.05
0510	Custom		96	.083		16.50	2.82		19.32	23
0520	Maple, economy		96	.083		11	2.82		13.82	16.85
0530	Average		96	.083		13.50	2.82		16.32	19.60
0540	Custom		96	.083		16.95	2.82		19.77	23.50
0550	Oak, for bending rail, economy		48	.167		23.50	5.65		29.15	35.50
0560	Average		48	.167		26	5.65		31.65	38
0570	Custom		48	.167		29.50	5.65		35.15	41.50
0580	Maple, for bending rail, economy		48	.167		32	5.65		37.65	44.50
0590	Average		48	.167		32	5.65		37.65	44.50
0600	Custom		48	.167		32	5.65		37.65	44.50
0610	Risers, oak, 3/4" x 8", 36" long		80	.100	Ea.	13	3.39		16.39	20
0620	42" long		70	.114		15.15	3.87		19.02	23
0630	48" long		63	.127		17.30	4.30		21.60	26.50
0640	54" long		56	.143		19.50	4.84		24.34	29.50
0650	60" long		50	.160		21.50	5.40		26.90	33
0660	72" long		42	.190		26	6.45		32.45	39.50
0670	Poplar, 3/4" x 8", 36" long		80	.100		12.50	3.39		15.89	19.45
0680	42" long		71	.113		14.55	3.82		18.37	22.50
0690	48" long		63	.127		16.65	4.30		20.95	25.50
0700	54" long		56	.143		18.70	4.84		23.54	28.50
0710	60" long		50	.160		21	5.40		26.40	32
0720	72" long		42	.190		25	6.45		31.45	38.50
0730	Pine, 1" x 8", 36" long		80	.100		3.52	3.39		6.91	9.55
0740	42" long		70	.114		4.11	3.87		7.98	11
0750	48" long		63	.127		4.69	4.30		8.99	12.35
0760	54" long		56	.143		5.30	4.84		10.14	13.90
0770	60" long		50	.160		5.85	5.40		11.25	15.55
0780	72" long		42	.190		7.05	6.45		13.50	18.55
0790	Treads, oak, no returns, 1-1/32" x 11-1/2" x 36" long		32	.250		27	8.50		35.50	43.50

06 43 Wood Stairs and Railings

06 43 13 – Wood Stairs

06 43 13.40 Wood Stair Parts

		Crew	Daily Output	Labor-Hours	Unit	Material	2016 Bare Costs Labor	Equipment	Total	Total Incl O&P
0800	42" long	1 Carp	32	.250	Ea.	31.50	8.50		40	48.50
0810	48" long		32	.250		36	8.50		44.50	53.50
0820	54" long		32	.250		40.50	8.50		49	58.50
0830	60" long		32	.250		45	8.50		53.50	63.50
0840	72" long		32	.250		54	8.50		62.50	73.50
0850	Mitred return one end, 1-1/32" x 11-1/2" x 36" long		24	.333		36	11.30		47.30	58.50
0860	42" long		24	.333		42	11.30		53.30	65
0870	48" long		24	.333		48	11.30		59.30	72
0880	54" long		24	.333		54	11.30		65.30	78.50
0890	60" long		24	.333		60	11.30		71.30	85
0900	72" long		24	.333		72	11.30		83.30	98
0910	Mitred return two ends, 1-1/32" x 11-1/2" x 36" long		12	.667		46	22.50		68.50	88.50
0920	42" long		12	.667		53.50	22.50		76	97
0930	48" long		12	.667		61.50	22.50		84	106
0940	54" long		12	.667		69	22.50		91.50	114
0950	60" long		12	.667		76.50	22.50		99	123
0960	72" long		12	.667		92	22.50		114.50	139
0970	Starting step, oak, 48", bullnose		8	1		172	34		206	246
0980	Double end bullnose		8	1		254	34		288	335
1030	Skirt board, pine, 1" x 10"		55	.145	L.F.	1.53	4.93		6.46	9.95
1040	1" x 12"		52	.154	"	1.88	5.20		7.08	10.80
1050	Oak landing tread, 1-1/16" thick		54	.148	S.F.	9	5		14	18.30
1060	Oak cove molding		96	.083	L.F.	1	2.82		3.82	5.85
1070	Oak stringer molding		96	.083	"	4	2.82		6.82	9.15
1090	Rail bolt, 5/16" x 3-1/2"		48	.167	Ea.	2.75	5.65		8.40	12.50
1100	5/16" x 4-1/2"		48	.167		2.75	5.65		8.40	12.50
1120	Newel post anchor		16	.500		13	16.95		29.95	43
1130	Tapered plug, 1/2"		240	.033		1	1.13		2.13	2.99
1140	1"		240	.033		.99	1.13		2.12	2.98

06 43 16 – Wood Railings

06 43 16.10 Wood Handrails and Railings

		Crew	Daily Output	Labor-Hours	Unit	Material	Labor	Equipment	Total	Total Incl O&P
0010	**WOOD HANDRAILS AND RAILINGS**									
0020	Custom design, architectural grade, hardwood, plain	1 Carp	38	.211	L.F.	12.05	7.15		19.20	25
0100	Shaped		30	.267		62.50	9.05		71.55	84
0300	Stock interior railing with spindles 4" O.C., 4' long		40	.200		38.50	6.80		45.30	54
0400	8' long		48	.167		38.50	5.65		44.15	52

06 44 Ornamental Woodwork

06 44 19 – Wood Grilles

06 44 19.10 Grilles

		Crew	Daily Output	Labor-Hours	Unit	Material	Labor	Equipment	Total	Total Incl O&P
0010	**GRILLES** and panels, hardwood, sanded									
0020	2' x 4' to 4' x 8', custom designs, unfinished, economy	1 Carp	38	.211	S.F.	62	7.15		69.15	80
0050	Average		30	.267		70	9.05		79.05	92
0100	Custom		19	.421		72	14.25		86.25	103

06 44 33 – Wood Mantels

06 44 33.10 Fireplace Mantels

		Crew	Daily Output	Labor-Hours	Unit	Material	Labor	Equipment	Total	Total Incl O&P
0010	**FIREPLACE MANTELS**									
0015	6" molding, 6' x 3'-6" opening, plain, paint grade	1 Carp	5	1.600	Opng.	440	54		494	570
0100	Ornate, oak		5	1.600		590	54		644	740
0300	Prefabricated pine, colonial type, stock, deluxe		2	4		1,500	136		1,636	1,875

06 44 Ornamental Woodwork

06 44 33 – Wood Mantels

06 44 33.10 Fireplace Mantels

		Crew	Daily Output	Labor-Hours	Unit	Material	2016 Bare Costs Labor	Equipment	Total	Total Incl O&P
0400	Economy	1 Carp	3	2.667	Opng.	690	90.50		780.50	910

06 44 33.20 Fireplace Mantel Beam

		Crew	Daily Output	Labor-Hours	Unit	Material	Labor	Equipment	Total	Total Incl O&P
0010	**FIREPLACE MANTEL BEAM**									
0020	Rough texture wood, 4" x 8"	1 Carp	36	.222	L.F.	8.30	7.55		15.85	22
0100	4" x 10"		35	.229	"	10.80	7.75		18.55	25
0300	Laminated hardwood, 2-1/4" x 10-1/2" wide, 6' long		5	1.600	Ea.	125	54		179	229
0400	8' long		5	1.600	"	150	54		204	256
0600	Brackets for above, rough sawn		12	.667	Pr.	12	22.50		34.50	51
0700	Laminated		12	.667	"	16	22.50		38.50	55.50

06 44 39 – Wood Posts and Columns

06 44 39.10 Decorative Beams

		Crew	Daily Output	Labor-Hours	Unit	Material	Labor	Equipment	Total	Total Incl O&P
0010	**DECORATIVE BEAMS**									
0020	Rough sawn cedar, non-load bearing, 4" x 4"	2 Carp	180	.089	L.F.	1.25	3.01		4.26	6.45
0100	4" x 6"		170	.094		1.80	3.19		4.99	7.35
0200	4" x 8"		160	.100		2.40	3.39		5.79	8.35
0300	4" x 10"		150	.107		3.65	3.62		7.27	10.05
0400	4" x 12"		140	.114		4.61	3.87		8.48	11.55
0500	8" x 8"		130	.123		4.80	4.17		8.97	12.30
0600	Plastic beam, "hewn finish", 6" x 2"		240	.067		3.40	2.26		5.66	7.50
0601	6" x 4"		220	.073		3.75	2.47		6.22	8.25

06 44 39.20 Columns

		Crew	Daily Output	Labor-Hours	Unit	Material	Labor	Equipment	Total	Total Incl O&P
0010	**COLUMNS**									
0050	Aluminum, round colonial, 6" diameter	2 Carp	80	.200	V.L.F.	19	6.80		25.80	32.50
0100	8" diameter		62.25	.257		22	8.70		30.70	38.50
0200	10" diameter		55	.291		22.50	9.85		32.35	41.50
0250	Fir, stock units, hollow round, 6" diameter		80	.200		29.50	6.80		36.30	44
0300	8" diameter		80	.200		35.50	6.80		42.30	50.50
0350	10" diameter		70	.229		44.50	7.75		52.25	62
0360	12" diameter		65	.246		54.50	8.35		62.85	74
0400	Solid turned, to 8' high, 3-1/2" diameter		80	.200		9.60	6.80		16.40	22
0500	4-1/2" diameter		75	.213		11.90	7.25		19.15	25
0600	5-1/2" diameter		70	.229		16	7.75		23.75	30.50
0800	Square columns, built-up, 5" x 5"		65	.246		14.20	8.35		22.55	29.50
0900	Solid, 3-1/2" x 3-1/2"		130	.123		9.60	4.17		13.77	17.55
1600	Hemlock, tapered, T & G, 12" diam., 10' high		100	.160		39.50	5.40		44.90	52.50
1700	16' high		65	.246		73	8.35		81.35	94
1900	14" diameter, 10' high		100	.160		113	5.40		118.40	134
2000	18' high		65	.246		103	8.35		111.35	127
2200	18" diameter, 12' high		65	.246		165	8.35		173.35	196
2300	20' high		50	.320		118	10.85		128.85	148
2500	20" diameter, 14' high		40	.400		180	13.55		193.55	221
2600	20' high		35	.457		170	15.50		185.50	213
2800	For flat pilasters, deduct					33%				
3000	For splitting into halves, add				Ea.	106			106	117
4000	Rough sawn cedar posts, 4" x 4"	2 Carp	250	.064	V.L.F.	3.95	2.17		6.12	8
4100	4" x 6"		235	.068		6.80	2.31		9.11	11.35
4200	6" x 6"		220	.073		9.90	2.47		12.37	15.05
4300	8" x 8"		200	.080		19.20	2.71		21.91	25.50

06 48 Wood Frames

06 48 13 – Exterior Wood Door Frames

06 48 13.10 Exterior Wood Door Frames and Accessories

	06 48 13.10 Exterior Wood Door Frames and Accessories	Crew	Daily Output	Labor-Hours	Unit	Material	2016 Bare Costs Labor	Equipment	Total	Total Incl O&P
0010	**EXTERIOR WOOD DOOR FRAMES AND ACCESSORIES**									
0400	Exterior frame, incl. ext. trim, pine, 5/4 x 4-9/16" deep	2 Carp	375	.043	L.F.	6.60	1.45		8.05	9.70
0420	5-3/16" deep		375	.043		7.90	1.45		9.35	11.05
0440	6-9/16" deep		375	.043		8.80	1.45		10.25	12.05
0600	Oak, 5/4 x 4-9/16" deep		350	.046		19.75	1.55		21.30	24
0620	5-3/16" deep		350	.046		21.50	1.55		23.05	26.50
0640	6-9/16" deep		350	.046		19.50	1.55		21.05	24
1000	Sills, 8/4 x 8" deep, oak, no horns		100	.160		6.65	5.40		12.05	16.40
1020	2" horns		100	.160		20.50	5.40		25.90	31.50
1040	3" horns		100	.160		20.50	5.40		25.90	31.50
1100	8/4 x 10" deep, oak, no horns		90	.178		6.40	6.05		12.45	17.15
1120	2" horns		90	.178		26.50	6.05		32.55	39
1140	3" horns		90	.178		26.50	6.05		32.55	39
2000	Wood frame & trim, ext, colonial, 3' opng, fluted pilasters, flat head		22	.727	Ea.	505	24.50		529.50	595
2010	Dentil head		21	.762		580	26		606	680
2020	Ram's head		20	.800		695	27		722	805
2100	5'-4" opening, in-swing, fluted pilasters, flat head		17	.941		440	32		472	540
2120	Ram's head		15	1.067		1,400	36		1,436	1,575
2140	Out swing, fluted pilasters, flat head		17	.941		520	32		552	625
2160	Ram's head		15	1.067		1,475	36		1,511	1,675
2400	6'-0" opening, in-swing, fluted pilasters, flat head		16	1		520	34		554	625
2420	Ram's head		10	1.600		1,475	54		1,529	1,725
2460	Out-swing, fluted pilasters, flat head		16	1		520	34		554	625
2480	Ram's head		10	1.600		1,475	54		1,529	1,725
2600	For two sidelights, flat head, add		30	.533	Opng.	226	18.10		244.10	280
2620	Ram's head, add		20	.800	"	840	27		867	970
2700	Custom birch frame, 3'-0" opening		16	1	Ea.	240	34		274	320
2750	6'-0" opening		16	1		360	34		394	450
2900	Exterior, modern, plain trim, 3' opng., in-swing, FJP		26	.615		46.50	21		67.50	86
2920	Fir		24	.667		55	22.50		77.50	98
2940	Oak		22	.727		63	24.50		87.50	111

06 48 16 – Interior Wood Door Frames

06 48 16.10 Interior Wood Door Jamb and Frames

		Crew	Daily Output	Labor-Hours	Unit	Material	Labor	Equipment	Total	Total Incl O&P
0010	**INTERIOR WOOD DOOR JAMB AND FRAMES**									
3000	Interior frame, pine, 11/16" x 3-5/8" deep	2 Carp	375	.043	L.F.	4.44	1.45		5.89	7.30
3020	4-9/16" deep		375	.043		4.92	1.45		6.37	7.80
3040	5-3/16" deep		375	.043		4.66	1.45		6.11	7.55
3200	Oak, 11/16" x 3-5/8" deep		350	.046		9.85	1.55		11.40	13.40
3220	4-9/16" deep		350	.046		9.95	1.55		11.50	13.55
3240	5-3/16" deep		350	.046		13.90	1.55		15.45	17.85
3400	Walnut, 11/16" x 3-5/8" deep		350	.046		9	1.55		10.55	12.50
3420	4-9/16" deep		350	.046		9.45	1.55		11	13
3440	5-3/16" deep		350	.046		9.55	1.55		11.10	13.10
3600	Pocket door frame		16	1	Ea.	81	34		115	146
3800	Threshold, oak, 5/8" x 3-5/8" deep		200	.080	L.F.	3.64	2.71		6.35	8.55
3820	4-5/8" deep		190	.084		4.35	2.85		7.20	9.55
3840	5-5/8" deep		180	.089		6.75	3.01		9.76	12.50

06 49 Wood Screens and Exterior Wood Shutters

06 49 19 – Exterior Wood Shutters

06 49 19.10 Shutters, Exterior

		Crew	Daily Output	Labor-Hours	Unit	Material	2016 Bare Costs Labor	Equipment	Total	Total Incl O&P
0010	**SHUTTERS, EXTERIOR**									
0012	Aluminum, louvered, 1'-4" wide, 3'-0" long	1 Carp	10	.800	Pr.	200	27		227	266
0200	4'-0" long		10	.800		240	27		267	310
0300	5'-4" long		10	.800		280	27		307	355
0400	6'-8" long		9	.889		355	30		385	440
1000	Pine, louvered, primed, each 1'-2" wide, 3'-3" long		10	.800		220	27		247	288
1100	4'-7" long		10	.800		270	27		297	345
1250	Each 1'-4" wide, 3'-0" long		10	.800		220	27		247	288
1350	5'-3" long		10	.800		330	27		357	405
1500	Each 1'-6" wide, 3'-3" long		10	.800		244	27		271	315
1600	4'-7" long		10	.800		325	27		352	400
1620	Cedar, louvered, 1'-2" wide, 5'-7" long		10	.800		315	27		342	390
1630	Each 1'-4" wide, 2'-2" long		10	.800		175	27		202	239
1640	3'-0" long		10	.800		217	27		244	285
1650	3'-3" long		10	.800		228	27		255	297
1660	3'-11" long		10	.800		265	27		292	340
1670	4'-3" long		10	.800		275	27		302	350
1680	5'-3" long		10	.800		290	27		317	365
1690	5'-11" long		10	.800		350	27		377	430
1700	Door blinds, 6'-9" long, each 1'-3" wide		9	.889		380	30		410	465
1710	1'-6" wide		9	.889		435	30		465	530
1720	Cedar, solid raised panel, each 1'-4" wide, 3'-3" long		10	.800		315	27		342	390
1730	3'-11" long		10	.800		360	27		387	440
1740	4'-3" long		10	.800		365	27		392	445
1750	4'-7" long		10	.800		390	27		417	475
1760	4'-11" long		10	.800		415	27		442	505
1770	5'-11" long		10	.800		520	27		547	615
1800	Door blinds, 6'-9" long, each 1'-3" wide		9	.889		535	30		565	640
1900	1'-6" wide		9	.889		630	30		660	745
2500	Polystyrene, solid raised panel, each 1'-4" wide, 3'-3" long		10	.800		72	27		99	125
2600	3'-11" long		10	.800		93	27		120	148
2700	4'-7" long		10	.800		105	27		132	162
2800	5'-3" long		10	.800		120	27		147	178
2900	6'-8" long		9	.889		152	30		182	218
4500	Polystyrene, louvered, each 1'-2" wide, 3'-3" long		10	.800		36	27		63	85
4600	4'-7" long		10	.800		46	27		73	96
4750	5'-3" long		10	.800		59	27		86	111
4850	6'-8" long		9	.889		68	30		98	126
6000	Vinyl, louvered, each 1'-2" x 4'-7" long		10	.800		64	27		91	116
6200	Each 1'-4" x 6'-8" long		9	.889		76	30		106	134
8000	PVC exterior rolling shutters									
8100	including crank control	1 Carp	8	1	Ea.	540	34		574	645
8500	Insulative - 6' x 6'8" stock unit	"	8	1	"	765	34		799	900

06 51 Structural Plastic Shapes and Plates

06 51 13 – Plastic Lumber

06 51 13.10 Recycled Plastic Lumber

			Crew	Daily Output	Labor-Hours	Unit	Material	2016 Bare Costs Labor	Equipment	Total	Total Incl O&P
0010	**RECYCLED PLASTIC LUMBER**										
4000	Sheeting, recycled plastic, black or white, 4' x 8' x 1/8"	G	2 Carp	1100	.015	S.F.	1.26	.49		1.75	2.22
4010	4' x 8' x 3/16"	G		1100	.015		1.90	.49		2.39	2.92
4020	4' x 8' x 1/4"	G		950	.017		2.25	.57		2.82	3.44
4030	4' x 8' x 3/8"	G		950	.017		3.80	.57		4.37	5.15
4040	4' x 8' x 1/2"	G		900	.018		5	.60		5.60	6.50
4050	4' x 8' x 5/8"	G		900	.018		7.50	.60		8.10	9.25
4060	4' x 8' x 3/4"	G		850	.019		9.40	.64		10.04	11.40
4070	Add for colors	G				Ea.	5%				
8500	100% recycled plastic, var colors, NLB, 2" x 2"	G				L.F.	1.91			1.91	2.10
8510	2" x 4"	G					3.95			3.95	4.35
8520	2" x 6"	G					6.20			6.20	6.80
8530	2" x 8"	G					8.50			8.50	9.35
8540	2" x 10"	G					12.40			12.40	13.60
8550	5/4" x 4"	G					4.45			4.45	4.90
8560	5/4" x 6"	G					5.60			5.60	6.15
8570	1" x 6"	G					3.06			3.06	3.37
8580	1/2" x 8"	G					3.45			3.45	3.80
8590	2" x 10" T & G	G					12.20			12.20	13.40
8600	3" x 10" T & G	G					17.15			17.15	18.85
8610	Add for premium colors	G					20%				

06 51 13.12 Structural Plastic Lumber

		Crew	Daily Output	Labor-Hours	Unit	Material	Labor	Equipment	Total	Total Incl O&P
0010	**STRUCTURAL PLASTIC LUMBER**									
1320	Plastic lumber, posts or columns, 4" x 4"	2 Carp	390	.041	L.F.	9.25	1.39		10.64	12.50
1325	4" x 6"		275	.058		13.25	1.97		15.22	17.90
1330	4" x 8"		220	.073		19.35	2.47		21.82	25.50
1340	Girder, single, 4" x 4"		675	.024		9.25	.80		10.05	11.50
1345	4" x 6"		600	.027		13.25	.90		14.15	16.10
1350	4" x 8"		525	.030		19.35	1.03		20.38	23
1352	Double, 2" x 4"		625	.026		8.55	.87		9.42	10.85
1354	2" x 6"		600	.027		12.85	.90		13.75	15.65
1356	2" x 8"		575	.028		16.90	.94		17.84	20
1358	2" x 10"		550	.029		20.50	.99		21.49	24
1360	2" x 12"		525	.030		25	1.03		26.03	29
1362	Triple, 2" x 4"		575	.028		12.80	.94		13.74	15.70
1364	2" x 6"		550	.029		19.30	.99		20.29	22.50
1366	2" x 8"		525	.030		25.50	1.03		26.53	29.50
1368	2" x 10"		500	.032		31	1.08		32.08	36
1370	2" x 12"		475	.034		37.50	1.14		38.64	43
1372	Ledger, bolted 4' O.C., 2" x 4"		400	.040		4.42	1.36		5.78	7.15
1374	2" x 6"		550	.029		6.50	.99		7.49	8.80
1376	2" x 8"		390	.041		8.55	1.39		9.94	11.80
1378	2" x 10"		385	.042		10.40	1.41		11.81	13.80
1380	2" x 12"		380	.042		12.55	1.43		13.98	16.25
1382	Joists, 2" x 4"		1250	.013		4.27	.43		4.70	5.40
1384	2" x 6"		1250	.013		6.45	.43		6.88	7.80
1386	2" x 8"		1100	.015		8.45	.49		8.94	10.15
1388	2" x 10"		500	.032		10.40	1.08		11.48	13.25
1390	2" x 12"		875	.018		12.45	.62		13.07	14.75
1392	Railings and trim, 5/4" x 4"	1 Carp	300	.027		4.75	.90		5.65	6.75
1394	2" x 2"		300	.027		2.40	.90		3.30	4.15
1396	2" x 4"		300	.027		4.25	.90		5.15	6.20

06 51 Structural Plastic Shapes and Plates

06 51 13 – Plastic Lumber

06 51 13.12 Structural Plastic Lumber		Crew	Daily Output	Labor-Hours	Unit	Material	2016 Bare Costs Labor	2016 Bare Costs Equipment	Total	Total Incl O&P
1398	2" x 6"	1 Carp	300	.027	L.F.	6.40	.90		7.30	8.55

06 63 Plastic Railings

06 63 10 – Plastic (PVC) Railings

06 63 10.10 Plastic Railings

		Crew	Daily Output	Labor-Hours	Unit	Material	Labor	Equipment	Total	Incl O&P
0010	**PLASTIC RAILINGS**									
0100	Horizontal PVC handrail with balusters, 3-1/2" wide, 36" high	1 Carp	96	.083	L.F.	25.50	2.82		28.32	32.50
0150	42" high		96	.083		28.50	2.82		31.32	36
0200	Angled PVC handrail with balusters, 3-1/2" wide, 36" high		72	.111		29	3.77		32.77	38.50
0250	42" high		72	.111		32.50	3.77		36.27	42
0300	Post sleeve for 4 x 4 post		96	.083		13.10	2.82		15.92	19.15
0400	Post cap for 4 x 4 post, flat profile		48	.167	Ea.	13.20	5.65		18.85	24
0450	Newel post style profile		48	.167		24	5.65		29.65	36
0500	Raised corbeled profile		48	.167		36.50	5.65		42.15	49.50
0550	Post base trim for 4 x 4 post		96	.083		18.80	2.82		21.62	25

06 65 Plastic Trim

06 65 10 – PVC Trim

06 65 10.10 PVC Trim, Exterior

		Crew	Daily Output	Labor-Hours	Unit	Material	Labor	Equipment	Total	Incl O&P
0010	**PVC TRIM, EXTERIOR**									
0100	Cornerboards, 5/4" x 6" x 6"	1 Carp	240	.033	L.F.	8	1.13		9.13	10.70
0110	Door/window casing, 1" x 4"		200	.040		1.46	1.36		2.82	3.88
0120	1" x 6"		200	.040		2.20	1.36		3.56	4.69
0130	1" x 8"		195	.041		2.93	1.39		4.32	5.55
0140	1" x 10"		195	.041		3.81	1.39		5.20	6.50
0150	1" x 12"		190	.042		4.36	1.43		5.79	7.20
0160	5/4" x 4"		195	.041		2.19	1.39		3.58	4.74
0170	5/4" x 6"		195	.041		3.15	1.39		4.54	5.80
0180	5/4" x 8"		190	.042		4.19	1.43		5.62	7
0190	5/4" x 10"		190	.042		5.10	1.43		6.53	8
0200	5/4" x 12"		185	.043		5.60	1.47		7.07	8.60
0210	Fascia, 1" x 4"		250	.032		1.46	1.08		2.54	3.43
0220	1" x 6"		250	.032		2.20	1.08		3.28	4.24
0230	1" x 8"		225	.036		2.93	1.21		4.14	5.25
0240	1" x 10"		225	.036		3.81	1.21		5.02	6.20
0250	1" x 12"		200	.040		4.36	1.36		5.72	7.05
0260	5/4" x 4"		240	.033		2.19	1.13		3.32	4.30
0270	5/4" x 6"		240	.033		3.15	1.13		4.28	5.35
0280	5/4" x 8"		215	.037		4.19	1.26		5.45	6.70
0290	5/4" x 10"		215	.037		5.10	1.26		6.36	7.70
0300	5/4" x 12"		190	.042		5.60	1.43		7.03	8.55
0310	Frieze, 1" x 4"		250	.032		1.46	1.08		2.54	3.43
0320	1" x 6"		250	.032		2.20	1.08		3.28	4.24
0330	1" x 8"		225	.036		2.93	1.21		4.14	5.25
0340	1" x 10"		225	.036		3.81	1.21		5.02	6.20
0350	1" x 12"		200	.040		4.36	1.36		5.72	7.05
0360	5/4" x 4"		240	.033		2.19	1.13		3.32	4.30
0370	5/4" x 6"		240	.033		3.15	1.13		4.28	5.35
0380	5/4" x 8"		215	.037		4.19	1.26		5.45	6.70

06 65 Plastic Trim

06 65 10 – PVC Trim

06 65 10.10 PVC Trim, Exterior		Crew	Daily Output	Labor-Hours	Unit	Material	2016 Bare Costs Labor	Equipment	Total	Total Incl O&P
0390	5/4" x 10"	1 Carp	215	.037	L.F.	5.10	1.26		6.36	7.70
0400	5/4" x 12"		190	.042		5.60	1.43		7.03	8.55
0410	Rake, 1" x 4"		200	.040		1.46	1.36		2.82	3.88
0420	1" x 6"		200	.040		2.20	1.36		3.56	4.69
0430	1" x 8"		190	.042		2.93	1.43		4.36	5.60
0440	1" x 10"		190	.042		3.81	1.43		5.24	6.60
0450	1" x 12"		180	.044		4.36	1.51		5.87	7.30
0460	5/4" x 4"		195	.041		2.19	1.39		3.58	4.74
0470	5/4" x 6"		195	.041		3.15	1.39		4.54	5.80
0480	5/4" x 8"		185	.043		4.19	1.47		5.66	7.05
0490	5/4" x 10"		185	.043		5.10	1.47		6.57	8.05
0500	5/4" x 12"		175	.046		5.60	1.55		7.15	8.75
0510	Rake trim, 1" x 4"		225	.036		1.46	1.21		2.67	3.63
0520	1" x 6"		225	.036		2.20	1.21		3.41	4.44
0560	5/4" x 4"		220	.036		2.19	1.23		3.42	4.47
0570	5/4" x 6"		220	.036		3.15	1.23		4.38	5.55
0610	Soffit, 1" x 4"	2 Carp	420	.038		1.46	1.29		2.75	3.77
0620	1" x 6"		420	.038		2.20	1.29		3.49	4.58
0630	1" x 8"		420	.038		2.93	1.29		4.22	5.40
0640	1" x 10"		400	.040		3.81	1.36		5.17	6.45
0650	1" x 12"		400	.040		4.36	1.36		5.72	7.05
0660	5/4" x 4"		410	.039		2.19	1.32		3.51	4.62
0670	5/4" x 6"		410	.039		3.15	1.32		4.47	5.70
0680	5/4" x 8"		410	.039		4.19	1.32		5.51	6.80
0690	5/4" x 10"		390	.041		5.10	1.39		6.49	7.95
0700	5/4" x 12"		390	.041		5.60	1.39		6.99	8.50

06 80 Composite Fabrications

06 80 10 – Composite Decking

06 80 10.10 Woodgrained Composite Decking

		Crew	Daily Output	Labor-Hours	Unit	Material	2016 Bare Costs Labor	Equipment	Total	Total Incl O&P
0010	**WOODGRAINED COMPOSITE DECKING**									
0100	Woodgrained composite decking, 1" x 6"	2 Carp	640	.025	L.F.	3.65	.85		4.50	5.45
0110	Grooved edge		660	.024		3.79	.82		4.61	5.55
0120	2" x 6"		640	.025		3.78	.85		4.63	5.60
0130	Encased, 1" x 6"		640	.025		3.78	.85		4.63	5.60
0140	Grooved edge		660	.024		3.92	.82		4.74	5.70
0150	2" x 6"		640	.025		5.10	.85		5.95	7.05

06 81 Composite Railings

06 81 10 – Encased Railings

06 81 10.10 Encased Composite Railings

		Crew	Daily Output	Labor-Hours	Unit	Material	2016 Bare Costs Labor	Equipment	Total	Total Incl O&P
0010	**ENCASED COMPOSITE RAILINGS**									
0100	Encased composite railing, 6' long, 36" high, incl. balusters	1 Carp	16	.500	Ea.	161	16.95		177.95	206
0110	42" high, incl. balusters		16	.500		181	16.95		197.95	228
0120	8' long, 36" high, incl. balusters		12	.667		161	22.50		183.50	215
0130	42" high, incl. balusters		12	.667		181	22.50		203.50	237
0140	Accessories, post sleeve, 4" x 4", 39" long		32	.250		23.50	8.50		32	40
0150	96" long		24	.333		89	11.30		100.30	117
0160	6" x 6", 39" long		32	.250		48	8.50		56.50	67

06 81 Composite Railings

06 81 10 – Encased Railings

06 81 10.10 Encased Composite Railings	Crew	Daily Output	Labor-Hours	Unit	Material	2016 Bare Costs Labor	2016 Bare Costs Equipment	Total	Total Incl O&P
0170 96" long	1 Carp	24	.333	Ea.	141	11.30		152.30	174
0180 Accessories, post skirt, 4" x 4"		96	.083		4.25	2.82		7.07	9.40
0190 6" x 6"		96	.083		5	2.82		7.82	10.25
0200 Post cap, 4" x 4", flat		48	.167		6.30	5.65		11.95	16.40
0210 Pyramid		48	.167		6.30	5.65		11.95	16.40
0220 Post cap, 6" x 6", flat		48	.167		10.20	5.65		15.85	20.50
0230 Pyramid		48	.167		10.20	5.65		15.85	20.50

Division Notes

	CREW	DAILY OUTPUT	LABOR-HOURS	UNIT	BARE COSTS				TOTAL INCL O&P
					MAT.	LABOR	EQUIP.	TOTAL	

Division 7 Thermal & Moisture Protection

Estimating Tips

07 10 00 Dampproofing and Waterproofing

- Be sure of the job specifications before pricing this subdivision. The difference in cost between waterproofing and dampproofing can be great. Waterproofing will hold back standing water. Dampproofing prevents the transmission of water vapor. Also included in this section are vapor retarding membranes.

07 20 00 Thermal Protection

- Insulation and fireproofing products are measured by area, thickness, volume or R-value. Specifications may give only what the specific R-value should be in a certain situation. The estimator may need to choose the type of insulation to meet that R-value.

07 30 00 Steep Slope Roofing
07 40 00 Roofing and Siding Panels

- Many roofing and siding products are bought and sold by the square. One square is equal to an area that measures 100 square feet.

  This simple change in unit of measure could create a large error if the estimator is not observant. Accessories necessary for a complete installation must be figured into any calculations for both material and labor.

07 50 00 Membrane Roofing
07 60 00 Flashing and Sheet Metal
07 70 00 Roofing and Wall Specialties and Accessories

- The items in these subdivisions compose a roofing system. No one component completes the installation, and all must be estimated. Built-up or single-ply membrane roofing systems are made up of many products and installation trades. Wood blocking at roof perimeters or penetrations, parapet coverings, reglets, roof drains, gutters, downspouts, sheet metal flashing, skylights, smoke vents, and roof hatches all need to be considered along with the roofing material. Several different installation trades will need to work together on the roofing system. Inherent difficulties in the scheduling and coordination of various trades must be accounted for when estimating labor costs.

07 90 00 Joint Protection

- To complete the weather-tight shell, the sealants and caulkings must be estimated. Where different materials meet—at expansion joints, at flashing penetrations, and at hundreds of other locations throughout a construction project—caulking and sealants provide another line of defense against water penetration. Often, an entire system is based on the proper location and placement of caulking or sealants. The detailed drawings that are included as part of a set of architectural plans show typical locations for these materials. When caulking or sealants are shown at typical locations, this means the estimator must include them for all the locations where this detail is applicable. Be careful to keep different types of sealants separate, and remember to consider backer rods and primers if necessary.

Reference Numbers

Reference numbers are shown at the beginning of some major classifications. These numbers refer to related items in the Reference Section. The reference information may be an estimating procedure, an alternate pricing method, or technical information.

Note: Not all subdivisions listed here necessarily appear. ■

Did you know?

RSMeans Online gives you the same access to RSMeans' data with 24/7 access:
- Quickly locate costs in the searchable database.
- Build cost lists, estimates, and reports in minutes.
- Adjust costs to any location in the U.S. and Canada with the click of a button.

Start your free trial today at **www.RSMeansOnline.com**

RSMeans Online
FROM THE GORDIAN GROUP

No part of this cost data may be reproduced, stored in a retrieval system, or transmitted in any form or by any means without prior written permission of RSMeans.

07 01 Operation and Maint. of Thermal and Moisture Protection

07 01 50 – Maintenance of Membrane Roofing

07 01 50.10 Roof Coatings		Crew	Daily Output	Labor-Hours	Unit	Material	2016 Bare Costs Labor	2016 Bare Costs Equipment	Total	Total Incl O&P
0010	**ROOF COATINGS**									
0012	Asphalt, brush grade, material only				Gal.	8.95			8.95	9.80
0800	Glass fibered roof & patching cement, 5 gallon					8.25			8.25	9.10
1100	Roof patch & flashing cement, 5 gallon					8.75			8.75	9.60

07 05 Common Work Results for Thermal and Moisture Protection

07 05 05 – Selective Demolition for Thermal and Moisture Protection

07 05 05.10 Selective Demo., Thermal and Moist. Protection			Crew	Daily Output	Labor-Hours	Unit	Material	2016 Bare Costs Labor	2016 Bare Costs Equipment	Total	Total Incl O&P
0010	**SELECTIVE DEMO., THERMAL AND MOISTURE PROTECTION**										
0020	Caulking/sealant, to 1" x 1" joint	R024119-10	1 Clab	600	.013	L.F.		.33		.33	.55
0120	Downspouts, including hangers			350	.023	"		.56		.56	.94
0220	Flashing, sheet metal			290	.028	S.F.		.68		.68	1.14
0420	Gutters, aluminum or wood, edge hung			240	.033	L.F.		.82		.82	1.37
0520	Built-in			100	.080	"		1.97		1.97	3.30
0620	Insulation, air/vapor barrier			3500	.002	S.F.		.06		.06	.09
0670	Batts or blankets			1400	.006	C.F.		.14		.14	.24
0720	Foamed or sprayed in place		2 Clab	1000	.016	B.F.		.39		.39	.66
0770	Loose fitting		1 Clab	3000	.003	C.F.		.07		.07	.11
0870	Rigid board			3450	.002	B.F.		.06		.06	.10
1120	Roll roofing, cold adhesive			12	.667	Sq.		16.45		16.45	27.50
1170	Roof accessories, adjustable metal chimney flashing			9	.889	Ea.		22		22	36.50
1325	Plumbing vent flashing			32	.250	"		6.15		6.15	10.30
1375	Ridge vent strip, aluminum			310	.026	L.F.		.64		.64	1.06
1620	Skylight to 10 S.F.			8	1	Ea.		24.50		24.50	41.50
2120	Roof edge, aluminum soffit and fascia			570	.014	L.F.		.35		.35	.58
2170	Concrete coping, up to 12" wide		2 Clab	160	.100			2.47		2.47	4.13
2220	Drip edge		1 Clab	1000	.008			.20		.20	.33
2270	Gravel stop			950	.008			.21		.21	.35
2370	Sheet metal coping, up to 12" wide			240	.033			.82		.82	1.37
2470	Roof insulation board, over 2" thick		B-2	7800	.005	B.F.		.13		.13	.22
2520	Up to 2" thick		"	3900	.010	S.F.		.26		.26	.43
2620	Roof ventilation, louvered gable vent		1 Clab	16	.500	Ea.		12.35		12.35	20.50
2670	Remove, roof hatch		G-3	15	2.133			66.50		66.50	111
2675	Rafter vents		1 Clab	960	.008			.21		.21	.34
2720	Soffit vent and/or fascia vent			575	.014	L.F.		.34		.34	.57
2775	Soffit vent strip, aluminum, 3" to 4" wide			160	.050			1.23		1.23	2.06
2820	Roofing accessories, shingle moulding, to 1" x 4"			1600	.005			.12		.12	.21
2870	Cant strip		B-2	2000	.020			.50		.50	.84
2920	Concrete block walkway		1 Clab	230	.035			.86		.86	1.43
3070	Roofing, felt paper, 15#			70	.114	Sq.		2.82		2.82	4.71
3125	#30 felt			30	.267	"		6.55		6.55	11
3170	Asphalt shingles, 1 layer		B-2	3500	.011	S.F.		.29		.29	.48
3180	2 layers			1750	.023	"		.57		.57	.96
3370	Modified bitumen			26	1.538	Sq.		38.50		38.50	64.50
3420	Built-up, no gravel, 3 ply			25	1.600			40		40	67
3470	4 ply			21	1.905			47.50		47.50	80
3620	5 ply			1600	.025	S.F.		.63		.63	1.05
3720	5 ply, with gravel			890	.045			1.13		1.13	1.88
3725	Loose gravel removal			5000	.008			.20		.20	.34
3730	Embedded gravel removal			2000	.020			.50		.50	.84
3870	Fiberglass sheet			1200	.033			.83		.83	1.40

07 05 Common Work Results for Thermal and Moisture Protection

07 05 05 – Selective Demolition for Thermal and Moisture Protection

07 05 05.10 Selective Demo., Thermal and Moist. Protection

		Crew	Daily Output	Labor-Hours	Unit	Material	2016 Bare Costs Labor	Equipment	Total	Total Incl O&P
4120	Slate shingles	B-2	1900	.021	S.F.		.53		.53	.88
4170	Ridge shingles, clay or slate		2000	.020	L.F.		.50		.50	.84
4320	Single ply membrane, attached at seams		52	.769	Sq.		19.25		19.25	32.50
4370	Ballasted		75	.533			13.35		13.35	22.50
4420	Fully adhered		39	1.026			25.50		25.50	43
4550	Roof hatch, 2'-6" x 3'-0"	1 Clab	10	.800	Ea.		19.70		19.70	33
4670	Wood shingles	B-2	2200	.018	S.F.		.46		.46	.76
4820	Sheet metal roofing	"	2150	.019			.47		.47	.78
4970	Siding, horizontal wood clapboards	1 Clab	380	.021			.52		.52	.87
5025	Exterior insulation finish system	"	120	.067			1.64		1.64	2.75
5070	Tempered hardboard, remove and reset	1 Carp	380	.021			.71		.71	1.19
5120	Tempered hardboard sheet siding	"	375	.021			.72		.72	1.21
5170	Metal, corner strips	1 Clab	850	.009	L.F.		.23		.23	.39
5225	Horizontal strips		444	.018	S.F.		.44		.44	.74
5320	Vertical strips		400	.020			.49		.49	.83
5520	Wood shingles		350	.023			.56		.56	.94
5620	Stucco siding		360	.022			.55		.55	.92
5670	Textured plywood		725	.011			.27		.27	.46
5720	Vinyl siding		510	.016			.39		.39	.65
5770	Corner strips		900	.009	L.F.		.22		.22	.37
5870	Wood, boards, vertical		400	.020	S.F.		.49		.49	.83
5920	Waterproofing, protection/drain board	2 Clab	3900	.004	B.F.		.10		.10	.17
5970	Over 1/2" thick		1750	.009	S.F.		.23		.23	.38
6020	To 1/2" thick		2000	.008	"		.20		.20	.33

07 11 Dampproofing

07 11 13 – Bituminous Dampproofing

07 11 13.10 Bituminous Asphalt Coating

		Crew	Daily Output	Labor-Hours	Unit	Material	2016 Bare Costs Labor	Equipment	Total	Total Incl O&P
0010	**BITUMINOUS ASPHALT COATING**									
0030	Brushed on, below grade, 1 coat	1 Rofc	665	.012	S.F.	.22	.35		.57	.89
0100	2 coat		500	.016		.45	.46		.91	1.34
0300	Sprayed on, below grade, 1 coat		830	.010		.22	.28		.50	.76
0400	2 coat		500	.016		.44	.46		.90	1.33
0500	Asphalt coating, with fibers				Gal.	8.25			8.25	9.10
0600	Troweled on, asphalt with fibers, 1/16" thick	1 Rofc	500	.016	S.F.	.36	.46		.82	1.24
0700	1/8" thick		400	.020		.63	.58		1.21	1.76
1000	1/2" thick		350	.023		2.06	.66		2.72	3.48

07 11 16 – Cementitious Dampproofing

07 11 16.20 Cementitious Parging

		Crew	Daily Output	Labor-Hours	Unit	Material	2016 Bare Costs Labor	Equipment	Total	Total Incl O&P
0010	**CEMENTITIOUS PARGING**									
0020	Portland cement, 2 coats, 1/2" thick	D-1	250	.064	S.F.	.35	1.89		2.24	3.53
0100	Waterproofed Portland cement, 1/2" thick, 2 coats	"	250	.064	"	3.95	1.89		5.84	7.50

07 19 Water Repellents

07 19 19 – Silicone Water Repellents

07 19 19.10 Silicone Based Water Repellents		Crew	Daily Output	Labor-Hours	Unit	Material	2016 Bare Costs Labor	Equipment	Total	Total Incl O&P
0010	**SILICONE BASED WATER REPELLENTS**									
0020	Water base liquid, roller applied	2 Rofc	7000	.002	S.F.	.53	.07		.60	.70
0200	Silicone or stearate, sprayed on CMU, 1 coat	1 Rofc	4000	.002		.37	.06		.43	.51
0300	2 coats	"	3000	.003		.73	.08		.81	.95

07 21 Thermal Insulation

07 21 13 – Board Insulation

07 21 13.10 Rigid Insulation

			Crew	Daily Output	Labor-Hours	Unit	Material	Labor	Equipment	Total	Total Incl O&P
0010	**RIGID INSULATION**, for walls										
0040	Fiberglass, 1.5#/C.F., unfaced, 1" thick, R4.1	G	1 Carp	1000	.008	S.F.	.29	.27		.56	.77
0060	1-1/2" thick, R6.2	G		1000	.008		.38	.27		.65	.87
0080	2" thick, R8.3	G		1000	.008		.49	.27		.76	.99
0120	3" thick, R12.4	G		800	.010		.56	.34		.90	1.19
0370	3#/C.F., unfaced, 1" thick, R4.3	G		1000	.008		.52	.27		.79	1.02
0390	1-1/2" thick, R6.5	G		1000	.008		.78	.27		1.05	1.31
0400	2" thick, R8.7	G		890	.009		1.05	.30		1.35	1.67
0420	2-1/2" thick, R10.9	G		800	.010		1.10	.34		1.44	1.78
0440	3" thick, R13	G		800	.010		1.59	.34		1.93	2.32
0520	Foil faced, 1" thick, R4.3	G		1000	.008		.90	.27		1.17	1.44
0540	1-1/2" thick, R6.5	G		1000	.008		1.35	.27		1.62	1.94
0560	2" thick, R8.7	G		890	.009		1.69	.30		1.99	2.37
0580	2-1/2" thick, R10.9	G		800	.010		1.98	.34		2.32	2.75
0600	3" thick, R13	G		800	.010		2.18	.34		2.52	2.97
1600	Isocyanurate, 4' x 8' sheet, foil faced, both sides										
1610	1/2" thick	G	1 Carp	800	.010	S.F.	.31	.34		.65	.91
1620	5/8" thick	G		800	.010		.33	.34		.67	.93
1630	3/4" thick	G		800	.010		.38	.34		.72	.99
1640	1" thick	G		800	.010		.51	.34		.85	1.13
1650	1-1/2" thick	G		730	.011		.59	.37		.96	1.27
1660	2" thick	G		730	.011		.75	.37		1.12	1.45
1670	3" thick	G		730	.011		1.85	.37		2.22	2.66
1680	4" thick	G		730	.011		2.10	.37		2.47	2.93
1700	Perlite, 1" thick, R2.77	G		800	.010		.44	.34		.78	1.05
1750	2" thick, R5.55	G		730	.011		.77	.37		1.14	1.47
1900	Extruded polystyrene, 25 PSI compressive strength, 1" thick, R5	G		800	.010		.56	.34		.90	1.19
1940	2" thick R10	G		730	.011		1.13	.37		1.50	1.86
1960	3" thick, R15	G		730	.011		1.58	.37		1.95	2.36
2100	Expanded polystyrene, 1" thick, R3.85	G		800	.010		.26	.34		.60	.86
2120	2" thick, R7.69	G		730	.011		.52	.37		.89	1.19
2140	3" thick, R11.49	G		730	.011		.78	.37		1.15	1.48

07 21 13.13 Foam Board Insulation

			Crew	Daily Output	Labor-Hours	Unit	Material	Labor	Equipment	Total	Total Incl O&P
0010	**FOAM BOARD INSULATION**										
0600	Polystyrene, expanded, 1" thick, R4	G	1 Carp	680	.012	S.F.	.26	.40		.66	.96
0700	2" thick, R8	G	"	675	.012	"	.52	.40		.92	1.24

07 21 Thermal Insulation

07 21 16 – Blanket Insulation

07 21 16.10 Blanket Insulation for Floors/Ceilings		Crew	Daily Output	Labor-Hours	Unit	Material	2016 Bare Costs Labor	Equipment	Total	Total Incl O&P
0010	**BLANKET INSULATION FOR FLOORS/CEILINGS**									
0020	Including spring type wire fasteners									
2000	Fiberglass, blankets or batts, paper or foil backing									
2100	3-1/2" thick, R13	G	1 Carp	700	.011	S.F.	.40	.39		.79	1.09
2150	6-1/4" thick, R19	G		600	.013		.51	.45		.96	1.32
2210	9-1/2" thick, R30	G		500	.016		.73	.54		1.27	1.71
2220	12" thick, R38	G		475	.017		1.05	.57		1.62	2.12
3000	Unfaced, 3-1/2" thick, R13	G		600	.013		.33	.45		.78	1.12
3010	6-1/4" thick, R19	G		500	.016		.39	.54		.93	1.34
3020	9-1/2" thick, R30	G		450	.018		.60	.60		1.20	1.67
3030	12" thick, R38	G		425	.019		.75	.64		1.39	1.90

07 21 16.20 Blanket Insulation for Walls		Crew	Daily Output	Labor-Hours	Unit	Material	Labor	Equipment	Total	Total Incl O&P
0010	**BLANKET INSULATION FOR WALLS**									
0020	Kraft faced fiberglass, 3-1/2" thick, R11, 15" wide	G	1 Carp	1350	.006	S.F.	.31	.20		.51	.68
0030	23" wide	G		1600	.005		.31	.17		.48	.62
0060	R13, 11" wide	G		1150	.007		.33	.24		.57	.76
0080	15" wide	G		1350	.006		.33	.20		.53	.70
0100	23" wide	G		1600	.005		.33	.17		.50	.64
0110	R15, 11" wide	G		1150	.007		.49	.24		.73	.94
0120	15" wide	G		1350	.006		.49	.20		.69	.88
0140	6" thick, R19, 11" wide	G		1150	.007		.44	.24		.68	.88
0160	15" wide	G		1350	.006		.44	.20		.64	.82
0180	23" wide	G		1600	.005		.44	.17		.61	.76
0201	9" thick, R30, 15" wide	G		1350	.006		.73	.20		.93	1.14
0241	12" thick, R38, 15" wide	G		1350	.006		1.05	.20		1.25	1.50
0410	Foil faced fiberglass, 3-1/2" thick, R13, 11" wide	G		1150	.007		.47	.24		.71	.92
0420	15" wide	G		1350	.006		.47	.20		.67	.86
0442	R15, 11" wide	G		1150	.007		.48	.24		.72	.93
0444	15" wide	G		1350	.006		.48	.20		.68	.87
0461	6" thick, R19, 15" wide	G		1600	.005		.62	.17		.79	.96
0482	R21, 11" wide	G		1150	.007		.64	.24		.88	1.10
0501	9" thick, R30, 15" wide	G		1350	.006		.94	.20		1.14	1.37
0620	Unfaced fiberglass, 3-1/2" thick, R13, 11" wide	G		1150	.007		.33	.24		.57	.76
0821	3-1/2" thick, R13, 15" wide	G		1600	.005		.33	.17		.50	.64
0832	R15, 11" wide	G		1150	.007		.46	.24		.70	.91
0834	15" wide	G		1350	.006		.46	.20		.66	.85
0861	6" thick, R19, 15" wide	G		1350	.006		.39	.20		.59	.77
0901	9" thick, R30, 15" wide	G		1150	.007		.60	.24		.84	1.06
0941	12" thick, R38, 15" wide	G		1150	.007		.75	.24		.99	1.23
1300	Wall or ceiling insulation, mineral wool batts									
1320	3-1/2" thick, R15	G	1 Carp	1600	.005	S.F.	.67	.17		.84	1.02
1340	5-1/2" thick, R23	G		1600	.005		1.05	.17		1.22	1.44
1380	7-1/4" thick, R30	G		1350	.006		1.39	.20		1.59	1.87
1700	Non-rigid insul, recycled blue cotton fiber, unfaced batts, R13, 16" wide	G		1600	.005		1.01	.17		1.18	1.39
1710	R19, 16" wide	G		1600	.005		1.38	.17		1.55	1.80
1850	Friction fit wire insulation supports, 16" O.C.			960	.008	Ea.	.07	.28		.35	.55

07 21 19 – Foamed In Place Insulation

07 21 19.10 Masonry Foamed In Place Insulation		Crew	Daily Output	Labor-Hours	Unit	Material	Labor	Equipment	Total	Total Incl O&P
0010	**MASONRY FOAMED IN PLACE INSULATION**									
0100	Amino-plast foam, injected into block core, 6" block	G	G-2A	6000	.004	Ea.	.16	.10	.12	.38	.49
0110	8" block	G		5000	.005		.20	.12	.14	.46	.59

For customer support on your Residential Cost Data, call 877.759.4771.

07 21 Thermal Insulation

07 21 19 – Foamed In Place Insulation

07 21 19.10 Masonry Foamed In Place Insulation

			Crew	Daily Output	Labor-Hours	Unit	Material	2016 Bare Costs Labor	Equipment	Total	Total Incl O&P
0120		10" block	G	G-2A	4000	.006	Ea.	.25	.15	.18	.58	.74
0130		12" block	G		3000	.008		.33	.20	.24	.77	.98
0140		Injected into cavity wall	G		13000	.002	B.F.	.06	.05	.05	.16	.20
0150		Preparation, drill holes into mortar joint every 4 VLF, 5/8" dia		1 Clab	960	.008	Ea.		.21		.21	.34
0160		7/8" dia			680	.012			.29		.29	.49
0170		Patch drilled holes, 5/8" diameter			1800	.004		.04	.11		.15	.22
0180		7/8" diameter			1200	.007		.05	.16		.21	.34

07 21 23 – Loose-Fill Insulation

07 21 23.10 Poured Loose-Fill Insulation

				Crew	Daily Output	Labor-Hours	Unit	Material	Labor	Equipment	Total	Total Incl O&P
0010	**POURED LOOSE-FILL INSULATION**											
0020	Cellulose fiber, R3.8 per inch	G		1 Carp	200	.040	C.F.	.69	1.36		2.05	3.03
0021	4" thick	G			1000	.008	S.F.	.16	.27		.43	.63
0022	6" thick	G			800	.010	"	.28	.34		.62	.88
0080	Fiberglass wool, R4 per inch	G			200	.040	C.F.	.55	1.36		1.91	2.87
0081	4" thick	G			600	.013	S.F.	.19	.45		.64	.97
0082	6" thick	G			400	.020	"	.26	.68		.94	1.43
0100	Mineral wool, R3 per inch	G			200	.040	C.F.	.45	1.36		1.81	2.77
0101	4" thick	G			600	.013	S.F.	.15	.45		.60	.92
0102	6" thick	G			400	.020	"	.23	.68		.91	1.39
0300	Polystyrene, R4 per inch	G			200	.040	C.F.	1.60	1.36		2.96	4.03
0301	4" thick	G			600	.013	S.F.	.53	.45		.98	1.34
0302	6" thick	G			400	.020	"	.80	.68		1.48	2.02
0400	Perlite, R2.78 per inch	G			200	.040	C.F.	5.25	1.36		6.61	8.05
0401	4" thick	G			1000	.008	S.F.	1.75	.27		2.02	2.37
0402	6" thick	G			800	.010	"	2.63	.34		2.97	3.46

07 21 23.20 Masonry Loose-Fill Insulation

				Crew	Daily Output	Labor-Hours	Unit	Material	Labor	Equipment	Total	Total Incl O&P
0010	**MASONRY LOOSE-FILL INSULATION**, vermiculite or perlite											
0100	In cores of concrete block, 4" thick wall, .115 C.F./S.F.	G		D-1	4800	.003	S.F.	.60	.10		.70	.82
0700	Foamed in place, urethane in 2-5/8" cavity	G		G-2A	1035	.023		1.38	.58	.69	2.65	3.32
0800	For each 1" added thickness, add	G		"	2372	.010		.53	.25	.30	1.08	1.36

07 21 26 – Blown Insulation

07 21 26.10 Blown Insulation

				Crew	Daily Output	Labor-Hours	Unit	Material	Labor	Equipment	Total	Total Incl O&P
0010	**BLOWN INSULATION** Ceilings, with open access											
0020	Cellulose, 3-1/2" thick, R13	G		G-4	5000	.005	S.F.	.24	.12	.08	.44	.55
0030	5-3/16" thick, R19	G			3800	.006		.35	.16	.11	.62	.78
0050	6-1/2" thick, R22	G			3000	.008		.45	.20	.14	.79	.99
0100	8-11/16" thick, R30	G			2600	.009		.61	.23	.16	1	1.23
0120	10-7/8" thick, R38	G			1800	.013		.78	.34	.23	1.35	1.66
1000	Fiberglass, 5.5" thick, R11	G			3800	.006		.19	.16	.11	.46	.59
1050	6" thick, R12	G			3000	.008		.26	.20	.14	.60	.78
1100	8.8" thick, R19	G			2200	.011		.33	.28	.19	.80	1.03
1200	10" thick, R22	G			1800	.013		.38	.34	.23	.95	1.23
1300	11.5" thick, R26	G			1500	.016		.46	.41	.27	1.14	1.48
1350	13" thick, R30	G			1400	.017		.53	.43	.29	1.25	1.63
1450	16" thick, R38	G			1145	.021		.68	.53	.36	1.57	2.02
1500	20" thick, R49	G			920	.026		.89	.66	.45	2	2.58

07 21 27 – Reflective Insulation

07 21 27.10 Reflective Insulation Options

				Crew	Daily Output	Labor-Hours	Unit	Material	Labor	Equipment	Total	Total Incl O&P
0010	**REFLECTIVE INSULATION OPTIONS**											
0020	Aluminum foil on reinforced scrim	G		1 Carp	19	.421	C.S.F.	14.90	14.25		29.15	40.50
0100	Reinforced with woven polyolefin	G			19	.421		22	14.25		36.25	48

For customer support on your Residential Cost Data, call 877.759.4771.

07 21 Thermal Insulation

07 21 27 – Reflective Insulation

07 21 27.10 Reflective Insulation Options		Crew	Daily Output	Labor-Hours	Unit	Material	2016 Bare Costs Labor	Equipment	Total	Total Incl O&P
0500	With single bubble air space, R8.8	G	1 Carp	15	.533	C.S.F.	26	18.10		44.10	59
0600	With double bubble air space, R9.8	G		15	.533		31	18.10		49.10	64.50

07 21 29 – Sprayed Insulation

07 21 29.10 Sprayed-On Insulation

0010	**SPRAYED-ON INSULATION**										
0300	Closed cell, spray polyurethane foam, 2 pounds per cubic foot density										
0310	1" thick	G	G-2A	6000	.004	S.F.	.53	.10	.12	.75	.89
0320	2" thick	G		3000	.008		1.05	.20	.24	1.49	1.78
0330	3" thick	G		2000	.012		1.58	.30	.36	2.24	2.66
0335	3-1/2" thick	G		1715	.014		1.84	.35	.42	2.61	3.11
0340	4" thick	G		1500	.016		2.10	.40	.48	2.98	3.55
0350	5" thick	G		1200	.020		2.63	.50	.59	3.72	4.43
0355	5-1/2" thick	G		1090	.022		2.89	.55	.65	4.09	4.88
0360	6" thick	G		1000	.024		3.15	.60	.71	4.46	5.30

07 22 Roof and Deck Insulation

07 22 16 – Roof Board Insulation

07 22 16.10 Roof Deck Insulation

0010	**ROOF DECK INSULATION**, fastening excluded										
0016	Asphaltic cover board, fiberglass lined, 1/8" thick		1 Rofc	1400	.006	S.F.	.47	.16		.63	.82
0018	1/4" thick			1400	.006		.94	.16		1.10	1.33
0020	Fiberboard low density, 1/2" thick R1.39	G		1300	.006		.33	.18		.51	.69
0030	1" thick R2.78	G		1040	.008		.55	.22		.77	1.02
0080	1-1/2" thick R4.17	G		1040	.008		.84	.22		1.06	1.33
0100	2" thick R5.56	G		1040	.008		1.10	.22		1.32	1.62
0110	Fiberboard high density, 1/2" thick R1.3	G		1300	.006		.30	.18		.48	.66
0120	1" thick R2.5	G		1040	.008		.56	.22		.78	1.03
0130	1-1/2" thick R3.8	G		1040	.008		.86	.22		1.08	1.36
0200	Fiberglass, 3/4" thick R2.78	G		1300	.006		.61	.18		.79	1
0400	15/16" thick R3.70	G		1300	.006		.81	.18		.99	1.22
0460	1-1/16" thick R4.17	G		1300	.006		1.01	.18		1.19	1.44
0600	1-5/16" thick R5.26	G		1300	.006		1.37	.18		1.55	1.84
0650	2-1/16" thick R8.33	G		1040	.008		1.45	.22		1.67	2.01
0700	2-7/16" thick R10	G		1040	.008		1.68	.22		1.90	2.26
0800	Gypsum cover board, fiberglass mat facer, 1/4" thick			1400	.006		.47	.16		.63	.82
0810	1/2" thick			1300	.006		.57	.18		.75	.96
0820	5/8" thick			1200	.007		.61	.19		.80	1.02
0830	Primed fiberglass mat facer, 1/4" thick			1400	.006		.48	.16		.64	.83
0840	1/2" thick			1300	.006		.57	.18		.75	.96
0850	5/8" thick			1200	.007		.60	.19		.79	1.01
1650	Perlite, 1/2" thick R1.32	G		1365	.006		.26	.17		.43	.60
1655	3/4" thick R2.08	G		1040	.008		.37	.22		.59	.82
1660	1" thick R2.78	G		1040	.008		.53	.22		.75	.99
1670	1-1/2" thick R4.17	G		1040	.008		.78	.22		1	1.27
1680	2" thick R5.56	G		910	.009		1.06	.25		1.31	1.64
1685	2-1/2" thick R6.67	G		910	.009		1.40	.25		1.65	2.01
1690	Tapered for drainage	G		1040	.008	B.F.	1.05	.22		1.27	1.57
1700	Polyisocyanurate, 2#/C.F. density, 3/4" thick	G		1950	.004	S.F.	.46	.12		.58	.73
1705	1" thick	G		1820	.004		.47	.13		.60	.75
1715	1-1/2" thick	G		1625	.005		.64	.14		.78	.96

07 22 Roof and Deck Insulation

07 22 16 – Roof Board Insulation

07 22 16.10 Roof Deck Insulation		Crew	Daily Output	Labor-Hours	Unit	Material	2016 Bare Costs Labor	Equipment	Total	Total Incl O&P
1725	2" thick	G 1 Rofc	1430	.006	S.F.	.82	.16		.98	1.20
1735	2-1/2" thick	G	1365	.006		1.10	.17		1.27	1.52
1745	3" thick	G	1300	.006		1.22	.18		1.40	1.67
1755	3-1/2" thick	G	1300	.006		1.90	.18		2.08	2.42
1765	Tapered for drainage	G	1820	.004	B.F.	.59	.13		.72	.88
1900	Extruded Polystyrene									
1910	15 PSI compressive strength, 1" thick, R5	G 1 Rofc	1950	.004	S.F.	.60	.12		.72	.88
1920	2" thick, R10	G	1625	.005		.78	.14		.92	1.12
1930	3" thick, R15	G	1300	.006		1.56	.18		1.74	2.05
1932	4" thick, R20	G	1300	.006		2.10	.18		2.28	2.64
1934	Tapered for drainage	G	1950	.004	B.F.	.62	.12		.74	.90
1940	25 PSI compressive strength, 1" thick, R5	G	1950	.004	S.F.	.69	.12		.81	.98
1942	2" thick, R10	G	1625	.005		1.31	.14		1.45	1.70
1944	3" thick, R15	G	1300	.006		2	.18		2.18	2.53
1946	4" thick, R20	G	1300	.006		2.76	.18		2.94	3.37
1948	Tapered for drainage	G	1950	.004	B.F.	.61	.12		.73	.89
1950	40 psi compressive strength, 1" thick, R5	G	1950	.004	S.F.	.53	.12		.65	.80
1952	2" thick, R10	G	1625	.005		1.01	.14		1.15	1.37
1954	3" thick, R15	G	1300	.006		1.46	.18		1.64	1.93
1956	4" thick, R20	G	1300	.006		1.91	.18		2.09	2.43
1958	Tapered for drainage	G	1820	.004	B.F.	.76	.13		.89	1.07
1960	60 PSI compressive strength, 1" thick, R5	G	1885	.004	S.F.	.74	.12		.86	1.03
1962	2" thick, R10	G	1560	.005		1.41	.15		1.56	1.82
1964	3" thick, R15	G	1270	.006		2.29	.18		2.47	2.85
1966	4" thick, R20	G	1235	.006		2.85	.19		3.04	3.47
1968	Tapered for drainage	G	1820	.004	B.F.	.96	.13		1.09	1.29
2010	Expanded polystyrene, 1#/C.F. density, 3/4" thick, R2.89	G	1950	.004	S.F.	.20	.12		.32	.43
2020	1" thick, R3.85	G	1950	.004		.26	.12		.38	.51
2100	2" thick, R7.69	G	1625	.005		.52	.14		.66	.83
2110	3" thick, R11.49	G	1625	.005		.78	.14		.92	1.12
2120	4" thick, R15.38	G	1625	.005		1.04	.14		1.18	1.40
2130	5" thick, R19.23	G	1495	.005		1.30	.15		1.45	1.71
2140	6" thick, R23.26	G	1495	.005		1.56	.15		1.71	2
2150	Tapered for drainage	G	1950	.004	B.F.	.53	.12		.65	.80
2400	Composites with 2" EPS									
2410	1" fiberboard	G 1 Rofc	1325	.006	S.F.	1.40	.17		1.57	1.86
2420	7/16" oriented strand board	G	1040	.008		1.13	.22		1.35	1.65
2430	1/2" plywood	G	1040	.008		1.39	.22		1.61	1.94
2440	1" perlite	G	1040	.008		1.14	.22		1.36	1.66
2450	Composites with 1-1/2" polyisocyanurate									
2460	1" fiberboard	G 1 Rofc	1040	.008	S.F.	1.19	.22		1.41	1.72
2470	1" perlite	G	1105	.007		1.06	.21		1.27	1.55
2480	7/16" oriented strand board	G	1040	.008		.93	.22		1.15	1.43
3000	Fastening alternatives, coated screws, 2" long		3744	.002	Ea.	.06	.06		.12	.18
3010	4" long		3120	.003		.11	.07		.18	.26
3020	6" long		2675	.003		.18	.09		.27	.36
3030	8" long		2340	.003		.27	.10		.37	.48
3040	10" long		1872	.004		.45	.12		.57	.73
3050	Pre-drill and drive wedge spike, 2-1/2"		1248	.006		.39	.18		.57	.77
3060	3-1/2"		1101	.007		.52	.21		.73	.96
3070	4-1/2"		936	.009		.64	.25		.89	1.15
3075	3" galvanized deck plates		7488	.001		.08	.03		.11	.15
3080	Spot mop asphalt	G-1	295	.190	Sq.	5.55	5.15	1.96	12.66	17.70

07 22 Roof and Deck Insulation

07 22 16 – Roof Board Insulation

07 22 16.10 Roof Deck Insulation

		Crew	Daily Output	Labor-Hours	Unit	Material	2016 Bare Costs Labor	Equipment	Total	Total Incl O&P
3090	Full mop asphalt	G-1	192	.292	Sq.	11.10	7.90	3.01	22.01	30
3110	Low-rise polyurethane adhesive, from 5 gallon kit, 12" OC beads	1 Rofc	45	.178		33	5.10		38.10	46
3120	6" OC beads		32	.250		66	7.20		73.20	86
3130	4" OC beads		30	.267		99	7.65		106.65	123

07 24 Exterior Insulation and Finish Systems

07 24 13 – Polymer-Based Exterior Insulation and Finish System

07 24 13.10 Exterior Insulation and Finish Systems

			Crew	Daily Output	Labor-Hours	Unit	Material	Labor	Equipment	Total	Total Incl O&P
0010	**EXTERIOR INSULATION AND FINISH SYSTEMS**										
0095	Field applied, 1" EPS insulation	G	J-1	390	.103	S.F.	1.93	3.03	.36	5.32	7.45
0100	With 1/2" cement board sheathing	G		268	.149		2.71	4.41	.52	7.64	10.75
0105	2" EPS insulation	G		390	.103		2.19	3.03	.36	5.58	7.75
0110	With 1/2" cement board sheathing	G		268	.149		2.97	4.41	.52	7.90	11.05
0115	3" EPS insulation	G		390	.103		2.45	3.03	.36	5.84	8.05
0120	With 1/2" cement board sheathing	G		268	.149		3.23	4.41	.52	8.16	11.35
0125	4" EPS insulation	G		390	.103		2.71	3.03	.36	6.10	8.35
0130	With 1/2" cement board sheathing	G		268	.149		4.28	4.41	.52	9.21	12.50
0140	Premium finish add			1265	.032		.35	.93	.11	1.39	2.04
0150	Heavy duty reinforcement add			914	.044		.85	1.29	.15	2.29	3.22
0160	2.5#/S.Y. metal lath substrate add		1 Lath	75	.107	S.Y.	2.82	3.53		6.35	8.80
0170	3.4#/S.Y. metal lath substrate add		"	75	.107	"	4.18	3.53		7.71	10.30
0180	Color or texture change,		J-1	1265	.032	S.F.	.80	.93	.11	1.84	2.53
0190	With substrate leveling base coat		1 Plas	530	.015		.80	.47		1.27	1.66
0210	With substrate sealing base coat		1 Pord	1224	.007		.12	.19		.31	.44
0370	V groove shape in panel face					L.F.	.65			.65	.72
0380	U groove shape in panel face					"	.85			.85	.94

07 25 Weather Barriers

07 25 10 – Weather Barriers or Wraps

07 25 10.10 Weather Barriers

		Crew	Daily Output	Labor-Hours	Unit	Material	Labor	Equipment	Total	Total Incl O&P
0010	**WEATHER BARRIERS**									
0400	Asphalt felt paper, 15#	1 Carp	37	.216	Sq.	5.30	7.35		12.65	18.10
0401	Per square foot	"	3700	.002	S.F.	.05	.07		.12	.18
0450	Housewrap, exterior, spun bonded polypropylene									
0470	Small roll	1 Carp	3800	.002	S.F.	.15	.07		.22	.29
0480	Large roll	"	4000	.002	"	.14	.07		.21	.26
2100	Asphalt felt roof deck vapor barrier, class 1 metal decks	1 Rofc	37	.216	Sq.	21.50	6.20		27.70	35.50
2200	For all other decks	"	37	.216		16.40	6.20		22.60	29.50
2800	Asphalt felt, 50% recycled content, 15 lb., 4 sq. per roll	1 Carp	36	.222		5.50	7.55		13.05	18.65
2810	30 lb., 2 sq. per roll	"	36	.222		10.85	7.55		18.40	24.50
3000	Building wrap, spunbonded polyethylene	2 Carp	8000	.002	S.F.	.16	.07		.23	.29

07 26 Vapor Retarders

07 26 10 – Above-Grade Vapor Retarders

07 26 10.10 Building Paper

		Crew	Daily Output	Labor-Hours	Unit	Material	2016 Bare Costs Labor	Equipment	Total	Total Incl O&P
0011	**BUILDING PAPER** Aluminum and kraft laminated, foil 1 side	1 Carp	3700	.002	S.F.	.12	.07		.19	.25
0101	Foil 2 sides [G]		3700	.002		.14	.07		.21	.27
0601	Polyethylene vapor barrier, standard, 2 mil [G]		3700	.002		.02	.07		.09	.14
0701	4 mil [G]		3700	.002		.03	.07		.10	.15
0901	6 mil [G]		3700	.002		.04	.07		.11	.16
1201	10 mil [G]		3700	.002		.08	.07		.15	.21
1801	Reinf. waterproof, 2 mil polyethylene backing, 1 side		3700	.002		.06	.07		.13	.19
1901	2 sides		3700	.002		.08	.07		.15	.21

07 27 Air Barriers

07 27 13 – Modified Bituminous Sheet Air Barriers

07 27 13.10 Modified Bituminous Sheet Air Barrier

		Crew	Daily Output	Labor-Hours	Unit	Material	Labor	Equipment	Total	Total Incl O&P
0010	**MODIFIED BITUMINOUS SHEET AIR BARRIER**									
0100	SBS Modified sheet laminated to polyethylene sheet, 40 mils, 4" wide	1 Carp	1200	.007	L.F.	.37	.23		.60	.79
0120	6" wide		1100	.007		.51	.25		.76	.97
0140	9" wide		1000	.008		.71	.27		.98	1.23
0160	12" wide		900	.009		.91	.30		1.21	1.50
0180	18" wide	2 Carp	1700	.009	S.F.	.85	.32		1.17	1.46
0200	36" wide	"	1800	.009		.83	.30		1.13	1.41
0220	Adhesive for above	1 Carp	1400	.006		.35	.19		.54	.70

07 27 26 – Fluid-Applied Membrane Air Barriers

07 27 26.10 Fluid Applied Membrane Air Barrier

		Crew	Daily Output	Labor-Hours	Unit	Material	Labor	Equipment	Total	Total Incl O&P
0010	**FLUID APPLIED MEMBRANE AIR BARRIER**									
0100	Spray applied vapor barrier, 25 S.F./gallon	1 Pord	1375	.006	S.F.	.01	.17		.18	.29

07 31 Shingles and Shakes

07 31 13 – Asphalt Shingles

07 31 13.10 Asphalt Roof Shingles

		Crew	Daily Output	Labor-Hours	Unit	Material	Labor	Equipment	Total	Total Incl O&P
0010	**ASPHALT ROOF SHINGLES**									
0100	Standard strip shingles									
0150	Inorganic, class A, 25 year	1 Rofc	5.50	1.455	Sq.	81	42		123	166
0155	Pneumatic nailed		7	1.143		81	33		114	150
0200	30 year		5	1.600		96	46		142	191
0205	Pneumatic nailed		6.25	1.280		96	37		133	174
0250	Standard laminated multi-layered shingles									
0300	Class A, 240-260 lb./square	1 Rofc	4.50	1.778	Sq.	109	51		160	214
0305	Pneumatic nailed		5.63	1.422		109	41		150	196
0350	Class A, 250-270 lb./square		4	2		109	57.50		166.50	226
0355	Pneumatic nailed		5	1.600		109	46		155	205
0400	Premium, laminated multi-layered shingles									
0450	Class A, 260-300 lb./square	1 Rofc	3.50	2.286	Sq.	158	65.50		223.50	295
0455	Pneumatic nailed		4.37	1.831		158	52.50		210.50	271
0500	Class A, 300-385 lb./square		3	2.667		241	76.50		317.50	405
0505	Pneumatic nailed		3.75	2.133		241	61.50		302.50	380
0800	#15 felt underlayment		64	.125		5.30	3.59		8.89	12.50
0825	#30 felt underlayment		58	.138		10.30	3.97		14.27	18.65
0850	Self adhering polyethylene and rubberized asphalt underlayment		22	.364		75	10.45		85.45	102
0900	Ridge shingles		330	.024	L.F.	2.20	.70		2.90	3.70
0905	Pneumatic nailed		412.50	.019	"	2.20	.56		2.76	3.45

07 31 Shingles and Shakes

07 31 13 – Asphalt Shingles

07 31 13.10 Asphalt Roof Shingles

		Crew	Daily Output	Labor-Hours	Unit	Material	2016 Bare Costs Labor	Equipment	Total	Total Incl O&P
1000	For steep roofs (7 to 12 pitch or greater), add						50%			

07 31 26 – Slate Shingles

07 31 26.10 Slate Roof Shingles

			Crew	Daily Output	Labor-Hours	Unit	Material	Labor	Equipment	Total	Total Incl O&P
0010	**SLATE ROOF SHINGLES**	R073126-20									
0100	Buckingham Virginia black, 3/16" - 1/4" thick	G	1 Rots	1.75	4.571	Sq.	540	132		672	840
0200	1/4" thick	G		1.75	4.571		540	132		672	840
0900	Pennsylvania black, Bangor, #1 clear	G		1.75	4.571		490	132		622	785
1200	Vermont, unfading, green, mottled green	G		1.75	4.571		495	132		627	790
1300	Semi-weathering green & gray	G		1.75	4.571		355	132		487	635
1400	Purple	G		1.75	4.571		435	132		567	720
1500	Black or gray	G		1.75	4.571		475	132		607	770
2700	Ridge shingles, slate			200	.040	L.F.	10	1.15		11.15	13.10

07 31 29 – Wood Shingles and Shakes

07 31 29.13 Wood Shingles

		Crew	Daily Output	Labor-Hours	Unit	Material	Labor	Equipment	Total	Total Incl O&P
0010	**WOOD SHINGLES**									
0012	16" No. 1 red cedar shingles, 5" exposure, on roof	1 Carp	2.50	3.200	Sq.	281	108		389	490
0015	Pneumatic nailed		3.25	2.462		281	83.50		364.50	450
0200	7-1/2" exposure, on walls		2.05	3.902		187	132		319	425
0205	Pneumatic nailed		2.67	2.996		187	102		289	375
0300	18" No. 1 red cedar perfections, 5-1/2" exposure, on roof		2.75	2.909		248	98.50		346.50	440
0305	Pneumatic nailed		3.57	2.241		248	76		324	400
0500	7-1/2" exposure, on walls		2.25	3.556		182	121		303	405
0505	Pneumatic nailed		2.92	2.740		182	93		275	355
0600	Resquared, and rebutted, 5-1/2" exposure, on roof		3	2.667		282	90.50		372.50	460
0605	Pneumatic nailed		3.90	2.051		282	69.50		351.50	425
0900	7-1/2" exposure, on walls		2.45	3.265		207	111		318	415
0905	Pneumatic nailed		3.18	2.516		207	85.50		292.50	370
1000	Add to above for fire retardant shingles					56			56	61.50
1060	Preformed ridge shingles	1 Carp	400	.020	L.F.	5	.68		5.68	6.65
2000	White cedar shingles, 16" long, extras, 5" exposure, on roof		2.40	3.333	Sq.	193	113		306	400
2005	Pneumatic nailed		3.12	2.564		193	87		280	360
2050	5" exposure on walls		2	4		193	136		329	440
2055	Pneumatic nailed		2.60	3.077		193	104		297	385
2100	7-1/2" exposure, on walls		2	4		138	136		274	380
2105	Pneumatic nailed		2.60	3.077		138	104		242	325
2150	"B" grade, 5" exposure on walls		2	4		165	136		301	410
2155	Pneumatic nailed		2.60	3.077		165	104		269	355
2300	For 15# organic felt underlayment on roof, 1 layer, add		64	.125		5.30	4.24		9.54	12.95
2400	2 layers, add		32	.250		10.60	8.50		19.10	26
2600	For steep roofs (7/12 pitch or greater), add to above						50%			
2700	Panelized systems, No.1 cedar shingles on 5/16" CDX plywood									
2800	On walls, 8' strips, 7" or 14" exposure	2 Carp	700	.023	S.F.	6	.78		6.78	7.90
3000	Ridge shakes or shingle wood	1 Carp	280	.029	L.F.	4.50	.97		5.47	6.55

07 31 29.16 Wood Shakes

		Crew	Daily Output	Labor-Hours	Unit	Material	Labor	Equipment	Total	Total Incl O&P
0010	**WOOD SHAKES**									
1100	Hand-split red cedar shakes, 1/2" thick x 24" long, 10" exp. on roof	1 Carp	2.50	3.200	Sq.	287	108		395	495
1105	Pneumatic nailed		3.25	2.462		287	83.50		370.50	455
1110	3/4" thick x 24" long, 10" exp. on roof		2.25	3.556		287	121		408	515
1115	Pneumatic nailed		2.92	2.740		287	93		380	470
1200	1/2" thick, 18" long, 8-1/2" exp. on roof		2	4		179	136		315	425
1205	Pneumatic nailed		2.60	3.077		179	104		283	370

07 31 Shingles and Shakes

07 31 29 – Wood Shingles and Shakes

07 31 29.16 Wood Shakes		Crew	Daily Output	Labor-Hours	Unit	Material	2016 Bare Costs Labor	Equipment	Total	Total Incl O&P
1210	3/4" thick x 18" long, 8 1/2" exp. on roof	1 Carp	1.80	4.444	Sq.	179	151		330	450
1215	Pneumatic nailed		2.34	3.419		179	116		295	390
1255	10" exp. on walls		2	4		173	136		309	415
1260	10" exposure on walls, pneumatic nailed		2.60	3.077		173	104		277	365
1700	Add to above for fire retardant shakes, 24" long					56			56	61.50
1800	18" long					56			56	61.50
1810	Ridge shakes	1 Carp	350	.023	L.F.	5.50	.78		6.28	7.35

07 32 Roof Tiles

07 32 13 – Clay Roof Tiles

07 32 13.10 Clay Tiles

		Crew	Daily Output	Labor-Hours	Unit	Material	Labor	Equipment	Total	Total Incl O&P
0010	**CLAY TILES**, including accessories									
0300	Flat shingle, interlocking, 15", 166 pcs/sq, fireflashed blend	3 Rots	6	4	Sq.	465	115		580	720
0500	Terra cotta red		6	4		510	115		625	770
0600	Roman pan and top, 18", 102 pcs/sq, fireflashed blend		5.50	4.364		500	126		626	780
0640	Terra cotta red	1 Rots	2.40	3.333		555	96		651	785
1100	Barrel mission tile, 18", 166 pcs/sq, fireflashed blend	3 Rots	5.50	4.364		410	126		536	680
1140	Terra cotta red		5.50	4.364		410	126		536	680
1700	Scalloped edge flat shingle, 14", 145 pcs/sq, fireflashed blend		6	4		1,125	115		1,240	1,450
1800	Terra cotta red		6	4		1,025	115		1,140	1,325
3010	#15 felt underlayment	1 Rofc	64	.125		5.30	3.59		8.89	12.50
3020	#30 felt underlayment		58	.138		10.30	3.97		14.27	18.65
3040	Polyethylene and rubberized asph. underlayment		22	.364		75	10.45		85.45	102

07 32 16 – Concrete Roof Tiles

07 32 16.10 Concrete Tiles

		Crew	Daily Output	Labor-Hours	Unit	Material	Labor	Equipment	Total	Total Incl O&P
0010	**CONCRETE TILES**									
0020	Corrugated, 13" x 16-1/2", 90 per sq., 950 lb. per sq.									
0050	Earthtone colors, nailed to wood deck	1 Rots	1.35	5.926	Sq.	103	171		274	430
0150	Blues		1.35	5.926		103	171		274	430
0200	Greens		1.35	5.926		103	171		274	430
0250	Premium colors		1.35	5.926		103	171		274	430
0500	Shakes, 13" x 16-1/2", 90 per sq., 950 lb. per sq.									
0600	All colors, nailed to wood deck	1 Rots	1.50	5.333	Sq.	124	154		278	420
1500	Accessory pieces, ridge & hip, 10" x 16-1/2", 8 lb. each	"	120	.067	Ea.	3.70	1.92		5.62	7.60
1700	Rake, 6-1/2" x 16-3/4", 9 lb. each					3.70			3.70	4.07
1800	Mansard hip, 10" x 16-1/2", 9.2 lb. each					3.70			3.70	4.07
1900	Hip starter, 10" x 16-1/2", 10.5 lb. each					10.40			10.40	11.45
2000	3 or 4 way apex, 10" each side, 11.5 lb. each					12			12	13.20

07 33 Natural Roof Coverings

07 33 63 – Vegetated Roofing

07 33 63.10 Green Roof Systems

		Crew	Daily Output	Labor-Hours	Unit	Material	2016 Bare Costs Labor	Equipment	Total	Total Incl O&P
0010	**GREEN ROOF SYSTEMS**									
0020	Soil mixture for green roof 30% sand, 55% gravel, 15% soil									
0100	Hoist and spread soil mixture 4 inch depth up to five stories tall roof G	B-13B	4000	.014	S.F.	.25	.39	.28	.92	1.22
0150	6 inch depth		2667	.021		.37	.58	.42	1.37	1.84
0200	8 inch depth		2000	.028		.50	.77	.56	1.83	2.46
0250	10 inch depth		1600	.035		.62	.97	.70	2.29	3.06
0300	12 inch depth		1335	.042		.74	1.16	.84	2.74	3.68
0310	Alt. man-made soil mix, hoist & spread, 4" deep up to 5 stories tall roof G		4000	.014		1.90	.39	.28	2.57	3.04
0350	Mobilization 55 ton crane to site G	1 Eqhv	3.60	2.222	Ea.		80		80	132
0355	Hoisting cost to five stories per day (Avg. 28 picks per day) G	B-13B	1	56	Day		1,550	1,125	2,675	3,800
0360	Mobilization or demobilization, 100 ton crane to site driver & escort G	A-3E	2.50	6.400	Ea.		213	62.50	275.50	425
0365	Hoisting cost six to ten stories per day (Avg. 21 picks per day) G	B-13C	1	56	Day		1,550	1,675	3,225	4,400
0370	Hoist and spread soil mixture 4 inch depth six to ten stories tall roof G		4000	.014	S.F.	.25	.39	.42	1.06	1.37
0375	6 inch depth		2667	.021		.37	.58	.63	1.58	2.07
0380	8 inch depth		2000	.028		.50	.77	.83	2.10	2.76
0385	10 inch depth		1600	.035		.62	.97	1.04	2.63	3.44
0390	12 inch depth		1335	.042		.74	1.16	1.25	3.15	4.12
0400	Green roof edging treated lumber 4" x 4" no hoisting included G	2 Carp	400	.040	L.F.	1.35	1.36		2.71	3.75
0410	4" x 6"		400	.040		2.02	1.36		3.38	4.49
0420	4" x 8"		360	.044		3.87	1.51		5.38	6.75
0430	4" x 6" double stacked		300	.053		4.04	1.81		5.85	7.45
0500	Green roof edging redwood lumber 4" x 4" no hoisting included G		400	.040		6.45	1.36		7.81	9.35
0510	4" x 6"		400	.040		12.60	1.36		13.96	16.10
0520	4" x 8"		360	.044		23.50	1.51		25.01	28.50
0530	4" x 6" double stacked		300	.053		25	1.81		26.81	30.50
0550	Components, not including membrane or insulation:									
0560	Fluid applied rubber membrane, reinforced, 215 mil thick G	G-5	350	.114	S.F.	.30	3.01	.55	3.86	6.50
0570	Root barrier G	2 Rofc	775	.021		.68	.59		1.27	1.84
0580	Moisture retention barrier and reservoir G	"	900	.018		2.59	.51		3.10	3.79
0600	Planting sedum, light soil, potted, 2-1/4" diameter, two per S.F. G	1 Clab	420	.019		5.65	.47		6.12	7
0610	one per S.F. G	"	840	.010		2.82	.23		3.05	3.49
0630	Planting sedum mat per S.F. including shipping (4000 S.F. min) G	4 Clab	4000	.008		6.75	.20		6.95	7.80
0640	Installation sedum mat system (no soil required) per S.F. (4000 S.F. min) G	"	4000	.008		9.40	.20		9.60	10.70
0645	Note: pricing of sedum mats shipped in full truck loads (4000-5000 S.F.)									

07 41 Roof Panels

07 41 13 – Metal Roof Panels

07 41 13.10 Aluminum Roof Panels

		Crew	Daily Output	Labor-Hours	Unit	Material	Labor	Equipment	Total	Total Incl O&P
0010	**ALUMINUM ROOF PANELS**									
0020	Corrugated or ribbed, .0155" thick, natural	G-3	1200	.027	S.F.	1	.83		1.83	2.49
0300	Painted	"	1200	.027	"	1.45	.83		2.28	2.99

07 41 33 – Plastic Roof Panels

07 41 33.10 Fiberglass Panels

		Crew	Daily Output	Labor-Hours	Unit	Material	Labor	Equipment	Total	Total Incl O&P
0010	**FIBERGLASS PANELS**									
0012	Corrugated panels, roofing, 8 oz. per S.F.	G-3	1000	.032	S.F.	2.15	1		3.15	4.04
0100	12 oz. per S.F.		1000	.032		4.20	1		5.20	6.30
0300	Corrugated siding, 6 oz. per S.F.		880	.036		1.90	1.14		3.04	3.99
0400	8 oz. per S.F.		880	.036		2.15	1.14		3.29	4.27
0500	Fire retardant		880	.036		3.80	1.14		4.94	6.10
0600	12 oz. siding, textured		880	.036		3.75	1.14		4.89	6.05

07 41 Roof Panels

07 41 33 – Plastic Roof Panels

07 41 33.10 Fiberglass Panels

		Crew	Daily Output	Labor-Hours	Unit	Material	2016 Bare Costs Labor	Equipment	Total	Total Incl O&P
0700	Fire retardant	G-3	880	.036	S.F.	4.35	1.14		5.49	6.70
0900	Flat panels, 6 oz. per S.F., clear or colors		880	.036		2.40	1.14		3.54	4.54
1100	Fire retardant, class A		880	.036		3.45	1.14		4.59	5.70
1300	8 oz. per S.F., clear or colors	↓	880	.036	↓	2.50	1.14		3.64	4.65

07 42 Wall Panels

07 42 13 – Metal Wall Panels

07 42 13.20 Aluminum Siding

		Crew	Daily Output	Labor-Hours	Unit	Material	Labor	Equipment	Total	Total Incl O&P
0011	**ALUMINUM SIDING**									
6040	.024 thick smooth white single 8" wide	2 Carp	515	.031	S.F.	2.73	1.05		3.78	4.76
6060	Double 4" pattern		515	.031		2.80	1.05		3.85	4.84
6080	Double 5" pattern		550	.029		2.72	.99		3.71	4.64
6120	Embossed white, 8" wide		515	.031		2.72	1.05		3.77	4.75
6140	Double 4" pattern		515	.031		2.79	1.05		3.84	4.83
6160	Double 5" pattern		550	.029		2.77	.99		3.76	4.70
6170	Vertical, embossed white, 12" wide		590	.027		2.77	.92		3.69	4.59
6320	.019 thick, insulated, smooth white, 8" wide		515	.031		2.45	1.05		3.50	4.46
6340	Double 4" pattern		515	.031		2.42	1.05		3.47	4.42
6360	Double 5" pattern		550	.029		2.42	.99		3.41	4.31
6400	Embossed white, 8" wide		515	.031		2.82	1.05		3.87	4.86
6420	Double 4" pattern		515	.031		2.85	1.05		3.90	4.90
6440	Double 5" pattern		550	.029		2.85	.99		3.84	4.79
6500	Shake finish 10" wide white		550	.029		3.06	.99		4.05	5
6600	Vertical pattern, 12" wide, white	↓	590	.027	↓	2.33	.92		3.25	4.10
6640	For colors add					.15			.15	.17
6700	Accessories, white									
6720	Starter strip 2-1/8"	2 Carp	610	.026	L.F.	.48	.89		1.37	2.02
6740	Sill trim		450	.036		.65	1.21		1.86	2.74
6760	Inside corner		610	.026		1.70	.89		2.59	3.36
6780	Outside corner post		610	.026		3.60	.89		4.49	5.45
6800	Door & window trim	↓	440	.036		.62	1.23		1.85	2.74
6820	For colors add					.14			.14	.15
6900	Soffit & fascia 1' overhang solid	2 Carp	110	.145		4.45	4.93		9.38	13.15
6920	Vented		110	.145		4.48	4.93		9.41	13.20
6940	2' overhang solid		100	.160		6.50	5.40		11.90	16.25
6960	Vented	↓	100	.160	↓	6.50	5.40		11.90	16.25

07 42 13.30 Steel Siding

		Crew	Daily Output	Labor-Hours	Unit	Material	Labor	Equipment	Total	Total Incl O&P
0010	**STEEL SIDING**									
0020	Beveled, vinyl coated, 8" wide	1 Carp	265	.030	S.F.	1.85	1.02		2.87	3.75
0050	10" wide	"	275	.029		1.90	.99		2.89	3.74
0081	Galv., corrugated or ribbed, on steel frame, 30 ga.	G-3	775	.041		1.20	1.29		2.49	3.47
0101	28 ga.		775	.041		1.30	1.29		2.59	3.58
0301	26 ga.		775	.041		1.80	1.29		3.09	4.13
0401	24 ga.		775	.041		2	1.29		3.29	4.35
0601	22 ga.		775	.041		2.25	1.29		3.54	4.63
0701	Colored, corrugated/ribbed, on steel frame, 10 yr. finish, 28 ga.		775	.041		1.95	1.29		3.24	4.30
0901	26 ga.		775	.041		2.12	1.29		3.41	4.48
1001	24 ga.	↓	775	.041	↓	2.38	1.29		3.67	4.77

07 46 Siding

07 46 23 – Wood Siding

07 46 23.10 Wood Board Siding

		Crew	Daily Output	Labor-Hours	Unit	Material	2016 Bare Costs Labor	Equipment	Total	Total Incl O&P
0010	**WOOD BOARD SIDING**									
2000	Board & batten, cedar, "B" grade, 1" x 10"	1 Carp	375	.021	S.F.	5.15	.72		5.87	6.90
2200	Redwood, clear, vertical grain, 1" x 10"		375	.021		5.35	.72		6.07	7.10
2400	White pine, #2 & better, 1" x 10"		375	.021		3.18	.72		3.90	4.71
2410	Board & batten siding, white pine #2, 1" x 12"		420	.019		3.18	.65		3.83	4.58
3200	Wood, cedar bevel, A grade, 1/2" x 6"		295	.027		4.25	.92		5.17	6.20
3300	1/2" x 8"		330	.024		6.50	.82		7.32	8.55
3500	3/4" x 10", clear grade		375	.021		6.75	.72		7.47	8.65
3600	"B" grade		375	.021		3.79	.72		4.51	5.40
3800	Cedar, rough sawn, 1" x 4", A grade, natural		220	.036		6.55	1.23		7.78	9.30
3900	Stained		220	.036		6.70	1.23		7.93	9.40
4100	1" x 12", board & batten, #3 & Btr., natural		420	.019		4.48	.65		5.13	6
4200	Stained		420	.019		4.81	.65		5.46	6.40
4400	1" x 8" channel siding, #3 & Btr., natural		330	.024		4.57	.82		5.39	6.40
4500	Stained		330	.024		4.92	.82		5.74	6.80
4700	Redwood, clear, beveled, vertical grain, 1/2" x 4"		220	.036		4.15	1.23		5.38	6.65
4750	1/2" x 6"		295	.027		4.79	.92		5.71	6.80
4800	1/2" x 8"		330	.024		5.20	.82		6.02	7.10
5000	3/4" x 10"		375	.021		4.90	.72		5.62	6.60
5200	Channel siding, 1" x 10", B grade		375	.021		4.50	.72		5.22	6.15
5250	Redwood, T&G boards, B grade, 1" x 4"		220	.036		3.25	1.23		4.48	5.65
5270	1" x 8"		330	.024		5.20	.82		6.02	7.10
5400	White pine, rough sawn, 1" x 8", natural		330	.024		2.19	.82		3.01	3.79
5500	Stained		330	.024		2.19	.82		3.01	3.79
5600	Tongue and groove, 1" x 8"		330	.024		2.19	.82		3.01	3.79

07 46 29 – Plywood Siding

07 46 29.10 Plywood Siding Options

		Crew	Daily Output	Labor-Hours	Unit	Material	Labor	Equipment	Total	Total Incl O&P
0010	**PLYWOOD SIDING OPTIONS**									
0900	Plywood, medium density overlaid, 3/8" thick	2 Carp	750	.021	S.F.	1.29	.72		2.01	2.63
1000	1/2" thick		700	.023		1.50	.78		2.28	2.95
1100	3/4" thick		650	.025		1.99	.83		2.82	3.59
1600	Texture 1-11, cedar, 5/8" thick, natural		675	.024		2.70	.80		3.50	4.32
1700	Factory stained		675	.024		2.80	.80		3.60	4.43
1900	Texture 1-11, fir, 5/8" thick, natural		675	.024		1.46	.80		2.26	2.96
2000	Factory stained		675	.024		1.80	.80		2.60	3.33
2050	Texture 1-11, S.Y.P., 5/8" thick, natural		675	.024		1.37	.80		2.17	2.86
2100	Factory stained		675	.024		1.44	.80		2.24	2.93
2200	Rough sawn cedar, 3/8" thick, natural		675	.024		1.25	.80		2.05	2.73
2300	Factory stained		675	.024		1.50	.80		2.30	3
2500	Rough sawn fir, 3/8" thick, natural		675	.024		.90	.80		1.70	2.34
2600	Factory stained		675	.024		1.04	.80		1.84	2.49
2800	Redwood, textured siding, 5/8" thick		675	.024		1.92	.80		2.72	3.46

07 46 33 – Plastic Siding

07 46 33.10 Vinyl Siding

		Crew	Daily Output	Labor-Hours	Unit	Material	Labor	Equipment	Total	Total Incl O&P
0010	**VINYL SIDING**									
3995	Clapboard profile, woodgrain texture, .048 thick, double 4	2 Carp	495	.032	S.F.	1.06	1.10		2.16	3
4000	Double 5		550	.029		1.06	.99		2.05	2.82
4005	Single 8		495	.032		1.35	1.10		2.45	3.32
4010	Single 10		550	.029		1.62	.99		2.61	3.44
4015	.044 thick, double 4		495	.032		1.04	1.10		2.14	2.98
4020	Double 5		550	.029		1.04	.99		2.03	2.80
4025	.042 thick, double 4		495	.032		1.04	1.10		2.14	2.98

For customer support on your Residential Cost Data, call 877.759.4771.

07 46 Siding

07 46 33 – Plastic Siding

07 46 33.10 Vinyl Siding

		Crew	Daily Output	Labor-Hours	Unit	Material	2016 Bare Costs Labor	Equipment	Total	Total Incl O&P
4030	Double 5	2 Carp	550	.029	S.F.	1.04	.99		2.03	2.80
4035	Cross sawn texture, .040 thick, double 4		495	.032		.72	1.10		1.82	2.63
4040	Double 5		550	.029		.72	.99		1.71	2.45
4045	Smooth texture, .042 thick, double 4		495	.032		.79	1.10		1.89	2.70
4050	Double 5		550	.029		.79	.99		1.78	2.52
4055	Single 8		495	.032		.79	1.10		1.89	2.70
4060	Cedar texture, .044 thick, double 4		495	.032		1.12	1.10		2.22	3.07
4065	Double 6		600	.027		1.12	.90		2.02	2.75
4070	Dutch lap profile, woodgrain texture, .048 thick, double 5		550	.029		1.06	.99		2.05	2.82
4075	.044 thick, double 4.5		525	.030		1.06	1.03		2.09	2.90
4080	.042 thick, double 4.5		525	.030		.90	1.03		1.93	2.73
4085	.040 thick, double 4.5		525	.030		.72	1.03		1.75	2.53
4100	Shake profile, 10" wide		400	.040		3.61	1.36		4.97	6.25
4105	Vertical pattern, .046 thick, double 5		550	.029		1.56	.99		2.55	3.37
4110	.044 thick, triple 3		550	.029		1.74	.99		2.73	3.57
4115	.040 thick, triple 4		550	.029		1.61	.99		2.60	3.43
4120	.040 thick, triple 2.66		550	.029		1.74	.99		2.73	3.57
4125	Insulation, fan folded extruded polystyrene, 1/4"		2000	.008		.29	.27		.56	.77
4130	3/8"		2000	.008		.32	.27		.59	.81
4135	Accessories, J channel, 5/8" pocket		700	.023	L.F.	.50	.78		1.28	1.85
4140	3/4" pocket		695	.023		.53	.78		1.31	1.89
4145	1-1/4" pocket		680	.024		.84	.80		1.64	2.26
4150	Flexible, 3/4" pocket		600	.027		2.33	.90		3.23	4.07
4155	Under sill finish trim		500	.032		.54	1.08		1.62	2.41
4160	Vinyl starter strip		700	.023		.65	.78		1.43	2.01
4165	Aluminum starter strip		700	.023		.29	.78		1.07	1.62
4170	Window casing, 2-1/2" wide, 3/4" pocket		510	.031		1.68	1.06		2.74	3.63
4175	Outside corner, woodgrain finish, 4" face, 3/4" pocket		700	.023		2.08	.78		2.86	3.59
4180	5/8" pocket		700	.023		2.08	.78		2.86	3.59
4185	Smooth finish, 4" face, 3/4" pocket		700	.023		2.08	.78		2.86	3.59
4190	7/8" pocket		690	.023		1.96	.79		2.75	3.48
4195	1-1/4" pocket		700	.023		1.38	.78		2.16	2.82
4200	Soffit and fascia, 1' overhang, solid		120	.133		4.65	4.52		9.17	12.65
4205	Vented		120	.133		4.65	4.52		9.17	12.65
4207	18" overhang, solid		110	.145		5.45	4.93		10.38	14.25
4208	Vented		110	.145		5.45	4.93		10.38	14.25
4210	2' overhang, solid		100	.160		6.20	5.40		11.60	15.95
4215	Vented		100	.160		6.20	5.40		11.60	15.95
4217	3' overhang, solid		100	.160		7.75	5.40		13.15	17.65
4218	Vented		100	.160		7.75	5.40		13.15	17.65
4220	Colors for siding and soffits, add				S.F.	.14			.14	.15
4225	Colors for accessories and trim, add				L.F.	.30			.30	.33

07 46 33.20 Polypropylene Siding

		Crew	Daily Output	Labor-Hours	Unit	Material	Labor	Equipment	Total	Total Incl O&P
0010	**POLYPROPYLENE SIDING**									
4090	Shingle profile, random grooves, double 7	2 Carp	400	.040	S.F.	3.11	1.36		4.47	5.70
4092	Cornerpost for above	1 Carp	365	.022	L.F.	12.65	.74		13.39	15.15
4095	Triple 5	2 Carp	400	.040	S.F.	3.11	1.36		4.47	5.70
4097	Cornerpost for above	1 Carp	365	.022	L.F.	11.80	.74		12.54	14.25
5000	Staggered butt, double 7"	2 Carp	400	.040	S.F.	3.61	1.36		4.97	6.25
5002	Cornerpost for above	1 Carp	365	.022	L.F.	12.75	.74		13.49	15.30
5010	Half round, double 6-1/4"	2 Carp	360	.044	S.F.	3.61	1.51		5.12	6.50
5020	Shake profile, staggered butt, double 9"	"	510	.031	"	3.61	1.06		4.67	5.75

07 46 Siding

07 46 33 – Plastic Siding

07 46 33.20 Polypropylene Siding

		Crew	Daily Output	Labor-Hours	Unit	Material	2016 Bare Costs Labor	Equipment	Total	Total Incl O&P
5022	Cornerpost for above	1 Carp	365	.022	L.F.	9.60	.74		10.34	11.80
5030	Straight butt, double 7"	2 Carp	400	.040	S.F.	3.61	1.36		4.97	6.25
5032	Cornerpost for above	1 Carp	365	.022	L.F.	12.60	.74		13.34	15.15
6000	Accessories, J channel, 5/8" pocket	2 Carp	700	.023		.50	.78		1.28	1.85
6010	3/4" pocket		695	.023		.53	.78		1.31	1.89
6020	1-1/4" pocket		680	.024		.84	.80		1.64	2.26
6030	Aluminum starter strip		700	.023		.29	.78		1.07	1.62

07 46 46 – Fiber Cement Siding

07 46 46.10 Fiber Cement Siding

		Crew	Daily Output	Labor-Hours	Unit	Material	Labor	Equipment	Total	Total Incl O&P
0010	**FIBER CEMENT SIDING**									
0020	Lap siding, 5/16" thick, 6" wide, 4-3/4" exposure, smooth texture	2 Carp	415	.039	S.F.	1.29	1.31		2.60	3.60
0025	Woodgrain texture		415	.039		1.29	1.31		2.60	3.60
0030	7-1/2" wide, 6-1/4" exposure, smooth texture		425	.038		1.59	1.28		2.87	3.89
0035	Woodgrain texture		425	.038		1.59	1.28		2.87	3.89
0040	8" wide, 6-3/4" exposure, smooth texture		425	.038		1.19	1.28		2.47	3.45
0045	Rough sawn texture		425	.038		1.19	1.28		2.47	3.45
0050	9-1/2" wide, 8-1/4" exposure, smooth texture		440	.036		1.26	1.23		2.49	3.45
0055	Woodgrain texture		440	.036		1.26	1.23		2.49	3.45
0060	12" wide, 10-3/8" exposure, smooth texture		455	.035		2.01	1.19		3.20	4.21
0065	Woodgrain texture		455	.035		2.01	1.19		3.20	4.21
0070	Panel siding, 5/16" thick, smooth texture		750	.021		1.31	.72		2.03	2.65
0075	Stucco texture		750	.021		1.31	.72		2.03	2.65
0080	Grooved woodgrain texture		750	.021		1.31	.72		2.03	2.65
0085	V - grooved woodgrain texture		750	.021		1.31	.72		2.03	2.65
0088	Shingle siding, 48" x 15-1/4" panels, 7" exposure		700	.023		4.12	.78		4.90	5.85
0090	Wood starter strip		400	.040	L.F.	.41	1.36		1.77	2.72
0200	Fiber cement siding, accessories, fascia, 5/4" x 3-1/2"		275	.058		.82	1.97		2.79	4.20
0210	5/4 x 5-1/2"		250	.064		1.98	2.17		4.15	5.80
0220	5/4 x 7-1/2"		225	.071		2.75	2.41		5.16	7.05
0230	5/4" x 9-1/2"		210	.076		3.30	2.58		5.88	7.95

07 46 73 – Soffit

07 46 73.10 Soffit Options

		Crew	Daily Output	Labor-Hours	Unit	Material	Labor	Equipment	Total	Total Incl O&P
0010	**SOFFIT OPTIONS**									
0012	Aluminum, residential, .020" thick	1 Carp	210	.038	S.F.	2	1.29		3.29	4.36
0100	Baked enamel on steel, 16 or 18 ga.		105	.076		5.90	2.58		8.48	10.80
0300	Polyvinyl chloride, white, solid		230	.035		2.10	1.18		3.28	4.28
0400	Perforated		230	.035		2.10	1.18		3.28	4.28
0500	For colors, add					.15			.15	.17

07 51 Built-Up Bituminous Roofing

07 51 13 – Built-Up Asphalt Roofing

07 51 13.10 Built-Up Roofing Components

		Crew	Daily Output	Labor-Hours	Unit	Material	Labor	Equipment	Total	Total Incl O&P
0010	**BUILT-UP ROOFING COMPONENTS**									
0012	Asphalt saturated felt, #30, 2 square per roll	1 Rofc	58	.138	Sq.	10.30	3.97		14.27	18.65
0200	#15, 4 sq. per roll, plain or perforated, not mopped		58	.138		5.30	3.97		9.27	13.15
0250	Perforated		58	.138		5.30	3.97		9.27	13.15
0300	Roll roofing, smooth, #65		15	.533		10.10	15.35		25.45	39.50
0500	#90		12	.667		44	19.15		63.15	83.50
0520	Mineralized		12	.667		35	19.15		54.15	74
0540	D.C. (Double coverage), 19" selvage edge		10	.800		54.50	23		77.50	103

07 51 Built-Up Bituminous Roofing

07 51 13 – Built-Up Asphalt Roofing

07 51 13.10 Built-Up Roofing Components	Crew	Daily Output	Labor-Hours	Unit	Material	2016 Bare Costs Labor	Equipment	Total	Total Incl O&P
0580 Adhesive (lap cement)				Gal.	8.85			8.85	9.75
07 51 13.20 Built-Up Roofing Systems									
0010 BUILT-UP ROOFING SYSTEMS									
0120 Asphalt flood coat with gravel/slag surfacing, not including									
0140 Insulation, flashing or wood nailers									
0200 Asphalt base sheet, 3 plies #15 asphalt felt, mopped	G-1	22	2.545	Sq.	99.50	69	26.50	195	265
0350 On nailable decks		21	2.667		103	72	27.50	202.50	277
0500 4 plies #15 asphalt felt, mopped		20	2.800		137	75.50	29	241.50	320
0550 On nailable decks		19	2.947		121	79.50	30.50	231	315
2000 Asphalt flood coat, smooth surface									
2200 Asphalt base sheet & 3 plies #15 asphalt felt, mopped	G-1	24	2.333	Sq.	106	63	24	193	259
2400 On nailable decks		23	2.435		98	66	25	189	257
2600 4 plies #15 asphalt felt, mopped		24	2.333		124	63	24	211	279
2700 On nailable decks		23	2.435		116	66	25	207	277
4500 Coal tar pitch with gravel/slag surfacing									
4600 4 plies #15 tarred felt, mopped	G-1	21	2.667	Sq.	200	72	27.50	299.50	385
4800 3 plies glass fiber felt (type IV), mopped	"	19	2.947	"	165	79.50	30.50	275	360
07 51 13.30 Cants									
0010 CANTS									
0012 Lumber, treated, 4" x 4" cut diagonally	1 Rofc	325	.025	L.F.	1.80	.71		2.51	3.28
0300 Mineral or fiber, trapezoidal, 1" x 4" x 48"		325	.025		.30	.71		1.01	1.63
0400 1-1/2" x 5-5/8" x 48"		325	.025		.48	.71		1.19	1.83

07 52 Modified Bituminous Membrane Roofing

07 52 13 – Atactic-Polypropylene-Modified Bituminous Membrane Roofing

07 52 13.10 APP Modified Bituminous Membrane	Crew	Daily Output	Labor-Hours	Unit	Material	2016 Bare Costs Labor	Equipment	Total	Total Incl O&P
0010 APP MODIFIED BITUMINOUS MEMBRANE R075213-30									
0020 Base sheet, #15 glass fiber felt, nailed to deck	1 Rofc	58	.138	Sq.	11.85	3.97		15.82	20.50
0030 Spot mopped to deck	G-1	295	.190		16.40	5.15	1.96	23.51	29.50
0040 Fully mopped to deck	"	192	.292		22	7.90	3.01	32.91	42
0050 #15 organic felt, nailed to deck	1 Rofc	58	.138		6.30	3.97		10.27	14.20
0060 Spot mopped to deck	G-1	295	.190		10.85	5.15	1.96	17.96	23.50
0070 Fully mopped to deck	"	192	.292		16.40	7.90	3.01	27.31	36
2100 APP mod., smooth surf. cap sheet, poly. reinf., torched, 160 mils	G-5	2100	.019	S.F.	.74	.50	.09	1.33	1.83
2150 170 mils		2100	.019		.75	.50	.09	1.34	1.85
2200 Granule surface cap sheet, poly. reinf., torched, 180 mils		2000	.020		.93	.53	.10	1.56	2.10
2250 Smooth surface flashing, torched, 160 mils		1260	.032		.74	.84	.15	1.73	2.52
2300 170 mils		1260	.032		.75	.84	.15	1.74	2.54
2350 Granule surface flashing, torched, 180 mils		1260	.032		.93	.84	.15	1.92	2.73
2400 Fibrated aluminum coating	1 Rofc	3800	.002		.08	.06		.14	.20
2450 Seam heat welding	"	205	.039	L.F.	.08	1.12		1.20	2.16

07 52 16 – Styrene-Butadiene-Styrene Modified Bituminous Membrane Roofing

07 52 16.10 SBS Modified Bituminous Membrane	Crew	Daily Output	Labor-Hours	Unit	Material	2016 Bare Costs Labor	Equipment	Total	Total Incl O&P
0010 SBS MODIFIED BITUMINOUS MEMBRANE									
0080 Mod bit rfng, SBS mod, gran surf cap sheet, poly reinf									
0650 120 to 149 mils thick	G-1	2000	.028	S.F.	1.27	.76	.29	2.32	3.11
0750 150 to 160 mils	"	2000	.028		1.71	.76	.29	2.76	3.59
1150 For reflective granules, add									
1600 Smooth surface cap sheet, mopped, 145 mils	G-1	2100	.027		.80	.72	.28	1.80	2.51
1620 Lightweight base sheet, fiberglass reinforced, 35 to 47 mil		2100	.027		.27	.72	.28	1.27	1.93

07 52 Modified Bituminous Membrane Roofing

07 52 16 – Styrene-Butadiene-Styrene Modified Bituminous Membrane Roofing

	07 52 16.10 SBS Modified Bituminous Membrane	Crew	Daily Output	Labor-Hours	Unit	Material	2016 Bare Costs Labor	Equipment	Total	Total Incl O&P
1625	Heavyweight base/ply sheet, reinforced, 87 to 120 mil thick	G-1	2100	.027	S.F.	.88	.72	.28	1.88	2.60
1650	Granulated walkpad, 180 – 220 mils	1 Rofc	400	.020		1.77	.58		2.35	3
1700	Smooth surface flashing, 145 mils	G-1	1260	.044		.80	1.20	.46	2.46	3.59
1800	150 mils		1260	.044		.50	1.20	.46	2.16	3.26
1900	Granular surface flashing, 150 mils		1260	.044		.70	1.20	.46	2.36	3.48
2000	160 mils		1260	.044		.72	1.20	.46	2.38	3.50
2010	Elastomeric asphalt primer	1 Rofc	2600	.003		.16	.09		.25	.34
2015	Roofing asphalt, 30 lb. per square	G-1	19000	.003		.14	.08	.03	.25	.33
2020	Cold process adhesive, 20 – 30 mils thick	1 Rofc	750	.011		.23	.31		.54	.83
2025	Self adhering vapor retarder, 30 to 45 mils thick	G-5	2150	.019		1	.49	.09	1.58	2.10
2050	Seam heat welding	1 Rofc	205	.039	L.F.	.08	1.12		1.20	2.16

07 57 Coated Foamed Roofing

07 57 13 – Sprayed Polyurethane Foam Roofing

	07 57 13.10 Sprayed Polyurethane Foam Roofing (S.P.F.)	Crew	Daily Output	Labor-Hours	Unit	Material	2016 Bare Costs Labor	Equipment	Total	Total Incl O&P
0010	**SPRAYED POLYURETHANE FOAM ROOFING (S.P.F.)**									
0100	Primer for metal substrate (when required)	G-2A	3000	.008	S.F.	.48	.20	.24	.92	1.15
0200	Primer for non-metal substrate (when required)		3000	.008		.19	.20	.24	.63	.83
0300	Closed cell spray, polyurethane foam, 3 lb. per C.F. density, 1", R6.7		15000	.002		.66	.04	.05	.75	.85
0400	2", R13.4		13125	.002		1.32	.05	.05	1.42	1.59
0500	3", R18.6		11485	.002		1.98	.05	.06	2.09	2.34
0550	4", R24.8		10080	.002		2.64	.06	.07	2.77	3.10
0700	Spray-on silicone coating		2500	.010		1.21	.24	.29	1.74	2.08
0800	Warranty 5-20 year manufacturer's								.15	.15
0900	Warranty 20 year, no dollar limit								.20	.20

07 58 Roll Roofing

07 58 10 – Asphalt Roll Roofing

	07 58 10.10 Roll Roofing	Crew	Daily Output	Labor-Hours	Unit	Material	2016 Bare Costs Labor	Equipment	Total	Total Incl O&P
0010	**ROLL ROOFING**									
0100	Asphalt, mineral surface									
0200	1 ply #15 organic felt, 1 ply mineral surfaced									
0300	Selvage roofing, lap 19", nailed & mopped	G-1	27	2.074	Sq.	77	56	21.50	154.50	211
0400	3 plies glass fiber felt (type IV), 1 ply mineral surfaced									
0500	Selvage roofing, lapped 19", mopped	G-1	25	2.240	Sq.	132	60.50	23	215.50	283
0600	Coated glass fiber base sheet, 2 plies of glass fiber									
0700	Felt (type IV), 1 ply mineral surfaced selvage									
0800	Roofing, lapped 19", mopped	G-1	25	2.240	Sq.	139	60.50	23	222.50	291
0900	On nailable decks	"	24	2.333	"	128	63	24	215	283
1000	3 plies glass fiber felt (type III), 1 ply mineral surfaced									
1100	Selvage roofing, lapped 19", mopped	G-1	25	2.240	Sq.	132	60.50	23	215.50	283

07 61 Sheet Metal Roofing

07 61 13 – Standing Seam Sheet Metal Roofing

07 61 13.10 Standing Seam Sheet Metal Roofing, Field Fab.	Crew	Daily Output	Labor-Hours	Unit	Material	2016 Bare Costs Labor	Equipment	Total	Total Incl O&P
0010 **STANDING SEAM SHEET METAL ROOFING, FIELD FABRICATED**									
0400 Copper standing seam roofing, over 10 squares, 16 oz., 125 lb. per sq.	1 Shee	1.30	6.154	Sq.	985	233		1,218	1,475
0600 18 oz., 140 lb. per sq.	"	1.20	6.667		1,100	252		1,352	1,625
1200 For abnormal conditions or small areas, add					25%	100%			
1300 For lead-coated copper, add					25%				

07 61 16 – Batten Seam Sheet Metal Roofing

07 61 16.10 Batten Seam Sheet Metal Roofing, Field Fabricated

0010 **BATTEN SEAM SHEET METAL ROOFING, FIELD FABRICATED**									
0012 Copper batten seam roofing, over 10 sq., 16 oz., 130 lb. per sq.	1 Shee	1.10	7.273	Sq.	1,250	275		1,525	1,825
0100 Zinc/copper alloy batten seam roofing, .020 thick		1.20	6.667		1,275	252		1,527	1,825
0200 Copper roofing, batten seam, over 10 sq., 18 oz., 145 lb. per sq.		1	8		1,400	300		1,700	2,025
0800 Zinc, copper alloy roofing, batten seam, .027" thick		1.15	6.957		1,600	263		1,863	2,225
0900 .032" thick		1.10	7.273		1,850	275		2,125	2,475
1000 .040" thick		1.05	7.619		2,300	288		2,588	3,000

07 61 19 – Flat Seam Sheet Metal Roofing

07 61 19.10 Flat Seam Sheet Metal Roofing, Field Fabricated

0010 **FLAT SEAM SHEET METAL ROOFING, FIELD FABRICATED**									
0900 Copper flat seam roofing, over 10 squares, 16 oz., 115 lb./sq.	1 Shee	1.20	6.667	Sq.	915	252		1,167	1,425
0950 18 oz., 130 lb./sq.		1.15	6.957		1,025	263		1,288	1,575
1000 20 oz., 145 lb./sq.		1.10	7.273		1,200	275		1,475	1,775
1008 Zinc flat seam roofing, .020" thick		1.20	6.667		1,100	252		1,352	1,625
1010 .027" thick		1.15	6.957		1,375	263		1,638	1,950
1020 .032" thick		1.12	7.143		1,575	270		1,845	2,175
1030 .040" thick		1.05	7.619		1,975	288		2,263	2,650
1100 Lead flat seam roofing, 5 lb. per S.F.		1.30	6.154		1,150	233		1,383	1,675

07 62 Sheet Metal Flashing and Trim

07 62 10 – Sheet Metal Trim

07 62 10.10 Sheet Metal Cladding

0010 **SHEET METAL CLADDING**									
0100 Aluminum, up to 6 bends, .032" thick, window casing	1 Carp	180	.044	S.F.	1.65	1.51		3.16	4.34
0200 Window sill		72	.111	L.F.	1.65	3.77		5.42	8.10
0300 Door casing		180	.044	S.F.	1.65	1.51		3.16	4.34
0400 Fascia		250	.032		1.65	1.08		2.73	3.64
0500 Rake trim		225	.036		1.65	1.21		2.86	3.84
0700 .024" thick, window casing		180	.044		1.45	1.51		2.96	4.12
0800 Window sill		72	.111	L.F.	1.45	3.77		5.22	7.90
0900 Door casing		180	.044	S.F.	1.45	1.51		2.96	4.12
1000 Fascia		250	.032		1.45	1.08		2.53	3.42
1100 Rake trim		225	.036		1.45	1.21		2.66	3.62
1200 Vinyl coated aluminum, up to 6 bends, window casing		180	.044		1.54	1.51		3.05	4.21
1300 Window sill		72	.111	L.F.	1.54	3.77		5.31	8
1400 Door casing		180	.044	S.F.	1.54	1.51		3.05	4.21
1500 Fascia		250	.032		1.54	1.08		2.62	3.51
1600 Rake trim		225	.036		1.54	1.21		2.75	3.71

07 65 Flexible Flashing

07 65 10 – Sheet Metal Flashing

07 65 10.10 Sheet Metal Flashing and Counter Flashing		Crew	Daily Output	Labor-Hours	Unit	Material	2016 Bare Costs Labor	Equipment	Total	Total Incl O&P
0010	SHEET METAL FLASHING AND COUNTER FLASHING									
0011	Including up to 4 bends									
0020	Aluminum, mill finish, .013" thick	1 Rofc	145	.055	S.F.	.83	1.59		2.42	3.83
0030	.016" thick		145	.055		.95	1.59		2.54	3.97
0060	.019" thick		145	.055		1.25	1.59		2.84	4.30
0100	.032" thick		145	.055		1.35	1.59		2.94	4.41
0200	.040" thick		145	.055		2.31	1.59		3.90	5.45
0300	.050" thick		145	.055		2.64	1.59		4.23	5.80
0325	Mill finish 5" x 7" step flashing, .016" thick		1920	.004	Ea.	.15	.12		.27	.39
0350	Mill finish 12" x 12" step flashing, .016" thick		1600	.005	"	.55	.14		.69	.88
0400	Painted finish, add				S.F.	.32			.32	.35
1600	Copper, 16 oz., sheets, under 1000 lb.	1 Rofc	115	.070		7.85	2		9.85	12.35
1900	20 oz. sheets, under 1000 lb.		110	.073		10.30	2.09		12.39	15.20
2200	24 oz. sheets, under 1000 lb.		105	.076		14.60	2.19		16.79	20
2500	32 oz. sheets, under 1000 lb.		100	.080		18.50	2.30		20.80	24.50
2700	W shape for valleys, 16 oz., 24" wide		100	.080	L.F.	16.20	2.30		18.50	22
5800	Lead, 2.5 lb. per S.F., up to 12" wide		135	.059	S.F.	5.40	1.70		7.10	9.10
5900	Over 12" wide		135	.059		3.99	1.70		5.69	7.55
8900	Stainless steel sheets, 32 ga.,		155	.052		3.25	1.48		4.73	6.30
9000	28 ga.		155	.052		4.20	1.48		5.68	7.35
9100	26 ga.		155	.052		4.45	1.48		5.93	7.65
9200	24 ga.		155	.052		4.95	1.48		6.43	8.20
9290	For mechanically keyed flashing, add					40%				
9320	Steel sheets, galvanized, 20 ga.	1 Rofc	130	.062	S.F.	1.25	1.77		3.02	4.64
9322	22 ga.		135	.059		1.21	1.70		2.91	4.47
9324	24 ga.		140	.057		.92	1.64		2.56	4.04
9326	26 ga.		148	.054		.80	1.55		2.35	3.74
9328	28 ga.		155	.052		.69	1.48		2.17	3.50
9340	30 ga.		160	.050		.58	1.44		2.02	3.29
9400	Terne coated stainless steel, .015" thick, 28 ga		155	.052		8.05	1.48		9.53	11.60
9500	.018" thick, 26 ga		155	.052		9	1.48		10.48	12.65
9600	Zinc and copper alloy (brass), .020" thick		155	.052		10.20	1.48		11.68	13.95
9700	.027" thick		155	.052		12	1.48		13.48	15.95
9800	.032" thick		155	.052		15.40	1.48		16.88	19.70
9900	.040" thick		155	.052		20	1.48		21.48	24.50

07 65 13 – Laminated Sheet Flashing

07 65 13.10 Laminated Sheet Flashing

		Crew	Daily Output	Labor-Hours	Unit	Material	Labor	Equipment	Total	Total Incl O&P
0010	LAMINATED SHEET FLASHING, Including up to 4 bends									
8550	Shower pan, 3 ply copper and fabric, 3 oz.	1 Rofc	155	.052	S.F.	4	1.48		5.48	7.15
8600	7 oz.	"	155	.052	"	4.20	1.48		5.68	7.35

07 65 19 – Plastic Sheet Flashing

07 65 19.10 Plastic Sheet Flashing and Counter Flashing

		Crew	Daily Output	Labor-Hours	Unit	Material	Labor	Equipment	Total	Total Incl O&P
0010	PLASTIC SHEET FLASHING AND COUNTER FLASHING									
7300	Polyvinyl chloride, black, 10 mil	1 Rofc	285	.028	S.F.	.21	.81		1.02	1.72
7400	20 mil		285	.028		.26	.81		1.07	1.78
7600	30 mil		285	.028		.34	.81		1.15	1.86
7700	60 mil		285	.028		.82	.81		1.63	2.39
8060	PVC tape, 5" x 45 mils, for joint covers, 100 L.F./roll				Ea.	172			172	189

07 65 Flexible Flashing

07 65 23 – Rubber Sheet Flashing

07 65 23.10 Rubber Sheet Flashing and Counterflashing		Crew	Daily Output	Labor-Hours	Unit	Material	2016 Bare Costs Labor	Equipment	Total	Total Incl O&P
0010	**RUBBER SHEET FLASHING AND COUNTERFLASHING**									
4810	EPDM 90 mils, 1" diameter pipe flashing	1 Rofc	32	.250	Ea.	19.35	7.20		26.55	35
4820	2" diameter		30	.267		19.30	7.65		26.95	35
4830	3" diameter		28	.286		20.50	8.20		28.70	38
4840	4" diameter		24	.333		24	9.60		33.60	43.50
4850	6" diameter		22	.364		23.50	10.45		33.95	45.50
8100	Rubber, butyl, 1/32" thick		285	.028	S.F.	1.96	.81		2.77	3.65
8200	1/16" thick		285	.028		2.65	.81		3.46	4.41
8300	Neoprene, cured, 1/16" thick		285	.028		2.55	.81		3.36	4.30
8400	1/8" thick		285	.028		6.05	.81		6.86	8.15

07 65 26 – Self-Adhering Sheet Flashing

07 65 26.10 Self-Adhering Sheet or Roll Flashing		Crew	Daily Output	Labor-Hours	Unit	Material	Labor	Equipment	Total	Total Incl O&P
0010	**SELF-ADHERING SHEET OR ROLL FLASHING**									
0020	Self-adhered flashing, 25 Mil cross laminated HDPE, 4" wide	1 Rofc	960	.008	L.F.	.20	.24		.44	.66
0040	6" wide		896	.009		.30	.26		.56	.79
0060	9" wide		832	.010		.44	.28		.72	1
0080	12" wide		768	.010		.59	.30		.89	1.20

07 71 Roof Specialties

07 71 19 – Manufactured Gravel Stops and Fasciae

07 71 19.10 Gravel Stop		Crew	Daily Output	Labor-Hours	Unit	Material	Labor	Equipment	Total	Total Incl O&P
0010	**GRAVEL STOP**									
0020	Aluminum, .050" thick, 4" face height, mill finish	1 Shee	145	.055	L.F.	6.40	2.09		8.49	10.55
0080	Duranodic finish		145	.055		7.35	2.09		9.44	11.60
0100	Painted		145	.055		7.45	2.09		9.54	11.70
1200	Copper, 16 oz., 3" face height		145	.055		24	2.09		26.09	30
1300	6" face height		135	.059		33	2.24		35.24	40
1350	Galv steel, 24 ga., 4" leg, plain, with continuous cleat, 4" face		145	.055		6.25	2.09		8.34	10.40
1360	6" face height		145	.055		6.50	2.09		8.59	10.65
1500	Polyvinyl chloride, 6" face height		135	.059		5.70	2.24		7.94	10
1800	Stainless steel, 24 ga., 6" face height		135	.059		15.30	2.24		17.54	20.50

07 71 19.30 Fascia

		Crew	Daily Output	Labor-Hours	Unit	Material	Labor	Equipment	Total	Total Incl O&P
0010	**FASCIA**									
0100	Aluminum, reverse board and batten, .032" thick, colored, no furring incl	1 Shee	145	.055	S.F.	6.75	2.09		8.84	10.95
0200	Residential type, aluminum	1 Carp	200	.040	L.F.	1.95	1.36		3.31	4.42
0220	Vinyl	"	200	.040	"	2.10	1.36		3.46	4.58
0300	Steel, galv and enameled, stock, no furring, long panels	1 Shee	145	.055	S.F.	5.25	2.09		7.34	9.30
0600	Short panels	"	115	.070	"	5.85	2.63		8.48	10.85

07 71 23 – Manufactured Gutters and Downspouts

07 71 23.10 Downspouts

		Crew	Daily Output	Labor-Hours	Unit	Material	Labor	Equipment	Total	Total Incl O&P
0010	**DOWNSPOUTS**									
0020	Aluminum, embossed, .020" thick, 2" x 3"	1 Shee	190	.042	L.F.	1.05	1.59		2.64	3.82
0100	Enameled		190	.042		1.39	1.59		2.98	4.19
0300	.024" thick, 2' x 3"		180	.044		2.05	1.68		3.73	5.05
0400	3" x 4"		140	.057		2.37	2.16		4.53	6.20
0600	Round, corrugated aluminum, 3" diameter, .020" thick		190	.042		2	1.59		3.59	4.86
0700	4" diameter, .025" thick		140	.057		2.95	2.16		5.11	6.85
0900	Wire strainer, round, 2" diameter		155	.052	Ea.	1.84	1.95		3.79	5.25
1000	4" diameter		155	.052		2.40	1.95		4.35	5.90

07 71 Roof Specialties

07 71 23 – Manufactured Gutters and Downspouts

07 71 23.10 Downspouts		Crew	Daily Output	Labor-Hours	Unit	Material	2016 Bare Costs Labor	Equipment	Total	Total Incl O&P
1200	Rectangular, perforated, 2" x 3"	1 Shee	145	.055	Ea.	2.33	2.09		4.42	6.05
1300	3" x 4"		145	.055	↓	3.33	2.09		5.42	7.15
1500	Copper, round, 16 oz., stock, 2" diameter		190	.042	L.F.	7.60	1.59		9.19	11
1600	3" diameter		190	.042		8	1.59		9.59	11.45
1800	4" diameter		145	.055		9.70	2.09		11.79	14.15
1900	5" diameter		130	.062		15.75	2.33		18.08	21
2100	Rectangular, corrugated copper, stock, 2" x 3"		190	.042		8.75	1.59		10.34	12.25
2200	3" x 4"		145	.055		10.40	2.09		12.49	14.95
2400	Rectangular, plain copper, stock, 2" x 3"		190	.042		11.05	1.59		12.64	14.80
2500	3" x 4"		145	.055	↓	13.95	2.09		16.04	18.85
2700	Wire strainers, rectangular, 2" x 3"		145	.055	Ea.	17.05	2.09		19.14	22
2800	3" x 4"		145	.055		17.80	2.09		19.89	23
3000	Round, 2" diameter		145	.055		6.40	2.09		8.49	10.55
3100	3" diameter		145	.055		7.10	2.09		9.19	11.30
3300	4" diameter		145	.055		12.10	2.09		14.19	16.80
3400	5" diameter		115	.070	↓	22.50	2.63		25.13	29
3600	Lead-coated copper, round, stock, 2" diameter		190	.042	L.F.	22	1.59		23.59	27
3700	3" diameter		190	.042		22.50	1.59		24.09	27
3900	4" diameter		145	.055		23.50	2.09		25.59	29.50
4000	5" diameter, corrugated		130	.062		23	2.33		25.33	29
4200	6" diameter, corrugated		105	.076		30	2.88		32.88	38
4300	Rectangular, corrugated, stock, 2" x 3"		190	.042		15.70	1.59		17.29	19.90
4500	Plain, stock, 2" x 3"		190	.042		24	1.59		25.59	29
4600	3" x 4"		145	.055		33	2.09		35.09	40
4800	Steel, galvanized, round, corrugated, 2" or 3" diameter, 28 ga.		190	.042		2.10	1.59		3.69	4.97
4900	4" diameter, 28 ga.		145	.055		2.52	2.09		4.61	6.25
5700	Rectangular, corrugated, 28 ga., 2" x 3"		190	.042		1.94	1.59		3.53	4.79
5800	3" x 4"		145	.055		1.82	2.09		3.91	5.50
6000	Rectangular, plain, 28 ga., galvanized, 2" x 3"		190	.042		3.65	1.59		5.24	6.70
6100	3" x 4"		145	.055		4.06	2.09		6.15	7.95
6300	Epoxy painted, 24 ga., corrugated, 2" x 3"		190	.042		2.33	1.59		3.92	5.20
6400	3" x 4"		145	.055	↓	2.83	2.09		4.92	6.60
6600	Wire strainers, rectangular, 2" x 3"		145	.055	Ea.	17.05	2.09		19.14	22
6700	3" x 4"		145	.055		17.80	2.09		19.89	23
6900	Round strainers, 2" or 3" diameter		145	.055		3.86	2.09		5.95	7.75
7000	4" diameter		145	.055	↓	5.80	2.09		7.89	9.85
8200	Vinyl, rectangular, 2" x 3"		210	.038	L.F.	2.08	1.44		3.52	4.69
8300	Round, 2-1/2"		220	.036	"	1.32	1.37		2.69	3.74

07 71 23.20 Downspout Elbows

		Crew	Daily Output	Labor-Hours	Unit	Material	Labor	Equipment	Total	Total Incl O&P
0010	**DOWNSPOUT ELBOWS**									
0020	Aluminum, embossed, 2" x 3", .020" thick	1 Shee	100	.080	Ea.	.95	3.02		3.97	6.10
0100	Enameled		100	.080		1.75	3.02		4.77	7
0200	Embossed, 3" x 4", .025" thick		100	.080		4.51	3.02		7.53	10
0300	Enameled		100	.080		3.75	3.02		6.77	9.20
0400	Embossed, corrugated, 3" diameter, .020" thick		100	.080		2.74	3.02		5.76	8.05
0500	4" diameter, .025" thick		100	.080		5.80	3.02		8.82	11.40
0600	Copper, 16 oz., 2" diameter		100	.080		9.45	3.02		12.47	15.45
0700	3" diameter		100	.080		9	3.02		12.02	14.95
0800	4" diameter		100	.080		13.90	3.02		16.92	20.50
1000	Rectangular, 2" x 3" corrugated		100	.080		9.35	3.02		12.37	15.35
1100	3" x 4" corrugated		100	.080		13.25	3.02		16.27	19.60
1300	Vinyl, 2-1/2" diameter, 45 or 75 degree bend		100	.080	↓	3.88	3.02		6.90	9.30

07 71 Roof Specialties

07 71 23 – Manufactured Gutters and Downspouts

	07 71 23.20 Downspout Elbows	Crew	Daily Output	Labor-Hours	Unit	Material	2016 Bare Costs Labor	Equipment	Total	Total Incl O&P
1400	Tee Y junction	1 Shee	75	.107	Ea.	12.90	4.03		16.93	21
07 71 23.30 Gutters										
0010	**GUTTERS**									
0012	Aluminum, stock units, 5" K type, .027" thick, plain	1 Shee	125	.064	L.F.	2.81	2.42		5.23	7.15
0100	Enameled		125	.064		2.79	2.42		5.21	7.10
0300	5" K type, .032" thick, plain		125	.064		3.47	2.42		5.89	7.85
0400	Enameled		125	.064		3.59	2.42		6.01	8
0700	Copper, half round, 16 oz., stock units, 4" wide		125	.064		9.50	2.42		11.92	14.50
0900	5" wide		125	.064		7.25	2.42		9.67	12
1000	6" wide		118	.068		11.85	2.56		14.41	17.25
1200	K type, 16 oz., stock, 5" wide		125	.064		8.05	2.42		10.47	12.90
1300	6" wide		125	.064		8.50	2.42		10.92	13.40
1500	Lead coated copper, 16 oz., half round, stock, 4" wide		125	.064		14.70	2.42		17.12	20
1600	6" wide		118	.068		22	2.56		24.56	28.50
1800	K type, stock, 5" wide		125	.064		18.35	2.42		20.77	24
1900	6" wide		125	.064		19.65	2.42		22.07	25.50
2100	Copper clad stainless steel, K type, 5" wide		125	.064		7.50	2.42		9.92	12.30
2200	6" wide		125	.064		9.45	2.42		11.87	14.45
2400	Steel, galv, half round or box, 28 ga., 5" wide, plain		125	.064		2.10	2.42		4.52	6.35
2500	Enameled		125	.064		2.18	2.42		4.60	6.45
2700	26 ga., stock, 5" wide		125	.064		2.40	2.42		4.82	6.70
2800	6" wide		125	.064		2.78	2.42		5.20	7.10
3000	Vinyl, O.G., 4" wide	1 Carp	115	.070		1.30	2.36		3.66	5.40
3100	5" wide		115	.070		1.60	2.36		3.96	5.70
3200	4" half round, stock units		115	.070		1.35	2.36		3.71	5.45
3250	Joint connectors				Ea.	2.95			2.95	3.25
3300	Wood, clear treated cedar, fir or hemlock, 3" x 4"	1 Carp	100	.080	L.F.	10	2.71		12.71	15.55
3400	4" x 5"	"	100	.080	"	16.75	2.71		19.46	23
5000	Accessories, end cap, K type, aluminum 5"	1 Shee	625	.013	Ea.	.70	.48		1.18	1.58
5010	6"		625	.013		1.50	.48		1.98	2.46
5020	Copper, 5"		625	.013		3.22	.48		3.70	4.35
5030	6"		625	.013		3.54	.48		4.02	4.70
5040	Lead coated copper, 5"		625	.013		12.50	.48		12.98	14.55
5050	6"		625	.013		13.35	.48		13.83	15.50
5060	Copper clad stainless steel, 5"		625	.013		3	.48		3.48	4.11
5070	6"		625	.013		3.60	.48		4.08	4.77
5080	Galvanized steel, 5"		625	.013		1.28	.48		1.76	2.22
5090	6"		625	.013		2.18	.48		2.66	3.21
5100	Vinyl, 4"	1 Carp	625	.013		13.15	.43		13.58	15.25
5110	5"	"	625	.013		13.15	.43		13.58	15.25
5120	Half round, copper, 4"	1 Shee	625	.013		4.53	.48		5.01	5.80
5130	5"		625	.013		4.53	.48		5.01	5.80
5140	6"		625	.013		7.65	.48		8.13	9.25
5150	Lead coated copper, 5"		625	.013		14.15	.48		14.63	16.35
5160	6"		625	.013		21	.48		21.48	24.50
5170	Copper clad stainless steel, 5"		625	.013		4.50	.48		4.98	5.75
5180	6"		625	.013		4.50	.48		4.98	5.75
5190	Galvanized steel, 5"		625	.013		2.25	.48		2.73	3.29
5200	6"		625	.013		2.80	.48		3.28	3.89
5210	Outlet, aluminum, 2" x 3"		420	.019		.62	.72		1.34	1.88
5220	3" x 4"		420	.019		1.05	.72		1.77	2.36
5230	2-3/8" round		420	.019		.56	.72		1.28	1.82

07 71 Roof Specialties

07 71 23 – Manufactured Gutters and Downspouts

07 71 23.30 Gutters

		Crew	Daily Output	Labor-Hours	Unit	Material	2016 Bare Costs Labor	Equipment	Total	Total Incl O&P
5240	Copper, 2" x 3"	1 Shee	420	.019	Ea.	6.50	.72		7.22	8.35
5250	3" x 4"		420	.019		7.85	.72		8.57	9.85
5260	2-3/8" round		420	.019		4.55	.72		5.27	6.20
5270	Lead coated copper, 2" x 3"		420	.019		25.50	.72		26.22	29
5280	3" x 4"		420	.019		28.50	.72		29.22	32.50
5290	2-3/8" round		420	.019		25.50	.72		26.22	29
5300	Copper clad stainless steel, 2" x 3"		420	.019		6.50	.72		7.22	8.35
5310	3" x 4"		420	.019		7.85	.72		8.57	9.85
5320	2-3/8" round		420	.019		4.55	.72		5.27	6.20
5330	Galvanized steel, 2" x 3"		420	.019		3.18	.72		3.90	4.70
5340	3" x 4"		420	.019		4.90	.72		5.62	6.60
5350	2-3/8" round		420	.019		3.98	.72		4.70	5.60
5360	K type mitres, aluminum		65	.123		3	4.65		7.65	11.05
5370	Copper		65	.123		17.40	4.65		22.05	27
5380	Lead coated copper		65	.123		54.50	4.65		59.15	68
5390	Copper clad stainless steel		65	.123		27.50	4.65		32.15	38.50
5400	Galvanized steel		65	.123		22.50	4.65		27.15	33
5420	Half round mitres, copper		65	.123		63.50	4.65		68.15	78
5430	Lead coated copper		65	.123		89.50	4.65		94.15	106
5440	Copper clad stainless steel		65	.123		56	4.65		60.65	70
5450	Galvanized steel		65	.123		25.50	4.65		30.15	36
5460	Vinyl mitres and outlets		65	.123		9.75	4.65		14.40	18.50
5470	Sealant		940	.009	L.F.	.01	.32		.33	.55
5480	Soldering		96	.083	"	.20	3.15		3.35	5.45

07 71 23.35 Gutter Guard

		Crew	Daily Output	Labor-Hours	Unit	Material	Labor	Equipment	Total	Total Incl O&P
0010	**GUTTER GUARD**									
0020	6" wide strip, aluminum mesh	1 Carp	500	.016	L.F.	2.50	.54		3.04	3.66
0100	Vinyl mesh	"	500	.016	"	2.70	.54		3.24	3.88

07 71 43 – Drip Edge

07 71 43.10 Drip Edge, Rake Edge, Ice Belts

		Crew	Daily Output	Labor-Hours	Unit	Material	Labor	Equipment	Total	Total Incl O&P
0010	**DRIP EDGE, RAKE EDGE, ICE BELTS**									
0020	Aluminum, .016" thick, 5" wide, mill finish	1 Carp	400	.020	L.F.	.57	.68		1.25	1.77
0100	White finish		400	.020		.63	.68		1.31	1.83
0200	8" wide, mill finish		400	.020		1.45	.68		2.13	2.74
0300	Ice belt, 28" wide, mill finish		100	.080		7.65	2.71		10.36	13
0310	Vented, mill finish		400	.020		1.99	.68		2.67	3.33
0320	Painted finish		400	.020		2.24	.68		2.92	3.60
0400	Galvanized, 5" wide		400	.020		.59	.68		1.27	1.79
0500	8" wide, mill finish		400	.020		.82	.68		1.50	2.04
0510	Rake edge, aluminum, 1-1/2" x 1-1/2"		400	.020		.33	.68		1.01	1.50
0520	3-1/2" x 1-1/2"		400	.020		.45	.68		1.13	1.64

07 72 Roof Accessories

07 72 23 – Relief Vents

07 72 23.20 Vents

		Crew	Daily Output	Labor-Hours	Unit	Material	2016 Bare Costs Labor	Equipment	Total	Total Incl O&P
0010	**VENTS**									
0100	Soffit or eave, aluminum, mill finish, strips, 2-1/2" wide	1 Carp	200	.040	L.F.	.46	1.36		1.82	2.78
0200	3" wide		200	.040		.48	1.36		1.84	2.80
0300	Enamel finish, 3" wide		200	.040		.50	1.36		1.86	2.82
0400	Mill finish, rectangular, 4" x 16"		72	.111	Ea.	1.52	3.77		5.29	7.95
0500	8" x 16"		72	.111		2.33	3.77		6.10	8.85
2420	Roof ventilator	Q-9	16	1		54	34		88	117
2500	Vent, roof vent	1 Rofc	24	.333		27	9.60		36.60	47

07 72 26 – Ridge Vents

07 72 26.10 Ridge Vents and Accessories

		Crew	Daily Output	Labor-Hours	Unit	Material	2016 Bare Costs Labor	Equipment	Total	Total Incl O&P
0010	**RIDGE VENTS AND ACCESSORIES**									
0100	Aluminum strips, mill finish	1 Rofc	160	.050	L.F.	2.86	1.44		4.30	5.80
0150	Painted finish		160	.050	"	4.05	1.44		5.49	7.10
0200	Connectors		48	.167	Ea.	4.75	4.79		9.54	14.10
0300	End caps		48	.167	"	2.05	4.79		6.84	11.10
0400	Galvanized strips		160	.050	L.F.	3.60	1.44		5.04	6.60
0430	Molded polyethylene, shingles not included		160	.050	"	2.75	1.44		4.19	5.65
0440	End plugs		48	.167	Ea.	2.05	4.79		6.84	11.10
0450	Flexible roll, shingles not included		160	.050	L.F.	2.34	1.44		3.78	5.20
2300	Ridge vent strip, mill finish	1 Shee	155	.052	"	3.88	1.95		5.83	7.50

07 72 53 – Snow Guards

07 72 53.10 Snow Guard Options

		Crew	Daily Output	Labor-Hours	Unit	Material	2016 Bare Costs Labor	Equipment	Total	Total Incl O&P
0010	**SNOW GUARD OPTIONS**									
0100	Slate & asphalt shingle roofs, fastened with nails	1 Rofc	160	.050	Ea.	12.55	1.44		13.99	16.45
0200	Standing seam metal roofs, fastened with set screws		48	.167		16.95	4.79		21.74	27.50
0300	Surface mount for metal roofs, fastened with solder		48	.167		6.05	4.79		10.84	15.50
0400	Double rail pipe type, including pipe		130	.062	L.F.	34	1.77		35.77	41

07 72 80 – Vents

07 72 80.30 Vent Options

		Crew	Daily Output	Labor-Hours	Unit	Material	2016 Bare Costs Labor	Equipment	Total	Total Incl O&P
0010	**VENT OPTIONS**									
0800	Polystyrene baffles, 12" wide for 16" O.C. rafter spacing	1 Carp	90	.089	Ea.	.44	3.01		3.45	5.55
0900	For 24" O.C. rafter spacing	"	110	.073	"	.78	2.47		3.25	4.99

07 76 Roof Pavers

07 76 16 – Roof Decking Pavers

07 76 16.10 Roof Pavers and Supports

		Crew	Daily Output	Labor-Hours	Unit	Material	2016 Bare Costs Labor	Equipment	Total	Total Incl O&P
0010	**ROOF PAVERS AND SUPPORTS**									
1000	Roof decking pavers, concrete blocks, 2" thick, natural	1 Clab	115	.070	S.F.	3.37	1.71		5.08	6.60
1100	Colors		115	.070	"	3.73	1.71		5.44	6.95
1200	Support pedestal, bottom cap		960	.008	Ea.	3	.21		3.21	3.64
1300	Top cap		960	.008		4.80	.21		5.01	5.65
1400	Leveling shims, 1/16"		1920	.004		1.20	.10		1.30	1.49
1500	1/8"		1920	.004		1.20	.10		1.30	1.49
1600	Buffer pad		960	.008		2.50	.21		2.71	3.09
1700	PVC legs (4" SDR 35)		2880	.003	Inch	.12	.07		.19	.25
2000	Alternate pricing method, system in place		101	.079	S.F.	7	1.95		8.95	10.95

07 91 Preformed Joint Seals

07 91 13 – Compression Seals

07 91 13.10 Compression Seals		Crew	Daily Output	Labor-Hours	Unit	Material	2016 Bare Costs Labor	Equipment	Total	Total Incl O&P
0010	COMPRESSION SEALS									
4900	O-ring type cord, 1/4"	1 Bric	472	.017	L.F.	.43	.55		.98	1.39
4910	1/2"		440	.018		1.20	.59		1.79	2.30
4920	3/4"		424	.019		2.40	.61		3.01	3.66
4930	1"		408	.020		4.08	.64		4.72	5.55
4940	1-1/4"		384	.021		7.40	.67		8.07	9.25
4950	1-1/2"		368	.022		9.20	.70		9.90	11.25
4960	1-3/4"		352	.023		18.55	.74		19.29	21.50
4970	2"		344	.023		22	.75		22.75	26

07 91 16 – Joint Gaskets

07 91 16.10 Joint Gaskets		Crew	Daily Output	Labor-Hours	Unit	Material	Labor	Equipment	Total	Total Incl O&P
0010	JOINT GASKETS									
4400	Joint gaskets, neoprene, closed cell w/adh, 1/8" x 3/8"	1 Bric	240	.033	L.F.	.32	1.08		1.40	2.15
4500	1/4" x 3/4"		215	.037		.62	1.21		1.83	2.69
4700	1/2" x 1"		200	.040		1.45	1.30		2.75	3.76
4800	3/4" x 1-1/2"		165	.048		1.60	1.57		3.17	4.38

07 91 23 – Backer Rods

07 91 23.10 Backer Rods		Crew	Daily Output	Labor-Hours	Unit	Material	Labor	Equipment	Total	Total Incl O&P
0010	BACKER RODS									
0030	Backer rod, polyethylene, 1/4" diameter	1 Bric	4.60	1.739	C.L.F.	2.28	56.50		58.78	96.50
0050	1/2" diameter		4.60	1.739		4	56.50		60.50	98.50
0070	3/4" diameter		4.60	1.739		6.50	56.50		63	101
0090	1" diameter		4.60	1.739		11.50	56.50		68	107

07 91 26 – Joint Fillers

07 91 26.10 Joint Fillers		Crew	Daily Output	Labor-Hours	Unit	Material	Labor	Equipment	Total	Total Incl O&P
0010	JOINT FILLERS									
4360	Butyl rubber filler, 1/4" x 1/4"	1 Bric	290	.028	L.F.	.22	.89		1.11	1.74
4365	1/2" x 1/2"		250	.032		.90	1.04		1.94	2.72
4370	1/2" x 3/4"		210	.038		1.35	1.23		2.58	3.54
4375	3/4" x 3/4"		230	.035		2.02	1.13		3.15	4.10
4380	1" x 1"		180	.044		2.70	1.44		4.14	5.35
4390	For coloring, add					12%				
4980	Polyethylene joint backing, 1/4" x 2"	1 Bric	2.08	3.846	C.L.F.	12.90	125		137.90	222
4990	1/4" x 6"		1.28	6.250	"	28.50	203		231.50	370
5600	Silicone, room temp vulcanizing foam seal, 1/4" x 1/2"		1312	.006	L.F.	.46	.20		.66	.83
5610	1/2" x 1/2"		656	.012		.91	.40		1.31	1.66
5620	1/2" x 3/4"		442	.018		1.37	.59		1.96	2.48
5630	3/4" x 3/4"		328	.024		2.05	.79		2.84	3.57
5640	1/8" x 1"		1312	.006		.46	.20		.66	.83
5650	1/8" x 3"		442	.018		1.37	.59		1.96	2.48
5670	1/4" x 3"		295	.027		2.73	.88		3.61	4.46
5680	1/4" x 6"		148	.054		5.45	1.75		7.20	8.90
5690	1/2" x 6"		82	.098		10.90	3.16		14.06	17.25
5700	1/2" x 9"		52.50	.152		16.40	4.94		21.34	26.50
5710	1/2" x 12"		33	.242		22	7.85		29.85	37

07 92 Joint Sealants

07 92 13 – Elastomeric Joint Sealants

07 92 13.20 Caulking and Sealant Options

		Crew	Daily Output	Labor-Hours	Unit	Material	2016 Bare Costs Labor	Equipment	Total	Total Incl O&P
0010	**CAULKING AND SEALANT OPTIONS**									
0050	Latex acrylic based, bulk				Gal.	27			27	29.50
0055	Bulk in place 1/4" x 1/4" bead	1 Bric	300	.027	L.F.	.08	.86		.94	1.53
0060	1/4" x 3/8"		294	.027		.14	.88		1.02	1.62
0065	1/4" x 1/2"		288	.028		.19	.90		1.09	1.71
0075	3/8" x 3/8"		284	.028		.21	.91		1.12	1.75
0080	3/8" x 1/2"		280	.029		.28	.93		1.21	1.85
0085	3/8" x 5/8"		276	.029		.35	.94		1.29	1.96
0095	3/8" x 3/4"		272	.029		.42	.95		1.37	2.05
0100	1/2" x 1/2"		275	.029		.38	.94		1.32	1.98
0105	1/2" x 5/8"		269	.030		.47	.96		1.43	2.13
0110	1/2" x 3/4"		263	.030		.56	.99		1.55	2.26
0115	1/2" x 7/8"		256	.031		.66	1.01		1.67	2.41
0120	1/2" x 1"		250	.032		.75	1.04		1.79	2.56
0125	3/4" x 3/4"		244	.033		.85	1.06		1.91	2.70
0130	3/4" x 1"		225	.036		1.13	1.15		2.28	3.16
0135	1" x 1"		200	.040		1.50	1.30		2.80	3.81
0190	Cartridges				Gal.	32.50			32.50	36
0200	11 fl. oz. cartridge				Ea.	2.80			2.80	3.08
0500	1/4" x 1/2"	1 Bric	288	.028	L.F.	.23	.90		1.13	1.75
0600	1/2" x 1/2"		275	.029		.46	.94		1.40	2.07
0800	3/4" x 3/4"		244	.033		1.03	1.06		2.09	2.90
0900	3/4" x 1"		225	.036		1.37	1.15		2.52	3.43
1000	1" x 1"		200	.040		1.72	1.30		3.02	4.05
1400	Butyl based, bulk				Gal.	35			35	38.50
1500	Cartridges				"	38.50			38.50	42
1700	1/4" x 1/2", 154 L.F./gal.	1 Bric	288	.028	L.F.	.23	.90		1.13	1.75
1800	1/2" x 1/2", 77 L.F./gal.	"	275	.029	"	.45	.94		1.39	2.07
2300	Polysulfide compounds, 1 component, bulk				Gal.	82			82	90
2400	Cartridges				"	128			128	141
2600	1 or 2 component, in place, 1/4" x 1/4", 308 L.F./gal.	1 Bric	300	.027	L.F.	.27	.86		1.13	1.73
2700	1/2" x 1/4", 154 L.F./gal.		288	.028		.53	.90		1.43	2.09
2900	3/4" x 3/8", 68 L.F./gal.		272	.029		1.21	.95		2.16	2.92
3000	1" x 1/2", 38 L.F./gal.		250	.032		2.16	1.04		3.20	4.10
3200	Polyurethane, 1 or 2 component				Gal.	53			53	58.50
3300	Cartridges				"	65			65	71.50
3500	Bulk, in place, 1/4" x 1/4"	1 Bric	300	.027	L.F.	.17	.86		1.03	1.63
3655	1/2" x 1/4"		288	.028		.34	.90		1.24	1.88
3800	3/4" x 3/8"		272	.029		.78	.95		1.73	2.45
3900	1" x 1/2"		250	.032		1.38	1.04		2.42	3.25
4100	Silicone rubber, bulk				Gal.	54			54	59.50
4200	Cartridges				"	53			53	58.50

07 92 19 – Acoustical Joint Sealants

07 92 19.10 Acoustical Sealant

		Crew	Daily Output	Labor-Hours	Unit	Material	2016 Bare Costs Labor	Equipment	Total	Total Incl O&P
0010	**ACOUSTICAL SEALANT**									
0020	Acoustical sealant, elastomeric, cartridges				Ea.	8.50			8.50	9.35
0025	In place, 1/4" x 1/4"	1 Bric	300	.027	L.F.	.35	.86		1.21	1.82
0030	1/4" x 1/2"		288	.028		.69	.90		1.59	2.26
0035	1/2" x 1/2"		275	.029		1.39	.94		2.33	3.09
0040	1/2" x 3/4"		263	.030		2.08	.99		3.07	3.93
0045	3/4" x 3/4"		244	.033		3.12	1.06		4.18	5.20
0050	1" x 1"		200	.040		5.55	1.30		6.85	8.25

Estimating Tips

08 10 00 Doors and Frames
All exterior doors should be addressed for their energy conservation (insulation and seals).
- Most metal doors and frames look alike, but there may be significant differences among them. When estimating these items, be sure to choose the line item that most closely compares to the specification or door schedule requirements regarding:
  - type of metal
  - metal gauge
  - door core material
  - fire rating
  - finish
- Wood and plastic doors vary considerably in price. The primary determinant is the veneer material. Lauan, birch, and oak are the most common veneers. Other variables include the following:
  - hollow or solid core
  - fire rating
  - flush or raised panel
  - finish
- Door pricing includes bore for cylindrical lockset and mortise for hinges.

08 30 00 Specialty Doors and Frames
- There are many varieties of special doors, and they are usually priced per each. Add frames, hardware, or operators required for a complete installation.

08 40 00 Entrances, Storefronts, and Curtain Walls
- Glazed curtain walls consist of the metal tube framing and the glazing material. The cost data in this subdivision is presented for the metal tube framing alone or the composite wall. If your estimate requires a detailed takeoff of the framing, be sure to add the glazing cost and any tints.

08 50 00 Windows
- Most metal windows are delivered preglazed. However, some metal windows are priced without glass. Refer to 08 80 00 Glazing for glass pricing. The grade C indicates commercial grade windows, usually ASTM C-35.
- All wood windows and vinyl are priced preglazed. The glazing is insulating glass. Add the cost of screens and grills if required, and not already included.

08 70 00 Hardware
- Hardware costs add considerably to the cost of a door. The most efficient method to determine the hardware requirements for a project is to review the door and hardware schedule together. One type of door may have different hardware, depending on the door usage.
- Door hinges are priced by the pair, with most doors requiring 1-1/2 pairs per door. The hinge prices do not include installation labor, because it is included in door installation. Hinges are classified according to the frequency of use, base material, and finish.

08 80 00 Glazing
- Different openings require different types of glass. The most common types are:
  - float
  - tempered
  - insulating
  - impact-resistant
  - ballistic-resistant
- Most exterior windows are glazed with insulating glass. Entrance doors and window walls, where the glass is less than 18" from the floor, are generally glazed with tempered glass. Interior windows and some residential windows are glazed with float glass.
- Coastal communities require the use of impact-resistant glass, dependant on wind speed.
- The insulation or 'u' value is a strong consideration, along with solar heat gain, to determine total energy efficiency.

Reference Numbers
Reference numbers are shown at the beginning of some major classifications. These numbers refer to related items in the Reference Section. The reference information may be an estimating procedure, an alternate pricing method, or technical information.

Note: Not all subdivisions listed here necessarily appear. ■

Did you know?
RSMeans Online gives you the same access to RSMeans' data with 24/7 access:
- Quickly locate costs in the searchable database.
- Build cost lists, estimates, and reports in minutes.
- Adjust costs to any location in the U.S. and Canada with the click of a button.

Start your free trial today at **www.RSMeansOnline.com**

RSMeans Online
FROM THE GORDIAN GROUP®

No part of this cost data may be reproduced, stored in a retrieval system, or transmitted in any form or by any means without prior written permission of RSMeans.

08 01 Operation and Maintenance of Openings

08 01 53 – Operation and Maintenance of Plastic Windows

08 01 53.81 Solid Vinyl Replacement Windows

	08 01 53.81 Solid Vinyl Replacement Windows		Crew	Daily Output	Labor-Hours	Unit	Material	2016 Bare Costs Labor	Equipment	Total	Total Incl O&P
0010	**SOLID VINYL REPLACEMENT WINDOWS**	R085313-20									
0020	Double hung, insulated glass, up to 83 united inches	G	2 Carp	8	2	Ea.	345	68		413	495
0040	84 to 93	G		8	2		385	68		453	540
0060	94 to 101	G		6	2.667		385	90.50		475.50	575
0080	102 to 111	G		6	2.667		405	90.50		495.50	595
0100	112 to 120	G		6	2.667		440	90.50		530.50	635
0120	For each united inch over 120, add	G		800	.020	Inch	5.60	.68		6.28	7.30
0140	Casement windows, one operating sash, 42 to 60 united inches	G		8	2	Ea.	240	68		308	380
0160	61 to 70	G		8	2		270	68		338	410
0180	71 to 80	G		8	2		290	68		358	435
0200	81 to 96	G		8	2		310	68		378	455
0220	Two operating sash, 58 to 78 united inches	G		8	2		475	68		543	640
0240	79 to 88	G		8	2		505	68		573	670
0260	89 to 98	G		8	2		550	68		618	720
0280	99 to 108	G		6	2.667		575	90.50		665.50	785
0300	109 to 121	G		6	2.667		615	90.50		705.50	825
0320	Two operating, one fixed sash, 73 to 108 united inches	G		8	2		745	68		813	935
0340	109 to 118	G		8	2		785	68		853	980
0360	119 to 128	G		6	2.667		805	90.50		895.50	1,025
0380	129 to 138	G		6	2.667		870	90.50		960.50	1,100
0400	139 to 156	G		6	2.667		945	90.50		1,035.50	1,200
0420	Four operating sash, 98 to 118 united inches	G		8	2		1,075	68		1,143	1,300
0440	119 to 128	G		8	2		1,150	68		1,218	1,375
0460	129 to 138	G		6	2.667		1,200	90.50		1,290.50	1,475
0480	139 to 148	G		6	2.667		1,275	90.50		1,365.50	1,550
0500	149 to 168	G		6	2.667		1,350	90.50		1,440.50	1,625
0520	169 to 178	G		6	2.667		1,450	90.50		1,540.50	1,750
0560	Fixed picture window, up to 63 united inches	G		8	2		180	68		248	310
0580	64 to 83	G		8	2		206	68		274	340
0600	84 to 101	G		8	2		257	68		325	395
0620	For each united inch over 101, add	G		900	.018	Inch	3.10	.60		3.70	4.42
0800	Cellulose fiber insulation, poured into sash balance cavity	G	1 Carp	36	.222	C.F.	.69	7.55		8.24	13.35
0820	Silicone caulking at perimeter	G	"	800	.010	L.F.	.17	.34		.51	.76
2000	Impact resistant replacement windows										
2005	Laminated Glass, 120 mph rating, measure in united inches										
2010	Installation labor does not cover any rework of the window opening										
2020	Double hung, insulated glass, up to 101 united inches		2 Carp	8	2	Ea.	480	68		548	640
2025	For each united inch over 101, add			80	.200	Inch	3.25	6.80		10.05	14.95
2100	Casement windows, impact resistant, up to 60 united inches			8	2	Ea.	450	68		518	610
2120	61 to 70			8	2		480	68		548	640
2130	71 to 80			8	2		500	68		568	665
2140	81 to 100			8	2		515	68		583	680
2150	For each united inch over 100, add			80	.200	Inch	4.55	6.80		11.35	16.35
2200	Awning windows, impact resistant, up to 60 united inches			8	2	Ea.	470	68		538	630
2220	61 to 70			8	2		480	68		548	640
2230	71 to 80			8	2		500	68		568	665
2240	For each united inch over 80, add			80	.200	Inch	4.55	6.80		11.35	16.35
2300	Picture windows, impact resistant, up to 63 united inches			8	2	Ea.	345	68		413	495
2320	63 to 83			8	2		370	68		438	520
2330	84 to 101			8	2		415	68		483	570
2340	For each united inch over 101, add			80	.200	Inch	3.25	6.80		10.05	14.95

08 05 Common Work Results for Openings

08 05 05 – Selective Demolition for Openings

08 05 05.10 Selective Demolition Doors

		Crew	Daily Output	Labor-Hours	Unit	Material	2016 Bare Costs Labor	Equipment	Total	Total Incl O&P
0010	**SELECTIVE DEMOLITION DOORS** R024119-10									
0200	Doors, exterior, 1-3/4" thick, single, 3' x 7' high	1 Clab	16	.500	Ea.		12.35		12.35	20.50
0210	3' x 8' high		10	.800			19.70		19.70	33
0215	Double, 3' x 8' high		6	1.333			33		33	55
0220	Double, 6' x 7' high		12	.667			16.45		16.45	27.50
0500	Interior, 1-3/8" thick, single, 3' x 7' high		20	.400			9.85		9.85	16.50
0520	Double, 6' x 7' high		16	.500			12.35		12.35	20.50
0700	Bi-folding, 3' x 6'-8" high		20	.400			9.85		9.85	16.50
0720	6' x 6'-8" high		18	.444			10.95		10.95	18.35
0900	Bi-passing, 3' x 6'-8" high		16	.500			12.35		12.35	20.50
0940	6' x 6'-8" high		14	.571			14.10		14.10	23.50
1500	Remove and reset, hollow core	1 Carp	8	1			34		34	57
1520	Solid		6	1.333			45		45	75.50
2000	Frames, including trim, metal		8	1			34		34	57
2200	Wood	2 Carp	32	.500			16.95		16.95	28.50
2201	Alternate pricing method	1 Carp	200	.040	L.F.		1.36		1.36	2.27
3000	Special doors, counter doors	2 Carp	6	2.667	Ea.		90.50		90.50	151
3300	Glass, sliding, including frames		12	1.333			45		45	75.50
3400	Overhead, commercial, 12' x 12' high		4	4			136		136	227
3500	Residential, 9' x 7' high		8	2			68		68	114
3540	16' x 7' high		7	2.286			77.50		77.50	130
3600	Remove and reset, small		4	4			136		136	227
3620	Large		2.50	6.400			217		217	365
3660	Remove and reset elec. garage door opener	1 Carp	8	1			34		34	57
4000	Residential lockset, exterior		28	.286			9.70		9.70	16.20
4010	Residential lockset, exterior w/deadbolt		26	.308			10.45		10.45	17.45
4020	Residential lockset, interior		30	.267			9.05		9.05	15.15
4200	Deadbolt lock		32	.250			8.50		8.50	14.20
4224	Pocket door		8	1			34		34	57
5590	Remove mail slot	1 Clab	45	.178			4.38		4.38	7.35
5600	Remove door sidelight	1 Carp	6	1.333			45		45	75.50

08 05 05.20 Selective Demolition of Windows

		Crew	Daily Output	Labor-Hours	Unit	Material	2016 Bare Costs Labor	Equipment	Total	Total Incl O&P
0010	**SELECTIVE DEMOLITION OF WINDOWS** R024119-10									
0200	Aluminum, including trim, to 12 S.F.	1 Clab	16	.500	Ea.		12.35		12.35	20.50
0240	To 25 S.F.		11	.727			17.95		17.95	30
0280	To 50 S.F.		5	1.600			39.50		39.50	66
0320	Storm windows/screens, to 12 S.F.		27	.296			7.30		7.30	12.20
0360	To 25 S.F.		21	.381			9.40		9.40	15.70
0400	To 50 S.F.		16	.500			12.35		12.35	20.50
0500	Screens, incl. aluminum frame, small		20	.400			9.85		9.85	16.50
0510	Large		16	.500			12.35		12.35	20.50
0600	Glass, up to 10 S.F. per window		200	.040	S.F.		.99		.99	1.65
0620	Over 10 S.F. per window		150	.053	"		1.31		1.31	2.20
2000	Wood, including trim, to 12 S.F.		22	.364	Ea.		8.95		8.95	15
2020	To 25 S.F.		18	.444			10.95		10.95	18.35
2060	To 50 S.F.		13	.615			15.15		15.15	25.50
2065	To 180 S.F.		8	1			24.50		24.50	41.50
4300	Remove bay/bow window	2 Carp	6	2.667			90.50		90.50	151
4410	Remove skylight, plstc domes, flush/curb mtd	G-3	395	.081	S.F.		2.53		2.53	4.22
5020	Remove and reset window, up to a 2' x 2' widow	1 Carp	6	1.333	Ea.		45		45	75.50
5040	Up to a 3' x 3' window		4	2			68		68	114
5080	Up to a 4' x 5' window		2	4			136		136	227

08 05 Common Work Results for Openings

08 05 05 – Selective Demolition for Openings

08 05 05.20 Selective Demolition of Windows		Crew	Daily Output	Labor-Hours	Unit	Material	2016 Bare Costs Labor	2016 Bare Costs Equipment	Total	Total Incl O&P
6000	Screening only	1 Clab	4000	.002	S.F.		.05		.05	.08
9100	Window awning, residential	"	80	.100	L.F.		2.47		2.47	4.13

08 11 Metal Doors and Frames

08 11 63 – Metal Screen and Storm Doors and Frames

08 11 63.23 Aluminum Screen and Storm Doors and Frames

		Crew	Daily Output	Labor-Hours	Unit	Material	Labor	Equipment	Total	Total Incl O&P
0010	**ALUMINUM SCREEN AND STORM DOORS AND FRAMES**									
0020	Combination storm and screen									
0420	Clear anodic coating, 2'-8" wide	2 Carp	14	1.143	Ea.	207	38.50		245.50	293
0440	3'-0" wide	"	14	1.143	"	180	38.50		218.50	262
0500	For 7' door height, add					8%				
1020	Mill finish, 2'-8" wide	2 Carp	14	1.143	Ea.	240	38.50		278.50	330
1040	3'-0" wide	"	14	1.143		263	38.50		301.50	355
1100	For 7'-0" door, add					8%				
1520	White painted, 2'-8" wide	2 Carp	14	1.143		288	38.50		326.50	380
1540	3'-0" wide		14	1.143		310	38.50		348.50	405
1541	Storm door, painted, alum., insul., 6'-8" x 2'-6" wide		14	1.143		260	38.50		298.50	350
1545	2'-8" wide		14	1.143		279	38.50		317.50	370
1600	For 7'-0" door, add					8%				
1800	Aluminum screen door, 6'-8" x 2'-8" wide	2 Carp	14	1.143		155	38.50		193.50	236
1810	3'-0" wide	"	14	1.143		260	38.50		298.50	350
2000	Wood door & screen, see Section 08 14 33.20									

08 12 Metal Frames

08 12 13 – Hollow Metal Frames

08 12 13.13 Standard Hollow Metal Frames

			Crew	Daily Output	Labor-Hours	Unit	Material	Labor	Equipment	Total	Total Incl O&P
0010	**STANDARD HOLLOW METAL FRAMES**										
0020	16 ga., up to 5-3/4" jamb depth										
0025	3'-0" x 6'-8" single	G	2 Carp	16	1	Ea.	148	34		182	220
0028	3'-6" wide, single	G		16	1		160	34		194	233
0030	4'-0" wide, single	G		16	1		159	34		193	231
0040	6'-0" wide, double	G		14	1.143		204	38.50		242.50	289
0045	8'-0" wide, double	G		14	1.143		213	38.50		251.50	299
0100	3'-0" x 7'-0" single	G		16	1		155	34		189	228
0110	3'-6" wide, single	G		16	1		165	34		199	239
0112	4'-0" wide, single	G		16	1		165	34		199	239
0140	6'-0" wide, double	G		14	1.143		194	38.50		232.50	279
0145	8'-0" wide, double	G		14	1.143		229	38.50		267.50	315
1000	16 ga., up to 4-7/8" deep, 3'-0" x 7'-0" single	G		16	1		166	34		200	240
1140	6'-0" wide, double	G		14	1.143		188	38.50		226.50	272
1200	16 ga., 8-3/4" deep, 3'-0" x 7'-0" single	G		16	1		199	34		233	276
1240	6'-0" wide, double	G		14	1.143		234	38.50		272.50	320
2800	14 ga., up to 3-7/8" deep, 3'-0" x 7'-0" single	G		16	1		181	34		215	256
2840	6'-0" wide, double	G		14	1.143		217	38.50		255.50	305
3000	14 ga., up to 5-3/4" deep, 3'-0" x 6'-8" single	G		16	1		155	34		189	228
3002	3'-6" wide, single	G		16	1		204	34		238	282
3005	4'-0" wide, single	G		16	1		210	34		244	288
3600	up to 5-3/4" jamb depth, 4'-0" x 7'-0" single	G		15	1.067		185	36		221	264
3620	6'-0" wide, double	G		12	1.333		235	45		280	335

08 12 Metal Frames

08 12 13 - Hollow Metal Frames

08 12 13.13 Standard Hollow Metal Frames

		Crew	Daily Output	Labor-Hours	Unit	Material	2016 Bare Costs Labor	Equipment	Total	Total Incl O&P
3640	8'-0" wide, double	G 2 Carp	12	1.333	Ea.	246	45		291	345
3700	8'-0" high, 4'-0" wide, single	G	15	1.067		235	36		271	320
3740	8'-0" wide, double	G	12	1.333		290	45		335	395
4000	6-3/4" deep, 4'-0" x 7'-0" single	G	15	1.067		223	36		259	305
4020	6'-0" wide, double	G	12	1.333		278	45		323	380
4040	8'-0", wide double	G	12	1.333		287	45		332	390
4100	8'-0" high, 4'-0" wide, single	G	15	1.067		276	36		312	365
4140	8'-0" wide, double	G	12	1.333		320	45		365	430
4400	8-3/4" deep, 4'-0" x 7'-0", single	G	15	1.067		252	36		288	340
4440	8'-0" wide, double	G	12	1.333		315	45		360	420
4500	4'-0" x 8'-0", single	G	15	1.067		283	36		319	370
4540	8'-0" wide, double	G	12	1.333		350	45		395	460
4900	For welded frames, add					63			63	69.50
5400	14 ga., "B" label, up to 5-3/4" deep, 4'-0" x 7'-0" single	G 2 Carp	15	1.067		211	36		247	294
5440	8'-0" wide, double	G	12	1.333		274	45		319	375
5800	6-3/4" deep, 7'-0" high, 4'-0" wide, single	G	15	1.067		218	36		254	300
5840	8'-0" wide, double	G	12	1.333		390	45		435	500
6200	8-3/4" deep, 4'-0" x 7'-0" single	G	15	1.067		296	36		332	385
6240	8'-0" wide, double	G	12	1.333		380	45		425	495
6300	For "A" label use same price as "B" label									
6400	For baked enamel finish, add					30%	15%			
6500	For galvanizing, add					20%				
6600	For hospital stop, add				Ea.	295			295	325
6620	For hospital stop, stainless steel add				"	380			380	420
7900	Transom lite frames, fixed, add	2 Carp	155	.103	S.F.	54	3.50		57.50	65.50
8000	Movable, add	"	130	.123	"	68	4.17		72.17	82

08 13 Metal Doors

08 13 13 - Hollow Metal Doors

08 13 13.15 Metal Fire Doors

		Crew	Daily Output	Labor-Hours	Unit	Material	Labor	Equipment	Total	Total Incl O&P
0010	**METAL FIRE DOORS** R081313-20									
0015	Steel, flush, "B" label, 90 minute									
0020	Full panel, 20 ga., 2'-0" x 6'-8"	2 Carp	20	.800	Ea.	420	27		447	505
0040	2'-8" x 6'-8"		18	.889		435	30		465	530
0060	3'-0" x 6'-8"		17	.941		435	32		467	535
0080	3'-0" x 7'-0"		17	.941		455	32		487	555
0140	18 ga., 3'-0" x 6'-8"		16	1		495	34		529	600
0160	2'-8" x 7'-0"		17	.941		520	32		552	625
0180	3'-0" x 7'-0"		16	1		505	34		539	610
0200	4'-0" x 7'-0"		15	1.067		650	36		686	775
0220	For "A" label, 3 hour, 18 ga., use same price as "B" label									
0240	For vision lite, add				Ea.	156			156	172
0520	Flush, "B" label 90 min., egress core, 20 ga., 2'-0" x 6'-8"	2 Carp	18	.889		655	30		685	770
0540	2'-8" x 6'-8"		17	.941		665	32		697	785
0560	3'-0" x 6'-8"		16	1		665	34		699	785
0580	3'-0" x 7'-0"		16	1		685	34		719	805
0640	Flush, "A" label 3 hour, egress core, 18 ga., 3'-0" x 6'-8"		15	1.067		720	36		756	855
0660	2'-8" x 7'-0"		16	1		750	34		784	875
0680	3'-0" x 7'-0"		15	1.067		740	36		776	875
0700	4'-0" x 7'-0"		14	1.143		885	38.50		923.50	1,025

08 13 Metal Doors

08 13 13 – Hollow Metal Doors

08 13 13.20 Residential Steel Doors		Crew	Daily Output	Labor-Hours	Unit	Material	2016 Bare Costs Labor	Equipment	Total	Total Incl O&P
0010	**RESIDENTIAL STEEL DOORS**									
0020	Prehung, insulated, exterior									
0030	Embossed, full panel, 2'-8" x 6'-8"	G	2 Carp	17	.941	Ea.	315	32		347	400
0040	3'-0" x 6'-8"	G		15	1.067		273	36		309	360
0060	3'-0" x 7'-0"	G		15	1.067		345	36		381	440
0070	5'-4" x 6'-8", double	G		8	2		630	68		698	810
0220	Half glass, 2'-8" x 6'-8"	G		17	.941		320	32		352	405
0240	3'-0" x 6'-8"	G		16	1		320	34		354	405
0260	3'-0" x 7'-0"	G		16	1		370	34		404	460
0270	5'-4" x 6'-8", double	G		8	2		650	68		718	830
1320	Flush face, full panel, 2'-8" x 6'-8"	G		16	1		264	34		298	345
1340	3'-0" x 6'-8"	G		15	1.067		264	36		300	350
1360	3'-0" x 7'-0"	G		15	1.067		296	36		332	385
1380	5'-4" x 6'-8", double	G		8	2		560	68		628	730
1420	Half glass, 2'-8" x 6'-8"	G		17	.941		325	32		357	410
1440	3'-0" x 6'-8"	G		16	1		325	34		359	410
1460	3'-0" x 7'-0"	G		16	1		400	34		434	495
1480	5'-4" x 6'-8", double	G		8	2		650	68		718	830
1500	Sidelight, full lite, 1'-0" x 6'-8" with grille	G					252			252	277
1510	1'-0" x 6'-8", low e	G					274			274	300
1520	1'-0" x 6'-8", half lite	G					280			280	310
1530	1'-0" x 6'-8", half lite, low e	G					284			284	310
2300	Interior, residential, closet, bi-fold, 2'-0" x 6'-8"	G	2 Carp	16	1		161	34		195	234
2330	3'-0" wide	G		16	1		200	34		234	277
2360	4'-0" wide	G		15	1.067		260	36		296	345
2400	5'-0" wide	G		14	1.143		315	38.50		353.50	415
2420	6'-0" wide	G		13	1.231		335	41.50		376.50	440
2510	Bi-passing closet, incl. hardware, no frame or trim incl.										
2511	Mirrored, metal frame, 4'-0" x 6'-8"		2 Carp	10	1.600	Opng.	217	54		271	330
2512	5'-0" wide			10	1.600		247	54		301	360
2513	6'-0" wide			10	1.600		281	54		335	400
2514	7'-0" wide			9	1.778		269	60.50		329.50	395
2515	8'-0" wide			9	1.778		440	60.50		500.50	585
2611	Mirrored, metal, 4'-0" x 8'-0"			10	1.600		320	54		374	445
2612	5'-0" wide			10	1.600		330	54		384	455
2613	6'-0" wide			10	1.600		385	54		439	515
2614	7'-0" wide			9	1.778		400	60.50		460.50	540
2615	8'-0" wide			9	1.778		445	60.50		505.50	585

08 14 Wood Doors

08 14 13 – Carved Wood Doors

08 14 13.10 Types of Wood Doors, Carved

		Crew	Daily Output	Labor-Hours	Unit	Material	Labor	Equipment	Total	Total Incl O&P
0010	**TYPES OF WOOD DOORS, CARVED**									
3000	Solid wood, 1-3/4" thick stile and rail									
3020	Mahogany, 3'-0" x 7'-0", six panel		2 Carp	14	1.143	Ea.	1,275	38.50		1,313.50	1,475
3030	With two lites			10	1.600		1,950	54		2,004	2,250
3040	3'-6" x 8'-0", six panel			10	1.600		1,550	54		1,604	1,800
3050	With two lites			8	2		2,600	68		2,668	2,975
3100	Pine, 3'-0" x 7'-0", six panel			14	1.143		570	38.50		608.50	690
3110	With two lites			10	1.600		825	54		879	1,000

08 14 Wood Doors

08 14 13 – Carved Wood Doors

08 14 13.10 Types of Wood Doors, Carved		Crew	Daily Output	Labor-Hours	Unit	Material	2016 Bare Costs Labor	Equipment	Total	Total Incl O&P
3120	3'-6" x 8'-0", six panel	2 Carp	10	1.600	Ea.	950	54		1,004	1,150
3130	With two lites		8	2		1,825	68		1,893	2,150
3200	Red oak, 3'-0" x 7'-0", six panel		14	1.143		1,850	38.50		1,888.50	2,100
3210	With two lites		10	1.600		2,450	54		2,504	2,800
3220	3'-6" x 8'-0", six panel		10	1.600		2,800	54		2,854	3,175
3230	With two lites		8	2		3,400	68		3,468	3,875
4000	Hand carved door, mahogany									
4020	3'-0" x 7'-0", simple design	2 Carp	14	1.143	Ea.	1,750	38.50		1,788.50	2,000
4030	Intricate design		11	1.455		3,700	49.50		3,749.50	4,150
4040	3'-6" x 8'-0", simple design		10	1.600		3,000	54		3,054	3,400
4050	Intricate design		8	2		3,700	68		3,768	4,200
4400	For custom finish, add					475			475	525
4600	Side light, mahogany, 7'-0" x 1'-6" wide, 4 lites	2 Carp	18	.889		1,100	30		1,130	1,250
4610	6 lites		14	1.143		2,625	38.50		2,663.50	2,975
4620	8'-0" x 1'-6" wide, 4 lites		14	1.143		1,800	38.50		1,838.50	2,050
4630	6 lites		10	1.600		2,100	54		2,154	2,400
4640	Side light, oak, 7'-0" x 1'-6" wide, 4 lites		18	.889		1,200	30		1,230	1,375
4650	6 lites		14	1.143		2,100	38.50		2,138.50	2,375
4660	8'-0" x 1'-6" wide, 4 lites		14	1.143		1,100	38.50		1,138.50	1,275
4670	6 lites		10	1.600		2,100	54		2,154	2,400

08 14 16 – Flush Wood Doors

08 14 16.09 Smooth Wood Doors

		Crew	Daily Output	Labor-Hours	Unit	Material	Labor	Equipment	Total	Total Incl O&P
0010	**SMOOTH WOOD DOORS**									
0015	Flush, interior, hollow core									
0025	Lauan face, 1-3/8", 3'-0" x 6'-8"	2 Carp	17	.941	Ea.	54.50	32		86.50	114
0030	4'-0" x 6'-8"		16	1		126	34		160	196
0140	Birch face, 1-3/8", 2'-6" x 6'-8"		17	.941		85	32		117	147
0180	3'-0" x 6'-8"		17	.941		95.50	32		127.50	159
0200	4'-0" x 6'-8"		16	1		158	34		192	231
0202	1-3/4", 2'-0" x 6'-8"		17	.941		57	32		89	117
0204	2'-4" x 7'-0"		16	1		121	34		155	190
0206	2'-6" x 7'-0"		16	1		125	34		159	195
0208	2'-8" x 7'-0"		16	1		131	34		165	201
0210	3'-0" x 7'-0"		16	1		138	34		172	209
0212	3'-4" x 7'-0"		15	1.067		214	36		250	296
0214	Pair of 3'-0" x 7'-0"		9	1.778	Pr.	266	60.50		326.50	395
0480	For prefinishing, clear, add				Ea.	46			46	50.50
0500	For prefinishing, stain, add				"	57			57	62.50
0620	For dutch door with shelf, add					140%				
1320	M.D. overlay on hardboard, 1-3/8", 2'-0" x 6'-8"	2 Carp	17	.941	Ea.	115	32		147	181
1340	2'-6" x 6'-8"		17	.941		115	32		147	181
1380	3'-0" x 6'-8"		17	.941		126	32		158	193
1400	4'-0" x 6'-8"		16	1		178	34		212	253
1720	H.P. plastic laminate, 1-3/8", 2'-0" x 6'-8"		16	1		260	34		294	345
1740	2'-6" x 6'-8"		16	1		260	34		294	345
1780	3'-0" x 6'-8"		15	1.067		290	36		326	380
1785	Door, plastic laminate, 3'-0" x 6'-8"					290			290	320
1800	4'-0" x 6'-8"	2 Carp	14	1.143		385	38.50		423.50	485
2020	Particle core, lauan face, 1-3/8", 2'-6" x 6'-8"		15	1.067		92	36		128	162
2040	3'-0" x 6'-8"		14	1.143		94	38.50		132.50	168
2120	Birch face, 1-3/8", 2'-6" x 6'-8"		15	1.067		103	36		139	174
2140	3'-0" x 6'-8"		14	1.143		113	38.50		151.50	189

08 14 Wood Doors

08 14 16 – Flush Wood Doors

08 14 16.09 Smooth Wood Doors

		Crew	Daily Output	Labor-Hours	Unit	Material	2016 Bare Costs Labor	Equipment	Total	Total Incl O&P
3320	M.D. overlay on hardboard, 1-3/8", 2'-6" x 6'-8"	2 Carp	14	1.143	Ea.	106	38.50		144.50	182
3340	3'-0" x 6'-8"		13	1.231		115	41.50		156.50	197
4000	Exterior, flush, solid core, birch, 1-3/4" x 2'-6" x 7'-0"		15	1.067		173	36		209	251
4020	2'-8" wide		15	1.067		159	36		195	236
4040	3'-0" wide		14	1.143		209	38.50		247.50	295
4045	3'-0" x 8'-0"	1 Carp	8	1		430	34		464	530
4100	Oak faced 1-3/4" x 2'-6" x 7'-0"	2 Carp	15	1.067		215	36		251	298
4120	2'-8" wide		15	1.067		225	36		261	310
4140	3'-0" wide		14	1.143		230	38.50		268.50	320
4160	Walnut faced, 1-3/4" x 3'-0" x 6'-8"	1 Carp	17	.471		295	15.95		310.95	350
4180	3'-6" wide	"	17	.471		365	15.95		380.95	425
4200	Walnut faced, 1-3/4" x 2'-6" x 7'-0"	2 Carp	15	1.067		310	36		346	400
4220	2'-8" wide		15	1.067		315	36		351	405
4240	3'-0" wide		14	1.143		320	38.50		358.50	415
4250	3'-6" wide	1 Carp	14	.571		420	19.35		439.35	495
4260	3'-0" x 8'-0"		8	1		375	34		409	465
4270	3'-6" wide		8	1		450	34		484	550
4285	Cherry faced, flush, sc, 1-3/4" x 3'-0" x 8'-0" wide		8	1		465	34		499	565

08 14 16.10 Wood Doors Decorator

		Crew	Daily Output	Labor-Hours	Unit	Material	Labor	Equipment	Total	Total Incl O&P
0010	**WOOD DOORS DECORATOR**									
1800	Exterior, flush, solid wood core, birch 1-3/4" x 2'-6" x 7'-0"	2 Carp	15	1.067	Ea.	320	36		356	410
1820	2'-8" wide		15	1.067		325	36		361	420
1840	3'-0" wide		14	1.143		335	38.50		373.50	435
1900	Oak faced, 1-3/4" x 2'-6" x 7'-0"		15	1.067		320	36		356	410
1920	2'-8" wide		15	1.067		330	36		366	425
1940	3'-0" wide		14	1.143		340	38.50		378.50	440
2100	Walnut faced, 1-3/4" x 2'-6" x 7'-0"		15	1.067		385	36		421	485
2120	2'-8" wide		15	1.067		400	36		436	500
2140	3'-0" wide		14	1.143		430	38.50		468.50	540

08 14 33 – Stile and Rail Wood Doors

08 14 33.10 Wood Doors Paneled

		Crew	Daily Output	Labor-Hours	Unit	Material	Labor	Equipment	Total	Total Incl O&P
0010	**WOOD DOORS PANELED**									
0020	Interior, six panel, hollow core, 1-3/8" thick									
0040	Molded hardboard, 2'-0" x 6'-8"	2 Carp	17	.941	Ea.	62	32		94	122
0060	2'-6" x 6'-8"		17	.941		64	32		96	124
0070	2'-8" x 6'-8"		17	.941		67	32		99	127
0080	3'-0" x 6'-8"		17	.941		72	32		104	133
0140	Embossed print, molded hardboard, 2'-0" x 6'-8"		17	.941		64	32		96	124
0160	2'-6" x 6'-8"		17	.941		64	32		96	124
0180	3'-0" x 6'-8"		17	.941		72	32		104	133
0540	Six panel, solid, 1-3/8" thick, pine, 2'-0" x 6'-8"		15	1.067		155	36		191	232
0560	2'-6" x 6'-8"		14	1.143		170	38.50		208.50	252
0580	3'-0" x 6'-8"		13	1.231		145	41.50		186.50	230
1020	Two panel, bored rail, solid, 1-3/8" thick, pine, 1'-6" x 6'-8"		16	1		270	34		304	355
1040	2'-0" x 6'-8"		15	1.067		355	36		391	450
1060	2'-6" x 6'-8"		14	1.143		400	38.50		438.50	505
1340	Two panel, solid, 1-3/8" thick, fir, 2'-0" x 6'-8"		15	1.067		160	36		196	237
1360	2'-6" x 6'-8"		14	1.143		210	38.50		248.50	296
1380	3'-0" x 6'-8"		13	1.231		415	41.50		456.50	525
1740	Five panel, solid, 1-3/8" thick, fir, 2'-0" x 6'-8"		15	1.067		280	36		316	370
1760	2'-6" x 6'-8"		14	1.143		420	38.50		458.50	525
1780	3'-0" x 6'-8"		13	1.231		420	41.50		461.50	530

08 14 Wood Doors

08 14 33 - Stile and Rail Wood Doors

08 14 33.10 Wood Doors Paneled

		Crew	Daily Output	Labor-Hours	Unit	Material	2016 Bare Costs Labor	2016 Bare Costs Equipment	Total	Total Incl O&P
4190	Exterior, Knotty pine, paneled, 1-3/4", 3'-0" x 6'-8"	2 Carp	16	1	Ea.	780	34		814	915
4195	Double 1-3/4", 3'-0" x 6'-8"		16	1		1,550	34		1,584	1,775
4200	Ash, paneled, 1-3/4", 3'-0" x 6'-8"		16	1		890	34		924	1,025
4205	Double 1-3/4", 3'-0" x 6'-8"		16	1		1,775	34		1,809	2,000
4210	Cherry, paneled, 1-3/4", 3'-0" x 6'-8"		16	1		890	34		924	1,025
4215	Double 1-3/4", 3'-0" x 6'-8"		16	1		1,775	34		1,809	2,000
4230	Ash, paneled, 1-3/4", 3'-0" x 8'-0"		16	1		950	34		984	1,100
4235	Double 1-3/4", 3'-0" x 8'-0"		16	1		1,900	34		1,934	2,150
4240	Hard Maple, paneled, 1-3/4", 3'-0" x 8'-0"		16	1		950	34		984	1,100
4245	Double 1-3/4", 3'-0" x 8'-0"		16	1		1,900	34		1,934	2,150
4250	Cherry, paneled, 1-3/4", 3'-0" x 8'-0"		16	1		1,000	34		1,034	1,150
4255	Double 1-3/4", 3'-0" x 8'-0"		16	1		2,000	34		2,034	2,250

08 14 33.20 Wood Doors Residential

		Crew	Daily Output	Labor-Hours	Unit	Material	Labor	Equipment	Total	Total Incl O&P
0010	**WOOD DOORS RESIDENTIAL**									
0200	Exterior, combination storm & screen, pine									
0260	2'-8" wide	2 Carp	10	1.600	Ea.	300	54		354	420
0280	3'-0" wide		9	1.778		310	60.50		370.50	440
0300	7'-1" x 3'-0" wide		9	1.778		365	60.50		425.50	500
0400	Full lite, 6'-9" x 2'-6" wide		11	1.455		320	49.50		369.50	435
0420	2'-8" wide		10	1.600		320	54		374	440
0440	3'-0" wide		9	1.778		325	60.50		385.50	455
0500	7'-1" x 3'-0" wide		9	1.778		355	60.50		415.50	490
0604	Door, screen, plain full		12	1.333		375	45		420	485
0614	Divided		12	1.333		405	45		450	520
0634	Decor full		12	1.333		460	45		505	580
0700	Dutch door, pine, 1-3/4" x 2'-8" x 6'-8", 6 panel		12	1.333		825	45		870	985
0720	Half glass		10	1.600		900	54		954	1,075
0800	3'-0" wide, 6 panel		12	1.333		790	45		835	940
0820	Half glass		10	1.600		975	54		1,029	1,175
1000	Entrance door, colonial, 1-3/4" x 6'-8" x 2'-8" wide		16	1		550	34		584	660
1020	6 panel pine, 3'-0" wide		15	1.067		445	36		481	550
1100	8 panel pine, 2'-8" wide		16	1		600	34		634	715
1120	3'-0" wide		15	1.067		570	36		606	685
1200	For tempered safety glass lites, (min of 2) add					79			79	87
1300	Flush, birch, solid core, 1-3/4" x 6'-8" x 2'-8" wide	2 Carp	16	1		113	34		147	182
1320	3'-0" wide		15	1.067		119	36		155	191
1350	7'-0" x 2'-8" wide		16	1		123	34		157	192
1360	3'-0" wide		15	1.067		143	36		179	218
1420	6'-8" x 3'-0" wide, fir		16	1		470	34		504	570
1720	Carved mahogany 3'-0" x 6'-8"		15	1.067		1,375	36		1,411	1,575
1760	Mahogany, 3'-0" x 6'-8"		15	1.067		700	36		736	830
1930	For dutch door with shelf, add					140%				
2700	Interior, closet, bi-fold, w/hardware, no frame or trim incl.									
2720	Flush, birch, 2'-6" x 6'-8"	2 Carp	13	1.231	Ea.	67.50	41.50		109	145
2740	3'-0" wide		13	1.231		71.50	41.50		113	149
2760	4'-0" wide		12	1.333		109	45		154	196
2780	5'-0" wide		11	1.455		105	49.50		154.50	199
2800	6'-0" wide		10	1.600		128	54		182	232
2804	Flush lauan 2'-0" x 6'-8"		14	1.143		52.50	38.50		91	123
2810	8'-0" wide		9	1.778		195	60.50		255.50	315
2817	6'-0" wide		9	1.778		120	60.50		180.50	233
2820	Flush, hardboard, primed, 6'-8" x 2'-6" wide		13	1.231		67.50	41.50		109	145

08 14 Wood Doors

08 14 33 – Stile and Rail Wood Doors

08 14 33.20 Wood Doors Residential		Crew	Daily Output	Labor-Hours	Unit	Material	2016 Bare Costs Labor	Equipment	Total	Total Incl O&P
2840	3'-0" wide	2 Carp	13	1.231	Ea.	75	41.50		116.50	153
2860	4'-0" wide		12	1.333		150	45		195	241
2880	5'-0" wide		11	1.455		158	49.50		207.50	256
2900	6'-0" wide		10	1.600		159	54		213	266
2920	Hardboard, primed 7'-0" x 4'-0", wide		12	1.333		182	45		227	276
2930	6'-0" wide		10	1.600		177	54		231	286
3000	Raised panel pine, 6'-6" or 6'-8" x 2'-6" wide		13	1.231		205	41.50		246.50	295
3020	3'-0" wide		13	1.231		282	41.50		323.50	380
3040	4'-0" wide		12	1.333		305	45		350	410
3060	5'-0" wide		11	1.455		365	49.50		414.50	485
3080	6'-0" wide		10	1.600		400	54		454	530
3180	Louvered, pine, 6'-6" or 6'-8" x 1'-6" wide		13	1.231		184	41.50		225.50	273
3190	2'-0" wide		14	1.143		158	38.50		196.50	238
3200	Louvered, pine 6'-6" or 6'-8" x 2'-6" wide		13	1.231		144	41.50		185.50	228
3220	3'-0" wide		13	1.231		210	41.50		251.50	300
3225	Door, interior louvered bi-fold, pine, 3'-0" x 6'-8"					210			210	231
3240	4'-0" wide	2 Carp	12	1.333		235	45		280	335
3260	5'-0" wide		11	1.455		262	49.50		311.50	370
3280	6'-0" wide		10	1.600		289	54		343	410
3290	8'-0" wide		10	1.600		445	54		499	580
3300	7'-0" x 3'-0" wide		12	1.333		269	45		314	370
3320	6'-0" wide		10	1.600		335	54		389	455
4400	Bi-passing closet, incl. hardware and frame, no trim incl.									
4420	Flush, lauan, 6'-8" x 4'-0" wide	2 Carp	12	1.333	Opng.	176	45		221	270
4440	5'-0" wide		11	1.455		187	49.50		236.50	288
4460	6'-0" wide		10	1.600		209	54		263	320
4600	Flush, birch, 6'-8" x 4'-0" wide		12	1.333		242	45		287	340
4620	5'-0" wide		11	1.455		212	49.50		261.50	315
4640	6'-0" wide		10	1.600		283	54		337	400
4800	Louvered, pine, 6'-8" x 4'-0" wide		12	1.333		485	45		530	605
4820	5'-0" wide		11	1.455		525	49.50		574.50	665
4840	6'-0" wide		10	1.600		625	54		679	780
4900	Mirrored, 6'-8" x 4'-0" wide		12	1.333	Ea.	300	45		345	405
5000	Paneled, pine, 6'-8" x 4'-0" wide		12	1.333	Opng.	470	45		515	595
5020	5'-0" wide		11	1.455		545	49.50		594.50	685
5040	6'-0" wide		10	1.600		730	54		784	895
5042	8'-0" wide		12	1.333		940	45		985	1,100
5061	Hardboard, 6'-8" x 4'-0" wide		10	1.600		190	54		244	300
5062	5'-0" wide		10	1.600		196	54		250	305
5063	6'-0" wide		10	1.600		217	54		271	330
6100	Folding accordion, closet, including track and frame									
6200	Rigid PVC	2 Carp	10	1.600	Ea.	55.50	54		109.50	152
7310	Passage doors, flush, no frame included									
7320	Hardboard, hollow core, 1-3/8" x 6'-8" x 1'-6" wide	2 Carp	18	.889	Ea.	40.50	30		70.50	95
7330	2'-0" wide		18	.889		43	30		73	98
7340	2'-6" wide		18	.889		47.50	30		77.50	103
7350	2'-8" wide		18	.889		49	30		79	105
7360	3'-0" wide		17	.941		52	32		84	111
7420	Lauan, hollow core, 1-3/8" x 6'-8" x 1'-6" wide		18	.889		35.50	30		65.50	89.50
7440	2'-0" wide		18	.889		35	30		65	89
7450	2'-4" wide		18	.889		38.50	30		68.50	93
7460	2'-6" wide		18	.889		38.50	30		68.50	93
7480	2'-8" wide		18	.889		40	30		70	94.50

08 14 Wood Doors

08 14 33 – Stile and Rail Wood Doors

08 14 33.20 Wood Doors Residential		Crew	Daily Output	Labor-Hours	Unit	Material	2016 Bare Costs Labor	2016 Bare Costs Equipment	Total	Total Incl O&P
7500	3'-0" wide	2 Carp	17	.941	Ea.	42.50	32		74.50	100
7540	2'-6" wide		16	1		61.50	34		95.50	125
7560	2'-8" wide		16	1		65.50	34		99.50	129
7580	3'-0" wide		16	1		68.50	34		102.50	132
7595	Pair of 3'-0" wide		9	1.778	Pr.	137	60.50		197.50	252
7700	Birch, hollow core, 1-3/8" x 6'-8" x 1'-6" wide		18	.889	Ea.	41	30		71	96
7720	2'-0" wide		18	.889		42.50	30		72.50	97
7740	2'-6" wide		18	.889		50	30		80	106
7760	2'-8" wide		18	.889		50.50	30		80.50	106
7780	3'-0" wide		17	.941		52	32		84	111
7790	2'-6" ash/oak door with hinges		18	.889		90	30		120	150
7910	2'-8" wide		16	1		82	34		116	147
7920	3'-0" wide		16	1		88.50	34		122.50	155
7940	Pair of 3'-0" wide		9	1.778	Pr.	185	60.50		245.50	305
8000	Pine louvered, 1-3/8" x 6'-8" x 1'-6" wide		19	.842	Ea.	120	28.50		148.50	180
8020	2'-0" wide		18	.889		131	30		161	195
8040	2'-6" wide		18	.889		150	30		180	216
8060	2'-8" wide		18	.889		161	30		191	228
8080	3'-0" wide		17	.941		175	32		207	246
8300	Pine paneled, 1-3/8" x 6'-8" x 1'-6" wide		19	.842		128	28.50		156.50	189
8320	2'-0" wide		18	.889		153	30		183	219
8330	2'-4" wide		18	.889		182	30		212	251
8340	2'-6" wide		18	.889		182	30		212	251
8360	2'-8" wide		18	.889		185	30		215	255
8380	3'-0" wide		17	.941		204	32		236	278
8450	French door, pine, 15 lites, 1-3/8" x 6'-8" x 2'-6" wide		18	.889		195	30		225	266
8470	2'-8" wide		18	.889		251	30		281	325
8490	3'-0" wide		17	.941		276	32		308	360
8804	Pocket door, 6 panel pine, 2'-6" x 6'-8" with frame		10.50	1.524		345	51.50		396.50	465
8814	2'-8" x 6'-8"		10.50	1.524		350	51.50		401.50	470
8824	3'-0" x 6'-8"		10.50	1.524		355	51.50		406.50	475
9000	Passage doors, flush, no frame, birch, solid core, 1-3/8" x 2'-4" x 7'-0"		16	1		118	34		152	187
9020	2'-8" wide		16	1		125	34		159	195
9040	3'-0" wide		16	1		135	34		169	206
9060	3'-4" wide		15	1.067		156	36		192	233
9080	Pair of 3'-0" wide		9	1.778	Pr.	267	60.50		327.50	395
9100	Lauan, solid core, 1-3/8" x 7'-0" x 2'-4" wide		16	1	Ea.	137	34		171	207
9120	2'-8" wide		16	1		146	34		180	218
9140	3'-0" wide		16	1		155	34		189	227
9160	3'-4" wide		15	1.067		163	36		199	240
9180	Pair of 3'-0" wide		9	1.778	Pr.	281	60.50		341.50	410
9200	Hardboard, solid core, 1-3/8" x 7'-0" x 2'-4" wide		16	1	Ea.	169	34		203	243
9220	2'-8" wide		16	1		175	34		209	250
9240	3'-0" wide		16	1		180	34		214	255
9260	3'-4" wide		15	1.067		199	36		235	280

08 14 Wood Doors

08 14 35 – Torrified Doors

08 14 35.10 Torrified Exterior Doors

		Crew	Daily Output	Labor-Hours	Unit	Material	2016 Bare Costs Labor	Equipment	Total	Total Incl O&P
0010	**TORRIFIED EXTERIOR DOORS**									
0020	Wood doors made from torrified wood, exterior.									
0030	All doors require a finish be applied, all glass is insulated									
0040	All doors require pilot holes for all fasteners									
0100	6 panel, Paint grade poplar, 1-3/4" x 3'-0"x 6'-8"	2 Carp	12	1.333	Ea.	1,075	45		1,120	1,250
0120	Half glass 3'-0"x 6'-8"	"	12	1.333		1,175	45		1,220	1,375
0200	Side lite, full glass, 1-3/4" x 1'-2" x 6'-8"					905			905	995
0220	Side lite, half glass, 1-3/4" x 1'-2" x 6'-8"					905			905	995
0300	Raised Face, 2 Panel, Paint grade poplar, 1-3/4" x 3'-0"x 7'-0"	2 Carp	12	1.333		1,275	45		1,320	1,475
0320	Side lite, raised face, half glass, 1-3/4" x 1'-2" x 7'-0"					1,050			1,050	1,175
0500	6 panel, Fir, 1-3/4" x 3'-0"x 6'-8"	2 Carp	12	1.333		1,550	45		1,595	1,775
0520	Half glass 3'-0"x 6'-8"	"	12	1.333		1,650	45		1,695	1,900
0600	Side lite, full glass, 1-3/4" x 1'-2" x 6'-8"					1,150			1,150	1,275
0620	Side lite, half glass, 1-3/4" x 1'-2" x 6'-8"					1,450			1,450	1,600
0700	6 panel, Mahogany, 1-3/4" x 3'-0"x 6'-8"	2 Carp	12	1.333		1,625	45		1,670	1,850
0800	Side lite, full glass, 1-3/4" x 1'-2" x 6'-8"					1,225			1,225	1,350
0820	Side lite, half glass, 1-3/4" x 1'-2" x 6'-8"					1,200			1,200	1,325

08 14 40 – Interior Cafe Doors

08 14 40.10 Cafe Style Doors

		Crew	Daily Output	Labor-Hours	Unit	Material	Labor	Equipment	Total	Total Incl O&P
0010	**CAFE STYLE DOORS**									
6520	Interior cafe doors, 2'-6" opening, stock, panel pine	2 Carp	16	1	Ea.	218	34		252	296
6540	3'-0" opening	"	16	1	"	240	34		274	320
6550	Louvered pine									
6560	2'-6" opening	2 Carp	16	1	Ea.	180	34		214	255
8000	3'-0" opening		16	1		193	34		227	269
8010	2'-6" opening, hardwood		16	1		264	34		298	345
8020	3'-0" opening		16	1		281	34		315	365

08 16 Composite Doors

08 16 13 – Fiberglass Doors

08 16 13.10 Entrance Doors, Fibrous Glass

			Crew	Daily Output	Labor-Hours	Unit	Material	Labor	Equipment	Total	Total Incl O&P
0010	**ENTRANCE DOORS, FIBROUS GLASS**										
0020	Exterior, fiberglass, door, 2'-8" wide x 6'-8" high	G	2 Carp	15	1.067	Ea.	270	36		306	360
0040	3'-0" wide x 6'-8" high	G		15	1.067		270	36		306	360
0060	3'-0" wide x 7'-0" high	G		15	1.067		460	36		496	565
0080	3'-0" wide x 6'-8" high, with two lites	G		15	1.067		315	36		351	405
0100	3'-0" wide x 8'-0" high, with two lites	G		15	1.067		525	36		561	640
0110	Half glass, 3'-0" wide x 6'-8" high	G		15	1.067		435	36		471	540
0120	3'-0" wide x 6'-8" high, low e	G		15	1.067		465	36		501	570
0130	3'-0" wide x 8'-0" high	G		15	1.067		585	36		621	705
0140	3'-0" wide x 8'-0" high, low e	G		15	1.067		655	36		691	780
0150	Side lights, 1'-0" wide x 6'-8" high	G					266			266	293
0160	1'-0" wide x 6'-8" high, low e	G					281			281	310
0180	1'-0" wide x 6'-8" high, full glass	G					310			310	340
0190	1'-0" wide x 6'-8" high, low e	G					340			340	370

08 16 Composite Doors

08 16 14 – French Doors

08 16 14.10 Exterior Doors With Glass Lites

		Crew	Daily Output	Labor-Hours	Unit	Material	2016 Bare Costs Labor	Equipment	Total	Total Incl O&P
0010	**EXTERIOR DOORS WITH GLASS LITES**									
0020	French, Fir, 1-3/4", 3'-0" wide x 6'-8" high	2 Carp	12	1.333	Ea.	600	45		645	735
0025	Double		12	1.333		1,200	45		1,245	1,400
0030	Maple, 1-3/4", 3'-0" wide x 6'-8" high		12	1.333		675	45		720	820
0035	Double		12	1.333		1,350	45		1,395	1,550
0040	Cherry, 1-3/4", 3'-0" wide x 6'-8" high		12	1.333		790	45		835	940
0045	Double		12	1.333		1,575	45		1,620	1,800
0100	Mahogany, 1-3/4", 3'-0" wide x 8'-0" high		10	1.600		800	54		854	970
0105	Double		10	1.600		1,600	54		1,654	1,850
0110	Fir, 1-3/4", 3'-0" wide x 8'-0" high		10	1.600		1,200	54		1,254	1,425
0115	Double		10	1.600		2,400	54		2,454	2,750
0120	Oak, 1-3/4", 3'-0" wide x 8'-0" high		10	1.600		1,825	54		1,879	2,100
0125	Double	▼	10	1.600	▼	3,650	54		3,704	4,100

08 17 Integrated Door Opening Assemblies

08 17 23 – Integrated Wood Door Opening Assemblies

08 17 23.10 Pre-Hung Doors

		Crew	Daily Output	Labor-Hours	Unit	Material	Labor	Equipment	Total	Total Incl O&P
0010	**PRE-HUNG DOORS**									
0300	Exterior, wood, comb. storm & screen, 6'-9" x 2'-6" wide	2 Carp	15	1.067	Ea.	296	36		332	385
0320	2'-8" wide		15	1.067		296	36		332	385
0340	3'-0" wide	▼	15	1.067		305	36		341	395
0360	For 7'-0" high door, add				▼	30			30	33
1600	Entrance door, flush, birch, solid core									
1620	4-5/8" solid jamb, 1-3/4" x 6'-8" x 2'-8" wide	2 Carp	16	1	Ea.	289	34		323	375
1640	3'-0" wide		16	1		375	34		409	470
1642	5-5/8" jamb	▼	16	1		325	34		359	415
1680	For 7'-0" high door, add				▼	25			25	27.50
2000	Entrance door, colonial, 6 panel pine									
2020	4-5/8" solid jamb, 1-3/4" x 6'-8" x 2'-8" wide	2 Carp	16	1	Ea.	640	34		674	760
2040	3'-0" wide	"	16	1		675	34		709	795
2060	For 7'-0" high door, add					54			54	59
2200	For 5-5/8" solid jamb, add					41.50			41.50	46
2230	French style, exterior, 1 lite, 1-3/4" x 3'-0" x 6'-8"	1 Carp	14	.571		590	19.35		609.35	685
2235	9 lites	"	14	.571		660	19.35		679.35	760
2245	15 lites	2 Carp	14	1.143	▼	640	38.50		678.50	770
2250	Double, 15 lites, 2'-0" x 6'-8", 4'-0" opening		7	2.286	Pr.	1,200	77.50		1,277.50	1,450
2260	2'-6" x 6'-8", 5'-0" opening		7	2.286		1,325	77.50		1,402.50	1,575
2280	3'-0" x 6'-8", 6'-0" opening		7	2.286	▼	1,350	77.50		1,427.50	1,600
2430	3'-0" x 7'-0", 15 lites		14	1.143	Ea.	950	38.50		988.50	1,125
2432	Two 3'-0" x 7'-0"		7	2.286	Pr.	1,975	77.50		2,052.50	2,275
2435	3'-0" x 8'-0"		14	1.143	Ea.	1,000	38.50		1,038.50	1,175
2437	Two, 3'-0" x 8'-0"	▼	7	2.286	Pr.	2,075	77.50		2,152.50	2,425
2500	Exterior, metal face, insulated, incl. jamb, brickmold and									
2520	Threshold, flush, 2'-8" x 6'-8"	2 Carp	16	1	Ea.	289	34		323	375
2550	3'-0" x 6'-8"		16	1		289	34		323	375
3500	Embossed, 6 panel, 2'-8" x 6'-8"		16	1		320	34		354	410
3550	3'-0" x 6'-8"		16	1		325	34		359	410
3600	2 narrow lites, 2'-8" x 6'-8"		16	1		296	34		330	380
3650	3'-0" x 6'-8"		16	1		300	34		334	385
3700	Half glass, 2'-8" x 6'-8"	▼	16	1		325	34		359	415

08 17 Integrated Door Opening Assemblies

08 17 23 – Integrated Wood Door Opening Assemblies

08 17 23.10 Pre-Hung Doors		Crew	Daily Output	Labor-Hours	Unit	Material	2016 Bare Costs Labor	Equipment	Total	Total Incl O&P
3750	3'-0" x 6'-8"	2 Carp	16	1	Ea.	340	34		374	430
3800	2 top lites, 2'-8" x 6'-8"		16	1		300	34		334	385
3850	3'-0" x 6'-8"	↓	16	1	↓	340	34		374	430
4000	Interior, passage door, 4-5/8" solid jamb									
4370	Pine, louvered, 2'-8"x 6'-8"	2 Carp	17	.941	Ea.	193	32		225	266
4380	3'-0"		17	.941		203	32		235	277
4400	Lauan, flush, solid core, 1-3/8" x 6'-8" x 2'-6" wide		17	.941		186	32		218	259
4420	2'-8" wide		17	.941		186	32		218	259
4440	3'-0" wide		16	1		201	34		235	278
4600	Hollow core, 1-3/8" x 6'-8" x 2'-6" wide		17	.941		125	32		157	192
4620	2'-8" wide		17	.941		125	32		157	191
4640	3'-0" wide	↓	16	1		140	34		174	211
4700	For 7'-0" high door, add					35.50			35.50	39
5000	Birch, flush, solid core, 1-3/8" x 6'-8" x 2'-6" wide	2 Carp	17	.941		275	32		307	360
5020	2'-8" wide		17	.941		201	32		233	275
5040	3'-0" wide		16	1		300	34		334	385
5200	Hollow core, 1-3/8" x 6'-8" x 2'-6" wide		17	.941		221	32		253	297
5220	2'-8" wide		17	.941		241	32		273	320
5240	3'-0" wide	↓	16	1		242	34		276	325
5280	For 7'-0" high door, add					30.50			30.50	33.50
5500	Hardboard paneled, 1-3/8" x 6'-8" x 2'-6" wide	2 Carp	17	.941		145	32		177	213
5520	2'-8" wide		17	.941		153	32		185	222
5540	3'-0" wide		16	1		150	34		184	222
6000	Pine paneled, 1-3/8" x 6'-8" x 2'-6" wide		17	.941		255	32		287	335
6020	2'-8" wide		17	.941		274	32		306	355
6040	3'-0" wide	↓	16	1		282	34		316	365
7200	Prehung, bifold, mirrored, 6'-8" x 5'-0"	1 Carp	9	.889		380	30		410	470
7220	6'-8" x 6'-0"		9	.889		380	30		410	470
7240	6'-8" x 8'-0"		6	1.333		635	45		680	770
7600	Oak, 6 panel, 1-3/4" x 6'-8" x 3'-0"		17	.471		895	15.95		910.95	1,000
8500	Pocket door frame with lauan, flush, hollow core, 1-3/8" x 3'-0" x 6'-8"	↓	17	.471	↓	223	15.95		238.95	273

08 31 Access Doors and Panels

08 31 13 – Access Doors and Frames

08 31 13.20 Bulkhead/Cellar Doors

		Crew	Daily Output	Labor-Hours	Unit	Material	Labor	Equipment	Total	Total Incl O&P
0010	**BULKHEAD/CELLAR DOORS**									
0020	Steel, not incl. sides, 44" x 62"	1 Carp	5.50	1.455	Ea.	600	49.50		649.50	745
0100	52" x 73"		5.10	1.569		850	53		903	1,025
0500	With sides and foundation plates, 57" x 45" x 24"		4.70	1.702		910	57.50		967.50	1,100
0600	42" x 49" x 51"	↓	4.30	1.860	↓	925	63		988	1,125

08 31 13.40 Kennel Doors

		Crew	Daily Output	Labor-Hours	Unit	Material	Labor	Equipment	Total	Total Incl O&P
0010	**KENNEL DOORS**									
0020	2 way, swinging type, 13" x 19" opening	2 Carp	11	1.455	Opng.	90	49.50		139.50	182
0100	17" x 29" opening		11	1.455		110	49.50		159.50	204
0200	9" x 9" opening, electronic with accessories	↓	11	1.455	↓	144	49.50		193.50	241

08 32 Sliding Glass Doors

08 32 13 – Sliding Aluminum-Framed Glass Doors

08 32 13.10 Sliding Aluminum Doors

		Crew	Daily Output	Labor-Hours	Unit	Material	2016 Bare Costs Labor	Equipment	Total	Total Incl O&P
0010	**SLIDING ALUMINUM DOORS**									
0350	Aluminum, 5/8" tempered insulated glass, 6' wide									
0400	Premium	2 Carp	4	4	Ea.	1,625	136		1,761	2,000
0450	Economy		4	4		820	136		956	1,125
0500	8' wide, premium		3	5.333		1,675	181		1,856	2,125
0550	Economy		3	5.333		1,450	181		1,631	1,875
0600	12' wide, premium		2.50	6.400		3,000	217		3,217	3,675
0650	Economy		2.50	6.400		1,550	217		1,767	2,075
4000	Aluminum, baked on enamel, temp glass, 6'-8" x 10'-0" wide		4	4		1,100	136		1,236	1,425
4020	Insulating glass, 6'-8" x 6'-0" wide		4	4		955	136		1,091	1,275
4040	8'-0" wide		3	5.333		1,125	181		1,306	1,525
4060	10'-0" wide		2	8		1,375	271		1,646	1,950
4080	Anodized, temp glass, 6'-8" x 6'-0" wide		4	4		455	136		591	725
4100	8'-0" wide		3	5.333		575	181		756	940
4120	10'-0" wide		2	8		645	271		916	1,175

08 32 19 – Sliding Wood-Framed Glass Doors

08 32 19.10 Sliding Wood Doors

		Crew	Daily Output	Labor-Hours	Unit	Material	Labor	Equipment	Total	Total Incl O&P
0010	**SLIDING WOOD DOORS**									
0020	Wood, tempered insul. glass, 6' wide, premium	2 Carp	4	4	Ea.	1,450	136		1,586	1,825
0100	Economy		4	4		1,175	136		1,311	1,500
0150	8' wide, wood, premium		3	5.333		1,850	181		2,031	2,325
0200	Economy		3	5.333		1,500	181		1,681	1,950
0235	10' wide, wood, premium		2.50	6.400		2,600	217		2,817	3,225
0240	Economy		2.50	6.400		2,200	217		2,417	2,800
0250	12' wide, wood, premium		2.50	6.400		3,000	217		3,217	3,675
0300	Economy		2.50	6.400		2,450	217		2,667	3,075

08 32 19.15 Sliding Glass Vinyl-Clad Wood Doors

			Crew	Daily Output	Labor-Hours	Unit	Material	Labor	Equipment	Total	Total Incl O&P
0010	**SLIDING GLASS VINYL-CLAD WOOD DOORS**										
0020	Glass, sliding vinyl clad, insul. glass, 6'-0" x 6'-8"	G	2 Carp	4	4	Opng.	1,500	136		1,636	1,900
0025	6'-0" x 6'-10" high	G		4	4		1,650	136		1,786	2,050
0030	6'-0" x 8'-0" high	G		4	4		1,975	136		2,111	2,400
0050	5'-0" x 6'-8" high	G		4	4		1,500	136		1,636	1,875
0100	8'-0" x 6'-10" high	G		4	4		2,025	136		2,161	2,450
0104	8'-0" x 6'-8"	G		4	4		1,925	136		2,061	2,325
0150	8'-0" x 8'-0" high	G		4	4		2,225	136		2,361	2,675
0500	4 leaf, 9'-0" x 6'-10" high	G		3	5.333		3,250	181		3,431	3,875
0550	9'-0" x 8'-0" high	G		3	5.333		3,775	181		3,956	4,450
0600	12'-0" x 6'-10" high	G		3	5.333		3,875	181		4,056	4,575

08 36 Panel Doors

08 36 13 – Sectional Doors

08 36 13.20 Residential Garage Doors

		Crew	Daily Output	Labor-Hours	Unit	Material	Labor	Equipment	Total	Total Incl O&P
0010	**RESIDENTIAL GARAGE DOORS**									
0050	Hinged, wood, custom, double door, 9' x 7'	2 Carp	4	4	Ea.	815	136		951	1,125
0070	16' x 7'		3	5.333		1,225	181		1,406	1,650
0200	Overhead, sectional, incl. hardware, fiberglass, 9' x 7', standard		5	3.200		960	108		1,068	1,225
0220	Deluxe		5	3.200		1,150	108		1,258	1,425
0300	16' x 7', standard		6	2.667		1,575	90.50		1,665.50	1,875
0320	Deluxe		6	2.667		2,150	90.50		2,240.50	2,500
0500	Hardboard, 9' x 7', standard		8	2		685	68		753	870

08 36 Panel Doors

08 36 13 – Sectional Doors

08 36 13.20 Residential Garage Doors		Crew	Daily Output	Labor-Hours	Unit	Material	2016 Bare Costs Labor	Equipment	Total	Total Incl O&P
0520	Deluxe	2 Carp	8	2	Ea.	815	68		883	1,025
0600	16' x 7', standard		6	2.667		1,225	90.50		1,315.50	1,500
0620	Deluxe		6	2.667		1,425	90.50		1,515.50	1,725
0700	Metal, 9' x 7', standard		8	2		835	68		903	1,025
0720	Deluxe		6	2.667		960	90.50		1,050.50	1,200
0800	16' x 7', standard		6	2.667		1,025	90.50		1,115.50	1,275
0820	Deluxe		5	3.200		1,400	108		1,508	1,725
0900	Wood, 9' x 7', standard		8	2		970	68		1,038	1,200
0920	Deluxe		8	2		2,150	68		2,218	2,475
1000	16' x 7', standard		6	2.667		1,625	90.50		1,715.50	1,950
1020	Deluxe		6	2.667		3,025	90.50		3,115.50	3,475
1800	Door hardware, sectional	1 Carp	4	2		350	68		418	500
1810	Door tracks only		4	2		163	68		231	294
1820	One side only		7	1.143		120	38.50		158.50	197
3000	Swing-up, including hardware, fiberglass, 9' x 7', standard	2 Carp	8	2		1,000	68		1,068	1,225
3020	Deluxe		8	2		1,100	68		1,168	1,325
3100	16' x 7', standard		6	2.667		1,250	90.50		1,340.50	1,525
3120	Deluxe		6	2.667		1,600	90.50		1,690.50	1,900
3200	Hardboard, 9' x 7', standard		8	2		550	68		618	720
3220	Deluxe		8	2		650	68		718	830
3300	16' x 7', standard		6	2.667		670	90.50		760.50	885
3320	Deluxe		6	2.667		850	90.50		940.50	1,075
3400	Metal, 9' x 7', standard		8	2		600	68		668	775
3420	Deluxe		8	2		975	68		1,043	1,200
3500	16' x 7', standard		6	2.667		800	90.50		890.50	1,025
3520	Deluxe		6	2.667		1,100	90.50		1,190.50	1,350
3600	Wood, 9' x 7', standard		8	2		700	68		768	885
3620	Deluxe		8	2		1,125	68		1,193	1,375
3700	16' x 7', standard		6	2.667		900	90.50		990.50	1,150
3720	Deluxe		6	2.667		2,100	90.50		2,190.50	2,450
3900	Door hardware only, swing up	1 Carp	4	2		148	68		216	276
3920	One side only		7	1.143		90	38.50		128.50	164
4000	For electric operator, economy, add		8	1		425	34		459	525
4100	Deluxe, including remote control		8	1		610	34		644	730
4500	For transmitter/receiver control, add to operator				Total	110			110	121
4600	Transmitters, additional				"	60			60	66
6000	Replace section, on sectional door, fiberglass, 9' x 7'	1 Carp	4	2	Ea.	640	68		708	815
6020	16' x 7'		3.50	2.286		765	77.50		842.50	970
6200	Hardboard, 9' x 7'		4	2		157	68		225	287
6220	16' x 7'		3.50	2.286		230	77.50		307.50	385
6300	Metal, 9' x 7'		4	2		200	68		268	335
6320	16' x 7'		3.50	2.286		320	77.50		397.50	480
6500	Wood, 9' x 7'		4	2		120	68		188	246
6520	16' x 7'		3.50	2.286		240	77.50		317.50	395
7010	Garage doors, row of lites					120			120	132

08 51 Metal Windows

08 51 13 – Aluminum Windows

08 51 13.20 Aluminum Windows

		Crew	Daily Output	Labor-Hours	Unit	Material	2016 Bare Costs Labor	2016 Bare Costs Equipment	Total	Total Incl O&P
0010	**ALUMINUM WINDOWS**, incl. frame and glazing, commercial grade									
1000	Stock units, casement, 3'-1" x 3'-2" opening	2 Sswk	10	1.600	Ea.	375	57		432	515
1040	Insulating glass	"	10	1.600		515	57		572	670
1050	Add for storms					120			120	132
1600	Projected, with screen, 3'-1" x 3'-2" opening	2 Sswk	10	1.600		355	57		412	495
1650	Insulating glass	"	10	1.600		390	57		447	530
1700	Add for storms					117			117	129
2000	4'-5" x 5'-3" opening	2 Sswk	8	2		400	71.50		471.50	570
2050	Insulating glass	"	8	2		475	71.50		546.50	655
2100	Add for storms					126			126	139
2500	Enamel finish windows, 3'-1" x 3'-2"	2 Sswk	10	1.600		360	57		417	500
2550	Insulating glass		10	1.600		315	57		372	450
2600	4'-5" x 5'-3"		8	2		405	71.50		476.50	575
2700	Insulating glass		8	2		415	71.50		486.50	585
3000	Single hung, 2' x 3' opening, enameled, standard glazed		10	1.600		208	57		265	335
3100	Insulating glass		10	1.600		252	57		309	380
3300	2'-8" x 6'-8" opening, standard glazed		8	2		365	71.50		436.50	530
3400	Insulating glass		8	2		475	71.50		546.50	650
3700	3'-4" x 5'-0" opening, standard glazed		9	1.778		300	63.50		363.50	445
3800	Insulating glass		9	1.778		335	63.50		398.50	480
4000	Sliding aluminum, 3' x 2' opening, standard glazed		10	1.600		217	57		274	345
4100	Insulating glass		10	1.600		232	57		289	360
4300	5' x 3' opening, standard glazed		9	1.778		330	63.50		393.50	480
4400	Insulating glass		9	1.778		385	63.50		448.50	540
4600	8' x 4' opening, standard glazed		6	2.667		350	95		445	560
4700	Insulating glass		6	2.667		565	95		660	795
5000	9' x 5' opening, standard glazed		4	4		530	143		673	845
5100	Insulating glass		4	4		850	143		993	1,200
5500	Sliding, with thermal barrier and screen, 6' x 4', 2 track		8	2		725	71.50		796.50	925
5700	4 track		8	2		910	71.50		981.50	1,125
6000	For above units with bronze finish, add					15%				
6200	For installation in concrete openings, add					8%				

08 51 13.30 Impact Resistant Aluminum Windows

		Crew	Daily Output	Labor-Hours	Unit	Material	Labor	Equipment	Total	Total Incl O&P
0010	**IMPACT RESISTANT ALUMINUM WINDOWS**, incl. frame and glazing									
0100	Single hung, impact resistant, 2'-8" x 5'-0"	2 Carp	9	1.778	Ea.	1,250	60.50		1,310.50	1,475
0120	3'-0" x 5'-0"		9	1.778		1,300	60.50		1,360.50	1,525
0130	4'-0" x 5'-0"		9	1.778		1,300	60.50		1,360.50	1,525
0250	Horizontal slider, impact resistant, 5'-5" x 5'-2"		9	1.778		1,550	60.50		1,610.50	1,800

08 51 23 – Steel Windows

08 51 23.40 Basement Utility Windows

		Crew	Daily Output	Labor-Hours	Unit	Material	Labor	Equipment	Total	Total Incl O&P
0010	**BASEMENT UTILITY WINDOWS**									
0015	1'-3" x 2'-8"	1 Carp	16	.500	Ea.	128	16.95		144.95	169
1100	1'-7" x 2'-8"		16	.500		140	16.95		156.95	183
1200	1'-11" x 2'-8"		14	.571		143	19.35		162.35	191

08 51 66 – Metal Window Screens

08 51 66.10 Screens

		Crew	Daily Output	Labor-Hours	Unit	Material	Labor	Equipment	Total	Total Incl O&P
0010	**SCREENS**									
0020	For metal sash, aluminum or bronze mesh, flat screen	2 Sswk	1200	.013	S.F.	4.30	.48		4.78	5.60
0500	Wicket screen, inside window	"	1000	.016	"	6.65	.57		7.22	8.35
0600	Residential, aluminum mesh and frame, 2' x 3'	2 Carp	32	.500	Ea.	16.30	16.95		33.25	46.50
0610	Rescreen		50	.320		13.25	10.85		24.10	33

For customer support on your Residential Cost Data, call 877.759.4771.

08 51 Metal Windows

08 51 66 – Metal Window Screens

08 51 66.10 Screens

		Crew	Daily Output	Labor-Hours	Unit	Material	2016 Bare Costs Labor	Equipment	Total	Total Incl O&P
0620	3' x 5'	2 Carp	32	.500	Ea.	55	16.95		71.95	89
0630	Rescreen		45	.356		33.50	12.05		45.55	57
0640	4' x 8'		25	.640		85	21.50		106.50	130
0650	Rescreen		40	.400		50	13.55		63.55	77.50
0660	Patio door		25	.640		200	21.50		221.50	257
0680	Rescreening		1600	.010	S.F.	2.50	.34		2.84	3.32
1000	Screens for solar louvers	2 Sswk	160	.100	"	24.50	3.57		28.07	33.50

08 52 Wood Windows

08 52 10 – Plain Wood Windows

08 52 10.10 Wood Windows

		Crew	Daily Output	Labor-Hours	Unit	Material	2016 Bare Costs Labor	Equipment	Total	Total Incl O&P
0010	**WOOD WINDOWS**, including frame, screens and grilles									
0020	Residential, stock units									
0050	Awning type, double insulated glass, 2'-10" x 1'-9" opening	2 Carp	12	1.333	Opng.	230	45		275	330
0100	2'-10" x 6'-0" opening	1 Carp	8	1		555	34		589	665
0200	4'-0" x 3'-6" single pane		10	.800		375	27		402	460
0300	6' x 5' single pane		8	1	Ea.	510	34		544	615
1000	Casement, 2'-0" x 3'-4" high	2 Carp	20	.800		264	27		291	335
1020	2'-0" x 4'-0"		18	.889		281	30		311	360
1040	2'-0" x 5'-0"		17	.941		310	32		342	395
1060	2'-0" x 6'-0"		16	1		305	34		339	395
1080	4'-0" x 3'-4"		15	1.067		590	36		626	705
1100	4'-0" x 4'-0"		15	1.067		660	36		696	790
1120	4'-0" x 5'-0"		14	1.143		745	38.50		783.50	885
1140	4'-0" x 6'-0"		12	1.333		845	45		890	1,000
1600	Casement units, 8' x 5', with screens, double insulated glass		2.50	6.400	Opng.	1,500	217		1,717	2,025
1700	Low E glass		2.50	6.400		1,625	217		1,842	2,175
2300	Casements, including screens, 2'-0" x 3'-4", dbl. insulated glass		11	1.455		275	49.50		324.50	390
2400	Low E glass		11	1.455		275	49.50		324.50	390
2600	2 lite, 4'-0" x 4'-0", double insulated glass		9	1.778		500	60.50		560.50	650
2700	Low E glass		9	1.778		515	60.50		575.50	665
2900	3 lite, 5'-2" x 5'-0", double insulated glass		7	2.286		760	77.50		837.50	965
3000	Low E glass		7	2.286		805	77.50		882.50	1,025
3200	4 lite, 7'-0" x 5'-0", double insulated glass		6	2.667		1,075	90.50		1,165.50	1,350
3300	Low E glass		6	2.667		1,150	90.50		1,240.50	1,425
3500	5 lite, 8'-6" x 5'-0", double insulated glass		5	3.200		1,450	108		1,558	1,750
3600	Low E glass		5	3.200		1,450	108		1,558	1,750
3800	For removable wood grilles, diamond pattern, add				Leaf	39			39	43
3900	Rectangular pattern, add				"	39			39	43
4000	Bow, fixed lites, 8' x 5', double insulated glass	2 Carp	3	5.333	Opng.	1,450	181		1,631	1,875
4100	Low E glass	"	3	5.333	"	1,925	181		2,106	2,400
4150	6'-0" x 5'-0"	1 Carp	8	1	Ea.	1,275	34		1,309	1,450
4300	Fixed lites, 9'-9" x 5'-0", double insulated glass	2 Carp	2	8	Opng.	995	271		1,266	1,550
4400	Low E glass	"	2	8	"	1,075	271		1,346	1,650
5000	Bow, casement, 8'-1" x 4'-8" high	3 Carp	8	3	Ea.	1,600	102		1,702	1,925
5020	9'-6" x 4'-8"		8	3		1,700	102		1,802	2,050
5040	8'-1" x 5'-1"		8	3		1,925	102		2,027	2,300
5060	9'-6" x 5'-1"		6	4		1,925	136		2,061	2,350
5080	8'-1" x 6'-0"		6	4		1,925	136		2,061	2,350
5100	9'-6" x 6'-0"		6	4		2,025	136		2,161	2,450
5800	Skylights, hatches, vents, and sky roofs, see Section 08 62 13.00									

08 52 Wood Windows

08 52 10 – Plain Wood Windows

08 52 10.20 Awning Window

		Crew	Daily Output	Labor-Hours	Unit	Material	2016 Bare Costs Labor	2016 Bare Costs Equipment	Total	Total Incl O&P
0010	**AWNING WINDOW**, Including frame, screens and grilles									
0100	34" x 22", insulated glass	1 Carp	10	.800	Ea.	271	27		298	345
0200	Low E glass		10	.800		272	27		299	345
0300	40" x 28", insulated glass		9	.889		315	30		345	395
0400	Low E Glass		9	.889		340	30		370	425
0500	48" x 36", insulated glass		8	1		470	34		504	570
0600	Low E glass		8	1		495	34		529	595

08 52 10.30 Wood Windows

		Crew	Daily Output	Labor-Hours	Unit	Material	Labor	Equipment	Total	Total Incl O&P
0010	**WOOD WINDOWS**, double hung									
0020	Including frame, double insulated glass, screens and grilles									
0040	Double hung, 2'-2" x 3'-4" high	2 Carp	15	1.067	Ea.	212	36		248	294
0060	2'-2" x 4'-4"		14	1.143		231	38.50		269.50	320
0080	2'-6" x 3'-4"		13	1.231		221	41.50		262.50	315
0100	2'-6" x 4'-0"		12	1.333		230	45		275	330
0120	2'-6" x 4'-8"		12	1.333		253	45		298	355
0140	2'-10" x 3'-4"		10	1.600		226	54		280	340
0160	2'-10" x 4'-0"		10	1.600		248	54		302	365
0180	3'-7" x 3'-4"		9	1.778		256	60.50		316.50	385
0200	3'-7" x 5'-4"		9	1.778		286	60.50		346.50	415
0220	3'-10" x 5'-4"		8	2		515	68		583	680
3800	Triple glazing for above, add					25%				

08 52 10.40 Casement Window

			Crew	Daily Output	Labor-Hours	Unit	Material	Labor	Equipment	Total	Total Incl O&P
0010	**CASEMENT WINDOW**, including frame, screen and grilles	R085216-10									
0100	2'-0" x 3'-0" H, dbl. insulated glass	G	1 Carp	10	.800	Ea.	273	27		300	345
0150	Low E glass	G		10	.800		259	27		286	330
0200	2'-0" x 4'-6" high, double insulated glass	G		9	.889		370	30		400	455
0250	Low E glass	G		9	.889		370	30		400	455
0260	Casement 4'-2" x 4'-2" double insulated glass	G		11	.727		890	24.50		914.50	1,025
0270	4'-0" x 4'-0" Low E glass	G		11	.727		535	24.50		559.50	630
0290	6'-4" x 5'-7" Low E glass	G		9	.889		1,125	30		1,155	1,300
0300	2'-4" x 6'-0" high, double insulated glass	G		8	1		440	34		474	540
0350	Low E glass	G		8	1		440	34		474	540
0522	Vinyl clad, premium, double insulated glass, 2'-0" x 3'-0"	G		10	.800		271	27		298	345
0524	2'-0" x 4'-0"	G		9	.889		315	30		345	400
0525	2'-0" x 5'-0"	G		8	1		360	34		394	455
0528	2'-0" x 6'-0"	G		8	1		380	34		414	475
0600	3'-0" x 5'-0"	G		8	1		665	34		699	785
0700	4'-0" x 3'-0"	G		8	1		730	34		764	860
0710	4'-0" x 4'-0"	G		8	1		625	34		659	740
0720	4'-8" x 4'-0"	G		8	1		690	34		724	815
0730	4'-8" x 5'-0"	G		6	1.333		790	45		835	940
0740	4'-8" x 6'-0"	G		6	1.333		880	45		925	1,050
0750	6'-0" x 4'-0"	G		6	1.333		805	45		850	960
0800	6'-0" x 5'-0"	G		6	1.333		895	45		940	1,050
0900	5'-6" x 5'-6"	G	2 Carp	15	1.067		1,425	36		1,461	1,625
2000	Bay, casement units, 8' x 5', w/screens, dbl. insul. glass			2.50	6.400	Opng.	1,625	217		1,842	2,175
2100	Low E glass			2.50	6.400	"	1,725	217		1,942	2,250
3020	Vinyl clad, premium, double insulated glass, multiple leaf units										
3080	Single unit, 1'-6" x 5'-0"	G	2 Carp	20	.800	Ea.	315	27		342	390
3100	2'-0" x 2'-0"	G		20	.800		215	27		242	282
3140	2'-0" x 2'-6"	G		20	.800		271	27		298	345
3220	2'-0" x 3'-6"	G		20	.800		273	27		300	345

08 52 Wood Windows

08 52 10 - Plain Wood Windows

08 52 10.40 Casement Window

			Crew	Daily Output	Labor-Hours	Unit	Material	2016 Bare Costs Labor	Equipment	Total	Total Incl O&P
3260	2'-0" x 4'-0"	G	2 Carp	19	.842	Ea.	315	28.50		343.50	400
3300	2'-0" x 4'-6"	G		19	.842		315	28.50		343.50	400
3340	2'-0" x 5'-0"	G		18	.889		360	30		390	450
3460	2'-4" x 3'-0"	G		20	.800		275	27		302	345
3500	2'-4" x 4'-0"	G		19	.842		340	28.50		368.50	425
3540	2'-4" x 5'-0"	G		18	.889		405	30		435	495
3700	Double unit, 2'-8" x 5'-0"	G		18	.889		570	30		600	675
3740	2'-8" x 6'-0"	G		17	.941		660	32		692	780
3840	3'-0" x 4'-6"	G		18	.889		515	30		545	620
3860	3'-0" x 5'-0"	G		17	.941		665	32		697	785
3880	3'-0" x 6'-0"	G		17	.941		715	32		747	840
3980	3'-4" x 2'-6"	G		19	.842		435	28.50		463.50	530
4000	3'-4" x 3'-0"	G		12	1.333		435	45		480	555
4030	3'-4" x 4'-0"	G		18	.889		530	30		560	635
4050	3'-4" x 5'-0"	G		12	1.333		660	45		705	800
4100	3'-4" x 6'-0"	G		11	1.455		715	49.50		764.50	870
4200	3'-6" x 3'-0"	G		18	.889		520	30		550	620
4340	4'-0" x 3'-0"	G		18	.889		490	30		520	585
4380	4'-0" x 3'-6"	G		17	.941		530	32		562	635
4420	4'-0" x 4'-0"	G		16	1		625	34		659	740
4460	4'-0" x 4'-4"	G		16	1		625	34		659	740
4540	4'-0" x 5'-0"	G		16	1		675	34		709	800
4580	4'-0" x 6'-0"	G		15	1.067		765	36		801	900
4740	4'-8" x 3'-0"	G		18	.889		555	30		585	660
4780	4'-8" x 3'-6"	G		17	.941		595	32		627	705
4820	4'-8" x 4'-0"	G		16	1		690	34		724	815
4860	4'-8" x 5'-0"	G		15	1.067		790	36		826	925
4900	4'-8" x 6'-0"	G		15	1.067		880	36		916	1,025
5060	5'-0" x 5'-0"	G		15	1.067		1,025	36		1,061	1,175
5100	Triple unit, 5'-6" x 3'-0"	G		17	.941		700	32		732	825
5140	5'-6" x 3'-6"	G		16	1		760	34		794	890
5180	5'-6" x 4'-6"	G		15	1.067		855	36		891	1,000
5220	5'-6" x 5'-6"	G		15	1.067		1,125	36		1,161	1,275
5300	6'-0" x 4'-6"	G		15	1.067		835	36		871	975
5850	5'-0" x 3'-0"	G		12	1.333		590	45		635	725
5900	5'-0" x 4'-0"	G		11	1.455		910	49.50		959.50	1,075
6100	5'-0" x 5'-6"	G		10	1.600		1,050	54		1,104	1,250
6150	5'-0" x 6'-0"	G		10	1.600		1,175	54		1,229	1,400
6200	6'-0" x 3'-0"	G		12	1.333		1,150	45		1,195	1,350
6250	6'-0" x 3'-4"	G		12	1.333		740	45		785	890
6300	6'-0" x 4'-0"	G		11	1.455		825	49.50		874.50	995
6350	6'-0" x 5'-0"	G		10	1.600		900	54		954	1,075
6400	6'-0" x 6'-0"	G		10	1.600		1,150	54		1,204	1,350
6500	Quadruple unit, 7'-0" x 4'-0"	G		9	1.778		1,100	60.50		1,160.50	1,300
6700	8'-0" x 4'-6"	G		9	1.778		1,350	60.50		1,410.50	1,600
6950	6'-8" x 4'-0"	G		10	1.600		1,075	54		1,129	1,275
7000	6'-8" x 6'-0"	G		10	1.600		1,425	54		1,479	1,675
8190	For installation, add per leaf							15%			
8200	For multiple leaf units, deduct for stationary sash										
8220	2' high					Ea.	23.50			23.50	25.50
8240	4'-6" high						26.50			26.50	29
8260	6' high						35			35	38.50

08 52 Wood Windows

08 52 10 – Plain Wood Windows

08 52 10.50 Double Hung

		Crew	Daily Output	Labor-Hours	Unit	Material	2016 Bare Costs Labor	Equipment	Total	Total Incl O&P
0010	**DOUBLE HUNG**, Including frame, screens and grilles									
0100	2'-0" x 3'-0" high, low E insul. glass [G]	1 Carp	10	.800	Ea.	210	27		237	277
0200	3'-0" x 4'-0" high, double insulated glass [G]		9	.889		279	30		309	355
0300	4'-0" x 4'-6" high, low E insulated glass [G]		8	1		320	34		354	410

08 52 10.55 Picture Window

		Crew	Daily Output	Labor-Hours	Unit	Material	Labor	Equipment	Total	Total Incl O&P
0010	**PICTURE WINDOW**, Including frame and grilles									
0100	3'-6" x 4'-0" high, dbl. insulated glass	2 Carp	12	1.333	Ea.	420	45		465	540
0150	Low E glass		12	1.333		435	45		480	550
0200	4'-0" x 4'-6" high, double insulated glass		11	1.455		550	49.50		599.50	690
0250	Low E glass		11	1.455		530	49.50		579.50	670
0300	5'-0" x 4'-0" high, double insulated glass		11	1.455		580	49.50		629.50	725
0350	Low E glass		11	1.455		605	49.50		654.50	750
0400	6'-0" x 4'-6" high, double insulated glass		10	1.600		640	54		694	790
0450	Low E glass		10	1.600		640	54		694	795

08 52 10.65 Wood Sash

		Crew	Daily Output	Labor-Hours	Unit	Material	Labor	Equipment	Total	Total Incl O&P
0010	**WOOD SASH**, Including glazing but not trim									
0050	Custom, 5'-0" x 4'-0", 1" dbl. glazed, 3/16" thick lites	2 Carp	3.20	5	Ea.	230	170		400	535
0100	1/4" thick lites		5	3.200		245	108		353	450
0200	1" thick, triple glazed		5	3.200		415	108		523	635
0300	7'-0" x 4'-6" high, 1" double glazed, 3/16" thick lites		4.30	3.721		420	126		546	670
0400	1/4" thick lites		4.30	3.721		475	126		601	730
0500	1" thick, triple glazed		4.30	3.721		540	126		666	805
0600	8'-6" x 5'-0" high, 1" double glazed, 3/16" thick lites		3.50	4.571		565	155		720	885
0700	1/4" thick lites		3.50	4.571		620	155		775	940
0800	1" thick, triple glazed		3.50	4.571		625	155		780	945
0900	Window frames only, based on perimeter length				L.F.	4.02			4.02	4.42

08 52 10.70 Sliding Windows

		Crew	Daily Output	Labor-Hours	Unit	Material	Labor	Equipment	Total	Total Incl O&P
0010	**SLIDING WINDOWS**									
0100	3'-0" x 3'-0" high, double insulated [G]	1 Carp	10	.800	Ea.	284	27		311	355
0120	Low E glass [G]		10	.800		310	27		337	385
0200	4'-0" x 3'-6" high, double insulated [G]		9	.889		355	30		385	440
0220	Low E glass [G]		9	.889		360	30		390	450
0300	6'-0" x 5'-0" high, double insulated [G]		8	1		490	34		524	590
0320	Low E glass [G]		8	1		530	34		564	635
6000	Sliding, insulating glass, including screens,									
6100	3'-0" x 3'-0"	2 Carp	6.50	2.462	Ea.	330	83.50		413.50	505
6200	4'-0" x 3'-6"		6.30	2.540		335	86		421	515
6300	5'-0" x 4'-0"		6	2.667		415	90.50		505.50	610

08 52 13 – Metal-Clad Wood Windows

08 52 13.10 Awning Windows, Metal-Clad

		Crew	Daily Output	Labor-Hours	Unit	Material	Labor	Equipment	Total	Total Incl O&P
0010	**AWNING WINDOWS, METAL-CLAD**									
2000	Metal clad, awning deluxe, double insulated glass, 34" x 22"	1 Carp	9	.889	Ea.	249	30		279	325
2050	36" x 25"		9	.889		276	30		306	355
2100	40" x 22"		9	.889		292	30		322	370
2150	40" x 30"		9	.889		340	30		370	425
2200	48" x 28"		8	1		350	34		384	440
2250	60" x 36"		8	1		375	34		409	465

08 52 Wood Windows

08 52 13 – Metal-Clad Wood Windows

08 52 13.20 Casement Windows, Metal-Clad

		Crew	Daily Output	Labor-Hours	Unit	Material	2016 Bare Costs Labor	Equipment	Total	Total Incl O&P
0010	**CASEMENT WINDOWS, METAL-CLAD**									
0100	Metal clad, deluxe, dbl. insul. glass, 2'-0" x 3'-0" high	G	1 Carp	10	.800	Ea.	285	27		312	360
0120	2'-0" x 4'-0" high	G		9	.889		320	30		350	400
0130	2'-0" x 5'-0" high	G		8	1		330	34		364	420
0140	2'-0" x 6'-0" high	G		8	1		370	34		404	460
0150	Casement window, metal clad, double insul. glass, 3'-6" x 3'-6"	G		8.90	.899		490	30.50		520.50	590
0300	Metal clad, casement, bldrs mdl, 6'-0" x 4'-0", dbl. insltd gls, 3 panels		2 Carp	10	1.600		1,200	54		1,254	1,425
0310	9'-0" x 4'-0", 4 panels			8	2		1,550	68		1,618	1,825
0320	10'-0" x 5'-0", 5 panels			7	2.286		2,100	77.50		2,177.50	2,450
0330	12'-0" x 6'-0", 6 panels			6	2.667		2,700	90.50		2,790.50	3,100

08 52 13.30 Double-Hung Windows, Metal-clad

		Crew	Daily Output	Labor-Hours	Unit	Material	Labor	Equipment	Total	Total Incl O&P
0010	**DOUBLE-HUNG WINDOWS, METAL-CLAD**									
0100	Metal clad, deluxe, dbl. insul. glass, 2'-6" x 3'-0" high	G	1 Carp	10	.800	Ea.	280	27		307	355
0120	3'-0" x 3'-6" high	G		10	.800		320	27		347	395
0140	3'-0" x 4'-0" high	G		9	.889		335	30		365	420
0160	3'-0" x 4'-6" high	G		9	.889		350	30		380	435
0180	3'-0" x 5'-0" high	G		8	1		380	34		414	470
0200	3'-6" x 6'-0" high	G		8	1		455	34		489	560

08 52 13.35 Picture and Sliding Windows Metal-Clad

		Crew	Daily Output	Labor-Hours	Unit	Material	Labor	Equipment	Total	Total Incl O&P
0010	**PICTURE AND SLIDING WINDOWS METAL-CLAD**									
2000	Metal clad, dlx picture, dbl. insul. glass, 4'-0" x 4'-0" high		2 Carp	12	1.333	Ea.	380	45		425	495
2100	4'-0" x 6'-0" high			11	1.455		560	49.50		609.50	700
2200	5'-0" x 6'-0" high			10	1.600		620	54		674	770
2300	6'-0" x 6'-0" high			10	1.600		710	54		764	870
2400	Metal clad, dlx sliding, double insulated glass, 3'-0" x 3'-0" high	G	1 Carp	10	.800		330	27		357	405
2420	4'-0" x 3'-6" high	G		9	.889		400	30		430	490
2440	5'-0" x 4'-0" high	G		9	.889		480	30		510	575
2460	6'-0" x 5'-0" high	G		8	1		730	34		764	855

08 52 13.40 Bow and Bay Windows, Metal-Clad

		Crew	Daily Output	Labor-Hours	Unit	Material	Labor	Equipment	Total	Total Incl O&P
0010	**BOW AND BAY WINDOWS, METAL-CLAD**									
0100	Metal clad, deluxe, dbl. insul. glass, 8'-0" x 5'-0" high, 4 panels		2 Carp	10	1.600	Ea.	1,700	54		1,754	1,975
0120	10'-0" x 5'-0" high, 5 panels			8	2		1,825	68		1,893	2,125
0140	10'-0" x 6'-0" high, 5 panels			7	2.286		2,150	77.50		2,227.50	2,500
0160	12'-0" x 6'-0" high, 6 panels			6	2.667		2,925	90.50		3,015.50	3,375
0400	Double hung, bldrs. model, bay, 8' x 4' high, dbl. insulated glass			10	1.600		1,350	54		1,404	1,575
0440	Low E glass			10	1.600		1,450	54		1,504	1,700
0460	9'-0" x 5'-0" high, double insulated glass			6	2.667		1,450	90.50		1,540.50	1,750
0480	Low E glass			6	2.667		1,525	90.50		1,615.50	1,825
0500	Metal clad, deluxe, dbl. insul. glass, 7'-0" x 4'-0" high			10	1.600		1,300	54		1,354	1,525
0520	8'-0" x 4'-0" high			8	2		1,350	68		1,418	1,600
0540	8'-0" x 5'-0" high			7	2.286		1,375	77.50		1,452.50	1,650
0560	9'-0" x 5'-0" high			6	2.667		1,475	90.50		1,565.50	1,775

08 52 16 – Plastic-Clad Wood Windows

08 52 16.10 Bow Window

		Crew	Daily Output	Labor-Hours	Unit	Material	Labor	Equipment	Total	Total Incl O&P
0010	**BOW WINDOW** Including frames, screens, and grilles									
0020	End panels operable									
1000	Bow type, casement, wood, bldrs. mdl., 8' x 5' dbl. insltd glass, 4 panel		2 Carp	10	1.600	Ea.	1,500	54		1,554	1,750
1050	Low E glass			10	1.600		1,325	54		1,379	1,550
1100	10'-0" x 5'-0", double insulated glass, 6 panels			6	2.667		1,350	90.50		1,440.50	1,650
1200	Low E glass, 6 panels			6	2.667		1,450	90.50		1,540.50	1,750
1300	Vinyl clad, bldrs. model, double insulated glass, 6'-0" x 4'-0", 3 panel			10	1.600		1,025	54		1,079	1,225

08 52 Wood Windows

08 52 16 – Plastic-Clad Wood Windows

08 52 16.10 Bow Window

		Crew	Daily Output	Labor-Hours	Unit	Material	2016 Bare Costs Labor	Equipment	Total	Total Incl O&P
1340	9'-0" x 4'-0", 4 panel	2 Carp	8	2	Ea.	1,350	68		1,418	1,625
1380	10'-0" x 6'-0", 5 panels		7	2.286		2,250	77.50		2,327.50	2,600
1420	12'-0" x 6'-0", 6 panels		6	2.667		2,925	90.50		3,015.50	3,375
2000	Bay window, 8' x 5', dbl. insul glass		10	1.600		1,925	54		1,979	2,200
2050	Low E glass		10	1.600		2,325	54		2,379	2,650
2100	12'-0" x 6'-0", double insulated glass, 6 panels		6	2.667		2,400	90.50		2,490.50	2,775
2200	Low E glass		6	2.667		3,175	90.50		3,265.50	3,650
2280	6'-0" x 4'-0"		11	1.455		1,250	49.50		1,299.50	1,450
2300	Vinyl clad, premium, double insulated glass, 8'-0" x 5'-0"		10	1.600		1,750	54		1,804	2,025
2340	10'-0" x 5'-0"		8	2		2,300	68		2,368	2,650
2380	10'-0" x 6'-0"		7	2.286		2,625	77.50		2,702.50	3,000
2420	12'-0" x 6'-0"		6	2.667		3,200	90.50		3,290.50	3,675
2430	14'-0" x 3'-0"		7	2.286		2,950	77.50		3,027.50	3,375
2440	14'-0" x 6'-0"		5	3.200		5,000	108		5,108	5,675
3300	Vinyl clad, premium, double insulated glass, 7'-0" x 4'-6"		10	1.600		1,400	54		1,454	1,625
3340	8'-0" x 4'-6"		8	2		1,425	68		1,493	1,675
3380	8'-0" x 5'-0"		7	2.286		1,500	77.50		1,577.50	1,775
3420	9'-0" x 5'-0"		6	2.667		1,525	90.50		1,615.50	1,825

08 52 16.15 Awning Window Vinyl-Clad

		Crew	Daily Output	Labor-Hours	Unit	Material	Labor	Equipment	Total	Total Incl O&P
0010	**AWNING WINDOW VINYL-CLAD** Including frames, screens, and grilles									
0200	Vinyl clad, premium, double insulated glass, 24" x 17"	1 Carp	12	.667	Ea.	212	22.50		234.50	272
0210	24" x 28"		11	.727		257	24.50		281.50	325
0250	36" x 17"		9	.889		263	30		293	340
0280	36" x 28"		9	.889		305	30		335	385
0300	36" x 36"		9	.889		345	30		375	425
0320	36" x 40"		9	.889		395	30		425	485
0340	40" x 22"		10	.800		291	27		318	365
0360	48" x 28"		8	1		370	34		404	460
0380	60" x 36"		8	1		500	34		534	605

08 52 16.20 Half Round, Vinyl Clad

		Crew	Daily Output	Labor-Hours	Unit	Material	Labor	Equipment	Total	Total Incl O&P
0010	**HALF ROUND, VINYL CLAD**, double insulated glass, incl. grille									
0800	14" height x 24" base	2 Carp	9	1.778	Ea.	415	60.50		475.50	555
1040	15" height x 25" base		8	2		385	68		453	535
1060	16" height x 28" base		7	2.286		405	77.50		482.50	575
1080	17" height x 29" base		7	2.286		420	77.50		497.50	590
2000	19" height x 33" base	1 Carp	6	1.333		465	45		510	585
2100	20" height x 35" base		6	1.333		465	45		510	585
2200	21" height x 37" base		6	1.333		480	45		525	605
2250	23" height x 41" base	2 Carp	6	2.667		520	90.50		610.50	720
2300	26" height x 48" base		6	2.667		535	90.50		625.50	735
2350	30" height x 56" base		6	2.667		615	90.50		705.50	825
3000	36" height x 67" base	1 Carp	4	2		1,050	68		1,118	1,275
3040	38" height x 71" base	2 Carp	5	3.200		975	108		1,083	1,250
3050	40" height x 75" base	"	5	3.200		1,275	108		1,383	1,575
5000	Elliptical, 71" x 16"	1 Carp	11	.727		975	24.50		999.50	1,125
5100	95" x 21"	"	10	.800		1,375	27		1,402	1,575

08 52 16.30 Palladian Windows

		Crew	Daily Output	Labor-Hours	Unit	Material	Labor	Equipment	Total	Total Incl O&P
0010	**PALLADIAN WINDOWS**									
0020	Vinyl clad, double insulated glass, including frame and grilles									
0040	3'-2" x 2'-6" high	2 Carp	11	1.455	Ea.	1,250	49.50		1,299.50	1,450
0060	3'-2" x 4'-10"		11	1.455		1,750	49.50		1,799.50	1,975
0080	3'-2" x 6'-4"		10	1.600		1,700	54		1,754	1,975

08 52 Wood Windows

08 52 16 – Plastic-Clad Wood Windows

08 52 16.30 Palladian Windows

		Crew	Daily Output	Labor-Hours	Unit	Material	2016 Bare Costs Labor	Equipment	Total	Total Incl O&P
0100	4'-0" x 4'-0"	2 Carp	10	1.600	Ea.	1,525	54		1,579	1,775
0120	4'-0" x 5'-4"	3 Carp	10	2.400		1,875	81.50		1,956.50	2,175
0140	4'-0" x 6'-0"		9	2.667		1,950	90.50		2,040.50	2,300
0160	4'-0" x 7'-4"		9	2.667		2,125	90.50		2,215.50	2,475
0180	5'-5" x 4'-10"		9	2.667		2,275	90.50		2,365.50	2,650
0200	5'-5" x 6'-10"		9	2.667		2,575	90.50		2,665.50	3,000
0220	5'-5" x 7'-9"		9	2.667		2,800	90.50		2,890.50	3,225
0240	6'-0" x 7'-11"		8	3		3,475	102		3,577	4,000
0260	8'-0" x 6'-0"		8	3		3,075	102		3,177	3,550

08 52 16.35 Double-Hung Window

			Crew	Daily Output	Labor-Hours	Unit	Material	Labor	Equipment	Total	Total Incl O&P
0010	**DOUBLE-HUNG WINDOW** Including frames, screens, and grilles										
0300	Vinyl clad, premium, double insulated glass, 2'-6" x 3'-0"	G	1 Carp	10	.800	Ea.	310	27		337	385
0305	2'-6" x 4'-0"	G		10	.800		360	27		387	440
0400	3'-0" x 3'-6"	G		10	.800		335	27		362	415
0500	3'-0" x 4'-0"	G		9	.889		395	30		425	485
0600	3'-0" x 4'-6"	G		9	.889		410	30		440	500
0700	3'-0" x 5'-0"	G		8	1		445	34		479	540
0790	3'-4" x 5'-0"	G		8	1		455	34		489	555
0800	3'-6" x 6'-0"	G		8	1		490	34		524	595
0820	4'-0" x 5'-0"	G		7	1.143		560	38.50		598.50	680
0830	4'-0" x 6'-0"	G		7	1.143		700	38.50		738.50	830

08 52 16.40 Transom Windows

		Crew	Daily Output	Labor-Hours	Unit	Material	Labor	Equipment	Total	Total Incl O&P
0010	**TRANSOM WINDOWS**									
0050	Vinyl clad, premium, double insulated glass, 32" x 8"	1 Carp	16	.500	Ea.	188	16.95		204.95	235
0100	36" x 8"		16	.500		199	16.95		215.95	248
0110	36" x 12"		16	.500		212	16.95		228.95	262
1000	Vinyl clad, premium, dbl. insul. glass, 4'-0" x 4'-0"	2 Carp	12	1.333		520	45		565	645
1100	4'-0" x 6'-0"		11	1.455		955	49.50		1,004.50	1,125
1200	5'-0" x 6'-0"		10	1.600		1,050	54		1,104	1,275
1300	6'-0" x 6'-0"		10	1.600		1,125	54		1,179	1,350

08 52 16.45 Trapezoid Windows

		Crew	Daily Output	Labor-Hours	Unit	Material	Labor	Equipment	Total	Total Incl O&P
0010	**TRAPEZOID WINDOWS**									
0100	Vinyl clad, including frame and exterior trim									
0900	20" base x 44" leg x 53" leg	2 Carp	13	1.231	Ea.	390	41.50		431.50	500
1000	24" base x 90" leg x 102" leg		8	2		670	68		738	855
3000	36" base x 40" leg x 22" leg		12	1.333		430	45		475	545
3010	36" base x 44" leg x 25" leg		13	1.231		450	41.50		491.50	565
3050	36" base x 26" leg x 48" leg		9	1.778		455	60.50		515.50	600
3100	36" base x 42" legs, 50" peak		9	1.778		545	60.50		605.50	700
3200	36" base x 60" leg x 81" leg		11	1.455		710	49.50		759.50	870
4320	44" base x 23" leg x 56" leg		11	1.455		560	49.50		609.50	705
4350	44" base x 59" leg x 92" leg		10	1.600		840	54		894	1,025
4500	46" base x 15" leg x 46" leg		8	2		425	68		493	580
4550	46" base x 16" leg x 48" leg		8	2		450	68		518	610
4600	46" base x 50" leg x 80" leg		7	2.286		680	77.50		757.50	880
6600	66" base x 12" leg x 42" leg		8	2		575	68		643	750
6650	66" base x 12" legs, 28" peak		9	1.778		455	60.50		515.50	600
6700	68" base x 23" legs, 31" peak		8	2		575	68		643	750

08 52 16.70 Vinyl Clad, Premium, Dbl. Insulated Glass

			Crew	Daily Output	Labor-Hours	Unit	Material	Labor	Equipment	Total	Total Incl O&P
0010	**VINYL CLAD, PREMIUM, DBL. INSULATED GLASS**										
1000	Sliding, 3'-0" x 3'-0"	G	1 Carp	10	.800	Ea.	605	27		632	710
1020	4'-0" x 1'-11"	G		11	.727		585	24.50		609.50	680

08 52 Wood Windows

08 52 16 – Plastic-Clad Wood Windows

08 52 16.70 Vinyl Clad, Premium, Dbl. Insulated Glass		Crew	Daily Output	Labor-Hours	Unit	Material	2016 Bare Costs Labor	Equipment	Total	Total Incl O&P
1040	4'-0" x 3'-0"	G 1 Carp	10	.800	Ea.	700	27		727	810
1050	4'-0" x 3'-6"	G	9	.889		685	30		715	805
1090	4'-0" x 5'-0"	G	9	.889		915	30		945	1,050
1100	5'-0" x 4'-0"	G	9	.889		900	30		930	1,050
1120	5'-0" x 5'-0"	G	8	1		1,000	34		1,034	1,150
1140	6'-0" x 4'-0"	G	8	1		1,025	34		1,059	1,175
1150	6'-0" x 5'-0"	G	8	1		1,125	34		1,159	1,300

08 52 50 – Window Accessories

08 52 50.10 Window Grille or Muntin

		Crew	Daily Output	Labor-Hours	Unit	Material	Labor	Equipment	Total	Total Incl O&P
0010	**WINDOW GRILLE OR MUNTIN**, snap in type									
0020	Standard pattern interior grilles									
2000	Wood, awning window, glass size 28" x 16" high	1 Carp	30	.267	Ea.	28	9.05		37.05	46
2060	44" x 24" high		32	.250		40	8.50		48.50	58
2100	Casement, glass size, 20" x 36" high		30	.267		32	9.05		41.05	50
2180	20" x 56" high		32	.250		43	8.50		51.50	61.50
2200	Double hung, glass size, 16" x 24" high		24	.333	Set	51	11.30		62.30	75.50
2280	32" x 32" high		34	.235	"	131	8		139	157
2500	Picture, glass size, 48" x 48" high		30	.267	Ea.	120	9.05		129.05	147
2580	60" x 68" high		28	.286	"	183	9.70		192.70	218
2600	Sliding, glass size, 14" x 36" high		24	.333	Set	35.50	11.30		46.80	58
2680	36" x 36" high		22	.364	"	43.50	12.35		55.85	68

08 52 66 – Wood Window Screens

08 52 66.10 Wood Screens

		Crew	Daily Output	Labor-Hours	Unit	Material	Labor	Equipment	Total	Total Incl O&P
0010	**WOOD SCREENS**									
0020	Over 3 S.F., 3/4" frames	2 Carp	375	.043	S.F.	4.89	1.45		6.34	7.80
0100	1-1/8" frames	"	375	.043	"	8.20	1.45		9.65	11.40

08 52 69 – Wood Storm Windows

08 52 69.10 Storm Windows

			Crew	Daily Output	Labor-Hours	Unit	Material	Labor	Equipment	Total	Total Incl O&P
0010	**STORM WINDOWS**, aluminum residential										
0300	Basement, mill finish, incl. fiberglass screen										
0320	1'-10" x 1'-0" high	G	2 Carp	30	.533	Ea.	36	18.10		54.10	70
0340	2'-9" x 1'-6" high	G		30	.533		39	18.10		57.10	73.50
0360	3'-4" x 2'-0" high	G		30	.533		46	18.10		64.10	81
1600	Double-hung, combination, storm & screen										
1700	Custom, clear anodic coating, 2'-0" x 3'-5" high		2 Carp	30	.533	Ea.	97	18.10		115.10	138
1720	2'-6" x 5'-0" high			28	.571		117	19.35		136.35	162
1740	4'-0" x 6'-0" high			25	.640		235	21.50		256.50	296
1800	White painted, 2'-0" x 3'-5" high			30	.533		112	18.10		130.10	154
1820	2'-6" x 5'-0" high			28	.571		173	19.35		192.35	223
1840	4'-0" x 6'-0" high			25	.640		286	21.50		307.50	350
2000	Clear anodic coating, 2'-0" x 3'-5" high	G		30	.533		97	18.10		115.10	138
2020	2'-6" x 5'-0" high	G		28	.571		121	19.35		140.35	166
2040	4'-0" x 6'-0" high	G		25	.640		133	21.50		154.50	183
2400	White painted, 2'-0" x 3'-5" high	G		30	.533		92	18.10		110.10	132
2420	2'-6" x 5'-0" high	G		28	.571		97	19.35		116.35	140
2440	4'-0" x 6'-0" high	G		25	.640		112	21.50		133.50	160
2600	Mill finish, 2'-0" x 3'-5" high	G		30	.533		87	18.10		105.10	126
2620	2'-6" x 5'-0" high	G		28	.571		92	19.35		111.35	134
2640	4'-0" x 6'-8" high	G		25	.640		112	21.50		133.50	160
4000	Picture window, storm, 1 lite, white or bronze finish										
4020	4'-6" x 4'-6" high		2 Carp	25	.640	Ea.	133	21.50		154.50	183

08 52 Wood Windows

08 52 69 – Wood Storm Windows

	08 52 69.10 Storm Windows	Crew	Daily Output	Labor-Hours	Unit	Material	2016 Bare Costs Labor	Equipment	Total	Total Incl O&P
4040	5'-8" x 4'-6" high	2 Carp	20	.800	Ea.	150	27		177	211
4400	Mill finish, 4'-6" x 4'-6" high		25	.640		133	21.50		154.50	183
4420	5'-8" x 4'-6" high	↓	20	.800	↓	154	27		181	215
4600	3 lite, white or bronze finish									
4620	4'-6" x 4'-6" high	2 Carp	25	.640	Ea.	163	21.50		184.50	216
4640	5'-8" x 4'-6" high		20	.800		173	27		200	236
4800	Mill finish, 4'-6" x 4'-6" high		25	.640		154	21.50		175.50	206
4820	5'-8" x 4'-6" high	↓	20	.800	↓	163	27		190	225
6000	Sliding window, storm, 2 lite, white or bronze finish									
6020	3'-4" x 2'-7" high	2 Carp	28	.571	Ea.	135	19.35		154.35	182
6040	4'-4" x 3'-3" high		25	.640		150	21.50		171.50	202
6060	5'-4" x 6'-0" high	↓	20	.800	↓	240	27		267	310
9000	Interior storm window									
9100	Storm window interior glass	1 Glaz	107	.075	S.F.	6.20	2.48		8.68	10.95

08 53 Plastic Windows

08 53 13 – Vinyl Windows

08 53 13.10 Solid Vinyl Windows

		Crew	Daily Output	Labor-Hours	Unit	Material	Labor	Equipment	Total	Total Incl O&P
0010	**SOLID VINYL WINDOWS**									
0020	Double hung, including frame and screen, 2'-0" x 2'-6"	2 Carp	15	1.067	Ea.	201	36		237	282
0040	2'-0" x 3'-6"		14	1.143		214	38.50		252.50	300
0060	2'-6" x 4'-6"		13	1.231		253	41.50		294.50	350
0080	3'-0" x 4'-0"		10	1.600		225	54		279	340
0100	3'-0" x 4'-6"		9	1.778		289	60.50		349.50	415
0120	3'-6" x 4'-6"		8	2		310	68		378	455
0140	3'-6" x 6'-0"	↓	7	2.286	↓	380	77.50		457.50	550

08 53 13.20 Vinyl Single Hung Windows

			Crew	Daily Output	Labor-Hours	Unit	Material	Labor	Equipment	Total	Total Incl O&P
0010	**VINYL SINGLE HUNG WINDOWS**, insulated glass										
0020	Grids, low e, J fin, extension jambs										
0130	25" x 41"	G	2 Carp	20	.800	Ea.	200	27		227	266
0140	25" x 49"	G		18	.889		210	30		240	282
0150	25" x 57"	G		17	.941		220	32		252	296
0160	25" x 65"	G		16	1		240	34		274	320
0170	29" x 41"	G		18	.889		205	30		235	277
0180	29" x 53"	G		18	.889		220	30		250	293
0190	29" x 57"	G		17	.941		230	32		262	305
0200	29" x 65"	G		16	1		240	34		274	320
0210	33" x 41"	G		20	.800		215	27		242	283
0220	33" x 53"	G		18	.889		230	30		260	305
0230	33" x 57"	G		17	.941		240	32		272	320
0240	33" x 65"	G		16	1		250	34		284	330
0250	37" x 41"	G		20	.800		242	27		269	310
0260	37" x 53"	G		18	.889		255	30		285	330
0270	37" x 57"	G		17	.941		265	32		297	345
0280	37" x 65"	G		16	1		290	34		324	375
0500	Vinyl clad, premium, double insulated glass, circle, 24" diameter		1 Carp	6	1.333		460	45		505	580
1000	2'-4" diameter			6	1.333		540	45		585	670
1500	2'-11" diameter		↓	6	1.333	↓	640	45		685	775

08 53 Plastic Windows

08 53 13 – Vinyl Windows

08 53 13.30 Vinyl Double Hung Windows

		Crew	Daily Output	Labor-Hours	Unit	Material	2016 Bare Costs Labor	2016 Bare Costs Equipment	Total	Total Incl O&P
0010	**VINYL DOUBLE HUNG WINDOWS**, insulated glass									
0100	Grids, low E, J fin, ext. jambs, 21" x 53" G	2 Carp	18	.889	Ea.	215	30		245	288
0102	21" x 37" G		18	.889		223	30		253	296
0104	21" x 41" G		18	.889		232	30		262	305
0106	21" x 49" G		18	.889		247	30		277	325
0110	21" x 57" G		17	.941		263	32		295	345
0120	21" x 65" G		16	1		279	34		313	360
0128	25" x 37" G		20	.800		234	27		261	305
0130	25" x 41" G		20	.800		242	27		269	310
0140	25" x 49" G		18	.889		258	30		288	335
0145	25" x 53" G		18	.889		265	30		295	345
0150	25" x 57" G		17	.941		274	32		306	355
0160	25" x 65" G		16	1		290	34		324	375
0162	25" x 69" G		16	1		297	34		331	380
0164	25" x 77" G		16	1		325	34		359	410
0168	29" x 37" G		18	.889		242	30		272	315
0170	29" x 41" G		18	.889		250	30		280	325
0172	29" x 49" G		18	.889		267	30		297	345
0180	29" x 53" G		18	.889		271	30		301	350
0190	29" x 57" G		17	.941		280	32		312	365
0200	29" x 65" G		16	1		300	34		334	385
0202	29" x 69" G		16	1		310	34		344	395
0205	29" x 77" G		16	1		335	34		369	420
0208	33" x 37" G		20	.800		255	27		282	325
0210	33" x 41" G		20	.800		263	27		290	335
0215	33" x 49" G		20	.800		280	27		307	355
0220	33" x 53" G		18	.889		285	30		315	365
0230	33" x 57" G		17	.941		297	32		329	380
0240	33" x 65" G		16	1		315	34		349	400
0242	33" x 69" G		16	1		330	34		364	415
0246	33" x 77" G		16	1		350	34		384	440
0250	37" x 41" G		20	.800		289	27		316	365
0255	37" x 49" G		20	.800		282	27		309	355
0260	37" x 53" G		18	.889		315	30		345	395
0270	37" x 57" G		17	.941		325	32		357	415
0280	37" x 65" G		16	1		350	34		384	440
0282	37" x 69" G		16	1		360	34		394	450
0286	37" x 77" G		16	1		380	34		414	470
0300	Solid vinyl, average quality, double insulated glass, 2'-0" x 3'-0" G	1 Carp	10	.800		291	27		318	365
0310	3'-0" x 4'-0" G		9	.889		209	30		239	281
0330	Premium, double insulated glass, 2'-6" x 3'-0" G		10	.800		271	27		298	345
0340	3'-0" x 3'-6" G		9	.889		305	30		335	385
0350	3'-0" x 4'-0" G		9	.889		325	30		355	405
0360	3'-0" x 4'-6" G		9	.889		335	30		365	420
0370	3'-0" x 5'-0" G		8	1		360	34		394	450
0380	3'-6" x 6'-0" G		8	1		390	34		424	480

08 53 13.40 Vinyl Casement Windows

		Crew	Daily Output	Labor-Hours	Unit	Material	Labor	Equipment	Total	Total Incl O&P
0010	**VINYL CASEMENT WINDOWS**, insulated glass									
0015	Grids, low E, J fin, extension jambs, screens									
0100	One lite, 21" x 41" G	2 Carp	20	.800	Ea.	295	27		322	370
0110	21" x 47" G		20	.800		320	27		347	400
0120	21" x 53" G		20	.800		345	27		372	425

08 53 Plastic Windows

08 53 13 – Vinyl Windows

08 53 13.40 Vinyl Casement Windows

			Crew	Daily Output	Labor-Hours	Unit	Material	2016 Bare Costs Labor	Equipment	Total	Total Incl O&P
0128	24" x 35"	G	2 Carp	19	.842	Ea.	283	28.50		311.50	360
0130	24" x 41"	G		19	.842		310	28.50		338.50	390
0140	24" x 47"	G		19	.842		335	28.50		363.50	415
0150	24" x 53"	G		19	.842		360	28.50		388.50	445
0158	28" x 35"	G		19	.842		300	28.50		328.50	380
0160	28" x 41"	G		19	.842		330	28.50		358.50	415
0170	28" x 47"	G		19	.842		350	28.50		378.50	435
0180	28" x 53"	G		19	.842		385	28.50		413.50	475
0184	28" x 59"	G		19	.842		395	28.50		423.50	485
0188	Two lites, 33" x 35"	G		18	.889		470	30		500	565
0190	33" x 41"	G		18	.889		500	30		530	600
0200	33" x 47"	G		18	.889		540	30		570	640
0210	33" x 53"	G		18	.889		575	30		605	680
0212	33" x 59"	G		18	.889		610	30		640	720
0215	33" x 72"	G		18	.889		635	30		665	745
0220	41" x 41"	G		18	.889		545	30		575	650
0230	41" x 47"	G		18	.889		585	30		615	690
0240	41" x 53"	G		17	.941		620	32		652	735
0242	41" x 59"	G		17	.941		650	32		682	770
0246	41" x 72"	G		17	.941		685	32		717	810
0250	47" x 41"	G		17	.941		555	32		587	665
0260	47" x 47"	G		17	.941		590	32		622	705
0270	47" x 53"	G		17	.941		625	32		657	745
0272	47" x 59"	G		17	.941		680	32		712	805
0280	56" x 41"	G		15	1.067		595	36		631	715
0290	56" x 47"	G		15	1.067		625	36		661	750
0300	56" x 53"	G		15	1.067		680	36		716	810
0302	56" x 59"	G		15	1.067		710	36		746	840
0310	56" x 72"	G		15	1.067		770	36		806	910
0340	Solid vinyl, premium, double insulated glass, 2'-0" x 3'-0" high	G	1 Carp	10	.800		270	27		297	345
0360	2'-0" x 4'-0" high	G		9	.889		299	30		329	380
0380	2'-0" x 5'-0" high	G		8	1		335	34		369	425

08 53 13.50 Vinyl Picture Windows

		Crew	Daily Output	Labor-Hours	Unit	Material	Labor	Equipment	Total	Total Incl O&P
0010	**VINYL PICTURE WINDOWS**, insulated glass									
0120	Grids, low e, J fin, ext. jambs, 47" x 35"	2 Carp	12	1.333	Ea.	300	45		345	405
0130	47" x 41"		12	1.333		390	45		435	505
0140	47" x 47"		12	1.333		345	45		390	450
0150	47" x 53"		11	1.455		370	49.50		419.50	490
0160	71" x 35"		11	1.455		390	49.50		439.50	515
0170	71" x 41"		11	1.455		410	49.50		459.50	535
0180	71" x 47"		11	1.455		440	49.50		489.50	570

08 53 13.60 Vinyl Half Round Windows

		Crew	Daily Output	Labor-Hours	Unit	Material	Labor	Equipment	Total	Total Incl O&P
0010	**VINYL HALF ROUND WINDOWS**, Including grille, j fin low E, ext. jambs									
0100	10" height x 20" base	2 Carp	8	2	Ea.	475	68		543	635
0110	15" height x 30" base		8	2		445	68		513	605
0120	17" height x 34" base		7	2.286		325	77.50		402.50	490
0130	19" height x 38" base		7	2.286		355	77.50		432.50	520
0140	19" height x 33" base		7	2.286		555	77.50		632.50	745
0150	24" height x 48" base	1 Carp	6	1.333		370	45		415	480
0160	25" height x 50" base	"	6	1.333		755	45		800	905
0170	30" height x 60" base	2 Carp	6	2.667		665	90.50		755.50	880

08 54 Composite Windows

08 54 13 – Fiberglass Windows

08 54 13.10 Fiberglass Single Hung Windows

		Crew	Daily Output	Labor-Hours	Unit	Material	2016 Bare Costs Labor	Equipment	Total	Total Incl O&P
0010	**FIBERGLASS SINGLE HUNG WINDOWS**									
0100	Grids, low E, 18" x 24" G	2 Carp	18	.889	Ea.	335	30		365	420
0110	18" x 40" G		17	.941		340	32		372	430
0130	24" x 40" G		20	.800		360	27		387	440
0230	36" x 36" G		17	.941		370	32		402	460
0250	36" x 48" G		20	.800		405	27		432	490
0260	36" x 60" G		18	.889		445	30		475	540
0280	36" x 72" G		16	1		470	34		504	570
0290	48" x 40" G		16	1		470	34		504	570

08 54 13.30 Fiberglass Slider Windows

		Crew	Daily Output	Labor-Hours	Unit	Material	Labor	Equipment	Total	Total Incl O&P
0010	**FIBERGLASS SLIDER WINDOWS**									
0100	Grids, low E, 36" x 24" G	2 Carp	20	.800	Ea.	340	27		367	420
0110	36" x 36" G	"	20	.800	"	390	27		417	475

08 54 13.50 Fiberglass Bay Windows

		Crew	Daily Output	Labor-Hours	Unit	Material	Labor	Equipment	Total	Total Incl O&P
0010	**FIBERGLASS BAY WINDOWS**									
0150	48" x 36" G	2 Carp	11	1.455	Ea.	1,100	49.50		1,149.50	1,275

08 61 Roof Windows

08 61 13 – Metal Roof Windows

08 61 13.10 Roof Windows

		Crew	Daily Output	Labor-Hours	Unit	Material	Labor	Equipment	Total	Total Incl O&P
0010	**ROOF WINDOWS**, fixed high perf tmpd glazing, metallic framed									
0020	46" x 21-1/2", Flashed for shingled roof	1 Carp	8	1	Ea.	281	34		315	365
0100	46" x 28"		8	1		305	34		339	390
0125	57" x 44"		6	1.333		375	45		420	490
0130	72" x 28"		7	1.143		375	38.50		413.50	480
0150	Fixed, laminated tempered glazing, 46" x 21-1/2"		8	1		465	34		499	565
0175	46" x 28"		8	1		515	34		549	620
0200	57" x 44"		6	1.333		485	45		530	605
0500	Vented flashing set for shingled roof, 46" x 21-1/2"		7	1.143		465	38.50		503.50	575
0525	46" x 28"		6	1.333		515	45		560	640
0550	57" x 44"		5	1.600		645	54		699	800
0560	72" x 28"		5	1.600		645	54		699	800
0575	Flashing set for low pitched roof, 46" x 21-1/2"		7	1.143		535	38.50		573.50	655
0600	46" x 28"		7	1.143		585	38.50		623.50	710
0625	57" x 44"		5	1.600		730	54		784	890
0650	Flashing set for curb 46" x 21-1/2"		7	1.143		620	38.50		658.50	745
0675	46" x 28"		7	1.143		670	38.50		708.50	805
0700	57" x 44"		5	1.600		825	54		879	1,000

08 61 16 – Wood Roof Windows

08 61 16.16 Roof Windows, Wood Framed

		Crew	Daily Output	Labor-Hours	Unit	Material	Labor	Equipment	Total	Total Incl O&P
0010	**ROOF WINDOWS, WOOD FRAMED**									
5600	Roof window incl. frame, flashing, double insulated glass & screens,									
5610	complete unit, 22" x 38"	2 Carp	3	5.333	Ea.	750	181		931	1,125
5650	2'-5" x 3'-8"		3.20	5		930	170		1,100	1,300
5700	3'-5" x 4'-9"		3.40	4.706		1,050	160		1,210	1,425

08 62 Unit Skylights

08 62 13 – Domed Unit Skylights

08 62 13.10 Domed Skylights

		Crew	Daily Output	Labor-Hours	Unit	Material	2016 Bare Costs Labor	Equipment	Total	Total Incl O&P
0010	**DOMED SKYLIGHTS**									
0020	Skylight, fixed dome type, 22" x 22"	G	G-3	12	2.667	Ea.	216	83.50		299.50	375
0030	22" x 46"	G		10	3.200		269	100		369	465
0040	30" x 30"	G		12	2.667		286	83.50		369.50	455
0050	30" x 46"	G		10	3.200		380	100		480	585
0110	Fixed, double glazed, 22" x 27"	G		12	2.667		275	83.50		358.50	445
0120	22" x 46"	G		10	3.200		293	100		393	485
0130	44" x 46"	G		10	3.200		430	100		530	640
0210	Operable, double glazed, 22" x 27"	G		12	2.667		405	83.50		488.50	585
0220	22" x 46"	G		10	3.200		465	100		565	680
0230	44" x 46"	G		10	3.200		905	100		1,005	1,175

08 62 13.20 Skylights

		Crew	Daily Output	Labor-Hours	Unit	Material	Labor	Equipment	Total	Total Incl O&P
0010	**SKYLIGHTS**, flush or curb mounted									
2120	Ventilating insulated plexiglass dome with									
2130	curb mounting, 36" x 36"	G	G-3	12	2.667	Ea.	480	83.50		563.50	670
2150	52" x 52"	G		12	2.667		670	83.50		753.50	875
2160	28" x 52"	G		10	3.200		490	100		590	705
2170	36" x 52"	G		10	3.200		545	100		645	760
2180	For electric opening system, add	G					315			315	345
2210	Operating skylight, with thermopane glass, 24" x 48"	G	G-3	10	3.200		585	100		685	810
2220	32" x 48"	G	"	9	3.556		615	111		726	860
2310	Non venting insulated plexiglass dome skylight with									
2320	Flush mount 22" x 46"	G	G-3	15.23	2.101	Ea.	330	65.50		395.50	475
2330	30" x 30"	G		16	2		305	62.50		367.50	440
2340	46" x 46"	G		13.91	2.301		565	72		637	740
2350	Curb mount 22" x 46"	G		15.23	2.101		365	65.50		430.50	510
2360	30" x 30"	G		16	2		405	62.50		467.50	555
2370	46" x 46"	G		13.91	2.301		625	72		697	810
2381	Non-insulated flush mount 22" x 46"			15.23	2.101		225	65.50		290.50	360
2382	30" x 30"			16	2		202	62.50		264.50	325
2383	46" x 46"			13.91	2.301		380	72		452	540
2384	Curb mount 22" x 46"			15.23	2.101		189	65.50		254.50	320
2385	30" x 30"			16	2		188	62.50		250.50	310
4000	Skylight, solar tube kit, incl dome, flashing, diffuser, 1 pipe, 10" dia.	G	1 Carp	2	4		242	136		378	495
4010	14" diam.	G		2	4		335	136		471	595
4020	21" diam.	G		2	4		420	136		556	685
4030	Accessories for, 1' long x 9" dia pipe	G		24	.333		50	11.30		61.30	74
4040	2' long x 9" dia pipe	G		24	.333		43	11.30		54.30	66.50
4050	4' long x 9" dia pipe	G		20	.400		75	13.55		88.55	105
4060	1' long x 13" dia pipe	G		24	.333		70	11.30		81.30	96
4070	2' long x 13" dia pipe	G		24	.333		55	11.30		66.30	79.50
4080	4' long x 13" dia pipe	G		20	.400		102	13.55		115.55	135
4090	6.5" turret ext for 21" dia pipe	G		16	.500		110	16.95		126.95	150
4100	12' long x 21" dia flexible pipe	G		12	.667		88	22.50		110.50	135
4110	45 degree elbow, 10"	G		16	.500		152	16.95		168.95	196
4120	14"	G		16	.500		82	16.95		98.95	119
4130	Interior decorative ring, 9"	G		20	.400		40	13.55		53.55	66.50
4140	13"	G		20	.400		60	13.55		73.55	88.50

08 71 Door Hardware

08 71 20 – Hardware

08 71 20.15 Hardware

		Crew	Daily Output	Labor-Hours	Unit	Material	2016 Bare Costs Labor	Equipment	Total	Total Incl O&P
0010	**HARDWARE**									
0020	Average percentage for hardware, total job cost									
0500	Total hardware for building, average distribution				Job	85%	15%			
1000	Door hardware, apartment, interior	1 Carp	4	2	Door	465	68		533	625
1300	Average, door hardware, motel/hotel interior, with access card		4	2	"	600	68		668	775
2100	Pocket door		6	1.333	Ea.	100	45		145	186
4000	Door knocker, bright brass		32	.250		75	8.50		83.50	96.50
4100	Mail slot, bright brass, 2" x 11"		25	.320		60	10.85		70.85	84
4200	Peep hole, add to price of door					12			12	13.20

08 71 20.40 Lockset

		Crew	Daily Output	Labor-Hours	Unit	Material	Labor	Equipment	Total	Total Incl O&P
0010	**LOCKSET**, Standard duty									
0020	Non-keyed, passage, w/sect.trim	1 Carp	12	.667	Ea.	70	22.50		92.50	115
0100	Privacy		12	.667		75	22.50		97.50	121
0400	Keyed, single cylinder function		10	.800		108	27		135	164
0500	Lever handled, keyed, single cylinder function		10	.800		128	27		155	187

08 71 20.41 Dead Locks

		Crew	Daily Output	Labor-Hours	Unit	Material	Labor	Equipment	Total	Total Incl O&P
0010	**DEAD LOCKS**									
1203	Deadlock night latch	1 Carp	7.70	1.039	Ea.	45	35		80	109
1420	Deadbolt lock, single cylinder		10	.800		48.50	27		75.50	99
1440	Double cylinder		10	.800		63	27		90	115

08 71 20.42 Mortise Locksets

		Crew	Daily Output	Labor-Hours	Unit	Material	Labor	Equipment	Total	Total Incl O&P
0010	**MORTISE LOCKSETS**, Comm., wrought knobs & full escutcheon trim									
0015	Assumes mortise is cut									
0020	Non-keyed, passage, grade 3	1 Carp	9	.889	Ea.	149	30		179	215
0030	Grade 1		8	1		405	34		439	500
0040	Privacy set Grade 3		9	.889		169	30		199	237
0050	Grade 1		8	1		455	34		489	555
0100	Keyed, office/entrance/apartment, Grade 2		8	1		197	34		231	274
0110	Grade 1		7	1.143		515	38.50		553.50	635
0120	Single cylinder, typical, Grade 3		8	1		189	34		223	265
0130	Grade 1		7	1.143		500	38.50		538.50	615
0200	Hotel, room, Grade 3		7	1.143		190	38.50		228.50	274
0210	Grade 1 (see also Section 08 71 20.15)		6	1.333		510	45		555	635
0300	Double cylinder, Grade 3		8	1		225	34		259	305
0310	Grade 1		7	1.143		515	38.50		553.50	635
1000	Wrought knobs and sectional trim, non-keyed, passage, Grade 3		10	.800		130	27		157	189
1010	Grade 1		9	.889		405	30		435	495
1040	Privacy, Grade 3		10	.800		145	27		172	206
1050	Grade 1		9	.889		455	30		485	550
1100	Keyed, entrance, office/apartment, Grade 3		9	.889		220	30		250	293
1103	Install lockset		6.92	1.156		220	39		259	310
1110	Grade 1		8	1		520	34		554	625
1120	Single cylinder, Grade 3		9	.889		225	30		255	299
1130	Grade 1		8	1		500	34		534	605
2000	Cast knobs and full escutcheon trim									
2010	Non-keyed, passage, Grade 3	1 Carp	9	.889	Ea.	275	30		305	355
2020	Grade 1		8	1		380	34		414	475
2040	Privacy, Grade 3		9	.889		320	30		350	400
2050	Grade 1		8	1		440	34		474	540
2120	Keyed, single cylinder, Grade 3		8	1		330	34		364	420
2123	Mortise lock		6.15	1.301		330	44		374	440
2130	Grade 1		7	1.143		525	38.50		563.50	645

08 71 Door Hardware

08 71 20 – Hardware

08 71 20.50 Door Stops

		Crew	Daily Output	Labor-Hours	Unit	Material	2016 Bare Costs Labor	Equipment	Total	Total Incl O&P
0010	**DOOR STOPS**									
0020	Holder & bumper, floor or wall	1 Carp	32	.250	Ea.	35	8.50		43.50	53
1300	Wall bumper, 4" diameter, with rubber pad, aluminum		32	.250		12.30	8.50		20.80	27.50
1600	Door bumper, floor type, aluminum		32	.250		8.70	8.50		17.20	24
1620	Brass		32	.250		10	8.50		18.50	25
1630	Bronze		32	.250		18.35	8.50		26.85	34
1900	Plunger type, door mounted		32	.250		29.50	8.50		38	46.50
2520	Wall type, aluminum		32	.250		35.50	8.50		44	53
2540	Plunger type, aluminum		32	.250		29.50	8.50		38	46.50
2560	Brass		32	.250		45	8.50		53.50	63
3020	Floor mounted, US3		3	2.667		315	90.50		405.50	500

08 71 20.60 Entrance Locks

		Crew	Daily Output	Labor-Hours	Unit	Material	Labor	Equipment	Total	Total Incl O&P
0010	**ENTRANCE LOCKS**									
0015	Cylinder, grip handle deadlocking latch	1 Carp	9	.889	Ea.	175	30		205	244
0020	Deadbolt		8	1		175	34		209	250
0100	Push and pull plate, dead bolt		8	1		225	34		259	305
0900	For handicapped lever, add					150			150	165

08 71 20.65 Thresholds

		Crew	Daily Output	Labor-Hours	Unit	Material	Labor	Equipment	Total	Total Incl O&P
0010	**THRESHOLDS**									
0011	Threshold 3' long saddles aluminum	1 Carp	48	.167	L.F.	9.70	5.65		15.35	20
0100	Aluminum, 8" wide, 1/2" thick		12	.667	Ea.	49	22.50		71.50	91.50
0500	Bronze		60	.133	L.F.	43	4.52		47.52	55
0600	Bronze, panic threshold, 5" wide, 1/2" thick		12	.667	Ea.	158	22.50		180.50	211
0700	Rubber, 1/2" thick, 5-1/2" wide		20	.400		40	13.55		53.55	66.50
0800	2-3/4" wide		20	.400		45.50	13.55		59.05	72.50
1950	ADA Compliant Thresholds									
2000	Threshold, wood oak 3-1/2" wide x 24" long	1 Carp	12	.667	Ea.	11	22.50		33.50	50
2010	3-1/2" wide x 36" long		12	.667		15	22.50		37.50	54.50
2020	3-1/2" wide x 48" long		12	.667		20	22.50		42.50	60
2030	4-1/2" wide x 24" long		12	.667		13	22.50		35.50	52.50
2040	4-1/2" wide x 36" long		12	.667		19	22.50		41.50	59
2050	4-1/2" wide x 48" long		12	.667		25	22.50		47.50	65.50
2060	6-1/2" wide x 24" long		12	.667		18	22.50		40.50	58
2070	6-1/2" wide x 36" long		12	.667		27	22.50		49.50	67.50
2080	6-1/2" wide x 48" long		12	.667		36	22.50		58.50	77.50
2090	Threshold, wood cherry 3-1/2" wide x 24" long		12	.667		15	22.50		37.50	54.50
2100	3-1/2" wide x 36" long		12	.667		22	22.50		44.50	62
2110	3-1/2" wide x 48" long		12	.667		29	22.50		51.50	70
2120	4-1/2" wide x 24" long		12	.667		19	22.50		41.50	59
2130	4-1/2" wide x 36" long		12	.667		28	22.50		50.50	69
2140	4-1/2" wide x 48" long		12	.667		37	22.50		59.50	78.50
2150	6-1/2" wide x 24" long		12	.667		27	22.50		49.50	67.50
2160	6-1/2" wide x 36" long		12	.667		40	22.50		62.50	82
2170	6-1/2" wide x 48" long		12	.667		53	22.50		75.50	96.50
2180	Threshold, wood walnut 3-1/2" wide x 24" long		12	.667		18	22.50		40.50	58
2190	3-1/2" wide x 36" long		12	.667		27	22.50		49.50	67.50
2200	3-1/2" wide x 48" long		12	.667		36	22.50		58.50	77.50
2210	4-1/2" wide x 24" long		12	.667		26	22.50		48.50	66.50
2220	4-1/2" wide x 36" long		12	.667		38	22.50		60.50	80
2230	4-1/2" wide x 48" long		12	.667		50	22.50		72.50	93
2240	6-1/2" wide x 24" long		12	.667		43	22.50		65.50	85.50
2250	6-1/2" wide x 36" long		12	.667		63	22.50		85.50	108

08 71 Door Hardware

08 71 20 – Hardware

08 71 20.65 Thresholds

		Crew	Daily Output	Labor-Hours	Unit	Material	2016 Bare Costs Labor	Equipment	Total	Total Incl O&P
2260	6-1/2" wide x 48" long	1 Carp	12	.667	Ea.	84	22.50		106.50	131
2300	Threshold, aluminum 4" wide x 36" long		12	.667		33	22.50		55.50	74.50
2310	4" wide x 48" long		12	.667		41	22.50		63.50	83
2320	4" wide x 72" long		12	.667		66	22.50		88.50	111
2330	5" wide x 36" long		12	.667		45	22.50		67.50	87.50
2340	5" wide x 48" long		12	.667		56	22.50		78.50	99.50
2350	5" wide x 72" long		12	.667		89	22.50		111.50	136
2360	6" wide x 36" long		12	.667		53	22.50		75.50	96.50
2370	6" wide x 48" long		12	.667		68	22.50		90.50	113
2380	6" wide x 72" long		12	.667		106	22.50		128.50	155
2390	7" wide x 36" long		12	.667		69	22.50		91.50	114
2400	7" wide x 48" long		12	.667		90	22.50		112.50	137
2410	7" wide x 72" long		12	.667		137	22.50		159.50	189
2500	Threshold, ramp, aluminum or rubber 24" x 24"		12	.667		190	22.50		212.50	247

08 71 20.75 Door Hardware Accessories

0010	DOOR HARDWARE ACCESSORIES									
1000	Knockers, brass, standard	1 Carp	16	.500	Ea.	49	16.95		65.95	82.50
1100	Deluxe		10	.800		121	27		148	179
4500	Rubber door silencers		540	.015		.45	.50		.95	1.34

08 71 20.90 Hinges

0010	HINGES									
0012	Full mortise, avg. freq., steel base, USP, 4-1/2" x 4-1/2"				Pr.	37			37	40.50
0100	5" x 5", USP					57.50			57.50	63
0200	6" x 6", USP					119			119	131
0400	Brass base, 4-1/2" x 4-1/2", US10					60			60	66
0500	5" x 5", US10					86			86	94.50
0600	6" x 6", US10					162			162	178
0800	Stainless steel base, 4-1/2" x 4-1/2", US32					76.50			76.50	84
0900	For non removable pin, add (security item)				Ea.	5.25			5.25	5.80
0910	For floating pin, driven tips, add					3.30			3.30	3.63
0930	For hospital type tip on pin, add					13.95			13.95	15.35
0940	For steeple type tip on pin, add					18.95			18.95	21
0950	Full mortise, high frequency, steel base, 3-1/2" x 3-1/2", US26D				Pr.	30.50			30.50	33.50
1000	4-1/2" x 4-1/2", USP					66			66	73
1100	5" x 5", USP					52			52	57.50
1200	6" x 6", USP					138			138	152
1400	Brass base, 3-1/2" x 3-1/2", US4					51			51	56
1430	4-1/2" x 4-1/2", US10					75.50			75.50	83
1500	5" x 5", US10					126			126	138
1600	6" x 6", US10					169			169	186
1800	Stainless steel base, 4-1/2" x 4-1/2", US32					103			103	114
1810	5" x 4-1/2", US32					137			137	151
1930	For hospital type tip on pin, add				Ea.	14.60			14.60	16.05
1950	Full mortise, low frequency, steel base, 3-1/2" x 3-1/2", US26D				Pr.	25			25	27.50
2000	4-1/2" x 4-1/2", USP					23			23	25
2100	5" x 5", USP					48			48	53
2200	6" x 6", USP					93.50			93.50	103
2300	4-1/2" x 4-1/2", US3					17.35			17.35	19.05
2310	5" x 5", US3					42			42	46
2400	Brass bass, 4-1/2" x 4-1/2", US10					54.50			54.50	59.50
2500	5" x 5", US10					78			78	86
2800	Stainless steel base, 4-1/2" x 4-1/2", US32					76			76	84

08 71 Door Hardware

08 71 20 – Hardware

08 71 20.90 Hinges		Crew	Daily Output	Labor-Hours	Unit	Material	2016 Bare Costs Labor	Equipment	Total	Total Incl O&P
8000	Install hinge	1 Carp	34	.235	Pr.		8		8	13.35
08 71 20.91 Special Hinges										
0010	**SPECIAL HINGES**									
8000	Continuous hinges									
8010	Steel, piano, 2" x 72"	1 Carp	20	.400	Ea.	22	13.55		35.55	46.50
8020	Brass, piano, 1-1/16" x 30"		30	.267		8	9.05		17.05	24
8030	Acrylic, piano, 1-3/4" x 12"	↓	40	.200	↓	15	6.80		21.80	28
08 71 20.92 Mortised Hinges										
0010	**MORTISED HINGES**									
0200	Average frequency, steel plated, ball bearing, 3-1/2" x 3-1/2"				Pr.	29.50			29.50	32.50
0300	Bronze, ball bearing					34.50			34.50	38
0900	High frequency, steel plated, ball bearing					78			78	86
1100	Bronze, ball bearing					78			78	86
1300	Average frequency, steel plated, ball bearing, 4-1/2" x 4-1/2"					37			37	40.50
1500	Bronze, ball bearing, to 36" wide					39			39	42.50
1700	Low frequency, steel, plated, plain bearing					20.50			20.50	22.50
1900	Bronze, plain bearing				↓	28.50			28.50	31
08 71 20.95 Kick Plates										
0010	**KICK PLATES**									
0020	Stainless steel, .050, 16 ga., 8" x 28", US32	1 Carp	15	.533	Ea.	38	18.10		56.10	72.50
0080	Mop/Kick, 4" x 28"		15	.533		34	18.10		52.10	68
0090	4" x 30"		15	.533		36	18.10		54.10	70
0100	4" x 34"		15	.533		41	18.10		59.10	75.50
0110	6" x 28"		15	.533		43	18.10		61.10	78
0120	6" x 30"		15	.533		47	18.10		65.10	82
0130	6" x 34"	↓	15	.533	↓	53	18.10		71.10	89

08 71 21 – Astragals

08 71 21.10 Exterior Mouldings, Astragals		Crew	Daily Output	Labor-Hours	Unit	Material	Labor	Equipment	Total	Total Incl O&P
0010	**EXTERIOR MOULDINGS, ASTRAGALS**									
4170	Astragal for double doors, aluminum	1 Carp	4	2	Opng.	34.50	68		102.50	152
4174	Bronze	"	4	2	"	47	68		115	166

08 71 25 – Weatherstripping

08 71 25.10 Mechanical Seals, Weatherstripping		Crew	Daily Output	Labor-Hours	Unit	Material	Labor	Equipment	Total	Total Incl O&P
0010	**MECHANICAL SEALS, WEATHERSTRIPPING**									
1000	Doors, wood frame, interlocking, for 3' x 7' door, zinc	1 Carp	3	2.667	Opng.	44	90.50		134.50	200
1100	Bronze		3	2.667		56	90.50		146.50	213
1300	6' x 7' opening, zinc		2	4		54	136		190	287
1400	Bronze		2	4	↓	65	136		201	299
1500	Vinyl V strip	↓	6.40	1.250	Ea.	12.50	42.50		55	85
1700	Wood frame, spring type, bronze									
1800	3' x 7' door	1 Carp	7.60	1.053	Opng.	23.50	35.50		59	85
1900	6' x 7' door		7	1.143		31	38.50		69.50	99
1920	Felt, 3' x 7' door		14	.571		3.90	19.35		23.25	37
1930	6' x 7' door		13	.615		4.45	21		25.45	40
1950	Rubber, 3' x 7' door		7.60	1.053		9.15	35.50		44.65	69.50
1960	6' x 7' door	↓	7	1.143	↓	11.15	38.50		49.65	77.50
2200	Metal frame, spring type, bronze									
2300	3' x 7' door	1 Carp	3	2.667	Opng.	47.50	90.50		138	203
2400	6' x 7' door	"	2.50	3.200	"	52	108		160	239
2500	For stainless steel, spring type, add					133%				
2700	Metal frame, extruded sections, 3' x 7' door, aluminum	1 Carp	3	2.667	Opng.	28	90.50		118.50	182

08 71 Door Hardware

08 71 25 – Weatherstripping

08 71 25.10 Mechanical Seals, Weatherstripping	Crew	Daily Output	Labor-Hours	Unit	Material	2016 Bare Costs Labor	Equipment	Total	Total Incl O&P
2800 Bronze	1 Carp	3	2.667	Opng.	82	90.50		172.50	241
3100 6' x 7' door, aluminum		1.50	5.333		35	181		216	345
3200 Bronze		1.50	5.333		137	181		318	455
3500 Threshold weatherstripping									
3650 Door sweep, flush mounted, aluminum	1 Carp	25	.320	Ea.	19	10.85		29.85	39
3700 Vinyl		25	.320		18	10.85		28.85	38
5000 Garage door bottom weatherstrip, 12' aluminum, clear		14	.571		25	19.35		44.35	60
5010 Bronze		14	.571		90	19.35		109.35	132
5050 Bottom protection, Rubber		14	.571		37	19.35		56.35	73
5100 Threshold		14	.571		73	19.35		92.35	113

08 75 Window Hardware

08 75 10 – Window Handles and Latches

08 75 10.10 Handles and Latches	Crew	Daily Output	Labor-Hours	Unit	Material	2016 Bare Costs Labor	Equipment	Total	Total Incl O&P
0010 HANDLES AND LATCHES									
1000 Handles, surface mounted, aluminum	1 Carp	24	.333	Ea.	4.56	11.30		15.86	24
1020 Brass		24	.333		5.85	11.30		17.15	25.50
1040 Chrome		24	.333		7.30	11.30		18.60	27
1500 Recessed, aluminum		12	.667		2.64	22.50		25.14	41
1520 Brass		12	.667		4.14	22.50		26.64	42.50
1540 Chrome		12	.667		3.67	22.50		26.17	42
2000 Latches, aluminum		20	.400		3.48	13.55		17.03	26.50
2020 Brass		20	.400		4.65	13.55		18.20	27.50
2040 Chrome		20	.400		3.39	13.55		16.94	26

08 75 30 – Weatherstripping

08 75 30.10 Mechanical Weather Seals	Crew	Daily Output	Labor-Hours	Unit	Material	2016 Bare Costs Labor	Equipment	Total	Total Incl O&P
0010 MECHANICAL WEATHER SEALS, Window, double hung, 3' X 5'									
0020 Zinc	1 Carp	7.20	1.111	Opng.	20	37.50		57.50	85
0100 Bronze		7.20	1.111		40	37.50		77.50	107
0200 Vinyl V strip		7	1.143		9.50	38.50		48	75.50
0500 As above but heavy duty, zinc		4.60	1.739		20	59		79	121
0600 Bronze		4.60	1.739		70	59		129	176

08 79 Hardware Accessories

08 79 20 – Door Accessories

08 79 20.10 Door Hardware Accessories	Crew	Daily Output	Labor-Hours	Unit	Material	2016 Bare Costs Labor	Equipment	Total	Total Incl O&P
0010 DOOR HARDWARE ACCESSORIES									
0140 Door bolt, surface, 4"	1 Carp	32	.250	Ea.	13	8.50		21.50	28.50
0160 Door latch	"	12	.667	"	8.55	22.50		31.05	47.50
0200 Sliding closet door									
0220 Track and hanger, single	1 Carp	10	.800	Ea.	65	27		92	117
0240 Double		8	1		80	34		114	145
0260 Door guide, single		48	.167		30	5.65		35.65	42.50
0280 Double		48	.167		40	5.65		45.65	53.50
0600 Deadbolt and lock cover plate, brass or stainless steel		30	.267		28	9.05		37.05	46
0620 Hole cover plate, brass or chrome		35	.229		8	7.75		15.75	22
2240 Mortise lockset, passage, lever handle		9	.889		160	30		190	227
4000 Security chain, standard		18	.444		10	15.05		25.05	36
4100 Deluxe		18	.444		45	15.05		60.05	74.50

08 81 Glass Glazing

08 81 10 – Float Glass

08 81 10.10 Various Types and Thickness of Float Glass		Crew	Daily Output	Labor-Hours	Unit	Material	2016 Bare Costs Labor	Equipment	Total	Total Incl O&P
0010	**VARIOUS TYPES AND THICKNESS OF FLOAT GLASS**									
0020	3/16" Plain	2 Glaz	130	.123	S.F.	5.10	4.08		9.18	12.40
0200	Tempered, clear		130	.123		7	4.08		11.08	14.50
0300	Tinted		130	.123		8.10	4.08		12.18	15.70
0600	1/4" thick, clear, plain		120	.133		6.40	4.42		10.82	14.40
0700	Tinted		120	.133		9.10	4.42		13.52	17.35
0800	Tempered, clear		120	.133		9	4.42		13.42	17.25
0900	Tinted		120	.133		11.35	4.42		15.77	19.85
1600	3/8" thick, clear, plain		75	.213		10.40	7.05		17.45	23
1700	Tinted		75	.213		15.70	7.05		22.75	29
1800	Tempered, clear		75	.213		16.75	7.05		23.80	30
1900	Tinted		75	.213		19.10	7.05		26.15	33
2200	1/2" thick, clear, plain		55	.291		17.65	9.65		27.30	35.50
2300	Tinted		55	.291		27.50	9.65		37.15	46
2400	Tempered, clear		55	.291		25	9.65		34.65	43.50
2500	Tinted		55	.291		26	9.65		35.65	45
2800	5/8" thick, clear, plain		45	.356		27.50	11.80		39.30	49.50
2900	Tempered, clear		45	.356		31.50	11.80		43.30	54

08 81 25 – Glazing Variables

08 81 25.10 Applications of Glazing

		Crew	Daily Output	Labor-Hours	Unit	Material	Labor	Equipment	Total	Total Incl O&P
0010	**APPLICATIONS OF GLAZING**									
0600	For glass replacement, add				S.F.		100%			
0700	For gasket settings, add				L.F.	5.75			5.75	6.35
0900	For sloped glazing, add				S.F.		26%			
2000	Fabrication, polished edges, 1/4" thick				Inch	.55			.55	.61
2100	1/2" thick					1.30			1.30	1.43
2500	Mitered edges, 1/4" thick					1.30			1.30	1.43
2600	1/2" thick					2.15			2.15	2.37

08 81 30 – Insulating Glass

08 81 30.10 Reduce Heat Transfer Glass

			Crew	Daily Output	Labor-Hours	Unit	Material	Labor	Equipment	Total	Total Incl O&P
0010	**REDUCE HEAT TRANSFER GLASS**										
0015	2 lites 1/8" float, 1/2" thk under 15 S.F.										
0100	Tinted	G	2 Glaz	95	.168	S.F.	14.25	5.60		19.85	25
0280	Double glazed, 5/8" thk unit, 3/16" float, 15-30 S.F., clear			90	.178		13.70	5.90		19.60	25
0400	1" thk, dbl. glazed, 1/4" float, 30-70 S.F., clear	G		75	.213		17.15	7.05		24.20	30.50
0500	Tinted	G		75	.213		23.50	7.05		30.55	38
2000	Both lites, light & heat reflective	G		85	.188		31.50	6.25		37.75	45
2500	Heat reflective, film inside, 1" thick unit, clear	G		85	.188		27.50	6.25		33.75	41
2600	Tinted	G		85	.188		28.50	6.25		34.75	42
3000	Film on weatherside, clear, 1/2" thick unit	G		95	.168		19.60	5.60		25.20	31
3100	5/8" thick unit	G		90	.178		20	5.90		25.90	32
3200	1" thick unit	G		85	.188		27	6.25		33.25	40.50
5000	Spectrally selective film, on ext, blocks solar gain/allows 70% of light	G		95	.168		14.25	5.60		19.85	25

08 81 40 – Plate Glass

08 81 40.10 Plate Glass

		Crew	Daily Output	Labor-Hours	Unit	Material	Labor	Equipment	Total	Total Incl O&P
0010	**PLATE GLASS** Twin ground, polished,									
0015	3/16" thick, material				S.F.	5.35			5.35	5.90
0020	3/16" thick	2 Glaz	100	.160		5.35	5.30		10.65	14.70
0100	1/4" thick		94	.170		7.35	5.65		13	17.45
0200	3/8" thick		60	.267		12.60	8.85		21.45	28.50
0300	1/2" thick		40	.400		24.50	13.25		37.75	49

08 81 Glass Glazing

08 81 55 – Window Glass

08 81 55.10 Sheet Glass

		Crew	Daily Output	Labor-Hours	Unit	Material	2016 Bare Costs Labor	Equipment	Total	Total Incl O&P
0010	**SHEET GLASS** (window), clear float, stops, putty bed									
0015	1/8" thick, clear float	2 Glaz	480	.033	S.F.	3.71	1.10		4.81	5.90
0500	3/16" thick, clear		480	.033		6	1.10		7.10	8.45
0600	Tinted		480	.033		7.45	1.10		8.55	10.05
0700	Tempered		480	.033		9.20	1.10		10.30	11.95

08 83 Mirrors

08 83 13 – Mirrored Glass Glazing

08 83 13.10 Mirrors

		Crew	Daily Output	Labor-Hours	Unit	Material	Labor	Equipment	Total	Total Incl O&P
0010	**MIRRORS**, No frames, wall type, 1/4" plate glass, polished edge									
0100	Up to 5 S.F.	2 Glaz	125	.128	S.F.	9.50	4.24		13.74	17.50
0200	Over 5 S.F.		160	.100		9.25	3.32		12.57	15.70
0500	Door type, 1/4" plate glass, up to 12 S.F.		160	.100		8.80	3.32		12.12	15.20
1000	Float glass, up to 10 S.F., 1/8" thick		160	.100		5.90	3.32		9.22	12
1100	3/16" thick		150	.107		7.30	3.54		10.84	13.90
1500	12" x 12" wall tiles, square edge, clear		195	.082		2.43	2.72		5.15	7.20
1600	Veined		195	.082		6.20	2.72		8.92	11.35
2010	Bathroom, unframed, laminated		160	.100		14	3.32		17.32	21

08 87 Glazing Surface Films

08 87 53 – Security Films On Glass

08 87 53.10 Security Film Adhered On Glass

		Crew	Daily Output	Labor-Hours	Unit	Material	Labor	Equipment	Total	Total Incl O&P
0010	**SECURITY FILM ADHERED ON GLASS** (glass not included)									
0020	Security Film, clear, 7 mil	1 Glaz	200	.040	S.F.	1.03	1.33		2.36	3.33
0030	8 mil		200	.040		1.86	1.33		3.19	4.25
0040	9 mil		200	.040		2.07	1.33		3.40	4.48
0050	10 mil		200	.040		2.24	1.33		3.57	4.66
0060	12 mil		200	.040		1.96	1.33		3.29	4.36
0075	15 mil		200	.040		2.38	1.33		3.71	4.82
0100	Security Film, sealed with structural adhesive, 7 mil		180	.044		5.55	1.47		7.02	8.55
0110	8 mil		180	.044		6.50	1.47		7.97	9.60
0140	14 mil		180	.044		7.10	1.47		8.57	10.25

08 91 Louvers

08 91 19 – Fixed Louvers

08 91 19.10 Aluminum Louvers

		Crew	Daily Output	Labor-Hours	Unit	Material	Labor	Equipment	Total	Total Incl O&P
0010	**ALUMINUM LOUVERS**									
0020	Aluminum with screen, residential, 8" x 8"	1 Carp	38	.211	Ea.	20	7.15		27.15	34
0100	12" x 12"		38	.211		16	7.15		23.15	29.50
0200	12" x 18"		35	.229		20	7.75		27.75	35
0250	14" x 24"		30	.267		29	9.05		38.05	47
0300	18" x 24"		27	.296		32	10.05		42.05	52
0500	24" x 30"		24	.333		60	11.30		71.30	85
0700	Triangle, adjustable, small		20	.400		58	13.55		71.55	86.50
0800	Large		15	.533		79.50	18.10		97.60	118
2100	Midget, aluminum, 3/4" deep, 1" diameter		85	.094		.77	3.19		3.96	6.20
2150	3" diameter		60	.133		2.48	4.52		7	10.30

08 91 Louvers

08 91 19 – Fixed Louvers

08 91 19.10 Aluminum Louvers		Crew	Daily Output	Labor-Hours	Unit	Material	2016 Bare Costs Labor	Equipment	Total	Total Incl O&P
2200	4" diameter	1 Carp	50	.160	Ea.	4.95	5.40		10.35	14.55
2250	6" diameter	↓	30	.267	↓	4.10	9.05		13.15	19.65

08 95 Vents

08 95 13 – Soffit Vents

08 95 13.10 Wall Louvers

		Crew	Daily Output	Labor-Hours	Unit	Material	Labor	Equipment	Total	Total Incl O&P
0010	**WALL LOUVERS**									
2400	Under eaves vent, aluminum, mill finish, 16" x 4"	1 Carp	48	.167	Ea.	1.93	5.65		7.58	11.55
2500	16" x 8"	"	48	.167	"	2.23	5.65		7.88	11.90

08 95 16 – Wall Vents

08 95 16.10 Louvers

		Crew	Daily Output	Labor-Hours	Unit	Material	Labor	Equipment	Total	Total Incl O&P
0010	**LOUVERS**									
0020	Redwood, 2'-0" diameter, full circle	1 Carp	16	.500	Ea.	190	16.95		206.95	238
0100	Half circle		16	.500		175	16.95		191.95	222
0200	Octagonal		16	.500		142	16.95		158.95	185
0300	Triangular, 5/12 pitch, 5'-0" at base		16	.500		200	16.95		216.95	249
1000	Rectangular, 1'-4" x 1'-3"		16	.500		19.50	16.95		36.45	50
1100	Rectangular, 1'-4" x 1'-8"		16	.500		28	16.95		44.95	59.50
1200	1'-4" x 2'-2"		15	.533		31	18.10		49.10	65
1300	1'-9" x 2'-2"		15	.533		37.50	18.10		55.60	72
1400	2'-3" x 2'-2"		14	.571		49.50	19.35		68.85	87
1700	2'-4" x 2'-11"		13	.615		49.50	21		70.50	89.50
2000	Aluminum, 12" x 16"		25	.320		20	10.85		30.85	40
2010	16" x 20"		25	.320		27.50	10.85		38.35	48.50
2020	24" x 30"		25	.320		59	10.85		69.85	83
2100	6' triangle		12	.667		168	22.50		190.50	223
3100	Round, 2'-2" diameter		16	.500		124	16.95		140.95	165
7000	Vinyl gable vent, 8" x 8"		38	.211		14	7.15		21.15	27.50
7020	12" x 12"		38	.211		28	7.15		35.15	43
7080	12" x 18"		35	.229		37	7.75		44.75	53.50
7200	18" x 24"	↓	30	.267	↓	45	9.05		54.05	64.50

Estimating Tips
General
- Room Finish Schedule: A complete set of plans should contain a room finish schedule. If one is not available, it would be well worth the time and effort to obtain one.

09 20 00 Plaster and Gypsum Board
- Lath is estimated by the square yard plus a 5% allowance for waste. Furring, channels, and accessories are measured by the linear foot. An extra foot should be allowed for each accessory miter or stop.
- Plaster is also estimated by the square yard. Deductions for openings vary by preference, from zero deduction to 50% of all openings over 2 feet in width. The estimator should allow one extra square foot for each linear foot of horizontal interior or exterior angle located below the ceiling level. Also, double the areas of small radius work.
- Drywall accessories, studs, track, and acoustical caulking are all measured by the linear foot. Drywall taping is figured by the square foot. Gypsum wallboard is estimated by the square foot. No material deductions should be made for door or window openings under 32 S.F.

09 60 00 Flooring
- Tile and terrazzo areas are taken off on a square foot basis. Trim and base materials are measured by the linear foot. Accent tiles are listed per each. Two basic methods of installation are used. Mud set is approximately 30% more expensive than thin set. In terrazzo work, be sure to include the linear footage of embedded decorative strips, grounds, machine rubbing, and power cleanup.
- Wood flooring is available in strip, parquet, or block configuration. The latter two types are set in adhesives with quantities estimated by the square foot. The laying pattern will influence labor costs and material waste. In addition to the material and labor for laying wood floors, the estimator must make allowances for sanding and finishing these areas, unless the flooring is prefinished.
- Sheet flooring is measured by the square yard. Roll widths vary, so consideration should be given to use the most economical width, as waste must be figured into the total quantity. Consider also the installation methods available, direct glue down or stretched.

09 70 00 Wall Finishes
- Wall coverings are estimated by the square foot. The area to be covered is measured, length by height of wall above baseboards, to calculate the square footage of each wall. This figure is divided by the number of square feet in the single roll which is being used. Deduct, in full, the areas of openings such as doors and windows. Where a pattern match is required allow 25%–30% waste.

09 80 00 Acoustic Treatment
- Acoustical systems fall into several categories. The takeoff of these materials should be by the square foot of area with a 5% allowance for waste. Do not forget about scaffolding, if applicable, when estimating these systems.

09 90 00 Painting and Coating
- A major portion of the work in painting involves surface preparation. Be sure to include cleaning, sanding, filling, and masking costs in the estimate.
- Protection of adjacent surfaces is not included in painting costs. When considering the method of paint application, an important factor is the amount of protection and masking required. These must be estimated separately and may be the determining factor in choosing the method of application.

Reference Numbers
Reference numbers are shown at the beginning of some major classifications. These numbers refer to related items in the Reference Section. The reference information may be an estimating procedure, an alternate pricing method, or technical information.

Note: Not all subdivisions listed here necessarily appear. ■

Did you know?
RSMeans Online gives you the same access to RSMeans' data with 24/7 access:
- Quickly locate costs in the searchable database.
- Build cost lists, estimates, and reports in minutes.
- Adjust costs to any location in the U.S. and Canada with the click of a button.

Start your free trial today at **www.RSMeansOnline.com**

RSMeans Online
FROM THE GORDIAN GROUP®

No part of this cost data may be reproduced, stored in a retrieval system, or transmitted in any form or by any means without prior written permission of RSMeans.

09 01 Maintenance of Finishes

09 01 70 – Maintenance of Wall Finishes

09 01 70.10 Gypsum Wallboard Repairs		Crew	Daily Output	Labor-Hours	Unit	Material	2016 Bare Costs Labor	Equipment	Total	Total Incl O&P
0010	**GYPSUM WALLBOARD REPAIRS**									
0100	Fill and sand, pin/nail holes	1 Carp	960	.008	Ea.		.28		.28	.47
0110	Screw head pops		480	.017			.57		.57	.95
0120	Dents, up to 2" square		48	.167		.01	5.65		5.66	9.45
0130	2" to 4" square		24	.333		.03	11.30		11.33	18.95
0140	Cut square, patch, sand and finish, holes, up to 2" square		12	.667		.03	22.50		22.53	38
0150	2" to 4" square		11	.727		.09	24.50		24.59	41.50
0160	4" to 8" square		10	.800		.24	27		27.24	46
0170	8" to 12" square		8	1		.48	34		34.48	57.50
0180	12" to 32" square		6	1.333		1.58	45		46.58	77
0210	16" by 48"		5	1.600		2.71	54		56.71	94
0220	32" by 48"		4	2		4.43	68		72.43	119
0230	48" square		3.50	2.286		6.25	77.50		83.75	137
0240	60" square		3.20	2.500		9.75	85		94.75	153
0500	Skim coat surface with joint compound		1600	.005	S.F.	.03	.17		.20	.32
0510	Prepare, retape and refinish joints		60	.133	L.F.	.64	4.52		5.16	8.25

09 05 Common Work Results for Finishes

09 05 05 – Selective Demolition for Finishes

09 05 05.10 Selective Demolition, Ceilings

		Crew	Daily Output	Labor-Hours	Unit	Material	Labor	Equipment	Total	Total Incl O&P
0010	**SELECTIVE DEMOLITION, CEILINGS** R024119-10									
0200	Ceiling, drywall, furred and nailed or screwed	2 Clab	800	.020	S.F.		.49		.49	.83
1000	Plaster, lime and horse hair, on wood lath, incl. lath		700	.023			.56		.56	.94
1200	Suspended ceiling, mineral fiber, 2' x 2' or 2' x 4'		1500	.011			.26		.26	.44
1250	On suspension system, incl. system		1200	.013			.33		.33	.55
1500	Tile, wood fiber, 12" x 12", glued		900	.018			.44		.44	.73
1540	Stapled		1500	.011			.26		.26	.44
2000	Wood, tongue and groove, 1" x 4"		1000	.016			.39		.39	.66
2040	1" x 8"		1100	.015			.36		.36	.60
2400	Plywood or wood fiberboard, 4' x 8' sheets		1200	.013			.33		.33	.55

09 05 05.20 Selective Demolition, Flooring

		Crew	Daily Output	Labor-Hours	Unit	Material	Labor	Equipment	Total	Total Incl O&P
0010	**SELECTIVE DEMOLITION, FLOORING** R024119-10									
0200	Brick with mortar	2 Clab	475	.034	S.F.		.83		.83	1.39
0400	Carpet, bonded, including surface scraping		2000	.008			.20		.20	.33
0480	Tackless		9000	.002			.04		.04	.07
0550	Carpet tile, releasable adhesive		5000	.003			.08		.08	.13
0560	Permanent adhesive		1850	.009			.21		.21	.36
0800	Resilient, sheet goods		1400	.011			.28		.28	.47
0850	Vinyl or rubber cove base	1 Clab	1000	.008	L.F.		.20		.20	.33
0860	Vinyl or rubber cove base, molded corner	"	1000	.008	Ea.		.20		.20	.33
0870	For glued and caulked installation, add to labor						50%			
0900	Vinyl composition tile, 12" x 12"	2 Clab	1000	.016	S.F.		.39		.39	.66
2000	Tile, ceramic, thin set		675	.024			.58		.58	.98
2020	Mud set		625	.026			.63		.63	1.06
3000	Wood, block, on end	1 Carp	400	.020			.68		.68	1.14
3200	Parquet		450	.018			.60		.60	1.01
3400	Strip flooring, interior, 2-1/4" x 25/32" thick		325	.025			.83		.83	1.40
3500	Exterior, porch flooring, 1" x 4"		220	.036			1.23		1.23	2.06
3800	Subfloor, tongue and groove, 1" x 6"		325	.025			.83		.83	1.40
3820	1" x 8"		430	.019			.63		.63	1.06
3840	1" x 10"		520	.015			.52		.52	.87

09 05 Common Work Results for Finishes

09 05 05 – Selective Demolition for Finishes

09 05 05.20 Selective Demolition, Flooring

		Crew	Daily Output	Labor-Hours	Unit	Material	2016 Bare Costs Labor	Equipment	Total	Total Incl O&P
4000	Plywood, nailed	1 Carp	600	.013	S.F.		.45		.45	.76
4100	Glued and nailed		400	.020			.68		.68	1.14
4200	Hardboard, 1/4" thick		760	.011			.36		.36	.60

09 05 05.30 Selective Demolition, Walls and Partitions

		Crew	Daily Output	Labor-Hours	Unit	Material	2016 Bare Costs Labor	Equipment	Total	Total Incl O&P
0010	**SELECTIVE DEMOLITION, WALLS AND PARTITIONS** R024119-10									
0020	Walls, concrete, reinforced	B-39	120	.400	C.F.		10	1.90	11.90	18.80
0025	Plain	"	160	.300	"		7.50	1.43	8.93	14.10
1000	Drywall, nailed or screwed	1 Clab	1000	.008	S.F.		.20		.20	.33
1010	2 layers		400	.020			.49		.49	.83
1500	Fiberboard, nailed		900	.009			.22		.22	.37
1568	Plenum barrier, sheet lead		300	.027			.66		.66	1.10
2200	Metal or wood studs, finish 2 sides, fiberboard	B-1	520	.046			1.17		1.17	1.96
2250	Lath and plaster		260	.092			2.34		2.34	3.91
2300	Plasterboard (drywall)		520	.046			1.17		1.17	1.96
2350	Plywood		450	.053			1.35		1.35	2.26
2800	Paneling, 4' x 8' sheets	1 Clab	475	.017			.42		.42	.69
3000	Plaster, lime and horsehair, on wood lath		400	.020			.49		.49	.83
3020	On metal lath		335	.024			.59		.59	.99
3450	Plaster, interior gypsum, acoustic, or cement		60	.133	S.Y.		3.29		3.29	5.50
3500	Stucco, on masonry		145	.055			1.36		1.36	2.28
3510	Commercial 3-coat		80	.100			2.47		2.47	4.13
3520	Interior stucco		25	.320			7.90		7.90	13.20
3760	Tile, ceramic, on walls, thin set		300	.027	S.F.		.66		.66	1.10
3765	Mud set		250	.032	"		.79		.79	1.32

09 22 Supports for Plaster and Gypsum Board

09 22 03 – Fastening Methods for Finishes

09 22 03.20 Drilling Plaster/Drywall

		Crew	Daily Output	Labor-Hours	Unit	Material	2016 Bare Costs Labor	Equipment	Total	Total Incl O&P
0010	**DRILLING PLASTER/DRYWALL**									
1100	Drilling & layout for drywall/plaster walls, up to 1" deep, no anchor									
1200	Holes, 1/4" diameter	1 Carp	150	.053	Ea.	.01	1.81		1.82	3.04
1300	3/8" diameter		140	.057		.01	1.94		1.95	3.25
1400	1/2" diameter		130	.062		.01	2.09		2.10	3.50
1500	3/4" diameter		120	.067		.02	2.26		2.28	3.80
1600	1" diameter		110	.073		.02	2.47		2.49	4.15
1700	1-1/4" diameter		100	.080		.03	2.71		2.74	4.58
1800	1-1/2" diameter		90	.089		.05	3.01		3.06	5.10
1900	For ceiling installations, add						40%			

09 22 13 – Metal Furring

09 22 13.13 Metal Channel Furring

		Crew	Daily Output	Labor-Hours	Unit	Material	2016 Bare Costs Labor	Equipment	Total	Total Incl O&P
0010	**METAL CHANNEL FURRING**									
0030	Beams and columns, 7/8" channels, galvanized, 12" O.C.	1 Lath	155	.052	S.F.	.42	1.71		2.13	3.23
0050	16" O.C.		170	.047		.35	1.56		1.91	2.90
0070	24" O.C.		185	.043		.23	1.43		1.66	2.56
0100	Ceilings, on steel, 7/8" channels, galvanized, 12" O.C.		210	.038		.38	1.26		1.64	2.46
0300	16" O.C.		290	.028		.35	.91		1.26	1.86
0400	24" O.C.		420	.019		.23	.63		.86	1.27
0600	1-5/8" channels, galvanized, 12" O.C.		190	.042		.51	1.39		1.90	2.82
0700	16" O.C.		260	.031		.46	1.02		1.48	2.16
0900	24" O.C.		390	.021		.31	.68		.99	1.44

09 22 Supports for Plaster and Gypsum Board

09 22 13 – Metal Furring

09 22 13.13 Metal Channel Furring

		Crew	Daily Output	Labor-Hours	Unit	Material	2016 Bare Costs Labor	2016 Bare Costs Equipment	Total	Total Incl O&P
0930	7/8" channels with sound isolation clips, 12" O.C.	1 Lath	120	.067	S.F.	1.76	2.21		3.97	5.50
0940	16" O.C.		100	.080		1.32	2.65		3.97	5.75
0950	24" O.C.		165	.048		.88	1.60		2.48	3.56
0960	1-5/8" channels, galvanized, 12" O.C.		110	.073		1.89	2.41		4.30	5.95
0970	16" O.C.		100	.080		1.42	2.65		4.07	5.85
0980	24" O.C.		155	.052		.94	1.71		2.65	3.80
1000	Walls, 7/8" channels, galvanized, 12" O.C.		235	.034		.38	1.13		1.51	2.24
1200	16" O.C.		265	.030		.35	1		1.35	2
1300	24" O.C.		350	.023		.23	.76		.99	1.47
1500	1-5/8" channels, galvanized, 12" O.C.		210	.038		.51	1.26		1.77	2.61
1600	16" O.C.		240	.033		.46	1.10		1.56	2.29
1800	24" O.C.		305	.026		.31	.87		1.18	1.74
1920	7/8" channels with sound isolation clips, 12" O.C.		125	.064		1.76	2.12		3.88	5.35
1940	16" O.C.		100	.080		1.32	2.65		3.97	5.75
1950	24" O.C.		150	.053		.88	1.77		2.65	3.82
1960	1-5/8" channels, galvanized, 12" O.C.		115	.070		1.89	2.30		4.19	5.80
1970	16" O.C.		95	.084		1.42	2.79		4.21	6.05
1980	24" O.C.		140	.057		.94	1.89		2.83	4.10

09 22 16 – Non-Structural Metal Framing

09 22 16.13 Non-Structural Metal Stud Framing

		Crew	Daily Output	Labor-Hours	Unit	Material	2016 Bare Costs Labor	2016 Bare Costs Equipment	Total	Total Incl O&P
0010	**NON-STRUCTURAL METAL STUD FRAMING**									
1600	Non-load bearing, galv., 8' high, 25 ga. 1-5/8" wide, 16" O.C.	1 Carp	619	.013	S.F.	.26	.44		.70	1.02
1610	24" O.C.		950	.008		.20	.29		.49	.70
1620	2-1/2" wide, 16" O.C.		613	.013		.33	.44		.77	1.10
1630	24" O.C.		938	.009		.24	.29		.53	.75
1640	3-5/8" wide, 16" O.C.		600	.013		.39	.45		.84	1.19
1650	24" O.C.		925	.009		.29	.29		.58	.81
1660	4" wide, 16" O.C.		594	.013		.44	.46		.90	1.24
1670	24" O.C.		925	.009		.33	.29		.62	.85
1680	6" wide, 16" O.C.		588	.014		.52	.46		.98	1.34
1690	24" O.C.		906	.009		.39	.30		.69	.93
1700	20 ga. studs, 1-5/8" wide, 16" O.C.		494	.016		.33	.55		.88	1.29
1710	24" O.C.		763	.010		.25	.36		.61	.86
1720	2-1/2" wide, 16" O.C.		488	.016		.42	.56		.98	1.40
1730	24" O.C.		750	.011		.32	.36		.68	.96
1740	3-5/8" wide, 16" O.C.		481	.017		.49	.56		1.05	1.47
1750	24" O.C.		738	.011		.36	.37		.73	1.02
1760	4" wide, 16" O.C.		475	.017		.58	.57		1.15	1.60
1770	24" O.C.		738	.011		.43	.37		.80	1.10
1780	6" wide, 16" O.C.		469	.017		.67	.58		1.25	1.71
1790	24" O.C.		725	.011		.51	.37		.88	1.19
2000	Non-load bearing, galv., 10' high, 25 ga. 1-5/8" wide, 16" O.C.		495	.016		.25	.55		.80	1.19
2100	24" O.C.		760	.011		.18	.36		.54	.80
2200	2-1/2" wide, 16" O.C.		490	.016		.31	.55		.86	1.27
2250	24" O.C.		750	.011		.23	.36		.59	.86
2300	3-5/8" wide, 16" O.C.		480	.017		.37	.57		.94	1.36
2350	24" O.C.		740	.011		.27	.37		.64	.91
2400	4" wide, 16" O.C.		475	.017		.41	.57		.98	1.41
2450	24" O.C.		740	.011		.30	.37		.67	.94
2500	6" wide, 16" O.C.		470	.017		.49	.58		1.07	1.51
2550	24" O.C.		725	.011		.36	.37		.73	1.03
2600	20 ga. studs, 1-5/8" wide, 16" O.C.		395	.020		.32	.69		1.01	1.50

09 22 Supports for Plaster and Gypsum Board

09 22 16 – Non-Structural Metal Framing

09 22 16.13 Non-Structural Metal Stud Framing		Crew	Daily Output	Labor-Hours	Unit	Material	2016 Bare Costs Labor	Equipment	Total	Total Incl O&P
2650	24" O.C.	1 Carp	610	.013	S.F.	.23	.44		.67	1
2700	2-1/2" wide, 16" O.C.		390	.021		.40	.70		1.10	1.60
2750	24" O.C.		600	.013		.30	.45		.75	1.09
2800	3-5/8" wide, 16" OC		385	.021		.46	.70		1.16	1.69
2850	24" O.C.		590	.014		.34	.46		.80	1.14
2900	4" wide, 16" O.C.		380	.021		.55	.71		1.26	1.79
2950	24" O.C.		590	.014		.40	.46		.86	1.21
3000	6" wide, 16" O.C.		375	.021		.64	.72		1.36	1.91
3050	24" O.C.		580	.014		.47	.47		.94	1.30
3060	Non-load bearing, galv., 12' high, 25 ga. 1-5/8" wide, 16" O.C.		413	.019		.24	.66		.90	1.36
3070	24" O.C.		633	.013		.17	.43		.60	.91
3080	2-1/2" wide, 16" O.C.		408	.020		.29	.66		.95	1.43
3090	24" O.C.		625	.013		.22	.43		.65	.97
3100	3-5/8" wide, 16" O.C.		400	.020		.35	.68		1.03	1.53
3110	24" O.C.		617	.013		.26	.44		.70	1.02
3120	4" wide, 16" O.C.		396	.020		.39	.68		1.07	1.58
3130	24" O.C.		617	.013		.29	.44		.73	1.06
3140	6" wide, 16" O.C.		392	.020		.47	.69		1.16	1.68
3150	24" O.C.		604	.013		.34	.45		.79	1.13
3160	20 ga. studs, 1-5/8" wide, 16" O.C.		329	.024		.30	.82		1.12	1.71
3170	24" O.C.		508	.016		.22	.53		.75	1.13
3180	2-1/2" wide, 16" O.C.		325	.025		.38	.83		1.21	1.82
3190	24" O.C.		500	.016		.28	.54		.82	1.22
3200	3-5/8" wide, 16" O.C.		321	.025		.44	.84		1.28	1.89
3210	24" O.C.		492	.016		.32	.55		.87	1.27
3220	4" wide, 16" O.C.		317	.025		.52	.86		1.38	2.01
3230	24" O.C.		492	.016		.38	.55		.93	1.34
3240	6" wide, 16" O.C.		313	.026		.61	.87		1.48	2.12
3250	24" O.C.		483	.017		.44	.56		1	1.43
5000	Load bearing studs, see Section 05 41 13.30									

09 22 26 – Suspension Systems

09 22 26.13 Ceiling Suspension Systems

		Crew	Daily Output	Labor-Hours	Unit	Material	Labor	Equipment	Total	Total Incl O&P
0010	**CEILING SUSPENSION SYSTEMS** for gypsum board or plaster									
8000	Suspended ceilings, including carriers									
8200	1-1/2" carriers, 24" O.C. with:									
8300	7/8" channels, 16" O.C.	1 Lath	275	.029	S.F.	.57	.96		1.53	2.18
8320	24" O.C.		310	.026		.45	.85		1.30	1.87
8400	1-5/8" channels, 16" O.C.		205	.039		.68	1.29		1.97	2.84
8420	24" O.C.		250	.032		.53	1.06		1.59	2.29
8600	2" carriers, 24" O.C. with:									
8700	7/8" channels, 16" O.C.	1 Lath	250	.032	S.F.	.61	1.06		1.67	2.38
8720	24" O.C.		285	.028		.49	.93		1.42	2.04
8800	1-5/8" channels, 16" O.C.		190	.042		.72	1.39		2.11	3.04
8820	24" O.C.		225	.036		.57	1.18		1.75	2.53

09 22 36 – Lath

09 22 36.13 Gypsum Lath

		Crew	Daily Output	Labor-Hours	Unit	Material	Labor	Equipment	Total	Total Incl O&P
0011	**GYPSUM LATH** Plain or perforated, nailed, 3/8" thick	1 Lath	765	.010	S.F.	.34	.35		.69	.93
0101	1/2" thick, nailed		720	.011		.27	.37		.64	.89
0301	Clipped to steel studs, 3/8" thick		675	.012		.34	.39		.73	1
0401	1/2" thick		630	.013		.27	.42		.69	.98
0601	Firestop gypsum base, to steel studs, 3/8" thick		630	.013		.27	.42		.69	.98
0701	1/2" thick		585	.014		.33	.45		.78	1.09

09 22 Supports for Plaster and Gypsum Board

09 22 36 – Lath

09 22 36.13 Gypsum Lath

		Crew	Daily Output	Labor-Hours	Unit	Material	2016 Bare Costs Labor	Equipment	Total	Total Incl O&P
0901	Foil back, to steel studs, 3/8" thick	1 Lath	675	.012	S.F.	.38	.39		.77	1.05
1001	1/2" thick		630	.013		.44	.42		.86	1.16
1501	For ceiling installations, add		1950	.004			.14		.14	.22
1601	For columns and beams, add		1550	.005			.17		.17	.28

09 22 36.23 Metal Lath

		Crew	Daily Output	Labor-Hours	Unit	Material	Labor	Equipment	Total	Total Incl O&P
0010	**METAL LATH** R092000-50									
3601	2.5 lb. diamond painted, on wood framing, on walls	1 Lath	765	.010	S.F.	.41	.35		.76	1.01
3701	On ceilings		675	.012		.41	.39		.80	1.08
4201	3.4 lb. diamond painted, wired to steel framing, on walls		675	.012		.43	.39		.82	1.11
4301	On ceilings		540	.015		.43	.49		.92	1.27
5101	Rib lath, painted, wired to steel, on walls, 2.75 lb.		675	.012		.35	.39		.74	1.01
5201	3.4 lb.		630	.013		.45	.42		.87	1.18
5701	Suspended ceiling system, incl. 3.4 lb. diamond lath, painted		135	.059		.47	1.96		2.43	3.69
5801	Galvanized		135	.059		.47	1.96		2.43	3.69

09 22 36.83 Accessories, Plaster

		Crew	Daily Output	Labor-Hours	Unit	Material	Labor	Equipment	Total	Total Incl O&P
0010	**ACCESSORIES, PLASTER**									
0020	Casing bead, expanded flange, galvanized	1 Lath	2.70	2.963	C.L.F.	56.50	98		154.50	221
0200	Foundation weep screed, galvanized	"	2.70	2.963		54.50	98		152.50	219
0900	Channels, cold rolled, 16 ga., 3/4" deep, galvanized					38.50			38.50	42.50
1620	Corner bead, expanded bullnose, 3/4" radius, #10, galvanized	1 Lath	2.60	3.077		26	102		128	194
1650	#1, galvanized		2.55	3.137		49.50	104		153.50	223
1670	Expanded wing, 2-3/4" wide, #1, galvanized		2.65	3.019		39	100		139	205
1700	Inside corner (corner rite), 3" x 3", painted		2.60	3.077		20.50	102		122.50	188
1750	Strip-ex, 4" wide, painted		2.55	3.137		24	104		128	195
1800	Expansion joint, 3/4" grounds, limited expansion, galv., 1 piece		2.70	2.963		75.50	98		173.50	242
2100	Extreme expansion, galvanized, 2 piece		2.60	3.077		141	102		243	320

09 23 Gypsum Plastering

09 23 13 – Acoustical Gypsum Plastering

09 23 13.10 Perlite or Vermiculite Plaster

		Crew	Daily Output	Labor-Hours	Unit	Material	Labor	Equipment	Total	Total Incl O&P
0010	**PERLITE OR VERMICULITE PLASTER** R092000-50									
0020	In 100 lb. bags, under 200 bags				Bag	17.55			17.55	19.35
0301	2 coats, no lath included, on walls	J-1	830	.048	S.F.	.65	1.42	.17	2.24	3.24
0401	On ceilings		710	.056		.65	1.66	.20	2.51	3.66
0901	3 coats, no lath included, on walls		665	.060		.71	1.78	.21	2.70	3.91
1001	On ceilings		565	.071		.71	2.09	.25	3.05	4.47
1700	For irregular or curved surfaces, add to above				S.Y.		30%			
1800	For columns and beams, add to above						50%			
1900	For soffits, add to ceiling prices						40%			

09 23 20 – Gypsum Plaster

09 23 20.10 Gypsum Plaster On Walls and Ceilings

		Crew	Daily Output	Labor-Hours	Unit	Material	Labor	Equipment	Total	Total Incl O&P
0010	**GYPSUM PLASTER ON WALLS AND CEILINGS** R092000-50									
0020	80# bag, less than 1 ton				Bag	15.95			15.95	17.55
0302	2 coats, no lath included, on walls	J-1	750	.053	S.F.	.41	1.57	.19	2.17	3.23
0402	On ceilings		660	.061		.41	1.79	.21	2.41	3.61
0903	3 coats, no lath included, on walls		620	.065		.59	1.91	.23	2.73	4.01
1002	On ceilings		560	.071		1.05	2.11	.25	3.41	4.88
1600	For irregular or curved surfaces, add						30%			
1800	For columns & beams, add						50%			

09 24 Cement Plastering

09 24 23 – Cement Stucco

09 24 23.40 Stucco

		Crew	Daily Output	Labor-Hours	Unit	Material	2016 Bare Costs Labor	Equipment	Total	Total Incl O&P
0010	STUCCO R092000-50									
0011	3 coats 1" thick, float finish, with mesh, on wood frame	J-2	470	.102	S.F.	.96	3.08	.30	4.34	6.40
0101	On masonry construction	J-1	495	.081		.26	2.39	.28	2.93	4.49
0151	2 coats, 3/4" thick, float finish, no lath incl.	"	980	.041		.26	1.21	.14	1.61	2.41
0301	For trowel finish, add	1 Plas	1530	.005			.16		.16	.27
0600	For coloring, add	J-1	685	.058	S.Y.	.40	1.72	.20	2.32	3.49
0700	For special texture add		200	.200	"	1.41	5.90	.70	8.01	11.95
1001	Exterior stucco, with bonding agent, 3 coats, on walls		1800	.022	S.F.	.37	.66	.08	1.11	1.56
1201	Ceilings		1620	.025		.37	.73	.09	1.19	1.69
1301	Beams		720	.056		.37	1.64	.20	2.21	3.29
1501	Columns		900	.044		.37	1.31	.16	1.84	2.71
1601	Mesh, painted, nailed to wood, 1.8 lb.	1 Lath	540	.015		.72	.49		1.21	1.59
1801	3.6 lb.		495	.016		.43	.53		.96	1.33
1901	Wired to steel, painted, 1.8 lb.		477	.017		.72	.56		1.28	1.70
2101	3.6 lb.		450	.018		.43	.59		1.02	1.42

09 25 Other Plastering

09 25 23 – Lime Based Plastering

09 25 23.10 Venetian Plaster

		Crew	Daily Output	Labor-Hours	Unit	Material	2016 Bare Costs Labor	Equipment	Total	Total Incl O&P
0010	VENETIAN PLASTER									
0100	Walls, 1 coat primer, roller applied	1 Plas	950	.008	S.F.	.16	.26		.42	.61
0210	For pigment, light colors add per S.F. plaster					.02			.02	.02
0220	For pigment, dark colors add per S.F. plaster					.05			.05	.06
0300	For sealer/wax coat incl. burnishing, add	1 Plas	300	.027		.41	.84		1.25	1.82

09 26 Veneer Plastering

09 26 13 – Gypsum Veneer Plastering

09 26 13.20 Blueboard

		Crew	Daily Output	Labor-Hours	Unit	Material	2016 Bare Costs Labor	Equipment	Total	Total Incl O&P
0010	BLUEBOARD For use with thin coat									
0100	plaster application see Section 09 26 13.80									
1000	3/8" thick, on walls or ceilings, standard, no finish included	2 Carp	1900	.008	S.F.	.35	.29		.64	.87
1100	With thin coat plaster finish		875	.018		.46	.62		1.08	1.55
1400	On beams, columns, or soffits, standard, no finish included		675	.024		.40	.80		1.20	1.79
1450	With thin coat plaster finish		475	.034		.51	1.14		1.65	2.48
3000	1/2" thick, on walls or ceilings, standard, no finish included		1900	.008		.34	.29		.63	.85
3100	With thin coat plaster finish		875	.018		.45	.62		1.07	1.54
3300	Fire resistant, no finish included		1900	.008		.34	.29		.63	.85
3400	With thin coat plaster finish		875	.018		.45	.62		1.07	1.54
3450	On beams, columns, or soffits, standard, no finish included		675	.024		.39	.80		1.19	1.78
3500	With thin coat plaster finish		475	.034		.50	1.14		1.64	2.46
3700	Fire resistant, no finish included		675	.024		.39	.80		1.19	1.78
3800	With thin coat plaster finish		475	.034		.50	1.14		1.64	2.46
5000	5/8" thick, on walls or ceilings, fire resistant, no finish included		1900	.008		.34	.29		.63	.85
5100	With thin coat plaster finish		875	.018		.45	.62		1.07	1.54
5500	On beams, columns, or soffits, no finish included		675	.024		.39	.80		1.19	1.78
5600	With thin coat plaster finish		475	.034		.50	1.14		1.64	2.46
6000	For high ceilings, over 8' high, add		3060	.005			.18		.18	.30
6500	For over 3 stories high, add per story		6100	.003			.09		.09	.15

09 26 Veneer Plastering

09 26 13 – Gypsum Veneer Plastering

09 26 13.80 Thin Coat Plaster

09 26 13.80 Thin Coat Plaster		Crew	Daily Output	Labor-Hours	Unit	Material	2016 Bare Costs Labor	2016 Bare Costs Equipment	Total	Total Incl O&P
0010	THIN COAT PLASTER R092000-50									
0012	1 coat veneer, not incl. lath	J-1	3600	.011	S.F.	.11	.33	.04	.48	.70
1000	In 50 lb. bags				Bag	15.05			15.05	16.55

09 28 Backing Boards and Underlayments

09 28 13 – Cementitious Backing Boards

09 28 13.10 Cementitious Backerboard

		Crew	Daily Output	Labor-Hours	Unit	Material	Labor	Equipment	Total	Total Incl O&P
0010	CEMENTITIOUS BACKERBOARD									
0070	Cementitious backerboard, on floor, 3' x 4' x 1/2" sheets	2 Carp	525	.030	S.F.	.84	1.03		1.87	2.65
0080	3' x 5' x 1/2" sheets		525	.030		.81	1.03		1.84	2.62
0090	3' x 6' x 1/2" sheets		525	.030		.79	1.03		1.82	2.59
0100	3' x 4' x 5/8" sheets		525	.030		1.01	1.03		2.04	2.84
0110	3' x 5' x 5/8" sheets		525	.030		1.01	1.03		2.04	2.84
0120	3' x 6' x 5/8" sheets		525	.030		1	1.03		2.03	2.82
0150	On wall, 3' x 4' x 1/2" sheets		350	.046		.84	1.55		2.39	3.51
0160	3' x 5' x 1/2" sheets		350	.046		.81	1.55		2.36	3.48
0170	3' x 6' x 1/2" sheets		350	.046		.79	1.55		2.34	3.45
0180	3' x 4' x 5/8" sheets		350	.046		1.01	1.55		2.56	3.70
0190	3' x 5' x 5/8" sheets		350	.046		1.01	1.55		2.56	3.70
0200	3' x 6' x 5/8" sheets		350	.046		1	1.55		2.55	3.68
0250	On counter, 3' x 4' x 1/2" sheets		180	.089		.84	3.01		3.85	5.95
0260	3' x 5' x 1/2" sheets		180	.089		.81	3.01		3.82	5.95
0270	3' x 6' x 1/2" sheets		180	.089		.79	3.01		3.80	5.90
0300	3' x 4' x 5/8" sheets		180	.089		1.01	3.01		4.02	6.15
0310	3' x 5' x 5/8" sheets		180	.089		1.01	3.01		4.02	6.15
0320	3' x 6' x 5/8" sheets		180	.089		1	3.01		4.01	6.15

09 29 Gypsum Board

09 29 10 – Gypsum Board Panels

09 29 10.20 Taping and Finishing

		Crew	Daily Output	Labor-Hours	Unit	Material	Labor	Equipment	Total	Total Incl O&P
0010	TAPING AND FINISHING									
3600	For taping and finishing joints, add	2 Carp	2000	.008	S.F.	.05	.27		.32	.50
4500	For thin coat plaster instead of taping, add	J-1	3600	.011	"	.11	.33	.04	.48	.70

09 29 10.30 Gypsum Board

		Crew	Daily Output	Labor-Hours	Unit	Material	Labor	Equipment	Total	Total Incl O&P
0010	GYPSUM BOARD on walls & ceilings R092910-10									
0100	Nailed or screwed to studs unless otherwise noted									
0110	1/4" thick, on walls or ceilings, standard, no finish included	2 Carp	1330	.012	S.F.	.36	.41		.77	1.08
0115	1/4" thick, on walls or ceilings, flexible, no finish included		1050	.015		.51	.52		1.03	1.42
0117	1/4" thick, on columns or soffits, flexible, no finish included		1050	.015		.51	.52		1.03	1.42
0130	1/4" thick, standard, no finish included, less than 800 S.F.		510	.031		.36	1.06		1.42	2.18
0150	3/8" thick, on walls, standard, no finish included		2000	.008		.35	.27		.62	.84
0200	On ceilings, standard, no finish included		1800	.009		.35	.30		.65	.89
0250	On beams, columns, or soffits, no finish included		675	.024		.35	.80		1.15	1.74
0300	1/2" thick, on walls, standard, no finish included		2000	.008		.33	.27		.60	.81
0350	Taped and finished (level 4 finish)		965	.017		.38	.56		.94	1.36
0390	With compound skim coat (level 5 finish)		775	.021		.43	.70		1.13	1.64
0400	Fire resistant, no finish included		2000	.008		.36	.27		.63	.85
0450	Taped and finished (level 4 finish)		965	.017		.41	.56		.97	1.39

09 29 Gypsum Board

09 29 10 – Gypsum Board Panels

09 29 10.30 Gypsum Board		Crew	Daily Output	Labor-Hours	Unit	Material	2016 Bare Costs Labor	Equipment	Total	Total Incl O&P
0490	With compound skim coat (level 5 finish)	2 Carp	775	.021	S.F.	.46	.70		1.16	1.68
0500	Water resistant, no finish included		2000	.008		.41	.27		.68	.90
0550	Taped and finished (level 4 finish)		965	.017		.46	.56		1.02	1.44
0590	With compound skim coat (level 5 finish)		775	.021		.51	.70		1.21	1.73
0600	Prefinished, vinyl, clipped to studs		900	.018		.49	.60		1.09	1.55
0700	Mold resistant, no finish included		2000	.008		.43	.27		.70	.92
0710	Taped and finished (level 4 finish)		965	.017		.48	.56		1.04	1.47
0720	With compound skim coat (level 5 finish)		775	.021		.53	.70		1.23	1.75
1000	On ceilings, standard, no finish included		1800	.009		.33	.30		.63	.86
1050	Taped and finished (level 4 finish)		765	.021		.38	.71		1.09	1.61
1090	With compound skim coat (level 5 finish)		610	.026		.43	.89		1.32	1.96
1100	Fire resistant, no finish included		1800	.009		.36	.30		.66	.90
1150	Taped and finished (level 4 finish)		765	.021		.41	.71		1.12	1.64
1195	With compound skim coat (level 5 finish)		610	.026		.46	.89		1.35	2
1200	Water resistant, no finish included		1800	.009		.41	.30		.71	.95
1250	Taped and finished (level 4 finish)		765	.021		.46	.71		1.17	1.69
1290	With compound skim coat (level 5 finish)		610	.026		.51	.89		1.40	2.05
1310	Mold resistant, no finish included		1800	.009		.43	.30		.73	.97
1320	Taped and finished (level 4 finish)		765	.021		.48	.71		1.19	1.72
1330	With compound skim coat (level 5 finish)		610	.026		.53	.89		1.42	2.07
1350	Sag resistant, no finish included		1600	.010		.35	.34		.69	.96
1360	Taped and finished (level 4 finish)		765	.021		.40	.71		1.11	1.63
1370	With compound skim coat (level 5 finish)		610	.026		.45	.89		1.34	1.99
1500	On beams, columns, or soffits, standard, no finish included		675	.024		.38	.80		1.18	1.77
1550	Taped and finished (level 4 finish)		540	.030		.38	1		1.38	2.10
1590	With compound skim coat (level 5 finish)		475	.034		.43	1.14		1.57	2.38
1600	Fire resistant, no finish included		675	.024		.36	.80		1.16	1.75
1650	Taped and finished (level 4 finish)		540	.030		.41	1		1.41	2.13
1690	With compound skim coat (level 5 finish)		475	.034		.46	1.14		1.60	2.42
1700	Water resistant, no finish included		675	.024		.47	.80		1.27	1.87
1750	Taped and finished (level 4 finish)		540	.030		.46	1		1.46	2.18
1790	With compound skim coat (level 5 finish)		475	.034		.51	1.14		1.65	2.47
1800	Mold resistant, no finish included		675	.024		.49	.80		1.29	1.89
1810	Taped and finished (level 4 finish)		540	.030		.48	1		1.48	2.21
1820	With compound skim coat (level 5 finish)		475	.034		.53	1.14		1.67	2.49
1850	Sag resistant, no finish included		675	.024		.40	.80		1.20	1.79
1860	Taped and finished (level 4 finish)		540	.030		.40	1		1.40	2.12
1870	With compound skim coat (level 5 finish)		475	.034		.45	1.14		1.59	2.41
2000	5/8" thick, on walls, standard, no finish included		2000	.008		.34	.27		.61	.82
2050	Taped and finished (level 4 finish)		965	.017		.39	.56		.95	1.37
2090	With compound skim coat (level 5 finish)		775	.021		.44	.70		1.14	1.65
2100	Fire resistant, no finish included		2000	.008		.35	.27		.62	.84
2150	Taped and finished (level 4 finish)		965	.017		.40	.56		.96	1.38
2195	With compound skim coat (level 5 finish)		775	.021		.45	.70		1.15	1.67
2200	Water resistant, no finish included		2000	.008		.43	.27		.70	.92
2250	Taped and finished (level 4 finish)		965	.017		.48	.56		1.04	1.47
2290	With compound skim coat (level 5 finish)		775	.021		.53	.70		1.23	1.75
2300	Prefinished, vinyl, clipped to studs		900	.018		.76	.60		1.36	1.85
2510	Mold resistant, no finish included		2000	.008		.47	.27		.74	.97
2520	Taped and finished (level 4 finish)		965	.017		.52	.56		1.08	1.51
2530	With compound skim coat (level 5 finish)		775	.021		.57	.70		1.27	1.80
3000	On ceilings, standard, no finish included		1800	.009		.34	.30		.64	.87
3050	Taped and finished (level 4 finish)		765	.021		.39	.71		1.10	1.62

09 29 Gypsum Board

09 29 10 – Gypsum Board Panels

09 29 10.30 Gypsum Board		Crew	Daily Output	Labor-Hours	Unit	Material	2016 Bare Costs Labor	Equipment	Total	Total Incl O&P
3090	With compound skim coat (level 5 finish)	2 Carp	615	.026	S.F.	.44	.88		1.32	1.96
3100	Fire resistant, no finish included		1800	.009		.35	.30		.65	.89
3150	Taped and finished (level 4 finish)		765	.021		.40	.71		1.11	1.63
3190	With compound skim coat (level 5 finish)		615	.026		.45	.88		1.33	1.98
3200	Water resistant, no finish included		1800	.009		.43	.30		.73	.97
3250	Taped and finished (level 4 finish)		765	.021		.48	.71		1.19	1.72
3290	With compound skim coat (level 5 finish)		615	.026		.53	.88		1.41	2.06
3300	Mold resistant, no finish included		1800	.009		.47	.30		.77	1.02
3310	Taped and finished (level 4 finish)		765	.021		.52	.71		1.23	1.76
3320	With compound skim coat (level 5 finish)		615	.026		.57	.88		1.45	2.11
3500	On beams, columns, or soffits, no finish included		675	.024		.39	.80		1.19	1.78
3550	Taped and finished (level 4 finish)		475	.034		.45	1.14		1.59	2.40
3590	With compound skim coat (level 5 finish)		380	.042		.51	1.43		1.94	2.95
3600	Fire resistant, no finish included		675	.024		.40	.80		1.20	1.79
3650	Taped and finished (level 4 finish)		475	.034		.46	1.14		1.60	2.41
3690	With compound skim coat (level 5 finish)		380	.042		.45	1.43		1.88	2.89
3700	Water resistant, no finish included		675	.024		.49	.80		1.29	1.89
3750	Taped and finished (level 4 finish)		475	.034		.53	1.14		1.67	2.49
3790	With compound skim coat (level 5 finish)		380	.042		.55	1.43		1.98	2.99
3800	Mold resistant, no finish included		675	.024		.54	.80		1.34	1.94
3810	Taped and finished (level 4 finish)		475	.034		.57	1.14		1.71	2.54
3820	With compound skim coat (level 5 finish)		380	.042		.59	1.43		2.02	3.04
4000	Fireproofing, beams or columns, 2 layers, 1/2" thick, incl finish		330	.048		.81	1.64		2.45	3.65
4010	Mold resistant		330	.048		.95	1.64		2.59	3.80
4050	5/8" thick		300	.053		.79	1.81		2.60	3.90
4060	Mold resistant		300	.053		1.03	1.81		2.84	4.17
4100	3 layers, 1/2" thick		225	.071		1.22	2.41		3.63	5.40
4110	Mold resistant		225	.071		1.43	2.41		3.84	5.60
4150	5/8" thick		210	.076		1.19	2.58		3.77	5.65
4160	Mold resistant		210	.076		1.55	2.58		4.13	6.05
5200	For work over 8' high, add		3060	.005			.18		.18	.30
5270	For textured spray, add	2 Lath	1600	.010		.04	.33		.37	.58
5350	For finishing inner corners, add	2 Carp	950	.017	L.F.	.10	.57		.67	1.07
5355	For finishing outer corners, add	"	1250	.013		.23	.43		.66	.99
5500	For acoustical sealant, add per bead	1 Carp	500	.016		.04	.54		.58	.96
5550	Sealant, 1 quart tube				Ea.	7.05			7.05	7.80
6000	Gypsum sound dampening panels									
6010	1/2" thick on walls, multi-layer, light weight, no finish included	2 Carp	1500	.011	S.F.	2.02	.36		2.38	2.83
6015	Taped and finished (level 4 finish)		725	.022		2.07	.75		2.82	3.52
6020	With compound skim coat (level 5 finish)		580	.028		2.12	.94		3.06	3.90
6025	5/8" thick on walls, for wood studs, no finish included		1500	.011		2.23	.36		2.59	3.06
6030	Taped and finished (level 4 finish)		725	.022		2.28	.75		3.03	3.76
6035	With compound skim coat (level 5 finish)		580	.028		2.33	.94		3.27	4.13
6040	For metal stud, no finish included		1500	.011		2.18	.36		2.54	3.01
6045	Taped and finished (level 4 finish)		725	.022		2.23	.75		2.98	3.70
6050	With compound skim coat (level 5 finish)		580	.028		2.28	.94		3.22	4.08
6055	Abuse resist, no finish included		1500	.011		3.95	.36		4.31	4.96
6060	Taped and finished (level 4 finish)		725	.022		4	.75		4.75	5.65
6065	With compound skim coat (level 5 finish)		580	.028		4.05	.94		4.99	6.05
6070	Shear rated, no finish included		1500	.011		4.53	.36		4.89	5.60
6075	Taped and finished (level 4 finish)		725	.022		4.58	.75		5.33	6.30
6080	With compound skim coat (level 5 finish)		580	.028		4.63	.94		5.57	6.65
6085	For SCIF applications, no finish included		1500	.011		4.96	.36		5.32	6.05

09 29 Gypsum Board

09 29 10 – Gypsum Board Panels

09 29 10.30 Gypsum Board		Crew	Daily Output	Labor-Hours	Unit	Material	2016 Bare Costs Labor	Equipment	Total	Total Incl O&P
6090	Taped and finished (level 4 finish)	2 Carp	725	.022	S.F.	5	.75		5.75	6.75
6095	With compound skim coat (level 5 finish)		580	.028		5.05	.94		5.99	7.10
6100	1-3/8" thick on walls, THX Certified, no finish included		1500	.011		8.40	.36		8.76	9.85
6105	Taped and finished (level 4 finish)		725	.022		8.45	.75		9.20	10.55
6110	With compound skim coat (level 5 finish)		580	.028		8.50	.94		9.44	10.90
6115	5/8" thick on walls, score & snap installation, no finish included		2000	.008		1.79	.27		2.06	2.42
6120	Taped and finished (level 4 finish)		965	.017		1.84	.56		2.40	2.96
6125	With compound skim coat (level 5 finish)		775	.021		1.89	.70		2.59	3.25
7020	5/8" thick on ceilings, for wood joists, no finish included		1200	.013		2.23	.45		2.68	3.21
7025	Taped and finished (level 4 finish)		510	.031		2.28	1.06		3.34	4.29
7030	With compound skim coat (level 5 finish)		410	.039		2.33	1.32		3.65	4.77
7035	For metal joists, no finish included		1200	.013		2.18	.45		2.63	3.16
7040	Taped and finished (level 4 finish)		510	.031		2.23	1.06		3.29	4.23
7045	With compound skim coat (level 5 finish)		410	.039		2.28	1.32		3.60	4.72
7050	Abuse resist, no finish included		1200	.013		3.95	.45		4.40	5.10
7055	Taped and finished (level 4 finish)		510	.031		4	1.06		5.06	6.20
7060	With compound skim coat (level 5 finish)		410	.039		4.05	1.32		5.37	6.65
7065	Shear rated, no finish included		1200	.013		4.53	.45		4.98	5.75
7070	Taped and finished (level 4 finish)		510	.031		4.58	1.06		5.64	6.85
7075	With compound skim coat (level 5 finish)		410	.039		4.63	1.32		5.95	7.30
7080	For SCIF applications, no finish included		1200	.013		4.96	.45		5.41	6.20
7085	Taped and finished (level 4 finish)		510	.031		5	1.06		6.06	7.30
7090	With compound skim coat (level 5 finish)		410	.039		5.05	1.32		6.37	7.75
8010	5/8" thick on ceilings, score & snap installation, no finish included		1600	.010		1.79	.34		2.13	2.54
8015	Taped and finished (level 4 finish)		680	.024		1.84	.80		2.64	3.36
8020	With compound skim coat (level 5 finish)		545	.029		1.89	1		2.89	3.75

09 29 15 – Gypsum Board Accessories

09 29 15.10 Accessories, Drywall

		Crew	Daily Output	Labor-Hours	Unit	Material	Labor	Equipment	Total	Total Incl O&P
0011	ACCESSORIES, DRYWALL Casing bead, galvanized steel	1 Carp	290	.028	L.F.	.24	.94		1.18	1.83
0101	Vinyl		290	.028		.23	.94		1.17	1.82
0401	Corner bead, galvanized steel, 1-1/4" x 1-1/4"		350	.023		.17	.78		.95	1.48
0411	1-1/4" x 1-1/4", 10' long		35	.229	Ea.	1.66	7.75		9.41	14.80
0601	Vinyl corner bead		400	.020	L.F.	.20	.68		.88	1.36
0901	Furring channel, galv. steel, 7/8" deep, standard		260	.031		.33	1.04		1.37	2.12
1001	Resilient		260	.031		.25	1.04		1.29	2.02
1101	J trim, galvanized steel, 1/2" wide		300	.027			.90		.90	1.51
1121	5/8" wide		300	.027		.32	.90		1.22	1.86
1160	Screws #6 x 1" A				M	10.25			10.25	11.30
1170	#6 x 1-5/8" A				"	15.10			15.10	16.60
1501	Z stud, galvanized steel, 1-1/2" wide	1 Carp	260	.031	L.F.	.39	1.04		1.43	2.18

09 30 Tiling

09 30 13 – Ceramic Tiling

09 30 13.10 Ceramic Tile

		Crew	Daily Output	Labor-Hours	Unit	Material	2016 Bare Costs Labor	2016 Bare Costs Equipment	Total	Total Incl O&P
0010	**CERAMIC TILE**									
0020	Backsplash, thinset, average grade tiles	1 Tilf	50	.160	S.F.	2.70	4.98		7.68	11
0022	Custom grade tiles		50	.160		5.40	4.98		10.38	14
0024	Luxury grade tiles		50	.160		10.80	4.98		15.78	19.90
0026	Economy grade tiles		50	.160		2.46	4.98		7.44	10.75
0050	Base, using 1' x 4" high pc. with 1" x 1" tiles, mud set	D-7	82	.195	L.F.	5.10	5.40		10.50	14.35
0100	Thin set	"	128	.125		4.84	3.47		8.31	10.90
0300	For 6" high base, 1" x 1" tile face, add					.77			.77	.85
0400	For 2" x 2" tile face, add to above					.42			.42	.46
0600	Cove base, 4-1/4" x 4-1/4" high, mud set	D-7	91	.176		4.19	4.88		9.07	12.50
0700	Thin set		128	.125		4.07	3.47		7.54	10.10
0900	6" x 4-1/4" high, mud set		100	.160		4.37	4.44		8.81	12
1000	Thin set		137	.117		4.25	3.24		7.49	9.90
1200	Sanitary cove base, 6" x 4-1/4" high, mud set		93	.172		4.67	4.77		9.44	12.85
1300	Thin set		124	.129		4.55	3.58		8.13	10.80
1500	6" x 6" high, mud set		84	.190		5.45	5.30		10.75	14.55
1600	Thin set		117	.137		5.30	3.79		9.09	12
1800	Bathroom accessories, average (soap dish, tooth brush holder)		82	.195	Ea.	9.55	5.40		14.95	19.25
1900	Bathtub, 5', rec. 4-1/4" x 4-1/4" tile wainscot, adhesive set 6' high		2.90	5.517		174	153		327	440
2100	7' high wainscot		2.50	6.400		198	178		376	505
2200	8' high wainscot		2.20	7.273		211	202		413	555
2400	Bullnose trim, 4-1/4" x 4-1/4", mud set		82	.195	L.F.	4.15	5.40		9.55	13.30
2500	Thin set		128	.125		4.07	3.47		7.54	10.05
2700	2" x 6" bullnose trim, mud set		84	.190		4.38	5.30		9.68	13.35
2800	Thin set		124	.129		4.32	3.58		7.90	10.55
3000	Floors, natural clay, random or uniform, thin set, color group 1		183	.087	S.F.	4.15	2.43		6.58	8.50
3100	Color group 2		183	.087		5.75	2.43		8.18	10.30
3255	Floors, glazed, thin set, 6" x 6", color group 1		300	.053		4.66	1.48		6.14	7.50
3260	8" x 8" tile		300	.053		4.66	1.48		6.14	7.50
3270	12" x 12" tile		290	.055		6.05	1.53		7.58	9.20
3280	16" x 16" tile		280	.057		7.45	1.59		9.04	10.70
3281	18" x 18" tile		270	.059		6.80	1.64		8.44	10.15
3282	20" x 20" tile		260	.062		9.90	1.71		11.61	13.65
3283	24" x 24" tile		250	.064		11.25	1.78		13.03	15.20
3285	Border, 6" x 12" tile		200	.080		11.50	2.22		13.72	16.25
3290	3" x 12" tile		200	.080		40	2.22		42.22	47.50
3300	Porcelain type, 1 color, color group 2, 1" x 1"		183	.087		5.20	2.43		7.63	9.65
3310	2" x 2" or 2" x 1", thin set		190	.084		6.40	2.34		8.74	10.85
3350	For random blend, 2 colors, add					1			1	1.10
3360	4 colors, add					1.50			1.50	1.65
4300	Specialty tile, 4-1/4" x 4-1/4" x 1/2", decorator finish	D-7	183	.087		10.40	2.43		12.83	15.40
4500	Add for epoxy grout, 1/16" joint, 1" x 1" tile		800	.020		.67	.56		1.23	1.64
4600	2" x 2" tile		820	.020		.62	.54		1.16	1.56
4610	Add for epoxy grout, 1/8" joint, 8" x 8" x 3/8" tile, add		900	.018		1.39	.49		1.88	2.33
4800	Pregrouted sheets, walls, 4-1/4" x 4-1/4", 6" x 4-1/4"									
4810	and 8-1/2" x 4-1/4", 4 S.F. sheets, silicone grout	D-7	240	.067	S.F.	5.40	1.85		7.25	8.95
5100	Floors, unglazed, 2 S.F. sheets,									
5110	urethane adhesive	D-7	180	.089	S.F.	5.10	2.47		7.57	9.60
5400	Walls, interior, thin set, 4-1/4" x 4-1/4" tile		190	.084		2.49	2.34		4.83	6.50
5500	6" x 4-1/4" tile		190	.084		3.01	2.34		5.35	7.10
5700	8-1/2" x 4-1/4" tile		190	.084		5.25	2.34		7.59	9.60
5800	6" x 6" tile		175	.091		3.46	2.54		6	7.90
5810	8" x 8" tile		170	.094		4.75	2.61		7.36	9.45

09 30 Tiling

09 30 13 – Ceramic Tiling

09 30 13.10 Ceramic Tile

		Crew	Daily Output	Labor-Hours	Unit	Material	2016 Bare Costs Labor	Equipment	Total	Total Incl O&P
5820	12" x 12" tile	D-7	160	.100	S.F.	4.55	2.78		7.33	9.50
5830	16" x 16" tile		150	.107		5.05	2.96		8.01	10.40
6000	Decorated wall tile, 4-1/4" x 4-1/4", color group 1		270	.059		3.53	1.64		5.17	6.55
6100	Color group 4		180	.089		52	2.47		54.47	61
6600	Crystalline glazed, 4-1/4" x 4-1/4", mud set, plain		100	.160		4.52	4.44		8.96	12.15
6700	4-1/4" x 4-1/4", scored tile		100	.160		6	4.44		10.44	13.80
6900	6" x 6" plain		93	.172		6.80	4.77		11.57	15.20
7000	For epoxy grout, 1/16" joints, 4-1/4" tile, add		800	.020		.41	.56		.97	1.35
7200	For tile set in dry mortar, add		1735	.009			.26		.26	.41
7300	For tile set in Portland cement mortar, add		290	.055		.16	1.53		1.69	2.66
9300	Ceramic tiles, recycled glass, standard colors, 2" x 2" thru 6" x 6" G		190	.084		21.50	2.34		23.84	27.50
9310	6" x 6" G		175	.091		21.50	2.54		24.04	28
9320	8" x 8" G		170	.094		23	2.61		25.61	29.50
9330	12" x 12" G		160	.100		23	2.78		25.78	30
9340	Earthtones, 2" x 2" to 4" x 8" G		190	.084		25.50	2.34		27.84	32
9350	6" x 6" G		175	.091		25.50	2.54		28.04	32
9360	8" x 8" G		170	.094		26.50	2.61		29.11	33
9370	12" x 12" G		160	.100		26.50	2.78		29.28	33.50
9380	Deep colors, 2" x 2" to 4" x 8" G		190	.084		30	2.34		32.34	37
9390	6" x 6" G		175	.091		30	2.54		32.54	37
9400	8" x 8" G		170	.094		31.50	2.61		34.11	38.50
9410	12" x 12" G		160	.100		31.50	2.78		34.28	39

09 30 13.45 Ceramic Tile Accessories

		Crew	Daily Output	Labor-Hours	Unit	Material	Labor	Equipment	Total	Total Incl O&P
0010	**CERAMIC TILE ACCESSORIES**									
0100	Spacers, 1/8"				C	1.98			1.98	2.18
1310	Sealer for natural stone tile, installed	1 Tilf	650	.012	S.F.	.05	.38		.43	.68

09 30 16 – Quarry Tiling

09 30 16.10 Quarry Tile

		Crew	Daily Output	Labor-Hours	Unit	Material	Labor	Equipment	Total	Total Incl O&P
0010	**QUARRY TILE**									
0100	Base, cove or sanitary, mud set, to 5" high, 1/2" thick	D-7	110	.145	L.F.	5.85	4.04		9.89	13
0300	Bullnose trim, red, mud set, 6" x 6" x 1/2" thick		120	.133		4.70	3.70		8.40	11.15
0400	4" x 4" x 1/2" thick		110	.145		4.83	4.04		8.87	11.85
0600	4" x 8" x 1/2" thick, using 8" as edge		130	.123		4.79	3.42		8.21	10.80
0700	Floors, mud set, 1,000 S.F. lots, red, 4" x 4" x 1/2" thick		120	.133	S.F.	8.05	3.70		11.75	14.85
0900	6" x 6" x 1/2" thick		140	.114		7.95	3.17		11.12	13.90
1000	4" x 8" x 1/2" thick		130	.123		5	3.42		8.42	11.05
1300	For waxed coating, add					.76			.76	.84
1500	For non-standard colors, add					.46			.46	.51
1600	For abrasive surface, add					.52			.52	.57
1800	Brown tile, imported, 6" x 6" x 3/4"	D-7	120	.133		7.40	3.70		11.10	14.15
1900	8" x 8" x 1"		110	.145		8.70	4.04		12.74	16.10
2100	For thin set mortar application, deduct		700	.023			.63		.63	1.03
2700	Stair tread, 6" x 6" x 3/4", plain		50	.320		7.40	8.90		16.30	22.50
2800	Abrasive		47	.340		6.80	9.45		16.25	23
3000	Wainscot, 6" x 6" x 1/2", thin set, red		105	.152		5.55	4.23		9.78	12.95
3100	Non-standard colors		105	.152		5.50	4.23		9.73	12.90
3300	Window sill, 6" wide, 3/4" thick		90	.178	L.F.	5.45	4.93		10.38	14
3400	Corners		80	.200	Ea.	6.10	5.55		11.65	15.75

09 30 Tiling

09 30 23 – Glass Mosaic Tiling

09 30 23.10 Glass Mosaics

		Crew	Daily Output	Labor-Hours	Unit	Material	2016 Bare Costs Labor	2016 Bare Costs Equipment	Total	Total Incl O&P
0010	**GLASS MOSAICS** 3/4" tile on 12" sheets, standard grout									
1020	1" tile on 12" sheets, opalescent finish	D-7	73	.219	S.F.	17.50	6.10		23.60	29
1040	1" x 2" tile on 12" sheet, blend		73	.219		19.05	6.10		25.15	31
1060	2" tile on 12" sheet, blend		73	.219		16.50	6.10		22.60	28
1080	5/8" x random tile, linear, on 12" sheet, blend		73	.219		25	6.10		31.10	37.50
1600	Dots on 12" sheet		73	.219		25	6.10		31.10	37.50
1700	For glass mosaic tiles set in dry mortar, add		290	.055		.45	1.53		1.98	2.98
1720	For glass mosaic tile set in Portland cement mortar, add		290	.055		.01	1.53		1.54	2.49
1730	For polyblend sanded tile grout		96.15	.166	Lb.	2.19	4.62		6.81	9.85

09 30 29 – Metal Tiling

09 30 29.10 Metal Tile

		Crew	Daily Output	Labor-Hours	Unit	Material	Labor	Equipment	Total	Total Incl O&P
0010	**METAL TILE** 4' x 4' sheet, 24 ga., tile pattern, nailed									
0200	Stainless steel	2 Carp	512	.031	S.F.	28	1.06		29.06	33
0400	Aluminized steel	"	512	.031	"	15.10	1.06		16.16	18.35

09 30 95 – Tile & Stone Setting Materials and Specialties

09 30 95.10 Moisture Resistant, Anti-Fracture Membrane

		Crew	Daily Output	Labor-Hours	Unit	Material	Labor	Equipment	Total	Total Incl O&P
0010	**MOISTURE RESISTANT, ANTI-FRACTURE MEMBRANE**									
0200	Elastomeric membrane, 1/16" thick	D-7	275	.058	S.F.	1.07	1.61		2.68	3.79

09 34 Waterproofing-Membrane Tiling

09 34 13 – Waterproofing-Membrane Ceramic Tiling

09 34 13.10 Ceramic Tile Waterproofing Membrane

		Crew	Daily Output	Labor-Hours	Unit	Material	Labor	Equipment	Total	Total Incl O&P
0010	**CERAMIC TILE WATERPROOFING MEMBRANE**									
0020	On floors, including thinset									
0030	Fleece laminated polyethylene grid, 1/8" thick	D-7	250	.064	S.F.	2.28	1.78		4.06	5.35
0040	5/16" thick	"	250	.064	"	2.60	1.78		4.38	5.75
0050	On walls, including thinset									
0060	Fleece laminated polyethylene sheet, 8 mil thick	D-7	480	.033	S.F.	2.28	.92		3.20	4
0070	Accessories, including thinset									
0080	Joint and corner sheet, 4 mils thick, 5" wide	1 Tilf	240	.033	L.F.	1.35	1.04		2.39	3.16
0090	7-1/4" wide		180	.044		1.71	1.38		3.09	4.12
0100	10" wide		120	.067		2.08	2.08		4.16	5.65
0110	Pre-formed corners, inside		32	.250	Ea.	6.95	7.80		14.75	20.50
0120	Outside		32	.250		7.65	7.80		15.45	21
0130	2" flanged floor drain with 6" stainless steel grate		16	.500		370	15.60		385.60	435
0140	EPS, sloped shower floor		480	.017	S.F.	4.97	.52		5.49	6.30
0150	Curb		32	.250	L.F.	14.05	7.80		21.85	28

09 51 Acoustical Ceilings

09 51 23 – Acoustical Tile Ceilings

09 51 23.10 Suspended Acoustic Ceiling Tiles

		Crew	Daily Output	Labor-Hours	Unit	Material	2016 Bare Costs Labor	Equipment	Total	Total Incl O&P
0010	**SUSPENDED ACOUSTIC CEILING TILES**, not including									
0100	suspension system									
0300	Fiberglass boards, film faced, 2' x 2' or 2' x 4', 5/8" thick	1 Carp	625	.013	S.F.	1.25	.43		1.68	2.11
0400	3/4" thick		600	.013		2.63	.45		3.08	3.65
0500	3" thick, thermal, R11		450	.018		2.62	.60		3.22	3.89
0600	Glass cloth faced fiberglass, 3/4" thick		500	.016		2.72	.54		3.26	3.90
0700	1" thick		485	.016		3.31	.56		3.87	4.58
0820	1-1/2" thick, nubby face		475	.017		2.60	.57		3.17	3.82
1110	Mineral fiber tile, lay-in, 2' x 2' or 2' x 4', 5/8" thick, fine texture		625	.013		.96	.43		1.39	1.79
1115	Rough textured		625	.013		.86	.43		1.29	1.68
1125	3/4" thick, fine textured		600	.013		1.99	.45		2.44	2.95
1130	Rough textured		600	.013		1.60	.45		2.05	2.52
1135	Fissured		600	.013		1.85	.45		2.30	2.80
1150	Tegular, 5/8" thick, fine textured		470	.017		1.04	.58		1.62	2.11
1155	Rough textured		470	.017		1.23	.58		1.81	2.32
1165	3/4" thick, fine textured		450	.018		2.08	.60		2.68	3.30
1170	Rough textured		450	.018		1.41	.60		2.01	2.56
1175	Fissured		450	.018		2.02	.60		2.62	3.23
1185	For plastic film face, add					.79			.79	.87
1190	For fire rating, add					.45			.45	.50
1600	Solid alum. planks, 3-1/4"x12', open reveal	1 Carp	500	.016		2.35	.54		2.89	3.50
1650	Closed reveal		500	.016		3	.54		3.54	4.21
1700	7-1/4"x12', open reveal		500	.016		4	.54		4.54	5.30
1750	Closed reveal		500	.016		5.10	.54		5.64	6.50
1775	Metal, open cell, 2'x2', 6" cell		500	.016		8.25	.54		8.79	10
1800	8" cell		500	.016		9.15	.54		9.69	10.95
1825	2'x4', 6" cell		500	.016		5.35	.54		5.89	6.80
1850	8" cell		500	.016		5.35	.54		5.89	6.80
1870	Translucent lay-in panels, 2'x2'		500	.016		23	.54		23.54	26
1890	2'x6'		500	.016		17.20	.54		17.74	19.80
3750	Wood fiber in cementitious binder, 2' x 2' or 4', painted, 1" thick		600	.013		2	.45		2.45	2.96
3760	2" thick		550	.015		3.50	.49		3.99	4.68
3770	2-1/2" thick		500	.016		4.37	.54		4.91	5.70
3780	3" thick		450	.018		5.25	.60		5.85	6.80

09 51 23.30 Suspended Ceilings, Complete

		Crew	Daily Output	Labor-Hours	Unit	Material	Labor	Equipment	Total	Total Incl O&P
0010	**SUSPENDED CEILINGS, COMPLETE**, including standard									
0100	suspension system but not incl. 1-1/2" carrier channels									
0600	Fiberglass ceiling board, 2' x 4' x 5/8", plain faced	1 Carp	500	.016	S.F.	2.02	.54		2.56	3.13
0700	Offices, 2' x 4' x 3/4"		380	.021		3.40	.71		4.11	4.93
1800	Tile, Z bar suspension, 5/8" mineral fiber tile		150	.053		2.32	1.81		4.13	5.60
1900	3/4" mineral fiber tile		150	.053		2.35	1.81		4.16	5.60

09 51 53 – Direct-Applied Acoustical Ceilings

09 51 53.10 Ceiling Tile

		Crew	Daily Output	Labor-Hours	Unit	Material	Labor	Equipment	Total	Total Incl O&P
0010	**CEILING TILE**, stapled or cemented									
0100	12" x 12" or 12" x 24", not including furring									
0600	Mineral fiber, vinyl coated, 5/8" thick	1 Carp	300	.027	S.F.	2.23	.90		3.13	3.96
0700	3/4" thick		300	.027		2.93	.90		3.83	4.73
0900	Fire rated, 3/4" thick, plain faced		300	.027		1.31	.90		2.21	2.95
1000	Plastic coated face		300	.027		1.93	.90		2.83	3.63
1200	Aluminum faced, 5/8" thick, plain		300	.027		1.74	.90		2.64	3.42
3300	For flameproofing, add					.10			.10	.11
3400	For sculptured 3 dimensional, add					.31			.31	.34

09 51 Acoustical Ceilings

09 51 53 – Direct-Applied Acoustical Ceilings

09 51 53.10 Ceiling Tile

		Crew	Daily Output	Labor-Hours	Unit	Material	2016 Bare Costs Labor	Equipment	Total	Total Incl O&P
3900	For ceiling primer, add				S.F.	.13			.13	.14
4000	For ceiling cement, add					.40			.40	.44

09 53 Acoustical Ceiling Suspension Assemblies

09 53 23 – Metal Acoustical Ceiling Suspension Assemblies

09 53 23.30 Ceiling Suspension Systems

		Crew	Daily Output	Labor-Hours	Unit	Material	Labor	Equipment	Total	Total Incl O&P
0010	**CEILING SUSPENSION SYSTEMS** for boards and tile									
0050	Class A suspension system, 15/16" T bar, 2' x 4' grid	1 Carp	800	.010	S.F.	.77	.34		1.11	1.41
0300	2' x 2' grid	"	650	.012		.99	.42		1.41	1.79
0350	For 9/16" grid, add					.16			.16	.18
0360	For fire rated grid, add					.09			.09	.10
0370	For colored grid, add					.21			.21	.23
0400	Concealed Z bar suspension system, 12" module	1 Carp	520	.015		.86	.52		1.38	1.82
0600	1-1/2" carrier channels, 4' O.C., add		470	.017		.12	.58		.70	1.10
0650	1-1/2" x 3-1/2" channels		470	.017		.32	.58		.90	1.32
0700	Carrier channels for ceilings with									
0900	recessed lighting fixtures, add	1 Carp	460	.017	S.F.	.22	.59		.81	1.23
5000	Wire hangers, #12 wire	"	300	.027	Ea.	.06	.90		.96	1.58

09 61 Flooring Treatment

09 61 19 – Concrete Floor Staining

09 61 19.40 Floors, Interior

		Crew	Daily Output	Labor-Hours	Unit	Material	Labor	Equipment	Total	Total Incl O&P
0010	**FLOORS, INTERIOR**									
0300	Acid stain and sealer									
0310	Stain, one coat	1 Pord	650	.012	S.F.	.12	.35		.47	.72
0320	Two coats		570	.014		.25	.40		.65	.93
0330	Acrylic sealer, one coat		2600	.003		.23	.09		.32	.41
0340	Two coats		1400	.006		.47	.16		.63	.78

09 62 Specialty Flooring

09 62 19 – Laminate Flooring

09 62 19.10 Floating Floor

		Crew	Daily Output	Labor-Hours	Unit	Material	Labor	Equipment	Total	Total Incl O&P
0010	**FLOATING FLOOR**									
8300	Floating floor, laminate, wood pattern strip, complete	1 Clab	133	.060	S.F.	4.50	1.48		5.98	7.45
8310	Components, T & G wood composite strips					4.02			4.02	4.42
8320	Film					.17			.17	.18
8330	Foam					.26			.26	.29
8340	Adhesive					.71			.71	.79
8350	Installation kit					.19			.19	.21
8360	Trim, 2" wide x 3' long				L.F.	4.30			4.30	4.73
8370	Reducer moulding				"	5.70			5.70	6.25

09 62 23 – Bamboo Flooring

09 62 23.10 Flooring, Bamboo

			Crew	Daily Output	Labor-Hours	Unit	Material	Labor	Equipment	Total	Total Incl O&P
0010	**FLOORING, BAMBOO**										
8600	Flooring, wood, bamboo strips, unfinished, 5/8" x 4" x 3'	G	1 Carp	255	.031	S.F.	4.60	1.06		5.66	6.85
8610	5/8" x 4" x 4'	G		275	.029		4.77	.99		5.76	6.90
8620	5/8" x 4" x 6'	G		295	.027		5.25	.92		6.17	7.30

09 62 Specialty Flooring

09 62 23 – Bamboo Flooring

09 62 23.10 Flooring, Bamboo

			Crew	Daily Output	Labor-Hours	Unit	Material	2016 Bare Costs Labor	Equipment	Total	Total Incl O&P
8630	Finished, 5/8" x 4" x 3'	G	1 Carp	255	.031	S.F.	5.05	1.06		6.11	7.35
8640	5/8" x 4" x 4'	G		275	.029		5.30	.99		6.29	7.50
8650	5/8" x 4" x 6'	G		295	.027		4.94	.92		5.86	7
8660	Stair treads, unfinished, 1-1/16" x 11-1/2" x 4'	G		18	.444	Ea.	44	15.05		59.05	73.50
8670	Finished, 1-1/16" x 11-1/2" x 4'	G		18	.444		82	15.05		97.05	116
8680	Stair risers, unfinished, 5/8" x 7-1/2" x 4'	G		18	.444		16.30	15.05		31.35	43
8690	Finished, 5/8" x 7-1/2" x 4'	G		18	.444		31	15.05		46.05	59
8700	Stair nosing, unfinished, 6' long	G		16	.500		36	16.95		52.95	68
8710	Finished, 6' long	G		16	.500		42	16.95		58.95	74.50

09 62 29 – Cork Flooring

09 62 29.10 Cork Tile Flooring

			Crew	Daily Output	Labor-Hours	Unit	Material	Labor	Equipment	Total	Total Incl O&P
0010	**CORK TILE FLOORING**										
2200	Cork tile, standard finish, 1/8" thick		1 Tilf	315	.025	S.F.	6.95	.79		7.74	8.95
2250	3/16" thick	G		315	.025		5.55	.79		6.34	7.45
2300	5/16" thick	G		315	.025		8.35	.79		9.14	10.50
2350	1/2" thick	G		315	.025		10.95	.79		11.74	13.30
2500	Urethane finish, 1/8" thick	G		315	.025		8.10	.79		8.89	10.20
2550	3/16" thick	G		315	.025		8.25	.79		9.04	10.35
2600	5/16" thick	G		315	.025		8.85	.79		9.64	11.05
2650	1/2" thick	G		315	.025		12.15	.79		12.94	14.65

09 63 Masonry Flooring

09 63 13 – Brick Flooring

09 63 13.10 Miscellaneous Brick Flooring

			Crew	Daily Output	Labor-Hours	Unit	Material	Labor	Equipment	Total	Total Incl O&P
0010	**MISCELLANEOUS BRICK FLOORING**										
0020	Acid-proof shales, red, 8" x 3-3/4" x 1-1/4" thick		D-7	.43	37.209	M	705	1,025		1,730	2,450
0050	2-1/4" thick		D-1	.40	40		965	1,175		2,140	3,025
0200	Acid-proof clay brick, 8" x 3-3/4" x 2-1/4" thick	G	"	.40	40		925	1,175		2,100	3,000
0260	Cast ceramic, pressed, 4" x 8" x 1/2", unglazed		D-7	100	.160	S.F.	6.35	4.44		10.79	14.15
0270	Glazed			100	.160		8.45	4.44		12.89	16.50
0280	Hand molded flooring, 4" x 8" x 3/4", unglazed			95	.168		8.35	4.67		13.02	16.75
0290	Glazed			95	.168		10.50	4.67		15.17	19.10
0300	8" hexagonal, 3/4" thick, unglazed			85	.188		9.20	5.20		14.40	18.55
0310	Glazed			85	.188		16.60	5.20		21.80	26.50
0450	Acid-proof joints, 1/4" wide		D-1	65	.246		1.46	7.25		8.71	13.70
0500	Pavers, 8" x 4", 1" to 1-1/4" thick, red		D-7	95	.168		3.69	4.67		8.36	11.60
0510	Ironspot		"	95	.168		5.20	4.67		9.87	13.30
0540	1-3/8" to 1-3/4" thick, red		D-1	95	.168		3.56	4.97		8.53	12.20
0560	Ironspot			95	.168		5.15	4.97		10.12	14
0580	2-1/4" thick, red			90	.178		3.62	5.25		8.87	12.75
0590	Ironspot			90	.178		5.60	5.25		10.85	14.95
0700	Paver, adobe brick, 6" x 12", 1/2" joint	G		42	.381		1.32	11.25		12.57	20
0710	Mexican red, 12" x 12"	G	1 Tilf	48	.167		1.72	5.20		6.92	10.30
0720	Saltillo, 12" x 12"	G	"	48	.167		1.49	5.20		6.69	10.05
0800	For sidewalks and patios with pavers, see Section 32 14 16.10										
0870	For epoxy joints, add		D-1	600	.027	S.F.	2.81	.79		3.60	4.40
0880	For Furan underlayment, add		"	600	.027		2.33	.79		3.12	3.87
0890	For waxed surface, steam cleaned, add		A-1H	1000	.008		.20	.20	.08	.48	.64

09 63 Masonry Flooring

09 63 40 – Stone Flooring

09 63 40.10 Marble		Crew	Daily Output	Labor-Hours	Unit	Material	2016 Bare Costs Labor	Equipment	Total	Total Incl O&P
0010	**MARBLE**									
0020	Thin gauge tile, 12" x 6", 3/8", white Carara	D-7	60	.267	S.F.	14.95	7.40		22.35	28.50
0100	Travertine		60	.267		9.95	7.40		17.35	23
0200	12" x 12" x 3/8", thin set, floors		60	.267		10.30	7.40		17.70	23.50
0300	On walls		52	.308		9.85	8.55		18.40	24.50
1000	Marble threshold, 4" wide x 36" long x 5/8" thick, white	▼	60	.267	Ea.	10.30	7.40		17.70	23.50

09 63 40.20 Slate Tile

0010	**SLATE TILE**									
0020	Vermont, 6" x 6" x 1/4" thick, thin set	D-7	180	.089	S.F.	7.50	2.47		9.97	12.25

09 64 Wood Flooring

09 64 23 – Wood Parquet Flooring

09 64 23.10 Wood Parquet

		Crew	Daily Output	Labor-Hours	Unit	Material	Labor	Equipment	Total	Total Incl O&P
0010	**WOOD PARQUET** flooring									
5200	Parquetry, 5/16" thk, no finish, oak, plain pattern	1 Carp	160	.050	S.F.	5.40	1.70		7.10	8.80
5300	Intricate pattern		100	.080		9.65	2.71		12.36	15.20
5500	Teak, plain pattern		160	.050		6.30	1.70		8	9.80
5600	Intricate pattern		100	.080		10.70	2.71		13.41	16.30
5650	13/16" thick, select grade oak, plain pattern		160	.050		9.75	1.70		11.45	13.60
5700	Intricate pattern		100	.080		17.25	2.71		19.96	23.50
5800	Custom parquetry, including finish, plain pattern		100	.080		17.05	2.71		19.76	23.50
5900	Intricate pattern		50	.160		25	5.40		30.40	36.50
6700	Parquetry, prefinished white oak, 5/16" thick, plain pattern		160	.050		7.90	1.70		9.60	11.55
6800	Intricate pattern		100	.080		8.45	2.71		11.16	13.80
7000	Walnut or teak, parquetry, plain pattern		160	.050		8.40	1.70		10.10	12.10
7100	Intricate pattern	▼	100	.080	▼	13.50	2.71		16.21	19.40
7200	Acrylic wood parquet blocks, 12" x 12" x 5/16",									
7210	Irradiated, set in epoxy	1 Carp	160	.050	S.F.	10.20	1.70		11.90	14.05

09 64 29 – Wood Strip and Plank Flooring

09 64 29.10 Wood

		Crew	Daily Output	Labor-Hours	Unit	Material	Labor	Equipment	Total	Total Incl O&P
0010	**WOOD**									
0020	Fir, vertical grain, 1" x 4", not incl. finish, grade B & better	1 Carp	255	.031	S.F.	2.79	1.06		3.85	4.85
0100	C grade & better		255	.031		2.63	1.06		3.69	4.67
0300	Flat grain, 1" x 4", not incl. finish, B & better		255	.031		3.19	1.06		4.25	5.30
0400	C & better		255	.031		3.07	1.06		4.13	5.15
4000	Maple, strip, 25/32" x 2-1/4", not incl. finish, select		170	.047		4.90	1.60		6.50	8.05
4100	#2 & better		170	.047		4.42	1.60		6.02	7.55
4300	33/32" x 3-1/4", not incl. finish, #1 grade		170	.047		4.56	1.60		6.16	7.65
4400	#2 & better	▼	170	.047	▼	4.06	1.60		5.66	7.15
4600	Oak, white or red, 25/32" x 2-1/4", not incl. finish									
4700	#1 common	1 Carp	170	.047	S.F.	3.33	1.60		4.93	6.35
4900	Select quartered, 2-1/4" wide		170	.047		4.19	1.60		5.79	7.30
5000	Clear		170	.047		4.10	1.60		5.70	7.20
6100	Prefinished, white oak, prime grade, 2-1/4" wide		170	.047		4.83	1.60		6.43	7.95
6200	3-1/4" wide		185	.043		5.25	1.47		6.72	8.20
6400	Ranch plank		145	.055		7.15	1.87		9.02	11
6500	Hardwood blocks, 9" x 9", 25/32" thick		160	.050		6	1.70		7.70	9.45
7400	Yellow pine, 3/4" x 3-1/8", T & G, C & better, not incl. finish	▼	200	.040		1.53	1.36		2.89	3.95
7500	Refinish wood floor, sand, 2 coats poly, wax, soft wood	1 Clab	400	.020		.90	.49		1.39	1.82
7600	Hard wood	▼	130	.062		1.34	1.52		2.86	4.01

09 64 Wood Flooring

09 64 29 – Wood Strip and Plank Flooring

09 64 29.10 Wood		Crew	Daily Output	Labor-Hours	Unit	Material	2016 Bare Costs Labor	Equipment	Total	Total Incl O&P
7800	Sanding and finishing, 2 coats polyurethane	1 Clab	295	.027	S.F.	.90	.67		1.57	2.11
7900	Subfloor and underlayment, see Section 06 16									
8015	Transition molding, 2 1/4" wide, 5' long	1 Carp	19.20	.417	Ea.	10.85	14.15		25	35.50

09 65 Resilient Flooring

09 65 10 – Resilient Tile Underlayment

09 65 10.10 Latex Underlayment

		Crew	Daily Output	Labor-Hours	Unit	Material	Labor	Equipment	Total	Total Incl O&P
0010	**LATEX UNDERLAYMENT**									
3600	Latex underlayment, 1/8" thk., cementitious for resilient flooring	1 Tilf	160	.050	S.F.	1.22	1.56		2.78	3.86
4000	Liquid, fortified				Gal.	32			32	35

09 65 13 – Resilient Base and Accessories

09 65 13.13 Resilient Base

		Crew	Daily Output	Labor-Hours	Unit	Material	Labor	Equipment	Total	Total Incl O&P
0010	**RESILIENT BASE**									
0690	1/8" vinyl base, 2 1/2" H, straight or cove, standard colors	1 Tilf	315	.025	L.F.	.68	.79		1.47	2.03
0700	4" high		315	.025		1.21	.79		2	2.61
0710	6" high		315	.025		1.38	.79		2.17	2.80
0720	Corners, 2 1/2" high		315	.025	Ea.	2.14	.79		2.93	3.63
0730	4" high		315	.025		2.35	.79		3.14	3.87
0740	6" high		315	.025		2.65	.79		3.44	4.20
0800	1/8" rubber base, 2 1/2" H, straight or cove, standard colors		315	.025	L.F.	1.11	.79		1.90	2.50
1100	4" high		315	.025		1.14	.79		1.93	2.53
1110	6" high		315	.025		1.91	.79		2.70	3.39
1150	Corners, 2 1/2" high		315	.025	Ea.	2.25	.79		3.04	3.76
1153	4" high		315	.025		2.30	.79		3.09	3.81
1155	6" high		315	.025		2.86	.79		3.65	4.43
1450	For premium color/finish add					50%				
1500	Millwork profile	1 Tilf	315	.025	L.F.	6.15	.79		6.94	8.05

09 65 16 – Resilient Sheet Flooring

09 65 16.10 Rubber and Vinyl Sheet Flooring

			Crew	Daily Output	Labor-Hours	Unit	Material	Labor	Equipment	Total	Total Incl O&P
0010	**RUBBER AND VINYL SHEET FLOORING**										
5500	Linoleum, sheet goods	G	1 Tilf	360	.022	S.F.	3.59	.69		4.28	5.05
5900	Rubber, sheet goods, 36" wide, 1/8" thick			120	.067		7.10	2.08		9.18	11.15
5950	3/16" thick			100	.080		10.45	2.49		12.94	15.55
6000	1/4" thick			90	.089		12.30	2.77		15.07	18.05
8000	Vinyl sheet goods, backed, .065" thick, plain pattern/colors			250	.032		3.69	1		4.69	5.65
8050	Intricate pattern/colors			200	.040		4.34	1.25		5.59	6.80
8100	.080" thick, plain pattern/colors			230	.035		4.12	1.08		5.20	6.30
8150	Intricate pattern/colors			200	.040		6.45	1.25		7.70	9.10
8200	.125" thick, plain pattern/colors			230	.035		3.87	1.08		4.95	6
8250	Intricate pattern/colors			200	.040		7.50	1.25		8.75	10.25
8700	Adhesive cement, 1 gallon per 200 to 300 S.F.					Gal.	29.50			29.50	32
8800	Asphalt primer, 1 gallon per 300 S.F.						14			14	15.40
8900	Emulsion, 1 gallon per 140 S.F.						18.85			18.85	20.50

09 65 19 – Resilient Tile Flooring

09 65 19.19 Vinyl Composition Tile Flooring

		Crew	Daily Output	Labor-Hours	Unit	Material	Labor	Equipment	Total	Total Incl O&P
0010	**VINYL COMPOSITION TILE FLOORING**									
7000	Vinyl composition tile, 12" x 12", 1/16" thick	1 Tilf	500	.016	S.F.	1.19	.50		1.69	2.12
7050	Embossed		500	.016		2.21	.50		2.71	3.24
7100	Marbleized		500	.016		2.21	.50		2.71	3.24

09 65 Resilient Flooring

09 65 19 – Resilient Tile Flooring

09 65 19.19 Vinyl Composition Tile Flooring

		Crew	Daily Output	Labor-Hours	Unit	Material	2016 Bare Costs Labor	Equipment	Total	Total Incl O&P
7150	Solid	1 Tilf	500	.016	S.F.	2.85	.50		3.35	3.95
7200	3/32" thick, embossed		500	.016		1.43	.50		1.93	2.38
7250	Marbleized		500	.016		2.54	.50		3.04	3.60
7300	Solid		500	.016		2.36	.50		2.86	3.41
7350	1/8" thick, marbleized		500	.016		2.44	.50		2.94	3.49
7400	Solid		500	.016		1.55	.50		2.05	2.52
7450	Conductive		500	.016		6.05	.50		6.55	7.45

09 65 19.23 Vinyl Tile Flooring

		Crew	Daily Output	Labor-Hours	Unit	Material	Labor	Equipment	Total	Total Incl O&P
0010	**VINYL TILE FLOORING**									
7500	Vinyl tile, 12" x 12", 3/32" thick, standard colors/patterns	1 Tilf	500	.016	S.F.	3.71	.50		4.21	4.89
7550	1/8" thick, standard colors/patterns		500	.016		5.20	.50		5.70	6.50
7600	1/8" thick, premium colors/patterns		500	.016		7	.50		7.50	8.50
7650	Solid colors		500	.016		3.25	.50		3.75	4.39
7700	Marbleized or Travertine pattern		500	.016		5.90	.50		6.40	7.30
7750	Florentine pattern		500	.016		6.30	.50		6.80	7.75
7800	Premium colors/patterns		500	.016		6.25	.50		6.75	7.65

09 65 19.33 Rubber Tile Flooring

		Crew	Daily Output	Labor-Hours	Unit	Material	Labor	Equipment	Total	Total Incl O&P
0010	**RUBBER TILE FLOORING**									
6050	Rubber tile, marbleized colors, 12" x 12", 1/8" thick	1 Tilf	400	.020	S.F.	5.75	.62		6.37	7.30
6100	3/16" thick		400	.020		9.05	.62		9.67	10.95
6300	Special tile, plain colors, 1/8" thick		400	.020		7.95	.62		8.57	9.75
6350	3/16" thick		400	.020		9.05	.62		9.67	10.95

09 65 33 – Conductive Resilient Flooring

09 65 33.10 Conductive Rubber and Vinyl Flooring

		Crew	Daily Output	Labor-Hours	Unit	Material	Labor	Equipment	Total	Total Incl O&P
0010	**CONDUCTIVE RUBBER AND VINYL FLOORING**									
1700	Conductive flooring, rubber tile, 1/8" thick	1 Tilf	315	.025	S.F.	6.95	.79		7.74	8.95
1800	Homogeneous vinyl tile, 1/8" thick	"	315	.025	"	6.70	.79		7.49	8.70

09 66 Terrazzo Flooring

09 66 13 – Portland Cement Terrazzo Flooring

09 66 13.10 Portland Cement Terrazzo

		Crew	Daily Output	Labor-Hours	Unit	Material	Labor	Equipment	Total	Total Incl O&P
0010	**PORTLAND CEMENT TERRAZZO**, cast-in-place									
4300	Stone chips, onyx gemstone, per 50 lb. bag				Bag	16.80			16.80	18.50

09 66 16 – Terrazzo Floor Tile

09 66 16.13 Portland Cement Terrazzo Floor Tile

		Crew	Daily Output	Labor-Hours	Unit	Material	Labor	Equipment	Total	Total Incl O&P
0010	**PORTLAND CEMENT TERRAZZO FLOOR TILE**									
1200	Floor tiles, non-slip, 1" thick, 12" x 12"	D-1	60	.267	S.F.	20.50	7.85		28.35	35.50
1300	1-1/4" thick, 12" x 12"		60	.267		20.50	7.85		28.35	36
1500	16" x 16"		50	.320		22.50	9.45		31.95	40.50
1600	1-1/2" thick, 16" x 16"		45	.356		20.50	10.50		31	40

09 66 16.30 Terrazzo, Precast

		Crew	Daily Output	Labor-Hours	Unit	Material	Labor	Equipment	Total	Total Incl O&P
0010	**TERRAZZO, PRECAST**									
0020	Base, 6" high, straight	1 Mstz	70	.114	L.F.	12.10	3.57		15.67	19.10
0100	Cove		60	.133		13.20	4.16		17.36	21.50
0300	8" high, straight		60	.133		11.80	4.16		15.96	19.70
0400	Cove		50	.160		17.35	4.99		22.34	27
0600	For white cement, add					.46			.46	.51
0700	For 16 ga. zinc toe strip, add					1.81			1.81	1.99
0900	Curbs, 4" x 4" high	1 Mstz	40	.200		33.50	6.25		39.75	46.50

09 66 Terrazzo Flooring

09 66 16 – Terrazzo Floor Tile

09 66 16.30 Terrazzo, Precast		Crew	Daily Output	Labor-Hours	Unit	Material	2016 Bare Costs Labor	2016 Bare Costs Equipment	Total	Total Incl O&P
1000	8" x 8" high	1 Mstz	30	.267	L.F.	38.50	8.30		46.80	56
4800	Wainscot, 12" x 12" x 1" tiles		12	.667	S.F.	7.10	21		28.10	41.50
4900	16" x 16" x 1-1/2" tiles		8	1	"	14.25	31		45.25	66

09 68 Carpeting

09 68 05 – Carpet Accessories

09 68 05.11 Flooring Transition Strip

		Crew	Daily Output	Labor-Hours	Unit	Material	Labor	Equipment	Total	Total Incl O&P
0010	**FLOORING TRANSITION STRIP**									
0107	Clamp down brass divider, 12' strip, vinyl to carpet	1 Tilf	31.25	.256	Ea.	13.95	7.95		21.90	28.50
0117	Vinyl to hard surface	"	31.25	.256	"	13.95	7.95		21.90	28.50

09 68 10 – Carpet Pad

09 68 10.10 Commercial Grade Carpet Pad

		Crew	Daily Output	Labor-Hours	Unit	Material	Labor	Equipment	Total	Total Incl O&P
0010	**COMMERCIAL GRADE CARPET PAD**									
9001	Sponge rubber pad, 20 oz./sq. yd.	1 Tilf	1350	.006	S.F.	.51	.18		.69	.86
9101	40-62 oz./sq. yd.		1350	.006		.97	.18		1.15	1.37
9201	Felt pad, 20-32 oz./sq. yd.		1350	.006		.62	.18		.80	.99
9301	Maximum		1350	.006		1.03	.18		1.21	1.44
9401	Bonded urethane pad, 2.7 density		1350	.006		.66	.18		.84	1.03
9501	13.0 density		1350	.006		.89	.18		1.07	1.28
9601	Prime urethane pad, 2.7 density		1350	.006		.36	.18		.54	.70
9701	13.0 density		1350	.006		.55	.18		.73	.91

09 68 13 – Tile Carpeting

09 68 13.10 Carpet Tile

		Crew	Daily Output	Labor-Hours	Unit	Material	Labor	Equipment	Total	Total Incl O&P
0010	**CARPET TILE**									
0100	Tufted nylon, 18" x 18", hard back, 20 oz.	1 Tilf	80	.100	S.Y.	24	3.12		27.12	31.50
0110	26 oz.		80	.100		29	3.12		32.12	37
0200	Cushion back, 20 oz.		80	.100		28	3.12		31.12	35.50
0210	26 oz.		80	.100		43	3.12		46.12	52

09 68 16 – Sheet Carpeting

09 68 16.10 Sheet Carpet

		Crew	Daily Output	Labor-Hours	Unit	Material	Labor	Equipment	Total	Total Incl O&P
0010	**SHEET CARPET**									
0701	Nylon, level loop, 26 oz., light to medium traffic	1 Tilf	675	.012	S.F.	2.38	.37		2.75	3.21
0901	32 oz., medium traffic		675	.012		4.39	.37		4.76	5.45
1101	40 oz., medium to heavy traffic		675	.012		5	.37		5.37	6.10
2101	Nylon, plush, 20 oz., light traffic		675	.012		2.03	.37		2.40	2.84
2801	24 oz., light to medium traffic		675	.012		2.12	.37		2.49	2.93
2901	30 oz., medium traffic		675	.012		3.44	.37		3.81	4.38
3001	36 oz., medium traffic		675	.012		3.88	.37		4.25	4.87
3101	42 oz., medium to heavy traffic		630	.013		5	.40		5.40	6.15
3201	46 oz., medium to heavy traffic		630	.013		5.60	.40		6	6.85
3301	54 oz., heavy traffic		630	.013		6.15	.40		6.55	7.40
3501	Olefin, 15 oz., light traffic		675	.012		1.67	.37		2.04	2.44
3651	22 oz., light traffic		675	.012		1.59	.37		1.96	2.34
4501	50 oz., medium to heavy traffic, level loop		630	.013		11.15	.40		11.55	12.95
4701	32 oz., medium to heavy traffic, patterned		630	.013		10.10	.40		10.50	11.80
4901	48 oz., heavy traffic, patterned		630	.013		11.15	.40		11.55	12.95
5000	For less than full roll (approx. 1500 S.F.), add					25%				
5100	For small rooms, less than 12' wide, add						25%			
5200	For large open areas (no cuts), deduct						25%			

09 68 Carpeting

09 68 16 – Sheet Carpeting

09 68 16.10 Sheet Carpet

		Crew	Daily Output	Labor-Hours	Unit	Material	2016 Bare Costs Labor	2016 Bare Costs Equipment	Total	Total Incl O&P
5600	For bound carpet baseboard, add	1 Tilf	300	.027	L.F.	3	.83		3.83	4.64
5610	For stairs, not incl. price of carpet, add	"	30	.267	Riser		8.30		8.30	13.45
8950	For tackless, stretched installation, add padding from 09 68 10.10 to above									
9850	For brand-named specific fiber, add				S.Y.	25%				

09 68 20 – Athletic Carpet

09 68 20.10 Indoor Athletic Carpet

		Crew	Daily Output	Labor-Hours	Unit	Material	Labor	Equipment	Total	Total Incl O&P
0010	**INDOOR ATHLETIC CARPET**									
3700	Polyethylene, in rolls, no base incl., landscape surfaces	1 Tilf	275	.029	S.F.	4.16	.91		5.07	6.05
3800	Nylon action surface, 1/8" thick		275	.029		3.76	.91		4.67	5.60
3900	1/4" thick		275	.029		5.45	.91		6.36	7.40
4000	3/8" thick	↓	275	.029	↓	6.80	.91		7.71	8.95

09 72 Wall Coverings

09 72 19 – Textile Wall Coverings

09 72 19.10 Textile Wall Covering

		Crew	Daily Output	Labor-Hours	Unit	Material	Labor	Equipment	Total	Total Incl O&P
0010	**TEXTILE WALL COVERING**, including sizing; add 10-30% waste @ takeoff									
0020	Silk	1 Pape	640	.013	S.F.	4.23	.36		4.59	5.25
0030	Cotton		640	.013		6.75	.36		7.11	8.05
0040	Linen		640	.013		1.89	.36		2.25	2.67
0050	Blend	↓	640	.013	↓	3.03	.36		3.39	3.92

09 72 20 – Natural Fiber Wall Covering

09 72 20.10 Natural Fiber Wall Covering

		Crew	Daily Output	Labor-Hours	Unit	Material	Labor	Equipment	Total	Total Incl O&P
0010	**NATURAL FIBER WALL COVERING**, including sizing; add 10-30% waste @ takeoff									
0015	Bamboo	1 Pape	640	.013	S.F.	2.31	.36		2.67	3.13
0030	Burlap		640	.013		2.07	.36		2.43	2.87
0045	Jute		640	.013		1.32	.36		1.68	2.04
0060	Sisal	↓	640	.013	↓	1.49	.36		1.85	2.23

09 72 23 – Wallpapering

09 72 23.10 Wallpaper

		Crew	Daily Output	Labor-Hours	Unit	Material	Labor	Equipment	Total	Total Incl O&P
0010	**WALLPAPER** including sizing; add 10-30 percent waste @ takeoff R097223-10									
0050	Aluminum foil	1 Pape	275	.029	S.F.	1.04	.84		1.88	2.52
0100	Copper sheets, .025" thick, vinyl backing		240	.033		5.55	.96		6.51	7.70
0300	Phenolic backing		240	.033		7.20	.96		8.16	9.55
0600	Cork tiles, light or dark, 12" x 12" x 3/16"		240	.033		4.66	.96		5.62	6.75
0700	5/16" thick		235	.034		3.13	.98		4.11	5.05
0900	1/4" basketweave		240	.033		3.50	.96		4.46	5.45
1000	1/2" natural, non-directional pattern		240	.033		6.55	.96		7.51	8.80
1100	3/4" natural, non-directional pattern		240	.033		11.95	.96		12.91	14.75
1200	Granular surface, 12" x 36", 1/2" thick		385	.021		1.31	.60		1.91	2.43
1300	1" thick		370	.022		1.68	.62		2.30	2.88
1500	Polyurethane coated, 12" x 12" x 3/16" thick		240	.033		4.08	.96		5.04	6.05
1600	5/16" thick		235	.034		6.60	.98		7.58	8.85
1800	Cork wallpaper, paperbacked, natural		480	.017		2	.48		2.48	2.99
1900	Colors		480	.017		2.87	.48		3.35	3.95
2100	Flexible wood veneer, 1/32" thick, plain woods		100	.080		2.44	2.31		4.75	6.50
2200	Exotic woods	↓	95	.084	↓	3.69	2.43		6.12	8.05
2400	Gypsum-based, fabric-backed, fire resistant									
2500	for masonry walls, 21 oz./S.Y.	1 Pape	800	.010	S.F.	.77	.29		1.06	1.33
2600	Average		720	.011		1.27	.32		1.59	1.93

09 72 Wall Coverings

09 72 23 – Wallpapering

09 72 23.10 Wallpaper

		Crew	Daily Output	Labor-Hours	Unit	Material	2016 Bare Costs Labor	Equipment	Total	Total Incl O&P
2700	Small quantities	1 Pape	640	.013	S.F.	.70	.36		1.06	1.36
2750	Acrylic, modified, semi-rigid PVC, .028" thick	2 Carp	330	.048		1.28	1.64		2.92	4.16
2800	.040" thick	"	320	.050		1.85	1.70		3.55	4.88
3000	Vinyl wall covering, fabric-backed, lightweight, type 1 (12-15 oz./S.Y.)	1 Pape	640	.013		.97	.36		1.33	1.66
3300	Medium weight, type 2 (20-24 oz./S.Y.)		480	.017		.90	.48		1.38	1.78
3400	Heavy weight, type 3 (28 oz./S.Y.)	↓	435	.018		1.37	.53		1.90	2.38
3600	Adhesive, 5 gal. lots (18 S.Y./gal.)				Gal.	14			14	15.40
3700	Wallpaper, average workmanship, solid pattern, low cost paper	1 Pape	640	.013	S.F.	.59	.36		.95	1.24
3900	Basic patterns (matching required), avg. cost paper		535	.015		1.13	.43		1.56	1.95
4000	Paper at $85 per double roll, quality workmanship	↓	435	.018	↓	1.72	.53		2.25	2.76
4100	Linen wall covering, paper backed									
4150	Flame treatment				S.F.	1.08			1.08	1.19
4180	Stain resistance treatment					1.72			1.72	1.89
4200	Grass cloths with lining paper G	1 Pape	400	.020		.96	.58		1.54	2.01
4300	Premium texture/color G	"	350	.023	↓	2.85	.66		3.51	4.23

09 91 Painting

09 91 03 – Paint Restoration

09 91 03.20 Sanding

		Crew	Daily Output	Labor-Hours	Unit	Material	Labor	Equipment	Total	Total Incl O&P
0010	**SANDING** and puttying interior trim, compared to									
0100	Painting 1 coat, on quality work				L.F.		100%			
0300	Medium work						50%			
0400	Industrial grade				↓		25%			
0500	Surface protection, placement and removal									
0510	Basic drop cloths	1 Pord	6400	.001	S.F.		.04		.04	.06
0520	Masking with paper		800	.010		.07	.29		.36	.55
0530	Volume cover up (using plastic sheathing, or building paper)	↓	16000	.001			.01		.01	.02

09 91 03.30 Exterior Surface Preparation

		Crew	Daily Output	Labor-Hours	Unit	Material	Labor	Equipment	Total	Total Incl O&P
0010	**EXTERIOR SURFACE PREPARATION**									
0015	Doors, per side, not incl. frames or trim									
0020	Scrape & sand									
0030	Wood, flush	1 Pord	616	.013	S.F.		.37		.37	.61
0040	Wood, detail		496	.016			.46		.46	.76
0050	Wood, louvered		280	.029			.82		.82	1.35
0060	Wood, overhead	↓	616	.013	↓		.37		.37	.61
0070	Wire brush									
0080	Metal, flush	1 Pord	640	.013	S.F.		.36		.36	.59
0090	Metal, detail		520	.015			.44		.44	.73
0100	Metal, louvered		360	.022			.64		.64	1.05
0110	Metal or fibr., overhead		640	.013			.36		.36	.59
0120	Metal, roll up		560	.014			.41		.41	.67
0130	Metal, bulkhead	↓	640	.013	↓		.36		.36	.59
0140	Power wash, based on 2500 lb. operating pressure									
0150	Metal, flush	A-1H	2240	.004	S.F.		.09	.03	.12	.19
0160	Metal, detail		2120	.004			.09	.04	.13	.20
0170	Metal, louvered		2000	.004			.10	.04	.14	.21
0180	Metal or fibr., overhead		2400	.003			.08	.03	.11	.18
0190	Metal, roll up		2400	.003			.08	.03	.11	.18
0200	Metal, bulkhead	↓	2200	.004	↓		.09	.04	.13	.19
0400	Windows, per side, not incl. trim									
0410	Scrape & sand									

09 91 Painting

09 91 03 – Paint Restoration

09 91 03.30 Exterior Surface Preparation

		Crew	Daily Output	Labor-Hours	Unit	Material	2016 Bare Costs Labor	2016 Bare Costs Equipment	Total	Total Incl O&P
0420	Wood, 1-2 lite	1 Pord	320	.025	S.F.		.72		.72	1.18
0430	Wood, 3-6 lite		280	.029			.82		.82	1.35
0440	Wood, 7-10 lite		240	.033			.95		.95	1.57
0450	Wood, 12 lite		200	.040			1.15		1.15	1.89
0460	Wood, Bay/Bow		320	.025			.72		.72	1.18
0470	Wire brush									
0480	Metal, 1-2 lite	1 Pord	480	.017	S.F.		.48		.48	.79
0490	Metal, 3-6 lite		400	.020			.57		.57	.94
0500	Metal, Bay/Bow		480	.017			.48		.48	.79
0510	Power wash, based on 2500 lb. operating pressure									
0520	1-2 lite	A-1H	4400	.002	S.F.		.04	.02	.06	.10
0530	3-6 lite		4320	.002			.05	.02	.07	.10
0540	7-10 lite		4240	.002			.05	.02	.07	.10
0550	12 lite		4160	.002			.05	.02	.07	.10
0560	Bay/Bow		4400	.002			.04	.02	.06	.10
0600	Siding, scrape and sand, light=10-30%, med.=30-70%									
0610	Heavy=70-100% of surface to sand									
0650	Texture 1-11, light	1 Pord	480	.017	S.F.		.48		.48	.79
0660	Med.		440	.018			.52		.52	.86
0670	Heavy		360	.022			.64		.64	1.05
0680	Wood shingles, shakes, light		440	.018			.52		.52	.86
0690	Med.		360	.022			.64		.64	1.05
0700	Heavy		280	.029			.82		.82	1.35
0710	Clapboard, light		520	.015			.44		.44	.73
0720	Med.		480	.017			.48		.48	.79
0730	Heavy		400	.020			.57		.57	.94
0740	Wire brush									
0750	Aluminum, light	1 Pord	600	.013	S.F.		.38		.38	.63
0760	Med.		520	.015			.44		.44	.73
0770	Heavy		440	.018			.52		.52	.86
0780	Pressure wash, based on 2500 lb. operating pressure									
0790	Stucco	A-1H	3080	.003	S.F.		.06	.03	.09	.14
0800	Aluminum or vinyl		3200	.003			.06	.02	.08	.13
0810	Siding, masonry, brick & block		2400	.003			.08	.03	.11	.18
1300	Miscellaneous, wire brush									
1310	Metal, pedestrian gate	1 Pord	100	.080	S.F.		2.29		2.29	3.78

09 91 03.40 Interior Surface Preparation

		Crew	Daily Output	Labor-Hours	Unit	Material	Labor	Equipment	Total	Total Incl O&P
0010	**INTERIOR SURFACE PREPARATION**									
0020	Doors, per side, not incl. frames or trim									
0030	Scrape & sand									
0040	Wood, flush	1 Pord	616	.013	S.F.		.37		.37	.61
0050	Wood, detail		496	.016			.46		.46	.76
0060	Wood, louvered		280	.029			.82		.82	1.35
0070	Wire brush									
0080	Metal, flush	1 Pord	640	.013	S.F.		.36		.36	.59
0090	Metal, detail		520	.015			.44		.44	.73
0100	Metal, louvered		360	.022			.64		.64	1.05
0110	Hand wash									
0120	Wood, flush	1 Pord	2160	.004	S.F.		.11		.11	.17
0130	Wood, detailed		2000	.004			.11		.11	.19
0140	Wood, louvered		1360	.006			.17		.17	.28
0150	Metal, flush		2160	.004			.11		.11	.17

09 91 Painting

09 91 03 – Paint Restoration

09 91 03.40 Interior Surface Preparation		Crew	Daily Output	Labor-Hours	Unit	Material	2016 Bare Costs Labor	Equipment	Total	Total Incl O&P
0160	Metal, detail	1 Pord	2000	.004	S.F.		.11		.11	.19
0170	Metal, louvered		1360	.006			.17		.17	.28
0400	Windows, per side, not incl. trim									
0410	Scrape & sand									
0420	Wood, 1-2 lite	1 Pord	360	.022	S.F.		.64		.64	1.05
0430	Wood, 3-6 lite		320	.025			.72		.72	1.18
0440	Wood, 7-10 lite		280	.029			.82		.82	1.35
0450	Wood, 12 lite		240	.033			.95		.95	1.57
0460	Wood, Bay/Bow		360	.022			.64		.64	1.05
0470	Wire brush									
0480	Metal, 1-2 lite	1 Pord	520	.015	S.F.		.44		.44	.73
0490	Metal, 3-6 lite		440	.018			.52		.52	.86
0500	Metal, Bay/Bow		520	.015			.44		.44	.73
0600	Walls, sanding, light=10-30%, medium - 30-70%,									
0610	heavy=70-100% of surface to sand									
0650	Walls, sand									
0660	Gypsum board or plaster, light	1 Pord	3077	.003	S.F.		.07		.07	.12
0670	Gypsum board or plaster, medium		2160	.004			.11		.11	.17
0680	Gypsum board or plaster, heavy		923	.009			.25		.25	.41
0690	Wood, T&G, light		2400	.003			.10		.10	.16
0700	Wood, T&G, med.		1600	.005			.14		.14	.24
0710	Wood, T&G, heavy		800	.010			.29		.29	.47
0720	Walls, wash									
0730	Gypsum board or plaster	1 Pord	3200	.003	S.F.		.07		.07	.12
0740	Wood, T&G		3200	.003			.07		.07	.12
0750	Masonry, brick & block, smooth		2800	.003			.08		.08	.14
0760	Masonry, brick & block, coarse		2000	.004			.11		.11	.19
8000	For chemical washing, see Section 04 01 30									

09 91 03.41 Scrape After Fire Damage		Crew	Daily Output	Labor-Hours	Unit	Material	Labor	Equipment	Total	Total Incl O&P
0010	**SCRAPE AFTER FIRE DAMAGE**									
0050	Boards, 1" x 4"	1 Pord	336	.024	L.F.		.68		.68	1.12
0060	1" x 6"		260	.031			.88		.88	1.45
0070	1" x 8"		207	.039			1.11		1.11	1.82
0080	1" x 10"		174	.046			1.32		1.32	2.17
0500	Framing, 2" x 4"		265	.030			.86		.86	1.43
0510	2" x 6"		221	.036			1.04		1.04	1.71
0520	2" x 8"		190	.042			1.21		1.21	1.99
0530	2" x 10"		165	.048			1.39		1.39	2.29
0540	2" x 12"		144	.056			1.59		1.59	2.62
1000	Heavy framing, 3" x 4"		226	.035			1.01		1.01	1.67
1010	4" x 4"		210	.038			1.09		1.09	1.80
1020	4" x 6"		191	.042			1.20		1.20	1.98
1030	4" x 8"		165	.048			1.39		1.39	2.29
1040	4" x 10"		144	.056			1.59		1.59	2.62
1060	4" x 12"		131	.061			1.75		1.75	2.88
2900	For sealing, light damage		825	.010	S.F.	.14	.28		.42	.61
2920	Heavy damage		460	.017	"	.31	.50		.81	1.16

09 91 Painting

09 91 13 – Exterior Painting

09 91 13.30 Fences

		Crew	Daily Output	Labor-Hours	Unit	Material	2016 Bare Costs Labor	Equipment	Total	Total Incl O&P
0010	**FENCES** R099100-20									
0100	Chain link or wire metal, one side, water base									
0110	Roll & brush, first coat	1 Pord	960	.008	S.F.	.08	.24		.32	.47
0120	Second coat		1280	.006		.07	.18		.25	.38
0130	Spray, first coat		2275	.004		.08	.10		.18	.25
0140	Second coat		2600	.003		.08	.09		.17	.23
0150	Picket, water base									
0160	Roll & brush, first coat	1 Pord	865	.009	S.F.	.08	.27		.35	.53
0170	Second coat		1050	.008		.08	.22		.30	.45
0180	Spray, first coat		2275	.004		.08	.10		.18	.26
0190	Second coat		2600	.003		.08	.09		.17	.24
0200	Stockade, water base									
0210	Roll & brush, first coat	1 Pord	1040	.008	S.F.	.08	.22		.30	.45
0220	Second coat		1200	.007		.08	.19		.27	.40
0230	Spray, first coat		2275	.004		.08	.10		.18	.26
0240	Second coat		2600	.003		.08	.09		.17	.24

09 91 13.42 Miscellaneous, Exterior

		Crew	Daily Output	Labor-Hours	Unit	Material	Labor	Equipment	Total	Total Incl O&P
0010	**MISCELLANEOUS, EXTERIOR** R099100-20									
0100	Railing, ext., decorative wood, incl. cap & baluster									
0110	Newels & spindles @ 12" O.C.									
0120	Brushwork, stain, sand, seal & varnish									
0130	First coat	1 Pord	90	.089	L.F.	.85	2.55		3.40	5.15
0140	Second coat	"	120	.067	"	.85	1.91		2.76	4.08
0150	Rough sawn wood, 42" high, 2" x 2" verticals, 6" O.C.									
0160	Brushwork, stain, each coat	1 Pord	90	.089	L.F.	.26	2.55		2.81	4.49
0170	Wrought iron, 1" rail, 1/2" sq. verticals									
0180	Brushwork, zinc chromate, 60" high, bars 6" O.C.									
0190	Primer	1 Pord	130	.062	L.F.	.88	1.76		2.64	3.87
0200	Finish coat		130	.062		1.13	1.76		2.89	4.14
0210	Additional coat		190	.042		1.32	1.21		2.53	3.44
0220	Shutters or blinds, single panel, 2' x 4', paint all sides									
0230	Brushwork, primer	1 Pord	20	.400	Ea.	.67	11.45		12.12	19.65
0240	Finish coat, exterior latex		20	.400		.62	11.45		12.07	19.60
0250	Primer & 1 coat, exterior latex		13	.615		1.14	17.65		18.79	30.50
0260	Spray, primer		35	.229		.98	6.55		7.53	11.90
0270	Finish coat, exterior latex		35	.229		1.31	6.55		7.86	12.25
0280	Primer & 1 coat, exterior latex		20	.400		1.06	11.45		12.51	20
0290	For louvered shutters, add				S.F.	10%				
0300	Stair stringers, exterior, metal									
0310	Roll & brush, zinc chromate, to 14", each coat	1 Pord	320	.025	L.F.	.38	.72		1.10	1.59
0320	Rough sawn wood, 4" x 12"									
0330	Roll & brush, exterior latex, each coat	1 Pord	215	.037	L.F.	.09	1.07		1.16	1.86
0340	Trellis/lattice, 2" x 2" @ 3" O.C. with 2" x 8" supports									
0350	Spray, latex, per side, each coat	1 Pord	475	.017	S.F.	.09	.48		.57	.89
0450	Decking, ext., sealer, alkyd, brushwork, sealer coat		1140	.007		.10	.20		.30	.44
0460	1st coat		1140	.007		.10	.20		.30	.44
0470	2nd coat		1300	.006		.07	.18		.25	.37
0500	Paint, alkyd, brushwork, primer coat		1140	.007		.12	.20		.32	.46
0510	1st coat		1140	.007		.13	.20		.33	.47
0520	2nd coat		1300	.006		.09	.18		.27	.39
0600	Sand paint, alkyd, brushwork, 1 coat		150	.053		.13	1.53		1.66	2.67

09 91 Painting

09 91 13 – Exterior Painting

09 91 13.60 Siding Exterior

		Crew	Daily Output	Labor-Hours	Unit	Material	2016 Bare Costs Labor	Equipment	Total	Total Incl O&P
0010	**SIDING EXTERIOR**, Alkyd (oil base)									
0450	Steel siding, oil base, paint 1 coat, brushwork	2 Pord	2015	.008	S.F.	.10	.23		.33	.48
0500	Spray		4550	.004		.15	.10		.25	.34
0800	Paint 2 coats, brushwork		1300	.012		.20	.35		.55	.80
1000	Spray		2750	.006		.17	.17		.34	.46
1200	Stucco, rough, oil base, paint 2 coats, brushwork		1300	.012		.20	.35		.55	.80
1400	Roller		1625	.010		.21	.28		.49	.69
1600	Spray		2925	.005		.22	.16		.38	.51
1800	Texture 1-11 or clapboard, oil base, primer coat, brushwork		1300	.012		.15	.35		.50	.75
2000	Spray		4550	.004		.15	.10		.25	.34
2100	Paint 1 coat, brushwork		1300	.012		.15	.35		.50	.74
2200	Spray		4550	.004		.15	.10		.25	.33
2400	Paint 2 coats, brushwork		810	.020		.30	.57		.87	1.25
2600	Spray		2600	.006		.33	.18		.51	.65
3000	Stain 1 coat, brushwork		1520	.011		.09	.30		.39	.60
3200	Spray		5320	.003		.10	.09		.19	.25
3400	Stain 2 coats, brushwork		950	.017		.17	.48		.65	.98
4000	Spray		3050	.005		.19	.15		.34	.46
4200	Wood shingles, oil base primer coat, brushwork		1300	.012		.14	.35		.49	.73
4400	Spray		3900	.004		.13	.12		.25	.33
4600	Paint 1 coat, brushwork		1300	.012		.12	.35		.47	.72
4800	Spray		3900	.004		.15	.12		.27	.36
5000	Paint 2 coats, brushwork		810	.020		.25	.57		.82	1.20
5200	Spray		2275	.007		.23	.20		.43	.59
5800	Stain 1 coat, brushwork		1500	.011		.09	.31		.40	.60
6000	Spray		3900	.004		.09	.12		.21	.29
6500	Stain 2 coats, brushwork		950	.017		.17	.48		.65	.98
7000	Spray		2660	.006		.24	.17		.41	.54
8000	For latex paint, deduct					10%				
8100	For work over 12' H, from pipe scaffolding, add						15%			
8200	For work over 12' H, from extension ladder, add						25%			
8300	For work over 12' H, from swing staging, add						35%			

09 91 13.62 Siding, Misc.

		Crew	Daily Output	Labor-Hours	Unit	Material	Labor	Equipment	Total	Total Incl O&P
0010	**SIDING, MISC.**, latex paint R099100-10									
0100	Aluminum siding									
0110	Brushwork, primer	2 Pord	2275	.007	S.F.	.06	.20		.26	.40
0120	Finish coat, exterior latex		2275	.007		.06	.20		.26	.39
0130	Primer & 1 coat exterior latex		1300	.012		.13	.35		.48	.72
0140	Primer & 2 coats exterior latex		975	.016		.19	.47		.66	.98
0150	Mineral fiber shingles									
0160	Brushwork, primer	2 Pord	1495	.011	S.F.	.15	.31		.46	.68
0170	Finish coat, industrial enamel		1495	.011		.17	.31		.48	.69
0180	Primer & 1 coat enamel		810	.020		.32	.57		.89	1.28
0190	Primer & 2 coats enamel		540	.030		.48	.85		1.33	1.93
0200	Roll, primer		1625	.010		.17	.28		.45	.65
0210	Finish coat, industrial enamel		1625	.010		.18	.28		.46	.66
0220	Primer & 1 coat enamel		975	.016		.35	.47		.82	1.15
0230	Primer & 2 coats enamel		650	.025		.53	.71		1.24	1.74
0240	Spray, primer		3900	.004		.13	.12		.25	.33
0250	Finish coat, industrial enamel		3900	.004		.15	.12		.27	.35
0260	Primer & 1 coat enamel		2275	.007		.28	.20		.48	.64
0270	Primer & 2 coats enamel		1625	.010		.43	.28		.71	.93

09 91 Painting

09 91 13 – Exterior Painting

09 91 13.62 Siding, Misc.

		Crew	Daily Output	Labor-Hours	Unit	Material	2016 Bare Costs Labor	Equipment	Total	Total Incl O&P
0280	Waterproof sealer, first coat	2 Pord	4485	.004	S.F.	.10	.10		.20	.28
0290	Second coat		5235	.003		.09	.09		.18	.24
0300	Rough wood incl. shingles, shakes or rough sawn siding									
0310	Brushwork, primer	2 Pord	1280	.013	S.F.	.14	.36		.50	.74
0320	Finish coat, exterior latex		1280	.013		.10	.36		.46	.70
0330	Primer & 1 coat exterior latex		960	.017		.24	.48		.72	1.05
0340	Primer & 2 coats exterior latex		700	.023		.34	.65		.99	1.46
0350	Roll, primer		2925	.005		.18	.16		.34	.46
0360	Finish coat, exterior latex		2925	.005		.12	.16		.28	.40
0370	Primer & 1 coat exterior latex		1790	.009		.30	.26		.56	.76
0380	Primer & 2 coats exterior latex		1300	.012		.43	.35		.78	1.05
0390	Spray, primer		3900	.004		.15	.12		.27	.36
0400	Finish coat, exterior latex		3900	.004		.10	.12		.22	.30
0410	Primer & 1 coat exterior latex		2600	.006		.25	.18		.43	.56
0420	Primer & 2 coats exterior latex		2080	.008		.34	.22		.56	.74
0430	Waterproof sealer, first coat		4485	.004		.18	.10		.28	.36
0440	Second coat		4485	.004		.10	.10		.20	.28
0450	Smooth wood incl. butt, T&G, beveled, drop or B&B siding									
0460	Brushwork, primer	2 Pord	2325	.007	S.F.	.10	.20		.30	.43
0470	Finish coat, exterior latex		1280	.013		.10	.36		.46	.70
0480	Primer & 1 coat exterior latex		800	.020		.20	.57		.77	1.16
0490	Primer & 2 coats exterior latex		630	.025		.30	.73		1.03	1.53
0500	Roll, primer		2275	.007		.11	.20		.31	.45
0510	Finish coat, exterior latex		2275	.007		.11	.20		.31	.45
0520	Primer & 1 coat exterior latex		1300	.012		.22	.35		.57	.82
0530	Primer & 2 coats exterior latex		975	.016		.33	.47		.80	1.13
0540	Spray, primer		4550	.004		.08	.10		.18	.26
0550	Finish coat, exterior latex		4550	.004		.10	.10		.20	.28
0560	Primer & 1 coat exterior latex		2600	.006		.18	.18		.36	.49
0570	Primer & 2 coats exterior latex		1950	.008		.28	.24		.52	.69
0580	Waterproof sealer, first coat		5230	.003		.10	.09		.19	.25
0590	Second coat		5980	.003		.10	.08		.18	.24
0600	For oil base paint, add					10%				

09 91 13.70 Doors and Windows, Exterior

		Crew	Daily Output	Labor-Hours	Unit	Material	2016 Bare Costs Labor	Equipment	Total	Total Incl O&P
0010	**DOORS AND WINDOWS, EXTERIOR** R099100-10									
0100	Door frames & trim, only									
0110	Brushwork, primer R099100-20	1 Pord	512	.016	L.F.	.06	.45		.51	.81
0120	Finish coat, exterior latex		512	.016		.08	.45		.53	.82
0130	Primer & 1 coat, exterior latex		300	.027		.14	.76		.90	1.41
0140	Primer & 2 coats, exterior latex		265	.030		.22	.86		1.08	1.67
0150	Doors, flush, both sides, incl. frame & trim									
0160	Roll & brush, primer	1 Pord	10	.800	Ea.	4.65	23		27.65	43
0170	Finish coat, exterior latex		10	.800		5.85	23		28.85	44.50
0180	Primer & 1 coat, exterior latex		7	1.143		10.50	32.50		43	65.50
0190	Primer & 2 coats, exterior latex		5	1.600		16.35	46		62.35	93.50
0200	Brushwork, stain, sealer & 2 coats polyurethane		4	2		30	57.50		87.50	128
0210	Doors, French, both sides, 10-15 lite, incl. frame & trim									
0220	Brushwork, primer	1 Pord	6	1.333	Ea.	2.32	38		40.32	65.50
0230	Finish coat, exterior latex		6	1.333		2.93	38		40.93	66
0240	Primer & 1 coat, exterior latex		3	2.667		5.25	76.50		81.75	132
0250	Primer & 2 coats, exterior latex		2	4		8	115		123	198
0260	Brushwork, stain, sealer & 2 coats polyurethane		2.50	3.200		10.70	91.50		102.20	163

09 91 Painting

09 91 13 – Exterior Painting

09 91 13.70 Doors and Windows, Exterior

		Crew	Daily Output	Labor-Hours	Unit	Material	2016 Bare Costs Labor	Equipment	Total	Total Incl O&P
0270	Doors, louvered, both sides, incl. frame & trim									
0280	Brushwork, primer	1 Pord	7	1.143	Ea.	4.65	32.50		37.15	59
0290	Finish coat, exterior latex		7	1.143		5.85	32.50		38.35	60.50
0300	Primer & 1 coat, exterior latex		4	2		10.50	57.50		68	106
0310	Primer & 2 coats, exterior latex		3	2.667		16	76.50		92.50	144
0320	Brushwork, stain, sealer & 2 coats polyurethane		4.50	1.778		30	51		81	117
0330	Doors, panel, both sides, incl. frame & trim									
0340	Roll & brush, primer	1 Pord	6	1.333	Ea.	4.65	38		42.65	68
0350	Finish coat, exterior latex		6	1.333		5.85	38		43.85	69.50
0360	Primer & 1 coat, exterior latex		3	2.667		10.50	76.50		87	138
0370	Primer & 2 coats, exterior latex		2.50	3.200		16	91.50		107.50	169
0380	Brushwork, stain, sealer & 2 coats polyurethane		3	2.667		30	76.50		106.50	159
0400	Windows, per ext. side, based on 15 S.F.									
0410	1 to 6 lite									
0420	Brushwork, primer	1 Pord	13	.615	Ea.	.92	17.65		18.57	30
0430	Finish coat, exterior latex		13	.615		1.15	17.65		18.80	30.50
0440	Primer & 1 coat, exterior latex		8	1		2.07	28.50		30.57	49.50
0450	Primer & 2 coats, exterior latex		6	1.333		3.16	38		41.16	66.50
0460	Stain, sealer & 1 coat varnish		7	1.143		4.23	32.50		36.73	58.50
0470	7 to 10 lite									
0480	Brushwork, primer	1 Pord	11	.727	Ea.	.92	21		21.92	35.50
0490	Finish coat, exterior latex		11	.727		1.15	21		22.15	36
0500	Primer & 1 coat, exterior latex		7	1.143		2.07	32.50		34.57	56.50
0510	Primer & 2 coats, exterior latex		5	1.600		3.16	46		49.16	79
0520	Stain, sealer & 1 coat varnish		6	1.333		4.23	38		42.23	67.50
0530	12 lite									
0540	Brushwork, primer	1 Pord	10	.800	Ea.	.92	23		23.92	39
0550	Finish coat, exterior latex		10	.800		1.15	23		24.15	39.50
0560	Primer & 1 coat, exterior latex		6	1.333		2.07	38		40.07	65.50
0570	Primer & 2 coats, exterior latex		5	1.600		3.16	46		49.16	79
0580	Stain, sealer & 1 coat varnish		6	1.333		4.33	38		42.33	68
0590	For oil base paint, add					10%				

09 91 13.80 Trim, Exterior

		Crew	Daily Output	Labor-Hours	Unit	Material	2016 Bare Costs Labor	Equipment	Total	Total Incl O&P
0010	**TRIM, EXTERIOR** R099100-10									
0100	Door frames & trim (see Doors, interior or exterior)									
0110	Fascia, latex paint, one coat coverage									
0120	1" x 4", brushwork	1 Pord	640	.013	L.F.	.02	.36		.38	.62
0130	Roll		1280	.006		.03	.18		.21	.33
0140	Spray		2080	.004		.02	.11		.13	.20
0150	1" x 6" to 1" x 10", brushwork		640	.013		.08	.36		.44	.68
0160	Roll		1230	.007		.09	.19		.28	.40
0170	Spray		2100	.004		.07	.11		.18	.25
0180	1" x 12", brushwork		640	.013		.08	.36		.44	.68
0190	Roll		1050	.008		.09	.22		.31	.45
0200	Spray		2200	.004		.07	.10		.17	.24
0210	Gutters & downspouts, metal, zinc chromate paint									
0220	Brushwork, gutters, 5", first coat	1 Pord	640	.013	L.F.	.40	.36		.76	1.03
0230	Second coat		960	.008		.38	.24		.62	.80
0240	Third coat		1280	.006		.30	.18		.48	.64
0250	Downspouts, 4", first coat		640	.013		.40	.36		.76	1.03
0260	Second coat		960	.008		.38	.24		.62	.80
0270	Third coat		1280	.006		.30	.18		.48	.64

09 91 Painting

09 91 13 – Exterior Painting

09 91 13.80 Trim, Exterior

		Crew	Daily Output	Labor-Hours	Unit	Material	2016 Bare Costs Labor	Equipment	Total	Total Incl O&P
0280	Gutters & downspouts, wood									
0290	Brushwork, gutters, 5", primer	1 Pord	640	.013	L.F.	.06	.36		.42	.66
0300	Finish coat, exterior latex		640	.013		.07	.36		.43	.67
0310	Primer & 1 coat exterior latex		400	.020		.14	.57		.71	1.09
0320	Primer & 2 coats exterior latex		325	.025		.22	.71		.93	1.40
0330	Downspouts, 4", primer		640	.013		.06	.36		.42	.66
0340	Finish coat, exterior latex		640	.013		.07	.36		.43	.67
0350	Primer & 1 coat exterior latex		400	.020		.14	.57		.71	1.09
0360	Primer & 2 coats exterior latex		325	.025		.11	.71		.82	1.28
0370	Molding, exterior, up to 14" wide									
0380	Brushwork, primer	1 Pord	640	.013	L.F.	.07	.36		.43	.67
0390	Finish coat, exterior latex		640	.013		.08	.36		.44	.68
0400	Primer & 1 coat exterior latex		400	.020		.17	.57		.74	1.12
0410	Primer & 2 coats exterior latex		315	.025		.17	.73		.90	1.38
0420	Stain & fill		1050	.008		.10	.22		.32	.47
0430	Shellac		1850	.004		.14	.12		.26	.36
0440	Varnish		1275	.006		.10	.18		.28	.41

09 91 13.90 Walls, Masonry (CMU), Exterior

		Crew	Daily Output	Labor-Hours	Unit	Material	Labor	Equipment	Total	Total Incl O&P
0010	**WALLS, MASONRY (CMU), EXTERIOR**									
0360	Concrete masonry units (CMU), smooth surface									
0370	Brushwork, latex, first coat	1 Pord	640	.013	S.F.	.07	.36		.43	.67
0380	Second coat		960	.008		.06	.24		.30	.45
0390	Waterproof sealer, first coat		736	.011		.26	.31		.57	.79
0400	Second coat		1104	.007		.26	.21		.47	.62
0410	Roll, latex, paint, first coat		1465	.005		.09	.16		.25	.35
0420	Second coat		1790	.004		.06	.13		.19	.28
0430	Waterproof sealer, first coat		1680	.005		.26	.14		.40	.50
0440	Second coat		2060	.004		.26	.11		.37	.46
0450	Spray, latex, paint, first coat		1950	.004		.07	.12		.19	.26
0460	Second coat		2600	.003		.05	.09		.14	.21
0470	Waterproof sealer, first coat		2245	.004		.26	.10		.36	.45
0480	Second coat		2990	.003		.26	.08		.34	.41
0490	Concrete masonry unit (CMU), porous									
0500	Brushwork, latex, first coat	1 Pord	640	.013	S.F.	.14	.36		.50	.75
0510	Second coat		960	.008		.07	.24		.31	.47
0520	Waterproof sealer, first coat		736	.011		.26	.31		.57	.79
0530	Second coat		1104	.007		.26	.21		.47	.62
0540	Roll latex, first coat		1465	.005		.11	.16		.27	.38
0550	Second coat		1790	.004		.07	.13		.20	.29
0560	Waterproof sealer, first coat		1680	.005		.26	.14		.40	.50
0570	Second coat		2060	.004		.26	.11		.37	.46
0580	Spray latex, first coat		1950	.004		.08	.12		.20	.28
0590	Second coat		2600	.003		.05	.09		.14	.21
0600	Waterproof sealer, first coat		2245	.004		.26	.10		.36	.45
0610	Second coat		2990	.003		.26	.08		.34	.41

09 91 23 – Interior Painting

09 91 23.20 Cabinets and Casework

		Crew	Daily Output	Labor-Hours	Unit	Material	Labor	Equipment	Total	Total Incl O&P
0010	**CABINETS AND CASEWORK**									
1000	Primer coat, oil base, brushwork	1 Pord	650	.012	S.F.	.06	.35		.41	.65
2000	Paint, oil base, brushwork, 1 coat		650	.012		.11	.35		.46	.70
2500	2 coats		400	.020		.20	.57		.77	1.17
3000	Stain, brushwork, wipe off		650	.012		.09	.35		.44	.68

09 91 Painting

09 91 23 - Interior Painting

09 91 23.20 Cabinets and Casework		Crew	Daily Output	Labor-Hours	Unit	Material	2016 Bare Costs Labor	Equipment	Total	Total Incl O&P
4000	Shellac, 1 coat, brushwork	1 Pord	650	.012	S.F.	.12	.35		.47	.71
4500	Varnish, 3 coats, brushwork, sand after 1st coat	↓	325	.025	↓	.26	.71		.97	1.44
5000	For latex paint, deduct					10%				

09 91 23.33 Doors and Windows, Interior Alkyd (Oil Base)

		Crew	Daily Output	Labor-Hours	Unit	Material	Labor	Equipment	Total	Incl O&P
0010	**DOORS AND WINDOWS, INTERIOR ALKYD (OIL BASE)**									
0500	Flush door & frame, 3' x 7', oil, primer, brushwork	1 Pord	10	.800	Ea.	3.70	23		26.70	42
1000	Paint, 1 coat		10	.800		4.16	23		27.16	42.50
1200	2 coats		6	1.333		4.54	38		42.54	68
1400	Stain, brushwork, wipe off		18	.444		1.83	12.75		14.58	23
1600	Shellac, 1 coat, brushwork		25	.320		2.47	9.15		11.62	17.80
1800	Varnish, 3 coats, brushwork, sand after 1st coat		9	.889		5.40	25.50		30.90	48
2000	Panel door & frame, 3' x 7', oil, primer, brushwork		6	1.333		2.41	38		40.41	65.50
2200	Paint, 1 coat		6	1.333		4.16	38		42.16	67.50
2400	2 coats		3	2.667		10.75	76.50		87.25	138
2600	Stain, brushwork, panel door, 3' x 7', not incl. frame		16	.500		1.83	14.35		16.18	25.50
2800	Shellac, 1 coat, brushwork		22	.364		2.47	10.40		12.87	19.85
3000	Varnish, 3 coats, brushwork, sand after 1st coat	↓	7.50	1.067	↓	5.40	30.50		35.90	56.50
3020	French door, incl. 3' x 7', 6 lites, frame & trim									
3022	Paint, 1 coat, over existing paint	1 Pord	5	1.600	Ea.	8.30	46		54.30	84.50
3024	2 coats, over existing paint		5	1.600		16.15	46		62.15	93.50
3026	Primer & 1 coat		3.50	2.286		13.15	65.50		78.65	122
3028	Primer & 2 coats		3	2.667		21.50	76.50		98	150
3032	Varnish or polyurethane, 1 coat		5	1.600		8.40	46		54.40	85
3034	2 coats, sanding between	↓	3	2.667	↓	16.85	76.50		93.35	145
4400	Windows, including frame and trim, per side									
4600	Colonial type, 6/6 lites, 2' x 3', oil, primer, brushwork	1 Pord	14	.571	Ea.	.38	16.35		16.73	27.50
5800	Paint, 1 coat		14	.571		.66	16.35		17.01	27.50
6000	2 coats		9	.889		1.28	25.50		26.78	43.50
6200	3' x 5' opening, 6/6 lites, primer coat, brushwork		12	.667		.95	19.10		20.05	32.50
6400	Paint, 1 coat		12	.667		1.64	19.10		20.74	33.50
6600	2 coats		7	1.143		3.19	32.50		35.69	57.50
6800	4' x 8' opening, 6/6 lites, primer coat, brushwork		8	1		2.03	28.50		30.53	49
7000	Paint, 1 coat		8	1		3.50	28.50		32	51
7200	2 coats		5	1.600		6.80	46		52.80	83
8000	Single lite type, 2' x 3', oil base, primer coat, brushwork		33	.242		.38	6.95		7.33	11.85
8200	Paint, 1 coat		33	.242		.66	6.95		7.61	12.15
8400	2 coats		20	.400		1.28	11.45		12.73	20.50
8600	3' x 5' opening, primer coat, brushwork		20	.400		.95	11.45		12.40	19.95
8800	Paint, 1 coat		20	.400		1.64	11.45		13.09	20.50
8900	2 coats		13	.615		3.19	17.65		20.84	32.50
9200	4' x 8' opening, primer coat, brushwork		14	.571		2.03	16.35		18.38	29
9400	Paint, 1 coat		14	.571		3.50	16.35		19.85	31
9600	2 coats	↓	8	1	↓	6.80	28.50		35.30	54.50

09 91 23.35 Doors and Windows, Interior Latex

		Crew	Daily Output	Labor-Hours	Unit	Material	Labor	Equipment	Total	Incl O&P
0010	**DOORS & WINDOWS, INTERIOR LATEX** R099100-10									
0100	Doors, flush, both sides, incl. frame & trim									
0110	Roll & brush, primer	1 Pord	10	.800	Ea.	4.15	23		27.15	42.50
0120	Finish coat, latex		10	.800		5.20	23		28.20	44
0130	Primer & 1 coat latex		7	1.143		9.35	32.50		41.85	64.50
0140	Primer & 2 coats latex		5	1.600		14.25	46		60.25	91
0160	Spray, both sides, primer		20	.400		4.36	11.45		15.81	23.50
0170	Finish coat, latex	↓	20	.400	↓	5.45	11.45		16.90	25

09 91 Painting

09 91 23 – Interior Painting

09 91 23.35 Doors and Windows, Interior Latex

		Crew	Daily Output	Labor-Hours	Unit	Material	2016 Bare Costs Labor	2016 Bare Costs Equipment	Total	Total Incl O&P
0180	Primer & 1 coat latex	1 Pord	11	.727	Ea.	9.90	21		30.90	45.50
0190	Primer & 2 coats latex		8	1		15.10	28.50		43.60	63.50
0200	Doors, French, both sides, 10-15 lite, incl. frame & trim									
0210	Roll & brush, primer	1 Pord	6	1.333	Ea.	2.07	38		40.07	65.50
0220	Finish coat, latex		6	1.333		2.61	38		40.61	66
0230	Primer & 1 coat latex		3	2.667		4.68	76.50		81.18	131
0240	Primer & 2 coats latex		2	4		7.15	115		122.15	197
0260	Doors, louvered, both sides, incl. frame & trim									
0270	Roll & brush, primer	1 Pord	7	1.143	Ea.	4.15	32.50		36.65	58.50
0280	Finish coat, latex		7	1.143		5.20	32.50		37.70	60
0290	Primer & 1 coat, latex		4	2		9.10	57.50		66.60	105
0300	Primer & 2 coats, latex		3	2.667		14.55	76.50		91.05	142
0320	Spray, both sides, primer		20	.400		4.36	11.45		15.81	23.50
0330	Finish coat, latex		20	.400		5.45	11.45		16.90	25
0340	Primer & 1 coat, latex		11	.727		9.90	21		30.90	45.50
0350	Primer & 2 coats, latex		8	1		15.45	28.50		43.95	64
0360	Doors, panel, both sides, incl. frame & trim									
0370	Roll & brush, primer	1 Pord	6	1.333	Ea.	4.36	38		42.36	68
0380	Finish coat, latex		6	1.333		5.20	38		43.20	69
0390	Primer & 1 coat, latex		3	2.667		9.35	76.50		85.85	136
0400	Primer & 2 coats, latex		2.50	3.200		14.55	91.50		106.05	167
0420	Spray, both sides, primer		10	.800		4.36	23		27.36	43
0430	Finish coat, latex		10	.800		5.45	23		28.45	44
0440	Primer & 1 coat, latex		5	1.600		9.90	46		55.90	86.50
0450	Primer & 2 coats, latex		4	2		15.45	57.50		72.95	112
0460	Windows, per interior side, based on 15 S.F.									
0470	1 to 6 lite									
0480	Brushwork, primer	1 Pord	13	.615	Ea.	.82	17.65		18.47	30
0490	Finish coat, enamel		13	.615		1.03	17.65		18.68	30
0500	Primer & 1 coat enamel		8	1		1.85	28.50		30.35	49
0510	Primer & 2 coats enamel		6	1.333		2.88	38		40.88	66
0530	7 to 10 lite									
0540	Brushwork, primer	1 Pord	11	.727	Ea.	.82	21		21.82	35.50
0550	Finish coat, enamel		11	.727		1.03	21		22.03	35.50
0560	Primer & 1 coat enamel		7	1.143		1.85	32.50		34.35	56
0570	Primer & 2 coats enamel		5	1.600		2.88	46		48.88	78.50
0590	12 lite									
0600	Brushwork, primer	1 Pord	10	.800	Ea.	.82	23		23.82	39
0610	Finish coat, enamel		10	.800		1.03	23		24.03	39
0620	Primer & 1 coat enamel		6	1.333		1.85	38		39.85	65
0630	Primer & 2 coats enamel		5	1.600		2.88	46		48.88	78.50
0650	For oil base paint, add					10%				

09 91 23.39 Doors and Windows, Interior Latex, Zero Voc

			Crew	Daily Output	Labor-Hours	Unit	Material	Labor	Equipment	Total	Total Incl O&P
0010	**DOORS & WINDOWS, INTERIOR LATEX, ZERO VOC**										
0100	Doors flush, both sides, incl. frame & trim										
0110	Roll & brush, primer	G	1 Pord	10	.800	Ea.	5.10	23		28.10	43.50
0120	Finish coat, latex	G		10	.800		5.55	23		28.55	44
0130	Primer & 1 coat latex	G		7	1.143		10.65	32.50		43.15	65.50
0140	Primer & 2 coats latex	G		5	1.600		15.90	46		61.90	93
0160	Spray, both sides, primer	G		20	.400		5.40	11.45		16.85	25
0170	Finish coat, latex	G		20	.400		5.80	11.45		17.25	25.50
0180	Primer & 1 coat latex	G		11	.727		11.25	21		32.25	47

09 91 Painting

09 91 23 – Interior Painting

09 91 23.39 Doors and Windows, Interior Latex, Zero Voc

		Crew	Daily Output	Labor-Hours	Unit	Material	2016 Bare Costs Labor	Equipment	Total	Total Incl O&P
0190	Primer & 2 coats latex	[G] 1 Pord	8	1	Ea.	16.80	28.50		45.30	65.50
0200	Doors, French, both sides, 10-15 lite, incl. frame & trim									
0210	Roll & brush, primer	[G] 1 Pord	6	1.333	Ea.	2.56	38		40.56	66
0220	Finish coat, latex	[G]	6	1.333		2.77	38		40.77	66
0230	Primer & 1 coat latex	[G]	3	2.667		5.35	76.50		81.85	132
0240	Primer & 2 coats latex	[G]	2	4		7.95	115		122.95	198
0360	Doors, panel, both sides, incl. frame & trim									
0370	Roll & brush, primer	[G] 1 Pord	6	1.333	Ea.	5.40	38		43.40	69
0380	Finish coat, latex	[G]	6	1.333		5.55	38		43.55	69
0390	Primer & 1 coat, latex	[G]	3	2.667		10.65	76.50		87.15	138
0400	Primer & 2 coats, latex	[G]	2.50	3.200		16.20	91.50		107.70	169
0420	Spray, both sides, primer	[G]	10	.800		5.40	23		28.40	44
0430	Finish coat, latex	[G]	10	.800		5.80	23		28.80	44.50
0440	Primer & 1 coat, latex	[G]	5	1.600		11.25	46		57.25	88
0450	Primer & 2 coats, latex	[G]	4	2		17.15	57.50		74.65	113
0460	Windows, per interior side, based on 15 S.F.									
0470	1 to 6 lite									
0480	Brushwork, primer	[G] 1 Pord	13	.615	Ea.	1.01	17.65		18.66	30
0490	Finish coat, enamel	[G]	13	.615		1.09	17.65		18.74	30
0500	Primer & 1 coat enamel	[G]	8	1		2.10	28.50		30.60	49.50
0510	Primer & 2 coats enamel	[G]	6	1.333		3.20	38		41.20	66.50

09 91 23.40 Floors, Interior

		Crew	Daily Output	Labor-Hours	Unit	Material	Labor	Equipment	Total	Total Incl O&P
0010	**FLOORS, INTERIOR**									
0100	Concrete paint, latex									
0110	Brushwork									
0120	1st coat	1 Pord	975	.008	S.F.	.14	.24		.38	.54
0130	2nd coat		1150	.007		.09	.20		.29	.43
0140	3rd coat		1300	.006		.07	.18		.25	.37
0150	Roll									
0160	1st coat	1 Pord	2600	.003	S.F.	.19	.09		.28	.36
0170	2nd coat		3250	.002		.11	.07		.18	.24
0180	3rd coat		3900	.002		.08	.06		.14	.19
0190	Spray									
0200	1st coat	1 Pord	2600	.003	S.F.	.16	.09		.25	.33
0210	2nd coat		3250	.002		.09	.07		.16	.22
0220	3rd coat		3900	.002		.07	.06		.13	.18

09 91 23.52 Miscellaneous, Interior

		Crew	Daily Output	Labor-Hours	Unit	Material	Labor	Equipment	Total	Total Incl O&P
0010	**MISCELLANEOUS, INTERIOR**									
2400	Floors, conc./wood, oil base, primer/sealer coat, brushwork	2 Pord	1950	.008	S.F.	.10	.24		.34	.49
2450	Roller		5200	.003		.10	.09		.19	.26
2600	Spray		6000	.003		.10	.08		.18	.24
2650	Paint 1 coat, brushwork		1950	.008		.10	.24		.34	.50
2800	Roller		5200	.003		.10	.09		.19	.26
2850	Spray		6000	.003		.11	.08		.19	.25
3000	Stain, wood floor, brushwork, 1 coat		4550	.004		.09	.10		.19	.27
3200	Roller		5200	.003		.09	.09		.18	.25
3250	Spray		6000	.003		.09	.08		.17	.23
3400	Varnish, wood floor, brushwork		4550	.004		.09	.10		.19	.26
3450	Roller		5200	.003		.09	.09		.18	.25
3600	Spray		6000	.003		.10	.08		.18	.24
3800	Grilles, per side, oil base, primer coat, brushwork	1 Pord	520	.015		.13	.44		.57	.87
3850	Spray		1140	.007		.13	.20		.33	.48

09 91 Painting

09 91 23 – Interior Painting

09 91 23.52 Miscellaneous, Interior		Crew	Daily Output	Labor-Hours	Unit	Material	2016 Bare Costs Labor	Equipment	Total	Total Incl O&P
3880	Paint 1 coat, brushwork	1 Pord	520	.015	S.F.	.22	.44		.66	.97
3900	Spray		1140	.007		.24	.20		.44	.60
3920	Paint 2 coats, brushwork		325	.025		.43	.71		1.14	1.63
3940	Spray		650	.012		.49	.35		.84	1.12
4500	Louvers, one side, primer, brushwork		524	.015		.10	.44		.54	.82
4520	Paint one coat, brushwork		520	.015		.10	.44		.54	.83
4530	Spray		1140	.007		.11	.20		.31	.45
4540	Paint two coats, brushwork		325	.025		.18	.71		.89	1.36
4550	Spray		650	.012		.20	.35		.55	.80
4560	Paint three coats, brushwork		270	.030		.27	.85		1.12	1.70
4570	Spray		500	.016		.30	.46		.76	1.09
5000	Pipe, 1" - 4" diameter, primer or sealer coat, oil base, brushwork	2 Pord	1250	.013	L.F.	.10	.37		.47	.71
5100	Spray		2165	.007		.09	.21		.30	.45
5200	Paint 1 coat, brushwork		1250	.013		.11	.37		.48	.72
5300	Spray		2165	.007		.10	.21		.31	.46
5350	Paint 2 coats, brushwork		775	.021		.19	.59		.78	1.18
5400	Spray		1240	.013		.22	.37		.59	.85
5450	5" - 8" diameter, primer or sealer coat, brushwork		620	.026		.20	.74		.94	1.44
5500	Spray		1085	.015		.33	.42		.75	1.06
5550	Paint 1 coat, brushwork		620	.026		.29	.74		1.03	1.54
5600	Spray		1085	.015		.33	.42		.75	1.06
5650	Paint 2 coats, brushwork		385	.042		.39	1.19		1.58	2.39
5700	Spray		620	.026		.43	.74		1.17	1.69
6600	Radiators, per side, primer, brushwork	1 Pord	520	.015	S.F.	.10	.44		.54	.83
6620	Paint, one coat		520	.015		.09	.44		.53	.83
6640	Two coats		340	.024		.18	.67		.85	1.31
6660	Three coats		283	.028		.27	.81		1.08	1.63
7000	Trim, wood, incl. puttying, under 6" wide									
7200	Primer coat, oil base, brushwork	1 Pord	650	.012	L.F.	.03	.35		.38	.61
7250	Paint, 1 coat, brushwork		650	.012		.05	.35		.40	.64
7400	2 coats		400	.020		.11	.57		.68	1.06
7450	3 coats		325	.025		.16	.71		.87	1.33
7500	Over 6" wide, primer coat, brushwork		650	.012		.06	.35		.41	.65
7550	Paint, 1 coat, brushwork		650	.012		.11	.35		.46	.70
7600	2 coats		400	.020		.21	.57		.78	1.17
7650	3 coats		325	.025		.32	.71		1.03	1.51
8000	Cornice, simple design, primer coat, oil base, brushwork		650	.012	S.F.	.06	.35		.41	.65
8250	Paint, 1 coat		650	.012		.11	.35		.46	.70
8300	2 coats		400	.020		.21	.57		.78	1.17
8350	Ornate design, primer coat		350	.023		.06	.65		.71	1.15
8400	Paint, 1 coat		350	.023		.11	.65		.76	1.20
8450	2 coats		400	.020		.21	.57		.78	1.17
8600	Balustrades, primer coat, oil base, brushwork		520	.015		.06	.44		.50	.80
8650	Paint, 1 coat		520	.015		.11	.44		.55	.85
8700	2 coats		325	.025		.21	.71		.92	1.39
8900	Trusses and wood frames, primer coat, oil base, brushwork		800	.010		.06	.29		.35	.54
8950	Spray		1200	.007		.07	.19		.26	.38
9000	Paint 1 coat, brushwork		750	.011		.11	.31		.42	.62
9200	Spray		1200	.007		.12	.19		.31	.44
9220	Paint 2 coats, brushwork		500	.016		.21	.46		.67	.99
9240	Spray		600	.013		.24	.38		.62	.89
9260	Stain, brushwork, wipe off		600	.013		.09	.38		.47	.73
9280	Varnish, 3 coats, brushwork		275	.029		.26	.83		1.09	1.65

09 91 Painting

09 91 23 – Interior Painting

09 91 23.52 Miscellaneous, Interior

		Crew	Daily Output	Labor-Hours	Unit	Material	2016 Bare Costs Labor	2016 Bare Costs Equipment	Total	Total Incl O&P
9350	For latex paint, deduct				S.F.	10%				

09 91 23.72 Walls and Ceilings, Interior

		Crew	Daily Output	Labor-Hours	Unit	Material	Labor	Equipment	Total	Total Incl O&P
0010	**WALLS AND CEILINGS, INTERIOR**									
0100	Concrete, drywall or plaster, latex, primer or sealer coat									
0200	Smooth finish, brushwork	1 Pord	1150	.007	S.F.	.06	.20		.26	.40
0240	Roller		1350	.006		.06	.17		.23	.35
0280	Spray		2750	.003		.05	.08		.13	.20
0300	Sand finish, brushwork		975	.008		.06	.24		.30	.46
0340	Roller		1150	.007		.06	.20		.26	.40
0380	Spray		2275	.004		.05	.10		.15	.23
0400	Paint 1 coat, smooth finish, brushwork		1200	.007		.07	.19		.26	.39
0440	Roller		1300	.006		.07	.18		.25	.37
0480	Spray		2275	.004		.06	.10		.16	.23
0500	Sand finish, brushwork		1050	.008		.06	.22		.28	.43
0540	Roller		1600	.005		.07	.14		.21	.32
0580	Spray		2100	.004		.02	.11		.13	.20
0800	Paint 2 coats, smooth finish, brushwork		680	.012		.14	.34		.48	.71
0840	Roller		800	.010		.14	.29		.43	.62
0880	Spray		1625	.005		.13	.14		.27	.37
0900	Sand finish, brushwork		605	.013		.14	.38		.52	.77
0940	Roller		1020	.008		.14	.22		.36	.52
0980	Spray		1700	.005		.13	.13		.26	.36
1200	Paint 3 coats, smooth finish, brushwork		510	.016		.21	.45		.66	.97
1240	Roller		650	.012		.21	.35		.56	.81
1280	Spray		1625	.005		.19	.14		.33	.44
1300	Sand finish, brushwork		454	.018		.30	.50		.80	1.16
1340	Roller		680	.012		.31	.34		.65	.90
1380	Spray		1133	.007		.27	.20		.47	.63
1600	Glaze coating, 2 coats, spray, clear		1200	.007		.49	.19		.68	.85
1640	Multicolor		1200	.007		.97	.19		1.16	1.38
1660	Painting walls, complete, including surface prep, primer &									
1670	2 coats finish, on drywall or plaster, with roller	1 Pord	325	.025	S.F.	.21	.71		.92	1.39
1700	For oil base paint, add					10%				
1800	For ceiling installations, add						25%			
2000	Masonry or concrete block, primer/sealer, latex paint									
2100	Primer, smooth finish, brushwork	1 Pord	1000	.008	S.F.	.11	.23		.34	.50
2110	Roller		1150	.007		.11	.20		.31	.45
2180	Spray		2400	.003		.10	.10		.20	.27
2200	Sand finish, brushwork		850	.009		.11	.27		.38	.56
2210	Roller		975	.008		.11	.24		.35	.51
2280	Spray		2050	.004		.10	.11		.21	.29
2400	Finish coat, smooth finish, brush		1100	.007		.08	.21		.29	.43
2410	Roller		1300	.006		.08	.18		.26	.38
2480	Spray		2400	.003		.07	.10		.17	.24
2500	Sand finish, brushwork		950	.008		.08	.24		.32	.49
2510	Roller		1090	.007		.08	.21		.29	.44
2580	Spray		2040	.004		.07	.11		.18	.27
2800	Primer plus one finish coat, smooth brush		525	.015		.30	.44		.74	1.05
2810	Roller		615	.013		.19	.37		.56	.82
2880	Spray		1200	.007		.17	.19		.36	.49
2900	Sand finish, brushwork		450	.018		.19	.51		.70	1.05
2910	Roller		515	.016		.19	.44		.63	.94

09 91 Painting

09 91 23 – Interior Painting

09 91 23.72 Walls and Ceilings, Interior

		Crew	Daily Output	Labor-Hours	Unit	Material	2016 Bare Costs Labor	Equipment	Total	Total Incl O&P
2980	Spray	1 Pord	1025	.008	S.F.	.17	.22		.39	.55
3200	Primer plus 2 finish coats, smooth, brush		355	.023		.27	.65		.92	1.36
3210	Roller		415	.019		.27	.55		.82	1.21
3280	Spray		800	.010		.24	.29		.53	.73
3300	Sand finish, brushwork		305	.026		.27	.75		1.02	1.54
3310	Roller		350	.023		.27	.65		.92	1.38
3380	Spray		675	.012		.24	.34		.58	.82
3600	Glaze coating, 3 coats, spray, clear		900	.009		.70	.25		.95	1.19
3620	Multicolor		900	.009		1.14	.25		1.39	1.67
4000	Block filler, 1 coat, brushwork		425	.019		.12	.54		.66	1.03
4100	Silicone, water repellent, 2 coats, spray		2000	.004		.37	.11		.48	.60
4120	For oil base paint, add					10%				
8200	For work 8' - 15' H, add						10%			
8300	For work over 15' H, add						20%			
8400	For light textured surfaces, add						10%			
8410	Heavy textured, add						25%			

09 91 23.74 Walls and Ceilings, Interior, Zero VOC Latex

			Crew	Daily Output	Labor-Hours	Unit	Material	Labor	Equipment	Total	Total Incl O&P
0010	**WALLS AND CEILINGS, INTERIOR, ZERO VOC LATEX**										
0100	Concrete, dry wall or plaster, latex, primer or sealer coat										
0200	Smooth finish, brushwork	G	1 Pord	1150	.007	S.F.	.07	.20		.27	.40
0240	Roller	G		1350	.006		.06	.17		.23	.35
0280	Spray	G		2750	.003		.05	.08		.13	.19
0300	Sand finish, brushwork	G		975	.008		.06	.24		.30	.46
0340	Roller	G		1150	.007		.07	.20		.27	.41
0380	Spray	G		2275	.004		.05	.10		.15	.23
0400	Paint 1 coat, smooth finish, brushwork	G		1200	.007		.08	.19		.27	.40
0440	Roller	G		1300	.006		.08	.18		.26	.38
0480	Spray	G		2275	.004		.07	.10		.17	.24
0500	Sand finish, brushwork	G		1050	.008		.07	.22		.29	.44
0540	Roller	G		1600	.005		.08	.14		.22	.33
0580	Spray	G		2100	.004		.07	.11		.18	.25
0800	Paint 2 coats, smooth finish, brushwork	G		680	.012		.15	.34		.49	.73
0840	Roller	G		800	.010		.16	.29		.45	.64
0880	Spray	G		1625	.005		.14	.14		.28	.38
0900	Sand finish, brushwork	G		605	.013		.15	.38		.53	.78
0940	Roller	G		1020	.008		.16	.22		.38	.54
0980	Spray	G		1700	.005		.14	.13		.27	.37
1200	Paint 3 coats, smooth finish, brushwork	G		510	.016		.22	.45		.67	.99
1240	Roller	G		650	.012		.24	.35		.59	.84
1280	Spray	G		1625	.005		.20	.14		.34	.45
1800	For ceiling installations, add	G						25%			
8200	For work 8' - 15' H, add							10%			
8300	For work over 15' H, add							20%			

09 91 23.75 Dry Fall Painting

		Crew	Daily Output	Labor-Hours	Unit	Material	Labor	Equipment	Total	Total Incl O&P
0010	**DRY FALL PAINTING**									
0100	Sprayed on walls, gypsum board or plaster									
0220	One coat	1 Pord	2600	.003	S.F.	.07	.09		.16	.23
0250	Two coats		1560	.005		.15	.15		.30	.40
0280	Concrete or textured plaster, one coat		1560	.005		.07	.15		.22	.32
0310	Two coats		1300	.006		.15	.18		.33	.45
0340	Concrete block, one coat		1560	.005		.07	.15		.22	.32
0370	Two coats		1300	.006		.15	.18		.33	.45

09 91 Painting

09 91 23 – Interior Painting

09 91 23.75 Dry Fall Painting

		Crew	Daily Output	Labor-Hours	Unit	Material	2016 Bare Costs Labor	Equipment	Total	Total Incl O&P
0400	Wood, one coat	1 Pord	877	.009	S.F.	.07	.26		.33	.51
0430	Two coats		650	.012		.15	.35		.50	.74
0440	On ceilings, gypsum board or plaster									
0470	One coat	1 Pord	1560	.005	S.F.	.07	.15		.22	.32
0500	Two coats		1300	.006		.15	.18		.33	.45
0530	Concrete or textured plaster, one coat		1560	.005		.07	.15		.22	.32
0560	Two coats		1300	.006		.15	.18		.33	.45
0570	Structural steel, bar joists or metal deck, one coat		1560	.005		.07	.15		.22	.32
0580	Two coats		1040	.008		.15	.22		.37	.52

09 93 Staining and Transparent Finishing

09 93 23 – Interior Staining and Finishing

09 93 23.10 Varnish

		Crew	Daily Output	Labor-Hours	Unit	Material	Labor	Equipment	Total	Total Incl O&P
0010	**VARNISH**									
0012	1 coat + sealer, on wood trim, brush, no sanding included	1 Pord	400	.020	S.F.	.07	.57		.64	1.02
0020	1 coat + sealer, on wood trim, brush, no sanding included, no VOC		400	.020		.21	.57		.78	1.17
0100	Hardwood floors, 2 coats, no sanding included, roller		1890	.004		.15	.12		.27	.37

09 96 High-Performance Coatings

09 96 56 – Epoxy Coatings

09 96 56.20 Wall Coatings

		Crew	Daily Output	Labor-Hours	Unit	Material	Labor	Equipment	Total	Total Incl O&P
0010	**WALL COATINGS**									
0100	Acrylic glazed coatings, matte	1 Pord	525	.015	S.F.	.31	.44		.75	1.06
0200	Gloss		305	.026		.65	.75		1.40	1.96
0300	Epoxy coatings, solvent based		525	.015		.40	.44		.84	1.16
0400	Water based		170	.047		.27	1.35		1.62	2.52
0600	Exposed aggregate, troweled on, 1/16" to 1/4", solvent based		235	.034		.62	.98		1.60	2.29
0700	Water based (epoxy or polyacrylate)		130	.062		1.33	1.76		3.09	4.36
0900	1/2" to 5/8" aggregate, solvent based		130	.062		1.20	1.76		2.96	4.22
1000	Water based		80	.100		2.08	2.87		4.95	7
1200	1" aggregate size, solvent based		90	.089		2.12	2.55		4.67	6.55
1300	Water based		55	.145		3.23	4.17		7.40	10.40
1500	Exposed aggregate, sprayed on, 1/8" aggregate, solvent based		295	.027		.57	.78		1.35	1.91
1600	Water based		145	.055		1.05	1.58		2.63	3.76

Division Notes

		CREW	DAILY OUTPUT	LABOR-HOURS	UNIT	BARE COSTS				TOTAL INCL O&P
						MAT.	LABOR	EQUIP.	TOTAL	

Estimating Tips
General
- The items in this division are usually priced per square foot or each.
- Many items in Division 10 require some type of support system or special anchors that are not usually furnished with the item. The required anchors must be added to the estimate in the appropriate division.
- Some items in Division 10, such as lockers, may require assembly before installation. Verify the amount of assembly required. Assembly can often exceed installation time.

10 20 00 Interior Specialties
- Support angles and blocking are not included in the installation of toilet compartments, shower/dressing compartments, or cubicles. Appropriate line items from Divisions 5 or 6 may need to be added to support the installations.
- Toilet partitions are priced by the stall. A stall consists of a side wall, pilaster, and door with hardware. Toilet tissue holders and grab bars are extra.
- The required acoustical rating of a folding partition can have a significant impact on costs. Verify the sound transmission coefficient rating of the panel priced to the specification requirements.
- Grab bar installation does not include supplemental blocking or backing to support the required load. When grab bars are installed at an existing facility, provisions must be made to attach the grab bars to solid structure.

Reference Numbers
Reference numbers are shown at the beginning of some major classifications. These numbers refer to related items in the Reference Section. The reference information may be an estimating procedure, an alternate pricing method, or technical information.

Note: Not all subdivisions listed here necessarily appear. ∎

Did you know?
RSMeans Online gives you the same access to RSMeans' data with 24/7 access:
- Quickly locate costs in the searchable database.
- Build cost lists, estimates, and reports in minutes.
- Adjust costs to any location in the U.S. and Canada with the click of a button.

Start your free trial today at **www.RSMeansOnline.com**

RSMeans Online
FROM THE GORDIAN GROUP®

No part of this cost data may be reproduced, stored in a retrieval system, or transmitted in any form or by any means without prior written permission of RSMeans.

10 28 Toilet, Bath, and Laundry Accessories

10 28 13 – Toilet Accessories

10 28 13.13 Commercial Toilet Accessories

		Crew	Daily Output	Labor-Hours	Unit	Material	2016 Bare Costs Labor	Equipment	Total	Total Incl O&P
0010	**COMMERCIAL TOILET ACCESSORIES**									
0200	Curtain rod, stainless steel, 5' long, 1" diameter	1 Carp	13	.615	Ea.	28	21		49	66
0300	1-1/4" diameter		13	.615		30.50	21		51.50	69
0800	Grab bar, straight, 1-1/4" diameter, stainless steel, 18" long		24	.333		30.50	11.30		41.80	52.50
1100	36" long		20	.400		37	13.55		50.55	63
1105	42" long		20	.400		39	13.55		52.55	65.50
1120	Corner, 36" long		20	.400		90	13.55		103.55	122
3000	Mirror, with stainless steel 3/4" square frame, 18" x 24"		20	.400		49.50	13.55		63.05	77
3100	36" x 24"		15	.533		110	18.10		128.10	152
3300	72" x 24"		6	1.333		305	45		350	410
4300	Robe hook, single, regular		96	.083		19.40	2.82		22.22	26
4400	Heavy duty, concealed mounting		56	.143		22.50	4.84		27.34	33
6400	Towel bar, stainless steel, 18" long		23	.348		41.50	11.80		53.30	65.50
6500	30" long		21	.381		66	12.90		78.90	94.50
7400	Tumbler holder, for tumbler only		30	.267		18.75	9.05		27.80	35.50
7410	Tumbler holder, recessed		20	.400		8	13.55		21.55	31.50
7500	Soap, tumbler & toothbrush		30	.267		19.60	9.05		28.65	36.50
7510	Tumbler & toothbrush holder		20	.400		13.50	13.55		27.05	37.50

10 28 16 – Bath Accessories

10 28 16.20 Medicine Cabinets

		Crew	Daily Output	Labor-Hours	Unit	Material	Labor	Equipment	Total	Total Incl O&P
0010	**MEDICINE CABINETS**									
0020	With mirror, sst frame, 16" x 22", unlighted	1 Carp	14	.571	Ea.	99	19.35		118.35	142
0100	Wood frame		14	.571		140	19.35		159.35	187
0300	Sliding mirror doors, 20" x 16" x 4-3/4", unlighted		7	1.143		133	38.50		171.50	211
0400	24" x 19" x 8-1/2", lighted		5	1.600		184	54		238	293
0600	Triple door, 30" x 32", unlighted, plywood body		7	1.143		325	38.50		363.50	425
0700	Steel body		7	1.143		375	38.50		413.50	475
0900	Oak door, wood body, beveled mirror, single door		7	1.143		199	38.50		237.50	284
1000	Double door		6	1.333		415	45		460	530

10 28 19 – Tub and Shower Enclosures

10 28 19.10 Partitions, Shower

		Crew	Daily Output	Labor-Hours	Unit	Material	Labor	Equipment	Total	Total Incl O&P
0010	**PARTITIONS, SHOWER** floor mounted, no plumbing									
0400	Cabinet, one piece, fiberglass, 32" x 32"	2 Carp	5	3.200	Ea.	535	108		643	770
0420	36" x 36"		5	3.200		625	108		733	865
0440	36" x 48"		5	3.200		1,375	108		1,483	1,675
0460	Acrylic, 32" x 32"		5	3.200		340	108		448	555
0480	36" x 36"		5	3.200		1,050	108		1,158	1,325
0500	36" x 48"		5	3.200		1,400	108		1,508	1,725
0520	Shower door for above, clear plastic, 24" wide	1 Carp	8	1		188	34		222	264
0540	28" wide		8	1		225	34		259	305
0560	Tempered glass, 24" wide		8	1		220	34		254	299
0580	28" wide		8	1		252	34		286	335
2400	Glass stalls, with doors, no receptors, chrome on brass	2 Shee	3	5.333		1,600	202		1,802	2,100
2700	Anodized aluminum	"	4	4		1,125	151		1,276	1,475
3200	Receptors, precast terrazzo, 32" x 32"	2 Marb	14	1.143		335	36.50		371.50	430
3300	48" x 34"		9.50	1.684		440	54		494	575
3500	Plastic, simulated terrazzo receptor, 32" x 32"		14	1.143		148	36.50		184.50	225
3600	32" x 48"		12	1.333		222	43		265	315
3800	Precast concrete, colors, 32" x 32"		14	1.143		201	36.50		237.50	283
3900	48" x 48"		8	2		277	64.50		341.50	410
4100	Shower doors, economy plastic, 24" wide	1 Shee	9	.889		144	33.50		177.50	214
4200	Tempered glass door, economy		8	1		258	38		296	345

10 28 Toilet, Bath, and Laundry Accessories

10 28 19 – Tub and Shower Enclosures

10 28 19.10 Partitions, Shower

		Crew	Daily Output	Labor-Hours	Unit	Material	2016 Bare Costs Labor	Equipment	Total	Total Incl O&P
4400	Folding, tempered glass, aluminum frame	1 Shee	6	1.333	Ea.	420	50.50		470.50	545
4700	Deluxe, tempered glass, chrome on brass frame, 42" to 44"		8	1		415	38		453	520
4800	39" to 48" wide		1	8		675	300		975	1,250
4850	On anodized aluminum frame, obscure glass		2	4		575	151		726	880
4900	Clear glass		1	8		675	300		975	1,250
5100	Shower enclosure, tempered glass, anodized alum. frame									
5120	2 panel & door, corner unit, 32" x 32"	1 Shee	2	4	Ea.	1,100	151		1,251	1,475
5140	Neo-angle corner unit, 16" x 24" x 16"	"	2	4		1,125	151		1,276	1,475
5200	Shower surround, 3 wall, polypropylene, 32" x 32"	1 Carp	4	2		480	68		548	645
5220	PVC, 32" x 32"		4	2		380	68		448	535
5240	Fiberglass		4	2		420	68		488	575
5250	2 wall, polypropylene, 32" x 32"		4	2		315	68		383	465
5270	PVC		4	2		395	68		463	550
5290	Fiberglass		4	2		400	68		468	555
5300	Tub doors, tempered glass & frame, obscure glass	1 Shee	8	1		229	38		267	315
5400	Clear glass		6	1.333		535	50.50		585.50	675
5600	Chrome plated, brass frame, obscure glass		8	1		305	38		343	400
5700	Clear glass		6	1.333		745	50.50		795.50	900
5900	Tub/shower enclosure, temp. glass, alum. frame, obscure glass		2	4		410	151		561	700
6200	Clear glass		1.50	5.333		855	202		1,057	1,275
6500	On chrome-plated brass frame, obscure glass		2	4		565	151		716	875
6600	Clear glass		1.50	5.333		1,200	202		1,402	1,650
6800	Tub surround, 3 wall, polypropylene	1 Carp	4	2		256	68		324	395
6900	PVC		4	2		390	68		458	545
7000	Fiberglass, obscure glass		4	2		400	68		468	555
7100	Clear glass		3	2.667		680	90.50		770.50	900

10 28 23 – Laundry Accessories

10 28 23.13 Built-In Ironing Boards

		Crew	Daily Output	Labor-Hours	Unit	Material	Labor	Equipment	Total	Total Incl O&P
0010	**BUILT-IN IRONING BOARDS**									
0020	Including cabinet, board & light, 42"	1 Carp	2	4	Ea.	440	136		576	705

10 31 Manufactured Fireplaces

10 31 13 – Manufactured Fireplace Chimneys

10 31 13.10 Fireplace Chimneys

		Crew	Daily Output	Labor-Hours	Unit	Material	Labor	Equipment	Total	Total Incl O&P
0010	**FIREPLACE CHIMNEYS**									
0500	Chimney dbl. wall, all stainless, over 8'-6", 7" diam., add to fireplace	1 Carp	33	.242	V.L.F.	84	8.20		92.20	106
0600	10" diameter, add to fireplace		32	.250		112	8.50		120.50	138
0700	12" diameter, add to fireplace		31	.258		167	8.75		175.75	198
0800	14" diameter, add to fireplace		30	.267		227	9.05		236.05	265
1000	Simulated brick chimney top, 4' high, 16" x 16"		10	.800	Ea.	435	27		462	525
1100	24" x 24"		7	1.143	"	555	38.50		593.50	675

10 31 13.20 Chimney Accessories

		Crew	Daily Output	Labor-Hours	Unit	Material	Labor	Equipment	Total	Total Incl O&P
0010	**CHIMNEY ACCESSORIES**									
0020	Chimney screens, galv., 13" x 13" flue	1 Bric	8	1	Ea.	58.50	32.50		91	118
0050	24" x 24" flue		5	1.600		124	52		176	224
0200	Stainless steel, 13" x 13" flue		8	1		97.50	32.50		130	161
0250	20" x 20" flue		5	1.600		152	52		204	254
2400	Squirrel and bird screens, galvanized, 8" x 8" flue		16	.500		54	16.20		70.20	86.50
2450	13" x 13" flue		12	.667		55	21.50		76.50	96.50

10 31 Manufactured Fireplaces

10 31 16 – Manufactured Fireplace Forms

10 31 16.10 Fireplace Forms	Crew	Daily Output	Labor-Hours	Unit	Material	2016 Bare Costs Labor	Equipment	Total	Total Incl O&P
0010 **FIREPLACE FORMS**									
1800 Fireplace forms, no accessories, 32" opening	1 Bric	3	2.667	Ea.	705	86.50		791.50	920
1900 36" opening		2.50	3.200		895	104		999	1,150
2000 40" opening		2	4		1,200	130		1,330	1,525
2100 78" opening		1.50	5.333		1,725	173		1,898	2,200

10 31 23 – Prefabricated Fireplaces

10 31 23.10 Fireplace, Prefabricated

	Crew	Daily Output	Labor-Hours	Unit	Material	Labor	Equipment	Total	Total Incl O&P
0010 **FIREPLACE, PREFABRICATED**, free standing or wall hung									
0100 With hood & screen, painted	1 Carp	1.30	6.154	Ea.	1,475	209		1,684	1,975
0150 Average		1	8		1,700	271		1,971	2,325
0200 Stainless steel		.90	8.889		3,050	300		3,350	3,850
1500 Simulated logs, gas fired, 40,000 BTU, 2' long, manual safety pilot		7	1.143	Set	440	38.50		478.50	550
1600 Adjustable flame remote pilot		6	1.333		1,200	45		1,245	1,400
1700 Electric, 1,500 BTU, 1'-6" long, incandescent flame		7	1.143		205	38.50		243.50	290
1800 1,500 BTU, LED flame		6	1.333		320	45		365	425
2000 Fireplace, built-in, 36" hearth, radiant		1.30	6.154	Ea.	680	209		889	1,100
2100 Recirculating, small fan		1	8		900	271		1,171	1,450
2150 Large fan		.90	8.889		2,000	300		2,300	2,700
2200 42" hearth, radiant		1.20	6.667		915	226		1,141	1,375
2300 Recirculating, small fan		.90	8.889		1,200	300		1,500	1,825
2350 Large fan		.80	10		1,350	340		1,690	2,050
2400 48" hearth, radiant		1.10	7.273		2,175	247		2,422	2,800
2500 Recirculating, small fan		.80	10		2,475	340		2,815	3,300
2550 Large fan		.70	11.429		2,475	385		2,860	3,375
3000 See through, including doors		.80	10		2,200	340		2,540	2,975
3200 Corner (2 wall)		1	8		3,425	271		3,696	4,200

10 32 Fireplace Specialties

10 32 13 – Fireplace Dampers

10 32 13.10 Dampers

	Crew	Daily Output	Labor-Hours	Unit	Material	Labor	Equipment	Total	Total Incl O&P
0010 **DAMPERS**									
0800 Damper, rotary control, steel, 30" opening	1 Bric	6	1.333	Ea.	119	43		162	203
0850 Cast iron, 30" opening		6	1.333		125	43		168	209
1200 Steel plate, poker control, 60" opening		8	1		320	32.50		352.50	410
1250 84" opening, special order		5	1.600		585	52		637	730
1400 "Universal" type, chain operated, 32" x 20" opening		8	1		250	32.50		282.50	330
1450 48" x 24" opening		5	1.600		375	52		427	495

10 32 23 – Fireplace Doors

10 32 23.10 Doors

	Crew	Daily Output	Labor-Hours	Unit	Material	Labor	Equipment	Total	Total Incl O&P
0010 **DOORS**									
0400 Cleanout doors and frames, cast iron, 8" x 8"	1 Bric	12	.667	Ea.	42.50	21.50		64	82.50
0450 12" x 12"		10	.800		97	26		123	150
0500 18" x 24"		8	1		150	32.50		182.50	219
0550 Cast iron frame, steel door, 24" x 30"		5	1.600		315	52		367	430
1600 Dutch Oven door and frame, cast iron, 12" x 15" opening		13	.615		131	19.95		150.95	178
1650 Copper plated, 12" x 15" opening		13	.615		257	19.95		276.95	315

10 35 Stoves

10 35 13 – Heating Stoves

10 35 13.10 Woodburning Stoves

		Crew	Daily Output	Labor-Hours	Unit	Material	2016 Bare Costs Labor	Equipment	Total	Total Incl O&P
0010	**WOODBURNING STOVES**									
0015	Cast iron, °1500 S.F.	2 Carp	1.30	12.308	Ea.	1,300	415		1,715	2,125
0020	1500-2000 S.F.		1	16		2,050	540		2,590	3,150
0030	2,000 S.F.		.80	20		2,800	680		3,480	4,200
0050	For gas log lighter, add					46.50			46.50	51

10 44 Fire Protection Specialties

10 44 16 – Fire Extinguishers

10 44 16.13 Portable Fire Extinguishers

		Crew	Daily Output	Labor-Hours	Unit	Material	2016 Bare Costs Labor	Equipment	Total	Total Incl O&P
0010	**PORTABLE FIRE EXTINGUISHERS**									
0140	CO_2, with hose and "H" horn, 10 lb.				Ea.	288			288	315
1000	Dry chemical, pressurized									
1040	Standard type, portable, painted, 2-1/2 lb.				Ea.	38.50			38.50	42.50
1080	10 lb.					82			82	90
1100	20 lb.					135			135	149
1120	30 lb.					435			435	480
2000	ABC all purpose type, portable, 2-1/2 lb.					21			21	23.50
2080	9-1/2 lb.					44			44	48

10 55 Postal Specialties

10 55 23 – Mail Boxes

10 55 23.10 Mail Boxes

		Crew	Daily Output	Labor-Hours	Unit	Material	2016 Bare Costs Labor	Equipment	Total	Total Incl O&P
0011	**MAIL BOXES**									
1900	Letter slot, residential	1 Carp	20	.400	Ea.	80	13.55		93.55	111
2400	Residential, galv. steel, small 20" x 7" x 9"	1 Clab	16	.500		200	12.35		212.35	241
2410	With galv. steel post, 54" long		6	1.333		250	33		283	330
2420	Large, 24" x 12" x 15"		16	.500		200	12.35		212.35	241
2430	With galv. steel post, 54" long		6	1.333		225	33		258	305
2440	Decorative, polyethylene, 22" x 10" x 10"		16	.500		60	12.35		72.35	86.50
2450	With alum. post, decorative, 54" long		6	1.333		190	33		223	264

10 56 Storage Assemblies

10 56 13 – Metal Storage Shelving

10 56 13.10 Shelving

		Crew	Daily Output	Labor-Hours	Unit	Material	2016 Bare Costs Labor	Equipment	Total	Total Incl O&P
0010	**SHELVING**									
0020	Metal, industrial, cross-braced, 3' wide, 12" deep	1 Sswk	175	.046	SF Shlf	7.20	1.63		8.83	10.90
0100	24" deep		330	.024		5.20	.86		6.06	7.30
2200	Wide span, 1600 lb. capacity per shelf, 6' wide, 24" deep		380	.021		7.65	.75		8.40	9.80
2400	36" deep		440	.018		6.75	.65		7.40	8.60
3000	Residential, vinyl covered wire, wardrobe, 12" deep	1 Carp	195	.041	L.F.	14.95	1.39		16.34	18.75
3100	16" deep		195	.041		10.70	1.39		12.09	14.10
3200	Standard, 6" deep		195	.041		4.25	1.39		5.64	7
3300	9" deep		195	.041		5.80	1.39		7.19	8.70
3400	12" deep		195	.041		7.70	1.39		9.09	10.80
3500	16" deep		195	.041		18	1.39		19.39	22
3600	20" deep		195	.041		21.50	1.39		22.89	26
3700	Support bracket		80	.100	Ea.	6.40	3.39		9.79	12.75

10 57 Wardrobe and Closet Specialties

10 57 23 – Closet and Utility Shelving

10 57 23.19 Wood Closet and Utility Shelving	Crew	Daily Output	Labor-Hours	Unit	Material	2016 Bare Costs Labor	Equipment	Total	Total Incl O&P
0010 **WOOD CLOSET AND UTILITY SHELVING**									
0020 Pine, clear grade, no edge band, 1" x 8"	1 Carp	115	.070	L.F.	3.47	2.36		5.83	7.75
0100 1" x 10"		110	.073		4.32	2.47		6.79	8.90
0200 1" x 12"		105	.076		5.20	2.58		7.78	10.05
0450 1" x 18"		95	.084		7.80	2.85		10.65	13.40
0460 1" x 24"		85	.094		10.40	3.19		13.59	16.80
0600 Plywood, 3/4" thick with lumber edge, 12" wide		75	.107		1.91	3.62		5.53	8.15
0700 24" wide		70	.114	↓	3.39	3.87		7.26	10.25
0900 Bookcase, clear grade pine, shelves 12" O.C., 8" deep, per S.F. shelf		70	.114	S.F.	11.30	3.87		15.17	18.90
1000 12" deep shelves		65	.123	"	16.95	4.17		21.12	25.50
1200 Adjustable closet rod and shelf, 12" wide, 3' long		20	.400	Ea.	11.60	13.55		25.15	35.50
1300 8' long		15	.533	"	21	18.10		39.10	54
1500 Prefinished shelves with supports, stock, 8" wide		75	.107	L.F.	4.89	3.62		8.51	11.45
1600 10" wide	↓	70	.114	"	5.70	3.87		9.57	12.75

10 73 Protective Covers

10 73 16 – Canopies

10 73 16.10 Canopies, Residential

	Crew	Daily Output	Labor-Hours	Unit	Material	Labor	Equipment	Total	Total Incl O&P
0010 **CANOPIES, RESIDENTIAL** Prefabricated									
0500 Carport, free standing, baked enamel, alum., .032", 40 psf									
0520 16' x 8', 4 posts	2 Carp	3	5.333	Ea.	4,725	181		4,906	5,500
0600 20' x 10', 6 posts		2	8		4,825	271		5,096	5,775
0605 30' x 10', 8 posts	↓	2	8		7,250	271		7,521	8,425
1000 Door canopies, extruded alum., .032", 42" projection, 4' wide	1 Carp	8	1		203	34		237	280
1020 6' wide	"	6	1.333		287	45		332	390
1040 8' wide	2 Carp	9	1.778		380	60.50		440.50	515
1060 10' wide		7	2.286		405	77.50		482.50	575
1080 12' wide	↓	5	3.200		635	108		743	880
1200 54" projection, 4' wide	1 Carp	8	1		251	34		285	335
1220 6' wide	"	6	1.333		294	45		339	400
1240 8' wide	2 Carp	9	1.778		355	60.50		415.50	490
1260 10' wide		7	2.286		530	77.50		607.50	715
1280 12' wide	↓	5	3.200		575	108		683	810
1300 Painted, add					20%				
1310 Bronze anodized, add					50%				
3000 Window awnings, aluminum, window 3' high, 4' wide	1 Carp	10	.800		130	27		157	189
3020 6' wide	"	8	1		185	34		219	261
3040 9' wide	2 Carp	9	1.778		310	60.50		370.50	445
3060 12' wide	"	5	3.200		310	108		418	525
3100 Window, 4' high, 4' wide	1 Carp	10	.800		164	27		191	226
3120 6' wide	"	8	1		236	34		270	315
3140 9' wide	2 Carp	9	1.778		385	60.50		445.50	525
3160 12' wide	"	5	3.200		385	108		493	605
3200 Window, 6' high, 4' wide	1 Carp	10	.800		295	27		322	370
3220 6' wide	"	8	1		340	34		374	430
3240 9' wide	2 Carp	9	1.778		935	60.50		995.50	1,125
3260 12' wide	"	5	3.200		1,275	108		1,383	1,575
3400 Roll-up aluminum, 2'-6" wide	1 Carp	14	.571		190	19.35		209.35	242
3420 3' wide		12	.667		195	22.50		217.50	252
3440 4' wide		10	.800		244	27		271	315
3460 6' wide	↓	8	1		290	34		324	375

10 73 Protective Covers

10 73 16 – Canopies

	10 73 16.10 Canopies, Residential	Crew	Daily Output	Labor-Hours	Unit	Material	2016 Bare Costs Labor	Equipment	Total	Total Incl O&P
3480	9' wide	2 Carp	9	1.778	Ea.	400	60.50		460.50	540
3500	12' wide	"	5	3.200	↓	535	108		643	770
3600	Window awnings, canvas, 24" drop, 3' wide	1 Carp	30	.267	L.F.	50	9.05		59.05	70
3620	4' wide		40	.200		43	6.80		49.80	59
3700	30" drop, 3' wide		30	.267		51.50	9.05		60.55	72
3720	4' wide		40	.200		44.50	6.80		51.30	60.50
3740	5' wide		45	.178		40	6.05		46.05	54
3760	6' wide		48	.167		37	5.65		42.65	50
3780	8' wide		48	.167		34.50	5.65		40.15	47.50
3800	10' wide	↓	50	.160	↓	42	5.40		47.40	55

10 74 Manufactured Exterior Specialties

10 74 23 – Cupolas

10 74 23.10 Wood Cupolas

		Crew	Daily Output	Labor-Hours	Unit	Material	Labor	Equipment	Total	Total Incl O&P
0010	**WOOD CUPOLAS**									
0020	Stock units, pine, painted, 18" sq., 28" high, alum. roof	1 Carp	4.10	1.951	Ea.	184	66		250	315
0100	Copper roof		3.80	2.105		255	71.50		326.50	400
0300	23" square, 33" high, aluminum roof		3.70	2.162		385	73.50		458.50	545
0400	Copper roof		3.30	2.424		505	82		587	700
0600	30" square, 37" high, aluminum roof		3.70	2.162		560	73.50		633.50	740
0700	Copper roof		3.30	2.424		670	82		752	875
0900	Hexagonal, 31" wide, 46" high, copper roof		4	2		930	68		998	1,150
1000	36" wide, 50" high, copper roof	↓	3.50	2.286		1,525	77.50		1,602.50	1,800
1200	For deluxe stock units, add to above					25%				
1400	For custom built units, add to above				↓	50%	50%			

10 74 33 – Weathervanes

10 74 33.10 Residential Weathervanes

		Crew	Daily Output	Labor-Hours	Unit	Material	Labor	Equipment	Total	Total Incl O&P
0010	**RESIDENTIAL WEATHERVANES**									
0020	Residential types, 18" to 24"	1 Carp	8	1	Ea.	96	34		130	163
0100	24" to 48"	"	2	4	"	1,700	136		1,836	2,100

10 74 46 – Window Wells

10 74 46.10 Area Window Wells

		Crew	Daily Output	Labor-Hours	Unit	Material	Labor	Equipment	Total	Total Incl O&P
0010	**AREA WINDOW WELLS**, Galvanized steel									
0020	20 ga., 3'-2" wide, 1' deep	1 Sswk	29	.276	Ea.	17.25	9.85		27.10	37
0100	2' deep		23	.348		31.50	12.40		43.90	57.50
0300	16 ga., 3'-2" wide, 1' deep		29	.276		23	9.85		32.85	43.50
0400	3' deep		23	.348		48	12.40		60.40	75.50
0600	Welded grating for above, 15 lb., painted		45	.178		91.50	6.35		97.85	113
0700	Galvanized		45	.178		124	6.35		130.35	149
0900	Translucent plastic cap for above	↓	60	.133	↓	20.50	4.75		25.25	31

10 75 Flagpoles

10 75 16 – Ground-Set Flagpoles

10 75 16.10 Flagpoles	Crew	Daily Output	Labor-Hours	Unit	Material	2016 Bare Costs Labor	Equipment	Total	Total Incl O&P
0010 **FLAGPOLES**, ground set									
0050 Not including base or foundation									
0100 Aluminum, tapered, ground set 20' high	K-1	2	8	Ea.	1,050	255	154	1,459	1,775
0200 25' high	↓	1.70	9.412		1,175	300	181	1,656	1,975
0300 30' high		1.50	10.667		1,350	340	205	1,895	2,300
0500 40' high	↓	1.20	13.333	↓	2,950	425	256	3,631	4,200

Estimating Tips
General
- The items in this division are usually priced per square foot or each. Many of these items are purchased by the owner for installation by the contractor. Check the specifications for responsibilities and include time for receiving, storage, installation, and mechanical and electrical hookups in the appropriate divisions.
- Many items in Division 11 require some type of support system that is not usually furnished with the item. Examples of these systems include blocking for the attachment of casework and support angles for ceiling-hung projection screens. The required blocking or supports must be added to the estimate in the appropriate division.
- Some items in Division 11 may require assembly or electrical hookups. Verify the amount of assembly required or the need for a hard electrical connection and add the appropriate costs.

Reference Numbers
Reference numbers are shown at the beginning of some major classifications. These numbers refer to related items in the Reference Section. The reference information may be an estimating procedure, an alternate pricing method, or technical information.

Note: Not all subdivisions listed here necessarily appear. ■

Did you know?
RSMeans Online gives you the same access to RSMeans' data with 24/7 access:
- Quickly locate costs in the searchable database.
- Build cost lists, estimates, and reports in minutes.
- Adjust costs to any location in the U.S. and Canada with the click of a button.

Start your free trial today at **www.RSMeansOnline.com**

No part of this cost data may be reproduced, stored in a retrieval system, or transmitted in any form or by any means without prior written permission of RSMeans.

11 30 Residential Equipment

11 30 13 – Residential Appliances

11 30 13.15 Cooking Equipment

		Crew	Daily Output	Labor-Hours	Unit	Material	2016 Bare Costs Labor	Equipment	Total	Total Incl O&P
0010	**COOKING EQUIPMENT**									
0020	Cooking range, 30" free standing, 1 oven, minimum	2 Clab	10	1.600	Ea.	470	39.50		509.50	580
0050	Maximum		4	4		1,900	98.50		1,998.50	2,250
0150	2 oven, minimum		10	1.600		1,250	39.50		1,289.50	1,450
0200	Maximum	↓	10	1.600		2,800	39.50		2,839.50	3,150
0350	Built-in, 30" wide, 1 oven, minimum	1 Elec	6	1.333		900	49		949	1,075
0400	Maximum	2 Carp	2	8		1,575	271		1,846	2,175
0500	2 oven, conventional, minimum		4	4		1,350	136		1,486	1,700
0550	1 conventional, 1 microwave, maximum	↓	2	8		2,525	271		2,796	3,225
0700	Free-standing, 1 oven, 21" wide range, minimum	2 Clab	10	1.600		505	39.50		544.50	620
0750	21" wide, maximum	"	4	4		580	98.50		678.50	805
0900	Countertop cooktops, 4 burner, standard, minimum	1 Elec	6	1.333		350	49		399	465
0950	Maximum		3	2.667		1,700	98.50		1,798.50	2,025
1050	As above, but with grill and griddle attachment, minimum		6	1.333		1,525	49		1,574	1,750
1100	Maximum		3	2.667		3,900	98.50		3,998.50	4,450
1250	Microwave oven, minimum		4	2		119	74		193	252
1300	Maximum		2	4		510	148		658	800
5380	Oven, built in, standard		4	2		900	74		974	1,100
5390	Deluxe	↓	2	4	↓	2,525	148		2,673	3,025

11 30 13.16 Refrigeration Equipment

		Crew	Daily Output	Labor-Hours	Unit	Material	Labor	Equipment	Total	Total Incl O&P
0010	**REFRIGERATION EQUIPMENT**									
2000	Deep freeze, 15 to 23 C.F., minimum	2 Clab	10	1.600	Ea.	570	39.50		609.50	690
2050	Maximum		5	3.200		835	79		914	1,050
2200	30 C.F., minimum		8	2		675	49.50		724.50	825
2250	Maximum	↓	3	5.333		900	131		1,031	1,200
5200	Icemaker, automatic, 20 lb. per day	1 Plum	7	1.143		1,100	44.50		1,144.50	1,275
5350	51 lb. per day	"	2	4		1,600	156		1,756	2,000
5450	Refrigerator, no frost, 6 C.F.	2 Clab	15	1.067		229	26.50		255.50	296
5500	Refrigerator, no frost, 10 C.F. to 12 C.F., minimum		10	1.600		475	39.50		514.50	585
5600	Maximum		6	2.667		555	65.50		620.50	725
5750	14 C.F. to 16 C.F., minimum		9	1.778		580	44		624	710
5800	Maximum		5	3.200		650	79		729	845
5950	18 C.F. to 20 C.F., minimum		8	2		735	49.50		784.50	895
6000	Maximum		4	4		1,700	98.50		1,798.50	2,050
6150	21 C.F. to 29 C.F., minimum		7	2.286		1,150	56.50		1,206.50	1,350
6200	Maximum	↓	3	5.333		3,850	131		3,981	4,450
6790	Energy-star qualified, 18 C.F., minimum [G]	2 Carp	4	4		650	136		786	940
6795	Maximum [G]		2	8		1,250	271		1,521	1,825
6797	21.7 C.F., minimum [G]		4	4		1,125	136		1,261	1,450
6799	Maximum [G]		4	4	↓	2,075	136		2,211	2,500

11 30 13.17 Kitchen Cleaning Equipment

		Crew	Daily Output	Labor-Hours	Unit	Material	Labor	Equipment	Total	Total Incl O&P
0010	**KITCHEN CLEANING EQUIPMENT**									
2750	Dishwasher, built-in, 2 cycles, minimum	L-1	4	2.500	Ea.	299	96.50		395.50	490
2800	Maximum		2	5		450	193		643	815
2950	4 or more cycles, minimum		4	2.500		425	96.50		521.50	630
2960	Average		4	2.500		565	96.50		661.50	785
3000	Maximum		2	5		1,225	193		1,418	1,650
3100	Energy-star qualified, minimum [G]		4	2.500		490	96.50		586.50	700
3110	Maximum [G]	↓	2	5	↓	1,650	193		1,843	2,150

11 30 Residential Equipment

11 30 13 – Residential Appliances

11 30 13.18 Waste Disposal Equipment		Crew	Daily Output	Labor-Hours	Unit	Material	2016 Bare Costs Labor	Equipment	Total	Total Incl O&P
0010	WASTE DISPOSAL EQUIPMENT									
1750	Compactor, residential size, 4 to 1 compaction, minimum	1 Carp	5	1.600	Ea.	690	54		744	850
1800	Maximum	"	3	2.667		1,150	90.50		1,240.50	1,425
3300	Garbage disposal, sink type, minimum	L-1	10	1		93	38.50		131.50	166
3350	Maximum	"	10	1		243	38.50		281.50	330

11 30 13.19 Kitchen Ventilation Equipment

		Crew	Daily Output	Labor-Hours	Unit	Material	Labor	Equipment	Total	Total Incl O&P
0010	KITCHEN VENTILATION EQUIPMENT									
4150	Hood for range, 2 speed, vented, 30" wide, minimum	L-3	5	2	Ea.	83	69		152	207
4200	Maximum		3	3.333		855	115		970	1,125
4300	42" wide, minimum		5	2		166	69		235	297
4330	Custom		5	2		1,625	69		1,694	1,925
4350	Maximum		3	3.333		1,975	115		2,090	2,375
4500	For ventless hood, 2 speed, add					20			20	22
4650	For vented 1 speed, deduct from maximum					55			55	60.50

11 30 13.24 Washers

			Crew	Daily Output	Labor-Hours	Unit	Material	Labor	Equipment	Total	Total Incl O&P
0010	WASHERS										
6650	Washing machine, automatic, minimum		1 Plum	3	2.667	Ea.	545	104		649	770
6700	Maximum			1	8		1,650	310		1,960	2,350
6750	Energy star qualified, front loading, minimum	G		3	2.667		700	104		804	940
6760	Maximum	G		1	8		1,600	310		1,910	2,275
6764	Top loading, minimum	G		3	2.667		570	104		674	795
6766	Maximum	G		3	2.667		1,400	104		1,504	1,725

11 30 13.25 Dryers

			Crew	Daily Output	Labor-Hours	Unit	Material	Labor	Equipment	Total	Total Incl O&P
0010	DRYERS										
6770	Electric, front loading, energy-star qualified, minimum	G	L-2	3	5.333	Ea.	475	156		631	790
6780	Maximum	G	"	2	8		1,600	235		1,835	2,150
7450	Vent kits for dryers		1 Carp	10	.800		40	27		67	89.50

11 30 15 – Miscellaneous Residential Appliances

11 30 15.13 Sump Pumps

		Crew	Daily Output	Labor-Hours	Unit	Material	Labor	Equipment	Total	Total Incl O&P
0010	SUMP PUMPS									
6400	Cellar drainer, pedestal, 1/3 H.P., molded PVC base	1 Plum	3	2.667	Ea.	140	104		244	325
6450	Solid brass	"	2	4	"	297	156		453	585
6460	Sump pump, see also Section 22 14 29.16									

11 30 15.23 Water Heaters

		Crew	Daily Output	Labor-Hours	Unit	Material	Labor	Equipment	Total	Total Incl O&P
0010	WATER HEATERS									
6900	Electric, glass lined, 30 gallon, minimum	L-1	5	2	Ea.	815	77		892	1,025
6950	Maximum		3	3.333		1,125	129		1,254	1,450
7100	80 gallon, minimum		2	5		1,475	193		1,668	1,950
7150	Maximum		1	10		2,050	385		2,435	2,875
7180	Gas, glass lined, 30 gallon, minimum	2 Plum	5	3.200		1,450	125		1,575	1,800
7220	Maximum		3	5.333		2,025	208		2,233	2,575
7260	50 gallon, minimum		2.50	6.400		1,525	250		1,775	2,075
7300	Maximum		1.50	10.667		2,125	415		2,540	3,025
7310	Water heater, see also Section 22 33 30.13									

11 30 15.43 Air Quality

		Crew	Daily Output	Labor-Hours	Unit	Material	Labor	Equipment	Total	Total Incl O&P
0010	AIR QUALITY									
2450	Dehumidifier, portable, automatic, 15 pint	1 Elec	4	2	Ea.	200	74		274	340
2550	40 pint		3.75	2.133		214	78.50		292.50	365
3550	Heater, electric, built-in, 1250 watt, ceiling type, minimum		4	2		117	74		191	249
3600	Maximum		3	2.667		204	98.50		302.50	385

11 30 Residential Equipment

11 30 15 – Miscellaneous Residential Appliances

11 30 15.43 Air Quality

		Crew	Daily Output	Labor-Hours	Unit	Material	2016 Bare Costs Labor	Equipment	Total	Total Incl O&P
3700	Wall type, minimum	1 Elec	4	2	Ea.	191	74		265	330
3750	Maximum		3	2.667		195	98.50		293.50	375
3900	1500 watt wall type, with blower		4	2		173	74		247	310
3950	3000 watt		3	2.667		390	98.50		488.50	590
4850	Humidifier, portable, 8 gallons per day					152			152	167
5000	15 gallons per day					222			222	244

11 30 33 – Retractable Stairs

11 30 33.10 Disappearing Stairway

		Crew	Daily Output	Labor-Hours	Unit	Material	Labor	Equipment	Total	Total Incl O&P
0010	**DISAPPEARING STAIRWAY** No trim included									
0020	One piece, yellow pine, 8'-0" ceiling	2 Carp	4	4	Ea.	250	136		386	500
0030	9'-0" ceiling		4	4		284	136		420	535
0040	10'-0" ceiling		3	5.333		254	181		435	585
0050	11'-0" ceiling		3	5.333		285	181		466	620
0060	12'-0" ceiling		3	5.333		360	181		541	700
0100	Custom grade, pine, 8'-6" ceiling, minimum	1 Carp	4	2		195	68		263	330
0150	Average		3.50	2.286		244	77.50		321.50	400
0200	Maximum		3	2.667		305	90.50		395.50	485
0500	Heavy duty, pivoted, from 7'-7" to 12'-10" floor to floor		3	2.667		1,025	90.50		1,115.50	1,275
0600	16'-0" ceiling		2	4		1,600	136		1,736	1,975
0800	Economy folding, pine, 8'-6" ceiling		4	2		176	68		244	310
0900	9'-6" ceiling		4	2		196	68		264	330
1100	Automatic electric, aluminum, floor to floor height, 8' to 9'	2 Carp	1	16		8,700	540		9,240	10,500

11 32 Unit Kitchens

11 32 13 – Metal Unit Kitchens

11 32 13.10 Commercial Unit Kitchens

		Crew	Daily Output	Labor-Hours	Unit	Material	Labor	Equipment	Total	Total Incl O&P
0010	**COMMERCIAL UNIT KITCHENS**									
1500	Combination range, refrigerator and sink, 30" wide, minimum	L-1	2	5	Ea.	1,100	193		1,293	1,550
1550	Maximum		1	10		1,250	385		1,635	2,000
1570	60" wide, average		1.40	7.143		1,425	276		1,701	2,025
1590	72" wide, average		1.20	8.333		1,500	320		1,820	2,175

11 41 Foodservice Storage Equipment

11 41 13 – Refrigerated Food Storage Cases

11 41 13.30 Wine Cellar

		Crew	Daily Output	Labor-Hours	Unit	Material	Labor	Equipment	Total	Total Incl O&P
0010	**WINE CELLAR**, refrigerated, Redwood interior, carpeted, walk-in type									
0020	6'-8" high, including racks									
0200	80" W x 48" D for 900 bottles	2 Carp	1.50	10.667	Ea.	4,250	360		4,610	5,275
0250	80" W x 72" D for 1300 bottles		1.33	12.030		5,175	410		5,585	6,375
0300	80" W x 94" D for 1900 bottles		1.17	13.675		6,275	465		6,740	7,675

11 81 Facility Maintenance Equipment

11 81 19 – Vacuum Cleaning Systems

11 81 19.10 Vacuum Cleaning	Crew	Daily Output	Labor-Hours	Unit	Material	2016 Bare Costs Labor	Equipment	Total	Total Incl O&P
0010 **VACUUM CLEANING**									
0020 Central, 3 inlet, residential	1 Skwk	.90	8.889	Total	1,100	305		1,405	1,700
0400 5 inlet system, residential		.50	16		1,550	545		2,095	2,625
0600 7 inlet system, commercial		.40	20		1,750	685		2,435	3,075
0800 9 inlet system, residential	↓	.30	26.667		3,875	910		4,785	5,775
4010 Rule of thumb: First 1200 S.F., installed								1,425	1,575
4020 For each additional S.F., add				S.F.				.26	.26

Division Notes

Estimating Tips
General
- The items in this division are usually priced per square foot or each. Most of these items are purchased by the owner and installed by the contractor. Do not assume the items in Division 12 will be purchased and installed by the contractor. Check the specifications for responsibilities and include receiving, storage, installation, and mechanical and electrical hookups in the appropriate divisions.
- Some items in this division require some type of support system that is not usually furnished with the item. Examples of these systems include blocking for the attachment of casework and heavy drapery rods. The required blocking must be added to the estimate in the appropriate division.

Reference Numbers
Reference numbers are shown at the beginning of some major classifications. These numbers refer to related items in the Reference Section. The reference information may be an estimating procedure, an alternate pricing method, or technical information.

Note: Not all subdivisions listed here necessarily appear. ■

Did you know?

RSMeans Online gives you the same access to RSMeans' data with 24/7 access:
- Quickly locate costs in the searchable database.
- Build cost lists, estimates, and reports in minutes.
- Adjust costs to any location in the U.S. and Canada with the click of a button.

Start your free trial today at **www.RSMeansOnline.com**

No part of this cost data may be reproduced, stored in a retrieval system, or transmitted in any form or by any means without prior written permission of RSMeans.

12 21 Window Blinds

12 21 13 – Horizontal Louver Blinds

12 21 13.13 Metal Horizontal Louver Blinds	Crew	Daily Output	Labor-Hours	Unit	Material	2016 Bare Costs Labor	Equipment	Total	Total Incl O&P
0010 **METAL HORIZONTAL LOUVER BLINDS**									
0020 Horizontal, 1" aluminum slats, solid color, stock	1 Carp	590	.014	S.F.	4.69	.46		5.15	5.90

12 21 13.33 Vinyl Horizontal Louver Blinds	Crew	Daily Output	Labor-Hours	Unit	Material	Labor	Equipment	Total	Total Incl O&P
0010 **VINYL HORIZONTAL LOUVER BLINDS**									
0015 1" composite, 48" wide, 48" high	1 Carp	30	.267	Ea.	37.50	9.05		46.55	56
0020 72" high		30	.267		45.50	9.05		54.55	65
0030 60" wide, 60" high		30	.267		54.50	9.05		63.55	75
0040 72" high		30	.267		54.50	9.05		63.55	75
0050 72" wide x 72" high		30	.267		65	9.05		74.05	86.50

12 22 Curtains and Drapes

12 22 16 – Drapery Track and Accessories

12 22 16.10 Drapery Hardware	Crew	Daily Output	Labor-Hours	Unit	Material	Labor	Equipment	Total	Total Incl O&P
0010 **DRAPERY HARDWARE**									
0030 Standard traverse, per foot, minimum	1 Carp	59	.136	L.F.	7.40	4.60		12	15.80
0100 Maximum		51	.157	"	10.55	5.30		15.85	20.50
0200 Decorative traverse, 28"-48", minimum		22	.364	Ea.	26.50	12.35		38.85	49.50
0220 Maximum		21	.381		58.50	12.90		71.40	85.50
0300 48"-84", minimum		20	.400		28.50	13.55		42.05	54
0320 Maximum		19	.421		74.50	14.25		88.75	106
0400 66"-120", minimum		18	.444		39	15.05		54.05	68
0420 Maximum		17	.471		114	15.95		129.95	152
0500 84"-156", minimum		16	.500		50	16.95		66.95	83.50
0520 Maximum		15	.533		155	18.10		173.10	202
0600 130"-240", minimum		14	.571		73	19.35		92.35	113
0620 Maximum		13	.615		215	21		236	272
0700 Slide rings, each, minimum					1.08			1.08	1.19
0720 Maximum					2.11			2.11	2.32
4000 Traverse rods, adjustable, 28" to 48"	1 Carp	22	.364		30	12.35		42.35	53.50
4020 48" to 84"		20	.400		38	13.55		51.55	64.50
4040 66" to 120"		18	.444		46	15.05		61.05	76
4060 84" to 156"		16	.500		52.50	16.95		69.45	86.50
4080 100" to 180"		14	.571		61	19.35		80.35	100
4100 228" to 312"		13	.615		85	21		106	129
4500 Curtain rod, 28" to 48", single		22	.364		6	12.35		18.35	27
4510 Double		22	.364		10.20	12.35		22.55	31.50
4520 48" to 86", single		20	.400		10.25	13.55		23.80	34
4530 Double		20	.400		17.10	13.55		30.65	41.50
4540 66" to 120", single		18	.444		17.15	15.05		32.20	44
4550 Double		18	.444		27	15.05		42.05	54.50
4600 Valance, pinch pleated fabric, 12" deep, up to 54" long, minimum					39.50			39.50	43.50
4610 Maximum					98			98	108
4620 Up to 77" long, minimum					69.50			69.50	76.50
4630 Maximum					158			158	174
5000 Stationary rods, first 2'					8.40			8.40	9.25

12 23 Interior Shutters

12 23 10 – Wood Interior Shutters

12 23 10.10 Wood Interior Shutters

		Crew	Daily Output	Labor-Hours	Unit	Material	2016 Bare Costs Labor	Equipment	Total	Total Incl O&P
0010	**WOOD INTERIOR SHUTTERS**, louvered									
0200	Two panel, 27" wide, 36" high	1 Carp	5	1.600	Set	158	54		212	265
0300	33" wide, 36" high		5	1.600		204	54		258	315
0500	47" wide, 36" high		5	1.600		273	54		327	390
1000	Four panel, 27" wide, 36" high		5	1.600		158	54		212	265
1100	33" wide, 36" high		5	1.600		202	54		256	315
1300	47" wide, 36" high		5	1.600		270	54		324	390
1400	Plantation shutters, 16" x 48"		5	1.600	Ea.	182	54		236	291
1450	16" x 96"		4	2		299	68		367	445
1460	36" x 96"		3	2.667		660	90.50		750.50	875

12 23 10.13 Wood Panels

		Crew	Daily Output	Labor-Hours	Unit	Material	2016 Bare Costs Labor	Equipment	Total	Total Incl O&P
0010	**WOOD PANELS**									
3000	Wood folding panels with movable louvers, 7" x 20" each	1 Carp	17	.471	Pr.	82.50	15.95		98.45	117
3300	8" x 28" each		17	.471		82.50	15.95		98.45	117
3450	9" x 36" each		17	.471		95	15.95		110.95	131
3600	10" x 40" each		17	.471		103	15.95		118.95	141
4000	Fixed louver type, stock units, 8" x 20" each		17	.471		94	15.95		109.95	130
4150	10" x 28" each		17	.471		79.50	15.95		95.45	114
4300	12" x 36" each		17	.471		94	15.95		109.95	130
4450	18" x 40" each		17	.471		134	15.95		149.95	175
5000	Insert panel type, stock, 7" x 20" each		17	.471		21	15.95		36.95	49.50
5150	8" x 28" each		17	.471		38	15.95		53.95	68.50
5300	9" x 36" each		17	.471		48.50	15.95		64.45	79.50
5450	10" x 40" each		17	.471		52	15.95		67.95	83.50
5600	Raised panel type, stock, 10" x 24" each		17	.471		132	15.95		147.95	172
5650	12" x 26" each		17	.471		132	15.95		147.95	172
5700	14" x 30" each		17	.471		146	15.95		161.95	187
5750	16" x 36" each		17	.471		161	15.95		176.95	204
6000	For custom built pine, add					22%				
6500	For custom built hardwood blinds, add					42%				

12 24 Window Shades

12 24 13 – Roller Window Shades

12 24 13.10 Shades

		Crew	Daily Output	Labor-Hours	Unit	Material	2016 Bare Costs Labor	Equipment	Total	Total Incl O&P
0010	**SHADES**									
0020	Basswood, roll-up, stain finish, 3/8" slats	1 Carp	300	.027	S.F.	17.15	.90		18.05	20.50
5011	Insulative shades	G	125	.064		12.15	2.17		14.32	17
6011	Solar screening, fiberglass	G	85	.094		6.90	3.19		10.09	12.95
8011	Interior insulative shutter									
8111	Stock unit, 15" x 60"	G	1 Carp	17	.471	Pr.	12.25	15.95	28.20	40

12 32 Manufactured Wood Casework

12 32 23 – Hardwood Casework

12 32 23.10 Manufactured Wood Casework, Stock Units		Crew	Daily Output	Labor-Hours	Unit	Material	2016 Bare Costs Labor	Equipment	Total	Total Incl O&P
0010	**MANUFACTURED WOOD CASEWORK, STOCK UNITS**									
0700	Kitchen base cabinets, hardwood, not incl. counter tops,									
0710	24" deep, 35" high, prefinished									
0800	One top drawer, one door below, 12" wide	2 Carp	24.80	.645	Ea.	271	22		293	335
0820	15" wide		24	.667		282	22.50		304.50	350
0840	18" wide		23.30	.687		305	23.50		328.50	380
0860	21" wide		22.70	.705		315	24		339	390
0880	24" wide		22.30	.717		370	24.50		394.50	450
1000	Four drawers, 12" wide		24.80	.645		285	22		307	350
1020	15" wide		24	.667		289	22.50		311.50	360
1040	18" wide		23.30	.687		320	23.50		343.50	390
1060	24" wide		22.30	.717		350	24.50		374.50	425
1200	Two top drawers, two doors below, 27" wide		22	.727		400	24.50		424.50	480
1220	30" wide		21.40	.748		440	25.50		465.50	525
1240	33" wide		20.90	.766		455	26		481	545
1260	36" wide		20.30	.788		470	26.50		496.50	565
1280	42" wide		19.80	.808		495	27.50		522.50	590
1300	48" wide		18.90	.847		535	28.50		563.50	635
1500	Range or sink base, two doors below, 30" wide		21.40	.748		365	25.50		390.50	445
1520	33" wide		20.90	.766		390	26		416	475
1540	36" wide		20.30	.788		405	26.50		431.50	495
1560	42" wide		19.80	.808		430	27.50		457.50	515
1580	48" wide		18.90	.847		450	28.50		478.50	545
1800	For sink front units, deduct					161			161	177
2000	Corner base cabinets, 36" wide, standard	2 Carp	18	.889		670	30		700	790
2100	Lazy Susan with revolving door	"	16.50	.970		855	33		888	1,000
4000	Kitchen wall cabinets, hardwood, 12" deep with two doors									
4050	12" high, 30" wide	2 Carp	24.80	.645	Ea.	245	22		267	305
4100	36" wide		24	.667		291	22.50		313.50	360
4400	15" high, 30" wide		24	.667		249	22.50		271.50	310
4420	33" wide		23.30	.687		310	23.50		333.50	380
4440	36" wide		22.70	.705		299	24		323	370
4450	42" wide		22.70	.705		335	24		359	410
4700	24" high, 30" wide		23.30	.687		335	23.50		358.50	405
4720	36" wide		22.70	.705		365	24		389	445
4740	42" wide		22.30	.717		415	24.50		439.50	500
5000	30" high, one door, 12" wide		22	.727		234	24.50		258.50	300
5020	15" wide		21.40	.748		246	25.50		271.50	315
5040	18" wide		20.90	.766		270	26		296	340
5060	24" wide		20.30	.788		315	26.50		341.50	395
5300	Two doors, 27" wide		19.80	.808		350	27.50		377.50	430
5320	30" wide		19.30	.829		370	28		398	450
5340	36" wide		18.80	.851		420	29		449	510
5360	42" wide		18.50	.865		455	29.50		484.50	550
5380	48" wide		18.40	.870		510	29.50		539.50	615
6000	Corner wall, 30" high, 24" wide		18	.889		360	30		390	445
6050	30" wide		17.20	.930		380	31.50		411.50	475

12 32 Manufactured Wood Casework

12 32 23 – Hardwood Casework

12 32 23.10 Manufactured Wood Casework, Stock Units

		Crew	Daily Output	Labor-Hours	Unit	Material	2016 Bare Costs Labor	Equipment	Total	Total Incl O&P
6100	36" wide	2 Carp	16.50	.970	Ea.	435	33		468	535
6500	Revolving Lazy Susan		15.20	1.053		490	35.50		525.50	600
7000	Broom cabinet, 84" high, 24" deep, 18" wide		10	1.600		670	54		724	830
7500	Oven cabinets, 84" high, 24" deep, 27" wide		8	2		1,025	68		1,093	1,250
7750	Valance board trim		396	.040	L.F.	16.25	1.37		17.62	20
7780	Toe kick trim	1 Carp	256	.031	"	3.10	1.06		4.16	5.20
7790	Base cabinet corner filler		16	.500	Ea.	43.50	16.95		60.45	76
7800	Cabinet filler, 3" x 24"		20	.400		17.10	13.55		30.65	41.50
7810	3" x 30"		20	.400		21.50	13.55		35.05	46
7820	3" x 42"		18	.444		30	15.05		45.05	58
7830	3" x 80"		16	.500		57	16.95		73.95	91.50
7850	Cabinet panel		50	.160	S.F.	9.30	5.40		14.70	19.35
9000	For deluxe models of all cabinets, add					40%				
9500	For custom built in place, add					25%	10%			
9558	Rule of thumb, kitchen cabinets not including									
9560	appliances & counter top, minimum	2 Carp	30	.533	L.F.	182	18.10		200.10	231
9600	Maximum	"	25	.640	"	405	21.50		426.50	480

12 32 23.30 Manufactured Wood Casework Vanities

		Crew	Daily Output	Labor-Hours	Unit	Material	Labor	Equipment	Total	Total Incl O&P
0010	**MANUFACTURED WOOD CASEWORK VANITIES**									
8000	Vanity bases, 2 doors, 30" high, 21" deep, 24" wide	2 Carp	20	.800	Ea.	315	27		342	390
8050	30" wide		16	1		375	34		409	470
8100	36" wide		13.33	1.200		365	40.50		405.50	470
8150	48" wide		11.43	1.400		480	47.50		527.50	605
9000	For deluxe models of all vanities, add to above					40%				
9500	For custom built in place, add to above					25%	10%			

12 32 23.35 Manufactured Wood Casework Hardware

		Crew	Daily Output	Labor-Hours	Unit	Material	Labor	Equipment	Total	Total Incl O&P
0010	**MANUFACTURED WOOD CASEWORK HARDWARE**									
1000	Catches, minimum	1 Carp	235	.034	Ea.	1.30	1.15		2.45	3.36
1020	Average		119.40	.067		4	2.27		6.27	8.20
1040	Maximum		80	.100		7.75	3.39		11.14	14.25
2000	Door/drawer pulls, handles									
2200	Handles and pulls, projecting, metal, minimum	1 Carp	48	.167	Ea.	5.20	5.65		10.85	15.15
2220	Average		42	.190		7.90	6.45		14.35	19.50
2240	Maximum		36	.222		10.65	7.55		18.20	24.50
2300	Wood, minimum		48	.167		5.30	5.65		10.95	15.30
2320	Average		42	.190		7.10	6.45		13.55	18.60
2340	Maximum		36	.222		9.80	7.55		17.35	23.50
2600	Flush, metal, minimum		48	.167		5.30	5.65		10.95	15.30
2620	Average		42	.190		7.25	6.45		13.70	18.80
2640	Maximum		36	.222		9.90	7.55		17.45	23.50
3000	Drawer tracks/glides, minimum		48	.167	Pr.	8.95	5.65		14.60	19.30
3020	Average		32	.250		16	8.50		24.50	32
3040	Maximum		24	.333		26	11.30		37.30	47.50
4000	Cabinet hinges, minimum		160	.050		3.10	1.70		4.80	6.25
4020	Average		95.24	.084		7.60	2.85		10.45	13.15
4040	Maximum		68	.118		12.40	3.99		16.39	20.50

12 34 Manufactured Plastic Casework

12 34 16 – Manufactured Solid-Plastic Casework

12 34 16.10 Outdoor Casework

		Crew	Daily Output	Labor-Hours	Unit	Material	2016 Bare Costs Labor	2016 Bare Costs Equipment	Total	Total Incl O&P
0010	**OUTDOOR CASEWORK**									
0020	Cabinet, base, sink/range, 36"	2 Carp	20.30	.788	Ea.	1,625	26.50		1,651.50	1,850
0100	Base, 36"		20.30	.788		1,500	26.50		1,526.50	1,700
0200	Filler strip, 1" x 30"	↓	158	.101	↓	43	3.43		46.43	53

12 36 Countertops

12 36 16 – Metal Countertops

12 36 16.10 Stainless Steel Countertops

		Crew	Daily Output	Labor-Hours	Unit	Material	Labor	Equipment	Total	Total Incl O&P
0010	**STAINLESS STEEL COUNTERTOPS**									
3200	Stainless steel, custom	1 Carp	24	.333	S.F.	155	11.30		166.30	190

12 36 19 – Wood Countertops

12 36 19.10 Maple Countertops

		Crew	Daily Output	Labor-Hours	Unit	Material	Labor	Equipment	Total	Total Incl O&P
0010	**MAPLE COUNTERTOPS**									
2900	Solid, laminated, 1-1/2" thick, no splash	1 Carp	28	.286	L.F.	76.50	9.70		86.20	101
3000	With square splash		28	.286	"	91	9.70		100.70	116
3400	Recessed cutting block with trim, 16" x 20" x 1"		8	1	Ea.	93	34		127	160
3411	Replace cutting block only	↓	16	.500	"	70	16.95		86.95	106

12 36 23 – Plastic Countertops

12 36 23.13 Plastic-Laminate-Clad Countertops

		Crew	Daily Output	Labor-Hours	Unit	Material	Labor	Equipment	Total	Total Incl O&P
0010	**PLASTIC-LAMINATE-CLAD COUNTERTOPS**									
0020	Stock, 24" wide w/backsplash, minimum	1 Carp	30	.267	L.F.	17.25	9.05		26.30	34
0100	Maximum		25	.320		35	10.85		45.85	56.50
0300	Custom plastic, 7/8" thick, aluminum molding, no splash		30	.267		30.50	9.05		39.55	48.50
0400	Cove splash		30	.267		29.50	9.05		38.55	47.50
0600	1-1/4" thick, no splash		28	.286		36	9.70		45.70	55.50
0700	Square splash		28	.286		43	9.70		52.70	63.50
0900	Square edge, plastic face, 7/8" thick, no splash		30	.267		33.50	9.05		42.55	52
1000	With splash	↓	30	.267		40	9.05		49.05	59
1200	For stainless channel edge, 7/8" thick, add					3.17			3.17	3.49
1300	1-1/4" thick, add					3.78			3.78	4.16
1500	For solid color suede finish, add				↓	4.14			4.14	4.55
1700	For end splash, add				Ea.	18.65			18.65	20.50
1901	For cut outs, standard, add, minimum	1 Carp	32	.250			8.50		8.50	14.20
2000	Maximum		8	1		6.20	34		40.20	64
2010	Cut out in backsplash for elec. wall outlet		38	.211			7.15		7.15	11.95
2020	Cut out for sink		20	.400			13.55		13.55	22.50
2030	Cut out for stove top		18	.444	↓		15.05		15.05	25
2100	Postformed, including backsplash and front edge		30	.267	L.F.	10.35	9.05		19.40	26.50
2110	Mitred, add		12	.667	Ea.		22.50		22.50	38
2200	Built-in place, 25" wide, plastic laminate	↓	25	.320	L.F.	42	10.85		52.85	64

12 36 33 – Tile Countertops

12 36 33.10 Ceramic Tile Countertops

		Crew	Daily Output	Labor-Hours	Unit	Material	Labor	Equipment	Total	Total Incl O&P
0010	**CERAMIC TILE COUNTERTOPS**									
2300	Ceramic tile mosaic	1 Carp	25	.320	L.F.	34	10.85		44.85	55.50

12 36 Countertops

12 36 40 – Stone Countertops

12 36 40.10 Natural Stone Countertops

		Crew	Daily Output	Labor-Hours	Unit	Material	2016 Bare Costs Labor	2016 Bare Costs Equipment	Total	Total Incl O&P
0010	**NATURAL STONE COUNTERTOPS**									
2500	Marble, stock, with splash, 1/2" thick, minimum	1 Bric	17	.471	L.F.	42.50	15.25		57.75	72
2700	3/4" thick, maximum		13	.615		107	19.95		126.95	150
2800	Granite, average, 1-1/4" thick, 24" wide, no splash		13.01	.615		140	19.90		159.90	187

12 36 61 – Simulated Stone Countertops

12 36 61.16 Solid Surface Countertops

		Crew	Daily Output	Labor-Hours	Unit	Material	Labor	Equipment	Total	Incl O&P
0010	**SOLID SURFACE COUNTERTOPS**, Acrylic polymer									
2000	Pricing for order of 1 – 50 L.F.									
2100	25" wide, solid colors	2 Carp	20	.800	L.F.	73.50	27		100.50	127
2200	Patterned colors		20	.800		93.50	27		120.50	149
2300	Premium patterned colors		20	.800		117	27		144	175
2400	With silicone attached 4" backsplash, solid colors		19	.842		81	28.50		109.50	137
2500	Patterned colors		19	.842		102	28.50		130.50	161
2600	Premium patterned colors		19	.842		128	28.50		156.50	188
2700	With hard seam attached 4" backsplash, solid colors		15	1.067		81	36		117	150
2800	Patterned colors		15	1.067		102	36		138	174
2900	Premium patterned colors		15	1.067		128	36		164	201
3800	Sinks, pricing for order of 1 – 50 units									
3900	Single bowl, hard seamed, solid colors, 13" x 17"	1 Carp	2	4	Ea.	500	136		636	775
4000	10" x 15"		4.55	1.758		230	59.50		289.50	355
4100	Cutouts for sinks		5.25	1.524			51.50		51.50	86.50

12 36 61.17 Solid Surface Vanity Tops

		Crew	Daily Output	Labor-Hours	Unit	Material	Labor	Equipment	Total	Incl O&P
0010	**SOLID SURFACE VANITY TOPS**									
0015	Solid surface, center bowl, 17" x 19"	1 Carp	12	.667	Ea.	190	22.50		212.50	247
0020	19" x 25"		12	.667		194	22.50		216.50	252
0030	19" x 31"		12	.667		227	22.50		249.50	288
0040	19" x 37"		12	.667		264	22.50		286.50	330
0050	22" x 25"		10	.800		345	27		372	425
0060	22" x 31"		10	.800		405	27		432	490
0070	22" x 37"		10	.800		470	27		497	565
0080	22" x 43"		10	.800		535	27		562	635
0090	22" x 49"		10	.800		595	27		622	700
0110	22" x 55"		8	1		675	34		709	800
0120	22" x 61"		8	1		770	34		804	905
0220	Double bowl, 22" x 61"		8	1		870	34		904	1,025
0230	Double bowl, 22" x 73"		8	1		950	34		984	1,100
0240	For aggregate colors, add					35%				
0250	For faucets and fittings, see Section 22 41 39.10									

12 36 61.19 Quartz Agglomerate Countertops

		Crew	Daily Output	Labor-Hours	Unit	Material	Labor	Equipment	Total	Incl O&P
0010	**QUARTZ AGGLOMERATE COUNTERTOPS**									
0100	25" wide, 4" backsplash, color group A, minimum	2 Carp	15	1.067	L.F.	65	36		101	132
0110	Maximum		15	1.067		91	36		127	161
0120	Color group B, minimum		15	1.067		67	36		103	134
0130	Maximum		15	1.067		95	36		131	166
0140	Color group C, minimum		15	1.067		78	36		114	147
0150	Maximum		15	1.067		107	36		143	179
0160	Color group D, minimum		15	1.067		85	36		121	154
0170	Maximum		15	1.067		115	36		151	188

Division Notes

	CREW	DAILY OUTPUT	LABOR-HOURS	UNIT	BARE COSTS				TOTAL INCL O&P
					MAT.	LABOR	EQUIP.	TOTAL	

Estimating Tips
General
- The items and systems in this division are usually estimated, purchased, supplied, and installed as a unit by one or more subcontractors. The estimator must ensure that all parties are operating from the same set of specifications and assumptions, and that all necessary items are estimated and will be provided. Many times the complex items and systems are covered, but the more common ones, such as excavation or a crane, are overlooked for the very reason that everyone assumes nobody could miss them. The estimator should be the central focus and be able to ensure that all systems are complete.
- Another area where problems can develop in this division is at the interface between systems. The estimator must ensure, for instance, that anchor bolts, nuts, and washers are estimated and included for the air-supported structures and pre-engineered buildings to be bolted to their foundations. Utility supply is a common area where essential items or pieces of equipment can be missed or overlooked, because each subcontractor may feel it is another's responsibility. The estimator should also be aware of certain items which may be supplied as part of a package but installed by others, and ensure that the installing contractor's estimate includes the cost of installation. Conversely, the estimator must also ensure that items are not costed by two different subcontractors, resulting in an inflated overall estimate.

13 30 00 Special Structures
- The foundations and floor slab, as well as rough mechanical and electrical, should be estimated, as this work is required for the assembly and erection of the structure. Generally, as noted in the data set, the pre-engineered building comes as a shell. Pricing is based on the size and structural design parameters stated in the reference section. Additional features, such as windows and doors with their related structural framing, must also be included by the estimator. Here again, the estimator must have a clear understanding of the scope of each portion of the work and all the necessary interfaces.

Reference Numbers
Reference numbers are shown at the beginning of some major classifications. These numbers refer to related items in the Reference Section. The reference information may be an estimating procedure, an alternate pricing method, or technical information.

Note: Not all subdivisions listed here necessarily appear. ■

No part of this cost data may be reproduced, stored in a retrieval system, or transmitted in any form or by any means without prior written permission of RSMeans.

Did you know?
RSMeans Online gives you the same access to RSMeans' data with 24/7 access:
- Quickly locate costs in the searchable database.
- Build cost lists, estimates, and reports in minutes.
- Adjust costs to any location in the U.S. and Canada with the click of a button.

Start your free trial today at **www.RSMeansOnline.com**

RSMeans Online
FROM THE GORDIAN GROUP®

13 11 Swimming Pools

13 11 13 – Below-Grade Swimming Pools

13 11 13.50 Swimming Pools

		Crew	Daily Output	Labor-Hours	Unit	Material	2016 Bare Costs Labor	Equipment	Total	Total Incl O&P
0010	**SWIMMING POOLS** Residential in-ground, vinyl lined									
0020	Concrete sides, w/equip, sand bottom	B-52	300	.187	SF Surf	25.50	5.20	2	32.70	39.50
0100	Metal or polystyrene sides R131113-20	B-14	410	.117		21.50	3.11	.89	25.50	29.50
0200	Add for vermiculite bottom					1.64			1.64	1.80
0500	Gunite bottom and sides, white plaster finish									
0600	12' x 30' pool	B-52	145	.386	SF Surf	48	10.75	4.13	62.88	75
0720	16' x 32' pool		155	.361		43	10.05	3.86	56.91	68.50
0750	20' x 40' pool	↓	250	.224	↓	38.50	6.25	2.39	47.14	55.50
0810	Concrete bottom and sides, tile finish									
0820	12' x 30' pool	B-52	80	.700	SF Surf	48.50	19.45	7.50	75.45	94
0830	16' x 32' pool		95	.589		40	16.40	6.30	62.70	78.50
0840	20' x 40' pool	↓	130	.431		32	11.95	4.60	48.55	60
1600	For water heating system, see Section 23 52 28.10									
1700	Filtration and deck equipment only, as % of total				Total				20%	20%
1800	Deck equipment, rule of thumb, 20' x 40' pool				SF Pool				1.18	1.30
3000	Painting pools, preparation + 3 coats, 20' x 40' pool, epoxy	2 Pord	.33	48.485	Total	1,725	1,400		3,125	4,200
3100	Rubber base paint, 18 gallons	"	.33	48.485	"	1,250	1,400		2,650	3,675

13 11 23 – On-Grade Swimming Pools

13 11 23.50 Swimming Pools

		Crew	Daily Output	Labor-Hours	Unit	Material	Labor	Equipment	Total	Total Incl O&P
0010	**SWIMMING POOLS** Residential above ground, steel construction									
0100	Round, 15' dia.	B-80A	3	8	Ea.	845	197	102	1,144	1,375
0120	18' dia.		2.50	9.600		915	237	123	1,275	1,525
0140	21' dia.		2	12		1,050	296	154	1,500	1,850
0160	24' dia.		1.80	13.333		1,150	330	171	1,651	2,025
0180	27' dia.		1.50	16		1,350	395	205	1,950	2,350
0200	30' dia.		1	24		1,500	590	305	2,395	2,975
0220	Oval, 12' x 24'		2.30	10.435		1,575	257	134	1,966	2,300
0240	15' x 30'		1.80	13.333		1,750	330	171	2,251	2,675
0260	18' x 33'	↓	1	24	↓	1,975	590	305	2,870	3,500

13 11 46 – Swimming Pool Accessories

13 11 46.50 Swimming Pool Equipment

		Crew	Daily Output	Labor-Hours	Unit	Material	Labor	Equipment	Total	Total Incl O&P
0010	**SWIMMING POOL EQUIPMENT**									
0020	Diving stand, stainless steel, 3 meter	2 Carp	.40	40	Ea.	15,600	1,350		16,950	19,500
0300	1 meter		2.70	5.926		9,500	201		9,701	10,800
0600	Diving boards, 16' long, aluminum		2.70	5.926		4,075	201		4,276	4,800
0700	Fiberglass		2.70	5.926		3,275	201		3,476	3,950
0800	14' long, aluminum		2.70	5.926		3,750	201		3,951	4,450
0850	Fiberglass	↓	2.70	5.926		3,250	201		3,451	3,900
1100	Bulkhead, movable, PVC, 8'-2" wide	2 Clab	8	2		2,375	49.50		2,424.50	2,700
1120	7'-9" wide		8	2		2,200	49.50		2,249.50	2,500
1140	7'-3" wide		8	2		1,975	49.50		2,024.50	2,250
1160	6'-9" wide	↓	8	2		1,975	49.50		2,024.50	2,250
1200	Ladders, heavy duty, stainless steel, 2 tread	2 Carp	7	2.286		855	77.50		932.50	1,075
1500	4 tread	"	6	2.667		1,125	90.50		1,215.50	1,400
2100	Lights, underwater, 12 volt, with transformer, 300 watt	1 Elec	1	8		305	295		600	825
2200	110 volt, 500 watt, standard	"	1	8	↓	293	295		588	805
3000	Pool covers, reinforced vinyl	3 Clab	1800	.013	S.F.	1.19	.33		1.52	1.86
3100	Vinyl, for winter, 400 S.F. max pool surface		3200	.008		.26	.18		.44	.60
3200	With water tubes, 400 S.F. max pool surface	↓	3000	.008	↓	.31	.20		.51	.67
3300	Slides, tubular, fiberglass, aluminum handrails & ladder, 5'-0", straight	2 Carp	1.60	10	Ea.	3,600	340		3,940	4,550
3320	8'-0", curved	"	3	5.333	"	7,225	181		7,406	8,250

13 12 Fountains

13 12 13 – Exterior Fountains

13 12 13.10 Outdoor Fountains

		Crew	Daily Output	Labor-Hours	Unit	Material	2016 Bare Costs Labor	2016 Bare Costs Equipment	Total	Total Incl O&P
0010	**OUTDOOR FOUNTAINS**									
0100	Outdoor fountain, 48" high with bowl and figures	2 Clab	2	8	Ea.	570	197		767	955
0200	Commercial, concrete or cast stone, 40-60" H, simple		2	8		760	197		957	1,175
0220	Average		2	8		1,300	197		1,497	1,750
0240	Ornate		2	8		2,475	197		2,672	3,050
0260	Metal, 72" high		2	8		965	197		1,162	1,375
0280	90" High		2	8		1,250	197		1,447	1,725
0300	120" High		2	8		5,000	197		5,197	5,825
0320	Resin or fiberglass, 40-60" H, wall type		2	8		475	197		672	850
0340	Waterfall type		2	8		1,200	197		1,397	1,625

13 12 23 – Interior Fountains

13 12 23.10 Indoor Fountains

		Crew	Daily Output	Labor-Hours	Unit	Material	Labor	Equipment	Total	Total Incl O&P
0010	**INDOOR FOUNTAINS**									
0100	Commercial, floor type, resin or fiberglass, lighted, cascade type	2 Clab	2	8	Ea.	292	197		489	650
0120	Tiered type		2	8		287	197		484	645
0140	Waterfall type		2	8		330	197		527	695

13 17 Tubs and Pools

13 17 13 – Hot Tubs

13 17 13.10 Redwood Hot Tub System

		Crew	Daily Output	Labor-Hours	Unit	Material	Labor	Equipment	Total	Total Incl O&P
0010	**REDWOOD HOT TUB SYSTEM**									
7050	4' diameter x 4' deep	Q-1	1	16	Ea.	3,225	560		3,785	4,475
7150	6' diameter x 4' deep		.80	20		4,950	705		5,655	6,575
7200	8' diameter x 4' deep		.80	20		7,250	705		7,955	9,125

13 17 33 – Whirlpool Tubs

13 17 33.10 Whirlpool Bath

		Crew	Daily Output	Labor-Hours	Unit	Material	Labor	Equipment	Total	Total Incl O&P
0010	**WHIRLPOOL BATH**									
6000	Whirlpool, bath with vented overflow, molded fiberglass									
6100	66" x 36" x 24"	Q-1	1	16	Ea.	3,475	560		4,035	4,750
6400	72" x 36" x 21"		1	16		1,800	560		2,360	2,925
6500	60" x 34" x 21"		1	16		1,800	560		2,360	2,900
6600	72" x 42" x 23"		1	16		2,175	560		2,735	3,325

13 24 Special Activity Rooms

13 24 16 – Saunas

13 24 16.50 Saunas and Heaters

		Crew	Daily Output	Labor-Hours	Unit	Material	Labor	Equipment	Total	Total Incl O&P
0010	**SAUNAS AND HEATERS**									
0020	Prefabricated, incl. heater & controls, 7' high, 6' x 4', C/C	L-7	2.20	11.818	Ea.	5,325	335		5,660	6,425
0050	6' x 4', C/P		2	13		4,800	370		5,170	5,900
0400	6' x 5', C/C		2	13		5,975	370		6,345	7,200
0450	6' x 5', C/P		2	13		5,425	370		5,795	6,575
0600	6' x 6', C/C		1.80	14.444		6,325	410		6,735	7,675
0650	6' x 6', C/P		1.80	14.444		5,775	410		6,185	7,050
0800	6' x 9', C/C		1.60	16.250		8,400	465		8,865	10,000
0850	6' x 9', C/P		1.60	16.250		7,650	465		8,115	9,175
1000	8' x 12', C/C		1.10	23.636		12,800	675		13,475	15,100
1050	8' x 12', C/P		1.10	23.636		11,600	675		12,275	13,800
1200	8' x 8', C/C		1.40	18.571		9,625	530		10,155	11,500

13 24 Special Activity Rooms

13 24 16 – Saunas

13 24 16.50 Saunas and Heaters

		Crew	Daily Output	Labor-Hours	Unit	Material	2016 Bare Costs Labor	Equipment	Total	Total Incl O&P
1250	8' x 8', C/P	L-7	1.40	18.571	Ea.	8,875	530		9,405	10,700
1400	8' x 10', C/C		1.20	21.667		10,600	620		11,220	12,700
1450	8' x 10', C/P		1.20	21.667		9,700	620		10,320	11,700
1600	10' x 12', C/C		1	26		12,800	740		13,540	15,400
1650	10' x 12', C/P		1	26		11,600	740		12,340	14,000
1700	Door only, cedar, 2' x 6', with 1' x 4' tempered insulated glass window	2 Carp	3.40	4.706		780	160		940	1,125
1800	Prehung, incl. jambs, pulls & hardware	"	12	1.333		760	45		805	915
2500	Heaters only (incl. above), wall mounted, to 200 C.F.					730			730	805
2750	To 300 C.F.					945			945	1,050
3000	Floor standing, to 720 C.F., 10,000 watts, w/controls	1 Elec	3	2.667		2,950	98.50		3,048.50	3,400
3250	To 1,000 C.F., 16,000 watts	"	3	2.667		3,900	98.50		3,998.50	4,425

13 24 26 – Steam Baths

13 24 26.50 Steam Baths and Components

		Crew	Daily Output	Labor-Hours	Unit	Material	Labor	Equipment	Total	Total Incl O&P
0010	**STEAM BATHS AND COMPONENTS**									
0020	Heater, timer & head, single, to 140 C.F.	1 Plum	1.20	6.667	Ea.	2,200	260		2,460	2,825
0500	To 300 C.F.	"	1.10	7.273		2,425	284		2,709	3,150
2700	Conversion unit for residential tub, including door					3,550			3,550	3,925

13 34 Fabricated Engineered Structures

13 34 13 – Glazed Structures

13 34 13.13 Greenhouses

		Crew	Daily Output	Labor-Hours	Unit	Material	Labor	Equipment	Total	Total Incl O&P
0010	**GREENHOUSES**, Shell only, stock units, not incl. 2' stub walls,									
0020	foundation, floors, heat or compartments									
0300	Residential type, free standing, 8'-6" long x 7'-6" wide	2 Carp	59	.271	SF Flr.	25	9.20		34.20	43
0400	10'-6" wide		85	.188		41	6.40		47.40	55.50
0600	13'-6" wide		108	.148		43.50	5		48.50	56
0700	17'-0" wide		160	.100		47	3.39		50.39	57
0900	Lean-to type, 3'-10" wide		34	.471		43.50	15.95		59.45	74
1000	6'-10" wide		58	.276		29	9.35		38.35	47.50
1100	Wall mounted to existing window, 3' x 3'	1 Carp	4	2	Ea.	1,600	68		1,668	1,875
1120	4' x 5'	"	3	2.667	"	1,950	90.50		2,040.50	2,300
1200	Deluxe quality, free standing, 7'-6" wide	2 Carp	55	.291	SF Flr.	85	9.85		94.85	110
1220	10'-6" wide		81	.198		63.50	6.70		70.20	81
1240	13'-6" wide		104	.154		53	5.20		58.20	67.50
1260	17'-0" wide		150	.107		49	3.62		52.62	59.50
1400	Lean-to type, 3'-10" wide		31	.516		105	17.50		122.50	146
1420	6'-10" wide		55	.291		70.50	9.85		80.35	94
1440	8'-0" wide		97	.165		60	5.60		65.60	75.50

13 34 13.19 Swimming Pool Enclosures

		Crew	Daily Output	Labor-Hours	Unit	Material	Labor	Equipment	Total	Total Incl O&P
0010	**SWIMMING POOL ENCLOSURES** Translucent, free standing									
0020	not including foundations, heat or light									
0200	Economy	2 Carp	200	.080	SF Hor.	33.50	2.71		36.21	41.50
0600	Deluxe	"	70	.229	"	93	7.75		100.75	115

13 34 63 – Natural Fiber Construction

13 34 63.50 Straw Bale Construction

			Crew	Daily Output	Labor-Hours	Unit	Material	Labor	Equipment	Total	Total Incl O&P
0010	**STRAW BALE CONSTRUCTION**										
2020	Straw bales in walls w/modified post and beam frame	G	2 Carp	320	.050	S.F.	6.55	1.70		8.25	10.05

Estimating Tips
General
- Many products in Division 14 will require some type of support or blocking for installation not included with the item itself. Examples are supports for conveyors or tube systems, attachment points for lifts, and footings for hoists or cranes. Add these supports in the appropriate division.

14 10 00 Dumbwaiters
14 20 00 Elevators
- Dumbwaiters and elevators are estimated and purchased in a method similar to buying a car. The manufacturer has a base unit with standard features. Added to this base unit price will be whatever options the owner or specifications require. Increased load capacity, additional vertical travel, additional stops, higher speed, and cab finish options are items to be considered. When developing an estimate for dumbwaiters and elevators, remember that some items needed by the installers may have to be included as part of the general contract.

Examples are:
- shaftway
- rail support brackets
- machine room
- electrical supply
- sill angles
- electrical connections
- pits
- roof penthouses
- pit ladders

Check the job specifications and drawings before pricing.
- Installation of elevators and handicapped lifts in historic structures can require significant additional costs. The associated structural requirements may involve cutting into and repairing finishes, moldings, flooring, etc. The estimator must account for these special conditions.

14 30 00 Escalators and Moving Walks
- Escalators and moving walks are specialty items installed by specialty contractors. There are numerous options associated with these items. For specific options, contact a manufacturer or contractor. In a method similar to estimating dumbwaiters and elevators, you should verify the extent of general contract work and add items as necessary.

14 40 00 Lifts
14 90 00 Other Conveying Equipment
- Products such as correspondence lifts, chutes, and pneumatic tube systems, as well as other items specified in this subdivision, may require trained installers. The general contractor might not have any choice as to who will perform the installation, or when it will be performed. Long lead times are often required for these products, making early decisions in scheduling necessary.

Reference Numbers
Reference numbers are shown at the beginning of some major classifications. These numbers refer to related items in the Reference Section. The reference information may be an estimating procedure, an alternate pricing method, or technical information.

Note: Not all subdivisions listed here necessarily appear. ■

No part of this cost data may be reproduced, stored in a retrieval system, or transmitted in any form or by any means without prior written permission of RSMeans.

Did you know?
RSMeans Online gives you the same access to RSMeans' data with 24/7 access:
- Quickly locate costs in the searchable database.
- Build cost lists, estimates, and reports in minutes.
- Adjust costs to any location in the U.S. and Canada with the click of a button.

Start your free trial today at **www.RSMeansOnline.com**

RSMeans Online
FROM THE GORDIAN GROUP®

14 21 Electric Traction Elevators

14 21 33 – Electric Traction Residential Elevators

14 21 33.20 Residential Elevators	Crew	Daily Output	Labor-Hours	Unit	Material	2016 Bare Costs Labor	2016 Bare Costs Equipment	Total	Total Incl O&P
0010 **RESIDENTIAL ELEVATORS**									
7000　Residential, cab type, 1 floor, 2 stop, economy model	2 Elev	.20	80	Ea.	11,400	4,250		15,650	19,500
7100　　Custom model		.10	160		19,300	8,475		27,775	35,000
7200　　2 floor, 3 stop, economy model		.12	133		17,000	7,075		24,075	30,200
7300　　Custom model	↓	.06	266	↓	27,700	14,100		41,800	53,500

14 42 Wheelchair Lifts

14 42 13 – Inclined Wheelchair Lifts

14 42 13.10 Inclined Wheelchair Lifts and Stairclimbers

	Crew	Daily Output	Labor-Hours	Unit	Material	Labor	Equipment	Total	Total Incl O&P
0010 **INCLINED WHEELCHAIR LIFTS AND STAIRCLIMBERS**									
7700　Stair climber (chair lift), single seat, minimum	2 Elev	1	16	Ea.	5,325	850		6,175	7,250
7800　　Maximum	"	.20	80	"	7,350	4,250		11,600	15,000

Estimating Tips

Pipe for fire protection and all uses is located in Subdivisions 21 11 13 and 22 11 13.

The labor adjustment factors listed in Subdivision 22 01 02.20 also apply to Division 21.

Many, but not all, areas in the U.S. require backflow protection in the fire system. It is advisable to check local building codes for specific requirements.

For your reference, the following is a list of the most applicable Fire Codes and Standards which may be purchased from the NFPA, 1 Batterymarch Park, Quincy, MA 02169-7471.
- NFPA 1: Uniform Fire Code
- NFPA 10: Portable Fire Extinguishers
- NFPA 11: Low-, Medium-, and High-Expansion Foam
- NFPA 12: Carbon Dioxide Extinguishing Systems (Also companion 12A)
- NFPA 13: Installation of Sprinkler Systems (Also companion 13D, 13E, and 13R)
- NFPA 14: Installation of Standpipe and Hose Systems
- NFPA 15: Water Spray Fixed Systems for Fire Protection
- NFPA 16: Installation of Foam-Water Sprinkler and Foam-Water Spray Systems
- NFPA 17: Dry Chemical Extinguishing Systems (Also companion 17A)
- NFPA 18: Wetting Agents
- NFPA 20: Installation of Stationary Pumps for Fire Protection
- NFPA 22: Water Tanks for Private Fire Protection
- NFPA 24: Installation of Private Fire Service Mains and their Appurtenances
- NFPA 25: Inspection, Testing and Maintenance of Water-Based Fire Protection

Reference Numbers

Reference numbers are shown at the beginning of some major classifications. These numbers refer to related items in the Reference Section. The reference information may be an estimating procedure, an alternate pricing method, or technical information.

Note: Not all subdivisions listed here necessarily appear. ■

No part of this cost data may be reproduced, stored in a retrieval system, or transmitted in any form or by any means without prior written permission of RSMeans.

Did you know?
RSMeans Online gives you the same access to RSMeans' data with 24/7 access:
- Quickly locate costs in the searchable database.
- Build cost lists, estimates, and reports in minutes.
- Adjust costs to any location in the U.S. and Canada with the click of a button.

Start your free trial today at www.RSMeansOnline.com

RSMeans Online
FROM THE GORDIAN GROUP

21 05 Common Work Results for Fire Suppression

21 05 23 – General-Duty Valves for Water-Based Fire-Suppression Piping

21 05 23.50 General-Duty Valves	Crew	Daily Output	Labor-Hours	Unit	Material	2016 Bare Costs Labor	Equipment	Total	Total Incl O&P
0010 **GENERAL-DUTY VALVES,** for water-based fire suppression									
6200 Valves and components									
6210 Alarm, includes									
6220 retard chamber, trim, gauges, alarm line strainer									
6260 3" size	Q-12	3	5.333	Ea.	1,725	182		1,907	2,200
6280 4" size	"	2	8		1,800	274		2,074	2,425
6300 6" size	Q-13	4	8	↓	2,000	274		2,274	2,650
6500 Check, swing, C.I. body, brass fittings, auto. ball drip									
6520 4" size	Q-12	3	5.333	Ea.	370	182		552	705
6800 Check, wafer, butterfly type, C.I. body, bronze fittings									
6820 4" size	Q-12	4	4	Ea.	1,050	137		1,187	1,375
8800 Flow control valve, includes trim and gauges, 2" size		2	8		4,925	274		5,199	5,875
8820 3" size	↓	1.50	10.667		5,425	365		5,790	6,575
8840 4" size	Q-13	2.80	11.429		6,050	390		6,440	7,325
8860 6" size	"	2	16		7,025	545		7,570	8,625
9200 Pressure operated relief valve, brass body	1 Spri	18	.444	↓	605	16.90		621.90	700
9600 Waterflow indicator, with recycling retard and									
9610 two single pole retard switches, 2" thru 6" pipe size	1 Spri	8	1	Ea.	144	38		182	221

21 11 Facility Fire-Suppression Water-Service Piping

21 11 13 – Facility Water Distribution Piping

21 11 13.16 Pipe, Plastic

		Crew	Daily Output	Labor-Hours	Unit	Material	Labor	Equipment	Total	Total Incl O&P
0010	**PIPE, PLASTIC**									
0020	CPVC, fire suppression (C-UL-S, FM, NFPA 13, 13D & 13R)									
0030	Socket joint, no couplings or hangers									
0100	SDR 13.5 (ASTM F442)									
0120	3/4" diameter	Q-12	420	.038	L.F.	1.67	1.30		2.97	4
0130	1" diameter		340	.047		2.58	1.61		4.19	5.50
0140	1-1/4" diameter		260	.062		4.10	2.10		6.20	8
0150	1-1/2" diameter		190	.084		5.65	2.88		8.53	11
0160	2" diameter		140	.114		9	3.91		12.91	16.35
0170	2-1/2" diameter		130	.123		16.65	4.21		20.86	25.50
0180	3" diameter	↓	120	.133	↓	25.50	4.56		30.06	35.50

21 11 13.18 Pipe Fittings, Plastic

		Crew	Daily Output	Labor-Hours	Unit	Material	Labor	Equipment	Total	Total Incl O&P
0010	**PIPE FITTINGS, PLASTIC**									
0020	CPVC, fire suppression (C-UL-S, FM, NFPA 13, 13D & 13R)									
0030	Socket joint									
0100	90° Elbow									
0120	3/4"	1 Plum	26	.308	Ea.	1.78	12		13.78	22
0130	1"		22.70	.352		3.91	13.75		17.66	27
0140	1-1/4"		20.20	.396		4.94	15.45		20.39	31
0150	1-1/2"	↓	18.20	.440		7	17.15		24.15	36
0160	2"	Q-1	33.10	.483		8.70	17		25.70	37.50
0170	2-1/2"		24.20	.661		16.75	23		39.75	57
0180	3"	↓	20.80	.769	↓	23	27		50	69.50
0200	45° Elbow									
0210	3/4"	1 Plum	26	.308	Ea.	2.45	12		14.45	22.50
0220	1"		22.70	.352		2.87	13.75		16.62	25.50
0230	1-1/4"		20.20	.396		4.15	15.45		19.60	30
0240	1-1/2"	↓	18.20	.440		5.80	17.15		22.95	35
0250	2"	Q-1	33.10	.483	↓	7.20	17		24.20	36

21 11 Facility Fire-Suppression Water-Service Piping

21 11 13 – Facility Water Distribution Piping

21 11 13.18 Pipe Fittings, Plastic		Crew	Daily Output	Labor-Hours	Unit	Material	2016 Bare Costs Labor	Equipment	Total	Total Incl O&P
0260	2-1/2"	Q-1	24.20	.661	Ea.	12.95	23		35.95	52.50
0270	3"		20.80	.769		18.55	27		45.55	65
0300	Tee									
0310	3/4"	1 Plum	17.30	.462	Ea.	2.45	18.05		20.50	32.50
0320	1"		15.20	.526		4.82	20.50		25.32	39.50
0330	1-1/4"		13.50	.593		7.25	23		30.25	46
0340	1-1/2"		12.10	.661		10.70	26		36.70	54.50
0350	2"	Q-1	20	.800		15.80	28		43.80	64
0360	2-1/2"		16.20	.988		25.50	34.50		60	85.50
0370	3"		13.90	1.151		40	40.50		80.50	111
0400	Tee, reducing x any size									
0420	1"	1 Plum	15.20	.526	Ea.	4.09	20.50		24.59	38.50
0430	1-1/4"		13.50	.593		7.50	23		30.50	46.50
0440	1-1/2"		12.10	.661		9.10	26		35.10	52.50
0450	2"	Q-1	20	.800		16.80	28		44.80	65
0460	2-1/2"		16.20	.988		19.95	34.50		54.45	79.50
0470	3"		13.90	1.151		23	40.50		63.50	92
0500	Coupling									
0510	3/4"	1 Plum	26	.308	Ea.	1.71	12		13.71	21.50
0520	1"		22.70	.352		2.26	13.75		16.01	25
0530	1-1/4"		20.20	.396		3.29	15.45		18.74	29
0540	1-1/2"		18.20	.440		4.70	17.15		21.85	33.50
0550	2"	Q-1	33.10	.483		6.35	17		23.35	35
0560	2-1/2"		24.20	.661		9.70	23		32.70	49
0570	3"		20.80	.769		12.60	27		39.60	58.50
0600	Coupling, reducing									
0610	1" x 3/4"	1 Plum	22.70	.352	Ea.	2.26	13.75		16.01	25
0620	1-1/4" x 1"		20.20	.396		3.42	15.45		18.87	29.50
0630	1-1/2" x 3/4"		18.20	.440		5.15	17.15		22.30	34
0640	1-1/2" x 1"		18.20	.440		4.94	17.15		22.09	34
0650	1-1/2" x 1-1/4"		18.20	.440		4.70	17.15		21.85	33.50
0660	2" x 1"	Q-1	33.10	.483		6.60	17		23.60	35.50
0670	2" x 1-1/2"	"	33.10	.483		6.35	17		23.35	35
0700	Cross									
0720	3/4"	1 Plum	13	.615	Ea.	3.84	24		27.84	43.50
0730	1"		11.30	.708		4.82	27.50		32.32	51
0740	1-1/4"		10.10	.792		6.65	31		37.65	58.50
0750	1-1/2"		9.10	.879		9.20	34.50		43.70	66.50
0760	2"	Q-1	16.60	.964		15	34		49	72.50
0770	2-1/2"	"	12.10	1.322		33	46.50		79.50	113
0800	Cap									
0820	3/4"	1 Plum	52	.154	Ea.	1.04	6		7.04	11.05
0830	1"		45	.178		1.47	6.95		8.42	13.05
0840	1-1/4"		40	.200		2.38	7.80		10.18	15.50
0850	1-1/2"		36.40	.220		3.29	8.60		11.89	17.75
0860	2"	Q-1	66	.242		4.94	8.50		13.44	19.50
0870	2-1/2"		48.40	.331		7.15	11.60		18.75	27
0880	3"		41.60	.385		11.50	13.50		25	35
0900	Adapter, sprinkler head, female w/metal thd. insert (s x FNPT)									
0920	3/4" x 1/2"	1 Plum	52	.154	Ea.	4.61	6		10.61	14.95
0930	1" x 1/2"		45	.178		4.86	6.95		11.81	16.80
0940	1" x 3/4"		45	.178		7.65	6.95		14.60	19.90

21 11 Facility Fire-Suppression Water-Service Piping

21 11 16 – Facility Fire Hydrants

21 11 16.50 Fire Hydrants for Buildings

		Crew	Daily Output	Labor-Hours	Unit	Material	2016 Bare Costs Labor	2016 Bare Costs Equipment	Total	Total Incl O&P
0010	**FIRE HYDRANTS FOR BUILDINGS**									
3750	Hydrants, wall, w/caps, single, flush, polished brass									
3800	2-1/2" x 2-1/2"	Q-12	5	3.200	Ea.	224	109		333	425
3840	2-1/2" x 3"		5	3.200		450	109		559	675
3860	3" x 3"		4.80	3.333		365	114		479	590
3900	For polished chrome, add					20%				
3950	Double, flush, polished brass									
4000	2-1/2" x 2-1/2" x 4"	Q-12	5	3.200	Ea.	600	109		709	840
4040	2-1/2" x 2-1/2" x 6"		4.60	3.478		865	119		984	1,150
4080	3" x 3" x 4"		4.90	3.265		1,275	112		1,387	1,575
4120	3" x 3" x 6"		4.50	3.556		1,325	122		1,447	1,650
4200	For polished chrome, add					10%				
4350	Double, projecting, polished brass									
4400	2-1/2" x 2-1/2" x 4"	Q-12	5	3.200	Ea.	267	109		376	475
4450	2-1/2" x 2-1/2" x 6"	"	4.60	3.478	"	545	119		664	795
4460	Valve control, dbl. flush/projecting hydrant, cap &									
4470	chain, extension rod & cplg., escutcheon, polished brass	Q-12	8	2	Ea.	315	68.50		383.50	465
4480	Four-way square, flush, polished brass									
4540	2-1/2"(4) x 6"	Q-12	3.60	4.444	Ea.	3,225	152		3,377	3,800

21 11 19 – Fire-Department Connections

21 11 19.50 Connections for the Fire-Department

		Crew	Daily Output	Labor-Hours	Unit	Material	2016 Bare Costs Labor	2016 Bare Costs Equipment	Total	Total Incl O&P
0010	**CONNECTIONS FOR THE FIRE-DEPARTMENT**									
6000	Roof manifold, horiz., brass, without valves & caps									
6040	2-1/2" x 2-1/2" x 4"	Q-12	4.80	3.333	Ea.	186	114		300	395
6060	2-1/2" x 2-1/2" x 6"		4.60	3.478		203	119		322	420
6080	2-1/2" x 2-1/2" x 2-1/2" x 4"		4.60	3.478		298	119		417	525
6090	2-1/2" x 2-1/2" x 2-1/2" x 6"		4.60	3.478		310	119		429	535
7000	Sprinkler line tester, cast brass					33			33	36.50
7140	Standpipe connections, wall, w/plugs & chains									
7160	Single, flush, brass, 2-1/2" x 2-1/2"	Q-12	5	3.200	Ea.	167	109		276	365
7180	2-1/2" x 3"	"	5	3.200	"	172	109		281	370
7240	For polished chrome, add					15%				
7280	Double, flush, polished brass									
7300	2-1/2" x 2-1/2" x 4"	Q-12	5	3.200	Ea.	545	109		654	780
7330	2-1/2" x 2-1/2" x 6"		4.60	3.478		760	119		879	1,025
7340	3" x 3" x 4"		4.90	3.265		990	112		1,102	1,275
7370	3" x 3" x 6"		4.50	3.556		1,200	122		1,322	1,500
7400	For polished chrome, add					15%				
7440	For sill cock combination, add				Ea.	95			95	104
7580	Double projecting, polished brass									
7600	2-1/2" x 2-1/2" x 4"	Q-12	5	3.200	Ea.	500	109		609	730
7630	2-1/2" x 2-1/2" x 6"	"	4.60	3.478	"	840	119		959	1,125
7680	For polished chrome, add					15%				
7900	Three way, flush, polished brass									
7920	2-1/2" (3) x 4"	Q-12	4.80	3.333	Ea.	1,750	114		1,864	2,125
7930	2-1/2" (3) x 6"	"	4.60	3.478		1,750	119		1,869	2,125
8000	For polished chrome, add					9%				
8020	Three way, projecting, polished brass									
8040	2-1/2"(3) x 4"	Q-12	4.80	3.333	Ea.	1,125	114		1,239	1,450
8070	2-1/2" (3) x 6"	"	4.60	3.478		1,625	119		1,744	1,975
8100	For polished chrome, add					12%				
8200	Four way, square, flush, polished brass,									

21 11 Facility Fire-Suppression Water-Service Piping

21 11 19 – Fire-Department Connections

21 11 19.50 Connections for the Fire-Department

		Crew	Daily Output	Labor-Hours	Unit	Material	2016 Bare Costs Labor	2016 Bare Costs Equipment	Total	Total Incl O&P
8240	2-1/2"(4) x 6"	Q-12	3.60	4.444	Ea.	1,250	152		1,402	1,625
8300	For polished chrome, add				"	10%				
8550	Wall, vertical, flush, cast brass									
8600	Two way, 2-1/2" x 2-1/2" x 4"	Q-12	5	3.200	Ea.	395	109		504	615
8660	Four way, 2-1/2"(4) x 6"		3.80	4.211		1,300	144		1,444	1,675
8680	Six way, 2-1/2"(6) x 6"		3.40	4.706		1,550	161		1,711	2,000
8700	For polished chrome, add					10%				
8800	Sidewalk siamese unit, polished brass, two way									
8820	2-1/2" x 2-1/2" x 4"	Q-12	2.50	6.400	Ea.	665	219		884	1,100
8850	2-1/2" x 2-1/2" x 6"		2	8		860	274		1,134	1,400
8860	3" x 3" x 4"		2.50	6.400		945	219		1,164	1,425
8890	3" x 3" x 6"		2	8		1,275	274		1,549	1,850
8940	For polished chrome, add					12%				
9100	Sidewalk siamese unit, polished brass, three way									
9120	2-1/2" x 2-1/2" x 2-1/2" x 6"	Q-12	2	8	Ea.	1,050	274		1,324	1,600
9160	For polished chrome, add				"	15%				

21 12 Fire-Suppression Standpipes

21 12 19 – Fire-Suppression Hose Racks

21 12 19.50 Fire Hose Racks

		Crew	Daily Output	Labor-Hours	Unit	Material	Labor	Equipment	Total	Total Incl O&P
0010	**FIRE HOSE RACKS**									
2600	Hose rack, swinging, for 1-1/2" diameter hose,									
2620	Enameled steel, 50' & 75' lengths of hose	Q-12	20	.800	Ea.	64	27.50		91.50	116
2640	100' and 125' lengths of hose		20	.800		64.50	27.50		92	117
2680	Chrome plated, 50' and 75' lengths of hose		20	.800		103	27.50		130.50	160
2700	100' and 125' lengths of hose		20	.800		110	27.50		137.50	166
2780	For hose rack nipple, 1-1/2" polished brass, add					29.50			29.50	32.50
2820	2-1/2" polished brass, add					54.50			54.50	60
2840	1-1/2" polished chrome, add					41.50			41.50	45.50
2860	2-1/2" polished chrome, add					59			59	64.50

21 12 23 – Fire-Suppression Hose Valves

21 12 23.70 Fire Hose Valves

		Crew	Daily Output	Labor-Hours	Unit	Material	Labor	Equipment	Total	Total Incl O&P
0010	**FIRE HOSE VALVES**									
0020	Angle, combination pressure adjust/restricting, rough brass									
0030	1-1/2"	1 Spri	12	.667	Ea.	89.50	25.50		115	141
0040	2-1/2"	"	7	1.143	"	184	43.50		227.50	274
0042	Nonpressure adjustable/restricting, rough brass									
0044	1-1/2"	1 Spri	12	.667	Ea.	56	25.50		81.50	104
0046	2-1/2"	"	7	1.143	"	94.50	43.50		138	176
0050	For polished brass, add					30%				
0060	For polished chrome, add					40%				
1000	Ball drip, automatic, rough brass, 1/2"	1 Spri	20	.400	Ea.	16.90	15.20		32.10	43.50
1010	3/4"	"	20	.400	"	18.75	15.20		33.95	45.50
1100	Ball, 175 lb., sprinkler system, FM/UL, threaded, bronze									
1120	Slow close									
1150	1" size	1 Spri	19	.421	Ea.	246	16		262	297
1160	1-1/4" size		15	.533		266	20.50		286.50	325
1170	1-1/2" size		13	.615		450	23.50		473.50	535
1180	2" size		11	.727		425	27.50		452.50	515
1190	2-1/2" size	Q-12	15	1.067		575	36.50		611.50	690

21 12 Fire-Suppression Standpipes

21 12 23 – Fire-Suppression Hose Valves

21 12 23.70 Fire Hose Valves

		Crew	Daily Output	Labor-Hours	Unit	Material	2016 Bare Costs Labor	Equipment	Total	Total Incl O&P
1230	For supervisory switch kit, all sizes									
1240	One circuit, add	1 Spri	48	.167	Ea.	169	6.35		175.35	197
1280	Quarter turn for trim									
1300	1/2" size	1 Spri	22	.364	Ea.	32.50	13.80		46.30	59
1310	3/4" size		20	.400		35	15.20		50.20	63.50
1320	1" size		19	.421		39	16		55	69
1330	1-1/4" size		15	.533		63.50	20.50		84	104
1340	1-1/2" size		13	.615		80	23.50		103.50	127
1350	2" size	↓	11	.727		95	27.50		122.50	151
1400	Caps, polished brass with chain, 3/4"					44			44	48.50
1420	1"					56			56	62
1440	1-1/2"					15.05			15.05	16.60
1460	2-1/2"					23			23	25
1480	3"					35.50			35.50	39
1900	Escutcheon plate, for angle valves, polished brass, 1-1/2"					15.20			15.20	16.70
1920	2-1/2"					25			25	27.50
1940	3"					31.50			31.50	34.50
1980	For polished chrome, add					15%				
3000	Gate, hose, wheel handle, N.R.S., rough brass, 1-1/2"	1 Spri	12	.667		141	25.50		166.50	197
3040	2-1/2", 300 lb.	"	7	1.143		199	43.50		242.50	291
3080	For polished brass, add					40%				
3090	For polished chrome, add					50%				
5000	Pressure reducing rough brass, 1-1/2"	1 Spri	12	.667		315	25.50		340.50	390
5020	2-1/2"	"	7	1.143		370	43.50		413.50	480
5080	For polished brass, add					105%				
5090	For polished chrome, add	↓				140%				

21 13 Fire-Suppression Sprinkler Systems

21 13 13 – Wet-Pipe Sprinkler Systems

21 13 13.50 Wet-Pipe Sprinkler System Components

		Crew	Daily Output	Labor-Hours	Unit	Material	Labor	Equipment	Total	Total Incl O&P
0010	**WET-PIPE SPRINKLER SYSTEM COMPONENTS**									
1100	Alarm, electric pressure switch (circuit closer)	1 Spri	26	.308	Ea.	93	11.70		104.70	121
1140	For explosion proof, max 20 PSI, contacts close or open		26	.308		560	11.70		571.70	635
1220	Water motor, complete with gong	↓	4	2	↓	435	76		511	605
1900	Flexible sprinkler head connectors									
1910	Braided stainless steel hose with mounting bracket									
1920	1/2" and 3/4" outlet size									
1940	40" length	1 Spri	30	.267	Ea.	74	10.15		84.15	98.50
1960	60" length	"	22	.364	"	85	13.80		98.80	117
1982	May replace hard-pipe armovers									
1984	For wet, pre-action, deluge or dry pipe systems									
2000	Release, emergency, manual, for hydraulic or pneumatic system	1 Spri	12	.667	Ea.	228	25.50		253.50	293
2060	Release, thermostatic, for hydraulic or pneumatic release line		20	.400		705	15.20		720.20	805
2200	Sprinkler cabinets, 6 head capacity		16	.500		83	19		102	123
2260	12 head capacity		16	.500		87.50	19		106.50	128
2340	Sprinkler head escutcheons, standard, brass tone, 1" size		40	.200		3	7.60		10.60	15.90
2360	Chrome, 1" size		40	.200		3.19	7.60		10.79	16.10
2400	Recessed type, bright brass		40	.200		10.60	7.60		18.20	24.50
2440	Chrome or white enamel	↓	40	.200		3.52	7.60		11.12	16.45
2600	Sprinkler heads, not including supply piping									
3700	Standard spray, pendent or upright, brass, 135°F to 286°F									

21 13 Fire-Suppression Sprinkler Systems

21 13 13 – Wet-Pipe Sprinkler Systems

21 13 13.50 Wet-Pipe Sprinkler System Components

		Crew	Daily Output	Labor-Hours	Unit	Material	2016 Bare Costs Labor	2016 Bare Costs Equipment	Total	Total Incl O&P
3720	1/2" NPT, 3/8" orifice	1 Spri	16	.500	Ea.	16.15	19		35.15	49.50
3730	1/2" NPT, 7/16" orifice		16	.500		16	19		35	49
3732	1/2" NPT, 7/16" orifice, chrome		16	.500		16.80	19		35.80	50
3740	1/2" NPT, 1/2" orifice		16	.500		10.40	19		29.40	43
3760	1/2" NPT, 17/32" orifice		16	.500		13.55	19		32.55	46.50
3780	3/4" NPT, 17/32" orifice		16	.500		12.45	19		31.45	45
3840	For chrome, add					3.63			3.63	3.99
4200	Sidewall, vertical brass, 135°F to 286°F									
4240	1/2" NPT, 1/2" orifice	1 Spri	16	.500	Ea.	26.50	19		45.50	61
4280	3/4" NPT, 17/32" orifice	"	16	.500		74	19		93	113
4360	For satin chrome, add					3			3	3.30
4500	Sidewall, horizontal, brass, 135°F to 286°F									
4520	1/2" NPT, 1/2" orifice	1 Spri	16	.500	Ea.	26.50	19		45.50	61
5600	Concealed, complete with cover plate									
5620	1/2" NPT, 1/2" orifice, 135°F to 212°F	1 Spri	9	.889	Ea.	25.50	34		59.50	84
6025	Residential sprinkler components, (one and two family)									
6026	Water motor alarm with strainer	1 Spri	4	2	Ea.	435	76		511	605
6027	Fast response, glass bulb, 135°F to 155°F									
6028	1/2" NPT, pendent, brass	1 Spri	16	.500	Ea.	28.50	19		47.50	63
6029	1/2" NPT, sidewall, brass		16	.500		28.50	19		47.50	63
6030	1/2" NPT, pendent, brass, extended coverage		16	.500		25	19		44	59
6031	1/2" NPT, sidewall, brass, extended coverage		16	.500		23	19		42	57
6032	3/4" NPT sidewall, brass, extended coverage		16	.500		25.50	19		44.50	59.50
6033	For chrome, add					15%				
6034	For polyester/teflon coating add					20%				
6100	Sprinkler head wrenches, standard head				Ea.	26			26	28.50
6120	Recessed head					39.50			39.50	43.50
6160	Tamper switch (valve supervisory switch)	1 Spri	16	.500		222	19		241	276

21 13 16 – Dry-Pipe Sprinkler Systems

21 13 16.50 Dry-Pipe Sprinkler System Components

		Crew	Daily Output	Labor-Hours	Unit	Material	Labor	Equipment	Total	Total Incl O&P
0010	**DRY-PIPE SPRINKLER SYSTEM COMPONENTS**									
0600	Accelerator	1 Spri	8	1	Ea.	795	38		833	940
0800	Air compressor for dry pipe system, automatic, complete									
0820	30 gal. system capacity, 3/4 HP	1 Spri	1.30	6.154	Ea.	810	234		1,044	1,275
0860	30 gal. system capacity, 1 HP		1.30	6.154		840	234		1,074	1,300
0960	Air pressure maintenance control		24	.333		360	12.65		372.65	420
1600	Dehydrator package, incl. valves and nipples		12	.667		745	25.50		770.50	860
2600	Sprinkler heads, not including supply piping									
2640	Dry, pendent, 1/2" orifice, 3/4" or 1" NPT									
2660	1/2" to 6" length	1 Spri	14	.571	Ea.	132	21.50		153.50	181
2670	6-1/4" to 8" length		14	.571		137	21.50		158.50	186
2680	8-1/4" to 12" length		14	.571		143	21.50		164.50	193
2800	For each inch or fraction, add					3.25			3.25	3.58
6330	Valves and components									
6340	Alarm test/shut off valve, 1/2"	1 Spri	20	.400	Ea.	22.50	15.20		37.70	50
8000	Dry pipe air check valve, 3" size	Q-12	2	8		1,850	274		2,124	2,500
8200	Dry pipe valve, incl. trim and gauges, 3" size		2	8		2,700	274		2,974	3,425
8220	4" size		1	16		3,000	545		3,545	4,200
8240	6" size	Q-13	2	16		3,525	545		4,070	4,775
8280	For accelerator trim with gauges, add	1 Spri	8	1		251	38		289	340

Division Notes

Estimating Tips

22 10 00 Plumbing Piping and Pumps

This subdivision is primarily basic pipe and related materials. The pipe may be used by any of the mechanical disciplines, i.e., plumbing, fire protection, heating, and air conditioning.

Note: CPVC plastic piping approved for fire protection is located in 21 11 13.

- The labor adjustment factors listed in Subdivision 22 01 02.20 apply throughout Divisions 21, 22, and 23. CAUTION: the correct percentage may vary for the same items. For example, the percentage add for the basic pipe installation should be based on the maximum height that the craftsman must install for that particular section. If the pipe is to be located 14' above the floor but it is suspended on threaded rod from beams, the bottom flange of which is 18' high (4' rods), then the height is actually 18' and the add is 20%. The pipe coverer, however, does not have to go above the 14', and so the add should be 10%.
- Most pipe is priced first as straight pipe with a joint (coupling, weld, etc.) every 10' and a hanger usually every 10'. There are exceptions with hanger spacing such as for cast iron pipe (5') and plastic pipe (3 per 10'). Following each type of pipe there are several lines listing sizes and the amount to be subtracted to delete couplings and hangers. This is for pipe that is to be buried or supported together on trapeze hangers. The reason that the couplings are deleted is that these runs are usually long, and frequently longer lengths of pipe are used. By deleting the couplings, the estimator is expected to look up and add back the correct reduced number of couplings.
- When preparing an estimate, it may be necessary to approximate the fittings. Fittings usually run between 25% and 50% of the cost of the pipe. The lower percentage is for simpler runs, and the higher number is for complex areas, such as mechanical rooms.
- For historic restoration projects, the systems must be as invisible as possible, and pathways must be sought for pipes, conduit, and ductwork. While installations in accessible spaces (such as basements and attics) are relatively straightforward to estimate, labor costs may be more difficult to determine when delivery systems must be concealed.

22 40 00 Plumbing Fixtures

- Plumbing fixture costs usually require two lines: the fixture itself and its "rough-in, supply, and waste."
- In the Assemblies Section (Plumbing D2010) for the desired fixture, the System Components Group at the center of the page shows the fixture on the first line. The rest of the list (fittings, pipe, tubing, etc.) will total up to what we refer to in the Unit Price section as "Rough-in, supply, waste, and vent." Note that for most fixtures we allow a nominal 5' of tubing to reach from the fixture to a main or riser.
- Remember that gas- and oil-fired units need venting.

Reference Numbers

Reference numbers are shown at the beginning of some major classifications. These numbers refer to related items in the Reference Section. The reference information may be an estimating procedure, an alternate pricing method, or technical information.

Note: Not all subdivisions listed here necessarily appear. ■

Did you know?

RSMeans Online gives you the same access to RSMeans' data with 24/7 access:

- Quickly locate costs in the searchable database.
- Build cost lists, estimates, and reports in minutes.
- Adjust costs to any location in the U.S. and Canada with the click of a button.

Start your free trial today at **www.RSMeansOnline.com**

No part of this cost data may be reproduced, stored in a retrieval system, or transmitted in any form or by any means without prior written permission of RSMeans.

22 05 Common Work Results for Plumbing

22 05 05 – Selective Demolition for Plumbing

22 05 05.10 Plumbing Demolition		Crew	Daily Output	Labor-Hours	Unit	Material	2016 Bare Costs Labor	Equipment	Total	Total Incl O&P
0010	**PLUMBING DEMOLITION**									
1020	Fixtures, including 10' piping									
1101	Bathtubs, cast iron	1 Clab	4	2	Ea.		49.50		49.50	82.50
1121	Fiberglass		6	1.333			33		33	55
1141	Steel	↓	5	1.600			39.50		39.50	66
1200	Lavatory, wall hung	1 Plum	10	.800			31		31	51.50
1221	Counter top	1 Clab	16	.500			12.35		12.35	20.50
1301	Sink, single compartment		16	.500			12.35		12.35	20.50
1321	Double	↓	10	.800			19.70		19.70	33
1400	Water closet, floor mounted	1 Plum	8	1			39		39	64.50
1421	Wall mounted	1 Clab	7	1.143	↓		28		28	47
2001	Piping, metal, to 1-1/2" diameter		200	.040	L.F.		.99		.99	1.65
2051	2" thru 3-1/2" diameter		150	.053			1.31		1.31	2.20
2101	4" thru 6" diameter	↓	100	.080			1.97		1.97	3.30
2160	Plastic pipe with fittings, up thru 1-1/2" diameter	1 Plum	250	.032			1.25		1.25	2.06
2162	2" thru 3" diameter	"	200	.040			1.56		1.56	2.58
2164	4" thru 6" diameter	Q-1	200	.080			2.81		2.81	4.64
2166	8" thru 14" diameter		150	.107			3.75		3.75	6.20
2168	16" diameter	↓	100	.160	↓		5.60		5.60	9.30
3000	Submersible sump pump	1 Plum	24	.333	Ea.		13		13	21.50
6000	Remove and reset fixtures, easy access		6	1.333			52		52	86
6100	Difficult access	↓	4	2	↓		78		78	129

22 05 23 – General-Duty Valves for Plumbing Piping

22 05 23.20 Valves, Bronze

		Crew	Daily Output	Labor-Hours	Unit	Material	Labor	Equipment	Total	Total Incl O&P
0010	**VALVES, BRONZE**									
1750	Check, swing, class 150, regrinding disc, threaded									
1860	3/4"	1 Plum	20	.400	Ea.	105	15.60		120.60	142
1870	1"	"	19	.421	"	152	16.45		168.45	194
2850	Gate, N.R.S., soldered, 125 psi									
2940	3/4"	1 Plum	20	.400	Ea.	66	15.60		81.60	98.50
2950	1"	"	19	.421	"	81.50	16.45		97.95	117
5600	Relief, pressure & temperature, self-closing, ASME, threaded									
5640	3/4"	1 Plum	28	.286	Ea.	206	11.15		217.15	245
5650	1"		24	.333		330	13		343	385
5660	1-1/4"	↓	20	.400	↓	655	15.60		670.60	745
6400	Pressure, water, ASME, threaded									
6440	3/4"	1 Plum	28	.286	Ea.	181	11.15		192.15	217
6450	1"	"	24	.333	"	278	13		291	325
6900	Reducing, water pressure									
6920	300 psi to 25-75 psi, threaded or sweat									
6940	1/2"	1 Plum	24	.333	Ea.	425	13		438	485
6950	3/4"		20	.400		435	15.60		450.60	500
6960	1"	↓	19	.421	↓	670	16.45		686.45	765
8350	Tempering, water, sweat connections									
8400	1/2"	1 Plum	24	.333	Ea.	102	13		115	134
8440	3/4"	"	20	.400	"	136	15.60		151.60	176
8650	Threaded connections									
8700	1/2"	1 Plum	24	.333	Ea.	151	13		164	188
8740	3/4"	"	20	.400	"	825	15.60		840.60	935
8800	Water heater water & gas safety shut off									
8810	Protection against a leaking water heater									
8814	Shut off valve	1 Plum	16	.500	Ea.	171	19.55		190.55	220

22 05 Common Work Results for Plumbing

22 05 23 – General-Duty Valves for Plumbing Piping

22 05 23.20 Valves, Bronze		Crew	Daily Output	Labor-Hours	Unit	Material	2016 Bare Costs Labor	Equipment	Total	Total Incl O&P
8818	Water heater dam	1 Plum	32	.250	Ea.	25	9.75		34.75	43.50
8822	Gas control wiring harness	↓	32	.250	↓	21.50	9.75		31.25	40
8830	Whole house flood safety shut off									
8834	Connections									
8838	3/4" NPT	1 Plum	12	.667	Ea.	960	26		986	1,100
8842	1" NPT		11	.727		985	28.50		1,013.50	1,125
8846	1-1/4" NPT	↓	10	.800	↓	1,025	31		1,056	1,175

22 05 29 – Hangers and Supports for Plumbing Piping and Equipment

22 05 29.10 Hangers & Supp. for Plumb'g/HVAC Pipe/Equip.

		Crew	Daily Output	Labor-Hours	Unit	Material	Labor	Equipment	Total	Total Incl O&P
0010	**HANGERS AND SUPPORTS FOR PLUMB'G/HVAC PIPE/EQUIP.**									
8000	Pipe clamp, plastic, 1/2" CTS	1 Plum	80	.100	Ea.	.24	3.91		4.15	6.70
8010	3/4" CTS		73	.110		.25	4.28		4.53	7.35
8020	1" CTS		68	.118		.56	4.59		5.15	8.20
8080	Economy clamp, 1/4" CTS		175	.046		.04	1.79		1.83	2.98
8090	3/8" CTS		168	.048		.04	1.86		1.90	3.11
8100	1/2" CTS		160	.050		.04	1.95		1.99	3.26
8110	3/4" CTS		145	.055		.05	2.15		2.20	3.61
8200	Half clamp, 1/2" CTS		80	.100		.08	3.91		3.99	6.55
8210	3/4" CTS		73	.110		.12	4.28		4.40	7.20
8300	Suspension clamp, 1/2" CTS		80	.100		.24	3.91		4.15	6.70
8310	3/4" CTS		73	.110		.25	4.28		4.53	7.35
8320	1" CTS		68	.118		.56	4.59		5.15	8.20
8400	Insulator, 1/2" CTS		80	.100		.39	3.91		4.30	6.90
8410	3/4" CTS		73	.110		.40	4.28		4.68	7.50
8420	1" CTS		68	.118		.41	4.59		5	8.05
8500	J hook clamp with nail, 1/2" CTS		240	.033		.13	1.30		1.43	2.29
8501	3/4" CTS	↓	240	.033		.13	1.30		1.43	2.29

22 05 48 – Vibration and Seismic Controls for Plumbing Piping and Equipment

22 05 48.10 Seismic Bracing Supports

		Crew	Daily Output	Labor-Hours	Unit	Material	Labor	Equipment	Total	Total Incl O&P
0010	**SEISMIC BRACING SUPPORTS**									
0020	Clamps									
0030	C-clamp, for mounting on steel beam									
0040	3/8" threaded rod	1 Skwk	160	.050	Ea.	2.07	1.71		3.78	5.15
0050	1/2" threaded rod		160	.050		2.22	1.71		3.93	5.30
0060	5/8" threaded rod		160	.050		3.74	1.71		5.45	7
0070	3/4" threaded rod	↓	160	.050	↓	4.60	1.71		6.31	7.90
0100	Brackets									
0110	Beam side or wall malleable iron									
0120	3/8" threaded rod	1 Skwk	48	.167	Ea.	3.14	5.70		8.84	13.05
0130	1/2" threaded rod		48	.167		4.43	5.70		10.13	14.45
0140	5/8" threaded rod		48	.167		8.30	5.70		14	18.75
0150	3/4" threaded rod		48	.167		11.40	5.70		17.10	22
0160	7/8" threaded rod	↓	48	.167	↓	11.80	5.70		17.50	22.50
0170	For concrete installation, add						30%			
0180	Wall, welded steel									
0190	0 size 12" wide 18" deep	1 Skwk	34	.235	Ea.	173	8.05		181.05	205
0200	1 size 18" wide 24" deep		34	.235		213	8.05		221.05	248
0210	2 size 24" wide 30" deep	↓	34	.235		293	8.05		301.05	335
0300	Rod, carbon steel									
0310	Continuous thread									
0320	1/4" thread	1 Skwk	144	.056	L.F.	1.78	1.90		3.68	5.15
0330	3/8" thread	↓	144	.056		1.90	1.90		3.80	5.30

22 05 Common Work Results for Plumbing

22 05 48 – Vibration and Seismic Controls for Plumbing Piping and Equipment

22 05 48.10 Seismic Bracing Supports		Crew	Daily Output	Labor-Hours	Unit	Material	2016 Bare Costs Labor	Equipment	Total	Total Incl O&P
0340	1/2" thread	1 Skwk	144	.056	L.F.	3	1.90		4.90	6.50
0350	5/8" thread		144	.056		4.26	1.90		6.16	7.90
0360	3/4" thread		144	.056		7.50	1.90		9.40	11.45
0370	7/8" thread	▼	144	.056	▼	9.40	1.90		11.30	13.55
0380	For galvanized, add					30%				
0400	Channel, steel									
0410	3/4" x 1-1/2"	1 Skwk	80	.100	L.F.	2.47	3.42		5.89	8.45
0420	1-1/2" x 1-1/2"		70	.114		3.16	3.91		7.07	10.05
0430	1-7/8" x 1-1/2"		60	.133		23	4.56		27.56	33
0440	3" x 1-1/2"	▼	50	.160	▼	39.50	5.45		44.95	52.50
0450	Spring nuts									
0460	3/8"	1 Skwk	100	.080	Ea.	1.49	2.74		4.23	6.25
0470	1/2"	"	80	.100	"	1.69	3.42		5.11	7.60
0500	Welding, field									
0510	Cleaning and welding plates, bars, or rods									
0520	To existing beams, columns, or trusses									
0530	1" weld	1 Skwk	144	.056	Ea.	.23	1.90		2.13	3.44
0540	2" weld		72	.111		.40	3.80		4.20	6.85
0550	3" weld		54	.148		.62	5.05		5.67	9.20
0560	4" weld		36	.222		.84	7.60		8.44	13.65
0570	5" weld		30	.267		1.07	9.10		10.17	16.50
0580	6" weld	▼	24	.333	▼	1.20	11.40		12.60	20.50
0600	Vibration absorbers									
0610	Hangers, neoprene flex									
0620	10-120 lb. capacity	1 Skwk	8	1	Ea.	27	34		61	87
0630	75-550 lb. capacity		8	1		38	34		72	99.50
0640	250-1100 lb. capacity		6	1.333		77.50	45.50		123	162
0650	1000-4000 lb. capacity	▼	6	1.333	▼	145	45.50		190.50	237

22 05 76 – Facility Drainage Piping Cleanouts

22 05 76.10 Cleanouts

		Crew	Daily Output	Labor-Hours	Unit	Material	Labor	Equipment	Total	Total Incl O&P
0010	**CLEANOUTS**									
0060	Floor type									
0080	Round or square, scoriated nickel bronze top									
0100	2" pipe size	1 Plum	10	.800	Ea.	197	31		228	269
0120	3" pipe size		8	1		295	39		334	390
0140	4" pipe size	▼	6	1.333	▼	295	52		347	410

22 05 76.20 Cleanout Tees

		Crew	Daily Output	Labor-Hours	Unit	Material	Labor	Equipment	Total	Total Incl O&P
0010	**CLEANOUT TEES**									
0100	Cast iron, B&S, with countersunk plug									
0220	3" pipe size	1 Plum	3.60	2.222	Ea.	297	87		384	470
0240	4" pipe size	"	3.30	2.424	"	370	94.50		464.50	560
0500	For round smooth access cover, same price									
4000	Plastic, tees and adapters. Add plugs									
4010	ABS, DWV									
4020	Cleanout tee, 1-1/2" pipe size	1 Plum	15	.533	Ea.	8.65	21		29.65	44

22 07 Plumbing Insulation

22 07 16 – Plumbing Equipment Insulation

22 07 16.10 Insulation for Plumbing Equipment		Crew	Daily Output	Labor-Hours	Unit	Material	2016 Bare Costs Labor	Equipment	Total	Total Incl O&P
0010	**INSULATION FOR PLUMBING EQUIPMENT**									
2900	Domestic water heater wrap kit									
2920	1-1/2" with vinyl jacket, 20-60 gal.	G	1 Plum	8	1	Ea.	15.10	39		54.10	81

22 07 19 – Plumbing Piping Insulation

22 07 19.10 Piping Insulation

			Crew	Daily Output	Labor-Hours	Unit	Material	Labor	Equipment	Total	Total Incl O&P
0010	**PIPING INSULATION**										
0230	Insulated protectors (ADA)										
0235	For exposed piping under sinks or lavatories										
0240	Vinyl coated foam, velcro tabs										
0245	P Trap, 1-1/4" or 1-1/2"		1 Plum	32	.250	Ea.	16.85	9.75		26.60	34.50
0260	Valve and supply cover										
0265	1/2", 3/8", and 7/16" pipe size		1 Plum	32	.250	Ea.	16.45	9.75		26.20	34
0285	1-1/4" pipe size		"	32	.250	"	13.65	9.75		23.40	31
0600	Pipe covering (price copper tube one size less than IPS)										
6600	Fiberglass, with all service jacket										
6840	1" wall, 1/2" iron pipe size	G	Q-14	240	.067	L.F.	.83	2.12		2.95	4.50
6860	3/4" iron pipe size	G		230	.070		.90	2.21		3.11	4.74
6870	1" iron pipe size	G		220	.073		.97	2.31		3.28	4.99
6900	2" iron pipe size	G		200	.080		1.22	2.54		3.76	5.65
7879	Rubber tubing, flexible closed cell foam										
8100	1/2" wall, 1/4" iron pipe size	G	1 Asbe	90	.089	L.F.	.54	3.13		3.67	5.90
8130	1/2" iron pipe size	G		89	.090		.67	3.17		3.84	6.15
8140	3/4" iron pipe size	G		89	.090		.75	3.17		3.92	6.25
8150	1" iron pipe size	G		88	.091		.82	3.20		4.02	6.35
8170	1-1/2" iron pipe size	G		87	.092		1.15	3.24		4.39	6.75
8180	2" iron pipe size	G		86	.093		1.47	3.28		4.75	7.15
8300	3/4" wall, 1/4" iron pipe size	G		90	.089		.85	3.13		3.98	6.25
8330	1/2" iron pipe size	G		89	.090		1.10	3.17		4.27	6.60
8340	3/4" iron pipe size	G		89	.090		1.35	3.17		4.52	6.90
8350	1" iron pipe size	G		88	.091		1.54	3.20		4.74	7.15
8380	2" iron pipe size	G		86	.093		2.68	3.28		5.96	8.50
8444	1" wall, 1/2" iron pipe size	G		86	.093		2.05	3.28		5.33	7.80
8445	3/4" iron pipe size	G		84	.095		2.48	3.36		5.84	8.45
8446	1" iron pipe size	G		84	.095		2.89	3.36		6.25	8.90
8447	1-1/4" iron pipe size	G		82	.098		3.23	3.44		6.67	9.40
8448	1-1/2" iron pipe size	G		82	.098		3.76	3.44		7.20	10
8449	2" iron pipe size	G		80	.100		4.95	3.53		8.48	11.45
8450	2-1/2" iron pipe size	G		80	.100		6.45	3.53		9.98	13.10
8456	Rubber insulation tape, 1/8" x 2" x 30'	G				Ea.	21			21	23

22 11 Facility Water Distribution

22 11 13 – Facility Water Distribution Piping

22 11 13.23 Pipe/Tube, Copper

		Crew	Daily Output	Labor-Hours	Unit	Material	Labor	Equipment	Total	Total Incl O&P
0010	**PIPE/TUBE, COPPER**, Solder joints									
1000	Type K tubing, couplings & clevis hanger assemblies 10' O.C.									
1180	3/4" diameter	1 Plum	74	.108	L.F.	8.15	4.22		12.37	15.90
1200	1" diameter	"	66	.121	"	10.60	4.73		15.33	19.45
2000	Type L tubing, couplings & clevis hanger assemblies 10' O.C.									
2140	1/2" diameter	1 Plum	81	.099	L.F.	3.68	3.86		7.54	10.40
2160	5/8" diameter		79	.101		5.30	3.95		9.25	12.35

22 11 Facility Water Distribution

22 11 13 – Facility Water Distribution Piping

22 11 13.23 Pipe/Tube, Copper

		Crew	Daily Output	Labor-Hours	Unit	Material	2016 Bare Costs Labor	Equipment	Total	Total Incl O&P
2180	3/4" diameter	1 Plum	76	.105	L.F.	5.25	4.11		9.36	12.55
2200	1" diameter		68	.118		7.40	4.59		11.99	15.75
2220	1-1/4" diameter	↓	58	.138	↓	10.05	5.40		15.45	19.95
3000	Type M tubing, couplings & clevis hanger assemblies 10' O.C.									
3140	1/2" diameter	1 Plum	84	.095	L.F.	3.21	3.72		6.93	9.70
3180	3/4" diameter		78	.103		4.39	4.01		8.40	11.40
3200	1" diameter		70	.114		6.60	4.46		11.06	14.60
3220	1-1/4" diameter		60	.133		9.15	5.20		14.35	18.70
3240	1-1/2" diameter		54	.148		12.10	5.80		17.90	23
3260	2" diameter	↓	44	.182	↓	18.15	7.10		25.25	31.50
4000	Type DWV tubing, couplings & clevis hanger assemblies 10' O.C.									
4100	1-1/4" diameter	1 Plum	60	.133	L.F.	9.20	5.20		14.40	18.70
4120	1-1/2" diameter		54	.148		11.30	5.80		17.10	22
4140	2" diameter	↓	44	.182		14.75	7.10		21.85	28
4160	3" diameter	Q-1	58	.276		26.50	9.70		36.20	45
4180	4" diameter	"	40	.400	↓	42.50	14.05		56.55	69.50

22 11 13.25 Pipe/Tube Fittings, Copper

		Crew	Daily Output	Labor-Hours	Unit	Material	2016 Bare Costs Labor	Equipment	Total	Total Incl O&P
0010	**PIPE/TUBE FITTINGS, COPPER**, Wrought unless otherwise noted									
0040	Solder joints, copper x copper									
0070	90° elbow, 1/4"	1 Plum	22	.364	Ea.	3.28	14.20		17.48	27
0100	1/2"		20	.400		1.13	15.60		16.73	27
0120	3/4"		19	.421		2.53	16.45		18.98	30
0250	45° elbow, 1/4"		22	.364		5.85	14.20		20.05	30
0280	1/2"		20	.400		2.06	15.60		17.66	28.50
0290	5/8"		19	.421		8.75	16.45		25.20	36.50
0300	3/4"		19	.421		3.52	16.45		19.97	31
0310	1"		16	.500		8.85	19.55		28.40	41.50
0320	1-1/4"		15	.533		11.85	21		32.85	47.50
0450	Tee, 1/4"		14	.571		6.55	22.50		29.05	44
0480	1/2"		13	.615		1.92	24		25.92	41.50
0490	5/8"		12	.667		12.05	26		38.05	56.50
0500	3/4"		12	.667		4.64	26		30.64	48
0510	1"		10	.800		13.95	31		44.95	67
0520	1-1/4"		9	.889		19.25	34.50		53.75	78
0612	Tee, reducing on the outlet, 1/4"		15	.533		10.75	21		31.75	46.50
0613	3/8"		15	.533		10.40	21		31.40	46
0614	1/2"		14	.571		9.65	22.50		32.15	47.50
0615	5/8"		13	.615		19.40	24		43.40	61
0616	3/4"		12	.667		6.45	26		32.45	50
0617	1"		11	.727		21	28.50		49.50	70
0618	1-1/4"		10	.800		24	31		55	78
0619	1-1/2"		9	.889		25	34.50		59.50	84.50
0620	2"	↓	8	1		40	39		79	109
0621	2-1/2"	Q-1	9	1.778		99	62.50		161.50	212
0622	3"		8	2		103	70.50		173.50	229
0623	4"		6	2.667		220	93.50		313.50	395
0624	5"	↓	5	3.200		1,125	112		1,237	1,425
0625	6"	Q-2	7	3.429		1,550	116		1,666	1,900
0626	8"	"	6	4		5,950	135		6,085	6,775
0630	Tee, reducing on the run, 1/4"	1 Plum	15	.533		13.90	21		34.90	50
0631	3/8"		15	.533		19.60	21		40.60	56
0632	1/2"	↓	14	.571		13.25	22.50		35.75	51.50

22 11 Facility Water Distribution

22 11 13 – Facility Water Distribution Piping

22 11 13.25 Pipe/Tube Fittings, Copper

		Crew	Daily Output	Labor-Hours	Unit	Material	2016 Bare Costs Labor	Equipment	Total	Total Incl O&P
0633	5/8"	1 Plum	13	.615	Ea.	19.55	24		43.55	61
0634	3/4"		12	.667		14.30	26		40.30	58.50
0635	1"		11	.727		16.95	28.50		45.45	65.50
0636	1-1/4"		10	.800		27.50	31		58.50	81.50
0637	1-1/2"		9	.889		47	34.50		81.50	109
0638	2"		8	1		60	39		99	131
0639	2-1/2"	Q-1	9	1.778		128	62.50		190.50	244
0640	3"		8	2		187	70.50		257.50	320
0641	4"		6	2.667		380	93.50		473.50	575
0642	5"		5	3.200		1,075	112		1,187	1,350
0643	6"	Q-2	7	3.429		1,625	116		1,741	2,000
0644	8"	"	6	4		5,950	135		6,085	6,775
0650	Coupling, 1/4"	1 Plum	24	.333		.77	13		13.77	22.50
0680	1/2"		22	.364		.85	14.20		15.05	24.50
0690	5/8"		21	.381		2.50	14.90		17.40	27.50
0700	3/4"		21	.381		1.72	14.90		16.62	26.50
0710	1"		18	.444		3.39	17.35		20.74	32
0715	1-1/4"		17	.471		5.85	18.40		24.25	37
2000	DWV, solder joints, copper x copper									
2030	90° Elbow, 1-1/4"	1 Plum	13	.615	Ea.	12.05	24		36.05	53
2050	1-1/2"		12	.667		16	26		42	60.50
2070	2"		10	.800		25.50	31		56.50	79.50
2090	3"	Q-1	10	1.600		62	56		118	161
2100	4"	"	9	1.778		300	62.50		362.50	435
2250	Tee, Sanitary, 1-1/4"	1 Plum	9	.889		21	34.50		55.50	80
2270	1-1/2"		8	1		26.50	39		65.50	93.50
2290	2"		7	1.143		36	44.50		80.50	113
2310	3"	Q-1	7	2.286		138	80.50		218.50	284
2330	4"	"	6	2.667		330	93.50		423.50	520
2400	Coupling, 1-1/4"	1 Plum	14	.571		5	22.50		27.50	42.50
2420	1-1/2"		13	.615		6.20	24		30.20	46.50
2440	2"		11	.727		8.65	28.50		37.15	56.50
2460	3"	Q-1	11	1.455		20	51		71	107
2480	4"	"	10	1.600		44	56		100	142

22 11 13.44 Pipe, Steel

		Crew	Daily Output	Labor-Hours	Unit	Material	Labor	Equipment	Total	Total Incl O&P
0010	**PIPE, STEEL**									
0050	Schedule 40, threaded, with couplings, and clevis hanger									
0060	assemblies sized for covering, 10' O.C.									
0540	Black, 1/4" diameter	1 Plum	66	.121	L.F.	4.96	4.73		9.69	13.25
0560	1/2" diameter		63	.127		2.90	4.96		7.86	11.40
0570	3/4" diameter		61	.131		3.35	5.10		8.45	12.15
0580	1" diameter		53	.151		4.30	5.90		10.20	14.45
0590	1-1/4" diameter	Q-1	89	.180		5.20	6.30		11.50	16.10
0600	1-1/2" diameter		80	.200		5.90	7.05		12.95	18.05
0610	2" diameter		64	.250		7.45	8.80		16.25	22.50

22 11 13.45 Pipe Fittings, Steel, Threaded

		Crew	Daily Output	Labor-Hours	Unit	Material	Labor	Equipment	Total	Total Incl O&P
0010	**PIPE FITTINGS, STEEL, THREADED**									
5000	Malleable iron, 150 lb.									
5020	Black									
5040	90° elbow, straight									
5090	3/4"	1 Plum	14	.571	Ea.	4.17	22.50		26.67	41.50
5100	1"	"	13	.615		7.25	24		31.25	47.50

22 11 Facility Water Distribution

22 11 13 – Facility Water Distribution Piping

22 11 13.45 Pipe Fittings, Steel, Threaded

		Crew	Daily Output	Labor-Hours	Unit	Material	2016 Bare Costs Labor	Equipment	Total	Total Incl O&P
5120	1-1/2"	Q-1	20	.800	Ea.	15.70	28		43.70	64
5130	2"	"	18	.889		28	31		59	82
5450	Tee, straight									
5500	3/4"	1 Plum	9	.889	Ea.	6.65	34.50		41.15	64.50
5510	1"	"	8	1		11.35	39		50.35	77
5520	1-1/4"	Q-1	14	1.143		18.40	40		58.40	86.50
5530	1-1/2"		13	1.231		23	43.50		66.50	96.50
5540	2"		11	1.455		39	51		90	128
5650	Coupling									
5700	3/4"	1 Plum	18	.444	Ea.	5.60	17.35		22.95	34.50
5710	1"	"	15	.533		8.40	21		29.40	43.50
5720	1-1/4"	Q-1	26	.615		10.85	21.50		32.35	47.50
5730	1-1/2"		24	.667		14.65	23.50		38.15	54.50
5740	2"		21	.762		21.50	27		48.50	68

22 11 13.74 Pipe, Plastic

		Crew	Daily Output	Labor-Hours	Unit	Material	2016 Bare Costs Labor	Equipment	Total	Total Incl O&P
0010	**PIPE, PLASTIC**									
1800	PVC, couplings 10' O.C., clevis hanger assemblies, 3 per 10'									
1820	Schedule 40									
1860	1/2" diameter	1 Plum	54	.148	L.F.	4.96	5.80		10.76	15
1870	3/4" diameter		51	.157		5.30	6.15		11.45	15.95
1880	1" diameter		46	.174		5.95	6.80		12.75	17.75
1890	1-1/4" diameter		42	.190		6.70	7.45		14.15	19.65
1900	1-1/2" diameter		36	.222		7	8.70		15.70	22
1910	2" diameter	Q-1	59	.271		8.25	9.55		17.80	25
1920	2-1/2" diameter		56	.286		10.65	10.05		20.70	28.50
1930	3" diameter		53	.302		13	10.60		23.60	32
1940	4" diameter		48	.333		16.60	11.70		28.30	37.50
4100	DWV type, schedule 40, couplings 10' O.C., clevis hanger assy's, 3 per 10'									
4210	ABS, schedule 40, foam core type									
4212	Plain end black									
4214	1-1/2" diameter	1 Plum	39	.205	L.F.	5.30	8		13.30	19
4216	2" diameter	Q-1	62	.258		5.70	9.05		14.75	21
4218	3" diameter		56	.286		8.15	10.05		18.20	25.50
4220	4" diameter		51	.314		10.45	11.05		21.50	29.50
4222	6" diameter		42	.381		18.80	13.40		32.20	42.50
4240	To delete coupling & hangers, subtract									
4244	1-1/2" diam. to 6" diam.					43%	48%			
4400	PVC									
4410	1-1/4" diameter	1 Plum	42	.190	L.F.	5.50	7.45		12.95	18.30
4420	1-1/2" diameter	"	36	.222		5.35	8.70		14.05	20
4460	2" diameter	Q-1	59	.271		5.70	9.55		15.25	22
4470	3" diameter		53	.302		8.20	10.60		18.80	26.50
4480	4" diameter		48	.333		10.25	11.70		21.95	30.50
5300	CPVC, socket joint, couplings 10' O.C., clevis hanger assemblies, 3 per 10'									
5302	Schedule 40									
5304	1/2" diameter	1 Plum	54	.148	L.F.	6.05	5.80		11.85	16.20
5305	3/4" diameter		51	.157		6.95	6.15		13.10	17.75
5306	1" diameter		46	.174		8.35	6.80		15.15	20.50
5307	1-1/4" diameter		42	.190		10	7.45		17.45	23.50
5308	1-1/2" diameter		36	.222		11.20	8.70		19.90	26.50
5309	2" diameter	Q-1	59	.271		13.30	9.55		22.85	30.50
5360	CPVC, threaded, couplings 10' O.C., clevis hanger assemblies, 3 per 10'									

22 11 Facility Water Distribution

22 11 13 – Facility Water Distribution Piping

22 11 13.74 Pipe, Plastic

		Crew	Daily Output	Labor-Hours	Unit	Material	2016 Bare Costs Labor	Equipment	Total	Total Incl O&P
5380	Schedule 40									
5460	1/2" diameter	1 Plum	54	.148	L.F.	6.90	5.80		12.70	17.15
5470	3/4" diameter		51	.157		8.35	6.15		14.50	19.25
5480	1" diameter		46	.174		9.85	6.80		16.65	22
5490	1-1/4" diameter		42	.190		11.15	7.45		18.60	24.50
5500	1-1/2" diameter		36	.222		12.20	8.70		20.90	28
5510	2" diameter	Q-1	59	.271		14.55	9.55		24.10	31.50
6500	Residential installation, plastic pipe									
6510	Couplings 10' O.C., strap hangers 3 per 10'									
6520	PVC, Schedule 40									
6530	1/2" diameter	1 Plum	138	.058	L.F.	1.05	2.26		3.31	4.89
6540	3/4" diameter		128	.063		1.26	2.44		3.70	5.40
6550	1" diameter		119	.067		1.86	2.63		4.49	6.40
6560	1-1/4" diameter		111	.072		2.27	2.81		5.08	7.15
6570	1-1/2" diameter		104	.077		2.51	3		5.51	7.70
6580	2" diameter	Q-1	197	.081		3.20	2.85		6.05	8.25
6590	2-1/2" diameter		162	.099		5.25	3.47		8.72	11.55
6600	4" diameter		123	.130		8.90	4.57		13.47	17.30
6700	PVC, DWV, Schedule 40									
6720	1-1/4" diameter	1 Plum	100	.080	L.F.	1.93	3.12		5.05	7.30
6730	1-1/2" diameter	"	94	.085		1.79	3.32		5.11	7.45
6740	2" diameter	Q-1	178	.090		2.12	3.16		5.28	7.55
6760	4" diameter	"	110	.145		5.85	5.10		10.95	14.90
7280	PEX, flexible, no couplings or hangers									
7282	Note: For labor costs add 25% to the couplings and fittings labor total.									
7285	For fittings see section 23 83 16.10 7000									
7300	Non-barrier type, hot/cold tubing rolls									
7310	1/4" diameter x 100'				L.F.	.49			.49	.54
7350	3/8" diameter x 100'					.55			.55	.61
7360	1/2" diameter x 100'					.61			.61	.67
7370	1/2" diameter x 500'					.61			.61	.67
7380	1/2" diameter x 1000'					.61			.61	.67
7400	3/4" diameter x 100'					1.11			1.11	1.22
7410	3/4" diameter x 500'					1.11			1.11	1.22
7420	3/4" diameter x 1000'					1.11			1.11	1.22
7460	1" diameter x 100'					1.90			1.90	2.09
7470	1" diameter x 300'					1.90			1.90	2.09
7480	1" diameter x 500'					1.90			1.90	2.09
7500	1-1/4" diameter x 100'					3.23			3.23	3.55
7510	1-1/4" diameter x 300'					3.23			3.23	3.55
7540	1-1/2" diameter x 100'					4.40			4.40	4.84
7550	1-1/2" diameter x 300'					4.40			4.40	4.84
7596	Most sizes available in red or blue									
7700	Non-barrier type, hot/cold tubing straight lengths									
7710	1/2" diameter x 20'				L.F.	.61			.61	.67
7750	3/4" diameter x 20'					1.11			1.11	1.22
7760	1" diameter x 20'					1.90			1.90	2.09
7770	1-1/4" diameter x 20'					3.23			3.23	3.55
7780	1-1/2" diameter x 20'					4.40			4.40	4.84
7790	2" diameter					8.60			8.60	9.45
7796	Most sizes available in red or blue									

22 11 Facility Water Distribution

22 11 13 – Facility Water Distribution Piping

22 11 13.76 Pipe Fittings, Plastic

		Crew	Daily Output	Labor-Hours	Unit	Material	2016 Bare Costs Labor	Equipment	Total	Total Incl O&P
0010	**PIPE FITTINGS, PLASTIC**									
2700	PVC (white), schedule 40, socket joints									
2760	90° elbow, 1/2"	1 Plum	33.30	.240	Ea.	.50	9.40		9.90	16
2770	3/4"		28.60	.280		.57	10.90		11.47	18.65
2780	1"		25	.320		1.01	12.50		13.51	21.50
2790	1-1/4"		22.20	.360		1.79	14.05		15.84	25
2800	1-1/2"		20	.400		1.94	15.60		17.54	28
2810	2"	Q-1	36.40	.440		3.03	15.45		18.48	29
2820	2-1/2"		26.70	.599		9.25	21		30.25	44.50
2830	3"		22.90	.699		11.05	24.50		35.55	52.50
2840	4"		18.20	.879		19.75	31		50.75	72.50
3180	Tee, 1/2"	1 Plum	22.20	.360		.63	14.05		14.68	23.50
3190	3/4"		19	.421		.73	16.45		17.18	28
3200	1"		16.70	.479		1.36	18.70		20.06	32.50
3210	1-1/4"		14.80	.541		2.11	21		23.11	37.50
3220	1-1/2"		13.30	.602		2.57	23.50		26.07	41.50
3230	2"	Q-1	24.20	.661		3.74	23		26.74	42.50
3240	2-1/2"		17.80	.899		12.35	31.50		43.85	65.50
3250	3"		15.20	1.053		16.20	37		53.20	79
3260	4"		12.10	1.322		29.50	46.50		76	109
3380	Coupling, 1/2"	1 Plum	33.30	.240		.33	9.40		9.73	15.80
3390	3/4"		28.60	.280		.46	10.90		11.36	18.50
3400	1"		25	.320		.80	12.50		13.30	21.50
3410	1-1/4"		22.20	.360		1.11	14.05		15.16	24
3420	1-1/2"		20	.400		1.19	15.60		16.79	27.50
3430	2"	Q-1	36.40	.440		1.80	15.45		17.25	27.50
3440	2-1/2"		26.70	.599		4	21		25	39
3450	3"		22.90	.699		6.25	24.50		30.75	47.50
3460	4"		18.20	.879		9.10	31		40.10	61
4500	DWV, ABS, non pressure, socket joints									
4540	1/4 Bend, 1-1/4"	1 Plum	20.20	.396	Ea.	3.17	15.45		18.62	29
4560	1-1/2"	"	18.20	.440		2.36	17.15		19.51	31
4570	2"	Q-1	33.10	.483		3.75	17		20.75	32
4650	1/8 Bend, same as 1/4 Bend									
4800	Tee, sanitary									
4820	1-1/4"	1 Plum	13.50	.593	Ea.	4.21	23		27.21	42.50
4830	1-1/2"	"	12.10	.661		3.60	26		29.60	46.50
4840	2"	Q-1	20	.800		5.55	28		33.55	52.50
5000	DWV, PVC, schedule 40, socket joints									
5040	1/4 bend, 1-1/4"	1 Plum	20.20	.396	Ea.	6.40	15.45		21.85	32.50
5060	1-1/2"	"	18.20	.440		1.82	17.15		18.97	30.50
5070	2"	Q-1	33.10	.483		2.87	17		19.87	31
5080	3"		20.80	.769		8.45	27		35.45	54
5090	4"		16.50	.970		16.60	34		50.60	74.50
5110	1/4 bend, long sweep, 1-1/2"	1 Plum	18.20	.440		4.21	17.15		21.36	33
5112	2"	Q-1	33.10	.483		4.70	17		21.70	33
5114	3"		20.80	.769		10.75	27		37.75	56.50
5116	4"		16.50	.970		20.50	34		54.50	78.50
5250	Tee, sanitary 1-1/4"	1 Plum	13.50	.593		6.65	23		29.65	45.50
5254	1-1/2"	"	12.10	.661		3.17	26		29.17	46
5255	2"	Q-1	20	.800		4.67	28		32.67	51.50
5256	3"		13.90	1.151		12.25	40.50		52.75	80

22 11 Facility Water Distribution

22 11 13 – Facility Water Distribution Piping

22 11 13.76 Pipe Fittings, Plastic

		Crew	Daily Output	Labor-Hours	Unit	Material	2016 Bare Costs Labor	Equipment	Total	Total Incl O&P
5257	4"	Q-1	11	1.455	Ea.	22.50	51		73.50	110
5259	6"	↓	6.70	2.388		91	84		175	238
5261	8"	Q-2	6.20	3.871		199	131		330	435
5264	2" x 1-1/2"	Q-1	22	.727		4.13	25.50		29.63	46.50
5266	3" x 1-1/2"		15.50	1.032		8.95	36.50		45.45	70
5268	4" x 3"		12.10	1.322		26.50	46.50		73	106
5271	6" x 4"	↓	6.90	2.319		88	81.50		169.50	231
5314	Combination Y & 1/8 bend, 1-1/2"	1 Plum	12.10	.661		7.70	26		33.70	51
5315	2"	Q-1	20	.800		8.20	28		36.20	55.50
5317	3"		13.90	1.151		21	40.50		61.50	89.50
5318	4"	↓	11	1.455	↓	42	51		93	131
5324	Combination Y & 1/8 bend, reducing									
5325	2" x 2" x 1-1/2"	Q-1	22	.727	Ea.	10.80	25.50		36.30	54
5327	3" x 3" x 1-1/2"		15.50	1.032		19.25	36.50		55.75	81
5328	3" x 3" x 2"		15.30	1.046		14.40	37		51.40	76.50
5329	4" x 4" x 2"	↓	12.20	1.311		22	46		68	100
5331	Wye, 1-1/4"	1 Plum	13.50	.593		8.50	23		31.50	47.50
5332	1-1/2"	"	12.10	.661		5.80	26		31.80	49
5333	2"	Q-1	20	.800		5.70	28		33.70	53
5334	3"		13.90	1.151		15.35	40.50		55.85	83.50
5335	4"		11	1.455		28	51		79	115
5336	6"	↓	6.70	2.388		81.50	84		165.50	228
5337	8"	Q-2	6.20	3.871		144	131		275	375
5341	2" x 1-1/2"	Q-1	22	.727		7	25.50		32.50	49.50
5342	3" x 1-1/2"		15.50	1.032		10.40	36.50		46.90	71.50
5343	4" x 3"		12.10	1.322		22.50	46.50		69	102
5344	6" x 4"	↓	6.90	2.319		62	81.50		143.50	202
5345	8" x 6"	Q-2	6.40	3.750		135	127		262	360
5347	Double wye, 1-1/2"	1 Plum	9.10	.879		13	34.50		47.50	71
5348	2"	Q-1	16.60	.964		14.50	34		48.50	72
5349	3"		10.40	1.538		30.50	54		84.50	123
5350	4"	↓	8.25	1.939	↓	62	68		130	180
5353	Double wye, reducing									
5354	2" x 2" x 1-1/2" x 1-1/2"	Q-1	16.80	.952	Ea.	13.25	33.50		46.75	69.50
5355	3" x 3" x 2" x 2"		10.60	1.509		22.50	53		75.50	113
5356	4" x 4" x 3" x 3"		8.45	1.893		49	66.50		115.50	164
5357	6" x 6" x 4" x 4"	↓	7.25	2.207		170	77.50		247.50	315
5374	Coupling, 1-1/4"	1 Plum	20.20	.396		4	15.45		19.45	30
5376	1-1/2"	"	18.20	.440		.85	17.15		18	29.50
5378	2"	Q-1	33.10	.483		1.17	17		18.17	29.50
5380	3"		20.80	.769		4.05	27		31.05	49
5390	4"		16.50	.970		6.95	34		40.95	63.50
5410	Reducer bushing, 2" x 1-1/4"		36.50	.438		3.35	15.40		18.75	29
5412	3" x 1-1/2"		27.30	.586		7.20	20.50		27.70	42
5414	4" x 2"		18.20	.879		12.55	31		43.55	65
5416	6" x 4"	↓	11.10	1.441		32	50.50		82.50	119
5418	8" x 6"	Q-2	10.20	2.353	↓	63	79.50		142.50	201
5500	CPVC, Schedule 80, threaded joints									
5540	90° Elbow, 1/4"	1 Plum	32	.250	Ea.	12.75	9.75		22.50	30
5560	1/2"		30.30	.264		7.40	10.30		17.70	25
5570	3/4"		26	.308		11.05	12		23.05	32
5580	1"		22.70	.352		15.55	13.75		29.30	39.50
5590	1-1/4"	↓	20.20	.396		30	15.45		45.45	58.50

22 11 Facility Water Distribution

22 11 13 – Facility Water Distribution Piping

22 11 13.76 Pipe Fittings, Plastic

		Crew	Daily Output	Labor-Hours	Unit	Material	2016 Bare Costs Labor	Equipment	Total	Total Incl O&P
5600	1-1/2"	1 Plum	18.20	.440	Ea.	32.50	17.15		49.65	64
5610	2"	Q-1	33.10	.483		43	17		60	75.50
5730	Coupling, 1/4"	1 Plum	32	.250		16.25	9.75		26	34
5732	1/2"		30.30	.264		13.35	10.30		23.65	31.50
5734	3/4"		26	.308		21	12		33	43
5736	1"		22.70	.352		24.50	13.75		38.25	49.50
5738	1-1/4"		20.20	.396		26	15.45		41.45	54
5740	1-1/2"	▼	18.20	.440		28	17.15		45.15	59
5742	2"	Q-1	33.10	.483	▼	33	17		50	64
5900	CPVC, Schedule 80, socket joints									
5904	90° Elbow, 1/4"	1 Plum	32	.250	Ea.	12.15	9.75		21.90	29.50
5906	1/2"		30.30	.264		4.76	10.30		15.06	22.50
5908	3/4"		26	.308		6.10	12		18.10	26.50
5910	1"		22.70	.352		9.65	13.75		23.40	33
5912	1-1/4"		20.20	.396		21	15.45		36.45	48.50
5914	1-1/2"	▼	18.20	.440		23.50	17.15		40.65	54
5916	2"	Q-1	33.10	.483		28	17		45	59
5930	45° Elbow, 1/4"	1 Plum	32	.250		18.05	9.75		27.80	36
5932	1/2"		30.30	.264		5.80	10.30		16.10	23.50
5934	3/4"		26	.308		8.35	12		20.35	29
5936	1"		22.70	.352		13.35	13.75		27.10	37
5938	1-1/4"		20.20	.396		26.50	15.45		41.95	54.50
5940	1-1/2"	▼	18.20	.440		27	17.15		44.15	58
5942	2"	Q-1	33.10	.483		30.50	17		47.50	61.50
5990	Coupling, 1/4"	1 Plum	32	.250		12.90	9.75		22.65	30.50
5992	1/2"		30.30	.264		5.05	10.30		15.35	22.50
5994	3/4"		26	.308		7.05	12		19.05	27.50
5996	1"		22.70	.352		9.45	13.75		23.20	33
5998	1-1/4"		20.20	.396		14.15	15.45		29.60	41
6000	1-1/2"	▼	18.20	.440		17.85	17.15		35	48
6002	2"	Q-1	33.10	.483	▼	20.50	17		37.50	51

22 11 19 – Domestic Water Piping Specialties

22 11 19.38 Water Supply Meters

		Crew	Daily Output	Labor-Hours	Unit	Material	Labor	Equipment	Total	Total Incl O&P
0010	**WATER SUPPLY METERS**									
2000	Domestic/commercial, bronze									
2020	Threaded									
2060	5/8" diameter, to 20 GPM	1 Plum	16	.500	Ea.	50	19.55		69.55	87
2080	3/4" diameter, to 30 GPM		14	.571		91	22.50		113.50	137
2100	1" diameter, to 50 GPM	▼	12	.667	▼	138	26		164	195

22 11 19.42 Backflow Preventers

		Crew	Daily Output	Labor-Hours	Unit	Material	Labor	Equipment	Total	Total Incl O&P
0010	**BACKFLOW PREVENTERS**, Includes valves									
0020	and four test cocks, corrosion resistant, automatic operation									
4000	Reduced pressure principle									
4100	Threaded, bronze, valves are ball									
4120	3/4" pipe size	1 Plum	16	.500	Ea.	450	19.55		469.55	525

22 11 19.50 Vacuum Breakers

		Crew	Daily Output	Labor-Hours	Unit	Material	Labor	Equipment	Total	Total Incl O&P
0010	**VACUUM BREAKERS**									
0013	See also backflow preventers Section 22 11 19.42									
1000	Anti-siphon continuous pressure type									
1010	Max. 150 PSI - 210°F									
1020	Bronze body									
1030	1/2" size	1 Stpi	24	.333	Ea.	161	13.35		174.35	199

22 11 Facility Water Distribution

22 11 19 – Domestic Water Piping Specialties

22 11 19.50 Vacuum Breakers

		Crew	Daily Output	Labor-Hours	Unit	Material	2016 Bare Costs Labor	Equipment	Total	Total Incl O&P
1040	3/4" size	1 Stpi	20	.400	Ea.	161	16		177	204
1050	1" size		19	.421		166	16.85		182.85	211
1060	1-1/4" size		15	.533		325	21.50		346.50	395
1070	1-1/2" size		13	.615		405	24.50		429.50	485
1080	2" size		11	.727		415	29		444	505
1200	Max. 125 PSI with atmospheric vent									
1210	Brass, in-line construction									
1220	1/4" size	1 Stpi	24	.333	Ea.	126	13.35		139.35	161
1230	3/8" size	"	24	.333		126	13.35		139.35	161
1260	For polished chrome finish, add					13%				
2000	Anti-siphon, non-continuous pressure type									
2010	Hot or cold water 125 PSI - 210°F									
2020	Bronze body									
2030	1/4" size	1 Stpi	24	.333	Ea.	69	13.35		82.35	98
2040	3/8" size		24	.333		69	13.35		82.35	98
2050	1/2" size		24	.333		78	13.35		91.35	108
2060	3/4" size		20	.400		93	16		109	129
2070	1" size		19	.421		144	16.85		160.85	186
2080	1-1/4" size		15	.533		252	21.50		273.50	310
2090	1-1/2" size		13	.615		296	24.50		320.50	365
2100	2" size		11	.727		460	29		489	555
2110	2-1/2" size		8	1		1,325	40		1,365	1,525
2120	3" size		6	1.333		1,750	53.50		1,803.50	2,025
2150	For polished chrome finish, add					50%				

22 11 19.54 Water Hammer Arresters/Shock Absorbers

		Crew	Daily Output	Labor-Hours	Unit	Material	Labor	Equipment	Total	Total Incl O&P
0010	**WATER HAMMER ARRESTERS/SHOCK ABSORBERS**									
0490	Copper									
0500	3/4" male I.P.S. For 1 to 11 fixtures	1 Plum	12	.667	Ea.	28	26		54	74

22 13 Facility Sanitary Sewerage

22 13 16 – Sanitary Waste and Vent Piping

22 13 16.20 Pipe, Cast Iron

		Crew	Daily Output	Labor-Hours	Unit	Material	Labor	Equipment	Total	Total Incl O&P
0010	**PIPE, CAST IRON**, Soil, on clevis hanger assemblies, 5' O.C. R221113-50									
0020	Single hub, service wt., lead & oakum joints 10' O.C.									
2120	2" diameter	Q-1	63	.254	L.F.	11.70	8.95		20.65	27.50
2140	3" diameter		60	.267		15.45	9.35		24.80	32.50
2160	4" diameter		55	.291		19.65	10.25		29.90	38.50
4000	No hub, couplings 10' O.C.									
4100	1-1/2" diameter	Q-1	71	.225	L.F.	11.50	7.90		19.40	25.50
4120	2" diameter		67	.239		12.05	8.40		20.45	27
4140	3" diameter		64	.250		15.45	8.80		24.25	31.50
4160	4" diameter		58	.276		19.70	9.70		29.40	37.50

22 13 16.30 Pipe Fittings, Cast Iron

		Crew	Daily Output	Labor-Hours	Unit	Material	Labor	Equipment	Total	Total Incl O&P
0010	**PIPE FITTINGS, CAST IRON**, Soil									
0040	Hub and spigot, service weight, lead & oakum joints									
0080	1/4 bend, 2"	Q-1	16	1	Ea.	20.50	35		55.50	80.50
0120	3"		14	1.143		27.50	40		67.50	96.50
0140	4"		13	1.231		43	43.50		86.50	119
0340	1/8 bend, 2"		16	1		14.65	35		49.65	74
0350	3"		14	1.143		23	40		63	91.50

22 13 Facility Sanitary Sewerage

22 13 16 – Sanitary Waste and Vent Piping

22 13 16.30 Pipe Fittings, Cast Iron

		Crew	Daily Output	Labor-Hours	Unit	Material	2016 Bare Costs Labor	Equipment	Total	Total Incl O&P
0360	4"	Q-1	13	1.231	Ea.	33.50	43.50		77	109
0500	Sanitary tee, 2"		10	1.600		28.50	56		84.50	125
0540	3"		9	1.778		46.50	62.50		109	154
0620	4"	↓	8	2	↓	57	70.50		127.50	179
5990	No hub									
6000	Cplg. & labor required at joints not incl. in fitting									
6010	price. Add 1 coupling per joint for installed price									
6020	1/4 Bend, 1-1/2"				Ea.	10.55			10.55	11.65
6060	2"					11.55			11.55	12.70
6080	3"					16.10			16.10	17.70
6120	4"					24			24	26
6184	1/4 Bend, long sweep, 1-1/2"					27			27	29.50
6186	2"					25			25	27.50
6188	3"					30.50			30.50	33.50
6189	4"					48.50			48.50	53.50
6190	5"					92.50			92.50	102
6191	6"					107			107	118
6192	8"					291			291	320
6193	10"					585			585	645
6200	1/8 Bend, 1-1/2"					8.90			8.90	9.80
6210	2"					9.95			9.95	10.95
6212	3"					13.30			13.30	14.65
6214	4"					17.45			17.45	19.15
6380	Sanitary tee, tapped, 1-1/2"					21			21	23
6382	2" x 1-1/2"					18.60			18.60	20.50
6384	2"					19.95			19.95	22
6386	3" x 2"					29.50			29.50	32.50
6388	3"					51			51	56.50
6390	4" x 1-1/2"					26.50			26.50	29
6392	4" x 2"					30			30	33
6393	4"					30			30	33
6394	6" x 1-1/2"					68.50			68.50	75.50
6396	6" x 2"					70			70	77
6459	Sanitary tee, 1-1/2"					14.85			14.85	16.30
6460	2"					15.90			15.90	17.50
6470	3"					19.60			19.60	21.50
6472	4"				↓	37			37	41
8000	Coupling, standard (by CISPI Mfrs.)									
8020	1-1/2"	Q-1	48	.333	Ea.	13.95	11.70		25.65	34.50
8040	2"		44	.364		15.25	12.80		28.05	38
8080	3"		38	.421		17.10	14.80		31.90	43.50
8120	4"		33	.485		19.90	17.05		36.95	50

22 13 16.50 Shower Drains

		Crew	Daily Output	Labor-Hours	Unit	Material	Labor	Equipment	Total	Total Incl O&P
0010	**SHOWER DRAINS**									
2780	Shower, with strainer, uniform diam. trap, bronze top									
2800	2" and 3" pipe size	Q-1	8	2	Ea.	335	70.50		405.50	480
2820	4" pipe size	"	7	2.286		405	80.50		485.50	580
2840	For galvanized body, add				↓	189			189	208

22 13 Facility Sanitary Sewerage

22 13 16 – Sanitary Waste and Vent Piping

22 13 16.60 Traps

		Crew	Daily Output	Labor-Hours	Unit	Material	2016 Bare Costs Labor	2016 Bare Costs Equipment	Total	Total Incl O&P
0010	**TRAPS**									
0030	Cast iron, service weight									
0050	Running P trap, without vent									
1100	2"	Q-1	16	1	Ea.	148	35		183	221
1150	4"	"	13	1.231		148	43.50		191.50	235
1160	6"	Q-2	17	1.412		655	48		703	800
3000	P trap, B&S, 2" pipe size	Q-1	16	1		35	35		70	96.50
3040	3" pipe size	"	14	1.143		52.50	40		92.50	124
4700	Copper, drainage, drum trap									
4840	3" x 6" swivel, 1-1/2" pipe size	1 Plum	16	.500	Ea.	160	19.55		179.55	208
5100	P trap, standard pattern									
5200	1-1/4" pipe size	1 Plum	18	.444	Ea.	78.50	17.35		95.85	115
5240	1-1/2" pipe size		17	.471		72	18.40		90.40	110
5260	2" pipe size		15	.533		111	21		132	157
5280	3" pipe size		11	.727		281	28.50		309.50	355
6710	ABS DWV P trap, solvent weld joint									
6720	1-1/2" pipe size	1 Plum	18	.444	Ea.	7.75	17.35		25.10	37
6722	2" pipe size		17	.471		10.20	18.40		28.60	41.50
6724	3" pipe size		15	.533		40	21		61	78.50
6726	4" pipe size		14	.571		80	22.50		102.50	125
6732	PVC DWV P trap, solvent weld joint									
6733	1-1/2" pipe size	1 Plum	18	.444	Ea.	6.10	17.35		23.45	35
6734	2" pipe size		17	.471		8.20	18.40		26.60	39.50
6735	3" pipe size		15	.533		28	21		49	65
6736	4" pipe size		14	.571		61.50	22.50		84	105
6860	PVC DWV hub x hub, basin trap, 1-1/4" pipe size		18	.444		41.50	17.35		58.85	74
6870	Sink P trap, 1-1/2" pipe size		18	.444		11.20	17.35		28.55	41
6880	Tubular S trap, 1-1/2" pipe size		17	.471		26.50	18.40		44.90	59.50
6890	PVC sch. 40 DWV, drum trap									
6900	1-1/2" pipe size	1 Plum	16	.500	Ea.	27.50	19.55		47.05	62
6910	P trap, 1-1/2" pipe size		18	.444		6.75	17.35		24.10	36
6920	2" pipe size		17	.471		9.15	18.40		27.55	40.50
6930	3" pipe size		15	.533		31	21		52	68.50
6940	4" pipe size		14	.571		70.50	22.50		93	115
6950	P trap w/clean out, 1-1/2" pipe size		18	.444		11.20	17.35		28.55	41
6960	2" pipe size		17	.471		18.80	18.40		37.20	51

22 13 16.80 Vent Flashing and Caps

		Crew	Daily Output	Labor-Hours	Unit	Material	Labor	Equipment	Total	Total Incl O&P
0010	**VENT FLASHING AND CAPS**									
0120	Vent caps									
0140	Cast iron									
0160	1-1/4" - 1-1/2" pipe	1 Plum	23	.348	Ea.	34	13.60		47.60	60
0170	2" - 2-1/8" pipe		22	.364		39.50	14.20		53.70	67
0180	2-1/2" - 3-5/8" pipe		21	.381		45	14.90		59.90	74
0190	4" - 4-1/8" pipe		19	.421		55	16.45		71.45	87.50
0200	5" - 6" pipe		17	.471		83.50	18.40		101.90	123
0300	PVC									
0320	1-1/4" - 1-1/2" pipe	1 Plum	24	.333	Ea.	9.20	13		22.20	31.50
0330	2" - 2-1/8" pipe	"	23	.348	"	9.30	13.60		22.90	32.50
0900	Vent flashing									
1350	Copper with neoprene ring									
1400	1-1/4" pipe	1 Plum	20	.400	Ea.	68	15.60		83.60	101
1430	1-1/2" pipe		20	.400		68	15.60		83.60	101

22 13 Facility Sanitary Sewerage

22 13 16 – Sanitary Waste and Vent Piping

22 13 16.80 Vent Flashing and Caps

		Crew	Daily Output	Labor-Hours	Unit	Material	2016 Bare Costs Labor	2016 Bare Costs Equipment	Total	Total Incl O&P
1440	2" pipe	1 Plum	18	.444	Ea.	68	17.35		85.35	103
1450	3" pipe		17	.471		82	18.40		100.40	121
1460	4" pipe		16	.500		82	19.55		101.55	122
2980	Neoprene, one piece									
3000	1-1/4" pipe	1 Plum	24	.333	Ea.	2.93	13		15.93	24.50
3030	1-1/2" pipe		24	.333		2.95	13		15.95	25
3040	2" pipe		23	.348		2.95	13.60		16.55	26
3050	3" pipe		21	.381		3.39	14.90		18.29	28
3060	4" pipe		20	.400		5	15.60		20.60	31.50

22 13 19 – Sanitary Waste Piping Specialties

22 13 19.13 Sanitary Drains

		Crew	Daily Output	Labor-Hours	Unit	Material	2016 Bare Costs Labor	2016 Bare Costs Equipment	Total	Total Incl O&P
0010	**SANITARY DRAINS**									
2000	Floor, medium duty, C.I., deep flange, 7" diam. top									
2040	2" and 3" pipe size	Q-1	12	1.333	Ea.	232	47		279	335
2080	For galvanized body, add					96.50			96.50	106
2120	With polished bronze top					315			315	345

22 14 Facility Storm Drainage

22 14 26 – Facility Storm Drains

22 14 26.13 Roof Drains

		Crew	Daily Output	Labor-Hours	Unit	Material	2016 Bare Costs Labor	2016 Bare Costs Equipment	Total	Total Incl O&P
0010	**ROOF DRAINS**									
3860	Roof, flat metal deck, C.I. body, 12" C.I. dome									
3890	3" pipe size	Q-1	14	1.143	Ea.	425	40		465	530

22 14 29 – Sump Pumps

22 14 29.16 Submersible Sump Pumps

		Crew	Daily Output	Labor-Hours	Unit	Material	2016 Bare Costs Labor	2016 Bare Costs Equipment	Total	Total Incl O&P
0010	**SUBMERSIBLE SUMP PUMPS**									
7000	Sump pump, automatic									
7100	Plastic, 1-1/4" discharge, 1/4 HP	1 Plum	6.40	1.250	Ea.	150	49		199	246
7500	Cast iron, 1-1/4" discharge, 1/4 HP	"	6	1.333	"	211	52		263	320

22 31 Domestic Water Softeners

22 31 13 – Residential Domestic Water Softeners

22 31 13.10 Residential Water Softeners

		Crew	Daily Output	Labor-Hours	Unit	Material	2016 Bare Costs Labor	2016 Bare Costs Equipment	Total	Total Incl O&P
0010	**RESIDENTIAL WATER SOFTENERS**									
7350	Water softener, automatic, to 30 grains per gallon	2 Plum	5	3.200	Ea.	400	125		525	645
7400	To 100 grains per gallon	"	4	4	"	850	156		1,006	1,200

22 33 Electric Domestic Water Heaters

22 33 30 – Residential, Electric Domestic Water Heaters

22 33 30.13 Residential, Small-Capacity Elec. Water Heaters

		Crew	Daily Output	Labor-Hours	Unit	Material	2016 Bare Costs Labor	Equipment	Total	Total Incl O&P
0010	RESIDENTIAL, SMALL-CAPACITY ELECTRIC DOMESTIC WATER HEATERS									
1000	Residential, electric, glass lined tank, 5 yr., 10 gal., single element	1 Plum	2.30	3.478	Ea.	395	136		531	660
1060	30 gallon, double element		2.20	3.636		905	142		1,047	1,225
1080	40 gallon, double element		2	4		965	156		1,121	1,325
1100	52 gallon, double element		2	4		1,075	156		1,231	1,450
1120	66 gallon, double element		1.80	4.444		1,475	174		1,649	1,900
1140	80 gallon, double element		1.60	5		1,650	195		1,845	2,125

22 34 Fuel-Fired Domestic Water Heaters

22 34 13 – Instantaneous, Tankless, Gas Domestic Water Heaters

22 34 13.10 Instantaneous, Tankless, Gas Water Heaters

			Crew	Daily Output	Labor-Hours	Unit	Material	Labor	Equipment	Total	Total Incl O&P
0010	INSTANTANEOUS, TANKLESS, GAS WATER HEATERS										
9410	Natural gas/propane, 3.2 GPM	G	1 Plum	2	4	Ea.	460	156		616	765
9420	6.4 GPM	G		1.90	4.211		625	164		789	955
9430	8.4 GPM	G		1.80	4.444		730	174		904	1,075
9440	9.5 GPM	G		1.60	5		920	195		1,115	1,325

22 34 30 – Residential Gas Domestic Water Heaters

22 34 30.13 Residential, Atmos, Gas Domestic Wtr Heaters

		Crew	Daily Output	Labor-Hours	Unit	Material	Labor	Equipment	Total	Total Incl O&P
0010	RESIDENTIAL, ATMOSPHERIC, GAS DOMESTIC WATER HEATERS									
2000	Gas fired, foam lined tank, 10 yr., vent not incl.									
2040	30 gallon	1 Plum	2	4	Ea.	1,625	156		1,781	2,025
2100	75 gallon	"	1.50	5.333	"	2,425	208		2,633	3,025
3000	Tank leak safety, water & gas shut off see 22 05 23.20 8800									

22 34 46 – Oil-Fired Domestic Water Heaters

22 34 46.10 Residential Oil-Fired Water Heaters

		Crew	Daily Output	Labor-Hours	Unit	Material	Labor	Equipment	Total	Total Incl O&P
0010	RESIDENTIAL OIL-FIRED WATER HEATERS									
3000	Oil fired, glass lined tank, 5 yr., vent not included, 30 gallon	1 Plum	2	4	Ea.	1,175	156		1,331	1,550
3040	50 gallon	"	1.80	4.444	"	1,400	174		1,574	1,800

22 41 Residential Plumbing Fixtures

22 41 13 – Residential Water Closets, Urinals, and Bidets

22 41 13.13 Water Closets

			Crew	Daily Output	Labor-Hours	Unit	Material	Labor	Equipment	Total	Total Incl O&P
0010	WATER CLOSETS										
0150	Tank type, vitreous china, incl. seat, supply pipe w/stop, 1.6 gpf or noted										
0200	Wall hung										
0400	Two piece, close coupled		Q-1	5.30	3.019	Ea.	525	106		631	755
0960	For rough-in, supply, waste, vent and carrier		"	2.73	5.861	"	945	206		1,151	1,400
0999	Floor mounted										
1020	One piece, low profile		Q-1	5.30	3.019	Ea.	735	106		841	980
1100	Two piece, close coupled			5.30	3.019		230	106		336	430
1102	Economy			5.30	3.019		125	106		231	315
1110	Two piece, close coupled, dual flush			5.30	3.019		320	106		426	525
1140	Two piece, close coupled, 1.28 gpf, ADA	G		5.30	3.019		310	106		416	515
1960	For color, add						30%				
1980	For rough-in, supply, waste and vent		Q-1	3.05	5.246	Ea.	325	184		509	665

22 41 Residential Plumbing Fixtures

22 41 16 – Residential Lavatories and Sinks

22 41 16.13 Lavatories

		Crew	Daily Output	Labor-Hours	Unit	Material	2016 Bare Costs Labor	Equipment	Total	Total Incl O&P
0010	**LAVATORIES**, With trim, white unless noted otherwise									
0500	Vanity top, porcelain enamel on cast iron									
0600	20" x 18"	Q-1	6.40	2.500	Ea.	315	88		403	490
0640	33" x 19" oval		6.40	2.500		475	88		563	665
0720	19" round		6.40	2.500		415	88		503	605
0860	For color, add					25%				
1000	Cultured marble, 19" x 17", single bowl	Q-1	6.40	2.500	Ea.	151	88		239	310
1120	25" x 22", single bowl		6.40	2.500		190	88		278	355
1160	37" x 22", single bowl		6.40	2.500		221	88		309	390
1580	For color, same price									
1900	Stainless steel, self-rimming, 25" x 22", single bowl, ledge	Q-1	6.40	2.500	Ea.	340	88		428	520
1960	17" x 22", single bowl		6.40	2.500		330	88		418	510
2600	Steel, enameled, 20" x 17", single bowl		5.80	2.759		150	97		247	325
2900	Vitreous china, 20" x 16", single bowl		5.40	2.963		236	104		340	430
3200	22" x 13", single bowl		5.40	2.963		243	104		347	440
3580	Rough-in, supply, waste and vent for all above lavatories		2.30	6.957		218	245		463	645
4000	Wall hung									
4040	Porcelain enamel on cast iron, 16" x 14", single bowl	Q-1	8	2	Ea.	500	70.50		570.50	665
4180	20" x 18", single bowl	"	8	2	"	265	70.50		335.50	405
4580	For color, add					30%				
6000	Vitreous china, 18" x 15", single bowl with backsplash	Q-1	7	2.286	Ea.	186	80.50		266.50	335
6060	19" x 17", single bowl		7	2.286		148	80.50		228.50	295
6960	Rough-in, supply, waste and vent for above lavatories		1.66	9.639		435	340		775	1,050
7000	Pedestal type									
7600	Vitreous china, 27" x 21", white	Q-1	6.60	2.424	Ea.	680	85		765	890
7610	27" x 21", colored		6.60	2.424		860	85		945	1,075
7620	27" x 21", premium color		6.60	2.424		975	85		1,060	1,225
7660	26" x 20", white		6.60	2.424		670	85		755	875
7670	26" x 20", colored		6.60	2.424		845	85		930	1,075
7680	26" x 20", premium color		6.60	2.424		960	85		1,045	1,200
7700	24" x 20", white		6.60	2.424		395	85		480	575
7710	24" x 20", colored		6.60	2.424		490	85		575	680
7720	24" x 20", premium color		6.60	2.424		550	85		635	750
7760	21" x 18", white		6.60	2.424		277	85		362	445
7770	21" x 18", colored		6.60	2.424		335	85		420	510
7990	Rough-in, supply, waste and vent for pedestal lavatories		1.66	9.639		435	340		775	1,050

22 41 16.16 Sinks

		Crew	Daily Output	Labor-Hours	Unit	Material	Labor	Equipment	Total	Total Incl O&P
0010	**SINKS**, With faucets and drain									
2000	Kitchen, counter top style, P.E. on C.I., 24" x 21" single bowl	Q-1	5.60	2.857	Ea.	297	100		397	490
2100	31" x 22" single bowl		5.60	2.857		550	100		650	765
2200	32" x 21" double bowl		4.80	3.333		370	117		487	600
3000	Stainless steel, self rimming, 19" x 18" single bowl		5.60	2.857		605	100		705	830
3100	25" x 22" single bowl		5.60	2.857		675	100		775	905
3200	33" x 22" double bowl		4.80	3.333		975	117		1,092	1,275
3300	43" x 22" double bowl		4.80	3.333		1,125	117		1,242	1,450
4000	Steel, enameled, with ledge, 24" x 21" single bowl		5.60	2.857		525	100		625	745
4100	32" x 21" double bowl		4.80	3.333		520	117		637	765
4960	For color sinks except stainless steel, add					10%				
4980	For rough-in, supply, waste and vent, counter top sinks	Q-1	2.14	7.477		247	263		510	705
5000	Kitchen, raised deck, P.E. on C.I.									
5100	32" x 21", dual level, double bowl	Q-1	2.60	6.154	Ea.	440	216		656	840
5790	For rough-in, supply, waste & vent, sinks	"	1.85	8.649	"	247	305		552	770

22 41 Residential Plumbing Fixtures

22 41 19 – Residential Bathtubs

22 41 19.10 Baths

		Crew	Daily Output	Labor-Hours	Unit	Material	2016 Bare Costs Labor	Equipment	Total	Total Incl O&P
0010	**BATHS**									
0100	Tubs, recessed porcelain enamel on cast iron, with trim									
0180	48" x 42"	Q-1	4	4	Ea.	2,675	141		2,816	3,150
0220	72" x 36"	"	3	5.333	"	2,750	187		2,937	3,325
0300	Mat bottom									
0380	5' long	Q-1	4.40	3.636	Ea.	1,150	128		1,278	1,450
0480	Above floor drain, 5' long		4	4		840	141		981	1,150
0560	Corner 48" x 44"		4.40	3.636		2,675	128		2,803	3,125
2000	Enameled formed steel, 4'-6" long		5.80	2.759		485	97		582	695
4600	Module tub & showerwall surround, molded fiberglass									
4610	5' long x 34" wide x 76" high	Q-1	4	4	Ea.	825	141		966	1,150
9600	Rough-in, supply, waste and vent, for all above tubs, add	"	2.07	7.729	"	355	272		627	840

22 41 23 – Residential Showers

22 41 23.20 Showers

		Crew	Daily Output	Labor-Hours	Unit	Material	Labor	Equipment	Total	Total Incl O&P
0010	**SHOWERS**									
1500	Stall, with drain only. Add for valve and door/curtain									
1520	32" square	Q-1	5	3.200	Ea.	1,175	112		1,287	1,450
1530	36" square		4.80	3.333		2,925	117		3,042	3,400
1540	Terrazzo receptor, 32" square		5	3.200		1,350	112		1,462	1,650
1560	36" square		4.80	3.333		1,475	117		1,592	1,825
1580	36" corner angle		4.80	3.333		1,725	117		1,842	2,100
3000	Fiberglass, one piece, with 3 walls, 32" x 32" square		5.50	2.909		335	102		437	540
3100	36" x 36" square		5.50	2.909		385	102		487	590
4200	Rough-in, supply, waste and vent for above showers		2.05	7.805		345	274		619	835

22 41 36 – Residential Laundry Trays

22 41 36.10 Laundry Sinks

		Crew	Daily Output	Labor-Hours	Unit	Material	Labor	Equipment	Total	Total Incl O&P
0010	**LAUNDRY SINKS**, With trim									
0020	Porcelain enamel on cast iron, black iron frame									
0050	24" x 21", single compartment	Q-1	6	2.667	Ea.	590	93.50		683.50	800
0100	26" x 21", single compartment	"	6	2.667	"	615	93.50		708.50	830
3000	Plastic, on wall hanger or legs									
3020	18" x 23", single compartment	Q-1	6.50	2.462	Ea.	154	86.50		240.50	310
3100	20" x 24", single compartment		6.50	2.462		169	86.50		255.50	330
3200	36" x 23", double compartment		5.50	2.909		199	102		301	390
3300	40" x 24", double compartment		5.50	2.909		295	102		397	495
5000	Stainless steel, counter top, 22" x 17" single compartment		6	2.667		76.50	93.50		170	239
5200	33" x 22", double compartment		5	3.200		91	112		203	286
9600	Rough-in, supply, waste and vent, for all laundry sinks		2.14	7.477		247	263		510	705

22 41 39 – Residential Faucets, Supplies and Trim

22 41 39.10 Faucets and Fittings

		Crew	Daily Output	Labor-Hours	Unit	Material	Labor	Equipment	Total	Total Incl O&P
0010	**FAUCETS AND FITTINGS**									
0150	Bath, faucets, diverter spout combination, sweat	1 Plum	8	1	Ea.	86	39		125	159
0200	For integral stops, IPS unions, add					109			109	120
0420	Bath, press-bal mix valve w/diverter, spout, shower head, arm/flange	1 Plum	8	1		173	39		212	255
0500	Drain, central lift, 1-1/2" IPS male		20	.400		49	15.60		64.60	80
0600	Trip lever, 1-1/2" IPS male		20	.400		55	15.60		70.60	86.50
1000	Kitchen sink faucets, top mount, cast spout		10	.800		77.50	31		108.50	137
1100	For spray, add		24	.333		16.95	13		29.95	40
1300	Single control lever handle									
1310	With pull out spray									

22 41 Residential Plumbing Fixtures

22 41 39 – Residential Faucets, Supplies and Trim

22 41 39.10 Faucets and Fittings		Crew	Daily Output	Labor-Hours	Unit	Material	2016 Bare Costs Labor	Equipment	Total	Total Incl O&P
1320	Polished chrome	1 Plum	10	.800	Ea.	200	31		231	271
2000	Laundry faucets, shelf type, IPS or copper unions		12	.667		61.50	26		87.50	111
2100	Lavatory faucet, centerset, without drain		10	.800		63.50	31		94.50	122
2120	With pop-up drain		6.66	1.201		83	47		130	169
2210	Porcelain cross handles and pop-up drain									
2220	Polished chrome	1 Plum	6.66	1.201	Ea.	198	47		245	296
2230	Polished brass	"	6.66	1.201	"	305	47		352	415
2260	Single lever handle and pop-up drain									
2280	Satin nickel	1 Plum	6.66	1.201	Ea.	283	47		330	390
2290	Polished chrome		6.66	1.201		198	47		245	296
2800	Self-closing, center set		10	.800		148	31		179	215
4000	Shower by-pass valve with union		18	.444		57	17.35		74.35	91.50
4200	Shower thermostatic mixing valve, concealed, with shower head trim kit		8	1		345	39		384	445
4220	Shower pressure balancing mixing valve									
4230	With shower head, arm, flange and diverter tub spout									
4240	Chrome	1 Plum	6.14	1.303	Ea.	375	51		426	500
4250	Satin nickel		6.14	1.303		505	51		556	640
4260	Polished graphite		6.14	1.303		505	51		556	640
5000	Sillcock, compact, brass, IPS or copper to hose		24	.333		10.20	13		23.20	33

22 41 39.70 Washer/Dryer Accessories

		Crew	Daily Output	Labor-Hours	Unit	Material	Labor	Equipment	Total	Total Incl O&P
0010	**WASHER/DRYER ACCESSORIES**									
1020	Valves ball type single lever									
1030	1/2" diam., IPS	1 Plum	21	.381	Ea.	58.50	14.90		73.40	89
1040	1/2" diam., solder	"	21	.381	"	58.50	14.90		73.40	89
1050	Recessed box, 16 ga., two hose valves and drain									
1060	1/2" size, 1-1/2" drain	1 Plum	18	.444	Ea.	133	17.35		150.35	175
1070	1/2" size, 2" drain	"	17	.471	"	119	18.40		137.40	162
1080	With grounding electric receptacle									
1090	1/2" size, 1-1/2" drain	1 Plum	18	.444	Ea.	145	17.35		162.35	189
1100	1/2" size, 2" drain	"	17	.471	"	157	18.40		175.40	204
1110	With grounding and dryer receptacle									
1120	1/2" size, 1-1/2" drain	1 Plum	18	.444	Ea.	180	17.35		197.35	227
1130	1/2" size, 2" drain	"	17	.471	"	182	18.40		200.40	231
1140	Recessed box 16 ga., ball valves with single lever and drain									
1150	1/2" size, 1-1/2" drain	1 Plum	19	.421	Ea.	251	16.45		267.45	305
1160	1/2" size, 2" drain	"	18	.444	"	218	17.35		235.35	268
1170	With grounding electric receptacle									
1180	1/2" size, 1-1/2" drain	1 Plum	19	.421	Ea.	249	16.45		265.45	300
1190	1/2" size, 2" drain	"	18	.444	"	241	17.35		258.35	295
1200	With grounding and dryer receptacles									
1210	1/2" size, 1-1/2" drain	1 Plum	19	.421	Ea.	238	16.45		254.45	288
1220	1/2" size, 2" drain	"	18	.444	"	264	17.35		281.35	320
1300	Recessed box, 20 ga., two hose valves and drain (economy type)									
1310	1/2" size, 1-1/2" drain	1 Plum	19	.421	Ea.	104	16.45		120.45	141
1320	1/2" size, 2" drain		18	.444		97	17.35		114.35	136
1330	Box with drain only		24	.333		60	13		73	87.50
1340	1/2" size, 1-1/2" ABS/PVC drain		19	.421		115	16.45		131.45	153
1350	1/2" size, 2" ABS/PVC drain		18	.444		112	17.35		129.35	152
1352	Box with drain and 15 A receptacle		24	.333		110	13		123	143
1360	1/2" size, 2" drain ABS/PVC, 15 A receptacle		24	.333		133	13		146	168
1400	Wall mounted									
1410	1/2" size, 1-1/2" plastic drain	1 Plum	19	.421	Ea.	30.50	16.45		46.95	60.50

22 41 Residential Plumbing Fixtures

22 41 39 – Residential Faucets, Supplies and Trim

22 41 39.70 Washer/Dryer Accessories

		Crew	Daily Output	Labor-Hours	Unit	Material	2016 Bare Costs Labor	Equipment	Total	Total Incl O&P
1420	1/2" size, 2" plastic drain	1 Plum	18	.444	Ea.	17.30	17.35		34.65	47.50
1500	Dryer vent kit									
1510	8' flex duct, clamps and outside hood	1 Plum	20	.400	Ea.	12.95	15.60		28.55	40.50
1980	Rough-in, supply, waste, and vent for washer boxes		3.46	2.310		250	90		340	425
9605	Washing machine valve assembly, hot & cold water supply, recessed		8	1		73.50	39		112.50	146
9610	Washing machine valve assembly, hot & cold water supply, mounted		8	1		58.50	39		97.50	129

22 42 Commercial Plumbing Fixtures

22 42 13 – Commercial Water Closets, Urinals, and Bidets

22 42 13.13 Water Closets

		Crew	Daily Output	Labor-Hours	Unit	Material	2016 Bare Costs Labor	Equipment	Total	Total Incl O&P
0010	**WATER CLOSETS**									
3000	Bowl only, with flush valve, seat, 1.6 gpf unless noted									
3100	Wall hung	Q-1	5.80	2.759	Ea.	965	97		1,062	1,200
3200	For rough-in, supply, waste and vent, single WC		2.56	6.250		980	220		1,200	1,425
3300	Floor mounted		5.80	2.759		315	97		412	505
3370	For rough-in, supply, waste and vent, single WC		2.84	5.634		365	198		563	725
3390	Floor mounted children's size, 10-3/4" high									
3392	With automatic flush sensor, 1.6 gpf	Q-1	6.20	2.581	Ea.	625	90.50		715.50	835
3396	With automatic flush sensor, 1.28 gpf		6.20	2.581		625	90.50		715.50	835
3400	For rough-in, supply, waste and vent, single WC		2.84	5.634		365	198		563	725

22 42 16 – Commercial Lavatories and Sinks

22 42 16.13 Lavatories

		Crew	Daily Output	Labor-Hours	Unit	Material	2016 Bare Costs Labor	Equipment	Total	Total Incl O&P
0010	**LAVATORIES**, With trim, white unless noted otherwise									
0020	Commercial lavatories same as residential. See Section 22 41 16									

22 42 16.40 Service Sinks

		Crew	Daily Output	Labor-Hours	Unit	Material	2016 Bare Costs Labor	Equipment	Total	Total Incl O&P
0010	**SERVICE SINKS**									
6650	Service, floor, corner, P.E. on C.I., 28" x 28"	Q-1	4.40	3.636	Ea.	1,025	128		1,153	1,325
6750	Vinyl coated rim guard, add					61.50			61.50	68
6760	Mop sink, molded stone, 24" x 36"	1 Plum	3.33	2.402		273	94		367	455
6770	Mop sink, molded stone, 24" x 36", w/rim 3 sides	"	3.33	2.402		278	94		372	460
6790	For rough-in, supply, waste & vent, floor service sinks	Q-1	1.64	9.756		715	345		1,060	1,350

22 42 39 – Commercial Faucets, Supplies, and Trim

22 42 39.10 Faucets and Fittings

		Crew	Daily Output	Labor-Hours	Unit	Material	2016 Bare Costs Labor	Equipment	Total	Total Incl O&P
0010	**FAUCETS AND FITTINGS**									
2790	Faucets for lavatories									
2800	Self-closing, center set	1 Plum	10	.800	Ea.	148	31		179	215
2810	Automatic sensor and operator, with faucet head		6.15	1.301		455	51		506	585
3000	Service sink faucet, cast spout, pail hook, hose end		14	.571		76	22.50		98.50	121

22 42 39.30 Carriers and Supports

		Crew	Daily Output	Labor-Hours	Unit	Material	2016 Bare Costs Labor	Equipment	Total	Total Incl O&P
0010	**CARRIERS AND SUPPORTS**, For plumbing fixtures									
0600	Plate type with studs, top back plate	1 Plum	7	1.143	Ea.	95	44.50		139.50	179
3000	Lavatory, concealed arm									
3050	Floor mounted, single									
3100	High back fixture	1 Plum	6	1.333	Ea.	570	52		622	710
3200	Flat slab fixture	"	6	1.333	"	455	52		507	585
8200	Water closet, residential									
8220	Vertical centerline, floor mount									
8240	Single, 3" caulk, 2" or 3" vent	1 Plum	6	1.333	Ea.	655	52		707	810
8260	4" caulk, 2" or 4" vent	"	6	1.333	"	845	52		897	1,025

22 51 Swimming Pool Plumbing Systems

22 51 19 – Swimming Pool Water Treatment Equipment

22 51 19.50 Swimming Pool Filtration Equipment	Crew	Daily Output	Labor-Hours	Unit	Material	2016 Bare Costs Labor	Equipment	Total	Total Incl O&P
0010 **SWIMMING POOL FILTRATION EQUIPMENT**									
0900 Filter system, sand or diatomite type, incl. pump, 6,000 gal./hr.	2 Plum	1.80	8.889	Total	2,025	345		2,370	2,800
1020 Add for chlorination system, 800 S.F. pool	"	3	5.333	Ea.	185	208		393	550

Estimating Tips
The labor adjustment factors listed in Subdivision 22 01 02.20 also apply to Division 23.

23 10 00 Facility Fuel Systems
- The prices in this subdivision for above- and below-ground storage tanks do not include foundations or hold-down slabs, unless noted. The estimator should refer to Divisions 3 and 31 for foundation system pricing. In addition to the foundations, required tank accessories, such as tank gauges, leak detection devices, and additional manholes and piping, must be added to the tank prices.

23 50 00 Central Heating Equipment
- When estimating the cost of an HVAC system, check to see who is responsible for providing and installing the temperature control system. It is possible to overlook controls, assuming that they would be included in the electrical estimate.

- When looking up a boiler, be careful on specified capacity. Some manufacturers rate their products on output while others use input.
- Include HVAC insulation for pipe, boiler, and duct (wrap and liner).
- Be careful when looking up mechanical items to get the correct pressure rating and connection type (thread, weld, flange).

23 70 00 Central HVAC Equipment
- Combination heating and cooling units are sized by the air conditioning requirements. (See Reference No. R236000-20 for preliminary sizing guide.)
- A ton of air conditioning is nominally 400 CFM.
- Rectangular duct is taken off by the linear foot for each size, but its cost is usually estimated by the pound. Remember that SMACNA standards now base duct on internal pressure.
- Prefabricated duct is estimated and purchased like pipe: straight sections and fittings.

- Note that cranes or other lifting equipment are not included on any lines in Division 23. For example, if a crane is required to lift a heavy piece of pipe into place high above a gym floor, or to put a rooftop unit on the roof of a four-story building, etc., it must be added. Due to the potential for extreme variation—from nothing additional required to a major crane or helicopter—we feel that including a nominal amount for "lifting contingency" would be useless and detract from the accuracy of the estimate. When using equipment rental cost data from RSMeans, do not forget to include the cost of the operator(s).

Reference Numbers
Reference numbers are shown at the beginning of some major classifications. These numbers refer to related items in the Reference Section. The reference information may be an estimating procedure, an alternate pricing method, or technical information.

Note: Not all subdivisions listed here necessarily appear. ■

No part of this cost data may be reproduced, stored in a retrieval system, or transmitted in any form or by any means without prior written permission of RSMeans.

Note: Trade Service, in part, has been used as a reference source for some of the material prices used in Division 23.

Did you know?
RSMeans Online gives you the same access to RSMeans' data with 24/7 access:
- Quickly locate costs in the searchable database.
- Build cost lists, estimates, and reports in minutes.
- Adjust costs to any location in the U.S. and Canada with the click of a button.

Start your free trial today at **www.RSMeansOnline.com**

RSMeans Online
FROM THE GORDIAN GROUP®

23 05 Common Work Results for HVAC

23 05 05 – Selective Demolition for HVAC

23 05 05.10 HVAC Demolition	Crew	Daily Output	Labor-Hours	Unit	Material	2016 Bare Costs Labor	Equipment	Total	Total Incl O&P
0010 **HVAC DEMOLITION**									
0100 Air conditioner, split unit, 3 ton	Q-5	2	8	Ea.		288		288	475
0150 Package unit, 3 ton	Q-6	3	8	"		278		278	460
0260 Baseboard, hydronic fin tube, 1/2"	Q-5	117	.137	L.F.		4.93		4.93	8.15
0298 Boilers									
0300 Electric, up thru 148 kW	Q-19	2	12	Ea.		435		435	715
0310 150 thru 518 kW	"	1	24			870		870	1,425
0320 550 thru 2000 kW	Q-21	.40	80			2,975		2,975	4,900
0330 2070 kW and up	"	.30	106			3,975		3,975	6,550
0340 Gas and/or oil, up thru 150 MBH	Q-7	2.20	14.545			525		525	865
0350 160 thru 2000 MBH		.80	40			1,450		1,450	2,375
0360 2100 thru 4500 MBH		.50	64			2,300		2,300	3,800
0370 4600 thru 7000 MBH		.30	106			3,850		3,850	6,350
0390 12,200 thru 25,000 MBH		.12	266			9,625		9,625	15,900
1000 Ductwork, 4" high, 8" wide	1 Clab	200	.040	L.F.		.99		.99	1.65
1100 6" high, 8" wide		165	.048			1.20		1.20	2
1200 10" high, 12" wide		125	.064			1.58		1.58	2.64
1300 12"-14" high, 16"-18" wide		85	.094			2.32		2.32	3.88
1500 30" high, 36" wide		56	.143			3.52		3.52	5.90
2200 Furnace, electric	Q-20	2	10	Ea.		345		345	575
2300 Gas or oil, under 120 MBH	Q-9	4	4			136		136	227
2340 Over 120 MBH	"	3	5.333			181		181	305
2800 Heat pump, package unit, 3 ton	Q-5	2.40	6.667			240		240	395
2840 Split unit, 3 ton		2	8			288		288	475
2950 Tank, steel, oil, 275 gal., above ground		10	1.600			57.50		57.50	95
2960 Remove and reset		3	5.333			192		192	315
5090 Remove refrigerant from system	1 Stpi	40	.200	Lb.		8		8	13.20

23 07 HVAC Insulation

23 07 13 – Duct Insulation

23 07 13.10 Duct Thermal Insulation

	Crew	Daily Output	Labor-Hours	Unit	Material	2016 Bare Costs Labor	Equipment	Total	Total Incl O&P
0010 **DUCT THERMAL INSULATION**									
3000 Ductwork									
3020 Blanket type, fiberglass, flexible									
3030 Fire rated for grease and hazardous exhaust ducts									
3060 1-1/2" thick	Q-14	84	.190	S.F.	4.54	6.05		10.59	15.25
3090 Fire rated for plenums									
3100 1/2" x 24" x 25'	Q-14	1.94	8.247	Roll	167	262		429	630
3110 1/2" x 24" x 25'		98	.163	S.F.	3.35	5.20		8.55	12.50
3120 1/2" x 48" x 25'		1.04	15.385	Roll	335	490		825	1,200
3126 1/2" x 48" x 25'		104	.154	S.F.	3.35	4.88		8.23	12
3140 FSK vapor barrier wrap, .75 lb. density									
3160 1" thick [G]	Q-14	350	.046	S.F.	.18	1.45		1.63	2.66
3170 1-1/2" thick [G]	"	320	.050	"	.22	1.59		1.81	2.93
3210 Vinyl jacket, same as FSK									

23 09 Instrumentation and Control for HVAC

23 09 53 – Pneumatic and Electric Control System for HVAC

23 09 53.10 Control Components		Crew	Daily Output	Labor-Hours	Unit	Material	2016 Bare Costs Labor	Equipment	Total	Total Incl O&P
0010	CONTROL COMPONENTS									
5000	Thermostats									
5030	Manual	1 Stpi	8	1	Ea.	53	40		93	125
5040	1 set back, electric, timed	G	8	1		38	40		78	108
5050	2 set back, electric, timed	G	8	1		190	40		230	275

23 13 Facility Fuel-Storage Tanks

23 13 13 – Facility Underground Fuel-Oil, Storage Tanks

23 13 13.09 Single-Wall Steel Fuel-Oil Tanks

		Crew	Daily Output	Labor-Hours	Unit	Material	Labor	Equipment	Total	Total Incl O&P
0010	SINGLE-WALL STEEL FUEL-OIL TANKS									
5000	Tanks, steel ugnd., sti-p3, not incl. hold-down bars									
5500	Excavation, pad, pumps and piping not included									
5510	Single wall, 500 gallon capacity, 7 ga. shell	Q-5	2.70	5.926	Ea.	2,050	214		2,264	2,600
5520	1,000 gallon capacity, 7 ga. shell	"	2.50	6.400		3,825	231		4,056	4,575
5530	2,000 gallon capacity, 1/4" thick shell	Q-7	4.60	6.957		6,200	251		6,451	7,250
5535	2,500 gallon capacity, 7 ga. shell	Q-5	3	5.333		6,850	192		7,042	7,850
5610	25,000 gallon capacity, 3/8" thick shell	Q-7	1.30	24.615		37,900	885		38,785	43,200
5630	40,000 gallon capacity, 3/8" thick shell		.90	35.556		42,000	1,275		43,275	48,300
5640	50,000 gallon capacity, 3/8" thick shell		.80	40		46,600	1,450		48,050	54,000

23 13 13.23 Glass-Fiber-Reinfcd-Plastic, Fuel-Oil, Storage

		Crew	Daily Output	Labor-Hours	Unit	Material	Labor	Equipment	Total	Total Incl O&P
0010	GLASS-FIBER-REINFCD-PLASTIC, UNDERGRND. FUEL-OIL, STORAGE									
0210	Fiberglass, underground, single wall, U.L. listed, not including									
0220	manway or hold-down strap									
0230	1,000 gallon capacity	Q-5	2.46	6.504	Ea.	4,325	234		4,559	5,125
0240	2,000 gallon capacity	Q-7	4.57	7.002		6,250	252		6,502	7,300
0245	3,000 gallon capacity		3.90	8.205		7,475	296		7,771	8,725
0255	5,000 gallon capacity		3.20	10		9,650	360		10,010	11,200
0500	For manway, fittings and hold-downs, add					20%	15%			
2210	Fiberglass, underground, single wall, U.L. listed, including									
2220	hold-down straps, no manways									
2230	1,000 gallon capacity	Q-5	1.88	8.511	Ea.	4,800	305		5,105	5,775
2240	2,000 gallon capacity	Q-7	3.55	9.014	"	6,725	325		7,050	7,925

23 13 23 – Facility Aboveground Fuel-Oil, Storage Tanks

23 13 23.16 Steel

		Crew	Daily Output	Labor-Hours	Unit	Material	Labor	Equipment	Total	Total Incl O&P
3001	STEEL, storage, above ground, including supports, coating									
3020	fittings, not including foundation, pumps or piping									
3040	Single wall, 275 gallon	Q-5	5	3.200	Ea.	490	115		605	730
3060	550 gallon	"	2.70	5.926		3,925	214		4,139	4,675
3080	1,000 gallon	Q-7	5	6.400		6,450	231		6,681	7,475
3320	Double wall, 500 gallon capacity	Q-5	2.40	6.667		2,625	240		2,865	3,275
3330	2000 gallon capacity	Q-7	4.15	7.711		9,975	278		10,253	11,500
3340	4000 gallon capacity		3.60	8.889		17,800	320		18,120	20,100
3350	6000 gallon capacity		2.40	13.333		21,000	480		21,480	23,900
3360	8000 gallon capacity		2	16		27,000	575		27,575	30,700
3370	10000 gallon capacity		1.80	17.778		30,200	640		30,840	34,300
3380	15000 gallon capacity		1.50	21.333		45,900	770		46,670	52,000
3390	20000 gallon capacity		1.30	24.615		52,500	885		53,385	59,000
3400	25000 gallon capacity		1.15	27.826		63,500	1,000		64,500	71,500
3410	30000 gallon capacity		1	32		69,500	1,150		70,650	78,500

23 13 Facility Fuel-Storage Tanks

23 13 23 – Facility Aboveground Fuel-Oil, Storage Tanks

23 13 23.26 Horizontal, Conc., Abvgrd Fuel-Oil, Stor. Tanks	Crew	Daily Output	Labor-Hours	Unit	Material	2016 Bare Costs Labor	Equipment	Total	Total Incl O&P
0010 **HORIZONTAL, CONCRETE, ABOVEGROUND FUEL-OIL, STORAGE TANKS**									
0050 Concrete, storage, aboveground, including pad & pump									
0100 500 gallon	F-3	2	20	Ea.	10,000	615	330	10,945	12,400
0200 1,000 gallon	"	2	20	"	14,000	615	330	14,945	16,800

23 21 Hydronic Piping and Pumps

23 21 20 – Hydronic HVAC Piping Specialties

23 21 20.46 Expansion Tanks

	Crew	Daily Output	Labor-Hours	Unit	Material	Labor	Equipment	Total	Total Incl O&P
0010 **EXPANSION TANKS**									
1507 Underground fuel-oil storage tanks, see Section 23 13 13									
2000 Steel, liquid expansion, ASME, painted, 15 gallon capacity	Q-5	17	.941	Ea.	665	34		699	785
2040 30 gallon capacity		12	1.333		740	48		788	895
3000 Steel ASME expansion, rubber diaphragm, 19 gal. cap. accept.		12	1.333		2,525	48		2,573	2,850
3020 31 gallon capacity		8	2		2,800	72		2,872	3,225

23 21 23 – Hydronic Pumps

23 21 23.13 In-Line Centrifugal Hydronic Pumps

	Crew	Daily Output	Labor-Hours	Unit	Material	Labor	Equipment	Total	Total Incl O&P
0010 **IN-LINE CENTRIFUGAL HYDRONIC PUMPS**									
0600 Bronze, sweat connections, 1/40 HP, in line									
0640 3/4" size	Q-1	16	1	Ea.	225	35		260	305
1000 Flange connection, 3/4" to 1-1/2" size									
1040 1/12 HP	Q-1	6	2.667	Ea.	580	93.50		673.50	795
1060 1/8 HP		6	2.667		1,000	93.50		1,093.50	1,250
2101 Pumps, circulating, 3/4" to 1-1/2" size, 1/3 HP		6	2.667		815	93.50		908.50	1,050

23 23 Refrigerant Piping

23 23 16 – Refrigerant Piping Specialties

23 23 16.16 Refrigerant Line Sets

	Crew	Daily Output	Labor-Hours	Unit	Material	Labor	Equipment	Total	Total Incl O&P
0010 **REFRIGERANT LINE SETS**									
0100 Copper tube									
0110 1/2" insulation, both tubes									
0120 Combination 1/4" and 1/2" tubes									
0130 10' set	Q-5	42	.381	Ea.	70.50	13.75		84.25	100
0140 20' set		40	.400		110	14.40		124.40	145
0150 30' set		37	.432		148	15.60		163.60	188
0160 40' set		35	.457		193	16.50		209.50	240
0170 50' set		32	.500		229	18.05		247.05	282
0180 100' set		22	.727		520	26		546	620
0300 Combination 1/4" and 3/4" tubes									
0310 10' set	Q-5	40	.400	Ea.	78.50	14.40		92.90	111
0320 20' set		38	.421		139	15.20		154.20	178
0330 30' set		35	.457		208	16.50		224.50	256
0340 40' set		33	.485		272	17.50		289.50	330
0350 50' set		30	.533		345	19.25		364.25	405
0380 100' set		20	.800		780	29		809	910
0500 Combination 3/8" & 3/4" tubes									
0510 10' set	Q-5	28	.571	Ea.	90.50	20.50		111	134
0520 20' set		36	.444		141	16		157	182
0530 30' set		34	.471		191	16.95		207.95	239

23 23 Refrigerant Piping

23 23 16 – Refrigerant Piping Specialties

23 23 16.16 Refrigerant Line Sets

		Crew	Daily Output	Labor-Hours	Unit	Material	2016 Bare Costs Labor	Equipment	Total	Total Incl O&P
0540	40' set	Q-5	31	.516	Ea.	252	18.60		270.60	310
0550	50' set		28	.571		296	20.50		316.50	360
0580	100' set		18	.889		875	32		907	1,025
0700	Combination 3/8" & 1-1/8" tubes									
0710	10' set	Q-5	36	.444	Ea.	158	16		174	201
0720	20' set		33	.485		265	17.50		282.50	320
0730	30' set		31	.516		355	18.60		373.60	420
0740	40' set		28	.571		535	20.50		555.50	620
0750	50' set		26	.615		615	22		637	715
0900	Combination 1/2" & 3/4" tubes									
0910	10' set	Q-5	37	.432	Ea.	95.50	15.60		111.10	131
0920	20' set		35	.457		169	16.50		185.50	213
0930	30' set		33	.485		255	17.50		272.50	310
0940	40' set		30	.533		335	19.25		354.25	400
0950	50' set		27	.593		420	21.50		441.50	500
0980	100' set		17	.941		950	34		984	1,100
2100	Combination 1/2" & 1-1/8" tubes									
2110	10' set	Q-5	35	.457	Ea.	163	16.50		179.50	206
2120	20' set		31	.516		279	18.60		297.60	335
2130	30' set		29	.552		420	19.90		439.90	495
2140	40' set		25	.640		565	23		588	660
2150	50' set		14	1.143		625	41		666	755
2300	For 1" thick insulation add					30%	15%			

23 31 HVAC Ducts and Casings

23 31 13 – Metal Ducts

23 31 13.13 Rectangular Metal Ducts

		Crew	Daily Output	Labor-Hours	Unit	Material	2016 Bare Costs Labor	Equipment	Total	Total Incl O&P
0010	**RECTANGULAR METAL DUCTS**									
0020	Fabricated rectangular, includes fittings, joints, supports,									
0021	allowance for flexible connections and field sketches.									
0030	Does not include "as-built dwgs." or insulation.									
0031	NOTE: Fabrication and installation are combined									
0040	as LABOR cost. Approx. 25% fittings assumed.									
0042	Fabrication/Inst. is to commercial quality standards									
0043	(SMACNA or equiv.) for structure, sealing, leak testing, etc.									
0100	Aluminum, alloy 3003-H14, under 100 lb.	Q-10	75	.320	Lb.	3.20	11.30		14.50	22.50
0110	100 to 500 lb.		80	.300		1.88	10.60		12.48	19.70
0120	500 to 1,000 lb.		95	.253		1.82	8.90		10.72	16.85
0140	1,000 to 2,000 lb.		120	.200		1.77	7.05		8.82	13.70
0500	Galvanized steel, under 200 lb.		235	.102		.57	3.60		4.17	6.65
0520	200 to 500 lb.		245	.098		.56	3.46		4.02	6.35
0540	500 to 1,000 lb.		255	.094		.55	3.32		3.87	6.15

23 33 Air Duct Accessories

23 33 13 – Dampers

23 33 13.13 Volume-Control Dampers		Crew	Daily Output	Labor-Hours	Unit	Material	2016 Bare Costs Labor	Equipment	Total	Total Incl O&P
0010	**VOLUME-CONTROL DAMPERS**									
6000	12" x 12"	1 Shee	21	.381	Ea.	29.50	14.40		43.90	56.50
8000	Multi-blade dampers, parallel blade									
8100	8" x 8"	1 Shee	24	.333	Ea.	82	12.60		94.60	111

23 33 13.16 Fire Dampers										
0010	**FIRE DAMPERS**									
3000	Fire damper, curtain type, 1-1/2 hr. rated, vertical, 6" x 6"	1 Shee	24	.333	Ea.	24	12.60		36.60	47.50
3020	8" x 6"	"	22	.364	"	24	13.75		37.75	49.50

23 33 46 – Flexible Ducts

23 33 46.10 Flexible Air Ducts

0010	**FLEXIBLE AIR DUCTS**									
1300	Flexible, coated fiberglass fabric on corr. resist. metal helix									
1400	pressure to 12" (WG) UL-181									
1500	Noninsulated, 3" diameter	Q-9	400	.040	L.F.	1.12	1.36		2.48	3.50
1540	5" diameter		320	.050		1.30	1.70		3	4.27
1560	6" diameter		280	.057		1.50	1.94		3.44	4.89
1580	7" diameter		240	.067		1.53	2.27		3.80	5.45
1900	Insulated, 1" thick, PE jacket, 3" diameter G		380	.042		2.60	1.43		4.03	5.25
1910	4" diameter G		340	.047		2.60	1.60		4.20	5.55
1920	5" diameter G		300	.053		2.60	1.81		4.41	5.90
1940	6" diameter G		260	.062		2.94	2.09		5.03	6.70
1960	7" diameter G		220	.073		3.20	2.48		5.68	7.65
1980	8" diameter G		180	.089		3.49	3.02		6.51	8.90
2040	12" diameter G		100	.160		4.90	5.45		10.35	14.50

23 33 53 – Duct Liners

23 33 53.10 Duct Liner Board

0010	**DUCT LINER BOARD**									
3490	Board type, fiberglass liner, 3 lb. density									
3940	Board type, non-fibrous foam									
3950	Temperature, bacteria and fungi resistant									
3960	1" thick G	Q-14	150	.107	S.F.	2.38	3.38		5.76	8.35
3970	1-1/2" thick G		130	.123		3.28	3.91		7.19	10.25
3980	2" thick G		120	.133		3.98	4.23		8.21	11.60

23 34 HVAC Fans

23 34 23 – HVAC Power Ventilators

23 34 23.10 HVAC Power Circulators and Ventilators

0010	**HVAC POWER CIRCULATORS AND VENTILATORS**									
6650	Residential, bath exhaust, grille, back draft damper									
6660	50 CFM	Q-20	24	.833	Ea.	65	29		94	120
6670	110 CFM		22	.909		100	31.50		131.50	163
6900	Kitchen exhaust, grille, complete, 160 CFM		22	.909		111	31.50		142.50	175
6910	180 CFM		20	1		93.50	34.50		128	161
6920	270 CFM		18	1.111		176	38.50		214.50	258
6930	350 CFM		16	1.250		133	43.50		176.50	218
6940	Residential roof jacks and wall caps									
6944	Wall cap with back draft damper									
6946	3" & 4" diam. round duct	1 Shee	11	.727	Ea.	26	27.50		53.50	74.50
6948	6" diam. round duct	"	11	.727	"	65	27.50		92.50	117
6958	Roof jack with bird screen and back draft damper									

23 34 HVAC Fans

23 34 23 – HVAC Power Ventilators

23 34 23.10 HVAC Power Circulators and Ventilators

		Crew	Daily Output	Labor-Hours	Unit	Material	2016 Bare Costs Labor	2016 Bare Costs Equipment	Total	Total Incl O&P
6960	3" & 4" diam. round duct	1 Shee	11	.727	Ea.	26	27.50		53.50	75
6962	3-1/4" x 10" rectangular duct	"	10	.800	"	49	30		79	105
8020	Attic, roof type									
8030	Aluminum dome, damper & curb									
8080	12" diameter, 1000 CFM (gravity)	1 Elec	10	.800	Ea.	595	29.50		624.50	705
8090	16" diameter, 1500 CFM (gravity)		9	.889		720	33		753	845
8100	20" diameter, 2500 CFM (gravity)		8	1		880	37		917	1,025
8160	Plastic, ABS dome									
8180	1050 CFM	1 Elec	14	.571	Ea.	175	21		196	228
8200	1600 CFM	"	12	.667	"	262	24.50		286.50	330
8240	Attic, wall type, with shutter, one speed									
8250	12" diameter, 1000 CFM	1 Elec	14	.571	Ea.	410	21		431	485
8260	14" diameter, 1500 CFM		12	.667		445	24.50		469.50	530
8270	16" diameter, 2000 CFM		9	.889		505	33		538	610
8290	Whole house, wall type, with shutter, one speed									
8300	30" diameter, 4800 CFM	1 Elec	7	1.143	Ea.	1,075	42		1,117	1,250
8310	36" diameter, 7000 CFM		6	1.333		1,175	49		1,224	1,375
8320	42" diameter, 10,000 CFM		5	1.600		1,325	59		1,384	1,550
8330	48" diameter, 16,000 CFM		4	2		1,625	74		1,699	1,925
8340	For two speed, add					98			98	108
8350	Whole house, lay-down type, with shutter, one speed									
8360	30" diameter, 4500 CFM	1 Elec	8	1	Ea.	1,150	37		1,187	1,300
8370	36" diameter, 6500 CFM		7	1.143		1,225	42		1,267	1,425
8380	42" diameter, 9000 CFM		6	1.333		1,350	49		1,399	1,575
8390	48" diameter, 12,000 CFM		5	1.600		1,550	59		1,609	1,800
8440	For two speed, add					74			74	81
8450	For 12 hour timer switch, add	1 Elec	32	.250		74	9.25		83.25	96

23 37 Air Outlets and Inlets

23 37 13 – Diffusers, Registers, and Grilles

23 37 13.10 Diffusers

		Crew	Daily Output	Labor-Hours	Unit	Material	2016 Bare Costs Labor	2016 Bare Costs Equipment	Total	Total Incl O&P
0010	**DIFFUSERS**, Aluminum, opposed blade damper unless noted									
0100	Ceiling, linear, also for sidewall									
0120	2" wide	1 Shee	32	.250	L.F.	16.80	9.45		26.25	34
0160	4" wide		26	.308	"	22.50	11.65		34.15	44
0500	Perforated, 24" x 24" lay-in panel size, 6" x 6"		16	.500	Ea.	157	18.90		175.90	204
0520	8" x 8"		15	.533		165	20		185	215
0530	9" x 9"		14	.571		167	21.50		188.50	220
0590	16" x 16"		11	.727		195	27.50		222.50	261
1000	Rectangular, 1 to 4 way blow, 6" x 6"		16	.500		50	18.90		68.90	86.50
1010	8" x 8"		15	.533		58	20		78	97
1014	9" x 9"		15	.533		66	20		86	106
1016	10" x 10"		15	.533		80	20		100	121
1020	12" x 6"		15	.533		71.50	20		91.50	112
1040	12" x 9"		14	.571		75.50	21.50		97	120
1060	12" x 12"		12	.667		84.50	25		109.50	135
1070	14" x 6"		13	.615		77.50	23.50		101	125
1074	14" x 14"		12	.667		128	25		153	182
1150	18" x 18"		9	.889		138	33.50		171.50	207
1170	24" x 12"		10	.800		163	30		193	231
1180	24" x 24"		7	1.143		270	43		313	370

For customer support on your Residential Cost Data, call 877.759.4771.

23 37 Air Outlets and Inlets

23 37 13 – Diffusers, Registers, and Grilles

23 37 13.10 Diffusers

		Crew	Daily Output	Labor-Hours	Unit	Material	2016 Bare Costs Labor	Equipment	Total	Total Incl O&P
1500	Round, butterfly damper, steel, diffuser size, 6" diameter	1 Shee	18	.444	Ea.	9.40	16.80		26.20	38.50
1520	8" diameter		16	.500		10	18.90		28.90	42.50
2000	T bar mounting, 24" x 24" lay-in frame, 6" x 6"		16	.500		72	18.90		90.90	111
2020	8" x 8"		14	.571		72.50	21.50		94	116
2040	12" x 12"		12	.667		88.50	25		113.50	140
2060	16" x 16"		11	.727		119	27.50		146.50	177
2080	18" x 18"		10	.800		133	30		163	197
6000	For steel diffusers instead of aluminum, deduct					10%				

23 37 13.30 Grilles

		Crew	Daily Output	Labor-Hours	Unit	Material	Labor	Equipment	Total	Total Incl O&P
0010	**GRILLES**									
0020	Aluminum, unless noted otherwise									
1000	Air return, steel, 6" x 6"	1 Shee	26	.308	Ea.	18.70	11.65		30.35	40
1020	10" x 6"		24	.333		18.70	12.60		31.30	41.50
1080	16" x 8"		22	.364		26.50	13.75		40.25	52
1100	12" x 12"		22	.364		26.50	13.75		40.25	52
1120	24" x 12"		18	.444		36	16.80		52.80	67.50
1180	16" x 16"		22	.364		34	13.75		47.75	60.50

23 37 13.60 Registers

		Crew	Daily Output	Labor-Hours	Unit	Material	Labor	Equipment	Total	Total Incl O&P
0010	**REGISTERS**									
0980	Air supply									
3000	Baseboard, hand adj. damper, enameled steel									
3012	8" x 6"	1 Shee	26	.308	Ea.	5.05	11.65		16.70	25
3020	10" x 6"		24	.333		5.50	12.60		18.10	27
3040	12" x 5"		23	.348		6	13.15		19.15	28.50
3060	12" x 6"		23	.348		6.75	13.15		19.90	29.50
4000	Floor, toe operated damper, enameled steel									
4020	4" x 8"	1 Shee	32	.250	Ea.	9.40	9.45		18.85	26
4040	4" x 12"	"	26	.308	"	11	11.65		22.65	31.50
4300	Spiral pipe supply register									
4310	Steel, with air scoop									
4320	4" x 12", for 8" thru 13" diameter duct	1 Shee	25	.320	Ea.	74.50	12.10		86.60	102
4330	4" x 18", for 8" thru 13" diameter duct		18	.444		87	16.80		103.80	124
4340	6" x 12", for 14" thru 21" diameter duct		19	.421		78.50	15.90		94.40	113
4350	6" x 16", for 14" thru 21" diameter duct		18	.444		89	16.80		105.80	126
4360	6" x 20", for 14" thru 21" diameter duct		17	.471		98.50	17.80		116.30	138
4370	6" x 24", for 14" thru 21" diameter duct		16	.500		116	18.90		134.90	159
4380	8" x 16", for 22" thru 31" diameter duct		19	.421		94	15.90		109.90	130
4390	8" x 18", for 22" thru 31" diameter duct		18	.444		98.50	16.80		115.30	136
4400	8" x 24", for 22" thru 31" diameter duct		15	.533		122	20		142	168

23 41 Particulate Air Filtration

23 41 13 – Panel Air Filters

23 41 13.10 Panel Type Air Filters

			Crew	Daily Output	Labor-Hours	Unit	Material	Labor	Equipment	Total	Total Incl O&P
0010	**PANEL TYPE AIR FILTERS**										
2950	Mechanical media filtration units										
3000	High efficiency type, with frame, non-supported	G				MCFM	35			35	38.50
3100	Supported type	G				"	50			50	55
5500	Throwaway glass or paper media type, 12" x 36" x 1"					Ea.	2.39			2.39	2.63

23 41 Particulate Air Filtration

23 41 16 – Renewable-Media Air Filters

23 41 16.10 Disposable Media Air Filters

		Crew	Daily Output	Labor-Hours	Unit	Material	2016 Bare Costs Labor	Equipment	Total	Total Incl O&P
0010	**DISPOSABLE MEDIA AIR FILTERS**									
5000	Renewable disposable roll				C.S.F.	5.25			5.25	5.75

23 41 19 – Washable Air Filters

23 41 19.10 Permanent Air Filters

0010	**PERMANENT AIR FILTERS**									
4500	Permanent washable	G			MCFM	20			20	22

23 41 23 – Extended Surface Filters

23 41 23.10 Expanded Surface Filters

0010	**EXPANDED SURFACE FILTERS**									
4000	Medium efficiency, extended surface	G			MCFM	6			6	6.60

23 42 Gas-Phase Air Filtration

23 42 13 – Activated-Carbon Air Filtration

23 42 13.10 Charcoal Type Air Filtration

0010	**CHARCOAL TYPE AIR FILTRATION**									
0050	Activated charcoal type, full flow				MCFM	600			600	660
0060	Full flow, impregnated media 12" deep					225			225	248
0070	HEPA filter & frame for field erection					400			400	440
0080	HEPA filter-diffuser, ceiling install.					300			300	330

23 43 Electronic Air Cleaners

23 43 13 – Washable Electronic Air Cleaners

23 43 13.10 Electronic Air Cleaners

0010	**ELECTRONIC AIR CLEANERS**									
2000	Electronic air cleaner, duct mounted									
2150	1000 CFM	1 Shee	4	2	Ea.	420	75.50		495.50	590
2200	1200 CFM		3.80	2.105		485	79.50		564.50	670
2250	1400 CFM		3.60	2.222		500	84		584	690

23 51 Breechings, Chimneys, and Stacks

23 51 23 – Gas Vents

23 51 23.10 Gas Chimney Vents

0010	**GAS CHIMNEY VENTS**, Prefab metal, U.L. listed									
0020	Gas, double wall, galvanized steel									
0080	3" diameter	Q-9	72	.222	V.L.F.	6.60	7.55		14.15	19.85
0100	4" diameter	"	68	.235	"	8.15	8		16.15	22.50

23 52 Heating Boilers

23 52 13 – Electric Boilers

23 52 13.10 Electric Boilers, ASME

		Crew	Daily Output	Labor-Hours	Unit	Material	2016 Bare Costs Labor	Equipment	Total	Total Incl O&P
0010	**ELECTRIC BOILERS, ASME**, Standard controls and trim									
1000	Steam, 6 KW, 20.5 MBH	Q-19	1.20	20	Ea.	4,100	725		4,825	5,700
1160	60 KW, 205 MBH		1	24		6,925	870		7,795	9,025
2000	Hot water, 7.5 KW, 25.6 MBH		1.30	18.462		4,950	670		5,620	6,550
2040	30 KW, 102 MBH		1.20	20		5,325	725		6,050	7,050
2060	45 KW, 164 MBH		1.20	20		5,425	725		6,150	7,150

23 52 23 – Cast-Iron Boilers

23 52 23.20 Gas-Fired Boilers

		Crew	Daily Output	Labor-Hours	Unit	Material	Labor	Equipment	Total	Total Incl O&P
0010	**GAS-FIRED BOILERS**, Natural or propane, standard controls, packaged									
1000	Cast iron, with insulated jacket									
3000	Hot water, gross output, 80 MBH	Q-7	1.46	21.918	Ea.	1,975	790		2,765	3,475
3020	100 MBH	"	1.35	23.704		2,550	855		3,405	4,225
7000	For tankless water heater, add					10%				
7050	For additional zone valves up to 312 MBH, add					194			194	214

23 52 23.30 Gas/Oil Fired Boilers

		Crew	Daily Output	Labor-Hours	Unit	Material	Labor	Equipment	Total	Total Incl O&P
0010	**GAS/OIL FIRED BOILERS**, Combination with burners and controls, packaged									
1000	Cast iron with insulated jacket									
2000	Steam, gross output, 720 MBH	Q-7	.43	74.074	Ea.	14,700	2,675		17,375	20,600
2900	Hot water, gross output									
2910	200 MBH	Q-6	.62	39.024	Ea.	10,600	1,350		11,950	13,900
2920	300 MBH		.49	49.080		10,600	1,700		12,300	14,500
2930	400 MBH		.41	57.971		12,400	2,025		14,425	17,000
2940	500 MBH		.36	67.039		13,400	2,325		15,725	18,600
3000	584 MBH	Q-7	.44	72.072		14,800	2,600		17,400	20,600

23 52 23.40 Oil-Fired Boilers

		Crew	Daily Output	Labor-Hours	Unit	Material	Labor	Equipment	Total	Total Incl O&P
0010	**OIL-FIRED BOILERS**, Standard controls, flame retention burner, packaged									
1000	Cast iron, with insulated flush jacket									
2000	Steam, gross output, 109 MBH	Q-7	1.20	26.667	Ea.	2,325	960		3,285	4,125
2060	207 MBH	"	.90	35.556	"	3,175	1,275		4,450	5,625
3000	Hot water, same price as steam									

23 52 26 – Steel Boilers

23 52 26.40 Oil-Fired Boilers

		Crew	Daily Output	Labor-Hours	Unit	Material	Labor	Equipment	Total	Total Incl O&P
0010	**OIL-FIRED BOILERS**, Standard controls, flame retention burner									
5000	Steel, with insulated flush jacket									
7000	Hot water, gross output, 103 MBH	Q-6	1.60	15	Ea.	1,825	520		2,345	2,850
7020	122 MBH		1.45	16.506		2,000	575		2,575	3,150
7060	168 MBH		1.30	18.405		2,250	640		2,890	3,525
7080	225 MBH		1.22	19.704		3,250	685		3,935	4,700

23 52 28 – Swimming Pool Boilers

23 52 28.10 Swimming Pool Heaters

		Crew	Daily Output	Labor-Hours	Unit	Material	Labor	Equipment	Total	Total Incl O&P
0010	**SWIMMING POOL HEATERS**, Not including wiring, external									
0020	piping, base or pad									
0160	Gas fired, input, 155 MBH	Q-6	1.50	16	Ea.	2,100	555		2,655	3,250
0200	199 MBH		1	24		2,225	835		3,060	3,825
0280	500 MBH		.40	60		9,325	2,075		11,400	13,700
2000	Electric, 12 KW, 4,800 gallon pool	Q-19	3	8		2,100	291		2,391	2,800
2020	15 KW, 7,200 gallon pool		2.80	8.571		2,150	310		2,460	2,850
2040	24 KW, 9,600 gallon pool		2.40	10		2,475	365		2,840	3,325
2100	57 KW, 24,000 gallon pool		1.20	20		3,650	725		4,375	5,200

23 52 Heating Boilers

23 52 88 – Burners

23 52 88.10 Replacement Type Burners		Crew	Daily Output	Labor-Hours	Unit	Material	2016 Bare Costs Labor	Equipment	Total	Total Incl O&P
0010	**REPLACEMENT TYPE BURNERS**									
0990	Residential, conversion, gas fired, LP or natural									
1000	Gun type, atmospheric input 50 to 225 MBH	Q-1	2.50	6.400	Ea.	915	225		1,140	1,375
1020	100 to 400 MBH	"	2	8		1,525	281		1,806	2,150
1025	Burner, gas, 100 – 400 MBH [G]					1,525			1,525	1,675
1040	300 to 1000 MBH	Q-1	1.70	9.412		4,600	330		4,930	5,625

23 54 Furnaces

23 54 13 – Electric-Resistance Furnaces

23 54 13.10 Electric Furnaces

		Crew	Daily Output	Labor-Hours	Unit	Material	Labor	Equipment	Total	Total Incl O&P
0010	**ELECTRIC FURNACES**, Hot air, blowers, std. controls									
0011	not including gas, oil or flue piping									
1000	Electric, UL listed									
1100	34.1 MBH	Q-20	4.40	4.545	Ea.	455	157		612	760

23 54 16 – Fuel-Fired Furnaces

23 54 16.13 Gas-Fired Furnaces

		Crew	Daily Output	Labor-Hours	Unit	Material	Labor	Equipment	Total	Total Incl O&P
0010	**GAS-FIRED FURNACES**									
3000	Gas, AGA certified, upflow, direct drive models									
3020	45 MBH input	Q-9	4	4	Ea.	580	136		716	860
3040	60 MBH input		3.80	4.211		595	143		738	895
3060	75 MBH input		3.60	4.444		660	151		811	975
3100	100 MBH input		3.20	5		690	170		860	1,050
3120	125 MBH input		3	5.333		725	181		906	1,100
3130	150 MBH input		2.80	5.714		1,375	194		1,569	1,825
3140	200 MBH input		2.60	6.154		2,950	209		3,159	3,600
4000	For starter plenum, add		16	1		91	34		125	157

23 54 16.16 Oil-Fired Furnaces

		Crew	Daily Output	Labor-Hours	Unit	Material	Labor	Equipment	Total	Total Incl O&P
0010	**OIL-FIRED FURNACES**									
6000	Oil, UL listed, atomizing gun type burner									
6020	56 MBH output	Q-9	3.60	4.444	Ea.	1,700	151		1,851	2,125
6030	84 MBH output		3.50	4.571		2,500	156		2,656	3,000
6040	95 MBH output		3.40	4.706		2,575	160		2,735	3,100
6060	134 MBH output		3.20	5		2,625	170		2,795	3,150
6080	151 MBH output		3	5.333		2,725	181		2,906	3,300
6100	200 MBH input		2.60	6.154		3,000	209		3,209	3,675

23 54 16.21 Solid Fuel-Fired Furnaces

			Crew	Daily Output	Labor-Hours	Unit	Material	Labor	Equipment	Total	Total Incl O&P
0010	**SOLID FUEL-FIRED FURNACES**										
6020	Wood fired furnaces										
6030	Includes hot water coil, thermostat, and auto draft control										
6040	24" long firebox	[G]	Q-9	4	4	Ea.	4,300	136		4,436	4,950
6050	30" long firebox	[G]		3.60	4.444		5,025	151		5,176	5,775
6060	With fireplace glass doors	[G]		3.20	5		6,350	170		6,520	7,275
6200	Wood/oil fired furnaces, includes two thermostats										
6210	Includes hot water coil and auto draft control										
6240	24" long firebox	[G]	Q-9	3.40	4.706	Ea.	5,150	160		5,310	5,950
6250	30" long firebox	[G]		3	5.333		5,775	181		5,956	6,650
6260	With fireplace glass doors	[G]		2.80	5.714		7,075	194		7,269	8,100
6400	Wood/gas fired furnaces, includes two thermostats										
6410	Includes hot water coil and auto draft control										

23 54 Furnaces

23 54 16 – Fuel-Fired Furnaces

23 54 16.21 Solid Fuel-Fired Furnaces		Crew	Daily Output	Labor-Hours	Unit	Material	2016 Bare Costs Labor	Equipment	Total	Total Incl O&P
6440	24" long firebox	Q-9 [G]	2.80	5.714	Ea.	5,925	194		6,119	6,850
6450	30" long firebox	[G]	2.40	6.667		6,625	227		6,852	7,675
6460	With fireplace glass doors	[G]	2	8		7,900	272		8,172	9,150
6600	Wood/oil/gas fired furnaces, optional accessories									
6610	Hot air plenum	Q-9	16	1	Ea.	108	34		142	176
6620	Safety heat dump		24	.667		94	22.50		116.50	141
6630	Auto air intake		18	.889		148	30.50		178.50	214
6640	Cold air return package		14	1.143		159	39		198	240
6650	Wood fork					51			51	56
6700	Wood fired outdoor furnace									
6740	24" long firebox	Q-9 [G]	3.80	4.211	Ea.	4,500	143		4,643	5,200
6760	Wood fired outdoor furnace, optional accessories									
6770	Chimney section, stainless steel, 6" ID x 3' lg.	Q-9 [G]	36	.444	Ea.	165	15.10		180.10	207
6780	Chimney cap, stainless steel	"	40	.400	"	95	13.60		108.60	128
6800	Wood fired hot water furnace									
6820	Includes 200 gal. hot water storage, thermostat, and auto draft control									
6840	30" long firebox	Q-9 [G]	2.10	7.619	Ea.	8,275	259		8,534	9,525
6850	Water to air heat exchanger									
6870	Includes mounting kit and blower relay									
6880	140 MBH, 18.75" W x 18.75" L	Q-9	7.50	2.133	Ea.	370	72.50		442.50	525
6890	200 MBH, 24" W x 24" L	"	7	2.286	"	525	78		603	710
6900	Water to water heat exchanger									
6940	100 MBH	Q-9	6.50	2.462	Ea.	370	84		454	550
6960	290 MBH	"	6	2.667	"	535	91		626	740
7000	Optional accessories									
7010	Large volume circulation pump, (2 Included)	Q-9	14	1.143	Ea.	242	39		281	330
7020	Air bleed fittings, (package)		24	.667		50.50	22.50		73	93.50
7030	Domestic water preheater		6	2.667		219	91		310	390
7040	Smoke pipe kit		4	4		88	136		224	325

23 54 24 – Furnace Components for Cooling

23 54 24.10 Furnace Components and Combinations

		Crew	Daily Output	Labor-Hours	Unit	Material	Labor	Equipment	Total	Total Incl O&P
0010	**FURNACE COMPONENTS AND COMBINATIONS**									
0080	Coils, A.C. evaporator, for gas or oil furnaces									
0090	Add-on, with holding charge									
0100	Upflow									
0120	1-1/2 ton cooling	Q-5	4	4	Ea.	222	144		366	480
0130	2 ton cooling		3.70	4.324		241	156		397	520
0140	3 ton cooling		3.30	4.848		400	175		575	730
0150	4 ton cooling		3	5.333		530	192		722	900
0160	5 ton cooling		2.70	5.926		550	214		764	955
0300	Downflow									
0330	2-1/2 ton cooling	Q-5	3	5.333	Ea.	330	192		522	680
0340	3-1/2 ton cooling		2.60	6.154		505	222		727	920
0350	5 ton cooling		2.20	7.273		550	262		812	1,025
0600	Horizontal									
0630	2 ton cooling	Q-5	3.90	4.103	Ea.	269	148		417	540
0640	3 ton cooling		3.50	4.571		380	165		545	690
0650	4 ton cooling		3.20	5		445	180		625	785
0660	5 ton cooling		2.90	5.517		445	199		644	820
2000	Cased evaporator coils for air handlers									
2100	1-1/2 ton cooling	Q-5	4.40	3.636	Ea.	251	131		382	490
2110	2 ton cooling		4.10	3.902		330	141		471	595

23 54 Furnaces

23 54 24 – Furnace Components for Cooling

23 54 24.10 Furnace Components and Combinations		Crew	Daily Output	Labor-Hours	Unit	Material	2016 Bare Costs Labor	Equipment	Total	Total Incl O&P
2120	2-1/2 ton cooling	Q-5	3.90	4.103	Ea.	340	148		488	620
2130	3 ton cooling		3.70	4.324		365	156		521	655
2140	3-1/2 ton cooling		3.50	4.571		455	165		620	775
2150	4 ton cooling		3.20	5		495	180		675	840
2160	5 ton cooling		2.90	5.517		500	199		699	880
3010	Air handler, modular									
3100	With cased evaporator cooling coil									
3120	1-1/2 ton cooling	Q-5	3.80	4.211	Ea.	855	152		1,007	1,200
3130	2 ton cooling		3.50	4.571		945	165		1,110	1,325
3140	2-1/2 ton cooling		3.30	4.848		1,000	175		1,175	1,400
3150	3 ton cooling		3.10	5.161		1,150	186		1,336	1,575
3160	3-1/2 ton cooling		2.90	5.517		1,325	199		1,524	1,800
3170	4 ton cooling		2.50	6.400		1,425	231		1,656	1,950
3180	5 ton cooling		2.10	7.619		1,700	275		1,975	2,325
3500	With no cooling coil									
3520	1-1/2 ton coil size	Q-5	12	1.333	Ea.	720	48		768	875
3530	2 ton coil size		10	1.600		720	57.50		777.50	890
3540	2-1/2 ton coil size		10	1.600		730	57.50		787.50	900
3554	3 ton coil size		9	1.778		750	64		814	930
3560	3-1/2 ton coil size		9	1.778		770	64		834	950
3570	4 ton coil size		8.50	1.882		780	68		848	970
3580	5 ton coil size		8	2		790	72		862	990
4000	With heater									
4120	5 kW, 17.1 MBH	Q-5	16	1	Ea.	455	36		491	560
4130	7.5 kW, 25.6 MBH		15.60	1.026		1,100	37		1,137	1,250
4140	10 kW, 34.2 MBH		15.20	1.053		1,175	38		1,213	1,375
4150	12.5 KW, 42.7 MBH		14.80	1.081		1,550	39		1,589	1,775
4160	15 KW, 51.2 MBH		14.40	1.111		1,675	40		1,715	1,925
4170	25 KW, 85.4 MBH		14	1.143		2,225	41		2,266	2,525
4180	30 KW, 102 MBH		13	1.231		2,575	44.50		2,619.50	2,925

23 62 Packaged Compressor and Condenser Units

23 62 13 – Packaged Air-Cooled Refrigerant Compressor and Condenser Units

23 62 13.10 Packaged Air-Cooled Refrig. Condensing Units

		Crew	Daily Output	Labor-Hours	Unit	Material	Labor	Equipment	Total	Total Incl O&P
0010	**PACKAGED AIR-COOLED REFRIGERANT CONDENSING UNITS**									
0020	Condensing unit									
0030	Air cooled, compressor, standard controls									
0050	1.5 ton	Q-5	2.50	6.400	Ea.	1,275	231		1,506	1,775
0100	2 ton		2.10	7.619		1,300	275		1,575	1,875
0200	2.5 ton		1.70	9.412		1,425	340		1,765	2,125
0300	3 ton		1.30	12.308		1,475	445		1,920	2,350
0350	3.5 ton		1.10	14.545		1,700	525		2,225	2,750
0400	4 ton		.90	17.778		1,950	640		2,590	3,175
0500	5 ton		.60	26.667		2,325	960		3,285	4,150

23 74 Packaged Outdoor HVAC Equipment

23 74 33 – Dedicated Outdoor-Air Units

23 74 33.10 Rooftop Air Conditioners		Crew	Daily Output	Labor-Hours	Unit	Material	2016 Bare Costs Labor	2016 Bare Costs Equipment	Total	Total Incl O&P
0010	**ROOFTOP AIR CONDITIONERS**, Standard controls, curb, economizer									
1000	Single zone, electric cool, gas heat									
1140	5 ton cooling, 112 MBH heating	Q-5	.56	28.521	Ea.	4,425	1,025		5,450	6,550
1150	7.5 ton cooling, 170 MBH heating		.50	32.258		5,875	1,175		7,050	8,400
1156	8.5 ton cooling, 170 MBH heating	↓	.46	34.783		7,100	1,250		8,350	9,900
1160	10 ton cooling, 200 MBH heating	Q-6	.67	35.982	↓	9,100	1,250		10,350	12,100

23 81 Decentralized Unitary HVAC Equipment

23 81 13 – Packaged Terminal Air-Conditioners

23 81 13.10 Packaged Cabinet Type Air-Conditioners

		Crew	Daily Output	Labor-Hours	Unit	Material	Labor	Equipment	Total	Total Incl O&P
0010	**PACKAGED CABINET TYPE AIR-CONDITIONERS**, Cabinet, wall sleeve,									
0100	louver, electric heat, thermostat, manual changeover, 208 V									
0200	6,000 BTUH cooling, 8800 BTU heat	Q-5	6	2.667	Ea.	825	96		921	1,075
0220	9,000 BTUH cooling, 13,900 BTU heat		5	3.200		1,225	115		1,340	1,550
0240	12,000 BTUH cooling, 13,900 BTU heat		4	4		1,350	144		1,494	1,750
0260	15,000 BTUH cooling, 13,900 BTU heat	↓	3	5.333	↓	1,475	192		1,667	1,950

23 81 19 – Self-Contained Air-Conditioners

23 81 19.10 Window Unit Air Conditioners

		Crew	Daily Output	Labor-Hours	Unit	Material	Labor	Equipment	Total	Total Incl O&P
0010	**WINDOW UNIT AIR CONDITIONERS**									
4000	Portable/window, 15 amp, 125 V grounded receptacle required									
4060	5000 BTUH	1 Carp	8	1	Ea.	330	34		364	415
4340	6000 BTUH		8	1		380	34		414	475
4480	8000 BTUH		6	1.333		475	45		520	600
4500	10,000 BTUH	↓	6	1.333		590	45		635	725
4520	12,000 BTUH	L-2	8	2	↓	740	58.50		798.50	915
4600	Window/thru-the-wall, 15 amp, 230 V grounded receptacle required									
4780	18,000 BTUH	L-2	6	2.667	Ea.	1,000	78		1,078	1,225
4940	25,000 BTUH		4	4		1,275	117		1,392	1,600
4960	29,000 BTUH	↓	4	4	↓	1,400	117		1,517	1,725

23 81 19.20 Self-Contained Single Package

		Crew	Daily Output	Labor-Hours	Unit	Material	Labor	Equipment	Total	Total Incl O&P
0010	**SELF-CONTAINED SINGLE PACKAGE**									
0100	Air cooled, for free blow or duct, not incl. remote condenser									
0110	Constant volume									
0200	3 ton cooling	Q-5	1	16	Ea.	3,700	575		4,275	5,025
0210	4 ton cooling	"	.80	20	"	4,025	720		4,745	5,625
1000	Water cooled for free blow or duct, not including tower									
1010	Constant volume									
1100	3 ton cooling	Q-6	1	24	Ea.	3,575	835		4,410	5,300
1300	For hot water or steam heat coils, add				"	12%	10%			

23 81 26 – Split-System Air-Conditioners

23 81 26.10 Split Ductless Systems

		Crew	Daily Output	Labor-Hours	Unit	Material	Labor	Equipment	Total	Total Incl O&P
0010	**SPLIT DUCTLESS SYSTEMS**									
0100	Cooling only, single zone									
0110	Wall mount									
0120	3/4 ton cooling	Q-5	2	8	Ea.	1,100	288		1,388	1,700
0130	1 ton cooling		1.80	8.889		1,225	320		1,545	1,875
0140	1-1/2 ton cooling		1.60	10		1,975	360		2,335	2,775
0150	2 ton cooling	↓	1.40	11.429	↓	2,250	410		2,660	3,150
1000	Ceiling mount									
1020	2 ton cooling	Q-5	1.40	11.429	Ea.	2,000	410		2,410	2,875

23 81 Decentralized Unitary HVAC Equipment

23 81 26 – Split-System Air-Conditioners

23 81 26.10 Split Ductless Systems		Crew	Daily Output	Labor-Hours	Unit	Material	2016 Bare Costs Labor	Equipment	Total	Total Incl O&P
1030	3 ton cooling	Q-5	1.20	13.333	Ea.	2,550	480		3,030	3,600
3000	Multizone									
3010	Wall mount									
3020	2 @ 3/4 ton cooling	Q-5	1.80	8.889	Ea.	3,375	320		3,695	4,225
5000	Cooling/Heating									
5010	Wall mount									
5110	1 ton cooling	Q-5	1.70	9.412	Ea.	1,300	340		1,640	1,975
5120	1-1/2 ton cooling	"	1.50	10.667	"	1,975	385		2,360	2,800
7000	Accessories for all split ductless systems									
7010	Add for ambient frost control	Q-5	8	2	Ea.	126	72		198	258
7020	Add for tube/wiring kit, (Line sets)									
7030	15' kit	Q-5	32	.500	Ea.	93	18.05		111.05	132
7036	25' kit		28	.571		186	20.50		206.50	238
7040	35' kit		24	.667		196	24		220	256
7050	50' kit		20	.800		237	29		266	310

23 81 43 – Air-Source Unitary Heat Pumps

23 81 43.10 Air-Source Heat Pumps		Crew	Daily Output	Labor-Hours	Unit	Material	Labor	Equipment	Total	Total Incl O&P
0010	**AIR-SOURCE HEAT PUMPS**, Not including interconnecting tubing									
1000	Air to air, split system, not including curbs, pads, fan coil and ductwork									
1012	Outside condensing unit only, for fan coil see Section 23 82 19.10									
1020	2 ton cooling, 8.5 MBH heat @ 0°F	Q-5	2	8	Ea.	1,600	288		1,888	2,225
1054	4 ton cooling, 24 MBH heat @ 0°F	"	.80	20	"	2,375	720		3,095	3,825
1500	Single package, not including curbs, pads, or plenums									
1520	2 ton cooling, 6.5 MBH heat @ 0°F	Q-5	1.50	10.667	Ea.	3,100	385		3,485	4,050
1580	4 ton cooling, 13 MBH heat @ 0°F	"	.96	16.667	"	4,200	600		4,800	5,625

23 81 46 – Water-Source Unitary Heat Pumps

23 81 46.10 Water Source Heat Pumps		Crew	Daily Output	Labor-Hours	Unit	Material	Labor	Equipment	Total	Total Incl O&P
0010	**WATER SOURCE HEAT PUMPS**, Not incl. connecting tubing or water source									
2000	Water source to air, single package									
2100	1 ton cooling, 13 MBH heat @ 75°F	Q-5	2	8	Ea.	1,775	288		2,063	2,425
2200	4 ton cooling, 31 MBH heat @ 75°F	"	1.20	13.333	"	2,975	480		3,455	4,075

23 82 Convection Heating and Cooling Units

23 82 19 – Fan Coil Units

23 82 19.10 Fan Coil Air Conditioning		Crew	Daily Output	Labor-Hours	Unit	Material	Labor	Equipment	Total	Total Incl O&P
0010	**FAN COIL AIR CONDITIONING**									
0030	Fan coil AC, cabinet mounted, filters and controls									
0100	Chilled water, 1/2 ton cooling	Q-5	8	2	Ea.	460	72		532	625
0110	3/4 ton cooling		7	2.286		565	82.50		647.50	755
0120	1 ton cooling		6	2.667		685	96		781	915

23 82 29 – Radiators

23 82 29.10 Hydronic Heating		Crew	Daily Output	Labor-Hours	Unit	Material	Labor	Equipment	Total	Total Incl O&P
0010	**HYDRONIC HEATING**, Terminal units, not incl. main supply pipe									
1000	Radiation									
1100	Panel, baseboard, C.I., including supports, no covers	Q-5	46	.348	L.F.	38	12.55		50.55	62
3000	Radiators, cast iron									
3100	Free standing or wall hung, 6 tube, 25" high	Q-5	96	.167	Section	47.50	6		53.50	62
3200	4 tube, 19" high	"	96	.167	"	30.50	6		36.50	44
3250	Adj. brackets, 2 per wall radiator up to 30 sections	1 Stpi	32	.250	Ea.	54.50	10		64.50	76.50

23 82 Convection Heating and Cooling Units

23 82 36 – Finned-Tube Radiation Heaters

23 82 36.10 Finned Tube Radiation	Crew	Daily Output	Labor-Hours	Unit	Material	2016 Bare Costs Labor	Equipment	Total	Total Incl O&P
0010 **FINNED TUBE RADIATION**, Terminal units, not incl. main supply pipe									
1310 Baseboard, pkgd, 1/2" copper tube, alum. fin, 7" high	Q-5	60	.267	L.F.	11.20	9.60		20.80	28
1320 3/4" copper tube, alum. fin, 7" high		58	.276		7.35	9.95		17.30	24.50
1340 1" copper tube, alum. fin, 8-7/8" high		56	.286		18.85	10.30		29.15	37.50
1360 1-1/4" copper tube, alum. fin, 8-7/8" high	↓	54	.296	↓	28	10.70		38.70	48
1500 Note: fin tube may also require corners, caps, etc.									

23 83 Radiant Heating Units

23 83 16 – Radiant-Heating Hydronic Piping

23 83 16.10 Radiant Floor Heating	Crew	Daily Output	Labor-Hours	Unit	Material	2016 Bare Costs Labor	Equipment	Total	Total Incl O&P
0010 **RADIANT FLOOR HEATING**									
0100 Tubing, PEX (cross-linked polyethylene)									
0110 Oxygen barrier type for systems with ferrous materials									
0120 1/2"	Q-5	800	.020	L.F.	1.07	.72		1.79	2.37
0130 3/4"		535	.030		1.51	1.08		2.59	3.44
0140 1"	↓	400	.040	↓	2.36	1.44		3.80	4.98
0200 Non barrier type for ferrous free systems									
0210 1/2"	Q-5	800	.020	L.F.	.61	.72		1.33	1.86
0220 3/4"		535	.030		1.11	1.08		2.19	3
0230 1"	↓	400	.040	↓	1.90	1.44		3.34	4.47
1000 Manifolds									
1110 Brass									
1120 With supply and return valves, flow meter, thermometer,									
1122 auto air vent and drain/fill valve.									
1130 1", 2 circuit	Q-5	14	1.143	Ea.	257	41		298	350
1140 1", 3 circuit		13.50	1.185		294	42.50		336.50	395
1150 1", 4 circuit		13	1.231		320	44.50		364.50	425
1154 1", 5 circuit		12.50	1.280		380	46		426	495
1158 1", 6 circuit		12	1.333		415	48		463	535
1162 1", 7 circuit		11.50	1.391		455	50		505	585
1166 1", 8 circuit		11	1.455		500	52.50		552.50	635
1172 1", 9 circuit		10.50	1.524		540	55		595	685
1174 1", 10 circuit		10	1.600		580	57.50		637.50	735
1178 1", 11 circuit		9.50	1.684		605	60.50		665.50	765
1182 1", 12 circuit	↓	9	1.778	↓	670	64		734	840
1610 Copper manifold header (cut to size)									
1620 1" header, 12 circuit 1/2" sweat outlets	Q-5	3.33	4.805	Ea.	91	173		264	385
1630 1-1/4" header, 12 circuit 1/2" sweat outlets		3.20	5		106	180		286	415
1640 1-1/4" header, 12 circuit 3/4" sweat outlets		3	5.333		114	192		306	440
1650 1-1/2" header, 12 circuit 3/4" sweat outlets		3.10	5.161		137	186		323	455
1660 2" header, 12 circuit 3/4" sweat outlets	↓	2.90	5.517	↓	201	199		400	550
3000 Valves									
3110 Thermostatic zone valve actuator with end switch	Q-5	40	.400	Ea.	38.50	14.40		52.90	66.50
3114 Thermostatic zone valve actuator	"	36	.444	"	81	16		97	116
3120 Motorized straight zone valve with operator complete									
3130 3/4"	Q-5	35	.457	Ea.	130	16.50		146.50	170
3140 1"		32	.500		141	18.05		159.05	185
3150 1-1/4"	↓	29.60	.541	↓	179	19.50		198.50	229
3500 4 way mixing valve, manual, brass									
3530 1"	Q-5	13.30	1.203	Ea.	180	43.50		223.50	270

23 83 Radiant Heating Units

23 83 16 – Radiant-Heating Hydronic Piping

23 83 16.10 Radiant Floor Heating

		Crew	Daily Output	Labor-Hours	Unit	Material	2016 Bare Costs Labor	Equipment	Total	Total Incl O&P
3540	1-1/4"	Q-5	11.40	1.404	Ea.	195	50.50		245.50	298
3550	1-1/2"		11	1.455		249	52.50		301.50	360
3560	2"		10.60	1.509		350	54.50		404.50	480
3800	Mixing valve motor, 4 way for valves, 1" and 1-1/4"		34	.471		355	16.95		371.95	420
3810	Mixing valve motor, 4 way for valves, 1-1/2" and 2"	↓	30	.533	↓	365	19.25		384.25	435
5000	Radiant floor heating, zone control panel									
5120	4 zone actuator valve control, expandable	Q-5	20	.800	Ea.	163	29		192	227
5130	6 zone actuator valve control, expandable		18	.889		226	32		258	300
6070	Thermal track, straight panel for long continuous runs, 5.333 S.F.		40	.400		26.50	14.40		40.90	53.50
6080	Thermal track, utility panel, for direction reverse at run end, 5.333 S.F.		40	.400		26.50	14.40		40.90	53.50
6090	Combination panel, for direction reverse plus straight run, 5.333 S.F.	↓	40	.400	↓	26.50	14.40		40.90	53.50
7000	PEX tubing fittings									
7100	Compression type									
7116	Coupling									
7120	1/2" x 1/2"	1 Stpi	27	.296	Ea.	6.70	11.85		18.55	27
7124	3/4" x 3/4"	"	23	.348	"	10.60	13.95		24.55	34.50
7130	Adapter									
7132	1/2" x female sweat 1/2"	1 Stpi	27	.296	Ea.	4.36	11.85		16.21	24.50
7134	1/2" x female sweat 3/4"		26	.308		4.88	12.30		17.18	26
7136	5/8" x female sweat 3/4"	↓	24	.333		7	13.35		20.35	29.50
7140	Elbow									
7142	1/2" x female sweat 1/2"	1 Stpi	27	.296	Ea.	6.65	11.85		18.50	27
7144	1/2" x female sweat 3/4"		26	.308		7.80	12.30		20.10	29
7146	5/8" x female sweat 3/4"	↓	24	.333		8.75	13.35		22.10	31.50
7200	Insert type									
7206	PEX x male NPT									
7210	1/2" x 1/2"	1 Stpi	29	.276	Ea.	2.59	11.05		13.64	21
7220	3/4" x 3/4"		27	.296		3.81	11.85		15.66	23.50
7230	1" x 1"	↓	26	.308	↓	6.45	12.30		18.75	27.50
7300	PEX coupling									
7310	1/2" x 1/2"	1 Stpi	30	.267	Ea.	.98	10.70		11.68	18.70
7320	3/4" x 3/4"		29	.276		1.22	11.05		12.27	19.55
7330	1" x 1"	↓	28	.286		2.29	11.45		13.74	21.50
7400	PEX stainless crimp ring									
7410	1/2" x 1/2"	1 Stpi	86	.093	Ea.	.37	3.73		4.10	6.55
7420	3/4" x 3/4"		84	.095		.51	3.81		4.32	6.85
7430	1" x 1"	↓	82	.098	↓	.73	3.91		4.64	7.25

23 83 33 – Electric Radiant Heaters

23 83 33.10 Electric Heating

		Crew	Daily Output	Labor-Hours	Unit	Material	Labor	Equipment	Total	Total Incl O&P
0010	**ELECTRIC HEATING**, not incl. conduit or feed wiring									
1100	Rule of thumb: Baseboard units, including control	1 Elec	4.40	1.818	kW	105	67		172	226
1300	Baseboard heaters, 2' long, 350 watt		8	1	Ea.	26	37		63	89
1400	3' long, 750 watt		8	1		31.50	37		68.50	95
1600	4' long, 1000 watt		6.70	1.194		35.50	44		79.50	111
1800	5' long, 935 watt		5.70	1.404		46.50	52		98.50	136
2000	6' long, 1500 watt		5	1.600		51	59		110	153
2400	8' long, 2000 watt		4	2		56	74		130	183
2800	10' long, 1875 watt	↓	3.30	2.424		180	89.50		269.50	345
2950	Wall heaters with fan, 120 to 277 volt									
3170	1000 watt	1 Elec	6	1.333	Ea.	83.50	49		132.50	173
3180	1250 watt		5	1.600		83.50	59		142.50	189
3190	1500 watt	↓	4	2	↓	83.50	74		157.50	213

For customer support on your Residential Cost Data, call 877.759.4771.

23 83 Radiant Heating Units

23 83 33 – Electric Radiant Heaters

23 83 33.10 Electric Heating		Crew	Daily Output	Labor-Hours	Unit	Material	2016 Bare Costs Labor	Equipment	Total	Total Incl O&P
3600	Thermostats, integral	1 Elec	16	.500	Ea.	26.50	18.45		44.95	59
3800	Line voltage, 1 pole		8	1		12.85	37		49.85	74.50
5000	Radiant heating ceiling panels, 2' x 4', 500 watt		16	.500		410	18.45		428.45	485
5050	750 watt		16	.500		455	18.45		473.45	530
5300	Infrared quartz heaters, 120 volts, 1000 watts		6.70	1.194		345	44		389	450
5350	1500 watt		5	1.600		345	59		404	475
5400	240 volts, 1500 watt		5	1.600		345	59		404	475
5450	2000 watt		4	2		345	74		419	500
5500	3000 watt		3	2.667		380	98.50		478.50	580

Estimating Tips
26 05 00 Common Work Results for Electrical

- Conduit should be taken off in three main categories—power distribution, branch power, and branch lighting—so the estimator can concentrate on systems and components, therefore making it easier to ensure all items have been accounted for.
- For cost modifications for elevated conduit installation, add the percentages to labor according to the height of installation, and only to the quantities exceeding the different height levels, not to the total conduit quantities.
- Remember that aluminum wiring of equal ampacity is larger in diameter than copper and may require larger conduit.
- If more than three wires at a time are being pulled, deduct percentages from the labor hours of that grouping of wires.
- When taking off grounding systems, identify separately the type and size of wire, and list each unique type of ground connection.
- The estimator should take the weights of materials into consideration when completing a takeoff. Topics to consider include: How will the materials be supported? What methods of support are available? How high will the support structure have to reach? Will the final support structure be able to withstand the total burden? Is the support material included or separate from the fixture, equipment, and material specified?
- Do not overlook the costs for equipment used in the installation. If scaffolding or highlifts are available in the field, contractors may use them in lieu of the proposed ladders and rolling staging.

26 20 00 Low-Voltage Electrical Transmission

- Supports and concrete pads may be shown on drawings for the larger equipment, or the support system may be only a piece of plywood for the back of a panelboard. In either case, it must be included in the costs.

26 40 00 Electrical and Cathodic Protection

- When taking off cathodic protections systems, identify the type and size of cable, and list each unique type of anode connection.

26 50 00 Lighting

- Fixtures should be taken off room by room, using the fixture schedule, specifications, and the ceiling plan. For large concentrations of lighting fixtures in the same area, deduct the percentages from labor hours.

Reference Numbers

Reference numbers are shown at the beginning of some major classifications. These numbers refer to related items in the Reference Section. The reference information may be an estimating procedure, an alternate pricing method, or technical information.

Note: Not all subdivisions listed here necessarily appear. ■

No part of this cost data may be reproduced, stored in a retrieval system, or transmitted in any form or by any means without prior written permission of RSMeans.

Note: Trade Service, in part, has been used as a reference source for some of the material prices used in Division 26.

Did you know?
RSMeans Online gives you the same access to RSMeans' data with 24/7 access:
- Quickly locate costs in the searchable database.
- Build cost lists, estimates, and reports in minutes.
- Adjust costs to any location in the U.S. and Canada with the click of a button.

Start your free trial today at **www.RSMeansOnline.com**

RSMeans Online
FROM THE GORDIAN GROUP

26 05 Common Work Results for Electrical

26 05 05 – Selective Demolition for Electrical

26 05 05.10 Electrical Demolition		Crew	Daily Output	Labor-Hours	Unit	Material	2016 Bare Costs Labor	Equipment	Total	Total Incl O&P
0010	**ELECTRICAL DEMOLITION**									
0020	Conduit to 15' high, including fittings & hangers									
0100	Rigid galvanized steel, 1/2" to 1" diameter	1 Elec	242	.033	L.F.		1.22		1.22	2
0120	1-1/4" to 2"	"	200	.040	"		1.48		1.48	2.41
0270	Armored cable (BX) avg. 50' runs									
0280	#14, 2 wire	1 Elec	690	.012	L.F.		.43		.43	.70
0290	#14, 3 wire		571	.014			.52		.52	.85
0300	#12, 2 wire		605	.013			.49		.49	.80
0310	#12, 3 wire		514	.016			.57		.57	.94
0320	#10, 2 wire		514	.016			.57		.57	.94
0330	#10, 3 wire		425	.019			.69		.69	1.14
0340	#8, 3 wire		342	.023			.86		.86	1.41
0350	Non metallic sheathed cable (Romex)									
0360	#14, 2 wire	1 Elec	720	.011	L.F.		.41		.41	.67
0370	#14, 3 wire		657	.012			.45		.45	.74
0380	#12, 2 wire		629	.013			.47		.47	.77
0390	#10, 3 wire		450	.018			.66		.66	1.07
0400	Wiremold raceway, including fittings & hangers									
0420	No. 3000	1 Elec	250	.032	L.F.		1.18		1.18	1.93
0440	No. 4000		217	.037			1.36		1.36	2.23
0460	No. 6000		166	.048			1.78		1.78	2.91
0462	Plugmold with receptacle		114	.070			2.59		2.59	4.24
0465	Telephone/power pole		12	.667	Ea.		24.50		24.50	40
0470	Non-metallic, straight section		480	.017	L.F.		.62		.62	1.01
0500	Channels, steel, including fittings & hangers									
0520	3/4" x 1-1/2"	1 Elec	308	.026	L.F.		.96		.96	1.57
0540	1-1/2" x 1-1/2"		269	.030			1.10		1.10	1.79
0560	1-1/2" x 1-7/8"		229	.035			1.29		1.29	2.11
1210	Panel boards, incl. removal of all breakers,									
1220	conduit terminations & wire connections									
1230	3 wire, 120/240 V, 100A, to 20 circuits	1 Elec	2.60	3.077	Ea.		114		114	186
1240	200 amps, to 42 circuits	2 Elec	2.60	6.154			227		227	370
1260	4 wire, 120/208 V, 125A, to 20 circuits	1 Elec	2.40	3.333			123		123	201
1270	200 amps, to 42 circuits	2 Elec	2.40	6.667			246		246	400
1720	Junction boxes, 4" sq. & oct.	1 Elec	80	.100			3.69		3.69	6.05
1740	Handy box		107	.075			2.76		2.76	4.51
1760	Switch box		107	.075			2.76		2.76	4.51
1780	Receptacle & switch plates		257	.031			1.15		1.15	1.88
1800	Wire, THW-THWN-THHN, removed from									
1810	in place conduit, to 15' high									
1830	#14	1 Elec	65	.123	C.L.F.		4.54		4.54	7.45
1840	#12		55	.145			5.35		5.35	8.80
1850	#10		45.50	.176			6.50		6.50	10.60
2000	Interior fluorescent fixtures, incl. supports									
2010	& whips, to 15' high									
2100	Recessed drop-in 2' x 2', 2 lamp	2 Elec	35	.457	Ea.		16.85		16.85	27.50
2140	2' x 4', 4 lamp	"	30	.533	"		19.70		19.70	32
2180	Surface mount, acrylic lens & hinged frame									
2220	2' x 2', 2 lamp	2 Elec	44	.364	Ea.		13.40		13.40	22
2260	2' x 4', 4 lamp	"	33	.485	"		17.90		17.90	29.50
2300	Strip fixtures, surface mount									
2320	4' long, 1 lamp	2 Elec	53	.302	Ea.		11.15		11.15	18.20
2380	8' long, 2 lamp	"	40	.400	"		14.75		14.75	24

26 05 Common Work Results for Electrical

26 05 05 – Selective Demolition for Electrical

26 05 05.10 Electrical Demolition		Crew	Daily Output	Labor-Hours	Unit	Material	2016 Bare Costs Labor	Equipment	Total	Total Incl O&P
2460	Interior incandescent, surface, ceiling									
2470	or wall mount, to 12' high									
2480	Metal cylinder type, 75 Watt	2 Elec	62	.258	Ea.		9.50		9.50	15.55
2600	Exterior fixtures, incandescent, wall mount									
2620	100 Watt	2 Elec	50	.320	Ea.		11.80		11.80	19.30
3000	Ceiling fan, tear out and remove	1 Elec	24	.333	"		12.30		12.30	20

26 05 19 – Low-Voltage Electrical Power Conductors and Cables

26 05 19.20 Armored Cable

		Crew	Daily Output	Labor-Hours	Unit	Material	Labor	Equipment	Total	Total Incl O&P
0010	**ARMORED CABLE**									
0051	600 volt, copper (BX), #14, 2 conductor, solid	1 Elec	240	.033	L.F.	.46	1.23		1.69	2.51
0101	3 conductor, solid		200	.040		.73	1.48		2.21	3.21
0151	#12, 2 conductor, solid		210	.038		.46	1.41		1.87	2.80
0201	3 conductor, solid		180	.044		.75	1.64		2.39	3.50
0251	#10, 2 conductor, solid		180	.044		.87	1.64		2.51	3.64
0301	3 conductor, solid		150	.053		1.21	1.97		3.18	4.56
0351	3 conductor, stranded		120	.067		2.29	2.46		4.75	6.55

26 05 19.55 Non-Metallic Sheathed Cable

		Crew	Daily Output	Labor-Hours	Unit	Material	Labor	Equipment	Total	Total Incl O&P
0010	**NON-METALLIC SHEATHED CABLE** 600 volt									
0100	Copper with ground wire (Romex)									
0151	#14, 2 wire	1 Elec	250	.032	L.F.	.21	1.18		1.39	2.17
0201	3 wire		230	.035		.31	1.28		1.59	2.44
0251	#12, 2 wire		220	.036		.33	1.34		1.67	2.55
0301	3 wire		200	.040		.47	1.48		1.95	2.92
0351	#10, 2 wire		200	.040		.50	1.48		1.98	2.97
0401	3 wire		140	.057		.74	2.11		2.85	4.26
0451	#8, 3 conductor		130	.062		1.18	2.27		3.45	5
0501	#6, 3 wire		120	.067		1.92	2.46		4.38	6.15
0550	SE type SER aluminum cable, 3 RHW and									
0601	1 bare neutral, 3 #8 & 1 #8	1 Elec	150	.053	L.F.	.69	1.97		2.66	3.98
0651	3 #6 & 1 #6	"	130	.062		.78	2.27		3.05	4.57
0701	3 #4 & 1 #6	2 Elec	220	.073		.87	2.68		3.55	5.35
0751	3 #2 & 1 #4		200	.080		1.29	2.95		4.24	6.25
0801	3 #1/0 & 1 #2		180	.089		1.95	3.28		5.23	7.50
0851	3 #2/0 & 1 #1		160	.100		2.29	3.69		5.98	8.55
0901	3 #4/0 & 1 #2/0		140	.114		3.27	4.22		7.49	10.50
2401	SEU service entrance cable, copper 2 conductors, #8 + #8 neutral	1 Elec	150	.053		1.07	1.97		3.04	4.40
2601	#6 + #8 neutral		130	.062		1.58	2.27		3.85	5.45
2801	#6 + #6 neutral		130	.062		1.80	2.27		4.07	5.70
3001	#4 + #6 neutral	2 Elec	220	.073		2.48	2.68		5.16	7.10
3201	#4 + #4 neutral		220	.073		2.80	2.68		5.48	7.45
3401	#3 + #5 neutral		210	.076		3.47	2.81		6.28	8.40
6500	Service entrance cap for copper SEU									
6600	100 amp	1 Elec	12	.667	Ea.	8.10	24.50		32.60	49
6700	150 amp		10	.800		13.80	29.50		43.30	63.50
6800	200 amp		8	1		21	37		58	83.50

26 05 19.90 Wire

		Crew	Daily Output	Labor-Hours	Unit	Material	Labor	Equipment	Total	Total Incl O&P
0010	**WIRE**									
0021	600 volt, copper type THW, solid, #14	1 Elec	1300	.006	L.F.	.07	.23		.30	.45
0031	#12		1100	.007		.11	.27		.38	.56
0041	#10		1000	.008		.17	.30		.47	.66
0161	#6		650	.012		.55	.45		1	1.35
0181	#4	2 Elec	1060	.015		.85	.56		1.41	1.85

26 05 Common Work Results for Electrical

26 05 19 – Low-Voltage Electrical Power Conductors and Cables

26 05 19.90 Wire		Crew	Daily Output	Labor-Hours	Unit	Material	2016 Bare Costs Labor	Equipment	Total	Total Incl O&P
0201	#3	2 Elec	1000	.016	L.F.	1.06	.59		1.65	2.13
0221	#2		900	.018		1.33	.66		1.99	2.53
0241	#1		800	.020		1.75	.74		2.49	3.14
0261	1/0		660	.024		2.18	.89		3.07	3.86
0281	2/0		580	.028		2.74	1.02		3.76	4.69
0301	3/0		500	.032		3.44	1.18		4.62	5.70
0351	4/0		440	.036		4.38	1.34		5.72	7

26 05 26 – Grounding and Bonding for Electrical Systems

26 05 26.80 Grounding

		Crew	Daily Output	Labor-Hours	Unit	Material	Labor	Equipment	Total	Total Incl O&P
0010	**GROUNDING**									
0030	Rod, copper clad, 8' long, 1/2" diameter	1 Elec	5.50	1.455	Ea.	20	53.50		73.50	110
0050	3/4" diameter		5.30	1.509		34	55.50		89.50	129
0080	10' long, 1/2" diameter		4.80	1.667		22.50	61.50		84	126
0100	3/4" diameter		4.40	1.818		39	67		106	153
0261	Wire, ground bare armored, #8-1 conductor		200	.040	L.F.	.70	1.48		2.18	3.18
0271	#6-1 conductor		180	.044	"	.84	1.64		2.48	3.60
0390	Bare copper wire, stranded, #8		11	.727	C.L.F.	25	27		52	71.50
0401	Bare copper, #6 wire		1000	.008	L.F.	.44	.30		.74	.97
0601	#2 stranded	2 Elec	1000	.016	"	1.20	.59		1.79	2.29
1800	Water pipe ground clamps, heavy duty									
2000	Bronze, 1/2" to 1" diameter	1 Elec	8	1	Ea.	24	37		61	86.50

26 05 33 – Raceway and Boxes for Electrical Systems

26 05 33.13 Conduit

		Crew	Daily Output	Labor-Hours	Unit	Material	Labor	Equipment	Total	Total Incl O&P
0010	**CONDUIT** To 15' high, includes 2 terminations, 2 elbows,									
0020	11 beam clamps, and 11 couplings per 100 L.F.									
1750	Rigid galvanized steel, 1/2" diameter	1 Elec	90	.089	L.F.	2.50	3.28		5.78	8.10
1770	3/4" diameter		80	.100		2.82	3.69		6.51	9.15
1800	1" diameter		65	.123		3.89	4.54		8.43	11.75
1830	1-1/4" diameter		60	.133		5.10	4.92		10.02	13.65
1850	1-1/2" diameter		55	.145		6.05	5.35		11.40	15.45
1870	2" diameter		45	.178		7.80	6.55		14.35	19.35
5000	Electric metallic tubing (EMT), 1/2" diameter		170	.047		.70	1.74		2.44	3.61
5020	3/4" diameter		130	.062		1	2.27		3.27	4.81
5040	1" diameter		115	.070		1.67	2.57		4.24	6.05
5060	1-1/4" diameter		100	.080		2.69	2.95		5.64	7.80
5080	1-1/2" diameter		90	.089		3.43	3.28		6.71	9.15
9100	PVC, schedule 40, 1/2" diameter		190	.042		.86	1.55		2.41	3.49
9110	3/4" diameter		145	.055		.95	2.04		2.99	4.37
9120	1" diameter		125	.064		2.06	2.36		4.42	6.15
9130	1-1/4" diameter		110	.073		2.53	2.68		5.21	7.15
9140	1-1/2" diameter		100	.080		2.86	2.95		5.81	8
9150	2" diameter		90	.089		3.73	3.28		7.01	9.45
9995	Do not include labor when adding couplings to a fitting installation									

26 05 33.16 Outlet Boxes

		Crew	Daily Output	Labor-Hours	Unit	Material	Labor	Equipment	Total	Total Incl O&P
0010	**OUTLET BOXES**									
0021	Pressed steel, octagon, 4"	1 Elec	18	.444	Ea.	2.22	16.40		18.62	29.50
0060	Covers, blank		64	.125		.93	4.61		5.54	8.55
0100	Extension rings		40	.200		3.68	7.40		11.08	16.10
0151	Square 4"		18	.444		2.88	16.40		19.28	30
0200	Extension rings		40	.200		3.72	7.40		11.12	16.15
0250	Covers, blank		64	.125		1	4.61		5.61	8.65

26 05 Common Work Results for Electrical

26 05 33 – Raceway and Boxes for Electrical Systems

26 05 33.16 Outlet Boxes		Crew	Daily Output	Labor-Hours	Unit	Material	2016 Bare Costs Labor	Equipment	Total	Total Incl O&P
0300	Plaster rings	1 Elec	64	.125	Ea.	2.04	4.61		6.65	9.80
0651	Switchbox		24	.333		3.55	12.30		15.85	24
1100	Concrete, floor, 1 gang		5.30	1.509		89.50	55.50		145	190

26 05 33.17 Outlet Boxes, Plastic										
0010	**OUTLET BOXES, PLASTIC**									
0051	4" diameter, round, with 2 mounting nails	1 Elec	23	.348	Ea.	2.39	12.85		15.24	23.50
0101	Bar hanger mounted		23	.348		5.10	12.85		17.95	26.50
0201	Square with 2 mounting nails		23	.348		4.63	12.85		17.48	26
0300	Plaster ring		64	.125		1.98	4.61		6.59	9.75
0401	Switch box with 2 mounting nails, 1 gang		27	.296		4.39	10.95		15.34	22.50
0501	2 gang		23	.348		3.33	12.85		16.18	24.50
0601	3 gang		18	.444		4.19	16.40		20.59	31.50

26 05 33.18 Pull Boxes										
0010	**PULL BOXES**									
0100	Steel, pull box, NEMA 1, type SC, 6" W x 6" H x 4" D	1 Elec	8	1	Ea.	9.65	37		46.65	71
0200	8" W x 8" H x 4" D		8	1		13.20	37		50.20	75
0300	10" W x 12" H x 6" D		5.30	1.509		22.50	55.50		78	116

26 05 33.25 Conduit Fittings for Rigid Galvanized Steel										
0010	**CONDUIT FITTINGS FOR RIGID GALVANIZED STEEL**									
2280	LB, LR or LL fittings & covers, 1/2" diameter	1 Elec	16	.500	Ea.	8.85	18.45		27.30	40
2290	3/4" diameter		13	.615		11	22.50		33.50	49
2300	1" diameter		11	.727		16.05	27		43.05	61.50
2330	1-1/4" diameter		8	1		25	37		62	88
2350	1-1/2" diameter		6	1.333		33	49		82	117
2370	2" diameter		5	1.600		54	59		113	156
5280	Service entrance cap, 1/2" diameter		16	.500		4.99	18.45		23.44	35.50
5300	3/4" diameter		13	.615		5.95	22.50		28.45	43.50
5320	1" diameter		10	.800		7.10	29.50		36.60	56.50
5340	1-1/4" diameter		8	1		8.95	37		45.95	70.50
5360	1-1/2" diameter		6.50	1.231		17.80	45.50		63.30	94
5380	2" diameter		5.50	1.455		26.50	53.50		80	118

26 05 33.35 Flexible Metallic Conduit										
0010	**FLEXIBLE METALLIC CONDUIT**									
0050	Steel, 3/8" diameter	1 Elec	200	.040	L.F.	.37	1.48		1.85	2.82
0100	1/2" diameter		200	.040		.41	1.48		1.89	2.86
0200	3/4" diameter		160	.050		.58	1.85		2.43	3.66
0250	1" diameter		100	.080		1.03	2.95		3.98	5.95
0300	1-1/4" diameter		70	.114		1.34	4.22		5.56	8.35
0350	1-1/2" diameter		50	.160		2.17	5.90		8.07	12.05
0370	2" diameter		40	.200		2.64	7.40		10.04	14.95

26 05 39 – Underfloor Raceways for Electrical Systems

26 05 39.30 Conduit In Concrete Slab										
0010	**CONDUIT IN CONCRETE SLAB** Including terminations,									
0020	fittings and supports									
3230	PVC, schedule 40, 1/2" diameter	1 Elec	270	.030	L.F.	.48	1.09		1.57	2.32
3250	3/4" diameter		230	.035		.56	1.28		1.84	2.71
3270	1" diameter		200	.040		.76	1.48		2.24	3.25
3300	1-1/4" diameter		170	.047		1	1.74		2.74	3.93
3330	1-1/2" diameter		140	.057		1.21	2.11		3.32	4.78
3350	2" diameter		120	.067		1.52	2.46		3.98	5.70
4350	Rigid galvanized steel, 1/2" diameter		200	.040		2.35	1.48		3.83	5

26 05 Common Work Results for Electrical

26 05 39 – Underfloor Raceways for Electrical Systems

26 05 39.30 Conduit In Concrete Slab

		Crew	Daily Output	Labor-Hours	Unit	Material	2016 Bare Costs Labor	Equipment	Total	Total Incl O&P
4400	3/4" diameter	1 Elec	170	.047	L.F.	2.48	1.74		4.22	5.55
4450	1" diameter		130	.062		3.44	2.27		5.71	7.50
4500	1-1/4" diameter		110	.073		4.78	2.68		7.46	9.65
4600	1-1/2" diameter		100	.080		5.35	2.95		8.30	10.70
4800	2" diameter		90	.089		6.60	3.28		9.88	12.60

26 05 39.40 Conduit In Trench

		Crew	Daily Output	Labor-Hours	Unit	Material	Labor	Equipment	Total	Total Incl O&P
0010	**CONDUIT IN TRENCH** Includes terminations and fittings									
0020	Does not include excavation or backfill, see Section 31 23 16									
0200	Rigid galvanized steel, 2" diameter	1 Elec	150	.053	L.F.	6.25	1.97		8.22	10.10
0400	2-1/2" diameter	"	100	.080		12.15	2.95		15.10	18.20
0600	3" diameter	2 Elec	160	.100		14.15	3.69		17.84	21.50
0800	3-1/2" diameter	"	140	.114		18.40	4.22		22.62	27.50

26 05 80 – Wiring Connections

26 05 80.10 Motor Connections

		Crew	Daily Output	Labor-Hours	Unit	Material	Labor	Equipment	Total	Total Incl O&P
0010	**MOTOR CONNECTIONS**									
0020	Flexible conduit and fittings, 115 volt, 1 phase, up to 1 HP motor	1 Elec	8	1	Ea.	5.55	37		42.55	66.50

26 05 90 – Residential Applications

26 05 90.10 Residential Wiring

		Crew	Daily Output	Labor-Hours	Unit	Material	Labor	Equipment	Total	Total Incl O&P
0010	**RESIDENTIAL WIRING**									
0020	20' avg. runs and #14/2 wiring incl. unless otherwise noted									
1000	Service & panel, includes 24' SE-AL cable, service eye, meter,									
1010	Socket, panel board, main bkr., ground rod, 15 or 20 amp									
1020	1-pole circuit breakers, and misc. hardware									
1100	100 amp, with 10 branch breakers	1 Elec	1.19	6.723	Ea.	485	248		733	940
1110	With PVC conduit and wire		.92	8.696		525	320		845	1,100
1120	With RGS conduit and wire		.73	10.959		685	405		1,090	1,425
1150	150 amp, with 14 branch breakers		1.03	7.767		705	287		992	1,250
1170	With PVC conduit and wire		.82	9.756		785	360		1,145	1,450
1180	With RGS conduit and wire		.67	11.940		1,100	440		1,540	1,925
1200	200 amp, with 18 branch breakers	2 Elec	1.80	8.889		1,050	330		1,380	1,675
1220	With PVC conduit and wire		1.46	10.959		1,125	405		1,530	1,900
1230	With RGS conduit and wire		1.24	12.903		1,525	475		2,000	2,450
1800	Lightning surge suppressor	1 Elec	32	.250		50.50	9.25		59.75	71
2000	Switch devices									
2100	Single pole, 15 amp, Ivory, with a 1-gang box, cover plate,									
2110	Type NM (Romex) cable	1 Elec	17.10	.468	Ea.	18.70	17.25		35.95	48.50
2120	Type MC (BX) cable		14.30	.559		27.50	20.50		48	64
2130	EMT & wire		5.71	1.401		36	51.50		87.50	125
2150	3-way, #14/3, type NM cable		14.55	.550		12.70	20.50		33.20	47
2170	Type MC cable		12.31	.650		25	24		49	66.50
2180	EMT & wire		5	1.600		30	59		89	130
2200	4-way, #14/3, type NM cable		14.55	.550		20.50	20.50		41	55.50
2220	Type MC cable		12.31	.650		32.50	24		56.50	75
2230	EMT & wire		5	1.600		37.50	59		96.50	138
2250	S.P., 20 amp, #12/2, type NM cable		13.33	.600		14.75	22		36.75	52
2270	Type MC cable		11.43	.700		21	26		47	65
2280	EMT & wire		4.85	1.649		33.50	61		94.50	137
2290	S.P. rotary dimmer, 600W, no wiring		17	.471		31.50	17.35		48.85	63
2300	S.P. rotary dimmer, 600W, type NM cable		14.55	.550		35.50	20.50		56	72
2320	Type MC cable		12.31	.650		44.50	24		68.50	87.50
2330	EMT & wire		5	1.600		54.50	59		113.50	157

26 05 Common Work Results for Electrical

26 05 90 – Residential Applications

26 05 90.10 Residential Wiring		Crew	Daily Output	Labor-Hours	Unit	Material	2016 Bare Costs Labor	Equipment	Total	Total Incl O&P
2350	3-way rotary dimmer, type NM cable	1 Elec	13.33	.600	Ea.	19.85	22		41.85	58
2370	Type MC cable		11.43	.700		28.50	26		54.50	73.50
2380	EMT & wire		4.85	1.649		39	61		100	142
2400	Interval timer wall switch, 20 amp, 1-30 min., #12/2									
2410	Type NM cable	1 Elec	14.55	.550	Ea.	57	20.50		77.50	96
2420	Type MC cable		12.31	.650		62	24		86	107
2430	EMT & wire		5	1.600		76	59		135	180
2500	Decorator style									
2510	S.P., 15 amp, type NM cable	1 Elec	17.10	.468	Ea.	22.50	17.25		39.75	53
2520	Type MC cable		14.30	.559		31.50	20.50		52	68.50
2530	EMT & wire		5.71	1.401		40	51.50		91.50	129
2550	3-way, #14/3, type NM cable		14.55	.550		16.60	20.50		37.10	51.50
2570	Type MC cable		12.31	.650		29	24		53	71
2580	EMT & wire		5	1.600		33.50	59		92.50	134
2600	4-way, #14/3, type NM cable		14.55	.550		24.50	20.50		45	59.50
2620	Type MC cable		12.31	.650		36.50	24		60.50	79.50
2630	EMT & wire		5	1.600		41.50	59		100.50	142
2650	S.P., 20 amp, #12/2, type NM cable		13.33	.600		18.60	22		40.60	56.50
2670	Type MC cable		11.43	.700		25	26		51	69.50
2680	EMT & wire		4.85	1.649		37.50	61		98.50	141
2700	S.P., slide dimmer, type NM cable		17.10	.468		24.50	17.25		41.75	55
2720	Type MC cable		14.30	.559		33	20.50		53.50	70.50
2730	EMT & wire		5.71	1.401		43.50	51.50		95	132
2750	S.P., touch dimmer, type NM cable		17.10	.468		41	17.25		58.25	73
2770	Type MC cable		14.30	.559		50	20.50		70.50	88.50
2780	EMT & wire		5.71	1.401		60	51.50		111.50	151
2800	3-way touch dimmer, type NM cable		13.33	.600		51.50	22		73.50	92.50
2820	Type MC cable		11.43	.700		60	26		86	108
2830	EMT & wire		4.85	1.649		70	61		131	177
3000	Combination devices									
3100	S.P. switch/15 amp recpt., Ivory, 1-gang box, plate									
3110	Type NM cable	1 Elec	11.43	.700	Ea.	24	26		50	68.50
3120	Type MC cable		10	.800		33	29.50		62.50	84.50
3130	EMT & wire		4.40	1.818		43	67		110	158
3150	S.P. switch/pilot light, type NM cable		11.43	.700		24.50	26		50.50	69
3170	Type MC cable		10	.800		33	29.50		62.50	85
3180	EMT & wire		4.43	1.806		43	66.50		109.50	157
3190	2-S.P. switches, 2-#14/2, no wiring		14	.571		13.70	21		34.70	49.50
3200	2-S.P. switches, 2-#14/2, type NM cables		10	.800		26.50	29.50		56	78
3220	Type MC cable		8.89	.900		40.50	33		73.50	99
3230	EMT & wire		4.10	1.951		45.50	72		117.50	168
3250	3-way switch/15 amp recpt., #14/3, type NM cable		10	.800		31.50	29.50		61	83
3270	Type MC cable		8.89	.900		44	33		77	103
3280	EMT & wire		4.10	1.951		48.50	72		120.50	172
3300	2-3 way switches, 2-#14/3, type NM cables		8.89	.900		40.50	33		73.50	99
3320	Type MC cable		8	1		61.50	37		98.50	128
3330	EMT & wire		4	2		55.50	74		129.50	182
3350	S.P. switch/20 amp recpt., #12/2, type NM cable		10	.800		31	29.50		60.50	83
3370	Type MC cable		8.89	.900		36	33		69	94
3380	EMT & wire		4.10	1.951		50	72		122	173
3400	Decorator style									
3410	S.P. switch/15 amp recpt., type NM cable	1 Elec	11.43	.700	Ea.	28	26		54	73
3420	Type MC cable		10	.800		36.50	29.50		66	89

26 05 Common Work Results for Electrical

26 05 90 – Residential Applications

26 05 90.10 Residential Wiring		Crew	Daily Output	Labor-Hours	Unit	Material	2016 Bare Costs Labor	Equipment	Total	Total Incl O&P
3430	EMT & wire	1 Elec	4.40	1.818	Ea.	47	67		114	162
3450	S.P. switch/pilot light, type NM cable		11.43	.700		28	26		54	73
3470	Type MC cable		10	.800		37	29.50		66.50	89
3480	EMT & wire		4.40	1.818		47	67		114	162
3500	2-S.P. switches, 2-#14/2, type NM cables		10	.800		30.50	29.50		60	82
3520	Type MC cable		8.89	.900		44	33		77	103
3530	EMT & wire		4.10	1.951		49.50	72		121.50	173
3550	3-way/15 amp recpt., #14/3, type NM cable		10	.800		35.50	29.50		65	87.50
3570	Type MC cable		8.89	.900		48	33		81	107
3580	EMT & wire		4.10	1.951		52.50	72		124.50	176
3650	2-3 way switches, 2-#14/3, type NM cables		8.89	.900		44.50	33		77.50	104
3670	Type MC cable		8	1		65	37		102	132
3680	EMT & wire		4	2		59.50	74		133.50	187
3700	S.P. switch/20 amp recpt., #12/2, type NM cable		10	.800		35	29.50		64.50	87
3720	Type MC cable		8.89	.900		40	33		73	98.50
3730	EMT & wire		4.10	1.951		54	72		126	177
4000	Receptacle devices									
4010	Duplex outlet, 15 amp recpt., Ivory, 1-gang box, plate									
4015	Type NM cable	1 Elec	14.55	.550	Ea.	10.30	20.50		30.80	44.50
4020	Type MC cable		12.31	.650		19	24		43	60
4030	EMT & wire		5.33	1.501		28	55.50		83.50	121
4050	With #12/2, type NM cable		12.31	.650		12.65	24		36.65	53
4070	Type MC cable		10.67	.750		19	27.50		46.50	66.50
4080	EMT & wire		4.71	1.699		31.50	62.50		94	138
4100	20 amp recpt., #12/2, type NM cable		12.31	.650		19.30	24		43.30	60.50
4120	Type MC cable		10.67	.750		25.50	27.50		53	73.50
4130	EMT & wire		4.71	1.699		38	62.50		100.50	145
4140	For GFI see Section 26 05 90.10 line 4300 below									
4150	Decorator style, 15 amp recpt., type NM cable	1 Elec	14.55	.550	Ea.	14.20	20.50		34.70	48.50
4170	Type MC cable		12.31	.650		23	24		47	64
4180	EMT & wire		5.33	1.501		31.50	55.50		87	126
4200	With #12/2, type NM cable		12.31	.650		16.55	24		40.55	57
4220	Type MC cable		10.67	.750		23	27.50		50.50	70.50
4230	EMT & wire		4.71	1.699		35.50	62.50		98	142
4250	20 amp recpt., #12/2, type NM cable		12.31	.650		23	24		47	64.50
4270	Type MC cable		10.67	.750		29.50	27.50		57	78
4280	EMT & wire		4.71	1.699		42	62.50		104.50	149
4300	GFI, 15 amp recpt., type NM cable		12.31	.650		23	24		47	64
4320	Type MC cable		10.67	.750		31.50	27.50		59	80
4330	EMT & wire		4.71	1.699		40.50	62.50		103	148
4350	GFI with #12/2, type NM cable		10.67	.750		25	27.50		52.50	73
4370	Type MC cable		9.20	.870		31.50	32		63.50	87
4380	EMT & wire		4.21	1.900		44	70		114	164
4400	20 amp recpt., #12/2, type NM cable		10.67	.750		51.50	27.50		79	102
4420	Type MC cable		9.20	.870		58	32		90	116
4430	EMT & wire		4.21	1.900		70.50	70		140.50	193
4500	Weather-proof cover for above receptacles, add		32	.250		2.94	9.25		12.19	18.35
4550	Air conditioner outlet, 20 amp-240 volt recpt.									
4560	30' of #12/2, 2 pole circuit breaker									
4570	Type NM cable	1 Elec	10	.800	Ea.	62	29.50		91.50	117
4580	Type MC cable		9	.889		70	33		103	131
4590	EMT & wire		4	2		81	74		155	210
4600	Decorator style, type NM cable		10	.800		67	29.50		96.50	122

26 05 Common Work Results for Electrical

26 05 90 – Residential Applications

26 05 90.10 Residential Wiring		Crew	Daily Output	Labor-Hours	Unit	Material	2016 Bare Costs Labor	Equipment	Total	Total Incl O&P
4620	Type MC cable	1 Elec	9	.889	Ea.	74.50	33		107.50	136
4630	EMT & wire	↓	4	2	↓	85.50	74		159.50	215
4650	Dryer outlet, 30 amp-240 volt recpt., 20' of #10/3									
4660	2 pole circuit breaker									
4670	Type NM cable	1 Elec	6.41	1.248	Ea.	58	46		104	140
4680	Type MC cable		5.71	1.401		64	51.50		115.50	155
4690	EMT & wire	↓	3.48	2.299	↓	75	85		160	222
4700	Range outlet, 50 amp-240 volt recpt., 30' of #8/3									
4710	Type NM cable	1 Elec	4.21	1.900	Ea.	85	70		155	209
4720	Type MC cable		4	2		126	74		200	259
4730	EMT & wire		2.96	2.703		104	99.50		203.50	278
4750	Central vacuum outlet, Type NM cable		6.40	1.250		53.50	46		99.50	135
4770	Type MC cable		5.71	1.401		66.50	51.50		118	158
4780	EMT & wire	↓	3.48	2.299	↓	81	85		166	228
4800	30 amp-110 volt locking recpt., #10/2 circ. bkr.									
4810	Type NM cable	1 Elec	6.20	1.290	Ea.	61.50	47.50		109	146
4820	Type MC cable		5.40	1.481		78.50	54.50		133	176
4830	EMT & wire	↓	3.20	2.500	↓	91.50	92.50		184	252
4900	Low voltage outlets									
4910	Telephone recpt., 20' of 4/C phone wire	1 Elec	26	.308	Ea.	8.40	11.35		19.75	28
4920	TV recpt., 20' of RG59U coax wire, F type connector	"	16	.500	"	16	18.45		34.45	47.50
4950	Door bell chime, transformer, 2 buttons, 60' of bellwire									
4970	Economy model	1 Elec	11.50	.696	Ea.	50	25.50		75.50	97
4980	Custom model		11.50	.696		99.50	25.50		125	151
4990	Luxury model, 3 buttons	↓	9.50	.842	↓	281	31		312	360
6000	Lighting outlets									
6050	Wire only (for fixture), type NM cable	1 Elec	32	.250	Ea.	6.35	9.25		15.60	22
6070	Type MC cable		24	.333		12.55	12.30		24.85	34
6080	EMT & wire		10	.800		20.50	29.50		50	71
6100	Box (4"), and wire (for fixture), type NM cable		25	.320		13.90	11.80		25.70	34.50
6120	Type MC cable		20	.400		20	14.75		34.75	46
6130	EMT & wire	↓	11	.727	↓	28	27		55	75
6200	Fixtures (use with lines 6050 or 6100 above)									
6210	Canopy style, economy grade	1 Elec	40	.200	Ea.	27	7.40		34.40	41.50
6220	Custom grade		40	.200		53.50	7.40		60.90	71
6250	Dining room chandelier, economy grade		19	.421		67.50	15.55		83.05	99.50
6260	Custom grade		19	.421		315	15.55		330.55	370
6270	Luxury grade		15	.533		820	19.70		839.70	935
6310	Kitchen fixture (fluorescent), economy grade		30	.267		71.50	9.85		81.35	95
6320	Custom grade		25	.320		221	11.80		232.80	263
6350	Outdoor, wall mounted, economy grade		30	.267		25.50	9.85		35.35	44
6360	Custom grade		30	.267		120	9.85		129.85	148
6370	Luxury grade		25	.320		248	11.80		259.80	292
6410	Outdoor PAR floodlights, 1 lamp, 150 watt		20	.400		27.50	14.75		42.25	54
6420	2 lamp, 150 watt each		20	.400		45.50	14.75		60.25	74
6425	Motion sensing, 2 lamp, 150 watt each		20	.400		75.50	14.75		90.25	107
6430	For infrared security sensor, add		32	.250		132	9.25		141.25	160
6450	Outdoor, quartz-halogen, 300 watt flood		20	.400		33	14.75		47.75	60.50
6600	Recessed downlight, round, pre-wired, 50 or 75 watt trim		30	.267		68	9.85		77.85	90.50
6610	With shower light trim		30	.267		82.50	9.85		92.35	107
6620	With wall washer trim		28	.286		85	10.55		95.55	111
6630	With eye-ball trim		28	.286		84.50	10.55		95.05	110
6700	Porcelain lamp holder	↓	40	.200	↓	2.52	7.40		9.92	14.80

For customer support on your Residential Cost Data, call 877.759.4771.

26 05 Common Work Results for Electrical

26 05 90 – Residential Applications

26 05 90.10 Residential Wiring		Crew	Daily Output	Labor-Hours	Unit	Material	2016 Bare Costs Labor	Equipment	Total	Total Incl O&P
6710	With pull switch	1 Elec	40	.200	Ea.	7.55	7.40		14.95	20.50
6750	Fluorescent strip, 2-20 watt tube, wrap around diffuser, 24"		24	.333		42	12.30		54.30	66.50
6760	1-34 watt tube, 48"		24	.333		100	12.30		112.30	130
6770	2-34 watt tubes, 48"		20	.400		118	14.75		132.75	154
6800	Bathroom heat lamp, 1-250 watt		28	.286		37.50	10.55		48.05	58.50
6810	2-250 watt lamps		28	.286		59.50	10.55		70.05	83
6820	For timer switch, see Section 26 05 90.10 line 2400									
6900	Outdoor post lamp, incl. post, fixture, 35' of #14/2									
6910	Type NM cable	1 Elec	3.50	2.286	Ea.	285	84.50		369.50	455
6920	Photo-eye, add		27	.296		27.50	10.95		38.45	48
6950	Clock dial time switch, 24 hr., w/enclosure, type NM cable		11.43	.700		73	26		99	123
6970	Type MC cable		11	.727		81.50	27		108.50	134
6980	EMT & wire		4.85	1.649		90.50	61		151.50	199
7000	Alarm systems									
7050	Smoke detectors, box, #14/3, type NM cable	1 Elec	14.55	.550	Ea.	33.50	20.50		54	70
7070	Type MC cable		12.31	.650		44.50	24		68.50	88
7080	EMT & wire		5	1.600		49	59		108	151
7090	For relay output to security system, add					10.20			10.20	11.20
8000	Residential equipment									
8050	Disposal hook-up, incl. switch, outlet box, 3' of flex									
8060	20 amp-1 pole circ. bkr., and 25' of #12/2									
8070	Type NM cable	1 Elec	10	.800	Ea.	36.50	29.50		66	88.50
8080	Type MC cable		8	1		43.50	37		80.50	109
8090	EMT & wire		5	1.600		58.50	59		117.50	161
8100	Trash compactor or dishwasher hook-up, incl. outlet box,									
8110	3' of flex, 15 amp-1 pole circ. bkr., and 25' of #14/2									
8130	Type MC cable	1 Elec	8	1	Ea.	31	37		68	94.50
8140	EMT & wire	"	5	1.600	"	43.50	59		102.50	144
8150	Hot water sink dispenser hook-up, use line 8100									
8200	Vent/exhaust fan hook-up, type NM cable	1 Elec	32	.250	Ea.	6.35	9.25		15.60	22
8220	Type MC cable		24	.333		12.55	12.30		24.85	34
8230	EMT & wire		10	.800		20.50	29.50		50	71
8250	Bathroom vent fan, 50 CFM (use with above hook-up)									
8260	Economy model	1 Elec	15	.533	Ea.	19.65	19.70		39.35	53.50
8270	Low noise model		15	.533		42	19.70		61.70	78.50
8280	Custom model		12	.667		131	24.50		155.50	184
8300	Bathroom or kitchen vent fan, 110 CFM									
8310	Economy model	1 Elec	15	.533	Ea.	67.50	19.70		87.20	107
8320	Low noise model	"	15	.533	"	92	19.70		111.70	133
8350	Paddle fan, variable speed (w/o lights)									
8360	Economy model (AC motor)	1 Elec	10	.800	Ea.	109	29.50		138.50	169
8362	With light kit		10	.800		150	29.50		179.50	214
8370	Custom model (AC motor)		10	.800		262	29.50		291.50	335
8372	With light kit		10	.800		300	29.50		329.50	385
8380	Luxury model (DC motor)		8	1		288	37		325	375
8382	With light kit		8	1		330	37		367	420
8390	Remote speed switch for above, add		12	.667		37.50	24.50		62	81
8500	Whole house exhaust fan, ceiling mount, 36", variable speed									
8510	Remote switch, incl. shutters, 20 amp-1 pole circ. bkr.									
8520	30' of #12/2, type NM cable	1 Elec	4	2	Ea.	1,325	74		1,399	1,600
8530	Type MC cable		3.50	2.286		1,350	84.50		1,434.50	1,625
8540	EMT & wire		3	2.667		1,350	98.50		1,448.50	1,650
8600	Whirlpool tub hook-up, incl. timer switch, outlet box									

26 05 Common Work Results for Electrical

26 05 90 – Residential Applications

26 05 90.10 Residential Wiring

		Crew	Daily Output	Labor-Hours	Unit	Material	2016 Bare Costs Labor	Equipment	Total	Total Incl O&P
8610	3' of flex, 20 amp-1 pole GFI circ. bkr.									
8620	30' of #12/2, type NM cable	1 Elec	5	1.600	Ea.	157	59		216	269
8630	Type MC cable		4.20	1.905		160	70.50		230.50	291
8640	EMT & wire	↓	3.40	2.353	↓	172	87		259	330
8650	Hot water heater hook-up, incl. 1-2 pole circ. bkr., box;									
8660	3' of flex, 20' of #10/2, type NM cable	1 Elec	5	1.600	Ea.	31.50	59		90.50	132
8670	Type MC cable		4.20	1.905		42.50	70.50		113	162
8680	EMT & wire	↓	3.40	2.353	↓	47	87		134	194
9000	Heating/air conditioning									
9050	Furnace/boiler hook-up, incl. firestat, local on-off switch									
9060	Emergency switch, and 40' of type NM cable	1 Elec	4	2	Ea.	52	74		126	178
9070	Type MC cable		3.50	2.286		65.50	84.50		150	210
9080	EMT & wire	↓	1.50	5.333	↓	83.50	197		280.50	410
9100	Air conditioner hook-up, incl. local 60 amp disc. switch									
9110	3' sealtite, 40 amp, 2 pole circuit breaker									
9130	40' of #8/2, type NM cable	1 Elec	3.50	2.286	Ea.	144	84.50		228.50	296
9140	Type MC cable		3	2.667		204	98.50		302.50	385
9150	EMT & wire	↓	1.30	6.154		185	227		412	575
9200	Heat pump hook-up, 1-40 & 1-100 amp 2 pole circ. bkr.									
9210	Local disconnect switch, 3' sealtite									
9220	40' of #8/2 & 30' of #3/2									
9230	Type NM cable	1 Elec	1.30	6.154	Ea.	460	227		687	875
9240	Type MC cable		1.08	7.407		510	273		783	1,000
9250	EMT & wire	↓	.94	8.511	↓	520	315		835	1,075
9500	Thermostat hook-up, using low voltage wire									
9520	Heating only, 25' of #18-3	1 Elec	24	.333	Ea.	8	12.30		20.30	29
9530	Heating/cooling, 25' of #18-4	"	20	.400	"	10.50	14.75		25.25	35.50

26 24 Switchboards and Panelboards

26 24 16 – Panelboards

26 24 16.10 Load Centers

		Crew	Daily Output	Labor-Hours	Unit	Material	2016 Bare Costs Labor	Equipment	Total	Total Incl O&P
0010	**LOAD CENTERS** (residential type)									
0100	3 wire, 120/240 V, 1 phase, including 1 pole plug-in breakers									
0200	100 amp main lugs, indoor, 8 circuits	1 Elec	1.40	5.714	Ea.	184	211		395	545
0300	12 circuits		1.20	6.667		93	246		339	500
0400	Rainproof, 8 circuits		1.40	5.714		220	211		431	585
0500	12 circuits	↓	1.20	6.667		310	246		556	740
0600	200 amp main lugs, indoor, 16 circuits	R-1A	1.80	8.889		281	295		576	795
0700	20 circuits		1.50	10.667		430	355		785	1,050
0800	24 circuits		1.30	12.308		465	410		875	1,175
1200	Rainproof, 16 circuits		1.80	8.889		405	295		700	930
1300	20 circuits		1.50	10.667		400	355		755	1,025
1400	24 circuits	↓	1.30	12.308		730	410		1,140	1,475

26 24 16.20 Panelboard and Load Center Circuit Breakers

		Crew	Daily Output	Labor-Hours	Unit	Material	2016 Bare Costs Labor	Equipment	Total	Total Incl O&P
0010	**PANELBOARD AND LOAD CENTER CIRCUIT BREAKERS**									
2000	Plug-in panel or load center, 120/240 volt, to 60 amp, 1 pole	1 Elec	12	.667	Ea.	9.70	24.50		34.20	50.50
2004	Circuit breaker, 120/240 volt, 20 A, 1 pole with NM cable		6.50	1.231		16.35	45.50		61.85	92.50
2006	30 A, 1 pole with NM cable		6.50	1.231		16.35	45.50		61.85	92.50
2010	2 pole		9	.889		26	33		59	82
2014	50 A, 2 pole with NM cable		5.50	1.455		32.50	53.50		86	124
2020	3 pole	↓	7.50	1.067	↓	76	39.50		115.50	148

26 24 Switchboards and Panelboards

26 24 16 – Panelboards

26 24 16.20 Panelboard and Load Center Circuit Breakers	Crew	Daily Output	Labor-Hours	Unit	Material	2016 Bare Costs Labor	Equipment	Total	Total Incl O&P
2030 100 amp, 2 pole	1 Elec	6	1.333	Ea.	102	49		151	193
2040 3 pole		4.50	1.778		113	65.50		178.50	231
2050 150 to 200 amp, 2 pole		3	2.667		218	98.50		316.50	400
2060 Plug-in tandem, 120/240 V, 2-15 A, 1 pole		11	.727		54.50	27		81.50	104
2070 1-15 A & 1-20 A		11	.727		54.50	27		81.50	104
2080 2-20 A		11	.727		54.50	27		81.50	104
2300 Ground fault, 240 volt, 30 amp, 1 pole		7	1.143		117	42		159	198
2310 2 pole		6	1.333		209	49		258	310

26 27 Low-Voltage Distribution Equipment

26 27 13 – Electricity Metering

26 27 13.10 Meter Centers and Sockets

		Crew	Daily Output	Labor-Hours	Unit	Material	2016 Bare Costs Labor	Equipment	Total	Total Incl O&P
0010	**METER CENTERS AND SOCKETS**									
0100	Sockets, single position, 4 terminal, 100 amp	1 Elec	3.20	2.500	Ea.	43	92.50		135.50	199
0200	150 amp		2.30	3.478		48	128		176	263
0300	200 amp		1.90	4.211		92	155		247	355
0500	Double position, 4 terminal, 100 amp		2.80	2.857		204	105		309	395
0600	150 amp		2.10	3.810		240	141		381	495
0700	200 amp		1.70	4.706		450	174		624	780
1100	Meter centers and sockets, three phase, single pos, 7 terminal, 100 amp		2.80	2.857		164	105		269	355
1200	200 amp		2.10	3.810		295	141		436	555
1400	400 amp		1.70	4.706		720	174		894	1,075
2590	Basic meter device									
2600	1P 3W 120/240 V 4 jaw 125A sockets, 3 meter	2 Elec	1	16	Ea.	340	590		930	1,350
2610	4 meter		.90	17.778		405	655		1,060	1,525
2620	5 meter		.80	20		490	740		1,230	1,725
2630	6 meter		.60	26.667		640	985		1,625	2,300
2640	7 meter		.56	28.571		1,400	1,050		2,450	3,275
2660	10 meter		.48	33.333		1,925	1,225		3,150	4,125
2680	Rainproof 1P 3W 120/240 V 4 jaw 125A sockets									
2690	3 meter	2 Elec	1	16	Ea.	640	590		1,230	1,675
2710	6 meter		.60	26.667		1,100	985		2,085	2,825
2730	8 meter		.52	30.769		1,525	1,125		2,650	3,550
2750	1P 3W 120/240 V 4 jaw sockets									
2760	with 125A circuit breaker, 3 meter	2 Elec	1	16	Ea.	1,200	590		1,790	2,300
2780	5 meter		.80	20		1,900	740		2,640	3,275
2800	7 meter		.56	28.571		2,725	1,050		3,775	4,700
2820	10 meter		.48	33.333		3,800	1,225		5,025	6,175
2830	Rainproof 1P 3W 120/240 V 4 jaw sockets									
2840	with 125A circuit breaker, 3 meter	2 Elec	1	16	Ea.	1,200	590		1,790	2,300
2870	6 meter		.60	26.667		2,225	985		3,210	4,050
2890	8 meter		.52	30.769		3,025	1,125		4,150	5,175
3250	1P 3W 120/240 V 4 jaw sockets									
3260	with 200A circuit breaker, 3 meter	2 Elec	1	16	Ea.	1,800	590		2,390	2,950
3290	6 meter		.60	26.667		3,600	985		4,585	5,550
3310	8 meter		.56	28.571		4,875	1,050		5,925	7,075
3330	Rainproof 1P 3W 120/240 V 4 jaw sockets									
3350	with 200A circuit breaker, 3 meter	2 Elec	1	16	Ea.	1,800	590		2,390	2,950
3380	6 meter		.60	26.667		3,600	985		4,585	5,550
3400	8 meter		.52	30.769		4,875	1,125		6,000	7,200

26 27 Low-Voltage Distribution Equipment

26 27 23 – Indoor Service Poles

26 27 23.40 Surface Raceway		Crew	Daily Output	Labor-Hours	Unit	Material	2016 Bare Costs Labor	Equipment	Total	Total Incl O&P
0010	**SURFACE RACEWAY**									
0090	Metal, straight section									
0100	No. 500	1 Elec	100	.080	L.F.	1.04	2.95		3.99	5.95
0110	No. 700		100	.080		1.17	2.95		4.12	6.10
0200	No. 1000		90	.089		1.45	3.28		4.73	6.95
0400	No. 1500, small pancake		90	.089		2.15	3.28		5.43	7.70
0600	No. 2000, base & cover, blank		90	.089		2.19	3.28		5.47	7.75
0800	No. 3000, base & cover, blank		75	.107		4.18	3.94		8.12	11.05
2400	Fittings, elbows, No. 500		40	.200	Ea.	1.90	7.40		9.30	14.15
2800	Elbow cover, No. 2000		40	.200		3.57	7.40		10.97	16
2880	Tee, No. 500		42	.190		3.66	7.05		10.71	15.55
2900	No. 2000		27	.296		11.85	10.95		22.80	31
3000	Switch box, No. 500		16	.500		11	18.45		29.45	42
3400	Telephone outlet, No. 1500		16	.500		14.10	18.45		32.55	45.50
3600	Junction box, No. 1500		16	.500		9.65	18.45		28.10	40.50
3800	Plugmold wired sections, No. 2000									
4000	1 circuit, 6 outlets, 3 ft. long	1 Elec	8	1	Ea.	36	37		73	100
4100	2 circuits, 8 outlets, 6 ft. long	"	5.30	1.509	"	59	55.50		114.50	156

26 27 26 – Wiring Devices

26 27 26.10 Low Voltage Switching

		Crew	Daily Output	Labor-Hours	Unit	Material	Labor	Equipment	Total	Total Incl O&P
0010	**LOW VOLTAGE SWITCHING**									
3600	Relays, 120 V or 277 V standard	1 Elec	12	.667	Ea.	41.50	24.50		66	86
3800	Flush switch, standard		40	.200		11.50	7.40		18.90	24.50
4000	Interchangeable		40	.200		15.05	7.40		22.45	28.50
4100	Surface switch, standard		40	.200		8.15	7.40		15.55	21
4200	Transformer 115 V to 25 V		12	.667		130	24.50		154.50	183
4400	Master control, 12 circuit, manual		4	2		126	74		200	260
4500	25 circuit, motorized		4	2		140	74		214	275
4600	Rectifier, silicon		12	.667		45.50	24.50		70	90
4800	Switchplates, 1 gang, 1, 2 or 3 switch, plastic		80	.100		5	3.69		8.69	11.55
5000	Stainless steel		80	.100		11.35	3.69		15.04	18.55
5400	2 gang, 3 switch, stainless steel		53	.151		23	5.55		28.55	34
5500	4 switch, plastic		53	.151		10.35	5.55		15.90	20.50
5600	2 gang, 4 switch, stainless steel		53	.151		21.50	5.55		27.05	32.50
5700	6 switch, stainless steel		53	.151		43	5.55		48.55	56.50
5800	3 gang, 9 switch, stainless steel		32	.250		64.50	9.25		73.75	86

26 27 26.20 Wiring Devices Elements

		Crew	Daily Output	Labor-Hours	Unit	Material	Labor	Equipment	Total	Total Incl O&P
0010	**WIRING DEVICES ELEMENTS**									
0200	Toggle switch, quiet type, single pole, 15 amp	1 Elec	40	.200	Ea.	9.50	7.40		16.90	22.50
0600	3 way, 15 amp		23	.348		1.90	12.85		14.75	23
0900	4 way, 15 amp		15	.533		7.45	19.70		27.15	40
1650	Dimmer switch, 120 volt, incandescent, 600 watt, 1 pole [G]		16	.500		22.50	18.45		40.95	55
2460	Receptacle, duplex, 120 volt, grounded, 15 amp		40	.200		1.26	7.40		8.66	13.45
2470	20 amp		27	.296		7.95	10.95		18.90	26.50
2490	Dryer, 30 amp		15	.533		4.46	19.70		24.16	37
2500	Range, 50 amp		11	.727		12.15	27		39.15	57.50
2600	Wall plates, stainless steel, 1 gang		80	.100		2.56	3.69		6.25	8.85
2800	2 gang		53	.151		4.33	5.55		9.88	13.85
3200	Lampholder, keyless		26	.308		16.85	11.35		28.20	37
3400	Pullchain with receptacle		22	.364		20.50	13.40		33.90	44.50

26 27 Low-Voltage Distribution Equipment

26 27 73 – Door Chimes

26 27 73.10 Doorbell System		Crew	Daily Output	Labor-Hours	Unit	Material	2016 Bare Costs Labor	Equipment	Total	Total Incl O&P
0010	**DOORBELL SYSTEM**, incl. transformer, button & signal									
1000	Door chimes, 2 notes	1 Elec	16	.500	Ea.	23.50	18.45		41.95	55.50
1020	with ambient light		12	.667		113	24.50		137.50	164
1100	Tube type, 3 tube system		12	.667		193	24.50		217.50	252
1180	4 tube system		10	.800		425	29.50		454.50	520
1900	For transformer & button, add		5	1.600		11.45	59		70.45	109
3000	For push button only		24	.333		2.26	12.30		14.56	22.50

26 28 Low-Voltage Circuit Protective Devices

26 28 16 – Enclosed Switches and Circuit Breakers

26 28 16.10 Circuit Breakers

		Crew	Daily Output	Labor-Hours	Unit	Material	Labor	Equipment	Total	Total Incl O&P
0010	**CIRCUIT BREAKERS** (in enclosure)									
0100	Enclosed (NEMA 1), 600 volt, 3 pole, 30 amp	1 Elec	3.20	2.500	Ea.	520	92.50		612.50	720
0200	60 amp		2.80	2.857		635	105		740	870
0400	100 amp		2.30	3.478		730	128		858	1,000

26 28 16.20 Safety Switches

		Crew	Daily Output	Labor-Hours	Unit	Material	Labor	Equipment	Total	Total Incl O&P
0010	**SAFETY SWITCHES**									
0100	General duty 240 volt, 3 pole NEMA 1, fusible, 30 amp	1 Elec	3.20	2.500	Ea.	62.50	92.50		155	220
0200	60 amp		2.30	3.478		106	128		234	325
0300	100 amp		1.90	4.211		181	155		336	455
0400	200 amp		1.30	6.154		385	227		612	795
0500	400 amp	2 Elec	1.80	8.889		980	330		1,310	1,600
9010	Disc. switch, 600 V 3 pole fusible, 30 amp, to 10 HP motor	1 Elec	3.20	2.500		355	92.50		447.50	540
9050	60 amp, to 30 HP motor		2.30	3.478		815	128		943	1,100
9070	100 amp, to 60 HP motor		1.90	4.211		815	155		970	1,150

26 28 16.40 Time Switches

		Crew	Daily Output	Labor-Hours	Unit	Material	Labor	Equipment	Total	Total Incl O&P
0010	**TIME SWITCHES**									
0100	Single pole, single throw, 24 hour dial	1 Elec	4	2	Ea.	145	74		219	281
0200	24 hour dial with reserve power		3.60	2.222		615	82		697	810
0300	Astronomic dial		3.60	2.222		240	82		322	400
0400	Astronomic dial with reserve power		3.30	2.424		665	89.50		754.50	875
0500	7 day calendar dial		3.30	2.424		146	89.50		235.50	305
0600	7 day calendar dial with reserve power		3.20	2.500		285	92.50		377.50	465
0700	Photo cell 2000 watt		8	1		28.50	37		65.50	92

26 32 Packaged Generator Assemblies

26 32 13 – Engine Generators

26 32 13.16 Gas-Engine-Driven Generator Sets

		Crew	Daily Output	Labor-Hours	Unit	Material	Labor	Equipment	Total	Total Incl O&P
0010	**GAS-ENGINE-DRIVEN GENERATOR SETS**									
0020	Gas or gasoline operated, includes battery,									
0050	charger, & muffler									
0200	3 phase 4 wire, 277/480 volt, 7.5 kW	R-3	.83	24.096	Ea.	8,550	890	169	9,609	11,100
0300	11.5 kW		.71	28.169		12,100	1,050	198	13,348	15,200
0400	20 kW		.63	31.746		14,300	1,175	223	15,698	17,900
0500	35 kW		.55	36.364		17,000	1,350	255	18,605	21,200

26 33 Battery Equipment

26 33 43 – Battery Chargers

26 33 43.55 Electric Vehicle Charging

		Crew	Daily Output	Labor-Hours	Unit	Material	2016 Bare Costs Labor	Equipment	Total	Total Incl O&P
0010	**ELECTRIC VEHICLE CHARGING**									
0020	Level 2, wall mounted									
2100	Light duty, hard wired	[G] 1 Elec	20.48	.391	Ea.	1,125	14.40		1,139.40	1,250
2110	plug in	[G] "	30.72	.260	"	1,125	9.60		1,134.60	1,250

26 36 Transfer Switches

26 36 23 – Automatic Transfer Switches

26 36 23.10 Automatic Transfer Switch Devices

		Crew	Daily Output	Labor-Hours	Unit	Material	Labor	Equipment	Total	Total Incl O&P
0010	**AUTOMATIC TRANSFER SWITCH DEVICES**									
0100	Switches, enclosed 480 volt, 3 pole, 30 amp	1 Elec	2.30	3.478	Ea.	2,775	128		2,903	3,250
0200	60 amp	"	1.90	4.211	"	2,775	155		2,930	3,300

26 41 Facility Lightning Protection

26 41 13 – Lightning Protection for Structures

26 41 13.13 Lightning Protection for Buildings

		Crew	Daily Output	Labor-Hours	Unit	Material	Labor	Equipment	Total	Total Incl O&P
0010	**LIGHTNING PROTECTION FOR BUILDINGS**									
0200	Air terminals & base, copper									
0400	3/8" diameter x 10" (to 75' high)	1 Elec	8	1	Ea.	19	37		56	81.50
1000	Aluminum, 1/2" diameter x 12" (to 75' high)		8	1		12.85	37		49.85	74.50
1020	1/2" diameter x 24"		7.30	1.096		14.55	40.50		55.05	82
1040	1/2" diameter x 60"		6.70	1.194		21.50	44		65.50	96
2000	Cable, copper, 220 lb. per thousand ft. (to 75' high)		320	.025	L.F.	3.03	.92		3.95	4.84
2500	Aluminum, 101 lb. per thousand ft. (to 75' high)		280	.029	"	.95	1.05		2	2.77
3000	Arrester, 175 volt AC to ground		8	1	Ea.	149	37		186	225

26 51 Interior Lighting

26 51 13 – Interior Lighting Fixtures, Lamps, and Ballasts

26 51 13.50 Interior Lighting Fixtures

		Crew	Daily Output	Labor-Hours	Unit	Material	Labor	Equipment	Total	Total Incl O&P
0010	**INTERIOR LIGHTING FIXTURES** Including lamps, mounting									
0030	hardware and connections									
0100	Fluorescent, C.W. lamps, troffer, recess mounted in grid, RS									
0130	Grid ceiling mount									
0200	Acrylic lens, 1'W x 4'L, two 40 watt	1 Elec	5.70	1.404	Ea.	50	52		102	140
0300	2'W x 2'L, two U40 watt		5.70	1.404		54	52		106	144
0600	2'W x 4'L, four 40 watt		4.70	1.702		60.50	63		123.50	170
1000	Surface mounted, RS									
1030	Acrylic lens with hinged & latched door frame									
1100	1'W x 4'L, two 40 watt	1 Elec	7	1.143	Ea.	64	42		106	140
1200	2'W x 2'L, two U40 watt		7	1.143		68.50	42		110.50	145
1500	2'W x 4'L, four 40 watt		5.30	1.509		81	55.50		136.50	180
2100	Strip fixture									
2200	4' long, one 40 watt, RS	1 Elec	8.50	.941	Ea.	23	34.50		57.50	82.50
2300	4' long, two 40 watt, RS	"	8	1		41	37		78	106
2600	8' long, one 75 watt, SL	2 Elec	13.40	1.194		50.50	44		94.50	128
2700	8' long, two 75 watt, SL	"	12.40	1.290		61	47.50		108.50	145
4450	Incandescent, high hat can, round alzak reflector, prewired									
4470	100 watt	1 Elec	8	1	Ea.	71.50	37		108.50	139

26 51 Interior Lighting

26 51 13 – Interior Lighting Fixtures, Lamps, and Ballasts

26 51 13.50 Interior Lighting Fixtures		Crew	Daily Output	Labor-Hours	Unit	Material	2016 Bare Costs Labor	Equipment	Total	Total Incl O&P
4480	150 watt	1 Elec	8	1	Ea.	104	37		141	175
5200	Ceiling, surface mounted, opal glass drum									
5300	8", one 60 watt lamp	1 Elec	10	.800	Ea.	44.50	29.50		74	97.50
5400	10", two 60 watt lamps		8	1		50	37		87	116
5500	12", four 60 watt lamps		6.70	1.194		70.50	44		114.50	150
6900	Mirror light, fluorescent, RS, acrylic enclosure, two 40 watt		8	1		128	37		165	201
6910	One 40 watt		8	1		106	37		143	177
6920	One 20 watt		12	.667		81.50	24.50		106	130

26 51 13.55 Interior LED Fixtures			Crew	Daily Output	Labor-Hours	Unit	Material	Labor	Equipment	Total	Total Incl O&P
0010	**INTERIOR LED FIXTURES** Incl. lamps, and mounting hardware										
0100	Downlight, recess mounted, 7.5" diameter, 25 watt	G	1 Elec	8	1	Ea.	335	37		372	425
0120	10" diameter, 36 watt	G		8	1		360	37		397	455
0160	cylinder, 10 watts	G		8	1		102	37		139	173
0180	20 watts	G		8	1		166	37		203	244
1000	Troffer, recess mounted, 2' x 4', 3200 Lumens	G		5.30	1.509		138	55.50		193.50	242
1010	4800 Lumens	G		5	1.600		179	59		238	293
1020	6400 Lumens	G		4.70	1.702		198	63		261	320
1100	Troffer retrofit lamp, 38 watt	G		21	.381		238	14.05		252.05	284
1110	60 watt	G		20	.400		340	14.75		354.75	400
1120	100 watt	G		18	.444		510	16.40		526.40	585
1200	Troffer, volumetric recess mounted, 2' x 2'	G		5.70	1.404		279	52		331	390
2000	Strip, surface mounted, one light bar 4' long, 3500K	G		8.50	.941		260	34.50		294.50	345
2010	5000K	G		8	1		260	37		297	345
2020	Two light bar 4' long, 5000K	G		7	1.143		410	42		452	520
3000	Linear, suspended mounted, one light bar 4' long, 37 watt	G		6.70	1.194		154	44		198	242
3010	One light bar 8' long, 74 watt	G	2 Elec	12.20	1.311		286	48.50		334.50	395
3020	Two light bar 4' long, 74 watt	G	1 Elec	5.70	1.404		305	52		357	425
3030	Two light bar 8' long, 148 watt	G	2 Elec	8.80	1.818		355	67		422	500
4000	High bay, surface mounted, round, 150 watts	G		5.41	2.959		480	109		589	710
4010	2 bars, 164 watts	G		5.41	2.959		450	109		559	675
4020	3 bars, 246 watts	G		5.01	3.197		580	118		698	835
4030	4 bars, 328 watts	G		4.60	3.478		710	128		838	990
4040	5 bars, 410 watts	G	3 Elec	4.20	5.716		830	211		1,041	1,250
4050	6 bars, 492 watts	G		3.80	6.324		935	233		1,168	1,400
4060	7 bars, 574 watts	G		3.39	7.075		1,025	261		1,286	1,550
4070	8 bars, 656 watts	G		2.99	8.029		1,150	296		1,446	1,725
5000	Track, lighthead, 6 watt	G	1 Elec	32	.250		54.50	9.25		63.75	75
5010	9 watt	G	"	32	.250		61.50	9.25		70.75	82.50
6000	Garage, surface mounted, 103 watts	G	2 Elec	6.50	2.462		970	91		1,061	1,225
6100	pendent mounted, 80 watts	G		6.50	2.462		695	91		786	915
6200	95 watts	G		6.50	2.462		800	91		891	1,025
6300	125 watts	G		6.50	2.462		865	91		956	1,100

26 51 13.70 Residential Fixtures		Crew	Daily Output	Labor-Hours	Unit	Material	Labor	Equipment	Total	Total Incl O&P
0010	**RESIDENTIAL FIXTURES**									
0400	Fluorescent, interior, surface, circline, 32 watt & 40 watt	1 Elec	20	.400	Ea.	141	14.75		155.75	179
0500	2' x 2', two U-tube 32 watt T8		8	1		165	37		202	242
0700	Shallow under cabinet, two 20 watt		16	.500		71.50	18.45		89.95	109
0900	Wall mounted, 4'L, two 32 watt T8, with baffle		10	.800		133	29.50		162.50	195
2000	Incandescent, exterior lantern, wall mounted, 60 watt		16	.500		60	18.45		78.45	96
2100	Post light, 150W, with 7' post		4	2		270	74		344	420
2500	Lamp holder, weatherproof with 150W PAR		16	.500		32	18.45		50.45	65
2550	With reflector and guard		12	.667		56	24.50		80.50	102

26 51 Interior Lighting

26 51 13 – Interior Lighting Fixtures, Lamps, and Ballasts

26 51 13.70 Residential Fixtures		Crew	Daily Output	Labor-Hours	Unit	Material	2016 Bare Costs Labor	Equipment	Total	Total Incl O&P
2600	Interior pendent, globe with shade, 150 watt	1 Elec	20	.400	Ea.	201	14.75		215.75	245

26 55 Special Purpose Lighting

26 55 59 – Display Lighting

26 55 59.10 Track Lighting

		Crew	Daily Output	Labor-Hours	Unit	Material	Labor	Equipment	Total	Total Incl O&P
0010	**TRACK LIGHTING**									
0100	8' section	2 Elec	10.60	1.509	Ea.	69	55.50		124.50	167
0300	3 circuits, 4' section	1 Elec	6.70	1.194		92.50	44		136.50	174
0400	8' section	2 Elec	10.60	1.509		109	55.50		164.50	211
0500	12' section	"	8.80	1.818		159	67		226	285
1000	Feed kit, surface mounting	1 Elec	16	.500		13	18.45		31.45	44.50
1100	End cover		24	.333		5.55	12.30		17.85	26
1200	Feed kit, stem mounting, 1 circuit		16	.500		42	18.45		60.45	76
1300	3 circuit		16	.500		42	18.45		60.45	76
2000	Electrical joiner, for continuous runs, 1 circuit		32	.250		24	9.25		33.25	41.50
2100	3 circuit		32	.250		59	9.25		68.25	80
2200	Fixtures, spotlight, 75W PAR halogen		16	.500		55.50	18.45		73.95	91
2210	50W MR16 halogen		16	.500		171	18.45		189.45	218
3000	Wall washer, 250 watt tungsten halogen		16	.500		132	18.45		150.45	175
3100	Low voltage, 25/50 watt, 1 circuit		16	.500		133	18.45		151.45	176
3120	3 circuit		16	.500		198	18.45		216.45	248

26 56 Exterior Lighting

26 56 13 – Lighting Poles and Standards

26 56 13.10 Lighting Poles

		Crew	Daily Output	Labor-Hours	Unit	Material	Labor	Equipment	Total	Total Incl O&P
0010	**LIGHTING POLES**									
6420	Wood pole, 4-1/2" x 5-1/8", 8' high	1 Elec	6	1.333	Ea.	335	49		384	450
6440	12' high		5.70	1.404		485	52		537	620
6460	20' high		4	2		685	74		759	875

26 56 23 – Area Lighting

26 56 23.10 Exterior Fixtures

		Crew	Daily Output	Labor-Hours	Unit	Material	Labor	Equipment	Total	Total Incl O&P
0010	**EXTERIOR FIXTURES** With lamps									
0400	Quartz, 500 watt	1 Elec	5.30	1.509	Ea.	52	55.50		107.50	149
1100	Wall pack, low pressure sodium, 35 watt		4	2		234	74		308	380
1150	55 watt		4	2		278	74		352	425

26 56 26 – Landscape Lighting

26 56 26.20 Landscape Fixtures

		Crew	Daily Output	Labor-Hours	Unit	Material	Labor	Equipment	Total	Total Incl O&P
0010	**LANDSCAPE FIXTURES**									
7380	Landscape recessed uplight, incl. housing, ballast, transformer									
7390	& reflector, not incl. conduit, wire, trench									
7420	Incandescent, 250 watt	1 Elec	5	1.600	Ea.	600	59		659	755
7440	Quartz, 250 watt	"	5	1.600	"	570	59		629	720

26 56 33 – Walkway Lighting

26 56 33.10 Walkway Luminaire

		Crew	Daily Output	Labor-Hours	Unit	Material	Labor	Equipment	Total	Total Incl O&P
0010	**WALKWAY LUMINAIRE**									
6500	Bollard light, lamp & ballast, 42" high with polycarbonate lens									
7200	Incandescent, 150 watt	1 Elec	3	2.667	Ea.	610	98.50		708.50	830

26 61 Lighting Systems and Accessories

26 61 23 – Lamps Applications

26 61 23.10 Lamps

			Crew	Daily Output	Labor-Hours	Unit	Material	2016 Bare Costs Labor	Equipment	Total	Total Incl O&P
0010	**LAMPS**										
0081	Fluorescent, rapid start, cool white, 2' long, 20 watt		1 Elec	100	.080	Ea.	3.60	2.95		6.55	8.80
0101	4' long, 40 watt			90	.089		2.75	3.28		6.03	8.40
1351	High pressure sodium, 70 watt			30	.267		26.50	9.85		36.35	45
1371	150 watt			30	.267		28.50	9.85		38.35	47

26 61 23.55 LED Lamps

			Crew	Daily Output	Labor-Hours	Unit	Material	Labor	Equipment	Total	Total Incl O&P
0010	**LED LAMPS**										
0100	LED lamp, interior, shape A60, equal to 60 watt	G	1 Elec	160	.050	Ea.	20	1.85		21.85	25
0200	Globe frosted A60, equal to 60 watt	G		160	.050		12	1.85		13.85	16.20
0300	Globe earth, equal to 100 watt	G		160	.050		29	1.85		30.85	35
1100	MR16, 3 watt, replacement of halogen lamp 25 watt	G		130	.062		20	2.27		22.27	25.50
1200	6 watt replacement of halogen lamp 45 watt	G		130	.062		21	2.27		23.27	27
2100	10 watt, PAR20, equal to 60 watt	G		130	.062		28	2.27		30.27	34.50
2200	15 watt, PAR30, equal to 100 watt	G		130	.062		73.50	2.27		75.77	84.50

Estimating Tips
27 20 00 Data Communications
27 30 00 Voice Communications
27 40 00 Audio-Video Communications

When estimating material costs for special systems, it is always prudent to obtain manufacturers' quotations for equipment prices and special installation requirements which will affect the total costs.

Reference Numbers

Reference numbers are shown at the beginning of some major classifications. These numbers refer to related items in the Reference Section. The reference information may be an estimating procedure, an alternate pricing method, or technical information.

Note: Not all subdivisions listed here necessarily appear. ■

No part of this cost data may be reproduced, stored in a retrieval system, or transmitted in any form or by any means without prior written permission of RSMeans.

Note: Trade Service, in part, has been used as a reference source for some of the material prices used in Division 27.

Did you know?
RSMeans Online gives you the same access to RSMeans' data with 24/7 access:
- Quickly locate costs in the searchable database.
- Build cost lists, estimates, and reports in minutes.
- Adjust costs to any location in the U.S. and Canada with the click of a button.

Start your free trial today at **www.RSMeansOnline.com**

RSMeans Online
FROM THE GORDIAN GROUP®

27 41 Audio-Video Systems

27 41 33 – Master Antenna Television Systems

27 41 33.10 T.V. Systems		Crew	Daily Output	Labor-Hours	Unit	Material	2016 Bare Costs Labor	2016 Bare Costs Equipment	Total	Total Incl O&P
0010	**T.V. SYSTEMS**, not including rough-in wires, cables & conduits									
0100	Master TV antenna system									
0200	VHF reception & distribution, 12 outlets	1 Elec	6	1.333	Outlet	139	49		188	234
0800	VHF & UHF reception & distribution, 12 outlets		6	1.333	"	218	49		267	320
5000	Antenna, small		6	1.333	Ea.	48	49		97	134
5100	Large		4	2		202	74		276	345
5110	Rotor unit		8	1		50	37		87	116
5120	Single booster		8	1		25	37		62	88
5130	Antenna pole, 10'		3.20	2.500		19.50	92.50		112	173
6100	Satellite TV system	2 Elec	1	16		2,200	590		2,790	3,400
6110	Dish, mesh, 10' diam.	"	2.40	6.667		1,675	246		1,921	2,250
6111	Two way RF/IF tapeoff	1 Elec	36	.222		3.59	8.20		11.79	17.35
6112	Two way RF/IF splitter		24	.333		11.60	12.30		23.90	33
6113	Line amplifier		24	.333		15.15	12.30		27.45	36.50
6114	Line splitters		36	.222		4.69	8.20		12.89	18.55
6115	Line multi switches		8	1		700	37		737	830
6120	Motor unit		2.40	3.333		259	123		382	485
7000	Home theater, widescreen, 42", high definition, TV		10.24	.781		400	29		429	485
7050	Flat wall mount bracket		10.24	.781		99	29		128	156
7100	7 channel home theater receiver		10.24	.781		385	29		414	470
7200	Home theater speakers		10.24	.781	Set	216	29		245	284
7300	Home theater programmable remote		10.24	.781	Ea.	249	29		278	320
8000	Main video splitter		4	2		3,200	74		3,274	3,650
8010	Video distribution units		4	2		140	74		214	275

Estimating Tips

- When estimating material costs for electronic safety and security systems, it is always prudent to obtain manufacturers' quotations for equipment prices and special installation requirements that affect the total cost.
- Fire alarm systems consist of control panels, annunciator panels, battery with rack, charger, and fire alarm actuating and indicating devices. Some fire alarm systems include speakers, telephone lines, door closer controls, and other components. Be careful not to overlook the costs related to installation for these items. Also be aware of costs for integrated automation instrumentation and terminal devices, control equipment, control wiring, and programming.
- Security equipment includes items such as CCTV, access control, and other detection and identification systems to perform alert and alarm functions. Be sure to consider the costs related to installation for this security equipment, such as for integrated automation instrumentation and terminal devices, control equipment, control wiring, and programming.

Reference Numbers

Reference numbers are shown at the beginning of some major classifications. These numbers refer to related items in the Reference Section. The reference information may be an estimating procedure, an alternate pricing method, or technical information.

Note: Not all subdivisions listed here necessarily appear. ■

No part of this cost data may be reproduced, stored in a retrieval system, or transmitted in any form or by any means without prior written permission of RSMeans.

Note: Trade Service, in part, has been used as a reference source for some of the material prices used in Division 28.

Did you know?

RSMeans Online gives you the same access to RSMeans' data with 24/7 access:
- Quickly locate costs in the searchable database.
- Build cost lists, estimates, and reports in minutes.
- Adjust costs to any location in the U.S. and Canada with the click of a button.

Start your free trial today at **www.RSMeansOnline.com**

RSMeans Online
FROM THE GORDIAN GROUP®

28 16 Intrusion Detection

28 16 16 – Intrusion Detection Systems Infrastructure

28 16 16.50 Intrusion Detection

		Crew	Daily Output	Labor-Hours	Unit	Material	2016 Bare Costs Labor	2016 Bare Costs Equipment	Total	Total Incl O&P
0010	**INTRUSION DETECTION**, not including wires & conduits									
0100	Burglar alarm, battery operated, mechanical trigger	1 Elec	4	2	Ea.	278	74		352	425
0200	Electrical trigger		4	2		330	74		404	485
0400	For outside key control, add		8	1		86	37		123	155
0600	For remote signaling circuitry, add		8	1		137	37		174	211
0800	Card reader, flush type, standard		2.70	2.963		790	109		899	1,050
1000	Multi-code		2.70	2.963		1,200	109		1,309	1,500
1200	Door switches, hinge switch		5.30	1.509		62.50	55.50		118	160
1400	Magnetic switch		5.30	1.509		77	55.50		132.50	176
2800	Ultrasonic motion detector, 12 volt		2.30	3.478		210	128		338	440
3000	Infrared photoelectric detector		4	2		144	74		218	280
3200	Passive infrared detector		4	2		234	74		308	380
3420	Switchmats, 30" x 5'		5.30	1.509		104	55.50		159.50	205
3440	30" x 25'		4	2		203	74		277	345
3460	Police connect panel		4	2		269	74		343	415
3480	Telephone dialer		5.30	1.509		385	55.50		440.50	515
3500	Alarm bell		4	2		104	74		178	236
3520	Siren		4	2		146	74		220	282

28 31 Fire Detection and Alarm

28 31 23 – Fire Detection and Alarm Annunciation Panels and Fire Stations

28 31 23.50 Alarm Panels and Devices

		Crew	Daily Output	Labor-Hours	Unit	Material	Labor	Equipment	Total	Total Incl O&P
0010	**ALARM PANELS AND DEVICES**, not including wires & conduits									
5600	Strobe and horn	1 Elec	5.30	1.509	Ea.	152	55.50		207.50	258
5800	Fire alarm horn		6.70	1.194		61	44		105	139
6600	Drill switch		8	1		370	37		407	470
6800	Master box		2.70	2.963		6,400	109		6,509	7,225
7800	Remote annunciator, 8 zone lamp		1.80	4.444		209	164		373	500
8000	12 zone lamp	2 Elec	2.60	6.154		335	227		562	740
8200	16 zone lamp	"	2.20	7.273		420	268		688	905

28 31 46 – Smoke Detection Sensors

28 31 46.50 Smoke Detectors

		Crew	Daily Output	Labor-Hours	Unit	Material	Labor	Equipment	Total	Total Incl O&P
0010	**SMOKE DETECTORS**									
5200	Smoke detector, ceiling type	1 Elec	6.20	1.290	Ea.	110	47.50		157.50	199

Estimating Tips
31 05 00 Common Work Results for Earthwork

- Estimating the actual cost of performing earthwork requires careful consideration of the variables involved. This includes items such as type of soil, whether water will be encountered, dewatering, whether banks need bracing, disposal of excavated earth, and length of haul to fill or spoil sites, etc. If the project has large quantities of cut or fill, consider raising or lowering the site to reduce costs, while paying close attention to the effect on site drainage and utilities.
- If the project has large quantities of fill, creating a borrow pit on the site can significantly lower the costs.
- It is very important to consider what time of year the project is scheduled for completion. Bad weather can create large cost overruns from dewatering, site repair, and lost productivity from cold weather.

Reference Numbers
Reference numbers are shown at the beginning of some major classifications. These numbers refer to related items in the Reference Section. The reference information may be an estimating procedure, an alternate pricing method, or technical information.

Note: Not all subdivisions listed here necessarily appear. ■

Did you know?
RSMeans Online gives you the same access to RSMeans' data with 24/7 access:
- Quickly locate costs in the searchable database.
- Build cost lists, estimates, and reports in minutes.
- Adjust costs to any location in the U.S. and Canada with the click of a button.

Start your free trial today at **www.RSMeansOnline.com**

RSMeans Online
FROM THE GORDIAN GROUP

No part of this cost data may be reproduced, stored in a retrieval system, or transmitted in any form or by any means without prior written permission of RSMeans.

31 05 Common Work Results for Earthwork

31 05 13 – Soils for Earthwork

31 05 13.10 Borrow

		Crew	Daily Output	Labor-Hours	Unit	Material	2016 Bare Costs Labor	Equipment	Total	Total Incl O&P
0010	**BORROW**									
0020	Spread, 200 H.P. dozer, no compaction, 2 mi. RT haul									
0200	Common borrow	B-15	600	.047	C.Y.	12.40	1.45	4.63	18.48	21

31 05 16 – Aggregates for Earthwork

31 05 16.10 Borrow

		Crew	Daily Output	Labor-Hours	Unit	Material	2016 Bare Costs Labor	Equipment	Total	Total Incl O&P
0010	**BORROW**									
0020	Spread, with 200 H.P. dozer, no compaction, 2 mi. RT haul									
0100	Bank run gravel	B-15	600	.047	L.C.Y.	20.50	1.45	4.63	26.58	30
0300	Crushed stone (1.40 tons per CY), 1-1/2"		600	.047		24	1.45	4.63	30.08	34
0320	3/4"		600	.047		24	1.45	4.63	30.08	34
0340	1/2"		600	.047		28	1.45	4.63	34.08	38.50
0360	3/8"		600	.047		29.50	1.45	4.63	35.58	40
0400	Sand, washed, concrete		600	.047		37.50	1.45	4.63	43.58	49
0500	Dead or bank sand		600	.047		20	1.45	4.63	26.08	29.50

31 11 Clearing and Grubbing

31 11 10 – Clearing and Grubbing Land

31 11 10.10 Clear and Grub Site

		Crew	Daily Output	Labor-Hours	Unit	Material	2016 Bare Costs Labor	Equipment	Total	Total Incl O&P
0010	**CLEAR AND GRUB SITE**									
0020	Cut & chip light trees to 6" diam.	B-7	1	48	Acre		1,275	1,675	2,950	4,000
0150	Grub stumps and remove	B-30	2	12			385	1,200	1,585	1,950
0200	Cut & chip medium, trees to 12" diam.	B-7	.70	68.571			1,825	2,400	4,225	5,675
0250	Grub stumps and remove	B-30	1	24			770	2,400	3,170	3,925
0300	Cut & chip heavy, trees to 24" diam.	B-7	.30	160			4,275	5,575	9,850	13,300
0350	Grub stumps and remove	B-30	.50	48			1,550	4,825	6,375	7,850
0400	If burning is allowed, deduct cut & chip								40%	40%

31 13 Selective Tree and Shrub Removal and Trimming

31 13 13 – Selective Tree and Shrub Removal

31 13 13.10 Selective Clearing

		Crew	Daily Output	Labor-Hours	Unit	Material	2016 Bare Costs Labor	Equipment	Total	Total Incl O&P
0010	**SELECTIVE CLEARING**									
0020	Clearing brush with brush saw	A-1C	.25	32	Acre		790	126	916	1,475
0100	By hand	1 Clab	.12	66.667			1,650		1,650	2,750
0300	With dozer, ball and chain, light clearing	B-11A	2	8			240	700	940	1,175
0400	Medium clearing	"	1.50	10.667			320	930	1,250	1,550

31 14 Earth Stripping and Stockpiling

31 14 13 – Soil Stripping and Stockpiling

31 14 13.23 Topsoil Stripping and Stockpiling

		Crew	Daily Output	Labor-Hours	Unit	Material	2016 Bare Costs Labor	Equipment	Total	Total Incl O&P
0010	**TOPSOIL STRIPPING AND STOCKPILING**									
1400	Loam or topsoil, remove and stockpile on site									
1420	6" deep, 200' haul	B-10B	865	.009	C.Y.		.33	1.62	1.95	2.32
1430	300' haul		520	.015			.54	2.69	3.23	3.85
1440	500' haul		225	.036			1.25	6.20	7.45	8.90
1450	Alternate method: 6" deep, 200' haul		5090	.002	S.Y.		.06	.27	.33	.39
1460	500' haul		1325	.006	"		.21	1.05	1.26	1.51
1500	Loam or topsoil, remove/stockpile on site									

31 14 Earth Stripping and Stockpiling

31 14 13 – Soil Stripping and Stockpiling

31 14 13.23 Topsoil Stripping and Stockpiling	Crew	Daily Output	Labor-Hours	Unit	Material	2016 Bare Costs Labor	Equipment	Total	Total Incl O&P
1510 By hand, 6" deep, 50' haul, less than 100 S.Y.	B-1	100	.240	S.Y.		6.10		6.10	10.15
1520 By skid steer, 6" deep, 100' haul, 101-500 S.Y.	B-62	500	.048			1.33	.35	1.68	2.60
1530 100' haul, 501-900 S.Y.	"	900	.027			.74	.19	.93	1.44
1540 200' haul, 901-1100 S.Y.	B-63	1000	.040			.99	.17	1.16	1.84
1550 By dozer, 200' haul, 1101-4000 S.Y.	B-10B	4000	.002	↓		.07	.35	.42	.50

31 22 Grading

31 22 16 – Fine Grading

31 22 16.10 Finish Grading

	Crew	Daily Output	Labor-Hours	Unit	Material	Labor	Equipment	Total	Total Incl O&P
0010 **FINISH GRADING**									
0012 Finish grading area to be paved with grader, small area	B-11L	400	.040	S.Y.		1.20	1.79	2.99	3.95
0100 Large area		2000	.008			.24	.36	.60	.79
0200 Grade subgrade for base course, roadways		3500	.005			.14	.20	.34	.45
1020 For large parking lots	B-32C	5000	.010			.29	.47	.76	1
1050 For small irregular areas	"	2000	.024			.73	1.18	1.91	2.50
1100 Fine grade for slab on grade, machine	B-11L	1040	.015			.46	.69	1.15	1.53
1150 Hand grading	B-18	700	.034			.87	.07	.94	1.52
1200 Fine grade granular base for sidewalks and bikeways	B-62	1200	.020	↓		.55	.14	.69	1.08
2550 Hand grade select gravel	2 Clab	60	.267	C.S.F.		6.55		6.55	11
3000 Hand grade select gravel, including compaction, 4" deep	B-18	555	.043	S.Y.		1.09	.08	1.17	1.92
3100 6" deep		400	.060			1.52	.12	1.64	2.67
3120 8" deep	↓	300	.080			2.03	.16	2.19	3.56
3300 Finishing grading slopes, gentle	B-11L	8900	.002			.05	.08	.13	.18
3310 Steep slopes	"	7100	.002	↓		.07	.10	.17	.22

31 23 Excavation and Fill

31 23 16 – Excavation

31 23 16.13 Excavating, Trench

	Crew	Daily Output	Labor-Hours	Unit	Material	Labor	Equipment	Total	Total Incl O&P
0010 **EXCAVATING, TRENCH**									
0011 Or continuous footing									
0050 1' to 4' deep, 3/8 C.Y. excavator	B-11C	150	.107	B.C.Y.		3.19	2.44	5.63	8
0060 1/2 C.Y. excavator	B-11M	200	.080			2.40	1.98	4.38	6.15
0090 4' to 6' deep, 1/2 C.Y. excavator	"	200	.080			2.40	1.98	4.38	6.15
0100 5/8 C.Y. excavator	B-12Q	250	.064			1.94	2.36	4.30	5.80
0300 1/2 C.Y. excavator, truck mounted	B-12J	200	.080			2.43	4.41	6.84	8.90
1352 4' to 6' deep, 1/2 C.Y. excavator w/trench box	B-13H	188	.085			2.58	5.10	7.68	9.95
1354 5/8 C.Y. excavator	"	235	.068			2.07	4.10	6.17	7.95
1400 By hand with pick and shovel 2' to 6' deep, light soil	1 Clab	8	1			24.50		24.50	41.50
1500 Heavy soil	"	4	2	↓		49.50		49.50	82.50
5020 Loam & sandy clay with no sheeting or dewatering included									
5050 1' to 4' deep, 3/8 C.Y. tractor loader/backhoe	B-11C	162	.099	B.C.Y.		2.96	2.26	5.22	7.40
5060 1/2 C.Y. excavator	B-11M	216	.074			2.22	1.83	4.05	5.70
5080 4' to 6' deep, 1/2 C.Y. excavator	"	216	.074			2.22	1.83	4.05	5.70
5090 5/8 C.Y. excavator	B-12Q	276	.058			1.76	2.14	3.90	5.25
5130 1/2 C.Y. excavator, truck mounted	B-12J	216	.074			2.25	4.08	6.33	8.25
5352 4' to 6' deep, 1/2 C.Y. excavator w/trench box	B-13H	205	.078			2.37	4.70	7.07	9.10
5354 5/8 C.Y. excavator	"	257	.062	↓		1.89	3.75	5.64	7.25
6020 Sand & gravel with no sheeting or dewatering included									
6050 1' to 4' deep, 3/8 C.Y. excavator	B-11C	165	.097	B.C.Y.		2.90	2.21	5.11	7.25

31 23 Excavation and Fill

31 23 16 – Excavation

31 23 16.13 Excavating, Trench

		Crew	Daily Output	Labor-Hours	Unit	Material	Labor	Equipment	Total	Total Incl O&P
6060	1/2 C.Y. excavator	B-11M	220	.073	B.C.Y.		2.18	1.80	3.98	5.60
6080	4' to 6' deep, 1/2 C.Y. excavator	"	220	.073			2.18	1.80	3.98	5.60
6090	5/8 C.Y. excavator	B-12Q	275	.058			1.77	2.14	3.91	5.30
6130	1/2 C.Y. excavator, truck mounted	B-12J	220	.073			2.21	4.01	6.22	8.10
6352	4' to 6' deep, 1/2 C.Y. excavator w/trench box	B-13H	209	.077			2.32	4.61	6.93	8.90
6354	5/8 C.Y. excavator	"	261	.061			1.86	3.69	5.55	7.15
7020	Dense hard clay with no sheeting or dewatering included									
7050	1' to 4' deep, 3/8 C.Y. excavator	B-11C	132	.121	B.C.Y.		3.63	2.77	6.40	9.10
7060	1/2 C.Y. excavator	B-11M	176	.091			2.72	2.25	4.97	7
7080	4' to 6' deep, 1/2 C.Y. excavator	"	176	.091			2.72	2.25	4.97	7
7090	5/8 C.Y. excavator	B-12Q	220	.073			2.21	2.68	4.89	6.60
7130	1/2 C.Y. excavator, truck mounted	B-12J	176	.091			2.76	5	7.76	10.10

31 23 16.14 Excavating, Utility Trench

		Crew	Daily Output	Labor-Hours	Unit	Material	Labor	Equipment	Total	Total Incl O&P
0010	**EXCAVATING, UTILITY TRENCH**									
0011	Common earth									
0050	Trenching with chain trencher, 12 H.P., operator walking									
0100	4" wide trench, 12" deep	B-53	800	.010	L.F.		.25	.09	.34	.50
1000	Backfill by hand including compaction, add									
1050	4" wide trench, 12" deep	A-1G	800	.010	L.F.		.25	.07	.32	.49

31 23 16.16 Structural Excavation for Minor Structures

		Crew	Daily Output	Labor-Hours	Unit	Material	Labor	Equipment	Total	Total Incl O&P
0010	**STRUCTURAL EXCAVATION FOR MINOR STRUCTURES**									
0015	Hand, pits to 6' deep, sandy soil	1 Clab	8	1	B.C.Y.		24.50		24.50	41.50
0100	Heavy soil or clay		4	2			49.50		49.50	82.50
1100	Hand loading trucks from stock pile, sandy soil		12	.667			16.45		16.45	27.50
1300	Heavy soil or clay		8	1			24.50		24.50	41.50
1500	For wet or muck hand excavation, add to above						50%			50%

31 23 16.42 Excavating, Bulk Bank Measure

		Crew	Daily Output	Labor-Hours	Unit	Material	Labor	Equipment	Total	Total Incl O&P
0010	**EXCAVATING, BULK BANK MEASURE**									
0011	Common earth piled									
0020	For loading onto trucks, add								15%	15%
0200	Excavator, hydraulic, crawler mtd., 1 C.Y. cap. = 100 C.Y./hr.	B-12A	800	.020	B.C.Y.		.61	1.02	1.63	2.13
0310	Wheel mounted, 1/2 C.Y. cap. = 40 C.Y./hr.	B-12E	320	.050			1.52	1.39	2.91	4.05
1200	Front end loader, track mtd., 1-1/2 C.Y. cap. = 70 C.Y./hr.	B-10N	560	.014			.50	.94	1.44	1.86
1500	Wheel mounted, 3/4 C.Y. cap. = 45 C.Y./hr.	B-10R	360	.022			.78	.83	1.61	2.20
5000	Excavating, bulk bank measure, sandy clay & loam piled									
5020	For loading onto trucks, add								15%	15%
5100	Excavator, hydraulic, crawler mtd., 1 C.Y. cap. = 120 C.Y./hr.	B-12A	960	.017	B.C.Y.		.51	.85	1.36	1.77
5610	Wheel mounted, 1/2 C.Y. cap. = 44 C.Y./hr.	B-12E	352	.045	"		1.38	1.27	2.65	3.68
8000	For hauling excavated material, see Section 31 23 23.20									

31 23 23 – Fill

31 23 23.13 Backfill

		Crew	Daily Output	Labor-Hours	Unit	Material	Labor	Equipment	Total	Total Incl O&P
0010	**BACKFILL**									
0015	By hand, no compaction, light soil	1 Clab	14	.571	L.C.Y.		14.10		14.10	23.50
0100	Heavy soil		11	.727	"		17.95		17.95	30
0300	Compaction in 6" layers, hand tamp, add to above		20.60	.388	E.C.Y.		9.55		9.55	16
0400	Roller compaction operator walking, add	B-10A	100	.080			2.82	1.85	4.67	6.70
0500	Air tamp, add	B-9D	190	.211			5.25	1.35	6.60	10.35
0600	Vibrating plate, add	A-1D	60	.133			3.29	.60	3.89	6.15
0800	Compaction in 12" layers, hand tamp, add to above	1 Clab	34	.235			5.80		5.80	9.70
1300	Dozer backfilling, bulk, up to 300' haul, no compaction	B-10B	1200	.007	L.C.Y.		.24	1.16	1.40	1.67
1400	Air tamped, add	B-11B	80	.200	E.C.Y.		5.85	3.66	9.51	13.80

31 23 Excavation and Fill

31 23 23 – Fill

31 23 23.16 Fill By Borrow and Utility Bedding

		Crew	Daily Output	Labor-Hours	Unit	Material	2016 Bare Costs Labor	Equipment	Total	Total Incl O&P
0010	**FILL BY BORROW AND UTILITY BEDDING**									
0049	Utility bedding, for pipe & conduit, not incl. compaction									
0050	Crushed or screened bank run gravel	B-6	150	.160	L.C.Y.	23	4.44	2.44	29.88	35.50
0100	Crushed stone 3/4" to 1/2"		150	.160		24	4.44	2.44	30.88	36.50
0200	Sand, dead or bank	↓	150	.160	↓	20	4.44	2.44	26.88	32
0500	Compacting bedding in trench	A-1D	90	.089	E.C.Y.		2.19	.40	2.59	4.11
0600	If material source exceeds 2 miles, add for extra mileage.									
0610	See Section 31 23 23.20 for hauling mileage add.									

31 23 23.17 General Fill

		Crew	Daily Output	Labor-Hours	Unit	Material	Labor	Equipment	Total	Total Incl O&P
0010	**GENERAL FILL**									
0011	Spread dumped material, no compaction									
0020	By dozer	B-10B	1000	.008	L.C.Y.		.28	1.40	1.68	2.01
0100	By hand	1 Clab	12	.667	"		16.45		16.45	27.50
0500	Gravel fill, compacted, under floor slabs, 4" deep	B-37	10000	.005	S.F.	.44	.13	.02	.59	.71
0600	6" deep		8600	.006		.66	.15	.02	.83	1
0700	9" deep		7200	.007		1.10	.18	.02	1.30	1.53
0800	12" deep		6000	.008	↓	1.54	.21	.03	1.78	2.07
1000	Alternate pricing method, 4" deep		120	.400	E.C.Y.	33	10.60	1.36	44.96	55.50
1100	6" deep		160	.300		33	7.95	1.02	41.97	51
1200	9" deep		200	.240		33	6.35	.81	40.16	48
1300	12" deep	↓	220	.218	↓	33	5.80	.74	39.54	47

31 23 23.20 Hauling

		Crew	Daily Output	Labor-Hours	Unit	Material	Labor	Equipment	Total	Total Incl O&P
0010	**HAULING**									
0011	Excavated or borrow, loose cubic yards									
0012	no loading equipment, including hauling, waiting, loading/dumping									
0013	time per cycle (wait, load, travel, unload or dump & return)									
0014	8 C.Y. truck, 15 MPH ave, cycle 0.5 miles, 10 min. wait/Ld./Uld.	B-34A	320	.025	L.C.Y.		.77	1.30	2.07	2.69
0016	cycle 1 mile		272	.029			.90	1.53	2.43	3.18
0018	cycle 2 miles		208	.038			1.18	2.01	3.19	4.15
0020	cycle 4 miles		144	.056			1.70	2.90	4.60	6
0022	cycle 6 miles		112	.071			2.19	3.73	5.92	7.70
0024	cycle 8 miles		88	.091			2.79	4.74	7.53	9.80
0026	20 MPH ave, cycle 0.5 mile		336	.024			.73	1.24	1.97	2.57
0028	cycle 1 mile		296	.027			.83	1.41	2.24	2.92
0030	cycle 2 miles		240	.033			1.02	1.74	2.76	3.59
0032	cycle 4 miles		176	.045			1.39	2.37	3.76	4.91
0034	cycle 6 miles		136	.059			1.80	3.07	4.87	6.35
0036	cycle 8 miles		112	.071			2.19	3.73	5.92	7.70
0044	25 MPH ave, cycle 4 miles		192	.042			1.28	2.17	3.45	4.50
0046	cycle 6 miles		160	.050			1.53	2.61	4.14	5.40
0048	cycle 8 miles		128	.063			1.92	3.26	5.18	6.75
0050	30 MPH ave, cycle 4 miles		216	.037			1.14	1.93	3.07	4
0052	cycle 6 miles		176	.045			1.39	2.37	3.76	4.91
0054	cycle 8 miles		144	.056			1.70	2.90	4.60	6
0114	15 MPH ave, cycle 0.5 mile, 15 min. wait/Ld./Uld.		224	.036			1.09	1.86	2.95	3.86
0116	cycle 1 mile		200	.040			1.23	2.09	3.32	4.32
0118	cycle 2 miles		168	.048			1.46	2.48	3.94	5.15
0120	cycle 4 miles		120	.067			2.04	3.48	5.52	7.20
0122	cycle 6 miles		96	.083			2.55	4.35	6.90	9
0124	cycle 8 miles		80	.100			3.07	5.20	8.27	10.80
0126	20 MPH ave, cycle 0.5 mile		232	.034			1.06	1.80	2.86	3.72
0128	cycle 1 mile	↓	208	.038	↓		1.18	2.01	3.19	4.15

31 23 Excavation and Fill

31 23 23 – Fill

31 23 23.20	Hauling	Crew	Daily Output	Labor-Hours	Unit	Material	2016 Bare Costs Labor	Equipment	Total	Total Incl O&P
0130	cycle 2 miles	B-34A	184	.043	L.C.Y.		1.33	2.27	3.60	4.70
0132	cycle 4 miles		144	.056			1.70	2.90	4.60	6
0134	cycle 6 miles		112	.071			2.19	3.73	5.92	7.70
0136	cycle 8 miles		96	.083			2.55	4.35	6.90	9
0144	25 MPH ave, cycle 4 miles		152	.053			1.61	2.75	4.36	5.70
0146	cycle 6 miles		128	.063			1.92	3.26	5.18	6.75
0148	cycle 8 miles		112	.071			2.19	3.73	5.92	7.70
0150	30 MPH ave, cycle 4 miles		168	.048			1.46	2.48	3.94	5.15
0152	cycle 6 miles		144	.056			1.70	2.90	4.60	6
0154	cycle 8 miles		120	.067			2.04	3.48	5.52	7.20
0214	15 MPH ave, cycle 0.5 mile, 20 min wait/Ld./Uld.		176	.045			1.39	2.37	3.76	4.91
0216	cycle 1 mile		160	.050			1.53	2.61	4.14	5.40
0218	cycle 2 miles		136	.059			1.80	3.07	4.87	6.35
0220	cycle 4 miles		104	.077			2.36	4.01	6.37	8.30
0222	cycle 6 miles		88	.091			2.79	4.74	7.53	9.80
0224	cycle 8 miles		72	.111			3.41	5.80	9.21	12
0226	20 MPH ave, cycle 0.5 mile		176	.045			1.39	2.37	3.76	4.91
0228	cycle 1 mile		168	.048			1.46	2.48	3.94	5.15
0230	cycle 2 miles		144	.056			1.70	2.90	4.60	6
0232	cycle 4 miles		120	.067			2.04	3.48	5.52	7.20
0234	cycle 6 miles		96	.083			2.55	4.35	6.90	9
0236	cycle 8 miles		88	.091			2.79	4.74	7.53	9.80
0244	25 MPH ave, cycle 4 miles		128	.063			1.92	3.26	5.18	6.75
0246	cycle 6 miles		112	.071			2.19	3.73	5.92	7.70
0248	cycle 8 miles		96	.083			2.55	4.35	6.90	9
0250	30 MPH ave, cycle 4 miles		136	.059			1.80	3.07	4.87	6.35
0252	cycle 6 miles		120	.067			2.04	3.48	5.52	7.20
0254	cycle 8 miles		104	.077			2.36	4.01	6.37	8.30
0314	15 MPH ave, cycle 0.5 mile, 25 min wait/Ld./Uld.		144	.056			1.70	2.90	4.60	6
0316	cycle 1 mile		128	.063			1.92	3.26	5.18	6.75
0318	cycle 2 miles		112	.071			2.19	3.73	5.92	7.70
0320	cycle 4 miles		96	.083			2.55	4.35	6.90	9
0322	cycle 6 miles		80	.100			3.07	5.20	8.27	10.80
0324	cycle 8 miles		64	.125			3.83	6.50	10.33	13.45
0326	20 MPH ave, cycle 0.5 mile		144	.056			1.70	2.90	4.60	6
0328	cycle 1 mile		136	.059			1.80	3.07	4.87	6.35
0330	cycle 2 miles		120	.067			2.04	3.48	5.52	7.20
0332	cycle 4 miles		104	.077			2.36	4.01	6.37	8.30
0334	cycle 6 miles		88	.091			2.79	4.74	7.53	9.80
0336	cycle 8 miles		80	.100			3.07	5.20	8.27	10.80
0344	25 MPH ave, cycle 4 miles		112	.071			2.19	3.73	5.92	7.70
0346	cycle 6 miles		96	.083			2.55	4.35	6.90	9
0348	cycle 8 miles		88	.091			2.79	4.74	7.53	9.80
0350	30 MPH ave, cycle 4 miles		112	.071			2.19	3.73	5.92	7.70
0352	cycle 6 miles		104	.077			2.36	4.01	6.37	8.30
0354	cycle 8 miles		96	.083			2.55	4.35	6.90	9
0414	15 MPH ave, cycle 0.5 mile, 30 min wait/Ld./Uld.		120	.067			2.04	3.48	5.52	7.20
0416	cycle 1 mile		112	.071			2.19	3.73	5.92	7.70
0418	cycle 2 miles		96	.083			2.55	4.35	6.90	9
0420	cycle 4 miles		80	.100			3.07	5.20	8.27	10.80
0422	cycle 6 miles		72	.111			3.41	5.80	9.21	12
0424	cycle 8 miles		64	.125			3.83	6.50	10.33	13.45
0426	20 MPH ave, cycle 0.5 mile		120	.067			2.04	3.48	5.52	7.20

31 23 Excavation and Fill

31 23 23 – Fill

31 23 23.20 Hauling		Crew	Daily Output	Labor-Hours	Unit	Material	2016 Bare Costs Labor	Equipment	Total	Total Incl O&P
0428	cycle 1 mile	B-34A	112	.071	L.C.Y.		2.19	3.73	5.92	7.70
0430	cycle 2 miles		104	.077			2.36	4.01	6.37	8.30
0432	cycle 4 miles		88	.091			2.79	4.74	7.53	9.80
0434	cycle 6 miles		80	.100			3.07	5.20	8.27	10.80
0436	cycle 8 miles		72	.111			3.41	5.80	9.21	12
0444	25 MPH ave, cycle 4 miles		96	.083			2.55	4.35	6.90	9
0446	cycle 6 miles		88	.091			2.79	4.74	7.53	9.80
0448	cycle 8 miles		80	.100			3.07	5.20	8.27	10.80
0450	30 MPH ave, cycle 4 miles		96	.083			2.55	4.35	6.90	9
0452	cycle 6 miles		88	.091			2.79	4.74	7.53	9.80
0454	cycle 8 miles		80	.100			3.07	5.20	8.27	10.80
0514	15 MPH ave, cycle 0.5 mile, 35 min wait/Ld./Uld.		104	.077			2.36	4.01	6.37	8.30
0516	cycle 1 mile		96	.083			2.55	4.35	6.90	9
0518	cycle 2 miles		88	.091			2.79	4.74	7.53	9.80
0520	cycle 4 miles		72	.111			3.41	5.80	9.21	12
0522	cycle 6 miles		64	.125			3.83	6.50	10.33	13.45
0524	cycle 8 miles		56	.143			4.38	7.45	11.83	15.40
0526	20 MPH ave, cycle 0.5 mile		104	.077			2.36	4.01	6.37	8.30
0528	cycle 1 mile		96	.083			2.55	4.35	6.90	9
0530	cycle 2 miles		96	.083			2.55	4.35	6.90	9
0532	cycle 4 miles		80	.100			3.07	5.20	8.27	10.80
0534	cycle 6 miles		72	.111			3.41	5.80	9.21	12
0536	cycle 8 miles		64	.125			3.83	6.50	10.33	13.45
0544	25 MPH ave, cycle 4 miles		88	.091			2.79	4.74	7.53	9.80
0546	cycle 6 miles		80	.100			3.07	5.20	8.27	10.80
0548	cycle 8 miles		72	.111			3.41	5.80	9.21	12
0550	30 MPH ave, cycle 4 miles		88	.091			2.79	4.74	7.53	9.80
0552	cycle 6 miles		80	.100			3.07	5.20	8.27	10.80
0554	cycle 8 miles		72	.111			3.41	5.80	9.21	12
1014	12 C.Y. truck, cycle 0.5 mile, 15 MPH ave, 15 min. wait/Ld./Uld.	B-34B	336	.024			.73	2.06	2.79	3.46
1016	cycle 1 mile		300	.027			.82	2.30	3.12	3.88
1018	cycle 2 miles		252	.032			.97	2.74	3.71	4.63
1020	cycle 4 miles		180	.044			1.36	3.84	5.20	6.45
1022	cycle 6 miles		144	.056			1.70	4.80	6.50	8.10
1024	cycle 8 miles		120	.067			2.04	5.75	7.79	9.70
1025	cycle 10 miles		96	.083			2.55	7.20	9.75	12.10
1026	20 MPH ave, cycle 0.5 mile		348	.023			.70	1.99	2.69	3.34
1028	cycle 1 mile		312	.026			.79	2.21	3	3.74
1030	cycle 2 miles		276	.029			.89	2.50	3.39	4.22
1032	cycle 4 miles		216	.037			1.14	3.20	4.34	5.40
1034	cycle 6 miles		168	.048			1.46	4.11	5.57	6.95
1036	cycle 8 miles		144	.056			1.70	4.80	6.50	8.10
1038	cycle 10 miles		120	.067			2.04	5.75	7.79	9.70
1040	25 MPH ave, cycle 4 miles		228	.035			1.08	3.03	4.11	5.10
1042	cycle 6 miles		192	.042			1.28	3.60	4.88	6.05
1044	cycle 8 miles		168	.048			1.46	4.11	5.57	6.95
1046	cycle 10 miles		144	.056			1.70	4.80	6.50	8.10
1050	30 MPH ave, cycle 4 miles		252	.032			.97	2.74	3.71	4.63
1052	cycle 6 miles		216	.037			1.14	3.20	4.34	5.40
1054	cycle 8 miles		180	.044			1.36	3.84	5.20	6.45
1056	cycle 10 miles		156	.051			1.57	4.43	6	7.45
1060	35 MPH ave, cycle 4 miles		264	.030			.93	2.62	3.55	4.41
1062	cycle 6 miles		228	.035			1.08	3.03	4.11	5.10

31 23 Excavation and Fill

31 23 23 – Fill

31 23 23.20 Hauling

		Crew	Daily Output	Labor-Hours	Unit	Material	2016 Bare Costs Labor	Equipment	Total	Total Incl O&P
1064	cycle 8 miles	B-34B	204	.039	L.C.Y.		1.20	3.39	4.59	5.70
1066	cycle 10 miles		180	.044			1.36	3.84	5.20	6.45
1068	cycle 20 miles		120	.067			2.04	5.75	7.79	9.70
1069	cycle 30 miles		84	.095			2.92	8.20	11.12	13.85
1070	cycle 40 miles		72	.111			3.41	9.60	13.01	16.15
1072	40 MPH ave, cycle 6 miles		240	.033			1.02	2.88	3.90	4.85
1074	cycle 8 miles		216	.037			1.14	3.20	4.34	5.40
1076	cycle 10 miles		192	.042			1.28	3.60	4.88	6.05
1078	cycle 20 miles		120	.067			2.04	5.75	7.79	9.70
1080	cycle 30 miles		96	.083			2.55	7.20	9.75	12.10
1082	cycle 40 miles		72	.111			3.41	9.60	13.01	16.15
1084	cycle 50 miles		60	.133			4.09	11.50	15.59	19.40
1094	45 MPH ave, cycle 8 miles		216	.037			1.14	3.20	4.34	5.40
1096	cycle 10 miles		204	.039			1.20	3.39	4.59	5.70
1098	cycle 20 miles		132	.061			1.86	5.25	7.11	8.80
1100	cycle 30 miles		108	.074			2.27	6.40	8.67	10.80
1102	cycle 40 miles		84	.095			2.92	8.20	11.12	13.85
1104	cycle 50 miles		72	.111			3.41	9.60	13.01	16.15
1106	50 MPH ave, cycle 10 miles		216	.037			1.14	3.20	4.34	5.40
1108	cycle 20 miles		144	.056			1.70	4.80	6.50	8.10
1110	cycle 30 miles		108	.074			2.27	6.40	8.67	10.80
1112	cycle 40 miles		84	.095			2.92	8.20	11.12	13.85
1114	cycle 50 miles		72	.111			3.41	9.60	13.01	16.15
1214	15 MPH ave, cycle 0.5 mile, 20 min. wait/Ld./Uld.		264	.030			.93	2.62	3.55	4.41
1216	cycle 1 mile		240	.033			1.02	2.88	3.90	4.85
1218	cycle 2 miles		204	.039			1.20	3.39	4.59	5.70
1220	cycle 4 miles		156	.051			1.57	4.43	6	7.45
1222	cycle 6 miles		132	.061			1.86	5.25	7.11	8.80
1224	cycle 8 miles		108	.074			2.27	6.40	8.67	10.80
1225	cycle 10 miles		96	.083			2.55	7.20	9.75	12.10
1226	20 MPH ave, cycle 0.5 mile		264	.030			.93	2.62	3.55	4.41
1228	cycle 1 mile		252	.032			.97	2.74	3.71	4.63
1230	cycle 2 miles		216	.037			1.14	3.20	4.34	5.40
1232	cycle 4 miles		180	.044			1.36	3.84	5.20	6.45
1234	cycle 6 miles		144	.056			1.70	4.80	6.50	8.10
1236	cycle 8 miles		132	.061			1.86	5.25	7.11	8.80
1238	cycle 10 miles		108	.074			2.27	6.40	8.67	10.80
1240	25 MPH ave, cycle 4 miles		192	.042			1.28	3.60	4.88	6.05
1242	cycle 6 miles		168	.048			1.46	4.11	5.57	6.95
1244	cycle 8 miles		144	.056			1.70	4.80	6.50	8.10
1246	cycle 10 miles		132	.061			1.86	5.25	7.11	8.80
1250	30 MPH ave, cycle 4 miles		204	.039			1.20	3.39	4.59	5.70
1252	cycle 6 miles		180	.044			1.36	3.84	5.20	6.45
1254	cycle 8 miles		156	.051			1.57	4.43	6	7.45
1256	cycle 10 miles		144	.056			1.70	4.80	6.50	8.10
1260	35 MPH ave, cycle 4 miles		216	.037			1.14	3.20	4.34	5.40
1262	cycle 6 miles		192	.042			1.28	3.60	4.88	6.05
1264	cycle 8 miles		168	.048			1.46	4.11	5.57	6.95
1266	cycle 10 miles		156	.051			1.57	4.43	6	7.45
1268	cycle 20 miles		108	.074			2.27	6.40	8.67	10.80
1269	cycle 30 miles		72	.111			3.41	9.60	13.01	16.15
1270	cycle 40 miles		60	.133			4.09	11.50	15.59	19.40
1272	40 MPH ave, cycle 6 miles		192	.042			1.28	3.60	4.88	6.05

31 23 Excavation and Fill

31 23 23 – Fill

31 23 23.20 Hauling

		Crew	Daily Output	Labor-Hours	Unit	Material	2016 Bare Costs Labor	Equipment	Total	Total Incl O&P
1274	cycle 8 miles	B-34B	180	.044	L.C.Y.		1.36	3.84	5.20	6.45
1276	cycle 10 miles		156	.051			1.57	4.43	6	7.45
1278	cycle 20 miles		108	.074			2.27	6.40	8.67	10.80
1280	cycle 30 miles		84	.095			2.92	8.20	11.12	13.85
1282	cycle 40 miles		72	.111			3.41	9.60	13.01	16.15
1284	cycle 50 miles		60	.133			4.09	11.50	15.59	19.40
1294	45 MPH ave, cycle 8 miles		180	.044			1.36	3.84	5.20	6.45
1296	cycle 10 miles		168	.048			1.46	4.11	5.57	6.95
1298	cycle 20 miles		120	.067			2.04	5.75	7.79	9.70
1300	cycle 30 miles		96	.083			2.55	7.20	9.75	12.10
1302	cycle 40 miles		72	.111			3.41	9.60	13.01	16.15
1304	cycle 50 miles		60	.133			4.09	11.50	15.59	19.40
1306	50 MPH ave, cycle 10 miles		180	.044			1.36	3.84	5.20	6.45
1308	cycle 20 miles		132	.061			1.86	5.25	7.11	8.80
1310	cycle 30 miles		96	.083			2.55	7.20	9.75	12.10
1312	cycle 40 miles		84	.095			2.92	8.20	11.12	13.85
1314	cycle 50 miles		72	.111			3.41	9.60	13.01	16.15
1414	15 MPH ave, cycle 0.5 mile, 25 min. wait/Ld./Uld.		204	.039			1.20	3.39	4.59	5.70
1416	cycle 1 mile		192	.042			1.28	3.60	4.88	6.05
1418	cycle 2 miles		168	.048			1.46	4.11	5.57	6.95
1420	cycle 4 miles		132	.061			1.86	5.25	7.11	8.80
1422	cycle 6 miles		120	.067			2.04	5.75	7.79	9.70
1424	cycle 8 miles		96	.083			2.55	7.20	9.75	12.10
1425	cycle 10 miles		84	.095			2.92	8.20	11.12	13.85
1426	20 MPH ave, cycle 0.5 mile		216	.037			1.14	3.20	4.34	5.40
1428	cycle 1 mile		204	.039			1.20	3.39	4.59	5.70
1430	cycle 2 miles		180	.044			1.36	3.84	5.20	6.45
1432	cycle 4 miles		156	.051			1.57	4.43	6	7.45
1434	cycle 6 miles		132	.061			1.86	5.25	7.11	8.80
1436	cycle 8 miles		120	.067			2.04	5.75	7.79	9.70
1438	cycle 10 miles		96	.083			2.55	7.20	9.75	12.10
1440	25 MPH ave, cycle 4 miles		168	.048			1.46	4.11	5.57	6.95
1442	cycle 6 miles		144	.056			1.70	4.80	6.50	8.10
1444	cycle 8 miles		132	.061			1.86	5.25	7.11	8.80
1446	cycle 10 miles		108	.074			2.27	6.40	8.67	10.80
1450	30 MPH ave, cycle 4 miles		168	.048			1.46	4.11	5.57	6.95
1452	cycle 6 miles		156	.051			1.57	4.43	6	7.45
1454	cycle 8 miles		132	.061			1.86	5.25	7.11	8.80
1456	cycle 10 miles		120	.067			2.04	5.75	7.79	9.70
1460	35 MPH ave, cycle 4 miles		180	.044			1.36	3.84	5.20	6.45
1462	cycle 6 miles		156	.051			1.57	4.43	6	7.45
1464	cycle 8 miles		144	.056			1.70	4.80	6.50	8.10
1466	cycle 10 miles		132	.061			1.86	5.25	7.11	8.80
1468	cycle 20 miles		96	.083			2.55	7.20	9.75	12.10
1469	cycle 30 miles		72	.111			3.41	9.60	13.01	16.15
1470	cycle 40 miles		60	.133			4.09	11.50	15.59	19.40
1472	40 MPH ave, cycle 6 miles		168	.048			1.46	4.11	5.57	6.95
1474	cycle 8 miles		156	.051			1.57	4.43	6	7.45
1476	cycle 10 miles		144	.056			1.70	4.80	6.50	8.10
1478	cycle 20 miles		96	.083			2.55	7.20	9.75	12.10
1480	cycle 30 miles		84	.095			2.92	8.20	11.12	13.85
1482	cycle 40 miles		60	.133			4.09	11.50	15.59	19.40
1484	cycle 50 miles		60	.133			4.09	11.50	15.59	19.40

31 23 Excavation and Fill

31 23 23 – Fill

31 23 23.20 Hauling		Crew	Daily Output	Labor-Hours	Unit	Material	2016 Bare Costs Labor	Equipment	Total	Total Incl O&P
1494	45 MPH ave, cycle 8 miles	B-34B	156	.051	L.C.Y.		1.57	4.43	6	7.45
1496	cycle 10 miles		144	.056			1.70	4.80	6.50	8.10
1498	cycle 20 miles		108	.074			2.27	6.40	8.67	10.80
1500	cycle 30 miles		84	.095			2.92	8.20	11.12	13.85
1502	cycle 40 miles		72	.111			3.41	9.60	13.01	16.15
1504	cycle 50 miles		60	.133			4.09	11.50	15.59	19.40
1506	50 MPH ave, cycle 10 miles		156	.051			1.57	4.43	6	7.45
1508	cycle 20 miles		120	.067			2.04	5.75	7.79	9.70
1510	cycle 30 miles		96	.083			2.55	7.20	9.75	12.10
1512	cycle 40 miles		72	.111			3.41	9.60	13.01	16.15
1514	cycle 50 miles		60	.133			4.09	11.50	15.59	19.40
1614	15 MPH ave, cycle 0.5 mile, 30 min. wait/Ld./Uld.		180	.044			1.36	3.84	5.20	6.45
1616	cycle 1 mile		168	.048			1.46	4.11	5.57	6.95
1618	cycle 2 miles		144	.056			1.70	4.80	6.50	8.10
1620	cycle 4 miles		120	.067			2.04	5.75	7.79	9.70
1622	cycle 6 miles		108	.074			2.27	6.40	8.67	10.80
1624	cycle 8 miles		84	.095			2.92	8.20	11.12	13.85
1625	cycle 10 miles		84	.095			2.92	8.20	11.12	13.85
1626	20 MPH ave, cycle 0.5 mile		180	.044			1.36	3.84	5.20	6.45
1628	cycle 1 mile		168	.048			1.46	4.11	5.57	6.95
1630	cycle 2 miles		156	.051			1.57	4.43	6	7.45
1632	cycle 4 miles		132	.061			1.86	5.25	7.11	8.80
1634	cycle 6 miles		120	.067			2.04	5.75	7.79	9.70
1636	cycle 8 miles		108	.074			2.27	6.40	8.67	10.80
1638	cycle 10 miles		96	.083			2.55	7.20	9.75	12.10
1640	25 MPH ave, cycle 4 miles		144	.056			1.70	4.80	6.50	8.10
1642	cycle 6 miles		132	.061			1.86	5.25	7.11	8.80
1644	cycle 8 miles		108	.074			2.27	6.40	8.67	10.80
1646	cycle 10 miles		108	.074			2.27	6.40	8.67	10.80
1650	30 MPH ave, cycle 4 miles		144	.056			1.70	4.80	6.50	8.10
1652	cycle 6 miles		132	.061			1.86	5.25	7.11	8.80
1654	cycle 8 miles		120	.067			2.04	5.75	7.79	9.70
1656	cycle 10 miles		108	.074			2.27	6.40	8.67	10.80
1660	35 MPH ave, cycle 4 miles		156	.051			1.57	4.43	6	7.45
1662	cycle 6 miles		144	.056			1.70	4.80	6.50	8.10
1664	cycle 8 miles		132	.061			1.86	5.25	7.11	8.80
1666	cycle 10 miles		120	.067			2.04	5.75	7.79	9.70
1668	cycle 20 miles		84	.095			2.92	8.20	11.12	13.85
1669	cycle 30 miles		72	.111			3.41	9.60	13.01	16.15
1670	cycle 40 miles		60	.133			4.09	11.50	15.59	19.40
1672	40 MPH ave, cycle 6 miles		144	.056			1.70	4.80	6.50	8.10
1674	cycle 8 miles		132	.061			1.86	5.25	7.11	8.80
1676	cycle 10 miles		120	.067			2.04	5.75	7.79	9.70
1678	cycle 20 miles		96	.083			2.55	7.20	9.75	12.10
1680	cycle 30 miles		72	.111			3.41	9.60	13.01	16.15
1682	cycle 40 miles		60	.133			4.09	11.50	15.59	19.40
1684	cycle 50 miles		48	.167			5.10	14.40	19.50	24.50
1694	45 MPH ave, cycle 8 miles		144	.056			1.70	4.80	6.50	8.10
1696	cycle 10 miles		132	.061			1.86	5.25	7.11	8.80
1698	cycle 20 miles		96	.083			2.55	7.20	9.75	12.10
1700	cycle 30 miles		84	.095			2.92	8.20	11.12	13.85
1702	cycle 40 miles		60	.133			4.09	11.50	15.59	19.40
1704	cycle 50 miles		60	.133			4.09	11.50	15.59	19.40

31 23 Excavation and Fill

31 23 23 – Fill

31 23 23.20 Hauling

		Crew	Daily Output	Labor-Hours	Unit	Material	2016 Bare Costs Labor	Equipment	Total	Total Incl O&P
1706	50 MPH ave, cycle 10 miles	B-34B	132	.061	L.C.Y.		1.86	5.25	7.11	8.80
1708	cycle 20 miles		108	.074			2.27	6.40	8.67	10.80
1710	cycle 30 miles		84	.095			2.92	8.20	11.12	13.85
1712	cycle 40 miles		72	.111			3.41	9.60	13.01	16.15
1714	cycle 50 miles		60	.133			4.09	11.50	15.59	19.40
2000	Hauling, 8 C.Y. truck, small project cost per hour	B-34A	8	1	Hr.		30.50	52	82.50	108
2100	12 C.Y. Truck	B-34B	8	1			30.50	86.50	117	146
2150	16.5 C.Y. Truck	B-34C	8	1			30.50	90.50	121	151
2175	18 C.Y. 8 wheel Truck	B-34I	8	1			30.50	109	139.50	171
2200	20 C.Y. Truck	B-34D	8	1			30.50	92.50	123	153
9014	18 C.Y. truck, 8 wheels,15 min. wait/Ld./Uld.,15 MPH, cycle 0.5 mi.	B-34I	504	.016	L.C.Y.		.49	1.72	2.21	2.70
9016	cycle 1 mile		450	.018			.55	1.93	2.48	3.03
9018	cycle 2 miles		378	.021			.65	2.30	2.95	3.60
9020	cycle 4 miles		270	.030			.91	3.22	4.13	5.05
9022	cycle 6 miles		216	.037			1.14	4.03	5.17	6.30
9024	cycle 8 miles		180	.044			1.36	4.83	6.19	7.55
9025	cycle 10 miles		144	.056			1.70	6.05	7.75	9.45
9026	20 MPH ave, cycle 0.5 mile		522	.015			.47	1.67	2.14	2.60
9028	cycle 1 mile		468	.017			.52	1.86	2.38	2.90
9030	cycle 2 miles		414	.019			.59	2.10	2.69	3.29
9032	cycle 4 miles		324	.025			.76	2.68	3.44	4.20
9034	cycle 6 miles		252	.032			.97	3.45	4.42	5.40
9036	cycle 8 miles		216	.037			1.14	4.03	5.17	6.30
9038	cycle 10 miles		180	.044			1.36	4.83	6.19	7.55
9040	25 MPH ave, cycle 4 miles		342	.023			.72	2.54	3.26	3.98
9042	cycle 6 miles		288	.028			.85	3.02	3.87	4.72
9044	cycle 8 miles		252	.032			.97	3.45	4.42	5.40
9046	cycle 10 miles		216	.037			1.14	4.03	5.17	6.30
9050	30 MPH ave, cycle 4 miles		378	.021			.65	2.30	2.95	3.60
9052	cycle 6 miles		324	.025			.76	2.68	3.44	4.20
9054	cycle 8 miles		270	.030			.91	3.22	4.13	5.05
9056	cycle 10 miles		234	.034			1.05	3.72	4.77	5.80
9060	35 MPH ave, cycle 4 miles		396	.020			.62	2.20	2.82	3.43
9062	cycle 6 miles		342	.023			.72	2.54	3.26	3.98
9064	cycle 8 miles		288	.028			.85	3.02	3.87	4.72
9066	cycle 10 miles		270	.030			.91	3.22	4.13	5.05
9068	cycle 20 miles		162	.049			1.51	5.35	6.86	8.40
9070	cycle 30 miles		126	.063			1.95	6.90	8.85	10.80
9072	cycle 40 miles		90	.089			2.72	9.65	12.37	15.15
9074	40 MPH ave, cycle 6 miles		360	.022			.68	2.41	3.09	3.78
9076	cycle 8 miles		324	.025			.76	2.68	3.44	4.20
9078	cycle 10 miles		288	.028			.85	3.02	3.87	4.72
9080	cycle 20 miles		180	.044			1.36	4.83	6.19	7.55
9082	cycle 30 miles		144	.056			1.70	6.05	7.75	9.45
9084	cycle 40 miles		108	.074			2.27	8.05	10.32	12.60
9086	cycle 50 miles		90	.089			2.72	9.65	12.37	15.15
9094	45 MPH ave, cycle 8 miles		324	.025			.76	2.68	3.44	4.20
9096	cycle 10 miles		306	.026			.80	2.84	3.64	4.44
9098	cycle 20 miles		198	.040			1.24	4.39	5.63	6.85
9100	cycle 30 miles		144	.056			1.70	6.05	7.75	9.45
9102	cycle 40 miles		126	.063			1.95	6.90	8.85	10.80
9104	cycle 50 miles		108	.074			2.27	8.05	10.32	12.60
9106	50 MPH ave, cycle 10 miles		324	.025			.76	2.68	3.44	4.20

31 23 Excavation and Fill

31 23 23 – Fill

31 23 23.20 Hauling		Crew	Daily Output	Labor-Hours	Unit	Material	2016 Bare Costs Labor	Equipment	Total	Total Incl O&P
9108	cycle 20 miles	B-34I	216	.037	L.C.Y.		1.14	4.03	5.17	6.30
9110	cycle 30 miles		162	.049			1.51	5.35	6.86	8.40
9112	cycle 40 miles		126	.063			1.95	6.90	8.85	10.80
9114	cycle 50 miles		108	.074			2.27	8.05	10.32	12.60
9214	20 min. wait/Ld./Uld.,15 MPH, cycle 0.5 mi.		396	.020			.62	2.20	2.82	3.43
9216	cycle 1 mile		360	.022			.68	2.41	3.09	3.78
9218	cycle 2 miles		306	.026			.80	2.84	3.64	4.44
9220	cycle 4 miles		234	.034			1.05	3.72	4.77	5.80
9222	cycle 6 miles		198	.040			1.24	4.39	5.63	6.85
9224	cycle 8 miles		162	.049			1.51	5.35	6.86	8.40
9225	cycle 10 miles		144	.056			1.70	6.05	7.75	9.45
9226	20 MPH ave, cycle 0.5 mile		396	.020			.62	2.20	2.82	3.43
9228	cycle 1 mile		378	.021			.65	2.30	2.95	3.60
9230	cycle 2 miles		324	.025			.76	2.68	3.44	4.20
9232	cycle 4 miles		270	.030			.91	3.22	4.13	5.05
9234	cycle 6 miles		216	.037			1.14	4.03	5.17	6.30
9236	cycle 8 miles		198	.040			1.24	4.39	5.63	6.85
9238	cycle 10 miles		162	.049			1.51	5.35	6.86	8.40
9240	25 MPH ave, cycle 4 miles		288	.028			.85	3.02	3.87	4.72
9242	cycle 6 miles		252	.032			.97	3.45	4.42	5.40
9244	cycle 8 miles		216	.037			1.14	4.03	5.17	6.30
9246	cycle 10 miles		198	.040			1.24	4.39	5.63	6.85
9250	30 MPH ave, cycle 4 miles		306	.026			.80	2.84	3.64	4.44
9252	cycle 6 miles		270	.030			.91	3.22	4.13	5.05
9254	cycle 8 miles		234	.034			1.05	3.72	4.77	5.80
9256	cycle 10 miles		216	.037			1.14	4.03	5.17	6.30
9260	35 MPH ave, cycle 4 miles		324	.025			.76	2.68	3.44	4.20
9262	cycle 6 miles		288	.028			.85	3.02	3.87	4.72
9264	cycle 8 miles		252	.032			.97	3.45	4.42	5.40
9266	cycle 10 miles		234	.034			1.05	3.72	4.77	5.80
9268	cycle 20 miles		162	.049			1.51	5.35	6.86	8.40
9270	cycle 30 miles		108	.074			2.27	8.05	10.32	12.60
9272	cycle 40 miles		90	.089			2.72	9.65	12.37	15.15
9274	40 MPH ave, cycle 6 miles		288	.028			.85	3.02	3.87	4.72
9276	cycle 8 miles		270	.030			.91	3.22	4.13	5.05
9278	cycle 10 miles		234	.034			1.05	3.72	4.77	5.80
9280	cycle 20 miles		162	.049			1.51	5.35	6.86	8.40
9282	cycle 30 miles		126	.063			1.95	6.90	8.85	10.80
9284	cycle 40 miles		108	.074			2.27	8.05	10.32	12.60
9286	cycle 50 miles		90	.089			2.72	9.65	12.37	15.15
9294	45 MPH ave, cycle 8 miles		270	.030			.91	3.22	4.13	5.05
9296	cycle 10 miles		252	.032			.97	3.45	4.42	5.40
9298	cycle 20 miles		180	.044			1.36	4.83	6.19	7.55
9300	cycle 30 miles		144	.056			1.70	6.05	7.75	9.45
9302	cycle 40 miles		108	.074			2.27	8.05	10.32	12.60
9304	cycle 50 miles		90	.089			2.72	9.65	12.37	15.15
9306	50 MPH ave, cycle 10 miles		270	.030			.91	3.22	4.13	5.05
9308	cycle 20 miles		198	.040			1.24	4.39	5.63	6.85
9310	cycle 30 miles		144	.056			1.70	6.05	7.75	9.45
9312	cycle 40 miles		126	.063			1.95	6.90	8.85	10.80
9314	cycle 50 miles		108	.074			2.27	8.05	10.32	12.60
9414	25 min. wait/Ld./Uld.,15 MPH, cycle 0.5 mi.		306	.026			.80	2.84	3.64	4.44
9416	cycle 1 mile		288	.028			.85	3.02	3.87	4.72

31 23 Excavation and Fill

31 23 23 – Fill

31 23 23.20 Hauling		Crew	Daily Output	Labor-Hours	Unit	Material	2016 Bare Costs Labor	Equipment	Total	Total Incl O&P
9418	cycle 2 miles	B-34I	252	.032	L.C.Y.		.97	3.45	4.42	5.40
9420	cycle 4 miles		198	.040			1.24	4.39	5.63	6.85
9422	cycle 6 miles		180	.044			1.36	4.83	6.19	7.55
9424	cycle 8 miles		144	.056			1.70	6.05	7.75	9.45
9425	cycle 10 miles		126	.063			1.95	6.90	8.85	10.80
9426	20 MPH ave, cycle 0.5 mile		324	.025			.76	2.68	3.44	4.20
9428	cycle 1 mile		306	.026			.80	2.84	3.64	4.44
9430	cycle 2 miles		270	.030			.91	3.22	4.13	5.05
9432	cycle 4 miles		234	.034			1.05	3.72	4.77	5.80
9434	cycle 6 miles		198	.040			1.24	4.39	5.63	6.85
9436	cycle 8 miles		180	.044			1.36	4.83	6.19	7.55
9438	cycle 10 miles		144	.056			1.70	6.05	7.75	9.45
9440	25 MPH ave, cycle 4 miles		252	.032			.97	3.45	4.42	5.40
9442	cycle 6 miles		216	.037			1.14	4.03	5.17	6.30
9444	cycle 8 miles		198	.040			1.24	4.39	5.63	6.85
9446	cycle 10 miles		180	.044			1.36	4.83	6.19	7.55
9450	30 MPH ave, cycle 4 miles		252	.032			.97	3.45	4.42	5.40
9452	cycle 6 miles		234	.034			1.05	3.72	4.77	5.80
9454	cycle 8 miles		198	.040			1.24	4.39	5.63	6.85
9456	cycle 10 miles		180	.044			1.36	4.83	6.19	7.55
9460	35 MPH ave, cycle 4 miles		270	.030			.91	3.22	4.13	5.05
9462	cycle 6 miles		234	.034			1.05	3.72	4.77	5.80
9464	cycle 8 miles		216	.037			1.14	4.03	5.17	6.30
9466	cycle 10 miles		198	.040			1.24	4.39	5.63	6.85
9468	cycle 20 miles		144	.056			1.70	6.05	7.75	9.45
9470	cycle 30 miles		108	.074			2.27	8.05	10.32	12.60
9472	cycle 40 miles		90	.089			2.72	9.65	12.37	15.15
9474	40 MPH ave, cycle 6 miles		252	.032			.97	3.45	4.42	5.40
9476	cycle 8 miles		234	.034			1.05	3.72	4.77	5.80
9478	cycle 10 miles		216	.037			1.14	4.03	5.17	6.30
9480	cycle 20 miles		144	.056			1.70	6.05	7.75	9.45
9482	cycle 30 miles		126	.063			1.95	6.90	8.85	10.80
9484	cycle 40 miles		90	.089			2.72	9.65	12.37	15.15
9486	cycle 50 miles		90	.089			2.72	9.65	12.37	15.15
9494	45 MPH ave, cycle 8 miles		234	.034			1.05	3.72	4.77	5.80
9496	cycle 10 miles		216	.037			1.14	4.03	5.17	6.30
9498	cycle 20 miles		162	.049			1.51	5.35	6.86	8.40
9500	cycle 30 miles		126	.063			1.95	6.90	8.85	10.80
9502	cycle 40 miles		108	.074			2.27	8.05	10.32	12.60
9504	cycle 50 miles		90	.089			2.72	9.65	12.37	15.15
9506	50 MPH ave, cycle 10 miles		234	.034			1.05	3.72	4.77	5.80
9508	cycle 20 miles		180	.044			1.36	4.83	6.19	7.55
9510	cycle 30 miles		144	.056			1.70	6.05	7.75	9.45
9512	cycle 40 miles		108	.074			2.27	8.05	10.32	12.60
9514	cycle 50 miles		90	.089			2.72	9.65	12.37	15.15
9614	30 min. wait/Ld./Uld.,15 MPH, cycle 0.5 mi.		270	.030			.91	3.22	4.13	5.05
9616	cycle 1 mile		252	.032			.97	3.45	4.42	5.40
9618	cycle 2 miles		216	.037			1.14	4.03	5.17	6.30
9620	cycle 4 miles		180	.044			1.36	4.83	6.19	7.55
9622	cycle 6 miles		162	.049			1.51	5.35	6.86	8.40
9624	cycle 8 miles		126	.063			1.95	6.90	8.85	10.80
9625	cycle 10 miles		126	.063			1.95	6.90	8.85	10.80
9626	20 MPH ave, cycle 0.5 mile		270	.030			.91	3.22	4.13	5.05

31 23 Excavation and Fill

31 23 23 – Fill

31 23 23.20 Hauling

		Crew	Daily Output	Labor-Hours	Unit	Material	2016 Bare Costs Labor	Equipment	Total	Total Incl O&P
9628	cycle 1 mile	B-34I	252	.032	L.C.Y.		.97	3.45	4.42	5.40
9630	cycle 2 miles		234	.034			1.05	3.72	4.77	5.80
9632	cycle 4 miles		198	.040			1.24	4.39	5.63	6.85
9634	cycle 6 miles		180	.044			1.36	4.83	6.19	7.55
9636	cycle 8 miles		162	.049			1.51	5.35	6.86	8.40
9638	cycle 10 miles		144	.056			1.70	6.05	7.75	9.45
9640	25 MPH ave, cycle 4 miles		216	.037			1.14	4.03	5.17	6.30
9642	cycle 6 miles		198	.040			1.24	4.39	5.63	6.85
9644	cycle 8 miles		180	.044			1.36	4.83	6.19	7.55
9646	cycle 10 miles		162	.049			1.51	5.35	6.86	8.40
9650	30 MPH ave, cycle 4 miles		216	.037			1.14	4.03	5.17	6.30
9652	cycle 6 miles		198	.040			1.24	4.39	5.63	6.85
9654	cycle 8 miles		180	.044			1.36	4.83	6.19	7.55
9656	cycle 10 miles		162	.049			1.51	5.35	6.86	8.40
9660	35 MPH ave, cycle 4 miles		234	.034			1.05	3.72	4.77	5.80
9662	cycle 6 miles		216	.037			1.14	4.03	5.17	6.30
9664	cycle 8 miles		198	.040			1.24	4.39	5.63	6.85
9666	cycle 10 miles		180	.044			1.36	4.83	6.19	7.55
9668	cycle 20 miles		126	.063			1.95	6.90	8.85	10.80
9670	cycle 30 miles		108	.074			2.27	8.05	10.32	12.60
9672	cycle 40 miles		90	.089			2.72	9.65	12.37	15.15
9674	40 MPH ave, cycle 6 miles		216	.037			1.14	4.03	5.17	6.30
9676	cycle 8 miles		198	.040			1.24	4.39	5.63	6.85
9678	cycle 10 miles		180	.044			1.36	4.83	6.19	7.55
9680	cycle 20 miles		144	.056			1.70	6.05	7.75	9.45
9682	cycle 30 miles		108	.074			2.27	8.05	10.32	12.60
9684	cycle 40 miles		90	.089			2.72	9.65	12.37	15.15
9686	cycle 50 miles		72	.111			3.41	12.10	15.51	18.90
9694	45 MPH ave, cycle 8 miles		216	.037			1.14	4.03	5.17	6.30
9696	cycle 10 miles		198	.040			1.24	4.39	5.63	6.85
9698	cycle 20 miles		144	.056			1.70	6.05	7.75	9.45
9700	cycle 30 miles		126	.063			1.95	6.90	8.85	10.80
9702	cycle 40 miles		108	.074			2.27	8.05	10.32	12.60
9704	cycle 50 miles		90	.089			2.72	9.65	12.37	15.15
9706	50 MPH ave, cycle 10 miles		198	.040			1.24	4.39	5.63	6.85
9708	cycle 20 miles		162	.049			1.51	5.35	6.86	8.40
9710	cycle 30 miles		126	.063			1.95	6.90	8.85	10.80
9712	cycle 40 miles		108	.074			2.27	8.05	10.32	12.60
9714	cycle 50 miles	↓	90	.089	↓		2.72	9.65	12.37	15.15

31 23 23.24 Compaction, Structural

		Crew	Daily Output	Labor-Hours	Unit	Material	Labor	Equipment	Total	Total Incl O&P
0010	**COMPACTION, STRUCTURAL**									
0020	Steel wheel tandem roller, 5 tons	B-10E	8	1	Hr.		35.50	20.50	56	81
0050	Air tamp, 6" to 8" lifts, common fill	B-9	250	.160	E.C.Y.		4.01	.92	4.93	7.70
0060	Select fill	"	300	.133			3.34	.76	4.10	6.45
0600	Vibratory plate, 8" lifts, common fill	A-1D	200	.040			.99	.18	1.17	1.85
0700	Select fill	"	216	.037	↓		.91	.17	1.08	1.71

31 25 Erosion and Sedimentation Controls

31 25 14 – Stabilization Measures for Erosion and Sedimentation Control

31 25 14.16 Rolled Erosion Control Mats and Blankets

		Crew	Daily Output	Labor-Hours	Unit	Material	2016 Bare Costs Labor	Equipment	Total	Total Incl O&P
0010	**ROLLED EROSION CONTROL MATS AND BLANKETS**									
0020	Jute mesh, 100 S.Y. per roll, 4' wide, stapled G	B-80A	2400	.010	S.Y.	.99	.25	.13	1.37	1.64
0100	Plastic netting, stapled, 2" x 1" mesh, 20 mil G	B-1	2500	.010		.47	.24		.71	.93
0120	Revegetation mat, webbed G	2 Clab	1000	.016		1.57	.39		1.96	2.39
0200	Polypropylene mesh, stapled, 6.5 oz./S.Y. G	B-1	2500	.010		1.65	.24		1.89	2.23
0300	Tobacco netting, or jute mesh #2, stapled G	"	2500	.010		.21	.24		.45	.64
1000	Silt fence, install and maintain, remove G	B-62	1300	.018	L.F.	.25	.51	.13	.89	1.28
1100	Allow 25% per month for maintenance; 6-month max life									

31 31 Soil Treatment

31 31 16 – Termite Control

31 31 16.13 Chemical Termite Control

		Crew	Daily Output	Labor-Hours	Unit	Material	2016 Bare Costs Labor	Equipment	Total	Total Incl O&P
0010	**CHEMICAL TERMITE CONTROL**									
0020	Slab and walls, residential	1 Skwk	1200	.007	SF Flr.	.32	.23		.55	.73
0030	SS mesh, no chemicals, avg 1400 S.F. home, min G		1000	.008	"	.32	.27		.59	.81
0400	Insecticides for termite control, minimum		14.20	.563	Gal.	68.50	19.25		87.75	108
0500	Maximum		11	.727	"	117	25		142	171

Division Notes

		CREW	DAILY OUTPUT	LABOR-HOURS	UNIT	BARE COSTS				TOTAL INCL O&P
						MAT.	LABOR	EQUIP.	TOTAL	

Estimating Tips

32 01 00 Operations and Maintenance of Exterior Improvements

- Recycling of asphalt pavement is becoming very popular and is an alternative to removal and replacement. It can be a good value engineering proposal if removed pavement can be recycled, either at the project site or at another site that is reasonably close to the project site. Sections on repair of flexible and rigid pavement are included.

32 10 00 Bases, Ballasts, and Paving

- When estimating paving, keep in mind the project schedule. Also note that prices for asphalt and concrete are generally higher in the cold seasons. Lines for pavement markings, including tactile warning systems and fence lines, are included.

32 90 00 Planting

- The timing of planting and guarantee specifications often dictate the costs for establishing tree and shrub growth and a stand of grass or ground cover. Establish the work performance schedule to coincide with the local planting season. Maintenance and growth guarantees can add from 20%–100% to the total landscaping cost and can be contractually cumbersome. The cost to replace trees and shrubs can be as high as 5% of the total cost, depending on the planting zone, soil conditions, and time of year.

Reference Numbers

Reference numbers are shown at the beginning of some major classifications. These numbers refer to related items in the Reference Section. The reference information may be an estimating procedure, an alternate pricing method, or technical information.

Note: Not all subdivisions listed here necessarily appear. ■

Did you know?

RSMeans Online gives you the same access to RSMeans' data with 24/7 access:
- Quickly locate costs in the searchable database.
- Build cost lists, estimates, and reports in minutes.
- Adjust costs to any location in the U.S. and Canada with the click of a button.

Start your free trial today at **www.RSMeansOnline.com**

RSMeans Online
FROM THE GORDIAN GROUP®

No part of this cost data may be reproduced, stored in a retrieval system, or transmitted in any form or by any means without prior written permission of RSMeans.

32 01 Operation and Maintenance of Exterior Improvements

32 01 13 – Flexible Paving Surface Treatment

32 01 13.66 Fog Seal

		Crew	Daily Output	Labor-Hours	Unit	Material	2016 Bare Costs Labor	2016 Bare Costs Equipment	Total	Total Incl O&P
0010	FOG SEAL									
0012	Sealcoating, 2 coat coal tar pitch emulsion over 10,000 S.Y.	B-45	5000	.003	S.Y.	.93	.09	.06	1.08	1.24
0030	1000 to 10,000 S.Y.	"	3000	.005		.93	.15	.11	1.19	1.38
0100	Under 1000 S.Y.	B-1	1050	.023		.93	.58		1.51	1.99
0300	Petroleum resistant, over 10,000 S.Y.	B-45	5000	.003		1.34	.09	.06	1.49	1.69
0320	1000 to 10,000 S.Y.	"	3000	.005		1.34	.15	.11	1.60	1.83
0400	Under 1000 S.Y.	B-1	1050	.023		1.34	.58		1.92	2.44

32 06 Schedules for Exterior Improvements

32 06 10 – Schedules for Bases, Ballasts, and Paving

32 06 10.10 Sidewalks, Driveways and Patios

		Crew	Daily Output	Labor-Hours	Unit	Material	Labor	Equipment	Total	Total Incl O&P
0010	SIDEWALKS, DRIVEWAYS AND PATIOS No base									
0021	Asphaltic concrete, 2" thick	B-37	6480	.007	S.F.	.81	.20	.03	1.04	1.25
0101	2-1/2" thick	"	5950	.008	"	1.02	.21	.03	1.26	1.51
0300	Concrete, 3000 psi, CIP, 6 x 6 - W1.4 x W1.4 mesh,									
0310	broomed finish, no base, 4" thick	B-24	600	.040	S.F.	1.79	1.21		3	3.96
0350	5" thick		545	.044		2.38	1.33		3.71	4.82
0400	6" thick		510	.047		2.78	1.42		4.20	5.40
0450	For bank run gravel base, 4" thick, add	B-18	2500	.010		.47	.24	.02	.73	.95
0520	8" thick, add	"	1600	.015		.96	.38	.03	1.37	1.72
1000	Crushed stone, 1" thick, white marble	2 Clab	1700	.009		.42	.23		.65	.85
1050	Bluestone	"	1700	.009		.12	.23		.35	.52
1700	Redwood, prefabricated, 4' x 4' sections	2 Carp	316	.051		4.82	1.72		6.54	8.15
1750	Redwood planks, 1" thick, on sleepers	"	240	.067		4.82	2.26		7.08	9.10
2250	Stone dust, 4" thick	B-62	900	.027	S.Y.	3.56	.74	.19	4.49	5.35

32 06 10.20 Steps

		Crew	Daily Output	Labor-Hours	Unit	Material	Labor	Equipment	Total	Total Incl O&P
0010	STEPS									
0011	Incl. excav., borrow & concrete base as required									
0100	Brick steps	B-24	35	.686	LF Riser	15.75	20.50		36.25	52
0200	Railroad ties	2 Clab	25	.640		3.59	15.80		19.39	30.50
0300	Bluestone treads, 12" x 2" or 12" x 1-1/2"	B-24	30	.800		33.50	24		57.50	77
0600	Precast concrete, see Section 03 41 23.50									
4025	Steel edge strips, incl. stakes, 1/4" x 5"	B-1	390	.062	L.F.	5	1.56		6.56	8.10
4050	Edging, landscape timber or railroad ties, 6" x 8"	2 Carp	170	.094	"	2.11	3.19		5.30	7.65

32 11 Base Courses

32 11 23 – Aggregate Base Courses

32 11 23.23 Base Course Drainage Layers

		Crew	Daily Output	Labor-Hours	Unit	Material	Labor	Equipment	Total	Total Incl O&P
0010	BASE COURSE DRAINAGE LAYERS									
0011	For roadways and large areas									
0051	3/4" stone compacted to 3" deep	B-36	36000	.001	S.F.	.26	.03	.05	.34	.39
0101	6" deep		35100	.001		.52	.03	.05	.60	.69
0201	9" deep		25875	.002		.76	.05	.06	.87	.99
0305	12" deep		21150	.002		1.35	.06	.08	1.49	1.66
0306	Crushed 1-1/2" stone base, compacted to 4" deep		47000	.001		.05	.02	.04	.11	.13
0307	6" deep		35100	.001		.67	.03	.05	.75	.85
0308	8" deep		27000	.001		.90	.04	.06	1	1.13
0309	12" deep		16200	.002		1.35	.07	.10	1.52	1.71
0350	Bank run gravel, spread and compacted									

32 11 Base Courses

32 11 23 – Aggregate Base Courses

32 11 23.23 Base Course Drainage Layers

		Crew	Daily Output	Labor-Hours	Unit	Material	2016 Bare Costs Labor	Equipment	Total	Total Incl O&P
0371	6" deep	B-32	54000	.001	S.F.	.44	.02	.04	.50	.56
0391	9" deep		39600	.001		.64	.03	.06	.73	.82
0401	12" deep		32400	.001		.88	.03	.07	.98	1.10
6900	For small and irregular areas, add						50%	50%		

32 11 26 – Asphaltic Base Courses

32 11 26.19 Bituminous-Stabilized Base Courses

		Crew	Daily Output	Labor-Hours	Unit	Material	Labor	Equipment	Total	Total Incl O&P
0010	**BITUMINOUS-STABILIZED BASE COURSES**									
0020	And large paved areas									
0700	Liquid application to gravel base, asphalt emulsion	B-45	6000	.003	Gal.	4.50	.07	.05	4.62	5.15
0800	Prime and seal, cut back asphalt		6000	.003	"	5.30	.07	.05	5.42	6.05
1000	Macadam penetration crushed stone, 2 gal. per S.Y., 4" thick		6000	.003	S.Y.	9	.07	.05	9.12	10.10
1100	6" thick, 3 gal. per S.Y.		4000	.004		13.50	.11	.08	13.69	15.10
1200	8" thick, 4 gal. per S.Y.		3000	.005		18	.15	.11	18.26	20
8900	For small and irregular areas, add						50%	50%		

32 12 Flexible Paving

32 12 16 – Asphalt Paving

32 12 16.14 Paving Asphaltic Concrete

		Crew	Daily Output	Labor-Hours	Unit	Material	Labor	Equipment	Total	Total Incl O&P
0010	**PAVING ASPHALTIC CONCRETE**									
0020	6" stone base, 2" binder course, 1" topping	B-25C	9000	.005	S.F.	1.83	.15	.27	2.25	2.56
0025	2" binder course, 2" topping		9000	.005		2.28	.15	.27	2.70	3.06
0030	3" binder course, 2" topping		9000	.005		2.70	.15	.27	3.12	3.52
0035	4" binder course, 2" topping		9000	.005		3.11	.15	.27	3.53	3.97
0040	1.5" binder course, 1" topping		9000	.005		1.62	.15	.27	2.04	2.33
0042	3" binder course, 1" topping		9000	.005		2.24	.15	.27	2.66	3.02
0045	3" binder course, 3" topping		9000	.005		3.15	.15	.27	3.57	4.02
0050	4" binder course, 3" topping		9000	.005		3.57	.15	.27	3.99	4.47
0055	4" binder course, 4" topping		9000	.005		4.02	.15	.27	4.44	4.97
0300	Binder course, 1-1/2" thick		35000	.001		.62	.04	.07	.73	.84
0400	2" thick		25000	.002		.81	.05	.10	.96	1.09
0500	3" thick		15000	.003		1.25	.09	.16	1.50	1.70
0600	4" thick		10800	.004		1.63	.13	.23	1.99	2.25
0800	Sand finish course, 3/4" thick		41000	.001		.33	.03	.06	.42	.49
0900	1" thick		34000	.001		.40	.04	.07	.51	.59
1000	Fill pot holes, hot mix, 2" thick	B-16	4200	.008		.85	.20	.16	1.21	1.46
1100	4" thick		3500	.009		1.25	.24	.20	1.69	2
1120	6" thick		3100	.010		1.67	.28	.22	2.17	2.55
1140	Cold patch, 2" thick	B-51	3000	.016		.97	.41	.08	1.46	1.85
1160	4" thick		2700	.018		1.85	.46	.09	2.40	2.90
1180	6" thick		1900	.025		2.87	.65	.13	3.65	4.39

32 13 Rigid Paving

32 13 13 – Concrete Paving

32 13 13.23 Concrete Paving Surface Treatment

		Crew	Daily Output	Labor-Hours	Unit	Material	2016 Bare Costs Labor	Equipment	Total	Total Incl O&P
0010	**CONCRETE PAVING SURFACE TREATMENT**									
0015	Including joints, finishing and curing									
0021	Fixed form, 12' pass, unreinforced, 6" thick	B-26	18000	.005	S.F.	3.34	.14	.20	3.68	4.12
0101	8" thick	"	13500	.007		4.50	.19	.26	4.95	5.55
0701	Finishing, broom finish small areas	2 Cefi	1215	.013			.42		.42	.68

32 14 Unit Paving

32 14 13 – Precast Concrete Unit Paving

32 14 13.18 Precast Concrete Plantable Pavers

		Crew	Daily Output	Labor-Hours	Unit	Material	Labor	Equipment	Total	Total Incl O&P
0010	**PRECAST CONCRETE PLANTABLE PAVERS** (50% grass)									
0300	3/4" crushed stone base for plantable pavers, 6 inch depth	B-62	1000	.024	S.Y.	4.71	.67	.17	5.55	6.50
0400	8 inch depth		900	.027		6.30	.74	.19	7.23	8.40
0500	10 inch depth		800	.030		7.85	.83	.22	8.90	10.30
0600	12 inch depth		700	.034		9.45	.95	.25	10.65	12.20
0700	Hydro seeding plantable pavers	B-81A	20	.800	M.S.F.	10.65	22	22	54.65	71.50
0800	Apply fertilizer and seed to plantable pavers	1 Clab	8	1	"	50.50	24.50		75	97

32 14 16 – Brick Unit Paving

32 14 16.10 Brick Paving

		Crew	Daily Output	Labor-Hours	Unit	Material	Labor	Equipment	Total	Total Incl O&P
0010	**BRICK PAVING**									
0012	4" x 8" x 1-1/2", without joints (4.5 brick/S.F.)	D-1	110	.145	S.F.	2.24	4.29		6.53	9.60
0100	Grouted, 3/8" joint (3.9 brick/S.F.)		90	.178		2.02	5.25		7.27	10.95
0200	4" x 8" x 2-1/4", without joints (4.5 bricks/S.F.)		110	.145		2.33	4.29		6.62	9.70
0300	Grouted, 3/8" joint (3.9 brick/S.F.)		90	.178		2.02	5.25		7.27	10.95
0455	Pervious brick paving, 4" x 8" x 3-1/4", without joints (4.5 bricks/S.F.)		110	.145		3.49	4.29		7.78	11
0500	Bedding, asphalt, 3/4" thick	B-25	5130	.017		.69	.48	.55	1.72	2.14
0540	Course washed sand bed, 1" thick	B-18	5000	.005		.35	.12	.01	.48	.60
0580	Mortar, 1" thick	D-1	300	.053		.80	1.57		2.37	3.50
0620	2" thick		200	.080		1.60	2.36		3.96	5.70
1500	Brick on 1" thick sand bed laid flat, 4.5 per S.F.		100	.160		2.85	4.72		7.57	11
2000	Brick pavers, laid on edge, 7.2 per S.F.		70	.229		3.93	6.75		10.68	15.60

32 14 23 – Asphalt Unit Paving

32 14 23.10 Asphalt Blocks

		Crew	Daily Output	Labor-Hours	Unit	Material	Labor	Equipment	Total	Total Incl O&P
0010	**ASPHALT BLOCKS**									
0020	Rectangular, 6" x 12" x 1-1/4", w/bed & neopr. adhesive	D-1	135	.119	S.F.	5.60	3.50		9.10	12
0100	3" thick		130	.123		7.80	3.63		11.43	14.65
0300	Hexagonal tile, 8" wide, 1-1/4" thick		135	.119		5.60	3.50		9.10	12
0400	2" thick		130	.123		7.80	3.63		11.43	14.65
0500	Square, 8" x 8", 1-1/4" thick		135	.119		5.60	3.50		9.10	12
0600	2" thick		130	.123		7.80	3.63		11.43	14.65

32 14 40 – Stone Paving

32 14 40.10 Stone Pavers

		Crew	Daily Output	Labor-Hours	Unit	Material	Labor	Equipment	Total	Total Incl O&P
0010	**STONE PAVERS**									
1100	Flagging, bluestone, irregular, 1" thick,	D-1	81	.198	S.F.	9	5.85		14.85	19.60
1150	Snapped random rectangular, 1" thick		92	.174		13.65	5.15		18.80	23.50
1200	1-1/2" thick		85	.188		16.40	5.55		21.95	27.50
1250	2" thick		83	.193		19.10	5.70		24.80	30.50
1300	Slate, natural cleft, irregular, 3/4" thick		92	.174		9	5.15		14.15	18.45
1310	1" thick		85	.188		10.50	5.55		16.05	21
1351	Random rectangular, gauged, 1/2" thick		105	.152		19.50	4.50		24	29

32 14 Unit Paving

32 14 40 – Stone Paving

32 14 40.10 Stone Pavers		Crew	Daily Output	Labor-Hours	Unit	Material	2016 Bare Costs Labor	Equipment	Total	Total Incl O&P
1400	Random rectangular, butt joint, gauged, 1/4" thick	D-1	150	.107	S.F.	21	3.15		24.15	28.50
1450	For sand rubbed finish, add					9			9	9.90
1500	For interior setting, add								25%	25%
1550	Granite blocks, 3-1/2" x 3-1/2" x 3-1/2"	D-1	92	.174	S.F.	15	5.15		20.15	25

32 16 Curbs, Gutters, Sidewalks, and Driveways

32 16 13 – Curbs and Gutters

32 16 13.13 Cast-in-Place Concrete Curbs and Gutters

		Crew	Daily Output	Labor-Hours	Unit	Material	Labor	Equipment	Total	Total Incl O&P
0010	**CAST-IN-PLACE CONCRETE CURBS AND GUTTERS**									
0290	Forms only, no concrete									
0300	Concrete, wood forms, 6" x 18", straight	C-2	500	.096	L.F.	3.13	2.85		5.98	8.25
0400	6" x 18", radius	"	200	.240		3.25	7.10		10.35	15.50
0404	Concrete, wood forms, 6" x 18", straight & concrete	C-2A	500	.096		6.15	3.11		9.26	11.95
0406	6" x 18", radius	"	200	.240		6.25	7.75		14	19.80

32 16 13.23 Precast Concrete Curbs and Gutters

0010	**PRECAST CONCRETE CURBS AND GUTTERS**									
0550	Precast, 6" x 18", straight	B-29	700	.069	L.F.	9	1.84	1.26	12.10	14.35
0600	6" x 18", radius	"	325	.148	"	10	3.97	2.71	16.68	20.50

32 16 13.33 Asphalt Curbs

0010	**ASPHALT CURBS**									
0012	Curbs, asphaltic, machine formed, 8" wide, 6" high, 40 L.F./ton	B-27	1000	.032	L.F.	1.73	.80	.30	2.83	3.58
0100	8" wide, 8" high, 30 L.F. per ton		900	.036		2.30	.89	.33	3.52	4.40
0150	Asphaltic berm, 12" W, 3"-6" H, 35 L.F./ton, before pavement		700	.046		.04	1.15	.42	1.61	2.42
0200	12" W, 1-1/2" to 4" H, 60 L.F. per ton, laid with pavement	B-2	1050	.038		.02	.95		.97	1.63

32 16 13.43 Stone Curbs

0010	**STONE CURBS**									
1000	Granite, split face, straight, 5" x 16"	D-13	275	.175	L.F.	14	5.50	1.74	21.24	26.50
1100	6" x 18"	"	250	.192		18.40	6.05	1.92	26.37	32
1300	Radius curbing, 6" x 18", over 10' radius	B-29	260	.185		22.50	4.96	3.39	30.85	37
1400	Corners, 2' radius	"	80	.600	Ea.	75.50	16.15	11.05	102.70	122
1600	Edging, 4-1/2" x 12", straight	D-13	300	.160	L.F.	7	5.05	1.60	13.65	17.80
1800	Curb inlets, (guttermouth) straight	B-29	41	1.171	Ea.	168	31.50	21.50	221	261
2000	Indian granite (belgian block)									
2100	Jumbo, 10-1/2" x 7-1/2" x 4", grey	D-1	150	.107	L.F.	7.90	3.15		11.05	13.95
2150	Pink		150	.107		8.60	3.15		11.75	14.70
2200	Regular, 9" x 4-1/2" x 4-1/2", grey		160	.100		4.67	2.95		7.62	10.05
2250	Pink		160	.100		6.15	2.95		9.10	11.65
2300	Cubes, 4" x 4" x 4", grey		175	.091		3.63	2.70		6.33	8.50
2350	Pink		175	.091		3.81	2.70		6.51	8.70
2400	6" x 6" x 6", pink		155	.103		12.60	3.05		15.65	18.95
2500	Alternate pricing method for indian granite									
2550	Jumbo, 10-1/2" x 7-1/2" x 4" (30 lb.), grey				Ton	450			450	495
2600	Pink					500			500	550
2650	Regular, 9" x 4-1/2" x 4-1/2" (20 lb.), grey					330			330	365
2700	Pink					430			430	475
2750	Cubes, 4" x 4" x 4" (5 lb.), grey					440			440	485
2800	Pink					490			490	540
2850	6" x 6" x 6" (25 lb.), pink					490			490	540
2900	For pallets, add					22			22	24

32 31 Fences and Gates

32 31 13 – Chain Link Fences and Gates

32 31 13.15 Chain Link Fence

		Crew	Daily Output	Labor-Hours	Unit	Material	2016 Bare Costs Labor	Equipment	Total	Total Incl O&P
0010	**CHAIN LINK FENCE**									
0020	1-5/8" post 10' O.C., 1-3/8" top rail, 2" corner post galv. stl., 3' high	B-1	185	.130	L.F.	9.10	3.28		12.38	15.55
0050	4' high		170	.141		8.70	3.57		12.27	15.60
0100	6' high		115	.209	↓	12.15	5.30		17.45	22
0150	Add for gate 3' wide, 1-3/8" frame 3' high		12	2	Ea.	94	50.50		144.50	188
0170	4' high		10	2.400		100	61		161	212
0190	6' high		10	2.400		120	61		181	234
0200	Add for gate 4' wide, 1-3/8" frame 3' high		9	2.667		95.50	67.50		163	218
0220	4' high		9	2.667		104	67.50		171.50	227
0240	6' high		8	3		126	76		202	266
0350	Aluminized steel, 9 ga. wire, 3' high		185	.130	L.F.	7.60	3.28		10.88	13.90
0380	4' high		170	.141		8	3.57		11.57	14.80
0400	6' high		115	.209	↓	10.40	5.30		15.70	20.50
0450	Add for gate 3' wide, 1-3/8" frame 3' high		12	2	Ea.	127	50.50		177.50	225
0470	4' high		10	2.400		137	61		198	252
0490	6' high		10	2.400		173	61		234	292
0500	Add for gate 4' wide, 1-3/8" frame 3' high		10	2.400		132	61		193	247
0520	4' high		9	2.667		136	67.50		203.50	263
0540	6' high		8	3	↓	184	76		260	330
0620	Vinyl covered 9 ga. wire, 3' high		185	.130	L.F.	7.40	3.28		10.68	13.65
0640	4' high		170	.141		7.25	3.57		10.82	14
0660	6' high		115	.209	↓	9.70	5.30		15	19.50
0720	Add for gate 3' wide, 1-3/8" frame 3' high		12	2	Ea.	102	50.50		152.50	197
0740	4' high		10	2.400		107	61		168	220
0760	6' high		10	2.400		133	61		194	248
0780	Add for gate 4' wide, 1-3/8" frame 3' high		10	2.400		105	61		166	217
0800	4' high		9	2.667		110	67.50		177.50	234
0820	6' high	↓	8	3	↓	142	76		218	283
0860	Tennis courts, 11 ga. wire, 2 1/2" post 10' O.C., 1-5/8" top rail									
0900	2-1/2" corner post, 10' high	B-1	95	.253	L.F.	7.70	6.40		14.10	19.20
0920	12' high		80	.300	"	7.95	7.60		15.55	21.50
1000	Add for gate 3' wide, 1-5/8" frame 10' high		10	2.400	Ea.	220	61		281	345
1040	Aluminized, 11 ga. wire 10' high		95	.253	L.F.	9.30	6.40		15.70	21
1100	12' high		80	.300	"	10.20	7.60		17.80	24
1140	Add for gate 3' wide, 1-5/8" frame, 10' high		10	2.400	Ea.	142	61		203	258
1250	Vinyl covered 11 ga. wire, 10' high		95	.253	L.F.	8.05	6.40		14.45	19.55
1300	12' high		80	.300	"	9.60	7.60		17.20	23.50
1400	Add for gate 3' wide, 1-3/8" frame, 10' high	↓	10	2.400	Ea.	283	61		344	410

32 31 23 – Plastic Fences and Gates

32 31 23.10 Fence, Vinyl

		Crew	Daily Output	Labor-Hours	Unit	Material	2016 Bare Costs Labor	Equipment	Total	Total Incl O&P
0010	**FENCE, VINYL**									
0011	White, steel reinforced, stainless steel fasteners									
0020	Picket, 4" x 4" posts @ 6' - 0" OC, 3' high	B-1	140	.171	L.F.	22.50	4.34		26.84	32.50
0030	4' high		130	.185		25	4.67		29.67	35.50
0040	5' high		120	.200		27.50	5.05		32.55	39
0100	Board (semi-privacy), 5" x 5" posts @ 7' - 6" OC, 5' high		130	.185		25.50	4.67		30.17	36
0120	6' high		125	.192		28.50	4.86		33.36	39.50
0200	Basketweave, 5" x 5" posts @ 7' - 6" OC, 5' high		160	.150		25.50	3.80		29.30	34.50
0220	6' high		150	.160		28.50	4.05		32.55	38.50
0300	Privacy, 5" x 5" posts @ 7' - 6" OC, 5' high		130	.185		25	4.67		29.67	35.50
0320	6' high		150	.160	↓	28.50	4.05		32.55	38.50
0350	Gate, 5' high	↓	9	2.667	Ea.	315	67.50		382.50	460

32 31 Fences and Gates

32 31 23 – Plastic Fences and Gates

32 31 23.10 Fence, Vinyl

		Crew	Daily Output	Labor-Hours	Unit	Material	2016 Bare Costs Labor	Equipment	Total	Total Incl O&P
0360	6' high	B-1	9	2.667	Ea.	360	67.50		427.50	510
0400	For posts set in concrete, add		25	.960	↓	8.90	24.50		33.40	50.50
0500	Post and rail fence, 2 rail		150	.160	L.F.	5.95	4.05		10	13.35
0510	3 rail		150	.160		7.70	4.05		11.75	15.25
0515	4 rail		150	.160	↓	9.95	4.05		14	17.75

32 31 26 – Wire Fences and Gates

32 31 26.10 Fences, Misc. Metal

		Crew	Daily Output	Labor-Hours	Unit	Material	Labor	Equipment	Total	Total Incl O&P
0010	**FENCES, MISC. METAL**									
0012	Chicken wire, posts @ 4', 1" mesh, 4' high	B-80C	410	.059	L.F.	3.37	1.54	.63	5.54	6.95
0100	2" mesh, 6' high		350	.069		3.78	1.81	.73	6.32	8
0200	Galv. steel, 12 ga., 2" x 4" mesh, posts 5' O.C., 3' high		300	.080		2.75	2.11	.86	5.72	7.50
0300	5' high		300	.080		3.35	2.11	.86	6.32	8.15
0400	14 ga., 1" x 2" mesh, 3' high		300	.080		3.39	2.11	.86	6.36	8.20
0500	5' high	↓	300	.080	↓	4.52	2.11	.86	7.49	9.40
1000	Kennel fencing, 1-1/2" mesh, 6' long, 3'-6" wide, 6'-2" high	2 Clab	4	4	Ea.	505	98.50		603.50	720
1050	12' long		4	4		710	98.50		808.50	945
1200	Top covers, 1-1/2" mesh, 6' long		15	1.067		135	26.50		161.50	192
1250	12' long	↓	12	1.333	↓	189	33		222	263

32 31 29 – Wood Fences and Gates

32 31 29.10 Fence, Wood

		Crew	Daily Output	Labor-Hours	Unit	Material	Labor	Equipment	Total	Total Incl O&P
0010	**FENCE, WOOD**									
0011	Basket weave, 3/8" x 4" boards, 2" x 4"									
0020	stringers on spreaders, 4" x 4" posts									
0050	No. 1 cedar, 6' high	B-80C	160	.150	L.F.	26	3.96	1.60	31.56	37
0070	Treated pine, 6' high	↓	150	.160		34.50	4.22	1.71	40.43	47
0090	Vertical weave 6' high		145	.166	↓	22.50	4.36	1.77	28.63	34
0200	Board fence, 1" x 4" boards, 2" x 4" rails, 4" x 4" post									
0220	Preservative treated, 2 rail, 3' high	B-80C	145	.166	L.F.	9.45	4.36	1.77	15.58	19.60
0240	4' high		135	.178		11.10	4.69	1.90	17.69	22
0260	3 rail, 5' high		130	.185		12.10	4.87	1.97	18.94	23.50
0300	6' high		125	.192		13.65	5.05	2.05	20.75	25.50
0320	No. 2 grade western cedar, 2 rail, 3' high		145	.166		10.40	4.36	1.77	16.53	20.50
0340	4' high		135	.178		11.45	4.69	1.90	18.04	22.50
0360	3 rail, 5' high		130	.185		12.85	4.87	1.97	19.69	24.50
0400	6' high		125	.192		13.75	5.05	2.05	20.85	26
0420	No. 1 grade cedar, 2 rail, 3' high		145	.166		13.20	4.36	1.77	19.33	23.50
0440	4' high		135	.178		14.25	4.69	1.90	20.84	25.50
0460	3 rail, 5' high		130	.185		17.35	4.87	1.97	24.19	29.50
0500	6' high	↓	125	.192	↓	19.10	5.05	2.05	26.20	31.50
0540	Shadow box, 1" x 6" board, 2" x 4" rail, 4" x 4"post									
0560	Pine, pressure treated, 3 rail, 6' high	B-80C	150	.160	L.F.	17.55	4.22	1.71	23.48	28
0600	Gate, 3'-6" wide		8	3	Ea.	109	79	32	220	288
0620	No. 1 cedar, 3 rail, 4' high		130	.185	L.F.	18.35	4.87	1.97	25.19	30.50
0640	6' high		125	.192		23	5.05	2.05	30.10	36
0860	Open rail fence, split rails, 2 rail 3' high, no. 1 cedar		160	.150		12.80	3.96	1.60	18.36	22.50
0870	No. 2 cedar		160	.150		11.75	3.96	1.60	17.31	21.50
0880	3 rail, 4' high, no. 1 cedar		150	.160		13.20	4.22	1.71	19.13	23.50
0890	No. 2 cedar		150	.160		9.35	4.22	1.71	15.28	19.15
0920	Rustic rails, 2 rail 3' high, no. 1 cedar		160	.150		10.20	3.96	1.60	15.76	19.55
0930	No. 2 cedar		160	.150		9.55	3.96	1.60	15.11	18.85
0940	3 rail, 4' high		150	.160		9.75	4.22	1.71	15.68	19.65
0950	No. 2 cedar	↓	150	.160		7.95	4.22	1.71	13.88	17.65

32 31 Fences and Gates

32 31 29 – Wood Fences and Gates

32 31 29.10 Fence, Wood

		Crew	Daily Output	Labor-Hours	Unit	Material	2016 Bare Costs Labor	Equipment	Total	Total Incl O&P
0960	Picket fence, gothic, pressure treated pine									
1000	2 rail, 3' high	B-80C	140	.171	L.F.	7.80	4.52	1.83	14.15	18.10
1020	3 rail, 4' high		130	.185	"	8.90	4.87	1.97	15.74	20
1040	Gate, 3'-6" wide		9	2.667	Ea.	74	70.50	28.50	173	230
1060	No. 2 cedar, 2 rail, 3' high		140	.171	L.F.	9	4.52	1.83	15.35	19.40
1100	3 rail, 4' high		130	.185	"	9.10	4.87	1.97	15.94	20.50
1120	Gate, 3'-6" wide		9	2.667	Ea.	79	70.50	28.50	178	236
1140	No. 1 cedar, 2 rail 3' high		140	.171	L.F.	13.70	4.52	1.83	20.05	24.50
1160	3 rail, 4' high		130	.185		17.85	4.87	1.97	24.69	30
1200	Rustic picket, molded pine, 2 rail, 3' high		140	.171		8.45	4.52	1.83	14.80	18.75
1220	No. 1 cedar, 2 rail, 3' high		140	.171		9.75	4.52	1.83	16.10	20
1240	Stockade fence, no. 1 cedar, 3-1/4" rails, 6' high		160	.150		13.15	3.96	1.60	18.71	23
1260	8' high		155	.155		18.10	4.08	1.66	23.84	28.50
1300	No. 2 cedar, treated wood rails, 6' high		160	.150		13.45	3.96	1.60	19.01	23
1320	Gate, 3'-6" wide		8	3	Ea.	89.50	79	32	200.50	266
1360	Treated pine, treated rails, 6' high		160	.150	L.F.	13.95	3.96	1.60	19.51	23.50
1400	8' high		150	.160	"	19.40	4.22	1.71	25.33	30.50

32 31 29.20 Fence, Wood Rail

		Crew	Daily Output	Labor-Hours	Unit	Material	Labor	Equipment	Total	Total Incl O&P
0010	**FENCE, WOOD RAIL**									
0012	Picket, No. 2 cedar, Gothic, 2 rail, 3' high	B-1	160	.150	L.F.	8	3.80		11.80	15.15
0050	Gate, 3'-6" wide	B-80C	9	2.667	Ea.	77	70.50	28.50	176	234
0400	3 rail, 4' high		150	.160	L.F.	9	4.22	1.71	14.93	18.80
0500	Gate, 3'-6" wide		9	2.667	Ea.	94	70.50	28.50	193	252
1200	Stockade, No. 2 cedar, treated wood rails, 6' high		160	.150	L.F.	9	3.96	1.60	14.56	18.25
1250	Gate, 3' wide		9	2.667	Ea.	96	70.50	28.50	195	255
1300	No. 1 cedar, 3-1/4" cedar rails, 6' high		160	.150	L.F.	20.50	3.96	1.60	26.06	31
1500	Gate, 3' wide		9	2.667	Ea.	220	70.50	28.50	319	390
2700	Prefabricated redwood or cedar, 4' high		160	.150	L.F.	15.75	3.96	1.60	21.31	25.50
2800	6' high		150	.160		23.50	4.22	1.71	29.43	34.50
3300	Board, shadow box, 1" x 6", treated pine, 6' high		160	.150		12.50	3.96	1.60	18.06	22
3400	No. 1 cedar, 6' high		150	.160		25	4.22	1.71	30.93	36
3900	Basket weave, No. 1 cedar, 6' high		160	.150		34	3.96	1.60	39.56	45.50
4200	Gate, 3'-6" wide	B-1	9	2.667	Ea.	176	67.50		243.50	305
5000	Fence rail, redwood, 2" x 4", merch. grade 8'	"	2400	.010	L.F.	2.50	.25		2.75	3.17

32 32 Retaining Walls

32 32 13 – Cast-in-Place Concrete Retaining Walls

32 32 13.10 Retaining Walls, Cast Concrete

		Crew	Daily Output	Labor-Hours	Unit	Material	Labor	Equipment	Total	Total Incl O&P
0010	**RETAINING WALLS, CAST CONCRETE**									
1800	Concrete gravity wall with vertical face including excavation & backfill									
1850	No reinforcing									
1900	6' high, level embankment	C-17C	36	2.306	L.F.	78.50	80	16.95	175.45	239
2000	33° slope embankment	"	32	2.594	"	91	90	19.10	200.10	272
2800	Reinforced concrete cantilever, incl. excavation, backfill & reinf.									
2900	6' high, 33° slope embankment	C-17C	35	2.371	L.F.	71.50	82	17.45	170.95	236

32 32 23 – Segmental Retaining Walls

32 32 23.13 Segmental Conc. Unit Masonry Retaining Walls

0010	**SEGMENTAL CONC. UNIT MASONRY RETAINING WALLS**									
7100	Segmental Retaining Wall system, incl. pins, and void fill									
7120	base and backfill not included									

32 32 Retaining Walls

32 32 23 – Segmental Retaining Walls

32 32 23.13 Segmental Conc. Unit Masonry Retaining Walls		Crew	Daily Output	Labor-Hours	Unit	Material	2016 Bare Costs Labor	Equipment	Total	Total Incl O&P
7140	Large unit, 8" high x 18" wide x 20" deep, 3 plane split	B-62	300	.080	S.F.	12.15	2.22	.58	14.95	17.70
7150	Straight split		300	.080		12.25	2.22	.58	15.05	17.80
7160	Medium, lt. wt., 8" high x 18" wide x 12" deep, 3 plane split		400	.060		9.45	1.66	.43	11.54	13.65
7170	Straight split		400	.060		9.35	1.66	.43	11.44	13.55
7180	Small unit, 4" x 18" x 10" deep, 3 plane split		400	.060		12.05	1.66	.43	14.14	16.50
7190	Straight split		400	.060		11.90	1.66	.43	13.99	16.30
7200	Cap unit, 3 plane split		300	.080		12.40	2.22	.58	15.20	18
7210	Cap unit, straight split		300	.080		12.40	2.22	.58	15.20	18
7250	Geo-grid soil reinforcement 4' x 50'	2 Clab	22500	.001		.76	.02		.78	.87
7255	Geo-grid soil reinforcement 6' x 150'	"	22500	.001		.60	.02		.62	.69
8000	For higher walls, add components as necessary									

32 32 26 – Metal Crib Retaining Walls

32 32 26.10 Metal Bin Retaining Walls

		Crew	Daily Output	Labor-Hours	Unit	Material	Labor	Equipment	Total	Total Incl O&P
0010	**METAL BIN RETAINING WALLS**									
0011	Aluminized steel bin, excavation									
0020	and backfill not included, 10' wide									
0100	4' high, 5.5' deep	B-13	650	.074	S.F.	26.50	1.99	1.15	29.64	33.50
0200	8' high, 5.5' deep		615	.078		30.50	2.10	1.22	33.82	38.50
0300	10' high, 7.7' deep		580	.083		34	2.22	1.29	37.51	42
0400	12' high, 7.7' deep		530	.091		36.50	2.43	1.41	40.34	45.50
0500	16' high, 7.7' deep		515	.093		38.50	2.51	1.45	42.46	48.50

32 32 29 – Timber Retaining Walls

32 32 29.10 Landscape Timber Retaining Walls

		Crew	Daily Output	Labor-Hours	Unit	Material	Labor	Equipment	Total	Total Incl O&P
0010	**LANDSCAPE TIMBER RETAINING WALLS**									
0100	Treated timbers, 6" x 6"	1 Clab	265	.030	L.F.	3	.74		3.74	4.55
0110	6" x 8"	"	200	.040	"	3.75	.99		4.74	5.80
0120	Drilling holes in timbers for fastening, 1/2"	1 Carp	450	.018	Inch		.60		.60	1.01
0130	5/8"	"	450	.018	"		.60		.60	1.01
0140	Reinforcing rods for fastening, 1/2"	1 Clab	312	.026	L.F.	.35	.63		.98	1.45
0150	5/8"	"	312	.026	"	.55	.63		1.18	1.67
0160	Reinforcing fabric	2 Clab	2500	.006	S.Y.	1.95	.16		2.11	2.41
0170	Gravel backfill		28	.571	C.Y.	18.50	14.10		32.60	44
0180	Perforated pipe, 4" diameter with silt sock		1200	.013	L.F.	1.25	.33		1.58	1.93
0190	Galvanized 60d common nails	1 Clab	625	.013	Ea.	.17	.32		.49	.71
0200	20d common nails	"	3800	.002	"	.04	.05		.09	.13

32 32 53 – Stone Retaining Walls

32 32 53.10 Retaining Walls, Stone

		Crew	Daily Output	Labor-Hours	Unit	Material	Labor	Equipment	Total	Total Incl O&P
0010	**RETAINING WALLS, STONE**									
0015	Including excavation, concrete footing and									
0020	stone 3' below grade. Price is exposed face area.									
0200	Decorative random stone, to 6' high, 1'-6" thick, dry set	D-1	35	.457	S.F.	58.50	13.50		72	87
0300	Mortar set		40	.400		60.50	11.80		72.30	86
0500	Cut stone, to 6' high, 1'-6" thick, dry set		35	.457		60.50	13.50		74	89.50
0600	Mortar set		40	.400		61.50	11.80		73.30	87
0800	Random stone, 6' to 10' high, 2' thick, dry set		45	.356		66.50	10.50		77	90.50
0900	Mortar set		50	.320		69	9.45		78.45	91.50
1100	Cut stone, 6' to 10' high, 2' thick, dry set		45	.356		67	10.50		77.50	91
1200	Mortar set		50	.320		69	9.45		78.45	92

32 33 Site Furnishings

32 33 33 – Site Manufactured Planters

32 33 33.10 Planters

		Crew	Daily Output	Labor-Hours	Unit	Material	2016 Bare Costs Labor	Equipment	Total	Total Incl O&P
0010	**PLANTERS**									
0012	Concrete, sandblasted, precast, 48" diameter, 24" high	2 Clab	15	1.067	Ea.	585	26.50		611.50	690
0300	Fiberglass, circular, 36" diameter, 24" high		15	1.067		655	26.50		681.50	770
1200	Wood, square, 48" side, 24" high		15	1.067		1,400	26.50		1,426.50	1,600
1300	Circular, 48" diameter, 30" high		10	1.600		1,025	39.50		1,064.50	1,200
1600	Planter/bench, 72"		5	3.200		3,450	79		3,529	3,900

32 33 43 – Site Seating and Tables

32 33 43.13 Site Seating

		Crew	Daily Output	Labor-Hours	Unit	Material	Labor	Equipment	Total	Total Incl O&P
0010	**SITE SEATING**									
0012	Seating, benches, park, precast conc., w/backs, wood rails, 4' long	2 Clab	5	3.200	Ea.	605	79		684	795
0100	8' long		4	4		1,025	98.50		1,123.50	1,300
0500	Steel barstock pedestals w/backs, 2" x 3" wood rails, 4' long		10	1.600		1,150	39.50		1,189.50	1,350
0510	8' long		7	2.286		1,500	56.50		1,556.50	1,750
0800	Cast iron pedestals, back & arms, wood slats, 4' long		8	2		395	49.50		444.50	520
0820	8' long		5	3.200		1,200	79		1,279	1,450
1700	Steel frame, fir seat, 10' long		10	1.600		375	39.50		414.50	480

32 84 Planting Irrigation

32 84 23 – Underground Sprinklers

32 84 23.10 Sprinkler Irrigation System

		Crew	Daily Output	Labor-Hours	Unit	Material	Labor	Equipment	Total	Total Incl O&P
0010	**SPRINKLER IRRIGATION SYSTEM**									
0011	For lawns									
0800	Residential system, custom, 1" supply	B-20	2000	.012	S.F.	.26	.30		.56	.80
0900	1-1/2" supply	"	1800	.013	"	.49	.34		.83	1.10

32 91 Planting Preparation

32 91 13 – Soil Preparation

32 91 13.16 Mulching

		Crew	Daily Output	Labor-Hours	Unit	Material	Labor	Equipment	Total	Total Incl O&P
0010	**MULCHING**									
0100	Aged barks, 3" deep, hand spread	1 Clab	100	.080	S.Y.	3.60	1.97		5.57	7.25
0150	Skid steer loader	B-63	13.50	2.963	M.S.F.	400	73	12.85	485.85	575
0200	Hay, 1" deep, hand spread	1 Clab	475	.017	S.Y.	.50	.42		.92	1.24
0250	Power mulcher, small	B-64	180	.089	M.S.F.	55.50	2.42	2.22	60.14	67.50
0350	Large	B-65	530	.030	"	55.50	.82	1.07	57.39	63.50
0400	Humus peat, 1" deep, hand spread	1 Clab	700	.011	S.Y.	2.60	.28		2.88	3.33
0450	Push spreader	"	2500	.003	"	2.60	.08		2.68	2.99
0550	Tractor spreader	B-66	700	.011	M.S.F.	289	.39	.38	289.77	320
0600	Oat straw, 1" deep, hand spread	1 Clab	475	.017	S.Y.	.50	.42		.92	1.24
0650	Power mulcher, small	B-64	180	.089	M.S.F.	55.50	2.42	2.22	60.14	67.50
0700	Large	B-65	530	.030	"	55.50	.82	1.07	57.39	63.50
0750	Add for asphaltic emulsion	B-45	1770	.009	Gal.	5.80	.25	.18	6.23	7
0800	Peat moss, 1" deep, hand spread	1 Clab	900	.009	S.Y.	2.60	.22		2.82	3.23
0850	Push spreader	"	2500	.003	"	2.60	.08		2.68	2.99
0950	Tractor spreader	B-66	700	.011	M.S.F.	289	.39	.38	289.77	320
1000	Polyethylene film, 6 mil	2 Clab	2000	.008	S.Y.	.41	.20		.61	.78
1100	Redwood nuggets, 3" deep, hand spread	1 Clab	150	.053	"	2.77	1.31		4.08	5.25
1150	Skid steer loader	B-63	13.50	2.963	M.S.F.	310	73	12.85	395.85	475
1200	Stone mulch, hand spread, ceramic chips, economy	1 Clab	125	.064	S.Y.	6.85	1.58		8.43	10.20
1250	Deluxe	"	95	.084	"	10.60	2.08		12.68	15.10

32 91 Planting Preparation

32 91 13 – Soil Preparation

32 91 13.16 Mulching

		Crew	Daily Output	Labor-Hours	Unit	Material	2016 Bare Costs Labor	Equipment	Total	Total Incl O&P
1300	Granite chips	B-1	10	2.400	C.Y.	60.50	61		121.50	169
1400	Marble chips	"	10	2.400		189	61		250	310
1600	Pea gravel		28	.857		109	21.50		130.50	157
1700	Quartz		10	2.400		190	61		251	310
1800	Tar paper, 15 lb. felt	1 Clab	800	.010	S.Y.	.48	.25		.73	.93
1900	Wood chips, 2" deep, hand spread	"	220	.036	"	1.80	.90		2.70	3.48
1950	Skid steer loader	B-63	20.30	1.970	M.S.F.	200	48.50	8.55	257.05	310

32 91 13.26 Planting Beds

		Crew	Daily Output	Labor-Hours	Unit	Material	Labor	Equipment	Total	Total Incl O&P
0010	**PLANTING BEDS**									
0100	Backfill planting pit, by hand, on site topsoil	2 Clab	18	.889	C.Y.		22		22	36.50
0200	Prepared planting mix, by hand	"	24	.667			16.45		16.45	27.50
0300	Skid steer loader, on site topsoil	B-62	340	.071			1.96	.51	2.47	3.82
0400	Prepared planting mix	"	410	.059			1.62	.42	2.04	3.16
1000	Excavate planting pit, by hand, sandy soil	2 Clab	16	1			24.50		24.50	41.50
1100	Heavy soil or clay	"	8	2			49.50		49.50	82.50
1200	1/2 C.Y. backhoe, sandy soil	B-11C	150	.107			3.19	2.44	5.63	8
1300	Heavy soil or clay	"	115	.139			4.17	3.18	7.35	10.40
2000	Mix planting soil, incl. loam, manure, peat, by hand	2 Clab	60	.267		42.50	6.55		49.05	58
2100	Skid steer loader	B-62	150	.160		42.50	4.44	1.16	48.10	55.50
3000	Pile sod, skid steer loader	"	2800	.009	S.Y.		.24	.06	.30	.47
3100	By hand	2 Clab	400	.040			.99		.99	1.65
4000	Remove sod, F.E. loader	B-10S	2000	.004			.14	.19	.33	.44
4100	Sod cutter	B-12K	3200	.005			.15	.31	.46	.59
4200	By hand	2 Clab	240	.067			1.64		1.64	2.75

32 91 19 – Landscape Grading

32 91 19.13 Topsoil Placement and Grading

		Crew	Daily Output	Labor-Hours	Unit	Material	Labor	Equipment	Total	Total Incl O&P
0010	**TOPSOIL PLACEMENT AND GRADING**									
0300	Fine grade, base course for paving, see Section 32 11 23.23									
0701	Furnish and place, truck dumped, unscreened, 4" deep	B-10S	12000	.001	S.F.	.41	.02	.03	.46	.52
0801	6" deep	"	7400	.001	"	.48	.04	.05	.57	.65
0900	Fine grading and seeding, incl. lime, fertilizer & seed,									
1001	With equipment	B-14	9000	.005	S.F.	.07	.14	.04	.25	.36

32 92 Turf and Grasses

32 92 19 – Seeding

32 92 19.13 Mechanical Seeding

		Crew	Daily Output	Labor-Hours	Unit	Material	Labor	Equipment	Total	Total Incl O&P
0010	**MECHANICAL SEEDING**									
0020	Mechanical seeding, 215 lb./acre	B-66	1.50	5.333	Acre	560	181	176	917	1,100
0100	44 lb./M.S.Y.	"	2500	.003	S.Y.	.20	.11	.11	.42	.52
0101	44 lb./M.S.Y.	1 Clab	13950	.001	S.F.	.02	.01		.03	.04
0300	Fine grading and seeding incl. lime, fertilizer & seed,									
0310	with equipment	B-14	1000	.048	S.Y.	.45	1.27	.37	2.09	3.03
0400	Fertilizer hand push spreader, 35 lb. per M.S.F.	1 Clab	200	.040	M.S.F.	9.75	.99		10.74	12.40
0600	Limestone hand push spreader, 50 lb. per M.S.F.		180	.044		5.40	1.10		6.50	7.75
0800	Grass seed hand push spreader, 4.5 lb. per M.S.F.		180	.044		22	1.10		23.10	26.50
1000	Hydro or air seeding for large areas, incl. seed and fertilizer	B-81	8900	.002	S.Y.	.50	.05	.08	.63	.72
1100	With wood fiber mulch added	"	8900	.002	"	2	.05	.08	2.13	2.37
1300	Seed only, over 100 lb., field seed, minimum				Lb.	1.80			1.80	1.98
1400	Maximum					1.70			1.70	1.87
1500	Lawn seed, minimum					1.40			1.40	1.54

32 92 Turf and Grasses

32 92 19 – Seeding

32 92 19.13 Mechanical Seeding

		Crew	Daily Output	Labor-Hours	Unit	Material	2016 Bare Costs Labor	Equipment	Total	Total Incl O&P
1600	Maximum				Lb.	2.50			2.50	2.75
1800	Aerial operations, seeding only, field seed	B-58	50	.480	Acre	585	13.30	63	661.30	735
1900	Lawn seed		50	.480		455	13.30	63	531.30	590
2100	Seed and liquid fertilizer, field seed		50	.480		725	13.30	63	801.30	890
2200	Lawn seed		50	.480		595	13.30	63	671.30	745

32 92 23 – Sodding

32 92 23.10 Sodding Systems

		Crew	Daily Output	Labor-Hours	Unit	Material	Labor	Equipment	Total	Total Incl O&P
0010	**SODDING SYSTEMS**									
0020	Sodding, 1" deep, bluegrass sod, on level ground, over 8 M.S.F.	B-63	22	1.818	M.S.F.	244	45	7.85	296.85	350
0200	4 M.S.F.		17	2.353		260	58	10.20	328.20	395
0300	1000 S.F.		13.50	2.963		295	73	12.85	380.85	460
0500	Sloped ground, over 8 M.S.F.		6	6.667		244	164	29	437	575
0600	4 M.S.F.		5	8		260	197	34.50	491.50	655
0700	1000 S.F.		4	10		295	247	43.50	585.50	790
1000	Bent grass sod, on level ground, over 6 M.S.F.		20	2		253	49.50	8.65	311.15	370
1100	3 M.S.F.		18	2.222		267	55	9.60	331.60	395
1200	Sodding 1000 S.F. or less		14	2.857		290	70.50	12.35	372.85	450
1500	Sloped ground, over 6 M.S.F.		15	2.667		253	65.50	11.55	330.05	400
1600	3 M.S.F.		13.50	2.963		267	73	12.85	352.85	430
1700	1000 S.F.		12	3.333		290	82	14.45	386.45	475

32 93 Plants

32 93 13 – Ground Covers

32 93 13.10 Ground Cover Plants

		Crew	Daily Output	Labor-Hours	Unit	Material	Labor	Equipment	Total	Total Incl O&P
0010	**GROUND COVER PLANTS**									
0012	Plants, pachysandra, in prepared beds	B-1	15	1.600	C	83.50	40.50		124	160
0200	Vinca minor, 1 yr., bare root, in prepared beds		12	2	"	71.50	50.50		122	163
0600	Stone chips, in 50 lb. bags, Georgia marble		520	.046	Bag	3.63	1.17		4.80	5.95
0700	Onyx gemstone		260	.092		17	2.34		19.34	22.50
0800	Quartz		260	.092		17	2.34		19.34	22.50
0900	Pea gravel, truckload lots		28	.857	Ton	27.50	21.50		49	66.50

32 93 33 – Shrubs

32 93 33.10 Shrubs and Trees

		Crew	Daily Output	Labor-Hours	Unit	Material	Labor	Equipment	Total	Total Incl O&P
0010	**SHRUBS AND TREES**									
0011	Evergreen, in prepared beds, B & B									
0100	Arborvitae pyramidal, 4'-5'	B-17	30	1.067	Ea.	101	30.50	26	157.50	191
0150	Globe, 12"-15"	B-1	96	.250		24.50	6.35		30.85	37.50
0300	Cedar, blue, 8'-10'	B-17	18	1.778		235	50.50	43.50	329	390
0500	Hemlock, Canadian, 2-1/2'-3'	B-1	36	.667		32	16.90		48.90	63.50
0550	Holly, Savannah, 8' - 10' H		9.68	2.479		262	63		325	395
0600	Juniper, andorra, 18"-24"		80	.300		37	7.60		44.60	53.50
0620	Wiltoni, 15"-18"		80	.300		26	7.60		33.60	41.50
0640	Skyrocket, 4-1/2'-5'	B-17	55	.582		108	16.55	14.25	138.80	162
0660	Blue pfitzer, 2'-2-1/2'	B-1	44	.545		39	13.80		52.80	65.50
0680	Ketleerie, 2-1/2'-3'		50	.480		53	12.15		65.15	78.50
0700	Pine, black, 2-1/2'-3'		50	.480		60.50	12.15		72.65	87
0720	Mugo, 18"-24"		60	.400		60.50	10.15		70.65	83.50
0740	White, 4'-5'	B-17	75	.427		52	12.15	10.45	74.60	88.50
0800	Spruce, blue, 18"-24"	B-1	60	.400		68	10.15		78.15	92
0840	Norway, 4'-5'	B-17	75	.427		84.50	12.15	10.45	107.10	125

32 93 Plants

32 93 33 – Shrubs

32 93 33.10 Shrubs and Trees

	32 93 33.10 Shrubs and Trees	Crew	Daily Output	Labor-Hours	Unit	Material	2016 Bare Costs Labor	Equipment	Total	Total Incl O&P
0900	Yew, denisforma, 12"-15"	B-1	60	.400	Ea.	35.50	10.15		45.65	56.50
1000	Capitata, 18"-24"		30	.800		33.50	20.50		54	70.50
1100	Hicksi, 2'-2-1/2'		30	.800		79.50	20.50		100	121

32 93 33.20 Shrubs

	32 93 33.20 Shrubs	Crew	Daily Output	Labor-Hours	Unit	Material	Labor	Equipment	Total	Total Incl O&P
0010	**SHRUBS**									
0011	Broadleaf Evergreen, planted in prepared beds									
0100	Andromeda, 15"-18", container	B-1	96	.250	Ea.	33	6.35		39.35	47
0200	Azalea, 15" - 18", container		96	.250		30	6.35		36.35	43.50
0300	Barberry, 9"-12", container		130	.185		17.95	4.67		22.62	27.50
0400	Boxwood, 15"-18", B&B		96	.250		43.50	6.35		49.85	58.50
0500	Euonymus, emerald gaiety, 12" to 15", container		115	.209		24.50	5.30		29.80	35.50
0600	Holly, 15"-18", B & B		96	.250		37	6.35		43.35	51
0900	Mount laurel, 18" - 24", B & B		80	.300		72	7.60		79.60	91.50
1000	Paxistema, 9" - 12" high		130	.185		21.50	4.67		26.17	32
1100	Rhododendron, 18"-24", container		48	.500		38.50	12.65		51.15	63
1200	Rosemary, 1 gal. container		600	.040		18.05	1.01		19.06	21.50
2000	Deciduous, planted in prepared beds, amelanchier, 2'-3', B & B		57	.421		121	10.65		131.65	152
2100	Azalea, 15"-18", B & B		96	.250		30	6.35		36.35	43
2300	Bayberry, 2'-3', B & B		57	.421		24.50	10.65		35.15	45
2600	Cotoneaster, 15"-18", B & B		80	.300		27	7.60		34.60	42
2800	Dogwood, 3'-4', B & B	B-17	40	.800		32.50	23	19.55	75.05	95.50
2900	Euonymus, alatus compacta, 15" to 18", container	B-1	80	.300		26.50	7.60		34.10	42
3200	Forsythia, 2'-3', container	"	60	.400		18.85	10.15		29	37.50
3300	Hibiscus, 3'-4', B & B	B-17	75	.427		49	12.15	10.45	71.60	85.50
3400	Honeysuckle, 3'-4', B & B	B-1	60	.400		27	10.15		37.15	46.50
3500	Hydrangea, 2'-3', B & B	"	57	.421		30.50	10.65		41.15	51.50
3600	Lilac, 3'-4', B & B	B-17	40	.800		28.50	23	19.55	71.05	91
3900	Privet, bare root, 18"-24"	B-1	80	.300		14.80	7.60		22.40	29
4100	Quince, 2'-3', B & B	"	57	.421		28.50	10.65		39.15	49.50
4200	Russian olive, 3'-4', B & B	B-17	75	.427		29.50	12.15	10.45	52.10	64
4400	Spirea, 3'-4', B & B	B-1	70	.343		21	8.70		29.70	37.50
4500	Viburnum, 3'-4', B & B	B-17	40	.800		25.50	23	19.55	68.05	87.50

32 93 43 – Trees

32 93 43.20 Trees

	32 93 43.20 Trees		Crew	Daily Output	Labor-Hours	Unit	Material	Labor	Equipment	Total	Total Incl O&P
0010	**TREES**										
0011	Deciduous, in prep. beds, balled & burlapped (B&B)										
0100	Ash, 2" caliper	G	B-17	8	4	Ea.	180	114	98	392	495
0200	Beech, 5'-6'	G		50	.640		181	18.20	15.65	214.85	248
0300	Birch, 6'-8', 3 stems	G		20	1.600		163	45.50	39	247.50	299
0500	Crabapple, 6'-8'	G		20	1.600		134	45.50	39	218.50	267
0600	Dogwood, 4'-5'	G		40	.800		127	23	19.55	169.55	200
0700	Eastern redbud 4'-5'	G		40	.800		145	23	19.55	187.55	219
0800	Elm, 8'-10'	G		20	1.600		250	45.50	39	334.50	395
0900	Ginkgo, 6'-7'	G		24	1.333		144	38	32.50	214.50	257
1000	Hawthorn, 8'-10', 1" caliper	G		20	1.600		157	45.50	39	241.50	292
1100	Honeylocust, 10'-12', 1-1/2" caliper	G		10	3.200		200	91	78.50	369.50	455
1300	Larch, 8'	G		32	1		123	28.50	24.50	176	211
1400	Linden, 8'-10', 1" caliper	G		20	1.600		139	45.50	39	223.50	272
1500	Magnolia, 4'-5'	G		20	1.600		98	45.50	39	182.50	227
1600	Maple, red, 8'-10', 1-1/2" caliper	G		10	3.200		196	91	78.50	365.50	455
1700	Mountain ash, 8'-10', 1" caliper	G		16	2		171	57	49	277	335
1800	Oak, 2-1/2"-3" caliper	G		6	5.333		315	152	130	597	740

32 93 Plants

32 93 43 – Trees

32 93 43.20 Trees		Crew	Daily Output	Labor-Hours	Unit	Material	2016 Bare Costs Labor	Equipment	Total	Total Incl O&P
2100	Planetree, 9'-11', 1-1/4" caliper	G B-17	10	3.200	Ea.	229	91	78.50	398.50	490
2200	Plum, 6'-8', 1" caliper	G	20	1.600		79.50	45.50	39	164	206
2300	Poplar, 9'-11', 1-1/4" caliper	G	10	3.200		140	91	78.50	309.50	390
2500	Sumac, 2'-3'	G	75	.427		42.50	12.15	10.45	65.10	78.50
2700	Tulip, 5'-6'	G	40	.800		45.50	23	19.55	88.05	110
2800	Willow, 6'-8', 1" caliper	G	20	1.600		93	45.50	39	177.50	221

32 94 Planting Accessories

32 94 13 – Landscape Edging

32 94 13.20 Edging

		Crew	Daily Output	Labor-Hours	Unit	Material	Labor	Equipment	Total	Total Incl O&P
0010	**EDGING**									
0050	Aluminum alloy, including stakes, 1/8" x 4", mill finish	B-1	390	.062	L.F.	4	1.56		5.56	7
0051	Black paint		390	.062		4.64	1.56		6.20	7.70
0052	Black anodized		390	.062		5.35	1.56		6.91	8.50
0100	Brick, set horizontally, 1-1/2 bricks per L.F.	D-1	370	.043		1.29	1.28		2.57	3.54
0150	Set vertically, 3 bricks per L.F.	"	135	.119		4	3.50		7.50	10.25
0200	Corrugated aluminum, roll, 4" wide	1 Carp	650	.012		2.20	.42		2.62	3.12
0250	6" wide	"	550	.015		2.75	.49		3.24	3.86
0600	Railroad ties, 6" x 8"	2 Carp	170	.094		2.11	3.19		5.30	7.65
0650	7" x 9"		136	.118		2.34	3.99		6.33	9.30
0750	Redwood 2" x 4"		330	.048		2.26	1.64		3.90	5.25
0800	Steel edge strips, incl. stakes, 1/4" x 5"	B-1	390	.062		5	1.56		6.56	8.10
0850	3/16" x 4"	"	390	.062		3.95	1.56		5.51	6.95

32 94 50 – Tree Guying

32 94 50.10 Tree Guying Systems

		Crew	Daily Output	Labor-Hours	Unit	Material	Labor	Equipment	Total	Total Incl O&P
0010	**TREE GUYING SYSTEMS**									
0015	Tree guying including stakes, guy wire and wrap									
0100	Less than 3" caliper, 2 stakes	2 Clab	35	.457	Ea.	13.10	11.25		24.35	33.50
0200	3" to 4" caliper, 3 stakes	"	21	.762	"	19.55	18.80		38.35	53
1000	Including arrowhead anchor, cable, turnbuckles and wrap									
1100	Less than 3" caliper, 3" anchors	2 Clab	20	.800	Ea.	28	19.70		47.70	64
1200	3" to 6" caliper, 4" anchors		15	1.067		34	26.50		60.50	81.50
1300	6" caliper, 6" anchors		12	1.333		28.50	33		61.50	86
1400	8" caliper, 8" anchors		9	1.778		114	44		158	199

32 96 Transplanting

32 96 23 – Plant and Bulb Transplanting

32 96 23.23 Planting

		Crew	Daily Output	Labor-Hours	Unit	Material	Labor	Equipment	Total	Total Incl O&P
0010	**PLANTING**									
0012	Moving shrubs on site, 12" ball	B-62	28	.857	Ea.	24	6.20		30.20	46.50
0100	24" ball	"	22	1.091	"	30.50	7.90		38.40	59

32 96 23.43 Moving Trees

		Crew	Daily Output	Labor-Hours	Unit	Material	Labor	Equipment	Total	Total Incl O&P
0010	**MOVING TREES**, On site									
0300	Moving trees on site, 36" ball	B-6	3.75	6.400	Ea.		177	97.50	274.50	405
0400	60" ball	"	1	24	"		665	365	1,030	1,500

Estimating Tips
33 10 00 Water Utilities
33 30 00 Sanitary Sewerage Utilities
33 40 00 Storm Drainage Utilities

- Never assume that the water, sewer, and drainage lines will go in at the early stages of the project. Consider the site access needs before dividing the site in half with open trenches, loose pipe, and machinery obstructions. Always inspect the site to establish that the site drawings are complete. Check off all existing utilities on your drawings as you locate them. Be especially careful with underground utilities because appurtenances are sometimes buried during regrading or repaving operations. If you find any discrepancies, mark up the site plan for further research. Differing site conditions can be very costly if discovered later in the project.

- See also Section 33 01 00 for restoration of pipe where removal/replacement may be undesirable. Use of new types of piping materials can reduce the overall project cost. Owners/design engineers should consider the installing contractor as a valuable source of current information on utility products and local conditions that could lead to significant cost savings.

Reference Numbers
Reference numbers are shown at the beginning of some major classifications. These numbers refer to related items in the Reference Section. The reference information may be an estimating procedure, an alternate pricing method, or technical information.

Note: Not all subdivisions listed here necessarily appear. ∎

No part of this cost data may be reproduced, stored in a retrieval system, or transmitted in any form or by any means without prior written permission of RSMeans.

Note: Trade Service, in part, has been used as a reference source for some of the material prices used in Division 33.

Did you know?
RSMeans Online gives you the same access to RSMeans' data with 24/7 access:
- Quickly locate costs in the searchable database.
- Build cost lists, estimates, and reports in minutes.
- Adjust costs to any location in the U.S. and Canada with the click of a button.

Start your free trial today at **www.RSMeansOnline.com**

RSMeans Online
FROM THE GORDIAN GROUP

33 05 Common Work Results for Utilities

33 05 16 – Utility Structures

33 05 16.13 Precast Concrete Utility Boxes	Crew	Daily Output	Labor-Hours	Unit	Material	2016 Bare Costs Labor	Equipment	Total	Total Incl O&P
0010 **PRECAST CONCRETE UTILITY BOXES**, 6" thick									
0050 5' x 10' x 6' high, I.D.	B-13	2	24	Ea.	1,775	645	375	2,795	3,425
0350 Hand hole, precast concrete, 1-1/2" thick									
0400 1'-0" x 2'-0" x 1'-9", I.D., light duty	B-1	4	6	Ea.	410	152		562	705
0450 4'-6" x 3'-2" x 2'-0", O.D., heavy duty	B-6	3	8	"	1,475	222	122	1,819	2,100

33 05 23 – Trenchless Utility Installation

33 05 23.19 Microtunneling

				Unit				Total	Total Incl O&P
0010 **MICROTUNNELING**									
0011 Not including excavation, backfill, shoring,									
0020 or dewatering, average 50'/day, slurry method									
0100 24" to 48" outside diameter, minimum				L.F.				875	965
0110 Adverse conditions, add				%				50%	50%
1000 Rent microtunneling machine, average monthly lease				Month				97,500	107,000
1010 Operating technician				Day				630	690
1100 Mobilization and demobilization, minimum				Job				41,200	45,900
1110 Maximum				"				445,500	490,500

33 11 Water Utility Distribution Piping

33 11 13 – Public Water Utility Distribution Piping

33 11 13.15 Water Supply, Ductile Iron Pipe

	Crew	Daily Output	Labor-Hours	Unit	Material	Labor	Equipment	Total	Total Incl O&P
0010 **WATER SUPPLY, DUCTILE IRON PIPE**									
0020 Not including excavation or backfill									
2000 Pipe, class 50 water piping, 18' lengths									
2020 Mechanical joint, 4" diameter	B-21A	200	.200	L.F.	30.50	6.30	2.40	39.20	46.50
2040 6" diameter		160	.250		32	7.90	3	42.90	51.50
3000 Push-on joint, 4" diameter		400	.100		15.50	3.15	1.20	19.85	23.50
3020 6" diameter		333.33	.120		18.25	3.78	1.44	23.47	28
8000 Piping, fittings, mechanical joint, AWWA C110									
8006 90° bend, 4" diameter	B-20A	16	2	Ea.	155	61		216	271
8020 6" diameter		12.80	2.500		229	76		305	380
8200 Wye or tee, 4" diameter		10.67	2.999		375	91		466	565
8220 6" diameter		8.53	3.751		565	114		679	810
8398 45° bends, 4" diameter		16	2		255	61		316	380
8400 6" diameter		12.80	2.500		199	76		275	345
8405 8" diameter		10.67	2.999		291	91		382	470
8450 Decreaser, 6" x 4" diameter		14.22	2.250		207	68.50		275.50	340
8460 8" x 6" diameter		11.64	2.749		310	83.50		393.50	480
8550 Piping, butterfly valves, cast iron									
8560 4" diameter	B-20	6	4	Ea.	470	101		571	685
8700 Joint restraint, ductile iron mechanical joints									
8710 4" diameter	B-20A	32	1	Ea.	17.15	30.50		47.65	69.50
8720 6" diameter		25.60	1.250		21.50	38		59.50	86.50
8730 8" diameter		21.33	1.500		30	45.50		75.50	109
8740 10" diameter		18.28	1.751		38.50	53		91.50	131
8750 12" diameter		16.84	1.900		50	58		108	151
8760 14" diameter		16	2		53	61		114	160
8770 16" diameter		11.64	2.749		58	83.50		141.50	203
8780 18" diameter		11.03	2.901		73	88		161	226
8785 20" diameter		9.14	3.501		78	106		184	263
8790 24" diameter		7.53	4.250		104	129		233	330

33 11 Water Utility Distribution Piping

33 11 13 – Public Water Utility Distribution Piping

33 11 13.15 Water Supply, Ductile Iron Pipe

		Crew	Daily Output	Labor-Hours	Unit	Material	2016 Bare Costs Labor	Equipment	Total	Total Incl O&P
9600	Steel sleeve with tap, 4" diameter	B-20	3	8	Ea.	405	203		608	785
9620	6" diameter	"	2	12	↓	470	305		775	1,025

33 11 13.25 Water Supply, Polyvinyl Chloride Pipe

		Crew	Daily Output	Labor-Hours	Unit	Material	Labor	Equipment	Total	Total Incl O&P
0010	**WATER SUPPLY, POLYVINYL CHLORIDE PIPE**									
2100	PVC pipe, Class 150, 1-1/2" diameter	Q-1A	750	.013	L.F.	.50	.53		1.03	1.42
2120	2" diameter		686	.015		.73	.58		1.31	1.75
2140	2-1/2" diameter		500	.020		1.25	.79		2.04	2.68
2160	3" diameter	B-20	430	.056	↓	1.40	1.41		2.81	3.90
8700	PVC pipe, joint restraint									
8710	4" diameter	B-20A	32	1	Ea.	45	30.50		75.50	100
8720	6" diameter		25.60	1.250		54.50	38		92.50	123
8730	8" diameter		21.33	1.500		81	45.50		126.50	165
8740	10" diameter		18.28	1.751		152	53		205	256
8750	12" diameter		16.84	1.900		159	58		217	271
8760	14" diameter		16	2		237	61		298	360
8770	16" diameter		11.64	2.749		320	83.50		403.50	490
8780	18" diameter		11.03	2.901		395	88		483	580
8785	20" diameter		9.14	3.501		480	106		586	700
8790	24" diameter	↓	7.53	4.250	↓	565	129		694	835

33 12 Water Utility Distribution Equipment

33 12 13 – Water Service Connections

33 12 13.15 Tapping, Crosses and Sleeves

		Crew	Daily Output	Labor-Hours	Unit	Material	Labor	Equipment	Total	Total Incl O&P
0010	**TAPPING, CROSSES AND SLEEVES**									
4000	Drill and tap pressurized main (labor only)									
4100	6" main, 1" to 2" service	Q-1	3	5.333	Ea.		187		187	310
4150	8" main, 1" to 2" service	"	2.75	5.818	"		205		205	335
4500	Tap and insert gate valve									
4600	8" main, 4" branch	B-21	3.20	8.750	Ea.		235	44	279	440
4650	6" branch		2.70	10.370			278	52	330	520
4700	10" Main, 4" branch		2.70	10.370			278	52	330	520
4750	6" branch		2.35	11.915			320	59.50	379.50	600
4800	12" main, 6" branch	↓	2.35	11.915	↓		320	59.50	379.50	600

33 21 Water Supply Wells

33 21 13 – Public Water Supply Wells

33 21 13.10 Wells and Accessories

		Crew	Daily Output	Labor-Hours	Unit	Material	Labor	Equipment	Total	Total Incl O&P
0010	**WELLS & ACCESSORIES**									
0011	Domestic									
0100	Drilled, 4" to 6" diameter	B-23	120	.333	L.F.	8.35		24	32.35	40
1500	Pumps, installed in wells to 100' deep, 4" submersible									
1520	3/4 H.P.	Q-1	2.66	6.015	Ea.	855	211		1,066	1,300
1600	1 H.P.	"	2.29	6.987	"	905	246		1,151	1,400

33 31 Sanitary Utility Sewerage Piping

33 31 13 – Public Sanitary Utility Sewerage Piping

33 31 13.15 Sewage Collection, Concrete Pipe

		Crew	Daily Output	Labor-Hours	Unit	Material	2016 Bare Costs Labor	Equipment	Total	Total Incl O&P
0010	**SEWAGE COLLECTION, CONCRETE PIPE**									
0020	See Section 33 41 13.60 for sewage/drainage collection, concrete pipe									

33 31 13.25 Sewage Collection, Polyvinyl Chloride Pipe

		Crew	Daily Output	Labor-Hours	Unit	Material	Labor	Equipment	Total	Total Incl O&P
0010	**SEWAGE COLLECTION, POLYVINYL CHLORIDE PIPE**									
0020	Not including excavation or backfill									
2000	20' lengths, SDR 35, B&S, 4" diameter	B-20	375	.064	L.F.	1.49	1.62		3.11	4.35
2040	6" diameter		350	.069		3.28	1.74		5.02	6.50
2080	13' lengths, SDR 35, B&S, 8" diameter		335	.072		6.15	1.81		7.96	9.80
2120	10" diameter	B-21	330	.085		11.10	2.28	.43	13.81	16.45
4000	Piping, DWV PVC, no exc./bkfill., 10' L, Sch 40, 4" diameter	B-20	375	.064		3.97	1.62		5.59	7.10
4010	6" diameter		350	.069		8	1.74		9.74	11.70
4020	8" diameter		335	.072		12.50	1.81		14.31	16.80
4030	Fittings, 1/4 bend DWV PVC, 4" diameter		19	1.263	Ea.	16.60	32		48.60	72
4040	6" diameter		12	2		75.50	50.50		126	168
4050	8" diameter		11	2.182		75.50	55		130.50	176
4060	1/8 bend DWV PVC, 4" diameter		19	1.263		13.25	32		45.25	68
4070	6" diameter		12	2		51.50	50.50		102	142
4080	8" diameter		11	2.182		165	55		220	275
4090	Tee DWV PVC, 4" diameter		12	2		22.50	50.50		73	110
4100	6" diameter		10	2.400		91	61		152	202
4110	8" diameter		9	2.667		199	67.50		266.50	330

33 36 Utility Septic Tanks

33 36 13 – Utility Septic Tank and Effluent Wet Wells

33 36 13.13 Concrete Utility Septic Tank

		Crew	Daily Output	Labor-Hours	Unit	Material	Labor	Equipment	Total	Total Incl O&P
0010	**CONCRETE UTILITY SEPTIC TANK**									
0011	Not including excavation or piping									
0015	Septic tanks, precast, 1,000 gallon	B-21	8	3.500	Ea.	1,025	94	17.55	1,136.55	1,300
0060	1,500 gallon		7	4		1,575	107	20	1,702	1,950
0100	2,000 gallon		5	5.600		2,200	150	28	2,378	2,700
0900	Concrete riser 24" X 12" with standard lid	1 Clab	6	1.333		85	33		118	149
0905	24" X 12" with heavy duty lid		6	1.333		108	33		141	174
0910	24" X 8" with standard lid		6	1.333		79.50	33		112.50	143
0915	24" X 8" with heavy duty lid		6	1.333		102	33		135	168
0917	24" X 12" extension		12	.667		57.50	16.45		73.95	91
0920	24" X 8" extension		12	.667		52	16.45		68.45	84.50
0950	HDPE riser 20" X 12" with standard lid		8	1		118	24.50		142.50	171
0951	HDPE 20" X 12" with heavy duty lid		8	1		131	24.50		155.50	186
0955	HDPE 24" X 12" with standard lid		8	1		115	24.50		139.50	168
0956	HDPE 24" X 12" with heavy duty lid		8	1		137	24.50		161.50	192
0960	HDPE 20" X 6" extension		48	.167		26	4.11		30.11	35.50
0962	HDPE 20" X 12" extension		48	.167		46	4.11		50.11	57.50
0965	HDPE 24" X 6" extension		48	.167		27.50	4.11		31.61	37.50
0967	HDPE 24" X 12" extension		48	.167		45.50	4.11		49.61	57
1150	Leaching field chambers, 13' x 3'-7" x 1'-4", standard	B-13	16	3		485	80.50	47	612.50	715
1420	Leaching pit, precast concrete, 6', dia, 3' deep	B-21	4.70	5.957		885	160	30	1,075	1,275

33 36 13.19 Polyethylene Utility Septic Tank

		Crew	Daily Output	Labor-Hours	Unit	Material	Labor	Equipment	Total	Total Incl O&P
0010	**POLYETHYLENE UTILITY SEPTIC TANK**									
0015	High density polyethylene, 1,000 gallon	B-21	8	3.500	Ea.	1,225	94	17.55	1,336.55	1,525
0020	1,250 gallon		8	3.500		1,400	94	17.55	1,511.55	1,725

33 36 Utility Septic Tanks

33 36 13 – Utility Septic Tank and Effluent Wet Wells

33 36 13.19 Polyethylene Utility Septic Tank

		Crew	Daily Output	Labor-Hours	Unit	Material	2016 Bare Costs Labor	Equipment	Total	Total Incl O&P
0025	1,500 gallon	B-21	7	4	Ea.	1,550	107	20	1,677	1,900

33 36 19 – Utility Septic Tank Effluent Filter

33 36 19.13 Utility Septic Tank Effluent Tube Filter

0010	**UTILITY SEPTIC TANK EFFLUENT TUBE FILTER**									
3000	Effluent filter, 4" diameter	1 Skwk	8	1	Ea.	43.50	34		77.50	105
3020	6" diameter	"	7	1.143	"	254	39		293	345

33 36 33 – Utility Septic Tank Drainage Field

33 36 33.13 Utility Septic Tank Tile Drainage Field

		Crew	Daily Output	Labor-Hours	Unit	Material	Labor	Equipment	Total	Total Incl O&P
0010	**UTILITY SEPTIC TANK TILE DRAINAGE FIELD**									
0015	Distribution box, concrete, 5 outlets	2 Clab	20	.800	Ea.	82	19.70		101.70	123
0020	7 outlets		16	1		82	24.50		106.50	132
0025	9 outlets		8	2		500	49.50		549.50	635
0115	Distribution boxes, HDPE, 5 outlets		20	.800		64.50	19.70		84.20	104
0117	6 outlets		15	1.067		65	26.50		91.50	116
0118	7 outlets		15	1.067		65	26.50		91.50	116
0120	8 outlets		10	1.600		68.50	39.50		108	142
0240	Distribution boxes, Outlet Flow Leveler	1 Clab	50	.160		2.35	3.94		6.29	9.20
0300	Precast concrete, galley, 4' x 4' x 4'	B-21	16	1.750		250	47	8.75	305.75	365
0350	HDPE infiltration chamber 12" H X 15" W	2 Clab	300	.053	L.F.	7.35	1.31		8.66	10.30
0351	12" H X 15" W End Cap	1 Clab	32	.250	Ea.	19.75	6.15		25.90	32
0355	chamber 12" H X 22" W	2 Clab	300	.053	L.F.	6.80	1.31		8.11	9.70
0356	12" H X 22" W End Cap	1 Clab	32	.250	Ea.	17.30	6.15		23.45	29.50
0360	chamber 13" H X 34" W	2 Clab	300	.053	L.F.	15	1.31		16.31	18.70
0361	13" H X 34" W End Cap	1 Clab	32	.250	Ea.	53.50	6.15		59.65	69
0365	chamber 16" H X 34" W	2 Clab	300	.053	L.F.	10.75	1.31		12.06	14
0366	16" H X 34" W End Cap	1 Clab	32	.250	Ea.	8	6.15		14.15	19.10
0370	chamber 8" H X 16" W	2 Clab	300	.053	L.F.	9.65	1.31		10.96	12.80
0371	8" H X 16" W End Cap	1 Clab	32	.250	Ea.	10.80	6.15		16.95	22

33 36 50 – Drainage Field Systems

33 36 50.10 Drainage Field Excavation and Fill

		Crew	Daily Output	Labor-Hours	Unit	Material	Labor	Equipment	Total	Total Incl O&P
0010	**DRAINAGE FIELD EXCAVATION AND FILL**									
2200	Septic tank & drainage field excavation with 3/4 C.Y. backhoe	B-12F	145	.110	C.Y.		3.35	4.51	7.86	10.50
2400	4' trench for disposal field, 3/4 C.Y. backhoe	"	335	.048	L.F.		1.45	1.95	3.40	4.56
2600	Gravel fill, run of bank	B-6	150	.160	C.Y.	18.50	4.44	2.44	25.38	30.50
2800	Crushed stone, 3/4"	"	150	.160	"	38.50	4.44	2.44	45.38	52

33 41 Storm Utility Drainage Piping

33 41 13 – Public Storm Utility Drainage Piping

33 41 13.60 Sewage/Drainage Collection, Concrete Pipe

		Crew	Daily Output	Labor-Hours	Unit	Material	Labor	Equipment	Total	Total Incl O&P
0010	**SEWAGE/DRAINAGE COLLECTION, CONCRETE PIPE**									
0020	Not including excavation or backfill									
1020	8" diameter	B-14	224	.214	L.F.	8	5.70	1.63	15.33	20
1030	10" diameter	"	216	.222	"	8.85	5.90	1.69	16.44	21.50
3780	Concrete slotted pipe, class 4 mortar joint									
3800	12" diameter	B-21	168	.167	L.F.	28	4.48	.84	33.32	39.50
3840	18" diameter	"	152	.184	"	32	4.95	.92	37.87	44.50
3900	Concrete slotted pipe, Class 4 O-ring joint									
3940	12" diameter	B-21	168	.167	L.F.	28	4.48	.84	33.32	39.50
3960	18" diameter	"	152	.184	"	32	4.95	.92	37.87	44.50

33 44 Storm Utility Water Drains

33 44 13 – Utility Area Drains

33 44 13.13 Catch Basins	Crew	Daily Output	Labor-Hours	Unit	Material	2016 Bare Costs Labor	Equipment	Total	Total Incl O&P
0010 **CATCH BASINS**									
0011 Not including footing & excavation									
1600 Frames & grates, C.I., 24" square, 500 lb.	B-6	7.80	3.077	Ea.	340	85.50	47	472.50	570

33 46 Subdrainage

33 46 16 – Subdrainage Piping

33 46 16.25 Piping, Subdrainage, Corrugated Metal

	Crew	Daily Output	Labor-Hours	Unit	Material	Labor	Equipment	Total	Total Incl O&P
0010 **PIPING, SUBDRAINAGE, CORRUGATED METAL**									
0021 Not including excavation and backfill									
2010 Aluminum, perforated									
2020 6" diameter, 18 ga.	B-20	380	.063	L.F.	6.50	1.60		8.10	9.85
2200 8" diameter, 16 ga.	"	370	.065		8.55	1.64		10.19	12.15
2220 10" diameter, 16 ga.	B-21	360	.078		10.70	2.09	.39	13.18	15.65
3000 Uncoated galvanized, perforated									
3020 6" diameter, 18 ga.	B-20	380	.063	L.F.	6.05	1.60		7.65	9.35
3200 8" diameter, 16 ga.	"	370	.065		8.30	1.64		9.94	11.90
3220 10" diameter, 16 ga.	B-21	360	.078		8.80	2.09	.39	11.28	13.60
3240 12" diameter, 16 ga.	"	285	.098		9.80	2.64	.49	12.93	15.70
4000 Steel, perforated, asphalt coated									
4020 6" diameter, 18 ga.	B-20	380	.063	L.F.	6.55	1.60		8.15	9.90
4030 8" diameter, 18 ga.	"	370	.065		8.60	1.64		10.24	12.20
4040 10" diameter, 16 ga.	B-21	360	.078		10.25	2.09	.39	12.73	15.20
4050 12" diameter, 16 ga.		285	.098		11.30	2.64	.49	14.43	17.40
4060 18" diameter, 16 ga.		205	.137		17.40	3.67	.68	21.75	26

33 49 Storm Drainage Structures

33 49 13 – Storm Drainage Manholes, Frames, and Covers

33 49 13.10 Storm Drainage Manholes, Frames and Covers

	Crew	Daily Output	Labor-Hours	Unit	Material	Labor	Equipment	Total	Total Incl O&P
0010 **STORM DRAINAGE MANHOLES, FRAMES & COVERS**									
0020 Excludes footing, excavation, backfill (See line items for frame & cover)									
0050 Brick, 4' inside diameter, 4' deep	D-1	1	16	Ea.	535	470		1,005	1,375
1110 Precast, 4' I.D., 4' deep	B-22	4.10	7.317	"	740	201	51.50	992.50	1,200

33 51 Natural-Gas Distribution

33 51 13 – Natural-Gas Piping

33 51 13.10 Piping, Gas Service and Distribution, P.E.

	Crew	Daily Output	Labor-Hours	Unit	Material	Labor	Equipment	Total	Total Incl O&P
0010 **PIPING, GAS SERVICE AND DISTRIBUTION, POLYETHYLENE**									
0020 Not including excavation or backfill									
1000 60 psi coils, compression coupling @ 100', 1/2" diameter, SDR 11	B-20A	608	.053	L.F.	.50	1.60		2.10	3.21
1010 1" diameter, SDR 11		544	.059		1.13	1.79		2.92	4.21
1040 1-1/4" diameter, SDR 11		544	.059		1.62	1.79		3.41	4.75
1100 2" diameter, SDR 11		488	.066		2.25	1.99		4.24	5.80
1160 3" diameter, SDR 11		408	.078		4.70	2.38		7.08	9.10
1500 60 PSI 40' joints with coupling, 3" diameter, SDR 11	B-21A	408	.098		9.40	3.09	1.17	13.66	16.75
1540 4" diameter, SDR 11		352	.114		13.50	3.58	1.36	18.44	22.50
1600 6" diameter, SDR 11		328	.122		33.50	3.85	1.46	38.81	45
1640 8" diameter, SDR 11		272	.147		46.50	4.64	1.76	52.90	60.50

33 51 Natural-Gas Distribution

33 51 13 – Natural-Gas Piping

33 51 13.20 Piping, Gas Service and Distribution, Steel	Crew	Daily Output	Labor-Hours	Unit	Material	2016 Bare Costs Labor	Equipment	Total	Total Incl O&P
0010 **PIPING, GAS SERVICE & DISTRIBUTION, STEEL**									
0020 Not including excavation or backfill, tar coated and wrapped									
4000 Pipe schedule 40, plain end									
4040 1" diameter	Q-4	300	.107	L.F.	5	3.96	.19	9.15	12.25
4080 2" diameter	"	280	.114	"	7.85	4.24	.21	12.30	15.90

Division Notes

	CREW	DAILY OUTPUT	LABOR-HOURS	UNIT	BARE COSTS				TOTAL INCL O&P
					MAT.	LABOR	EQUIP.	TOTAL	

Division 48 — Electrical Power Generation

Estimating Tips

- When estimating costs for the installation of electrical power generation equipment, factors to review include access to the job site, access and setting at the installation site, required connections, uncrating pads, anchors, leveling, final assembly of the components, and temporary protection from physical damage, including from exposure to the environment.

- Be aware of the cost of equipment supports, concrete pads, and vibration isolators, and cross-reference to other trades' specifications. Also, review site and structural drawings for items that must be included in the estimates.

- It is important to include items that are not documented in the plans and specifications but must be priced. These items include, but are not limited to, testing, dust protection, roof penetration, core drilling concrete floors and walls, patching, cleanup, and final adjustments. Add a contingency or allowance for utility company fees for power hookups, if needed.

- The project size and scope of electrical power generation equipment will have a significant impact on cost. The intent of RSMeans cost data is to provide a benchmark cost so that owners, engineers, and electrical contractors will have a comfortable number with which to start a project. Additionally, there are many websites available to use for research and to obtain a vendor's quote to finalize costs.

Reference Numbers

Reference numbers are shown at the beginning of some major classifications. These numbers refer to related items in the Reference Section. The reference information may be an estimating procedure, an alternate pricing method, or technical information.

Note: Not all subdivisions listed here necessarily appear. ■

No part of this cost data may be reproduced, stored in a retrieval system, or transmitted in any form or by any means without prior written permission of RSMeans.

Note: Trade Service, in part, has been used as a reference source for some of the material prices used in Division 48.

Did you know?

RSMeans Online gives you the same access to RSMeans' data with 24/7 access:

- Quickly locate costs in the searchable database.
- Build cost lists, estimates, and reports in minutes.
- Adjust costs to any location in the U.S. and Canada with the click of a button.

Start your free trial today at **www.RSMeansOnline.com**

RSMeans Online
FROM THE GORDIAN GROUP®

48 15 Wind Energy Electrical Power Generation Equipment

48 15 13 – Wind Turbines

48 15 13.50 Wind Turbines and Components		Crew	Daily Output	Labor-Hours	Unit	Material	2016 Bare Costs Labor	Equipment	Total	Total Incl O&P
0010	**WIND TURBINES & COMPONENTS**									
0500	Complete system, grid connected									
1000	20 kW, 31' dia, incl. labor & material	G			System				49,900	49,900
1010	Enhanced	G							55,000	55,000
1500	10 kW, 23' dia, incl. labor & material	G							74,000	74,000
2000	2.4 kW, 12' dia, incl. labor & material	G			▼				18,000	18,000
2900	Component system									
3000	Turbine, 400 watt, 3' dia	G	1 Elec	3.41	2.346	Ea.	720	86.50		806.50	930
3100	600 watt, 3' dia	G		2.56	3.125		895	115		1,010	1,175
3200	1000 watt, 9' dia	G	▼	2.05	3.902	▼	3,700	144		3,844	4,300
3400	Mounting hardware										
3500	30' guyed tower kit	G	2 Clab	5.12	3.125	Ea.	460	77		537	635
3505	3' galvanized helical earth screw	G	1 Clab	8	1		53	24.50		77.50	99.50
3510	Attic mount kit	G	1 Rofc	2.56	3.125		166	90		256	350
3520	Roof mount kit	G	1 Clab	3.41	2.346	▼	152	58		210	264
8900	Equipment										
9100	DC to AC inverter for, 48 V, 4,000 watt	G	1 Elec	2	4	Ea.	2,300	148		2,448	2,775

Reference Section

All the reference information is in one section, making it easy to find what you need to know ... and easy to use the data set on a daily basis. This section is visually identified by a vertical black bar on the page edges.

In this Reference Section, we've included Equipment Rental Costs, a listing of rental and operating costs; Crew Listings, a full listing of all crews, equipment, and their costs; Location Factors for adjusting costs to the region you are in; Reference Tables, where you will find explanations, estimating information and procedures, or technical data; an explanation of all the Abbreviations in the data set; and sample Estimating Forms.

Table of Contents

Construction Equipment Rental Costs	659
Crew Listings	671
Location Factors	705
Reference Tables	711
R01 General Requirements	711
R02 Existing Conditions	719
R04 Masonry	720
R05 Metals	722
R06 Wood, Plastics & Composites	723

Reference Tables (cont.)

R07 Thermal & Moisture Protection	723
R08 Openings	724
R09 Finishes	726
R13 Special Construction	729
R22 Plumbing	730
R32 Exterior Improvements	730
R33 Utilities	731
Abbreviations	732
Estimating Forms	736

Equipment Rental Costs

Estimating Tips
- This section contains the average costs to rent and operate hundreds of pieces of construction equipment. This is useful information when estimating the time and material requirements of any particular operation in order to establish a unit or total cost. Bare equipment costs shown on a unit cost line include not only rental, but also operating costs for equipment under normal use.

Rental Costs
- Equipment rental rates are obtained from the following industry sources throughout North America: contractors, suppliers, dealers, manufacturers, and distributors.
- Rental rates vary throughout the country, with larger cities generally having lower rates. Lease plans for new equipment are available for periods in excess of six months, with a percentage of payments applying toward purchase.
- Monthly rental rates vary from 2% to 5% of the purchase price of the equipment depending on the anticipated life of the equipment and its wearing parts.
- Weekly rental rates are about 1/3 the monthly rates, and daily rental rates are about 1/3 the weekly rate.
- Rental rates can also be treated as reimbursement costs for contractor-owned equipment.

Owned equipment costs include depreciation, loan payments, interest, taxes, insurance, storage, and major repairs.

Operating Costs
- The operating costs include parts and labor for routine servicing, such as repair and replacement of pumps, filters and worn lines. Normal operating expendables, such as fuel, lubricants, tires and electricity (where applicable), are also included.
- Extraordinary operating expendables with highly variable wear patterns, such as diamond bits and blades, are excluded. These costs can be found as material costs in the Unit Price section.
- The hourly operating costs listed do not include the operator's wages.

Equipment Cost/Day
- Any power equipment required by a crew is shown in the Crew Listings with a daily cost.
- This daily cost of equipment needed by a crew includes both the rental cost and the operating cost and is based on dividing the weekly rental rate by 5 (number of working days in the week), and then adding the hourly operating cost times 8 (the number of hours in a day). This "Equipment Cost/ Day" is shown in the far right column of the Equipment Rental pages.

- If equipment is needed for only one or two days, it is best to develop your own cost by including components for daily rent and hourly operating cost. This is important when the listed Crew for a task does not contain the equipment needed, such as a crane for lifting mechanical heating/cooling equipment up onto a roof.
- If the quantity of work is less than the crew's Daily Output shown for a Unit Price line item that includes a bare unit equipment cost, it is recommended to estimate one day's rental cost and operating cost for equipment shown in the Crew Listing for that line item.

Mobilization/ Demobilization
- The cost to move construction equipment from an equipment yard or rental company to the job site and back again is not included in equipment rental costs listed in the Reference Section, nor in the bare equipment cost of any unit price line item, nor in any equipment costs shown in the Crew listings.
- Mobilization (to the site) and demobilization (from the site) costs can be found in the Unit Price Section.
- If a piece of equipment is already at the job site, it is not appropriate to utilize mobilization/demobilization costs again in an estimate. ■

Did you know?
RSMeans Online gives you the same access to RSMeans' data with 24/7 access:
- Quickly locate costs in the searchable database.
- Build cost lists, estimates, and reports in minutes.
- Adjust costs to any location in the U.S. and Canada with the click of a button.

Start your free trial today at **www.RSMeansOnline.com**

No part of this cost data may be reproduced, stored in a retrieval system, or transmitted in any form or by any means without prior written permission of RSMeans.

01 54 | Construction Aids

01 54 33 | Equipment Rental

		UNIT	HOURLY OPER. COST	RENT PER DAY	RENT PER WEEK	RENT PER MONTH	EQUIPMENT COST/DAY	
10	**0010 CONCRETE EQUIPMENT RENTAL** without operators R015433-10							**10**
	0200 Bucket, concrete lightweight, 1/2 C.Y.	Ea.	.80	23.50	70	210	20.40	
	0300 1 C.Y.		.85	28	84	252	23.60	
	0400 1-1/2 C.Y.		1.10	38.50	115	345	31.80	
	0500 2 C.Y.		1.20	45	135	405	36.60	
	0580 8 C.Y.		6.00	257	770	2,300	202	
	0600 Cart, concrete, self-propelled, operator walking, 10 C.F.		3.25	56.50	170	510	60	
	0700 Operator riding, 18 C.F.		5.35	95	285	855	99.80	
	0800 Conveyer for concrete, portable, gas, 16" wide, 26' long		12.60	123	370	1,100	174.80	
	0900 46' long		13.00	148	445	1,325	193	
	1000 56' long		13.15	157	470	1,400	199.20	
	1100 Core drill, electric, 2-1/2 H.P., 1" to 8" bit diameter		2.15	78.50	235	705	64.20	
	1150 11 H.P., 8" to 18" cores		6.00	115	345	1,025	117	
	1200 Finisher, concrete floor, gas, riding trowel, 96" wide		12.90	145	435	1,300	190.20	
	1300 Gas, walk-behind, 3 blade, 36" trowel		2.15	20	60	180	29.20	
	1400 4 blade, 48" trowel		4.25	27.50	83	249	50.60	
	1500 Float, hand-operated (Bull float) 48" wide		.08	13.35	40	120	8.65	
	1570 Curb builder, 14 H.P., gas, single screw		14.40	248	745	2,225	264.20	
	1590 Double screw		15.10	292	875	2,625	295.80	
	1600 Floor grinder, concrete and terrazzo, electric, 22" path		2.70	167	500	1,500	121.60	
	1700 Edger, concrete, electric, 7" path		1.04	51.50	155	465	39.30	
	1750 Vacuum pick-up system for floor grinders, wet/dry		1.49	81.50	245	735	60.90	
	1800 Mixer, powered, mortar and concrete, gas, 6 C.F., 18 H.P.		8.65	118	355	1,075	140.20	
	1900 10 C.F., 25 H.P.		10.85	143	430	1,300	172.80	
	2000 16 C.F.		11.15	165	495	1,475	188.20	
	2100 Concrete, stationary, tilt drum, 2 C.Y.		6.80	232	695	2,075	193.40	
	2120 Pump, concrete, truck mounted 4" line 80' boom		24.70	885	2,650	7,950	727.60	
	2140 5" line, 110' boom		32.05	1,175	3,510	10,500	958.40	
	2160 Mud jack, 50 C.F. per hr.		7.25	128	385	1,150	135	
	2180 225 C.F. per hr.		9.55	147	440	1,325	164.40	
	2190 Shotcrete pump rig, 12 C.Y./hr.		15.00	223	670	2,000	254	
	2200 35 C.Y./hr.		17.50	240	720	2,150	284	
	2600 Saw, concrete, manual, gas, 18 H.P.		6.80	45	135	405	81.40	
	2650 Self-propelled, gas, 30 H.P.		13.45	102	305	915	168.60	
	2675 V-groove crack chaser, manual, gas, 6 H.P.		2.35	17.35	52	156	29.20	
	2700 Vibrators, concrete, electric, 60 cycle, 2 H.P.		.41	8.65	26	78	8.50	
	2800 3 H.P.		.60	11.65	35	105	11.80	
	2900 Gas engine, 5 H.P.		2.05	16.35	49	147	26.20	
	3000 8 H.P.		2.80	15.35	46	138	31.60	
	3050 Vibrating screed, gas engine, 8 H.P.		2.58	71.50	215	645	63.65	
	3120 Concrete transit mixer, 6 x 4, 250 H.P., 8 C.Y., rear discharge		61.15	570	1,715	5,150	832.20	
	3200 Front discharge		72.05	700	2,100	6,300	996.40	
	3300 6 x 6, 285 H.P., 12 C.Y., rear discharge		71.10	660	1,980	5,950	964.80	
	3400 Front discharge		74.30	710	2,125	6,375	1,019	
20	**0010 EARTHWORK EQUIPMENT RENTAL** without operators R015433-10							**20**
	0040 Aggregate spreader, push type 8' to 12' wide	Ea.	3.05	25.50	76	228	39.60	
	0045 Tailgate type, 8' wide		2.90	32.50	98	294	42.80	
	0055 Earth auger, truck-mounted, for fence & sign posts, utility poles		19.10	445	1,340	4,025	420.80	
	0060 For borings and monitoring wells		46.50	690	2,070	6,200	786	
	0070 Portable, trailer mounted		3.25	33	99	297	45.80	
	0075 Truck-mounted, for caissons, water wells		98.70	2,950	8,815	26,400	2,553	
	0080 Horizontal boring machine, 12" to 36" diameter, 45 H.P.		24.40	195	585	1,750	312.20	
	0090 12" to 48" diameter, 65 H.P.		34.30	340	1,015	3,050	477.40	
	0095 Auger, for fence posts, gas engine, hand held		.60	6	18	54	8.40	
	0100 Excavator, diesel hydraulic, crawler mounted, 1/2 C.Y. cap.		24.45	415	1,250	3,750	445.60	
	0120 5/8 C.Y. capacity		32.30	550	1,655	4,975	589.40	
	0140 3/4 C.Y. capacity		35.25	620	1,860	5,575	654	
	0150 1 C.Y. capacity		49.35	695	2,090	6,275	812.80	

01 54 | Construction Aids

01 54 33 | Equipment Rental

		UNIT	HOURLY OPER. COST	RENT PER DAY	RENT PER WEEK	RENT PER MONTH	EQUIPMENT COST/DAY
0200	1-1/2 C.Y. capacity	Ea.	59.40	925	2,770	8,300	1,029
0300	2 C.Y. capacity		66.90	1,075	3,205	9,625	1,176
0320	2-1/2 C.Y. capacity		102.00	1,275	3,830	11,500	1,582
0325	3-1/2 C.Y. capacity		143.70	2,150	6,460	19,400	2,442
0330	4-1/2 C.Y. capacity		172.45	2,650	7,970	23,900	2,974
0335	6 C.Y. capacity		218.85	2,925	8,750	26,300	3,501
0340	7 C.Y. capacity		221.10	3,050	9,140	27,400	3,597
0342	Excavator attachments, bucket thumbs		3.15	245	735	2,200	172.20
0345	Grapples		2.70	193	580	1,750	137.60
0346	Hydraulic hammer for boom mounting, 4000 ft lb.		12.15	350	1,045	3,125	306.20
0347	5000 ft lb.		14.10	425	1,275	3,825	367.80
0348	8000 ft lb.		20.90	625	1,870	5,600	541.20
0349	12,000 ft lb.		22.90	750	2,245	6,725	632.20
0350	Gradall type, truck mounted, 3 ton @ 15' radius, 5/8 C.Y.		48.75	820	2,460	7,375	882
0370	1 C.Y. capacity		53.85	940	2,825	8,475	995.80
0400	Backhoe-loader, 40 to 45 H.P., 5/8 C.Y. capacity		14.85	242	725	2,175	263.80
0450	45 H.P. to 60 H.P., 3/4 C.Y. capacity		23.30	298	895	2,675	365.40
0460	80 H.P., 1-1/4 C.Y. capacity		25.90	315	945	2,825	396.20
0470	112 H.P., 1-1/2 C.Y. capacity		40.25	640	1,915	5,750	705
0482	Backhoe-loader attachment, compactor, 20,000 lb.		5.70	142	425	1,275	130.60
0485	Hydraulic hammer, 750 ft lb.		3.20	95	285	855	82.60
0486	Hydraulic hammer, 1200 ft lb.		6.20	217	650	1,950	179.60
0500	Brush chipper, gas engine, 6" cutter head, 35 H.P.		11.10	103	310	930	150.80
0550	Diesel engine, 12" cutter head, 130 H.P.		27.45	282	845	2,525	388.60
0600	15" cutter head, 165 H.P.		33.00	340	1,020	3,050	468
0750	Bucket, clamshell, general purpose, 3/8 C.Y.		1.25	38.50	115	345	33
0800	1/2 C.Y.		1.35	45	135	405	37.80
0850	3/4 C.Y.		1.50	55	165	495	45
0900	1 C.Y.		1.55	58.50	175	525	47.40
0950	1-1/2 C.Y.		2.50	80	240	720	68
1000	2 C.Y.		2.65	90	270	810	75.20
1010	Bucket, dragline, medium duty, 1/2 C.Y.		.70	23.50	70	210	19.60
1020	3/4 C.Y.		.75	24.50	73	219	20.60
1030	1 C.Y.		.75	26	78	234	21.60
1040	1-1/2 C.Y.		1.20	40	120	360	33.60
1050	2 C.Y.		1.25	45	135	405	37
1070	3 C.Y.		1.95	61.50	185	555	52.60
1200	Compactor, manually guided 2-drum vibratory smooth roller, 7.5 H.P.		7.35	210	630	1,900	184.80
1250	Rammer/tamper, gas, 8"		2.75	46.50	140	420	50
1260	15"		3.10	53.50	160	480	56.80
1300	Vibratory plate, gas, 18" plate, 3000 lb. blow		2.70	24.50	73	219	36.20
1350	21" plate, 5000 lb. blow		3.35	33	99	297	46.60
1370	Curb builder/extruder, 14 H.P., gas, single screw		14.40	248	745	2,225	264.20
1390	Double screw		15.10	292	875	2,625	295.80
1500	Disc harrow attachment, for tractor		.45	75	225	675	48.60
1810	Feller buncher, shearing & accumulating trees, 100 H.P.		47.95	755	2,270	6,800	837.60
1860	Grader, self-propelled, 25,000 lb.		39.40	665	1,990	5,975	713.20
1910	30,000 lb.		43.05	615	1,850	5,550	714.40
1920	40,000 lb.		64.75	1,125	3,385	10,200	1,195
1930	55,000 lb.		84.00	1,775	5,345	16,000	1,741
1950	Hammer, pavement breaker, self-propelled, diesel, 1000 to 1250 lb.		29.55	350	1,055	3,175	447.40
2000	1300 to 1500 lb.		44.33	705	2,110	6,325	776.65
2050	Pile driving hammer, steam or air, 4150 ft lb. @ 225 bpm		10.25	480	1,435	4,300	369
2100	8750 ft lb. @ 145 bpm		12.25	665	2,000	6,000	498
2150	15,000 ft lb. @ 60 bpm		13.80	800	2,405	7,225	591.40
2200	24,450 ft lb. @ 111 bpm		14.85	890	2,670	8,000	652.80
2250	Leads, 60' high for pile driving hammers up to 20,000 ft lb.		3.25	80	240	720	74
2300	90' high for hammers over 20,000 ft lb.		4.95	141	424	1,275	124.40

01 54 | Construction Aids

01 54 33 | Equipment Rental

		UNIT	HOURLY OPER. COST	RENT PER DAY	RENT PER WEEK	RENT PER MONTH	EQUIPMENT COST/DAY
2350	Diesel type hammer, 22,400 ft lb.	Ea.	18.00	390	1,170	3,500	378
2400	41,300 ft lb.		27.45	505	1,515	4,550	522.60
2450	141,000 ft lb.		47.40	880	2,640	7,925	907.20
2500	Vib. elec. hammer/extractor, 200 kW diesel generator, 34 H.P.		54.65	645	1,930	5,800	823.20
2550	80 H.P.		100.35	945	2,830	8,500	1,369
2600	150 H.P.		190.65	1,800	5,435	16,300	2,612
2800	Log chipper, up to 22" diameter, 600 H.P.		63.25	640	1,920	5,750	890
2850	Logger, for skidding & stacking logs, 150 H.P.		52.35	830	2,490	7,475	916.80
2860	Mulcher, diesel powered, trailer mounted		23.95	213	640	1,925	319.60
2900	Rake, spring tooth, with tractor		18.08	345	1,042	3,125	353.05
3000	Roller, vibratory, tandem, smooth drum, 20 H.P.		8.95	152	455	1,375	162.60
3050	35 H.P.		11.05	252	755	2,275	239.40
3100	Towed type vibratory compactor, smooth drum, 50 H.P.		25.70	325	980	2,950	401.60
3150	Sheepsfoot, 50 H.P.		27.05	350	1,050	3,150	426.40
3170	Landfill compactor, 220 H.P.		88.90	1,550	4,640	13,900	1,639
3200	Pneumatic tire roller, 80 H.P.		16.35	360	1,085	3,250	347.80
3250	120 H.P.		24.55	615	1,840	5,525	564.40
3300	Sheepsfoot vibratory roller, 240 H.P.		67.15	1,125	3,355	10,100	1,208
3320	340 H.P.		97.60	1,650	4,960	14,900	1,773
3350	Smooth drum vibratory roller, 75 H.P.		24.75	615	1,840	5,525	566
3400	125 H.P.		32.10	725	2,175	6,525	691.80
3410	Rotary mower, brush, 60", with tractor		22.30	320	955	2,875	369.40
3420	Rototiller, walk-behind, gas, 5 H.P.		1.76	51.50	155	465	45.10
3422	8 H.P.		2.75	80	240	720	70
3440	Scrapers, towed type, 7 C.Y. capacity		5.65	113	340	1,025	113.20
3450	10 C.Y. capacity		6.50	155	465	1,400	145
3500	15 C.Y. capacity		6.95	180	540	1,625	163.60
3525	Self-propelled, single engine, 14 C.Y. capacity		130.55	1,875	5,640	16,900	2,172
3550	Dual engine, 21 C.Y. capacity		173.20	2,225	6,645	19,900	2,715
3600	31 C.Y. capacity		230.80	3,150	9,445	28,300	3,735
3640	44 C.Y. capacity		286.85	4,075	12,220	36,700	4,739
3650	Elevating type, single engine, 11 C.Y. capacity		71.80	1,000	2,995	8,975	1,173
3700	22 C.Y. capacity		138.90	2,400	7,185	21,600	2,548
3710	Screening plant 110 H.P. w/5' x 10' screen		19.62	375	1,125	3,375	381.95
3720	5' x 16' screen		24.98	485	1,450	4,350	489.85
3850	Shovel, crawler-mounted, front-loading, 7 C.Y. capacity		257.95	3,575	10,740	32,200	4,212
3855	12 C.Y. capacity		419.50	4,950	14,860	44,600	6,328
3860	Shovel/backhoe bucket, 1/2 C.Y.		2.40	66.50	200	600	59.20
3870	3/4 C.Y.		2.45	73.50	220	660	63.60
3880	1 C.Y.		2.55	83.50	250	750	70.40
3890	1-1/2 C.Y.		2.70	96.50	290	870	79.60
3910	3 C.Y.		3.05	130	390	1,175	102.40
3950	Stump chipper, 18" deep, 30 H.P.		6.83	198	593	1,775	173.25
4110	Dozer, crawler, torque converter, diesel 80 H.P.		28.80	400	1,205	3,625	471.40
4150	105 H.P.		35.30	535	1,605	4,825	603.40
4200	140 H.P.		50.95	790	2,365	7,100	880.60
4260	200 H.P.		76.95	1,300	3,905	11,700	1,397
4310	300 H.P.		100.25	1,800	5,425	16,300	1,887
4360	410 H.P.		132.10	2,225	6,675	20,000	2,392
4370	500 H.P.		170.65	3,275	9,820	29,500	3,329
4380	700 H.P.		280.40	5,175	15,520	46,600	5,347
4400	Loader, crawler, torque conv., diesel, 1-1/2 C.Y., 80 H.P.		31.25	460	1,375	4,125	525
4450	1-1/2 to 1-3/4 C.Y., 95 H.P.		34.40	600	1,795	5,375	634.20
4510	1-3/4 to 2-1/4 C.Y., 130 H.P.		53.50	895	2,680	8,050	964
4530	2-1/2 to 3-1/4 C.Y., 190 H.P.		65.50	1,125	3,365	10,100	1,197
4560	3-1/2 to 5 C.Y., 275 H.P.		87.45	1,450	4,330	13,000	1,566
4610	Front end loader, 4WD, articulated frame, diesel, 1 to 1-1/4 C.Y., 70 H.P.		19.75	235	705	2,125	299
4620	1-1/2 to 1-3/4 C.Y., 95 H.P.		24.70	300	905	2,725	378.60

01 54 | Construction Aids

01 54 33 | Equipment Rental

			UNIT	HOURLY OPER. COST	RENT PER DAY	RENT PER WEEK	RENT PER MONTH	EQUIPMENT COST/DAY	
20	4650	1-3/4 to 2 C.Y., 130 H.P.	Ea.	29.25	365	1,095	3,275	453	20
	4710	2-1/2 to 3-1/2 C.Y., 145 H.P.		33.25	415	1,250	3,750	516	
	4730	3 to 4-1/2 C.Y., 185 H.P.		42.60	535	1,605	4,825	661.80	
	4760	5-1/4 to 5-3/4 C.Y., 270 H.P.		67.20	925	2,770	8,300	1,092	
	4810	7 to 9 C.Y., 475 H.P.		113.75	1,750	5,270	15,800	1,964	
	4870	9 - 11 C.Y., 620 H.P.		160.25	3,000	9,015	27,000	3,085	
	4880	Skid steer loader, wheeled, 10 C.F., 30 H.P. gas		10.40	150	450	1,350	173.20	
	4890	1 C.Y., 78 H.P., diesel		20.20	257	770	2,300	315.60	
	4892	Skid-steer attachment, auger		.47	77.50	233	700	50.35	
	4893	Backhoe		.66	110	329	985	71.10	
	4894	Broom		.70	117	351	1,050	75.80	
	4895	Forks		.22	37	111	335	23.95	
	4896	Grapple		.53	88	264	790	57.05	
	4897	Concrete hammer		1.00	167	502	1,500	108.40	
	4898	Tree spade		.93	156	467	1,400	100.85	
	4899	Trencher		.59	98.50	295	885	63.70	
	4900	Trencher, chain, boom type, gas, operator walking, 12 H.P.		5.00	46.50	140	420	68	
	4910	Operator riding, 40 H.P.		20.05	290	870	2,600	334.40	
	5000	Wheel type, diesel, 4' deep, 12" wide		85.15	835	2,510	7,525	1,183	
	5100	6' deep, 20" wide		91.45	1,925	5,795	17,400	1,891	
	5150	Chain type, diesel, 5' deep, 8" wide		32.25	470	1,405	4,225	539	
	5200	Diesel, 8' deep, 16" wide		133.50	2,975	8,960	26,900	2,860	
	5202	Rock trencher, wheel type, 6" wide x 18" deep		33.65	575	1,730	5,200	615.20	
	5206	Chain type, 18" wide x 7' deep		110.75	2,850	8,525	25,600	2,591	
	5210	Tree spade, self-propelled		12.30	267	800	2,400	258.40	
	5250	Truck, dump, 2-axle, 12 ton, 8 C.Y. payload, 220 H.P.		34.30	238	715	2,150	417.40	
	5300	Three axle dump, 16 ton, 12 C.Y. payload, 400 H.P.		60.85	340	1,020	3,050	690.80	
	5310	Four axle dump, 25 ton, 18 C.Y. payload, 450 H.P.		71.55	495	1,485	4,450	869.40	
	5350	Dump trailer only, rear dump, 16-1/2 C.Y.		5.35	138	415	1,250	125.80	
	5400	20 C.Y.		5.80	157	470	1,400	140.40	
	5450	Flatbed, single axle, 1-1/2 ton rating		25.90	68.50	205	615	248.20	
	5500	3 ton rating		31.05	98.50	295	885	307.40	
	5550	Off highway rear dump, 25 ton capacity		73.60	1,300	3,905	11,700	1,370	
	5600	35 ton capacity		82.20	1,450	4,355	13,100	1,529	
	5610	50 ton capacity		102.35	1,725	5,210	15,600	1,861	
	5620	65 ton capacity		105.35	1,775	5,360	16,100	1,915	
	5630	100 ton capacity		150.95	2,900	8,710	26,100	2,950	
	6000	Vibratory plow, 25 H.P., walking		8.50	61.50	185	555	105	
40	0010	**GENERAL EQUIPMENT RENTAL** without operators	R015433 -10						40
	0150	Aerial lift, scissor type, to 15' high, 1000 lb. cap., electric	Ea.	3.15	51.50	155	465	56.20	
	0160	To 25' high, 2000 lb. capacity		3.60	68.50	205	615	69.80	
	0170	Telescoping boom to 40' high, 500 lb. capacity, diesel		13.50	335	1,005	3,025	309	
	0180	To 45' high, 500 lb. capacity		14.75	365	1,090	3,275	336	
	0190	To 60' high, 600 lb. capacity		17.30	515	1,550	4,650	448.40	
	0195	Air compressor, portable, 6.5 CFM, electric		.79	13	39	117	14.10	
	0196	Gasoline		.70	19.65	59	177	17.40	
	0200	Towed type, gas engine, 60 CFM		14.10	50	150	450	142.80	
	0300	160 CFM		16.40	51.50	155	465	162.20	
	0400	Diesel engine, rotary screw, 250 CFM		16.35	112	335	1,000	197.80	
	0500	365 CFM		22.05	137	410	1,225	258.40	
	0550	450 CFM		27.95	170	510	1,525	325.60	
	0600	600 CFM		49.10	235	705	2,125	533.80	
	0700	750 CFM		49.25	245	735	2,200	541	
	0800	For silenced models, small sizes, add to rent		3%	5%	5%	5%		
	0900	Large sizes, add to rent		5%	7%	7%	7%		
	0930	Air tools, breaker, pavement, 60 lb.	Ea.	.50	10.35	31	93	10.20	
	0940	80 lb.		.50	10.35	31	93	10.20	
	0950	Drills, hand (jackhammer) 65 lb.		.60	17.65	53	159	15.40	

01 54 | Construction Aids

01 54 33 | Equipment Rental

		Unit	Hourly Oper. Cost	Rent Per Day	Rent Per Week	Rent Per Month	Equipment Cost/Day
0960	Track or wagon, swing boom, 4" drifter	Ea.	62.25	910	2,735	8,200	1,045
0970	5" drifter		73.80	1,100	3,290	9,875	1,248
0975	Track mounted quarry drill, 6" diameter drill		126.40	1,675	4,990	15,000	2,009
0980	Dust control per drill		1.04	25	75	225	23.30
0990	Hammer, chipping, 12 lb.		.55	26	78	234	20
1000	Hose, air with couplings, 50' long, 3/4" diameter		.03	4.67	14	42	3.05
1100	1" diameter		.03	5.65	17	51	3.65
1200	1-1/2" diameter		.05	8	24	72	5.20
1300	2" diameter		.06	10.65	32	96	6.90
1400	2-1/2" diameter		.11	19	57	171	12.30
1410	3" diameter		.13	21.50	64	192	13.85
1450	Drill, steel, 7/8" x 2'		.06	9.35	28	84	6.10
1460	7/8" x 6'		.05	8.65	26	78	5.60
1520	Moil points		.02	4	12	36	2.55
1525	Pneumatic nailer w/accessories		.55	36.50	109	325	26.20
1530	Sheeting driver for 60 lb. breaker		.04	6.65	20	60	4.30
1540	For 90 lb. breaker		.14	9	27	81	6.50
1550	Spade, 25 lb.		.50	7	21	63	8.20
1560	Tamper, single, 35 lb.		.50	33.50	100	300	24
1570	Triple, 140 lb.		.75	50	150	450	36
1580	Wrenches, impact, air powered, up to 3/4" bolt		.40	12.65	38	114	10.80
1590	Up to 1-1/4" bolt		.50	23.50	71	213	18.20
1600	Barricades, barrels, reflectorized, 1 to 99 barrels		.03	4.60	13.80	41.50	3
1610	100 to 200 barrels		.02	3.53	10.60	32	2.30
1620	Barrels with flashers, 1 to 99 barrels		.03	5.25	15.80	47.50	3.40
1630	100 to 200 barrels		.03	4.20	12.60	38	2.75
1640	Barrels with steady burn type C lights		.04	7	21	63	4.50
1650	Illuminated board, trailer mounted, with generator		3.50	130	390	1,175	106
1670	Portable barricade, stock, with flashers, 1 to 6 units		.03	5.25	15.80	47.50	3.40
1680	25 to 50 units		.03	4.90	14.70	44	3.20
1685	Butt fusion machine, wheeled, 1.5 HP electric, 2" - 8" diameter pipe		2.63	167	500	1,500	121.05
1690	Tracked, 20 HP diesel, 4"-12" diameter pipe		10.39	490	1,465	4,400	376.10
1695	83 HP diesel, 8" - 24" diameter pipe		27.35	975	2,930	8,800	804.80
1700	Carts, brick, hand powered, 1000 lb. capacity		.41	68.50	205	615	44.30
1800	Gas engine, 1500 lb., 7-1/2' lift		3.61	108	324	970	93.70
1822	Dehumidifier, medium, 6 lb./hr., 150 CFM		1.08	67.50	202.50	610	49.15
1824	Large, 18 lb./hr., 600 CFM		2.06	129	385.50	1,150	93.60
1830	Distributor, asphalt, trailer mounted, 2000 gal., 38 H.P. diesel		10.25	345	1,035	3,100	289
1840	3000 gal., 38 H.P. diesel		11.80	375	1,125	3,375	319.40
1850	Drill, rotary hammer, electric		1.07	26.50	80	240	24.55
1860	Carbide bit, 1-1/2" diameter, add to electric rotary hammer		.03	4.44	13.33	40	2.90
1865	Rotary, crawler, 250 H.P.		147.20	2,150	6,480	19,400	2,474
1870	Emulsion sprayer, 65 gal., 5 H.P. gas engine		2.79	101	303	910	82.90
1880	200 gal., 5 H.P. engine		8.05	168	505	1,525	165.40
1900	Floor auto-scrubbing machine, walk-behind, 28" path		5.13	335	1,000	3,000	241.05
1930	Floodlight, mercury vapor, or quartz, on tripod, 1000 watt		.44	20.50	62	186	15.90
1940	2000 watt		.83	41.50	124	370	31.45
1950	Floodlights, trailer mounted with generator, 1 - 300 watt light		3.60	71.50	215	645	71.80
1960	2 - 1000 watt lights		4.80	96.50	290	870	96.40
2000	4 - 300 watt lights		4.50	91.50	275	825	91
2005	Foam spray rig, incl. box trailer, compressor, generator, proportioner		33.75	495	1,485	4,450	567
2020	Forklift, straight mast, 12' lift, 5000 lb., 2 wheel drive, gas		26.25	207	620	1,850	334
2040	21' lift, 5000 lb., 4 wheel drive, diesel		20.45	250	750	2,250	313.60
2050	For rough terrain, 42' lift, 35' reach, 9000 lb., 110 H.P.		29.40	500	1,495	4,475	534.20
2060	For plant, 4 ton capacity, 80 H.P., 2 wheel drive, gas		16.05	95	285	855	185.40
2080	10 ton capacity, 120 H.P., 2 wheel drive, diesel		23.20	165	495	1,475	284.60
2100	Generator, electric, gas engine, 1.5 kW to 3 kW		3.70	11.65	35	105	36.60
2200	5 kW		4.80	15.35	46	138	47.60

01 54 | Construction Aids

01 54 33 | Equipment Rental

		UNIT	HOURLY OPER. COST	RENT PER DAY	RENT PER WEEK	RENT PER MONTH	EQUIPMENT COST/DAY
2300	10 kW	Ea.	9.05	38.50	115	345	95.40
2400	25 kW		10.25	86.50	260	780	134
2500	Diesel engine, 20 kW		11.80	70	210	630	136.40
2600	50 kW		22.50	105	315	945	243
2700	100 kW		41.50	135	405	1,225	413
2800	250 kW		81.70	257	770	2,300	807.60
2850	Hammer, hydraulic, for mounting on boom, to 500 ft lb.		2.50	75	225	675	65
2860	1000 ft lb.		4.25	125	375	1,125	109
2900	Heaters, space, oil or electric, 50 MBH		1.59	7.65	23	69	17.30
3000	100 MBH		2.93	10.65	32	96	29.85
3100	300 MBH		8.54	38.50	115	345	91.30
3150	500 MBH		14.00	45	135	405	139
3200	Hose, water, suction with coupling, 20' long, 2" diameter		.02	3	9	27	1.95
3210	3" diameter		.03	4.33	13	39	2.85
3220	4" diameter		.03	5	15	45	3.25
3230	6" diameter		.11	17.65	53	159	11.50
3240	8" diameter		.32	53.50	161	485	34.75
3250	Discharge hose with coupling, 50' long, 2" diameter		.01	1.33	4	12	.90
3260	3" diameter		.01	2.33	7	21	1.50
3270	4" diameter		.02	3.67	11	33	2.35
3280	6" diameter		.06	9.35	28	84	6.10
3290	8" diameter		.20	33	99	297	21.40
3295	Insulation blower		.78	6	18	54	9.85
3300	Ladders, extension type, 16' to 36' long		.14	22.50	68	204	14.70
3400	40' to 60' long		.19	31	93	279	20.10
3405	Lance for cutting concrete		2.28	64	192	575	56.65
3407	Lawn mower, rotary, 22", 5 H.P.		1.55	39.50	118	355	36
3408	48" self propelled		3.36	109	327	980	92.30
3410	Level, electronic, automatic, with tripod and leveling rod		.91	60.50	182	545	43.70
3430	Laser type, for pipe and sewer line and grade		.73	48.50	145	435	34.85
3440	Rotating beam for interior control		.78	52	156	470	37.45
3460	Builder's optical transit, with tripod and rod		.09	14.65	44	132	9.50
3500	Light towers, towable, with diesel generator, 2000 watt		4.50	91.50	275	825	91
3600	4000 watt		4.80	96.50	290	870	96.40
3700	Mixer, powered, plaster and mortar, 6 C.F., 7 H.P.		2.70	20	60	180	33.60
3800	10 C.F., 9 H.P.		2.85	31.50	95	285	41.80
3850	Nailer, pneumatic		.55	36.50	109	325	26.20
3900	Paint sprayers complete, 8 CFM		1.05	70	210	630	50.40
4000	17 CFM		1.98	132	395	1,175	94.85
4020	Pavers, bituminous, rubber tires, 8' wide, 50 H.P., diesel		31.90	510	1,535	4,600	562.20
4030	10' wide, 150 H.P.		106.55	1,875	5,635	16,900	1,979
4050	Crawler, 8' wide, 100 H.P., diesel		90.15	1,900	5,670	17,000	1,855
4060	10' wide, 150 H.P.		113.75	2,300	6,890	20,700	2,288
4070	Concrete paver, 12' to 24' wide, 250 H.P.		100.45	1,550	4,685	14,100	1,741
4080	Placer-spreader-trimmer, 24' wide, 300 H.P.		149.30	2,675	8,010	24,000	2,796
4100	Pump, centrifugal gas pump, 1-1/2" diam., 65 GPM		4.05	51.50	155	465	63.40
4200	2" diameter, 130 GPM		5.45	56.50	170	510	77.60
4300	3" diameter, 250 GPM		5.75	58.50	175	525	81
4400	6" diameter, 1500 GPM		30.35	180	540	1,625	350.80
4500	Submersible electric pump, 1-1/4" diameter, 55 GPM		.40	17	51	153	13.40
4600	1-1/2" diameter, 83 GPM		.44	19.35	58	174	15.10
4700	2" diameter, 120 GPM		1.55	24.50	73	219	27
4800	3" diameter, 300 GPM		2.79	43.50	130	390	48.30
4900	4" diameter, 560 GPM		14.32	162	485	1,450	211.55
5000	6" diameter, 1590 GPM		21.51	217	650	1,950	302.10
5100	Diaphragm pump, gas, single, 1-1/2" diameter		1.08	49.50	148	445	38.25
5200	2" diameter		4.35	61.50	185	555	71.80
5300	3" diameter		4.35	65	195	585	73.80

01 54 | Construction Aids

01 54 33 | Equipment Rental

		UNIT	HOURLY OPER. COST	RENT PER DAY	RENT PER WEEK	RENT PER MONTH	EQUIPMENT COST/DAY	
40	5400	Double, 4" diameter	Ea.	6.50	110	330	990	118	40
	5450	Pressure washer 5 GPM, 3000 psi		4.95	53.50	160	480	71.60	
	5460	7 GPM, 3000 psi		6.45	61.50	185	555	88.60	
	5500	Trash pump, self-priming, gas, 2" diameter		4.65	22	66	198	50.40	
	5600	Diesel, 4" diameter		8.95	91.50	275	825	126.60	
	5650	Diesel, 6" diameter		24.25	153	460	1,375	286	
	5655	Grout Pump		26.25	272	815	2,450	373	
	5700	Salamanders, L.P. gas fired, 100,000 Btu		3.13	13	39	117	32.85	
	5705	50,000 Btu		1.74	10.35	31	93	20.10	
	5720	Sandblaster, portable, open top, 3 C.F. capacity		.60	26.50	80	240	20.80	
	5730	6 C.F. capacity		.95	40	120	360	31.60	
	5740	Accessories for above		.13	22	66	198	14.25	
	5750	Sander, floor		.98	27.50	83	249	24.45	
	5760	Edger		.53	15.65	47	141	13.65	
	5800	Saw, chain, gas engine, 18" long		2.30	22	66	198	31.60	
	5900	Hydraulic powered, 36" long		.75	65	195	585	45	
	5950	60" long		.75	66.50	200	600	46	
	6000	Masonry, table mounted, 14" diameter, 5 H.P.		1.32	56.50	170	510	44.55	
	6050	Portable cut-off, 8 H.P.		2.50	33.50	100	300	40	
	6100	Circular, hand held, electric, 7-1/4" diameter		.23	4.67	14	42	4.65	
	6200	12" diameter		.23	8	24	72	6.65	
	6250	Wall saw, w/hydraulic power, 10 H.P.		9.75	61.50	185	555	115	
	6275	Shot blaster, walk-behind, 20" wide		4.80	295	885	2,650	215.40	
	6280	Sidewalk broom, walk-behind		2.04	65	195	585	55.30	
	6300	Steam cleaner, 100 gallons per hour		3.80	78.50	235	705	77.40	
	6310	200 gallons per hour		5.40	95	285	855	100.20	
	6340	Tar Kettle/Pot, 400 gallons		16.30	76.50	230	690	176.40	
	6350	Torch, cutting, acetylene-oxygen, 150' hose, excludes gases		.30	15	45	135	11.40	
	6360	Hourly operating cost includes tips and gas		21.00				168	
	6410	Toilet, portable chemical		.13	21.50	64	192	13.85	
	6420	Recycle flush type		.15	25.50	77	231	16.60	
	6430	Toilet, fresh water flush, garden hose,		.19	31	93	279	20.10	
	6440	Hoisted, non-flush, for high rise		.15	25	75	225	16.20	
	6465	Tractor, farm with attachment		21.35	305	910	2,725	352.80	
	6480	Trailers, platform, flush deck, 2 axle, 3 ton capacity		1.45	19.65	59	177	23.40	
	6500	25 ton capacity		5.40	117	350	1,050	113.20	
	6600	40 ton capacity		6.95	165	495	1,475	154.60	
	6700	3 axle, 50 ton capacity		7.50	183	550	1,650	170	
	6800	75 ton capacity		9.30	240	720	2,150	218.40	
	6810	Trailer mounted cable reel for high voltage line work		5.57	265	795	2,375	203.55	
	6820	Trailer mounted cable tensioning rig		11.06	525	1,580	4,750	404.50	
	6830	Cable pulling rig		71.33	2,950	8,850	26,600	2,341	
	6850	Portable cable/wire puller, 8000 lb max pulling capacity		3.71	167	502	1,500	130.10	
	6900	Water tank trailer, engine driven discharge, 5000 gallons		6.95	143	430	1,300	141.60	
	6925	10,000 gallons		9.45	198	595	1,775	194.60	
	6950	Water truck, off highway, 6000 gallons		87.60	780	2,335	7,000	1,168	
	7010	Tram car for high voltage line work, powered, 2 conductor		6.66	144	431	1,300	139.50	
	7020	Transit (builder's level) with tripod		.09	14.65	44	132	9.50	
	7030	Trench box, 3000 lb., 6' x 8'		.56	93.50	280	840	60.50	
	7040	7200 lb., 6' x 20'		.75	125	375	1,125	81	
	7050	8000 lb., 8' x 16'		1.08	180	540	1,625	116.65	
	7060	9500 lb., 8' x 20'		1.21	201	603	1,800	130.30	
	7065	11,000 lb., 8' x 24'		1.27	211	633	1,900	136.75	
	7070	12,000 lb., 10' x 20'		1.50	249	748	2,250	161.60	
	7100	Truck, pickup, 3/4 ton, 2 wheel drive		13.85	58.50	175	525	145.80	
	7200	4 wheel drive		14.10	73.50	220	660	156.80	
	7250	Crew carrier, 9 passenger		19.60	86.50	260	780	208.80	
	7290	Flat bed truck, 20,000 lb. GVW		21.60	125	375	1,125	247.80	

01 54 | Construction Aids

01 54 33 | Equipment Rental

			UNIT	HOURLY OPER. COST	RENT PER DAY	RENT PER WEEK	RENT PER MONTH	EQUIPMENT COST/DAY	
40	7300	Tractor, 4 x 2, 220 H.P.	Ea.	30.10	197	590	1,775	358.80	40
	7410	330 H.P.		44.65	272	815	2,450	520.20	
	7500	6 x 4, 380 H.P.		51.25	315	950	2,850	600	
	7600	450 H.P.		62.10	385	1,150	3,450	726.80	
	7610	Tractor, with A frame, boom and winch, 225 H.P.		33.00	273	820	2,450	428	
	7620	Vacuum truck, hazardous material, 2500 gallons		11.85	292	875	2,625	269.80	
	7625	5,000 gallons		13.20	410	1,230	3,700	351.60	
	7650	Vacuum, HEPA, 16 gallon, wet/dry		.85	18	54	162	17.60	
	7655	55 gallon, wet/dry		.80	27	81	243	22.60	
	7660	Water tank, portable		.16	26.50	80	240	17.30	
	7690	Sewer/catch basin vacuum, 14 C.Y., 1500 gallons		17.54	615	1,850	5,550	510.30	
	7700	Welder, electric, 200 amp		3.88	16.35	49	147	40.85	
	7800	300 amp		5.76	20	60	180	58.10	
	7900	Gas engine, 200 amp		14.45	24.50	73	219	130.20	
	8000	300 amp		16.55	24.50	74	222	147.20	
	8100	Wheelbarrow, any size		.08	13	39	117	8.45	
	8200	Wrecking ball, 4000 lb.		2.30	68.50	205	615	59.40	
50	0010	**HIGHWAY EQUIPMENT RENTAL** without operators							50
	0050	Asphalt batch plant, portable drum mixer, 100 ton/hr.	Ea.	80.81	1,450	4,380	13,100	1,522	
	0060	200 ton/hr.		92.38	1,550	4,655	14,000	1,670	
	0070	300 ton/hr.		109.99	1,825	5,465	16,400	1,973	
	0100	Backhoe attachment, long stick, up to 185 H.P., 10.5' long		.36	23.50	71	213	17.10	
	0140	Up to 250 H.P., 12' long		.39	25.50	77	231	18.50	
	0180	Over 250 H.P., 15' long		.54	35.50	107	320	25.70	
	0200	Special dipper arm, up to 100 H.P., 32' long		1.10	73	219	655	52.60	
	0240	Over 100 H.P., 33' long		1.37	91	273	820	65.55	
	0280	Catch basin/sewer cleaning truck, 3 ton, 9 C.Y., 1000 gal.		42.90	390	1,170	3,500	577.20	
	0300	Concrete batch plant, portable, electric, 200 C.Y./hr.		22.52	515	1,545	4,625	489.15	
	0520	Grader/dozer attachment, ripper/scarifier, rear mounted, up to 135 H.P.		2.90	58.50	175	525	58.20	
	0540	Up to 180 H.P.		3.85	88.50	265	795	83.80	
	0580	Up to 250 H.P.		5.50	142	425	1,275	129	
	0700	Pvmt. removal bucket, for hyd. excavator, up to 90 H.P.		1.85	51.50	155	465	45.80	
	0740	Up to 200 H.P.		2.05	68.50	205	615	57.40	
	0780	Over 200 H.P.		2.20	85	255	765	68.60	
	0900	Aggregate spreader, self-propelled, 187 H.P.		52.46	685	2,050	6,150	829.70	
	1000	Chemical spreader, 3 C.Y.		3.25	43.50	130	390	52	
	1900	Hammermill, traveling, 250 H.P.		70.05	2,125	6,360	19,100	1,832	
	2000	Horizontal borer, 3" diameter, 13 H.P. gas driven		6.40	55	165	495	84.20	
	2150	Horizontal directional drill, 20,000 lb. thrust, 78 H.P. diesel		31.00	680	2,045	6,125	657	
	2160	30,000 lb. thrust, 115 H.P.		38.60	1,050	3,135	9,400	935.80	
	2170	50,000 lb. thrust, 170 H.P.		55.35	1,325	4,005	12,000	1,244	
	2190	Mud trailer for HDD, 1500 gallons, 175 H.P., gas		32.25	153	460	1,375	350	
	2200	Hydromulcher, diesel, 3000 gallon, for truck mounting		23.30	247	740	2,225	334.40	
	2300	Gas, 600 gallon		8.60	98.50	295	885	127.80	
	2400	Joint & crack cleaner, walk behind, 25 H.P.		4.00	50	150	450	62	
	2500	Filler, trailer mounted, 400 gallons, 20 H.P.		9.45	212	635	1,900	202.60	
	3000	Paint striper, self-propelled, 40 gallon, 22 H.P.		7.25	157	470	1,400	152	
	3100	120 gallon, 120 H.P.		23.15	400	1,195	3,575	424.20	
	3200	Post drivers, 6" I-Beam frame, for truck mounting		19.00	385	1,155	3,475	383	
	3400	Road sweeper, self-propelled, 8' wide, 90 H.P.		39.40	585	1,755	5,275	666.20	
	3450	Road sweeper, vacuum assisted, 4 C.Y., 220 gallons		75.35	635	1,905	5,725	983.80	
	4000	Road mixer, self-propelled, 130 H.P.		47.70	785	2,350	7,050	851.60	
	4100	310 H.P.		83.25	2,125	6,395	19,200	1,945	
	4220	Cold mix paver, incl. pug mill and bitumen tank, 165 H.P.		98.45	2,400	7,190	21,600	2,226	
	4240	Pavement brush, towed		3.15	93.50	280	840	81.20	
	4250	Paver, asphalt, wheel or crawler, 130 H.P., diesel		98.15	2,375	7,125	21,400	2,210	
	4300	Paver, road widener, gas 1' to 6', 67 H.P.		49.10	915	2,750	8,250	942.80	

For customer support on your Residential Cost Data, call 877.759.4771.

01 54 | Construction Aids

01 54 33 | Equipment Rental

		UNIT	HOURLY OPER. COST	RENT PER DAY	RENT PER WEEK	RENT PER MONTH	EQUIPMENT COST/DAY	
50	4400 Diesel, 2' to 14', 88 H.P.	Ea.	62.75	1,075	3,205	9,625	1,143	**50**
	4600 Slipform pavers, curb and gutter, 2 track, 75 H.P.		61.85	1,000	3,035	9,100	1,102	
	4700 4 track, 165 H.P.		41.15	750	2,255	6,775	780.20	
	4800 Median barrier, 215 H.P.		66.45	1,225	3,670	11,000	1,266	
	4901 Trailer, low bed, 75 ton capacity		10.05	238	715	2,150	223.40	
	5000 Road planer, walk behind, 10" cutting width, 10 H.P.		3.70	33.50	100	300	49.60	
	5100 Self-propelled, 12" cutting width, 64 H.P.		10.50	113	340	1,025	152	
	5120 Traffic line remover, metal ball blaster, truck mounted, 115 H.P.		48.45	765	2,295	6,875	846.60	
	5140 Grinder, truck mounted, 115 H.P.		54.85	830	2,490	7,475	936.80	
	5160 Walk-behind, 11 H.P.		4.20	53.50	160	480	65.60	
	5200 Pavement profiler, 4' to 6' wide, 450 H.P.		252.75	3,400	10,165	30,500	4,055	
	5300 8' to 10' wide, 750 H.P.		395.65	4,450	13,350	40,100	5,835	
	5400 Roadway plate, steel, 1" x 8' x 20'		.08	13.65	41	123	8.85	
	5600 Stabilizer, self-propelled, 150 H.P.		48.85	625	1,870	5,600	764.80	
	5700 310 H.P.		96.75	1,775	5,295	15,900	1,833	
	5800 Striper, truck mounted, 120 gallon paint, 460 H.P.		64.90	485	1,455	4,375	810.20	
	5900 Thermal paint heating kettle, 115 gallons		8.29	25.50	77	231	81.70	
	6000 Tar kettle, 330 gallon, trailer mounted		12.54	58.50	175	525	135.30	
	7000 Tunnel locomotive, diesel, 8 to 12 ton		32.15	595	1,780	5,350	613.20	
	7005 Electric, 10 ton		26.90	680	2,035	6,100	622.20	
	7010 Muck cars, 1/2 C.Y. capacity		2.10	24.50	74	222	31.60	
	7020 1 C.Y. capacity		2.30	33	99	297	38.20	
	7030 2 C.Y. capacity		2.45	38.50	115	345	42.60	
	7040 Side dump, 2 C.Y. capacity		2.65	45	135	405	48.20	
	7050 3 C.Y. capacity		3.60	51.50	155	465	59.80	
	7060 5 C.Y. capacity		5.15	65	195	585	80.20	
	7100 Ventilating blower for tunnel, 7-1/2 H.P.		2.10	51.50	155	465	47.80	
	7110 10 H.P.		2.28	53.50	160	480	50.25	
	7120 20 H.P.		3.54	69.50	208	625	69.90	
	7140 40 H.P.		5.81	98.50	295	885	105.50	
	7160 60 H.P.		8.92	153	460	1,375	163.35	
	7175 75 H.P.		11.71	210	630	1,900	219.70	
	7180 200 H.P.		23.75	305	920	2,750	374	
	7800 Windrow loader, elevating		55.85	1,325	4,000	12,000	1,247	
60	0010 **LIFTING AND HOISTING EQUIPMENT RENTAL** without operators R015433-10							**60**
	0120 Aerial lift truck, 2 person, to 80'	Ea.	26.40	745	2,240	6,725	659.20	
	0140 Boom work platform, 40' snorkel		11.75	285	855	2,575	265	
	0150 Crane, flatbed mounted, 3 ton capacity R015433-15		16.15	198	595	1,775	248.20	
	0200 Crane, climbing, 106' jib, 6000 lb. capacity, 410 fpm		39.85	1,675	5,060	15,200	1,331	
	0300 101' jib, 10,250 lb. capacity, 270 fpm R312316-45		46.60	2,125	6,410	19,200	1,655	
	0500 Tower, static, 130' high, 106' jib, 6200 lb. capacity at 400 fpm		43.75	1,950	5,840	17,500	1,518	
	0600 Crawler mounted, lattice boom, 1/2 C.Y., 15 tons at 12' radius		36.96	665	2,000	6,000	695.70	
	0700 3/4 C.Y., 20 tons at 12' radius		49.28	835	2,510	7,525	896.25	
	0800 1 C.Y., 25 tons at 12' radius		65.70	1,125	3,340	10,000	1,194	
	0900 1-1/2 C.Y., 40 tons at 12' radius		65.75	1,125	3,385	10,200	1,203	
	1000 2 C.Y., 50 tons at 12' radius		69.75	1,325	3,940	11,800	1,346	
	1100 3 C.Y., 75 tons at 12' radius		74.30	1,525	4,565	13,700	1,507	
	1200 100 ton capacity, 60' boom		84.05	1,650	4,975	14,900	1,667	
	1300 165 ton capacity, 60' boom		106.80	2,050	6,170	18,500	2,088	
	1400 200 ton capacity, 70' boom		129.75	2,575	7,740	23,200	2,586	
	1500 350 ton capacity, 80' boom		181.25	3,850	11,575	34,700	3,765	
	1600 Truck mounted, lattice boom, 6 x 4, 20 tons at 10' radius		35.42	1,150	3,420	10,300	967.35	
	1700 25 tons at 10' radius		38.28	1,250	3,720	11,200	1,050	
	1800 8 x 4, 30 tons at 10' radius		41.49	1,325	3,960	11,900	1,124	
	1900 40 tons at 12' radius		44.28	1,375	4,140	12,400	1,182	
	2000 60 tons at 15' radius		49.79	1,450	4,380	13,100	1,274	
	2050 82 tons at 15' radius		55.71	1,550	4,680	14,000	1,382	
	2100 90 tons at 15' radius		62.48	1,700	5,100	15,300	1,520	

01 54 | Construction Aids

01 54 33 | Equipment Rental

		UNIT	HOURLY OPER. COST	RENT PER DAY	RENT PER WEEK	RENT PER MONTH	EQUIPMENT COST/DAY
60	2200	115 tons at 15' radius	Ea.	70.50	1,900	5,700	17,100	1,704
	2300	150 tons at 18' radius		83.50	2,000	6,000	18,000	1,868
	2350	165 tons at 18' radius		82.77	2,125	6,360	19,100	1,934
	2400	Truck mounted, hydraulic, 12 ton capacity		41.55	545	1,630	4,900	658.40
	2500	25 ton capacity		44.10	660	1,980	5,950	748.80
	2550	33 ton capacity		44.65	675	2,030	6,100	763.20
	2560	40 ton capacity		57.95	795	2,390	7,175	941.60
	2600	55 ton capacity		74.95	870	2,615	7,850	1,123
	2700	80 ton capacity		98.50	1,400	4,210	12,600	1,630
	2720	100 ton capacity		92.65	1,475	4,425	13,300	1,626
	2740	120 ton capacity		107.65	1,625	4,845	14,500	1,830
	2760	150 ton capacity		126.15	2,100	6,320	19,000	2,273
	2800	Self-propelled, 4 x 4, with telescoping boom, 5 ton		17.35	237	710	2,125	280.80
	2900	12-1/2 ton capacity		31.65	375	1,130	3,400	479.20
	3000	15 ton capacity		32.35	395	1,190	3,575	496.80
	3050	20 ton capacity		35.10	455	1,370	4,100	554.80
	3100	25 ton capacity		36.70	520	1,565	4,700	606.60
	3150	40 ton capacity		45.10	585	1,755	5,275	711.80
	3200	Derricks, guy, 20 ton capacity, 60' boom, 75' mast		23.97	415	1,238	3,725	439.35
	3300	100' boom, 115' mast		37.86	710	2,130	6,400	728.90
	3400	Stiffleg, 20 ton capacity, 70' boom, 37' mast		26.57	535	1,610	4,825	534.55
	3500	100' boom, 47' mast		41.01	860	2,580	7,750	844.10
	3550	Helicopter, small, lift to 1250 lb. maximum, w/pilot		96.80	3,325	10,000	30,000	2,774
	3600	Hoists, chain type, overhead, manual, 3/4 ton		.15	.33	1	3	1.40
	3900	10 ton		.75	6	18	54	9.60
	4000	Hoist and tower, 5000 lb. cap., portable electric, 40' high		5.02	238	713	2,150	182.75
	4100	For each added 10' section, add		.11	18.65	56	168	12.10
	4200	Hoist and single tubular tower, 5000 lb. electric, 100' high		6.80	330	996	3,000	253.60
	4300	For each added 6'-6" section, add		.19	32.50	97	291	20.90
	4400	Hoist and double tubular tower, 5000 lb., 100' high		7.30	365	1,097	3,300	277.80
	4500	For each added 6'-6" section, add		.21	35.50	107	320	23.10
	4550	Hoist and tower, mast type, 6000 lb., 100' high		7.87	380	1,137	3,400	290.35
	4570	For each added 10' section, add		.13	22	66	198	14.25
	4600	Hoist and tower, personnel, electric, 2000 lb., 100' @ 125 fpm		16.61	1,000	3,030	9,100	738.90
	4700	3000 lb., 100' @ 200 fpm		18.97	1,150	3,430	10,300	837.75
	4800	3000 lb., 150' @ 300 fpm		21.02	1,275	3,840	11,500	936.15
	4900	4000 lb., 100' @ 300 fpm		21.78	1,300	3,920	11,800	958.25
	5000	6000 lb., 100' @ 275 fpm		23.46	1,375	4,110	12,300	1,010
	5100	For added heights up to 500', add	L.F.	.01	1.67	5	15	1.10
	5200	Jacks, hydraulic, 20 ton	Ea.	.05	2	6	18	1.60
	5500	100 ton		.40	11.65	35	105	10.20
	6100	Jacks, hydraulic, climbing w/50' jackrods, control console, 30 ton cap.		2.05	137	410	1,225	98.40
	6150	For each added 10' jackrod section, add		.05	3.33	10	30	2.40
	6300	50 ton capacity		3.30	220	659	1,975	158.20
	6350	For each added 10' jackrod section, add		.06	4	12	36	2.90
	6500	125 ton capacity		8.60	575	1,720	5,150	412.80
	6550	For each added 10' jackrod section, add		.59	39	117	350	28.10
	6600	Cable jack, 10 ton capacity with 200' cable		1.72	114	343	1,025	82.35
	6650	For each added 50' of cable, add		.21	13.65	41	123	9.90
70	0010	**WELLPOINT EQUIPMENT RENTAL** without operators	R015433 -10					
	0020	Based on 2 months rental						
	0100	Combination jetting & wellpoint pump, 60 H.P. diesel	Ea.	16.29	340	1,016	3,050	333.50
	0200	High pressure gas jet pump, 200 H.P., 300 psi	"	36.68	289	868	2,600	467.05
	0300	Discharge pipe, 8" diameter	L.F.	.01	.55	1.65	4.95	.40
	0350	12" diameter		.01	.81	2.43	7.30	.55
	0400	Header pipe, flows up to 150 GPM, 4" diameter		.01	.50	1.50	4.50	.40
	0500	400 GPM, 6" diameter		.01	.58	1.75	5.25	.45

01 54 | Construction Aids

01 54 33 | Equipment Rental

			UNIT	HOURLY OPER. COST	RENT PER DAY	RENT PER WEEK	RENT PER MONTH	EQUIPMENT COST/DAY	
70	0600	800 GPM, 8" diameter	L.F.	.01	.81	2.43	7.30	.55	70
	0700	1500 GPM, 10" diameter		.01	.85	2.56	7.70	.60	
	0800	2500 GPM, 12" diameter		.02	1.61	4.83	14.50	1.15	
	0900	4500 GPM, 16" diameter		.03	2.06	6.19	18.55	1.50	
	0950	For quick coupling aluminum and plastic pipe, add		.03	2.14	6.41	19.25	1.50	
	1100	Wellpoint, 25' long, with fittings & riser pipe, 1-1/2" or 2" diameter	Ea.	.06	4.26	12.79	38.50	3.05	
	1200	Wellpoint pump, diesel powered, 4" suction, 20 H.P.		7.16	195	585	1,750	174.30	
	1300	6" suction, 30 H.P.		9.67	242	726	2,175	222.55	
	1400	8" suction, 40 H.P.		13.09	330	996	3,000	303.90	
	1500	10" suction, 75 H.P.		19.62	390	1,164	3,500	389.75	
	1600	12" suction, 100 H.P.		28.25	615	1,850	5,550	596	
	1700	12" suction, 175 H.P.		41.13	685	2,050	6,150	739.05	
80	0010	**MARINE EQUIPMENT RENTAL** without operators R015433-10							80
	0200	Barge, 400 Ton, 30' wide x 90' long	Ea.	17.40	1,075	3,200	9,600	779.20	
	0240	800 Ton, 45' wide x 90' long		21.05	1,300	3,880	11,600	944.40	
	2000	Tugboat, diesel, 100 H.P.		37.80	217	650	1,950	432.40	
	2040	250 H.P.		78.95	395	1,185	3,550	868.60	
	2080	380 H.P.		155.50	1,175	3,535	10,600	1,951	
	3000	Small work boat, gas, 16-foot, 50 H.P.		17.35	60	180	540	174.80	
	4000	Large, diesel, 48-foot, 200 H.P.		88.50	1,250	3,735	11,200	1,455	

Crews - Residential

Crew No.	Bare Costs		Incl. Subs O&P		Cost Per Labor-Hour	
Crew A-1	Hr.	Daily	Hr.	Daily	Bare Costs	Incl. O&P
1 Building Laborer	$24.65	$197.20	$41.25	$330.00	$24.65	$41.25
1 Concrete Saw, Gas Manual		81.40		89.54	10.18	11.19
8 L.H., Daily Totals		$278.60		$419.54	$34.83	$52.44
Crew A-1A	Hr.	Daily	Hr.	Daily	Bare Costs	Incl. O&P
1 Skilled Worker	$34.20	$273.60	$57.45	$459.60	$34.20	$57.45
1 Shot Blaster, 20"		215.40		236.94	26.93	29.62
8 L.H., Daily Totals		$489.00		$696.54	$61.13	$87.07
Crew A-1B	Hr.	Daily	Hr.	Daily	Bare Costs	Incl. O&P
1 Building Laborer	$24.65	$197.20	$41.25	$330.00	$24.65	$41.25
1 Concrete Saw		168.60		185.46	21.07	23.18
8 L.H., Daily Totals		$365.80		$515.46	$45.73	$64.43
Crew A-1C	Hr.	Daily	Hr.	Daily	Bare Costs	Incl. O&P
1 Building Laborer	$24.65	$197.20	$41.25	$330.00	$24.65	$41.25
1 Chain Saw, Gas, 18"		31.60		34.76	3.95	4.34
8 L.H., Daily Totals		$228.80		$364.76	$28.60	$45.59
Crew A-1D	Hr.	Daily	Hr.	Daily	Bare Costs	Incl. O&P
1 Building Laborer	$24.65	$197.20	$41.25	$330.00	$24.65	$41.25
1 Vibrating Plate, Gas, 18"		36.20		39.82	4.53	4.98
8 L.H., Daily Totals		$233.40		$369.82	$29.18	$46.23
Crew A-1E	Hr.	Daily	Hr.	Daily	Bare Costs	Incl. O&P
1 Building Laborer	$24.65	$197.20	$41.25	$330.00	$24.65	$41.25
1 Vibrating Plate, Gas, 21"		46.60		51.26	5.83	6.41
8 L.H., Daily Totals		$243.80		$381.26	$30.48	$47.66
Crew A-1F	Hr.	Daily	Hr.	Daily	Bare Costs	Incl. O&P
1 Building Laborer	$24.65	$197.20	$41.25	$330.00	$24.65	$41.25
1 Rammer/Tamper, Gas, 8"		50.00		55.00	6.25	6.88
8 L.H., Daily Totals		$247.20		$385.00	$30.90	$48.13
Crew A-1G	Hr.	Daily	Hr.	Daily	Bare Costs	Incl. O&P
1 Building Laborer	$24.65	$197.20	$41.25	$330.00	$24.65	$41.25
1 Rammer/Tamper, Gas, 15"		56.80		62.48	7.10	7.81
8 L.H., Daily Totals		$254.00		$392.48	$31.75	$49.06
Crew A-1H	Hr.	Daily	Hr.	Daily	Bare Costs	Incl. O&P
1 Building Laborer	$24.65	$197.20	$41.25	$330.00	$24.65	$41.25
1 Exterior Steam Cleaner		77.40		85.14	9.68	10.64
8 L.H., Daily Totals		$274.60		$415.14	$34.33	$51.89
Crew A-1J	Hr.	Daily	Hr.	Daily	Bare Costs	Incl. O&P
1 Building Laborer	$24.65	$197.20	$41.25	$330.00	$24.65	$41.25
1 Cultivator, Walk-Behind, 5 H.P.		45.10		49.61	5.64	6.20
8 L.H., Daily Totals		$242.30		$379.61	$30.29	$47.45
Crew A-1K	Hr.	Daily	Hr.	Daily	Bare Costs	Incl. O&P
1 Building Laborer	$24.65	$197.20	$41.25	$330.00	$24.65	$41.25
1 Cultivator, Walk-Behind, 8 H.P.		70.00		77.00	8.75	9.63
8 L.H., Daily Totals		$267.20		$407.00	$33.40	$50.88
Crew A-1M	Hr.	Daily	Hr.	Daily	Bare Costs	Incl. O&P
1 Building Laborer	$24.65	$197.20	$41.25	$330.00	$24.65	$41.25
1 Snow Blower, Walk-Behind		55.30		60.83	6.91	7.60
8 L.H., Daily Totals		$252.50		$390.83	$31.56	$48.85

Crew No.	Bare Costs		Incl. Subs O&P		Cost Per Labor-Hour	
Crew A-2	Hr.	Daily	Hr.	Daily	Bare Costs	Incl. O&P
2 Laborers	$24.65	$394.40	$41.25	$660.00	$26.37	$43.88
1 Truck Driver (light)	29.80	238.40	49.15	393.20		
1 Flatbed Truck, Gas, 1.5 Ton		248.20		273.02	10.34	11.38
24 L.H., Daily Totals		$881.00		$1326.22	$36.71	$55.26
Crew A-2A	Hr.	Daily	Hr.	Daily	Bare Costs	Incl. O&P
2 Laborers	$24.65	$394.40	$41.25	$660.00	$26.37	$43.88
1 Truck Driver (light)	29.80	238.40	49.15	393.20		
1 Flatbed Truck, Gas, 1.5 Ton		248.20		273.02		
1 Concrete Saw		168.60		185.46	17.37	19.10
24 L.H., Daily Totals		$1049.60		$1511.68	$43.73	$62.99
Crew A-2B	Hr.	Daily	Hr.	Daily	Bare Costs	Incl. O&P
1 Truck Driver (light)	$29.80	$238.40	$49.15	$393.20	$29.80	$49.15
1 Flatbed Truck, Gas, 1.5 Ton		248.20		273.02	31.02	34.13
8 L.H., Daily Totals		$486.60		$666.22	$60.83	$83.28
Crew A-3A	Hr.	Daily	Hr.	Daily	Bare Costs	Incl. O&P
1 Equip. Oper. (light)	$33.90	$271.20	$56.05	$448.40	$33.90	$56.05
1 Pickup Truck, 4x4, 3/4 Ton		156.80		172.48	19.60	21.56
8 L.H., Daily Totals		$428.00		$620.88	$53.50	$77.61
Crew A-3B	Hr.	Daily	Hr.	Daily	Bare Costs	Incl. O&P
1 Equip. Oper. (medium)	$35.25	$282.00	$58.25	$466.00	$32.95	$54.40
1 Truck Driver (heavy)	30.65	245.20	50.55	404.40		
1 Dump Truck, 12 C.Y., 400 H.P.		690.80		759.88		
1 F.E. Loader, W.M.,2.5 C.Y.		516.00		567.60	75.42	82.97
16 L.H., Daily Totals		$1734.00		$2197.88	$108.38	$137.37
Crew A-3C	Hr.	Daily	Hr.	Daily	Bare Costs	Incl. O&P
1 Equip. Oper. (light)	$33.90	$271.20	$56.05	$448.40	$33.90	$56.05
1 Loader, Skid Steer, 78 H.P.		315.60		347.16	39.45	43.40
8 L.H., Daily Totals		$586.80		$795.56	$73.35	$99.44
Crew A-3D	Hr.	Daily	Hr.	Daily	Bare Costs	Incl. O&P
1 Truck Driver (light)	$29.80	$238.40	$49.15	$393.20	$29.80	$49.15
1 Pickup Truck, 4x4, 3/4 Ton		156.80		172.48		
1 Flatbed Trailer, 25 Ton		113.20		124.52	33.75	37.13
8 L.H., Daily Totals		$508.40		$690.20	$63.55	$86.28
Crew A-3E	Hr.	Daily	Hr.	Daily	Bare Costs	Incl. O&P
1 Equip. Oper. (crane)	$36.05	$288.40	$59.60	$476.80	$33.35	$55.08
1 Truck Driver (heavy)	30.65	245.20	50.55	404.40		
1 Pickup Truck, 4x4, 3/4 Ton		156.80		172.48	9.80	10.78
16 L.H., Daily Totals		$690.40		$1053.68	$43.15	$65.86
Crew A-3F	Hr.	Daily	Hr.	Daily	Bare Costs	Incl. O&P
1 Equip. Oper. (crane)	$36.05	$288.40	$59.60	$476.80	$33.35	$55.08
1 Truck Driver (heavy)	30.65	245.20	50.55	404.40		
1 Pickup Truck, 4x4, 3/4 Ton		156.80		172.48		
1 Truck Tractor, 6x4, 380 H.P.		600.00		660.00		
1 Lowbed Trailer, 75 Ton		223.40		245.74	61.26	67.39
16 L.H., Daily Totals		$1513.80		$1959.42	$94.61	$122.46

Crews - Residential

Crew No.	Bare Costs		Incl. Subs O&P		Cost Per Labor-Hour	
Crew A-3G	Hr.	Daily	Hr.	Daily	Bare Costs	Incl. O&P
1 Equip. Oper. (crane)	$36.05	$288.40	$59.60	$476.80	$33.35	$55.08
1 Truck Driver (heavy)	30.65	245.20	50.55	404.40		
1 Pickup Truck, 4x4, 3/4 Ton		156.80		172.48		
1 Truck Tractor, 6x4, 450 H.P.		726.80		799.48		
1 Lowbed Trailer, 75 Ton		223.40		245.74	69.19	76.11
16 L.H., Daily Totals		$1640.60		$2098.90	$102.54	$131.18

Crew A-3H	Hr.	Daily	Hr.	Daily	Bare Costs	Incl. O&P
1 Equip. Oper. (crane)	$36.05	$288.40	$59.60	$476.80	$36.05	$59.60
1 Hyd. Crane, 12 Ton (Daily)		877.40		965.14	109.68	120.64
8 L.H., Daily Totals		$1165.80		$1441.94	$145.72	$180.24

Crew A-3I	Hr.	Daily	Hr.	Daily	Bare Costs	Incl. O&P
1 Equip. Oper. (crane)	$36.05	$288.40	$59.60	$476.80	$36.05	$59.60
1 Hyd. Crane, 25 Ton (Daily)		1013.00		1114.30	126.63	139.29
8 L.H., Daily Totals		$1301.40		$1591.10	$162.68	$198.89

Crew A-3J	Hr.	Daily	Hr.	Daily	Bare Costs	Incl. O&P
1 Equip. Oper. (crane)	$36.05	$288.40	$59.60	$476.80	$36.05	$59.60
1 Hyd. Crane, 40 Ton (Daily)		1259.00		1384.90	157.38	173.11
8 L.H., Daily Totals		$1547.40		$1861.70	$193.43	$232.71

Crew A-3K	Hr.	Daily	Hr.	Daily	Bare Costs	Incl. O&P
1 Equip. Oper. (crane)	$36.05	$288.40	$59.60	$476.80	$33.90	$56.05
1 Equip. Oper. (oiler)	31.75	254.00	52.50	420.00		
1 Hyd. Crane, 55 Ton (Daily)		1470.00		1617.00		
1 P/U Truck, 3/4 Ton (Daily)		170.80		187.88	102.55	112.81
16 L.H., Daily Totals		$2183.20		$2701.68	$136.45	$168.85

Crew A-3L	Hr.	Daily	Hr.	Daily	Bare Costs	Incl. O&P
1 Equip. Oper. (crane)	$36.05	$288.40	$59.60	$476.80	$33.90	$56.05
1 Equip. Oper. (oiler)	31.75	254.00	52.50	420.00		
1 Hyd. Crane, 80 Ton (Daily)		2193.00		2412.30		
1 P/U Truck, 3/4 Ton (Daily)		170.80		187.88	147.74	162.51
16 L.H., Daily Totals		$2906.20		$3496.98	$181.64	$218.56

Crew A-3M	Hr.	Daily	Hr.	Daily	Bare Costs	Incl. O&P
1 Equip. Oper. (crane)	$36.05	$288.40	$59.60	$476.80	$33.90	$56.05
1 Equip. Oper. (oiler)	31.75	254.00	52.50	420.00		
1 Hyd. Crane, 100 Ton (Daily)		2216.00		2437.60		
1 P/U Truck, 3/4 Ton (Daily)		170.80		187.88	149.18	164.09
16 L.H., Daily Totals		$2929.20		$3522.28	$183.07	$220.14

Crew A-3N	Hr.	Daily	Hr.	Daily	Bare Costs	Incl. O&P
1 Equip. Oper. (crane)	$36.05	$288.40	$59.60	$476.80	$36.05	$59.60
1 Tower Cane (monthly)		1145.00		1259.50	143.13	157.44
8 L.H., Daily Totals		$1433.40		$1736.30	$179.18	$217.04

Crew A-3P	Hr.	Daily	Hr.	Daily	Bare Costs	Incl. O&P
1 Equip. Oper. (light)	$33.90	$271.20	$56.05	$448.40	$33.90	$56.05
1 A.T. Forklift, 42' lift		534.20		587.62	66.78	73.45
8 L.H., Daily Totals		$805.40		$1036.02	$100.68	$129.50

Crew A-3Q	Hr.	Daily	Hr.	Daily	Bare Costs	Incl. O&P
1 Equip. Oper. (light)	$33.90	$271.20	$56.05	$448.40	$33.90	$56.05
1 Pickup Truck, 4x4, 3/4 Ton		156.80		172.48		
1 Flatbed trailer, 3 Ton		23.40		25.74	22.52	24.78
8 L.H., Daily Totals		$451.40		$646.62	$56.42	$80.83

Crew A-4	Hr.	Daily	Hr.	Daily	Bare Costs	Incl. O&P
2 Carpenters	$33.90	$542.40	$56.75	$908.00	$32.15	$53.57
1 Painter, Ordinary	28.65	229.20	47.20	377.60		
24 L.H., Daily Totals		$771.60		$1285.60	$32.15	$53.57

Crew A-5	Hr.	Daily	Hr.	Daily	Bare Costs	Incl. O&P
2 Laborers	$24.65	$394.40	$41.25	$660.00	$25.22	$42.13
.25 Truck Driver (light)	29.80	59.60	49.15	98.30		
.25 Flatbed Truck, Gas, 1.5 Ton		62.05		68.25	3.45	3.79
18 L.H., Daily Totals		$516.05		$826.55	$28.67	$45.92

Crew A-6	Hr.	Daily	Hr.	Daily	Bare Costs	Incl. O&P
1 Instrument Man	$34.20	$273.60	$57.45	$459.60	$33.67	$56.25
1 Rodman/Chainman	33.15	265.20	55.05	440.40		
1 Level, Electronic		43.70		48.07	2.73	3.00
16 L.H., Daily Totals		$582.50		$948.07	$36.41	$59.25

Crew A-7	Hr.	Daily	Hr.	Daily	Bare Costs	Incl. O&P
1 Chief of Party	$40.70	$325.60	$67.55	$540.40	$36.02	$60.02
1 Instrument Man	34.20	273.60	57.45	459.60		
1 Rodman/Chainman	33.15	265.20	55.05	440.40		
1 Level, Electronic		43.70		48.07	1.82	2.00
24 L.H., Daily Totals		$908.10		$1488.47	$37.84	$62.02

Crew A-8	Hr.	Daily	Hr.	Daily	Bare Costs	Incl. O&P
1 Chief of Party	$40.70	$325.60	$67.55	$540.40	$35.30	$58.77
1 Instrument Man	34.20	273.60	57.45	459.60		
2 Rodmen/Chainmen	33.15	530.40	55.05	880.80		
1 Level, Electronic		43.70		48.07	1.37	1.50
32 L.H., Daily Totals		$1173.30		$1928.87	$36.67	$60.28

Crew A-9	Hr.	Daily	Hr.	Daily	Bare Costs	Incl. O&P
1 Asbestos Foreman	$35.75	$286.00	$60.65	$485.20	$35.31	$59.91
7 Asbestos Workers	35.25	1974.00	59.80	3348.80		
64 L.H., Daily Totals		$2260.00		$3834.00	$35.31	$59.91

Crew A-10A	Hr.	Daily	Hr.	Daily	Bare Costs	Incl. O&P
1 Asbestos Foreman	$35.75	$286.00	$60.65	$485.20	$35.42	$60.08
2 Asbestos Workers	35.25	564.00	59.80	956.80		
24 L.H., Daily Totals		$850.00		$1442.00	$35.42	$60.08

Crew A-10B	Hr.	Daily	Hr.	Daily	Bare Costs	Incl. O&P
1 Asbestos Foreman	$35.75	$286.00	$60.65	$485.20	$35.38	$60.01
3 Asbestos Workers	35.25	846.00	59.80	1435.20		
32 L.H., Daily Totals		$1132.00		$1920.40	$35.38	$60.01

Crew A-10C	Hr.	Daily	Hr.	Daily	Bare Costs	Incl. O&P
3 Asbestos Workers	$35.25	$846.00	$59.80	$1435.20	$35.25	$59.80
1 Flatbed Truck, Gas, 1.5 Ton		248.20		273.02	10.34	11.38
24 L.H., Daily Totals		$1094.20		$1708.22	$45.59	$71.18

Crew A-10D	Hr.	Daily	Hr.	Daily	Bare Costs	Incl. O&P
2 Asbestos Workers	$35.25	$564.00	$59.80	$956.80	$34.58	$57.92
1 Equip. Oper. (crane)	36.05	288.40	59.60	476.80		
1 Equip. Oper. (oiler)	31.75	254.00	52.50	420.00		
1 Hydraulic Crane, 33 Ton		763.20		839.52	23.85	26.23
32 L.H., Daily Totals		$1869.60		$2693.12	$58.42	$84.16

Crews - Residential

Crew No.	Bare Costs		Incl. Subs O&P		Cost Per Labor-Hour	
Crew A-11	Hr.	Daily	Hr.	Daily	Bare Costs	Incl. O&P
1 Asbestos Foreman	$35.75	$286.00	$60.65	$485.20	$35.31	$59.91
7 Asbestos Workers	35.25	1974.00	59.80	3348.80		
2 Chip. Hammers, 12 Lb., Elec.		40.00		44.00	0.63	0.69
64 L.H., Daily Totals		$2300.00		$3878.00	$35.94	$60.59
Crew A-12	Hr.	Daily	Hr.	Daily	Bare Costs	Incl. O&P
1 Asbestos Foreman	$35.75	$286.00	$60.65	$485.20	$35.31	$59.91
7 Asbestos Workers	35.25	1974.00	59.80	3348.80		
1 Trk-Mtd Vac, 14 CY, 1500 Gal.		510.30		561.33		
1 Flatbed Truck, 20,000 GVW		247.80		272.58	11.85	13.03
64 L.H., Daily Totals		$3018.10		$4667.91	$47.16	$72.94
Crew A-13	Hr.	Daily	Hr.	Daily	Bare Costs	Incl. O&P
1 Equip. Oper. (light)	$33.90	$271.20	$56.05	$448.40	$33.90	$56.05
1 Trk-Mtd Vac, 14 CY, 1500 Gal.		510.30		561.33		
1 Flatbed Truck, 20,000 GVW		247.80		272.58	94.76	104.24
8 L.H., Daily Totals		$1029.30		$1282.31	$128.66	$160.29
Crew B-1	Hr.	Daily	Hr.	Daily	Bare Costs	Incl. O&P
1 Labor Foreman (outside)	$26.65	$213.20	$44.60	$356.80	$25.32	$42.37
2 Laborers	24.65	394.40	41.25	660.00		
24 L.H., Daily Totals		$607.60		$1016.80	$25.32	$42.37
Crew B-1A	Hr.	Daily	Hr.	Daily	Bare Costs	Incl. O&P
1 Labor Foreman (outside)	$26.65	$213.20	$44.60	$356.80	$25.32	$42.37
2 Laborers	24.65	394.40	41.25	660.00		
2 Cutting Torches		22.80		25.08		
2 Sets of Gases		336.00		369.60	14.95	16.45
24 L.H., Daily Totals		$966.40		$1411.48	$40.27	$58.81
Crew B-1B	Hr.	Daily	Hr.	Daily	Bare Costs	Incl. O&P
1 Labor Foreman (outside)	$26.65	$213.20	$44.60	$356.80	$28.00	$46.67
2 Laborers	24.65	394.40	41.25	660.00		
1 Equip. Oper. (crane)	36.05	288.40	59.60	476.80		
2 Cutting Torches		22.80		25.08		
2 Sets of Gases		336.00		369.60		
1 Hyd. Crane, 12 Ton		658.40		724.24	31.79	34.97
32 L.H., Daily Totals		$1913.20		$2612.52	$59.79	$81.64
Crew B-1C	Hr.	Daily	Hr.	Daily	Bare Costs	Incl. O&P
1 Labor Foreman (outside)	$26.65	$213.20	$44.60	$356.80	$25.32	$42.37
2 Laborers	24.65	394.40	41.25	660.00		
1 Aerial Lift Truck, 60' Boom		448.40		493.24	18.68	20.55
24 L.H., Daily Totals		$1056.00		$1510.04	$44.00	$62.92
Crew B-1D	Hr.	Daily	Hr.	Daily	Bare Costs	Incl. O&P
2 Laborers	$24.65	$394.40	$41.25	$660.00	$24.65	$41.25
1 Small Work Boat, Gas, 50 H.P.		174.80		192.28		
1 Pressure Washer, 7 GPM		88.60		97.46	16.46	18.11
16 L.H., Daily Totals		$657.80		$949.74	$41.11	$59.36
Crew B-1E	Hr.	Daily	Hr.	Daily	Bare Costs	Incl. O&P
1 Labor Foreman (outside)	$26.65	$213.20	$44.60	$356.80	$25.15	$42.09
3 Laborers	24.65	591.60	41.25	990.00		
1 Work Boat, Diesel, 200 H.P.		1455.00		1600.50		
2 Pressure Washers, 7 GPM		177.20		194.92	51.01	56.11
32 L.H., Daily Totals		$2437.00		$3142.22	$76.16	$98.19

Crew No.	Bare Costs		Incl. Subs O&P		Cost Per Labor-Hour	
Crew B-1F	Hr.	Daily	Hr.	Daily	Bare Costs	Incl. O&P
2 Skilled Workers	$34.20	$547.20	$57.45	$919.20	$31.02	$52.05
1 Laborer	24.65	197.20	41.25	330.00		
1 Small Work Boat, Gas, 50 H.P.		174.80		192.28		
1 Pressure Washer, 7 GPM		88.60		97.46	10.98	12.07
24 L.H., Daily Totals		$1007.80		$1538.94	$41.99	$64.12
Crew B-1G	Hr.	Daily	Hr.	Daily	Bare Costs	Incl. O&P
2 Laborers	$24.65	$394.40	$41.25	$660.00	$24.65	$41.25
1 Small Work Boat, Gas, 50 H.P.		174.80		192.28	10.93	12.02
16 L.H., Daily Totals		$569.20		$852.28	$35.58	$53.27
Crew B-1H	Hr.	Daily	Hr.	Daily	Bare Costs	Incl. O&P
2 Skilled Workers	$34.20	$547.20	$57.45	$919.20	$31.02	$52.05
1 Laborer	24.65	197.20	41.25	330.00		
1 Small Work Boat, Gas, 50 H.P.		174.80		192.28	7.28	8.01
24 L.H., Daily Totals		$919.20		$1441.48	$38.30	$60.06
Crew B-1J	Hr.	Daily	Hr.	Daily	Bare Costs	Incl. O&P
1 Labor Foreman (inside)	$25.15	$201.20	$42.10	$336.80	$24.90	$41.67
1 Laborer	24.65	197.20	41.25	330.00		
16 L.H., Daily Totals		$398.40		$666.80	$24.90	$41.67
Crew B-1K	Hr.	Daily	Hr.	Daily	Bare Costs	Incl. O&P
1 Carpenter Foreman (inside)	$34.40	$275.20	$57.60	$460.80	$34.15	$57.17
1 Carpenter	33.90	271.20	56.75	454.00		
16 L.H., Daily Totals		$546.40		$914.80	$34.15	$57.17
Crew B-2	Hr.	Daily	Hr.	Daily	Bare Costs	Incl. O&P
1 Labor Foreman (outside)	$26.65	$213.20	$44.60	$356.80	$25.05	$41.92
4 Laborers	24.65	788.80	41.25	1320.00		
40 L.H., Daily Totals		$1002.00		$1676.80	$25.05	$41.92
Crew B-2A	Hr.	Daily	Hr.	Daily	Bare Costs	Incl. O&P
1 Labor Foreman (outside)	$26.65	$213.20	$44.60	$356.80	$25.32	$42.37
2 Laborers	24.65	394.40	41.25	660.00		
1 Aerial Lift Truck, 60' Boom		448.40		493.24	18.68	20.55
24 L.H., Daily Totals		$1056.00		$1510.04	$44.00	$62.92
Crew B-3	Hr.	Daily	Hr.	Daily	Bare Costs	Incl. O&P
1 Labor Foreman (outside)	$26.65	$213.20	$44.60	$356.80	$28.75	$47.74
2 Laborers	24.65	394.40	41.25	660.00		
1 Equip. Oper. (medium)	35.25	282.00	58.25	466.00		
2 Truck Drivers (heavy)	30.65	490.40	50.55	808.80		
1 Crawler Loader, 3 C.Y.		1197.00		1316.70		
2 Dump Trucks, 12 C.Y., 400 H.P.		1381.60		1519.76	53.72	59.09
48 L.H., Daily Totals		$3958.60		$5128.06	$82.47	$106.83
Crew B-3A	Hr.	Daily	Hr.	Daily	Bare Costs	Incl. O&P
4 Laborers	$24.65	$788.80	$41.25	$1320.00	$26.77	$44.65
1 Equip. Oper. (medium)	35.25	282.00	58.25	466.00		
1 Hyd. Excavator, 1.5 C.Y.		1029.00		1131.90	25.73	28.30
40 L.H., Daily Totals		$2099.80		$2917.90	$52.49	$72.95
Crew B-3B	Hr.	Daily	Hr.	Daily	Bare Costs	Incl. O&P
2 Laborers	$24.65	$394.40	$41.25	$660.00	$28.80	$47.83
1 Equip. Oper. (medium)	35.25	282.00	58.25	466.00		
1 Truck Driver (heavy)	30.65	245.20	50.55	404.40		
1 Backhoe Loader, 80 H.P.		396.20		435.82		
1 Dump Truck, 12 C.Y., 400 H.P.		690.80		759.88	33.97	37.37
32 L.H., Daily Totals		$2008.60		$2726.10	$62.77	$85.19

Crews - Residential

Crew No.	Bare Costs		Incl. Subs O&P		Cost Per Labor-Hour	
Crew B-3C	**Hr.**	**Daily**	**Hr.**	**Daily**	**Bare Costs**	**Incl. O&P**
3 Laborers	$24.65	$591.60	$41.25	$990.00	$27.30	$45.50
1 Equip. Oper. (medium)	35.25	282.00	58.25	466.00		
1 Crawler Loader, 4 C.Y.		1566.00		1722.60	48.94	53.83
32 L.H., Daily Totals		$2439.60		$3178.60	$76.24	$99.33
Crew B-4	**Hr.**	**Daily**	**Hr.**	**Daily**	**Bare Costs**	**Incl. O&P**
1 Labor Foreman (outside)	$26.65	$213.20	$44.60	$356.80	$25.98	$43.36
4 Laborers	24.65	788.80	41.25	1320.00		
1 Truck Driver (heavy)	30.65	245.20	50.55	404.40		
1 Truck Tractor, 220 H.P.		358.80		394.68		
1 Flatbed Trailer, 40 Ton		154.60		170.06	10.70	11.77
48 L.H., Daily Totals		$1760.60		$2645.94	$36.68	$55.12
Crew B-5	**Hr.**	**Daily**	**Hr.**	**Daily**	**Bare Costs**	**Incl. O&P**
1 Labor Foreman (outside)	$26.65	$213.20	$44.60	$356.80	$27.17	$45.32
3 Laborers	24.65	591.60	41.25	990.00		
1 Equip. Oper. (medium)	35.25	282.00	58.25	466.00		
1 Air Compressor, 250 cfm		197.80		217.58		
2 Breakers, Pavement, 60 lb.		20.40		22.44		
2 -50' Air Hoses, 1.5"		10.40		11.44		
1 Crawler Loader, 3 C.Y.		1197.00		1316.70	35.64	39.20
40 L.H., Daily Totals		$2512.40		$3380.96	$62.81	$84.52
Crew B-5A	**Hr.**	**Daily**	**Hr.**	**Daily**	**Bare Costs**	**Incl. O&P**
1 Labor Foreman (outside)	$26.65	$213.20	$44.60	$356.80	$28.35	$47.15
6 Laborers	24.65	1183.20	41.25	1980.00		
2 Equip. Opers. (medium)	35.25	564.00	58.25	932.00		
1 Equip. Oper. (light)	33.90	271.20	56.05	448.40		
2 Truck Drivers (heavy)	30.65	490.40	50.55	808.80		
1 Air Compressor, 365 cfm		258.40		284.24		
2 Breakers, Pavement, 60 lb.		20.40		22.44		
8 -50' Air Hoses, 1"		29.20		32.12		
2 Dump Trucks, 8 C.Y., 220 H.P.		834.80		918.28	11.90	13.09
96 L.H., Daily Totals		$3864.80		$5783.08	$40.26	$60.24
Crew B-5B	**Hr.**	**Daily**	**Hr.**	**Daily**	**Bare Costs**	**Incl. O&P**
1 Powderman	$34.20	$273.60	$57.45	$459.60	$32.77	$54.27
2 Equip. Opers. (medium)	35.25	564.00	58.25	932.00		
3 Truck Drivers (heavy)	30.65	735.60	50.55	1213.20		
1 F.E. Loader, W.M., 2.5 C.Y.		516.00		567.60		
3 Dump Trucks, 12 C.Y., 400 H.P.		2072.40		2279.64		
1 Air Compressor, 365 CFM		258.40		284.24	59.31	65.24
48 L.H., Daily Totals		$4420.00		$5736.28	$92.08	$119.51
Crew B-5C	**Hr.**	**Daily**	**Hr.**	**Daily**	**Bare Costs**	**Incl. O&P**
3 Laborers	$24.65	$591.60	$41.25	$990.00	$29.79	$49.40
1 Equip. Oper. (medium)	35.25	282.00	58.25	466.00		
2 Truck Drivers (heavy)	30.65	490.40	50.55	808.80		
1 Equip. Oper. (crane)	36.05	288.40	59.60	476.80		
1 Equip. Oper. (oiler)	31.75	254.00	52.50	420.00		
2 Dump Trucks, 12 C.Y., 400 H.P.		1381.60		1519.76		
1 Crawler Loader, 4 C.Y.		1566.00		1722.60		
1 S.P. Crane, 4x4, 25 Ton		606.60		667.26	55.53	61.09
64 L.H., Daily Totals		$5460.60		$7071.22	$85.32	$110.49

Crew No.	Bare Costs		Incl. Subs O&P		Cost Per Labor-Hour	
Crew B-5D	**Hr.**	**Daily**	**Hr.**	**Daily**	**Bare Costs**	**Incl. O&P**
1 Labor Foreman (outside)	$26.65	$213.20	$44.60	$356.80	$27.75	$46.19
3 Laborers	24.65	591.60	41.25	990.00		
1 Equip. Oper. (medium)	35.25	282.00	58.25	466.00		
1 Truck Driver (heavy)	30.65	245.20	50.55	404.40		
1 Air Compressor, 250 cfm		197.80		217.58		
2 Breakers, Pavement, 60 lb.		20.40		22.44		
2 -50' Air Hoses, 1.5"		10.40		11.44		
1 Crawler Loader, 3 C.Y.		1197.00		1316.70		
1 Dump Truck, 12 C.Y., 400 H.P.		690.80		759.88	44.09	48.50
48 L.H., Daily Totals		$3448.40		$4545.24	$71.84	$94.69
Crew B-6	**Hr.**	**Daily**	**Hr.**	**Daily**	**Bare Costs**	**Incl. O&P**
2 Laborers	$24.65	$394.40	$41.25	$660.00	$27.73	$46.18
1 Equip. Oper. (light)	33.90	271.20	56.05	448.40		
1 Backhoe Loader, 48 H.P.		365.40		401.94	15.23	16.75
24 L.H., Daily Totals		$1031.00		$1510.34	$42.96	$62.93
Crew B-6B	**Hr.**	**Daily**	**Hr.**	**Daily**	**Bare Costs**	**Incl. O&P**
2 Labor Foremen (outside)	$26.65	$426.40	$44.60	$713.60	$25.32	$42.37
4 Laborers	24.65	788.80	41.25	1320.00		
1 S.P. Crane, 4x4, 5 Ton		280.80		308.88		
1 Flatbed Truck, Gas, 1.5 Ton		248.20		273.02		
1 Butt Fusion Mach., 4"-12" diam.		376.10		413.71	18.86	20.74
48 L.H., Daily Totals		$2120.30		$3029.21	$44.17	$63.11
Crew B-6C	**Hr.**	**Daily**	**Hr.**	**Daily**	**Bare Costs**	**Incl. O&P**
2 Labor Foremen (outside)	$26.65	$426.40	$44.60	$713.60	$25.32	$42.37
4 Laborers	24.65	788.80	41.25	1320.00		
1 S.P. Crane, 4x4, 12 Ton		479.20		527.12		
1 Flatbed Truck, Gas, 3 Ton		307.40		338.14		
1 Butt Fusion Mach., 8"-24" diam.		804.80		885.28	33.15	36.47
48 L.H., Daily Totals		$2806.60		$3784.14	$58.47	$78.84
Crew B-7	**Hr.**	**Daily**	**Hr.**	**Daily**	**Bare Costs**	**Incl. O&P**
1 Labor Foreman (outside)	$26.65	$213.20	$44.60	$356.80	$26.75	$44.64
4 Laborers	24.65	788.80	41.25	1320.00		
1 Equip. Oper. (medium)	35.25	282.00	58.25	466.00		
1 Brush Chipper, 12", 130 H.P.		388.60		427.46		
1 Crawler Loader, 3 C.Y.		1197.00		1316.70		
2 Chain Saws, Gas, 36" Long		90.00		99.00	34.91	38.40
48 L.H., Daily Totals		$2959.60		$3985.96	$61.66	$83.04
Crew B-7A	**Hr.**	**Daily**	**Hr.**	**Daily**	**Bare Costs**	**Incl. O&P**
2 Laborers	$24.65	$394.40	$41.25	$660.00	$27.73	$46.18
1 Equip. Oper. (light)	33.90	271.20	56.05	448.40		
1 Rake w/Tractor		353.05		388.36		
2 Chain Saws, Gas, 18"		63.20		69.52	17.34	19.08
24 L.H., Daily Totals		$1081.85		$1566.28	$45.08	$65.26
Crew B-7B	**Hr.**	**Daily**	**Hr.**	**Daily**	**Bare Costs**	**Incl. O&P**
1 Labor Foreman (outside)	$26.65	$213.20	$44.60	$356.80	$27.31	$45.49
4 Laborers	24.65	788.80	41.25	1320.00		
1 Equip. Oper. (medium)	35.25	282.00	58.25	466.00		
1 Truck Driver (heavy)	30.65	245.20	50.55	404.40		
1 Brush Chipper, 12", 130 H.P.		388.60		427.46		
1 Crawler Loader, 3 C.Y.		1197.00		1316.70		
2 Chain Saws, Gas, 36" Long		90.00		99.00		
1 Dump Truck, 8 C.Y., 220 H.P.		417.40		459.14	37.38	41.11
56 L.H., Daily Totals		$3622.20		$4849.50	$64.68	$86.60

Crews - Residential

Crew No.	Bare Costs		Incl. Subs O&P		Cost Per Labor-Hour	
Crew B-7C	Hr.	Daily	Hr.	Daily	Bare Costs	Incl. O&P
1 Labor Foreman (outside)	$26.65	$213.20	$44.60	$356.80	$27.31	$45.49
4 Laborers	24.65	788.80	41.25	1320.00		
1 Equip. Oper. (medium)	35.25	282.00	58.25	466.00		
1 Truck Driver (heavy)	30.65	245.20	50.55	404.40		
1 Brush Chipper, 12", 130 H.P.		388.60		427.46		
1 Crawler Loader, 3 C.Y.		1197.00		1316.70		
2 Chain Saws, Gas, 36" Long		90.00		99.00		
1 Dump Truck, 12 C.Y., 400 H.P.		690.80		759.88	42.26	46.48
56 L.H., Daily Totals		$3895.60		$5150.24	$69.56	$91.97
Crew B-8	Hr.	Daily	Hr.	Daily	Bare Costs	Incl. O&P
1 Labor Foreman (outside)	$26.65	$213.20	$44.60	$356.80	$29.68	$49.24
2 Laborers	24.65	394.40	41.25	660.00		
2 Equip. Opers. (medium)	35.25	564.00	58.25	932.00		
2 Truck Drivers (heavy)	30.65	490.40	50.55	808.80		
1 Hyd. Crane, 25 Ton		748.80		823.68		
1 Crawler Loader, 3 C.Y.		1197.00		1316.70		
2 Dump Trucks, 12 C.Y., 400 H.P.		1381.60		1519.76	59.42	65.36
56 L.H., Daily Totals		$4989.40		$6417.74	$89.10	$114.60
Crew B-9	Hr.	Daily	Hr.	Daily	Bare Costs	Incl. O&P
1 Labor Foreman (outside)	$26.65	$213.20	$44.60	$356.80	$25.05	$41.92
4 Laborers	24.65	788.80	41.25	1320.00		
1 Air Compressor, 250 cfm		197.80		217.58		
2 Breakers, Pavement, 60 lb.		20.40		22.44		
2 -50' Air Hoses, 1.5"		10.40		11.44	5.71	6.29
40 L.H., Daily Totals		$1230.60		$1928.26	$30.77	$48.21
Crew B-9A	Hr.	Daily	Hr.	Daily	Bare Costs	Incl. O&P
2 Laborers	$24.65	$394.40	$41.25	$660.00	$26.65	$44.35
1 Truck Driver (heavy)	30.65	245.20	50.55	404.40		
1 Water Tank Trailer, 5000 Gal.		141.60		155.76		
1 Truck Tractor, 220 H.P.		358.80		394.68		
2 -50' Discharge Hoses, 3"		3.00		3.30	20.98	23.07
24 L.H., Daily Totals		$1143.00		$1618.14	$47.63	$67.42
Crew B-9B	Hr.	Daily	Hr.	Daily	Bare Costs	Incl. O&P
2 Laborers	$24.65	$394.40	$41.25	$660.00	$26.65	$44.35
1 Truck Driver (heavy)	30.65	245.20	50.55	404.40		
2 -50' Discharge Hoses, 3"		3.00		3.30		
1 Water Tank Trailer, 5000 Gal.		141.60		155.76		
1 Truck Tractor, 220 H.P.		358.80		394.68		
1 Pressure Washer		71.60		78.76	23.96	26.35
24 L.H., Daily Totals		$1214.60		$1696.90	$50.61	$70.70
Crew B-9D	Hr.	Daily	Hr.	Daily	Bare Costs	Incl. O&P
1 Labor Foreman (outside)	$26.65	$213.20	$44.60	$356.80	$25.05	$41.92
4 Common Laborers	24.65	788.80	41.25	1320.00		
1 Air Compressor, 250 cfm		197.80		217.58		
2 -50' Air Hoses, 1.5"		10.40		11.44		
2 Air Powered Tampers		48.00		52.80	6.41	7.05
40 L.H., Daily Totals		$1258.20		$1958.62	$31.45	$48.97
Crew B-10	Hr.	Daily	Hr.	Daily	Bare Costs	Incl. O&P
1 Equip. Oper. (medium)	35.25	282.00	58.25	466.00	35.25	58.25
8 L.H., Daily Totals		$282.00		$466.00	$35.25	$58.25
Crew B-10A	Hr.	Daily	Hr.	Daily	Bare Costs	Incl. O&P
1 Equip. Oper. (medium)	$35.25	$282.00	$58.25	$466.00	$35.25	$58.25
1 Roller, 2-Drum, W.B., 7.5 H.P.		184.80		203.28	23.10	25.41
8 L.H., Daily Totals		$466.80		$669.28	$58.35	$83.66
Crew B-10B	Hr.	Daily	Hr.	Daily	Bare Costs	Incl. O&P
1 Equip. Oper. (medium)	$35.25	$282.00	$58.25	$466.00	$35.25	$58.25
1 Dozer, 200 H.P.		1397.00		1536.70	174.63	192.09
8 L.H., Daily Totals		$1679.00		$2002.70	$209.88	$250.34
Crew B-10C	Hr.	Daily	Hr.	Daily	Bare Costs	Incl. O&P
1 Equip. Oper. (medium)	$35.25	$282.00	$58.25	$466.00	$35.25	$58.25
1 Dozer, 200 H.P.		1397.00		1536.70		
1 Vibratory Roller, Towed, 23 Ton		401.60		441.76	224.82	247.31
8 L.H., Daily Totals		$2080.60		$2444.46	$260.07	$305.56
Crew B-10D	Hr.	Daily	Hr.	Daily	Bare Costs	Incl. O&P
1 Equip. Oper. (medium)	$35.25	$282.00	$58.25	$466.00	$35.25	$58.25
1 Dozer, 200 H.P.		1397.00		1536.70		
1 Sheepsft. Roller, Towed		426.40		469.04	227.93	250.72
8 L.H., Daily Totals		$2105.40		$2471.74	$263.18	$308.97
Crew B-10E	Hr.	Daily	Hr.	Daily	Bare Costs	Incl. O&P
1 Equip. Oper. (medium)	$35.25	$282.00	$58.25	$466.00	$35.25	$58.25
1 Tandem Roller, 5 Ton		162.60		178.86	20.32	22.36
8 L.H., Daily Totals		$444.60		$644.86	$55.58	$80.61
Crew B-10F	Hr.	Daily	Hr.	Daily	Bare Costs	Incl. O&P
1 Equip. Oper. (medium)	$35.25	$282.00	$58.25	$466.00	$35.25	$58.25
1 Tandem Roller, 10 Ton		239.40		263.34	29.93	32.92
8 L.H., Daily Totals		$521.40		$729.34	$65.17	$91.17
Crew B-10G	Hr.	Daily	Hr.	Daily	Bare Costs	Incl. O&P
1 Equip. Oper. (medium)	$35.25	$282.00	$58.25	$466.00	$35.25	$58.25
1 Sheepsfoot Roller, 240 H.P.		1208.00		1328.80	151.00	166.10
8 L.H., Daily Totals		$1490.00		$1794.80	$186.25	$224.35
Crew B-10H	Hr.	Daily	Hr.	Daily	Bare Costs	Incl. O&P
1 Equip. Oper. (medium)	$35.25	$282.00	$58.25	$466.00	$35.25	$58.25
1 Diaphragm Water Pump, 2"		71.80		78.98		
1 -20' Suction Hose, 2"		1.95		2.15		
2 -50' Discharge Hoses, 2"		1.80		1.98	9.44	10.39
8 L.H., Daily Totals		$357.55		$549.11	$44.69	$68.64
Crew B-10I	Hr.	Daily	Hr.	Daily	Bare Costs	Incl. O&P
1 Equip. Oper. (medium)	$35.25	$282.00	$58.25	$466.00	$35.25	$58.25
1 Diaphragm Water Pump, 4"		118.00		129.80		
1 -20' Suction Hose, 4"		3.25		3.58		
2 -50' Discharge Hoses, 4"		4.70		5.17	15.74	17.32
8 L.H., Daily Totals		$407.95		$604.54	$50.99	$75.57
Crew B-10J	Hr.	Daily	Hr.	Daily	Bare Costs	Incl. O&P
1 Equip. Oper. (medium)	$35.25	$282.00	$58.25	$466.00	$35.25	$58.25
1 Centrifugal Water Pump, 3"		81.00		89.10		
1 -20' Suction Hose, 3"		2.85		3.13		
2 -50' Discharge Hoses, 3"		3.00		3.30	10.86	11.94
8 L.H., Daily Totals		$368.85		$561.53	$46.11	$70.19
Crew B-10K	Hr.	Daily	Hr.	Daily	Bare Costs	Incl. O&P
1 Equip. Oper. (medium)	$35.25	$282.00	$58.25	$466.00	$35.25	$58.25
1 Centr. Water Pump, 6"		350.80		385.88		
1 -20' Suction Hose, 6"		11.50		12.65		
2 -50' Discharge Hoses, 6"		12.20		13.42	46.81	51.49
8 L.H., Daily Totals		$656.50		$877.95	$82.06	$109.74

Crews - Residential

Crew No.	Bare Costs Hr.	Daily	Incl. Subs O&P Hr.	Daily	Cost Per Labor-Hour Bare Costs	Incl. O&P
Crew B-10L					Bare Costs	Incl. O&P
1 Equip. Oper. (medium)	$35.25	$282.00	$58.25	$466.00	$35.25	$58.25
1 Dozer, 80 H.P.		471.40		518.54	58.92	64.82
8 L.H., Daily Totals		$753.40		$984.54	$94.17	$123.07
Crew B-10M					Bare Costs	Incl. O&P
1 Equip. Oper. (medium)	$35.25	$282.00	$58.25	$466.00	$35.25	$58.25
1 Dozer, 300 H.P.		1887.00		2075.70	235.88	259.46
8 L.H., Daily Totals		$2169.00		$2541.70	$271.13	$317.71
Crew B-10N					Bare Costs	Incl. O&P
1 Equip. Oper. (medium)	$35.25	$282.00	$58.25	$466.00	$35.25	$58.25
1 F.E. Loader, T.M., 1.5 C.Y		525.00		577.50	65.63	72.19
8 L.H., Daily Totals		$807.00		$1043.50	$100.88	$130.44
Crew B-10O					Bare Costs	Incl. O&P
1 Equip. Oper. (medium)	$35.25	$282.00	$58.25	$466.00	$35.25	$58.25
1 F.E. Loader, T.M., 2.25 C.Y.		964.00		1060.40	120.50	132.55
8 L.H., Daily Totals		$1246.00		$1526.40	$155.75	$190.80
Crew B-10P					Bare Costs	Incl. O&P
1 Equip. Oper. (medium)	$35.25	$282.00	$58.25	$466.00	$35.25	$58.25
1 Crawler Loader, 3 C.Y.		1197.00		1316.70	149.63	164.59
8 L.H., Daily Totals		$1479.00		$1782.70	$184.88	$222.84
Crew B-10Q					Bare Costs	Incl. O&P
1 Equip. Oper. (medium)	$35.25	$282.00	$58.25	$466.00	$35.25	$58.25
1 Crawler Loader, 4 C.Y.		1566.00		1722.60	195.75	215.32
8 L.H., Daily Totals		$1848.00		$2188.60	$231.00	$273.57
Crew B-10R					Bare Costs	Incl. O&P
1 Equip. Oper. (medium)	$35.25	$282.00	$58.25	$466.00	$35.25	$58.25
1 F.E. Loader, W.M., 1 C.Y.		299.00		328.90	37.38	41.11
8 L.H., Daily Totals		$581.00		$794.90	$72.63	$99.36
Crew B-10S					Bare Costs	Incl. O&P
1 Equip. Oper. (medium)	$35.25	$282.00	$58.25	$466.00	$35.25	$58.25
1 F.E. Loader, W.M., 1.5 C.Y.		378.60		416.46	47.33	52.06
8 L.H., Daily Totals		$660.60		$882.46	$82.58	$110.31
Crew B-10T					Bare Costs	Incl. O&P
1 Equip. Oper. (medium)	$35.25	$282.00	$58.25	$466.00	$35.25	$58.25
1 F.E. Loader, W.M., 2.5 C.Y.		516.00		567.60	64.50	70.95
8 L.H., Daily Totals		$798.00		$1033.60	$99.75	$129.20
Crew B-10U					Bare Costs	Incl. O&P
1 Equip. Oper. (medium)	$35.25	$282.00	$58.25	$466.00	$35.25	$58.25
1 F.E. Loader, W.M., 5.5 C.Y.		1092.00		1201.20	136.50	150.15
8 L.H., Daily Totals		$1374.00		$1667.20	$171.75	$208.40
Crew B-10V					Bare Costs	Incl. O&P
1 Equip. Oper. (medium)	$35.25	$282.00	$58.25	$466.00	$35.25	$58.25
1 Dozer, 700 H.P.		5347.00		5881.70	668.38	735.21
8 L.H., Daily Totals		$5629.00		$6347.70	$703.63	$793.46
Crew B-10W					Bare Costs	Incl. O&P
1 Equip. Oper. (medium)	$35.25	$282.00	$58.25	$466.00	$35.25	$58.25
1 Dozer, 105 H.P.		603.40		663.74	75.42	82.97
8 L.H., Daily Totals		$885.40		$1129.74	$110.68	$141.22

Crew No.	Bare Costs Hr.	Daily	Incl. Subs O&P Hr.	Daily	Cost Per Labor-Hour Bare Costs	Incl. O&P
Crew B-10X					Bare Costs	Incl. O&P
1 Equip. Oper. (medium)	$35.25	$282.00	$58.25	$466.00	$35.25	$58.25
1 Dozer, 410 H.P.		2392.00		2631.20	299.00	328.90
8 L.H., Daily Totals		$2674.00		$3097.20	$334.25	$387.15
Crew B-10Y					Bare Costs	Incl. O&P
1 Equip. Oper. (medium)	$35.25	$282.00	$58.25	$466.00	$35.25	$58.25
1 Vibr. Roller, Towed, 12 Ton		566.00		622.60	70.75	77.83
8 L.H., Daily Totals		$848.00		$1088.60	$106.00	$136.07
Crew B-11A					Bare Costs	Incl. O&P
1 Equipment Oper. (med.)	$35.25	$282.00	$58.25	$466.00	$29.95	$49.75
1 Laborer	24.65	197.20	41.25	330.00		
1 Dozer, 200 H.P.		1397.00		1536.70	87.31	96.04
16 L.H., Daily Totals		$1876.20		$2332.70	$117.26	$145.79
Crew B-11B					Bare Costs	Incl. O&P
1 Equipment Oper. (light)	$33.90	$271.20	$56.05	$448.40	$29.27	$48.65
1 Laborer	24.65	197.20	41.25	330.00		
1 Air Powered Tamper		24.00		26.40		
1 Air Compressor, 365 cfm		258.40		284.24		
2 -50' Air Hoses, 1.5"		10.40		11.44	18.30	20.13
16 L.H., Daily Totals		$761.20		$1100.48	$47.58	$68.78
Crew B-11C					Bare Costs	Incl. O&P
1 Equipment Oper. (med.)	$35.25	$282.00	$58.25	$466.00	$29.95	$49.75
1 Laborer	24.65	197.20	41.25	330.00		
1 Backhoe Loader, 48 H.P.		365.40		401.94	22.84	25.12
16 L.H., Daily Totals		$844.60		$1197.94	$52.79	$74.87
Crew B-11K					Bare Costs	Incl. O&P
1 Equipment Oper. (med.)	$35.25	$282.00	$58.25	$466.00	$29.95	$49.75
1 Laborer	24.65	197.20	41.25	330.00		
1 Trencher, Chain Type, 8' D		2860.00		3146.00	178.75	196.63
16 L.H., Daily Totals		$3339.20		$3942.00	$208.70	$246.38
Crew B-11L					Bare Costs	Incl. O&P
1 Equipment Oper. (med.)	$35.25	$282.00	$58.25	$466.00	$29.95	$49.75
1 Laborer	24.65	197.20	41.25	330.00		
1 Grader, 30,000 Lbs.		714.40		785.84	44.65	49.12
16 L.H., Daily Totals		$1193.60		$1581.84	$74.60	$98.86
Crew B-11M					Bare Costs	Incl. O&P
1 Equipment Oper. (med.)	$35.25	$282.00	$58.25	$466.00	$29.95	$49.75
1 Laborer	24.65	197.20	41.25	330.00		
1 Backhoe Loader, 80 H.P.		396.20		435.82	24.76	27.24
16 L.H., Daily Totals		$875.40		$1231.82	$54.71	$76.99
Crew B-11W					Bare Costs	Incl. O&P
1 Equipment Operator (med.)	$35.25	$282.00	$58.25	$466.00	$30.53	$50.42
1 Common Laborer	24.65	197.20	41.25	330.00		
10 Truck Drivers (heavy)	30.65	2452.00	50.55	4044.00		
1 Dozer, 200 H.P.		1397.00		1536.70		
1 Vibratory Roller, Towed, 23 Ton		401.60		441.76		
10 Dump Trucks, 8 C.Y., 220 H.P.		4174.00		4591.40	62.21	68.44
96 L.H., Daily Totals		$8903.80		$11409.86	$92.75	$118.85

Crews - Residential

Crew No.	Bare Costs		Incl. Subs O&P		Cost Per Labor-Hour	
Crew B-11Y	Hr.	Daily	Hr.	Daily	Bare Costs	Incl. O&P
1 Labor Foreman (outside)	$26.65	$213.20	$44.60	$356.80	$28.41	$47.29
5 Common Laborers	24.65	986.00	41.25	1650.00		
3 Equipment Operators (med.)	35.25	846.00	58.25	1398.00		
1 Dozer, 80 H.P.		471.40		518.54		
2 Rollers, 2-Drum, W.B., 7.5 H.P.		369.60		406.56		
4 Vibrating Plates, Gas, 21"		186.40		205.04	14.27	15.70
72 L.H., Daily Totals		$3072.60		$4534.94	$42.67	$62.99

Crew B-12A	Hr.	Daily	Hr.	Daily	Bare Costs	Incl. O&P
1 Equip. Oper. (crane)	$36.05	$288.40	$59.60	$476.80	$30.35	$50.42
1 Laborer	24.65	197.20	41.25	330.00		
1 Hyd. Excavator, 1 C.Y.		812.80		894.08	50.80	55.88
16 L.H., Daily Totals		$1298.40		$1700.88	$81.15	$106.31

Crew B-12B	Hr.	Daily	Hr.	Daily	Bare Costs	Incl. O&P
1 Equip. Oper. (crane)	$36.05	$288.40	$59.60	$476.80	$30.35	$50.42
1 Laborer	24.65	197.20	41.25	330.00		
1 Hyd. Excavator, 1.5 C.Y.		1029.00		1131.90	64.31	70.74
16 L.H., Daily Totals		$1514.60		$1938.70	$94.66	$121.17

Crew B-12C	Hr.	Daily	Hr.	Daily	Bare Costs	Incl. O&P
1 Equip. Oper. (crane)	$36.05	$288.40	$59.60	$476.80	$30.35	$50.42
1 Laborer	24.65	197.20	41.25	330.00		
1 Hyd. Excavator, 2 C.Y.		1176.00		1293.60	73.50	80.85
16 L.H., Daily Totals		$1661.60		$2100.40	$103.85	$131.28

Crew B-12D	Hr.	Daily	Hr.	Daily	Bare Costs	Incl. O&P
1 Equip. Oper. (crane)	$36.05	$288.40	$59.60	$476.80	$30.35	$50.42
1 Laborer	24.65	197.20	41.25	330.00		
1 Hyd. Excavator, 3.5 C.Y.		2442.00		2686.20	152.63	167.89
16 L.H., Daily Totals		$2927.60		$3493.00	$182.97	$218.31

Crew B-12E	Hr.	Daily	Hr.	Daily	Bare Costs	Incl. O&P
1 Equip. Oper. (crane)	$36.05	$288.40	$59.60	$476.80	$30.35	$50.42
1 Laborer	24.65	197.20	41.25	330.00		
1 Hyd. Excavator, .5 C.Y.		445.60		490.16	27.85	30.64
16 L.H., Daily Totals		$931.20		$1296.96	$58.20	$81.06

Crew B-12F	Hr.	Daily	Hr.	Daily	Bare Costs	Incl. O&P
1 Equip. Oper. (crane)	$36.05	$288.40	$59.60	$476.80	$30.35	$50.42
1 Laborer	24.65	197.20	41.25	330.00		
1 Hyd. Excavator, .75 C.Y.		654.00		719.40	40.88	44.96
16 L.H., Daily Totals		$1139.60		$1526.20	$71.22	$95.39

Crew B-12G	Hr.	Daily	Hr.	Daily	Bare Costs	Incl. O&P
1 Equip. Oper. (crane)	$36.05	$288.40	$59.60	$476.80	$30.35	$50.42
1 Laborer	24.65	197.20	41.25	330.00		
1 Crawler Crane, 15 Ton		695.70		765.27		
1 Clamshell Bucket, .5 C.Y.		37.80		41.58	45.84	50.43
16 L.H., Daily Totals		$1219.10		$1613.65	$76.19	$100.85

Crew B-12H	Hr.	Daily	Hr.	Daily	Bare Costs	Incl. O&P
1 Equip. Oper. (crane)	$36.05	$288.40	$59.60	$476.80	$30.35	$50.42
1 Laborer	24.65	197.20	41.25	330.00		
1 Crawler Crane, 25 Ton		1194.00		1313.40		
1 Clamshell Bucket, 1 C.Y.		47.40		52.14	77.59	85.35
16 L.H., Daily Totals		$1727.00		$2172.34	$107.94	$135.77

Crew B-12I	Hr.	Daily	Hr.	Daily	Bare Costs	Incl. O&P
1 Equip. Oper. (crane)	$36.05	$288.40	$59.60	$476.80	$30.35	$50.42
1 Laborer	24.65	197.20	41.25	330.00		
1 Crawler Crane, 20 Ton		896.25		985.88		
1 Dragline Bucket, .75 C.Y.		20.60		22.66	57.30	63.03
16 L.H., Daily Totals		$1402.45		$1815.34	$87.65	$113.46

Crew B-12J	Hr.	Daily	Hr.	Daily	Bare Costs	Incl. O&P
1 Equip. Oper. (crane)	$36.05	$288.40	$59.60	$476.80	$30.35	$50.42
1 Laborer	24.65	197.20	41.25	330.00		
1 Gradall, 5/8 C.Y.		882.00		970.20	55.13	60.64
16 L.H., Daily Totals		$1367.60		$1777.00	$85.47	$111.06

Crew B-12K	Hr.	Daily	Hr.	Daily	Bare Costs	Incl. O&P
1 Equip. Oper. (crane)	$36.05	$288.40	$59.60	$476.80	$30.35	$50.42
1 Laborer	24.65	197.20	41.25	330.00		
1 Gradall, 3 Ton, 1 C.Y.		995.80		1095.38	62.24	68.46
16 L.H., Daily Totals		$1481.40		$1902.18	$92.59	$118.89

Crew B-12L	Hr.	Daily	Hr.	Daily	Bare Costs	Incl. O&P
1 Equip. Oper. (crane)	$36.05	$288.40	$59.60	$476.80	$30.35	$50.42
1 Laborer	24.65	197.20	41.25	330.00		
1 Crawler Crane, 15 Ton		695.70		765.27		
1 F.E. Attachment, .5 C.Y.		59.20		65.12	47.18	51.90
16 L.H., Daily Totals		$1240.50		$1637.19	$77.53	$102.32

Crew B-12M	Hr.	Daily	Hr.	Daily	Bare Costs	Incl. O&P
1 Equip. Oper. (crane)	$36.05	$288.40	$59.60	$476.80	$30.35	$50.42
1 Laborer	24.65	197.20	41.25	330.00		
1 Crawler Crane, 20 Ton		896.25		985.88		
1 F.E. Attachment, .75 C.Y.		63.60		69.96	59.99	65.99
16 L.H., Daily Totals		$1445.45		$1862.64	$90.34	$116.41

Crew B-12N	Hr.	Daily	Hr.	Daily	Bare Costs	Incl. O&P
1 Equip. Oper. (crane)	$36.05	$288.40	$59.60	$476.80	$30.35	$50.42
1 Laborer	24.65	197.20	41.25	330.00		
1 Crawler Crane, 25 Ton		1194.00		1313.40		
1 F.E. Attachment, 1 C.Y.		70.40		77.44	79.03	86.93
16 L.H., Daily Totals		$1750.00		$2197.64	$109.38	$137.35

Crew B-12O	Hr.	Daily	Hr.	Daily	Bare Costs	Incl. O&P
1 Equip. Oper. (crane)	$36.05	$288.40	$59.60	$476.80	$30.35	$50.42
1 Laborer	24.65	197.20	41.25	330.00		
1 Crawler Crane, 40 Ton		1203.00		1323.30		
1 F.E. Attachment, 1.5 C.Y.		79.60		87.56	80.16	88.18
16 L.H., Daily Totals		$1768.20		$2217.66	$110.51	$138.60

Crew B-12P	Hr.	Daily	Hr.	Daily	Bare Costs	Incl. O&P
1 Equip. Oper. (crane)	$36.05	$288.40	$59.60	$476.80	$30.35	$50.42
1 Laborer	24.65	197.20	41.25	330.00		
1 Crawler Crane, 40 Ton		1203.00		1323.30		
1 Dragline Bucket, 1.5 C.Y.		33.60		36.96	77.29	85.02
16 L.H., Daily Totals		$1722.20		$2167.06	$107.64	$135.44

Crew B-12Q	Hr.	Daily	Hr.	Daily	Bare Costs	Incl. O&P
1 Equip. Oper. (crane)	$36.05	$288.40	$59.60	$476.80	$30.35	$50.42
1 Laborer	24.65	197.20	41.25	330.00		
1 Hyd. Excavator, 5/8 C.Y.		589.40		648.34	36.84	40.52
16 L.H., Daily Totals		$1075.00		$1455.14	$67.19	$90.95

Crews - Residential

Crew No.	Bare Costs		Incl. Subs O&P		Cost Per Labor-Hour	
Crew B-12S	Hr.	Daily	Hr.	Daily	Bare Costs	Incl. O&P
1 Equip. Oper. (crane)	$36.05	$288.40	$59.60	$476.80	$30.35	$50.42
1 Laborer	24.65	197.20	41.25	330.00		
1 Hyd. Excavator, 2.5 C.Y.		1582.00		1740.20	98.88	108.76
16 L.H., Daily Totals		$2067.60		$2547.00	$129.22	$159.19
Crew B-12T	Hr.	Daily	Hr.	Daily	Bare Costs	Incl. O&P
1 Equip. Oper. (crane)	$36.05	$288.40	$59.60	$476.80	$30.35	$50.42
1 Laborer	24.65	197.20	41.25	330.00		
1 Crawler Crane, 75 Ton		1507.00		1657.70		
1 F.E. Attachment, 3 C.Y.		102.40		112.64	100.59	110.65
16 L.H., Daily Totals		$2095.00		$2577.14	$130.94	$161.07
Crew B-12V	Hr.	Daily	Hr.	Daily	Bare Costs	Incl. O&P
1 Equip. Oper. (crane)	$36.05	$288.40	$59.60	$476.80	$30.35	$50.42
1 Laborer	24.65	197.20	41.25	330.00		
1 Crawler Crane, 75 Ton		1507.00		1657.70		
1 Dragline Bucket, 3 C.Y.		52.60		57.86	97.47	107.22
16 L.H., Daily Totals		$2045.20		$2522.36	$127.83	$157.65
Crew B-12Y	Hr.	Daily	Hr.	Daily	Bare Costs	Incl. O&P
1 Equip. Oper. (crane)	$36.05	$288.40	$59.60	$476.80	$28.45	$47.37
2 Laborers	24.65	394.40	41.25	660.00		
1 Hyd. Excavator, 3.5 C.Y.		2442.00		2686.20	101.75	111.93
24 L.H., Daily Totals		$3124.80		$3823.00	$130.20	$159.29
Crew B-12Z	Hr.	Daily	Hr.	Daily	Bare Costs	Incl. O&P
1 Equip. Oper. (crane)	$36.05	$288.40	$59.60	$476.80	$28.45	$47.37
2 Laborers	24.65	394.40	41.25	660.00		
1 Hyd. Excavator, 2.5 C.Y.		1582.00		1740.20	65.92	72.51
24 L.H., Daily Totals		$2264.80		$2877.00	$94.37	$119.88
Crew B-13	Hr.	Daily	Hr.	Daily	Bare Costs	Incl. O&P
1 Labor Foreman (outside)	$26.65	$213.20	$44.60	$356.80	$26.88	$44.87
4 Laborers	24.65	788.80	41.25	1320.00		
1 Equip. Oper. (crane)	36.05	288.40	59.60	476.80		
1 Hyd. Crane, 25 Ton		748.80		823.68	15.60	17.16
48 L.H., Daily Totals		$2039.20		$2977.28	$42.48	$62.03
Crew B-13A	Hr.	Daily	Hr.	Daily	Bare Costs	Incl. O&P
1 Labor Foreman (outside)	$26.65	$213.20	$44.60	$356.80	$29.68	$49.24
2 Laborers	24.65	394.40	41.25	660.00		
2 Equipment Operators (med.)	35.25	564.00	58.25	932.00		
2 Truck Drivers (heavy)	30.65	490.40	50.55	808.80		
1 Crawler Crane, 75 Ton		1507.00		1657.70		
1 Crawler Loader, 4 C.Y.		1566.00		1722.60		
2 Dump Trucks, 8 C.Y., 220 H.P.		834.80		918.28	69.78	76.76
56 L.H., Daily Totals		$5569.80		$7056.18	$99.46	$126.00
Crew B-13B	Hr.	Daily	Hr.	Daily	Bare Costs	Incl. O&P
1 Labor Foreman (outside)	$26.65	$213.20	$44.60	$356.80	$27.58	$45.96
4 Laborers	24.65	788.80	41.25	1320.00		
1 Equip. Oper. (crane)	36.05	288.40	59.60	476.80		
1 Equip. Oper. (oiler)	31.75	254.00	52.50	420.00		
1 Hyd. Crane, 55 Ton		1123.00		1235.30	20.05	22.06
56 L.H., Daily Totals		$2667.40		$3808.90	$47.63	$68.02

Crew No.	Bare Costs		Incl. Subs O&P		Cost Per Labor-Hour	
Crew B-13C	Hr.	Daily	Hr.	Daily	Bare Costs	Incl. O&P
1 Labor Foreman (outside)	$26.65	$213.20	$44.60	$356.80	$27.58	$45.96
4 Laborers	24.65	788.80	41.25	1320.00		
1 Equip. Oper. (crane)	36.05	288.40	59.60	476.80		
1 Equip. Oper. (oiler)	31.75	254.00	52.50	420.00		
1 Crawler Crane, 100 Ton		1667.00		1833.70	29.77	32.74
56 L.H., Daily Totals		$3211.40		$4407.30	$57.35	$78.70
Crew B-13D	Hr.	Daily	Hr.	Daily	Bare Costs	Incl. O&P
1 Laborer	$24.65	$197.20	$41.25	$330.00	$30.35	$50.42
1 Equip. Oper. (crane)	36.05	288.40	59.60	476.80		
1 Hyd. Excavator, 1 C.Y.		812.80		894.08		
1 Trench Box		81.00		89.10	55.86	61.45
16 L.H., Daily Totals		$1379.40		$1789.98	$86.21	$111.87
Crew B-13E	Hr.	Daily	Hr.	Daily	Bare Costs	Incl. O&P
1 Laborer	$24.65	$197.20	$41.25	$330.00	$30.35	$50.42
1 Equip. Oper. (crane)	36.05	288.40	59.60	476.80		
1 Hyd. Excavator, 1.5 C.Y.		1029.00		1131.90		
1 Trench Box		81.00		89.10	69.38	76.31
16 L.H., Daily Totals		$1595.60		$2027.80	$99.72	$126.74
Crew B-13F	Hr.	Daily	Hr.	Daily	Bare Costs	Incl. O&P
1 Laborer	$24.65	$197.20	$41.25	$330.00	$30.35	$50.42
1 Equip. Oper. (crane)	36.05	288.40	59.60	476.80		
1 Hyd. Excavator, 3.5 C.Y.		2442.00		2686.20		
1 Trench Box		81.00		89.10	157.69	173.46
16 L.H., Daily Totals		$3008.60		$3582.10	$188.04	$223.88
Crew B-13G	Hr.	Daily	Hr.	Daily	Bare Costs	Incl. O&P
1 Laborer	$24.65	$197.20	$41.25	$330.00	$30.35	$50.42
1 Equip. Oper. (crane)	36.05	288.40	59.60	476.80		
1 Hyd. Excavator, .75 C.Y.		654.00		719.40		
1 Trench Box		81.00		89.10	45.94	50.53
16 L.H., Daily Totals		$1220.60		$1615.30	$76.29	$100.96
Crew B-13H	Hr.	Daily	Hr.	Daily	Bare Costs	Incl. O&P
1 Laborer	$24.65	$197.20	$41.25	$330.00	$30.35	$50.42
1 Equip. Oper. (crane)	36.05	288.40	59.60	476.80		
1 Gradall, 5/8 C.Y.		882.00		970.20		
1 Trench Box		81.00		89.10	60.19	66.21
16 L.H., Daily Totals		$1448.60		$1866.10	$90.54	$116.63
Crew B-13I	Hr.	Daily	Hr.	Daily	Bare Costs	Incl. O&P
1 Laborer	$24.65	$197.20	$41.25	$330.00	$30.35	$50.42
1 Equip. Oper. (crane)	36.05	288.40	59.60	476.80		
1 Gradall, 3 Ton, 1 C.Y.		995.80		1095.38		
1 Trench Box		81.00		89.10	67.30	74.03
16 L.H., Daily Totals		$1562.40		$1991.28	$97.65	$124.46
Crew B-13J	Hr.	Daily	Hr.	Daily	Bare Costs	Incl. O&P
1 Laborer	$24.65	$197.20	$41.25	$330.00	$30.35	$50.42
1 Equip. Oper. (crane)	36.05	288.40	59.60	476.80		
1 Hyd. Excavator, 2.5 C.Y.		1582.00		1740.20		
1 Trench Box		81.00		89.10	103.94	114.33
16 L.H., Daily Totals		$2148.60		$2636.10	$134.29	$164.76

Crews - Residential

Crew No.		Bare Costs		Incl. Subs O&P		Cost Per Labor-Hour	
Crew B-13K	Hr.	Daily	Hr.	Daily	Bare Costs	Incl. O&P
2 Equip. Opers. (crane)	$36.05	$576.80	$59.60	$953.60	$36.05	$59.60
1 Hyd. Excavator, .75 C.Y.		654.00		719.40		
1 Hyd. Hammer, 4000 ft-lb		306.20		336.82		
1 Hyd. Excavator, .75 C.Y.		654.00		719.40	100.89	110.98
16 L.H., Daily Totals		$2191.00		$2729.22	$136.94	$170.58
Crew B-13L	Hr.	Daily	Hr.	Daily	Bare Costs	Incl. O&P
2 Equip. Opers. (crane)	$36.05	$576.80	$59.60	$953.60	$36.05	$59.60
1 Hyd. Excavator, 1.5 C.Y.		1029.00		1131.90		
1 Hyd. Hammer, 5000 ft-lb		367.80		404.58		
1 Hyd. Excavator, .75 C.Y.		654.00		719.40	128.18	140.99
16 L.H., Daily Totals		$2627.60		$3209.48	$164.22	$200.59
Crew B-13M	Hr.	Daily	Hr.	Daily	Bare Costs	Incl. O&P
2 Equip. Opers. (crane)	$36.05	$576.80	$59.60	$953.60	$36.05	$59.60
1 Hyd. Excavator, 2.5 C.Y.		1582.00		1740.20		
1 Hyd. Hammer, 8000 ft-lb		541.20		595.32		
1 Hyd. Excavator, 1.5 C.Y.		1029.00		1131.90	197.01	216.71
16 L.H., Daily Totals		$3729.00		$4421.02	$233.06	$276.31
Crew B-13N	Hr.	Daily	Hr.	Daily	Bare Costs	Incl. O&P
2 Equip. Opers. (crane)	$36.05	$576.80	$59.60	$953.60	$36.05	$59.60
1 Hyd. Excavator, 3.5 C.Y.		2442.00		2686.20		
1 Hyd. Hammer, 12,000 ft-lb		632.20		695.42		
1 Hyd. Excavator, 1.5 C.Y.		1029.00		1131.90	256.45	282.10
16 L.H., Daily Totals		$4680.00		$5467.12	$292.50	$341.69
Crew B-14	Hr.	Daily	Hr.	Daily	Bare Costs	Incl. O&P
1 Labor Foreman (outside)	$26.65	$213.20	$44.60	$356.80	$26.52	$44.27
4 Laborers	24.65	788.80	41.25	1320.00		
1 Equip. Oper. (light)	33.90	271.20	56.05	448.40		
1 Backhoe Loader, 48 H.P.		365.40		401.94	7.61	8.37
48 L.H., Daily Totals		$1638.60		$2527.14	$34.14	$52.65
Crew B-14A	Hr.	Daily	Hr.	Daily	Bare Costs	Incl. O&P
1 Equip. Oper. (crane)	$36.05	$288.40	$59.60	$476.80	$32.25	$53.48
.5 Laborer	24.65	98.60	41.25	165.00		
1 Hyd. Excavator, 4.5 C.Y.		2974.00		3271.40	247.83	272.62
12 L.H., Daily Totals		$3361.00		$3913.20	$280.08	$326.10
Crew B-14B	Hr.	Daily	Hr.	Daily	Bare Costs	Incl. O&P
1 Equip. Oper. (crane)	$36.05	$288.40	$59.60	$476.80	$32.25	$53.48
.5 Laborer	24.65	98.60	41.25	165.00		
1 Hyd. Excavator, 6 C.Y.		3501.00		3851.10	291.75	320.93
12 L.H., Daily Totals		$3888.00		$4492.90	$324.00	$374.41
Crew B-14C	Hr.	Daily	Hr.	Daily	Bare Costs	Incl. O&P
1 Equip. Oper. (crane)	$36.05	$288.40	$59.60	$476.80	$32.25	$53.48
.5 Laborer	24.65	98.60	41.25	165.00		
1 Hyd. Excavator, 7 C.Y.		3597.00		3956.70	299.75	329.73
12 L.H., Daily Totals		$3984.00		$4598.50	$332.00	$383.21
Crew B-14F	Hr.	Daily	Hr.	Daily	Bare Costs	Incl. O&P
1 Equip. Oper. (crane)	$36.05	$288.40	$59.60	$476.80	$32.25	$53.48
.5 Laborer	24.65	98.60	41.25	165.00		
1 Hyd. Shovel, 7 C.Y.		4212.00		4633.20	351.00	386.10
12 L.H., Daily Totals		$4599.00		$5275.00	$383.25	$439.58

Crew No.		Bare Costs		Incl. Subs O&P		Cost Per Labor-Hour	
Crew B-14G	Hr.	Daily	Hr.	Daily	Bare Costs	Incl. O&P
1 Equip. Oper. (crane)	$36.05	$288.40	$59.60	$476.80	$32.25	$53.48
.5 Laborer	24.65	98.60	41.25	165.00		
1 Hyd. Shovel, 12 C.Y.		6328.00		6960.80	527.33	580.07
12 L.H., Daily Totals		$6715.00		$7602.60	$559.58	$633.55
Crew B-14J	Hr.	Daily	Hr.	Daily	Bare Costs	Incl. O&P
1 Equip. Oper. (medium)	$35.25	$282.00	$58.25	$466.00	$31.72	$52.58
.5 Laborer	24.65	98.60	41.25	165.00		
1 F.E. Loader, 8 C.Y.		1964.00		2160.40	163.67	180.03
12 L.H., Daily Totals		$2344.60		$2791.40	$195.38	$232.62
Crew B-14K	Hr.	Daily	Hr.	Daily	Bare Costs	Incl. O&P
1 Equip. Oper. (medium)	$35.25	$282.00	$58.25	$466.00	$31.72	$52.58
.5 Laborer	24.65	98.60	41.25	165.00		
1 F.E. Loader, 10 C.Y.		3085.00		3393.50	257.08	282.79
12 L.H., Daily Totals		$3465.60		$4024.50	$288.80	$335.38
Crew B-15	Hr.	Daily	Hr.	Daily	Bare Costs	Incl. O&P
1 Equipment Oper. (med.)	$35.25	$282.00	$58.25	$466.00	$31.11	$51.42
.5 Laborer	24.65	98.60	41.25	165.00		
2 Truck Drivers (heavy)	30.65	490.40	50.55	808.80		
2 Dump Trucks, 12 C.Y., 400 H.P.		1381.60		1519.76		
1 Dozer, 200 H.P.		1397.00		1536.70	99.24	109.16
28 L.H., Daily Totals		$3649.60		$4496.26	$130.34	$160.58
Crew B-16	Hr.	Daily	Hr.	Daily	Bare Costs	Incl. O&P
1 Labor Foreman (outside)	$26.65	$213.20	$44.60	$356.80	$26.65	$44.41
2 Laborers	24.65	394.40	41.25	660.00		
1 Truck Driver (heavy)	30.65	245.20	50.55	404.40		
1 Dump Truck, 12 C.Y., 400 H.P.		690.80		759.88	21.59	23.75
32 L.H., Daily Totals		$1543.60		$2181.08	$48.24	$68.16
Crew B-17	Hr.	Daily	Hr.	Daily	Bare Costs	Incl. O&P
2 Laborers	$24.65	$394.40	$41.25	$660.00	$28.46	$47.27
1 Equip. Oper. (light)	33.90	271.20	56.05	448.40		
1 Truck Driver (heavy)	30.65	245.20	50.55	404.40		
1 Backhoe Loader, 48 H.P.		365.40		401.94		
1 Dump Truck, 8 C.Y., 220 H.P.		417.40		459.14	24.46	26.91
32 L.H., Daily Totals		$1693.60		$2373.88	$52.92	$74.18
Crew B-17A	Hr.	Daily	Hr.	Daily	Bare Costs	Incl. O&P
2 Labor Foremen (outside)	$26.65	$426.40	$44.60	$713.60	$27.16	$45.49
6 Laborers	24.65	1183.20	41.25	1980.00		
1 Skilled Worker Foreman (out)	36.20	289.60	60.80	486.40		
1 Skilled Worker	34.20	273.60	57.45	459.60		
80 L.H., Daily Totals		$2172.80		$3639.60	$27.16	$45.49
Crew B-17B	Hr.	Daily	Hr.	Daily	Bare Costs	Incl. O&P
2 Laborers	$24.65	$394.40	$41.25	$660.00	$28.46	$47.27
1 Equip. Oper. (light)	33.90	271.20	56.05	448.40		
1 Truck Driver (heavy)	30.65	245.20	50.55	404.40		
1 Backhoe Loader, 48 H.P.		365.40		401.94		
1 Dump Truck, 12 C.Y., 400 H.P.		690.80		759.88	33.01	36.31
32 L.H., Daily Totals		$1967.00		$2674.62	$61.47	$83.58
Crew B-18	Hr.	Daily	Hr.	Daily	Bare Costs	Incl. O&P
1 Labor Foreman (outside)	$26.65	$213.20	$44.60	$356.80	$25.32	$42.37
2 Laborers	24.65	394.40	41.25	660.00		
1 Vibrating Plate, Gas, 21"		46.60		51.26	1.94	2.14
24 L.H., Daily Totals		$654.20		$1068.06	$27.26	$44.50

For customer support on your Residential Cost Data, call 877.759.4771.

Crews - Residential

Crew No.		Bare Costs		Incl. Subs O&P		Cost Per Labor-Hour	
Crew B-19	Hr.	Daily	Hr.	Daily	Bare Costs	Incl. O&P
1 Pile Driver Foreman (outside)	$35.90	$287.20	$61.95	$495.60	$33.17	$56.69
4 Pile Drivers	33.90	1084.80	58.50	1872.00		
1 Equip. Oper. (crane)	36.05	288.40	59.60	476.80		
1 Building Laborer	24.65	197.20	41.25	330.00		
1 Crawler Crane, 40 Ton		1203.00		1323.30		
1 Lead, 90' High		124.40		136.84		
1 Hammer, Diesel, 22k ft-lb		378.00		415.80	30.45	33.50
56 L.H., Daily Totals		$3563.00		$5050.34	$63.63	$90.18

Crew B-19A	Hr.	Daily	Hr.	Daily	Bare Costs	Incl. O&P
1 Pile Driver Foreman (outside)	$35.90	$287.20	$61.95	$495.60	$33.17	$56.69
4 Pile Drivers	33.90	1084.80	58.50	1872.00		
1 Equip. Oper. (crane)	36.05	288.40	59.60	476.80		
1 Common Laborer	24.65	197.20	41.25	330.00		
1 Crawler Crane, 75 Ton		1507.00		1657.70		
1 Lead, 90' high		124.40		136.84		
1 Hammer, Diesel, 41k ft-lb		522.60		574.86	38.46	42.31
56 L.H., Daily Totals		$4011.60		$5543.80	$71.64	$99.00

Crew B-19B	Hr.	Daily	Hr.	Daily	Bare Costs	Incl. O&P
1 Pile Driver Foreman (outside)	$35.90	$287.20	$61.95	$495.60	$33.17	$56.69
4 Pile Drivers	33.90	1084.80	58.50	1872.00		
1 Equip. Oper. (crane)	36.05	288.40	59.60	476.80		
1 Common Laborer	24.65	197.20	41.25	330.00		
1 Crawler Crane, 40 Ton		1203.00		1323.30		
1 Lead, 90' High		124.40		136.84		
1 Hammer, Diesel, 22k ft-lb		378.00		415.80		
1 Barge, 400 Ton		779.20		857.12	44.37	48.80
56 L.H., Daily Totals		$4342.20		$5907.46	$77.54	$105.49

Crew B-19C	Hr.	Daily	Hr.	Daily	Bare Costs	Incl. O&P
1 Pile Driver Foreman (outside)	$35.90	$287.20	$61.95	$495.60	$33.17	$56.69
4 Pile Drivers	33.90	1084.80	58.50	1872.00		
1 Equip. Oper. (crane)	36.05	288.40	59.60	476.80		
1 Common Laborer	24.65	197.20	41.25	330.00		
1 Crawler Crane, 75 Ton		1507.00		1657.70		
1 Lead, 90' High		124.40		136.84		
1 Hammer, Diesel, 41k ft-lb		522.60		574.86		
1 Barge, 400 Ton		779.20		857.12	52.38	57.62
56 L.H., Daily Totals		$4790.80		$6400.92	$85.55	$114.30

Crew B-20	Hr.	Daily	Hr.	Daily	Bare Costs	Incl. O&P
1 Labor Foreman (outside)	$26.65	$213.20	$44.60	$356.80	$25.32	$42.37
2 Laborers	24.65	394.40	41.25	660.00		
24 L.H., Daily Totals		$607.60		$1016.80	$25.32	$42.37

Crew B-20A	Hr.	Daily	Hr.	Daily	Bare Costs	Incl. O&P
1 Labor Foreman (outside)	$26.65	$213.20	$44.60	$356.80	$30.40	$50.45
1 Laborer	24.65	197.20	41.25	330.00		
1 Plumber	39.05	312.40	64.40	515.20		
1 Plumber Apprentice	31.25	250.00	51.55	412.40		
32 L.H., Daily Totals		$972.80		$1614.40	$30.40	$50.45

Crew B-21	Hr.	Daily	Hr.	Daily	Bare Costs	Incl. O&P
1 Labor Foreman (outside)	$26.65	$213.20	$44.60	$356.80	$26.85	$44.83
2 Laborers	24.65	394.40	41.25	660.00		
.5 Equip. Oper. (crane)	36.05	144.20	59.60	238.40		
.5 S.P. Crane, 4x4, 5 Ton		140.40		154.44	5.01	5.52
28 L.H., Daily Totals		$892.20		$1409.64	$31.86	$50.34

Crew B-21A	Hr.	Daily	Hr.	Daily	Bare Costs	Incl. O&P
1 Labor Foreman (outside)	$26.65	$213.20	$44.60	$356.80	$31.53	$52.28
1 Laborer	24.65	197.20	41.25	330.00		
1 Plumber	39.05	312.40	64.40	515.20		
1 Plumber Apprentice	31.25	250.00	51.55	412.40		
1 Equip. Oper. (crane)	36.05	288.40	59.60	476.80		
1 S.P. Crane, 4x4, 12 Ton		479.20		527.12	11.98	13.18
40 L.H., Daily Totals		$1740.40		$2618.32	$43.51	$65.46

Crew B-21B	Hr.	Daily	Hr.	Daily	Bare Costs	Incl. O&P
1 Labor Foreman (outside)	$26.65	$213.20	$44.60	$356.80	$27.33	$45.59
3 Laborers	24.65	591.60	41.25	990.00		
1 Equip. Oper. (crane)	36.05	288.40	59.60	476.80		
1 Hyd. Crane, 12 Ton		658.40		724.24	16.46	18.11
40 L.H., Daily Totals		$1751.60		$2547.84	$43.79	$63.70

Crew B-21C	Hr.	Daily	Hr.	Daily	Bare Costs	Incl. O&P
1 Labor Foreman (outside)	$26.65	$213.20	$44.60	$356.80	$27.58	$45.96
4 Laborers	24.65	788.80	41.25	1320.00		
1 Equip. Oper. (crane)	36.05	288.40	59.60	476.80		
1 Equip. Oper. (oiler)	31.75	254.00	52.50	420.00		
2 Cutting Torches		22.80		25.08		
2 Sets of Gases		336.00		369.60		
1 Lattice Boom Crane, 90 Ton		1520.00		1672.00	33.55	36.91
56 L.H., Daily Totals		$3423.20		$4640.28	$61.13	$82.86

Crew B-22	Hr.	Daily	Hr.	Daily	Bare Costs	Incl. O&P
1 Labor Foreman (outside)	$26.65	$213.20	$44.60	$356.80	$27.46	$45.81
2 Laborers	24.65	394.40	41.25	660.00		
.75 Equip. Oper. (crane)	36.05	216.30	59.60	357.60		
.75 S.P. Crane, 4x4, 5 Ton		210.60		231.66	7.02	7.72
30 L.H., Daily Totals		$1034.50		$1606.06	$34.48	$53.54

Crew B-22A	Hr.	Daily	Hr.	Daily	Bare Costs	Incl. O&P
1 Labor Foreman (outside)	$26.65	$213.20	$44.60	$356.80	$29.24	$48.83
1 Skilled Worker	34.20	273.60	57.45	459.60		
2 Laborers	24.65	394.40	41.25	660.00		
1 Equipment Operator, Crane	36.05	288.40	59.60	476.80		
1 S.P. Crane, 4x4, 5 Ton		280.80		308.88		
1 Butt Fusion Mach., 4"-12" diam.		376.10		413.71	16.42	18.06
40 L.H., Daily Totals		$1826.50		$2675.79	$45.66	$66.89

Crew B-22B	Hr.	Daily	Hr.	Daily	Bare Costs	Incl. O&P
1 Labor Foreman (outside)	$26.65	$213.20	$44.60	$356.80	$29.24	$48.83
1 Skilled Worker	34.20	273.60	57.45	459.60		
2 Laborers	24.65	394.40	41.25	660.00		
1 Equip. Oper. (crane)	36.05	288.40	59.60	476.80		
1 S.P. Crane, 4x4, 5 Ton		280.80		308.88		
1 Butt Fusion Mach., 8"-24" diam.		804.80		885.28	27.14	29.85
40 L.H., Daily Totals		$2255.20		$3147.36	$56.38	$78.68

Crew B-22C	Hr.	Daily	Hr.	Daily	Bare Costs	Incl. O&P
1 Skilled Worker	$34.20	$273.60	$57.45	$459.60	$29.43	$49.35
1 Laborer	24.65	197.20	41.25	330.00		
1 Butt Fusion Mach., 2"-8" diam.		121.05		133.16	7.57	8.32
16 L.H., Daily Totals		$591.85		$922.76	$36.99	$57.67

Crews - Residential

Crew No.	Bare Costs		Incl. Subs O&P		Cost Per Labor-Hour	
Crew B-23	Hr.	Daily	Hr.	Daily	Bare Costs	Incl. O&P
1 Labor Foreman (outside)	$26.65	$213.20	$44.60	$356.80	$25.05	$41.92
4 Laborers	24.65	788.80	41.25	1320.00		
1 Drill Rig, Truck-Mounted		2553.00		2808.30		
1 Flatbed Truck, Gas, 3 Ton		307.40		338.14	71.51	78.66
40 L.H., Daily Totals		$3862.40		$4823.24	$96.56	$120.58
Crew B-23A	Hr.	Daily	Hr.	Daily	Bare Costs	Incl. O&P
1 Labor Foreman (outside)	$26.65	$213.20	$44.60	$356.80	$28.85	$48.03
1 Laborer	24.65	197.20	41.25	330.00		
1 Equip. Oper. (medium)	35.25	282.00	58.25	466.00		
1 Drill Rig, Truck-Mounted		2553.00		2808.30		
1 Pickup Truck, 3/4 Ton		145.80		160.38	112.45	123.69
24 L.H., Daily Totals		$3391.20		$4121.48	$141.30	$171.73
Crew B-23B	Hr.	Daily	Hr.	Daily	Bare Costs	Incl. O&P
1 Labor Foreman (outside)	$26.65	$213.20	$44.60	$356.80	$28.85	$48.03
1 Laborer	24.65	197.20	41.25	330.00		
1 Equip. Oper. (medium)	35.25	282.00	58.25	466.00		
1 Drill Rig, Truck-Mounted		2553.00		2808.30		
1 Pickup Truck, 3/4 Ton		145.80		160.38		
1 Centr. Water Pump, 6"		350.80		385.88	127.07	139.77
24 L.H., Daily Totals		$3742.00		$4507.36	$155.92	$187.81
Crew B-24	Hr.	Daily	Hr.	Daily	Bare Costs	Incl. O&P
1 Cement Finisher	$31.95	$255.60	$51.85	$414.80	$30.17	$49.95
1 Laborer	24.65	197.20	41.25	330.00		
1 Carpenter	33.90	271.20	56.75	454.00		
24 L.H., Daily Totals		$724.00		$1198.80	$30.17	$49.95
Crew B-25	Hr.	Daily	Hr.	Daily	Bare Costs	Incl. O&P
1 Labor Foreman (outside)	$26.65	$213.20	$44.60	$356.80	$27.72	$46.19
7 Laborers	24.65	1380.40	41.25	2310.00		
3 Equip. Opers. (medium)	35.25	846.00	58.25	1398.00		
1 Asphalt Paver, 130 H.P.		2210.00		2431.00		
1 Tandem Roller, 10 Ton		239.40		263.34		
1 Roller, Pneum. Whl., 12 Ton		347.80		382.58	31.79	34.97
88 L.H., Daily Totals		$5236.80		$7141.72	$59.51	$81.16
Crew B-25B	Hr.	Daily	Hr.	Daily	Bare Costs	Incl. O&P
1 Labor Foreman (outside)	$26.65	$213.20	$44.60	$356.80	$28.35	$47.20
7 Laborers	24.65	1380.40	41.25	2310.00		
4 Equip. Opers. (medium)	35.25	1128.00	58.25	1864.00		
1 Asphalt Paver, 130 H.P.		2210.00		2431.00		
2 Tandem Rollers, 10 Ton		478.80		526.68		
1 Roller, Pneum. Whl., 12 Ton		347.80		382.58	31.63	34.79
96 L.H., Daily Totals		$5758.20		$7871.06	$59.98	$81.99
Crew B-25C	Hr.	Daily	Hr.	Daily	Bare Costs	Incl. O&P
1 Labor Foreman (outside)	$26.65	$213.20	$44.60	$356.80	$28.52	$47.48
3 Laborers	24.65	591.60	41.25	990.00		
2 Equip. Opers. (medium)	35.25	564.00	58.25	932.00		
1 Asphalt Paver, 130 H.P.		2210.00		2431.00		
1 Tandem Roller, 10 Ton		239.40		263.34	51.03	56.13
48 L.H., Daily Totals		$3818.20		$4973.14	$79.55	$103.61

Crew No.	Bare Costs		Incl. Subs O&P		Cost Per Labor-Hour	
Crew B-25D	Hr.	Daily	Hr.	Daily	Bare Costs	Incl. O&P
1 Labor Foreman (outside)	$26.65	$213.20	$44.60	$356.80	$28.69	$47.75
3 Laborers	24.65	591.60	41.25	990.00		
2.125 Equip. Opers. (medium)	35.25	599.25	58.25	990.25		
.125 Truck Driver (heavy)	30.65	30.65	50.55	50.55		
.125 Truck Tractor, 6x4, 380 H.P.		75.00		82.50		
.125 Dist. Tanker, 3000 Gallon		39.92		43.92		
1 Asphalt Paver, 130 H.P.		2210.00		2431.00		
1 Tandem Roller, 10 Ton		239.40		263.34	51.29	56.42
50 L.H., Daily Totals		$3999.03		$5208.36	$79.98	$104.17
Crew B-25E	Hr.	Daily	Hr.	Daily	Bare Costs	Incl. O&P
1 Labor Foreman (outside)	$26.65	$213.20	$44.60	$356.80	$28.86	$48.01
3 Laborers	24.65	591.60	41.25	990.00		
2.250 Equip. Oper. (medium)	35.25	634.50	58.25	1048.50		
.25 Truck Driver (heavy)	30.65	61.30	50.55	101.10		
.25 Truck Tractor, 6x4, 380 H.P.		150.00		165.00		
.25 Dist. Tanker, 3000 Gallon		79.85		87.83		
1 Asphalt Paver, 130 H.P.		2210.00		2431.00		
1 Tandem Roller, 10 Ton		239.40		263.34	51.52	56.68
52 L.H., Daily Totals		$4179.85		$5443.57	$80.38	$104.68
Crew B-26	Hr.	Daily	Hr.	Daily	Bare Costs	Incl. O&P
1 Labor Foreman (outside)	$26.65	$213.20	$44.60	$356.80	$28.41	$47.31
6 Laborers	24.65	1183.20	41.25	1980.00		
2 Equip. Opers. (medium)	35.25	564.00	58.25	932.00		
1 Rodman (reinf.)	35.50	284.00	59.95	479.60		
1 Cement Finisher	31.95	255.60	51.85	414.80		
1 Grader, 30,000 Lbs.		714.40		785.84		
1 Paving Mach. & Equip.		2796.00		3075.60	39.89	43.88
88 L.H., Daily Totals		$6010.40		$8024.64	$68.30	$91.19
Crew B-26A	Hr.	Daily	Hr.	Daily	Bare Costs	Incl. O&P
1 Labor Foreman (outside)	$26.65	$213.20	$44.60	$356.80	$28.41	$47.31
6 Laborers	24.65	1183.20	41.25	1980.00		
2 Equip. Opers. (medium)	35.25	564.00	58.25	932.00		
1 Rodman (reinf.)	35.50	284.00	59.95	479.60		
1 Cement Finisher	31.95	255.60	51.85	414.80		
1 Grader, 30,000 Lbs.		714.40		785.84		
1 Paving Mach. & Equip.		2796.00		3075.60		
1 Concrete Saw		168.60		185.46	41.81	45.99
88 L.H., Daily Totals		$6179.00		$8210.10	$70.22	$93.30
Crew B-26B	Hr.	Daily	Hr.	Daily	Bare Costs	Incl. O&P
1 Labor Foreman (outside)	$26.65	$213.20	$44.60	$356.80	$28.98	$48.22
6 Laborers	24.65	1183.20	41.25	1980.00		
3 Equip. Opers. (medium)	35.25	846.00	58.25	1398.00		
1 Rodman (reinf.)	35.50	284.00	59.95	479.60		
1 Cement Finisher	31.95	255.60	51.85	414.80		
1 Grader, 30,000 Lbs.		714.40		785.84		
1 Paving Mach. & Equip.		2796.00		3075.60		
1 Concrete Pump, 110' Boom		958.40		1054.24	46.55	51.20
96 L.H., Daily Totals		$7250.80		$9544.88	$75.53	$99.43
Crew B-26C	Hr.	Daily	Hr.	Daily	Bare Costs	Incl. O&P
1 Labor Foreman (outside)	$26.65	$213.20	$44.60	$356.80	$27.73	$46.22
6 Laborers	24.65	1183.20	41.25	1980.00		
1 Equip. Oper. (medium)	35.25	282.00	58.25	466.00		
1 Rodman (reinf.)	35.50	284.00	59.95	479.60		
1 Cement Finisher	31.95	255.60	51.85	414.80		
1 Paving Mach. & Equip.		2796.00		3075.60		
1 Concrete Saw		168.60		185.46	37.06	40.76
80 L.H., Daily Totals		$5182.60		$6958.26	$64.78	$86.98

For customer support on your Residential Cost Data, call 877.759.4771.

Crews - Residential

Crew No.	Bare Costs		Incl. Subs O&P		Cost Per Labor-Hour	
	Hr.	Daily	Hr.	Daily	Bare Costs	Incl. O&P
Crew B-27					$25.15	$42.09
1 Labor Foreman (outside)	$26.65	$213.20	$44.60	$356.80		
3 Laborers	24.65	591.60	41.25	990.00		
1 Berm Machine		295.80		325.38	9.24	10.17
32 L.H., Daily Totals		$1100.60		$1672.18	$34.39	$52.26
Crew B-28					$30.82	$51.58
2 Carpenters	$33.90	$542.40	$56.75	$908.00		
1 Laborer	24.65	197.20	41.25	330.00		
24 L.H., Daily Totals		$739.60		$1238.00	$30.82	$51.58
Crew B-29					$26.88	$44.87
1 Labor Foreman (outside)	$26.65	$213.20	$44.60	$356.80		
4 Laborers	24.65	788.80	41.25	1320.00		
1 Equip. Oper. (crane)	36.05	288.40	59.60	476.80		
1 Gradall, 5/8 C.Y.		882.00		970.20	18.38	20.21
48 L.H., Daily Totals		$2172.40		$3123.80	$45.26	$65.08
Crew B-30					$32.18	$53.12
1 Equip. Oper. (medium)	$35.25	$282.00	$58.25	$466.00		
2 Truck Drivers (heavy)	30.65	490.40	50.55	808.80		
1 Hyd. Excavator, 1.5 C.Y.		1029.00		1131.90		
2 Dump Trucks, 12 C.Y., 400 H.P.		1381.60		1519.76	100.44	110.49
24 L.H., Daily Totals		$3183.00		$3926.46	$132.63	$163.60
Crew B-31					$25.05	$41.92
1 Labor Foreman (outside)	$26.65	$213.20	$44.60	$356.80		
4 Laborers	24.65	788.80	41.25	1320.00		
1 Air Compressor, 250 cfm		197.80		217.58		
1 Sheeting Driver		6.50		7.15		
2 -50' Air Hoses, 1.5"		10.40		11.44	5.37	5.90
40 L.H., Daily Totals		$1216.70		$1912.97	$30.42	$47.82
Crew B-32					$32.60	$54.00
1 Laborer	$24.65	$197.20	$41.25	$330.00		
3 Equip. Opers. (medium)	35.25	846.00	58.25	1398.00		
1 Grader, 30,000 Lbs.		714.40		785.84		
1 Tandem Roller, 10 Ton		239.40		263.34		
1 Dozer, 200 H.P.		1397.00		1536.70	73.46	80.81
32 L.H., Daily Totals		$3394.00		$4313.88	$106.06	$134.81
Crew B-32A					$31.72	$52.58
1 Laborer	$24.65	$197.20	$41.25	$330.00		
2 Equip. Opers. (medium)	35.25	564.00	58.25	932.00		
1 Grader, 30,000 Lbs.		714.40		785.84		
1 Roller, Vibratory, 25 Ton		691.80		760.98	58.59	64.45
24 L.H., Daily Totals		$2167.40		$2808.82	$90.31	$117.03
Crew B-32B					$31.72	$52.58
1 Laborer	$24.65	$197.20	$41.25	$330.00		
2 Equip. Opers. (medium)	35.25	564.00	58.25	932.00		
1 Dozer, 200 H.P.		1397.00		1536.70		
1 Roller, Vibratory, 25 Ton		691.80		760.98	87.03	95.74
24 L.H., Daily Totals		$2850.00		$3559.68	$118.75	$148.32

Crew No.	Bare Costs		Incl. Subs O&P		Cost Per Labor-Hour	
	Hr.	Daily	Hr.	Daily	Bare Costs	Incl. O&P
Crew B-32C					$30.28	$50.31
1 Labor Foreman (outside)	$26.65	$213.20	$44.60	$356.80		
2 Laborers	24.65	394.40	41.25	660.00		
3 Equip. Opers. (medium)	35.25	846.00	58.25	1398.00		
1 Grader, 30,000 Lbs.		714.40		785.84		
1 Tandem Roller, 10 Ton		239.40		263.34		
1 Dozer, 200 H.P.		1397.00		1536.70	48.98	53.87
48 L.H., Daily Totals		$3804.40		$5000.68	$79.26	$104.18
Crew B-33A					$35.25	$58.25
1 Equip. Oper. (medium)	$35.25	$282.00	$58.25	$466.00		
.25 Equip. Oper. (medium)	35.25	70.50	58.25	116.50		
1 Scraper, Towed, 7 C.Y.		113.20		124.52		
1.250 Dozers, 300 H.P.		2358.75		2594.63	247.19	271.91
10 L.H., Daily Totals		$2824.45		$3301.65	$282.44	$330.16
Crew B-33B					$35.25	$58.25
1 Equip. Oper. (medium)	$35.25	$282.00	$58.25	$466.00		
.25 Equip. Oper. (medium)	35.25	70.50	58.25	116.50		
1 Scraper, Towed, 10 C.Y.		145.00		159.50		
1.250 Dozers, 300 H.P.		2358.75		2594.63	250.38	275.41
10 L.H., Daily Totals		$2856.25		$3336.63	$285.63	$333.66
Crew B-33C					$35.25	$58.25
1 Equip. Oper. (medium)	$35.25	$282.00	$58.25	$466.00		
.25 Equip. Oper. (medium)	35.25	70.50	58.25	116.50		
1 Scraper, Towed, 15 C.Y.		163.60		179.96		
1.250 Dozers, 300 H.P.		2358.75		2594.63	252.24	277.46
10 L.H., Daily Totals		$2874.85		$3357.09	$287.49	$335.71
Crew B-33D					$35.25	$58.25
1 Equip. Oper. (medium)	$35.25	$282.00	$58.25	$466.00		
.25 Equip. Oper. (medium)	35.25	70.50	58.25	116.50		
1 S.P. Scraper, 14 C.Y.		2172.00		2389.20		
.25 Dozer, 300 H.P.		471.75		518.92	264.38	290.81
10 L.H., Daily Totals		$2996.25		$3490.63	$299.63	$349.06
Crew B-33E					$35.25	$58.25
1 Equip. Oper. (medium)	$35.25	$282.00	$58.25	$466.00		
.25 Equip. Oper. (medium)	35.25	70.50	58.25	116.50		
1 S.P. Scraper, 21 C.Y.		2715.00		2986.50		
.25 Dozer, 300 H.P.		471.75		518.92	318.68	350.54
10 L.H., Daily Totals		$3539.25		$4087.93	$353.93	$408.79
Crew B-33F					$35.25	$58.25
1 Equip. Oper. (medium)	$35.25	$282.00	$58.25	$466.00		
.25 Equip. Oper. (medium)	35.25	70.50	58.25	116.50		
1 Elev. Scraper, 11 C.Y.		1173.00		1290.30		
.25 Dozer, 300 H.P.		471.75		518.92	164.47	180.92
10 L.H., Daily Totals		$1997.25		$2391.72	$199.72	$239.17
Crew B-33G					$35.25	$58.25
1 Equip. Oper. (medium)	$35.25	$282.00	$58.25	$466.00		
.25 Equip. Oper. (medium)	35.25	70.50	58.25	116.50		
1 Elev. Scraper, 22 C.Y.		2548.00		2802.80		
.25 Dozer, 300 H.P.		471.75		518.92	301.98	332.17
10 L.H., Daily Totals		$3372.25		$3904.22	$337.23	$390.42

Crews - Residential

Crew No.	Bare Costs Hr.	Bare Costs Daily	Incl. Subs O&P Hr.	Incl. Subs O&P Daily	Bare Costs	Incl. O&P	Cost Per Labor-Hour Bare Costs	Cost Per Labor-Hour Incl. O&P
Crew B-33K								
1 Equipment Operator (med.)	$35.25	$282.00	$58.25	$466.00	$32.22	$53.39		
.25 Equipment Operator (med.)	35.25	70.50	58.25	116.50				
.5 Laborer	24.65	98.60	41.25	165.00				
1 S.P. Scraper, 31 C.Y.		3735.00		4108.50				
.25 Dozer, 410 H.P.		598.00		657.80	309.50	340.45		
14 L.H., Daily Totals		$4784.10		$5513.80	$341.72	$393.84		
Crew B-34A								
1 Truck Driver (heavy)	$30.65	$245.20	$50.55	$404.40	$30.65	$50.55		
1 Dump Truck, 8 C.Y., 220 H.P.		417.40		459.14	52.17	57.39		
8 L.H., Daily Totals		$662.60		$863.54	$82.83	$107.94		
Crew B-34B								
1 Truck Driver (heavy)	$30.65	$245.20	$50.55	$404.40	$30.65	$50.55		
1 Dump Truck, 12 C.Y., 400 H.P.		690.80		759.88	86.35	94.98		
8 L.H., Daily Totals		$936.00		$1164.28	$117.00	$145.54		
Crew B-34C								
1 Truck Driver (heavy)	$30.65	$245.20	$50.55	$404.40	$30.65	$50.55		
1 Truck Tractor, 6x4, 380 H.P.		600.00		660.00				
1 Dump Trailer, 16.5 C.Y.		125.80		138.38	90.72	99.80		
8 L.H., Daily Totals		$971.00		$1202.78	$121.38	$150.35		
Crew B-34D								
1 Truck Driver (heavy)	$30.65	$245.20	$50.55	$404.40	$30.65	$50.55		
1 Truck Tractor, 6x4, 380 H.P.		600.00		660.00				
1 Dump Trailer, 20 C.Y.		140.40		154.44	92.55	101.81		
8 L.H., Daily Totals		$985.60		$1218.84	$123.20	$152.35		
Crew B-34E								
1 Truck Driver (heavy)	$30.65	$245.20	$50.55	$404.40	$30.65	$50.55		
1 Dump Truck, Off Hwy., 25 Ton		1370.00		1507.00	171.25	188.38		
8 L.H., Daily Totals		$1615.20		$1911.40	$201.90	$238.93		
Crew B-34F								
1 Truck Driver (heavy)	$30.65	$245.20	$50.55	$404.40	$30.65	$50.55		
1 Dump Truck, Off Hwy., 35 Ton		1529.00		1681.90	191.13	210.24		
8 L.H., Daily Totals		$1774.20		$2086.30	$221.78	$260.79		
Crew B-34G								
1 Truck Driver (heavy)	$30.65	$245.20	$50.55	$404.40	$30.65	$50.55		
1 Dump Truck, Off Hwy., 50 Ton		1861.00		2047.10	232.63	255.89		
8 L.H., Daily Totals		$2106.20		$2451.50	$263.27	$306.44		
Crew B-34H								
1 Truck Driver (heavy)	$30.65	$245.20	$50.55	$404.40	$30.65	$50.55		
1 Dump Truck, Off Hwy., 65 Ton		1915.00		2106.50	239.38	263.31		
8 L.H., Daily Totals		$2160.20		$2510.90	$270.02	$313.86		
Crew B-34I								
1 Truck Driver (heavy)	$30.65	$245.20	$50.55	$404.40	$30.65	$50.55		
1 Dump Truck, 18 C.Y., 450 H.P.		869.40		956.34	108.68	119.54		
8 L.H., Daily Totals		$1114.60		$1360.74	$139.32	$170.09		
Crew B-34J								
1 Truck Driver (heavy)	$30.65	$245.20	$50.55	$404.40	$30.65	$50.55		
1 Dump Truck, Off Hwy., 100 Ton		2950.00		3245.00	368.75	405.63		
8 L.H., Daily Totals		$3195.20		$3649.40	$399.40	$456.18		
Crew B-34K								
1 Truck Driver (heavy)	$30.65	$245.20	$50.55	$404.40	$30.65	$50.55		
1 Truck Tractor, 6x4, 450 H.P.		726.80		799.48				
1 Lowbed Trailer, 75 Ton		223.40		245.74	118.78	130.65		
8 L.H., Daily Totals		$1195.40		$1449.62	$149.43	$181.20		
Crew B-34L								
1 Equip. Oper. (light)	$33.90	$271.20	$56.05	$448.40	$33.90	$56.05		
1 Flatbed Truck, Gas, 1.5 Ton		248.20		273.02	31.02	34.13		
8 L.H., Daily Totals		$519.40		$721.42	$64.92	$90.18		
Crew B-34M								
1 Equip. Oper. (light)	$33.90	$271.20	$56.05	$448.40	$33.90	$56.05		
1 Flatbed Truck, Gas, 3 Ton		307.40		338.14	38.42	42.27		
8 L.H., Daily Totals		$578.60		$786.54	$72.33	$98.32		
Crew B-34N								
1 Truck Driver (heavy)	$30.65	$245.20	$50.55	$404.40	$32.95	$54.40		
1 Equip. Oper. (medium)	35.25	282.00	58.25	466.00				
1 Truck Tractor, 6x4, 380 H.P.		600.00		660.00				
1 Flatbed Trailer, 40 Ton		154.60		170.06	47.16	51.88		
16 L.H., Daily Totals		$1281.80		$1700.46	$80.11	$106.28		
Crew B-34P								
1 Pipe Fitter	$40.05	$320.40	$66.05	$528.40	$35.03	$57.82		
1 Truck Driver (light)	29.80	238.40	49.15	393.20				
1 Equip. Oper. (medium)	35.25	282.00	58.25	466.00				
1 Flatbed Truck, Gas, 3 Ton		307.40		338.14				
1 Backhoe Loader, 48 H.P.		365.40		401.94	28.03	30.84		
24 L.H., Daily Totals		$1513.60		$2127.68	$63.07	$88.65		
Crew B-34Q								
1 Pipe Fitter	$40.05	$320.40	$66.05	$528.40	$35.30	$58.27		
1 Truck Driver (light)	29.80	238.40	49.15	393.20				
1 Equip. Oper. (crane)	36.05	288.40	59.60	476.80				
1 Flatbed Trailer, 25 Ton		113.20		124.52				
1 Dump Truck, 8 C.Y., 220 H.P.		417.40		459.14				
1 Hyd. Crane, 25 Ton		748.80		823.68	53.31	58.64		
24 L.H., Daily Totals		$2126.60		$2805.74	$88.61	$116.91		
Crew B-34R								
1 Pipe Fitter	$40.05	$320.40	$66.05	$528.40	$35.30	$58.27		
1 Truck Driver (light)	29.80	238.40	49.15	393.20				
1 Equip. Oper. (crane)	36.05	288.40	59.60	476.80				
1 Flatbed Trailer, 25 Ton		113.20		124.52				
1 Dump Truck, 8 C.Y., 220 H.P.		417.40		459.14				
1 Hyd. Crane, 25 Ton		748.80		823.68				
1 Hyd. Excavator, 1 C.Y.		812.80		894.08	87.17	95.89		
24 L.H., Daily Totals		$2939.40		$3699.82	$122.47	$154.16		
Crew B-34S								
2 Pipe Fitters	$40.05	$640.80	$66.05	$1056.80	$36.70	$60.56		
1 Truck Driver (heavy)	30.65	245.20	50.55	404.40				
1 Equip. Oper. (crane)	36.05	288.40	59.60	476.80				
1 Flatbed Trailer, 40 Ton		154.60		170.06				
1 Truck Tractor, 6x4, 380 H.P.		600.00		660.00				
1 Hyd. Crane, 80 Ton		1630.00		1793.00				
1 Hyd. Excavator, 2 C.Y.		1176.00		1293.60	111.27	122.40		
32 L.H., Daily Totals		$4735.00		$5854.66	$147.97	$182.96		

Crews - Residential

Crew B-34T	Hr.	Daily	Hr.	Daily	Bare Costs	Incl. O&P
2 Pipe Fitters	$40.05	$640.80	$66.05	$1056.80	$36.70	$60.56
1 Truck Driver (heavy)	30.65	245.20	50.55	404.40		
1 Equip. Oper. (crane)	36.05	288.40	59.60	476.80		
1 Flatbed Trailer, 40 Ton		154.60		170.06		
1 Truck Tractor, 6x4, 380 H.P.		600.00		660.00		
1 Hyd. Crane, 80 Ton		1630.00		1793.00	74.52	81.97
32 L.H., Daily Totals		$3559.00		$4561.06	$111.22	$142.53

Crew B-34U	Hr.	Daily	Hr.	Daily	Bare Costs	Incl. O&P
1 Truck Driver (heavy)	$30.65	$245.20	$50.55	$404.40	$32.27	$53.30
1 Equip. Oper. (light)	33.90	271.20	56.05	448.40		
1 Truck Tractor, 220 H.P.		358.80		394.68		
1 Flatbed Trailer, 25 Ton		113.20		124.52	29.50	32.45
16 L.H., Daily Totals		$988.40		$1372.00	$61.77	$85.75

Crew B-34V	Hr.	Daily	Hr.	Daily	Bare Costs	Incl. O&P
1 Truck Driver (heavy)	$30.65	$245.20	$50.55	$404.40	$33.53	$55.40
1 Equip. Oper. (crane)	36.05	288.40	59.60	476.80		
1 Equip. Oper. (light)	33.90	271.20	56.05	448.40		
1 Truck Tractor, 6x4, 450 H.P.		726.80		799.48		
1 Equipment Trailer, 50 Ton		170.00		187.00		
1 Pickup Truck, 4x4, 3/4 Ton		156.80		172.48	43.90	48.29
24 L.H., Daily Totals		$1858.40		$2488.56	$77.43	$103.69

Crew B-34W	Hr.	Daily	Hr.	Daily	Bare Costs	Incl. O&P
5 Truck Drivers (heavy)	$30.65	$1226.00	$50.55	$2022.00	$31.81	$52.57
2 Equip. Opers. (crane)	36.05	576.80	59.60	953.60		
1 Equip. Oper. (mechanic)	36.25	290.00	59.90	479.20		
1 Laborer	24.65	197.20	41.25	330.00		
4 Truck Tractors, 6x4, 380 H.P.		2400.00		2640.00		
2 Equipment Trailers, 50 Ton		340.00		374.00		
2 Flatbed Trailers, 40 Ton		309.20		340.12		
1 Pickup Truck, 4x4, 3/4 Ton		156.80		172.48		
1 S.P. Crane, 4x4, 20 Ton		554.80		610.28	52.23	57.46
72 L.H., Daily Totals		$6050.80		$7921.68	$84.04	$110.02

Crew B-35	Hr.	Daily	Hr.	Daily	Bare Costs	Incl. O&P
1 Labor Foreman (outside)	$26.65	$213.20	$44.60	$356.80	$32.12	$53.46
1 Skilled Worker	34.20	273.60	57.45	459.60		
1 Welder (plumber)	39.05	312.40	64.40	515.20		
1 Laborer	24.65	197.20	41.25	330.00		
1 Equip. Oper. (crane)	36.05	288.40	59.60	476.80		
1 Welder, Electric, 300 amp		58.10		63.91		
1 Hyd. Excavator, .75 C.Y.		654.00		719.40	17.80	19.58
40 L.H., Daily Totals		$1996.90		$2921.71	$49.92	$73.04

Crew B-35A	Hr.	Daily	Hr.	Daily	Bare Costs	Incl. O&P
1 Labor Foreman (outside)	$26.65	$213.20	$44.60	$356.80	$31.00	$51.58
2 Laborers	24.65	394.40	41.25	660.00		
1 Skilled Worker	34.20	273.60	57.45	459.60		
1 Welder (plumber)	39.05	312.40	64.40	515.20		
1 Equip. Oper. (crane)	36.05	288.40	59.60	476.80		
1 Equip. Oper. (oiler)	31.75	254.00	52.50	420.00		
1 Welder, Gas Engine, 300 amp		147.20		161.92		
1 Crawler Crane, 75 Ton		1507.00		1657.70	29.54	32.49
56 L.H., Daily Totals		$3390.20		$4708.02	$60.54	$84.07

Crew B-36	Hr.	Daily	Hr.	Daily	Bare Costs	Incl. O&P
1 Labor Foreman (outside)	$26.65	$213.20	$44.60	$356.80	$29.29	$48.72
2 Laborers	24.65	394.40	41.25	660.00		
2 Equip. Opers. (medium)	35.25	564.00	58.25	932.00		
1 Dozer, 200 H.P.		1397.00		1536.70		
1 Aggregate Spreader		39.60		43.56		
1 Tandem Roller, 10 Ton		239.40		263.34	41.90	46.09
40 L.H., Daily Totals		$2847.60		$3792.40	$71.19	$94.81

Crew B-36A	Hr.	Daily	Hr.	Daily	Bare Costs	Incl. O&P
1 Labor Foreman (outside)	$26.65	$213.20	$44.60	$356.80	$30.99	$51.44
2 Laborers	24.65	394.40	41.25	660.00		
4 Equip. Opers. (medium)	35.25	1128.00	58.25	1864.00		
1 Dozer, 200 H.P.		1397.00		1536.70		
1 Aggregate Spreader		39.60		43.56		
1 Tandem Roller, 10 Ton		239.40		263.34		
1 Roller, Pneum. Whl., 12 Ton		347.80		382.58	36.14	39.75
56 L.H., Daily Totals		$3759.40		$5106.98	$67.13	$91.20

Crew B-36B	Hr.	Daily	Hr.	Daily	Bare Costs	Incl. O&P
1 Labor Foreman (outside)	$26.65	$213.20	$44.60	$356.80	$30.95	$51.33
2 Laborers	24.65	394.40	41.25	660.00		
4 Equip. Opers. (medium)	35.25	1128.00	58.25	1864.00		
1 Truck Driver (heavy)	30.65	245.20	50.55	404.40		
1 Grader, 30,000 Lbs.		714.40		785.84		
1 F.E. Loader, Crl, 1.5 C.Y.		634.20		697.62		
1 Dozer, 300 H.P.		1887.00		2075.70		
1 Roller, Vibratory, 25 Ton		691.80		760.98		
1 Truck Tractor, 6x4, 450 H.P.		726.80		799.48		
1 Water Tank Trailer, 5000 Gal.		141.60		155.76	74.93	82.43
64 L.H., Daily Totals		$6776.60		$8560.58	$105.88	$133.76

Crew B-36C	Hr.	Daily	Hr.	Daily	Bare Costs	Incl. O&P
1 Labor Foreman (outside)	$26.65	$213.20	$44.60	$356.80	$32.61	$53.98
3 Equip. Opers. (medium)	35.25	846.00	58.25	1398.00		
1 Truck Driver (heavy)	30.65	245.20	50.55	404.40		
1 Grader, 30,000 Lbs.		714.40		785.84		
1 Dozer, 300 H.P.		1887.00		2075.70		
1 Roller, Vibratory, 25 Ton		691.80		760.98		
1 Truck Tractor, 6x4, 450 H.P.		726.80		799.48		
1 Water Tank Trailer, 5000 Gal.		141.60		155.76	104.04	114.44
40 L.H., Daily Totals		$5466.00		$6736.96	$136.65	$168.42

Crew B-36E	Hr.	Daily	Hr.	Daily	Bare Costs	Incl. O&P
1 Labor Foreman (outside)	$26.65	$213.20	$44.60	$356.80	$33.05	$54.69
4 Equip. Opers. (medium)	35.25	1128.00	58.25	1864.00		
1 Truck Driver (heavy)	30.65	245.20	50.55	404.40		
1 Grader, 30,000 Lbs.		714.40		785.84		
1 Dozer, 300 H.P.		1887.00		2075.70		
1 Roller, Vibratory, 25 Ton		691.80		760.98		
1 Truck Tractor, 6x4, 380 H.P.		600.00		660.00		
1 Dist. Tanker, 3000 Gallon		319.40		351.34	87.76	96.54
48 L.H., Daily Totals		$5799.00		$7259.06	$120.81	$151.23

Crew B-37	Hr.	Daily	Hr.	Daily	Bare Costs	Incl. O&P
1 Labor Foreman (outside)	$26.65	$213.20	$44.60	$356.80	$26.52	$44.27
4 Laborers	24.65	788.80	41.25	1320.00		
1 Equip. Oper. (light)	33.90	271.20	56.05	448.40		
1 Tandem Roller, 5 Ton		162.60		178.86	3.39	3.73
48 L.H., Daily Totals		$1435.80		$2304.06	$29.91	$48.00

Crews - Residential

Crew No.	Bare Costs		Incl. Subs O&P		Cost Per Labor-Hour	

Crew B-37A	Hr.	Daily	Hr.	Daily	Bare Costs	Incl. O&P
2 Laborers	$24.65	$394.40	$41.25	$660.00	$26.37	$43.88
1 Truck Driver (light)	29.80	238.40	49.15	393.20		
1 Flatbed Truck, Gas, 1.5 Ton		248.20		273.02		
1 Tar Kettle, T.M.		135.30		148.83	15.98	17.58
24 L.H., Daily Totals		$1016.30		$1475.05	$42.35	$61.46

Crew B-37B	Hr.	Daily	Hr.	Daily	Bare Costs	Incl. O&P
3 Laborers	$24.65	$591.60	$41.25	$990.00	$25.94	$43.23
1 Truck Driver (light)	29.80	238.40	49.15	393.20		
1 Flatbed Truck, Gas, 1.5 Ton		248.20		273.02		
1 Tar Kettle, T.M.		135.30		148.83	11.98	13.18
32 L.H., Daily Totals		$1213.50		$1805.05	$37.92	$56.41

Crew B-37C	Hr.	Daily	Hr.	Daily	Bare Costs	Incl. O&P
2 Laborers	$24.65	$394.40	$41.25	$660.00	$27.23	$45.20
2 Truck Drivers (light)	29.80	476.80	49.15	786.40		
2 Flatbed Trucks, Gas, 1.5 Ton		496.40		546.04		
1 Tar Kettle, T.M.		135.30		148.83	19.74	21.71
32 L.H., Daily Totals		$1502.90		$2141.27	$46.97	$66.91

Crew B-37D	Hr.	Daily	Hr.	Daily	Bare Costs	Incl. O&P
1 Laborer	$24.65	$197.20	$41.25	$330.00	$27.23	$45.20
1 Truck Driver (light)	29.80	238.40	49.15	393.20		
1 Pickup Truck, 3/4 Ton		145.80		160.38	9.11	10.02
16 L.H., Daily Totals		$581.40		$883.58	$36.34	$55.22

Crew B-37E	Hr.	Daily	Hr.	Daily	Bare Costs	Incl. O&P
3 Laborers	$24.65	$591.60	$41.25	$990.00	$28.96	$48.05
1 Equip. Oper. (light)	33.90	271.20	56.05	448.40		
1 Equip. Oper. (medium)	35.25	282.00	58.25	466.00		
2 Truck Drivers (light)	29.80	476.80	49.15	786.40		
4 Barrels w/ Flasher		13.60		14.96		
1 Concrete Saw		168.60		185.46		
1 Rotary Hammer Drill		24.55		27.00		
1 Hammer Drill Bit		2.90		3.19		
1 Loader, Skid Steer, 30 H.P.		173.20		190.52		
1 Conc. Hammer Attach.		108.40		119.24		
1 Vibrating Plate, Gas, 18"		36.20		39.82		
2 Flatbed Trucks, Gas, 1.5 Ton		496.40		546.04	18.28	20.11
56 L.H., Daily Totals		$2645.45		$3817.03	$47.24	$68.16

Crew B-37F	Hr.	Daily	Hr.	Daily	Bare Costs	Incl. O&P
3 Laborers	$24.65	$591.60	$41.25	$990.00	$25.94	$43.23
1 Truck Driver (light)	29.80	238.40	49.15	393.20		
4 Barrels w/ Flasher		13.60		14.96		
1 Concrete Mixer, 10 C.F.		172.80		190.08		
1 Air Compressor, 60 cfm		142.80		157.08		
1 -50' Air Hose, 3/4"		3.05		3.36		
1 Spade (Chipper)		8.20		9.02		
1 Flatbed Truck, Gas, 1.5 Ton		248.20		273.02	18.40	20.23
32 L.H., Daily Totals		$1418.65		$2030.71	$44.33	$63.46

Crew B-37G	Hr.	Daily	Hr.	Daily	Bare Costs	Incl. O&P
1 Labor Foreman (outside)	$26.65	$213.20	$44.60	$356.80	$26.52	$44.27
4 Laborers	24.65	788.80	41.25	1320.00		
1 Equip. Oper. (light)	33.90	271.20	56.05	448.40		
1 Berm Machine		295.80		325.38		
1 Tandem Roller, 5 Ton		162.60		178.86	9.55	10.51
48 L.H., Daily Totals		$1731.60		$2629.44	$36.08	$54.78

Crew B-37H	Hr.	Daily	Hr.	Daily	Bare Costs	Incl. O&P
1 Labor Foreman (outside)	$26.65	$213.20	$44.60	$356.80	$26.52	$44.27
4 Laborers	24.65	788.80	41.25	1320.00		
1 Equip. Oper. (light)	33.90	271.20	56.05	448.40		
1 Tandem Roller, 5 Ton		162.60		178.86		
1 Flatbed Truck, Gas, 1.5 Ton		248.20		273.02		
1 Tar Kettle, T.M.		135.30		148.83	11.38	12.51
48 L.H., Daily Totals		$1819.30		$2725.91	$37.90	$56.79

Crew B-37I	Hr.	Daily	Hr.	Daily	Bare Costs	Incl. O&P
3 Laborers	$24.65	$591.60	$41.25	$990.00	$28.96	$48.05
1 Equip. Oper. (light)	33.90	271.20	56.05	448.40		
1 Equip. Oper. (medium)	35.25	282.00	58.25	466.00		
2 Truck Drivers (light)	29.80	476.80	49.15	786.40		
4 Barrels w/ Flasher		13.60		14.96		
1 Concrete Saw		168.60		185.46		
1 Rotary Hammer Drill		24.55		27.00		
1 Hammer Drill Bit		2.90		3.19		
1 Air Compressor, 60 cfm		142.80		157.08		
1 -50' Air Hose, 3/4"		3.05		3.36		
1 Spade (Chipper)		8.20		9.02		
1 Loader, Skid Steer, 30 H.P.		173.20		190.52		
1 Conc. Hammer Attach.		108.40		119.24		
1 Concrete Mixer, 10 C.F.		172.80		190.08		
1 Vibrating Plate, Gas, 18"		36.20		39.82		
2 Flatbed Trucks, Gas, 1.5 Ton		496.40		546.04	24.12	26.53
56 L.H., Daily Totals		$2972.30		$4176.57	$53.08	$74.58

Crew B-37J	Hr.	Daily	Hr.	Daily	Bare Costs	Incl. O&P
1 Labor Foreman (outside)	$26.65	$213.20	$44.60	$356.80	$26.52	$44.27
4 Laborers	24.65	788.80	41.25	1320.00		
1 Equip. Oper. (light)	33.90	271.20	56.05	448.40		
1 Air Compressor, 60 cfm		142.80		157.08		
1 -50' Air Hose, 3/4"		3.05		3.36		
2 Concrete Mixers, 10 C.F.		345.60		380.16		
2 Flatbed Trucks, Gas, 1.5 Ton		496.40		546.04		
1 Shot Blaster, 20"		215.40		236.94	25.07	27.57
48 L.H., Daily Totals		$2476.45		$3448.78	$51.59	$71.85

Crew B-37K	Hr.	Daily	Hr.	Daily	Bare Costs	Incl. O&P
1 Labor Foreman (outside)	$26.65	$213.20	$44.60	$356.80	$26.52	$44.27
4 Laborers	24.65	788.80	41.25	1320.00		
1 Equip. Oper. (light)	33.90	271.20	56.05	448.40		
1 Air Compressor, 60 cfm		142.80		157.08		
1 -50' Air Hose, 3/4"		3.05		3.36		
2 Flatbed Trucks, Gas, 1.5 Ton		496.40		546.04		
1 Shot Blaster, 20"		215.40		236.94	17.87	19.65
48 L.H., Daily Totals		$2130.85		$3068.61	$44.39	$63.93

Crew B-38	Hr.	Daily	Hr.	Daily	Bare Costs	Incl. O&P
2 Laborers	$24.65	$394.40	$41.25	$660.00	$27.73	$46.18
1 Equip. Oper. (light)	33.90	271.20	56.05	448.40		
1 Backhoe Loader, 48 H.P.		365.40		401.94		
1 Hyd. Hammer, (1200 lb.)		179.60		197.56	22.71	24.98
24 L.H., Daily Totals		$1210.60		$1707.90	$50.44	$71.16

Crew B-39	Hr.	Daily	Hr.	Daily	Bare Costs	Incl. O&P
1 Labor Foreman (outside)	$26.65	$213.20	$44.60	$356.80	$24.98	$41.81
5 Laborers	24.65	986.00	41.25	1650.00		
1 Air Compressor, 250 cfm		197.80		217.58		
2 Breakers, Pavement, 60 lb.		20.40		22.44		
2 -50' Air Hoses, 1.5"		10.40		11.44	4.76	5.24
48 L.H., Daily Totals		$1427.80		$2258.26	$29.75	$47.05

Crews - Residential

Crew No.	Bare Costs		Incl. Subs O&P		Cost Per Labor-Hour	
	Hr.	Daily	Hr.	Daily	Bare Costs	Incl. O&P
Crew B-40						
1 Pile Driver Foreman (outside)	$35.90	$287.20	$61.95	$495.60	$33.17	$56.69
4 Pile Drivers	33.90	1084.80	58.50	1872.00		
1 Building Laborer	24.65	197.20	41.25	330.00		
1 Equip. Oper. (crane)	36.05	288.40	59.60	476.80		
1 Crawler Crane, 40 Ton		1203.00		1323.30		
1 Vibratory Hammer & Gen.		2612.00		2873.20	68.13	74.94
56 L.H., Daily Totals		$5672.60		$7370.90	$101.30	$131.62
Crew B-40B						
1 Labor Foreman (outside)	$26.65	$213.20	$44.60	$356.80	$28.07	$46.74
3 Laborers	24.65	591.60	41.25	990.00		
1 Equip. Oper. (crane)	36.05	288.40	59.60	476.80		
1 Equip. Oper. (oiler)	31.75	254.00	52.50	420.00		
1 Lattice Boom Crane, 40 Ton		1182.00		1300.20	24.63	27.09
48 L.H., Daily Totals		$2529.20		$3543.80	$52.69	$73.83
Crew B-41						
1 Labor Foreman (outside)	$26.65	$213.20	$44.60	$356.80	$25.85	$43.20
4 Laborers	24.65	788.80	41.25	1320.00		
.25 Equip. Oper. (crane)	36.05	72.10	59.60	119.20		
.25 Equip. Oper. (oiler)	31.75	63.50	52.50	105.00		
.25 Crawler Crane, 40 Ton		300.75		330.82	6.84	7.52
44 L.H., Daily Totals		$1438.35		$2231.82	$32.69	$50.72
Crew B-42						
1 Labor Foreman (outside)	$26.65	$213.20	$44.60	$356.80	$28.62	$47.66
4 Laborers	24.65	788.80	41.25	1320.00		
1 Equip. Oper. (crane)	36.05	288.40	59.60	476.80		
1 Welder	39.05	312.40	64.40	515.20		
1 Hyd. Crane, 25 Ton		748.80		823.68		
1 Welder, Gas Engine, 300 amp		147.20		161.92		
1 Horz. Boring Csg. Mch.		477.40		525.14	24.52	26.98
56 L.H., Daily Totals		$2976.20		$4179.54	$53.15	$74.63
Crew B-43						
1 Labor Foreman (outside)	$26.65	$213.20	$44.60	$356.80	$25.05	$41.92
4 Laborers	24.65	788.80	41.25	1320.00		
1 Drill Rig, Truck-Mounted		2553.00		2808.30	63.83	70.21
40 L.H., Daily Totals		$3555.00		$4485.10	$88.88	$112.13
Crew B-44						
1 Pile Driver Foreman (outside)	$35.90	$287.20	$61.95	$495.60	$32.11	$54.76
4 Pile Drivers	33.90	1084.80	58.50	1872.00		
1 Equip. Oper. (crane)	36.05	288.40	59.60	476.80		
2 Laborers	24.65	394.40	41.25	660.00		
1 Crawler Crane, 40 Ton		1203.00		1323.30		
1 Lead, 60' High		74.00		81.40		
1 Hammer, Diesel, 15K ft-lbs.		591.40		650.54	29.19	32.11
64 L.H., Daily Totals		$3923.20		$5559.64	$61.30	$86.87
Crew B-45						
1 Building Laborer	$24.65	$197.20	$41.25	$330.00	$27.65	$45.90
1 Truck Driver (heavy)	30.65	245.20	50.55	404.40		
1 Dist. Tanker, 3000 Gallon		319.40		351.34	19.96	21.96
16 L.H., Daily Totals		$761.80		$1085.74	$47.61	$67.86
Crew B-46						
1 Pile Driver Foreman (outside)	$35.90	$287.20	$61.95	$495.60	$29.61	$50.45
2 Pile Drivers	33.90	542.40	58.50	936.00		
3 Laborers	24.65	591.60	41.25	990.00		
1 Chain Saw, Gas, 36" Long		45.00		49.50	0.94	1.03
48 L.H., Daily Totals		$1466.20		$2471.10	$30.55	$51.48
Crew B-47						
1 Blast Foreman (outside)	$26.65	$213.20	$44.60	$356.80	$25.65	$42.92
1 Driller	24.65	197.20	41.25	330.00		
1 Air Track Drill, 4"		1045.00		1149.50		
1 Air Compressor, 600 cfm		533.80		587.18		
2 -50' Air Hoses, 3"		27.70		30.47	100.41	110.45
16 L.H., Daily Totals		$2016.90		$2453.95	$126.06	$153.37
Crew B-47A						
1 Drilling Foreman (outside)	$26.65	$213.20	$44.60	$356.80	$31.48	$52.23
1 Equip. Oper. (heavy)	36.05	288.40	59.60	476.80		
1 Equip. Oper. (oiler)	31.75	254.00	52.50	420.00		
1 Air Track Drill, 5"		1248.00		1372.80	52.00	57.20
24 L.H., Daily Totals		$2003.60		$2626.40	$83.48	$109.43
Crew B-47C						
1 Laborer	$24.65	$197.20	$41.25	$330.00	$29.27	$48.65
1 Equip. Oper. (light)	33.90	271.20	56.05	448.40		
1 Air Compressor, 750 cfm		541.00		595.10		
2 -50' Air Hoses, 3"		27.70		30.47		
1 Air Track Drill, 4"		1045.00		1149.50	100.86	110.94
16 L.H., Daily Totals		$2082.10		$2553.47	$130.13	$159.59
Crew B-47E						
1 Labor Foreman (outside)	$26.65	$213.20	$44.60	$356.80	$25.15	$42.09
3 Laborers	24.65	591.60	41.25	990.00		
1 Flatbed Truck, Gas, 3 Ton		307.40		338.14	9.61	10.57
32 L.H., Daily Totals		$1112.20		$1684.94	$34.76	$52.65
Crew B-47G						
1 Labor Foreman (outside)	$26.65	$213.20	$44.60	$356.80	$25.32	$42.37
2 Laborers	24.65	394.40	41.25	660.00		
1 Air Track Drill, 4"		1045.00		1149.50		
1 Air Compressor, 600 cfm		533.80		587.18		
2 -50' Air Hoses, 3"		27.70		30.47		
1 Gunite Pump Rig		373.00		410.30	82.48	90.73
24 L.H., Daily Totals		$2587.10		$3194.25	$107.80	$133.09
Crew B-47H						
1 Skilled Worker Foreman (out)	$36.20	$289.60	$60.80	$486.40	$34.70	$58.29
3 Skilled Workers	34.20	820.80	57.45	1378.80		
1 Flatbed Truck, Gas, 3 Ton		307.40		338.14	9.61	10.57
32 L.H., Daily Totals		$1417.80		$2203.34	$44.31	$68.85
Crew B-48						
1 Labor Foreman (outside)	$26.65	$213.20	$44.60	$356.80	$26.88	$44.87
4 Laborers	24.65	788.80	41.25	1320.00		
1 Equip. Oper. (crane)	36.05	288.40	59.60	476.80		
1 Centr. Water Pump, 6"		350.80		385.88		
1 -20' Suction Hose, 6"		11.50		12.65		
1 -50' Discharge Hose, 6"		6.10		6.71		
1 Drill Rig, Truck-Mounted		2553.00		2808.30	60.86	66.95
48 L.H., Daily Totals		$4211.80		$5367.14	$87.75	$111.82

Crews - Residential

Crew No.		Bare Costs		Incl. Subs O&P		Cost Per Labor-Hour	
Crew B-49	Hr.	Daily	Hr.	Daily	Bare Costs	Incl. O&P
1 Labor Foreman (outside)	$26.65	$213.20	$44.60	$356.80	$28.19	$47.49
5 Laborers	24.65	986.00	41.25	1650.00		
1 Equip. Oper. (crane)	36.05	288.40	59.60	476.80		
2 Pile Drivers	33.90	542.40	58.50	936.00		
1 Hyd. Crane, 25 Ton		748.80		823.68		
1 Centr. Water Pump, 6"		350.80		385.88		
1 -20' Suction Hose, 6"		11.50		12.65		
1 -50' Discharge Hose, 6"		6.10		6.71		
1 Drill Rig, Truck-Mounted		2553.00		2808.30	50.98	56.07
72 L.H., Daily Totals		$5700.20		$7456.82	$79.17	$103.57
Crew B-50	Hr.	Daily	Hr.	Daily	Bare Costs	Incl. O&P
1 Pile Driver Foreman (outside)	$35.90	$287.20	$61.95	$495.60	$30.66	$52.22
6 Pile Drivers	33.90	1627.20	58.50	2808.00		
1 Equip. Oper. (crane)	36.05	288.40	59.60	476.80		
5 Laborers	24.65	986.00	41.25	1650.00		
1 Crawler Crane, 40 Ton		1203.00		1323.30		
1 Lead, 60' High		74.00		81.40		
1 Hammer, Diesel, 15K ft.-lbs.		591.40		650.54		
1 Air Compressor, 600 cfm		533.80		587.18		
2 -50' Air Hoses, 3"		27.70		30.47		
1 Chain Saw, Gas, 36" Long		45.00		49.50	23.80	26.18
104 L.H., Daily Totals		$5663.70		$8152.79	$54.46	$78.39
Crew B-51	Hr.	Daily	Hr.	Daily	Bare Costs	Incl. O&P
1 Labor Foreman (outside)	$26.65	$213.20	$44.60	$356.80	$25.84	$43.13
4 Laborers	24.65	788.80	41.25	1320.00		
1 Truck Driver (light)	29.80	238.40	49.15	393.20		
1 Flatbed Truck, Gas, 1.5 Ton		248.20		273.02	5.17	5.69
48 L.H., Daily Totals		$1488.60		$2343.02	$31.01	$48.81
Crew B-52	Hr.	Daily	Hr.	Daily	Bare Costs	Incl. O&P
1 Labor Foreman (outside)	$26.65	$213.20	$44.60	$356.80	$27.79	$46.49
1 Carpenter	33.90	271.20	56.75	454.00		
4 Laborers	24.65	788.80	41.25	1320.00		
.5 Rodman (reinf.)	35.50	142.00	59.95	239.80		
.5 Equip. Oper. (medium)	35.25	141.00	58.25	233.00		
.5 Crawler Loader, 3 C.Y.		598.50		658.35	10.69	11.76
56 L.H., Daily Totals		$2154.70		$3261.95	$38.48	$58.25
Crew B-53	Hr.	Daily	Hr.	Daily	Bare Costs	Incl. O&P
1 Building Laborer	$24.65	$197.20	$41.25	$330.00	$24.65	$41.25
1 Trencher, Chain, 12 H.P.		68.00		74.80	8.50	9.35
8 L.H., Daily Totals		$265.20		$404.80	$33.15	$50.60
Crew B-54	Hr.	Daily	Hr.	Daily	Bare Costs	Incl. O&P
1 Equip. Oper. (light)	$33.90	$271.20	$56.05	$448.40	$33.90	$56.05
1 Trencher, Chain, 40 H.P.		334.40		367.84	41.80	45.98
8 L.H., Daily Totals		$605.60		$816.24	$75.70	$102.03
Crew B-54A	Hr.	Daily	Hr.	Daily	Bare Costs	Incl. O&P
.17 Labor Foreman (outside)	$26.65	$36.24	$44.60	$60.66	$34.00	$56.27
1 Equipment Operator (med.)	35.25	282.00	58.25	466.00		
1 Wheel Trencher, 67 H.P.		1183.00		1301.30	126.39	139.03
9.36 L.H., Daily Totals		$1501.24		$1827.96	$160.39	$195.29
Crew B-54B	Hr.	Daily	Hr.	Daily	Bare Costs	Incl. O&P
.25 Labor Foreman (outside)	$26.65	$53.30	$44.60	$89.20	$33.53	$55.52
1 Equipment Operator (med.)	35.25	282.00	58.25	466.00		
1 Wheel Trencher, 150 H.P.		1891.00		2080.10	189.10	208.01
10 L.H., Daily Totals		$2226.30		$2635.30	$222.63	$263.53

Crew No.		Bare Costs		Incl. Subs O&P		Cost Per Labor-Hour	
Crew B-54D	Hr.	Daily	Hr.	Daily	Bare Costs	Incl. O&P
1 Laborer	$24.65	$197.20	$41.25	$330.00	$29.95	$49.75
1 Equipment Operator (med.)	35.25	282.00	58.25	466.00		
1 Rock Trencher, 6" Width		615.20		676.72	38.45	42.30
16 L.H., Daily Totals		$1094.40		$1472.72	$68.40	$92.05
Crew B-54E	Hr.	Daily	Hr.	Daily	Bare Costs	Incl. O&P
1 Laborer	$24.65	$197.20	$41.25	$330.00	$29.95	$49.75
1 Equipment Operator (med.)	35.25	282.00	58.25	466.00		
1 Rock Trencher, 18" Width		2591.00		2850.10	161.94	178.13
16 L.H., Daily Totals		$3070.20		$3646.10	$191.89	$227.88
Crew B-55	Hr.	Daily	Hr.	Daily	Bare Costs	Incl. O&P
1 Laborer	$24.65	$197.20	$41.25	$330.00	$27.23	$45.20
1 Truck Driver (light)	29.80	238.40	49.15	393.20		
1 Truck-Mounted Earth Auger		786.00		864.60		
1 Flatbed Truck, Gas, 3 Ton		307.40		338.14	68.34	75.17
16 L.H., Daily Totals		$1529.00		$1925.94	$95.56	$120.37
Crew B-56	Hr.	Daily	Hr.	Daily	Bare Costs	Incl. O&P
2 Laborers	$24.65	$394.40	$41.25	$660.00	$24.65	$41.25
1 Air Track Drill, 4"		1045.00		1149.50		
1 Air Compressor, 600 cfm		533.80		587.18		
1 -50' Air Hose, 3"		13.85		15.23	99.54	109.49
16 L.H., Daily Totals		$1987.05		$2411.92	$124.19	$150.74
Crew B-57	Hr.	Daily	Hr.	Daily	Bare Costs	Incl. O&P
1 Labor Foreman (outside)	$26.65	$213.20	$44.60	$356.80	$27.33	$45.59
3 Laborers	24.65	591.60	41.25	990.00		
1 Equip. Oper. (crane)	36.05	288.40	59.60	476.80		
1 Barge, 400 Ton		779.20		857.12		
1 Crawler Crane, 25 Ton		1194.00		1313.40		
1 Clamshell Bucket, 1 C.Y.		47.40		52.14		
1 Centr. Water Pump, 6"		350.80		385.88		
1 -20' Suction Hose, 6"		11.50		12.65		
20 -50' Discharge Hoses, 6"		122.00		134.20	62.62	68.88
40 L.H., Daily Totals		$3598.10		$4578.99	$89.95	$114.47
Crew B-58	Hr.	Daily	Hr.	Daily	Bare Costs	Incl. O&P
2 Laborers	$24.65	$394.40	$41.25	$660.00	$27.73	$46.18
1 Equip. Oper. (light)	33.90	271.20	56.05	448.40		
1 Backhoe Loader, 48 H.P.		365.40		401.94		
1 Small Helicopter, w/ Pilot		2774.00		3051.40	130.81	143.89
24 L.H., Daily Totals		$3805.00		$4561.74	$158.54	$190.07
Crew B-59	Hr.	Daily	Hr.	Daily	Bare Costs	Incl. O&P
1 Truck Driver (heavy)	$30.65	$245.20	$50.55	$404.40	$30.65	$50.55
1 Truck Tractor, 220 H.P.		358.80		394.68		
1 Water Tank Trailer, 5000 Gal.		141.60		155.76	62.55	68.81
8 L.H., Daily Totals		$745.60		$954.84	$93.20	$119.36
Crew B-60	Hr.	Daily	Hr.	Daily	Bare Costs	Incl. O&P
1 Labor Foreman (outside)	$26.65	$213.20	$44.60	$356.80	$28.43	$47.33
3 Laborers	24.65	591.60	41.25	990.00		
1 Equip. Oper. (crane)	36.05	288.40	59.60	476.80		
1 Equip. Oper. (light)	33.90	271.20	56.05	448.40		
1 Crawler Crane, 40 Ton		1203.00		1323.30		
1 Lead, 60' High		74.00		81.40		
1 Hammer, Diesel, 15K ft.-lbs.		591.40		650.54		
1 Backhoe Loader, 48 H.P.		365.40		401.94	46.54	51.19
48 L.H., Daily Totals		$3598.20		$4729.18	$74.96	$98.52

Crews - Residential

Crew B-61

Crew B-61	Bare Costs Hr.	Bare Costs Daily	Incl. Subs O&P Hr.	Incl. Subs O&P Daily	Cost Per Labor-Hour Bare Costs	Cost Per Labor-Hour Incl. O&P
1 Labor Foreman (outside)	$26.65	$213.20	$44.60	$356.80	$25.05	$41.92
4 Laborers	24.65	788.80	41.25	1320.00		
1 Cement Mixer, 2 C.Y.		193.40		212.74		
1 Air Compressor, 160 cfm		162.20		178.42	8.89	9.78
40 L.H., Daily Totals		$1357.60		$2067.96	$33.94	$51.70

Crew B-62

Crew B-62	Hr.	Daily	Hr.	Daily	Bare Costs	Incl. O&P
2 Laborers	$24.65	$394.40	$41.25	$660.00	$27.73	$46.18
1 Equip. Oper. (light)	33.90	271.20	56.05	448.40		
1 Loader, Skid Steer, 30 H.P.		173.20		190.52	7.22	7.94
24 L.H., Daily Totals		$838.80		$1298.92	$34.95	$54.12

Crew B-62A

Crew B-62A	Hr.	Daily	Hr.	Daily	Bare Costs	Incl. O&P
2 Laborers	$24.65	$394.40	$41.25	$660.00	$27.73	$46.18
1 Equip. Oper. (light)	33.90	271.20	56.05	448.40		
1 Loader, Skid Steer, 30 H.P.		173.20		190.52		
1 Trencher Attachment		63.70		70.07	9.87	10.86
24 L.H., Daily Totals		$902.50		$1368.99	$37.60	$57.04

Crew B-63

Crew B-63	Hr.	Daily	Hr.	Daily	Bare Costs	Incl. O&P
5 Laborers	$24.65	$986.00	$41.25	$1650.00	$24.65	$41.25
1 Loader, Skid Steer, 30 H.P.		173.20		190.52	4.33	4.76
40 L.H., Daily Totals		$1159.20		$1840.52	$28.98	$46.01

Crew B-63B

Crew B-63B	Hr.	Daily	Hr.	Daily	Bare Costs	Incl. O&P
1 Labor Foreman (inside)	$25.15	$201.20	$42.10	$336.80	$27.09	$45.16
2 Laborers	24.65	394.40	41.25	660.00		
1 Equip. Oper. (light)	33.90	271.20	56.05	448.40		
1 Loader, Skid Steer, 78 H.P.		315.60		347.16	9.86	10.85
32 L.H., Daily Totals		$1182.40		$1792.36	$36.95	$56.01

Crew B-64

Crew B-64	Hr.	Daily	Hr.	Daily	Bare Costs	Incl. O&P
1 Laborer	$24.65	$197.20	$41.25	$330.00	$27.23	$45.20
1 Truck Driver (light)	29.80	238.40	49.15	393.20		
1 Power Mulcher (small)		150.80		165.88		
1 Flatbed Truck, Gas, 1.5 Ton		248.20		273.02	24.94	27.43
16 L.H., Daily Totals		$834.60		$1162.10	$52.16	$72.63

Crew B-65

Crew B-65	Hr.	Daily	Hr.	Daily	Bare Costs	Incl. O&P
1 Laborer	$24.65	$197.20	$41.25	$330.00	$27.23	$45.20
1 Truck Driver (light)	29.80	238.40	49.15	393.20		
1 Power Mulcher (Large)		319.60		351.56		
1 Flatbed Truck, Gas, 1.5 Ton		248.20		273.02	35.49	39.04
16 L.H., Daily Totals		$1003.40		$1347.78	$62.71	$84.24

Crew B-66

Crew B-66	Hr.	Daily	Hr.	Daily	Bare Costs	Incl. O&P
1 Equip. Oper. (light)	$33.90	$271.20	$56.05	$448.40	$33.90	$56.05
1 Loader-Backhoe, 40 H.P.		263.80		290.18	32.98	36.27
8 L.H., Daily Totals		$535.00		$738.58	$66.88	$92.32

Crew B-67

Crew B-67	Hr.	Daily	Hr.	Daily	Bare Costs	Incl. O&P
1 Millwright	$35.75	$286.00	$57.40	$459.20	$34.83	$56.73
1 Equip. Oper. (light)	33.90	271.20	56.05	448.40		
1 Forklift, R/T, 4,000 Lb.		313.60		344.96	19.60	21.56
16 L.H., Daily Totals		$870.80		$1252.56	$54.42	$78.28

Crew B-67B

Crew B-67B	Hr.	Daily	Hr.	Daily	Bare Costs	Incl. O&P
1 Millwright Foreman (inside)	$36.25	$290.00	$58.20	$465.60	$36.00	$57.80
1 Millwright	35.75	286.00	57.40	459.20		
16 L.H., Daily Totals		$576.00		$924.80	$36.00	$57.80

Crew B-68

Crew B-68	Hr.	Daily	Hr.	Daily	Bare Costs	Incl. O&P
2 Millwrights	$35.75	$572.00	$57.40	$918.40	$35.13	$56.95
1 Equip. Oper. (light)	33.90	271.20	56.05	448.40		
1 Forklift, R/T, 4,000 Lb.		313.60		344.96	13.07	14.37
24 L.H., Daily Totals		$1156.80		$1711.76	$48.20	$71.32

Crew B-68A

Crew B-68A	Hr.	Daily	Hr.	Daily	Bare Costs	Incl. O&P
1 Millwright Foreman (inside)	$36.25	$290.00	$58.20	$465.60	$35.92	$57.67
2 Millwrights	35.75	572.00	57.40	918.40		
1 Forklift, 8,000 Lb.		185.40		203.94	7.72	8.50
24 L.H., Daily Totals		$1047.40		$1587.94	$43.64	$66.16

Crew B-68B

Crew B-68B	Hr.	Daily	Hr.	Daily	Bare Costs	Incl. O&P
1 Millwright Foreman (inside)	$36.25	$290.00	$58.20	$465.60	$37.09	$60.36
2 Millwrights	35.75	572.00	57.40	918.40		
2 Electricians	36.90	590.40	60.35	965.60		
2 Plumbers	39.05	624.80	64.40	1030.40		
1 Forklift, 5,000 Lb.		334.00		367.40	5.96	6.56
56 L.H., Daily Totals		$2411.20		$3747.40	$43.06	$66.92

Crew B-68C

Crew B-68C	Hr.	Daily	Hr.	Daily	Bare Costs	Incl. O&P
1 Millwright Foreman (inside)	$36.25	$290.00	$58.20	$465.60	$36.99	$60.09
1 Millwright	35.75	286.00	57.40	459.20		
1 Electrician	36.90	295.20	60.35	482.80		
1 Plumber	39.05	312.40	64.40	515.20		
1 Forklift, 5,000 Lb.		334.00		367.40	10.44	11.48
32 L.H., Daily Totals		$1517.60		$2290.20	$47.42	$71.57

Crew B-68D

Crew B-68D	Hr.	Daily	Hr.	Daily	Bare Costs	Incl. O&P
1 Labor Foreman (inside)	$25.15	$201.20	$42.10	$336.80	$27.90	$46.47
1 Laborer	24.65	197.20	41.25	330.00		
1 Equip. Oper. (light)	33.90	271.20	56.05	448.40		
1 Forklift, 5,000 Lb.		334.00		367.40	13.92	15.31
24 L.H., Daily Totals		$1003.60		$1482.60	$41.82	$61.77

Crew B-68E

Crew B-68E	Hr.	Daily	Hr.	Daily	Bare Costs	Incl. O&P
1 Struc. Steel Foreman (inside)	$36.15	$289.20	$66.30	$530.40	$35.75	$65.58
3 Struc. Steel Workers	35.65	855.60	65.40	1569.60		
1 Welder	35.65	285.20	65.40	523.20		
1 Forklift, 8,000 Lb.		185.40		203.94	4.63	5.10
40 L.H., Daily Totals		$1615.40		$2827.14	$40.38	$70.68

Crew B-68F

Crew B-68F	Hr.	Daily	Hr.	Daily	Bare Costs	Incl. O&P
1 Skilled Worker Foreman (out)	$36.20	$289.60	$60.80	$486.40	$34.87	$58.57
2 Skilled Workers	34.20	547.20	57.45	919.20		
1 Forklift, 5,000 Lb.		334.00		367.40	13.92	15.31
24 L.H., Daily Totals		$1170.80		$1773.00	$48.78	$73.88

Crew B-68G

Crew B-68G	Hr.	Daily	Hr.	Daily	Bare Costs	Incl. O&P
2 Structural Steel Workers	$35.65	$570.40	$65.40	$1046.40	$35.65	$65.40
1 Forklift, 5,000 Lb.		334.00		367.40	20.88	22.96
16 L.H., Daily Totals		$904.40		$1413.80	$56.52	$88.36

Crew B-69

Crew B-69	Hr.	Daily	Hr.	Daily	Bare Costs	Incl. O&P
1 Labor Foreman (outside)	$26.65	$213.20	$44.60	$356.80	$28.07	$46.74
3 Laborers	24.65	591.60	41.25	990.00		
1 Equip. Oper. (crane)	36.05	288.40	59.60	476.80		
1 Equip. Oper. (oiler)	31.75	254.00	52.50	420.00		
1 Hyd. Crane, 80 Ton		1630.00		1793.00	33.96	37.35
48 L.H., Daily Totals		$2977.20		$4036.60	$62.02	$84.10

Crews - Residential

Crew No.	Bare Costs		Incl. Subs O&P		Cost Per Labor-Hour	
Crew B-69A	**Hr.**	**Daily**	**Hr.**	**Daily**	**Bare Costs**	**Incl. O&P**
1 Labor Foreman (outside)	$26.65	$213.20	$44.60	$356.80	$27.97	$46.41
3 Laborers	24.65	591.60	41.25	990.00		
1 Equip. Oper. (medium)	35.25	282.00	58.25	466.00		
1 Concrete Finisher	31.95	255.60	51.85	414.80		
1 Curb/Gutter Paver, 2-Track		1102.00		1212.20	22.96	25.25
48 L.H., Daily Totals		$2444.40		$3439.80	$50.92	$71.66
Crew B-69B	**Hr.**	**Daily**	**Hr.**	**Daily**	**Bare Costs**	**Incl. O&P**
1 Labor Foreman (outside)	$26.65	$213.20	$44.60	$356.80	$27.97	$46.41
3 Laborers	24.65	591.60	41.25	990.00		
1 Equip. Oper. (medium)	35.25	282.00	58.25	466.00		
1 Cement Finisher	31.95	255.60	51.85	414.80		
1 Curb/Gutter Paver, 4-Track		780.20		858.22	16.25	17.88
48 L.H., Daily Totals		$2122.60		$3085.82	$44.22	$64.29
Crew B-70	**Hr.**	**Daily**	**Hr.**	**Daily**	**Bare Costs**	**Incl. O&P**
1 Labor Foreman (outside)	$26.65	$213.20	$44.60	$356.80	$29.48	$49.01
3 Laborers	24.65	591.60	41.25	990.00		
3 Equip. Opers. (medium)	35.25	846.00	58.25	1398.00		
1 Grader, 30,000 Lbs.		714.40		785.84		
1 Ripper, Beam & 1 Shank		83.80		92.18		
1 Road Sweeper, S.P., 8' wide		666.20		732.82		
1 F.E. Loader, W.M., 1.5 C.Y.		378.60		416.46	32.91	36.20
56 L.H., Daily Totals		$3493.80		$4772.10	$62.39	$85.22
Crew B-71	**Hr.**	**Daily**	**Hr.**	**Daily**	**Bare Costs**	**Incl. O&P**
1 Labor Foreman (outside)	$26.65	$213.20	$44.60	$356.80	$29.48	$49.01
3 Laborers	24.65	591.60	41.25	990.00		
3 Equip. Opers. (medium)	35.25	846.00	58.25	1398.00		
1 Pvmt. Profiler, 750 H.P.		5835.00		6418.50		
1 Road Sweeper, S.P., 8' wide		666.20		732.82		
1 F.E. Loader, W.M., 1.5 C.Y.		378.60		416.46	122.85	135.14
56 L.H., Daily Totals		$8530.60		$10312.58	$152.33	$184.15
Crew B-72	**Hr.**	**Daily**	**Hr.**	**Daily**	**Bare Costs**	**Incl. O&P**
1 Labor Foreman (outside)	$26.65	$213.20	$44.60	$356.80	$30.20	$50.17
3 Laborers	24.65	591.60	41.25	990.00		
4 Equip. Opers. (medium)	35.25	1128.00	58.25	1864.00		
1 Pvmt. Profiler, 750 H.P.		5835.00		6418.50		
1 Hammermill, 250 H.P.		1832.00		2015.20		
1 Windrow Loader		1247.00		1371.70		
1 Mix Paver, 165 H.P.		2226.00		2448.60		
1 Roller, Pneum. Whl., 12 Ton		347.80		382.58	179.50	197.45
64 L.H., Daily Totals		$13420.60		$15847.38	$209.70	$247.62
Crew B-73	**Hr.**	**Daily**	**Hr.**	**Daily**	**Bare Costs**	**Incl. O&P**
1 Labor Foreman (outside)	$26.65	$213.20	$44.60	$356.80	$31.52	$52.29
2 Laborers	24.65	394.40	41.25	660.00		
5 Equip. Opers. (medium)	35.25	1410.00	58.25	2330.00		
1 Road Mixer, 310 H.P.		1945.00		2139.50		
1 Tandem Roller, 10 Ton		239.40		263.34		
1 Hammermill, 250 H.P.		1832.00		2015.20		
1 Grader, 30,000 Lbs.		714.40		785.84		
.5 F.E. Loader, W.M., 1.5 C.Y.		189.30		208.23		
.5 Truck Tractor, 220 H.P.		179.40		197.34		
.5 Water Tank Trailer, 5000 Gal.		70.80		77.88	80.79	88.86
64 L.H., Daily Totals		$7187.90		$9034.13	$112.31	$141.16

Crew No.	Bare Costs		Incl. Subs O&P		Cost Per Labor-Hour	
Crew B-74	**Hr.**	**Daily**	**Hr.**	**Daily**	**Bare Costs**	**Incl. O&P**
1 Labor Foreman (outside)	$26.65	$213.20	$44.60	$356.80	$31.70	$52.49
1 Laborer	24.65	197.20	41.25	330.00		
4 Equip. Opers. (medium)	35.25	1128.00	58.25	1864.00		
2 Truck Drivers (heavy)	30.65	490.40	50.55	808.80		
1 Grader, 30,000 Lbs.		714.40		785.84		
1 Ripper, Beam & 1 Shank		83.80		92.18		
2 Stabilizers, 310 H.P.		3666.00		4032.60		
1 Flatbed Truck, Gas, 3 Ton		307.40		338.14		
1 Chem. Spreader, Towed		52.00		57.20		
1 Roller, Vibratory, 25 Ton		691.80		760.98		
1 Water Tank Trailer, 5000 Gal.		141.60		155.76		
1 Truck Tractor, 220 H.P.		358.80		394.68	94.00	103.40
64 L.H., Daily Totals		$8044.60		$9976.98	$125.70	$155.89
Crew B-75	**Hr.**	**Daily**	**Hr.**	**Daily**	**Bare Costs**	**Incl. O&P**
1 Labor Foreman (outside)	$26.65	$213.20	$44.60	$356.80	$31.85	$52.77
1 Laborer	24.65	197.20	41.25	330.00		
4 Equip. Oper. (medium)	35.25	1128.00	58.25	1864.00		
1 Truck Driver (heavy)	30.65	245.20	50.55	404.40		
1 Grader, 30,000 Lbs.		714.40		785.84		
1 Ripper, Beam & 1 Shank		83.80		92.18		
2 Stabilizers, 310 H.P.		3666.00		4032.60		
1 Dist. Tanker, 3000 Gallon		319.40		351.34		
1 Truck Tractor, 6x4, 380 H.P.		600.00		660.00		
1 Roller, Vibratory, 25 Ton		691.80		760.98	108.49	119.34
56 L.H., Daily Totals		$7859.00		$9638.14	$140.34	$172.11
Crew B-76	**Hr.**	**Daily**	**Hr.**	**Daily**	**Bare Costs**	**Incl. O&P**
1 Dock Builder Foreman (outside)	$35.90	$287.20	$61.95	$495.60	$34.36	$58.46
5 Dock Builders	33.90	1356.00	58.50	2340.00		
2 Equip. Opers. (crane)	36.05	576.80	59.60	953.60		
1 Equip. Oper. (oiler)	31.75	254.00	52.50	420.00		
1 Crawler Crane, 50 Ton		1346.00		1480.60		
1 Barge, 400 Ton		779.20		857.12		
1 Hammer, Diesel, 15K ft.-lbs.		591.40		650.54		
1 Lead, 60' High		74.00		81.40		
1 Air Compressor, 600 cfm		533.80		587.18		
2 -50' Air Hoses, 3"		27.70		30.47	46.56	51.21
72 L.H., Daily Totals		$5826.10		$7896.51	$80.92	$109.67
Crew B-76A	**Hr.**	**Daily**	**Hr.**	**Daily**	**Bare Costs**	**Incl. O&P**
1 Labor Foreman (outside)	$26.65	$213.20	$44.60	$356.80	$27.21	$45.37
5 Laborers	24.65	986.00	41.25	1650.00		
1 Equip. Oper. (crane)	36.05	288.40	59.60	476.80		
1 Equip. Oper. (oiler)	31.75	254.00	52.50	420.00		
1 Crawler Crane, 50 Ton		1346.00		1480.60		
1 Barge, 400 Ton		779.20		857.12	33.21	36.53
64 L.H., Daily Totals		$3866.80		$5241.32	$60.42	$81.90
Crew B-77	**Hr.**	**Daily**	**Hr.**	**Daily**	**Bare Costs**	**Incl. O&P**
1 Labor Foreman (outside)	$26.65	$213.20	$44.60	$356.80	$26.08	$43.50
3 Laborers	24.65	591.60	41.25	990.00		
1 Truck Driver (light)	29.80	238.40	49.15	393.20		
1 Crack Cleaner, 25 H.P.		62.00		68.20		
1 Crack Filler, Trailer Mtd.		202.60		222.86		
1 Flatbed Truck, Gas, 3 Ton		307.40		338.14	14.30	15.73
40 L.H., Daily Totals		$1615.20		$2369.20	$40.38	$59.23

Crews - Residential

Crew No.	Bare Costs		Incl. Subs O&P		Cost Per Labor-Hour	
Crew B-78	Hr.	Daily	Hr.	Daily	Bare Costs	Incl. O&P
1 Labor Foreman (outside)	$26.65	$213.20	$44.60	$356.80	$25.05	$41.92
4 Laborers	24.65	788.80	41.25	1320.00		
1 Paint Striper, S.P., 40 Gallon		152.00		167.20		
1 Flatbed Truck, Gas, 3 Ton		307.40		338.14		
1 Pickup Truck, 3/4 Ton		145.80		160.38	15.13	16.64
40 L.H., Daily Totals		$1607.20		$2342.52	$40.18	$58.56
Crew B-78B	Hr.	Daily	Hr.	Daily	Bare Costs	Incl. O&P
2 Laborers	$24.65	$394.40	$41.25	$660.00	$25.68	$42.89
.25 Equip. Oper. (light)	33.90	67.80	56.05	112.10		
1 Pickup Truck, 3/4 Ton		145.80		160.38		
1 Line Rem.,11 H.P.,Walk Behind		65.60		72.16		
.25 Road Sweeper, S.P., 8' wide		166.55		183.21	21.00	23.10
18 L.H., Daily Totals		$840.15		$1187.85	$46.67	$65.99
Crew B-78C	Hr.	Daily	Hr.	Daily	Bare Costs	Incl. O&P
1 Labor Foreman (outside)	$26.65	$213.20	$44.60	$356.80	$25.84	$43.13
4 Laborers	24.65	788.80	41.25	1320.00		
1 Truck Driver (light)	29.80	238.40	49.15	393.20		
1 Paint Striper, T.M., 120 Gal.		810.20		891.22		
1 Flatbed Truck, Gas, 3 Ton		307.40		338.14		
1 Pickup Truck, 3/4 Ton		145.80		160.38	26.32	28.95
48 L.H., Daily Totals		$2503.80		$3459.74	$52.16	$72.08
Crew B-78D	Hr.	Daily	Hr.	Daily	Bare Costs	Incl. O&P
2 Labor Foremen (outside)	$26.65	$426.40	$44.60	$713.60	$25.57	$42.71
7 Laborers	24.65	1380.40	41.25	2310.00		
1 Truck Driver (light)	29.80	238.40	49.15	393.20		
1 Paint Striper, T.M., 120 Gal.		810.20		891.22		
1 Flatbed Truck, Gas, 3 Ton		307.40		338.14		
3 Pickup Trucks, 3/4 Ton		437.40		481.14		
1 Air Compressor, 60 cfm		142.80		157.08		
1 -50' Air Hose, 3/4"		3.05		3.36		
1 Breaker, Pavement, 60 lb.		10.20		11.22	21.39	23.53
80 L.H., Daily Totals		$3756.25		$5298.95	$46.95	$66.24
Crew B-78E	Hr.	Daily	Hr.	Daily	Bare Costs	Incl. O&P
2 Labor Foremen (outside)	$26.65	$426.40	$44.60	$713.60	$25.41	$42.47
9 Laborers	24.65	1774.80	41.25	2970.00		
1 Truck Driver (light)	29.80	238.40	49.15	393.20		
1 Paint Striper, T.M., 120 Gal.		810.20		891.22		
1 Flatbed Truck, Gas, 3 Ton		307.40		338.14		
4 Pickup Trucks, 3/4 Ton		583.20		641.52		
2 Air Compressors, 60 cfm		285.60		314.16		
2 -50' Air Hoses, 3/4"		6.10		6.71		
2 Breakers, Pavement, 60 lb.		20.40		22.44	20.97	23.06
96 L.H., Daily Totals		$4452.50		$6290.99	$46.38	$65.53
Crew B-78F	Hr.	Daily	Hr.	Daily	Bare Costs	Incl. O&P
2 Labor Foremen (outside)	$26.65	$426.40	$44.60	$713.60	$25.30	$42.29
11 Laborers	24.65	2169.20	41.25	3630.00		
1 Truck Driver (light)	29.80	238.40	49.15	393.20		
1 Paint Striper, T.M., 120 Gal.		810.20		891.22		
1 Flatbed Truck, Gas, 3 Ton		307.40		338.14		
7 Pickup Trucks, 3/4 Ton		1020.60		1122.66		
3 Air Compressors, 60 cfm		428.40		471.24		
3 -50' Air Hoses, 3/4"		9.15		10.07		
3 Breakers, Pavement, 60 lb.		30.60		33.66	23.27	25.60
112 L.H., Daily Totals		$5440.35		$7603.78	$48.57	$67.89

Crew No.	Bare Costs		Incl. Subs O&P		Cost Per Labor-Hour	
Crew B-79	Hr.	Daily	Hr.	Daily	Bare Costs	Incl. O&P
1 Labor Foreman (outside)	$26.65	$213.20	$44.60	$356.80	$26.08	$43.50
3 Laborers	24.65	591.60	41.25	990.00		
1 Truck Driver (light)	29.80	238.40	49.15	393.20		
1 Paint Striper, T.M., 120 Gal.		810.20		891.22		
1 Heating Kettle, 115 Gallon		81.70		89.87		
1 Flatbed Truck, Gas, 3 Ton		307.40		338.14		
2 Pickup Trucks, 3/4 Ton		291.60		320.76	37.27	41.00
40 L.H., Daily Totals		$2534.10		$3379.99	$63.35	$84.50
Crew B-79B	Hr.	Daily	Hr.	Daily	Bare Costs	Incl. O&P
1 Laborer	$24.65	$197.20	$41.25	$330.00	$24.65	$41.25
1 Set of Gases		168.00		184.80	21.00	23.10
8 L.H., Daily Totals		$365.20		$514.80	$45.65	$64.35
Crew B-79C	Hr.	Daily	Hr.	Daily	Bare Costs	Incl. O&P
1 Labor Foreman (outside)	$26.65	$213.20	$44.60	$356.80	$25.67	$42.86
5 Laborers	24.65	986.00	41.25	1650.00		
1 Truck Driver (light)	29.80	238.40	49.15	393.20		
1 Paint Striper, T.M., 120 Gal.		810.20		891.22		
1 Heating Kettle, 115 Gallon		81.70		89.87		
1 Flatbed Truck, Gas, 3 Ton		307.40		338.14		
3 Pickup Trucks, 3/4 Ton		437.40		481.14		
1 Air Compressor, 60 cfm		142.80		157.08		
1 -50' Air Hose, 3/4"		3.05		3.36		
1 Breaker, Pavement, 60 lb.		10.20		11.22	32.01	35.21
56 L.H., Daily Totals		$3230.35		$4372.02	$57.68	$78.07
Crew B-79D	Hr.	Daily	Hr.	Daily	Bare Costs	Incl. O&P
2 Labor Foremen (outside)	$26.65	$426.40	$44.60	$713.60	$25.79	$43.08
5 Laborers	24.65	986.00	41.25	1650.00		
1 Truck Driver (light)	29.80	238.40	49.15	393.20		
1 Paint Striper, T.M., 120 Gal.		810.20		891.22		
1 Heating Kettle, 115 Gallon		81.70		89.87		
1 Flatbed Truck, Gas, 3 Ton		307.40		338.14		
4 Pickup Trucks, 3/4 Ton		583.20		641.52		
1 Air Compressor, 60 cfm		142.80		157.08		
1 -50' Air Hose, 3/4"		3.05		3.36		
1 Breaker, Pavement, 60 lb.		10.20		11.22	30.29	33.32
64 L.H., Daily Totals		$3589.35		$4889.20	$56.08	$76.39
Crew B-79E	Hr.	Daily	Hr.	Daily	Bare Costs	Incl. O&P
2 Labor Foremen (outside)	$26.65	$426.40	$44.60	$713.60	$25.57	$42.71
7 Laborers	24.65	1380.40	41.25	2310.00		
1 Truck Driver (light)	29.80	238.40	49.15	393.20		
1 Paint Striper, T.M., 120 Gal.		810.20		891.22		
1 Heating Kettle, 115 Gallon		81.70		89.87		
1 Flatbed Truck, Gas, 3 Ton		307.40		338.14		
5 Pickup Trucks, 3/4 Ton		729.00		801.90		
2 Air Compressors, 60 cfm		285.60		314.16		
2 -50' Air Hoses, 3/4"		6.10		6.71		
2 Breakers, Pavement, 60 lb.		20.40		22.44	28.00	30.81
80 L.H., Daily Totals		$4285.60		$5881.24	$53.57	$73.52
Crew B-80	Hr.	Daily	Hr.	Daily	Bare Costs	Incl. O&P
1 Labor Foreman (outside)	$26.65	$213.20	$44.60	$356.80	$25.32	$42.37
2 Laborers	24.65	394.40	41.25	660.00		
1 Flatbed Truck, Gas, 3 Ton		307.40		338.14		
1 Earth Auger, Truck-Mtd.		420.80		462.88	30.34	33.38
24 L.H., Daily Totals		$1335.80		$1817.82	$55.66	$75.74

Crews - Residential

Crew No.	Bare Costs Hr.	Bare Costs Daily	Incl. Subs O&P Hr.	Incl. Subs O&P Daily	Cost Per Labor-Hour Bare Costs	Cost Per Labor-Hour Incl. O&P
Crew B-80A					Bare Costs	Incl. O&P
3 Laborers	$24.65	$591.60	$41.25	$990.00	$24.65	$41.25
1 Flatbed Truck, Gas, 3 Ton		307.40		338.14	12.81	14.09
24 L.H., Daily Totals		$899.00		$1328.14	$37.46	$55.34
Crew B-80B	Hr.	Daily	Hr.	Daily	Bare Costs	Incl. O&P
3 Laborers	$24.65	$591.60	$41.25	$990.00	$26.96	$44.95
1 Equip. Oper. (light)	33.90	271.20	56.05	448.40		
1 Crane, Flatbed Mounted, 3 Ton		248.20		273.02	7.76	8.53
32 L.H., Daily Totals		$1111.00		$1711.42	$34.72	$53.48
Crew B-80C	Hr.	Daily	Hr.	Daily	Bare Costs	Incl. O&P
2 Laborers	$24.65	$394.40	$41.25	$660.00	$26.37	$43.88
1 Truck Driver (light)	29.80	238.40	49.15	393.20		
1 Flatbed Truck, Gas, 1.5 Ton		248.20		273.02		
1 Manual Fence Post Auger, Gas		8.40		9.24	10.69	11.76
24 L.H., Daily Totals		$889.40		$1335.46	$37.06	$55.64
Crew B-81	Hr.	Daily	Hr.	Daily	Bare Costs	Incl. O&P
1 Laborer	$24.65	$197.20	$41.25	$330.00	$27.65	$45.90
1 Truck Driver (heavy)	30.65	245.20	50.55	404.40		
1 Hydromulcher, T.M., 3000 Gal.		334.40		367.84		
1 Truck Tractor, 220 H.P.		358.80		394.68	43.33	47.66
16 L.H., Daily Totals		$1135.60		$1496.92	$70.97	$93.56
Crew B-81A	Hr.	Daily	Hr.	Daily	Bare Costs	Incl. O&P
1 Laborer	$24.65	$197.20	$41.25	$330.00	$27.23	$45.20
1 Truck Driver (light)	29.80	238.40	49.15	393.20		
1 Hydromulcher, T.M., 600 Gal.		127.80		140.58		
1 Flatbed Truck, Gas, 3 Ton		307.40		338.14	27.20	29.92
16 L.H., Daily Totals		$870.80		$1201.92	$54.42	$75.12
Crew B-82	Hr.	Daily	Hr.	Daily	Bare Costs	Incl. O&P
1 Laborer	$24.65	$197.20	$41.25	$330.00	$29.27	$48.65
1 Equip. Oper. (light)	33.90	271.20	56.05	448.40		
1 Horiz. Borer, 6 H.P.		84.20		92.62	5.26	5.79
16 L.H., Daily Totals		$552.60		$871.02	$34.54	$54.44
Crew B-82A	Hr.	Daily	Hr.	Daily	Bare Costs	Incl. O&P
2 Laborers	$24.65	$394.40	$41.25	$660.00	$29.27	$48.65
2 Equip. Opers. (light)	33.90	542.40	56.05	896.80		
2 Dump Trucks, 8 C.Y., 220 H.P.		834.80		918.28		
1 Flatbed Trailer, 25 Ton		113.20		124.52		
1 Horiz. Dir. Drill, 20k lb. Thrust		657.00		722.70		
1 Mud Trailer for HDD, 1500 Gal.		350.00		385.00		
1 Pickup Truck, 4x4, 3/4 Ton		156.80		172.48		
1 Flatbed trailer, 3 Ton		23.40		25.74		
1 Loader, Skid Steer, 78 H.P.		315.60		347.16	76.59	84.25
32 L.H., Daily Totals		$3387.60		$4252.68	$105.86	$132.90
Crew B-82B	Hr.	Daily	Hr.	Daily	Bare Costs	Incl. O&P
2 Laborers	$24.65	$394.40	$41.25	$660.00	$29.27	$48.65
2 Equip. Opers. (light)	33.90	542.40	56.05	896.80		
2 Dump Trucks, 8 C.Y., 220 H.P.		834.80		918.28		
1 Flatbed Trailer, 25 Ton		113.20		124.52		
1 Horiz. Dir. Drill, 30k lb. Thrust		935.80		1029.38		
1 Mud Trailer for HDD, 1500 Gal.		350.00		385.00		
1 Pickup Truck, 4x4, 3/4 Ton		156.80		172.48		
1 Flatbed trailer, 3 Ton		23.40		25.74		
1 Loader, Skid Steer, 78 H.P.		315.60		347.16	85.30	93.83
32 L.H., Daily Totals		$3666.40		$4559.36	$114.58	$142.48

Crew No.	Bare Costs Hr.	Bare Costs Daily	Incl. Subs O&P Hr.	Incl. Subs O&P Daily	Cost Per Labor-Hour Bare Costs	Cost Per Labor-Hour Incl. O&P
Crew B-82C	Hr.	Daily	Hr.	Daily	Bare Costs	Incl. O&P
2 Laborers	$24.65	$394.40	$41.25	$660.00	$29.27	$48.65
2 Equip. Opers. (light)	33.90	542.40	56.05	896.80		
2 Dump Trucks, 8 C.Y., 220 H.P.		834.80		918.28		
1 Flatbed Trailer, 25 Ton		113.20		124.52		
1 Horiz. Dir. Drill, 50k lb. Thrust		1244.00		1368.40		
1 Mud Trailer for HDD, 1500 Gal.		350.00		385.00		
1 Pickup Truck, 4x4, 3/4 Ton		156.80		172.48		
1 Flatbed trailer, 3 Ton		23.40		25.74		
1 Loader, Skid Steer, 78 H.P.		315.60		347.16	94.93	104.42
32 L.H., Daily Totals		$3974.60		$4898.38	$124.21	$153.07
Crew B-82D	Hr.	Daily	Hr.	Daily	Bare Costs	Incl. O&P
1 Equip. Oper. (light)	$33.90	$271.20	$56.05	$448.40	$33.90	$56.05
1 Mud Trailer for HDD, 1500 Gal.		350.00		385.00	43.75	48.13
8 L.H., Daily Totals		$621.20		$833.40	$77.65	$104.18
Crew B-83	Hr.	Daily	Hr.	Daily	Bare Costs	Incl. O&P
1 Tugboat Captain	$35.25	$282.00	$58.25	$466.00	$29.95	$49.75
1 Tugboat Hand	24.65	197.20	41.25	330.00		
1 Tugboat, 250 H.P.		868.60		955.46	54.29	59.72
16 L.H., Daily Totals		$1347.80		$1751.46	$84.24	$109.47
Crew B-84	Hr.	Daily	Hr.	Daily	Bare Costs	Incl. O&P
1 Equip. Oper. (medium)	$35.25	$282.00	$58.25	$466.00	$35.25	$58.25
1 Rotary Mower/Tractor		369.40		406.34	46.17	50.79
8 L.H., Daily Totals		$651.40		$872.34	$81.42	$109.04
Crew B-85	Hr.	Daily	Hr.	Daily	Bare Costs	Incl. O&P
3 Laborers	$24.65	$591.60	$41.25	$990.00	$27.97	$46.51
1 Equip. Oper. (medium)	35.25	282.00	58.25	466.00		
1 Truck Driver (heavy)	30.65	245.20	50.55	404.40		
1 Aerial Lift Truck, 80'		659.20		725.12		
1 Brush Chipper, 12", 130 H.P.		388.60		427.46		
1 Pruning Saw, Rotary		6.65		7.32	26.36	29.00
40 L.H., Daily Totals		$2173.25		$3020.30	$54.33	$75.51
Crew B-86	Hr.	Daily	Hr.	Daily	Bare Costs	Incl. O&P
1 Equip. Oper. (medium)	$35.25	$282.00	$58.25	$466.00	$35.25	$58.25
1 Stump Chipper, S.P.		173.25		190.57	21.66	23.82
8 L.H., Daily Totals		$455.25		$656.58	$56.91	$82.07
Crew B-86A	Hr.	Daily	Hr.	Daily	Bare Costs	Incl. O&P
1 Equip. Oper. (medium)	$35.25	$282.00	$58.25	$466.00	$35.25	$58.25
1 Grader, 30,000 Lbs.		714.40		785.84	89.30	98.23
8 L.H., Daily Totals		$996.40		$1251.84	$124.55	$156.48
Crew B-86B	Hr.	Daily	Hr.	Daily	Bare Costs	Incl. O&P
1 Equip. Oper. (medium)	$35.25	$282.00	$58.25	$466.00	$35.25	$58.25
1 Dozer, 200 H.P.		1397.00		1536.70	174.63	192.09
8 L.H., Daily Totals		$1679.00		$2002.70	$209.88	$250.34
Crew B-87	Hr.	Daily	Hr.	Daily	Bare Costs	Incl. O&P
1 Laborer	$24.65	$197.20	$41.25	$330.00	$33.13	$54.85
4 Equip. Opers. (medium)	35.25	1128.00	58.25	1864.00		
2 Feller Bunchers, 100 H.P.		1675.20		1842.72		
1 Log Chipper, 22" Tree		890.00		979.00		
1 Dozer, 105 H.P.		603.40		663.74		
1 Chain Saw, Gas, 36" Long		45.00		49.50	80.34	88.37
40 L.H., Daily Totals		$4538.80		$5728.96	$113.47	$143.22

Crews - Residential

Crew No.	Bare Costs		Incl. Subs O&P		Cost Per Labor-Hour	
Crew B-88	Hr.	Daily	Hr.	Daily	Bare Costs	Incl. O&P
1 Laborer	$24.65	$197.20	$41.25	$330.00	$33.74	$55.82
6 Equip. Opers. (medium)	35.25	1692.00	58.25	2796.00		
2 Feller Bunchers, 100 H.P.		1675.20		1842.72		
1 Log Chipper, 22" Tree		890.00		979.00		
2 Log Skidders, 50 H.P.		1833.60		2016.96		
1 Dozer, 105 H.P.		603.40		663.74		
1 Chain Saw, Gas, 36" Long		45.00		49.50	90.13	99.14
56 L.H., Daily Totals		$6936.40		$8677.92	$123.86	$154.96
Crew B-89	Hr.	Daily	Hr.	Daily	Bare Costs	Incl. O&P
1 Skilled Worker	$34.20	$273.60	$57.45	$459.60	$29.43	$49.35
1 Building Laborer	24.65	197.20	41.25	330.00		
1 Flatbed Truck, Gas, 3 Ton		307.40		338.14		
1 Concrete Saw		168.60		185.46		
1 Water Tank, 65 Gal.		17.30		19.03	30.83	33.91
16 L.H., Daily Totals		$964.10		$1332.23	$60.26	$83.26
Crew B-89A	Hr.	Daily	Hr.	Daily	Bare Costs	Incl. O&P
1 Skilled Worker	$34.20	$273.60	$57.45	$459.60	$29.43	$49.35
1 Laborer	24.65	197.20	41.25	330.00		
1 Core Drill (Large)		117.00		128.70	7.31	8.04
16 L.H., Daily Totals		$587.80		$918.30	$36.74	$57.39
Crew B-89B	Hr.	Daily	Hr.	Daily	Bare Costs	Incl. O&P
1 Equip. Oper. (light)	$33.90	$271.20	$56.05	$448.40	$31.85	$52.60
1 Truck Driver (light)	29.80	238.40	49.15	393.20		
1 Wall Saw, Hydraulic, 10 H.P.		115.00		126.50		
1 Generator, Diesel, 100 kW		413.00		454.30		
1 Water Tank, 65 Gal.		17.30		19.03		
1 Flatbed Truck, Gas, 3 Ton		307.40		338.14	53.29	58.62
16 L.H., Daily Totals		$1362.30		$1779.57	$85.14	$111.22
Crew B-90	Hr.	Daily	Hr.	Daily	Bare Costs	Incl. O&P
1 Labor Foreman (outside)	$26.65	$213.20	$44.60	$356.80	$28.71	$47.69
3 Laborers	24.65	591.60	41.25	990.00		
2 Equip. Opers. (light)	33.90	542.40	56.05	896.80		
2 Truck Drivers (heavy)	30.65	490.40	50.55	808.80		
1 Road Mixer, 310 H.P.		1945.00		2139.50		
1 Dist. Truck, 2000 Gal.		289.00		317.90	34.91	38.40
64 L.H., Daily Totals		$4071.60		$5509.80	$63.62	$86.09
Crew B-90A	Hr.	Daily	Hr.	Daily	Bare Costs	Incl. O&P
1 Labor Foreman (outside)	$26.65	$213.20	$44.60	$356.80	$30.99	$51.44
2 Laborers	24.65	394.40	41.25	660.00		
4 Equip. Opers. (medium)	35.25	1128.00	58.25	1864.00		
2 Graders, 30,000 Lbs.		1428.80		1571.68		
1 Tandem Roller, 10 Ton		239.40		263.34		
1 Roller, Pneum. Whl., 12 Ton		347.80		382.58	36.00	39.60
56 L.H., Daily Totals		$3751.60		$5098.40	$66.99	$91.04
Crew B-90B	Hr.	Daily	Hr.	Daily	Bare Costs	Incl. O&P
1 Labor Foreman (outside)	$26.65	$213.20	$44.60	$356.80	$30.28	$50.31
2 Laborers	24.65	394.40	41.25	660.00		
3 Equip. Opers. (medium)	35.25	846.00	58.25	1398.00		
1 Roller, Pneum. Whl., 12 Ton		347.80		382.58		
1 Road Mixer, 310 H.P.		1945.00		2139.50	47.77	52.54
48 L.H., Daily Totals		$3746.40		$4936.88	$78.05	$102.85

Crew No.	Bare Costs		Incl. Subs O&P		Cost Per Labor-Hour	
Crew B-90C	Hr.	Daily	Hr.	Daily	Bare Costs	Incl. O&P
1 Labor Foreman (outside)	$26.65	$213.20	$44.60	$356.80	$29.36	$48.73
4 Laborers	24.65	788.80	41.25	1320.00		
3 Equip. Opers. (medium)	35.25	846.00	58.25	1398.00		
3 Truck Drivers (heavy)	30.65	735.60	50.55	1213.20		
3 Road Mixers, 310 H.P.		5835.00		6418.50	66.31	72.94
88 L.H., Daily Totals		$8418.60		$10706.50	$95.67	$121.66
Crew B-90D	Hr.	Daily	Hr.	Daily	Bare Costs	Incl. O&P
1 Labor Foreman (outside)	$26.65	$213.20	$44.60	$356.80	$28.63	$47.58
6 Laborers	24.65	1183.20	41.25	1980.00		
3 Equip. Opers. (medium)	35.25	846.00	58.25	1398.00		
3 Truck Drivers (heavy)	30.65	735.60	50.55	1213.20		
3 Road Mixers, 310 H.P.		5835.00		6418.50	56.11	61.72
104 L.H., Daily Totals		$8813.00		$11366.50	$84.74	$109.29
Crew B-90E	Hr.	Daily	Hr.	Daily	Bare Costs	Incl. O&P
1 Labor Foreman (outside)	$26.65	$213.20	$44.60	$356.80	$29.07	$48.32
4 Laborers	24.65	788.80	41.25	1320.00		
3 Equip. Opers. (medium)	35.25	846.00	58.25	1398.00		
1 Truck Driver (heavy)	30.65	245.20	50.55	404.40		
1 Road Mixer, 310 H.P.		1945.00		2139.50	27.01	29.72
72 L.H., Daily Totals		$4038.20		$5618.70	$56.09	$78.04
Crew B-91	Hr.	Daily	Hr.	Daily	Bare Costs	Incl. O&P
1 Labor Foreman (outside)	$26.65	$213.20	$44.60	$356.80	$30.95	$51.33
2 Laborers	24.65	394.40	41.25	660.00		
4 Equip. Opers. (medium)	35.25	1128.00	58.25	1864.00		
1 Truck Driver (heavy)	30.65	245.20	50.55	404.40		
1 Dist. Tanker, 3000 Gallon		319.40		351.34		
1 Truck Tractor, 6x4, 380 H.P.		600.00		660.00		
1 Aggreg. Spreader, S.P.		829.70		912.67		
1 Roller, Pneum. Whl., 12 Ton		347.80		382.58		
1 Tandem Roller, 10 Ton		239.40		263.34	36.50	40.16
64 L.H., Daily Totals		$4317.10		$5855.13	$67.45	$91.49
Crew B-91B	Hr.	Daily	Hr.	Daily	Bare Costs	Incl. O&P
1 Laborer	$24.65	$197.20	$41.25	$330.00	$29.95	$49.75
1 Equipment Oper. (med.)	35.25	282.00	58.25	466.00		
1 Road Sweeper, Vac. Assist.		983.80		1082.18	61.49	67.64
16 L.H., Daily Totals		$1463.00		$1878.18	$91.44	$117.39
Crew B-91C	Hr.	Daily	Hr.	Daily	Bare Costs	Incl. O&P
1 Laborer	$24.65	$197.20	$41.25	$330.00	$27.23	$45.20
1 Truck Driver (light)	29.80	238.40	49.15	393.20		
1 Catch Basin Cleaning Truck		577.20		634.92	36.08	39.68
16 L.H., Daily Totals		$1012.80		$1358.12	$63.30	$84.88
Crew B-91D	Hr.	Daily	Hr.	Daily	Bare Costs	Incl. O&P
1 Labor Foreman (outside)	$26.65	$213.20	$44.60	$356.80	$29.80	$49.48
5 Laborers	24.65	986.00	41.25	1650.00		
5 Equip. Opers. (medium)	35.25	1410.00	58.25	2330.00		
2 Truck Drivers (heavy)	30.65	490.40	50.55	808.80		
1 Aggreg. Spreader, S.P.		829.70		912.67		
2 Truck Tractors, 6x4, 380 H.P.		1200.00		1320.00		
2 Dist. Tankers, 3000 Gallon		638.80		702.68		
2 Pavement Brushes, Towed		162.40		178.64		
2 Rollers Pneum. Whl., 12 Ton		695.60		765.16	33.91	37.30
104 L.H., Daily Totals		$6626.10		$9024.75	$63.71	$86.78

Crews - Residential

Crew No.	Bare Costs		Incl. Subs O&P		Cost Per Labor-Hour	
Crew B-92	Hr.	Daily	Hr.	Daily	Bare Costs	Incl. O&P
1 Labor Foreman (outside)	$26.65	$213.20	$44.60	$356.80	$25.15	$42.09
3 Laborers	24.65	591.60	41.25	990.00		
1 Crack Cleaner, 25 H.P.		62.00		68.20		
1 Air Compressor, 60 cfm		142.80		157.08		
1 Tar Kettle, T.M.		135.30		148.83		
1 Flatbed Truck, Gas, 3 Ton		307.40		338.14	20.23	22.26
32 L.H., Daily Totals		$1452.30		$2059.05	$45.38	$64.35
Crew B-93	Hr.	Daily	Hr.	Daily	Bare Costs	Incl. O&P
1 Equip. Oper. (medium)	$35.25	$282.00	$58.25	$466.00	$35.25	$58.25
1 Feller Buncher, 100 H.P.		837.60		921.36	104.70	115.17
8 L.H., Daily Totals		$1119.60		$1387.36	$139.95	$173.42
Crew B-94A	Hr.	Daily	Hr.	Daily	Bare Costs	Incl. O&P
1 Laborer	$24.65	$197.20	$41.25	$330.00	$24.65	$41.25
1 Diaphragm Water Pump, 2"		71.80		78.98		
1 -20' Suction Hose, 2"		1.95		2.15		
2 -50' Discharge Hoses, 2"		1.80		1.98	9.44	10.39
8 L.H., Daily Totals		$272.75		$413.11	$34.09	$51.64
Crew B-94B	Hr.	Daily	Hr.	Daily	Bare Costs	Incl. O&P
1 Laborer	$24.65	$197.20	$41.25	$330.00	$24.65	$41.25
1 Diaphragm Water Pump, 4"		118.00		129.80		
1 -20' Suction Hose, 4"		3.25		3.58		
2 -50' Discharge Hoses, 4"		4.70		5.17	15.74	17.32
8 L.H., Daily Totals		$323.15		$468.55	$40.39	$58.57
Crew B-94C	Hr.	Daily	Hr.	Daily	Bare Costs	Incl. O&P
1 Laborer	$24.65	$197.20	$41.25	$330.00	$24.65	$41.25
1 Centrifugal Water Pump, 3"		81.00		89.10		
1 -20' Suction Hose, 3"		2.85		3.13		
2 -50' Discharge Hoses, 3"		3.00		3.30	10.86	11.94
8 L.H., Daily Totals		$284.05		$425.54	$35.51	$53.19
Crew B-94D	Hr.	Daily	Hr.	Daily	Bare Costs	Incl. O&P
1 Laborer	$24.65	$197.20	$41.25	$330.00	$24.65	$41.25
1 Centr. Water Pump, 6"		350.80		385.88		
1 -20' Suction Hose, 6"		11.50		12.65		
2 -50' Discharge Hoses, 6"		12.20		13.42	46.81	51.49
8 L.H., Daily Totals		$571.70		$741.95	$71.46	$92.74
Crew C-1	Hr.	Daily	Hr.	Daily	Bare Costs	Incl. O&P
2 Carpenters	$33.90	$542.40	$56.75	$908.00	$29.30	$49.14
1 Carpenter Helper	24.75	198.00	41.80	334.40		
1 Laborer	24.65	197.20	41.25	330.00		
32 L.H., Daily Totals		$937.60		$1572.40	$29.30	$49.14
Crew C-2	Hr.	Daily	Hr.	Daily	Bare Costs	Incl. O&P
1 Carpenter Foreman (outside)	$35.90	$287.20	$60.10	$480.80	$29.64	$49.74
2 Carpenters	33.90	542.40	56.75	908.00		
2 Carpenter Helpers	24.75	396.00	41.80	668.80		
1 Laborer	24.65	197.20	41.25	330.00		
48 L.H., Daily Totals		$1422.80		$2387.60	$29.64	$49.74
Crew C-2A	Hr.	Daily	Hr.	Daily	Bare Costs	Incl. O&P
1 Carpenter Foreman (outside)	$35.90	$287.20	$60.10	$480.80	$32.37	$53.91
3 Carpenters	33.90	813.60	56.75	1362.00		
1 Cement Finisher	31.95	255.60	51.85	414.80		
1 Laborer	24.65	197.20	41.25	330.00		
48 L.H., Daily Totals		$1553.60		$2587.60	$32.37	$53.91

Crew No.	Bare Costs		Incl. Subs O&P		Cost Per Labor-Hour	
Crew C-3	Hr.	Daily	Hr.	Daily	Bare Costs	Incl. O&P
1 Rodman Foreman (outside)	$37.50	$300.00	$63.35	$506.80	$31.48	$52.88
3 Rodmen (reinf.)	35.50	852.00	59.95	1438.80		
1 Equip. Oper. (light)	33.90	271.20	56.05	448.40		
3 Laborers	24.65	591.60	41.25	990.00		
3 Stressing Equipment		30.60		33.66		
.5 Grouting Equipment		82.20		90.42	1.76	1.94
64 L.H., Daily Totals		$2127.60		$3508.08	$33.24	$54.81
Crew C-4	Hr.	Daily	Hr.	Daily	Bare Costs	Incl. O&P
1 Rodman Foreman (outside)	$37.50	$300.00	$63.35	$506.80	$33.29	$56.13
2 Rodmen (reinf.)	35.50	568.00	59.95	959.20		
1 Building Laborer	24.65	197.20	41.25	330.00		
3 Stressing Equipment		30.60		33.66	0.96	1.05
32 L.H., Daily Totals		$1095.80		$1829.66	$34.24	$57.18
Crew C-4A	Hr.	Daily	Hr.	Daily	Bare Costs	Incl. O&P
2 Rodmen (reinf.)	$35.50	$568.00	$59.95	$959.20	$35.50	$59.95
4 Stressing Equipment		40.80		44.88	2.55	2.81
16 L.H., Daily Totals		$608.80		$1004.08	$38.05	$62.76
Crew C-5	Hr.	Daily	Hr.	Daily	Bare Costs	Incl. O&P
1 Rodman Foreman (outside)	$37.50	$300.00	$63.35	$506.80	$32.31	$54.23
2 Rodmen (reinf.)	35.50	568.00	59.95	959.20		
1 Equip. Oper. (crane)	36.05	288.40	59.60	476.80		
2 Building Laborers	24.65	394.40	41.25	660.00		
1 Hyd. Crane, 25 Ton		748.80		823.68	15.60	17.16
48 L.H., Daily Totals		$2299.60		$3426.48	$47.91	$71.39
Crew C-6	Hr.	Daily	Hr.	Daily	Bare Costs	Incl. O&P
1 Labor Foreman (outside)	$26.65	$213.20	$44.60	$356.80	$26.20	$43.58
4 Laborers	24.65	788.80	41.25	1320.00		
1 Cement Finisher	31.95	255.60	51.85	414.80		
2 Gas Engine Vibrators		63.20		69.52	1.32	1.45
48 L.H., Daily Totals		$1320.80		$2161.12	$27.52	$45.02
Crew C-7	Hr.	Daily	Hr.	Daily	Bare Costs	Incl. O&P
1 Labor Foreman (outside)	$26.65	$213.20	$44.60	$356.80	$27.65	$45.94
5 Laborers	24.65	986.00	41.25	1650.00		
1 Cement Finisher	31.95	255.60	51.85	414.80		
1 Equip. Oper. (medium)	35.25	282.00	58.25	466.00		
1 Equip. Oper. (oiler)	31.75	254.00	52.50	420.00		
2 Gas Engine Vibrators		63.20		69.52		
1 Concrete Bucket, 1 C.Y.		23.60		25.96		
1 Hyd. Crane, 55 Ton		1123.00		1235.30	16.80	18.48
72 L.H., Daily Totals		$3200.60		$4638.38	$44.45	$64.42
Crew C-8	Hr.	Daily	Hr.	Daily	Bare Costs	Incl. O&P
1 Labor Foreman (outside)	$26.65	$213.20	$44.60	$356.80	$28.54	$47.19
3 Laborers	24.65	591.60	41.25	990.00		
2 Cement Finishers	31.95	511.20	51.85	829.60		
1 Equip. Oper. (medium)	35.25	282.00	58.25	466.00		
1 Concrete Pump (Small)		727.60		800.36	12.99	14.29
56 L.H., Daily Totals		$2325.60		$3442.76	$41.53	$61.48
Crew C-8A	Hr.	Daily	Hr.	Daily	Bare Costs	Incl. O&P
1 Labor Foreman (outside)	$26.65	$213.20	$44.60	$356.80	$27.42	$45.34
3 Laborers	24.65	591.60	41.25	990.00		
2 Cement Finishers	31.95	511.20	51.85	829.60		
48 L.H., Daily Totals		$1316.00		$2176.40	$27.42	$45.34

Crews - Residential

Crew No.	Bare Costs		Incl. Subs O&P		Cost Per Labor-Hour	
Crew C-8B	Hr.	Daily	Hr.	Daily	Bare Costs	Incl. O&P
1 Labor Foreman (outside)	$26.65	$213.20	$44.60	$356.80	$27.17	$45.32
3 Laborers	24.65	591.60	41.25	990.00		
1 Equip. Oper. (medium)	35.25	282.00	58.25	466.00		
1 Vibrating Power Screed		63.65		70.02		
1 Roller, Vibratory, 25 Ton		691.80		760.98		
1 Dozer, 200 H.P.		1397.00		1536.70	53.81	59.19
40 L.H., Daily Totals		$3239.25		$4180.49	$80.98	$104.51
Crew C-8C	Hr.	Daily	Hr.	Daily	Bare Costs	Incl. O&P
1 Labor Foreman (outside)	$26.65	$213.20	$44.60	$356.80	$27.97	$46.41
3 Laborers	24.65	591.60	41.25	990.00		
1 Cement Finisher	31.95	255.60	51.85	414.80		
1 Equip. Oper. (medium)	35.25	282.00	58.25	466.00		
1 Shotcrete Rig, 12 C.Y./hr		254.00		279.40		
1 Air Compressor, 160 cfm		162.20		178.42		
4 -50' Air Hoses, 1"		14.60		16.06		
4 -50' Air Hoses, 2"		27.60		30.36	9.55	10.51
48 L.H., Daily Totals		$1800.80		$2731.84	$37.52	$56.91
Crew C-8D	Hr.	Daily	Hr.	Daily	Bare Costs	Incl. O&P
1 Labor Foreman (outside)	$26.65	$213.20	$44.60	$356.80	$29.29	$48.44
1 Laborer	24.65	197.20	41.25	330.00		
1 Cement Finisher	31.95	255.60	51.85	414.80		
1 Equipment Oper. (light)	33.90	271.20	56.05	448.40		
1 Air Compressor, 250 cfm		197.80		217.58		
2 -50' Air Hoses, 1"		7.30		8.03	6.41	7.05
32 L.H., Daily Totals		$1142.30		$1775.61	$35.70	$55.49
Crew C-8E	Hr.	Daily	Hr.	Daily	Bare Costs	Incl. O&P
1 Labor Foreman (outside)	$26.65	$213.20	$44.60	$356.80	$27.74	$46.04
3 Laborers	24.65	591.60	41.25	990.00		
1 Cement Finisher	31.95	255.60	51.85	414.80		
1 Equipment Oper. (light)	33.90	271.20	56.05	448.40		
1 Shotcrete Rig, 35 C.Y./hr		284.00		312.40		
1 Air Compressor, 250 cfm		197.80		217.58		
4 -50' Air Hoses, 1"		14.60		16.06		
4 -50' Air Hoses, 2"		27.60		30.36	10.92	12.01
48 L.H., Daily Totals		$1855.60		$2786.40	$38.66	$58.05
Crew C-10	Hr.	Daily	Hr.	Daily	Bare Costs	Incl. O&P
1 Laborer	$24.65	$197.20	$41.25	$330.00	$29.52	$48.32
2 Cement Finishers	31.95	511.20	51.85	829.60		
24 L.H., Daily Totals		$708.40		$1159.60	$29.52	$48.32
Crew C-10B	Hr.	Daily	Hr.	Daily	Bare Costs	Incl. O&P
3 Laborers	$24.65	$591.60	$41.25	$990.00	$27.57	$45.49
2 Cement Finishers	31.95	511.20	51.85	829.60		
1 Concrete Mixer, 10 C.F.		172.80		190.08		
2 Trowels, 48" Walk-Behind		101.20		111.32	6.85	7.54
40 L.H., Daily Totals		$1376.80		$2121.00	$34.42	$53.02
Crew C-10C	Hr.	Daily	Hr.	Daily	Bare Costs	Incl. O&P
1 Laborer	$24.65	$197.20	$41.25	$330.00	$29.52	$48.32
2 Cement Finishers	31.95	511.20	51.85	829.60		
1 Trowel, 48" Walk-Behind		50.60		55.66	2.11	2.32
24 L.H., Daily Totals		$759.00		$1215.26	$31.63	$50.64

Crew No.	Bare Costs		Incl. Subs O&P		Cost Per Labor-Hour	
Crew C-10D	Hr.	Daily	Hr.	Daily	Bare Costs	Incl. O&P
1 Laborer	$24.65	$197.20	$41.25	$330.00	$29.52	$48.32
2 Cement Finishers	31.95	511.20	51.85	829.60		
1 Vibrating Power Screed		63.65		70.02		
1 Trowel, 48" Walk-Behind		50.60		55.66	4.76	5.24
24 L.H., Daily Totals		$822.65		$1285.28	$34.28	$53.55
Crew C-10E	Hr.	Daily	Hr.	Daily	Bare Costs	Incl. O&P
1 Laborer	$24.65	$197.20	$41.25	$330.00	$29.52	$48.32
2 Cement Finishers	31.95	511.20	51.85	829.60		
1 Vibrating Power Screed		63.65		70.02		
1 Cement Trowel, 96" Ride-On		190.20		209.22	10.58	11.63
24 L.H., Daily Totals		$962.25		$1438.84	$40.09	$59.95
Crew C-10F	Hr.	Daily	Hr.	Daily	Bare Costs	Incl. O&P
1 Laborer	$24.65	$197.20	$41.25	$330.00	$29.52	$48.32
2 Cement Finishers	31.95	511.20	51.85	829.60		
1 Aerial Lift Truck, 60' Boom		448.40		493.24	18.68	20.55
24 L.H., Daily Totals		$1156.80		$1652.84	$48.20	$68.87
Crew C-11	Hr.	Daily	Hr.	Daily	Bare Costs	Incl. O&P
1 Skilled Worker Foreman	$36.20	$289.60	$60.80	$486.40	$34.75	$58.24
5 Skilled Workers	34.20	1368.00	57.45	2298.00		
1 Equip. Oper. (crane)	36.05	288.40	59.60	476.80		
1 Lattice Boom Crane, 150 Ton		1868.00		2054.80	33.36	36.69
56 L.H., Daily Totals		$3814.00		$5316.00	$68.11	$94.93
Crew C-12	Hr.	Daily	Hr.	Daily	Bare Costs	Incl. O&P
1 Carpenter Foreman (outside)	$35.90	$287.20	$60.10	$480.80	$33.05	$55.20
3 Carpenters	33.90	813.60	56.75	1362.00		
1 Laborer	24.65	197.20	41.25	330.00		
1 Equip. Oper. (crane)	36.05	288.40	59.60	476.80		
1 Hyd. Crane, 12 Ton		658.40		724.24	13.72	15.09
48 L.H., Daily Totals		$2244.80		$3373.84	$46.77	$70.29
Crew C-13	Hr.	Daily	Hr.	Daily	Bare Costs	Incl. O&P
2 Struc. Steel Workers	$35.65	$570.40	$65.40	$1046.40	$35.07	$62.52
1 Carpenter	33.90	271.20	56.75	454.00		
1 Welder, Gas Engine, 300 amp		147.20		161.92	6.13	6.75
24 L.H., Daily Totals		$988.80		$1662.32	$41.20	$69.26
Crew C-14	Hr.	Daily	Hr.	Daily	Bare Costs	Incl. O&P
1 Carpenter Foreman (outside)	$35.90	$287.20	$60.10	$480.80	$29.77	$49.75
3 Carpenters	33.90	813.60	56.75	1362.00		
2 Carpenter Helpers	24.75	396.00	41.80	668.80		
4 Laborers	24.65	788.80	41.25	1320.00		
2 Rodmen (reinf.)	35.50	568.00	59.95	959.20		
2 Rodman Helpers	24.75	396.00	41.80	668.80		
2 Cement Finishers	31.95	511.20	51.85	829.60		
1 Equip. Oper. (crane)	36.05	288.40	59.60	476.80		
1 Hyd. Crane, 80 Ton		1630.00		1793.00	11.99	13.18
136 L.H., Daily Totals		$5679.20		$8559.00	$41.76	$62.93

Crews - Residential

Crew No.	Bare Costs		Incl. Subs O&P		Cost Per Labor-Hour	
Crew C-14A	**Hr.**	**Daily**	**Hr.**	**Daily**	**Bare Costs**	**Incl. O&P**
1 Carpenter Foreman (outside)	$35.90	$287.20	$60.10	$480.80	$33.47	$56.02
16 Carpenters	33.90	4339.20	56.75	7264.00		
4 Rodmen (reinf.)	35.50	1136.00	59.95	1918.40		
2 Laborers	24.65	394.40	41.25	660.00		
1 Cement Finisher	31.95	255.60	51.85	414.80		
1 Equip. Oper. (medium)	35.25	282.00	58.25	466.00		
1 Gas Engine Vibrator		31.60		34.76		
1 Concrete Pump (Small)		727.60		800.36	3.80	4.18
200 L.H., Daily Totals		$7453.60		$12039.12	$37.27	$60.20
Crew C-14B	**Hr.**	**Daily**	**Hr.**	**Daily**	**Bare Costs**	**Incl. O&P**
1 Carpenter Foreman (outside)	$35.90	$287.20	$60.10	$480.80	$33.41	$55.86
16 Carpenters	33.90	4339.20	56.75	7264.00		
4 Rodmen (reinf.)	35.50	1136.00	59.95	1918.40		
2 Laborers	24.65	394.40	41.25	660.00		
2 Cement Finishers	31.95	511.20	51.85	829.60		
1 Equip. Oper. (medium)	35.25	282.00	58.25	466.00		
1 Gas Engine Vibrator		31.60		34.76		
1 Concrete Pump (Small)		727.60		800.36	3.65	4.01
208 L.H., Daily Totals		$7709.20		$12453.92	$37.06	$59.87
Crew C-14C	**Hr.**	**Daily**	**Hr.**	**Daily**	**Bare Costs**	**Incl. O&P**
1 Carpenter Foreman (outside)	$35.90	$287.20	$60.10	$480.80	$31.49	$52.67
6 Carpenters	33.90	1627.20	56.75	2724.00		
2 Rodmen (reinf.)	35.50	568.00	59.95	959.20		
4 Laborers	24.65	788.80	41.25	1320.00		
1 Cement Finisher	31.95	255.60	51.85	414.80		
1 Gas Engine Vibrator		31.60		34.76	0.28	0.31
112 L.H., Daily Totals		$3558.40		$5933.56	$31.77	$52.98
Crew C-14D	**Hr.**	**Daily**	**Hr.**	**Daily**	**Bare Costs**	**Incl. O&P**
1 Carpenter Foreman (outside)	$35.90	$287.20	$60.10	$480.80	$33.34	$55.76
18 Carpenters	33.90	4881.60	56.75	8172.00		
2 Rodmen (reinf.)	35.50	568.00	59.95	959.20		
2 Laborers	24.65	394.40	41.25	660.00		
1 Cement Finisher	31.95	255.60	51.85	414.80		
1 Equip. Oper. (medium)	35.25	282.00	58.25	466.00		
1 Gas Engine Vibrator		31.60		34.76		
1 Concrete Pump (Small)		727.60		800.36	3.80	4.18
200 L.H., Daily Totals		$7428.00		$11987.92	$37.14	$59.94
Crew C-14E	**Hr.**	**Daily**	**Hr.**	**Daily**	**Bare Costs**	**Incl. O&P**
1 Carpenter Foreman (outside)	$35.90	$287.20	$60.10	$480.80	$31.96	$53.55
2 Carpenters	33.90	542.40	56.75	908.00		
4 Rodmen (reinf.)	35.50	1136.00	59.95	1918.40		
3 Laborers	24.65	591.60	41.25	990.00		
1 Cement Finisher	31.95	255.60	51.85	414.80		
1 Gas Engine Vibrator		31.60		34.76	0.36	0.40
88 L.H., Daily Totals		$2844.40		$4746.76	$32.32	$53.94
Crew C-14F	**Hr.**	**Daily**	**Hr.**	**Daily**	**Bare Costs**	**Incl. O&P**
1 Labor Foreman (outside)	$26.65	$213.20	$44.60	$356.80	$29.74	$48.69
2 Laborers	24.65	394.40	41.25	660.00		
6 Cement Finishers	31.95	1533.60	51.85	2488.80		
1 Gas Engine Vibrator		31.60		34.76	0.44	0.48
72 L.H., Daily Totals		$2172.80		$3540.36	$30.18	$49.17
Crew C-14G	**Hr.**	**Daily**	**Hr.**	**Daily**	**Bare Costs**	**Incl. O&P**
1 Labor Foreman (outside)	$26.65	$213.20	$44.60	$356.80	$29.11	$47.79
2 Laborers	24.65	394.40	41.25	660.00		
4 Cement Finishers	31.95	1022.40	51.85	1659.20		
1 Gas Engine Vibrator		31.60		34.76	0.56	0.62
56 L.H., Daily Totals		$1661.60		$2710.76	$29.67	$48.41
Crew C-14H	**Hr.**	**Daily**	**Hr.**	**Daily**	**Bare Costs**	**Incl. O&P**
1 Carpenter Foreman (outside)	$35.90	$287.20	$60.10	$480.80	$32.63	$54.44
2 Carpenters	33.90	542.40	56.75	908.00		
1 Rodman (reinf.)	35.50	284.00	59.95	479.60		
1 Laborer	24.65	197.20	41.25	330.00		
1 Cement Finisher	31.95	255.60	51.85	414.80		
1 Gas Engine Vibrator		31.60		34.76	0.66	0.72
48 L.H., Daily Totals		$1598.00		$2647.96	$33.29	$55.17
Crew C-14L	**Hr.**	**Daily**	**Hr.**	**Daily**	**Bare Costs**	**Incl. O&P**
1 Carpenter Foreman (outside)	$35.90	$287.20	$60.10	$480.80	$30.82	$51.45
6 Carpenters	33.90	1627.20	56.75	2724.00		
4 Laborers	24.65	788.80	41.25	1320.00		
1 Cement Finisher	31.95	255.60	51.85	414.80		
1 Gas Engine Vibrator		31.60		34.76	0.33	0.36
96 L.H., Daily Totals		$2990.40		$4974.36	$31.15	$51.82
Crew C-14M	**Hr.**	**Daily**	**Hr.**	**Daily**	**Bare Costs**	**Incl. O&P**
1 Carpenter Foreman (outside)	$35.90	$287.20	$60.10	$480.80	$31.96	$53.27
2 Carpenters	33.90	542.40	56.75	908.00		
1 Rodman (reinf.)	35.50	284.00	59.95	479.60		
2 Laborers	24.65	394.40	41.25	660.00		
1 Cement Finisher	31.95	255.60	51.85	414.80		
1 Equip. Oper. (medium)	35.25	282.00	58.25	466.00		
1 Gas Engine Vibrator		31.60		34.76		
1 Concrete Pump (Small)		727.60		800.36	11.86	13.05
64 L.H., Daily Totals		$2804.80		$4244.32	$43.83	$66.32
Crew C-15	**Hr.**	**Daily**	**Hr.**	**Daily**	**Bare Costs**	**Incl. O&P**
1 Carpenter Foreman (outside)	$35.90	$287.20	$60.10	$480.80	$30.78	$51.22
2 Carpenters	33.90	542.40	56.75	908.00		
3 Laborers	24.65	591.60	41.25	990.00		
2 Cement Finishers	31.95	511.20	51.85	829.60		
1 Rodman (reinf.)	35.50	284.00	59.95	479.60		
72 L.H., Daily Totals		$2216.40		$3688.00	$30.78	$51.22
Crew C-16	**Hr.**	**Daily**	**Hr.**	**Daily**	**Bare Costs**	**Incl. O&P**
1 Labor Foreman (outside)	$26.65	$213.20	$44.60	$356.80	$28.54	$47.19
3 Laborers	24.65	591.60	41.25	990.00		
2 Cement Finishers	31.95	511.20	51.85	829.60		
1 Equip. Oper. (medium)	35.25	282.00	58.25	466.00		
1 Gunite Pump Rig		373.00		410.30		
2 -50' Air Hoses, 3/4"		6.10		6.71		
2 -50' Air Hoses, 2"		13.80		15.18	7.02	7.72
56 L.H., Daily Totals		$1990.90		$3074.59	$35.55	$54.90
Crew C-16A	**Hr.**	**Daily**	**Hr.**	**Daily**	**Bare Costs**	**Incl. O&P**
1 Laborer	$24.65	$197.20	$41.25	$330.00	$30.95	$50.80
2 Cement Finishers	31.95	511.20	51.85	829.60		
1 Equip. Oper. (medium)	35.25	282.00	58.25	466.00		
1 Gunite Pump Rig		373.00		410.30		
2 -50' Air Hoses, 3/4"		6.10		6.71		
2 -50' Air Hoses, 2"		13.80		15.18		
1 Aerial Lift Truck, 60' Boom		448.40		493.24	26.29	28.92
32 L.H., Daily Totals		$1831.70		$2551.03	$57.24	$79.72

Crews - Residential

Crew No.	Bare Costs		Incl. Subs O&P		Cost Per Labor-Hour	
Crew C-17	Hr.	Daily	Hr.	Daily	Bare Costs	Incl. O&P
2 Skilled Worker Foremen (out)	$36.20	$579.20	$60.80	$972.80	$34.60	$58.12
8 Skilled Workers	34.20	2188.80	57.45	3676.80		
80 L.H., Daily Totals		$2768.00		$4649.60	$34.60	$58.12
Crew C-17A	Hr.	Daily	Hr.	Daily	Bare Costs	Incl. O&P
2 Skilled Worker Foremen (out)	$36.20	$579.20	$60.80	$972.80	$34.62	$58.14
8 Skilled Workers	34.20	2188.80	57.45	3676.80		
.125 Equip. Oper. (crane)	36.05	36.05	59.60	59.60		
.125 Hyd. Crane, 80 Ton		203.75		224.13	2.52	2.77
81 L.H., Daily Totals		$3007.80		$4933.32	$37.13	$60.91
Crew C-17B	Hr.	Daily	Hr.	Daily	Bare Costs	Incl. O&P
2 Skilled Worker Foremen (out)	$36.20	$579.20	$60.80	$972.80	$34.64	$58.16
8 Skilled Workers	34.20	2188.80	57.45	3676.80		
.25 Equip. Oper. (crane)	36.05	72.10	59.60	119.20		
.25 Hyd. Crane, 80 Ton		407.50		448.25		
.25 Trowel, 48" Walk-Behind		12.65		13.91	5.12	5.64
82 L.H., Daily Totals		$3260.25		$5230.97	$39.76	$63.79
Crew C-17C	Hr.	Daily	Hr.	Daily	Bare Costs	Incl. O&P
2 Skilled Worker Foremen (out)	$36.20	$579.20	$60.80	$972.80	$34.65	$58.17
8 Skilled Workers	34.20	2188.80	57.45	3676.80		
.375 Equip. Oper. (crane)	36.05	108.15	59.60	178.80		
.375 Hyd. Crane, 80 Ton		611.25		672.38	7.36	8.10
83 L.H., Daily Totals		$3487.40		$5500.77	$42.02	$66.27
Crew C-17D	Hr.	Daily	Hr.	Daily	Bare Costs	Incl. O&P
2 Skilled Worker Foremen (out)	$36.20	$579.20	$60.80	$972.80	$34.67	$58.19
8 Skilled Workers	34.20	2188.80	57.45	3676.80		
.5 Equip. Oper. (crane)	36.05	144.20	59.60	238.40		
.5 Hyd. Crane, 80 Ton		815.00		896.50	9.70	10.67
84 L.H., Daily Totals		$3727.20		$5784.50	$44.37	$68.86
Crew C-17E	Hr.	Daily	Hr.	Daily	Bare Costs	Incl. O&P
2 Skilled Worker Foremen (out)	$36.20	$579.20	$60.80	$972.80	$34.60	$58.12
8 Skilled Workers	34.20	2188.80	57.45	3676.80		
1 Hyd. Jack with Rods		98.40		108.24	1.23	1.35
80 L.H., Daily Totals		$2866.40		$4757.84	$35.83	$59.47
Crew C-18	Hr.	Daily	Hr.	Daily	Bare Costs	Incl. O&P
.125 Labor Foreman (outside)	$26.65	$26.65	$44.60	$44.60	$24.87	$41.62
1 Laborer	24.65	197.20	41.25	330.00		
1 Concrete Cart, 10 C.F.		60.00		66.00	6.67	7.33
9 L.H., Daily Totals		$283.85		$440.60	$31.54	$48.96
Crew C-19	Hr.	Daily	Hr.	Daily	Bare Costs	Incl. O&P
.125 Labor Foreman (outside)	$26.65	$26.65	$44.60	$44.60	$24.87	$41.62
1 Laborer	24.65	197.20	41.25	330.00		
1 Concrete Cart, 18 C.F.		99.80		109.78	11.09	12.20
9 L.H., Daily Totals		$323.65		$484.38	$35.96	$53.82
Crew C-20	Hr.	Daily	Hr.	Daily	Bare Costs	Incl. O&P
1 Labor Foreman (outside)	$26.65	$213.20	$44.60	$356.80	$27.14	$45.12
5 Laborers	24.65	986.00	41.25	1650.00		
1 Cement Finisher	31.95	255.60	51.85	414.80		
1 Equip. Oper. (medium)	35.25	282.00	58.25	466.00		
2 Gas Engine Vibrators		63.20		69.52		
1 Concrete Pump (Small)		727.60		800.36	12.36	13.59
64 L.H., Daily Totals		$2527.60		$3757.48	$39.49	$58.71

Crew No.	Bare Costs		Incl. Subs O&P		Cost Per Labor-Hour	
Crew C-21	Hr.	Daily	Hr.	Daily	Bare Costs	Incl. O&P
1 Labor Foreman (outside)	$26.65	$213.20	$44.60	$356.80	$27.14	$45.12
5 Laborers	24.65	986.00	41.25	1650.00		
1 Cement Finisher	31.95	255.60	51.85	414.80		
1 Equip. Oper. (medium)	35.25	282.00	58.25	466.00		
2 Gas Engine Vibrators		63.20		69.52		
1 Concrete Conveyer		199.20		219.12	4.10	4.51
64 L.H., Daily Totals		$1999.20		$3176.24	$31.24	$49.63
Crew C-22	Hr.	Daily	Hr.	Daily	Bare Costs	Incl. O&P
1 Rodman Foreman (outside)	$37.50	$300.00	$63.35	$506.80	$35.80	$60.41
4 Rodmen (reinf.)	35.50	1136.00	59.95	1918.40		
.125 Equip. Oper. (crane)	36.05	36.05	59.60	59.60		
.125 Equip. Oper. (oiler)	31.75	31.75	52.50	52.50		
.125 Hyd. Crane, 25 Ton		93.60		102.96	2.23	2.45
42 L.H., Daily Totals		$1597.40		$2640.26	$38.03	$62.86
Crew C-23	Hr.	Daily	Hr.	Daily	Bare Costs	Incl. O&P
2 Skilled Worker Foremen (out)	$36.20	$579.20	$60.80	$972.80	$34.54	$57.84
6 Skilled Workers	34.20	1641.60	57.45	2757.60		
1 Equip. Oper. (crane)	36.05	288.40	59.60	476.80		
1 Equip. Oper. (oiler)	31.75	254.00	52.50	420.00		
1 Lattice Boom Crane, 90 Ton		1520.00		1672.00	19.00	20.90
80 L.H., Daily Totals		$4283.20		$6299.20	$53.54	$78.74
Crew C-24	Hr.	Daily	Hr.	Daily	Bare Costs	Incl. O&P
2 Skilled Worker Foremen (out)	$36.20	$579.20	$60.80	$972.80	$34.54	$57.84
6 Skilled Workers	34.20	1641.60	57.45	2757.60		
1 Equip. Oper. (crane)	36.05	288.40	59.60	476.80		
1 Equip. Oper. (oiler)	31.75	254.00	52.50	420.00		
1 Lattice Boom Crane, 150 Ton		1868.00		2054.80	23.35	25.68
80 L.H., Daily Totals		$4631.20		$6682.00	$57.89	$83.53
Crew C-25	Hr.	Daily	Hr.	Daily	Bare Costs	Incl. O&P
2 Rodmen (reinf.)	$35.50	$568.00	$59.95	$959.20	$28.57	$49.92
2 Rodmen Helpers	21.65	346.40	39.90	638.40		
32 L.H., Daily Totals		$914.40		$1597.60	$28.57	$49.92
Crew C-27	Hr.	Daily	Hr.	Daily	Bare Costs	Incl. O&P
2 Cement Finishers	$31.95	$511.20	$51.85	$829.60	$31.95	$51.85
1 Concrete Saw		168.60		185.46	10.54	11.59
16 L.H., Daily Totals		$679.80		$1015.06	$42.49	$63.44
Crew C-28	Hr.	Daily	Hr.	Daily	Bare Costs	Incl. O&P
1 Cement Finisher	$31.95	$255.60	$51.85	$414.80	$31.95	$51.85
1 Portable Air Compressor, Gas		17.40		19.14	2.17	2.39
8 L.H., Daily Totals		$273.00		$433.94	$34.13	$54.24
Crew C-29	Hr.	Daily	Hr.	Daily	Bare Costs	Incl. O&P
1 Laborer	$24.65	$197.20	$41.25	$330.00	$24.65	$41.25
1 Pressure Washer		71.60		78.76	8.95	9.85
8 L.H., Daily Totals		$268.80		$408.76	$33.60	$51.09
Crew C-30	Hr.	Daily	Hr.	Daily	Bare Costs	Incl. O&P
1 Laborer	$24.65	$197.20	$41.25	$330.00	$24.65	$41.25
1 Concrete Mixer, 10 C.F.		172.80		190.08	21.60	23.76
8 L.H., Daily Totals		$370.00		$520.08	$46.25	$65.01

Crews - Residential

Crew No.	Bare Costs		Incl. Subs O&P		Cost Per Labor-Hour	
Crew C-31	Hr.	Daily	Hr.	Daily	Bare Costs	Incl. O&P
1 Cement Finisher	$31.95	$255.60	$51.85	$414.80	$31.95	$51.85
1 Grout Pump		373.00		410.30	46.63	51.29
8 L.H., Daily Totals		$628.60		$825.10	$78.58	$103.14
Crew C-32	Hr.	Daily	Hr.	Daily	Bare Costs	Incl. O&P
1 Cement Finisher	$31.95	$255.60	$51.85	$414.80	$28.30	$46.55
1 Laborer	24.65	197.20	41.25	330.00		
1 Crack Chaser Saw, Gas, 6 H.P.		29.20		32.12		
1 Vacuum Pick-Up System		60.90		66.99	5.63	6.19
16 L.H., Daily Totals		$542.90		$843.91	$33.93	$52.74
Crew D-1	Hr.	Daily	Hr.	Daily	Bare Costs	Incl. O&P
1 Bricklayer	$32.40	$259.20	$54.00	$432.00	$29.50	$49.17
1 Bricklayer Helper	26.60	212.80	44.35	354.80		
16 L.H., Daily Totals		$472.00		$786.80	$29.50	$49.17
Crew D-2	Hr.	Daily	Hr.	Daily	Bare Costs	Incl. O&P
3 Bricklayers	$32.40	$777.60	$54.00	$1296.00	$30.08	$50.14
2 Bricklayer Helpers	26.60	425.60	44.35	709.60		
40 L.H., Daily Totals		$1203.20		$2005.60	$30.08	$50.14
Crew D-3	Hr.	Daily	Hr.	Daily	Bare Costs	Incl. O&P
3 Bricklayers	$32.40	$777.60	$54.00	$1296.00	$30.26	$50.45
2 Bricklayer Helpers	26.60	425.60	44.35	709.60		
.25 Carpenter	33.90	67.80	56.75	113.50		
42 L.H., Daily Totals		$1271.00		$2119.10	$30.26	$50.45
Crew D-4	Hr.	Daily	Hr.	Daily	Bare Costs	Incl. O&P
1 Bricklayer	$32.40	$259.20	$54.00	$432.00	$27.37	$45.66
3 Bricklayer Helpers	26.60	638.40	44.35	1064.40		
1 Building Laborer	24.65	197.20	41.25	330.00		
1 Grout Pump, 50 C.F./hr.		135.00		148.50	3.38	3.71
40 L.H., Daily Totals		$1229.80		$1974.90	$30.75	$49.37
Crew D-5	Hr.	Daily	Hr.	Daily	Bare Costs	Incl. O&P
1 Block Mason Helper	26.60	212.80	44.35	354.80	26.60	44.35
8 L.H., Daily Totals		$212.80		$354.80	$26.60	$44.35
Crew D-6	Hr.	Daily	Hr.	Daily	Bare Costs	Incl. O&P
3 Bricklayers	$32.40	$777.60	$54.00	$1296.00	$29.50	$49.17
3 Bricklayer Helpers	26.60	638.40	44.35	1064.40		
48 L.H., Daily Totals		$1416.00		$2360.40	$29.50	$49.17
Crew D-7	Hr.	Daily	Hr.	Daily	Bare Costs	Incl. O&P
1 Tile Layer	$31.15	$249.20	$50.40	$403.20	$27.75	$44.90
1 Tile Layer Helper	24.35	194.80	39.40	315.20		
16 L.H., Daily Totals		$444.00		$718.40	$27.75	$44.90
Crew D-8	Hr.	Daily	Hr.	Daily	Bare Costs	Incl. O&P
3 Bricklayers	$32.40	$777.60	$54.00	$1296.00	$30.08	$50.14
2 Bricklayer Helpers	26.60	425.60	44.35	709.60		
40 L.H., Daily Totals		$1203.20		$2005.60	$30.08	$50.14
Crew D-9	Hr.	Daily	Hr.	Daily	Bare Costs	Incl. O&P
3 Bricklayers	$32.40	$777.60	$54.00	$1296.00	$29.50	$49.17
3 Bricklayer Helpers	26.60	638.40	44.35	1064.40		
48 L.H., Daily Totals		$1416.00		$2360.40	$29.50	$49.17
Crew D-10	Hr.	Daily	Hr.	Daily	Bare Costs	Incl. O&P
1 Bricklayer Foreman (outside)	$34.40	$275.20	$57.35	$458.80	$32.36	$53.83
1 Bricklayer	32.40	259.20	54.00	432.00		
1 Bricklayer Helper	26.60	212.80	44.35	354.80		
1 Equip. Oper. (crane)	36.05	288.40	59.60	476.80		
1 S.P. Crane, 4x4, 12 Ton		479.20		527.12	14.98	16.47
32 L.H., Daily Totals		$1514.80		$2249.52	$47.34	$70.30
Crew D-11	Hr.	Daily	Hr.	Daily	Bare Costs	Incl. O&P
2 Bricklayers	$32.40	$518.40	$54.00	$864.00	$30.47	$50.78
1 Bricklayer Helper	26.60	212.80	44.35	354.80		
24 L.H., Daily Totals		$731.20		$1218.80	$30.47	$50.78
Crew D-12	Hr.	Daily	Hr.	Daily	Bare Costs	Incl. O&P
2 Bricklayers	$32.40	$518.40	$54.00	$864.00	$29.50	$49.17
2 Bricklayer Helpers	26.60	425.60	44.35	709.60		
32 L.H., Daily Totals		$944.00		$1573.60	$29.50	$49.17
Crew D-13	Hr.	Daily	Hr.	Daily	Bare Costs	Incl. O&P
1 Bricklayer Foreman (outside)	$34.40	$275.20	$57.35	$458.80	$31.41	$52.27
2 Bricklayers	32.40	518.40	54.00	864.00		
2 Bricklayer Helpers	26.60	425.60	44.35	709.60		
1 Equip. Oper. (crane)	36.05	288.40	59.60	476.80		
1 S.P. Crane, 4x4, 12 Ton		479.20		527.12	9.98	10.98
48 L.H., Daily Totals		$1986.80		$3036.32	$41.39	$63.26
Crew E-1	Hr.	Daily	Hr.	Daily	Bare Costs	Incl. O&P
2 Struc. Steel Workers	$35.65	$570.40	$65.40	$1046.40	$35.65	$65.40
1 Welder, Gas Engine, 300 amp		147.20		161.92	9.20	10.12
16 L.H., Daily Totals		$717.60		$1208.32	$44.85	$75.52
Crew E-2	Hr.	Daily	Hr.	Daily	Bare Costs	Incl. O&P
1 Struc. Steel Foreman (outside)	$37.65	$301.20	$69.05	$552.40	$36.05	$65.04
4 Struc. Steel Workers	35.65	1140.80	65.40	2092.80		
1 Equip. Oper. (crane)	36.05	288.40	59.60	476.80		
1 Lattice Boom Crane, 90 Ton		1520.00		1672.00	31.67	34.83
48 L.H., Daily Totals		$3250.40		$4794.00	$67.72	$99.88
Crew E-3	Hr.	Daily	Hr.	Daily	Bare Costs	Incl. O&P
1 Struc. Steel Foreman (outside)	$37.65	$301.20	$69.05	$552.40	$36.32	$66.62
2 Struc. Steel Workers	35.65	570.40	65.40	1046.40		
1 Welder, Gas Engine, 300 amp		147.20		161.92	6.13	6.75
24 L.H., Daily Totals		$1018.80		$1760.72	$42.45	$73.36
Crew E-3A	Hr.	Daily	Hr.	Daily	Bare Costs	Incl. O&P
1 Struc. Steel Foreman (outside)	$37.65	$301.20	$69.05	$552.40	$36.32	$66.62
2 Struc. Steel Workers	35.65	570.40	65.40	1046.40		
1 Welder, Gas Engine, 300 amp		147.20		161.92		
1 Aerial Lift Truck, 40' Boom		309.00		339.90	19.01	20.91
24 L.H., Daily Totals		$1327.80		$2100.62	$55.33	$87.53
Crew E-4	Hr.	Daily	Hr.	Daily	Bare Costs	Incl. O&P
1 Struc. Steel Foreman (outside)	$37.65	$301.20	$69.05	$552.40	$36.15	$66.31
3 Struc. Steel Workers	35.65	855.60	65.40	1569.60		
1 Welder, Gas Engine, 300 amp		147.20		161.92	4.60	5.06
32 L.H., Daily Totals		$1304.00		$2283.92	$40.75	$71.37

Crews - Residential

Crew No.	Bare Costs		Incl. Subs O&P		Cost Per Labor-Hour	
Crew E-5	Hr.	Daily	Hr.	Daily	Bare Costs	Incl. O&P
1 Struc. Steel Foreman (outside)	$37.65	$301.20	$69.05	$552.40	$35.92	$65.16
7 Struc. Steel Workers	35.65	1996.40	65.40	3662.40		
1 Equip. Oper. (crane)	36.05	288.40	59.60	476.80		
1 Lattice Boom Crane, 90 Ton		1520.00		1672.00		
1 Welder, Gas Engine, 300 amp		147.20		161.92	23.16	25.47
72 L.H., Daily Totals		$4253.20		$6525.52	$59.07	$90.63

Crew No.	Bare Costs		Incl. Subs O&P		Cost Per Labor-Hour	
Crew E-6	Hr.	Daily	Hr.	Daily	Bare Costs	Incl. O&P
1 Struc. Steel Foreman (outside)	$37.65	$301.20	$69.05	$552.40	$35.69	$64.63
12 Struc. Steel Workers	35.65	3422.40	65.40	6278.40		
1 Equip. Oper. (crane)	36.05	288.40	59.60	476.80		
1 Equip. Oper. (light)	33.90	271.20	56.05	448.40		
1 Lattice Boom Crane, 90 Ton		1520.00		1672.00		
1 Welder, Gas Engine, 300 amp		147.20		161.92		
1 Air Compressor, 160 cfm		162.20		178.42		
2 Impact Wrenches		36.40		40.04	15.55	17.10
120 L.H., Daily Totals		$6149.00		$9808.38	$51.24	$81.74

Crew No.	Bare Costs		Incl. Subs O&P		Cost Per Labor-Hour	
Crew E-7	Hr.	Daily	Hr.	Daily	Bare Costs	Incl. O&P
1 Struc. Steel Foreman (outside)	$37.65	$301.20	$69.05	$552.40	$35.92	$65.16
7 Struc. Steel Workers	35.65	1996.40	65.40	3662.40		
1 Equip. Oper. (crane)	36.05	288.40	59.60	476.80		
1 Lattice Boom Crane, 90 Ton		1520.00		1672.00		
2 Welders, Gas Engine, 300 amp		294.40		323.84	25.20	27.72
72 L.H., Daily Totals		$4400.40		$6687.44	$61.12	$92.88

Crew No.	Bare Costs		Incl. Subs O&P		Cost Per Labor-Hour	
Crew E-8	Hr.	Daily	Hr.	Daily	Bare Costs	Incl. O&P
1 Struc. Steel Foreman (outside)	$37.65	$301.20	$69.05	$552.40	$35.87	$65.20
9 Struc. Steel Workers	35.65	2566.80	65.40	4708.80		
1 Equip. Oper. (crane)	36.05	288.40	59.60	476.80		
1 Lattice Boom Crane, 90 Ton		1520.00		1672.00		
4 Welders, Gas Engine, 300 amp		588.80		647.68	23.96	26.36
88 L.H., Daily Totals		$5265.20		$8057.68	$59.83	$91.56

Crew No.	Bare Costs		Incl. Subs O&P		Cost Per Labor-Hour	
Crew E-9	Hr.	Daily	Hr.	Daily	Bare Costs	Incl. O&P
2 Struc. Steel Foremen (outside)	$37.65	$602.40	$69.05	$1104.80	$35.70	$64.33
5 Struc. Steel Workers	35.65	1426.00	65.40	2616.00		
1 Welder Foreman (outside)	37.65	301.20	69.05	552.40		
5 Welders	35.65	1426.00	65.40	2616.00		
1 Equip. Oper. (crane)	36.05	288.40	59.60	476.80		
1 Equip. Oper. (oiler)	31.75	254.00	52.50	420.00		
1 Equip. Oper. (light)	33.90	271.20	56.05	448.40		
1 Lattice Boom Crane, 90 Ton		1520.00		1672.00		
5 Welders, Gas Engine, 300 amp		736.00		809.60	17.63	19.39
128 L.H., Daily Totals		$6825.20		$10716.00	$53.32	$83.72

Crew No.	Bare Costs		Incl. Subs O&P		Cost Per Labor-Hour	
Crew E-10	Hr.	Daily	Hr.	Daily	Bare Costs	Incl. O&P
1 Struc. Steel Foreman (outside)	$37.65	$301.20	$69.05	$552.40	$36.32	$66.62
2 Struc. Steel Workers	35.65	570.40	65.40	1046.40		
1 Welder, Gas Engine, 300 amp		147.20		161.92		
1 Flatbed Truck, Gas, 3 Ton		307.40		338.14	18.94	20.84
24 L.H., Daily Totals		$1326.20		$2098.86	$55.26	$87.45

Crew No.	Bare Costs		Incl. Subs O&P		Cost Per Labor-Hour	
Crew E-11	Hr.	Daily	Hr.	Daily	Bare Costs	Incl. O&P
2 Painters, Struc. Steel	$29.30	$468.80	$54.40	$870.40	$29.29	$51.52
1 Building Laborer	24.65	197.20	41.25	330.00		
1 Equip. Oper. (light)	33.90	271.20	56.05	448.40		
1 Air Compressor, 250 cfm		197.80		217.58		
1 Sandblaster, Portable, 3 C.F.		20.80		22.88		
1 Set Sand Blasting Accessories		14.25		15.68	7.28	8.00
32 L.H., Daily Totals		$1170.05		$1904.93	$36.56	$59.53

Crew No.	Bare Costs		Incl. Subs O&P		Cost Per Labor-Hour	
Crew E-11A	Hr.	Daily	Hr.	Daily	Bare Costs	Incl. O&P
2 Painters, Struc. Steel	$29.30	$468.80	$54.40	$870.40	$29.29	$51.52
1 Building Laborer	24.65	197.20	41.25	330.00		
1 Equip. Oper. (light)	33.90	271.20	56.05	448.40		
1 Air Compressor, 250 cfm		197.80		217.58		
1 Sandblaster, Portable, 3 C.F.		20.80		22.88		
1 Set Sand Blasting Accessories		14.25		15.68		
1 Aerial Lift Truck, 60' Boom		448.40		493.24	21.29	23.42
32 L.H., Daily Totals		$1618.45		$2398.18	$50.58	$74.94

Crew No.	Bare Costs		Incl. Subs O&P		Cost Per Labor-Hour	
Crew E-11B	Hr.	Daily	Hr.	Daily	Bare Costs	Incl. O&P
2 Painters, Struc. Steel	$29.30	$468.80	$54.40	$870.40	$27.75	$50.02
1 Building Laborer	24.65	197.20	41.25	330.00		
2 Paint Sprayers, 8 C.F.M.		100.80		110.88		
1 Aerial Lift Truck, 60' Boom		448.40		493.24	22.88	25.17
24 L.H., Daily Totals		$1215.20		$1804.52	$50.63	$75.19

Crew No.	Bare Costs		Incl. Subs O&P		Cost Per Labor-Hour	
Crew E-12	Hr.	Daily	Hr.	Daily	Bare Costs	Incl. O&P
1 Welder Foreman (outside)	$37.65	$301.20	$69.05	$552.40	$35.77	$62.55
1 Equip. Oper. (light)	33.90	271.20	56.05	448.40		
1 Welder, Gas Engine, 300 amp		147.20		161.92	9.20	10.12
16 L.H., Daily Totals		$719.60		$1162.72	$44.98	$72.67

Crew No.	Bare Costs		Incl. Subs O&P		Cost Per Labor-Hour	
Crew E-13	Hr.	Daily	Hr.	Daily	Bare Costs	Incl. O&P
1 Welder Foreman (outside)	$37.65	$301.20	$69.05	$552.40	$36.40	$64.72
.5 Equip. Oper. (light)	33.90	135.60	56.05	224.20		
1 Welder, Gas Engine, 300 amp		147.20		161.92	12.27	13.49
12 L.H., Daily Totals		$584.00		$938.52	$48.67	$78.21

Crew No.	Bare Costs		Incl. Subs O&P		Cost Per Labor-Hour	
Crew E-14	Hr.	Daily	Hr.	Daily	Bare Costs	Incl. O&P
1 Struc. Steel Worker	$35.65	$285.20	$65.40	$523.20	$35.65	$65.40
1 Welder, Gas Engine, 300 amp		147.20		161.92	18.40	20.24
8 L.H., Daily Totals		$432.40		$685.12	$54.05	$85.64

Crew No.	Bare Costs		Incl. Subs O&P		Cost Per Labor-Hour	
Crew E-16	Hr.	Daily	Hr.	Daily	Bare Costs	Incl. O&P
1 Welder Foreman (outside)	$37.65	$301.20	$69.05	$552.40	$36.65	$67.22
1 Welder	35.65	285.20	65.40	523.20		
1 Welder, Gas Engine, 300 amp		147.20		161.92	9.20	10.12
16 L.H., Daily Totals		$733.60		$1237.52	$45.85	$77.34

Crew No.	Bare Costs		Incl. Subs O&P		Cost Per Labor-Hour	
Crew E-17	Hr.	Daily	Hr.	Daily	Bare Costs	Incl. O&P
1 Struc. Steel Foreman (outside)	$37.65	$301.20	$69.05	$552.40	$36.65	$67.22
1 Structural Steel Worker	35.65	285.20	65.40	523.20		
16 L.H., Daily Totals		$586.40		$1075.60	$36.65	$67.22

Crew No.	Bare Costs		Incl. Subs O&P		Cost Per Labor-Hour	
Crew E-18	Hr.	Daily	Hr.	Daily	Bare Costs	Incl. O&P
1 Struc. Steel Foreman (outside)	$37.65	$301.20	$69.05	$552.40	$35.97	$64.70
3 Structural Steel Workers	35.65	855.60	65.40	1569.60		
1 Equipment Operator (med.)	35.25	282.00	58.25	466.00		
1 Lattice Boom Crane, 20 Ton		967.35		1064.09	24.18	26.60
40 L.H., Daily Totals		$2406.15		$3652.09	$60.15	$91.30

Crew No.	Bare Costs		Incl. Subs O&P		Cost Per Labor-Hour	
Crew E-19	Hr.	Daily	Hr.	Daily	Bare Costs	Incl. O&P
1 Struc. Steel Foreman (outside)	$37.65	$301.20	$69.05	$552.40	$35.73	$63.50
1 Structural Steel Worker	35.65	285.20	65.40	523.20		
1 Equip. Oper. (light)	33.90	271.20	56.05	448.40		
1 Lattice Boom Crane, 20 Ton		967.35		1064.09	40.31	44.34
24 L.H., Daily Totals		$1824.95		$2588.09	$76.04	$107.84

Crews - Residential

Crew No.	Bare Costs		Incl. Subs O&P		Cost Per Labor-Hour	
Crew E-20	Hr.	Daily	Hr.	Daily	Bare Costs	Incl. O&P
1 Struc. Steel Foreman (outside)	$37.65	$301.20	$69.05	$552.40	$35.46	$63.52
5 Structural Steel Workers	35.65	1426.00	65.40	2616.00		
1 Equip. Oper. (crane)	36.05	288.40	59.60	476.80		
1 Equip. Oper. (oiler)	31.75	254.00	52.50	420.00		
1 Lattice Boom Crane, 40 Ton		1182.00		1300.20	18.47	20.32
64 L.H., Daily Totals		$3451.60		$5365.40	$53.93	$83.83
Crew E-22	Hr.	Daily	Hr.	Daily	Bare Costs	Incl. O&P
1 Skilled Worker Foreman (out)	$36.20	$289.60	$60.80	$486.40	$34.87	$58.57
2 Skilled Workers	34.20	547.20	57.45	919.20		
24 L.H., Daily Totals		$836.80		$1405.60	$34.87	$58.57
Crew E-24	Hr.	Daily	Hr.	Daily	Bare Costs	Incl. O&P
3 Structural Steel Workers	$35.65	$855.60	$65.40	$1569.60	$35.55	$63.61
1 Equipment Operator (med.)	35.25	282.00	58.25	466.00		
1 Hyd. Crane, 25 Ton		748.80		823.68	23.40	25.74
32 L.H., Daily Totals		$1886.40		$2859.28	$58.95	$89.35
Crew E-25	Hr.	Daily	Hr.	Daily	Bare Costs	Incl. O&P
1 Welder	$35.65	$285.20	$65.40	$523.20	$35.65	$65.40
1 Cutting Torch		11.40		12.54	1.43	1.57
8 L.H., Daily Totals		$296.60		$535.74	$37.08	$66.97
Crew E-26	Hr.	Daily	Hr.	Daily	Bare Costs	Incl. O&P
1 Struc. Steel Foreman (outside)	$37.65	$301.20	$69.05	$552.40	$36.55	$66.01
1 Struc. Steel Worker	35.65	285.20	65.40	523.20		
1 Welder	35.65	285.20	65.40	523.20		
.25 Electrician	36.90	73.80	60.35	120.70		
.25 Plumber	39.05	78.10	64.40	128.80		
1 Welder, Gas Engine, 300 amp		147.20		161.92	5.26	5.78
28 L.H., Daily Totals		$1170.70		$2010.22	$41.81	$71.79
Crew F-3	Hr.	Daily	Hr.	Daily	Bare Costs	Incl. O&P
2 Carpenters	$33.90	$542.40	$56.75	$908.00	$30.67	$51.34
2 Carpenter Helpers	24.75	396.00	41.80	668.80		
1 Equip. Oper. (crane)	36.05	288.40	59.60	476.80		
1 Hyd. Crane, 12 Ton		658.40		724.24	16.46	18.11
40 L.H., Daily Totals		$1885.20		$2777.84	$47.13	$69.45
Crew F-4	Hr.	Daily	Hr.	Daily	Bare Costs	Incl. O&P
2 Carpenters	$33.90	$542.40	$56.75	$908.00	$30.67	$51.34
2 Carpenter Helpers	24.75	396.00	41.80	668.80		
1 Equip. Oper. (crane)	36.05	288.40	59.60	476.80		
1 Hyd. Crane, 55 Ton		1123.00		1235.30	28.07	30.88
40 L.H., Daily Totals		$2349.80		$3288.90	$58.74	$82.22
Crew F-5	Hr.	Daily	Hr.	Daily	Bare Costs	Incl. O&P
2 Carpenters	$33.90	$542.40	$56.75	$908.00	$29.32	$49.27
2 Carpenter Helpers	24.75	396.00	41.80	668.80		
32 L.H., Daily Totals		$938.40		$1576.80	$29.32	$49.27
Crew F-6	Hr.	Daily	Hr.	Daily	Bare Costs	Incl. O&P
2 Carpenters	$33.90	$542.40	$56.75	$908.00	$30.63	$51.12
2 Building Laborers	24.65	394.40	41.25	660.00		
1 Equip. Oper. (crane)	36.05	288.40	59.60	476.80		
1 Hyd. Crane, 12 Ton		658.40		724.24	16.46	18.11
40 L.H., Daily Totals		$1883.60		$2769.04	$47.09	$69.23
Crew F-7	Hr.	Daily	Hr.	Daily	Bare Costs	Incl. O&P
2 Carpenters	$33.90	$542.40	$56.75	$908.00	$29.27	$49.00
2 Building Laborers	24.65	394.40	41.25	660.00		
32 L.H., Daily Totals		$936.80		$1568.00	$29.27	$49.00
Crew G-1	Hr.	Daily	Hr.	Daily	Bare Costs	Incl. O&P
1 Roofer Foreman (outside)	$30.75	$246.00	$56.70	$453.60	$27.01	$49.79
4 Roofers Composition	28.75	920.00	53.00	1696.00		
2 Roofer Helpers	21.65	346.40	39.90	638.40		
1 Application Equipment		193.00		212.30		
1 Tar Kettle/Pot		176.40		194.04		
1 Crew Truck		208.80		229.68	10.32	11.36
56 L.H., Daily Totals		$2090.60		$3424.02	$37.33	$61.14
Crew G-2	Hr.	Daily	Hr.	Daily	Bare Costs	Incl. O&P
1 Plasterer	$31.45	$251.60	$51.40	$411.20	$27.58	$45.40
1 Plasterer Helper	26.65	213.20	43.55	348.40		
1 Building Laborer	24.65	197.20	41.25	330.00		
1 Grout Pump, 50 C.F./hr.		135.00		148.50	5.63	6.19
24 L.H., Daily Totals		$797.00		$1238.10	$33.21	$51.59
Crew G-2A	Hr.	Daily	Hr.	Daily	Bare Costs	Incl. O&P
1 Roofer Composition	$28.75	$230.00	$53.00	$424.00	$25.02	$44.72
1 Roofer Helper	21.65	173.20	39.90	319.20		
1 Building Laborer	24.65	197.20	41.25	330.00		
1 Foam Spray Rig, Trailer-Mtd.		567.00		623.70		
1 Pickup Truck, 3/4 Ton		145.80		160.38	29.70	32.67
24 L.H., Daily Totals		$1313.20		$1857.28	$54.72	$77.39
Crew G-3	Hr.	Daily	Hr.	Daily	Bare Costs	Incl. O&P
2 Sheet Metal Workers	$37.80	$604.80	$63.05	$1008.80	$31.23	$52.15
2 Building Laborers	24.65	394.40	41.25	660.00		
32 L.H., Daily Totals		$999.20		$1668.80	$31.23	$52.15
Crew G-4	Hr.	Daily	Hr.	Daily	Bare Costs	Incl. O&P
1 Labor Foreman (outside)	$26.65	$213.20	$44.60	$356.80	$25.32	$42.37
2 Building Laborers	24.65	394.40	41.25	660.00		
1 Flatbed Truck, Gas, 1.5 Ton		248.20		273.02		
1 Air Compressor, 160 cfm		162.20		178.42	17.10	18.81
24 L.H., Daily Totals		$1018.00		$1468.24	$42.42	$61.18
Crew G-5	Hr.	Daily	Hr.	Daily	Bare Costs	Incl. O&P
1 Roofer Foreman (outside)	$30.75	$246.00	$56.70	$453.60	$26.31	$48.50
2 Roofers Composition	28.75	460.00	53.00	848.00		
2 Roofer Helpers	21.65	346.40	39.90	638.40		
1 Application Equipment		193.00		212.30	4.83	5.31
40 L.H., Daily Totals		$1245.40		$2152.30	$31.14	$53.81
Crew G-6A	Hr.	Daily	Hr.	Daily	Bare Costs	Incl. O&P
2 Roofers Composition	$28.75	$460.00	$53.00	$848.00	$28.75	$53.00
1 Small Compressor, Electric		14.10		15.51		
2 Pneumatic Nailers		52.40		57.64	4.16	4.57
16 L.H., Daily Totals		$526.50		$921.15	$32.91	$57.57
Crew G-7	Hr.	Daily	Hr.	Daily	Bare Costs	Incl. O&P
1 Carpenter	$33.90	$271.20	$56.75	$454.00	$33.90	$56.75
1 Small Compressor, Electric		14.10		15.51		
1 Pneumatic Nailer		26.20		28.82	5.04	5.54
8 L.H., Daily Totals		$311.50		$498.33	$38.94	$62.29

Crews - Residential

Crew No.	Bare Costs		Incl. Subs O&P		Cost Per Labor-Hour	
Crew H-1	Hr.	Daily	Hr.	Daily	Bare Costs	Incl. O&P
2 Glaziers	$33.15	$530.40	$55.05	$880.80	$34.40	$60.23
2 Struc. Steel Workers	35.65	570.40	65.40	1046.40		
32 L.H., Daily Totals		$1100.80		$1927.20	$34.40	$60.23
Crew H-2	Hr.	Daily	Hr.	Daily	Bare Costs	Incl. O&P
2 Glaziers	$33.15	$530.40	$55.05	$880.80	$30.32	$50.45
1 Building Laborer	24.65	197.20	41.25	330.00		
24 L.H., Daily Totals		$727.60		$1210.80	$30.32	$50.45
Crew H-3	Hr.	Daily	Hr.	Daily	Bare Costs	Incl. O&P
1 Glazier	$33.15	$265.20	$55.05	$440.40	$28.95	$48.42
1 Helper	24.75	198.00	41.80	334.40		
16 L.H., Daily Totals		$463.20		$774.80	$28.95	$48.42
Crew H-4	Hr.	Daily	Hr.	Daily	Bare Costs	Incl. O&P
1 Carpenter	$33.90	$271.20	$56.75	$454.00	$30.84	$51.49
1 Carpenter Helper	24.75	198.00	41.80	334.40		
.5 Electrician	36.90	147.60	60.35	241.40		
20 L.H., Daily Totals		$616.80		$1029.80	$30.84	$51.49
Crew J-1	Hr.	Daily	Hr.	Daily	Bare Costs	Incl. O&P
3 Plasterers	$31.45	$754.80	$51.40	$1233.60	$29.53	$48.26
2 Plasterer Helpers	26.65	426.40	43.55	696.80		
1 Mixing Machine, 6 C.F.		140.20		154.22	3.50	3.86
40 L.H., Daily Totals		$1321.40		$2084.62	$33.03	$52.12
Crew J-2	Hr.	Daily	Hr.	Daily	Bare Costs	Incl. O&P
3 Plasterers	$31.45	$754.80	$51.40	$1233.60	$30.13	$49.13
2 Plasterer Helpers	26.65	426.40	43.55	696.80		
1 Lather	33.10	264.80	53.50	428.00		
1 Mixing Machine, 6 C.F.		140.20		154.22	2.92	3.21
48 L.H., Daily Totals		$1586.20		$2512.62	$33.05	$52.35
Crew J-3	Hr.	Daily	Hr.	Daily	Bare Costs	Incl. O&P
1 Terrazzo Worker	$31.20	$249.60	$50.50	$404.00	$28.70	$46.45
1 Terrazzo Helper	26.20	209.60	42.40	339.20		
1 Floor Grinder, 22" Path		121.60		133.76		
1 Terrazzo Mixer		188.20		207.02	19.36	21.30
16 L.H., Daily Totals		$769.00		$1083.98	$48.06	$67.75
Crew J-4	Hr.	Daily	Hr.	Daily	Bare Costs	Incl. O&P
2 Cement Finishers	$31.95	$511.20	$51.85	$829.60	$29.52	$48.32
1 Laborer	24.65	197.20	41.25	330.00		
1 Floor Grinder, 22" Path		121.60		133.76		
1 Floor Edger, 7" Path		39.30		43.23		
1 Vacuum Pick-Up System		60.90		66.99	9.24	10.17
24 L.H., Daily Totals		$930.20		$1403.58	$38.76	$58.48
Crew J-4A	Hr.	Daily	Hr.	Daily	Bare Costs	Incl. O&P
2 Cement Finishers	$31.95	$511.20	$51.85	$829.60	$28.30	$46.55
2 Laborers	24.65	394.40	41.25	660.00		
1 Floor Grinder, 22" Path		121.60		133.76		
1 Floor Edger, 7" Path		39.30		43.23		
1 Vacuum Pick-Up System		60.90		66.99		
1 Floor Auto Scrubber		241.05		265.15	14.46	15.91
32 L.H., Daily Totals		$1368.45		$1998.73	$42.76	$62.46

Crew No.	Bare Costs		Incl. Subs O&P		Cost Per Labor-Hour	
Crew J-4B	Hr.	Daily	Hr.	Daily	Bare Costs	Incl. O&P
1 Laborer	$24.65	$197.20	$41.25	$330.00	$24.65	$41.25
1 Floor Auto Scrubber		241.05		265.15	30.13	33.14
8 L.H., Daily Totals		$438.25		$595.15	$54.78	$74.39
Crew J-6	Hr.	Daily	Hr.	Daily	Bare Costs	Incl. O&P
2 Painters	$28.65	$458.40	$47.20	$755.20	$28.96	$47.92
1 Building Laborer	24.65	197.20	41.25	330.00		
1 Equip. Oper. (light)	33.90	271.20	56.05	448.40		
1 Air Compressor, 250 cfm		197.80		217.58		
1 Sandblaster, Portable, 3 C.F.		20.80		22.88		
1 Set Sand Blasting Accessories		14.25		15.68	7.28	8.00
32 L.H., Daily Totals		$1159.65		$1789.73	$36.24	$55.93
Crew J-7	Hr.	Daily	Hr.	Daily	Bare Costs	Incl. O&P
2 Painters	$28.65	$458.40	$47.20	$755.20	$28.65	$47.20
1 Floor Belt Sander		24.45		26.90		
1 Floor Sanding Edger		13.65		15.02	2.38	2.62
16 L.H., Daily Totals		$496.50		$797.11	$31.03	$49.82
Crew K-1	Hr.	Daily	Hr.	Daily	Bare Costs	Incl. O&P
1 Carpenter	$33.90	$271.20	$56.75	$454.00	$31.85	$52.95
1 Truck Driver (light)	29.80	238.40	49.15	393.20		
1 Flatbed Truck, Gas, 3 Ton		307.40		338.14	19.21	21.13
16 L.H., Daily Totals		$817.00		$1185.34	$51.06	$74.08
Crew K-2	Hr.	Daily	Hr.	Daily	Bare Costs	Incl. O&P
1 Struc. Steel Foreman (outside)	$37.65	$301.20	$69.05	$552.40	$34.37	$61.20
1 Struc. Steel Worker	35.65	285.20	65.40	523.20		
1 Truck Driver (light)	29.80	238.40	49.15	393.20		
1 Flatbed Truck, Gas, 3 Ton		307.40		338.14	12.81	14.09
24 L.H., Daily Totals		$1132.20		$1806.94	$47.17	$75.29
Crew L-1	Hr.	Daily	Hr.	Daily	Bare Costs	Incl. O&P
.25 Electrician	$36.90	$73.80	$60.35	$120.70	$38.62	$63.59
1 Plumber	39.05	312.40	64.40	515.20		
10 L.H., Daily Totals		$386.20		$635.90	$38.62	$63.59
Crew L-2	Hr.	Daily	Hr.	Daily	Bare Costs	Incl. O&P
1 Carpenter	$33.90	$271.20	$56.75	$454.00	$29.32	$49.27
1 Carpenter Helper	24.75	198.00	41.80	334.40		
16 L.H., Daily Totals		$469.20		$788.40	$29.32	$49.27
Crew L-3	Hr.	Daily	Hr.	Daily	Bare Costs	Incl. O&P
1 Carpenter	$33.90	$271.20	$56.75	$454.00	$34.50	$57.47
.25 Electrician	36.90	73.80	60.35	120.70		
10 L.H., Daily Totals		$345.00		$574.70	$34.50	$57.47
Crew L-3A	Hr.	Daily	Hr.	Daily	Bare Costs	Incl. O&P
1 Carpenter Foreman (outside)	$35.90	$287.20	$60.10	$480.80	$36.53	$61.08
.5 Sheet Metal Worker	37.80	151.20	63.05	252.20		
12 L.H., Daily Totals		$438.40		$733.00	$36.53	$61.08
Crew L-4	Hr.	Daily	Hr.	Daily	Bare Costs	Incl. O&P
1 Skilled Worker	$34.20	$273.60	$57.45	$459.60	$29.48	$49.63
1 Helper	24.75	198.00	41.80	334.40		
16 L.H., Daily Totals		$471.60		$794.00	$29.48	$49.63

Crews - Residential

Crew No.	Bare Costs Hr.	Bare Costs Daily	Incl. Subs O&P Hr.	Incl. Subs O&P Daily	Cost Per Labor-Hour Bare Costs	Cost Per Labor-Hour Incl. O&P
Crew L-5						
1 Struc. Steel Foreman (outside)	$37.65	$301.20	$69.05	$552.40	$35.99	$65.09
5 Struc. Steel Workers	35.65	1426.00	65.40	2616.00		
1 Equip. Oper. (crane)	36.05	288.40	59.60	476.80		
1 Hyd. Crane, 25 Ton		748.80		823.68	13.37	14.71
56 L.H., Daily Totals		$2764.40		$4468.88	$49.36	$79.80
Crew L-5A						
1 Struc. Steel Foreman (outside)	$37.65	$301.20	$69.05	$552.40	$36.25	$64.86
2 Structural Steel Workers	35.65	570.40	65.40	1046.40		
1 Equip. Oper. (crane)	36.05	288.40	59.60	476.80		
1 S.P. Crane, 4x4, 25 Ton		606.60		667.26	18.96	20.85
32 L.H., Daily Totals		$1766.60		$2742.86	$55.21	$85.71
Crew L-5B						
1 Struc. Steel Foreman (outside)	$37.65	$301.20	$69.05	$552.40	$35.95	$61.50
2 Structural Steel Workers	35.65	570.40	65.40	1046.40		
2 Electricians	36.90	590.40	60.35	965.60		
2 Steamfitters/Pipefitters	40.05	640.80	66.05	1056.80		
1 Equip. Oper. (crane)	36.05	288.40	59.60	476.80		
1 Common Laborer	24.65	197.20	41.25	330.00		
1 Hyd. Crane, 80 Ton		1630.00		1793.00	22.64	24.90
72 L.H., Daily Totals		$4218.40		$6221.00	$58.59	$86.40
Crew L-6						
1 Plumber	$39.05	$312.40	$64.40	$515.20	$38.33	$63.05
.5 Electrician	36.90	147.60	60.35	241.40		
12 L.H., Daily Totals		$460.00		$756.60	$38.33	$63.05
Crew L-7						
1 Carpenter	$33.90	$271.20	$56.75	$454.00	$28.50	$47.83
2 Carpenter Helpers	24.75	396.00	41.80	668.80		
.25 Electrician	36.90	73.80	60.35	120.70		
26 L.H., Daily Totals		$741.00		$1243.50	$28.50	$47.83
Crew L-8						
1 Carpenter	$33.90	$271.20	$56.75	$454.00	$31.27	$52.30
1 Carpenter Helper	24.75	198.00	41.80	334.40		
.5 Plumber	39.05	156.20	64.40	257.60		
20 L.H., Daily Totals		$625.40		$1046.00	$31.27	$52.30
Crew L-9						
1 Skilled Worker Foreman	$36.20	$289.60	$60.80	$486.40	$30.74	$51.56
1 Skilled Worker	34.20	273.60	57.45	459.60		
2 Helpers	24.75	396.00	41.80	668.80		
.5 Electrician	36.90	147.60	60.35	241.40		
36 L.H., Daily Totals		$1106.80		$1856.20	$30.74	$51.56
Crew L-10						
1 Struc. Steel Foreman (outside)	$37.65	$301.20	$69.05	$552.40	$36.45	$64.68
1 Structural Steel Worker	35.65	285.20	65.40	523.20		
1 Equip. Oper. (crane)	36.05	288.40	59.60	476.80		
1 Hyd. Crane, 12 Ton		658.40		724.24	27.43	30.18
24 L.H., Daily Totals		$1533.20		$2276.64	$63.88	$94.86
Crew L-11						
2 Wreckers	$25.40	$406.40	$44.70	$715.20	$30.19	$51.26
1 Equip. Oper. (crane)	36.05	288.40	59.60	476.80		
1 Equip. Oper. (light)	33.90	271.20	56.05	448.40		
1 Hyd. Excavator, 2.5 C.Y.		1582.00		1740.20		
1 Loader, Skid Steer, 78 H.P.		315.60		347.16	59.30	65.23
32 L.H., Daily Totals		$2863.60		$3727.76	$89.49	$116.49
Crew M-1						
3 Elevator Constructors	$53.00	$1272.00	$86.50	$2076.00	$50.35	$82.17
1 Elevator Apprentice	42.40	339.20	69.20	553.60		
5 Hand Tools		48.00		52.80	1.50	1.65
32 L.H., Daily Totals		$1659.20		$2682.40	$51.85	$83.83
Crew M-3						
1 Electrician Foreman (outside)	$38.90	$311.20	$63.65	$509.20	$39.47	$64.74
1 Common Laborer	24.65	197.20	41.25	330.00		
.25 Equipment Operator (med.)	35.25	70.50	58.25	116.50		
1 Elevator Constructor	53.00	424.00	86.50	692.00		
1 Elevator Apprentice	42.40	339.20	69.20	553.60		
.25 S.P. Crane, 4x4, 20 Ton		138.70		152.57	4.08	4.49
34 L.H., Daily Totals		$1480.80		$2353.87	$43.55	$69.23
Crew M-4						
1 Electrician Foreman (outside)	$38.90	$311.20	$63.65	$509.20	$39.09	$64.14
1 Common Laborer	24.65	197.20	41.25	330.00		
.25 Equipment Operator, Crane	36.05	72.10	59.60	119.20		
.25 Equip. Oper. (oiler)	31.75	63.50	52.50	105.00		
1 Elevator Constructor	53.00	424.00	86.50	692.00		
1 Elevator Apprentice	42.40	339.20	69.20	553.60		
.25 S.P. Crane, 4x4, 40 Ton		177.95		195.75	4.94	5.44
36 L.H., Daily Totals		$1585.15		$2504.74	$44.03	$69.58
Crew Q-1						
1 Plumber	$39.05	$312.40	$64.40	$515.20	$35.15	$57.98
1 Plumber Apprentice	31.25	250.00	51.55	412.40		
16 L.H., Daily Totals		$562.40		$927.60	$35.15	$57.98
Crew Q-1A						
.25 Plumber Foreman (outside)	$41.05	$82.10	$67.70	$135.40	$39.45	$65.06
1 Plumber	39.05	312.40	64.40	515.20		
10 L.H., Daily Totals		$394.50		$650.60	$39.45	$65.06
Crew Q-1C						
1 Plumber	$39.05	$312.40	$64.40	$515.20	$35.18	$58.07
1 Plumber Apprentice	31.25	250.00	51.55	412.40		
1 Equip. Oper. (medium)	35.25	282.00	58.25	466.00		
1 Trencher, Chain Type, 8' D		2860.00		3146.00	119.17	131.08
24 L.H., Daily Totals		$3704.40		$4539.60	$154.35	$189.15
Crew Q-2						
1 Plumber	$39.05	$312.40	$64.40	$515.20	$33.85	$55.83
2 Plumber Apprentices	31.25	500.00	51.55	824.80		
24 L.H., Daily Totals		$812.40		$1340.00	$33.85	$55.83
Crew Q-3						
2 Plumbers	$39.05	$624.80	$64.40	$1030.40	$35.15	$57.98
2 Plumber Apprentices	31.25	500.00	51.55	824.80		
32 L.H., Daily Totals		$1124.80		$1855.20	$35.15	$57.98

Crews - Residential

Crew No.	Bare Costs Hr.	Bare Costs Daily	Incl. Subs O&P Hr.	Incl. Subs O&P Daily	Cost Per Labor-Hour Bare Costs	Cost Per Labor-Hour Incl. O&P
Crew Q-4					Bare Costs	Incl. O&P
2 Plumbers	$39.05	$624.80	$64.40	$1030.40	$37.10	$61.19
1 Welder (plumber)	39.05	312.40	64.40	515.20		
1 Plumber Apprentice	31.25	250.00	51.55	412.40		
1 Welder, Electric, 300 amp		58.10		63.91	1.82	2.00
32 L.H., Daily Totals		$1245.30		$2021.91	$38.92	$63.18
Crew Q-5	Hr.	Daily	Hr.	Daily	Bare Costs	Incl. O&P
1 Steamfitter	$40.05	$320.40	$66.05	$528.40	$36.05	$59.45
1 Steamfitter Apprentice	32.05	256.40	52.85	422.80		
16 L.H., Daily Totals		$576.80		$951.20	$36.05	$59.45
Crew Q-6	Hr.	Daily	Hr.	Daily	Bare Costs	Incl. O&P
1 Steamfitter	$40.05	$320.40	$66.05	$528.40	$34.72	$57.25
2 Steamfitter Apprentices	32.05	512.80	52.85	845.60		
24 L.H., Daily Totals		$833.20		$1374.00	$34.72	$57.25
Crew Q-7	Hr.	Daily	Hr.	Daily	Bare Costs	Incl. O&P
2 Steamfitters	$40.05	$640.80	$66.05	$1056.80	$36.05	$59.45
2 Steamfitter Apprentices	32.05	512.80	52.85	845.60		
32 L.H., Daily Totals		$1153.60		$1902.40	$36.05	$59.45
Crew Q-8	Hr.	Daily	Hr.	Daily	Bare Costs	Incl. O&P
2 Steamfitters	$40.05	$640.80	$66.05	$1056.80	$38.05	$62.75
1 Welder (steamfitter)	40.05	320.40	66.05	528.40		
1 Steamfitter Apprentice	32.05	256.40	52.85	422.80		
1 Welder, Electric, 300 amp		58.10		63.91	1.82	2.00
32 L.H., Daily Totals		$1275.70		$2071.91	$39.87	$64.75
Crew Q-9	Hr.	Daily	Hr.	Daily	Bare Costs	Incl. O&P
1 Sheet Metal Worker	$37.80	$302.40	$63.05	$504.40	$34.02	$56.75
1 Sheet Metal Apprentice	30.25	242.00	50.45	403.60		
16 L.H., Daily Totals		$544.40		$908.00	$34.02	$56.75
Crew Q-10	Hr.	Daily	Hr.	Daily	Bare Costs	Incl. O&P
2 Sheet Metal Workers	$37.80	$604.80	$63.05	$1008.80	$35.28	$58.85
1 Sheet Metal Apprentice	30.25	242.00	50.45	403.60		
24 L.H., Daily Totals		$846.80		$1412.40	$35.28	$58.85
Crew Q-11	Hr.	Daily	Hr.	Daily	Bare Costs	Incl. O&P
2 Sheet Metal Workers	$37.80	$604.80	$63.05	$1008.80	$34.02	$56.75
2 Sheet Metal Apprentices	30.25	484.00	50.45	807.20		
32 L.H., Daily Totals		$1088.80		$1816.00	$34.02	$56.75
Crew Q-12	Hr.	Daily	Hr.	Daily	Bare Costs	Incl. O&P
1 Sprinkler Installer	$38.00	$304.00	$62.95	$503.60	$34.20	$56.65
1 Sprinkler Apprentice	30.40	243.20	50.35	402.80		
16 L.H., Daily Totals		$547.20		$906.40	$34.20	$56.65
Crew Q-13	Hr.	Daily	Hr.	Daily	Bare Costs	Incl. O&P
2 Sprinkler Installers	$38.00	$608.00	$62.95	$1007.20	$34.20	$56.65
2 Sprinkler Apprentices	30.40	486.40	50.35	805.60		
32 L.H., Daily Totals		$1094.40		$1812.80	$34.20	$56.65
Crew Q-14	Hr.	Daily	Hr.	Daily	Bare Costs	Incl. O&P
1 Asbestos Worker	$35.25	$282.00	$59.80	$478.40	$31.73	$53.83
1 Asbestos Apprentice	28.20	225.60	47.85	382.80		
16 L.H., Daily Totals		$507.60		$861.20	$31.73	$53.83

Crew No.	Bare Costs Hr.	Bare Costs Daily	Incl. Subs O&P Hr.	Incl. Subs O&P Daily	Cost Per Labor-Hour Bare Costs	Cost Per Labor-Hour Incl. O&P
Crew Q-15	Hr.	Daily	Hr.	Daily	Bare Costs	Incl. O&P
1 Plumber	$39.05	$312.40	$64.40	$515.20	$35.15	$57.98
1 Plumber Apprentice	31.25	250.00	51.55	412.40		
1 Welder, Electric, 300 amp		58.10		63.91	3.63	3.99
16 L.H., Daily Totals		$620.50		$991.51	$38.78	$61.97
Crew Q-16	Hr.	Daily	Hr.	Daily	Bare Costs	Incl. O&P
2 Plumbers	$39.05	$624.80	$64.40	$1030.40	$36.45	$60.12
1 Plumber Apprentice	31.25	250.00	51.55	412.40		
1 Welder, Electric, 300 amp		58.10		63.91	2.42	2.66
24 L.H., Daily Totals		$932.90		$1506.71	$38.87	$62.78
Crew Q-17	Hr.	Daily	Hr.	Daily	Bare Costs	Incl. O&P
1 Steamfitter	$40.05	$320.40	$66.05	$528.40	$36.05	$59.45
1 Steamfitter Apprentice	32.05	256.40	52.85	422.80		
1 Welder, Electric, 300 amp		58.10		63.91	3.63	3.99
16 L.H., Daily Totals		$634.90		$1015.11	$39.68	$63.44
Crew Q-17A	Hr.	Daily	Hr.	Daily	Bare Costs	Incl. O&P
1 Steamfitter	$40.05	$320.40	$66.05	$528.40	$36.05	$59.50
1 Steamfitter Apprentice	32.05	256.40	52.85	422.80		
1 Equip. Oper. (crane)	36.05	288.40	59.60	476.80		
1 Hyd. Crane, 12 Ton		658.40		724.24		
1 Welder, Electric, 300 amp		58.10		63.91	29.85	32.84
24 L.H., Daily Totals		$1581.70		$2216.15	$65.90	$92.34
Crew Q-18	Hr.	Daily	Hr.	Daily	Bare Costs	Incl. O&P
2 Steamfitters	$40.05	$640.80	$66.05	$1056.80	$37.38	$61.65
1 Steamfitter Apprentice	32.05	256.40	52.85	422.80		
1 Welder, Electric, 300 amp		58.10		63.91	2.42	2.66
24 L.H., Daily Totals		$955.30		$1543.51	$39.80	$64.31
Crew Q-19	Hr.	Daily	Hr.	Daily	Bare Costs	Incl. O&P
1 Steamfitter	$40.05	$320.40	$66.05	$528.40	$36.33	$59.75
1 Steamfitter Apprentice	32.05	256.40	52.85	422.80		
1 Electrician	36.90	295.20	60.35	482.80		
24 L.H., Daily Totals		$872.00		$1434.00	$36.33	$59.75
Crew Q-20	Hr.	Daily	Hr.	Daily	Bare Costs	Incl. O&P
1 Sheet Metal Worker	$37.80	$302.40	$63.05	$504.40	$34.60	$57.47
1 Sheet Metal Apprentice	30.25	242.00	50.45	403.60		
.5 Electrician	36.90	147.60	60.35	241.40		
20 L.H., Daily Totals		$692.00		$1149.40	$34.60	$57.47
Crew Q-21	Hr.	Daily	Hr.	Daily	Bare Costs	Incl. O&P
2 Steamfitters	$40.05	$640.80	$66.05	$1056.80	$37.26	$61.33
1 Steamfitter Apprentice	32.05	256.40	52.85	422.80		
1 Electrician	36.90	295.20	60.35	482.80		
32 L.H., Daily Totals		$1192.40		$1962.40	$37.26	$61.33
Crew Q-22	Hr.	Daily	Hr.	Daily	Bare Costs	Incl. O&P
1 Plumber	$39.05	$312.40	$64.40	$515.20	$35.15	$57.98
1 Plumber Apprentice	31.25	250.00	51.55	412.40		
1 Hyd. Crane, 12 Ton		658.40		724.24	41.15	45.27
16 L.H., Daily Totals		$1220.80		$1651.84	$76.30	$103.24

Crews - Residential

Crew Q-22A	Hr.	Daily	Hr.	Daily	Bare Costs	Incl. O&P
1 Plumber	$39.05	$312.40	$64.40	$515.20	$32.75	$54.20
1 Plumber Apprentice	31.25	250.00	51.55	412.40		
1 Laborer	24.65	197.20	41.25	330.00		
1 Equip. Oper. (crane)	36.05	288.40	59.60	476.80		
1 Hyd. Crane, 12 Ton		658.40		724.24	20.57	22.63
32 L.H., Daily Totals		$1706.40		$2458.64	$53.33	$76.83

Crew Q-23	Hr.	Daily	Hr.	Daily	Bare Costs	Incl. O&P
1 Plumber Foreman (outside)	$41.05	$328.40	$67.70	$541.60	$38.45	$63.45
1 Plumber	39.05	312.40	64.40	515.20		
1 Equip. Oper. (medium)	35.25	282.00	58.25	466.00		
1 Lattice Boom Crane, 20 Ton		967.35		1064.09	40.31	44.34
24 L.H., Daily Totals		$1890.15		$2586.89	$78.76	$107.79

Crew R-1	Hr.	Daily	Hr.	Daily	Bare Costs	Incl. O&P
1 Electrician Foreman	$37.40	$299.20	$61.20	$489.60	$34.52	$56.46
3 Electricians	36.90	885.60	60.35	1448.40		
2 Electrician Apprentices	29.50	472.00	48.25	772.00		
48 L.H., Daily Totals		$1656.80		$2710.00	$34.52	$56.46

Crew R-1A	Hr.	Daily	Hr.	Daily	Bare Costs	Incl. O&P
1 Electrician	$36.90	$295.20	$60.35	$482.80	$33.20	$54.30
1 Electrician Apprentice	29.50	236.00	48.25	386.00		
16 L.H., Daily Totals		$531.20		$868.80	$33.20	$54.30

Crew R-1B	Hr.	Daily	Hr.	Daily	Bare Costs	Incl. O&P
1 Electrician	$36.90	$295.20	$60.35	$482.80	$31.97	$52.28
2 Electrician Apprentices	29.50	472.00	48.25	772.00		
24 L.H., Daily Totals		$767.20		$1254.80	$31.97	$52.28

Crew R-1C	Hr.	Daily	Hr.	Daily	Bare Costs	Incl. O&P
2 Electricians	$36.90	$590.40	$60.35	$965.60	$33.20	$54.30
2 Electrician Apprentices	29.50	472.00	48.25	772.00		
1 Portable cable puller, 8000 lb.		130.10		143.11	4.07	4.47
32 L.H., Daily Totals		$1192.50		$1880.71	$37.27	$58.77

Crew R-2	Hr.	Daily	Hr.	Daily	Bare Costs	Incl. O&P
1 Electrician Foreman	$37.40	$299.20	$61.20	$489.60	$34.74	$56.91
3 Electricians	36.90	885.60	60.35	1448.40		
2 Electrician Apprentices	29.50	472.00	48.25	772.00		
1 Equip. Oper. (crane)	36.05	288.40	59.60	476.80		
1 S.P. Crane, 4x4, 5 Ton		280.80		308.88	5.01	5.52
56 L.H., Daily Totals		$2226.00		$3495.68	$39.75	$62.42

Crew R-3	Hr.	Daily	Hr.	Daily	Bare Costs	Incl. O&P
1 Electrician Foreman	$37.40	$299.20	$61.20	$489.60	$36.93	$60.54
1 Electrician	36.90	295.20	60.35	482.80		
.5 Equip. Oper. (crane)	36.05	144.20	59.60	238.40		
.5 S.P. Crane, 4x4, 5 Ton		140.40		154.44	7.02	7.72
20 L.H., Daily Totals		$879.00		$1365.24	$43.95	$68.26

Crew R-4	Hr.	Daily	Hr.	Daily	Bare Costs	Incl. O&P
1 Struc. Steel Foreman (outside)	$37.65	$301.20	$69.05	$552.40	$36.30	$65.12
3 Struc. Steel Workers	35.65	855.60	65.40	1569.60		
1 Electrician	36.90	295.20	60.35	482.80		
1 Welder, Gas Engine, 300 amp		147.20		161.92	3.68	4.05
40 L.H., Daily Totals		$1599.20		$2766.72	$39.98	$69.17

Crew R-5	Hr.	Daily	Hr.	Daily	Bare Costs	Incl. O&P
1 Electrician Foreman	$37.40	$299.20	$61.20	$489.60	$32.53	$53.68
4 Electrician Linemen	36.90	1180.80	60.35	1931.20		
2 Electrician Operators	36.90	590.40	60.35	965.60		
4 Electrician Groundmen	24.75	792.00	41.80	1337.60		
1 Crew Truck		208.80		229.68		
1 Flatbed Truck, 20,000 GVW		247.80		272.58		
1 Pickup Truck, 3/4 Ton		145.80		160.38		
.2 Hyd. Crane, 55 Ton		224.60		247.06		
.2 Hyd. Crane, 12 Ton		131.68		144.85		
.2 Earth Auger, Truck-Mtd.		84.16		92.58		
1 Tractor w/Winch		428.00		470.80	16.71	18.39
88 L.H., Daily Totals		$4333.24		$6341.92	$49.24	$72.07

Crew R-6	Hr.	Daily	Hr.	Daily	Bare Costs	Incl. O&P
1 Electrician Foreman	$37.40	$299.20	$61.20	$489.60	$32.53	$53.68
4 Electrician Linemen	36.90	1180.80	60.35	1931.20		
2 Electrician Operators	36.90	590.40	60.35	965.60		
4 Electrician Groundmen	24.75	792.00	41.80	1337.60		
1 Crew Truck		208.80		229.68		
1 Flatbed Truck, 20,000 GVW		247.80		272.58		
1 Pickup Truck, 3/4 Ton		145.80		160.38		
.2 Hyd. Crane, 55 Ton		224.60		247.06		
.2 Hyd. Crane, 12 Ton		131.68		144.85		
.2 Earth Auger, Truck-Mtd.		84.16		92.58		
1 Tractor w/Winch		428.00		470.80		
3 Cable Trailers		610.65		671.72		
.5 Tensioning Rig		202.25		222.47		
.5 Cable Pulling Rig		1170.50		1287.55	39.25	43.18
88 L.H., Daily Totals		$6316.64		$8523.66	$71.78	$96.86

Crew R-7	Hr.	Daily	Hr.	Daily	Bare Costs	Incl. O&P
1 Electrician Foreman	$37.40	$299.20	$61.20	$489.60	$26.86	$45.03
5 Electrician Groundmen	24.75	990.00	41.80	1672.00		
1 Crew Truck		208.80		229.68	4.35	4.79
48 L.H., Daily Totals		$1498.00		$2391.28	$31.21	$49.82

Crew R-8	Hr.	Daily	Hr.	Daily	Bare Costs	Incl. O&P
1 Electrician Foreman	$37.40	$299.20	$61.20	$489.60	$32.93	$54.31
3 Electrician Linemen	36.90	885.60	60.35	1448.40		
2 Electrician Groundmen	24.75	396.00	41.80	668.80		
1 Pickup Truck, 3/4 Ton		145.80		160.38		
1 Crew Truck		208.80		229.68	7.39	8.13
48 L.H., Daily Totals		$1935.40		$2996.86	$40.32	$62.43

Crew R-9	Hr.	Daily	Hr.	Daily	Bare Costs	Incl. O&P
1 Electrician Foreman	$37.40	$299.20	$61.20	$489.60	$30.89	$51.18
1 Electrician Lineman	36.90	295.20	60.35	482.80		
2 Electrician Operators	36.90	590.40	60.35	965.60		
4 Electrician Groundmen	24.75	792.00	41.80	1337.60		
1 Pickup Truck, 3/4 Ton		145.80		160.38		
1 Crew Truck		208.80		229.68	5.54	6.09
64 L.H., Daily Totals		$2331.40		$3665.66	$36.43	$57.28

Crew R-10	Hr.	Daily	Hr.	Daily	Bare Costs	Incl. O&P
1 Electrician Foreman	$37.40	$299.20	$61.20	$489.60	$34.96	$57.40
4 Electrician Linemen	36.90	1180.80	60.35	1931.20		
1 Electrician Groundman	24.75	198.00	41.80	334.40		
1 Crew Truck		208.80		229.68		
3 Tram Cars		418.50		460.35	13.07	14.38
48 L.H., Daily Totals		$2305.30		$3445.23	$48.03	$71.78

Crews - Residential

Crew R-11	Hr.	Daily	Hr.	Daily	Bare Costs	Incl. O&P
1 Electrician Foreman	$37.40	$299.20	$61.20	$489.60	$35.10	$57.64
4 Electricians	36.90	1180.80	60.35	1931.20		
1 Equip. Oper. (crane)	36.05	288.40	59.60	476.80		
1 Common Laborer	24.65	197.20	41.25	330.00		
1 Crew Truck		208.80		229.68		
1 Hyd. Crane, 12 Ton		658.40		724.24	15.49	17.03
56 L.H., Daily Totals		$2832.80		$4181.52	$50.59	$74.67

Crew R-12	Hr.	Daily	Hr.	Daily	Bare Costs	Incl. O&P
1 Carpenter Foreman (inside)	$34.40	$275.20	$57.60	$460.80	$30.86	$52.11
4 Carpenters	33.90	1084.80	56.75	1816.00		
4 Common Laborers	24.65	788.80	41.25	1320.00		
1 Equip. Oper. (medium)	35.25	282.00	58.25	466.00		
1 Steel Worker	35.65	285.20	65.40	523.20		
1 Dozer, 200 H.P.		1397.00		1536.70		
1 Pickup Truck, 3/4 Ton		145.80		160.38	17.53	19.29
88 L.H., Daily Totals		$4258.80		$6283.08	$48.40	$71.40

Crew R-15	Hr.	Daily	Hr.	Daily	Bare Costs	Incl. O&P
1 Electrician Foreman	$37.40	$299.20	$61.20	$489.60	$36.48	$59.77
4 Electricians	36.90	1180.80	60.35	1931.20		
1 Equipment Oper. (light)	33.90	271.20	56.05	448.40		
1 Aerial Lift Truck, 40' Boom		309.00		339.90	6.44	7.08
48 L.H., Daily Totals		$2060.20		$3209.10	$42.92	$66.86

Crew R-15A	Hr.	Daily	Hr.	Daily	Bare Costs	Incl. O&P
1 Electrician Foreman	$37.40	$299.20	$61.20	$489.60	$32.40	$53.41
2 Electricians	36.90	590.40	60.35	965.60		
2 Common Laborers	24.65	394.40	41.25	660.00		
1 Equip. Oper. (light)	33.90	271.20	56.05	448.40		
1 Aerial Lift Truck, 40' Boom		309.00		339.90	6.44	7.08
48 L.H., Daily Totals		$1864.20		$2903.50	$38.84	$60.49

Crew R-18	Hr.	Daily	Hr.	Daily	Bare Costs	Incl. O&P
.25 Electrician Foreman	$37.40	$74.80	$61.20	$122.40	$32.38	$52.97
1 Electrician	36.90	295.20	60.35	482.80		
2 Electrician Apprentices	29.50	472.00	48.25	772.00		
26 L.H., Daily Totals		$842.00		$1377.20	$32.38	$52.97

Crew R-19	Hr.	Daily	Hr.	Daily	Bare Costs	Incl. O&P
.5 Electrician Foreman	$37.40	$149.60	$61.20	$244.80	$37.00	$60.52
2 Electricians	36.90	590.40	60.35	965.60		
20 L.H., Daily Totals		$740.00		$1210.40	$37.00	$60.52

Crew R-21	Hr.	Daily	Hr.	Daily	Bare Costs	Incl. O&P
1 Electrician Foreman	$37.40	$299.20	$61.20	$489.60	$36.98	$60.51
3 Electricians	36.90	885.60	60.35	1448.40		
.1 Equip. Oper. (medium)	35.25	28.20	58.25	46.60		
.1 S.P. Crane, 4x4, 25 Ton		60.66		66.73	1.85	2.03
32.8 L.H., Daily Totals		$1273.66		$2051.33	$38.83	$62.54

Crew R-22	Hr.	Daily	Hr.	Daily	Bare Costs	Incl. O&P
.66 Electrician Foreman	$37.40	$197.47	$61.20	$323.14	$33.79	$55.28
2 Electricians	36.90	590.40	60.35	965.60		
2 Electrician Apprentices	29.50	472.00	48.25	772.00		
37.28 L.H., Daily Totals		$1259.87		$2060.74	$33.79	$55.28

Crew R-30	Hr.	Daily	Hr.	Daily	Bare Costs	Incl. O&P
.25 Electrician Foreman (outside)	$38.90	$77.80	$63.65	$127.30	$29.52	$48.85
1 Electrician	36.90	295.20	60.35	482.80		
2 Laborers (Semi-Skilled)	24.65	394.40	41.25	660.00		
26 L.H., Daily Totals		$767.40		$1270.10	$29.52	$48.85

Location Factors - Residential

Costs shown in *RSMeans Residential Cost Data* are based on national averages for materials and installation. To adjust these costs to a specific location, simply multiply the base cost by the factor for that city. The data is arranged alphabetically by state and postal zip code numbers. For a city not listed, use the factor for a nearby city with similar economic characteristics.

STATE	CITY	Residential
ALABAMA		
350-352	Birmingham	.88
354	Tuscaloosa	.89
355	Jasper	.85
356	Decatur	.88
357-358	Huntsville	.87
359	Gadsden	.79
360-361	Montgomery	.88
362	Anniston	.87
363	Dothan	.88
364	Evergreen	.84
365-366	Mobile	.89
367	Selma	.85
368	Phenix City	.87
369	Butler	.86
ALASKA		
995-996	Anchorage	1.23
997	Fairbanks	1.27
998	Juneau	1.25
999	Ketchikan	1.27
ARIZONA		
850,853	Phoenix	.85
851,852	Mesa/Tempe	.85
855	Globe	.82
856-857	Tucson	.84
859	Show Low	.86
860	Flagstaff	.87
863	Prescott	.82
864	Kingman	.83
865	Chambers	.84
ARKANSAS		
716	Pine Bluff	.79
717	Camden	.71
718	Texarkana	.75
719	Hot Springs	.71
720-722	Little Rock	.82
723	West Memphis	.78
724	Jonesboro	.76
725	Batesville	.73
726	Harrison	.75
727	Fayetteville	.70
728	Russellville	.76
729	Fort Smith	.81
CALIFORNIA		
900-902	Los Angeles	1.08
903-905	Inglewood	1.06
906-908	Long Beach	1.05
910-912	Pasadena	1.05
913-916	Van Nuys	1.08
917-918	Alhambra	1.09
919-921	San Diego	1.05
922	Palm Springs	1.05
923-924	San Bernardino	1.06
925	Riverside	1.07
926-927	Santa Ana	1.07
928	Anaheim	1.07
930	Oxnard	1.07
931	Santa Barbara	1.07
932-933	Bakersfield	1.05
934	San Luis Obispo	1.08
935	Mojave	1.07
936-938	Fresno	1.11
939	Salinas	1.14
940-941	San Francisco	1.24
942,956-958	Sacramento	1.12
943	Palo Alto	1.17
944	San Mateo	1.21
945	Vallejo	1.16
946	Oakland	1.21
947	Berkeley	1.23
948	Richmond	1.23
949	San Rafael	1.22
950	Santa Cruz	1.17
951	San Jose	1.22
952	Stockton	1.13
953	Modesto	1.11

STATE	CITY	Residential
CALIFORNIA (CONT'D)		
954	Santa Rosa	1.20
955	Eureka	1.16
959	Marysville	1.12
960	Redding	1.18
961	Susanville	1.16
COLORADO		
800-802	Denver	.90
803	Boulder	.92
804	Golden	.88
805	Fort Collins	.89
806	Greeley	.86
807	Fort Morgan	.90
808-809	Colorado Springs	.86
810	Pueblo	.86
811	Alamosa	.85
812	Salida	.88
813	Durango	.89
814	Montrose	.87
815	Grand Junction	.92
816	Glenwood Springs	.88
CONNECTICUT		
060	New Britain	1.11
061	Hartford	1.10
062	Willimantic	1.12
063	New London	1.11
064	Meriden	1.11
065	New Haven	1.11
066	Bridgeport	1.12
067	Waterbury	1.11
068	Norwalk	1.13
069	Stamford	1.13
D.C.		
200-205	Washington	.97
DELAWARE		
197	Newark	1.02
198	Wilmington	1.01
199	Dover	1.01
FLORIDA		
320,322	Jacksonville	.83
321	Daytona Beach	.86
323	Tallahassee	.82
324	Panama City	.78
325	Pensacola	.88
326,344	Gainesville	.82
327-328,347	Orlando	.85
329	Melbourne	.86
330-332,340	Miami	.85
333	Fort Lauderdale	.84
334,349	West Palm Beach	.84
335-336,346	Tampa	.85
337	St. Petersburg	.84
338	Lakeland	.83
339,341	Fort Myers	.82
342	Sarasota	.86
GEORGIA		
300-303,399	Atlanta	.90
304	Statesboro	.77
305	Gainesville	.79
306	Athens	.78
307	Dalton	.78
308-309	Augusta	.87
310-312	Macon	.77
313-314	Savannah	.82
315	Waycross	.79
316	Valdosta	.73
317,398	Albany	.76
318-319	Columbus	.78
HAWAII		
967	Hilo	1.22
968	Honolulu	1.25

Location Factors - Residential

STATE	CITY	Residential	STATE	CITY	Residential
STATES & POSS.			**KENTUCKY (CONT'D)**		
969	Guam	.98	406	Frankfort	.86
IDAHO			407-409	Corbin	.81
832	Pocatello	.87	410	Covington	.90
833	Twin Falls	.81	411-412	Ashland	.93
834	Idaho Falls	.85	413-414	Campton	.86
835	Lewiston	.97	415-416	Pikeville	.90
836-837	Boise	.87	417-418	Hazard	.86
838	Coeur d'Alene	.98	420	Paducah	.88
			421-422	Bowling Green	.89
ILLINOIS			423	Owensboro	.90
600-603	North Suburban	1.20	424	Henderson	.89
604	Joliet	1.22	425-426	Somerset	.85
605	South Suburban	1.20	427	Elizabethtown	.86
606-608	Chicago	1.21			
609	Kankakee	1.13	**LOUISIANA**		
610-611	Rockford	1.10	700-701	New Orleans	.86
612	Rock Island	.98	703	Thibodaux	.82
613	La Salle	1.10	704	Hammond	.77
614	Galesburg	1.03	705	Lafayette	.84
615-616	Peoria	1.06	706	Lake Charles	.86
617	Bloomington	1.03	707-708	Baton Rouge	.84
618-619	Champaign	1.04	710-711	Shreveport	.78
620-622	East St. Louis	1.02	712	Monroe	.75
623	Quincy	1.02	713-714	Alexandria	.78
624	Effingham	1.04			
625	Decatur	1.03	**MAINE**		
626-627	Springfield	1.04	039	Kittery	.92
628	Centralia	1.03	040-041	Portland	.96
629	Carbondale	1.00	042	Lewiston	.96
			043	Augusta	.92
INDIANA			044	Bangor	.95
460	Anderson	.91	045	Bath	.91
461-462	Indianapolis	.93	046	Machias	.95
463-464	Gary	1.05	047	Houlton	.96
465-466	South Bend	.91	048	Rockland	.95
467-468	Fort Wayne	.89	049	Waterville	.90
469	Kokomo	.91			
470	Lawrenceburg	.86	**MARYLAND**		
471	New Albany	.85	206	Waldorf	.88
472	Columbus	.91	207-208	College Park	.86
473	Muncie	.91	209	Silver Spring	.88
474	Bloomington	.92	210-212	Baltimore	.92
475	Washington	.90	214	Annapolis	.89
476-477	Evansville	.90	215	Cumberland	.90
478	Terre Haute	.91	216	Easton	.85
479	Lafayette	.92	217	Hagerstown	.89
			218	Salisbury	.83
IOWA			219	Elkton	.92
500-503,509	Des Moines	.89			
504	Mason City	.75	**MASSACHUSETTS**		
505	Fort Dodge	.74	010-011	Springfield	1.06
506-507	Waterloo	.83	012	Pittsfield	1.06
508	Creston	.84	013	Greenfield	1.04
510-511	Sioux City	.86	014	Fitchburg	1.13
512	Sibley	.73	015-016	Worcester	1.15
513	Spencer	.75	017	Framingham	1.18
514	Carroll	.79	018	Lowell	1.18
515	Council Bluffs	.86	019	Lawrence	1.18
516	Shenandoah	.81	020-022, 024	Boston	1.23
520	Dubuque	.87	023	Brockton	1.16
521	Decorah	.79	025	Buzzards Bay	1.14
522-524	Cedar Rapids	.92	026	Hyannis	1.13
525	Ottumwa	.87	027	New Bedford	1.16
526	Burlington	.88			
527-528	Davenport	.97	**MICHIGAN**		
			480,483	Royal Oak	1.01
KANSAS			481	Ann Arbor	1.02
660-662	Kansas City	.98	482	Detroit	1.05
664-666	Topeka	.81	484-485	Flint	.95
667	Fort Scott	.89	486	Saginaw	.91
668	Emporia	.83	487	Bay City	.91
669	Belleville	.81	488-489	Lansing	.94
670-672	Wichita	.80	490	Battle Creek	.90
673	Independence	.89	491	Kalamazoo	.90
674	Salina	.80	492	Jackson	.92
675	Hutchinson	.80	493,495	Grand Rapids	.90
676	Hays	.82	494	Muskegon	.88
677	Colby	.84	496	Traverse City	.85
678	Dodge City	.81	497	Gaylord	.88
679	Liberal	.80	498-499	Iron Mountain	.89
KENTUCKY			**MINNESOTA**		
400-402	Louisville	.89	550-551	Saint Paul	1.11
403-405	Lexington	.86	553-555	Minneapolis	1.12
			556-558	Duluth	1.04

Location Factors - Residential

STATE	CITY	Residential
MINNESOTA (CONT'D)		
559	Rochester	1.03
560	Mankato	1.00
561	Windom	.92
562	Willmar	.95
563	St. Cloud	1.04
564	Brainerd	.95
565	Detroit Lakes	.94
566	Bemidji	.94
567	Thief River Falls	.93
MISSISSIPPI		
386	Clarksdale	.74
387	Greenville	.82
388	Tupelo	.75
389	Greenwood	.77
390-392	Jackson	.84
393	Meridian	.82
394	Laurel	.76
395	Biloxi	.80
396	McComb	.74
397	Columbus	.75
MISSOURI		
630-631	St. Louis	1.02
633	Bowling Green	.97
634	Hannibal	.92
635	Kirksville	.90
636	Flat River	.96
637	Cape Girardeau	.91
638	Sikeston	.88
639	Poplar Bluff	.89
640-641	Kansas City	1.03
644-645	St. Joseph	.98
646	Chillicothe	.96
647	Harrisonville	.97
648	Joplin	.90
650-651	Jefferson City	.93
652	Columbia	.93
653	Sedalia	.92
654-655	Rolla	.97
656-658	Springfield	.89
MONTANA		
590-591	Billings	.90
592	Wolf Point	.86
593	Miles City	.87
594	Great Falls	.90
595	Havre	.83
596	Helena	.86
597	Butte	.86
598	Missoula	.87
599	Kalispell	.86
NEBRASKA		
680-681	Omaha	.89
683-685	Lincoln	.89
686	Columbus	.86
687	Norfolk	.88
688	Grand Island	.87
689	Hastings	.89
690	McCook	.82
691	North Platte	.87
692	Valentine	.83
693	Alliance	.82
NEVADA		
889-891	Las Vegas	1.03
893	Ely	1.02
894-895	Reno	.91
897	Carson City	.91
898	Elko	.93
NEW HAMPSHIRE		
030	Nashua	.98
031	Manchester	.96
032-033	Concord	.97
034	Keene	.88
035	Littleton	.91
036	Charleston	.84
037	Claremont	.85
038	Portsmouth	.96

STATE	CITY	Residential
NEW JERSEY		
070-071	Newark	1.15
072	Elizabeth	1.17
073	Jersey City	1.14
074-075	Paterson	1.15
076	Hackensack	1.14
077	Long Branch	1.11
078	Dover	1.14
079	Summit	1.15
080,083	Vineland	1.12
081	Camden	1.13
082,084	Atlantic City	1.17
085-086	Trenton	1.15
087	Point Pleasant	1.12
088-089	New Brunswick	1.17
NEW MEXICO		
870-872	Albuquerque	.83
873	Gallup	.82
874	Farmington	.83
875	Santa Fe	.84
877	Las Vegas	.82
878	Socorro	.82
879	Truth or Consequences	.81
880	Las Cruces	.81
881	Clovis	.82
882	Roswell	.83
883	Carrizozo	.82
884	Tucumcari	.83
NEW YORK		
100-102	New York	1.35
103	Staten Island	1.30
104	Bronx	1.30
105	Mount Vernon	1.16
106	White Plains	1.19
107	Yonkers	1.22
108	New Rochelle	1.20
109	Suffern	1.17
110	Queens	1.30
111	Long Island City	1.33
112	Brooklyn	1.36
113	Flushing	1.32
114	Jamaica	1.31
115,117,118	Hicksville	1.23
116	Far Rockaway	1.31
119	Riverhead	1.23
120-122	Albany	1.00
123	Schenectady	1.02
124	Kingston	1.04
125-126	Poughkeepsie	1.21
127	Monticello	1.04
128	Glens Falls	.95
129	Plattsburgh	1.00
130-132	Syracuse	.98
133-135	Utica	.96
136	Watertown	.94
137-139	Binghamton	.98
140-142	Buffalo	1.07
143	Niagara Falls	1.03
144-146	Rochester	.99
147	Jamestown	.92
148-149	Elmira	.95
NORTH CAROLINA		
270,272-274	Greensboro	.90
271	Winston-Salem	.90
275-276	Raleigh	.89
277	Durham	.90
278	Rocky Mount	.89
279	Elizabeth City	.80
280	Gastonia	.94
281-282	Charlotte	.93
283	Fayetteville	.90
284	Wilmington	.91
285	Kinston	.91
286	Hickory	.89
287-288	Asheville	.91
289	Murphy	.83
NORTH DAKOTA		
580-581	Fargo	.83
582	Grand Forks	.78
583	Devils Lake	.81
584	Jamestown	.79
585	Bismarck	.81

Location Factors - Residential

STATE	CITY	Residential
NORTH DAKOTA (CONT'D)		
586	Dickinson	.79
587	Minot	.86
588	Williston	.79
OHIO		
430-432	Columbus	.93
433	Marion	.90
434-436	Toledo	.98
437-438	Zanesville	.90
439	Steubenville	.94
440	Lorain	.96
441	Cleveland	1.00
442-443	Akron	.98
444-445	Youngstown	.95
446-447	Canton	.94
448-449	Mansfield	.91
450	Hamilton	.92
451-452	Cincinnati	.92
453-454	Dayton	.93
455	Springfield	.94
456	Chillicothe	.97
457	Athens	.94
458	Lima	.92
OKLAHOMA		
730-731	Oklahoma City	.84
734	Ardmore	.79
735	Lawton	.82
736	Clinton	.80
737	Enid	.79
738	Woodward	.80
739	Guymon	.82
740-741	Tulsa	.80
743	Miami	.85
744	Muskogee	.77
745	McAlester	.76
746	Ponca City	.79
747	Durant	.77
748	Shawnee	.78
749	Poteau	.78
OREGON		
970-972	Portland	1.00
973	Salem	.99
974	Eugene	1.00
975	Medford	.98
976	Klamath Falls	.99
977	Bend	1.01
978	Pendleton	.99
979	Vale	.97
PENNSYLVANIA		
150-152	Pittsburgh	1.01
153	Washington	.97
154	Uniontown	.95
155	Bedford	.92
156	Greensburg	.98
157	Indiana	.96
158	Dubois	.92
159	Johnstown	.93
160	Butler	.94
161	New Castle	.94
162	Kittanning	.94
163	Oil City	.93
164-165	Erie	.95
166	Altoona	.89
167	Bradford	.92
168	State College	.91
169	Wellsboro	.93
170-171	Harrisburg	.96
172	Chambersburg	.90
173-174	York	.93
175-176	Lancaster	.94
177	Williamsport	.93
178	Sunbury	.94
179	Pottsville	.93
180	Lehigh Valley	1.00
181	Allentown	1.05
182	Hazleton	.94
183	Stroudsburg	.94
184-185	Scranton	.97
186-187	Wilkes-Barre	.95
188	Montrose	.92
189	Doylestown	1.09

STATE	CITY	Residential
PENNSYLVANIA (CONT'D)		
190-191	Philadelphia	1.17
193	Westchester	1.11
194	Norristown	1.11
195-196	Reading	.99
PUERTO RICO		
009	San Juan	.76
RHODE ISLAND		
028	Newport	1.09
029	Providence	1.09
SOUTH CAROLINA		
290-292	Columbia	.96
293	Spartanburg	.96
294	Charleston	.94
295	Florence	.93
296	Greenville	.95
297	Rock Hill	.83
298	Aiken	.93
299	Beaufort	.81
SOUTH DAKOTA		
570-571	Sioux Falls	.77
572	Watertown	.73
573	Mitchell	.74
574	Aberdeen	.76
575	Pierre	.78
576	Mobridge	.73
577	Rapid City	.78
TENNESSEE		
370-372	Nashville	.85
373-374	Chattanooga	.84
375,380-381	Memphis	.84
376	Johnson City	.71
377-379	Knoxville	.81
382	McKenzie	.73
383	Jackson	.77
384	Columbia	.78
385	Cookeville	.72
TEXAS		
750	McKinney	.81
751	Waxahackie	.81
752-753	Dallas	.84
754	Greenville	.82
755	Texarkana	.80
756	Longview	.80
757	Tyler	.82
758	Palestine	.79
759	Lufkin	.81
760-761	Fort Worth	.82
762	Denton	.85
763	Wichita Falls	.85
764	Eastland	.83
765	Temple	.81
766-767	Waco	.84
768	Brownwood	.79
769	San Angelo	.79
770-772	Houston	.84
773	Huntsville	.81
774	Wharton	.82
775	Galveston	.83
776-777	Beaumont	.86
778	Bryan	.80
779	Victoria	.83
780	Laredo	.81
781-782	San Antonio	.82
783-784	Corpus Christi	.85
785	McAllen	.85
786-787	Austin	.80
788	Del Rio	.82
789	Giddings	.81
790-791	Amarillo	.80
792	Childress	.82
793-794	Lubbock	.81
795-796	Abilene	.83
797	Midland	.85
798-799,885	El Paso	.80
UTAH		
840-841	Salt Lake City	.82
842,844	Ogden	.81
843	Logan	.81

Location Factors - Residential

STATE	CITY	Residential
UTAH (CONT'D)		
845	Price	.78
846-847	Provo	.82
VERMONT		
050	White River Jct.	.89
051	Bellows Falls	.96
052	Bennington	.99
053	Brattleboro	.95
054	Burlington	.94
056	Montpelier	.92
057	Rutland	.93
058	St. Johnsbury	.90
059	Guildhall	.90
VIRGINIA		
220-221	Fairfax	1.07
222	Arlington	1.10
223	Alexandria	1.11
224-225	Fredericksburg	1.08
226	Winchester	1.02
227	Culpeper	1.07
228	Harrisonburg	.86
229	Charlottesville	.89
230-232	Richmond	.92
233-235	Norfolk	.94
236	Newport News	.94
237	Portsmouth	.89
238	Petersburg	.92
239	Farmville	.83
240-241	Roanoke	.95
242	Bristol	.83
243	Pulaski	.82
244	Staunton	.86
245	Lynchburg	.92
246	Grundy	.80
WASHINGTON		
980-981,987	Seattle	1.02
982	Everett	1.03
983-984	Tacoma	1.00
985	Olympia	.98
986	Vancouver	.96
988	Wenatchee	.93
989	Yakima	.98
990-992	Spokane	.97
993	Richland	.97
994	Clarkston	.94
WEST VIRGINIA		
247-248	Bluefield	.94
249	Lewisburg	.93
250-253	Charleston	.97
254	Martinsburg	.90
255-257	Huntington	.98
258-259	Beckley	.94
260	Wheeling	.94
261	Parkersburg	.93
262	Buckhannon	.94
263-264	Clarksburg	.94
265	Morgantown	.94
266	Gassaway	.93
267	Romney	.91
268	Petersburg	.91
WISCONSIN		
530,532	Milwaukee	1.07
531	Kenosha	1.03
534	Racine	1.01
535	Beloit	1.00
537	Madison	1.00
538	Lancaster	.97
539	Portage	.96
540	New Richmond	.97
541-543	Green Bay	1.03
544	Wausau	.97
545	Rhinelander	.95
546	La Crosse	.96
547	Eau Claire	.99
548	Superior	.96
549	Oshkosh	.95
WYOMING		
820	Cheyenne	.82
821	Yellowstone Nat. Pk.	.82
822	Wheatland	.77

STATE	CITY	Residential
WYOMING (CONT'D)		
823	Rawlins	.85
824	Worland	.80
825	Riverton	.80
826	Casper	.79
827	Newcastle	.84
828	Sheridan	.83
829-831	Rock Springs	.86
CANADIAN FACTORS (reflect Canadian currency)		
ALBERTA		
	Calgary	1.07
	Edmonton	1.06
	Fort McMurray	1.10
	Lethbridge	1.08
	Lloydminster	1.03
	Medicine Hat	1.03
	Red Deer	1.03
BRITISH COLUMBIA		
	Kamloops	1.01
	Prince George	1.01
	Vancouver	1.02
	Victoria	1.02
MANITOBA		
	Brandon	1.08
	Portage la Prairie	.98
	Winnipeg	.95
NEW BRUNSWICK		
	Bathurst	.90
	Dalhousie	.92
	Fredericton	.95
	Moncton	.92
	Newcastle	.91
	St. John	.99
NEWFOUNDLAND		
	Corner Brook	1.02
	St. John's	1.03
NORTHWEST TERRITORIES		
	Yellowknife	1.12
NOVA SCOTIA		
	Bridgewater	.94
	Dartmouth	1.03
	Halifax	.97
	New Glasgow	1.02
	Sydney	1.01
	Truro	.94
	Yarmouth	1.02
ONTARIO		
	Barrie	1.11
	Brantford	1.10
	Cornwall	1.09
	Hamilton	1.06
	Kingston	1.09
	Kitchener	1.02
	London	1.05
	North Bay	1.17
	Oshawa	1.07
	Ottawa	1.06
	Owen Sound	1.10
	Peterborough	1.08
	Sarnia	1.10
	Sault Ste. Marie	1.04
	St. Catharines	1.04
	Sudbury	1.01
	Thunder Bay	1.06
	Timmins	1.07
	Toronto	1.08
	Windsor	1.05
PRINCE EDWARD ISLAND		
	Charlottetown	.90
	Summerside	.97
QUEBEC		
	Cap-de-la-Madeleine	1.08
	Charlesbourg	1.08
	Chicoutimi	1.11
	Gatineau	1.08
	Granby	1.08

Location Factors - Residential

STATE	CITY	Residential
QUEBEC (CONT'D)		
	Hull	1.08
	Joliette	1.09
	Laval	1.08
	Montreal	1.05
	Quebec City	1.06
	Rimouski	1.11
	Rouyn-Noranda	1.08
	Saint-Hyacinthe	1.08
	Sherbrooke	1.08
	Sorel	1.09
	Saint-Jerome	1.08
	Trois-Rivieres	1.19
SASKATCHEWAN		
	Moose Jaw	.91
	Prince Albert	.90
	Regina	1.05
	Saskatoon	1.03
YUKON		
	Whitehorse	1.03

General Requirements — R0111 Summary of Work

R011105-05 Tips for Accurate Estimating

1. Use pre-printed or columnar forms for orderly sequence of dimensions and locations and for recording telephone quotations.
2. Use only the front side of each paper or form except for certain pre-printed summary forms.
3. Be consistent in listing dimensions: For example, length x width x height. This helps in rechecking to ensure that, the total length of partitions is appropriate for the building area.
4. Use printed (rather than measured) dimensions where given.
5. Add up multiple printed dimensions for a single entry where possible.
6. Measure all other dimensions carefully.
7. Use each set of dimensions to calculate multiple related quantities.
8. Convert foot and inch measurements to decimal feet when listing. Memorize decimal equivalents to .01 parts of a foot (1/8" equals approximately .01').
9. Do not "round off" quantities until the final summary.
10. Mark drawings with different colors as items are taken off.
11. Keep similar items together, different items separate.
12. Identify location and drawing numbers to aid in future checking for completeness.
13. Measure or list everything on the drawings or mentioned in the specifications.
14. It may be necessary to list items not called for to make the job complete.
15. Be alert for: Notes on plans such as N.T.S. (not to scale); changes in scale throughout the drawings; reduced size drawings; discrepancies between the specifications and the drawings.
16. Develop a consistent pattern of performing an estimate. For example:
    a. Start the quantity takeoff at the lower floor and move to the next higher floor.
    b. Proceed from the main section of the building to the wings.
    c. Proceed from south to north or vice versa, clockwise or counterclockwise.
    d. Take off floor plan quantities first, elevations next, then detail drawings.
17. List all gross dimensions that can be either used again for different quantities, or used as a rough check of other quantities for verification (exterior perimeter, gross floor area, individual floor areas, etc.).
18. Utilize design symmetry or repetition (repetitive floors, repetitive wings, symmetrical design around a center line, similar room layouts, etc.). Note: Extreme caution is needed here so as not to omit or duplicate an area.
19. Do not convert units until the final total is obtained. For instance, when estimating concrete work, keep all units to the nearest cubic foot, then summarize and convert to cubic yards.
20. When figuring alternatives, it is best to total all items involved in the basic system, then total all items involved in the alternates. Therefore you work with positive numbers in all cases. When adds and deducts are used, it is often confusing whether to add or subtract a portion of an item; especially on a complicated or involved alternate.

General Requirements — R0111 Summary of Work

R011105-50 Metric Conversion Factors

Description: This table is primarily for converting customary U.S. units in the left hand column to SI metric units in the right hand column. In addition, conversion factors for some commonly encountered Canadian and non-SI metric units are included.

	If You Know		Multiply By		To Find
Length	Inches	x	25.4[a]	=	Millimeters
	Feet	x	0.3048[a]	=	Meters
	Yards	x	0.9144[a]	=	Meters
	Miles (statute)	x	1.609	=	Kilometers
Area	Square inches	x	645.2	=	Square millimeters
	Square feet	x	0.0929	=	Square meters
	Square yards	x	0.8361	=	Square meters
Volume (Capacity)	Cubic inches	x	16,387	=	Cubic millimeters
	Cubic feet	x	0.02832	=	Cubic meters
	Cubic yards	x	0.7646	=	Cubic meters
	Gallons (U.S. liquids)[b]	x	0.003785	=	Cubic meters[c]
	Gallons (Canadian liquid)[b]	x	0.004546	=	Cubic meters[c]
	Ounces (U.S. liquid)[b]	x	29.57	=	Milliliters[c, d]
	Quarts (U.S. liquid)[b]	x	0.9464	=	Liters[c, d]
	Gallons (U.S. liquid)[b]	x	3.785	=	Liters[c, d]
Force	Kilograms force[d]	x	9.807	=	Newtons
	Pounds force	x	4.448	=	Newtons
	Pounds force	x	0.4536	=	Kilograms force[d]
	Kips	x	4448	=	Newtons
	Kips	x	453.6	=	Kilograms force[d]
Pressure, Stress, Strength (Force per unit area)	Kilograms force per square centimeter[d]	x	0.09807	=	Megapascals
	Pounds force per square inch (psi)	x	0.006895	=	Megapascals
	Kips per square inch	x	6.895	=	Megapascals
	Pounds force per square inch (psi)	x	0.07031	=	Kilograms force per square centimeter[d]
	Pounds force per square foot	x	47.88	=	Pascals
	Pounds force per square foot	x	4.882	=	Kilograms force per square meter[d]
Flow	Cubic feet per minute	x	0.4719	=	Liters per second
	Gallons per minute	x	0.0631	=	Liters per second
	Gallons per hour	x	1.05	=	Milliliters per second
Bending Moment Or Torque	Inch-pounds force	x	0.01152	=	Meter-kilograms force[d]
	Inch-pounds force	x	0.1130	=	Newton-meters
	Foot-pounds force	x	0.1383	=	Meter-kilograms force[d]
	Foot-pounds force	x	1.356	=	Newton-meters
	Meter-kilograms force[d]	x	9.807	=	Newton-meters
Mass	Ounces (avoirdupois)	x	28.35	=	Grams
	Pounds (avoirdupois)	x	0.4536	=	Kilograms
	Tons (metric)	x	1000	=	Kilograms
	Tons, short (2000 pounds)	x	907.2	=	Kilograms
	Tons, short (2000 pounds)	x	0.9072	=	Megagrams[e]
Mass per Unit Volume	Pounds mass per cubic foot	x	16.02	=	Kilograms per cubic meter
	Pounds mass per cubic yard	x	0.5933	=	Kilograms per cubic meter
	Pounds mass per gallon (U.S. liquid)[b]	x	119.8	=	Kilograms per cubic meter
	Pounds mass per gallon (Canadian liquid)[b]	x	99.78	=	Kilograms per cubic meter
Temperature	Degrees Fahrenheit	(F-32)/1.8		=	Degrees Celsius
	Degrees Fahrenheit	(F+459.67)/1.8		=	Degrees Kelvin
	Degrees Celsius	C+273.15		=	Degrees Kelvin

[a] The factor given is exact
[b] One U.S. gallon = 0.8327 Canadian gallon
[c] 1 liter = 1000 milliliters = 1000 cubic centimeters
 1 cubic decimeter = 0.001 cubic meter
[d] Metric but not SI unit
[e] Called "tonne" in England and "metric ton" in other metric countries

General Requirements — R0111 Summary of Work

R011105-60 Weights and Measures

Measures of Length
1 Mile = 1760 Yards = 5280 Feet
1 Yard = 3 Feet = 36 inches
1 Foot = 12 Inches
1 Mil = 0.001 Inch
1 Fathom = 2 Yards = 6 Feet
1 Rod = 5.5 Yards = 16.5 Feet
1 Hand = 4 Inches
1 Span = 9 Inches
1 Micro-inch = One Millionth Inch or 0.000001 Inch
1 Micron = One Millionth Meter + 0.00003937 Inch

Surveyor's Measure
1 Mile = 8 Furlongs = 80 Chains
1 Furlong = 10 Chains = 220 Yards
1 Chain = 4 Rods = 22 Yards = 66 Feet = 100 Links
1 Link = 7.92 Inches

Square Measure
1 Square Mile = 640 Acres = 6400 Square Chains
1 Acre = 10 Square Chains = 4840 Square Yards = 43,560 Sq. Ft.
1 Square Chain = 16 Square Rods = 484 Square Yards = 4356 Sq. Ft.
1 Square Rod = 30.25 Square Yards = 272.25 Square Feet = 625 Square Lines
1 Square Yard = 9 Square Feet
1 Square Foot = 144 Square Inches
An Acre equals a Square 208.7 Feet per Side

Cubic Measure
1 Cubic Yard = 27 Cubic Feet
1 Cubic Foot = 1728 Cubic Inches
1 Cord of Wood = 4 x 4 x 8 Feet = 128 Cubic Feet
1 Perch of Masonry = 16½ x 1½ x 1 Foot = 24.75 Cubic Feet

Avoirdupois or Commercial Weight
1 Gross or Long Ton = 2240 Pounds
1 Net or Short Ton = 2000 Pounds
1 Pound = 16 Ounces = 7000 Grains
1 Ounce = 16 Drachms = 437.5 Grains
1 Stone = 14 Pounds

Power
1 British Thermal Unit per Hour = 0.2931 Watts
1 Ton (Refrigeration) = 3.517 Kilowatts
1 Horsepower (Boiler) = 9.81 Kilowatts
1 Horsepower (550 ft-lb/s) = 0.746 Kilowatts

Shipping Measure
For Measuring Internal Capacity of a Vessel:
 1 Register Ton = 100 Cubic Feet

For Measurement of Cargo:
 Approximately 40 Cubic Feet of Merchandise is considered a Shipping Ton, unless that bulk would weigh more than 2000 Pounds, in which case Freight Charge may be based upon weight.

40 Cubic Feet = 32.143 U.S. Bushels = 31.16 Imp. Bushels

Liquid Measure
1 Imperial Gallon = 1.2009 U.S. Gallon = 277.42 Cu. In.
1 Cubic Foot = 7.48 U.S. Gallons

R011110-10 Architectural Fees

Tabulated below are typical percentage fees by project size, for good professional architectural service. Fees may vary from those listed depending upon degree of design difficulty and economic conditions in any particular area.

Rates can be interpolated horizontally and vertically. Various portions of the same project requiring different rates should be adjusted proportionately. For alterations, add 50% to the fee for the first $500,000 of project cost and add 25% to the fee for project cost over $500,000.

Architectural fees tabulated below include Structural, Mechanical and Electrical Engineering Fees. They do not include the fees for special consultants such as kitchen planning, security, acoustical, interior design, etc.

Civil Engineering fees are included in the Architectural fee for project sites requiring minimal design such as city sites. However, separate Civil Engineering fees must be added when utility connections require design, drainage calculations are needed, stepped foundations are required, or provisions are required to protect adjacent wetlands.

Building Types	Total Project Size in Thousands of Dollars						
	100	250	500	1,000	5,000	10,000	50,000
Factories, garages, warehouses, repetitive housing	9.0%	8.0%	7.0%	6.2%	5.3%	4.9%	4.5%
Apartments, banks, schools, libraries, offices, municipal buildings	12.2	12.3	9.2	8.0	7.0	6.6	6.2
Churches, hospitals, homes, laboratories, museums, research	15.0	13.6	12.7	11.9	9.5	8.8	8.0
Memorials, monumental work, decorative furnishings	—	16.0	14.5	13.1	10.0	9.0	8.3

General Requirements — R0129 Payment Procedures

R012909-80 Sales Tax by State

State sales tax on materials is tabulated below (5 states have no sales tax). Many states allow local jurisdictions, such as a county or city, to levy additional sales tax.

Some projects may be sales tax exempt, particularly those constructed with public funds.

State	Tax (%)	State	Tax (%)	State	Tax (%)	State	Tax (%)
Alabama	4	Illinois	6.25	Montana	0	Rhode Island	7
Alaska	0	Indiana	7	Nebraska	5.5	South Carolina	6
Arizona	5.6	Iowa	6	Nevada	6.85	South Dakota	4
Arkansas	6	Kansas	6.15	New Hampshire	0	Tennessee	7
California	7.5	Kentucky	6	New Jersey	7	Texas	6.25
Colorado	2.9	Louisiana	4	New Mexico	5.125	Utah	5.95
Connecticut	6.35	Maine	5.5	New York	4	Vermont	6
Delaware	0	Maryland	6	North Carolina	4.75	Virginia	5.3
District of Columbia	5.75	Massachusetts	6.25	North Dakota	5	Washington	6.5
Florida	6	Michigan	6	Ohio	5.75	West Virginia	6
Georgia	4	Minnesota	6.875	Oklahoma	4.5	Wisconsin	5
Hawaii	4	Mississippi	7	Oregon	0	Wyoming	4
Idaho	6	Missouri	4.225	Pennsylvania	6	Average	5.08%

Sales Tax by Province (Canada)

GST - a value-added tax, which the government imposes on most goods and services provided in or imported into Canada. PST - a retail sales tax, which three of the provinces impose on the prices of most goods and some services. QST - a value-added tax, similar to the federal GST, which Quebec imposes. HST - Five provinces have combined their retail sales taxes with the federal GST into one harmonized tax.

Province	PST (%)	QST (%)	GST (%)	HST (%)
Alberta	0	0	5	0
British Columbia	7	0	5	0
Manitoba	8	0	5	0
New Brunswick	0	0	0	13
Newfoundland	0	0	0	13
Northwest Territories	0	0	5	0
Nova Scotia	0	0	0	15
Ontario	0	0	0	13
Prince Edward Island	0	0	0	14
Quebec	0	9.975	5	0
Saskatchewan	5	0	5	0
Yukon	0	0	5	0

R012909-85 Unemployment Taxes and Social Security Taxes

State unemployment tax rates vary not only from state to state, but also with the experience rating of the contractor. The federal unemployment tax rate is 6.0% of the first $7,000 of wages. This is reduced by a credit of up to 5.4% for timely payment to the state. The minimum federal unemployment tax is 0.6% after all credits.

Social security (FICA) for 2016 is estimated at time of publication to be 7.65% of wages up to $118,500.

General Requirements — R0131 Project Management & Coordination

R013113-40 Builder's Risk Insurance

Builder's Risk Insurance is insurance on a building during construction. Premiums are paid by the owner or the contractor. Blasting, collapse and underground insurance would raise total insurance costs above those listed. Floater policy for materials delivered to the job runs $.75 to $1.25 per $100 value. Contractor equipment insurance runs $.50 to $1.50 per $100 value. Insurance for miscellaneous tools to $1,500 value runs from $3.00 to $7.50 per $100 value.

Tabulated below are New England Builder's Risk insurance rates in dollars per $100 value for $1,000 deductible. For $25,000 deductible, rates can be reduced 13% to 34%. On contracts over $1,000,000, rates may be lower than those tabulated. Policies are written annually for the total completed value in place. For "all risk" insurance (excluding flood, earthquake and certain other perils) add $.025 to total rates below.

Coverage	Frame Construction (Class 1) Range	Frame Construction (Class 1) Average	Brick Construction (Class 4) Range	Brick Construction (Class 4) Average	Fire Resistive (Class 6) Range	Fire Resistive (Class 6) Average
Fire Insurance	$.350 to $.850	$.600	$.158 to $.189	$.174	$.052 to $.080	$.070
Extended Coverage	.115 to .200	.158	.080 to .105	.101	.081 to .105	.100
Vandalism	.012 to .016	.014	.008 to .011	.011	.008 to .011	.010
Total Annual Rate	$.477 to $1.066	$.772	$.246 to $.305	$.286	$.141 to $.196	$.180

R013113-50 General Contractor's Overhead

There are two distinct types of overhead on a construction project: Project Overhead and Main Office Overhead. Project Overhead includes those costs at a construction site not directly associated with the installation of construction materials. Examples of Project Overhead costs include the following:

1. Superintendent
2. Construction office and storage trailers
3. Temporary sanitary facilities
4. Temporary utilities
5. Security fencing
6. Photographs
7. Clean up
8. Performance and payment bonds

The above Project Overhead items are also referred to as General Requirements and therefore are estimated in Division 1. Division 1 is the first division listed in the CSI MasterFormat but it is usually the last division estimated. The sum of the costs in Divisions 1 through 49 is referred to as the sum of the direct costs.

All construction projects also include indirect costs. The primary components of indirect costs are the contractor's Main Office Overhead and profit. The amount of the Main Office Overhead expense varies depending on the the following:

1. Owner's compensation
2. Project managers and estimator's wages
3. Clerical support wages
4. Office rent and utilities
5. Corporate legal and accounting costs
6. Advertising
7. Automobile expenses
8. Association dues
9. Travel and entertainment expenses

These costs are usually calculated as a percentage of annual sales volume. This percentage can range from 35% for a small contractor doing less than $500,000 to 5% for a large contractor with sales in excess of $100 million.

General Requirements — R0131 Project Management & Coordination

R013113-60 Workers' Compensation Insurance Rates by Trade

The table below tabulates the national averages for workers' compensation insurance rates by trade and type of building. The average "Insurance Rate" is multiplied by the "% of Building Cost" for each trade. This produces the "Workers' Compensation" cost by % of total labor cost, to be added for each trade by building type to determine the weighted average workers' compensation rate for the building types analyzed.

Trade	Insurance Rate (% Labor Cost) Range	Insurance Rate (% Labor Cost) Average	% of Building Cost Office Bldgs.	% of Building Cost Schools & Apts.	% of Building Cost Mfg.	Workers' Compensation Office Bldgs.	Workers' Compensation Schools & Apts.	Workers' Compensation Mfg.
Excavation, Grading, etc.	3.1% to 21.5%	9.3%	4.8%	4.9%	4.5%	0.45%	0.46%	0.42%
Piles & Foundations	6.4 to 32.8	14.6	7.1	5.2	8.7	1.04	0.76	1.27
Concrete	4.3 to 32.5	12.6	5.0	14.8	3.7	0.63	1.86	0.47
Masonry	4.6 to 36.8	13.7	6.9	7.5	1.9	0.95	1.03	0.26
Structural Steel	6.5 to 95.8	27.5	10.7	3.9	17.6	2.94	1.07	4.84
Miscellaneous & Ornamental Metals	4.1 to 28.9	11.7	2.8	4.0	3.6	0.33	0.47	0.42
Carpentry & Millwork	5.0 to 42.8	14.4	3.7	4.0	0.5	0.53	0.58	0.07
Metal or Composition Siding	6.5 to 75.9	18.7	2.3	0.3	4.3	0.43	0.06	0.80
Roofing	6.5 to 107.1	31.4	2.3	2.6	3.1	0.72	0.82	0.97
Doors & Hardware	4.1 to 42.8	11.4	0.9	1.4	0.4	0.10	0.16	0.05
Sash & Glazing	4.8 to 28.0	13.1	3.5	4.0	1.0	0.46	0.52	0.13
Lath & Plaster	2.8 to 41.3	10.5	3.3	6.9	0.8	0.35	0.72	0.08
Tile, Marble & Floors	2.9 to 22.4	8.8	2.6	3.0	0.5	0.23	0.26	0.04
Acoustical Ceilings	3.1 to 33.8	8.6	2.4	0.2	0.3	0.21	0.02	0.03
Painting	4.1 to 34.3	11.7	1.5	1.6	1.6	0.18	0.19	0.19
Interior Partitions	5.0 to 42.8	14.4	3.9	4.3	4.4	0.56	0.62	0.63
Miscellaneous Items	2.5 to 105.9	12.6	5.2	3.7	9.7	0.65	0.47	1.22
Elevators	1.4 to 13.6	5.2	2.1	1.1	2.2	0.11	0.06	0.11
Sprinklers	2.8 to 25.9	7.6	0.5	—	2.0	0.04	—	0.15
Plumbing	1.9 to 19.5	6.9	4.9	7.2	5.2	0.34	0.50	0.36
Heat., Vent., Air Conditioning	3.8 to 18.4	8.8	13.5	11.0	12.9	1.19	0.97	1.14
Electrical	2.3 to 14.4	5.6	10.1	8.4	11.1	0.57	0.47	0.62
Total	1.4% to 107.1%	—	100.0%	100.0%	100.0%	13.01%	12.07%	14.27%

Overall Weighted Average 13.12%

Workers' Compensation Insurance Rates by States

The table below lists the weighted average workers' compensation base rate for each state with a factor comparing this with the national average of 13.0%.

State	Weighted Average	Factor	State	Weighted Average	Factor	State	Weighted Average	Factor
Alabama	19.5%	150	Kentucky	14.3%	110	North Dakota	9.2%	71
Alaska	13.7	105	Louisiana	18.6	143	Ohio	8.2	63
Arizona	12.6	97	Maine	12.2	94	Oklahoma	12.5	96
Arkansas	8.3	64	Maryland	14.1	108	Oregon	11.8	91
California	25.8	198	Massachusetts	11.6	89	Pennsylvania	17.4	134
Colorado	7.2	55	Michigan	16.3	125	Rhode Island	12.1	93
Connecticut	23.7	182	Minnesota	22.6	174	South Carolina	18.9	145
Delaware	11.9	92	Mississippi	13.2	102	South Dakota	14.9	115
District of Columbia	11.0	85	Missouri	14.8	114	Tennessee	12.0	92
Florida	10.7	82	Montana	8.5	65	Texas	9.1	70
Georgia	28.1	216	Nebraska	17.7	136	Utah	8.8	68
Hawaii	8.1	62	Nevada	8.7	67	Vermont	13.1	101
Idaho	10.3	79	New Hampshire	20.0	154	Virginia	9.5	73
Illinois	27.1	208	New Jersey	14.4	111	Washington	10.5	81
Indiana	5.5	42	New Mexico	14.4	111	West Virginia	8.0	62
Iowa	14.6	112	New York	17.2	132	Wisconsin	13.1	101
Kansas	8.4	65	North Carolina	16.7	128	Wyoming	6.1	47

Weighted Average for U.S. is 13.7% of payroll = 100%

The weighted average skilled worker rate for 35 trades is 13.0%. For bidding purposes, apply the full value of workers' compensation directly to total labor costs, or if labor is 38%, materials 42% and overhead and profit 20% of total cost, carry 38/80 x 13.0% = 6.2% of cost (before overhead and profit) into overhead. Rates vary not only from state to state but also with the experience rating of the contractor.

Rates are the most current available at the time of publication.

General Requirements — R0154 Construction Aids

R015423-10 Steel Tubular Scaffolding

On new construction, tubular scaffolding is efficient up to 60' high or five stories. Above this it is usually better to use a hung scaffolding if construction permits. Swing scaffolding operations may interfere with tenants. In this case, the tubular is more practical at all heights.

In repairing or cleaning the front of an existing building the cost of tubular scaffolding per S.F. of building front increases as the height increases above the first tier. The first tier cost is relatively high due to leveling and alignment.

The minimum efficient crew for erecting and dismantling is three workers. They can set up and remove 18 frame sections per day up to 5 stories high. For 6 to 12 stories high, a crew of four is most efficient. Use two or more on top and two on the bottom for handing up or hoisting. They can also set up and remove 18 frame sections per day. At 7' horizontal spacing, this will run about 800 S.F. per day of erecting and dismantling. Time for placing and removing planks must be added to the above. A crew of three can place and remove 72 planks per day up to 5 stories. For over 5 stories, a crew of four can place and remove 80 planks per day.

The table below shows the number of pieces required to erect tubular steel scaffolding for 1000 S.F. of building frontage. This area is made up of a scaffolding system that is 12 frames (11 bays) long by 2 frames high.

For jobs under twenty-five frames, add 50% to rental cost. Rental rates will be lower for jobs over three months duration. Large quantities for long periods can reduce rental rates by 20%.

Description of Component	Number of Pieces for 1000 S.F. of Building Front	Unit
5' Wide Standard Frame, 6'-4" High	24	Ea.
Leveling Jack & Plate	24	
Cross Brace	44	
Side Arm Bracket, 21"	12	
Guardrail Post	12	
Guardrail, 7' section	22	
Stairway Section	2	
Stairway Starter Bar	1	
Stairway Inside Handrail	2	
Stairway Outside Handrail	2	
Walk-Thru Frame Guardrail	2	

Scaffolding is often used as falsework over 15' high during construction of cast-in-place concrete beams and slabs. Two foot wide scaffolding is generally used for heavy beam construction. The span between frames depends upon the load to be carried with a maximum span of 5'.

Heavy duty shoring frames with a capacity of 10,000#/leg can be spaced up to 10' O.C. depending upon form support design and loading.

Scaffolding used as horizontal shoring requires less than half the material required with conventional shoring.

On new construction, erection is done by carpenters.

Rolling towers supporting horizontal shores can reduce labor and speed the job. For maintenance work, catwalks with spans up to 70' can be supported by the rolling towers.

R015423-20 Pump Staging

Pump staging is generally not available for rent. The table below shows the number of pieces required to erect pump staging for 2400 S.F. of building frontage. This area is made up of a pump jack system that is 3 poles (2 bays) wide by 2 poles high.

Item	Number of Pieces for 2400 S.F. of Building Front	Unit
Aluminum pole section, 24' long	6	Ea.
Aluminum splice joint, 6' long	3	
Aluminum foldable brace	3	
Aluminum pump jack	3	
Aluminum support for workbench/back safety rail	3	
Aluminum scaffold plank/workbench, 14" wide x 24' long	4	
Safety net, 22' long	2	
Aluminum plank end safety rail	2	

The cost in place for this 2400 S.F. will depend on how many uses are realized during the life of the equipment.

General Requirements — R0154 Construction Aids

R015436-50 Mobilization

Costs to move rented construction equipment to a job site from an equipment dealer's or contractor's yard (mobilization) or off the job site (demobilization) are not included in the rental or operating rates, nor in the equipment cost on a unit price line or in a crew listing. These costs can be found consolidated in the Mobilization section of the data and elsewhere in particular site work sections. If a piece of equipment is already on the job site, it is not appropriate to include mob/demob costs in a new estimate that requires use of that equipment. The following table identifies approximate sizes of rented construction equipment that would be hauled on a towed trailer. Because this listing is not all-encompassing, the user can infer as to what size trailer might be required for a piece of equipment not listed.

3-ton Trailer	20-ton Trailer	40-ton Trailer	50-ton Trailer
20 H.P. Excavator	110 H.P. Excavator	200 H.P. Excavator	270 H.P. Excavator
50 H.P. Skid Steer	165 H.P. Dozer	300 H.P. Dozer	Small Crawler Crane
35 H.P. Roller	150 H.P. Roller	400 H.P. Scraper	500 H.P. Scraper
40 H.P. Trencher	Backhoe	450 H.P. Art. Dump Truck	500 H.P. Art. Dump Truck

Existing Conditions   R0241 Demolition

R024119-10 Demolition Defined

Whole Building Demolition - Demolition of the whole building with no concern for any particular building element, component, or material type being demolished. This type of demolition is accomplished with large pieces of construction equipment that break up the structure, load it into trucks and haul it to a disposal site, but disposal or dump fees are not included. Demolition of below-grade foundation elements, such as footings, foundation walls, grade beams, slabs on grade, etc., is not included. Certain mechanical equipment containing flammable liquids or ozone-depleting refrigerants, electric lighting elements, communication equipment components, and other building elements may contain hazardous waste, and must be removed, either selectively or carefully, as hazardous waste before the building can be demolished.

Foundation Demolition - Demolition of below-grade foundation footings, foundation walls, grade beams, and slabs on grade. This type of demolition is accomplished by hand or pneumatic hand tools, and does not include saw cutting, or handling, loading, hauling, or disposal of the debris.

Gutting - Removal of building interior finishes and electrical/mechanical systems down to the load-bearing and sub-floor elements of the rough building frame, with no concern for any particular building element, component, or material type being demolished. This type of demolition is accomplished by hand or pneumatic hand tools, and includes loading into trucks, but not hauling, disposal or dump fees, scaffolding, or shoring. Certain mechanical equipment containing flammable liquids or ozone-depleting refrigerants, electric lighting elements, communication equipment components, and other building elements may contain hazardous waste, and must be removed, either selectively or carefully, as hazardous waste, before the building is gutted.

Selective Demolition - Demolition of a selected building element, component, or finish, with some concern for surrounding or adjacent elements, components, or finishes (see the first Subdivision (s) at the beginning of appropriate Divisions). This type of demolition is accomplished by hand or pneumatic hand tools, and does not include handling, loading, storing, hauling, or disposal of the debris, scaffolding, or shoring. "Gutting" methods may be used in order to save time, but damage that is caused to surrounding or adjacent elements, components, or finishes may have to be repaired at a later time.

Careful Removal - Removal of a piece of service equipment, building element or component, or material type, with great concern for both the removed item and surrounding or adjacent elements, components or finishes. The purpose of careful removal may be to protect the removed item for later re-use, preserve a higher salvage value of the removed item, or replace an item while taking care to protect surrounding or adjacent elements, components, connections, or finishes from cosmetic and/or structural damage. An approximation of the time required to perform this type of removal is 1/3 to 1/2 the time it would take to install a new item of like kind. This type of removal is accomplished by hand or pneumatic hand tools, and does not include loading, hauling, or storing the removed item, scaffolding, shoring, or lifting equipment.

Cutout Demolition - Demolition of a small quantity of floor, wall, roof, or other assembly, with concern for the appearance and structural integrity of the surrounding materials. This type of demolition is accomplished by hand or pneumatic hand tools, and does not include saw cutting, handling, loading, hauling, or disposal of debris, scaffolding, or shoring.

Rubbish Handling - Work activities that involve handling, loading or hauling of debris. Generally, the cost of rubbish handling must be added to the cost of all types of demolition, with the exception of whole building demolition.

Minor Site Demolition - Demolition of site elements outside the footprint of a building. This type of demolition is accomplished by hand or pneumatic hand tools, or with larger pieces of construction equipment, and may include loading a removed item onto a truck (check the Crew for equipment used). It does not include saw cutting, hauling or disposal of debris, and, sometimes, handling or loading.

R024119-20 Dumpsters

Dumpster rental costs on construction sites are presented in two ways. The cost per week rental includes the delivery of the dumpster; its pulling or emptying once per week, and its final removal. The assumption is made that the dumpster contractor could choose to empty a dumpster by simply bringing in an empty unit and removing the full one. These costs also include the disposal of the materials in the dumpster.

The Alternate Pricing can be used when actual planned conditions are not approximated by the weekly numbers. For example, these lines can be used when a dumpster is needed for 4 weeks and will need to be emptied 2 or 3 times per week. Conversely the Alternate Pricing lines can be used when a dumpster will be rented for several weeks or months but needs to be emptied only a few times over this period.

Masonry — R0401 Maintenance of Masonry

R040130-10 Cleaning Face Brick

On smooth brick a person can clean 70 S.F. an hour; on rough brick 50 S.F. per hour. Use one gallon muriatic acid to 20 gallons of water for 1000 S.F. Do not use acid solution until wall is at least seven days old, but a mild soap solution may be used after two days.

Time has been allowed for clean-up in brick prices.

Masonry — R0405 Common Work Results for Masonry

R040513-10 Cement Mortar (material only)

Type N - 1:1:6 mix by volume. Use everywhere above grade except as noted below. - 1:3 mix using conventional masonry cement which saves handling two separate bagged materials.

Type M - 1:1/4:3 mix by volume, or 1 part cement, 1/4 (10% by wt.) lime, 3 parts sand. Use for heavy loads and where earthquakes or hurricanes may occur. Also for reinforced brick, sewers, manholes and everywhere below grade.

Mix Proportions by Volume and Compressive Strength of Mortar

Where Used	Mortar Type	Portland Cement	Masonry Cement	Hydrated Lime	Masonry Sand	Compressive Strength @ 28 days
Plain Masonry	M	1	1	—	6	
		1	—	1/4	3	2500 psi
	S	1/2	1	—	4	
		1	—	1/4 to 1/2	4	1800 psi
	N	—	1	—	3	
		1	—	1/2 to 1-1/4	6	750 psi
	O	—	1	—	3	
		1	—	1-1/4 to 2-1/2	9	350 psi
	K	1	—	2-1/2 to 4	12	75 psi
Reinforced Masonry	PM	1	1	—	6	2500 psi
	PL	1	—	1/4 to 1/2	4	2500 psi

Note: The total aggregate should be between 2.25 to 3 times the sum of the cement and lime used.

The labor cost to mix the mortar is included in the productivity and labor cost of unit price lines in unit cost sections for brickwork, blockwork and stonework.

The material cost of mixed mortar is included in the material cost of those same unit price lines and includes the cost of renting and operating a 10 C.F. mixer at the rate of 200 C.F. per day.

There are two types of mortar color used. One type is the inert additive type with about 100 lbs. per M brick as the typical quantity required. These colors are also available in smaller-batch-sized bags (1 lb. to 15 lb.) which can be placed directly into the mixer without measuring. The other type is premixed and replaces the masonry cement. Dark green color has the highest cost.

R040519-50 Masonry Reinforcing

Horizontal joint reinforcing helps prevent wall cracks where wall movement may occur and in many locations is required by code. Horizontal joint reinforcing is generally not considered to be structural reinforcing and an unreinforced wall may still contain joint reinforcing.

Reinforcing strips come in 10' and 12' lengths and in truss and ladder shapes, with and without drips. Field labor runs between 2.7 to 5.3 hours per 1000 L.F. for wall thicknesses up to 12".

The wire meets ASTM A82 for cold drawn steel wire and the typical size is 9 ga. sides and ties with 3/16" diameter also available. Typical finish is mill galvanized with zinc coating at .10 oz. per S.F. Class I (.40 oz. per S.F.) and Class III (.80 oz. per S.F.) are also available, as is hot dipped galvanizing at 1.50 oz. per S.F.

Masonry   R0421 Clay Unit Masonry

R042110-10   Economy in Bricklaying

Have adequate supervision. Be sure bricklayers are always supplied with materials so there is no waiting. Place experienced bricklayers at corners and openings.

Use only screened sand for mortar. Otherwise, labor time will be wasted picking out pebbles. Use seamless metal tubs for mortar as they do not leak or catch the trowel. Locate stack and mortar for easy wheeling.

Have brick delivered for stacking. This makes for faster handling, reduces chipping and breakage, and requires less storage space. Many dealers will deliver select common in 2' x 3' x 4' pallets or face brick packaged. This affords quick handling with a crane or forklift and easy tonging in units of ten, which reduces waste.

Use wider bricks for one wythe wall construction. Keep scaffolding away from the wall to allow mortar to fall clear and not stain the wall.

On large jobs develop specialized crews for each type of masonry unit.

Consider designing for prefabricated panel construction on high rise projects.

Avoid excessive corners or openings. Each opening adds about 50% to the labor cost for area of opening.

Bolting stone panels and using window frames as stops reduce labor costs and speed up erection.

R042110-20   Common and Face Brick

Common building brick manufactured according to ASTM C62 and facing brick manufactured according to ASTM C216 are the two standard bricks available for general building use.

Building brick is made in three grades: SW, where high resistance to damage caused by cyclic freezing is required; MW, where moderate resistance to cyclic freezing is needed; and NW, where little resistance to cyclic freezing is needed. Facing brick is made in only the two grades SW and MW. Additionally, facing brick is available in three types: FBS, for general use; FBX, for general use where a higher degree of precision and lower permissible variation in size than FBS are needed; and FBA, for general use to produce characteristic architectural effects resulting from non-uniformity in size and texture of the units.

In figuring the material cost of brickwork, an allowance of 25% mortar waste and 3% brick breakage was included. If bricks are delivered palletized with 280 to 300 per pallet, or packaged, allow only 1-1/2% for breakage. Packaged or palletized delivery is practical when a job is big enough to have a crane or other equipment available to handle a package of brick. This is so on all industrial work but not always true on small commercial buildings.

The use of buff and gray face is increasing, and there is a continuing trend to the Norman, Roman, Jumbo and SCR brick.

Common red clay brick for backup is not used that often. Concrete block is the most usual backup material with occasional use of sand lime or cement brick. Building brick is commonly used in solid walls for strength and as a fire stop.

Brick panels built on the ground and then crane erected to the upper floors have proven to be economical. This allows the work to be done under cover and without scaffolding.

R042110-50   Brick, Block & Mortar Quantities

Type Brick	Nominal Size (incl. mortar) L H W	Modular Coursing	Number of Brick per S.F.	C.F. of Mortar per M Bricks, Waste Included 3/8" Joint	C.F. of Mortar per M Bricks, Waste Included 1/2" Joint	Bond Type	Description	Factor
Standard	8 x 2-2/3 x 4	3C=8"	6.75	10.3	12.9	Common	full header every fifth course	+20%
Economy	8 x 4 x 4	1C=4"	4.50	11.4	14.6		full header every sixth course	+16.7%
Engineer	8 x 3-1/5 x 4	5C=16"	5.63	10.6	13.6	English	full header every second course	+50%
Fire	9 x 2-1/2 x 4-1/2	2C=5"	6.40	550 # Fireclay	—	Flemish	alternate headers every course	+33.3%
Jumbo	12 x 4 x 6 or 8	1C=4"	3.00	23.8	30.8		every sixth course	+5.6%
Norman	12 x 2-2/3 x 4	3C=8"	4.50	14.0	17.9	Header = W x H exposed		+100%
Norwegian	12 x 3-1/5 x 4	5C=16"	3.75	14.6	18.6	Rowlock = H x W exposed		+100%
Roman	12 x 2 x 4	2C=4"	6.00	13.4	17.0	Rowlock stretcher = L x W exposed		+33.3%
SCR	12 x 2-2/3 x 6	3C=8"	4.50	21.8	28.0	Soldier = H x L exposed		—
Utility	12 x 4 x 4	1C=4"	3.00	15.4	19.6	Sailor = W x L exposed		-33.3%

Concrete Blocks Nominal Size	Approximate Weight per S.F. Standard	Approximate Weight per S.F. Lightweight	Blocks per 100 S.F.	Mortar per M block, waste included Partitions	Mortar per M block, waste included Back up
2" x 8" x 16"	20 PSF	15 PSF	113	27 C.F.	36 C.F.
4"	30	20		41	51
6"	42	30		56	66
8"	55	38		72	82
10"	70	47		87	97
12"	85	55		102	112

Brick & Mortar Quantities
©Brick Industry Association. 2009 Feb. Technical Notes on Brick Construction 10:
Dimensioning and Estimating Brick Masonry. Reston (VA): BIA. Table 1 Modular Brick Sizes and Table 4 Quantity Estimates for Brick Masonry.

Masonry — R0422 Concrete Unit Masonry

R042210-20 Concrete Block

The material cost of special block such as corner, jamb and head block can be figured at the same price as ordinary block of equal size. Labor on specials is about the same as equal-sized regular block.

Bond beams and 16" high lintel blocks are more expensive than regular units of equal size. Lintel blocks are 8" long and either 8" or 16" high.

Use of a motorized mortar spreader box will speed construction of continuous walls.

Hollow non-load-bearing units are made according to ASTM C129 and hollow load-bearing units according to ASTM C90.

Metals — R0531 Steel Decking

R053100-10 Decking Descriptions

General - All Deck Products

A steel deck is made by cold forming structural grade sheet steel into a repeating pattern of parallel ribs. The strength and stiffness of the panels are the result of the ribs and the material properties of the steel. Deck lengths can be varied to suit job conditions, but because of shipping considerations, are usually less than 40 feet. Standard deck width varies with the product used but full sheets are usually 12", 18", 24", 30", or 36". The deck is typically furnished in a standard width with the ends cut square. Any cutting for width, such as at openings or for angular fit, is done at the job site.

The deck is typically attached to the building frame with arc puddle welds, self-drilling screws, or powder or pneumatically driven pins. Sheet to sheet fastening is done with screws, button punching (crimping), or welds.

Composite Floor Deck

After installation and adequate fastening, a floor deck serves several purposes. It (a) acts as a working platform, (b) stabilizes the frame, (c) serves as a concrete form for the slab, and (d) reinforces the slab to carry the design loads applied during the life of the building. Composite decks are distinguished by the presence of shear connector devices as part of the deck. These devices are designed to mechanically lock the concrete and deck together so that the concrete and the deck work together to carry subsequent floor loads. These shear connector devices can be rolled-in embossments, lugs, holes, or wires welded to the panels. The deck profile can also be used to interlock concrete and steel.

Composite deck finishes are either galvanized (zinc coated) or phosphatized/painted. Galvanized deck has a zinc coating on both the top and bottom surfaces. The phosphatized/painted deck has a bare (phosphatized) top surface that will come into contact with the concrete. This bare top surface can be expected to develop rust before the concrete is placed. The bottom side of the deck has a primer coat of paint.

A composite floor deck is normally installed so the panel ends do not overlap on the supporting beams. Shear lugs or panel profile shapes often prevent a tight metal to metal fit if the panel ends overlap; the air gap caused by overlapping will prevent proper fusion with the structural steel supports when the panel end laps are shear stud welded.

Adequate end bearing of the deck must be obtained as shown on the drawings. If bearing is actually less in the field than shown on the drawings, further investigation is required.

Roof Deck

A roof deck is not designed to act compositely with other materials. A roof deck acts alone in transferring horizontal and vertical loads into the building frame. Roof deck rib openings are usually narrower than floor deck rib openings. This provides adequate support of the rigid thermal insulation board.

A roof deck is typically installed to endlap approximately 2" over supports. However, it can be butted (or lapped more than 2") to solve field fit problems. Since designers frequently use the installed deck system as part of the horizontal bracing system (the deck as a diaphragm), any fastening substitution or change should be approved by the designer. Continuous perimeter support of the deck is necessary to limit edge deflection in the finished roof and may be required for diaphragm shear transfer.

Standard roof deck finishes are galvanized or primer painted. The standard factory applied paint for roof decks is a primer paint and is not intended to weather for extended periods of time. Field painting or touching up of abrasions and deterioration of the primer coat or other protective finishes is the responsibility of the contractor.

Cellular Deck

A cellular deck is made by attaching a bottom steel sheet to a roof deck or composite floor deck panel. A cellular deck can be used in the same manner as a floor deck. Electrical, telephone, and data wires are easily run through the chase created between the deck panel and the bottom sheet.

When used as part of the electrical distribution system, the cellular deck must be installed so that the ribs line up and create a smooth cell transition at abutting ends. The joint that occurs at butting cell ends must be taped or otherwise sealed to prevent wet concrete from seeping into the cell. Cell interiors must be free of welding burrs, or other sharp intrusions, to prevent damage to wires.

When used as a roof deck, the bottom flat plate is usually left exposed to view. Care must be maintained during erection to keep good alignment and prevent damage.

A cellular deck is sometimes used with the flat plate on the top side to provide a flat working surface. Installation of the deck for this purpose requires special methods for attachment to the frame because the flat plate, now on the top, can prevent direct access to the deck material that is bearing on the structural steel. It may be advisable to treat the flat top surface to prevent slipping.

A cellular deck is always furnished galvanized or painted over galvanized.

Form Deck

A form deck can be any floor or roof deck product used as a concrete form. Connections to the frame are by the same methods used to anchor floor and roof decks. Welding washers are recommended when welding a deck that is less than 20 gauge thickness.

A form deck is furnished galvanized, prime painted, or uncoated. A galvanized deck must be used for those roof deck systems where a form deck is used to carry a lightweight insulating concrete fill.

Wood, Plastics & Comp. — R0611 Wood Framing

R061110-30   Lumber Product Material Prices

The price of forest products fluctuates widely from location to location and from season to season depending upon economic conditions. The bare material prices in the unit cost sections of the data set show the National Average material prices in effect Jan. 1 of this data year. It must be noted that lumber prices in general may change significantly during the year.

Availability of certain items depends upon geographic location and must be checked prior to firm-price bidding.

Wood, Plastics & Comp. — R0616 Sheathing

R061636-20   Plywood

There are two types of plywood used in construction: interior, which is moisture-resistant but not waterproofed, and exterior, which is waterproofed.

The grade of the exterior surface of the plywood sheets is designated by the first letter: A, for smooth surface with patches allowed; B, for solid surface with patches and plugs allowed; C, which may be surface plugged or may have knot holes up to 1″ wide; and D, which is used only for interior type plywood and may have knot holes up to 2-1/2″ wide. "Structural Grade" is specifically designed for engineered applications such as box beams. All CC & DD grades have roof and floor spans marked on them.

Underlayment-grade plywood runs from 1/4″ to 1-1/4″ thick. Thicknesses 5/8″ and over have optional tongue and groove joints which eliminate the need for blocking the edges. Underlayment 19/32″ and over may be referred to as Sturd-i-Floor.

The price of plywood can fluctuate widely due to geographic and economic conditions.

Typical uses for various plywood grades are as follows:

AA-AD Interior — cupboards, shelving, paneling, furniture
BB Plyform — concrete form plywood
CDX — wall and roof sheathing
Structural — box beams, girders, stressed skin panels
AA-AC Exterior — fences, signs, siding, soffits, etc.
Underlayment — base for resilient floor coverings
Overlaid HDO — high density for concrete forms & highway signs
Overlaid MDO — medium density for painting, siding, soffits & signs
303 Siding — exterior siding, textured, striated, embossed, etc.

Thermal & Moist. Protec. — R0731 Shingles & Shakes

R073126-20   Roof Slate

16″, 18″ and 20″ are standard lengths, and slate usually comes in random widths. For standard 3/16″ thickness use 1-1/2″ copper nails. Allow for 3% breakage.

Thermal & Moist. Protec. — R0752 Modified Bituminous Membrane Roofing

R075213-30   Modified Bitumen Roofing

The cost of modified bitumen roofing is highly dependent on the type of installation that is planned. Installation is based on the type of modifier used in the bitumen. The two most popular modifiers are atactic polypropylene (APP) and styrene butadiene styrene (SBS). The modifiers are added to heated bitumen during the manufacturing process to change its characteristics. A polyethylene, polyester or fiberglass reinforcing sheet is then sandwiched between layers of this bitumen. When completed, the result is a pre-assembled, built-up roof that has increased elasticity and weatherability. Some manufacturers include a surfacing material such as ceramic or mineral granules, metal particles or sand.

The preferred method of adhering SBS-modified bitumen roofing to the substrate is with hot-mopped asphalt (much the same as built-up roofing). This installation method requires a tar kettle/pot to heat the asphalt, as well as the labor, tools and equipment necessary to distribute and spread the hot asphalt.

The alternative method for applying APP and SBS modified bitumen is as follows. A skilled installer uses a torch to melt a small pool of bitumen off the membrane. This pool must form across the entire roll for proper adhesion. The installer must unroll the roofing at a pace slow enough to melt the bitumen, but fast enough to prevent damage to the rest of the membrane.

Modified bitumen roofing provides the advantages of both built-up and single-ply roofing. Labor costs are reduced over those of built-up roofing because only a single ply is necessary. The elasticity of single-ply roofing is attained with the reinforcing sheet and polymer modifiers. Modifieds have some self-healing characteristics and because of their multi-layer construction, they offer the reliability and safety of built-up roofing.

Openings — R0813 Metal Doors

R081313-20 Steel Door Selection Guide

Standard steel doors are classified into four levels, as recommended by the Steel Door Institute in the chart below. Each of the four levels offers a range of construction models and designs to meet architectural requirements for preference and appearance, including full flush, seamless, and stile & rail. Recommended minimum gauge requirements are also included.

For complete standard steel door construction specifications and available sizes, refer to the Steel Door Institute Technical Data Series, ANSI A250.8-98 (SDI-100), and ANSI A250.4-94 Test Procedure and Acceptance Criteria for Physical Endurance of Steel Door and Hardware Reinforcements.

Level		Model	Construction	For Full Flush or Seamless		
				Min. Gauge	Thickness (in)	Thickness (mm)
I	Standard Duty	1	Full Flush	20	0.032	0.8
		2	Seamless			
II	Heavy Duty	1	Full Flush	18	0.042	1.0
		2	Seamless			
III	Extra Heavy Duty	1	Full Flush	16	0.053	1.3
		2	Seamless			
		3	*Stile & Rail			
IV	Maximum Duty	1	Full Flush	14	0.067	1.6
		2	Seamless			

*Stiles & rails are 16 gauge; flush panels, when specified, are 18 gauge.

Openings — R0852 Wood Windows

R085216-10 Window Estimates

To ensure a complete window estimate, be sure to include the material and labor costs for each window, as well as the material and labor costs for an interior wood trim set.

Openings — R0853 Plastic Windows

R085313-20 Replacement Windows

Replacement windows are typically measured per United Inch.

United Inches are calculated by rounding the width and height of the window opening up to the nearest inch, then adding the two figures.

The labor cost for replacement windows includes removal of sash, existing sash balance or weights, parting bead where necessary and installation of new window.

Debris hauling and dump fees are not included.

Openings

R0871 Door Hardware

R087110-10 Hardware Finishes

This table describes hardware finishes used throughout the industry. It also shows the base metal and the respective symbols in the three predominate systems of identification. Many of these are used in pricing descriptions in Division Eight.

US″	BMHA*	CDN^	Base	Description
US P	600	CP	Steel	Primed for Painting
US 1B	601	C1B	Steel	Bright Black Japanned
US 2C	602	C2C	Steel	Zinc Plated
US 2G	603	C2G	Steel	Zinc Plated
US 3	605	C3	Brass	Bright Brass, Clear Coated
US 4	606	C4	Brass	Satin Brass, Clear Coated
US 5	609	C5	Brass	Satin Brass, Blackened, Satin Relieved, Clear Coated
US 7	610	C7	Brass	Satin Brass, Blackened, Bright Relieved, Clear Coated
US 9	611	C9	Bronze	Bright Bronze, Clear Coated
US 10	612	C10	Bronze	Satin Bronze, Clear Coated
US 10A	641	C10A	Steel	Antiqued Bronze, Oiled and Lacquered
US 10B	613	C10B	Bronze	Antiqued Bronze, Oiled
US 11	616	C11	Bronze	Satin Bronze, Blackened, Satin Relieved, Clear Coated
US 14	618	C14	Brass/Bronze	Bright Nickel Plated, Clear Coated
US 15	619	C15	Brass/Bronze	Satin Nickel, Clear Coated
US 15A	620	C15A	Brass/Bronze	Satin Nickel Plated, Blackened, Satin Relieved, Clear Coated
US 17A	621	C17A	Brass/Bronze	Nickel Plated, Blackened, Relieved, Clear Coated
US 19	622	C19	Brass/Bronze	Flat Black Coated
US 20	623	C20	Brass/Bronze	Statuary Bronze, Light
US 20A	624	C20A	Brass/Bronze	Statuary Bronze, Dark
US 26	625	C26	Brass/Bronze	Bright Chromium
US 26D	626	C26D	Brass/Bronze	Satin Chromium
US 20	627	C27	Aluminum	Satin Aluminum Clear
US 28	628	C28	Aluminum	Anodized Dull Aluminum
US 32	629	C32	Stainless Steel	Bright Stainless Steel
US 32D	630	C32D	Stainless Steel	Stainless Steel
US 3	632	C3	Steel	Bright Brass Plated, Clear Coated
US 4	633	C4	Steel	Satin Brass, Clear Coated
US 7	636	C7	Steel	Satin Brass Plated, Blackened, Bright Relieved, Clear Coated
US 9	637	C9	Steel	Bright Bronze Plated, Clear Coated
US 5	638	C5	Steel	Satin Brass Plated, Blackened, Bright Relieved, Clear Coated
US 10	639	C10	Steel	Satin Bronze Plated, Clear Coated
US 10B	640	C10B	Steel	Antique Bronze, Oiled
US 10A	641	C10A	Steel	Antiqued Bronze, Oiled and Lacquered
US 11	643	C11	Steel	Satin Bronze Plated, Blackened, Bright Relieved, Clear Coated
US 14	645	C14	Steel	Bright Nickel Plated, Clear Coated
US 15	646	C15	Steel	Satin Nickel Plated, Clear Coated
US 15A	647	C15A	Steel	Nickel Plated, Blackened, Bright Relieved, Clear Coated
US 17A	648	C17A	Steel	Nickel Plated, Blackened, Relieved, Clear Coated
US 20	649	C20	Steel	Statuary Bronze, Light
US 20A	650	C20A	Steel	Statuary Bronze, Dark
US 26	651	C26	Steel	Bright Chromium Plated
US 26D	652	C26D	Steel	Satin Chromium Plated

* - BMHA Builders Hardware Manufacturing Association
″ - US Equivalent
^ - Canadian Equivalent
Japanning is imitating Asian lacquer work

Finishes

R0920 Plaster & Gypsum Board

R092000-50 Lath, Plaster and Gypsum Board

Gypsum board lath is available in 3/8" thick x 16" wide x 4' long sheets as a base material for multi-layer plaster applications. It is also available as a base for either multi-layer or veneer plaster applications in 1/2" and 5/8" thick—4' wide x 8', 10' or 12' long sheets. Fasteners are screws or blued ring shank nails for wood framing and screws for metal framing.

Metal lath is available in diamond mesh patterns with flat or self-furring profiles. Paper backing is available for applications where excessive plaster waste needs to be avoided. A slotted mesh ribbed lath should be used in areas where the span between structural supports is greater than normal. Most metal lath comes in 27" x 96" sheets. Diamond mesh weighs 1.75, 2.5 or 3.4 pounds per square yard, slotted mesh lath weighs 2.75 or 3.4 pounds per square yard. Metal lath can be nailed, screwed or tied in place.

Many **accessories** are available. Corner beads, flat reinforcing strips, casing beads, control and expansion joints, furring brackets and channels are some examples. Note that accessories are not included in plaster or stucco line items.

Plaster is defined as a material or combination of materials that when mixed with a suitable amount of water, forms a plastic mass or paste. When applied to a surface, the paste adheres to it and subsequently hardens, preserving in a rigid state the form or texture imposed during the period of elasticity.

Gypsum plaster is made from ground calcined gypsum. It is mixed with aggregates and water for use as a base coat plaster.

Vermiculite plaster is a fire-retardant plaster covering used on steel beams, concrete slabs and other heavy construction materials. Vermiculite is a group name for certain clay minerals, hydrous silicates or aluminum, magnesium and iron that have been expanded by heat.

Perlite plaster is a plaster using perlite as an aggregate instead of sand. Perlite is a volcanic glass that has been expanded by heat.

Gauging plaster is a mix of gypsum plaster and lime putty that when applied produces a quick drying finish coat.

Veneer plaster is a one or two component gypsum plaster used as a thin finish coat over special gypsum board.

Keenes cement is a white cementitious material manufactured from gypsum that has been burned at a high temperature and ground to a fine powder. Alum is added to accelerate the set. The resulting plaster is hard and strong and accepts and maintains a high polish, hence it is used as a finishing plaster.

Stucco is a Portland cement based plaster used primarily as an exterior finish.

Plaster is used on both interior and exterior surfaces. Generally it is applied in multiple-coat systems. A three-coat system uses the terms scratch, brown and finish to identify each coat. A two-coat system uses base and finish to describe each coat. Each type of plaster and application system has attributes that are chosen by the designer to best fit the intended use.

Gypsum Plaster Quantities for 100 S.Y.	2 Coat, 5/8" Thick		3 Coat, 3/4" Thick		
	Base	Finish	Scratch	Brown	Finish
	1:3 Mix	2:1 Mix	1:2 Mix	1:3 Mix	2:1 Mix
Gypsum plaster	1,300 lb.		1,350 lb.	650 lb.	
Sand	1.75 C.Y.		1.85 C.Y.	1.35 C.Y.	
Finish hydrated lime		340 lb.			340 lb.
Gauging plaster		170 lb.			170 lb.

Vermiculite or Perlite Plaster Quantities for 100 S.Y.	2 Coat, 5/8" Thick		3 Coat, 3/4" Thick		
	Base	Finish	Scratch	Brown	Finish
Gypsum plaster	1,250 lb.		1,450 lb.	800 lb.	
Vermiculite or perlite	7.8 bags		8.0 bags	3.3 bags	
Finish hydrated lime		340 lb.			340 lb.
Gauging plaster		170 lb.			170 lb.

Stucco–Three-Coat System Quantities for 100 S.Y.	On Wood Frame	On Masonry
Portland cement	29 bags	21 bags
Sand	2.6 C.Y.	2.0 C.Y.
Hydrated lime	180 lb.	120 lb.

Finishes

R0929 Gypsum Board

R092910-10 Levels of Gypsum Board Finish

In the past, contract documents often used phrases such as "industry standard" and "workmanlike finish" to specify the expected quality of gypsum board wall and ceiling installations. The vagueness of these descriptions led to unacceptable work and disputes.

In order to resolve this problem, four major trade associations concerned with the manufacture, erection, finish and decoration of gypsum board wall and ceiling systems developed an industry-wide *Recommended Levels of Gypsum Board Finish*.

The finish of gypsum board walls and ceilings for specific final decoration is dependent on a number of factors. A primary consideration is the location of the surface and the degree of decorative treatment desired. Painted and unpainted surfaces in warehouses and other areas where appearance is normally not critical may simply require the taping of wallboard joints and 'spotting' of fastener heads. Blemish-free, smooth, monolithic surfaces often intended for painted and decorated walls and ceilings in habitated structures, ranging from single-family dwellings through monumental buildings, require additional finishing prior to the application of the final decoration.

Other factors to be considered in determining the level of finish of the gypsum board surface are (1) the type of angle of surface illumination (both natural and artificial lighting), and (2) the paint and method of application or the type and finish of wallcovering specified as the final decoration. Critical lighting conditions, gloss paints, and thin wall coverings require a higher level of gypsum board finish than heavily textured surfaces which are subsequently painted or surfaces which are to be decorated with heavy grade wall coverings.

The following descriptions were developed by the Association of the Wall and Ceiling Industries-International (AWCI), Ceiling & Interior Systems Construction Association (CISCA), Gypsum Association (GA), and Painting and Decorating Contractors of America (PDCA) as a guide.

Level 0: Used in temporary construction or wherever the final decoration has not been determined. Unfinished. No taping, finishing or corner beads are required. Also could be used where non-predecorated panels will be used in demountable-type partitions that are to be painted as a final finish.

Level 1: Frequently used in plenum areas above ceilings, in attics, in areas where the assembly would generally be concealed, or in building service corridors and other areas not normally open to public view. Some degree of sound and smoke control is provided; in some geographic areas, this level is referred to as "fire-taping," although this level of finish does not typically meet fire-resistant assembly requirements. Where a fire resistance rating is required for the gypsum board assembly, details of construction should be in accordance with reports of fire tests of assemblies that have met the requirements of the fire rating acceptable.

All joints and interior angles shall have tape embedded in joint compound. Accessories are optional at specifier discretion in corridors and other areas with pedestrian traffic. Tape and fastener heads need not be covered with joint compound. Surface shall be free of excess joint compound. Tool marks and ridges are acceptable.

Level 2: It may be specified for standard gypsum board surfaces in garages, warehouse storage, or other similar areas where surface appearance is not of primary importance.

All joints and interior angles shall have tape embedded in joint compound and shall be immediately wiped with a joint knife or trowel, leaving a thin coating of joint compound over all joints and interior angles. Fastener heads and accessories shall be covered with a coat of joint compound. Surface shall be free of excess joint compound. Tool marks and ridges are acceptable.

Level 3: Typically used in areas receiving heavy texture (spray or hand applied) finishes before final painting, or where commercial-grade (heavy duty) wall coverings are to be applied as the final decoration. This level of finish should not be used where smooth painted surfaces or where lighter weight wall coverings are specified. The prepared surface shall be coated with a drywall primer prior to the application of final finishes.

All joints and interior angles shall have tape embedded in joint compound and shall be immediately wiped with a joint knife or trowel, leaving a thin coating of joint compound over all joints and interior angles. One additional coat of joint compound shall be applied over all joints and interior angles. Fastener heads and accessories shall be covered with two separate coats of joint compound. All joint compounds shall be smooth and free of tool marks and ridges. The prepared surface shall be covered with a drywall primer prior to the application of the final decoration.

Level 4: This level should be used where residential grade (light duty) wall coverings, flat paints, or light textures are to be applied. The prepared surface shall be coated with a drywall primer prior to the application of final finishes. Release agents for wall coverings are specifically formulated to minimize damage if coverings are subsequently removed.

The weight, texture, and sheen level of the wall covering material selected should be taken into consideration when specifying wall coverings over this level of drywall treatment. Joints and fasteners must be sufficiently concealed if the wall covering material is lightweight, contains limited pattern, has a glossy finish, or has any combination of these features. In critical lighting areas, flat paints applied over light textures tend to reduce joint photographing. Gloss, semi-gloss, and enamel paints are not recommended over this level of finish.

All joints and interior angles shall have tape embedded in joint compound and shall be immediately wiped with a joint knife or trowel, leaving a thin coating of joint compound over all joints and interior angles. In addition, two separate coats of joint compound shall be applied over all flat joints and one separate coat of joint compound applied over interior angles. Fastener heads and accessories shall be covered with three separate coats of joint compound. All joint compounds shall be smooth and free of tool marks and ridges. The prepared surface shall be covered with a drywall primer like Sheetrock first coat prior to the application of the final decoration.

Level 5: The highest quality finish is the most effective method to provide a uniform surface and minimize the possibility of joint photographing and of fasteners showing through the final decoration. This level of finish is required where gloss, semi-gloss, or enamel is specified; when flat joints are specified over an untextured surface; or where critical lighting conditions occur. The prepared surface shall be coated with a drywall primer prior to the application of the final decoration.

All joints and interior angles shall have tape embedded in joint compound and be immediately wiped with a joint knife or trowel, leaving a thin coating of joint compound over all joints and interior angles. Two separate coats of joint compound shall be applied over all flat joints and one separate coat of joint compound applied over interior angles. Fastener heads and accessories shall be covered with three separate coats of joint compound.

A thin skim coat of joint compound shall be trowel applied to the entire surface. Excess compound is immediately troweled off, leaving a film or skim coating of compound completely covering the paper. As an alternative to a skim coat, a material manufactured especially for this purpose may be applied such as Sheetrock Tuff-Hide primer surfacer. The surface must be smooth and free of tool marks and ridges. The prepared surface shall be covered with a drywall primer prior to the application of the final decoration.

Finishes · R0972 Wall Coverings

R097223-10   Wall Covering

The table below lists the quantities required for 100 S.F. of wall covering.

Description	Medium-Priced Paper	Expensive Paper
Paper	1.6 dbl. rolls	1.6 dbl. rolls
Wall sizing	0.25 gallon	0.25 gallon
Vinyl wall paste	0.6 gallon	0.6 gallon
Apply sizing	0.3 hour	0.3 hour
Apply paper	1.2 hours	1.5 hours

Most wallpapers now come in double rolls only.
To remove old paper, allow 1.3 hours per 100 S.F.

Finishes · R0991 Painting

R099100-10   Painting Estimating Techniques

Proper estimating methodology is needed to obtain an accurate painting estimate. There is no known reliable shortcut or square foot method. The following steps should be followed:

- List all surfaces to be painted, with an accurate quantity (area) of each. Items having similar surface condition, finish, application method and accessibility may be grouped together.
- List all the tasks required for each surface to be painted, including surface preparation, masking, and protection of adjacent surfaces. Surface preparation may include minor repairs, washing, sanding and puttying.
- Select the proper Means line for each task. Review and consider all adjustments to labor and materials for type of paint and location of work. Apply the height adjustment carefully. For instance, when applying the adjustment for work over 8' high to a wall that is 12' high, apply the adjustment only to the area between 8' and 12' high, and not to the entire wall.

When applying more than one percent (%) adjustment, apply each to the base cost of the data, rather than applying one percentage adjustment on top of the other.

When estimating the cost of painting walls and ceilings remember to add the brushwork for all cut-ins at inside corners and around windows and doors as a LF measure. One linear foot of cut-in with a brush equals one square foot of painting.

All items for spray painting include the labor for roll-back.

Deduct for openings greater than 100 SF or openings that extend from floor to ceiling and are greater than 5' wide. Do not deduct small openings.

The cost of brushes, rollers, ladders and spray equipment are considered part of a painting contractor's overhead, and should not be added to the estimate. The cost of rented equipment such as scaffolding and swing staging should be added to the estimate.

R099100-20   Painting

Item	Coat	One Gallon Covers			In 8 Hours a Laborer Covers			Labor-Hours per 100 S.F.		
		Brush	Roller	Spray	Brush	Roller	Spray	Brush	Roller	Spray
Paint wood siding	prime	250 S.F.	225 S.F.	290 S.F.	1150 S.F.	1300 S.F.	2275 S.F.	.695	.615	.351
	others	270	250	290	1300	1625	2600	.615	.492	.307
Paint exterior trim	prime	400	—	—	650	—	—	1.230	—	—
	1st	475	—	—	800	—	—	1.000	—	—
	2nd	520	—	—	975	—	—	.820	—	—
Paint shingle siding	prime	270	255	300	650	975	1950	1.230	.820	.410
	others	360	340	380	800	1150	2275	1.000	.695	.351
Stain shingle siding	1st	180	170	200	750	1125	2250	1.068	.711	.355
	2nd	270	250	290	900	1325	2600	.888	.603	.307
Paint brick masonry	prime	180	135	160	750	800	1800	1.066	1.000	.444
	1st	270	225	290	815	975	2275	.981	.820	.351
	2nd	340	305	360	815	1150	2925	.981	.695	.273
Paint interior plaster or drywall	prime	400	380	495	1150	2000	3250	.695	.400	.246
	others	450	425	495	1300	2300	4000	.615	.347	.200
Paint interior doors and windows	prime	400	—	—	650	—	—	1.230	—	—
	1st	425	—	—	800	—	—	1.000	—	—
	2nd	450	—	—	975	—	—	.820	—	—

Special Construction — R1311 Swimming Pools

R131113-20 Swimming Pools

Pool prices given per square foot of surface area include pool structure, filter and chlorination equipment, pumps, related piping, ladders/steps, maintenance kit, skimmer and vacuum system. Decks and electrical service to equipment are not included.

Residential in-ground pool construction can be divided into two categories: vinyl lined and gunite. Vinyl lined pool walls are constructed of different materials including wood, concrete, plastic or metal. The bottom is often graded with sand over which the vinyl liner is installed. Vermiculite or soil cement bottoms may be substituted for an added cost.

Gunite pool construction is used both in residential and municipal installations. These structures are steel reinforced for strength and finished with a white cement limestone plaster.

Municipal pools will have a higher cost because plumbing codes require more expensive materials, chlorination equipment and higher filtration rates.

Municipal pools greater than 1,800 S.F. require gutter systems to control waves. This gutter may be formed into the concrete wall. Often a vinyl/stainless steel gutter or gutter/wall system is specified, which will raise the pool cost.

Competition pools usually require tile bottoms and sides with contrasting lane striping, which will also raise the pool cost.

Plumbing — R2211 Facility Water Distribution

R221113-50 Pipe Material Considerations

1. Malleable fittings should be used for gas service.
2. Malleable fittings are used where there are stresses/strains due to expansion and vibration.
3. Cast fittings may be broken as an aid to disassembling heating lines frozen by long use, temperature and minerals.
4. A cast iron pipe is extensively used for underground and submerged service.
5. Type M (light wall) copper tubing is available in hard temper only and is used for nonpressure and less severe applications than K and L.
6. Type L (medium wall) copper tubing, available hard or soft for interior service.
7. Type K (heavy wall) copper tubing, available in hard or soft temper for use where conditions are severe. For underground and interior service.
8. Hard drawn tubing requires fewer hangers or supports but should not be bent. Silver brazed fittings are recommended, but soft solder is normally used.
9. Type DMV (very light wall) copper tubing designed for drainage, waste and vent plus other non-critical pressure services.

Domestic/Imported Pipe and Fittings Costs

The prices shown in this publication for steel/cast iron pipe and steel, cast iron, and malleable iron fittings are based on domestic production sold at the normal trade discounts. The above listed items of foreign manufacture may be available at prices 1/3 to 1/2 of those shown. Some imported items after minor machining or finishing operations are being sold as domestic to further complicate the system.

Caution: Most pipe prices in this data set also include a coupling and pipe hangers which for the larger sizes can add significantly to the per foot cost and should be taken into account when comparing "book cost" with the quoted supplier's cost.

Exterior Improvements — R3292 Turf & Grasses

R329219-50 Seeding

The type of grass is determined by light, shade and moisture content of soil plus intended use. Fertilizer should be disked 4" before seeding. For steep slopes disk five tons of mulch and lay two tons of hay or straw on surface per acre after seeding. Surface mulch can be staked, lightly disked or tar emulsion sprayed. Material for mulch can be wood chips, peat moss, partially rotted hay or straw, wood fibers and sprayed emulsions. Hemp seed blankets with fertilizer are also available. For spring seeding, watering is necessary. Late fall seeding may have to be reseeded in the spring. Hydraulic seeding, power mulching, and aerial seeding can be used on large areas.

Utilities — R3311 Water Utility Distribution Piping

R331113-80 Piping Designations

There are several systems currently in use to describe pipe and fittings. The following paragraphs will help to identify and clarify classifications of piping systems used for water distribution.

Piping may be classified by schedule. Piping schedules include 5S, 10S, 10, 20, 30, Standard, 40, 60, Extra Strong, 80, 100, 120, 140, 160 and Double Extra Strong. These schedules are dependent upon the pipe wall thickness. The wall thickness of a particular schedule may vary with pipe size.

Ductile iron pipe for water distribution is classified by Pressure Classes such as Class 150, 200, 250, 300 and 350. These classes are actually the rated water working pressure of the pipe in pounds per square inch (psi). The pipe in these pressure classes is designed to withstand the rated water working pressure plus a surge allowance of 100 psi.

The American Water Works Association (AWWA) provides standards for various types of **plastic pipe.** C-900 is the specification for polyvinyl chloride (PVC) piping used for water distribution in sizes ranging from 4" through 12". C-901 is the specification for polyethylene (PE) pressure pipe, tubing and fittings used for water distribution in sizes ranging from 1/2" through 3". C-905 is the specification for PVC piping sizes 14" and greater.

PVC pressure-rated pipe is identified using the standard dimensional ratio (SDR) method. This method is defined by the American Society for Testing and Materials (ASTM) Standard D 2241. This pipe is available in SDR numbers 64, 41, 32.5, 26, 21, 17, and 13.5. A pipe with an SDR of 64 will have the thinnest wall while a pipe with an SDR of 13.5 will have the thickest wall. When the pressure rating (PR) of a pipe is given in psi, it is based on a line supplying water at 73 degrees F.

The National Sanitation Foundation (NSF) seal of approval is applied to products that can be used with potable water. These products have been tested to ANSI/NSF Standard 14.

Valves and strainers are classified by American National Standards Institute (ANSI) Classes. These Classes are 125, 150, 200, 250, 300, 400, 600, 900, 1500 and 2500. Within each class there is an operating pressure range dependent upon temperature. Design parameters should be compared to the appropriate material dependent, pressure-temperature rating chart for accurate valve selection.

Abbreviations

A	Area Square Feet; Ampere	Brk., brk	Brick	Csc	Cosecant
AAFES	Army and Air Force Exchange Service	brkt	Bracket	C.S.F.	Hundred Square Feet
		Brs.	Brass	CSI	Construction Specifications Institute
ABS	Acrylonitrile Butadiene Stryrene; Asbestos Bonded Steel	Brz.	Bronze		
		Bsn.	Basin	CT	Current Transformer
A.C., AC	Alternating Current; Air-Conditioning; Asbestos Cement; Plywood Grade A & C	Btr.	Better	CTS	Copper Tube Size
		BTU	British Thermal Unit	Cu	Copper, Cubic
		BTUH	BTU per Hour	Cu. Ft.	Cubic Foot
		Bu.	Bushels	cw	Continuous Wave
ACI	American Concrete Institute	BUR	Built-up Roofing	C.W.	Cool White; Cold Water
ACR	Air Conditioning Refrigeration	BX	Interlocked Armored Cable	Cwt.	100 Pounds
ADA	Americans with Disabilities Act	°C	Degree Centigrade	C.W.X.	Cool White Deluxe
AD	Plywood, Grade A & D	c	Conductivity, Copper Sweat	C.Y.	Cubic Yard (27 cubic feet)
Addit.	Additional	C	Hundred; Centigrade	C.Y./Hr.	Cubic Yard per Hour
Adj.	Adjustable	C/C	Center to Center, Cedar on Cedar	Cyl.	Cylinder
af	Audio-frequency	C-C	Center to Center	d	Penny (nail size)
AFUE	Annual Fuel Utilization Efficiency	Cab	Cabinet	D	Deep; Depth; Discharge
AGA	American Gas Association	Cair.	Air Tool Laborer	Dis., Disch.	Discharge
Agg.	Aggregate	Cal.	Caliper	Db	Decibel
A.H., Ah	Ampere Hours	Calc	Calculated	Dbl.	Double
A hr.	Ampere-hour	Cap.	Capacity	DC	Direct Current
A.H.U., AHU	Air Handling Unit	Carp.	Carpenter	DDC	Direct Digital Control
A.I.A.	American Institute of Architects	C.B.	Circuit Breaker	Demob.	Demobilization
AIC	Ampere Interrupting Capacity	C.C.A.	Chromate Copper Arsenate	d.f.t.	Dry Film Thickness
Allow.	Allowance	C.C.F.	Hundred Cubic Feet	d.f.u.	Drainage Fixture Units
alt., alt	Alternate	cd	Candela	D.H.	Double Hung
Alum.	Aluminum	cd/sf	Candela per Square Foot	DHW	Domestic Hot Water
a.m.	Ante Meridiem	CD	Grade of Plywood Face & Back	DI	Ductile Iron
Amp.	Ampere	CDX	Plywood, Grade C & D, exterior glue	Diag.	Diagonal
Anod.	Anodized			Diam., Dia	Diameter
ANSI	American National Standards Institute	Cefi.	Cement Finisher	Distrib.	Distribution
		Cem.	Cement	Div.	Division
APA	American Plywood Association	CF	Hundred Feet	Dk.	Deck
Approx.	Approximate	C.F.	Cubic Feet	D.L.	Dead Load; Diesel
Apt.	Apartment	CFM	Cubic Feet per Minute	DLH	Deep Long Span Bar Joist
Asb.	Asbestos	CFRP	Carbon Fiber Reinforced Plastic	dlx	Deluxe
A.S.B.C.	American Standard Building Code	c.g.	Center of Gravity	Do.	Ditto
Asbe.	Asbestos Worker	CHW	Chilled Water; Commercial Hot Water	DOP	Dioctyl Phthalate Penetration Test (Air Filters)
ASCE	American Society of Civil Engineers				
A.S.H.R.A.E.	American Society of Heating, Refrig. & AC Engineers	C.I., CI	Cast Iron	Dp., dp	Depth
		C.I.P., CIP	Cast in Place	D.P.S.T.	Double Pole, Single Throw
ASME	American Society of Mechanical Engineers	Circ.	Circuit	Dr.	Drive
		C.L.	Carload Lot	DR	Dimension Ratio
ASTM	American Society for Testing and Materials	CL	Chain Link	Drink.	Drinking
		Clab.	Common Laborer	D.S.	Double Strength
Attchmt.	Attachment	Clam	Common Maintenance Laborer	D.S.A.	Double Strength A Grade
Avg., Ave.	Average	C.L.F.	Hundred Linear Feet	D.S.B.	Double Strength B Grade
AWG	American Wire Gauge	CLF	Current Limiting Fuse	Dty.	Duty
AWWA	American Water Works Assoc.	CLP	Cross Linked Polyethylene	DWV	Drain Waste Vent
Bbl.	Barrel	cm	Centimeter	DX	Deluxe White, Direct Expansion
B&B, BB	Grade B and Better; Balled & Burlapped	CMP	Corr. Metal Pipe	dyn	Dyne
		CMU	Concrete Masonry Unit	e	Eccentricity
B&S	Bell and Spigot	CN	Change Notice	E	Equipment Only; East; Emissivity
B.&W.	Black and White	Col.	Column	Ea.	Each
b.c.c.	Body-centered Cubic	CO_2	Carbon Dioxide	EB	Encased Burial
B.C.Y.	Bank Cubic Yards	Comb.	Combination	Econ.	Economy
BE	Bevel End	comm.	Commercial, Communication	E.C.Y	Embankment Cubic Yards
B.F.	Board Feet	Compr.	Compressor	EDP	Electronic Data Processing
Bg. cem.	Bag of Cement	Conc.	Concrete	EIFS	Exterior Insulation Finish System
BHP	Boiler Horsepower; Brake Horsepower	Cont., cont	Continuous; Continued, Container	E.D.R.	Equiv. Direct Radiation
		Corr.	Corrugated	Eq.	Equation
B.I.	Black Iron	Cos	Cosine	EL	Elevation
bidir.	bidirectional	Cot	Cotangent	Elec.	Electrician; Electrical
Bit., Bitum.	Bituminous	Cov.	Cover	Elev.	Elevator; Elevating
Bit., Conc.	Bituminous Concrete	C/P	Control Point Adjustment	EMT	Electrical Metallic Conduit; Thin Wall Conduit
Bk.	Backed	CPA			
Bkrs.	Breakers	Cplg.	Coupling	Eng.	Engine, Engineered
Bldg., bldg	Building	CPM	Critical Path Method	EPDM	Ethylene Propylene Diene Monomer
Blk.	Block	CPVC	Chlorinated Polyvinyl Chloride		
Bm.	Beam	C.Pr.	Hundred Pair	EPS	Expanded Polystyrene
Boil.	Boilermaker	CRC	Cold Rolled Channel	Eqhv.	Equip. Oper., Heavy
bpm	Blows per Minute	Creos.	Creosote	Eqlt.	Equip. Oper., Light
BR	Bedroom	Crpt.	Carpet & Linoleum Layer	Eqmd.	Equip. Oper., Medium
Brg., brng.	Bearing	CRT	Cathode-ray Tube	Eqmm.	Equip. Oper., Master Mechanic
Brhe.	Bricklayer Helper	CS	Carbon Steel, Constant Shear Bar Joist	Eqol.	Equip. Oper., Oilers
Bric.	Bricklayer			Equip.	Equipment
				ERW	Electric Resistance Welded

Abbreviations

E.S.	Energy Saver	H	High Henry	Lath.	Lather
Est.	Estimated	HC	High Capacity	Lav.	Lavatory
esu	Electrostatic Units	H.D., HD	Heavy Duty; High Density	lb.; #	Pound
E.W.	Each Way	H.D.O.	High Density Overlaid	L.B., LB	Load Bearing; L Conduit Body
EWT	Entering Water Temperature	HDPE	High Density Polyethylene Plastic	L. & E.	Labor & Equipment
Excav.	Excavation	Hdr.	Header	lb./hr.	Pounds per Hour
excl	Excluding	Hdwe.	Hardware	lb./L.F.	Pounds per Linear Foot
Exp., exp	Expansion, Exposure	H.I.D., HID	High Intensity Discharge	lbf/sq.in.	Pound-force per Square Inch
Ext., ext	Exterior; Extension	Help.	Helper Average	L.C.L.	Less than Carload Lot
Extru.	Extrusion	HEPA	High Efficiency Particulate Air Filter	L.C.Y.	Loose Cubic Yard
f.	Fiber Stress			Ld.	Load
F	Fahrenheit; Female; Fill	Hg	Mercury	LE	Lead Equivalent
Fab, fab	Fabricated; Fabric	HIC	High Interrupting Capacity	LED	Light Emitting Diode
FBGS	Fiberglass	HM	Hollow Metal	L.F.	Linear Foot
F.C.	Footcandles	HMWPE	High Molecular Weight Polyethylene	L.F. Nose	Linear Foot of Stair Nosing
f.c.c.	Face-centered Cubic			L.F. Rsr	Linear Foot of Stair Riser
f'c.	Compressive Stress in Concrete; Extreme Compressive Stress	HO	High Output	Lg.	Long; Length; Large
		Horiz.	Horizontal	L & H	Light and Heat
F.E.	Front End	H.P., HP	Horsepower; High Pressure	LH	Long Span Bar Joist
FEP	Fluorinated Ethylene Propylene (Teflon)	H.P.F.	High Power Factor	L.H.	Labor Hours
		Hr.	Hour	L.L., LL	Live Load
F.G.	Flat Grain	Hrs./Day	Hours per Day	L.L.D.	Lamp Lumen Depreciation
F.H.A.	Federal Housing Administration	HSC	High Short Circuit	lm	Lumen
Fig.	Figure	Ht.	Height	lm/sf	Lumen per Square Foot
Fin.	Finished	Htg.	Heating	lm/W	Lumen per Watt
FIPS	Female Iron Pipe Size	Htrs.	Heaters	LOA	Length Over All
Fixt.	Fixture	HVAC	Heating, Ventilation & Air-Conditioning	log	Logarithm
FJP	Finger jointed and primed			L-O-L	Lateralolet
Fl. Oz.	Fluid Ounces	Hvy.	Heavy	long.	Longitude
Flr.	Floor	HW	Hot Water	L.P., LP	Liquefied Petroleum; Low Pressure
FM	Frequency Modulation; Factory Mutual	Hyd.; Hydr.	Hydraulic	L.P.F.	Low Power Factor
		Hz	Hertz (cycles)	LR	Long Radius
Fmg.	Framing	I.	Moment of Inertia	L.S.	Lump Sum
FM/UL	Factory Mutual/Underwriters Labs	IBC	International Building Code	Lt.	Light
Fdn.	Foundation	I.C.	Interrupting Capacity	Lt. Ga.	Light Gauge
FNPT	Female National Pipe Thread	ID	Inside Diameter	L.T.L.	Less than Truckload Lot
Fori.	Foreman, Inside	I.D.	Inside Dimension; Identification	Lt. Wt.	Lightweight
Foro.	Foreman, Outside	I.F.	Inside Frosted	L.V.	Low Voltage
Fount.	Fountain	I.M.C.	Intermediate Metal Conduit	M	Thousand; Material; Male; Light Wall Copper Tubing
fpm	Feet per Minute	In.	Inch		
FPT	Female Pipe Thread	Incan.	Incandescent	M²CA	Meters Squared Contact Area
Fr	Frame	Incl.	Included; Including	m/hr.; M.H.	Man-hour
F.R.	Fire Rating	Int.	Interior	mA	Milliampere
FRK	Foil Reinforced Kraft	Inst.	Installation	Mach.	Machine
FSK	Foil/Scrim/Kraft	Insul., insul	Insulation/Insulated	Mag. Str.	Magnetic Starter
FRP	Fiberglass Reinforced Plastic	I.P.	Iron Pipe	Maint.	Maintenance
FS	Forged Steel	I.P.S., IPS	Iron Pipe Size	Marb.	Marble Setter
FSC	Cast Body; Cast Switch Box	IPT	Iron Pipe Threaded	Mat; Mat'l.	Material
Ft., ft	Foot; Feet	I.W.	Indirect Waste	Max.	Maximum
Ftng.	Fitting	J	Joule	MBF	Thousand Board Feet
Ftg.	Footing	J.I.C.	Joint Industrial Council	MBH	Thousand BTU's per hr.
Ft lb.	Foot Pound	K	Thousand; Thousand Pounds; Heavy Wall Copper Tubing, Kelvin	MC	Metal Clad Cable
Furn.	Furniture			MCC	Motor Control Center
FVNR	Full Voltage Non-Reversing	K.A.H.	Thousand Amp. Hours	M.C.F.	Thousand Cubic Feet
FVR	Full Voltage Reversing	kcmil	Thousand Circular Mils	MCFM	Thousand Cubic Feet per Minute
FXM	Female by Male	KD	Knock Down	M.C.M.	Thousand Circular Mils
Fy.	Minimum Yield Stress of Steel	K.D.A.T.	Kiln Dried After Treatment	MCP	Motor Circuit Protector
g	Gram	kg	Kilogram	MD	Medium Duty
G	Gauss	kG	Kilogauss	MDF	Medium-density fibreboard
Ga.	Gauge	kgf	Kilogram Force	M.D.O.	Medium Density Overlaid
Gal., gal.	Gallon	kHz	Kilohertz	Med.	Medium
Galv., galv	Galvanized	Kip	1000 Pounds	MF	Thousand Feet
GC/MS	Gas Chromatograph/Mass Spectrometer	KJ	Kilojoule	M.F.B.M.	Thousand Feet Board Measure
		K.L.	Effective Length Factor	Mfg.	Manufacturing
Gen.	General	K.L.F.	Kips per Linear Foot	Mfrs.	Manufacturers
GFI	Ground Fault Interrupter	Km	Kilometer	mg	Milligram
GFRC	Glass Fiber Reinforced Concrete	KO	Knock Out	MGD	Million Gallons per Day
Glaz.	Glazier	K.S.F.	Kips per Square Foot	MGPH	Million Gallons per Hour
GPD	Gallons per Day	K.S.I.	Kips per Square Inch	MH, M.H.	Manhole; Metal Halide; Man-Hour
gpf	Gallon per Flush	kV	Kilovolt	MHz	Megahertz
GPH	Gallons per Hour	kVA	Kilovolt Ampere	Mi.	Mile
gpm, GPM	Gallons per Minute	kVAR	Kilovar (Reactance)	MI	Malleable Iron; Mineral Insulated
GR	Grade	KW	Kilowatt	MIPS	Male Iron Pipe Size
Gran.	Granular	KWh	Kilowatt-hour	mj	Mechanical Joint
Grnd.	Ground	L	Labor Only; Length; Long; Medium Wall Copper Tubing	m	Meter
GVW	Gross Vehicle Weight			mm	Millimeter
GWB	Gypsum Wall Board	Lab.	Labor	Mill.	Millwright
		lat	Latitude	Min., min.	Minimum, Minute

Abbreviations

Misc.	Miscellaneous	PDCA	Painting and Decorating Contractors of America	SC	Screw Cover
ml	Milliliter, Mainline			SCFM	Standard Cubic Feet per Minute
M.L.F.	Thousand Linear Feet	P.E., PE	Professional Engineer; Porcelain Enamel; Polyethylene; Plain End	Scaf.	Scaffold
Mo.	Month			Sch., Sched.	Schedule
Mobil.	Mobilization			S.C.R.	Modular Brick
Mog.	Mogul Base	P.E.C.I.	Porcelain Enamel on Cast Iron	S.D.	Sound Deadening
MPH	Miles per Hour	Perf.	Perforated	SDR	Standard Dimension Ratio
MPT	Male Pipe Thread	PEX	Cross Linked Polyethylene	S.E.	Surfaced Edge
MRGWB	Moisture Resistant Gypsum Wallboard	Ph.	Phase	Sel.	Select
		P.I.	Pressure Injected	SER, SEU	Service Entrance Cable
MRT	Mile Round Trip	Pile.	Pile Driver	S.F.	Square Foot
ms	Millisecond	Pkg.	Package	S.F.C.A.	Square Foot Contact Area
M.S.F.	Thousand Square Feet	Pl.	Plate	S.F. Flr.	Square Foot of Floor
Mstz.	Mosaic & Terrazzo Worker	Plah.	Plasterer Helper	S.F.G.	Square Foot of Ground
M.S.Y.	Thousand Square Yards	Plas.	Plasterer	S.F. Hor.	Square Foot Horizontal
Mtd., mtd., mtd	Mounted	plf	Pounds Per Linear Foot	SFR	Square Feet of Radiation
Mthe.	Mosaic & Terrazzo Helper	Pluh.	Plumber Helper	S.F. Shlf.	Square Foot of Shelf
Mtng.	Mounting	Plum.	Plumber	S4S	Surface 4 Sides
Mult.	Multi; Multiply	Ply.	Plywood	Shee.	Sheet Metal Worker
M.V.A.	Million Volt Amperes	p.m.	Post Meridiem	Sin.	Sine
M.V.A.R.	Million Volt Amperes Reactance	Pntd.	Painted	Skwk.	Skilled Worker
MV	Megavolt	Pord.	Painter, Ordinary	SL	Saran Lined
MW	Megawatt	pp	Pages	S.L.	Slimline
MXM	Male by Male	PP, PPL	Polypropylene	Sldr.	Solder
MYD	Thousand Yards	P.P.M.	Parts per Million	SLH	Super Long Span Bar Joist
N	Natural; North	Pr.	Pair	S.N.	Solid Neutral
nA	Nanoampere	P.E.S.B.	Pre-engineered Steel Building	SO	Stranded with oil resistant inside insulation
NA	Not Available; Not Applicable	Prefab.	Prefabricated		
N.B.C.	National Building Code	Prefin.	Prefinished	S-O-L	Socketolet
NC	Normally Closed	Prop.	Propelled	sp	Standpipe
NEMA	National Electrical Manufacturers Assoc.	PSF, psf	Pounds per Square Foot	S.P.	Static Pressure; Single Pole; Self-Propelled
		PSI, psi	Pounds per Square Inch		
NEHB	Bolted Circuit Breaker to 600V.	PSIG	Pounds per Square Inch Gauge	Spri.	Sprinkler Installer
NFPA	National Fire Protection Association	PSP	Plastic Sewer Pipe	spwg	Static Pressure Water Gauge
NLB	Non-Load-Bearing	Pspr.	Painter, Spray	S.P.D.T.	Single Pole, Double Throw
NM	Non-Metallic Cable	Psst.	Painter, Structural Steel	SPF	Spruce Pine Fir; Sprayed Polyurethane Foam
nm	Nanometer	P.T.	Potential Transformer		
No.	Number	P. & T.	Pressure & Temperature	S.P.S.T.	Single Pole, Single Throw
NO	Normally Open	Ptd.	Painted	SPT	Standard Pipe Thread
N.O.C.	Not Otherwise Classified	Ptns.	Partitions	Sq.	Square; 100 Square Feet
Nose.	Nosing	Pu	Ultimate Load	Sq. Hd.	Square Head
NPT	National Pipe Thread	PVC	Polyvinyl Chloride	Sq. In.	Square Inch
NQOD	Combination Plug-on/Bolt on Circuit Breaker to 240V.	Pvmt.	Pavement	S.S.	Single Strength; Stainless Steel
		PRV	Pressure Relief Valve	S.S.B.	Single Strength B Grade
N.R.C., NRC	Noise Reduction Coefficient/ Nuclear Regulator Commission	Pwr.	Power	sst, ss	Stainless Steel
		Q	Quantity Heat Flow	Sswk.	Structural Steel Worker
N.R.S.	Non Rising Stem	Qt.	Quart	Sswl.	Structural Steel Welder
ns	Nanosecond	Quan., Qty.	Quantity	St.; Stl.	Steel
nW	Nanowatt	Q.C.	Quick Coupling	STC	Sound Transmission Coefficient
OB	Opposing Blade	r	Radius of Gyration	Std.	Standard
OC	On Center	R	Resistance	Stg.	Staging
OD	Outside Diameter	R.C.P.	Reinforced Concrete Pipe	STK	Select Tight Knot
O.D.	Outside Dimension	Rect.	Rectangle	STP	Standard Temperature & Pressure
ODS	Overhead Distribution System	recpt.	Receptacle	Stpi.	Steamfitter, Pipefitter
O.G.	Ogee	Reg.	Regular	Str.	Strength; Starter; Straight
O.H.	Overhead	Reinf.	Reinforced	Strd.	Stranded
O&P	Overhead and Profit	Req'd.	Required	Struct.	Structural
Oper.	Operator	Res.	Resistant	Sty.	Story
Opng.	Opening	Resi.	Residential	Subj.	Subject
Orna.	Ornamental	RF	Radio Frequency	Subs.	Subcontractors
OSB	Oriented Strand Board	RFID	Radio-frequency Identification	Surf.	Surface
OS&Y	Outside Screw and Yoke	Rgh.	Rough	Sw.	Switch
OSHA	Occupational Safety and Health Act	RGS	Rigid Galvanized Steel	Swbd.	Switchboard
		RHW	Rubber, Heat & Water Resistant; Residential Hot Water	S.Y.	Square Yard
Ovhd.	Overhead			Syn.	Synthetic
OWG	Oil, Water or Gas	rms	Root Mean Square	S.Y.P.	Southern Yellow Pine
Oz.	Ounce	Rnd.	Round	Sys.	System
P.	Pole; Applied Load; Projection	Rodm.	Rodman	t.	Thickness
p.	Page	Rofc.	Roofer, Composition	T	Temperature; Ton
Pape.	Paperhanger	Rofp.	Roofer, Precast	Tan	Tangent
P.A.P.R.	Powered Air Purifying Respirator	Rohe.	Roofer Helpers (Composition)	T.C.	Terra Cotta
PAR	Parabolic Reflector	Rots.	Roofer, Tile & Slate	T & C	Threaded and Coupled
P.B., PB	Push Button	R.O.W.	Right of Way	T.D.	Temperature Difference
Pc., Pcs.	Piece, Pieces	RPM	Revolutions per Minute	TDD	Telecommunications Device for the Deaf
P.C.	Portland Cement; Power Connector	R.S.	Rapid Start		
P.C.F.	Pounds per Cubic Foot	Rsr	Riser	T.E.M.	Transmission Electron Microscope
PCM	Phase Contrast Microscopy	RT	Round Trip	temp	Temperature, Tempered, Temporary
		S.	Suction; Single Entrance; South	TFFN	Nylon Jacketed Wire
		SBS	Styrene Butadiene Styrene		

Abbreviations

TFE	Tetrafluoroethylene (Teflon)	U.L., UL	Underwriters Laboratory	w/	With
T. & G.	Tongue & Groove; Tar & Gravel	Uld.	Unloading	W.C., WC	Water Column; Water Closet
Th., Thk.	Thick	Unfin.	Unfinished	W.F.	Wide Flange
Thn.	Thin	UPS	Uninterruptible Power Supply	W.G.	Water Gauge
Thrded	Threaded	URD	Underground Residential Distribution	Wldg.	Welding
Tilf.	Tile Layer, Floor	US	United States	W. Mile	Wire Mile
Tilh.	Tile Layer, Helper	USGBC	U.S. Green Building Council	W-O-L	Weldolet
THHN	Nylon Jacketed Wire	USP	United States Primed	W.R.	Water Resistant
THW.	Insulated Strand Wire	UTMCD	Uniform Traffic Manual For Control Devices	Wrck.	Wrecker
THWN	Nylon Jacketed Wire			WSFU	Water Supply Fixture Unit
T.L., TL	Truckload	UTP	Unshielded Twisted Pair	W.S.P.	Water, Steam, Petroleum
T.M.	Track Mounted	V	Volt	WT., Wt.	Weight
Tot.	Total	VA	Volt Amperes	WWF	Welded Wire Fabric
T-O-L	Threadolet	VAT	Vinyl Asbestos Tile	XFER	Transfer
tmpd	Tempered	V.C.T.	Vinyl Composition Tile	XFMR	Transformer
TPO	Thermoplastic Polyolefin	VAV	Variable Air Volume	XHD	Extra Heavy Duty
T.S.	Trigger Start	VC	Veneer Core	XHHW	Cross-Linked Polyethylene Wire
Tr.	Trade	VDC	Volts Direct Current	XLPE	Insulation
Transf.	Transformer	Vent.	Ventilation	XLP	Cross-linked Polyethylene
Trhv.	Truck Driver, Heavy	Vert.	Vertical	Xport	Transport
Trlr	Trailer	V.F.	Vinyl Faced	Y	Wye
Trlt.	Truck Driver, Light	V.G.	Vertical Grain	yd	Yard
TTY	Teletypewriter	VHF	Very High Frequency	yr	Year
TV	Television	VHO	Very High Output	Δ	Delta
T.W.	Thermoplastic Water Resistant Wire	Vib.	Vibrating	%	Percent
		VLF	Vertical Linear Foot	~	Approximately
UCI	Uniform Construction Index	VOC	Volatile Organic Compound	Ø	Phase; diameter
UF	Underground Feeder	Vol.	Volume	@	At
UGND	Underground Feeder	VRP	Vinyl Reinforced Polyester	#	Pound; Number
UHF	Ultra High Frequency	W	Wire; Watt; Wide; West	<	Less Than
U.I.	United Inch			>	Greater Than
				Z	Zone

FORMS

RESIDENTIAL COST ESTIMATE

OWNER'S NAME: _____ APPRAISER: _____

RESIDENCE ADDRESS: _____ PROJECT: _____

CITY, STATE, ZIP CODE: _____ DATE: _____

CLASS OF CONSTRUCTION
- ☐ ECONOMY
- ☐ AVERAGE
- ☐ CUSTOM
- ☐ LUXURY

RESIDENCE TYPE
- ☐ 1 STORY
- ☐ 1-1/2 STORY
- ☐ 2 STORY
- ☐ 2-1/2 STORY
- ☐ 3 STORY
- ☐ BI-LEVEL
- ☐ TRI-LEVEL

CONFIGURATION
- ☐ DETACHED
- ☐ TOWN/ROW HOUSE
- ☐ SEMI-DETACHED

OCCUPANCY
- ☐ ONE STORY
- ☐ TWO FAMILY
- ☐ THREE FAMILY
- ☐ OTHER _____

EXTERIOR WALL SYSTEM
- ☐ WOOD SIDING—WOOD FRAME
- ☐ BRICK VENEER—WOOD FRAME
- ☐ STUCCO ON WOOD FRAME
- ☐ PAINTED CONCRETE BLOCK
- ☐ SOLID MASONRY (AVERAGE & CUSTOM)
- ☐ STONE VENEER—WOOD FRAME
- ☐ SOLID BRICK (LUXURY)
- ☐ SOLID STONE (LUXURY)

*LIVING AREA (Main Building)
- First Level _____ S.F.
- Second Level _____ S.F.
- Third Level _____ S.F.
- Total _____ S.F.

*LIVING AREA (Wing or Ell) (   )
- First Level _____ S.F.
- Second Level _____ S.F.
- Third Level _____ S.F.
- Total _____ S.F.

*LIVING AREA (Wing or Ell) (   )
- First Level _____ S.F.
- Second Level _____ S.F.
- Third Level _____ S.F.
- Total _____ S.F.

*Basement Area is not part of living area.

MAIN BUILDING	COSTS PER S.F. LIVING AREA
Cost per Square Foot of Living Area, from Page _____	$
Basement Addition: _____ % Finished, _____ % Unfinished	+
Roof Cover Adjustment: _____ Type, Page _____ (Add or Deduct)	( )
Central Air Conditioning: ☐ Separate Ducts ☐ Heating Ducts, Page _____	+
Heating System Adjustment: _____ Type, Page _____ (Add or Deduct)	( )
Main Building: Adjusted Cost per S.F. of Living Area	$

MAIN BUILDING TOTAL COST: $ _____ /S.F. (Cost per S.F. Living Area) × _____ S.F. (Living Area) × _____ (Town/Row House Multiplier, Use 1 for Detached) = $ _____ TOTAL COST

WING OR ELL ( ) _____ STORY	COSTS PER S.F. LIVING AREA
Cost per Square Foot of Living Area, from Page _____	$
Basement Addition: _____ % Finished, _____ % Unfinished	+
Roof Cover Adjustment: _____ Type, Page _____ (Add or Deduct)	( )
Central Air Conditioning: ☐ Separate Ducts ☐ Heating Ducts, Page _____	+
Heating System Adjustment: _____ Type, Page _____ (Add or Deduct)	( )
Wing or Ell ( ): Adjusted Cost per S.F. of Living Area	$

WING OR ELL (   ) TOTAL COST: $ _____ /S.F. (Cost per S.F. Living Area) × _____ S.F. (Living Area) = $ _____ TOTAL COST

WING OR ELL ( ) _____ STORY	COSTS PER S.F. LIVING AREA
Cost per Square Foot of Living Area, from Page _____	$
Basement Addition: _____ % Finished, _____ % Unfinished	+
Roof Cover Adjustment: _____ Type, Page _____ (Add or Deduct)	( )
Central Air Conditioning: ☐ Separate Ducts ☐ Heating Ducts, Page _____	+
Heating System Adjustment: _____ Type, Page _____ (Add or Deduct)	( )
Wing or Ell ( ): Adjusted Cost per S.F. of Living Area	$

WING OR ELL (   ) TOTAL COST: $ _____ /S.F. (Cost per S.F. Living Area) × _____ S.F. (Living Area) = $ _____ TOTAL COST

TOTAL THIS PAGE: _____

FORMS

RESIDENTIAL COST ESTIMATE

Total Page 1			$
	QUANTITY	UNIT COST	
Additional Bathrooms: _____ Full, _____ Half			
Finished Attic: _____ Ft. x _____ Ft.	S.F.		+
Breezeway: ☐ Open ☐ Enclosed _____ Ft. x _____ Ft.	S.F.		+
Covered Porch: ☐ Open ☐ Enclosed _____ Ft. x _____ Ft.	S.F.		+
Fireplace: ☐ Interior Chimney ☐ Exterior Chimney ☐ No. of Flues ☐ Additional Fireplaces			+
Appliances:			+
Kitchen Cabinets Adjustment: (+/−)			
☐ Garage ☐ Carport: _____ Car(s) Description _____ (+/−)			
Miscellaneous:			+

ADJUSTED TOTAL BUILDING COST $ _____

REPLACEMENT COST

ADJUSTED TOTAL BUILDING COST $ _____
Site Improvements
   (A) Paving & Sidewalks $ _____
   (B) Landscaping $ _____
   (C) Fences $ _____
   (D) Swimming Pool $ _____
   (E) Miscellaneous $ _____
TOTAL $ _____
Location Factor x _____
Location Replacement Cost $ _____
Depreciation −$ _____
LOCAL DEPRECIATED COST $ _____

INSURANCE COST

ADJUSTED TOTAL BUILDING COST $ _____
Insurance Exclusions
   (A) Footings, Site Work, Underground Piping −$ _____
   (B) Architects' Fees −$ _____
Total Building Cost Less Exclusion $ _____
Location Factor x _____
LOCAL INSURABLE REPLACEMENT COST $ _____

SKETCH AND ADDITIONAL CALCULATIONS

For customer support on your Residential Cost Data, call 877.759.4771.

Index

A

Abandon catch basin	300
ABC extinguisher portable	523
Aboveground storage tank concrete	580
Abrasive floor tile	493
tread	493
ABS DWV pipe	562
Absorber shock	567
A/C coils furnace	588
packaged terminal	590
Accelerator sprinkler system	553
Access door and panel	456
door basement	318
Accessory anchor bolt	311
bathroom	492, 520
door	477
drainage	437
drywall	491
duct	582
fireplace	521
gypsum board	491
plaster	486
toilet	520
Accordion door	452
Acid etch finish concrete	317
proof floor	497
Acoustic ceiling board	495
ceiling tile suspended	495
Acoustical ceiling	495
sealant	442, 490
Acrylic carpet	228
wall coating	517
wallcovering	503
wood block	498
Adhesive cement	499
roof	432
wallpaper	503
Adobe brick	329
Aerial lift	663
lift truck	668
seeding	644
Aggregate exposed	317
spreader	660, 667
stone	644
Aid staging	295
Air cleaner electronic	585
compressor	663
compressor portable	663
compressor sprinkler system	553
conditioner cooling & heating	590
conditioner fan coil	591
conditioner packaged terminal	590
conditioner portable	590
conditioner receptacle	602
conditioner removal	578
conditioner rooftop	590
conditioner self-contained	590
conditioner thru-wall	590
conditioner window	590
conditioner wiring	605
conditioning	268, 270
conditioning ventilating	583, 590
cooled condensing unit	589
filter	584, 585
filter mechanical media	585
filter permanent washable	585
filter roll type	585
filter washable	585
filtration charcoal type	585
handler heater	589
handler modular	589
hose	664
quality	529
return grille	584
spade	664
supply register	584
Air-handling unit	591
Air-source heat pump	591
Alarm fire	552
panel and device	616
residential	604
standpipe	552
valve sprinkler	548
water motor	552
Alteration fee	292
Alternative fastening	422
Aluminum ceiling tile	495
chain link	638
column	353, 407
column base	353
column cap	353
coping	333
diffuser perforated	583
downspout	436
drip edge	439
ductwork	581
edging	646
flagpole	526
flashing	211, 435
foil	420, 424
gravel stop	436
grille	584
gutter	438
louver	479
nail	361
plank scaffold	295
rivet	338
roof	427
roof panel	427
screen/storm door	446
service entrance cable	597
sheet metal	436
siding	166, 428
siding accessory	428
siding paint	507
sliding door	457
storm door	446
threshold	475
threshold ramp	475
tile	494
window	459
window demolition	445
window impact resistant	459
Anchor bolt	311, 324
bolt accessory	311
bolt sleeve	311
brick	324
buck	324
chemical	319, 336
epoxy	319, 336
expansion	336
framing	362
hollow wall	336
lead screw	336
masonry	324
metal nailing	336
nailing	336
nylon nailing	336
partition	324
plastic screw	337
rafter	362
rigid	324
roof safety	295
screw	336
sill	362
steel	324
stone	324
toggle-bolt	336
Antenna TV	614
Anti-siphon device	566
Apartment hardware	473
Appliance	528
residential	528, 604
Arch laminated	388
radial	388
Architectural fee	292, 713
Area window well	525
Armored cable	597
Arrester lightning	609
Ashlar stone	331
Asphalt base sheet	432
block	636
block floor	636
coating	417
curb	637
distributor	664
felt	431
flood coat	432
paver	665
paving	635
plant portable	667
primer	499
roof shingle	424
sheathing	386
shingle	424
sidewalk	634
tile	229
Asphaltic cover board	421
emulsion	642
Astragal	476
molding	401
one piece	476
Athletic carpet indoor	502
Attachment ripper	667
Attachments skid steer	663
Attic stair	530
ventilation fan	583
Auger earth	660
hole	300
Autoclave aerated concrete block	327
Automatic transfer switch	609
washing machine	529
Auto-scrub machine	664
Awning canvas	525
window	174, 461
window metal-clad	463
window vinyl clad	465

B

Backer rod	441
Backerboard cementitious	488
Backfill	620, 621
planting pit	643
trench	620
Backflow preventer	566
Backhoe	620, 661
bucket	662
extension	667
Backsplash countertop	538
Backup block	327
Baffle roof	440
Baked enamel frame	447
Ball valve slow close	551
wrecking	667
Baluster	231
birch	231
pine	231
wood stair	404
Balustrade painting	514
Bamboo flooring	496
Band joist framing	345
Bank run gravel	618
Bar grab	520
masonry reinforcing	325
towel	520
Zee	491
Barge construction	670
Bark mulch	642
mulch redwood	642
Barrels flasher	664
reflectorized	664
Barricade flasher	664
Barrier and enclosure	296
dust	296
slipform paver	668
weather	423
Base cabinet	536
carpet	502
ceramic tile	492
column	362
course	634, 635
course drainage layer	634
gravel	634
molding	391
quarry tile	493
resilient	499
sheet	432
sheet asphalt	432
sink	536
stabilizer	668
stone	331
terrazzo	500
vanity	537
wall	499
wood	391
Baseboard demolition	360
heat	591
heater electric	593
heating	279
register	584
Basement bulkhead stair	318
stair	231
Basement/utility window	459
Baseplate scaffold	294
shoring	294
Basic finish material	482
meter device	606
Basket weave fence vinyl	638
Basketweave fence	639
Bath accessory	237
spa	543
steam	544
whirlpool	543
Bathroom	248, 250, 252, 254, 256, 258, 260, 262, 264, 266, 573
accessory	492, 520
exhaust fan	582
faucet	573
fixture	571-573
Bathtub enclosure	521
removal	556
residential	573
Batt insulation	168, 419
Bay window	178, 460
wood window double hung	464
Bead casing	491
corner	486
parting	402
Beam & girder framing	371
bond	324
bondcrete	487
ceiling	407
decorative	407
drywall	488
hanger	362
laminated	388
load-bearing stud wall box	342
mantel	407
plaster	486
steel	340

738

Index

wood ... 371, 383
Bed molding ... 394
Bedding brick ... 636
pipe ... 621
Belgian block ... 637
Bell & spigot pipe ... 567
Bench ... 642
park ... 642
player ... 642
Berm pavement ... 637
road ... 637
Bevel siding ... 429
Bi-fold door ... 226, 448, 451
Bin retaining walls metal ... 641
Bi-passing closet door ... 452
door ... 226
Birch exterior door ... 455
hollow core door ... 453
molding ... 401
paneling ... 403
solid core door ... 451
wood frame ... 408
Bituminous block ... 636
coating ... 417
concrete curb ... 637
paver ... 665
Bituminous-stabilized base course ... 635
Blanket insulation ... 578
insulation wall ... 419
Blaster shot ... 666
Blind ... 191
exterior ... 409
vinyl ... 409, 534
window ... 534
wood ... 535
Block asphalt ... 636
autoclave concrete ... 327
backup ... 327
belgian ... 637
bituminous ... 636
Block, brick and mortar ... 721
Block concrete ... 327-329, 722
concrete bond beam ... 327
concrete exterior ... 328
filler ... 516
floor ... 498
glass ... 329
insulation ... 327
insulation insert concrete ... 327
lightweight ... 329
lintel ... 328
partition ... 329
profile ... 328
reflective ... 329
split rib ... 328
wall ... 158
wall removal ... 302
Blocking ... 370
carpentry ... 370
wood ... 374
Blower insulation ... 665
Blown-in cellulose ... 420
fiberglass ... 420
insulation ... 420
Blueboard ... 487
Bluegrass sod ... 644
Bluestone sidewalk ... 634
sill ... 332
step ... 634
Board & batten siding ... 429
batten fence ... 640
ceiling ... 495
fence ... 639
gypsum ... 488
insulation ... 418
insulation foam ... 418
paneling ... 404
sheathing ... 385
siding ... 162
siding wood ... 429
valance ... 537
verge ... 400
Boat work ... 670
Boiler ... 586
demolition ... 578
electric ... 586
electric steam ... 586
gas fired ... 586
gas/oil combination ... 586
hot water ... 586
oil-fired ... 586
removal ... 578
steam ... 586
Bollard ... 339
light ... 611
Bolt & hex nut steel ... 337
anchor ... 311, 324
steel ... 362
Bond beam ... 324
Bondcrete ... 487
Bookcase ... 524
Boom truck ... 668
Borer horizontal ... 667
Boring and exploratory drilling ... 300
cased ... 300
machine horizontal ... 660
Borrow ... 618
Boundary and survey marker ... 300
Bow & bay metal-clad window ... 464
window ... 178, 460, 464
Bowl toilet ... 571
Bowstring truss ... 388
Box & wiring device ... 599
beam load-bearing stud wall ... 342
distribution ... 651
electrical pull ... 599
fence shadow ... 639
flow leveler distribution ... 651
mail ... 523
pull ... 599
stair ... 404
trench ... 666
utility ... 648
Boxed header/beams ... 346
header/beams framing ... 346
Bracing ... 370
framing ... 348
let-in ... 370
load-bearing stud wall ... 341
metal ... 370
metal joist ... 344
roof rafter ... 348
shoring ... 294
support seismic ... 557
wood ... 370
Bracket scaffold ... 294
sidewall ... 295
Brass hinge ... 475
screw ... 361
Break ties concrete ... 317
Breaker circuit ... 608
pavement ... 663
vacuum ... 566
Brick adobe ... 329
anchor ... 324
bedding ... 636
Brick, block and mortar ... 721
Brick cart ... 664
chimney simulated ... 521
cleaning ... 322
common building ... 327
demolition ... 323, 482
edging ... 646
face ... 327, 721
floor ... 497
flooring miscellaneous ... 497
manhole ... 652
molding ... 398, 401
oversized ... 326
paving ... 636
removal ... 301
sidewalk ... 636
sill ... 332
stair ... 330
step ... 634
veneer ... 160, 325
veneer demolition ... 323
veneer thin ... 325
wall ... 160, 330
wash ... 322
Bricklaying ... 721
Bridging ... 371
framing ... 348
load-bearing stud wall ... 341
metal joist ... 345
roof rafter ... 348
Bronze swing check valve ... 556
valve ... 556
Broom cabinet ... 537
finish concrete ... 317
sidewalk ... 666
Brownstone ... 331
Brush chipper ... 661
clearing ... 618
cutter ... 661, 662
Buck anchor ... 324
rough ... 373
Bucket backhoe ... 662
clamshell ... 661
concrete ... 660
dragline ... 661
excavator ... 667
pavement ... 667
Buggy concrete ... 316, 660
Builder's risk insurance ... 715
Building demo footing & foundation ... 302
demolition ... 301
demolition selective ... 302
excavation ... 102, 104
greenhouse ... 544
hardware ... 473
insulation ... 420
paper ... 424
permit ... 293
prefabricated ... 544
Built-in range ... 528
Built-up roof ... 431, 432
roofing ... 210
roofing system ... 432
Bulk bank measure excavating ... 620
Bulkhead door ... 456
movable ... 542
Bulkhead/cellar door ... 456
Bulldozer ... 620, 662
Bullfloat concrete ... 660
Bumper door ... 474
wall ... 474
Buncher feller ... 661
Burlap curing ... 318
rubbing concrete ... 317
Burner gas conversion ... 587
gun type ... 587
residential ... 587
Bush hammer ... 317
hammer finish concrete ... 317
Butt fusion machine ... 664
Butterfly valve ... 648
Butyl caulking ... 442
flashing ... 436
rubber ... 441

C

Cabinet ... 234
& casework paint/coating ... 510
base ... 536
broom ... 537
corner base ... 536
corner wall ... 536
demolition ... 360
filler ... 537
hardboard ... 536
hardware ... 537
hinge ... 537
kitchen ... 536
medicine ... 520
outdoor ... 538
oven ... 537
shower ... 520
stain ... 510
varnish ... 511
wall ... 536
Cable armored ... 597
copper ... 597
electric ... 597
electrical ... 598
jack ... 669
pulling ... 666
sheathed non-metallic ... 597
sheathed Romex ... 597
tensioning ... 666
trailer ... 666
Cable/wire puller ... 666
Cafe door ... 454
Canopy door ... 524
framing ... 377
residential ... 524
wood ... 377
Cant roof ... 376, 432
Cantilever retaining wall ... 640
Canvas awning ... 525
Cap dowel ... 312
post ... 362
service entrance ... 597
vent ... 569
Car tram ... 666
Carpet ... 228
base ... 502
felt pad ... 501
floor ... 501
nylon ... 501
olefin ... 501
pad ... 501
pad commercial grade ... 501
padding ... 501
removal ... 482
sheet ... 501
stair ... 502
tile ... 501
tile removal ... 482
urethane pad ... 501
wool ... 501
Carrier ceiling ... 485
channel ... 496
Carrier/support fixture ... 575
Cart brick ... 664
concrete ... 316, 660
Cased boring ... 300
evaporator coil ... 588
Casement window ... 172, 461, 465
window metal-clad ... 464

Index

window vinyl		469
Casework custom		537
demolition		360
ground		381
painting		510
stain		510
varnish		511
Casing bead		491
molding		392
Cast in place concrete		314, 315
in place concrete curb		637
iron bench		642
iron damper		522
iron fitting		567
iron pipe		567
iron pipe fitting		567
iron radiator		591
iron trap		569
iron vent cap		569
trim lock		473
Caster scaffold		294
Catch basin		652
basin cleaner		667
basin precast		652
basin removal		300
basin vacuum		667
door		537
Caulking		442
polyurethane		442
sealant		442
Cavity wall insulation		420
wall reinforcing		325
Cedar closet		403
fence		639, 640
paneling		404
post		407
roof deck		384
roof plank		384
shake siding		164
shingle		425
siding		429
Ceiling		216, 220, 495
acoustical		495
beam		407
board		495
board acoustic		495
board fiberglass		495
bondcrete		487
carrier		485
demolition		482
diffuser		583
drywall		489
framing		372
furring		380, 483, 485
heater		529
insulation		419, 420
lath		486
luminous		612
molding		394
painting		515, 516
plaster		486
stair		530
support		339
suspended		485, 495
suspension system		496
tile		495
Cellar door		456
wine		530
Cellulose blown-in		420
insulation		420
Cement adhesive		499
flashing		416
grout		324
masonry		324
masonry unit		327-329
mortar		493, 494, 720
parging		417
terrazzo Portland		500
Cementitious backerboard		488
Central vacuum		531
Centrifugal pump		665
Ceramic mulch		642
tile		229, 492
tile accessories		493
tile countertop		538
tile demolition		482
tile floor		492
tile waterproofing membrane		494
tiling		492, 494
Chain fall hoist		669
link fence		114, 638
link fence paint		506
saw		666
trencher		620, 663
Chair lift		546
molding		401
rail demolition		360
Chamber infiltration		651
Chamfer strip		310
Channel carrier		496
furring		483, 491
siding		429
steel		486
Charcoal type air filter		585
Charge powder		338
Check valve		548
Chemical anchor		319, 336
dry extinguisher		523
spreader		667
termite control		631
toilet		666
Cherry threshold		474
Chime door		608
Chimney		326
accessories		521
brick		326
demolition		323
fireplace		521
flue		332
foundation		314
metal		521
screen		521
simulated brick		521
vent		585
Chip wood		643
Chipper brush		661
log		662
stump		662
Chipping hammer		664
Chloride accelerator concrete		316
Chlorination system		576
Circline fixture		610
Circuit wiring		280
Circuit breaker		608
Circular saw		666
Circulating pump		580
Cladding sheet metal		434
Clamp water pipe ground		598
Clamshell bucket		661
Clapboard painting		507
Clay tile		426
tile coping		333
Clean control joint		311
Cleaner catch basin		667
crack		667
steam		666
Cleaning brick		322
face brick		720
masonry		322
up		297
Cleanout door		522
floor		558
pipe		558
tee		558
Clear and grub site		618
grub		618
Clearing brush		618
selective		618
site		618
Climbing crane		668
hydraulic jack		669
Clip plywood		362
Clock timer		604
Closet cedar		403
door		226, 448, 452
pole		401
rod		524
water		571, 575
CMU		327
pointing		322
Coal tar pitch		432
Coat glaze		515
Coated roofing foamed		433
Coating bituminous		417
roof		416
rubber		418
silicone		418
spray		417
trowel		417
wall		517
water repellent		418
waterproofing		417
Cold mix paver		667
Collection concrete pipe sewage		650
sewage/drainage		651
Colonial door		451, 455
wood frame		408
Column		407
accessories formwork		310
aluminum		353, 407
base		362
base aluminum		353
bondcrete		487
brick		326
cap aluminum		353
demolition		323
drywall		488
formwork		308
framing		372
laminated wood		388
lath		486
lightweight		339
ornamental		353
plaster		486
plywood formwork		308
removal		323
round fiber tube formwork		308
steel		340
steel framed formwork		308
structural		339
wood		372, 383, 407
Combination device		601
storm door		451, 455
Commercial lavatory		575
toilet accessories		520
Common brick		327
building brick		327
nail		361
rafter roof framing		138
Communicating lockset		473
Compact fill		621
Compaction		630
soil		620
structural		630
Compactor earth		661
landfill		662
residential		529
Compensation workers'		293
Component control		579
furnace		588
Composite decking		375, 412
decking woodgrained		412
door		454
fabrication		412
insulation		422
rafter		376
railing		412
railing encased		412
Compression seal		441
Compressor air		663
Concrete acid etch finish		317
block		327-329, 722
block autoclave		327
block back-up		327
block bond beam		327
block decorative		328
block demolition		303
block exterior		328
block foundation		328
block grout		324
block insulation		420
block insulation insert		327
block lintel		328
block painting		515
block partition		329
block wall		158
break tie		317
broom finish		317
bucket		660
buggy		316, 660
bullfloat		660
burlap rubbing		317
bush hammer finish		317
cart		316, 660
cast-in-place		314, 315
chloride accelerator		316
conveyer		660
coping		333
curb		637
curing		318
cutout		302
demolition		302
distribution box		651
drilling		319
equipment rental		660
fiber reinforcing		316
float finish		317
floor edger		660
floor grinder		660
floor patching		308
floor staining		496
foundation		122
furring		380
grout		324
hand mix		315
hand trowel finish		317
impact drilling		319
lance		665
machine trowel finish		317
membrane curing		318
mixer		660
non-chloride accelerator		316
paver		665
paving surface treatment		636
pipe		651
placement footing		316
placement slab		316
placement wall		316
placing		316
plantable paver precast		636
pump		660
ready mix		315
reinforcing		312
removal		300

Index

retaining wall 640
retaining wall segmental 640
retarder 316
saw 660
septic tank 650
shingle 426
short load 316
sidewalk 634
slab 126
slab conduit 599
small quantity 315
spreader 665
stair 315
stair finish 317
tank aboveground storage 580
tile 426
trowel 660
truck 660
truck holding 316
utility vault 648
vibrator 660
volumetric mixed 315
wall demolition 302
water reducer 316
wheeling 316
winter 316
Condensing unit air cooled 589
Conditioner portable air 590
Conditioning ventilating air ... 590
Conductive floor 500
  rubber & vinyl flooring 500
Conductor 597
  & grounding 597
Conduit & fitting flexible 600
  concrete slab 599
  electrical 598
  fitting 599
  flexible metallic 599
  Greenfield 278
  in slab PVC 599
  in trench electrical 600
  in trench steel 600
  PVC 598
  rigid in slab 599
  rigid steel 598
Connection fire-department 550
  motor 600
  split system tubing 591
  standpipe 550
Connector joist 362
  timber 361
Construction barge 670
  management fee 292
  temporary 296
  time 292
Containment area lead 304
Continuous hinge 476
Control component 579
  damper volume 582
  joint 311
  joint clean 311
  joint sawn 311
  pressure sprinkler 553
Cooking equipment 528
  range 528
Cooling 268, 270
Coping 331, 333
  aluminum 333
  clay tile 333
  concrete 333
  removal 323
  terra cotta 333
Copper cable 597
  downspout 437
  drum trap 569
  DWV tubing 560

fitting 560
flashing 435, 569
gravel stop 436
gutter 438
pipe 559
rivet 338
roof 434
tube fitting 560
wall covering 502
wire 597
Core drill 660
Cork floor 497
  tile 497
  tile flooring 497
  wall tile 502
Corner base cabinet 536
  bead 486
  wall cabinet 536
Cornerboard PVC 411
Cornice 330
  molding 394, 397
  painting 514
Corrugated metal pipe 652
  roof tile 426
  siding 427, 428
  steel pipe 652
Counter top 538
  top demolition 360
Countertop backsplash 538
  ceramic tile 538
  engineered stone 539
  granite 539
  laminated 538
  maple 538
  marble 538, 539
  plastic-laminate-clad 538
  postformed 538
  sink 572
  solid surface 539
  stainless steel 538
Coupling plastic 565
Course drainage layers base .. 634
Cove base ceramic tile 492
  base terrazzo 500
  molding 394
Cover board asphaltic 421
  board gypsum 421
  pool 542
Covering wall 728
CPVC pipe 562
Crack cleaner 667
  filler 667
Crane climbing 668
  crawler 668
  crew daily 294
  hydraulic 669
  tower 668
  truck mounted 668
Crawler crane 668
  drill rotary 664
  shovel 662
Crew daily crane 294
  forklift 293
  survey 297
Crown molding 394, 401
Crushed stone 618
  stone sidewalk 634
Cubicle shower 573
Cultured stone 333
Cupola 525
Curb 637
  and gutter 637
  asphalt 637
  builder 660
  concrete 637
  edging 637

extruder 661
granite 637
inlet 637
precast 637
removal 300
roof 377
slipform paver 668
terrazzo 500
Curing concrete 318
  concrete membrane 318
  concrete water 318
  paper 423
Curtain damper fire 582
  rod 520
Custom casework 536
Cut stone curb 637
  stone trim 331
Cutout counter 538
  demolition 302
  slab 302
Cutter brush 661, 662
Cutting block 538
  torch 666
Cylinder lockset 473

D

Daily crane crew 294
Damper fire curtain 582
  fireplace 522
  multi-blade 582
Dead lock 473
Deadbolt 474
Deciduous shrub 645
Deck framing 374, 410
  roof 384, 386
  slab form 340
  wood 244, 384
Decking 722
  composite 375, 412
  form 340
  log roof 383
  roof 340
  woodgrained composite 412
Decorative beam 407
Decorator device 601
  switch 601
Deep freeze 528
Dehumidifier 664
Dehydrator package sprinkler . 553
Delivery charge 296
Demo footing & foundation
building 302
  thermal & moisture protection 416
Demolish decking 357
  remove pavement and curb .. 300
Demolition 719
  baseboard 360
  boiler 578
  brick 323, 482
  brick veneer 323
  cabinet 360
  casework 360
  ceiling 482
  ceramic tile 482
  chimney 323
  column 323
  concrete 302
  concrete block 303
  concrete wall 302
  door 445
  drywall 483
  ductwork 578
  electrical 596
  fencing 301

fireplace 323
flooring 482
framing 357
furnace 578
glass 445
granite 323
gutter 416
gutting 304
hammer 661
HVAC 578
joist 357
masonry 302, 323
metal stud 483
millwork 360
millwork & trim 360
mold contaminated area 306
paneling 360
partition 483
pavement 300
plaster 482, 483
plenum 483
plumbing 556
plywood 482, 483
post 358
rafter 358
railing 360
roofing 416
selective building 302
site 300
subfloor 482
tile 482
truss 360
walls and partitions 483
window 445
window & door 445
wood 482
wood framing 356
Dentil crown molding 395
Derrick crane guyed 669
  crane stiffleg 669
Detection intrusion 616
  system 616
Detector infrared 616
  motion 616
  smoke 616
Device anti-siphon 566
  combination 601
  decorator 601
  GFI 602
  receptacle 602
  residential 600
  wiring 607
Dewatering equipment 669
Diaphragm pump 665
Diesel hammer 662
  tugboat 670
Diffuser ceiling 583
  linear 583
  opposed blade damper 583
  perforated aluminum 583
  rectangular 583
  steel 584
  T-bar mount 584
Dimmer switch 600, 607
Directional drill horizontal 667
Disappearing stair 530
Disc harrow 661
Discharge hose 665
Disconnect safety switch 608
Dishwasher residential 528
Disposal 302
  field 650
  garbage 529
Disposer garbage 236
Distribution box 651
  box concrete 651

Index

box flow leveler 651	patio 457	track 664	scraper 662
box HDPE 651	plastic laminate 449	wood 362, 641	vibrator 630
pipe water 648, 649	porch 451	Drilling concrete 319	Earthwork 619
Distributor asphalt 664	prefinished 449	concrete impact 319	equipment 620
Diving board 542	pre-hung 454, 455	plaster 483	equipment rental 660
stand 542	removal 445	rig 660	Eave vent 440
Domed skylight 472	residential 408, 448, 451	steel 337	Edge drip 439
Door & window interior paint ... 511,	residential garage 457	Drinking fountain support 575	Edger concrete floor 660
512	residential steel 448	Drip edge 439	Edging 646
accessories 477	residential storm 446	edge aluminum 439	aluminum 646
accordion 452	rough buck 373	Driver post 667	curb 637
and panel access 456	sauna 544	sheeting 664	Effluent filter septic system 651
and window material 445	sectional 458	Driveway 110, 634	EIFS 423
bell residential 603	sectional overhead 457	removal 300	Elastomeric membrane 494
bi-fold 448, 451	shower 520	Drum trap copper 569	Elbow downspout 437
bi-passing closet 452	sidelight 448, 454	Dry fall painting 516	Electric appliance 528
birch exterior 455	silencer 475	pipe sprinkler head 553	baseboard heater 593
birch hollow core 453	sill 402, 408, 474	Dryer accessories washer 574	boiler 586
birch solid core 451	sliding aluminum 457	receptacle 603	cable 597
blind 409	sliding glass 457	vent 529, 575	fixture 609-611
bulkhead 456	sliding glass vinyl clad 457	wiring 280	furnace 587
bulkhead/cellar 456	sliding vinyl clad 457	Dry-pipe sprinkler system 553	generator 664
bumper 474	sliding wood 457	Drywall 214, 216, 488	heater 529
cafe 454	special 456	accessories 491	heating 279, 593
canopy 524	stain 511	column 488	lamp 612
casing PVC 411	steel 447, 724	cutout 303	log 522
catch 537	stop 474	demolition 483	metallic tubing 598
chime 608	storm 190, 446	finish 727	pool heater 586
cleanout 522	switch burglar alarm 616	frame 446	service 278, 608
closet 448, 452	threshold 408	gypsum 488	stair 530
colonial 451, 455	torrified 454	nail 361	switch 607, 608
combination storm ... 451, 455	varnish 511	painting 515, 516	vehicle charging 609
composite 454	weatherstrip 476	prefinished 489	water heater residential ... 571
decorator wood 450	weatherstrip garage 477	screw 491	Electrical cable 598
demolition 445	wood 224, 226, 448, 449	Duck tarpaulin 296	conduit 598
double 448	wood panel 450	Duct accessories 582	demolition 596
dutch 451	wood storm 455	fire rated blanket 578	wire 597
dutch oven 522	Doorbell system 608	flexible 582	Electronic air cleaner 585
embossed metal 455	Dormer framing 150, 152	flexible insulated 582	Elevator 546
entrance 451, 455	gable 380	flexible noninsulated 582	residential 546
exterior 182, 455	roofing 204, 206	HVAC 581	Embossed metal door 455
exterior pre-hung 455	Double hung bay wd. window .. 464	insulation 578	print door 450
fiberglass 454	hung window 466	liner 582	Employer liability 293
fireplace 522	hung wood window 463	liner non-fibrous 582	EMT 598
flush 449, 451	Double-hung window 170	mechanical 581	Emulsion adhesive 499
frame 446	Dowel cap 312	thermal insulation 578	asphaltic 642
frame exterior wood 408	reinforcing 312	Ductile iron fitting 648	pavement 634
frame interior 408	sleeve 312	iron fitting mech joint 648	penetration 635
french 455	Downspout 436, 437	iron pipe 648	sprayer 664
french exterior 455	aluminum 436	Ductless split system 590	Encased railing 412
garage 186, 457	copper 437	Ductwork aluminum 581	Enclosure bathtub 521
glass 457	demolition 416	demolition 578	shower 521
hand carved 449	elbow 437	fabric coated flexible 582	swimming pool 544
handle 537	lead coated copper 437	fabricated 581	Engineered stone countertop ... 539
hardboard 451	steel 437	galvanized 581	Entrance cable service 597
hardware 473	strainer 436	metal 581	door 451, 455
hardware accessory 475	Dozer 662	rectangular 581	door fibrous glass 454
interior pre-hung 456	Dragline bucket 661	rigid 581	frame 408
jamb molding 402	Drain 568	Dump charge 304	lock 474
kennel 456	floor 570	truck 663	weather cap 599
knob 473	roof 570	Dumpster 303	EPDM flashing 436
knocker 473, 475	sanitary 570	Dumpsters 719	Epoxy anchor 319, 336
labeled 447	shower 568	Dump truck off highway 663	grout 492
louvered 453, 456	Drainage accessories 437	Duplex receptacle 280, 607	wall coating 517
mahogany 449	field 651	Dust barrier 296	welded wire 312
metal 446	pipe 650, 651	Dutch door 451	Equipment 660
metal faced 455	trap 569	oven door 522	earthwork 620
metal fire 447	Drapery hardware 534	DWV pipe ABS 562	insurance 293
mirror 479	ring 534	PVC pipe 562, 650	pad 314
molding 402	Drawer tracks 537	tubing copper 560	rental 660, 718
moulded 450	Drill core 660		rental concrete 660
opener 458	main 649		rental earthwork 660
overhead 457	quarry 664	**E**	rental general 663
panel 450, 451	rig 300		rental highway 667
paneled 448	steel 664	Earth auger 660	rental lifting 668
passage 452, 456	tap main 649	compactor 661	rental marine 670

Index

rental wellpoint 669
swimming pool 542
Erosion control synthetic 631
Escutcheon plate 552
sprinkler 552
Estimate electrical heating 593
Estimates window 724
Estimating 711
Evap/condens unit line set 591
Evaporator coils cased 588
Evergreen shrub 644
tree 644
Excavating bulk bank measure .. 620
trench 619
utility trench 620
Excavation 619, 620
footing 102
foundation 104
hand 619, 620
planting pit 643
septic tank 651
structural 620
tractor 619
utility 106
Excavator bucket 667
hydraulic 660
Exhaust hood 529
Expansion anchor 336
joint 311, 486
shield 336
tank 580
tank steel 580
Exposed aggregate 317
aggregate coating 517
Extension backhoe 667
ladder 665
Exterior blind 409
concrete block 328
door 182, 455
door frame 408
fixture lighting 611
insulation 423
insulation finish system 423
light 281
lighting fixture 611
molding 395
paint doors & windows 508
plaster 487
pre-hung door 455
PVC molding 411
residential door 451
shutter 409
siding painting 507
surface preparation 503
trim 395
trim paint 509
wall framing 136
wood door frame 408
wood frame 408
Extinguisher chemical dry 523
fire 523
portable ABC 523
standard 523
Extruder curb 661

F

Fabric welded wire 312
Fabricated ductwork 581
Fabrication composite 412
Face brick 327, 721
brick cleaning 720
Facing stone 331
Fall painting dry 516
Fan attic ventilation 583

bathroom exhaust 582
house ventilation 583
kitchen exhaust 582
paddle 604
residential 604
ventilation 604
wiring 280, 604
Fascia aluminum 428
board 376
board demolition 357
metal 436
PVC 411
wood 398
Fastener timber 361
wood 361
Fastening alternative 422
Faucet & fitting 573
bathroom 573
laundry 574
lavatory 574
Fee architectural 292
Feller buncher 661
Felt asphalt 431
carpet pad 501
Fence 506
and gate 638
board 639
board batten 640
cedar 639, 640
chain link 638
metal 638
misc metal 639
open rail 639
picket paint 506
residential 638
vinyl 638
wire 639
wood 639
wood rail 640
Fencing demolition 301
Fertilizer 643
Fiber cement siding 431
reinforcing concrete 316
steel 313
synthetic 313
Fiberboard insulation 421
Fiberglass area wall cap 525
bay window 471
blown-in 420
ceiling board 495
door 454
insulation 418-420
panel 427
planter 642
single hung window 471
slider window 471
tank 579
window 471
wool 420
Field disposal 650
drainage 651
seeding 643
Fieldstone 330
Fill 620, 621
by borrow & utility bedding .. 621
gravel 618, 621
Filler block 516
cabinet 537
crack 667
joint 441
Film polyethylene 642
security 479
Filter air 585
mechanical media 584
swimming pool 576
Filtration equipment 542

Fine grade 643
Finish drywall 488
floor 499, 517
grading 619
nail 361
wall 502
Finishing floor concrete 317
wall concrete 317
Fir column 407
floor 498
molding 401
roof deck 384
roof plank 384
Fire alarm 552
call pullbox 616
damage repair 505
damper curtain type 582
door frame 447
door metal 447
escape stair 351
extinguisher 523
extinguisher portable 523
extinguishing system 548, 551
horn 616
hose gate valve 552
hose rack 551
hose valve 551
hydrant building 550
rated blanket duct 578
resistant drywall 489, 490
sprinkler head 553
Fire-department connection 550
Fireplace accessories 521
box 332
built-in 522
chimney 521
damper 522
demolition 323
door 522
form 522
free standing 522
mantel 406
mantel beam 407
masonry 238, 332
prefabricated 240, 521, 522
Firestop gypsum 485
wood 373
Fire-suppression pipe fitting 548
plastic pipe 548
Fitting cast iron 567
cast iron pipe 567
conduit 599
copper 560
copper pipe 560
ductile iron 648
DWV pipe 564
malleable iron 561
plastic 564
plastic pipe 548
PVC 564
steel 561
Fixed window 180
Fixture bathroom 571-573
carrier/support 575
electric 609
fluorescent 609, 610
incandescent 609
interior light 609
landscape 611
lantern 610
mirror light 610
plumbing 571, 580
removal 556
residential 603, 610
Flagging 497, 636
Flagpole 526

aluminum 526
Flasher barrels 664
barricade 664
Flashing 434, 435
aluminum 435
butyl 436
cement 416
copper 435, 569
EPDM 436
laminated sheet 435
masonry 435
neoprene 436, 570
plastic sheet 435
PVC 435
self-adhering 436
sheet metal 435
stainless 435
vent 569
Flat seam sheet metal roofing ... 434
Flatbed truck 663, 666
truck crane 668
Flexible conduit & fitting 600
duct 582
ductwork fabric coated 582
insulated duct 582
metallic conduit 599
noninsulated duct 582
sprinkler connector 552
Float finish concrete 317
glass 478
Floater equipment 293
Floating floor 496
pin 475
Floodlight trailer 664
tripod 664
Floor 497
acid proof 497
asphalt block 636
brick 497
carpet 501
ceramic tile 492
cleanout 558
concrete finishing 317
conductive 500
cork 497
drain 570
finish 517
flagging 636
framing 130, 132, 134
framing log 382
framing removal 303
heating radiant 592
insulation 419
marble 331
nail 361
oak 498
paint 513
paint & coating interior 513
parquet 498
plank 383
plywood 384
polyethylene 502
quarry tile 493
register 584
removal 302
rubber 499
rubber and vinyl sheet 499
sander 666
stain 513
subfloor 384
tile terrazzo 500
transition strip 501
underlayment 385
varnish 513, 517
vinyl 500
vinyl sheet 499

743

Index

wood	498
wood composition	498
Flooring	229
bamboo	496
conductive rubber & vinyl	500
demolition	482
masonry	497
miscellaneous brick	497
treatment	496
wood strip	498
Flue chimney	332
chimney metal	585
liner	326, 332
screen	521
tile	332
Fluid applied membrane	
air barrier	424
Fluorescent fixture	609, 610
Flush door	449, 451
wood door	450
Foam board insulation	418
core DWV ABS pipe	562
insulation	420
roofing	433
spray rig	664
Foamed coated roofing	433
in place insulation	419
Fog seal	634
Foil aluminum	420, 424
metallic	421
Folding door shower	521
Footing	118
concrete placement	316
excavation	102
formwork continuous	308
keyway	308
keyway formwork	308
reinforcing	312
removal	302
spread	314
Forest stewardship council	370
Forklift	664
crew	293
Form decking	340
fireplace	522
Formwork column	308
column accessories	310
column plywood	308
column round fiber tube	308
column steel framed	308
continuous footing	308
footing keyway	308
grade beam	308
insert concrete	311
insulating concrete	309
slab blockout	309
slab bulkhead	308
slab curb	309
slab edge	309
slab haunch	314
slab on grade	308
slab thickened edge	314
slab turndown	314
slab void	309
sleeve	310
snap-tie	310
spread footing	308
stair	310
stake	310
wall	309
wall accessories	310
wall boxout	309
wall brick shelf	309
wall bulkhead	309
wall plywood	309
wall prefab plywood	309
Foundation	118
chimney	314
concrete	122
concrete block	328
excavation	104
mat	314
mat concrete placement	316
removal	301
wall	328
wood	124
Fountain indoor	543
outdoor	543
yard	543
Frame baked enamel	447
door	446
drywall	446
entrance	408
exterior wood door	408
fire door	447
labeled	447
metal	446
metal butt	447
scaffold	294
steel	446
welded	447
window	463
Framing anchor	362
band joist	345
beam & girder	371
boxed headers/beams	346
bracing	348
bridging	348
canopy	377
ceiling	372
columns	372
deck	374, 410
demolition	357
dormer	150, 152
floor	130, 132, 134
heavy	383
joist	372
laminated	388
ledger & sill	377
load-bearing stud partition	342
load-bearing stud wall	342
log floor	382
log roof	383
metal joist	346
metal roof parapet	348
miscellaneous wood	373
NLB partition	484
open web joist wood	372
partition	154
porch	374
removal	356
roof	138, 140, 142, 144, 146, 148
roof metal rafter	349
roof metal truss	350
roof rafter	376
roof soffit	349
sill & ledger	377
sleeper	377
soffit & canopy	377
timber	383
treated lumber	377
wall	378
web stiffener	347
wood	370, 371
Freeze deep	528
French door	455
exterior door	455
Frieze PVC	411
Front end loader	620
Fuel furnace solid	587
Furnace A/C coils	588
components	588
demolition	578
electric	587
gas fired	587
hot air	587
oil fired	587
wiring	280
wood	587
Furring and lathing	483
ceiling	380, 483, 485
channel	483, 491
concrete	380
masonry	380
metal	483
steel	483
wall	380, 484
wood	380
Fusion machine butt	664

G

Gable dormer	380
dormer framing	150
dormer roofing	204
roof framing	138
Galley septic	651
Galvanized ductwork	581
welded wire	312
Gambrel roof framing	144
Garage door	186, 457
door residential	457
door weatherstrip	477
Garbage disposal	529
Gas conversion burner	587
fired boiler	586
fired furnace	587
fired heating	268
furnace wood	587
generator	608
generator set	608
heat air conditioner	590
heating	272
log	522
pipe	652, 653
service steel piping	653
vent	585
water heater instantaneous	571
water heater residential	571
water heater tankless	571
Gasket joint	441
neoprene	441
Gas/oil combination boiler	586
Gasoline generator	608
Gate metal	638
General contractor's overhead	715
equipment rental	663
fill	621
Generator electric	664
emergency	608
gas	608
gasoline	608
set	608
Geo-grid	641
GFI receptacle	602
Girder wood	371, 383
Glass	478, 479
block	329
demolition	445
door	457
door shower	521
door sliding	457
float	478
heat reflective	478
lined water heater	529
mirror	479
mosaic	493, 494
mosaic sheet	494
plate	478
reduce heat transfer	478
sheet	479
shower stall	520
tempered	478
tile	479
tinted	478
window	479
Glaze coat	515
Glazed ceramic tile	492
wall coating	517
Glazing	478
application	478
Glued laminated	388
Grab bar	520
Gradall	661
Grade beam formwork	308
fine	643
Grader motorized	661
Grading	618
and seeding	643
slope	619
Granite building	330
chip	643
countertop	539
curb	637
demolition	323
Indian	637
paver	330
reclaimed or antique	330
sidewalk	637
Grass cloth wallpaper	503
lawn	643
seed	643
Grating area wall	525
area way	525
Gravel base	634
fill	618, 621
pea	643
stop	436
Gravity retaining wall	640
Green roof membrane	427
roof soil mixture	427
roof system	427
Greenfield conduit	599
Greenhouse	242, 544
residential	544
Grid spike	362
Grille air return	584
aluminum	584
decorative wood	406
painting	513
window	467
Grinder concrete floor	660
Ground	381
clamp water pipe	598
cover plant	644
fault	280
rod	598
wire	598
Grounding	598
& conductor	597
Grout	324
cement	324
concrete	324
concrete block	324
epoxy	492
pump	666
tile	492
wall	324
Guard gutter	439
snow	440
Guardrail scaffold	294
Gun type burner	587
Gunite pool	243

Index

Gutter 438
   aluminum 438
   copper 438
   demolition 416
   guard 439
   lead coated copper 438
   stainless 438
   steel 438
   strainer 439
   vinyl 438
   wood 438
Gutting 304
   demolition 304
Guyed derrick crane 669
Gypsum block demolition 303
   board 488
   board accessory 491
   cover board 421
   drywall 488
   fabric wallcovering 502
   firestop 485
   lath 485
   lath nail 361
   plaster 486
   restoration 482
   sheathing 387
   sound dampening panel 490
   underlayment poured 319
   wallboard repair 482
   weatherproof 387

H

Half round molding 401
   round vinyl clad window 465
   round vinyl window 470
Hammer bush 317
   chipping 664
   demolition 661
   diesel 662
   drill rotary 664
   hydraulic 661, 665
   pile 661
   vibratory 662
Hammermill 667
Hand carved door 449
   clearing 618
   excavation 619, 620
   hole 648
   split shake 425
   trowel finish concrete 317
Handicap lever 474
   ramp 314
Handle door 537
   latch 477
Handrail oak 231
   wood 401, 406
Hand-split shake 164
Hanger beam 362
   joist 362
   pipe 557
   plastic pipe 557
Hardboard cabinet 536
   door 451
   molding 403
   overhead door 457
   paneling 402
   tempered 402
   underlayment 385
Hardware 473
   apartment 473
   cabinet 537
   door 473
   drapery 534
   finishes 725

   motel/hotel 473
   window 473
Hardwood floor 498
   grille 406
Harrow disc 661
Hauling 621
   cycle 621
Haunch slab 314
Hay 642
HDPE distribution box 651
   infiltration chamber 651
Head sprinkler 552
Header load-bearing stud wall 342
   pipe wellpoint 669
   wood 378
Headers/beams boxed 346
Hearth 332
Heat baseboard 591
   electric baseboard 593
   pump 591
   pump air-source 591
   pump residential 605
   reflective glass 478
   temporary 308
Heater air handler 589
   electric 529
   gas residential water 571
   infrared quartz 594
   quartz 594
   sauna 544
   space 665
   swimming pool 586
   water 529, 605
   wiring 280
Heating 268, 270, 272
   electric 279, 593
   estimate electrical 593
   hot air 587
   hydronic 586, 591
   insulation 559
   kettle 668
   panel radiant 594
Heavy duty shoring 294
   framing 383
   timber 383
Helicopter 669
Hemlock column 407
Hex bolt steel 337
Highway equipment rental 667
   paver 636
Hinge 475, 476
   brass 475
   cabinet 537
   continuous 476
   mortised 476
   paumelle 476
   residential 475
   stainless steel 475
   steel 475
Hip rafter 376
   roof framing 142
Hoist chain fall 669
   personnel 669
   tower 669
Holdown 362
Hollow core door 449, 453
   metal frame 446
   wall anchor 336
Home theater 614
Hood exhaust 529
   range 529
Hook robe 520
Horizontal aluminum siding 428
   borer 667
   boring machine 660
   directional drill 667

Horn fire 616
Hose air 664
   bibb sillcock 574
   discharge 665
   rack 551
   suction 665
   water 665
Hospital tip pin 475
Hot air furnace 587
   air heating 587
   tub 543
   water boiler 586
   water heating 586, 591
Hot-air heating 268, 270
Hotel lockset 473
Hot water heating 272
House flood safety shut-off 557
   ventilation fan 583
Housewrap 423
Humidifier 530
Humus peat 642
HVAC demolition 578
   duct 581
Hydrant building fire 550
Hydraulic crane 669
   excavator 660
   hammer 661, 665
   jack 669
   jack climbing 669
   seeding 643
Hydromulcher 667
Hydronic heating 586, 591

I

Icemaker 528
I-joist wood & composite 387
Impact wrench 664
Impact-resistant aluminum
   window 459
   replacement window 444
Incandescent fixture 609, 610
   light 281
Indian granite 637
Indoor athletic carpet 502
   fountain 543
   load center 605
Infiltration chamber 651
   chamber HDPE 651
Infrared detector 616
   quartz heater 594
Inlet curb 637
Insecticide 631
Insert concrete formwork 311
Instantaneous gas water heater 571
Insulated panel 423
   protector ADA 559
Insulating concrete formwork 309
   sheathing 384
Insulation 418, 420
   batt 419
   blanket 578
   blower 665
   blown-in 420
   board 211, 418
   building 420
   cavity wall 420
   ceiling 419, 420
   cellulose 420
   composite 422
   duct 578
   duct thermal 578
   exterior 423
   fiberglass 418-420
   finish system exterior 423

   floor 419
   foam 420
   foamed in place 419
   heating 559
   insert concrete block 327
   isocyanurate 418
   loose fill 420
   masonry 420
   mineral fiber 419
   pipe 559
   polystyrene 418, 420
   reflective 420
   removal 416
   rigid 418
   roof 421
   roof deck 421
   sprayed-on 421
   vapor barrier 424
   vermiculite 420
   wall 418
   wall blanket 419
   water heater 559
Insurance 293, 715
   builder risk 293
   equipment 293
   public liability 293
Interior door 224, 226
   door frame 408
   LED fixture 610
   light 281
   light fixture 609
   lighting 610
   paint doors & windows 511
   paint wall & ceiling 516
   partition framing 154
   pre-hung door 456
   residential door 451
   shutter 535
   wood door frame 408
Interval timer 601
Intrusion detection 616
   system 616
Ironing center 521
Ironspot brick 497
Irrigation system sprinkler 642
Isocyanurate insulation 418

J

J hook clamp w/nail 557
Jack cable 669
   hydraulic 669
   rafter 376
Jackhammer 663
Joint control 311
   expansion 311, 486
   filler 441
   gaskets 441
   push-on 648
   reinforcing 325
   restraint 648, 649
   sealer 441, 442
Joist connector 362
   demolition 357
   framing 372
   hanger 362
   metal framing 346
   removal 357
   sister 372
   web stiffener 347
   wood 372, 387
Jute mesh 631

Index

K

Kennel door	456
fence	639
Kettle heating	668
tar	666, 668
Keyway footing	308
Kick plate	476
Kitchen	234
appliance	528
cabinet	536
exhaust fan	582
sink	572
sink faucet	573
sink residential	572
unit commercial	530
Knob door	473
Knocker door	473, 475

L

Labeled door	447
frame	447
Ladder extension	665
swimming pool	542
Lag screw	338
screw shield	336
Laminated beam	388
countertop	538
framing	388
glued	388
roof deck	384
sheet flashing	435
veneer members	388
wood	388
Lamp electric	612
LED	612
post	352
Lampholder	607
Lance concrete	665
Landfill compactor	662
Landscape fixture	611
light	611
surface	502
Lantern fixture	610
Laser level	665
Latch handle	477
set	473
Latex caulking	442
underlayment	499
Lath gypsum	485
Lath, plaster and gypsum board	726
Lattice molding	401
Lauan door	452
Laundry faucet	574
sink	573
Lavatory	248
commercial	575
faucet	574
pedestal type	572
removal	556
residential	572, 575
sink	572
support	575
vanity top	572
wall hung	572
Lawn grass	643
mower	665
seed	643
Lazy Susan	536
Leaching pit	650
Lead coated copper downspout	437
coated copper gutter	438
coated downspout	437
containment area	304

flashing	435
paint encapsulation	305
paint removal	305
roof	434
screw anchor	336
Leads pile	661
Lean-to type greenhouse	544
LED fixture interior	610
lamp	612
Ledger & sill framing	377
Let-in bracing	370
Letter slot	523
Level laser	665
Leveling jack shoring	294
Lever handicap	474
Liability employer	293
insurance	715
Lift	546
aerial	663
scissor	663
truck aerial	668
Lifting equipment rental	668
Light bollard	611
fixture interior	609
landscape	611
post	610
support	339
tower	665
track	611
Lighting	280, 612
exterior fixtures	611
incandescent	610
interior	610
outlet	603
pole	611
residential	603, 610
strip	609
track	611
Lightning arrester	609
protection	609
suppressor	600
Lightweight block	329
column	339
Limestone	331, 643
coping	333
Line remover traffic	668
set	591
sets refrigerant	580
Linear diffuser	583
Linen wall covering	503
Liner duct	582
flue	326, 332
non-fibrous duct	582
Lintel	331
block	328
concrete block	328
precast concrete	318
steel	339
Load center indoor	605
center plug-in breaker	605
center rainproof	605
center residential	605
Load-bearing stud partition framing	342
stud wall bracing	341
stud wall bridging	341
stud wall framing	342
stud wall header	342
Loader front end	620
skidsteer	663
tractor	662
wheeled	662
windrow	668
Loam	618
Lock entrance	474
Locking receptacle	603

Lockset communicating	473
cylinder	473
hotel	473
mortise	473
Locomotive tunnel	668
Log chipper	662
electric	522
gas	522
skidder	662
structure	382
wall	382
Loose fill insulation	420
Louver	479
aluminum	479
midget	479
redwood	480
ventilation	480
wall	480
wood	406
Louvered blind	409
door	224, 226, 453, 456
Lowbed trailer	668
Low-voltage silicon rectifier	607
switching	607
switchplate	607
transformer	607
Lumber	371
core paneling	403
plastic	410
product prices	723
recycled plastic	410
structural plastic	410
Luminaire walkway	611
Luminous ceiling	612

M

Macadam	635
penetration	635
Machine auto-scrub	664
excavation	620
trowel finish concrete	317
welding	667
Mahogany door	449
Mail box	523
Malleable iron fitting	561
iron pipe fitting	561
Management fee construction	292
Manhole brick	652
removal	301
Manifold roof fire valve	550
Man-made soil mix	427
Mansard roof framing	146
Mantel beam	407
fireplace	406
Maple countertop	538
Marble	331
chip	643
coping	333
countertop	538, 539
floor	331
sill	332
synthetic	498
tile	498
Marine equipment rental	670
Marker boundary and survey	300
Mason scaffold	294
Masonry anchor	324
brick	326
cement	324
cleaning	322
cornice	330
demolition	302, 323
fireplace	238, 332
flashing	435

flooring	497
furring	380
insulation	420
nail	361
painting	515
pointing	322
reinforcing	325, 720
removal	301, 323
saw	666
selective demolition	322
sill	332
step	634
toothing	303
wall	158, 160, 327, 641
wall tie	324
Mat concrete placement foundation	316
foundation	314
Material door and window	445
Mechanical duct	581
media filter	584
seeding	643
Medicine cabinet	520
Membrane	211
elastomeric	494
roofing	431
Metal bin retaining wall	641
bracing	370
butt frame	447
chimney	521
clad window	170, 172, 174, 176, 178, 180
door	446
door residential	448
ductwork	581
faced door	455
fascia	436
fence	638
fireplace	240
flue chimney	585
frame	446
framing parapet	348
furring	483
gate	638
halide fixture	281
joist bracing	344
joist bridging	345
joist framing	346
nailing anchor	336
overhead door	457
rafter framing	349
roof	427
roof parapet framing	348
roof window	471
sheet	434
shelf	523
siding	166
soffit	431
stud demolition	483
stud NLB	484
support assemblies	483
threshold	474
tile	494
truss framing	350
window	459
Metal-clad double hung window	464
window bow & bay	464
window picture & sliding	464
Metallic foil	421
Meter center rainproof	606
device basic	606
socket	606
water supply	566
water supply domestic	566
Metric conversion factors	712
Microtunneling	648

Index

Microwave oven	528
Mill construction	383
Millwork	391
&trim demolition	360
demolition	360
Mineral fiber ceiling	495
fiber insulation	419
roof	433
Minor site demolition	301
Mirror	479
door	479
glass	479
light fixture	610
wall	479
Miscellaneous painting	505, 510
Mix planting pit	643
Mixer concrete	660
mortar	660, 665
plaster	665
road	667
Mixing valve	574
Mobilization or demobilization	296
Modification to cost	292
Modified bitumen roof	432
bitumen roofing	723
bituminous barrier sheet	424
bituminous membrane SBS	432
Modular air handler	589
Module tub-shower	573
Moil point	664
Mold contaminated area demolition	306
Molding base	391
bed	394
birch	401
brick	401
casing	392
ceiling	394
chair	401
cornice	394, 397
cove	394
crown	394
dentil crown	395
exterior	395
hardboard	403
pine	396
soffit	402
trim	401
window and door	402
wood transition	499
Monitor support	339
Monument survey	300
Mortar, brick and block	721
Mortar cement	493, 494, 720
mixer	660, 665
thinset	493
Mortise lockset	473
Mortised hinge	476
Mosaic glass	493, 494
Moss peat	642
Motel/hotel hardware	473
Motion detector	616
Motor connections	600
support	339
Motorized grader	661
Moulded door	450
Mounting board plywood	386
Movable louver blind	535
Movable bulkhead	542
Moving shrub	646
tree	646
Mower lawn	665
Muck car tunnel	668
Mud pump	660
trailer	667
Mulch bark	642
ceramic	642
stone	642
Mulcher power	662
Mulching	642
Multi-blade damper	582
Muntin	460
window	467

N

Nail	361
common	361
stake	310
Nailer pneumatic	664, 665
steel	373
wood	373
Nailing anchor	336
Natural fiber wall covering	502
Neoprene flashing	436, 570
gasket	441
Newel	231
wood stair	405
No hub pipe	567
hub pipe fitting	568
Non-chloride accelerator concrete	316
Non removable pin	475
Nylon carpet	228, 501
nailing anchor	336

O

Oak door frame	408
floor	229, 498
molding	401
paneling	403
threshold	408, 474
Off highway dump truck	663
Oil fired boiler	586
fired furnace	587
furnace wood	587
heating	270, 272
Oil-fired boiler	586
water heater residential	571
Olefin carpet	501
One piece astragal	476
One-way vent	440
Onyx	500
Open rail fence	639
Opener door	458
Ornamental column	353
OSB faced panel	381
subfloor	385
Outdoor cabinet	538
fountain	543
Outlet box plastic	599
box steel	598
lighting	603
Oven	528
cabinet	537
microwave	528
Overhaul	304
Overhead contractor	715
door	186, 457
Overlay face door	449
Oversized brick	326

P

P trap	569
trap running	569
Packaged terminal air-conditioner	590
Pad carpet	501
commercial grade carpet	501
equipment	314
Padding	228
carpet	501
Paddle fan	604
Paint & coating	503
& coating interior floor	513
aluminum siding	507
chain link fence	506
doors & windows exterior	508
doors & windows interior	511, 512
encapsulation lead	305
exterior miscellaneous	506
fence picket	506
floor	513
floor concrete	513
floor wood	513
interior miscellaneous	513
removal	305
removal lead	305
siding	507
sprayer	665
striper	667, 668
trim exterior	509
walls & ceilings interior	515, 516
wall masonry exterior	510
Paint/coating cabinet & casework	510
Painting	728
balustrade	514
casework	510
ceiling	515, 516
clapboard	507
concrete block	515
cornice	514
decking	506
drywall	515, 516
exterior siding	507
grille	513
masonry	515
miscellaneous	505, 510
pipe	514
plaster	515, 516
railing	506
shutter	506
siding	507
stair stringer	506
steel siding	507
stucco	507
swimming pool	542
trellis/lattice	506
trim	514
truss	514
wall	507, 516
window	511
Palladian window	465
Panel and device alarm	616
door	450, 451
fiberglass	427
insulated	423
OSB faced	381
prefabricated	423
radiant heat	594
shearwall	381
sound dampening gypsum	490
structural	381
structural insulated	381
system	403
wood folding	535
Panelboard w/circuit-breaker	605
Paneled door	224, 448
pine door	453
Paneling	402
birch	403
board	404
cedar	404
cutout	303
demolition	360
hardboard	402
plywood	403
redwood	404
wood	402
Panelized shingle	425
Paper building	424
sheathing	424
Paperhanging	502
Parapet metal framing	348
Parging cement	417
Park bench	642
Parquet floor	229, 498
wood	498
Particle board underlayment	385
core door	449
Parting bead	402
Partition	214, 218
anchor	324
block	329
concrete block	329
demolition	483
framing	154
framing load-bearing stud	342
framing NLB	484
shower	520
support	339
wood frame	373
Passage door	224, 452, 456
Patch roof	416
Patching concrete floor	308
Patio door	457
Paumelle hinge	476
Pavement	636
berm	637
breaker	663
bucket	667
demolition	300
emulsion	634
planer	668
profiler	668
widener	668
Paver asphalt	665
bituminous	665
cold mix	667
concrete	665
floor	497
highway	636
roof	440
shoulder	667
Paving	635
asphalt	635
brick	636
surface treatment concrete	636
Pea gravel	643
Peastone	644
Peat humus	642
moss	642
Pedestal type lavatory	572
Pegboard	402
Penetration macadam	635
Perforated aluminum pipe	652
pipe	652
Perlite insulation	418
plaster	486
Permit building	293
Personnel hoist	669
Pex pipe	563
tubing	563, 592
tubing fitting	593
Picket fence	640
fence vinyl	638
railing	352
Pickup truck	666

747

For customer support on your Residential Cost Data, call 877.759.4771.

Index

Pick-up vacuum 660
Picture window 180, 463
   window vinyl 470
Pier brick 326
Pilaster wood column 407
Pile driving 718
   hammer 661
   leads 661
   sod 643
Pin powder 338
Pine door 451
   door frame 408
   fireplace mantel 406
   floor 498
   molding 396
   roof deck 384
   shelving 524
   siding 429
   stair tread 404
Pipe & fitting backflow
preventer 566
   & fitting bronze 556
   & fitting copper 560
   & fitting DWV 564
   & fitting malleable iron .. 562
   & fitting PVC 562
   and fittings 730
   bedding 621
   cast iron 567
   clamp plastic 557
   cleanout 558
   concrete 651
   copper 559
   corrugated metal 652
   covering 559
   covering fiberglass 559
   CPVC 562
   drainage 650, 651
   ductile iron 648
   DWV PVC 562, 650
   fire-suppression plastic .. 548
   fitting cast iron 567
   fitting copper 560
   fitting DWV 564
   fitting fire-suppression .. 548
   fitting no hub 568
   fitting plastic 548, 564-566
   fitting soil 567
   foam core DWV ABS 562
   gas 652, 653
   hanger 557
   hanger plastic 557
   insulation 559
   no hub 567
   painting 514
   perforated aluminum 652
   PEX 563
   plastic 548, 562
   polyethylene 652
   PVC 562, 650
   rail aluminum 351
   rail galvanized 352
   rail stainless 352
   rail steel 352
   rail wall 352
   railing 351
   removal 301, 556
   removal plastic 556
   residential PVC 563
   sewage 650
   sewage collection PVC 650
   single hub 567
   sleeve plastic 310
   soil 567
   steel 653
   supply register spiral 584

Piping designations 731
Piping gas service polyethylene 652
   gas service steel 653
Pit leaching 650
Pitch coal tar 432
   emulsion tar 634
Placing concrete 316
Planer pavement 668
Plank floor 383
   roof 384
   scaffolding 295
Plant and bulb transplanting . 646
   bed preparation 643
   ground cover 644
   screening 662
Plantation shutters 535
Planter 642
   fiberglass 642
Planting 646
Plaster 218, 220
   accessories 486
   beam 486
   board 214, 216
   ceiling 486
   column 486
   cutout 303
   demolition 482, 483
   drilling 483
   ground 381
   gypsum 486
   mixer 665
   painting 515, 516
   perlite 486
   soffit 486
   thincoat 488
   venetian 487
   wall 486
Plasterboard 488
Plastic blind 191
   clad window .. 170, 172, 174, 176, 178, 180
   coupling 565
   faced hardboard 403
   fitting 564
   laminate door 449
   lumber 410
   lumber structural 410
   outlet box 599
   pipe 548, 562
   pipe clamp 557
   pipe fire-suppression 548
   pipe fitting 564-566
   pipe hanger 557
   pipe removal 556
   railing 411
   screw anchor 337
   sheet flashing 435
   siding 166
   skylight 472
   trap 569
   vent caps 569
   window 468
Plastic-laminate-clad countertop 538
Plate escutcheon 552
   glass 478
   roadway 668
   shear 362
   steel 340
   stiffener 340
   wall switch 607
Platform trailer 666
Plating zinc 361
Player bench 642
Plenum demolition 483
Plow vibrator 663
Plug in circuit breaker 605

in tandem circuit breaker ... 606
Plugmold raceway 607
Plumbing 566
   demolition 556
   fixture 571, 580
   fixtures removal 556
Plywood 723
   clip 362
   demolition 482, 483
   floor 384
   joist 387
   mounting board 386
   paneling 403
   sheathing roof & walls 386
   shelving 524
   siding 429
   soffit 402
   subfloor 385
   underlayment 385
Pneumatic nailer 664, 665
Pocket door 473
   door frame 408
Point moil 664
Pointing CMU 322
   masonry 322
Pole closet 401
   lighting 611
Police connect panel 616
Polyethylene film 642
   floor 502
   pipe 652
   pool cover 542
   septic tank 650
   tarpaulin 296
Polypropylene siding 430
Polystyrene blind 409
   insulation 418, 420
Polysulfide caulking 442
Polyurethane caulking 442
Polyvinyl soffit 402, 431
Pool accessories 542
   cover 542
   cover polyethylene 542
   filtration swimming 576
   heater electric 586
   swimming 542
Porcelain tile 492
Porch door 451
   framing 374
Portable air compressor 663
   asphalt plant 667
   fire extinguisher 523
Portland cement terrazzo 500
Post cap 362
   cedar 407
   demolition 358
   driver 667
   lamp 352
   light 610
   wood 372
Postformed countertop 538
Post-tensioned slab on grade . 313
Poured gypsum underlayment .. 319
   insulation 168
Powder actuated tool 338
   charge 338
   pin 338
Power mulcher 662
   trowel 660
   wiring 608
Precast catch basin 652
   concrete lintel 318
   concrete plantable paver .. 636
   concrete stairs 318
   concrete window sill 318
   coping 333

curb 637
receptor 520
terrazzo 500
Prefabricated building 544
   fireplace 240, 521
   panel 423
   wood stair 404
Prefinished door 449
   drywall 489
   floor 229, 498
   hardboard paneling 402
   shelving 524
Preformed roof panel 427
   roofing & siding 427
Pre-hung door 454, 455
Preparation exterior surface . 503
   interior surface 504
   plant bed 643
Pressure reducing valve water 556
   restricting valve 552
   valve relief 556
   wash 504
   washer 666
Prestressing steel 313
Preventer backflow 566
Prices lumber products 723
Prime coat 635
Primer asphalt 499
Privacy fence vinyl 638
Profile block 328
Profiler pavement 668
Projected window aluminum ... 459
Property line survey 300
Protection lightning 609
   slope 631
   temporary 297
   termite 631
   winter 308
Protector ADA insulated 559
P&T relief valve 556
PTAC unit 590
Pull box 599
   box electrical 599
   door 537
Pump 649
   centrifugal 665
   circulating 580
   concrete 660
   diaphragm 665
   grout 666
   heat 591
   mud 660
   shotcrete 660
   staging 294, 717
   submersible 570, 649, 665
   sump 529
   trash 666
   water 580, 649, 665
   wellpoint 669
Purlin log roof 383
   roof 383
Push-on joint 648
Putting 503
PVC conduit 598
   conduit in slab 599
   cornerboards 411
   door casing 411
   fascia 411
   fitting 564
   flashing 435
   frieze 411
   gravel stop 436
   molding exterior 411
   pipe 562, 650
   pipe residential 563
   rake 412

Index

soffit 412
trap 569
trim 411
waterstop 310

Q

Quarry drill 664
   tile 493
Quarter round molding 401
Quartz 643
   heater 594
Quoin 331

R

Raceway plugmold 607
   surface 607
   wiremold 607
Rack hose 551
Radial arch 388
Radiant floor heating 592
   heating panel 594
Radiator cast iron 591
Rafter 376
   anchor 362
   composite 376
   demolition 358
   framing metal 349
   metal bracing 348
   metal bridging 348
   roof framing 138
   wood 376
Rail aluminum pipe 351
   galvanized pipe 352
   stainless pipe 352
   steel pipe 352
   wall pipe 352
Railing composite 412
   demolition 360
   encased 412
   encased composite 412
   picket 352
   pipe 351
   plastic 411
   wood 401, 404, 406
   wood stair 405
Railroad tie 634, 646
   tie step 634
Rainproof load centers 605
   meter centers 606
Rake PVC 412
   tractor 662
Rammer/tamper 661
Ramp handicap 314
Ranch plank floor 498
Range circuit 280
   cooking 528
   hood 236, 529
   receptacle 603, 607
Ready mix concrete 315
Receptacle air conditioner 602
   device 602
   dryer 603
   duplex 607
   GFI 602
   locking 603
   range 603, 607
   telephone 603
   television 603
   weatherproof 602
Receptor precast 520
   shower 520
   terrazzo 520

Reclaimed or antique granite 330
Rectangular diffuser 583
   ductwork 581
Rectifier low-voltage silicon 607
Recycled plastic lumber 410
Reduce heat transfer glass 478
Redwood bark mulch 642
   cupola 525
   louver 480
   paneling 404
   siding 162, 429
   tub 543
   wine cellar 530
Refinish floor 498
Reflective block 329
   insulation 420
Reflectorized barrels 664
Refrigerant line sets 580
   removal 578
Refrigerated wine cellar 530
Refrigeration residential 528
Register air supply 584
   baseboard 584
Reinforcing concrete 312
   dowel 312
   footing 312
   joint 325
   masonry 325
   steel fiber 313
   synthetic fiber 313
   wall 312
Release emergency sprinkler 552
Relief valve P&T 556
   valve self-closing 556
   valve temperature 556
Removal air conditioner 578
   bathtub 556
   block wall 302
   boiler 578
   catch basin 300
   concrete 300
   curb 300
   driveway 300
   fixture 556
   floor 302
   foundation 301
   insulation 416
   lavatory 556
   masonry 301, 323
   paint 305
   pipe 301, 556
   plumbing fixtures 556
   refrigerant 578
   shingle 416
   sidewalk 301
   sink 556
   sod 643
   stone 301
   tree 618
   utility line 300
   water closet 556
   window 445
Rental equipment 660
Repair fire damage 505
Repellent water 516
Replacement windows 444
   windows impact-resistant 444
   windows 724
Residential alarm 604
   appliance 528, 604
   application 600
   bathtubs 573
   burner 587
   canopy 524
   closet door 448
   device 600

dishwasher 528
door 408, 448, 451
door bell 603
elevator 546
fan 604
fence 638
fixture 603, 610
gas water heater 571
greenhouse 544
gutting 304
heat pump 605
hinge 475
kitchen sinks 572
lavatory 572, 575
lighting 603, 610
load center 605
oil-fired water heater 571
overhead door 457
panel board 600
PVC pipe 563
refrigeration 528
roof jack 583
service 600
sinks 572
smoke detector 604
sprinkler 553
stair 404
storm door 446
switch 600
wall cap 582
wash bowl 572
water heater 529, 571, 605
water heater electric 571
wiring 600, 604
wood window 460
Resilient base 499
   flooring 229
Resquared shingle 425
Restoration gypsum 482
   window 306
Retaining wall 641
   wall cast concrete 640
   wall concrete segmental 640
   wall segmental 641
   wall stone 641
   wall timber 641
Retarder concrete 316
   vapor 423
Retractable stair 530
Ribbed waterstop 310
Ridge board 376
   shingle 425
   shingle slate 425
   vent 440
Rig drill 300
Rigid anchor 324
   conduit in trench 600
   in slab conduit 599
   insulation 168, 418
Ring drapery 534
   split 362
   toothed 362
Ripper attachment 667
Riser beech 231
   oak 231
   pipe wellpoint 670
   wood stair 405
Rivet 338
   aluminum 338
   copper 338
   stainless 338
   steel 338
   tool 338
Road berm 637
   mixer 667
   sweeper 667

Roadway plate 668
Robe hook 520
Rock trencher 663
Rod backer 441
   closet 524
   curtain 520
   ground 598
Roll roof 431
   roofing 433
   type air filter 584
Roller sheepsfoot 662
   tandem 662
   vibratory 662
Romex copper 597
Roof adhesive 432
   aluminum 427
   baffle 440
   beam 388
   built-up 431, 432
   cant 376, 432
   clay tile 426
   coating 416
   copper 434
   deck 384
   deck insulation 421
   deck laminated 384
   deck wood 384
   decking 340
   decking log 383
   drains 570
   fiberglass 427
   fire valve manifold 550
   framing 138, 140, 142, 144, 146, 148
   framing log 383
   framing removal 303
   hatch removal 417
   insulation 421
   jack residential 583
   lead 434
   membrane green 427
   metal 427
   metal rafter framing 349
   metal truss framing 350
   mineral 433
   modified bitumen 432
   nail 361
   panel aluminum 427
   panel preformed 427
   patch 416
   paver 440
   paver and support 440
   purlin 383
   purlin log 383
   rafter 376
   rafter bracing 348
   rafter bridging 348
   rafter framing 376
   roll 431
   safety anchor 295
   sheathing 386
   shingle asphalt 424
   slate 425, 723
   soffit framing 349
   soil mixture green 427
   specialty prefab 436
   system green 427
   truss 387, 388
   ventilator 440
   window 471
   window metal 471
   zinc 434
Roofing 194, 196, 198, 200, 202
   & siding preformed 427
   demolition 416
   flat seam sheet metal 434

Index

gambrel	198
hip roof	196
mansard	200
membrane	431
modified bitumen	432
roll	433
shed	202
SPF	433
system built-up	432
Rooftop air conditioner	590
Rope safety line	295
Rotary crawler drill	664
hammer drill	664
Rototiller	662
Rough buck	373
stone wall	330
Rough-in sink countertop	572
sink raised deck	572
sink service floor	575
tub	573
Round diffuser	584
Rubber and vinyl sheet floor	499
base	499
coating	418
floor	499
floor tile	500
sheet	499
threshold	474
tile	500
Rubbish handling	303
Run gravel bank	618

S

Safety flood shut-off house	557
line rope	295
shut-off flood whole house	557
switch	608
switch disconnect	608
water and gas shut-off	556
Salamander	666
Sales tax	292, 714
Sand fill	618
Sandblasting equipment	666
Sander floor	666
Sanding	503
floor	499
Sandstone	331
flagging	636
Sanitary base cove	492
drain	570
tee	568
Sash wood	463
Sauna	543
door	544
Saw chain	666
circular	666
concrete	660
masonry	666
Sawn control joint	311
SBS modified bituminous membrane	432
Scaffold aluminum plank	295
baseplate	294
bracket	294
caster	294
frame	294
guardrail	294
mason	294
stairway	294
wood plank	294
Scaffolding	717
plank	295
tubular	294
Scissor lift	663

Scrape after damage	505
Scraper earth	662
Screed, gas engine, 8HP vibrating	660
Screen chimney	521
fence	640
molding	401
security	459
squirrel and bird	521
window	459, 467
wood	467
Screening plant	662
Screen/storm door aluminum	446
Screw anchor	336
brass	361
drywall	491
lag	338
steel	361
wood	361
Seal compression	441
fog	634
pavement	634
Sealant	417
acoustical	442, 490
caulking	442
Sealcoat	634
Sealer joint	441, 442
Sectional door	458
overhead door	457
Security film	479
screen	459
Seeding	643, 730
Segmental concrete retaining wall	640
retaining wall	641
Seismic bracing supports	557
Selective clearing	618
demolition fencing	301
demolition masonry	322
Self-adhering flashing	436
Self-closing relief valve	556
Self-contained air conditioner	590
Septic galley	651
system	112, 650
system chamber	651
system effluent-filter	651
tank	650
tank concrete	650
tank polyethylene	650
Service electric	278, 608
entrance cable aluminum	597
entrance cap	597
residential	600
sink	575
sink faucet	575
Sewage collection concrete pipe	650
collection PVC pipe	650
pipe	650
Sewage/drainage collection	651
Shade	535
Shadow box fence	639
Shake siding	164
wood	425
Shear plate	362
wall	386
Shearwall panel	381
Sheathed nonmetallic cable	597
Romex cable	597
Sheathing	385, 386
asphalt	386
board	385
gypsum	387
insulating	384
paper	424
roof	386
roof & wall plywood	386

wall	386
Shed dormer framing	152
dormer roofing	206
roof framing	148
Sheepsfoot roller	662
Sheet carpet	501
glass	479
glass mosaic	494
metal	434
metal aluminum	436
metal cladding	434
metal flashing	435
modified bituminous barrier	424
rock	214, 216
Sheeting driver	664
Shelf metal	523
Shellac door	511
Shelter temporary	308
Shelving	524
pine	524
plywood	524
prefinished	524
storage	523
wood	524
Shield expansion	336
lag screw	336
Shingle	196, 198, 200, 202, 424
asphalt	424
cedar	425
concrete	426
panelized	425
removal	416
ridge	425
siding	164
slate	425
stain	507
strip	424
wood	425
Shock absorber	567
Shoring baseplate	294
bracing	294
heavy duty	294
leveling jack	294
Short load concrete	316
Shot blaster	666
Shotcrete pump	660
Shoulder paver	667
Shovel crawler	662
Shower by-pass valve	574
cabinet	520
cubicle	573
door	520
drain	568
enclosure	521
glass door	521
partition	520
receptor	520
stall	573
surround	521
Shower/tub control set	574
Shower-tub valve spout set	574
Shrub and tree	644
broadleaf evergreen	645
deciduous	645
evergreen	644
moving	646
Shut-off whole house flood safety	557
safety water and gas	556
water heater safety	556
Shutter	191, 535
exterior	409
interior	535
plantation	535
wood	409
wood interior	535

Sidelight	408
door	448, 454
Sidewalk	497
asphalt	634
brick	636
broom	666
concrete	634
driveway and patio	634
removal	301
Sidewall bracket	295
Siding aluminum	428
bevel	429
cedar	429
fiber cement	431
fiberglass	427
metal	166
nail	361
paint	507
painting	507
plywood	429
polypropylene	430
redwood	429
removal	417
shingle	164
stain	429, 507
steel	428
vinyl	429
wood	162, 429
wood board	429
Silencer door	475
Silicone	441
caulking	442
coating	418
water repellent	418
Sill	331
& ledger framing	377
anchor	362
door	402, 408, 474
masonry	332
precast concrete window	318
quarry tile	493
wood	377
Sillcock hose bibb	574
Silt fence	631
Simulated stone	333
Single hub pipe	567
hung aluminum window	459
zone rooftop unit	590
Sink	236
base	536
countertop	572
countertop rough-in	572
faucet kitchen	573
kitchen	572
laundry	573
lavatory	572
raised deck rough-in	572
removal	556
residential	572
service	575
service floor rough-in	575
slop	575
Siren	616
Sister joist	372
Site clear and grub	618
clearing	618
demolition	300
improvement	642
preparation	300
Sitework	102
Skidder log	662
Skidsteer attachments	663
loader	663
Skim coat	214, 216
Skylight	208, 471
domed	472

Index

plastic	472
removal	416
Skywindow	208
Slab blockout formwork	309
bulkhead formwork	308
concrete	126
concrete placement	316
curb formwork	309
cutout	302
edge formwork	309
haunch	314
haunch formwork	314
on grade	314
on grade formwork	308
on grade post-tensioned	313
on grade removal	301
stamping	317
textured	315
thickened edge	314
thickened edge formwork	314
turndown	314
turndown formwork	314
void formwork	309
Slate	331
roof	425, 723
shingle	425
sidewalk	636
sill	332
stair	331
tile	498
Slatwall	403
Sleeper	377
framing	377
wood	229
Sleeve anchor bolt	311
and tap	649
dowel	312
formwork	310
plastic pipe	310
Sliding aluminum door	457
door	185
glass door	457
glass vinyl-clad door	457
mirror	520
vinyl-clad door	457
window	176, 463
window aluminum	459
wood door	457
Slipform paver barrier	668
paver curb	668
Slop sink	575
Slope grading	619
protection	631
Slot letter	523
Slotted pipe	651
Slow close ball valve	551
Smoke detector	616
Snap-tie formwork	310
Snow guard	440
Soap holder	520
Socket meter	606
Sod	644
Sodding	644
Sodium low pressure fixture	611
Soffit	377
& canopy framing	377
aluminum	428
drywall	489
metal	431
molding	402
plaster	486
plywood	402
polyvinyl	402, 431
PVC	412
vent	440
wood	402
Softener water	570
Soil compaction	620
mix man-made	427
pipe	567
tamping	620
treatment	631
Solid fuel furnace	587
surface countertops	539
wood door	450
Sound dampening panel gypsum	490
Source heat pump water	591
Spa bath	543
Space heater	665
Spade air	664
tree	663
Spanish roof tile	426
Special door	456
stairway	231
SPF roofing	433
Spike grid	362
Spiral pipe supply register	584
stair	353, 404
Split rib block	328
ring	362
system ductless	590
Spray coating	417
rig foam	664
Sprayed-on insulation	421
Sprayer emulsion	664
paint	665
Spread footing	314
footing formwork	308
Spreader aggregate	660, 667
chemical	667
concrete	665
Sprinkler ball valve	551
cabinet	552
connector flexible	552
control pressure	553
dehydrator package	553
escutcheon	552
flow control valve	548
head	552
head wrench	553
irrigation system	642
line tester	550
release emergency	552
residential	553
system	552
system accelerator	553
system dry-pipe	553
trim valve	552
underground	642
valve	548, 553
Stabilizer base	668
Staging aids	295
pump	294
Stain cabinet	510
casework	510
door	511
floor	513
lumber	388
shingle	507
siding	429, 507
truss	514
Staining concrete floor	496
Stainless flashing	435
gutter	438
rivet	338
steel countertop	538
steel gravel stop	436
steel hinge	475
Stair basement	404
basement bulkhead	318
brick	330
carpet	502
ceiling	530
climber	546
concrete	315
disappearing	530
electric	530
finish concrete	317
fire escape	351
formwork	310
parts wood	404
precast concrete	318
prefabricated wood	404
railroad tie	634
removal	359
residential	404
retractable	530
slate	331
spiral	353, 404
stringer	373
stringer wood	373
tread	231
tread tile	493
wood	404
Stairway	230
scaffold	294
Stairwork handrails	406
Stake formwork	310
nail	310
Stall shower	573
Stamping slab	317
texture	317
Standard extinguisher	523
Standing seam	434
Standpipe alarm	552
connection	550
Steam bath	544
bath residential	544
boiler	586
boiler electric	586
cleaner	666
Steel anchor	324
beam	340
bin wall	641
bolt	362
bolt & hex nut	337
bridging	371
chain link	638
channel	486
column	340
conduit in slab	600
conduit in trench	600
conduit rigid	598
diffuser	584
door	447, 724
door residential	448
downspout	437
drill	664
drilling	337
edging	634, 646
expansion tank	580
fiber	313
fiber reinforcing	313
fireplace	240
fitting	561
flashing	435
frame	446
furring	483
gravel stop	436
gutter	438
hex bolt	337
hinge	475
lintel	339
member structural	340
nailer	373
outlet box	598
pipe	653
pipe corrugated	652
plate	340
prestressing	313
rivet	338
screw	361
siding	428
stud NLB	484
underground storage tank	579
window	459
Steeple tip pin	475
Step bluestone	634
brick	634
masonry	634
railroad tie	634
stone	331
Stiffener joist web	347
plate	340
Stiffleg derrick crane	669
Stockade fence	640
Stockpiling of soil	618
Stone aggregate	644
anchor	324
ashlar	331
base	331
crushed	618
cultured	333
curb cut	637
fill	618
floor	331
ground cover	644
mulch	642
paver	330, 636
removal	301
retaining walls	641
simulated	333
step	331
trim cut	331
wall	160, 330, 641
Stool cap	402
window	331, 332
Stop door	402
gravel	436
Storage shelving	523
tank	579
tank concrete aboveground	580
tank steel aboveground	579
tank steel underground	579
tank underground	579
Storm door	190, 446
door residential	446
drainage manhole frames	652
window	190, 467
window aluminum residential	467
Stove	523
woodburning	523
Strainer downspout	436
gutter	439
roof	437
wire	437
Strap tie	362
Straw	642
bale construction	544
Stringer	231
stair	373
Strip chamfer	310
floor	498
footing	314
lighting	609
shingle	424
soil	618
Striper paint	667, 668
Stripping topsoil	618
Structural columns	339
compaction	630
excavation	620
insulated panel	381
panel	381

Index

Structure log 382
Stucco 220, 487
   interior 218
   painting 507
   wall 158
Stud demolition 359
   NLB metal 484
   NLB steel 484
   NLB wall 484
   partition 373
   partition framing load-bearing 342
   wall 136, 373
   wall blocking load-bearing 341
   wall box beam load-bearing 342
   wall bracing load-bearing 341
   wall framing load bearing 342
   wall header load-bearing 342
Stump chipper 662
Subdrainage system 652
Subfloor 384
   adhesive 385
   demolition 482
   OSB 385
   plywood 229, 385
   wood 385
Submersible pump 570, 649, 665
   sump pump 570
Suction hose 665
Sump pump 529
   pump submersible 570
Supply ductile iron pipe water 648
Support ceiling 339
   drinking fountain 575
   lavatory 575
   light 339
   monitor 339
   motor 339
   partition 339
   X-ray 339
Suppressor lightning 600
Surface landscape 502
   preparation exterior 503
   preparation interior 504
   raceway 607
Surfacing 634
Surround shower 521
   tub 521
Survey crew 297
   monument 300
   property line 300
   topographic 300
Suspended acoustic ceiling tiles 495
   ceiling 485, 495
Suspension system ceiling 496
Sweeper road 667
Swimming pool 243, 542
   pool enclosure 544
   pool equipment 542
   pool filter 576
   pool filtration 576
   pool ladder 542
   pools 729
Swing check valve 556
   check valve bronze 556
Swing-up door 186
   overhead door 458
Switch box 599
   box plastic 599
   decorator 601
   dimmer 600, 607
   electric 607, 608
   general duty 608
   residential 600
   safety 608
   tamper 553
   time 608

toggle 607
   wiring 280
Switching low-voltage 607
Switchplate low-voltage 607
Synthetic erosion control 631
   fiber 313
   fiber reinforcing 313
   marble 498
System chamber septic 651
   fire extinguishing 548, 551
   septic 650
   sprinkler 548, 552, 553
   subdrainage 652
   T.V. 614
   VHF 614

T

T & G siding 162
Tamper 664
   switch 553
Tamping soil 620
Tandem roller 662
Tank aboveground storage concrete 580
   expansion 580
   fiberglass 579
   horizontal aboveground 579
   septic 650
   steel aboveground storage 579
   steel underground storage 579
   storage 579
   water 667
Tankless gas water heaters 571
Tap and sleeve 649
Taping and finishing gypsum board 488
Tapping crosses and sleeves 649
   main 649
Tar kettle 666, 668
   paper 643
   pitch emulsion 634
Tarpaulin 296
   duck 296
   polyethylene 296
Tax 292
   sales 292
   social security 292
   unemployment 292
T-bar mount diffuser 584
Teak floor 229, 498
   molding 401
   paneling 403
Tee cleanout 558
Telephone receptacle 603
Television receptacle 603
   system 614
Temperature relief valve 556
Tempered glass 478
   hardboard 402
Tempering valve 556
   valve water 556
Temporary construction 296
   heat 308
   protection 297
   shelter 308
   toilet 666
Tennis court fence 114, 638
Terminal A/C packaged 590
   air conditioner packaged 590
Termite control chemical 631
   protection 631
Terne coated flashing 435
Terra cotta coping 333
   cotta demolition 303

Terrazzo base 500
   cove base 500
   curb 500
   floor tile 500
   precast 500
   receptor 520
   wainscot 501
Tester sprinkler line 550
Textile wall covering 502
Texture stamping 317
Textured slab 315
Theater home 614
Thermal & moisture protection demo 416
Thermostat 579
   integral 594
   wire 605
Thickened edge slab 314
Thin brick veneer 325
Thincoat 214, 216
   plaster 488
Thinset ceramic tile 492
   mortar 493
Threshold 474
   aluminum 475
   cherry 474
   door 408
   oak 474
   ramp aluminum 475
   stone 331
   walnut 474
   wood 402
Thru-wall air conditioner 590
Tie rafter 376
   railroad 634, 646
   strap 362
   wall 324
Tile 492, 494, 499
   aluminum 494
   carpet 501
   ceiling 495
   ceramic 492
   clay 426
   concrete 426
   cork 497
   cork wall 502
   demolition 482
   flooring cork 497
   flue 332
   glass 479
   grout 492, 493
   marble 498
   metal 494
   porcelain 492
   quarry 493
   rubber 500
   slate 498
   stainless steel 494
   stair tread 493
   vinyl composition 499
   wall 492
   waterproofing membrane ceramic 494
   window sill 493
Tiling ceramic 492, 494
Timber connector 361
   fastener 361
   framing 383
   heavy 383
   laminated 388
   retaining wall 641
Time switch 608
Timer clock 604
   interval 601
   switch ventilator 583
Tinted glass 478

Toggle switch 607
Toggle bolt anchor 336
Toilet accessories 520
   accessories commercial 520
   bowl 571
   chemical 666
   temporary 666
Tongue and groove siding 162
Tool powder actuated 338
   rivet 338
Toothed ring 362
Toothing masonry 303
Top demolition counter 360
   vanity 539
Topographical surveys 300
Topsoil 618
   placement and grading 643
   stripping 618
Torch cutting 666
Torrified door 454
Towel bar 520
Tower crane 668, 718
   hoist 669
   light 665
Track drawer 537
   drill 664
   light 611
   lighting 611
   traverse 534
Tractor loader 662
   rake 662
   truck 667
Traffic line remover 668
Trailer floodlight 664
   lowbed 668
   mud 667
   platform 666
   truck 666
   water 666
Tram car 666
Transfer switch automatic 609
Transformer low-voltage 607
Transition molding wood 499
   strip floor 501
Transom lite frame 447
   windows 466
Transplanting plant and bulb 646
Trap cast iron 569
   drainage 569
   plastic 569
   PVC 569
Trapezoid windows 466
Trash pump 666
Traverse 534
   track 534
Travertine 331
Tread abrasive 493
   beech 231
   oak 231
   stone 332
   wood stair 405
Treated lumber framing 377
Treatment flooring 496
Tree deciduous 645
   evergreen 644
   guying systems 646
   moving 646
   removal 618
   spade 663
Trench backfill 620
   box 666
   disposal field 651
   excavating 619
   utility 620
Trencher chain 620, 663
   rock 663

Index

wheel 663
Trenching 106
Trim exterior 395, 396
   exterior paint 509
   molding 401
   painting 514
   PVC 411
   tile 492
   wood 398
Tripod floodlight 664
Trowel coating 417
   concrete 660
   power 660
Truck boom 668
   concrete 660
   crane flatbed 668
   dump 663
   flatbed 663, 666
   holding concrete 316
   loading 620
   mounted crane 668
   pickup 666
   tractor 667
   trailer 666
   vacuum 667
   winch 667
Truss bowstring 388
   demolition 360
   flat wood 387
   framing metal 350
   painting 514
   plate 362
   roof 387, 388
   roof framing 140
   stain 514
   varnish 514
Tub hot 543
   redwood 543
   rough-in 573
   surround 521
Tube and wire kit 591
   fittings copper 560
Tubing connection split system .. 591
   copper 559
   electric metallic 598
   fitting PEX 593
   PEX 563, 592
Tub-shower module 573
Tubular scaffolding 294
Tugboat diesel 670
Tumbler holder 520
Tunnel locomotive 668
   muck car 668
   ventilator 668
Turbine wind 656
Turndown slab 314
Turned column 407
TV antenna 614
   system 614

U

Undereave vent 480
Underground sprinkler ... 642
   storage tank 579
   storage tank steel 579
Underlayment hardboard .. 385
   latex 499
Unemployment tax 292
Unit air-handling 591
   commercial kitchen ... 530
   line set evap/condens .. 591
Utility 106
   boxes 648
   excavation 106

line removal 300
trench 620
trench excavating 620

V

Vacuum breaker 566
   catch basin 667
   central 531
   cleaning 531
   pick-up 660
   truck 667
   wet/dry 667
Valance board 537
Valley rafter 376
Valve assembly dry pipe sprinkler 553
   bronze 556
   butterfly 648
   cap fire 552
   check 548
   fire hose 551
   mixing 574
   relief pressure 556
   shower by-pass 574
   slow close ball 551
   spout set shower-tub .. 574
   sprinkler 548, 553
   sprinkler alarm 548
   sprinkler ball 551
   sprinkler flow control . 548
   sprinkler trim 552
   swing check 556
   tempering 556
   washing machine 574
   water pressure 556
Vanity base 537
   top 539
   top lavatory 572
Vapor barrier 424
   barrier sheathing 386
   retarder 423
Varnish cabinet 511
   casework 511
   door 511
   floor 513, 517
   truss 514
VCT removal 482
Vehicle charging electric .. 609
Veneer brick 160, 325
   core paneling 403
   member laminated ... 388
   removal 323
Venetian plaster 487
Vent cap 569
   cap cast iron 569
   cap plastic 569
   chimney 585
   dryer 529, 575
   eave 440
   flashing 569
   gas 585
   one-way 440
   ridge 440
   ridge strip 480
   soffit 440
Ventilating air conditioning . 583, 590
Ventilation fan 604
   louver 480
Ventilator roof 440
   timer switch 583
   tunnel 668
Verge board 400
Vermiculite insulation ... 420
Vertical aluminum siding .. 428

VHF system 614
Vibrating screed, gas engine, 8HP 660
Vibrator concrete 660
   earth 630
   plow 663
Vibratory hammer 662
   roller 662
Vinyl blind 191, 409, 534
   casement window ... 469
   chain link 638
   clad half round window .. 465
   clad premium window .. 466
   composition floor 499
   composition tile 499
   downspout 437
   faced wallboard 489
   fence 638
   floor 500
   flooring 229
   gutter 438
   replacement window .. 444
   sheet floor 499
   shutter 191
   siding 166, 429
   wallpaper 503
   window 469
   window double hung .. 469
   window half round ... 470
   window single hung .. 468
   window solid 468
Volume control damper .. 582

W

Wainscot ceramic tile 492
   molding 402
   quarry tile 493
   terrazzo 501
Walk 634
Walkway luminaire 611
Wall accessories formwork .. 310
   and partition demolition .. 483
   base 499
   blocking load-bearing stud .. 341
   box beam load-bearing stud .. 342
   boxout formwork 309
   bracing load-bearing stud ... 341
   brick 330
   brick shelf formwork .. 309
   bulkhead formwork .. 309
   bumper 474
   cabinet 536
   cap residential 582
   cast concrete retaining .. 640
   ceramic tile 492
   coating 517
   concrete finishing ... 317
   concrete placement .. 316
   concrete segmental retaining .. 640
   covering 728
   covering textile 502
   cutout 302
   drywall 488
   finish 502
   formwork 309
   foundation 328
   framing 136, 378
   framing load-bearing stud .. 342
   framing removal 303
   furring 380, 484
   grout 324
   header load-bearing stud .. 342
   heater 530
   hung lavatory 572

insulation 418
   interior 214, 218
   lath 485
   log 382
   louver 480
   masonry 158, 160, 327, 641
   mirror 479
   painting 507, 516
   paneling 403
   plaster 486
   plywood formwork ... 309
   prefab plywood formwork .. 309
   reinforcing 312
   retaining 641
   shear 386
   sheathing 386
   steel bin 641
   stone 330, 641
   stucco 487
   stud 373
   stud NLB 484
   switch plate 607
   tie 324
   tie masonry 324
   tile 492
   tile cork 502
Wallboard repair gypsum .. 482
Wallcovering 502
   acrylic 503
   gypsum fabric 502
Wall covering natural fiber .. 502
Wallpaper 502, 728
   grass cloth 503
   vinyl 503
Walnut floor 498
   threshold 474
Wardrobe 524
Wash bowl residential ... 572
   brick 322
   pressure 504
Washable air filter 585
   air filter permanent .. 585
Washer 362
   dryer accessories 574
   pressure 666
Washing machine automatic .. 529
   machine valve 574
Water closet 571, 575
   closet removal 556
   closet support 575
   curing concrete 318
   distribution pipe .. 648, 649
   flow indicator 548
   hammer arrester 567
   heater 529, 605
   heater gas residential .. 571
   heater insulation 559
   heater residential ... 529, 571, 605
   heater residential electric .. 571
   heater safety shut-off .. 556
   heater wrap kit 559
   heating hot 586, 591
   heating wiring 280
   hose 665
   motor alarm 552
   pipe ground clamp .. 598
   pressure reducing valve .. 556
   pressure relief valve .. 556
   pressure valve 556
   pump 580, 649, 665
   reducer concrete 316
   repellent 516
   repellent coating 418
   repellent silicone 418
   softener 570
   source heat pump ... 591

Index

supply domestic meter	566
supply ductile iron pipe	648
supply meter	566
supply PVC pipe	649
tank	667
tempering valve	556
trailer	666
well	649
Waterproofing coating	417
demolition	417
Waterstop PVC	310
ribbed	310
Weather barrier	423
barrier or wrap	423
cap entrance	599
Weatherproof receptacle	280, 602
Weatherstrip door	476
window	477
zinc	476
Weatherstripping	476
Weathervane residential	525
Web stiffener framing	347
Weights and measures	713
Welded frame	447
wire epoxy	312
wire fabric	312
wire galvanized	312
Welding machine	667
Wells & accessories	649
area window	525
water	649
Wellpoint discharge pipe	669
equipment rental	669
header pipe	669
pump	669
riser pipe	670
Wet/dry vacuum	667
Wheel trencher	663
Wheelbarrow	667
Wheeled loader	662
Whirlpool bath	543
Widener pavement	668
Winch truck	667
Wind turbine	656
Window	170, 172, 174, 176, 178, 180, 459, 472
& door demolition	445
air conditioner	590
aluminum	459
aluminum projected	459
aluminum residential storm	467
aluminum sliding	459
and door molding	402
awning	461
basement/utility	459
bay	460
blind	534
bow & bay metal-clad	464
casement	461, 465
casement wood	461
demolition	445
double hung	466
double hung bay wood	464
double hung vinyl	469
double hung wood	461, 463
double hung	170
estimates	724
fiberglass	471
fiberglass bay	471
fiberglass single hung	471
fiberglass slider	471
frame	463
glass	479
grille	467
half round vinyl	470
half round vinyl clad	465
hardware	473
impact resistant aluminum	459
impact resistant replacement	444
metal	459
metal-clad awning	463
metal-clad casement	464
metal-clad double hung	464
muntin	467
painting	511
palladian	465
picture	463
picture & sliding metal-clad	464
plastic	468
precast concrete sill	318
removal	445
replacement	444
restoration	306
roof	471
sash wood	463
screen	459, 467
sill	332
sill marble	331
sill tile	493
single hung aluminum	459
single hung vinyl	468
sliding	463
sliding wood	463
solid vinyl	468
steel	459
stool	331, 332
storm	190, 467
transom	466
trapezoid	466
trim set	402
vinyl	469
vinyl casement	469
vinyl clad awning	465
vinyl clad half round	465
vinyl clad premium	466
vinyl replacement	444
weatherstrip	477
well area	525
wood	460
wood awning	461
wood bow	465
wood double hung	463
wood residential	460
Windrow loader	668
Wine cellar	530
Winter concrete	316
protection	308
Wire copper	597
electrical	597
fence	639
ground	598
mesh	487
strainer	437
thermostat	605
THW	597
Wiremold raceway	607
Wiring air conditioner	605
device	607
fan	604
power	608
residential	600, 604
Wood & composite I-joists	387
awning window	461
base	391
beam	371, 383
blind	191, 535
block floor demolition	482
blocking	211, 374
bow window	465
bracing	370
canopy	377
chip	643
column	372, 383, 407
composition floor	498
cupola	525
deck	244, 384
demolition	482
door	224, 226, 448, 449
door decorator	450
door frame exterior	408
door frame interior	408
door residential	451
double hung window	463
fascia	398
fastener	361
fence	639
fiber ceiling	495
fiber sheathing	386
fiber underlayment	385
floor	130, 132, 134, 229, 498
floor demolition	482
folding panel	535
foundation	124
framing	370, 371
framing demolition	356
framing miscellaneous	373
framing open web joist	372
furnace	587
furring	380
gas furnace	587
girder	371
gutter	438
handrail	401, 406
interior shutter	535
joist	372, 387
laminated	388
louver	406
nailer	373
oil furnace	587
overhead door	458
panel door	450
paneling	402
parquet	498
partition	373
plank scaffold	294
planter	642
post	372
rafter	376
rail fence	640
railing	401, 404, 406
residential window	460
roof deck	384
sash	463
screen	467
screw	361
shake	425
shelving	524
shingle	425
shutter	191, 409
sidewalk	634
siding	162, 429
siding demolition	417
sill	377
soffit	402
stair	404
stair baluster	404
stair newel	405
stair part	404
stair railing	405
stair riser	405
stair stringer	373
stair tread	405
storm door	190, 455
storm window	190
strip flooring	498
subfloor	385
threshold	402
trim	398
truss	387
veneer wallpaper	502
wall framing	136
window	460
window casement	461
window demolition	445
window double hung	461
window double hung bay	464
window sash	463
window sliding	463
Woodburning stove	523
Wool carpet	501
fiberglass	420
Work boat	670
Workers' compensation	293, 716
Wrecking ball	667
Wrench impact	664
sprinkler head	553

X

X-ray support	339

Y

Yard fountain	543
Yellow pine floor	498

Z

Z bar suspension	496
Zee bar	491
Zinc plating	361
roof	434
weatherstrip	476

Notes

Notes

Notes

Notes

Notes

Division Notes

	CREW	DAILY OUTPUT	LABOR-HOURS	UNIT	BARE COSTS				TOTAL INCL O&P
					MAT.	LABOR	EQUIP.	TOTAL	

Division Notes

	CREW	DAILY OUTPUT	LABOR-HOURS	UNIT	BARE COSTS				TOTAL INCL O&P
					MAT.	LABOR	EQUIP.	TOTAL	

Division Notes

		CREW	DAILY OUTPUT	LABOR-HOURS	UNIT	BARE COSTS MAT.	LABOR	EQUIP.	TOTAL	TOTAL INCL O&P

Division Notes

Division Notes

	CREW	DAILY OUTPUT	LABOR-HOURS	UNIT	BARE COSTS				TOTAL INCL O&P
					MAT.	LABOR	EQUIP.	TOTAL	

Other RSMeans Products & Services

RSMeans—a tradition of excellence in construction cost information and services since 1942

Table of Contents
Annual Cost Guides
RSMeans Online
Seminars

For more information visit the RSMeans website at www.RSMeans.com

Unit prices according to the latest MasterFormat

Cost Data Selection Guide

The following table provides definitive information on the content of each cost data publication. The number of lines of data provided in each unit price or assemblies division, as well as the number of crews, is listed for each data set. The presence of other elements such as reference tables, square foot models, equipment rental costs, historical cost indexes, and city cost indexes, is also indicated. You can use the table to help select the RSMeans data set that has the quantity and type of information you most need in your work.

Unit Cost Divisions	Building Construction	Mechanical	Electrical	Commercial Renovation	Square Foot	Site Work Landsc.	Green Building	Interior	Concrete Masonry	Open Shop	Heavy Construction	Light Commercial	Facilities Construction	Plumbing	Residential
1	562	383	400	504		498	205	302	445	561	496	245	1040	393	177
2	776	279	85	732		991	206	399	213	775	733	480	1220	286	273
3	1691	340	230	1084		1483	986	354	2037	1691	1693	482	1791	316	389
4	960	21	0	920		725	180	615	1158	928	615	533	1175	0	445
5	1893	158	155	1090		844	1799	1098	720	1893	1037	979	1909	204	746
6	2453	18	18	2111		110	589	1528	281	2449	123	2141	2125	22	2661
7	1594	215	128	1633		580	763	531	523	1591	26	1328	1695	227	1048
8	2111	81	44	2645		257	1136	1757	105	2096	0	2219	2890	0	1534
9	2004	86	45	1828		310	451	2088	391	1944	15	1668	2252	54	1437
10	1049	17	10	645		216	27	864	157	1049	29	539	1141	233	224
11	1096	206	166	551		129	54	933	28	1071	0	231	1116	169	110
12	530	0	2	285		205	133	1533	14	500	0	259	1554	23	217
13	744	149	158	253		367	122	252	78	720	244	108	759	115	103
14	273	36	0	221		0	0	257	0	273	0	12	293	16	6
21	91	0	16	37		0	0	250	0	91	0	73	535	540	220
22	1168	7548	160	1201		1572	1066	848	20	1157	1681	874	7450	9370	718
23	1178	6975	580	929		157	903	778	38	1171	110	879	5213	1922	469
26	1380	458	10150	1033		793	630	1144	55	1372	562	1310	10099	399	619
27	72	0	339	34		13	0	71	0	72	39	52	321	0	22
28	96	58	136	71		0	21	78	0	99	0	40	151	44	25
31	1510	733	610	806		3263	288	7	1217	1455	3277	604	1569	660	613
32	813	49	8	881		4409	353	405	291	784	1826	421	1687	133	468
33	529	1076	538	252		2173	41	0	236	522	2157	128	1697	1283	154
34	107	0	47	4		190	0	0	31	62	214	0	136	0	0
35	18	0	0	0		327	0	0	0	18	442	0	84	0	0
41	60	0	0	33		8	0	22	0	61	31	0	68	14	0
44	75	79	0	0		0	0	0	0	0	0	0	75	75	0
46	23	16	0	0		274	261	0	0	23	264	0	33	33	0
48	12	0	25	0		0	25	0	0	12	17	12	12	0	12
Totals	24868	18981	14050	19783		19894	10239	16114	8038	24440	15631	15617	50090	16531	12690

Assem Div	Building Construction	Mechanical	Electrical	Commercial Renovation	Square Foot	Site Work Landscape	Assemblies	Green Building	Interior	Concrete Masonry	Heavy Construction	Light Commercial	Facilities Construction	Plumbing	Asm Div	Residential
A		15	0	188	150	577	598	0	0	536	571	154	24	0	1	378
B		0	0	848	2506	0	5666	56	329	1975	368	2098	174	0	2	211
C		0	0	647	928	0	1309	0	1628	146	0	818	249	0	3	588
D		1067	941	712	1859	72	2538	330	825	0	0	1345	1105	1088	4	851
E		0	0	86	261	0	301	0	5	0	0	258	5	0	5	390
F		0	0	0	114	0	114	0	0	0	0	114	3	0	6	357
G		527	447	318	312	3364	792	0	0	534	1349	205	293	677	7	307
															8	760
															9	80
															10	0
															11	0
															12	0
Totals		1609	1388	2799	6130	4013	11318	386	2787	3191	2288	4992	1853	1765		3922

Reference Section	Building Construction Costs	Mechanical	Electrical	Commercial Renovation	Square Foot	Site Work Landscape	Assem.	Green Building	Interior	Concrete Masonry	Open Shop	Heavy Construction	Light Commercial	Facilities Construction	Plumbing	Resi.
Reference Tables	yes	yes	yes	yes	no	yes	yes	yes	yes	yes	yes	yes	yes	yes	yes	yes
Models					111			25					50			28
Crews	575	575	575	553		575		575	575	575	551	575	551	553	575	551
Equipment Rental Costs	yes	yes	yes	yes		yes		yes	yes	yes	yes	yes	yes	yes	yes	yes
Historical Cost Indexes	yes	yes	yes	yes	yes	yes	yes	yes	yes	yes	yes	yes	yes	yes	yes	no
City Cost Indexes	yes	yes	yes	yes	yes	yes	yes	yes	yes	yes	yes	yes	yes	yes	yes	yes

Visit RSMeans.com/Online for more details on data titles in online format.

For more information visit our website at www.RSMeans.com

 Online   Book   CD   eBook Online add-on available for additional fee

Annual Cost Guides

Unit prices according to the latest MasterFormat

RSMeans Building Construction Cost Data 2016

Offers you unchallenged unit price reliability in an easy-to-use format. Whether used for verifying complete, finished estimates or for periodic checks, it supplies more cost facts better and faster than any comparable source. More than 24,700 unit prices have been updated for 2016. The City Cost Indexes and Location Factors cover more than 930 areas for indexing to any project location in North America. Order and get *RSMeans Quarterly Update Service* FREE.

RSMeans Green Building Cost Data 2016

Estimate, plan, and budget the costs of green building for both new commercial construction and renovation work with this sixth edition of *RSMeans Green Building Cost Data*. More than 10,000 unit costs for a wide array of green building products plus assemblies costs. Easily identified cross references to LEED and Green Globes building rating systems criteria.

RSMeans Mechanical Cost Data 2016

Total unit and systems price guidance for mechanical construction—materials, parts, fittings, and complete labor cost information. Includes prices for piping, heating, air conditioning, ventilation, and all related construction.

Plus new 2016 unit costs for:

- Thousands of installed HVAC/controls, sub-assemblies and assemblies
- "On-site" Location Factors for more than 930 cities and towns in the U.S. and Canada
- Crews, labor, and equipment

RSMeans Facilities Construction Cost Data 2016

For the maintenance and construction of commercial, industrial, municipal, and institutional properties. Costs are shown for new and remodeling construction and are broken down into materials, labor, equipment, and overhead and profit. Special emphasis is given to sections on mechanical, electrical, furnishings, site work, building maintenance, finish work, and demolition.

More than 49,800 unit costs, plus assemblies costs and a comprehensive Reference Section are included.

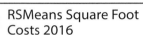

RSMeans Square Foot Costs 2016

Accurate and Easy-to-Use

- Updated price information based on nationwide figures from suppliers, estimators, labor experts, and contractors.
- Green building models
- Realistic graphics, offering true-to-life illustrations of building projects
- Extensive information on using square foot cost data, including sample estimates and alternate pricing methods

RSMeans Commercial Renovation Cost Data 2016

Commercial/Multi-family Residential

Use this valuable tool to estimate commercial and multi-family residential renovation and remodeling.

Includes: Updated costs for hundreds of unique methods, materials, and conditions that only come up in repair and remodeling, PLUS:

- Unit costs for more than 19,600 construction components
- Installed costs for more than 2,700 assemblies
- More than 930 "on-site" localization factors for the U.S. and Canada

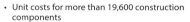

RSMeans Electrical Cost Data 2016

Pricing information for every part of electrical cost planning. More than 13,800 unit and systems costs with design tables; clear specifications and drawings; engineering guides; illustrated estimating procedures; complete labor-hour and materials costs for better scheduling and procurement; and the latest electrical products and construction methods.

- A variety of special electrical systems, including cathodic protection
- Costs for maintenance, demolition, HVAC/mechanical, specialties, equipment, and more

RSMeans Electrical Change Order Cost Data 2016

RSMeans Electrical Change Order Cost Data provides you with electrical unit prices exclusively for pricing change orders. Analyze and check your own change order estimates against those prepared when using RSMeans cost data. It also covers productivity analysis and change order cost justifications. With useful information for calculating the effects of change orders and dealing with their administration.

RSMeans Assemblies Cost Data 2016

RSMeans Assemblies Cost Data takes the guesswork out of preliminary or conceptual estimates. Now you don't have to try to calculate the assembled cost by working up individual component costs. We've done all the work for you.

Presents detailed illustrations, descriptions, specifications, and costs for every conceivable building assembly—over 350 types in all—arranged in the easy-to-use UNIFORMAT II system. Each illustrated "assembled" cost includes a complete grouping of materials and associated installation costs, including the installing contractor's overhead and profit.

RSMeans Open Shop Building Construction Cost Data 2016

The latest costs for accurate budgeting and estimating of new commercial and residential construction, renovation work, change orders, and cost engineering.

RSMeans Open Shop "BCCD" will assist you to:

- Develop benchmark prices for change orders.
- Plug gaps in preliminary estimates and budgets.
- Estimate complex projects.
- Substantiate invoices on contracts.
- Price ADA-related renovations.

For more information visit our website at www.RSMeans.com

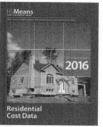

RSMeans Residential Cost Data 2016

Contains square foot costs for 28 basic home models with the look of today, plus hundreds of custom additions and modifications you can quote right off the page. Includes more than 3,900 costs for 89 residential systems. Complete with blank estimating forms, sample estimates, and step-by-step instructions.

Contains line items for cultured stone and brick, PVC trim, lumber, and TPO roofing.

RSMeans Site Work & Landscape Cost Data 2016

Includes unit and assemblies costs for earthwork, sewerage, piped utilities, site improvements, drainage, paving, trees and shrubs, street openings/repairs, underground tanks, and more. Contains more than 60 types of assemblies costs for accurate conceptual estimates.

Includes:

- Estimating for infrastructure improvements
- Environmentally-oriented construction
- ADA-mandated handicapped access
- Hazardous waste line items

RSMeans Facilities Maintenance & Repair Cost Data 2016

RSMeans Facilities Maintenance & Repair Cost Data gives you a complete system to manage and plan your facility repair and maintenance costs and budget efficiently. Guidelines for auditing a facility and developing an annual maintenance plan. Budgeting is included, along with reference tables on cost and management, and information on frequency and productivity of maintenance operations.

The only nationally recognized source of maintenance and repair costs. Developed in cooperation with the Civil Engineering Research Laboratory (CERL) of the Army Corps of Engineers.

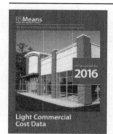

RSMeans Light Commercial Cost Data 2016

Specifically addresses the light commercial market, which is a specialized niche in the construction industry. Aids you, the owner/designer/contractor, in preparing all types of estimates—from budgets to detailed bids. Includes new advances in methods and materials.

The Assemblies section allows you to evaluate alternatives in the early stages of design/planning.

More than 15,500 unit costs ensure that you have the prices you need, when you need them.

RSMeans Concrete & Masonry Cost Data 2016

Provides you with cost facts for virtually all concrete/masonry estimating needs, from complicated formwork to various sizes and face finishes of brick and block—all in great detail. The comprehensive Unit Price Section contains more than 8,000 selected entries. Also contains an Assemblies [Cost] Section, and a detailed Reference Section that supplements the cost data.

RSMeans Labor Rates for the Construction Industry 2016

Complete information for estimating labor costs, making comparisons, and negotiating wage rates by trade for more than 300 U.S. and Canadian cities. With 46 construction trades in each city, and historical wage rates included for comparison. Each city chart lists the county and is alphabetically arranged with handy visual flip tabs for quick reference.

RSMeans Construction Cost Indexes 2016

What materials and labor costs will change unexpectedly this year? By how much?

- Breakdowns for 318 major cities
- National averages for 30 key cities
- Expanded five major city indexes
- Historical construction cost indexes

RSMeans Interior Cost Data 2016

Provides you with prices and guidance needed to make accurate interior work estimates. Contains costs on materials, equipment, hardware, custom installations, furnishings, and labor for new and remodel commercial and industrial interior construction, including updated information on office furnishings, as well as reference information.

RSMeans Heavy Construction Cost Data 2016

A comprehensive guide to heavy construction costs. Includes costs for highly specialized projects such as tunnels, dams, highways, airports, and waterways. Information on labor rates, equipment, and materials costs is included. Features unit price costs, systems costs, and numerous reference tables for costs and design.

RSMeans Plumbing Cost Data 2016

Comprehensive unit prices and assemblies for plumbing, irrigation systems, commercial and residential fire protection, point-of-use water heaters, and the latest approved materials. This publication and its companion, *RSMeans Mechanical Cost Data*, provide full-range cost estimating coverage for all the mechanical trades.

Contains updated costs for potable water, radiant heat systems, and high efficiency fixtures.

Visit RSMe

For more information visit our website at www.RSMeans.com

RSMeans Online

Competitive Cost Estimates Made Easy

RSMeans Online is a web-based service that provides accurate and up-to-date cost information to help you build competitive estimates or budgets in less time.

Quick, intuitive, easy to use, and automatically updated, RSMeans Online gives you instant access to hundreds of thousands of material, labor, and equipment costs from RSMeans' comprehensive database, delivering the information you need to build budgets and competitive estimates every time.

With RSMeans Online, you can perform quick searches to locate specific costs and adjust costs to reflect prices in your geographic area. Tag and store your favorites for fast access to your frequently used line items and assemblies and clone estimates to save time. System notifications will alert you as updated data becomes available. RSMeans Online is automatically updated throughout the year.

RSMeans Online's visual, interactive estimating features help you create, manage, save and share estimates with ease! It provides increased flexibility with customizable advanced reports. Easily edit custom report templates and import your company logo onto your estimates.

Unit prices according to the latest MasterFormat

	Standard	Advanced	Professional
Unit Prices	✓	✓	✓
Assemblies	X	✓	✓
Sq Ft Models	X	X	✓
Editable Sq Foot Models	X	X	✓
Editable Assembly Components	X	✓	✓
Custom Cost Data	X	✓	✓
User Defined Components	X	✓	✓
Advanced Reporting & Customization	X	✓	✓
Union Labor Type	✓	✓	✓
Open Shop*	Add on for additional fee	Add on for additional fee	✓
History (3-year lookback)**	Add on for additional fee	Add on for additional fee	Add on for additional fee
Full History (2007)	Add on for additional fee	Add on for additional fee	Add on for additional fee

*Open shop included in Complete Library & Bundles
** 3-year history lookback included with complete library

Estimate with Precision
Find everything you need to develop complete, accurate estimates.
- Verified costs for construction materials
- Equipment rental costs
- Crew sizing, labor hours and labor rates
- Localized costs for U.S. and Canada

Save Time & Increase Efficiency
Make cost estimating and calculating faster and easier than ever with secure, online estimating tools.
- Quickly locate costs in the searchable database
- Create estimates in minutes with RSMeans cost lines
- Tag and store favorites for fast access to frequently used items

Improve Planning & Decision-Making
Back your estimates with complete, accurate and up-to-date cost data for informed business decisions.
- Verify construction costs from third parties
- Check validity of subcontractor proposals
- Evaluate material and assembly alternatives

Increase Profits
Use RSMeans Online to estimate projects quickly and accurately, so you can gain an edge over your competition.
- Create accurate and competitive bids
- Minimize the risk of cost overruns
- Reduce variability
- Gain control over costs

30 Day Free Trial
Register for a free trial at www.RSMeansOnline.com
- Access to unit prices, building assemblies and facilities repair and remodeling items covering every category of construction
- Powerful tools to customize, save and share cost lists, estimates and reports with ease
- Access to square foot models for quick conceptual estimates
- Allows up to 10 users to evaluate the full range of collaborative functionality used items.

2016 RSMeans Seminar Schedule
Note: call for exact dates and details.

Location	Dates	Location	Dates
Seattle, WA	January and August	Bethesda, MD	June
Dallas/Ft. Worth, TX	January	El Segundo, CA	August
Austin, TX	February	Jacksonville, FL	September
Anchorage, AK	March and September	Dallas, TX	September
Las Vegas, NV	March	Philadelphia, PA	October
New Orleans, LA	March	Houston, TX	October
Washington, DC	April and September	Salt Lake City, UT	November
Phoenix, AZ	April	Baltimore, MD	November
Kansas City, MO	April	Orlando, FL	November
Toronto	May	San Diego, CA	December
Denver, CO	May	San Antonio, TX	December
San Francisco, CA	June	Raleigh, NC	December

☎ 877-620-6245

RSMeans beginning early 2016 will be offering a suite of online self paced offerings. Check our website at www.RSMeans.com for more information.

Professional Development

eLearning Training Sessions
Learn how to use *RSMeans Online*® or the *RSMeans CostWorks*® CD from the convenience of your home or office. Our eLearning training sessions let you join a training conference call and share the instructors' desktops, so you can view the presentation and step-by-step instructions on your own computer screen. The live webinars are held from 9 a.m. to 4 p.m. or 11 a.m. to 6 p.m. eastern standard time, with a one-hour break for lunch.

For these sessions, you must have a computer with high speed Internet access and a compatible Web browser. Learn more at www.RSMeansOnline.com or call for a schedule: 781-422-5115. Webinars are generally held on selected Wednesdays each month.

RSMeans Online	RSMeans CostWorks CD
$299 per person	$299 per person

Visit RSMe

For more information visit our website at www.RSMeans.com

Professional Development

RSMeans Online Training

Construction estimating is vital to the decision-making process at each state of every project. RSMeansOnline works the way you do. It's systematic, flexible and intuitive. In this one day class you will see how you can estimate any phase of any project faster and better.

Some of what you'll learn:
- Customizing RSMeansOnline
- Making the most of RSMeans "Circle Reference" numbers
- How to integrate your cost data
- Generate reports, exporting estimates to MS Excel, sharing, collaborating and more

Also available: RSMeans Online training webinar

Facilities Construction Estimating

In this *two-day* course, professionals working in facilities management can get help with their daily challenges to establish budgets for all phases of a project.

Some of what you'll learn:
- Determining the full scope of a project
- Identifying the scope of risks and opportunities
- Creative solutions to estimating issues
- Organizing estimates for presentation and discussion
- Special techniques for repair/remodel and maintenance projects
- Negotiating project change orders

Who should attend: facility managers, engineers, contractors, facility tradespeople, planners, and project managers.

Construction Cost Estimating: Concepts and Practice

This introductory course to improve estimating skills and effectiveness starts with the details of interpreting bid documents and ends with the summary of the estimate and bid submission.

Some of what you'll learn:
- Using the plans and specifications for creating estimates
- The takeoff process—deriving all tasks with correct quantities
- Developing pricing using various sources; how subcontractor pricing fits in
- Summarizing the estimate to arrive at the final number
- Formulas for area and cubic measure, adding waste and adjusting productivity to specific projects
- Evaluating subcontractors' proposals and prices
- Adding insurance and bonds
- Understanding how labor costs are calculated
- Submitting bids and proposals

Who should attend: project managers, architects, engineers, owners' representatives, contractors, and anyone who's responsible for budgeting or estimating construction projects.

Maintenance & Repair Estimating for Facilities

This *two-day* course teaches attendees how to plan, budget, and estimate the cost of ongoing and preventive maintenance and repair for existing buildings and grounds.

Some of what you'll learn:
- The most financially favorable maintenance, repair, and replacement scheduling and estimating
- Auditing and value engineering facilities
- Preventive planning and facilities upgrading
- Determining both in-house and contract-out service costs
- Annual, asset-protecting M&R plan

Who should attend: facility managers, maintenance supervisors, buildings and grounds superintendents, plant managers, planners, estimators, and others involved in facilities planning and budgeting.

Practical Project Management for Construction Professionals

In this *two-day* course, acquire the essential knowledge and develop the skills to effectively and efficiently execute the day-to-day responsibilities of the construction project manager.

Some of what you'll learn:
- General conditions of the construction contract
- Contract modifications: change orders and construction change directives
- Negotiations with subcontractors and vendors
- Effective writing: notification and communications
- Dispute resolution: claims and liens

Who should attend: architects, engineers, owners' representatives, and project managers.

Mechanical & Electrical Estimating

This *two-day* course teaches attendees how to prepare more accurate and complete mechanical/electrical estimates, avoid the pitfalls of omission and double-counting, and understand the composition and rationale within the RSMeans mechanical/electrical database.

Some of what you'll learn:
- The unique way mechanical and electrical systems are interrelated
- M&E estimates—conceptual, planning, budgeting, and bidding stages
- Order of magnitude, square foot, assemblies, and unit price estimating
- Comparative cost analysis of equipment and design alternatives

Who should attend: architects, engineers, facilities managers, mechanical and electrical contractors, and others who need a highly reliable method for developing, understanding, and evaluating mechanical and electrical contracts.

Unit Price Estimating

This interactive *two-day* seminar teaches attendees how to interpret project information and process it into final, detailed estimates with the greatest accuracy level.

The most important credential an estimator can take to the job is the ability to visualize construction and estimate accurately.

Some of what you'll learn:
- Interpreting the design in terms of cost
- The most detailed, time-tested methodology for accurate pricing
- Key cost drivers—material, labor, equipment, staging, and subcontracts
- Understanding direct and indirect costs for accurate job cost accounting and change order management

Who should attend: corporate and government estimators and purchasers, architects, engineers, and others who need to produce accurate project estimates.

Conceptual Estimating Using the RSMeans CostWorks CD

This *two-day* class uses the leading industry data and a powerful software package to develop highly accurate conceptual estimates for your construction projects. All attendees must bring a laptop computer loaded with the current year *Square Foot Costs* and the *Assemblies Cost Data* CostWorks titles.

Some of what you'll learn:
- Introduction to conceptual estimating
- Types of conceptual estimates
- Helpful hints
- Order of magnitude estimating
- Square foot estimating
- Assemblies estimating

Who should attend: architects, engineers, contractors, construction estimators, owners' representatives, and anyone looking for an electronic method for performing square foot estimating.

RSMeans CostWorks CD Training

This *one-day* course helps users become more familiar with the functionality of the *RSMeans CostWorks* program. Each menu, icon, screen, and function found in the program is explained in depth. Time is devoted to hands-on estimating exercises.

Some of what you'll learn:
- Searching the database using all navigation methods
- Exporting RSMeans data to your preferred spreadsheet format
- Viewing crews, assembly components, and much more
- Automatically regionalizing the database

This training session requires you to bring a laptop computer to class.

When you register for this course you will receive an outline for your laptop requirements.

Also offering web training for the RSMeans CostWorks CD!

Assessing Scope of Work for Facility Construction Estimating

This *two-day* practical training program addresses the vital importance of understanding the scope of projects in order to produce accurate cost estimates for facility repair and remodeling.

Some of what you'll learn:
- Discussions of site visits, plans/specs, record drawings of facilities, and site-specific lists
- Review of CSI divisions, including means, methods, materials, and the challenges of scoping each topic
- Exercises in scope identification and scope writing for accurate estimating of projects
- Hands-on exercises that require scope, take-off, and pricing

Who should attend: corporate and government estimators, planners, facility managers, and others who need to produce accurate project estimates.

Facilities Estimating Using the RSMeans CostWorks CD

Combines hands-on skill building with best estimating practices and real-life problems. Brings you up-to-date with key concepts and provides tips, pointers, and guidelines to save time and avoid cost oversights and errors.

Some of what you'll learn:
- Estimating process concepts
- Customizing and adapting RSMeans cost data
- Establishing scope of work to account for all known variables
- Budget estimating: when, why, and how
- Site visits: what to look for—what you can't afford to overlook
- How to estimate repair and remodeling variables

This training session requires you to bring a laptop computer to class.

Who should attend: facility managers, architects, engineers, contractors, facility tradespeople, planners, project managers and anyone involved with JOC, SABRE, or IDIQ.

Unit Price Estimating Using the RSMeans CostWorks CD

Step-by-step instructions and practice problems to identify and track key cost drivers—material, labor, equipment, staging, and subcontractors—for each specific task. Learn the most detailed, time-tested methodology for accurately "pricing" these variables, their impact on each other and on total cost.

Some of what you'll learn:
- Unit price cost estimating
- Order of magnitude, square foot, and assemblies estimating
- Quantity takeoff
- Direct and indirect construction costs
- Development of contractors' bill rates
- How to use *RSMeans Building Construction Cost Data*

This training session requires you to bring a laptop computer to class.

Who should attend: architects, engineers, corporate and government estimators, facility managers, and government procurement staff.

Visit RSMeans

For more information visit our website at www.RSMeans.com

Registration Information

Register early and save up to $100!
Register 30 days before the start date of a seminar and save $100 off your total fee. Note: This discount can be applied only once per order. It cannot be applied to team discount registrations or any other special offer.

How to register
Register by phone today! The RSMeans toll-free number for making reservations is 781-422-5115.

Two-day seminar registration fee - $935. One-day RSMeans CostWorks® or RSMeans Online training registration fee - $375.
To register by mail, complete the registration form and return, with your full fee, to: RSMeans Seminars, 1099 Hingham Street, Suite 201, Rockland, MA 02370.

One-day Construction Cost Estimating - $575.

Government pricing
All federal government employees save off the regular seminar price. Other promotional discounts cannot be combined with the government discount.

Team discount program
For over five attendee registrations. Call for pricing: 781-422-5115

Multiple course discounts
When signing up for two or more courses, call for pricing.

Refund policy
Cancellations will be accepted up to ten business days prior to the seminar start. There are no refunds for cancellations received later than ten working days prior to the first day of the seminar. A $150 processing fee will be applied for all cancellations. Written notice of the cancellation is required. Substitutions can be made at any time before the session starts. No-shows are subject to the full seminar fee.

AACE approved courses
Many seminars described and offered here have been approved for 14 hours (1.4 recertification credits) of credit by the AACE International Certification Board toward meeting the continuing education requirements for recertification as a Certified Cost Engineer/Certified Cost Consultant.

AIA Continuing Education
We are registered with the AIA Continuing Education System (AIA/CES) and are committed to developing quality learning activities in accordance with the CES criteria. Many seminars meet the AIA/CES criteria for Quality Level 2. AIA members may receive 14 learning units (LUs) for each two-day RSMeans course.

Daily course schedule
The first day of each seminar session begins at 8:30 a.m. and ends at 4:30 p.m. The second day begins at 8:00 a.m. and ends at 4:00 p.m. Participants are urged to bring a hand-held calculator since many actual problems will be worked out in each session.

Continental breakfast
Your registration includes the cost of a continental breakfast and a morning and afternoon refreshment break. These informal segments allow you to discuss topics of mutual interest with other seminar attendees. (You are free to make your own lunch and dinner arrangements.)

Hotel/transportation arrangements
RSMeans arranges to hold a block of rooms at most host hotels. To take advantage of special group rates when making your reservation, be sure to mention that you are attending the RSMeans seminar. You are, of course, free to stay at the lodging place of your choice. (Hotel reservations and transportation arrangements should be made directly by seminar attendees.)

Important
Class sizes are limited, so please register as soon as possible.

Note: Pricing subject to change.

Registration Form

ADDS-1000

Call 781-422-5115 to register or fax this form to 800-632-6732. Visit our website: www.RSMeans.com

Please register the following people for the RSMeans construction seminars as shown here. We understand that we must make our own hotel reservations if overnight stays are necessary.

☐ Full payment of $_____ enclosed.

☐ Bill me.

Please print name of registrant(s).
(To appear on certificate of completion)

P.O. #: _____
GOVERNMENT AGENCIES MUST SUPPLY PURCHASE ORDER NUMBER OR TRAINING FORM.

Firm name _____
Address _____
City/State/Zip _____
Telephone no. _____ Fax no. _____
E-mail address _____

Charge registration(s) to: ☐ MasterCard ☐ VISA ☐ American Express

Account no. _____ Exp. date _____
Cardholder's signature _____
Seminar name _____
Seminar City _____

Please mail check to: 1099 Hingham Street, Suite 201, Rockland, MA 02370 USA